Handbook of
PETROLEUM
REFINING

CHEMICAL INDUSTRIES
A Series of Reference Books and Textbooks

Founding Editor

HEINZ HEINEMANN
Berkeley, California

Series Editor

JAMES G. SPEIGHT
CD & W, Inc.
Laramie, Wyoming

MOST RECENTLY PUBLISHED

Handbook of Petroleum Refining, James G. Speight

Handbook of Refinery Desulfurization, Nour Shafik El-Gendy and James G. Speight

Petroleum and Gas Field Processing, Second Edition, Hussein K. Abdel-Aal, Mohamed A. Aggour, and Mohamed A. Fahim

Refining Used Lubricating Oils, James Speight and Douglas I. Exall

The Chemistry and Technology of Petroleum, Fifth Edition, James G. Speight

Transport Phenomena Fundamentals, Third Edition, Joel Plawsky

Synthetics, Mineral Oils, and Bio-Based Lubricants: Chemistry and Technology, Second Edition, Leslie R. Rudnick

Modeling of Processes and Reactors for Upgrading of Heavy Petroleum, Jorge Ancheyta

Synthetics, Mineral Oils, and Bio-Based Lubricants: Chemistry and Technology, Second Edition, Leslie R. Rudnick

Fundamentals of Automatic Process Control, Uttam Ray Chaudhuri and Utpal Ray Chaudhuri

The Chemistry and Technology of Coal, Third Edition, James G. Speight

Practical Handbook on Biodiesel Production and Properties, Mushtaq Ahmad, Mir Ajab Khan, Muhammad Zafar, and Shazia Sultana

Introduction to Process Control, Second Edition, Jose A. Romagnoli and Ahmet Palazoglu

Fundamentals of Petroleum and Petrochemical Engineering, Uttam Ray Chaudhuri

Advances in Fluid Catalytic Cracking: Testing, Characterization, and Environmental Regulations, edited by Mario L. Occelli

Advances in Fischer-Tropsch Synthesis, Catalysts, and Catalysis, edited by Burtron H. Davis and Mario L. Occelli

Transport Phenomena Fundamentals, Second Edition, Joel Plawsky

Asphaltenes: Chemical Transformation during Hydroprocessing of Heavy Oils, Jorge Ancheyta, Fernando Trejo, and Mohan Singh Rana

Handbook of
PETROLEUM REFINING

JAMES G. SPEIGHT

CRC Press
Taylor & Francis Group
Boca Raton London New York

CRC Press is an imprint of the
Taylor & Francis Group, an **informa** business

CRC Press
Taylor & Francis Group
6000 Broken Sound Parkway NW, Suite 300
Boca Raton, FL 33487-2742

First issued in paperback 2020

ISBN 13: 978-0-367-57440-6 (pbk)
ISBN 13: 978-1-4665-9160-8 (hbk)

Library of Congress Cataloging-in-Publication Data

Names: Speight, James G., author.
Title: Handbook of petroleum refining / James G. Speight.
Description: Boca Raton : Taylor & Francis, a CRC title, part of the Taylor & Francis imprint, a member of the Taylor & Francis Group, the academic division of T&F Informa, plc, [2017] | Series: Chemical industries ; 143 | Includes bibliographical references and index.
Identifiers: LCCN 2016008773 | ISBN 9781466591608 (alk. paper)
Subjects: LCSH: Petroleum--Refining--Handbooks, manuals, etc.
Classification: LCC TP690 .S743 2017 | DDC 622/.338--dc23
LC record available at https://lccn.loc.gov/2016008773

**Visit the Taylor & Francis Web site at
http://www.taylorandfrancis.com**

**and the CRC Press Web site at
http://www.crcpress.com**

Contents

SECTION II   Refining

Preface

Over the past three decades, the energy industry has experienced significant changes in resource availability, petropolitics, and technological advancements dictated by the changing quality of refinery feedstock. However, the dependence on fossil fuels as a primary energy source has remained unchanged. Advancements made in exploration, production, and refining technologies now allow for the utilization of resources that might have been considered unsuitable in the middle decades of the twentieth century.

It has been estimated that global energy consumption will grow approximately 50% by the first quarter of the twenty-first century, approximately 90% of which is projected to be supplied by fossil fuels such as oil, natural gas, and coal. In this supply-and-demand scenario, it is expected that the existing peak in conventional oil production will decline within the next two to three decades and production of oil from residua, heavy oil, and tar sand bitumen will increase significantly. For the purposes of this book, residua (resids), heavy oil, extra heavy oil, and tar sand bitumen are (for convenience) included in the term "heavy feedstocks."

Thus, just as there was a surge in upgrading technologies during the 1940s and 1950s to produce marketable products from residua, an equal surge in technologies, which is related to producing marketable products, will occur over the next two decades. So, the need for the development of upgrading processes continues in order to fulfill the product market demand as well as to satisfy environmental regulations. In one area in particular, the need for residuum conversion, technology has emerged as a result of the declining residual fuel oil market and the necessity to upgrade crude oil residua beyond the capabilities of the visbreaking, coking, and low-severity hydrodesulfurization processes.

In the meantime, the refining industry has entered a significant transition period with the arrival of the twenty-first century and the continued reassessment by various levels of government, and by various governments, of oil importing and oil exporting policies. Therefore, it is not surprising that refinery operations have evolved to include a range of *next-generation processes* as the demand for transportation fuels and fuel oil has shown a steady growth. These processes are different from one another in terms of the method and product slates and will find employment in refineries according to their respective features. The primary goal of these processes is to convert heavy feedstocks, such as residua, to lower-boiling products. Thus, these processes are given some consideration in this book.

There are, however, challenges when refining heavy feedstocks. The first challenge in heavy feedstock refining is the deposition of solids (phase separation), which is a direct consequence of high asphaltene and any inorganic solids. One of the most notorious effects of asphaltene constituents is the pronounced tendency to form aggregates in oil media, which under unfavorable solvent conditions leads to separation from the liquid medium. Inorganic fine solids are generally associated with asphaltene constituents, and as a result, the separated asphaltene constituents often contain a high concentration of inorganic fine solids. The separation of organic and/or inorganic solids during processing poses severe problems, which lead to coking in reactors and refinery lines, as well as catalyst deactivation.

Other challenges in heavy feedstock processing can be traced to the high content of heteroatoms (sulfur, nitrogen, oxygen) and heavy metals (particularly, nickel and vanadium). Although the concentration of these elements may be quite small, their impact is significant. For example, the presence of heteroatoms can also be responsible for objectionable characteristics in finished products, which can cause environmental concerns. Therefore, the levels of heteroatoms in finished products need to be reduced and follow more and more stringent environmental regulations. Also, the deposition of trace heavy metals (vanadium and nickel) and/or chemisorption of nitrogen-containing compounds on catalysts are the main reasons for catalyst passivation and/or poisoning in catalytic operations, and thus necessitate frequent regeneration of the catalyst or its replacement.

This book is designed to address the problems of solid deposition during heavy feedstock refining and is, accordingly, divided into two sections: (I) Feedstock Availability and Properties and (II) Refining. Each section will take the reader through the various steps that are necessary for crude oil evaluation and refining. Section I deals with the prerefining steps and outlines how feedstock can be evaluated prior to applying refining processes. Section II presents in detail, with relevant process data, the various processes that can be applied to a variety of feedstocks. All of the processes are described with sufficient detail to understand the process operations. The book brings the reader further up to date and adds more data as well as processing options that may be the processes of the twenty-first century.

By understanding the evolutionary changes that have occurred to date, this book will satisfy the needs of engineers and scientists at all levels from academia to the refinery and help them to understand refining and prepare for the new changes and evolution of the industry.

The target audience includes engineers, scientists, and students who require an update on petroleum processing and are interested in the direction of the industry in the next 50 years. Nontechnical readers, with help from the extensive Glossary at the end of the book, will also benefit from the book.

Dr. James G. Speight
Laramie, Wyoming

Author

James G. Speight, PhD, holds doctorate degrees in chemistry, geological sciences, and petroleum engineering and is the author of more than 60 books on petroleum science, petroleum engineering, and environmental sciences. He has served as an adjunct professor in the Department of Chemical and Fuels Engineering at the University of Utah and in the Departments of Chemistry and Chemical and Petroleum Engineering at the University of Wyoming. In addition, he has been a visiting professor in chemical engineering at the following universities: the University of Missouri–Columbia, the Technical University of Denmark, and the University of Trinidad and Tobago.

As a result of his work, Dr. Speight has been honored as the recipient of the following awards:

- Diploma of Honor, U.S. National Petroleum Engineering Society. *For Outstanding Contributions to the Petroleum Industry*, 1995.
- Gold Medal of the Russian Academy of Sciences. *For Outstanding Work in the Area of Petroleum Science*, 1996.
- Einstein Medal of the Russian Academy of Sciences. *In Recognition of Outstanding Contributions and Service in the Field of Geologic Sciences*, 2001.
- Gold Medal: Scientists without Frontiers, Russian Academy of Sciences. *In Recognition of His Continuous Encouragement of Scientists to Work Together across International Borders*, 2005.
- Methanex Distinguished Professor, University of Trinidad and Tobago. *In Recognition of Excellence in Research*, 2006.
- Gold Medal: Giants of Science and Engineering, Russian Academy of Sciences. *In Recognition of Continued Excellence in Science and Engineering*, 2006.

Section I

Feedstock: Availability and Evaluation

1 Crude Oil, Heavy Oil, and Tar Sand Bitumen

1.1 INTRODUCTION

Historically, petroleum and its derivatives have been known and used for millennia. Ancient workers recognized that certain derivatives of petroleum (asphalt) could be used for civic and decorative purposes, while others (naphtha) could provide certain advantages in warfare (Abraham, 1945; Pfeiffer, 1950; Van Nes and van Westen, 1951; Forbes, 1958a,b, 1959, 1964; Hoiberg, 1964; Speight, 2014a; Cobb and Goldwhite, 1995). Scientifically, petroleum is a carbon-based resource and is an extremely complex mixture of hydrocarbon compounds, usually with minor amounts of nitrogen-, oxygen-, and sulfur-containing compounds, as well as trace amounts of metal-containing compounds. Heavy oil is a subcategory of petroleum that contains a greater proportion of the higher-boiling constituents and heteroatom compounds. Tar sand bitumen is different to petroleum and heavy oil insofar as it cannot be recovered by any of the conventional (including enhanced recovery) methods (Speight, 2014a, 2015c). For the purposes of this book, residua (resids), heavy oil, extra heavy oil, and tar sand bitumen are (for convenience) included in the term "heavy feedstocks."

In the crude state, petroleum, heavy oil, and bitumen have minimal value, but when refined they provide high-value liquid fuels, solvents, lubricants, and many other products. The fuels derived from petroleum contribute approximately one-third to one-half of the total world energy supply and are used not only for transportation fuels (i.e., gasoline, diesel fuel, and aviation fuel, among others) but also to heat buildings. Petroleum products have a wide variety of uses that vary from gaseous and liquid fuels to near-solid machinery lubricants. In addition, asphalt (a once-maligned by-product and the residue of many refinery processes) is now a premium value product for highway surfaces, roofing materials, and miscellaneous waterproofing use (Speight, 2014a, 2015c).

In the context of this chapter, it is pertinent to note that throughout the millennia in which petroleum has been known and used, it is only in the last four decades that some attempts have been made to standardize petroleum nomenclature and terminology. But confusion may still exist. In fact, the *definition* of petroleum has been varied, unsystematic, diverse, and often archaic. Furthermore, the terminology of petroleum is a product of many years of growth. Thus, the long established use of an expression, however inadequate it may be, is altered with difficulty, and a new term, however precise, is at best adopted only slowly. Because of the need for a thorough understanding of petroleum and the associated technologies, it is essential that the definitions and the terminology of petroleum science and technology be given prime consideration. This will aid in a better understanding of petroleum, its constituents, and its various fractions. Of the many forms of terminology that have been used not all have survived, but the more commonly used are illustrated here. Particularly troublesome, and more confusing, are those terms that are applied to the more viscous materials, for example, the use of the terms "bitumen" and "asphalt." This part of the text attempts to alleviate much of the confusion that exists, but it must be remembered that the terminology of petroleum is still open to personal choice and historical usage.

Therefore, the purpose of this chapter is to provide some semblance of order into the disordered state that exists in the segment of petroleum technology that is known as "terminology." There is no effort here to define the individual processes since these will be defined in the relevant chapters. The purpose is to also define the various aspects of the feedstocks that are used in a refinery so that the reader can make a ready reference to any such word used in the text.

For the purposes of definitions and terminology, it is preferable to subdivide petroleum and related materials into three major classes: (1) materials that are of natural origin, (2) materials that are manufactured, and (3) materials that are integral fractions derived from the natural or manufactured products.

1.2 NATIVE MATERIALS

If there is to be a thorough understanding of petroleum and the associated technologies, it is essential that the definitions and the terminology of petroleum science and technology be given prime consideration (Meyer and DeWitt, 1990; Speight, 2014a). This will aid in a better understanding of petroleum, its constituents, and its various fractions. Of the many forms of terminology that have been used not all have survived, but the more commonly used are illustrated here. Particularly troublesome, and more confusing, are those terms that are applied to the more viscous materials, for example, the use of the terms "bitumen" and "asphalt." This part of the text attempts to alleviate much of the confusion that exists, but it must be remembered that the terminology of petroleum is still open to personal choice and historical usage.

1.2.1 Petroleum

"Petroleum" and the equivalent term "crude oil" cover a wide assortment of materials consisting of mixtures of hydrocarbons and other compounds containing variable amounts of sulfur, nitrogen, and oxygen, which may vary widely in API gravity and sulfur content (Table 1.1), as well as viscosity and the amount of residuum (the portion of crude oil boiling above 510°C, 950°F). Metal-containing constituents, notably those compounds that contain vanadium and nickel, usually occur in the more viscous crude oils in amounts up to several thousand parts per million and can have serious consequences during processing of these feedstocks (Speight, 1984). Because petroleum is a mixture of widely varying constituents and proportions, its physical properties also vary widely and the color from colorless to black. Thus, because petroleum is a mixture of widely varying constituents and proportions, its physical properties also vary widely and the color from colorless to black. The terms *heavy oil* and *bitumen* add further complexity to the nature of refinery feedstocks.

Petroleum is a mixture of gaseous, liquid, and solid hydrocarbon compounds that occur in sedimentary rock deposits throughout the world and also contains small quantities of nitrogen-, oxygen-, and sulfur-containing compounds, as well as trace amounts of metallic constituents (Speight, 2012a, 2014a).

Petroleum is a naturally occurring mixture of hydrocarbons, generally in a liquid state, which may also include compounds of sulfur, nitrogen, oxygen, metals, and other elements. Petroleum has also been defined as (1) any naturally occurring hydrocarbon, whether in a liquid, gaseous, or solid state; (2) any naturally occurring mixture of hydrocarbons, whether in a liquid, gaseous, or solid state; or (3) any naturally occurring mixture of one or more hydrocarbons, whether in a liquid, gaseous, or solid state and one or more of the following, that is to say, hydrogen sulfide, helium, and carbon dioxide. The definition also includes any petroleum as defined by paragraph by the earlier three categories, which has been returned to a natural reservoir (ITAA, 1936; ASTM D4175, 2015).

Crude petroleum is a mixture of compounds boiling at different temperatures that can be separated into a variety of different generic fractions by distillation. And the terminology of these fractions has been bound by utility and often bears little relationship to composition. Furthermore, there is a wide variation in the properties of crude petroleum because the proportions in which the different constituents occur vary with origin. Thus, some crude oils have higher proportions of the

TABLE 1.1
API Gravity and Sulfur Content of Selected Crude Oils

Country	Crude Oil	API	Sulfur (% w/w)
Abu Dhabi (UAE)	Abu Al Bu Khoosh	31.6	2.00
Abu Dhabi (UAE)	Abu Mubarras	38.1	0.93
Abu Dhabi (UAE)	El Bunduq	38.5	1.12
Abu Dhabi (UAE)	Murban	40.5	0.78
Abu Dhabi (UAE)	Umm Shaif	37.4	1.51
Abu Dhabi (UAE)	Zakum (Lower)	40.6	1.05
Abu Dhabi (UAE)	Zakum (Upper)	33.1	2.00
Algeria	Zarzaitine	43.0	0.07
Angola	Cabinda	31.7	0.17
Angola	Palanca	40.1	0.11
Angola	Takula	32.4	0.09
Australia	Airlie	44.7	0.01
Australia	Barrow Island	37.3	0.05
Australia	Challis	39.5	0.070
Australia	Cooper Basin	45.2	0.02
Australia	Gippsland	47.0	0.09
Australia	Griffin	55.0	0.03
Australia	Harriet	37.9	0.05
Australia	Jabiru	42.3	0.05
Australia	Jackson	43.8	0.03
Australia	Saladin	48.2	0.02
Australia	Skua	41.9	0.06
Brazil	Garoupa	30.0	0.68
Brazil	Sergipano Platforma	38.4	0.19
Brazil	Sergipano Terra	24.1	0.41
Brunei	Champion Export	23.9	0.12
Brunei	Seria	40.5	0.06
Cameroon	Kole Marine	32.6	0.33
Cameroon	Lokele	20.7	0.46
Canada (Alberta)	Bow River Heavy	26.7	2.10
Canada (Alberta)	Pembina	38.8	0.20
Canada (Alberta)	Rainbow	40.7	0.50
Canada (Alberta)	Rangeland South	39.5	0.75
Canada (Alberta)	Wainwright-Kinsella	23.1	2.58
China	Daqing (Taching)	32.6	0.09
China	Nanhai Light	40.6	0.06
China	Shengli	24.2	1.00
China	Weizhou	39.7	0.08
Colombia	Cano Limon	29.3	0.51
Congo (Brazzaville)	Emeraude	23.6	0.600
Dubai (UAE)	Fateh	31.1	2.000
Dubai (UAE)	Margham Light	50.3	0.040
Ecuador	Oriente	29.2	0.880
Egypt	Belayim	27.5	2.200
Egypt	Gulf of Suez	31.9	1.520
Egypt	Ras Gharib	21.5	3.640

(Continued)

TABLE 1.1 (*Continued*)
API Gravity and Sulfur Content of Selected Crude Oils

Country	Crude Oil	API	Sulfur (% w/w)
Gabon	Gamba	31.4	0.090
Gabon	Rabi-Kounga	33.5	0.070
Ghana	Salt Pond	37.4	0.097
India	Bombay High	39.2	0.150
Indonesia	Anoa	45.2	0.040
Indonesia	Ardjuna	35.2	0.105
Indonesia	Attaka	43.3	0.040
Indonesia	Badak	49.5	0.032
Indonesia	Bekapai	41.2	0.080
Indonesia	Belida	45.1	0.020
Indonesia	Bima	21.1	0.250
Indonesia	Cinta	33.4	0.080
Indonesia	Duri (Sumatran Heavy)	21.3	0.180
Indonesia	Ikan Pari	48.0	0.020
Indonesia	Kakap	51.5	0.050
Indonesia	Katapa	50.8	0.060
Indonesia	Lalang (Malacca Straits)	39.7	0.050
Indonesia	Minas (Sumatran Light)	34.5	0.081
Indonesia	Udang	38.0	0.050
Iran	Aboozar (Ardeshir)	26.9	2.480
Iran	Bahrgansar/Nowruz	27.1	2.450
Iran	Dorrood (Darius)	33.6	2.350
Iran	Foroozan (Fereidoon)	31.3	2.500
Iran	Iranian Heavy	30.9	1.730
Iran	Iranian Light	33.8	1.350
Iran	Rostam	35.9	1.550
Iran	Salmon (Sassan)	33.9	1.910
Iraq	Basrah Heavy	24.7	3.500
Iraq	Basrah Light	33.7	1.950
Iraq	Basrah Medium	31.1	2.580
Iraq	North Rumaila	33.7	1.980
Ivory Coast	Espoir	32.3	0.340
Kazakhstan	Kumkol	42.5	0.07
Kuwait	Kuwait Export	31.4	2.52
Libya	Amna	36.0	0.15
Libya	Brega	40.4	0.21
Libya	Bu Attifel	43.3	0.04
Libya	Buri	26.2	1.76
Libya	Es Sider	37.0	0.45
Libya	Sarir	38.4	0.16
Libya	Sirtica	41.3	0.45
Libya	Zueitina	41.3	0.28
Malaysia	Bintulu	28.1	0.08
Malaysia	Dulang	39.0	0.12
Malaysia	Labuan	32.2	0.07
Malaysia	Miri Light	32.6	0.04

(Continued)

TABLE 1.1 (*Continued*)
API Gravity and Sulfur Content of Selected Crude Oils

Country	Crude Oil	API	Sulfur (% w/w)
Malaysia	Tembungo	37.4	0.04
Mexico	Isthmus	33.3	1.49
Mexico	Maya	22.2	3.30
Mexico	Olmeca	39.8	0.80
Neutral Zone	Burgan	23.3	3.37
Neutral Zone	Eocene	18.6	4.55
Neutral Zone	Hout	32.8	1.91
Neutral Zone	Khafji	28.5	2.85
Neutral Zone	Ratawi	23.5	4.07
Nigeria	Antan	32.1	0.32
Nigeria	Bonny Light	33.9	0.14
Nigeria	Bonny Medium	25.2	0.23
Nigeria	Brass River	42.8	0.06
Nigeria	Escravos	36.4	0.12
Nigeria	Forcados	29.6	0.18
Nigeria	Pennington	36.6	0.07
Nigeria	Qua Iboe	35.8	0.12
North Sea (Denmark)	Danish North Sea	34.5	0.260
North Sea (Norway)	Ekofisk	39.2	0.169
North Sea (Norway)	Emerald	22.0	0.750
North Sea (Norway)	Oseberg	33.7	0.310
North Sea (United Kingdom)	Alba	20.0	1.330
North Sea (United Kingdom)	Duncan	38.5	0.180
North Sea (United Kingdom)	Forties Blend	40.5	0.350
North Sea (United Kingdom)	Innes	45.7	0.130
North Sea (United Kingdom)	Kittiwake	37.0	0.65
North Yemen	Alif	40.3	0.10
Oman	Oman Export	34.7	0.94
Papua New Guinea	Kubutu	44.0	0.04
Peru	Loreto Peruvian	33.1	0.23
Qatar	Dukhan (Qatar Land)	40.9	1.27
Qatar	Qatar Marine	36.0	1.42
Ras Al Khaiman (UAE)	Ras Al Khaiman	44.3	0.15
Russia	Siberian Light	37.8	0.42
Saudi Arabia	Arab Extra Light (Berri)	37.2	1.15
Saudi Arabia	Arab Heavy (Safaniya)	27.4	2.80
Saudi Arabia	Arab Light	33.4	1.77
Saudi Arabia	Arab Medium (Zuluf)	28.8	2.49
Sharjah (UAE)	Mubarek	37.0	0.62
Sumatra	Duri	20.3	0.21
Syria	Souedie	24.9	3.82
Timor Sea (Indonesia)	Hydra	37.5	0.08
Trinidad Tobago	Galeota Mix	32.8	0.27
Tunisia	Ashtart	30.0	0.99
United States (Alaska)	Alaskan North Slope	27.5	1.11
United States (Alaska)	Cook Inlet	35.0	0.10
United States (Alaska)	Drift River	35.3	0.09

(*Continued*)

TABLE 1.1 (*Continued*)
API Gravity and Sulfur Content of Selected Crude Oils

Country	Crude Oil	API	Sulfur (% w/w)
United States (Alaska)	Nikiski Terminal	34.6	0.10
United States (California)	Hondo Sandstone	35.2	0.21
United States (California)	Huntington Beach	20.7	1.38
United States (Florida)	Sunniland	24.9	3.25
United States (Louisiana)	Grand Isle	34.2	0.35
United States (Louisiana)	Lake Arthur	41.9	0.06
United States (Louisiana)	Louisiana Light Sweet	36.1	0.45
United States (Louisiana)	Ostrica	32.0	0.30
United States (Louisiana)	South Louisiana	32.8	0.28
United States (Michigan)	Lakehead Sweet	47.0	0.31
United States (New Mexico)	New Mexico Intermediate	37.6	0.17
United States (New Mexico)	New Mexico Light	43.3	0.07
United States (Oklahoma)	Basin Cushing Composite	34.0	1.95
United States (Texas)	Coastal B-2	32.2	0.22
United States (Texas)	East Texas	37.0	0.21
United States (Texas)	Sea Breeze	37.9	0.10
United States (Texas)	West Texas Intermediate	40.8	0.34
United States (Texas)	West Texas Semi-Sweet	39.0	0.27
United States (Texas)	West Texas Sour	34.1	1.640
United States (Wyoming)	Tom Brown	38.2	0.100
United States (Wyoming)	Wyoming Sweet	37.2	0.330
Venezuela	Lago Medio	32.2	1.010
Venezuela	Leona	24.4	1.510
Venezuela	Mesa	29.8	1.010
Venezuela	Ceuta Export	27.8	1.370
Venezuela	Guanipa	30.3	0.850
Venezuela	La Rosa Medium	25.3	1.730
Venezuela	LagoTreco	26.7	1.500
Venezuela	Oficina	33.3	0.780
Venezuela	Temblador	21.0	0.830
Venezuela	Tia Juana	25.8	1.630
Venezuela	Tia Juana Light	31.8	1.160
Venezuela	Tia Juana Medium 24	24.8	1.610
Venezuela	Tia Juana Medium 26	26.9	1.540
Viet Nam	Bach Ho (White Tiger)	38.6	0.030
Viet Nam	Dai Hung (Big Bear)	36.9	0.080
Yemen	Masila	30.5	0.670
Zaire	Zaire	31.7	0.130

lower-boiling components and others (such as heavy oil and bitumen) have higher proportions of higher-boiling components (asphaltic components and residuum).

The molecular boundaries of petroleum cover a wide range of boiling points and carbon numbers of hydrocarbon compounds and other compounds containing nitrogen, oxygen, and sulfur, as well as metallic (porphyrin) constituents that dictate the options to be used in a refinery (Long and Speight, 1998; Speight and Ozum, 2002; Parkash, 2003; Hsu and Robinson, 2006; Gary et al., 2007; Speight, 2014a). However, the actual boundaries of such a *petroleum map* can only be arbitrarily defined in terms of boiling point and carbon number. In fact, petroleum is so diverse that materials from different sources exhibit different boundary limits, and for this reason alone it is not surprising that petroleum has been difficult to *map* in a precise manner.

Petroleum occurs underground at various pressures depending on the depth. Because of the pressure, it contains considerable natural gas in solution. The oil underground is much more fluid than it is on the surface and is generally mobile under reservoir conditions because the elevated temperatures in subterranean formations (on the average, the temperature rises 1°C for every 100 ft, 33 m, of depth) decrease the viscosity.

Petroleum is derived from aquatic plants and animals that lived and died hundreds of millions of years ago. Their remains mixed with mud and sand in layered deposits that, over the millennia, were geologically transformed into sedimentary rock. Gradually, the organic matter decomposed and eventually formed petroleum (or a related precursor), which migrated from the original source beds to more porous and permeable rocks, such as *sandstone* and *siltstone*, where it finally became entrapped. Such entrapped accumulations of petroleum are called "reservoirs." A series of reservoirs within a common rock structure or a series of reservoirs in separate but neighboring formations are commonly referred to as an "oil field." A group of fields is often found in a single geologic environment known as a "sedimentary basin" or "province."

Petroleum reservoirs are generally classified according to their geologic structure and their production (drive) mechanism. Petroleum reservoirs exist in many different sizes and shapes of geologic structures. It is usually convenient to classify the reservoirs according to the conditions of their formation. For example, *dome-shaped* and *anticline reservoirs* are formed by the folding of the rock. Typically, the dome is circular in outline, and the anticline is long and narrow. Oil or gas moved or migrated upward through the porous strata where it was trapped by the sealing cap rock and the shape of the structure. On the other hand, *faulted reservoirs* are formed by shearing and offsetting of the strata (faulting). The movement of the nonporous rock opposite the porous formation containing the oil/gas creates the sealing. The tilt of the petroleum-bearing rock and the faulting trap the oil/gas in the reservoir. *Salt-dome reservoirs* take the shape of a dome and were formed due to the upward movement of a large, impermeable salt dome that deformed and lifted the overlying layers of rock.

Unconformities are formed as a result of an unconformity where the impermeable cap rock was laid down across the cutoff surfaces of the lower beds. In the *lens-type reservoir*, the petroleum-bearing porous formation is sealed by the surrounding, nonporous formation. Irregular deposition of sediments and shale at the time the formation was laid down is the probable cause for the change in the porosity of the formation. Finally, a *combination reservoir* is, as the name implies, a combination of folding, faulting, abrupt changes in porosity, or other conditions creating the trap that exists as this type of petroleum reservoir.

The major components of petroleum are *hydrocarbons*, compounds of hydrogen and carbon that display great variation in their molecular structure. The simplest hydrocarbons are a large group of chain-shaped molecules known as the "paraffins." This broad series extends from methane, which forms natural gas, through liquids that are refined into gasoline, to crystalline waxes. A series of saturated hydrocarbons containing a (usually six-membered) ring, known as the "naphthenes," range from volatile liquids such as *naphtha* to high-molecular-weight substances isolated as the *asphaltene* fraction. Another group of hydrocarbons containing a single or condensed aromatic ring system is known as the "aromatics"; the chief compound in this series is benzene, a popular raw material for making petrochemicals. *Nonhydrocarbon constituents* of petroleum include organic derivatives of nitrogen, oxygen, sulfur, and the metals nickel and vanadium. Most of these impurities are removed during refining.

Other members of the petroleum family that may invoke the need for enhanced recovery methods, including the application of hydraulic fracturing techniques (Speight, 2015a), and that are worthy of mention here include: (1) crude oil from tight formations, (2) opportunity crudes, (3) high-acid crude oils, and (4) foamy oil.

1.2.1.1 Crude Oil from Tight Formations

Generally, unconventional tight oil resources are found at considerable depths in sedimentary rock formations that are characterized by very low permeability. While some of the tight oil plays produce oil directly from shales, tight oil resources are also produced from low-permeability siltstone formations, sandstone formations, and carbonate formations that occur in close association with a shale source rock. Tight formations scattered throughout North America have the potential to produce crude oil (*tight oil*) (US EIA, 2011, 2013; Mayes, 2015). Such formations might be composted of shale sediments or sandstone sediments. In a conventional sandstone reservoir, the pores are interconnected so that gas and oil can flow easily from the rock to a wellbore. In tight sandstones, the pores are smaller and are poorly connected by very narrow capillaries, which results in low permeability. Tight oil occurs in sandstone sediments that have an effective permeability of less than 1 millidarcy (<1 mD). A shale play is a defined geographic area containing an organic-rich fine-grained sedimentary rock that underwent physical and chemical compaction during diagenesis to produce the following characteristics: (1) clay to silt-sized particles; (2) high percentage of silica, and sometimes carbonate minerals; (3) thermally mature; (4) hydrocarbon-filled porosity, on the order of 6%–14%; (5) low permeability, on the order of <0.1 mD; (6) large areal distribution; and (7) fracture stimulation required for economic production.

The most notable tight oil plays in North America include the Bakken shale, the Niobrara formation, the Barnett shale, the Eagle Ford shale, the Miocene Monterey play of California's San Joaquin Basin (California) and the Cardium play (Alberta, Canada). In many of these tight formations, the existence of large quantities of crude oil has been known for decades and efforts to commercially produce those resources have occurred sporadically with typically disappointing results. However, starting in the mid-2000s, advancements in well drilling and stimulation technologies combined with high oil prices have turned tight oil resources into one of the most actively explored and produced targets in North America.

Other known tight formations (on a worldwide basis) include the R'Mah Formation (Syria), the Sargelu Formation (northern Persian Gulf region), the Athel Formation (Oman), the Bazhenov Formation and Achimov Formation (West Siberia, Russia), the Coober Pedy Formation (Australia), the Chicontepec Formation (Mexico), and the Vaca Muerta field (Argentina) (U.S. EIA, 2011, 2013). However, tight oil formations are heterogeneous and vary widely over relatively short distances. Thus, even in a single horizontal production well, the amount of oil recovered may vary as may recovery within a field or even between adjacent wells. This makes evaluation of *shale plays* and decisions regarding the profitability of wells on a particular lease difficult. In addition, tight reservoirs that contain only crude oil (without natural gas as the pressurizing agent) cannot be economically produced (U.S. EIA, 2011, 2013).

Typical of the crude oil from tight formations (*tight oil*, *tight light oil*, and *tight shale oil* have been suggested as alternate terms) is the Bakken crude oil, which is a light, highly volatile crude oil. Briefly, Bakken crude oil is a light, sweet (low-sulfur) crude oil that has a relatively high proportion of volatile constituents. The production of the oil also yields a significant amount of volatile gases (including propane and butane) and low-boiling liquids (such as pentane and natural gasoline), which are often referred to collectively as (low-boiling or light) naphtha. By definition, natural gasoline (sometimes also referred to as *gas condensate*) is a mixture of low-boiling liquid hydrocarbons isolate from petroleum and natural gas wells suitable for blending with light naphtha and/or refinery gasoline (Mokhatab et al., 2006; Speight, 2007, 2014a). Because of the presence of low-boiling hydrocarbons, low-boiling naphtha (*light naphtha*) can become extremely explosive, even at relatively low ambient temperatures. Some of these gases may be burned off (flared) at the field wellhead, but others remain in the liquid products extracted from the well (Speight, 2014a).

Oil from tight shale formation is characterized by low asphaltene content, low sulfur content, and a significant molecular weight distribution of the paraffinic wax content (Speight, 2014a, 2015b).

Paraffin carbon chains of C_{10}–C_{60} have been found, with some shale oils containing carbon chains up to C_{72}. To control deposition and plugging in formations due to paraffins, the dispersants are commonly used. In upstream applications, these paraffin dispersants are applied as part of multi-functional additive packages where asphaltene stability and corrosion control are also addressed simultaneously (Speight, 2014a,b,c, 2015b,c). In addition, scale deposits of calcite ($CaCO_3$), other carbonate minerals (minerals containing the carbonate ion, $CO_3{}^{2-}$), and silicate minerals (minerals classified on the basis of the structure of the silicate group, which contains different ratios of silicon and oxygen) must be controlled during production or plugging problems arise. A wide range of scale additives are available, which can be highly effective when selected appropriately. Depending on the nature of the well and the operational conditions, a specific chemistry is recommended or blends of products are used to address scale deposition.

While the basic approach toward developing a tight oil resource is expected to be similar from area to area, the application of specific strategies, especially with respect to well completion and stimulation techniques, will almost certainly differ from play to play and often even within a given play. The differences depend on the geology (which can be very heterogeneous, even within a play) and reflect the evolution of technologies over time with increased experience and availability.

Finally, the properties of crude oils from tight formations are highly variable. Density and other properties can show wide variation, even within the same field. The Bakken crude is light and sweet with an API of 42° and a sulfur content of 0.19% w/w. Similarly, Eagle Ford is a light sweet feed, with a sulfur content of approximately 0.1% w/w and with published API gravity between 40° API and 62° API.

1.2.1.2 Opportunity Crudes

There is also the need for a refinery to be configured to accommodate *opportunity crude oils* and/or *high-acid crude oils* that, for many purposes, are often included with heavy feedstocks (Speight, 2014a,b; Yeung, 2014). *Opportunity crude oils* are either new crude oils with unknown or poorly understood properties relating to processing issues or are existing crude oils with well-known properties and processing concerns (Ohmes, 2014). Opportunity crude oils are often, but not always, heavy crude oils but in either case are more difficult to process due to high levels of solids (and other contaminants) produced with the oil, high levels of acidity, and high viscosity. These crude oils may also be incompatible with other oils in the refinery feedstock blend and cause excessive equipment fouling when processed either in a blend or separately (Speight, 2015b).

In addition to taking preventative measures, for the refinery to process these feedstocks without serious deleterious effects on the equipment, refiners need to develop programs for detailed and immediate feedstock evaluation so that they can understand the qualities of a crude oil very quickly and it can be valued appropriately and management of the crude processing can be planned meticulously (Babich and Moulijn, 2003; Speight, 2014a). For example, the compatibility of opportunity crudes with other opportunity crudes and with conventional crude oil, and heavy oil is a very important property to consider when making decisions regarding which crude to purchase. Blending crudes that are incompatible can lead to extensive fouling and processing difficulties due to unstable asphaltene constituents (Speight, 2014a, 2015b). These problems can quickly reduce the benefits of purchasing the opportunity crude in the first place. For example, extensive fouling in the crude preheat train may occur resulting in decreased energy efficiency, increased emissions of carbon dioxide, and increased frequency at which heat exchangers need to be cleaned. In a worst-case scenario, crude throughput may be reduced leading to significant financial losses.

Opportunity crude oils, while offering initial pricing advantages, may have composition problems that can cause severe problems at the refinery, harming infrastructure, yield, and profitability. Before refining, there is the need for comprehensive evaluations of opportunity crudes, giving the potential buyer and seller the needed data to make informed decisions regarding fair pricing and the suitability of a particular opportunity crude oil for a refinery. This will assist the refiner to manage the ever-changing crude oil quality input to a refinery—including quality and quantity requirements and situations, crude oil variations, contractual specifications, and risks associated with such opportunity crudes.

1.2.1.3 High-Acid Crude Oil

High-acid crude oils are crude oils that contain considerable proportions of naphthenic acids, which, as commonly used in the petroleum industry, refer collectively to all of the organic acids present in the crude oil (Shalaby, 2005; Speight, 2014b). In many instances, the high-acid crude oils are actually the heavier crude oils (Speight, 2014a,b). The total acid matrix is therefore complex, and it is unlikely that a simple titration, such as the traditional methods for the measurement of the total acid number, can give meaningful results to use in predictions of problems. An alternative way of defining the relative organic acid fraction of crude oils is therefore a real need in the oil industry, both upstream and downstream.

By the original definition, a naphthenic acid is a monobasic carboxyl group attached to a saturated cycloaliphatic structure. However, it has been a convention accepted in the oil industry that all organic acids in crude oil are called naphthenic acids. Naphthenic acids in crude oils are now known to be mixtures of low- to high-molecular-weight acids and the naphthenic acid fraction also contains other acidic species. Naphthenic acids, which are not *user friendly* in terms of refining (Kane and Cayard, 2002; Ghoshal and Sainik, 2013), can be either (or both) water soluble or oil soluble depending on their molecular weight, process temperatures, salinity of waters, and fluid pressures. In the water phase, naphthenic acids can cause stable reverse emulsions (oil droplets in a continuous water phase). In the oil phase with residual water, these acids have the potential to react with a host of minerals, which are capable of neutralizing the acids. The main reaction product found in practice is the calcium naphthenate soap (the calcium salt of naphthenic acids). The total acid matrix is therefore complex, and it is unlikely that a simple titration, such as the traditional methods for the measurement of the total acid number, can give meaningful results to use in predictions of problems. An alternative way of defining the relative organic acid fraction of crude oils is therefore a real need in the oil industry, both upstream and downstream.

Normally, the end result of formation of low-molecular-weight acidic species is treated in the overheads in refineries. A combined approach to front-end treating at crude inlet to heaters and preheat exchangers should be considered. It is commonly assumed that acidity in crude oils is related to carboxylic acid species, that is, components containing a –COOH functional group. While it is clear that carboxylic acid functionality is an important feature (60% of the ions have two or more oxygen atoms), a major portion (40%) of the acid types are not carboxylic acids. In fact, naphthenic acids are a mixture of different compounds that may be polycyclic and may have unsaturated bonds, aromatic rings, and hydroxyl groups. Even the carboxylic acids are more diverse than expected, with approximately 85% containing more heteroatoms than the two oxygen atoms needed to account for the carboxylic acid groups. Examining the distribution of component types in the acid fraction reveals that there is a broad distribution of species.

High-acid crude oils cause corrosion in the refinery—corrosion is predominant at temperatures in excess of 180°C (355°F) (Kane and Cayard, 2002; Ghoshal and Sainik, 2013; Speight, 2014c)—and occur particularly in the atmospheric distillation unit (the first point of entry of the high-acid crude oil) and also in the vacuum distillation units. In addition, overhead corrosion is caused by the mineral salts, magnesium, calcium, and sodium chloride, which are hydrolyzed to produce volatile hydrochloric acid, causing a highly corrosive condition in the overhead exchangers. Therefore, these salts present a significant contamination in opportunity crude oils. Other contaminants in opportunity crude oils that are shown to accelerate the hydrolysis reactions are inorganic clays and organic acids.

Corrosion by naphthenic acids typically has a localized pattern, particularly at areas of high velocity and, in some cases, where condensation of concentrated acid vapors can occur in crude distillation units. The attack is also described as lacking corrosion products. Damage is in the form of unexpected high corrosion rates on alloys that would normally be expected to resist sulfidic corrosion (particularly steels with more than 9% Cr). In some cases, even very highly alloyed materials (i.e., 12% Cr, type 316 stainless steel [SS] and type 317 SS) or, in severe cases, even 6% Mo stainless steel has been found to exhibit sensitivity to corrosion under these conditions.

The corrosion reaction processes involve the formation of iron naphthenates:

$$Fe + 2RCOOH = Fe(RCOO)_2 + H_2 (4)$$

$$Fe(RCOO)_2 + H_2S = FeS + 2RCOOH$$

The iron naphthenates are soluble in oil, and the surface is relatively film free. In the presence of hydrogen sulfide, a sulfide film is formed, which can offer some protection depending on the acid concentration. If the sulfur-containing compounds are reduced to hydrogen sulfide, the formation of a potentially protective layer of iron sulfide occurs on the unit walls and corrosion is reduced (Kane and Cayard, 2002). When the reduction product is water, coming from the reduction of sulfoxides, the naphthenic acid corrosion is enhanced.

Thermal decarboxylation can occur during the distillation process (during which the temperature of the crude oil in the distillation column can be as high as 400°C (750°F):

$$R{-}CO_2H \rightarrow R{-}H + CO_2$$

However, not all acidic species in petroleum are derivatives of carboxylic acids (–COOH), and some of the acidic species are resistant to high temperatures. For example, acidic species appear in the vacuum residue after having been subjected to the inlet temperatures of an atmospheric distillation tower and a vacuum distillation tower (Speight and Francisco, 1990). In addition, for the acid species that are volatile, naphthenic acids are most active at their boiling point, and the most severe corrosion generally occurs on condensation from the vapor phase back to the liquid phase.

1.2.1.4 Foamy Oil

Foamy oil is an oil-continuous foam that contains dispersed gas bubbles produced at the wellhead from heavy oil reservoirs under solution gas drive. The nature of the gas dispersions in oil distinguishes foamy oil behavior from conventional heavy oil. The gas that comes out of solution in the reservoir coalesces neither into large gas bubbles nor into a continuous flowing gas phase. Instead, it remains as small bubbles entrained in the crude oil, keeping the effective oil viscosity low while providing expansive energy that helps drive the oil toward the producing. Foamy oil accounts for unusually high production in heavy oil reservoirs under solution gas drive.

During primary production of heavy oil from solution gas drive reservoirs, the oil is pushed into the production wells by energy supplied by the dissolved gas. As fluid is withdrawn from the production wells, the pressure in the reservoir declines and the gas that was dissolved in the oil at high pressure starts to come out of solution (*foamy oil*). As pressure declines further with continued removal of fluids from the production wells, more gas is released from solution and the gas already released expands in volume. The expanding gas, which at this point is in the form of isolated bubbles, pushes the oil out of the pores and provides energy for the flow of oil into the production well. This process is very efficient until the isolated gas bubbles link up and the gas itself starts flowing into the production well. Once the gas flow starts, the oil has to compete with the gas for available flow energy. Thus, in some heavy oil reservoirs, due to the properties of the oil and the sand and also due to the production methods, the released gas forms foam with the oil and remains subdivided in the form of dispersed bubbles much longer.

Thus, foamy oil is formed in solution gas drive reservoirs when gas is released from solution with a decline in reservoir pressure. It has been noted that the oil at the wellhead of these heavy oil reservoirs resembles the form of foam, hence the term "foamy oil." The gas initially exists in the form of small bubbles within individual pores in the rock. As time passes and pressure continues to decline, the bubbles grow to fill the pores. With further declines in pressure, the bubbles created in different locations become large enough to coalesce into a continuous gas phase. Once the gas phase becomes continuous (i.e., when gas saturation exceeds the critical level—the minimum saturation at which a continuous gas phase exists in porous media) traditional two-phase (oil and gas) flow with classical relative permeability occurs. As a result, the gas-oil ratio (GOR) production increases rapidly after the critical gas saturation has been exceeded.

However, it has been observed that many heavy oil reservoirs in Alberta and Saskatchewan (Canada) exhibit foamy oil behavior, which is accompanied by sand production, leading to anomalously high oil recovery and lower gas-oil ratio (Chugh et al., 2000). These observations suggest that the foamy oil flow might be physically linked to sand production. It is apparent that some additional factors, which remain to be discovered, are involved in making the foamy solution gas possible at field rates of decline. One possible mechanism is the synergistic influence of sand influx into the production wells. Allowing 1%–3% w/w sand to enter the wellbore with the fluids can result in the propagation of a front of sharp pressure gradients away from the wellbore. These sharp pressure gradients occur at the advancing edge of the solution gas drive. It is still unknown how far from the wellbore the dilated zone can propagate.

However, the actual structure of foamy oil flow and its mathematical description are still not well understood. Much of the earlier discussion of such flows was based on the concept of microbubbles (i.e., bubbles that are much smaller than the average pore throat size and are thus free to move with the oil during flow) (Sheng et al., 1999). Dispersion of this type can be produced only by nucleation of a very large number of bubbles (explosive nucleation) and by the availability of a mechanism that prevents these bubbles from growing into larger bubbles with decline in pressure. Another hypothesis for the structure of foamy oil flow is that much larger bubbles migrating with the oil, with the dispersion created by the breakup of bubbles during migration. The major difference between conventional solution gas drive and foamy solution gas drive is that the pressure gradient in the latter is strong enough to mobilize gas clusters once they have grown to a certain size (Maini, 1999).

Reservoirs that exhibit foamy oil behavior are typically characterized by the appearance of an oil-continuous foam at the wellhead. When oil is produced as this nonequilibrium mixture, reservoirs can perform with higher than expected rates of production: up to 30 times that predicted by Darcy's law and lower than expected production gas-oil ratio (Poon and Kisman, 1992). Moreover, foamy oil flow is often accompanied by sand production along with the oil and gas—the presence of sand at the wellhead leads to sand dilation and the presence of high porosity, high-permeability zones (wormholes) in the reservoir (Maini, 1999, 2001). It is generally believed that in the field, the high rates and recoveries observed are the combination of the foamy oil mechanism and the presence of these wormholes.

1.2.2 Heavy Oil

There are also other *types* of petroleum that are different from the conventional petroleum insofar as they are much more difficult to recover from the subsurface reservoir. These materials have a higher viscosity (and lower API gravity) than conventional petroleum, and recovery of these petroleum types usually requires thermal stimulation of the reservoir leading to the application of various thermal methods (such as coking) in the refinery for suitable conversion to low-boiling distillates.

For example, petroleum and heavy oil have been arbitrarily defined in terms of physical properties. For example, heavy oil was considered to be the type of crude oil that had an API gravity less than 20°. For example, an API gravity equal to 12° signifies a heavy oil, while extra heavy oil and tar sand bitumen usually have an API gravity less than 10° (e.g., Athabasca bitumen = 8° API). Residua would vary depending upon the temperature at which distillation was terminated, but usually vacuum residua are in the range 2°–8° API (Speight, 2000, 2014a; Speight and Ozum, 2002; Parkash, 2003; Hsu and Robinson, 2006; Gary et al., 2007). The term *heavy oil* has also been used collectively to describe both the heavy oils that require thermal stimulation of recovery from the reservoir and the bitumen in bituminous sand formations from which the viscous bituminous material is recovered by a mining operation (Speight, 2013b,c, 2014a). Convenient as this may be, it is scientifically and technically incorrect.

Heavy oil is a *type* of petroleum that is different from conventional petroleum insofar as they are much more difficult to recover from the subsurface reservoir. Heavy oil, particularly heavy oil formed by the biodegradation of organic deposits, is found in shallow reservoirs, formed by unconsolidated sands. This characteristic, which causes difficulties during well drilling and completion operations, may become a production advantage due to higher permeability. In simple terms, heavy oil is a type of crude oil that is very viscous and does not flow easily. The common characteristic properties (relative to conventional crude oil) are high specific gravity, low hydrogen-to-carbon ratios, high carbon residues, and high contents of asphaltenes, heavy metal, sulfur, and nitrogen. Specialized refining processes are required to produce more useful fractions, such as naphtha, kerosene, and gas oil.

Thus, when petroleum occurs in a reservoir that allows the crude material to be recovered by pumping operations as a free-flowing dark to light colored liquid, it is often referred to as "conventional petroleum." Heavy oils are the other *types* of petroleum that are different from conventional petroleum insofar as they are much more difficult to recover from the subsurface reservoir. The definition of heavy oil is usually based on the API gravity or viscosity, and the definition is quite arbitrary although there have been attempts to rationalize the definition based upon viscosity, API gravity, and density.

There are large resources of *heavy oil* in Canada, Venezuela, Russia, the United States, and many other countries. The resources in North America alone provide a small percentage of current oil production (approximately 2%), existing commercial technologies could allow for significantly increased production. Under current economic conditions, heavy oil can be profitably produced, but at a smaller profit margin than for conventional oil, due to higher production costs and upgrading costs in conjunction with the lower market price for heavier crude oils. In fact, heavy oil accounts for more than double the resources of conventional oil in the world and heavy oil offers the potential to satisfy current and future oil demand. Not surprisingly, heavy oil has become an important theme in the petroleum industry with an increasing number of operators getting involved or expanding their plans in this market around the world.

However, heavy oil is more difficult to recover from the subsurface reservoir than conventional or light oil. A very general definition of heavy oils has been and remains based on the API gravity or viscosity, and the definition is quite arbitrary although there have been attempts to rationalize the definition based upon viscosity, API gravity, and density. For example, heavy oils were considered to be those crude oils that had gravity somewhat less than 20° API with the heavy oils falling into the API gravity range 10°–15°. For example, Cold Lake heavy crude oil has an API gravity equal to 12°, and tar sand bitumen usually has an API gravity in the range 5°–10° (Athabasca bitumen = 8° API).

Extra heavy oil is a nondescript term (related to viscosity) of little scientific meaning that is usually applied to tar sand bitumen, which is generally incapable of free flow under reservoir conditions. The general difference is that extra heavy oil, which may have properties similar to tar sand bitumen in the laboratory but, unlike tar sand bitumen in the deposit, has some degree of mobility in the reservoir or deposit (Table 1.2) (Delbianco and Montanari, 2009; Speight, 2014a). Extra heavy oils can flow at reservoir temperature and can be produced economically, without additional viscosity reduction techniques, through variants of conventional processes such as long horizontal wells or multilaterals. This is the case, for instance, in the Orinoco Basin (Venezuela) or in offshore reservoirs of the coast of Brazil but, once outside of the influence of the high reservoir temperature, these oils are too viscous at surface to be transported through conventional pipelines and require heated

TABLE 1.2

Simplified Differentiation between Conventional Crude Oil, Tight Oil, Heavy Crude Oil, Extra Heavy Crude Oil, and Tar Sand Bitumen[a]

Conventional crude oil
Mobile in the reservoir; API gravity, >25°
High-permeability reservoir
Primary recovery
Secondary recovery

Tight oil
Similar properties to the properties of conventional crude oil; API gravity, >25°
Immobile in the reservoir
Low-permeability reservoir
Horizontal drilling into reservoir
Fracturing (typically multifracturing) to release fluids/gases

Heavy crude oil
More viscous than conventional crude oil; API gravity, 10°–20°
Mobile in the reservoir
High-permeability reservoir
Secondary recovery
Tertiary recovery (enhanced oil recovery [EOR]; e.g., steam stimulation)

Extra heavy crude oil
Similar properties to the properties of tar sand bitumen; API gravity, <10°
Mobile in the reservoir
High-permeability reservoir
Secondary recovery
Tertiary recovery (EOR; e.g., steam stimulation)

Tar sand bitumen
Immobile in the deposit; API gravity, <10°
High-permeability reservoir
Mining (often preceded by explosive fracturing)
Steam-assisted gravity draining (SAGD)
Solvent methods (VAPEX)
Extreme heating methods
Innovative methods[b]

[a]   This list is not intended for use as a means of classification.
[b]   Innovative methods exclude tertiary recovery methods and methods such as steam-assisted gravity drainage (SAGD) and vapor-assisted extraction (VAPEX) methods but does include variants or hybrids thereof (Speight, 2016).

pipelines for transportation. Alternatively, the oil must be partially upgraded or fully upgraded or diluted with a light hydrocarbon (such as aromatic naphtha) to create a mix that is suitable for transportation (Speight, 2014a).

In the context of this book, the methods outlined in this book for heavy oil refining focus on heavy oil with an API gravity of less than 20 with a variable sulfur content (Table 1.3) (Speight, 2000). However, it must be recognized that some of the heavy oil sufficiently liquid to be recovered

TABLE 1.3
API Gravity and Sulfur Content of Selected Heavy Oils

Country	Crude Oil	API	Sulfur (% w/w)
Brazil	Albacor-Leste	18.9	0.66
Canada (Alberta)	Athabasca	8.0	4.8
Canada (Alberta)	Cold Lake	13.2	4.11
Canada (Alberta)	Lloydminster	16.0	2.60
Canada (Alberta)	Wabasca	19.6	3.90
Chad	Bolobo	16.8	0.14
Chad	Kome	18.5	0.20
China	Bozhong	16.7	0.30
China	Qinhuangdao	16.00	0.26
China	Zhao Dong	18.4	0.25
Colombia	Castilla	13.3	0.22
Colombia	Chichimene	19.8	1.12
Congo	Yombo	17.7	0.33
Ecuador	Ecuador Heavy	18.2	2.23
Ecuador	Napo	19.2	1.98
Guatemala	Xan-Coban	18.7	6.00
Indonesia	Kulin (South)	19.8	0.30
Iran	Soroosh (Cyrus)	18.1	3.30
Kuwait	Eocene	18.4	4.00
United States (California)	Arroyo Grande/Edna	14.9	2.03
United States (California)	Belridge (South)	13.7	1.00
United States (California)	Beta Offshore	15.9	3.60
United States (California)	Beta Offshore	16.9	3.30
United States (California)	Hondo Monterey	19.4	4.70
United States (California)	Hondo Monterey	17.2	4.70
United States (California)	Huntington Beach	19.4	2.00
United States (California)	Huntington Beach	14.4	0.90
United States (California)	Kern River	13.3	1.10
United States (California)	Kern River	14.4	1.02
United States (California	Lost Hills	18.4	1.00
United States (California)	Midway Sunset	12.6	1.60
United States (California)	Midway Sunset	11.0	1.55
United States (California)	Monterey	12.2	2.30
United States (California)	Mount Poso	16.0	0.70
United States (California)	Newport Beach	15.1	2.00
United States (California)	Point Arguello	19.5	3.50
United States (California)	Point Pedernales	15.9	5.10
United States (California)	San Ardo	12.2	2.30
United States (California)	San Joaquin Valley	15.7	1.20
United States (California)	Santa Maria	13.7	5.20
United States (California)	Sockeye	19.6	5.30
United States (California)	Sockeye	15.9	5.40
United States (California)	Torrance	18.2	1.80
United States (California)	Wilmington	18.6	1.59
United States (California)	Wilmington	16.9	1.70
United States (Mississippi)	Baxterville	16.3	3.02

(Continued)

TABLE 1.3 (*Continued*)
API Gravity and Sulfur Content of Selected Heavy Oils

Country	Crude Oil	API	Sulfur (% w/w)
Venezuela	Bachaquero	16.3	2.35
Venezuela	Bachaquero	12.2	2.80
Venezuela	Bachaquero	14.4	2.52
Venezuela	Bachaquero Heavy	10.7	2.78
Venezuela	Boscan	10.1	5.50
Venezuela	Hamaca	8.4	3.82
Venezuela	Jobo	9.2	4.10
Venezuela	Laguna	10.9	2.66
Venezuela	Lagunillas Heavy	17.0	2.19
Venezuela	Merey	18.0	2.28
Venezuela	Morichal	12.2	2.78
Venezuela	Pilon	14.1	1.91
Venezuela	Tia Juana Pesado	12.1	2.70
Venezuela	Tia Juana Heavy	18.2	2.24
Venezuela	Tia Juana Heavy	11.6	2.68
Venezuela	Tremblador	19.0	0.80
Venezuela	Zuata	15.7	2.69

Note: For reference, Athabasca tar sand bitumen has API = 8° and sulfur content = 4.8%–5.0% w/w.

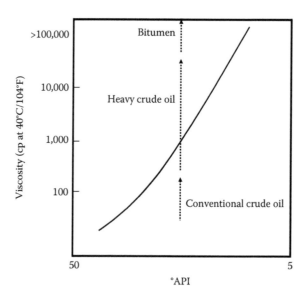

FIGURE 1.1 General relationship of viscosity to API gravity.

by pumping operations and are already being recovered by this method. Refining depends on the characteristics (properties) since these heavy oils fall into a range of high viscosity (Figure 1.1) and the viscosity is subject to temperature effects (Figure 1.2), which is the reason for the application of thermal conditions or dilution with a suitable solvent (such as aromatic naphtha) to enable heavy oil to flow in pipeline or within the refinery system.

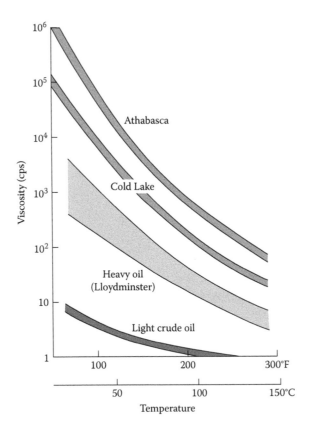

FIGURE 1.2 Variation of viscosity with temperature.

1.2.3 TAR SAND BITUMEN

In addition to conventional petroleum and heavy crude oil, there remains an even more viscous material that offers some relief to the potential shortfalls in supply (Meyer and De Witt, 1990; BP, 2015). This is the *bitumen* found in *tar sand* (*oil sand*) deposits. However, many of these reserves are only available with some difficulty and optional refinery scenarios will be necessary for conversion of these materials to liquid products (Speight, 2000, 2014a) because of the substantial differences in character between conventional petroleum and tar sand bitumen (Table 1.4). "Tar sands," also variously called "oil sands" or "bituminous sands," are a loose-to-consolidated sandstone or a porous carbonate rock, impregnated with bitumen, a heavy asphaltic crude oil with an extremely high viscosity under reservoir conditions.

The term *tar sand bitumen* (also, on occasion referred to as "extra heavy oil" and "native asphalt," although the latter term is incorrect) includes a wide variety of reddishbrown to black materials of near-solid to solid character that exist in nature either with no mineral impurity or with mineral matter contents that exceed 50% by weight. Bitumen is frequently found filling pores and crevices of sandstone, limestone, or argillaceous sediments, in which case the organic and associated mineral matrix is known as "rock asphalt."

Bitumen is also a naturally occurring material that is found in deposits that are incorrectly referred to as "tar sand" since tar is a product of the thermal processing of coal (Speight, 2013a). The permeability of a tar sand deposit low and passage of fluids through the deposit can only be achieved by prior application of fracturing techniques. Alternatively, bitumen recovery can be achieved by the conversion of the bitumen to a product *in situ* (*in situ* upgrading) followed by product recovery

TABLE 1.4

Comparison of the Properties of Crude Oil and Tar Sand Bitumen

Property	Bitumen	Conventional
Gravity, °API	8.6	25–37
Distillation, °F		
Vol%		
IBP	—	
5	430	
10	560	—
30	820	
50	1,010	<650
Viscosity		
SUS, 100°F (38°C)	35,000	<30
SUS, 210°F (99°C)	513	
Pour point, °F	+50	
Elemental analysis, wt%		
Carbon	83.1	86
Hydrogen	10.6	13.5
Sulfur	4.8	0.1–2.0
Nitrogen	0.4	0.2
Oxygen	1.1	
Hydrocarbon type, wt%		
Asphaltenes	19	<10
Resins	32	
Oils	49	>60
Metals, ppm		
Vanadium	250	
Nickel	100	
Iron	75	2–10
Copper	5	
Ash. wt%	0.75	0
Conradson carbon, wt%	13.5	1–2
Net heating value, Btu/lb	17,500	~19,500

from the deposit (Speight, 2013b,c, 2014a, 2016). Tar sand bitumen is a high-boiling material with little, if any, material boiling below 350°C (660°F), and the boiling range approximates the boiling range of an atmospheric residuum.

There have been many attempts to define tar sand deposits and the bitumen contained therein. In order to define conventional petroleum, heavy oil, and bitumen, the use of a single physical parameter such as viscosity is not sufficient. Other properties such as API gravity, elemental analysis, composition, and, most of all, the properties of the bulk deposit must also be included in any definition of these materials. Only then will it be possible to classify petroleum and its derivatives.

In fact, the most appropriate definition of *tar sands* is found in the writings of the U.S. government, namely:

> Tar sands are the several rock types that contain an extremely viscous hydrocarbon which is not recoverable in its natural state by conventional oil well production methods including currently used enhanced recovery techniques. The hydrocarbon-bearing rocks are variously known as bitumen-rocks oil, impregnated rocks, oil sands, and rock asphalt.

This definition speaks to the character of the bitumen through the method of recovery. Thus, the bitumen found in tar sand deposits is an extremely viscous material that is *immobile under reservoir conditions* and cannot be recovered through a well by the application of secondary or enhanced recovery techniques. Mining methods match the requirements of this definition (since mining is not one of the specified recovery methods), and the bitumen can be recovered by alteration of its natural state such as thermal conversion to a product that is then recovered. In this sense, changing the natural state (the chemical composition) as occurs during several thermal processes (such as some *in situ* combustion processes) also matches the requirements of the definition.

By inference and by omission, conventional petroleum and heavy oil are also included in this definition. Petroleum is the material that can be recovered by conventional oil well production methods, whereas heavy oil is the material that can be recovered by enhanced recovery methods. Tar sand is currently recovered by a mining process followed by separation of the bitumen by the hot water process. The bitumen is then used to produce hydrocarbons by a conversion process.

The only commercial operations for the recovery of bitumen from tar sand and its subsequent conversion to liquid fuels exist in the Canadian Province of Alberta where Suncor (initially as Great Canadian Oil Sands before the name change) went onstream in 1967 and Syncrude (a consortium of several companies) went onstream in 1977. Thus, throughout this text, there are frequent references to tar sand bitumen, but because commercial operations have been in place for approximately 50 years, it is not surprising that more is known about the Alberta (Canada) tar sand reserves than any other reserves in the world. Therefore, when there is a discussion on tar sand deposits, reference is made to the relevant deposit, but when the information is not available, the Alberta material is used for the purposes of the discussion.

Refining tar sand bitumen depends to a large degree on the composition of the bitumen (Tables 1.5 and 1.6) and the structure of the bitumen (Speight, 2014a). Generally, the bitumen found in tar sand deposits is an extremely viscous material that is *immobile* and cannot be refined by the application of conventional refinery. On the other hand, extra heavy oil, which is often likened to tar sand bitumen because of similarities in the properties of the two, has a degree of mobility under reservoir or deposit conditions but typically suffers from the same drawbacks as the immobile bitumen in refinery operations.

It is incorrect to refer to native bituminous materials as *tar* or *pitch*. Although the word tar is descriptive of the black, heavy bituminous material, it is best to avoid its use with respect to natural materials and to restrict the meaning to the volatile or near-volatile products produced in the destructive distillation of such organic substances as coal (Speight, 2013a). In the simplest sense, pitch is the distillation residue of the various types of tar.

Thus, alternative names, such as *bituminous sand* or *oil sand*, are gradually finding usage, with the former name (bituminous sands) more technically correct. The term *oil sand* is also used in the same way as the term *tar sand*, and these terms are used interchangeably throughout this text.

Bituminous rock and *bituminous sand* are those formations in which the bituminous material is found as a filling in veins and fissures in fractured rocks or impregnating relatively shallow sand, sandstone, and limestone strata. The deposits contain as much as 20% bituminous material, and if the organic material in the rock matrix is bitumen, it is usual (although chemically incorrect) to refer to the deposit as "rock asphalt" to distinguish it from bitumen that is relatively mineral free. A standard test (ASTM D4, 2015) is available for determining the bitumen content of various mixtures with inorganic materials, although the use of word "bitumen" as applied in this test might be questioned and it might be more appropriate to use the term *organic residues* to include *tar* and *pitch*. If the material is the asphaltite type or asphaltoid type, the corresponding terms should be used: "rock asphaltite" or "rock asphaltoid."

Bituminous rocks generally have a coarse, porous structure, with the bituminous material in the voids. A much more common situation is that in which the organic material is present as an inherent part of the rock composition insofar as it is a diagenetic residue of the organic material detritus that

TABLE 1.5
Properties of Bitumen from Different California Tar Sand Deposits

Deposit/Field	Reservoir Temperature (°F)	Reservoir Permeability (mD)	Reservoir Porosity	API	Viscosity (cP)					Sulfur (% w/w)
					Reservoir Conditions	80°F	87°F	100°F	200°F	
Arroyo Grande		700	38	8				15,000		3–5
Basal Foxen		300	25	9–17 (9.5)		47,000				4–5
Cat Canyon				0–12 (6)						
Brooks Sand	135	1400–5000	32	0–12	15,000					
S Sand	110	3450	37		12,000–1,000,000					
Casmalia		<1	48							
Zaca-Sisquoc			35	4–6						
Oxnard (Vaca)		6000	35	5				500,000	2,000	6–7
Paris Valley		3700	32							1.5
Upper Lobe							227,000			
Lower Lobe							23,000			
Midway Sunset								1,650		
Webster Sands	100	1300	28	14						

TABLE 1.6

Specific Gravity, API Gravity, and Viscosity of Various Bitumen Samples

Source	Specific Gravity	API Gravity	Viscosity (cP)	Viscosity (°F)
Athabasca (Canada)				
Mildred-Ruth Lakes	1.025	6.5	35,000	100
Abasand	1.027	6.3	500,000	100
	1.034	5.4	570,000	100
Ells River	1.008	8.9	25,000	
Utah (United States)				
Asphalt Ridge			610,000	140
Tar Sand Triangle			760,000	140
Sunnyside			1,650,000	100
California				
Arroyo Grande	1.055	2.6	1,300,000	220

was deposited with the sediment. The organic components of such rocks are usually refractory and are only slightly affected by most organic solvents.

Tar sand deposits occur throughout the world and the largest deposits occur in Alberta, Canada (the Athabasca, Wabasca, Cold Lake, and Peace River areas), and in Venezuela. Smaller deposits occur in the United States, with the larger individual deposits in Utah, California, New Mexico, and Kentucky. The term *tar sand*, also known as *oil sand* (in Canada), or bituminous sand, commonly describes sandstones or friable sand (quartz) impregnated with a viscous organic material known as *bitumen* (a hydrocarbonaceous material that is soluble in carbon disulfide). Significant amounts of fine material, usually largely or completely clay, are also present. The degree of porosity in tar sand varies from deposit to deposit and is an important characteristic in terms of recovery processes. The bitumen makes up the desirable fraction of the tar sands from which liquid fuels can be derived. However, the bitumen is usually not recoverable by conventional petroleum production techniques (Speight, 2013b,c, 2014a, 2016).

The properties and composition of the tar sands and the bitumen significantly influence the selection of recovery and the bitumen conversion processes and vary among the bitumen from different deposits. In the so-called wet sands or water-wet sands of the Athabasca deposit, a layer of water surrounds the sand grain, and the bitumen partially fills the voids between the wet grains. Utah tar sands lack the water layer; the bitumen is directly in contact with the sand grains without any intervening water; such tar sands are sometimes referred to as oil-wet sands. Typically, more than 99% w/w of mineral matter is composed of quartz and clays—the latter are detrimental to most refining processes. The general composition of typical deposits at the P.R. Spring Special Tar Sand Area showed a porosity of 8.4 vol% with the solid/liquid fraction being 90.5% w/w sand, 1.5% w/w fines, 7.5% w/w bitumen, and 0.5% w/w water. Utah deposits range from largely consolidated sands with low porosity and permeability to, in some cases, unconsolidated sand. High concentrations of heteroatoms (nitrogen, oxygen, sulfur, and metals) tend to increase viscosity, increase the bonding of bitumen with minerals, reduce yields, and make processing more difficult.

Additionally, the term *extra heavy oil* has been used to describe materials that occur in the solid or near-solid state in the deposit or reservoir and are generally incapable of free flow under ambient conditions (bitumen, *q.v.*). Whether such a material exists in the near-solid or solid state in the reservoir can be determined from the pour point and the reservoir temperature.

Briefly, *extra heavy oil is* a material that occurs in the solid or near-solid state and is generally has mobility under reservoir conditions. However, the term *extra heavy oil* is a recently evolved term (related to viscosity) of little scientific meaning. While this type of oil may resemble tar sand bitumen and does not flow easily, extra heavy oil is generally recognized as having mobility in the reservoir compared to tar sand bitumen, which is typically incapable of mobility (free flow) under reservoir conditions. For example, the tar sand bitumen located in Alberta, Canada is not mobile in the deposit and requires extreme methods of recovery to recover the bitumen. On the other hand, much of the extra heavy oil located in the Orinoco Basin of Venezuela requires recovery methods that are less extreme because of the mobility of the material in the reservoir. Whether the mobility of extra heavy oil is due to a high reservoir temperature (i.e., higher than the pour point of the extra heavy oil) or due to other factors is variable and subject to localized conditions in the reservoir. This may also be reflected in the choice of suitable extra heavy oil or bitumen conversion processes in the refinery.

In order to utilize the extra heavy oil produced in Venezuela, the government-run oil company PDVSA has developed Orimulsion fuel, which is a dispersion of extra heavy oil and approximately 30% water which is targeted for use as a boiler fuel in applications such as power generation and industrial use. However, concerns about the environmental impact of the use of Orimulsion have been raised owing to the relatively high levels of sulfur, nickel, and vanadium compared with other fuel oils. Technologies such as *selective catalytic reduction, flue gas desulfurization*, and *electrostatic precipitation* are suitable for cleanup of the exhaust emissions (Mokhatab et al., 2006; Speight, 2007).

Finally, because of the potential relevance to refining in the future (Chapter 17) (Speight, 2011a), there is the need to define *oil shale*, which is the term applied to a class of bituminous rocks that has achieved some importance. Oil shale does not contain oil (Scouten, 1990; Lee, 1991; Speight, 2012b). It is an argillaceous, laminated sediment of generally high organic content (*kerogen*) that can be thermally decomposed to yield appreciable amounts of a hydrocarbon-based oil that is commonly referred to as *shale oil*. Oil shale does not yield shale oil without the application of high temperatures and the ensuing thermal decomposition that is necessary to decompose the organic material (*kerogen*) in the shale.

1.2.4 Kerogen

Kerogen is the complex carbonaceous macromolecular (organic) material that occurs in sedimentary rocks and shale. It is not, by any definition, a member of the crude oil family of the bitumen family. Because kerogen produces oil on high-temperature (>500°C, >830°F) thermal decomposition (Scouten, 1990; Lee, 1991; Speight, 2014a), it is believed by some to be the precursor to petroleum but it may also be a by-product of the petroleum maturation process (Speight, 2014a). The production of a petroleum-type product on thermal decomposition of the source material is not always an indicator of petroleum source material. It is included here because kerogen is the source of shale oil (Scouten, 1990; Lee, 1991; Speight, 2012b) that could, at some future time, be a feedstock for refinery operations (Speight, 2011a).

Kerogen is, for the most part, insoluble in the common organic solvents and produces distillable hydrocarbons when subjected to a heat (Scouten, 1990; Lee, 1991; Speight, 2012b). When the kerogen occurs in shale, the whole material is often referred to as "oil shale." Kerogen is not the same as the bitumen that occurs in tar sand deposits (Scouten, 1990; Lee, 1991; Speight, 2012b, 2014a). A *synthetic crude oil* can be produced from oil shale kerogen by thermal decomposition, typically at temperatures above 500°C (930°F) at which the kerogen is thermally decomposed (cracked) to produce the lower-molecular-weight distillable products (Scouten, 1990; Lee, 1991; Speight, 2012b). Kerogen is also reputed to be a precursor of petroleum but this concept and the actual maturation pathway of the kerogen to petroleum is still the subject of considerable speculation and debate (Speight, 2014a).

For comparison with tar sand, *oil shale* is any fine-grained sedimentary rock containing solid organic matter (*kerogen*) that yields a hydrocarbon oil when heated. Oil shale varies in mineral composition. For example, clay minerals predominate in true shale, while other minerals (e.g., dolomite and calcite) occur in appreciable but subordinate amounts in the carbonates. In all shale types, layers of the constituent mineral alternate with layers of kerogen.

1.2.5 Biomass

A gasification refinery (Chapter 14) and the future refinery (Chapter 17) could well involve the incorporation of biomass into refinery feedstocks. "Biomass" (also referred to as "bio-feedstocks") refers to living and recently dead biological material that can be used as fuel or for industrial production (Wright et al., 2006; Speight, 2008, 2011b; Demirbaş, 2010; Lorenzini et al., 2010; Nersesian, 2010; Klaas et al., 2015; Syngellakis, 2015a,b).

Although it is projected (Speight, 2011a) that petroleum will be the refinery feedstock of choice for the next 50 years, reducing the national dependence on imported crude oil is of critical importance for long-term security and continued economic growth. Supplementing petroleum consumption with renewable biomass resources is a first step toward this goal. The realignment of the chemical industry from one of petrochemical refining to a biorefinery concept that is, given time, feasible has become a national goal of many oil importing countries. However, clearly defined goals are necessary for increasing the use of biomass-derived feedstocks in industrial chemical production, and it is important to keep the goal in perspective. In this context, the increased use of biofuels should be viewed as one of a range of possible measures for achieving self-sufficiency in energy, rather than a panacea (Crocker and Crofcheck, 2006; Langeveld at al., 2010), although there are arguments against the rush to the large-scale production of biofuels (Giampietro and Mayumi, 2009).

Biomass is carbon based and is composed of a mixture of organic molecules containing hydrogen, usually including atoms of oxygen, often nitrogen, and also small quantities of other atoms, including alkali metals, alkaline earth metals, and heavy metals. These metals are often found in functional molecules such as the porphyrin molecules, which include chlorophyll that contains magnesium. Unlike petroleum, heavy oil, and tar sand bitumen, biomass is a *renewable* energy source and, as a result, biomass has a high potential to play role in the production of liquid fuels in the future and could, more than likely, be an additional feedstock to a refinery.

Biomass is a term used to describe any material of recent biological origin, including plant materials such as trees, grasses, agricultural crops, and even animal manure. Other biomass components, which are generally present in minor amounts, include triglyceride derivatives, sterol derivatives, alkaloid derivatives, terpene derivatives, terpenoid derivatives, and waxes. This includes everything from *primary sources* of crops and residues harvested/collected directly from the land to *secondary sources* such as sawmill residuals, to *tertiary sources* of postconsumer residuals that often end up in landfills. A *fourth source*, although not usually categorized as such, includes the gases that result from anaerobic digestion of animal manures or organic materials in landfills (Wright et al., 2006). Generally, there are four distinct sources of biomass that can be converted to energy: (1) agricultural crops, (2) wood, (3) municipal and industrial wastes, and (4) landfill waste that produced landfill gas (Speight, 2008, 2011b).

Primary biomass is produced directly by photosynthesis and includes all terrestrial plants now used for food, feed, fiber, and fuel wood. All plants in natural and conservation areas (as well as algae and other aquatic plants growing in ponds, lakes, oceans, or artificial ponds and bioreactors) are also considered primary biomass. However, only a small portion of the primary biomass produced will ever be harvested as feedstock material for the production of bioenergy and by-products.

Secondary biomass feedstocks differ from primary biomass feedstocks in that the secondary feedstocks are a by-product of processing of the primary feedstocks. By *processing*, it is meant that there is substantial physical or chemical breakdown of the primary biomass and production of by-products; *processors* may be factories or animals. Field processes such as harvesting, bundling, chipping, or pressing do not cause a biomass resource that was produced by photosynthesis (e.g., tree tops and limbs) to be classified as secondary biomass. Specific examples of secondary biomass include sawdust from sawmills, black liquor (which is a by-product of papermaking), and cheese whey (which is a by-product of cheese-making processes). Manures from concentrated animal feeding operations are collectable secondary biomass resources. Vegetable oils used for biodiesel that are derived directly from the processing of oilseeds for various uses are also a secondary biomass resource.

Tertiary biomass feedstock includes postconsumer residues and wastes, such as fats, greases, oils, construction and demolition wood debris, other waste wood from the urban environments, as well as packaging wastes, municipal solid wastes, and landfill gases. A category *other wood waste from the urban environment* includes trimmings from urban trees, which technically fits the definition of primary biomass. However, because this material is normally handled as a waste stream along with other postconsumer wastes from urban environments (and included in those statistics), it makes the most sense to consider it to be part of the tertiary biomass stream.

As expected from the earlier descriptions, biomass feedstocks exhibit a wide range of physical, chemical, and agricultural/process engineering properties. Despite their wide range of possible sources, biomass feedstocks are remarkably uniform in many of their fuel properties, compared with competing feedstocks such as coal or petroleum. Approximately 6% of contiguous U.S. land area put into cultivation for biomass could supply all current demands for oil and gas. And this production would not add any net carbon dioxide to the atmosphere.

Dried biomass has a heating value of 5000–8000 Btu/lb. with virtually no ash or sulfur produced during combustion. However, nearly all kinds of biomass feedstocks destined for combustion fall in the range of 6450–8200 Btu/lb. For most agricultural residues, the heating values are even more uniform—approximately 6450–7300 Btu/lb; the values for woody materials are on the order of 7750–8200 Btu/lb. Moisture content is probably the most important determinant of heating value. Air-dried biomass typically has approximately 15%–20% moisture, whereas the moisture content for oven-dried biomass is around 0%. The bulk density (and hence energy density) of most biomass feedstocks is generally low, even after densification, approximately 10% and 40% of the bulk density of most fossil fuels. Liquid biofuels have comparable bulk densities to fossil fuels. Finally, many types of biomass do contain high amounts of mineral matter (translated as mineral ash during analysis), and the amount can vary from a fraction of 1% w/w to as much as 30% w/w. As is well known in the petroleum refining industry, mineral matter can be a *catalyst killer* in catalyst-based processes.

Biofuel is any fuel derived from biomass (Speight, 2008, 2011b) and has the potential to produce fuels that are more environmentally benign than petroleum-based fuels (Speight, 2008). Biofuel is a renewable energy source, unlike other natural resources, such as petroleum, and like petroleum, biomass is a form of stored energy. The production of biofuels to replace oil and natural gas is in active development, focusing on the use of cheap organic matter (usually cellulose, agricultural, and sewage waste) in the efficient production of liquid and gas biofuels that yield high net energy gain. One advantage of biofuel over most other fuel types is that it is biodegradable and is relatively harmless to the environment if spilled.

Direct biofuels are biofuels that can be used in existing unmodified petroleum engines. Because engine technology changes all the time, direct biofuel can be hard to define; a fuel that works well in one unmodified engine may not work in another. In general, newer engines are more sensitive to fuel than older engines, but new engines are also likely to be designed with some amount of biofuel in mind. Straight vegetable oil can be used in many older diesel engines (equipped with indirect injection system) but only in the warmest climates. In reality, small amounts of biofuel (such as bioethanol) are often blended with traditional fuels. The biofuel portion of these fuels is a direct replacement for the fuel they offset, but the total offset is small. For biodiesel, 5% or 20% v/v are commonly approved by various engine manufacturers.

The reserves of biomass are unlimited insofar as biomass resources are renewable and available each year unlike nonrenewable resources (such as the fossil fuels) that are eventually depleted

(Speight, 2008, 2011b). While forests and grasslands, say, could be depleted, if managed properly, they do represent a continuous supply of energy—compared to the fossil fuels that, once they have been exhausted, are no longer available.

The numbers illustrating the amount of available biomass are, at best, estimates and do not give any indications of the true (almost inexhaustible) amounts of material available. However, the production of biomass and biofuels brings up the food crops versus fuel crops concept, and there must be plans to ensure that local, regional, and national food needs will be met prior to shifting crop acreage into bioenergy feedstocks. This can place limitations on the amount of land available to produce energy crops but does leave the door open to use as much waste as possible for energy production.

1.3 NATURAL GAS

Natural gas is a gaseous mixture, which is predominantly methane but does contain other combustible hydrocarbon compounds as well as nonhydrocarbon compounds (Table 1.7) (Mokhatab et al., 2006; Speight, 2007). Natural gas is colorless, odorless, tasteless, shapeless, and lighter than air. In the natural state, it is not possible to see or smell natural gas. In addition to composition and thermal content (Btu/scf, Btu/ft^3), natural gas can also be characterized on the basis of the mode of the natural gas found in reservoirs where there is no or, at best only minimal amounts of, crude oil.

Other constituents are paraffinic hydrocarbons such as ethane (CH_3CH_3), propane ($CH_3CH_2CH_3$), and the butanes (C_4H_{10}). Many natural gases contain nitrogen (N_2) as well as carbon dioxide (CO_2) and hydrogen sulfide (H_2S). Trace quantities of argon, hydrogen, and helium may also be present. Generally, the hydrocarbons having a higher molecular weight than methane, carbon dioxide, and hydrogen sulfide are removed from natural gas prior to its use as a fuel. However, since the composition of natural gas and refinery gas is never constant, there are standard test methods that can be used to determine the suitability of natural gas (and refinery gas) for further use and indicate the processes by which the composition of natural gas can be prepared for use (Table 1.8) (Mokhatab et al., 2006; Speight, 2007, 2014a, 2015b).

TABLE 1.7
Range of Composition of Natural Gas

Category	Component	Amount (%)
Paraffinic	Methane (CH_4)	70–98
	Ethane (C_2H_6)	1–10
	Propane (C_3H_6)	Trace–5
	Butane (C_4H_{10})	Trace–2
	Pentane (C_5H_{12})	Trace–1
	Hexane (C_6H_{14})	Trace–0.5
	Heptane and higher (C_{7+})	None–trace
Cyclic	Cyclopropane (C_3H_6)	Traces
	Cyclohexane (C_6H_{12})	Traces
Aromatic	Benzene (C_6H_6), others	Traces
Nonhydrocarbon	Nitrogen (N_2)	Trace–15
	Carbon dioxide (CO_2)	Trace–1
	Hydrogen sulfide (H_2S)	Trace occasionally
	Helium (He)	Trace–5
	Other sulfur and nitrogen compounds	Trace occasionally
	Water (H_2O)	Trace–5

TABLE 1.8

Standard Test Methods That Can Be Applied to Determining Gas Properties[a]

ASTM D1070 Standard Test Methods for Relative Density of Gaseous Fuels

ASTM D1071 Standard Test Methods for Volumetric Measurement of Gaseous Fuel Samples

ASTM D1072 Standard Test Method for Total Sulfur in Fuel Gases

ASTM D1142 Standard Test Method for Water Vapor Content of Gaseous Fuels by Measurement of Dew-Point Temperature

ASTM D1826 Standard Test Method for Calorific (Heating) Value of Gases in Natural Gas ASTM D1945 Standard Test Method for Analysis of Natural Gas by Gas Chromatography

ASTM D1946 Standard Practice for Analysis of Reformed Gas by Gas Chromatography

ASTM D1988 Standard Test Method for Mercaptans in Natural Gas Using Length-of-Stain Detector Tubes

ASTM D3588 Standard Practice for Calculating Heat Value, Compressibility Factor, and Relative Density of Gaseous Fuels

ASTM D3956 Standard Specification for Methane Thermophysical Property Tables

ASTM D3984 Standard Specification for Ethane Thermophysical Property Tables

ASTM D4084 Standard Test Method for Analysis of Hydrogen Sulfide in Gaseous Fuels (Lead Acetate Reaction Rate Method)

ASTM D4150 Standard Terminology Relating to Gaseous Fuels

ASTM D4362 Standard Specification for Propane Thermophysical Property Tables

ASTM D4468 Standard Test Method for Total Sulfur in Gaseous Fuels by Hydrogenolysis and Rateometric Colorimetry

ASTM D4650 Standard Specification for Normal Butane Thermophysical Property Tables

ASTM D4651 Standard Specification for *Iso*-Butane Thermophysical Property Tables

ASTM D4784 Standard for LNG Density Calculation Models

ASTM D4810 Standard Test Method for Hydrogen Sulfide in Natural Gas Using Length-of-Stain Detector Tubes

ASTM D4888 Standard Test Method for Water Vapor in Natural Gas Using Length-of-Stain Detector Tubes

ASTM D4891 Standard Test Method for Heating Value of Gases in Natural Gas Range by Stoichiometric Combustion

ASTM D4984 Standard Test Method for Carbon Dioxide in Natural Gas Using Length-of-Stain Detector Tubes

ASTM D5287 Standard Practice for Automatic Sampling of Gaseous Fuels

ASTM D5454 Standard Test Method for Water Vapor Content of Gaseous Fuels Using Electronic Moisture Analyzers

ASTM D5503 Standard Practice for Natural Gas Sample-Handling and Conditioning Systems for Pipeline Instrumentation

ASTM D5504 Standard Test Method for Determination of Sulfur Compounds in Natural Gas and Gaseous Fuels by Gas Chromatography and Chemiluminescence

ASTM D5954 Standard Test Method for Mercury Sampling and Measurement in Natural Gas by Atomic Absorption Spectroscopy

ASTM D6228 Standard Test Method for Determination of Sulfur Compounds in Natural Gas and Gaseous Fuels by Gas Chromatography and Flame Photometric Detection

ASTM D6273 Standard Test Methods for Natural Gas Odor Intensity

ASTM D6350 Standard Test Method for Mercury Sampling and Analysis in Natural Gas by Atomic Fluorescence Spectroscopy

Note: Listed numerically rather than by order of importance.

[a] ASTM D4, *Test Method for Bitumen Content*, Annual Book of Standards, ASTM International, West Conshohocken, PA, 2015.

1.3.1 PETROLEUM-RELATED GAS

The generic term "natural gas" applies to gas commonly associated with petroliferous (petroleum-producing, petroleum-containing) geologic formations (Figure 1.3). Natural gas generally contains high proportions of methane (CH_4), and some of the higher-molecular-weight paraffins (C_nH_{2n+2}) generally containing up to six carbon atoms may also be present in small quantities.

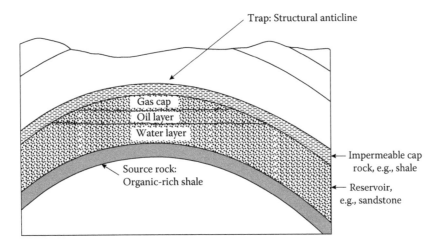

FIGURE 1.3 Schematic of a petroleum reservoir showing the gas cap.

The hydrocarbon constituents of natural gas are combustible, but nonflammable nonhydrocarbon components such as carbon dioxide, nitrogen, and helium are often present in the minority and are regarded as contaminants. In addition to the gas found associated with petroleum in reservoirs, there are also those reservoirs in which natural gas may be the sole occupant. And just as petroleum can vary in composition, so can natural gas.

Natural gas is often located in the same reservoir as with petroleum, but it can also be found trapped in gas reservoirs and within coal deposits. The occurrence of methane in coal seams is not a new discovery and methane (called "firedamp" by the miners because of its explosive nature) was known to coal miners for at least 150 years (or more) before it was *rediscovered* and developed as coalbed methane (Speight, 2013a). The natural gas can originate by thermogenic alteration of coal or by biogenic action of indigenous microbes on the coal. There are some horizontally drilled coalbed methane wells and some that receive hydraulic fracturing treatments. However, some coalbed methane reservoirs are also underground sources of drinking water, and as such, there are restrictions on hydraulic fracturing operations. The coalbed methane wells are mostly shallow, as the coal matrix does not have the strength to maintain porosity under the pressure of significant overburden thickness.

In addition to defining natural gas as *associated* and *nonassociated*, the types of natural gas vary according to composition. There is *dry gas* or *lean gas*, which is mostly methane, and *wet gas*, which contains considerable amounts of higher-molecular-weight and higher-boiling hydrocarbons (Table 1.9). *Sour gas* contains high proportions of much hydrogen sulfide, whereas *sweet gas* contains little or no hydrogen sulfide. *Residue gas* is the gas remaining (mostly methane) after the higher-molecular-weight paraffins have been extracted. *Casinghead gas* is the gas derived from an oil well by extraction at the surface. Natural gas has no distinct odor and the main use is for fuel, but it can also be used to make chemicals and liquefied petroleum gas (LPG).

Some natural gas wells also produce helium, which can occur in commercial quantities; nitrogen and carbon dioxide are also found in some natural gases. Gas is usually separated at as high a pressure as possible, reducing compression costs when the gas is to be used for gas lift or delivered to a pipeline. After gas removal, lighter hydrocarbons and hydrogen sulfide are removed as necessary to obtain a crude oil of suitable vapor pressure for transport yet retaining most of the natural gasoline constituents.

The nonhydrocarbon constituents of natural gas can be classified as two types of materials: (1) diluents, such as nitrogen, carbon dioxide, and water vapors, and (2) contaminants, such as hydrogen sulfide and/or other sulfur compounds. The diluents are noncombustible gases that reduce the

TABLE 1.9
**Range of Composition for *Wet* and *Dry*
Natural Gas**

Constituents	Composition (vol%)		
	Wet	Dry	Range
Hydrocarbons			
Methane	84.6	96.0	
Ethane	6.4	2.0	
Propane	5.3	0.6	
Iso-Butane	1.2	0.18	
n-Butane	1.4	0.12	
Isopentane	0.4	0.14	
n-Pentane	0.2	0.06	
Hexanes	0.4	0.10	
Heptanes	0.1	0.80	
Nonhydrocarbons			
Carbon dioxide			0–5
Helium			0–0.5
Hydrogen sulfide			0–5
Nitrogen			0–10
Argon			0–0.05
Radon, krypton, xenon			Traces

heating value of the gas and are on occasion used as *fillers* when it is necessary to reduce the heat content of the gas. On the other hand, the contaminants are detrimental to production and transportation equipment in addition to being obnoxious pollutants.

Thus, the primary reason for gas processing is to remove the unwanted constituents of natural gas such as (1) acid gas, which is predominantly hydrogen sulfide although carbon dioxide does occur to a lesser extent; (2) water, which includes all entrained free water or water in condensed forms; (3) liquids in the gas, such as higher-boiling hydrocarbons as well as pump lubricating oil, scrubber oil, and, on occasion, methanol; and (4) any solid matter that may be present, such as fine silica (sand) and scaling from the pipe. As with petroleum, natural gas from different wells varies widely in composition and analyses (Table 1.10), and the proportion of nonhydrocarbon constituents can vary over a very wide range. Thus, a particular natural gas field could require production, processing, and handling protocols different from those used for gas from another field.

Just as petroleum was used in antiquity, natural gas was also known in antiquity, although the use of petroleum was relatively better documented because of its use as a mastic for walls and roads as well as for its use in warfare (Abraham, 1945; Pfeiffer, 1950; Van Nes and van Westen, 1951; Forbes, 1958a,b, 1959, 1964; Hoiberg, 1964; Speight, 2014a; Cobb and Goldwhite, 1995). The use of natural gas in antiquity is somewhat less well documented, although historical records indicate that the use of natural gas (for other than religious purposes) dates back to approximately AD 250 when it was used as a fuel in China. The gas was obtained from shallow wells and was distributed through a piping system constructed from hollow bamboo stems. There is other fragmentary evidence for the use of natural gas in certain old texts, but the use is usually inferred since the gas is not named specifically. However, it is known that natural gas was used on a small scale for heating and lighting in northern Italy during the early seventeenth century. From this, it might be conjectured that natural gas found some use from the seventeenth century to the present day, recognizing that gas from coal would be a strong competitor.

TABLE 1.10

Variation in Natural Gas Composition with Source

Component	Type of Gas Field			Natural Gas Separated from Crude Oil, Ventura[a]		
	Dry Gas, Los Medanos[a] (mol%)	Sour Gas, Jumping Pound[b] (mol%)	Gas Condensate, Paloma[a] (mol%)	400 lb (mol%)	50 lb (mol%)	Vapor (mol%)
Hydrogen sulfide	0	3.3	0	0	0	0
Carbon dioxide	0	6.7	0.7	0.3	0.7	0.8
Nitrogen and air	0.8	0	0	0	—	2.2
Methane	95.8	84.0	74.5	89.6	81.8	69.1
Ethane	2.9	3.6	8.3	4.7	5.8	5.1
Propane	0.4	1.0	4.7	3.6	6.5	8.8
Iso-butane	0.1	0.3	0.9	0.5	0.9	2.1
n-Butane	Trace	0.4	1.9	0.9	2.3	5.0
Isopentane	0		0.8	0.2	0.5	1.4
n-Pentane	0		0.6	0.1	0.5	1.4
Hexane	0	0.7	1.3			
Heptane	0			0.1	1.0	4.1
Octane	0		6.3			
Nonane	0					
	100.0	100.0	100.0	100.0	100.0	100.0

[a] In California.

[b] In Canada.

Differences in natural gas composition occur between different reservoirs, and two wells in the same field may also yield gaseous products that are different in composition. Indeed, there is no single composition of components that might be termed *typical* natural gas. Methane and ethane constitute the bulk of the combustible components; carbon dioxide (CO_2) and nitrogen (N_2) are the major noncombustible (inert) components. Other constituents, such as hydrogen sulfide (H_2S), mercaptan derivatives (thiols; R-SH), as well as trace amounts of other sulfur-containing constituents may also be present.

Before the discovery of natural gas, the principal gaseous fuel source was the gas produced by the surface gasification of coal (Speight, 2013a). In fact, each town of any size had a plant for the gasification of coal (hence the use of the term "town gas"). Most of the natural gas produced at the petroleum fields was vented to the air or burned in a flare stack; only a small amount of the natural gas from the petroleum fields was pipelined to industrial areas for commercial use. It was only in the years after World War II that natural gas became a popular fuel commodity, leading to the recognition that it has at the present time.

There are several general definitions that have been applied to natural gas. Thus, *lean* gas is gas in which methane is the major constituent. *Wet* gas contains considerable amounts of the higher-molecular-weight hydrocarbons. *Sour* gas contains hydrogen sulfide, whereas *sweet* gas contains very little, if any, hydrogen sulfide. *Residue gas* is natural gas from which the higher-molecular-weight hydrocarbons have been extracted and *casinghead gas* is derived from petroleum but is separated at the separation facility at the wellhead.

To further define the terms "dry" and "wet" in quantitative measures, the term "dry natural gas" indicates that there is less than 0.1 gal (1 gal, U.S. = 264.2 m³) of gasoline vapor (higher-molecular-weight paraffins) per 1000 ft³ (1 ft³ = 0.028 m³). The term "wet natural gas" indicates that there are such paraffins present in the gas, in fact more than 0.1 gal/1000 ft³. *Associated* or *dissolved* natural gas occurs either as free gas or as gas in solution in the petroleum. Gas that occurs as a solution in the petroleum is *dissolved* gas, whereas the gas that exists in contact with the petroleum (*gas cap*) is *associated* gas.

1.3.2 Gas Hydrates

Methane hydrates, which consist of methane molecules trapped in a cage of water molecules, occur as crystalline solids in sediments in arctic regions and below the floor of the deep ocean. Although taking on the appearance of ice, methane hydrates will burn if ignited. Methane hydrates are the most abundant unconventional natural gas source and the most difficult to extract. Methane hydrates are conservatively estimated to hold twice the amount of energy found in all conventional fossil fuels, but the technical challenges of economically retrieving the resource are significant. There is also a significant risk that rising temperatures from global warming could destabilize the deposits, releasing the methane—a potent greenhouse gas—into the atmosphere and further exacerbating the problem.

Another product is *gas condensate*, which contains relatively high amounts of the higher-molecular-weight liquid hydrocarbons (up to and including octane, C_8H_{18}). These hydrocarbons may occur in the gas phase in the reservoir. On the other hand, natural gasoline (like refinery gasoline) consists mostly of pentane (C_5H_{12}) and higher-molecular-weight hydrocarbons. The term "natural gasoline" has also on occasion in the gas industry been applied to mixtures of liquefied petroleum gas, pentanes, and higher-molecular-weight hydrocarbons. Caution should be taken not to confuse *natural gasoline* with the term "straight-run gasoline" (often also incorrectly referred to as natural gasoline), which is the gasoline distilled unchanged from petroleum.

Liquefied petroleum gas (LPG) is composed of propane (C_3H_8), butanes (C_4H_{10}), and/or mixtures thereof; small amounts of ethane and pentane may also be present as impurities.

1.3.3 Coalbed Methane

In coalbeds (coal seams), methane (the primary component of natural gas) is generally adsorbed to the coal rather than contained in the pore space or structurally trapped in the formation. Pumping the injected and native water out of the coalbeds after fracturing serves to depressurize the coal, thereby allowing the methane to desorb and flow into the well and to the surface. Methane has traditionally posed a hazard to underground coal miners, as the highly flammable gas is released during mining activities. Otherwise inaccessible coal seams can also be tapped to collect this gas, known as coalbed methane, by employing similar well drilling and hydraulic fracturing techniques as are used in shale gas extraction.

Coalbed methane is a gas formed as part of the geological process of coal generation and is contained in varying quantities within all coal. Coalbed methane is exceptionally pure compared to conventional natural gas, containing only very small proportions of higher-molecular-weight hydrocarbons such as ethane and butane and other gases (such as hydrogen sulfide and carbon dioxide). Coalbed gas is over 90% methane and, subject to gas composition, may be suitable for introduction into a commercial pipeline with little or no treatment (Levine, 1993; Rice, 1993; Mokhatab et al., 2006; Speight, 2013a). Methane within coalbeds is not structurally trapped by overlying geologic strata, as in the geologic environments typical of conventional gas deposits (Speight, 2013a, 2014a). Only a small amount (on the order of 5%–10% v/v) of the coalbed methane is present as free gas within the joints and cleats of coalbeds. Most of the coalbed methane is contained within the coal itself (adsorbed to the sides of the small pores in the coal).

The primary (or natural) permeability of coal is very low, typically ranging from 0.1 to 30 mD, and because coal is a very weak (low modulus) material and cannot take much stress without fracturing, coal is almost always highly fractured and cleated. The resulting network of fractures commonly gives coalbeds a high secondary permeability (despite coal's typically low primary permeability).

Groundwater, hydraulic fracturing fluids, and methane gas can more easily flow through the network of fractures. Because hydraulic fracturing generally enlarges preexisting fractures in addition to creating new fractures, this network of natural fractures is very important to the extraction of methane from the coal.

1.3.4 BIOGENIC GAS

Biogenic gas (predominantly methane) is produced by certain types of bacteria (methanogens) during the process of breaking down organic matter in an oxygen-free environment (Speight, 2011b). Thus, biogenic gas is created by methanogenic organisms in marches, bogs, landfills, and shallow sediments. On the other hand, as a point of differentiation, thermogenic gas is created from buried organic material deeper in the earth, at greater temperature and pressure. Livestock manure, food waste, and sewage are all potential sources of biogenic gas, or biogas, which is usually considered a form of renewable energy. Small-scale biogas production is a well-established technology in parts of the developing world, particularly Asia, where farmers collect animal manure in vats and capture the methane given off while it decays.

Landfills offer another underutilized source of biogas (Speight, 2011b). When municipal waste is buried in a landfill, bacteria break down the organic material contained in garbage such as newspapers, cardboard, and food waste, producing gases such as carbon dioxide and methane. Rather than allowing these gases to go into the atmosphere, where they contribute to global warming, landfill gas facilities can capture them, separate the methane, and combust it to generate electricity, heat, or both.

1.4 MANUFACTURED MATERIALS

Manufactured materials are those materials that are produced as a result of petroleum refining (Speight and Ozum, 2002; Parkash, 2003; Hsu and Robinson, 2006; Gary et al., 2007; Speight, 2014a) and, for the purposes of this book, include products such as (1) residua, (2) asphalt, (3) synthetic crude oil, and (4) shale oil. In many cases, these products are subject to further refining after production and are worthy of mention here as optional refinery feedstocks either further refined as single feedstocks or as blends with other feedstocks, such as gas oil, as well as whole crudes, heavy crudes, and tar sand bitumen.

1.4.1 RESIDUUM

A *residuum* (*pl. residua*, also shortened to *resid*, *pl. resids*) is the residue obtained from petroleum after nondestructive distillation has removed all the volatile materials (Figure 1.4). The temperature of the distillation is usually maintained below 350°C (660°F) since the rate of thermal decomposition of petroleum constituents is minimal below this temperature, but the rate of thermal decomposition of petroleum constituents is substantial above 350°C (660°F). If the temperature of the distillation unit rises above 350°C (660°F) as happens in certain units where temperatures up to 395°C (740°F) are known to occur, cracking can be controlled by adjustment of the residence time.

Residua are black, viscous materials and are obtained by distillation of a crude oil under atmospheric pressure (atmospheric residuum) or under reduced pressure (vacuum residuum). They may be liquid at room temperature (generally atmospheric residua) or almost solid (generally vacuum residua) depending upon the nature of the crude oil.

When a residuum is obtained from crude oil and thermal decomposition has commenced, it is more usual to refer to this product as *pitch*. The differences between conventional petroleum and the related residua are due to the relative amounts of various constituents present, which are removed or remain by virtue of their relative volatility.

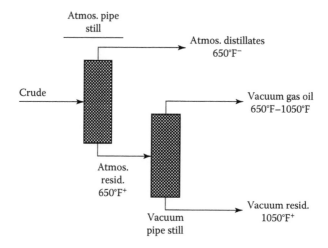

FIGURE 1.4 Simplified crude oil distillation scheme.

The chemical composition of a residuum is complex. Physical methods of fractionation usually indicate high proportions of asphaltenes and resins, even in amounts up to 50% (or higher) of the residuum. In addition, the presence of ash-forming metallic constituents, including such organometallic compounds as those of vanadium and nickel, is also a distinguishing feature of residua and the heavier oils. Furthermore, the deeper the *cut* into the crude oil, the greater is the concentration of sulfur and metals in the residuum and the greater the deterioration in physical properties.

1.4.2 ASPHALT

Asphalt is manufactured from petroleum (Figure 1.5) and is a black or brown material that has a consistency varying from a viscous liquid to a glassy solid (Speight, 2015c). To a point, asphalt may resemble tar sand bitumen, hence the tendency to refer to bitumen (incorrectly) as "native asphalt." It is recommended that there be differentiation between asphalt (manufactured) and bitumen (naturally occurring) other than by the use of the qualifying terms "petroleum" and "native" since the origins of the materials may be reflected in the resulting physicochemical properties of the

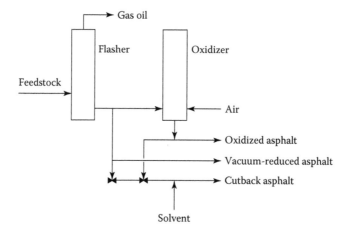

FIGURE 1.5 Simplified representation of asphalt production.

two types of materials. It is also necessary to distinguish between the asphalt that originates from petroleum by refining and the product in which the source of the asphalt is a material other than petroleum, for example, *wurtzilite asphalt*. In the absence of a qualifying word, it is assumed that the term "asphalt" refers to the product manufactured from petroleum.

When the asphalt is produced simply by distillation of an asphaltic crude oil, the product can be referred to as "residual asphalt" or "straight-run asphalt." If the asphalt is prepared by solvent extraction of residua or by light hydrocarbon (propane) precipitation, or if blown or otherwise treated, the term should be modified accordingly to qualify the product (e.g., *propane asphalt, blown asphalt*).

Asphalt softens when heated and is elastic under certain conditions. The mechanical properties of asphalt are of particular significance when it is used as a binder or adhesive. The principal application of asphalt is in road surfacing, which may be done in a variety of ways. Light oil *dust layer* treatments may be built up by repetition to form a hard surface, or a granular aggregate may be added to an asphalt coat, or earth materials from the road surface itself may be mixed with the asphalt.

Other important applications of asphalt include canal and reservoir linings, dam facings, and sea works. The asphalt so used may be a thin, sprayed membrane, covered with earth for protection against weathering and mechanical damage, or thicker surfaces, often including riprap (crushed rock). Asphalt is also used for roofs, coatings, floor tiles, soundproofing, waterproofing, and other building construction elements as well as in a number of industrial products, such as batteries. For certain applications, an asphaltic emulsion is prepared in which fine globules of asphalt are suspended in water.

1.4.3 Synthetic Crude Oil

Coal, oil shale, and tar sand bitumen can be upgraded (converted) through thermal decomposition by a variety of processes to produce a marketable and transportable product. Typically, the product of the conversion (synthetic crude oil) contains very few constituents that passed unchanged from the source (coal, oil shale, and tar sand bitumen) into the product (synthetic crude oil).

The synthetic crude oil so produced varies in nature, but the principal product is a hydrocarbon mixture that may resemble (but not always resembles) conventional crude oil, hence the use of the terms "synthetic crude oil" and "syncrude." However, the synthetic crude oil, although it may be produced from one of the less conventional conversion processes, can actually be refined by the usual refinery system, with or without some modification to the refinery process(es). For example, in the current and modern context, tar sand bitumen is recovered commercially from the tar sand deposits of northeastern Alberta (Canada) by mining followed by a hot water process for separation of the sand and bitumen after which the bitumen is upgraded by a combination of a thermal or hydrothermal process followed by product hydrotreating to produce a low-sulfur hydrocarbon-containing *synthetic crude oil* (Speight, 1995, 2013c, 2014a). By comparison, the unrefined synthetic crude oil from bitumen will generally resemble petroleum more closely than either the synthetic crude oil from coal or the synthetic crude oil from oil shale. Unrefined synthetic crude oil from coal can be identified by a high content of phenolic compounds whereas the unrefined synthetic crude oil from oil shale will contain high proportions of nitrogen-containing compounds (Scouten, 1990; Lee, 1991; Speight, 1995, 2013a,c).

Coal liquids, which are the products produced from the thermal decomposition of coal, also fall into the category of synthetic crude oil, whether the liquids have been hydrotreated or not (Speight, 2013a). The composition can vary from a mixture majority of hydrocarbon species to a mixture containing a majority of heteroatom species. Predominant heteroatom species contain oxygen, usually in the form of phenolic oxygen (e.g., C_6H_5OH) or ether oxygen (R_1OR_2).

The refined (hydrotreated) coal liquids, in which the heteroatom content has been reduced to acceptable levels, may also be referred to as *synthetic crude oil* that is sent to a refinery for further processing into various products.

1.4.4 Shale Oil

Shale oil is produced from a special class of bituminous rocks that has achieved some importance (Scouten, 1990; Lee, 1991; Speight, 2012b). These are argillaceous, laminated sediments of generally high organic content that can be thermally decomposed to yield appreciable amounts of oil, commonly referred to as "shale oil." Oil shale does not yield shale oil without the application of high temperatures and the ensuing thermal decomposition that is necessary to decompose the organic material (*kerogen*) in the shale. The kerogen produces a liquid product (shale oil) by thermal decomposition at high temperature (>500°C, >930°F). The raw oil shale can even be used directly as a fuel akin to a low-quality coal. Indeed, oil shale deposits have been exploited as such for several centuries and shale oil has been produced from oil shale since the nineteenth century.

Thus, "oil shale" is the term applied to a class of bituminous rocks that contain a complex heteroatomic molecule known as "kerogen (q.v.)" and oil shale does not contain oil. *Shale oil* is produced when *kerogen* is thermally decomposed to yield appreciable amounts of a hydrocarbon-based oil; it is this product that is commonly referred to as "shale oil." Oil shale does not yield shale oil without the application of high temperatures and the ensuing thermal decomposition that is necessary to decompose the organic material (*kerogen*) in the shale. Shale oil also contains heteroatom species, predominantly organic nitrogen-containing molecules. The refined (hydrotreated) shale oil, in which the heteroatom content has been reduced to acceptable levels, may also be referred to as "synthetic crude oil" that is sent to a refinery for further processing into various products.

1.5 DERIVED MATERIALS

Any feedstock or product mentioned in the various sections of this book is capable of being separated into several fractions that are sufficiently distinct in character to warrant the application of individual names (Figure 1.6). For example, distillation fractions that are separated by boiling point have been known and the names used for several decades. There may be some slight variation in the boiling ranges, but in general the names are recognized. On the other hand, fractions of feedstocks and product that are separated by other names are less well defined and in many cases can only be described with some difficulty.

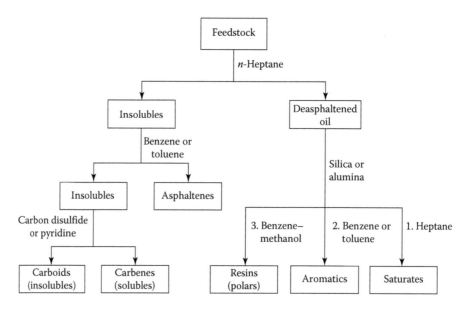

FIGURE 1.6   Fraction scheme for various feedstocks.

1.5.1 ASPHALTENE CONSTITUENTS

For example, treatment of petroleum, residua, heavy oil, or bitumen with a low-boiling liquid hydrocarbon results in the separation of brown to a black powdery material known as the "asphaltene fraction" (Figure 1.6) (Girdler, 1965; Mitchell and Speight, 1973). The reagents for effecting this separation are *n*-pentane and *n*-heptane, although other low-molecular-weight liquid hydrocarbons have been used. It must also be recognized that the character and yield of such a fraction vary with the liquid hydrocarbon used for the separation.

Asphaltene constituents separated from petroleum, residua, heavy oil, and bitumen dissolve readily in benzene, carbon disulfide, chloroform, or other chlorinated hydrocarbon solvents. However, in the case of the higher-molecular-weight native materials or petroleum residua that have been heated intensively or for prolonged periods, the *n*-pentane insoluble (or *n*-heptane insoluble) fraction may not dissolve completely in the solvents mentioned earlier. Definition of the asphaltene fraction has therefore been restricted to that of the *n*-pentane- or *n*-heptane-insoluble material that dissolves in such solvents as benzene.

The benzene- or toluene-insoluble material is collectively referred to as "carbenes" and "carboids," and the fraction soluble in carbon disulfide (or pyridine) but insoluble in benzene is defined as "carbenes" (and not the long-barreled gun used to shoot bison). By difference, *carboids* are insoluble in carbon disulfide (or pyridine). However, because of the different solvents that might be used in place of benzene, it is advisable to define the carbenes (or carboids) by prefixing them with the name of the solvent used for the separation.

1.5.2 NONASPHALTENE CONSTITUENTS

The portion of petroleum that is soluble in, for example, pentane or heptane is often referred to as the "maltene fraction" (*malthene fraction*). This fraction can be further subdivided by percolation through any surface-active material, such as fuller's earth or alumina to yield an *oil* fraction. A more strongly adsorbed, red to brown semisolid material known as "resins" remains on the adsorbent until desorbed by a solvent such as pyridine or a benzene/methanol mixed solvent. The *oil* fraction can be further subdivided into an *aromatics* fraction and a *saturates* fraction. Several other ways have been proposed for separating the resin fraction; for example, a common procedure in the refining industry that can also be used in laboratory practice involves precipitation by liquid propane.

The resin fraction and the *oil* (maltene fraction) fraction have also been referred to collectively as *petrolene fraction*, thereby adding further confusion to this system of nomenclature. However, it has been accepted by many workers in petroleum chemistry that the term *petrolene fractions* be applied to that part of the *n*-pentane-soluble (or *n*-heptane soluble) material that is low boiling (<300°C, <570°F, 760 mm) and can be distilled without thermal decomposition. Consequently, the term *maltene fraction* is now arbitrarily assigned to the pentane-soluble portion of petroleum that is relatively high boiling (>300°C, 760 mm).

Different feedstocks have different amounts of the asphaltene, resin, and oil fractions, which can also lead to different yields of thermal coke as produced in the Conradson carbon residue or Ramsbottom carbon residue tests. Such differences have effects on the methods chosen for refining the different feedstocks.

1.6 RESOURCES AND RESERVES

Fossil fuels (of which petroleum is one) are those fuels, namely, coal, petroleum (including heavy oil), tar sand bitumen, natural gas, and oil shale, that produced by the decay of plant remains over geological time (Scouten, 1990; Lee, 1991; Mokhatab et al., 2006; Speight, 2007, 2013a, 2014a). They are carbon-based resources and represent a vast source of energy. Resources such as heavy oil and bitumen (in tar sand formations) are discussed in this text and represent an unrealized potential,

TABLE 1.11

Nomenclature and Approximate Age of Geologic Strata

Era	Period	Epoch	Age (10⁶ Years)
Cenozoic	Quaternary	Recent	0.01
		Pleistocene	3
	Tertiary	Pliocene	12
		Miocene	25
		Oligocene	38
		Eocene	55
		Paleocene	65
Mesozoic	Cretaceous		135
	Jurassic		180
	Triassic		225
Paleozoic	Permian		275
	Carboniferous		
	Pennsylvanian		350
	Mississippian		
	Devonian		413
	Silurian		430
	Ordovician		500
	Cambrian		600

with liquid fuels from petroleum being only a fraction of those that could ultimately be produced from heavy oil and tar sand bitumen. At the present time, the majority of the energy consumed by humans is produced from fossil fuels with smaller amounts of energy coming from nuclear and hydroelectric sources, and the potential for nonfossil fuel resources to supplant fossil fuel resources as the predominant sources of energy is not yet possible. As a result, fossil fuels are projected to be the major sources of energy for the next 50 years (Speight, 2011a).

Petroleum occurs scattered throughout the earth's crust, which is divided into natural groups or strata, categorized in order of their antiquity (Table 1.11). In fact, the reservoir rocks that yield crude oil range in age from Precambrian to recent geological time, but rocks deposited during the Tertiary, Cretaceous, Permian, Pennsylvanian, Mississippian, Devonian, and Ordovician periods are particularly productive. In contrast, rocks of Jurassic, Triassic, Silurian, and Cambrian age are less productive but do contain some petroleum, while the rocks of Precambrian age yield petroleum only under exceptional circumstances. Most of the crude oil currently recovered and sent to refineries is produced from underground reservoirs, and it is these reserves that are the predominant source of petroleum and are discussed here. These divisions are recognized by the distinctive systems of debris (organic material, minerals, and fossils) that form a chronological time chart that indicates the relative ages of the strata. Petroleum, and other carbonaceous materials such as coal and oil shale, occurs in all of these geologic strata from the Precambrian to the recent, and the origin of petroleum within these formations is a question that remains open to conjecture and the basis for much research and speculation (Speight, 2014a). The answer cannot be given in this text, nor for that matter can it be presented in any advanced treatise. It is more pertinent to the present text that historical data be introduced to illustrate the reserves of fossil fuels decreasing the quality of crude oil and current prospects for the continuation and evolution of the refining industry.

Various definitions have been applied to energy reserves (Figures 1.7 and 1.8), but the crux of the matter is the amount of a resource that is recoverable using current technology. The definitions that are used to describe fossil fuel *reserves* are often varied and misunderstood because they are not

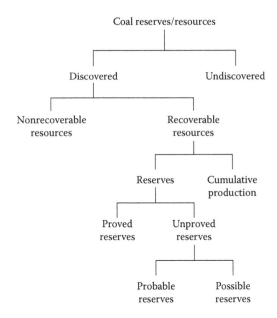

FIGURE 1.7 Subdivision of resources.

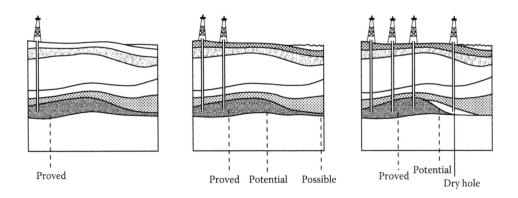

FIGURE 1.8 Representation of various reserves.

adequately defined at the time of use. Therefore, as a means of alleviating this problem, it is pertinent at this point to consider the definitions used to describe the amount of petroleum that remains in subterranean reservoirs.

By way of explanation, *proven reserves* are those reserves of petroleum that have been identified drilling operations and are recoverable by means of current technology. They have a high degree of accuracy since the reservoir or fields are often clearly demarcated and estimates of the oil in place have a high degree of accuracy. *Proven reserves* are frequently updated as recovery operations proceed by means of reservoir characteristics, such as production data and pressure transient analysis. The term "inferred reserves" is commonly used in addition to, or in place of, *potential reserves*. Inferred reserves are regarded as of a higher degree of accuracy than potential reserves, and the term is applied to those reserves that are estimated using an improved understanding of reservoir frameworks. The term also usually includes those reserves that can be recovered by further development of the various recovery technologies.

Probable reserves are those reserves of petroleum that are not fully certain and about which a slight doubt exists, and there is also an even greater degree of uncertainty about recovery but about which there is some information. An additional term, *potential reserves*, is also used on occasion; these reserves are based upon geological information about the types of sediments where such resources are likely to occur and they are considered to represent an educated guess. Then, there are the so-called undiscovered reserves, which can be, and often are, little more than figments of the imagination! The terms "undiscovered reserves" or "undiscovered resources" should be used with caution, especially when applied as a means of estimating reserves of petroleum reserves. The data are very speculative and are regarded by many energy scientists as having little value other than unbridled optimism.

The differences between the data obtained from these various estimates can be considerable, but it must be remembered that any data about the reserves of petroleum (and, for that matter, about any other fuel or mineral resource) will always be open to questions about the degree of certainty. Thus, in reality, and in spite of the use of unjustified word manipulation, *proven reserves* may be a very small part of the total hypothetical and/or speculative amounts of a resource.

At some time in the future, certain resources may become reserves. Such a reclassification can arise as a result of improvements in recovery techniques, which may either make the resource accessible or bring about a lowering of the recovery costs and render winning of the resource an economical proposition. In addition, other uses may also be found for a commodity, and the increased demand may result in an increase in price. Alternatively, a large deposit may become exhausted and unable to produce any more of the resource thus forcing production to focus on a resource that is lower grade but has a higher recovery cost.

Finally, it is rare that petroleum (the exception being tar sand deposits, from which most of the volatile material has disappeared over time) does not occur without an accompanying gas cap, referred to as *natural gas* (Figure 1.3). It is important, when describing reserves of petroleum, to also acknowledge the occurrence, properties, character, and reserves of *natural gas*, which also required refining before sales.

1.6.1 Conventional Crude Oil and Heavy Oil

At the present time, several countries are recognized as producers of petroleum (BP, 2015) and have available reserves which also include opportunity crude oils and high-acid crude oil and foamy oil.

On a worldwide basis, the proved reserves of conventional crude oil are estimated to be approximately 1.7 billion barrels (1.7×10^9 bbls) with approximately 232.5 million barrels of reserves in the United States and Canada (232.5×10^6 bbls) (BP, 2015). In addition to this substantial resource base, the crude oils available to the modern refinery are quite different in composition and properties to those available 40–50 years ago and require different refinery scenarios for conversion to liquid fuels (Speight, 2005, 2013d, 2014a). The current crude oils are somewhat heavier insofar as they have higher proportions of nonvolatile constituents, that is, a higher proportion of the nonvolatile residue. In fact, by the older and more arbitrary nonscientific definitions of the mid-twentieth century, many of the crude oils currently in use would have been classified as heavy crude oils. Changes in feedstock character, such as this tendency to contain higher proportions of nonvolatile residua, require adjustments to refinery operations to handle these heavier crude oils to reduce the amount of coke formed during processing and to balance the overall product slate (Speight 2011b, 2014a).

1.6.2 Tar Sand Bitumen

The utilization of tar sand resources for the production of liquid fuels was considered uneconomical during the first half of the twentieth century. The utilization of these resources became economical by the increase in the price of conventional crude and by the developments made in recovery, mining, extraction, and upgrading technologies. Thus, interest in the availability of these resources widened and formed an important part of the planning for the production of liquid fuels.

On an international note, the bitumen in tar sand deposits represents a potentially large supply of energy (Meyer and De Witt, 1990), which have (finally) been included in the resource base (BP, 2015). However, many of the reserves are available only with some difficulty (as is the case with some petroleum resources) but optional refinery scenarios are necessary for conversion of the bitumen to liquid products because of the substantial differences in character between conventional petroleum and bitumen (Speight, 2013d, 2014a).

Because of the diversity of available information and the continuing attempts to delineate the various world tar sand deposits, it is virtually impossible to present accurate numbers that reflect the extent of the reserves in terms of the barrel unit. Indeed, investigations into the extent of many of the world's deposits are continuing at such a rate that the numbers vary from 1 year to the next. Accordingly, the data quoted here must be recognized as approximate with the potential of being quite different at the time of publication.

Tar sand deposits are widely distributed throughout the world (Phizackerley and Scott, 1967; Demaison, 1977; Meyer and Dietzman, 1981; Speight, 1997). The various deposits have been described as belonging to two types: (1) materials that are found in stratigraphic traps and (2) deposits that are located in structural traps. Inevitably, there are gradations, and combinations of these two types of deposits and a broad pattern of deposit entrapment are believed to exist. In general terms, the entrapment characteristics for the very large tar sand deposits all involve a combination of stratigraphic and structural traps, and there are no very large ($>4 \times 10^9$ bbl) oil sand accumulations either in purely structural or in purely stratigraphic traps.

The potential reserves of bitumen that occur in tar sand deposits have been variously estimated on a world basis as being in excess of 3 trillion ($>3 \times 10^{12}$) barrels of petroleum equivalent. The reserves that have been estimated for the United States have been estimated to be in excess of 52 million ($>52 \times 10^6$) barrels. That commercialization has taken place in Canada does not mean that commercialization is imminent for other tar sand deposits. There are considerable differences between the Canadian and the U.S. deposits that could preclude across-the-board application of the Canadian principles to the U.S. sands (Speight, 1990, 2013c, 2014a).

Thus, in spite of the high estimations of the reserves of bitumen, the two conditions of vital concern for the economic development of tar sand deposits are the concentration of the resource, or the percent bitumen saturation, and its accessibility, usually measured by the overburden thickness. Recovery methods are based either on mining combined with some further processing or operation on the oil sands *in situ*. The mining methods are applicable to shallow deposits, characterized by an overburden ratio (i.e., overburden depth to thickness of tar sand deposit). For example, indications are that for the Athabasca deposit, no more than 10% of the in-place bitumen can be recovered by mining within current concepts of the economics and technology of open-pit mining; this 10% portion may be considered as the *proven reserves* of bitumen in the deposit.

1.6.3 NATURAL GAS

Natural gas occurs in the porous rock of the earth's crust either alone (*nonassociated natural gas*) or with accumulations of petroleum (*associated natural gas*). In the latter case, the gas forms the gas cap, which is the mass of gas trapped between the liquid petroleum and the impervious cap rock of the petroleum reservoir. When the pressure in the reservoir is sufficiently high, the natural gas may be dissolved in the petroleum and is released upon penetration of the reservoir as a result of drilling operations.

The proven reserves of natural gas are of the order of 1926.9 trillion cubic feet (1 Tcf = 1×10^{12}). Approximately 429 Tcf (429×10^{12} ft^3) exist in the United States and Canada (BP Statistical Review of World Energy, 2015) with demand in the United States continuing to increase. It should also be remembered that the total gas resource base (like any fossil fuel or mineral resource base) is dictated by economics. Therefore, when resource data are quoted, some attention must be given to the cost of recovering those resources and, most important, the economics of natural gas production must also include a cost factor that reflects the willingness to secure total, or a specific degree of, energy independence.

Thus, the overabundance of heavy crude oil and tar sand bitumen in the geographical vicinity of important markets—such as in the western hemisphere and the emerging Asiatic markets (China and India)—points to the need to find secure outlets for these reserves. This has led to strategic associations between producers and refiners, in which construction of new conversion units for such feedstocks is an essential part of future planning (Gembicki et al., 2007; Patel, 2007; Falkler and Sandu, 2010).

REFERENCES

Abraham, H. 1945. *Asphalts and Allied Substances.* Van Nostrand Scientific Publishers, New York.
ASTM D4. 2015. *Test Method for Bitumen Content.* Annual Book of Standards. ASTM International, West Conshohocken, PA.
ASTM D4175. 2015. *Standard Terminology Relating to Petroleum, Petroleum Products, and Lubricants.* Annual Book of Standards. ASTM International, Philadelphia, PA.
Babich, I.V. and Moulijn, J.A. 2003. Science and technology of novel processes for deep desulfurization of oil refinery streams: A review. *Fuel*, 82: 607–631.
BP. 2015. *BP Statistical Review of World Energy.* British Petroleum Company, London, UK.
Chen, Z., Huan, G., and Ma, Y. 2006. Computational methods for multiphase flows in porous media. In: *Computational Science and Engineering Series*, Volume 2. Society for Industrial and Applied Mathematics (SIAM), Philadelphia, PA.
Chugh, S., Baker, R., Telesford, A., and Zhang, E. 2000. Mainstream options for heavy oil: Part I—Cold production. *Journal of Canadian Petroleum Technology*, 39(4): 31–39.
Cobb, C. and Goldwhite, H. 1995. *Creations of Fire: Chemistry's Lively History from Alchemy to the Atomic Age.* Plenum Press, New York.
Crocker, M. and Crofcheck, C. 2006. Reducing national dependence on imported oil. In: *Energeia*, Volume 17, No. 6. Center for Applied Energy Research, University of Kentucky, Lexington, KY.
Delbianco, A. and Montanari, R. 2009. *Encyclopedia of Hydrocarbons, Volume III/New Developments: Energy, Transport, Sustainability.* EniS.p.A., Rome, Italy.
Demaison, G.J. 1977. Tars sand deposits of the world. In: *The Oil Sands of Canada-Venezuela.* D.A. Redford and A.G. Winestock (Editors). Canadian Institute of Mining and Metallurgy, Special Volume 17, p. 9.
Demirbaş, A. 2010. *Biorefineries for Upgrading Biomass Facilities.* Springer-Verlag, London, UK.
Falkler, T. and Sandu, C. 2010. Fine-tune processing heavy crudes in your facility. *Hydrocarbon Processing*, 89(9): 67–73.
Forbes, R.J. 1958a. *A History of Technology.* Oxford University Press, Oxford, UK.
Forbes, R.J. 1958b. *Studies in Early Petroleum Chemistry.* E.J. Brill Publishers, Leiden, the Netherlands.
Forbes, R.J. 1959. *More Studies in Early Petroleum Chemistry.* E.J. Brill Publishers, Leiden, the Netherlands.
Forbes, R.J. 1964. *Studies in Ancient Technology.* E.J. Brill Publishers, Leiden, the Netherlands.
Gary, J.G., Handwerk, G.E., and Kaiser, M.J. 2007. *Petroleum Refining: Technology and Economics*, 5th Edition. CRC Press/Taylor & Francis Group, Boca Raton, FL.
Gembicki, V.A., Cowan, T.M., and Brierley, G.R. 2007. Update processing operations to handle heavy feedstocks. *Hydrocarbon Processing*, 86(2): 41–53.
Ghoshal, S. and Sainik, V. 2013. Monitor and minimize corrosion in high-TAN crude processing. *Hydrocarbon Processing*, 92(3): 35–38.
Giampietro, M. and Mayumi, K., 2009. *The Biofuel Delusion: The Fallacy of Large-Scale Agro-Biofuel Production.* Earthscan, Washington, DC.
Girdler, R.B. 1965. Constitution of asphaltenes and related studies. *Proceedings of Association of Asphalt Paving Technologists*, 34: 45.
Hoiberg, A.J. 1964. *Bituminous Materials: Asphalts, Tars, and Pitches.* John Wiley & Sons, New York.
Hsu, C.S. and Robinson, P.R. 2006. *Practical Advances in Petroleum Processing*, Volumes 1 and 2. Springer, New York.

ITAA. 1936. Income Tax Assessment Act. Government of the Commonwealth of Australia.

Kane, R.D. and Cayard, M.S. 2002. A comprehensive study on naphthenic acid corrosion. Corrosion Paper No. 02555. NACE International, Houston, TX.

Klaas, M., Pischinger, S., and Schröder, W. (Editors). 2015. *Fuels from Biomass: An Interdisciplinary Approach*. Springer-Verlag, Berlin, Germany.

Koots, J.A. and Speight, J.G. 1975. The relation of petroleum resins to asphaltenes. *Fuel*, 54: 179–184.

Langeveld, H., Sanders, J., and Meeusen, M. (Editors). 2010. *The Biobased Economy*. Earthscan, Washington, DC.

Lee, S. 1991. *Oil Shale Technology*. CRC Press/Taylor & Francis Group, Boca Raton, FL.

Levine, J.R. 1993. Coalification: The evolution of coal as a source rock and reservoir rock for oil and gas. In: *Studies in Geology*, Volume 38. American Association of Petroleum Geologists, Tulsa, OK, pp. 39–77.

Long, R.B. and Speight, J.G. 1998. The composition of petroleum. In: *Petroleum Chemistry and Refining*. J.G. Speight (Editor). Taylor & Francis Group, Washington, DC, Chapter 2.

Lorenzini, G., Biserni, C., and Flacco, G. 2010. *Solar, Thermal, and Biomass Energy*. WIT Press, Boston, MA.

Maini, B.B. 1999. Foamy oil flow in primary production of heavy oil under solution gas drive. Paper No. SPE 56541. *Proceedings of Annual Technical Conference and Exhibition*. Houston, TX, October 3–6. Society of Petroleum Engineers, Richardson, TX.

Maini, B.B. 2001. Foamy oil flow. Paper No. SPE 68885. SPE J. Pet. Tech., Distinguished Authors Series. Society of Petroleum Engineers, Richardson, TX, pp. 54–64.

Mayes, J.M. 2015. What are the possible impacts on us refineries processing shale oils? *Hydrocarbon Processing*, 94(2): 67–70.

Meyer, R.F. and De Witt, W. Jr. 1990. *Definition and World Resources of Natural Bitumens*. Bulletin No. 1944. U.S. Geological Survey, Reston, VA.

Meyer, R.F. and Dietzman, W.D. 1981. World geography of heavy crude oils. In: *The Future of Heat Crude and Tar Sands*. R.F. Meyer and C.T. Steele (Editors). McGraw-Hill, New York, p. 16.

Mitchell, D.L. and Speight J.G. 1973. The solubility of asphaltenes in hydrocarbon solvents. *Fuel*, 52: 149–152.

Mokhatab, S., Poe, W.A., and Speight, J.G. 2006. *Handbook of Natural Gas Transmission and Processing*. Elsevier, Amsterdam, the Netherlands.

Nersesian, R.L. 2010. *Energy for the 21st Century: A Comprehensive Guide to Conventional and Alternative Energy Sources*. Earthscan, Washington, DC.

Ohmes, R. 2014. Characterizing and tracking contaminants in opportunity crudes. Digital refining. http://www.digitalrefining.com/article/1000893, Characterising_and_tracking_contaminants_in_opportunity_crudes_.html#.VJhFjV4AA; accessed November 1, 2014.

Parkash, S. 2003. *Refining Processes Handbook*. Gulf Professional Publishing, Elsevier, Amsterdam, the Netherlands.

Patel, S. 2007. Canadian oil sands—Opportunities, technologies, and challenges. *Hydrocarbon Processing*, 86(2): 65–74.

Pfeiffer, J.H. 1950. *The Properties of Asphaltic Bitumen*. Elsevier, Amsterdam, the Netherlands.

Phizackerley, P.H. and Scott, L.O. 1967. The major tar sand deposits of the world. *Proceedings of Seventh World Petroleum Congress*. Mexico City, Mexico, Volume 3, p. 551.

Poon, D. and Kisman, K. 1992. Non-Newtonian effects on the primary production of heavy oil reservoirs. *Journal of Canadian Petroleum Technology*, 31(7): 1–6.

Rice, D.D. 1993. Composition and origins of coalbed gas. In: *Studies in Geology*, Volume 38. American Association of Petroleum Geologists, Tulsa, OK, pp. 159–184.

Scouten, C.S. 1990. Oil shale. In: *Fuel Science and Technology Handbook*. J.G. Speight (Editor). Marcel Dekker, New York, Chapters 25–31.

Shalaby, H.M. 2005. Refining of Kuwait's heavy crude oil: Materials challenges. *Proceedings of Workshop on Corrosion and Protection of Metals*. Arab School for Science and Technology, Kuwait, Western Asia, December 3–7.

Sheng, J.J., Maini, B.B., Hayes, R.E., and Tortike, W.S. 1999. Critical review of foamy oil flow. *Transport in Porous Media*, 35: 157–187.

Speight, J.G. 1984. In: *Characterization of Heavy Crude Oils and Petroleum Residues*. S. Kaliaguine and A. Mahay (Editors). Elsevier, Amsterdam, the Netherlands, p. 515.

Speight, J.G. (Editor). 1990. *Fuel Science and Technology Handbook*. Marcel Dekker, New York.

Speight, J.G. 1995. Tar sands, recovery and processing. *Encyclopedia of Energy Technology and the Environment*. p. 2589.

Speight, J.G. 1997. Tar sands. *Kirk-Othmer Encyclopedia of Chemical Technology*, 4th Edition, Volume 23, p. 717.

Speight J.G. 2000. *The Desulfurization of Heavy Oils and Residua*, 2nd Edition. Marcel Dekker, New York.

Speight, J.G. 2005. Upgrading and refining of natural bitumen and heavy oil. In: *Coal, Oil Shale, Natural Bitumen, Heavy Oil and Peat. Encyclopedia of Life Support Systems (EOLSS)*. Developed under the Auspices of the UNESCO. EOLSS Publishers, Oxford, UK.

Speight, J.G. 2007. *Natural Gas: A Basic Handbook*. GPC Books, Gulf Publishing Company, Houston, TX.

Speight, J.G. 2008. *Synthetic Fuels Handbook: Properties, Processes, and Performance*. McGraw-Hill, New York.

Speight, J.G. 2011a. *The Refinery of the Future*. Gulf Professional Publishing, Elsevier, Oxford, UK.

Speight, J.G. (Editor). 2011b. *The Biofuels Handbook*. Royal Society of Chemistry, London, UK.

Speight, J.G. 2012a. *Crude Oil Assay Database*. Knovel, Elsevier, New York. http://www.knovel.com/web/portal/browse/display?_EXT_KNOVEL_DISPLAY_bookid=5485&VerticalID=0.

Speight, J.G. 2012b. *Shale Oil Production Processes*. Gulf Professional Publishing, Elsevier, Oxford, UK.

Speight, J.G. 2013a. *The Chemistry and Technology of Coal*, 3rd Edition. CRC Press/Taylor & Francis Group, Boca Raton, FL.

Speight, J.G. 2013b. *Heavy Oil Production Processes*. Gulf Professional Publishing, Elsevier, Oxford, UK.

Speight, J.G. 2013c. *Oil Sand Production Processes*. Gulf Professional Publishing, Elsevier, Oxford, UK.

Speight, J.G. 2013d. *Heavy and Extra Heavy Oil Upgrading Technologies*. Gulf Professional Publishing, Elsevier, Oxford, UK.

Speight, J.G. 2014a. *The Chemistry and Technology of Petroleum*, 5th Edition. CRC Press/Taylor & Francis Group, Boca Raton, FL.

Speight, J.G. 2014b. *High Acid Crudes*. Gulf Professional Publishing, Elsevier, Oxford, UK.

Speight, J.G. 2014c. *Oil and Gas Corrosion Prevention*. Gulf Professional Publishing, Elsevier, Oxford, UK.

Speight, J.G. 2015a. *Handbook of Hydraulic Fracturing*. John Wiley & Sons, Hoboken, NJ.

Speight, J.G. 2015b. *Handbook of Petroleum Product Analysis*, 2nd Edition. John Wiley & Sons, Hoboken, NJ.

Speight, J.G. 2015c. *Asphalt Materials Science and Technology*. Butterworth-Heinemann, Elsevier, Oxford, UK.

Speight, J.G. 2016. *Introduction to Enhanced Recovery Methods for Heavy Oil and Tar Sands*, 2nd Edition. Gulf Professional Publishing, Elsevier, Oxford, UK.

Speight, J.G. and Francisco, M.A. 1990. Studies in petroleum composition IV: Changes in the nature of chemical constituents during crude oil distillation. *Revue de l'InstitutFrançais du Pétrole*, 45: 733.

Speight, J.G. and Ozum, B. 2002. *Petroleum Refining Processes*. Marcel Dekker, New York.

Syngellakis, S. (Editor). 2015a. *Biomass to Biofuels*. WIT Press, Boston, MA.

Syngellakis, S. (Editor). 2015b. *Waste to Energy*. WIT Press, Boston, MA.

U.S. EIA. 2011. Review of emerging resources. U.S. Shale Gas and Shale Oil Plays. Energy Information Administration, U.S. Department of Energy, Washington, DC.

U.S. EIA. 2013. Technically recoverable shale oil and shale gas resources: An assessment of 137 shale formations in 41 countries outside the United States. Energy Information Administration, U.S. Department of Energy, Washington, DC.

Van Nes, K. and van Westen, H.A. 1951. *Aspects of the Constitution of Mineral Oils*. Elsevier, Amsterdam, the Netherlands.

Wright, L., Boundy, R., Perlack, R., Davis, S., and Saulsbury B. 2006. *Biomass Energy Data Book*, Edition 1. Office of Planning, Budget and Analysis, Energy Efficiency and Renewable Energy, U.S. Department of Energy. Contract No. DE-AC05-00OR22725. Oak Ridge National Laboratory, Oak Ridge, TN.

Yeung, T.W. 2014. Evaluating opportunity crude processing. Digital refining. http://www.digitalrefining.com/article/1000644; accessed October 25, 2014.

2 Feedstock Evaluation

2.1 INTRODUCTION

Petroleum evaluation is an important aspect of the prerefining examination of a refinery feedstock. Evaluation, in this context, is the determination of the physical and chemical characteristics of crude oil, heavy oil, and tar sand bitumen since the yields and properties of products or fractions produced from these feedstocks vary considerably and are dependent on the concentration of the various types of hydrocarbons as well as the amounts of the heteroatom compounds (i.e., molecular constituents contacting nitrogen and/or oxygen and/or sulfur and/or metals). Some types of feedstocks have economic advantages as sources of fuels and lubricants with highly restrictive characteristics because they require less specialized processing than that needed for the production of the same products from many types of crude oil. Others may contain unusually low concentrations of components that are desirable fuel or lubricant constituents, and the production of these products from such crude oils may not be economically feasible.

Since petroleum exhibits a wide range of physical properties, it is not surprising that the behavior of various feedstocks in these refinery operations is not simple. The atomic ratios from ultimate analysis (Figure 2.1) can give an indication of the nature of a feedstock and the generic hydrogen requirements to satisfy the refining chemistry, but it is not possible to predict with any degree of certainty how the feedstock will behave during refining. Any deductions made from such data are pure speculation and are open to much doubt.

In addition, the chemical composition of a feedstock is also an indicator of refining behavior (Speight and Ozum, 2002; Parkash, 2003; Gary et al., 2007; Speight, 2014, 2015). Whether the composition is represented in terms of compound types or in terms of generic compound classes, it can enable the refiner to determine the nature of the reactions. Hence, chemical composition can play a large part in determining the nature of the products that arise from the refining operations. It can also play a role in determining the means by which a particular feedstock should be processed (Speight and Ozum, 2002; Parkash, 2003; Gary et al., 2007; Speight, 2014).

Therefore, the judicious choice of a crude oil to produce any given product is just as important as the selection of the product for any given purpose. Thus, initial inspection of the nature of the petroleum will provide deductions related to the most logical means of refining. Indeed, careful evaluation of petroleum from physical property data is a major part of the initial study of any petroleum destined as a refinery feedstock. Proper interpretation of the data resulting from the inspection of crude oil requires an understanding of their significance.

Petroleum exhibits a wide range of physical properties and several relationships can be made between various physical properties (Speight, 2001). Whereas the properties such as viscosity, density, boiling point, and color of petroleum may vary widely, the ultimate or elemental analysis varies, as already noted, over a narrow range for a large number of petroleum samples. The carbon content is relatively constant, while the hydrogen and heteroatom contents are responsible for the major differences between petroleum. Coupled with the changes brought about to the feedstock constituents by refinery operations, it is not surprising that petroleum characterization is a monumental task.

Although it is possible to classify refinery operations using the three general terms, that is, (1) separation, (2) conversion, and (3) finishing, the chemical composition of a feedstock is a much truer indicator of refining behavior. Whether the composition is represented in terms of compound types or in terms of generic compound classes, it can enable the refiner to determine the nature of the reactions. Hence, chemical composition can play a large part in determining the nature of the

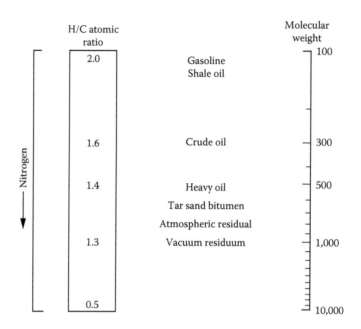

FIGURE 2.1 Atomic carbon/hydrogen ratio of various feedstocks.

products that arise from the refining operations. It can also play a role in determining the means by which a particular feedstock should be processed (Wallace, 1988; Wallace et al., 1988; Speight, 2001; Speight and Ozum, 2002; Parkash, 2003; Gary et al., 2007; Speight, 2014, 2015).

The physical and chemical characteristics of crude oils and the yields and properties of products or fractions prepared from them vary considerably and are dependent on the concentration of the various types of hydrocarbons and minor constituents present. Some types of petroleum have economic advantages as sources of fuels and lubricants with highly restrictive characteristics because they require less specialized processing than that needed for the production of the same products from many types of crude oil. Others may contain unusually low concentrations of components that are desirable fuel or lubricant constituents, and the production of these products from such crude oils may not be economically feasible.

Evaluation of petroleum for use as a feedstock usually involves an examination of one or more of the physical properties of the material. By this means, a set of basic characteristics can be obtained that can be correlated with utility. The physical properties of petroleum and petroleum products are often equated with those of the hydrocarbons for although petroleum is indeed a very complex mixture, there is gasoline, produced by nondestructive distillation, in which fewer than a dozen hydrocarbons make up at least 50% of the material (Speight, 2014, 2015).

To satisfy specific needs with regard to the type of petroleum to be processed, as well as to the nature of the product, most refiners have, through time, developed their own methods of petroleum analysis and evaluation. However, such methods are considered proprietary and are not normally available. Consequently, various standards organizations, such as the American Society for Testing and Materials in North America and the Institute of Petroleum in Britain, have devoted considerable time and effort to the correlation and standardization of methods for the inspection and evaluation of petroleum and petroleum products. A complete discussion of the large number of routine tests available for petroleum fills an entire book (Speight, 2015). However, it seems appropriate that in any discussion of the physical properties of petroleum and petroleum products reference be made to the corresponding test, and accordingly, the various test numbers have been included in the text.

The purpose of this chapter is to present an outline of the tests that may be applied to petroleum, heavy oil, and tar sand bitumen, including their respective products or even petroleum products as well as the resulting chemical properties and physical properties from which a feedstock or product can be evaluated (Speight, 2014, 2015). For these purposes, data relating to various chemical physical properties have been included as illustrative examples, but theoretical discussions of the physical properties of hydrocarbons were deemed irrelevant and are omitted.

2.2 PETROLEUM ASSAY

An efficient assay is derived from a series of test data that give an accurate description of petroleum quality and allow an indication of its behavior during refining. The first step is, of course, to assure adequate (correct) sampling by the use of the prescribed protocols (ASTM D4057).

Thus, analyses are performed to determine whether each batch of crude oil received at the refinery is suitable for refining purposes. The tests are also applied to determine if there has been any contamination during wellhead recovery, storage, or transportation that may increase the processing difficulty (cost). The information required is generally crude oil dependent or specific to a particular refinery and is also a function of refinery operations and desired product slate. To obtain the necessary information, two different analytical schemes are commonly used and these are as follows: (1) an inspection assay and (2) a comprehensive assay (Table 2.1).

Inspection assays usually involve determination of several key bulk properties of petroleum (e.g., API gravity, sulfur content, pour point, and distillation range) as a means of determining if

---

TABLE 2.1
Recommended Inspection Data Required for Petroleum and Heavy Feedstocks, Which Include Heavy Oil, Extra Heavy Oil, and Tar Sand Bitumen

Petroleum	Heavy Feedstocks
Density, specific gravity	Density, specific gravity
API gravity	API gravity
Carbon, wt%	Carbon, wt%
Hydrogen, wt%	Hydrogen, wt%
Nitrogen, wt%	Nitrogen, wt%
Sulfur, wt%	Sulfur, wt%
	Nickel, ppm
	Vanadium, ppm
	Iron, ppm
Pour point	Pour point
Wax content	
Wax appearance temperature	
Viscosity (various temperatures)	Viscosity (various temperatures)
Carbon residue of residuum	Carbon residue
	Ash, wt%
Distillation profile:	*Fractional composition:*
All fractions plus vacuum residue	Asphaltenes, wt%
	Resin constituents, wt%
	Aromatics, wt%
	Saturates, wt%

major changes in characteristics have occurred since the last comprehensive assay was performed. For example, a more detailed inspection assay might consist of the following tests: API gravity (or density or relative density), sulfur content, pour point, viscosity, salt content, water and sediment content, trace metals (or organic halides). The results from these tests with the archived data from a comprehensive assay provide an estimate of any changes that have occurred in the crude oil that may be critical to refinery operations. Inspection assays are performed routinely on all crude oils received at a refinery.

On the other hand, the comprehensive (or full) assay is more complex (as well as time-consuming and costly), and is usually only performed when a new field comes onstream or when the inspection assay indicates that significant changes in the composition of the crude oil have occurred. Except for these circumstances, a comprehensive assay of a particular crude oil stream may not (unfortunately) be updated for several years. A full petroleum assay may involve at least determinations of (1) carbon residue yield, (2) density (specific gravity), (3) sulfur content, (4) distillation profile (volatility), (5) metallic constituents, (6) viscosity, and (7) pour point, as well as any tests designated necessary to understand the properties and behavior of the crude oil under examination.

The inspection assay tests discussed earlier are not exhaustive but are the ones most commonly used and provide data on the impurities that are present, as well as a general idea of the products that may be recoverable. Other properties that are determined on an as-needed basis include, but are not limited to, the following: (1) vapor pressure (Reid method) (ASTM D323), (2) total acid number (ASTM D664), and (3) the aniline point (or mixed aniline point) (ASTM D611).

The *Reid vapor pressure* test method (ASTM D323) measures the vapor pressure of volatile petroleum. The Reid vapor pressure differs from the true vapor pressure of the sample due to some small sample vaporization and the presence of water vapor and air in the confined space. The *acid number* is the quantity of base, expressed in milligrams of potassium hydroxide per gram of sample that is required to titrate a sample in this solvent to a green/green-brown end point, using p-naph-tholbenzein indicator solution. The *strong acid number* is the quantity of base, expressed as milligrams of potassium hydroxide per gram of sample, required to titrate a sample in the solvent from its initial meter reading to a meter reading corresponding to a freshly prepared nonaqueous acidic buffer solution or a well-defined inflection point as specified in the test method (ASTM D664). The *aniline point* (or *mixed aniline point*) (ASTM D611) has been used for the characterization of crude oil, although it is more applicable to pure hydrocarbons and in their mixtures and is used to estimate the aromatic content of mixtures. Aromatic mixtures exhibit the lowest aniline points, and paraffin mixtures have the highest aniline points. Cycloparaffin derivatives and olefin derivatives exhibit values between these two extremes. In any hydrocarbon homologous series, the aniline point increases with increasing molecular weight.

Using the data derived from the test assay, it is possible to assess petroleum quality acquire a degree of predictability of performance during refining. However, knowledge of the basic concepts of refining will help the analyst understand the production and, to a large extent, the anticipated properties of the product, which in turn is related to storage, sampling, and handling of the products.

Therefore, the judicious choice of a crude oil to produce any given product is just as important as the selection of the product for any given purpose. Thus, initial inspection of petroleum using properties such as API gravity and sulfur (Tables 2.2 and 2.3) will provide information relative to the most logical means of refining or correlation of various properties to structural types present and hence attempted classification of the petroleum (Speight, 2014; Ghashghaee, 2015). Indeed, careful evaluation of petroleum from physical property data is a major part of the initial study of any petroleum destined as a refinery feedstock. Proper interpretation of the data resulting from the inspection of crude oil requires an understanding of their significance. In the following section, an indication of the physical properties that may be applied to petroleum, or even petroleum product, evaluation will be presented.

TABLE 2.2
API Gravity and Sulfur Content of Selected Crude Oils

Country	Crude Oil	API	Sulfur % w/w
Abu Dhabi (UAE)	Abu Al Bu Khoosh	31.6	2.00
Abu Dhabi (UAE)	Abu Mubarras	38.1	0.93
Abu Dhabi (UAE)	El Bunduq	38.5	1.12
Abu Dhabi (UAE)	Murban	40.5	0.78
Abu Dhabi (UAE)	Umm Shaif	37.4	1.51
Abu Dhabi (UAE)	Zakum (Lower)	40.6	1.05
Abu Dhabi (UAE)	Zakum (Upper)	33.1	2.00
Algeria	Zarzaitine	43.0	0.07
Angola	Cabinda	31.7	0.17
Angola	Palanca	40.1	0.11
Angola	Takula	32.4	0.09
Australia	Airlie	44.7	0.01
Australia	Barrow Island	37.3	0.05
Australia	Challis	39.5	0.070
Australia	Cooper Basin	45.2	0.02
Australia	Gippsland	47.0	0.09
Australia	Griffin	55.0	0.03
Australia	Harriet	37.9	0.05
Australia	Jabiru	42.3	0.05
Australia	Jackson	43.8	0.03
Australia	Saladin	48.2	0.02
Australia	Skua	41.9	0.06
Brazil	Garoupa	30.0	0.68
Brazil	Sergipano Platforma	38.4	0.19
Brazil	Sergipano Terra	24.1	0.41
Brunei	Champion Export	23.9	0.12
Brunei	Seria	40.5	0.06
Cameroon	Kole Marine	32.6	0.33
Cameroon	Lokele	20.7	0.46
Canada (Alberta)	Bow River Heavy	26.7	2.10
Canada (Alberta)	Pembina	38.8	0.20
Canada (Alberta)	Rainbow	40.7	0.50
Canada (Alberta)	Rangeland South	39.5	0.75
Canada (Alberta)	Wainwright-Kinsella	23.1	2.58
China	Daqing (Taching)	32.6	0.09
China	Nanhai Light	40.6	0.06
China	Shengli	24.2	1.00
China	Weizhou	39.7	0.08
Colombia	Cano Limon	29.3	0.51
Congo (Brazzaville)	Emeraude	23.6	0.600
Dubai (UAE)	Fateh	31.1	2.000
Dubai (UAE)	Margham Light	50.3	0.040
Ecuador	Oriente	29.2	0.880
Egypt	Belayim	27.5	2.200
Egypt	Gulf of Suez	31.9	1.520
Egypt	Ras Gharib	21.5	3.640
Gabon	Gamba	31.4	0.090

(*Continued*)

TABLE 2.2 (*Continued*)
API Gravity and Sulfur Content of Selected Crude Oils

Country	Crude Oil	API	Sulfur % w/w
Gabon	Rabi-Kounga	33.5	0.070
Ghana	Salt Pond	37.4	0.097
India	Bombay High	39.2	0.150
Indonesia	Anoa	45.2	0.040
Indonesia	Ardjuna	35.2	0.105
Indonesia	Attaka	43.3	0.040
Indonesia	Badak	49.5	0.032
Indonesia	Bekapai	41.2	0.080
Indonesia	Belida	45.1	0.020
Indonesia	Bima	21.1	0.250
Indonesia	Cinta	33.4	0.080
Indonesia	Duri (Sumatran Heavy)	21.3	0.180
Indonesia	Ikan Pari	48.0	0.020
Indonesia	Kakap	51.5	0.050
Indonesia	Katapa	50.8	0.060
Indonesia	Lalang (Malacca Straits)	39.7	0.050
Indonesia	Minas (Sumatran Light)	34.5	0.081
Indonesia	Udang	38.0	0.050
Iran	Aboozar (Ardeshir)	26.9	2.480
Iran	Bahrgansar/Nowruz	27.1	2.450
Iran	Dorrood (Darius)	33.6	2.350
Iran	Foroozan (Fereidoon)	31.3	2.500
Iran	Iranian Heavy	30.9	1.730
Iran	Iranian Light	33.8	1.350
Iran	Rostam	35.9	1.550
Iran	Salmon (Sassan)	33.9	1.910
Iraq	Basrah Heavy	24.7	3.500
Iraq	Basrah Light	33.7	1.950
Iraq	Basrah Medium	31.1	2.580
Iraq	North Rumaila	33.7	1.980
Ivory Coast	Espoir	32.3	0.340
Kazakhstan	Kumkol	42.5	0.07
Kuwait	Kuwait Export	31.4	2.52
Libya	Amna	36.0	0.15
Libya	Brega	40.4	0.21
Libya	Bu Attifel	43.3	0.04
Libya	Buri	26.2	1.76
Libya	Es Sider	37.0	0.45
Libya	Sarir	38.4	0.16
Libya	Sirtica	41.3	0.45
Libya	Zueitina	41.3	0.28
Malaysia	Bintulu	28.1	0.08
Malaysia	Dulang	39.0	0.12
Malaysia	Labuan	32.2	0.07
Malaysia	Miri Light	32.6	0.04
Malaysia	Tembungo	37.4	0.04
Mexico	Isthmus	33.3	1.49

(Continued)

TABLE 2.2 (*Continued*)
API Gravity and Sulfur Content of Selected Crude Oils

Country	Crude Oil	API	Sulfur % w/w
Mexico	Maya	22.2	3.30
Mexico	Olmeca	39.8	0.80
Neutral Zone	Burgan	23.3	3.37
Neutral Zone	Eocene	18.6	4.55
Neutral Zone	Hout	32.8	1.91
Neutral Zone	Khafji	28.5	2.85
Neutral Zone	Ratawi	23.5	4.07
Nigeria	Antan	32.1	0.32
Nigeria	Bonny Light	33.9	0.14
Nigeria	Bonny Medium	25.2	0.23
Nigeria	Brass River	42.8	0.06
Nigeria	Escravos	36.4	0.12
Nigeria	Forcados	29.6	0.18
Nigeria	Pennington	36.6	0.07
Nigeria	Qua Iboe	35.8	0.12
North Sea (Denmark)	Danish North Sea	34.5	0.260
North Sea (Norway)	Ekofisk	39.2	0.169
North Sea (Norway)	Emerald	22.0	0.750
North Sea (Norway)	Oseberg	33.7	0.310
North Sea (United Kingdom)	Alba	20.0	1.330
North Sea (United Kingdom)	Duncan	38.5	0.180
North Sea (United Kingdom)	Forties Blend	40.5	0.350
North Sea (United Kingdom)	Innes	45.7	0.130
North Sea (United Kingdom)	Kittiwake	37.0	0.65
North Yemen	Alif	40.3	0.10
Oman	Oman Export	34.7	0.94
Papua New Guinea	Kubutu	44.0	0.04
Peru	Loreto Peruvian	33.1	0.23
Qatar	Dukhan (Qatar Land)	40.9	1.27
Qatar	Qatar Marine	36.0	1.42
Ras Al Khaiman (UAE)	Ras Al Khaiman	44.3	0.15
Russia	Siberian Light	37.8	0.42
Saudi Arabia	Arab Extra Light (Berri)	37.2	1.15
Saudi Arabia	Arab Heavy (Safaniya)	27.4	2.80
Saudi Arabia	Arab Light	33.4	1.77
Saudi Arabia	Arab Medium (Zuluf	28.8	2.49
Sharjah (UAE)	Mubarek	37.0	0.62
Sumatra	Duri	20.3	0.21
Syria	Souedie	24.9	3.82
Timor Sea (Indonesia)	Hydra	37.5	0.08
Trinidad Tobago	Galeota Mix	32.8	0.27
Tunisia	Ashtart	30.0	0.99
United States (Alaska)	Alaskan North Slope	27.5	1.11
United States (Alaska)	Cook Inlet	35.0	0.10
United States (Alaska)	Drift River	35.3	0.09
United States (Alaska)	Nikiski Terminal	34.6	0.10
United States (California)	Hondo Sandstone	35.2	0.21

(*Continued*)

TABLE 2.2 (*Continued*)
API Gravity and Sulfur Content of Selected Crude Oils

Country	Crude Oil	API	Sulfur % w/w
United States (California)	Huntington Beach	20.7	1.38
United States (Florida)	Sunniland	24.9	3.25
United States (Louisiana)	Grand Isle	34.2	0.35
United States (Louisiana)	Lake Arthur	41.9	0.06
United States (Louisiana)	Louisiana Light Sweet	36.1	0.45
United States (Louisiana)	Ostrica	32.0	0.30
United States (Louisiana)	South Louisiana	32.8	0.28
United States (Michigan)	Lakehead Sweet	47.0	0.31
United States (New Mexico)	New Mexico Intermediate	37.6	0.17
United States (New Mexico)	New Mexico Light	43.3	0.07
United States (Oklahoma)	Basin-Cushing Composite	34.0	1.95
United States (Texas)	Coastal B-2	32.2	0.22
United States (Texas)	East Texas	37.0	0.21
United States (Texas)	West Texas Intermediate	40.8	0.34
United States (Texas)	Sea Breeze	37.9	0.10
United States (Texas)	West Texas Semi-sweet	39.0	0.27
United States (Texas)	West Texas Sour	34.1	1.640
United States (Wyoming)	Tom Brown	38.2	0.100
United States (Wyoming)	Wyoming Sweet	37.2	0.330
Venezuela	Lago Medio	32.2	1.010
Venezuela	Leona	24.4	1.510
Venezuela	Mesa	29.8	1.010
Venezuela	Ceuta Export	27.8	1.370
Venezuela	Guanipa	30.3	0.850
Venezuela	La Rosa Medium	25.3	1.730
Venezuela	Lago Treco	26.7	1.500
Venezuela	Oficina	33.3	0.780
Venezuela	Temblador	21.0	0.830
Venezuela	Tia Juana	25.8	1.630
Venezuela	Tia Juana Light	31.8	1.160
Venezuela	Tia Juana Medium 24	24.8	1.610
Venezuela	Tia Juana Medium 26	26.9	1.540
Vietnam	Bach Ho (White Tiger)	38.6	0.030
Vietnam	Dai Hung (Big Bear)	36.9	0.080
Yemen	Masila	30.5	0.670
Zaire	Zaire	31.7	0.130

2.3 PHYSICAL PROPERTIES

For the purposes of this text, a *physical property* is any property that is measurable and the value of which describes the physical state of petroleum that do not change the chemical nature of petroleum. The changes in the physical properties of a system can be used to describe its transformations (or evolutions between its momentary states). Physical properties are contrasted with chemical properties, which determine the way a material behaves in a chemical reaction.

Before any volatility tests are carried out, it must be recognized that the presence of more than 0.5% water in test samples of crude can cause several problems during various test procedures and produce erroneous results. For example, during various thermal tests, water (which has a high heat

TABLE 2.3
API Gravity and Sulfur Content of Selected Heavy Oils

Country	Crude Oil	API	Sulfur % w/w
Brazil	Albacor-Leste	18.9	0.66
Canada (Alberta)	Athabasca	8.0	4.8
Canada (Alberta)	Cold Lake	13.2	4.11
Canada (Alberta)	Lloydminster	16.0	2.60
Canada (Alberta)	Wabasca	19.6	3.90
Chad	Bolobo	16.8	0.14
Chad	Kome	18.5	0.20
China	Bozhong	16.7	0.30
China	Qinhuangdao	16.00	0.26
China	Zhao Dong	18.4	0.25
Colombia	Castilla	13.3	0.22
Colombia	Chichimene	19.8	1.12
Congo	Yombo	17.7	0.33
Ecuador	Ecuador Heavy	18.2	2.23
Ecuador	Napo	19.2	1.98
Guatemala	Xan-Coban	18.7	6.00
Indonesia	Kulin (South)	19.8	0.30
Iran	Soroosh (Cyrus)	18.1	3.30
Kuwait	Eocene	18.4	4.00
United States (California)	Arroyo Grande/Edna	14.9	2.03
United States (California)	Belridge (South)	13.7	1.00
United States (California)	Beta Offshore	15.9	3.60
United States (California)	Beta Offshore	16.9	3.30
United States (California)	Hondo Monterey	19.4	4.70
United States (California)	Hondo Monterey	17.2	4.70
United States (California)	Huntington Beach	19.4	2.00
United States (California)	Huntington Beach	14.4	0.90
United States (California)	Kern River	13.3	1.10
United States (California)	Kern River	14.4	1.02
United States (California	Lost Hills	18.4	1.00
United States (California)	Midway Sunset	12.6	1.60
United States (California)	Midway Sunset	11.0	1.55
United States (California)	Monterey	12.2	2.30
United States (California)	Mount Poso	16.0	0.70
United States (California)	Newport Beach	15.1	2.00
United States (California)	Point Arguello	19.5	3.50
United States (California)	Point Pedernales	15.9	5.10
United States (California)	San Ardo	12.2	2.30
United States (California)	San Joaquin Valley	15.7	1.20
United States (California)	Santa Maria	13.7	5.20
United States (California)	Sockeye	19.6	5.30
United States (California)	Sockeye	15.9	5.40
United States (California)	Torrance	18.2	1.80
United States (California)	Wilmington	18.6	1.59
United States (California)	Wilmington	16.9	1.70
United States (Mississippi)	Baxterville	16.3	3.02

(Continued)

TABLE 2.3 (*Continued*)
API Gravity and Sulfur Content of Selected Heavy Oils

Country	Crude Oil	API	Sulfur % w/w
Venezuela	Bachaquero	16.3	2.35
Venezuela	Bachaquero	12.2	2.80
Venezuela	Bachaquero	14.4	2.52
Venezuela	Bachaquero Heavy	10.7	2.78
Venezuela	Boscan	10.1	5.50
Venezuela	Hamaca	8.4	3.82
Venezuela	Jobo	9.2	4.10
Venezuela	Laguna	10.9	2.66
Venezuela	Lagunillas Heavy	17.0	2.19
Venezuela	Merey	18.0	2.28
Venezuela	Morichal	12.2	2.78
Venezuela	Pilon	14.1	1.91
Venezuela	Tia Juana Pesado	12.1	2.70
Venezuela	Tia Juana Heavy	18.2	2.24
Venezuela	Tia Juana Heavy	11.6	2.68
Venezuela	Tremblador	19.0	0.80
Venezuela	Zuata	15.7	2.69

Note: For reference, Athabasca tar sand bitumen has API = 8° and sulfur content = 4.8%–5.0% w/w.

of vaporization) requires the application of additional thermal energy to the distillation flask. In addition, water is relatively easily superheated and therefore excessive *bumping* can occur, leading to erroneous readings and the potential for the destruction of the glass equipment is real. Steam formed during distillation can act as a carrier gas, and high boiling point components may end up in the distillate (often referred to as *steam distillation*).

Removal of water (and sediment) can be achieved by centrifugation if the sample is not a tight emulsion. Other methods that are used to remove water include (1) heating in a pressure vessel to control loss of light ends, (2) addition of calcium chloride as recommended in ASTM D1160, (3) addition of an agent such as *iso*-propanol or *n*-butanol, (4) removal of water in a preliminary low efficiency or flash distillation followed by reblending the hydrocarbon that codistills with the water into the sample (see also IP 74), and (5) separation of the water from the hydrocarbon distillate by freezing.

Thus, the standard test methods described later in the various sections (presented in alphabetical order rather than order of preference, which is feedstock dependent) are based on the assumption that water has been reduced to an acceptable level, which is usually defined by each standard test.

2.3.1 Acid Number

Another characteristic of crude oil is the *TAN*, which represents a composite of acids present in the crude oil (ASTM D664) and is often also expressed as the *neutralization number*. Crude oils having a high acid number (high TAN number) account for an increasing percentage of the global crude oil market, and, in fact, the increase in world production of heavy, sour, and high-TAN crude oils will impact many world's (especially those of the United States) refineries (Shafizadeh et al., 2003; Sheridan, 2006).

High-acid crude oils are considered to be those with an acid content >1.0 mg KOH/g sample (many refiners consider TAN number greater than 0.5 mg KOH/g to be high) and refiners looking for discounted crude supplies will import and use greater volumes of high-total acid number (TAN) crude oils.

In the United States, using California as the example, Wilmington crude oil and Kern crude oil have a TAN ranging from 2.2 to 3.2 mg, respectively. However, some acids are relatively inert, and the TAN number does not always give a true reflection of the corrosive properties of the crude oil. Furthermore, different naphthenic acids (a broad group of organic acids in crude oil) will react at different temperatures—making it difficult to pinpoint the processing units within the refinery that will be affected by a particular high-TAN crude oil. Nonetheless, high-TAN crude oils contain naphthenic acids, and these acids corrode the distillation unit in the refinery and form sludge and gum that can block pipelines and pumps entering the refinery. The impact of corrosive, high-TAN crude oils can be overcome by blending higher- and lower-TAN crude oils, by installing or retrofitting equipment with anticorrosive materials, or by developing low-temperature catalytic decarboxylation processes using metal catalysts such as copper.

In the test method, the sample normally dissolved in toluene/isopropyl alcohol/water is titrated with potassium hydroxide, and the results are expressed as milligrams of potassium hydroxide per gram of sample (mg KOH/g). Crude oils having high acid numbers have a high potential to cause corrosion problems in the refineries, especially in the atmospheric and vacuum distillation units where the hot crude oil first comes into contact with hot metal surfaces. Crude oil typically has a total acid number (TAN value) value on the order of 0.05–6.0 mg KOH/g of sample.

Current methods for the determination of the acid content of hydrocarbon compositions are well established (ASTM D664), which includes potentiometric titration in nonaqueous conditions to clearly defined end points as detected by changes in millivolts readings versus volume of titrant used. A color indicator method (ASTM D974) is also available.

2.3.1.1 Potentiometric Titration

In this method (ASTM D664), the sample is normally dissolved in toluene and propanol with a little water and titrated with alcoholic potassium hydroxide (if sample is acidic). A glass electrode and reference electrode are immersed in the sample and connected to a voltmeter/potentiometer. The meter reading (in millivolts) is plotted against the volume of titrant. The end point is taken at the distinct inflection of the resulting titration curve corresponding to the basic buffer solution.

2.3.1.2 Color-Indicating Titration

In this test method (ASTM D947), an appropriate pH color indicator (such as phenolphthalein) is used. The titrant is added to the sample by means of a burette, and the volume of titrant used to cause a permanent color change in the sample is recorded from which the *total acid number* is calculated. It can be difficult to observe color changes in crude oil solutions. It is also possible that the results from the color indicator method may or may not be the same as the potentiometric results.

Test method ASTM D1534 is similar to ASTM D974 in that they both use a color change to indicate the end point. ASTM D1534 is designed for electric insulating oils (transformer oils), where the viscosity will not exceed 24 cSt at 40°C. The standard range of applications is for oils with an acid number between 0.05 and 0.50 mg KOH/g, which is applicable to the transformer oils. Test method ASTM D3339 is also similar to ASTM D974 but is designed for use on smaller oil samples. ASTM D974 and D664 roughly use a 20 g sample; ASTM D3339 uses a 2.0 g sample.

In terms of *repeatability* (*the difference between successive test results obtained by the same operator with the same apparatus under constant operating conditions on identical test material*), data acquired using D664 were found to be within ±7% of the mean 95% of the time for fresh oils using the inflection point method or ±12% of the mean for used oils with the buffer end point method. On the other hand, when using ASTM D974, a sample that having an acid number of 0.15 could vary from 0.10 to 0.20 and, when using ASTM D664, the acid number could vary from 0.17 to 0.13. In terms of *reproducibility* (*the difference between two single independent results obtained by different operators working in different laboratories on identical test material*), 95% of the time, the reproducibility of ASTM D664 is ±20% of the mean for fresh oils using the inflection point method or ±44% of the mean for used oil using the buffer end point method. For example, if a mean acid number was 0.10, the results could be expected to vary from an acid number of 0.14 to an acid number of 0.06, 95% of the time. When using ASTM D974, the analyses (on the same oil) from multiple laboratories could vary from 0.09 to 0.01. Furthermore, according to ASTM, the AN obtained by this standard (D664) may or may not be numerically the same as that obtained in accordance with test methods D974 and D3339.

In addition, the total acid number (TAN) values as conventionally analyzed in accordance with standard test methods (ASTM D664) do not correlate at all with their risk of forming naphthenates or other soaps during production in oilfields. The TAN of oil has frequently been used to quantify the presence of naphthenic acids, because the carboxylic acid components of oils are believed to be largely responsible for oil acidity. However, more recent research has begun to highlight deficiencies in relying upon this method for such a direct correlation, and the total acid number is no longer considered to be such a reliable indicator. Furthermore, the ASTM D974 test method is an older method and used for distillates, while the ASTM D664 test method is more accurate but measures acid gases and hydrolyzable salts in addition to organic acids. These differences are important on crude oils but less significant on distillates, and the Nalco NAT testing is more precise for quantifying the naphthenic acid content. Inorganic acids, esters, phenolic compounds, sulfur compounds, lactones, resins, salts, and additives such as inhibitors and detergents interfere with both methods. In addition, these ASTM methods do not differentiate between naphthenic acids, phenols, carbon dioxide, hydrogen sulfide, mercaptan derivatives, and other acidic compounds present in the oil.

2.3.2 Elemental Analysis

The elemental analysis (ultimate analysis) of petroleum for the percentages of carbon, hydrogen, nitrogen, oxygen, and sulfur is perhaps the first method used to examine the general nature, and perform an evaluation, of a feedstock. The atomic ratios of the various elements to carbon (i.e., H/C, N/C, O/C, and S/C) are frequently used for indications of the overall character of the feedstock. It is also of value to determine the amounts of trace elements, such as vanadium, nickel, and other metals, in a feedstock since these materials can have serious deleterious effects on catalyst performance during refining by catalytic processes.

However, it has become apparent, with the introduction of the heavier feedstocks into refinery operations, that these ratios are not the only requirement for predicting feedstock character before refining. The use of more complex feedstocks (in terms of chemical composition) has added a new dimension to refining operations. Thus, although atomic ratios, as determined by elemental analyses, may be used on a comparative basis between feedstocks, there is now no guarantee that a particular feedstock will behave as predicted from these data. Product slates cannot be predicted accurately, if at all, from these ratios.

The ultimate analysis (elemental composition) of petroleum is not reported to the same extent as for coal (Speight, 1994). Nevertheless, there are ASTM procedures for the ultimate analysis of petroleum and petroleum products but many such methods may have been designed for other materials (Speight, 2015). For example, *carbon content* can be determined by the method designated for coal and coke (ASTM D3178) or by the method designated for municipal solid waste (ASTM E777). There are also methods designated for the followings:

For example, *carbon content* can be determined by the method designated for coal and coke (ASTM D3178) or by the method designated for municipal solid waste (ASTM E777). There are also methods designated for

1. *Carbon* and *hydrogen content* (ASTM D1018, ASTM D3178, ASTM D3343, ASTM D3701, ASTM D5291, ASTM E777, IP 338)
2. *Nitrogen content* (ASTM D3179, ASTM D3228, ASTM E258, ASTM D5291, and ASTM E778)
3. *Oxygen content* (ASTM E385)

4. *Sulfur content* (ASTM D129, ASTM D139, ASTM D1266, ASTM D1552, ASTM D1757, ASTM D2622, ASTM D3120, ASTM D3177, ASTM D4045, and ASTM D4294)
5. *Metals content* (ASTM C1109, ASTM C1111, ASTM D482, ASTM D1318, ASTM D3340, ASTM D3341, ASTM D3605)

Of the data that are available, the proportions of the elements in petroleum vary only slightly over narrow limits:

Carbon	83.0%–87.0%
Hydrogen	10.0%–14.0%
Nitrogen	0.1%–2.0%
Oxygen	0.05%–1.5%
Sulfur	0.05%–6.0%
Metals (Ni and V)	<1000 ppm

And yet, there is a wide variation in physical properties from the lighter more mobile crude oil at one extreme to the heavier asphaltic crude oils at the other extreme. The majority of the more aromatic species and the heteroatoms occur in the higher-boiling fractions of feedstocks. The heavier feedstocks are relatively rich in these higher-boiling fractions (Speight, 2014, 2015).

Of the ultimate analytical data, more has been made of the sulfur content than any other property. For example, the sulfur content (ASTM D1552, ASTM D4294) and the API gravity represent the two properties that have, in the past, had the greatest influence on determining the value of petroleum as a feedstock. The sulfur content varies from approximately 0.1% w/w to approximately 3% w/w for the more conventional crude oils to as much as 5%–6% for heavy oil and bitumen. Residua, depending on the sulfur content of the crude oil feedstock, may be of the same order or even have higher sulfur content. Indeed, the very nature of the distillation process by which residua are produced, that is, removal of distillate without thermal decomposition, dictates that the majority of the sulfur, which is located predominantly in the higher-molecular-weight fractions, be concentrated in the residuum.

The sulfur content varies from approximately 0.1 wt% to approximately 3 wt% for the more conventional crude oils to as much as 5%–6% for heavy oil and bitumen. Residua, depending on the sulfur content of the crude oil feedstock, may be of the same order or even have a substantially higher sulfur content. Indeed, the very nature of the distillation process by which residua are produced, that is, removal of distillate without thermal decomposition, dictates that the majority of the sulfur, which is located predominantly in the higher-molecular-weight fractions, be concentrated in the residuum.

2.3.3 DENSITY AND SPECIFIC GRAVITY

The *density* and *specific gravity* of crude oil (ASTM D70, ASTM D71, ASTM D287, ASTM D1217, ASTM D1298, ASTM D1480, ASTM D1481, ASTM D1555, ASTM D1657, ASTM D4052, IP 235, IP 160, IP 249, IP 365) are two properties that have found wide use in the industry for preliminary assessment of the character and quality of crude oil.

Density is the mass of a unit volume of material at a specified temperature and has the dimensions of grams per cubic centimeter (a close approximation to grams per milliliter). *Specific gravity* is the ratio of the mass of a volume of the substance to the mass of the same volume of water and is dependent on two temperatures, those at which the masses of the sample and the water are measured. When the water temperature is 4°C (39°F), the specific gravity is equal to the density in the centimeter–gram–second (cgs) system, since the volume of 1 g of water at that temperature is, by definition, 1 mL. Thus, the density of water, for example, varies with temperature, and its specific gravity at equal temperatures is always unity. The standard temperatures for a specific gravity in the petroleum industry in North America are 60/60°F (15.6/15.6°C).

In the early years of the petroleum industry, density was the principal specification for petroleum and refinery products; it was used to give an estimation of the gasoline and, more particularly, the kerosene present in the crude oil. However, the derived relationships between the density of petroleum and its fractional composition were valid only if they were applied to a certain type of petroleum and lost some of their significance when applied to different types of petroleum. Nevertheless, density is still used to give a rough estimation of the nature of petroleum and petroleum products. Although density and specific gravity are used extensively, the API (American Petroleum Institute) gravity is the preferred property:

$$\text{Degrees API} = (141.5/\text{sp g at } 60/60°F) - 131.5$$

Specific gravity is influenced by the chemical composition of petroleum, but quantitative correlation is difficult to establish. Nevertheless, it is generally recognized that increased amounts of aromatic compounds result in an increase in density, whereas an increase in saturated compounds results in a decrease in density. Indeed, it is also possible to recognize certain preferred trends between the density of petroleum and one or another of the physical properties. For example, an approximate correlation exists between the density (API gravity) and sulfur content, Conradson carbon residue, viscosity, and nitrogen content (Speight, 2000).

The density or specific gravity of petroleum, petroleum products, heavy oil, and bitumen may be measured by means of a hydrometer (ASTM D287, ASTM D1298, ASTM D1657, IP 160), a pycnometer (ASTM D70, ASTM D1217, ASTM D1480, and ASTM D1481), or a digital density meter (ASTM D4052) and a digital density analyzer (ASTM D5002). Not all of these methods are suitable for measuring the density or specific gravity of heavy oil and bitumen although some methods lend themselves to adaptation. The API gravity of a feedstock (ASTM D287) is calculated directly from the specific gravity.

The pycnometer method (ASTM D70, ASTM D1217, ASTM D1480, ASTM D1481) for determining density is reliable, precise, and requires relatively small test samples. However, because of the time required, other methods such as using the hydrometer (ASTM D1298), the density meter (ASTM D4052), and the digital density analyzer (ASTM D5002) are often preferred. However, surface tension effects can affect the displacement method, and the density meter method loses some of its advantage when measuring the density of heavy oil and bitumen.

The pycnometer method (ASTM D70, ASTM D1217, ASTM D1480, and ASTM D1481) is routinely used to measure the density of samples being charged to a distillation flask, where volume charge is needed, but the volume is not conveniently measured. The volume may be found weighing the sample and determining the sample density. It is also used in routine measurements of material properties. It is worthy of note that even a small amount of solids in the sample will influence its measured density. For example, one per cent by weight solids in the sample can raise the density by 0.007 g/cm³.

The densimeter method (ASTM D4052) uses an instrument that measures the total mass of a tube by determining its natural frequency of vibration. This frequency is a function of the dimensions and the elastic properties of the tube and the weight of the tube and contents. Calibration with water and air provides data for the determination of the instrument constraints, which allow conversion of the natural frequency of vibration to sample density.

The density of petroleum usually ranges from approximately 0.8 (45.3° API) for the lighter crude oils to over 1.0 (less than 10° API) for heavy crude oil and bitumen. The variation of density with temperature (Table 2.4), effectively the coefficient of expansion, is a property of great technical importance, since most petroleum products are sold by volume and specific gravity is usually determined at the prevailing temperature (21°C, 70°F) rather than at the standard temperature (60°F, 15.6°C). The tables of gravity corrections (ASTM D1555) are based on an assumption that the coefficient of expansion of all petroleum products is a function (at fixed temperatures) of density only.

TABLE 2.4
Variation of Density and API Gravity of Selected Residua with Temperature and Pressure

Source Residuum	Temperature °C	°F							
			Pressure, psi	14.21	2,843	5,685	8,528	11,371	14,214
			Pressure, atmos.	0.97	193	387	580	774	967
			Pressure, MPa	0.098	19.6	39.2	58.8	78.4	98.0
California	25	77	Density, g/cc	1.014	1.023	1.031	1.038	1.045	1.051
			API gravity	8.0	6.8	5.7	4.8	3.9	3.3
	45	113	Density, g/cc	1.002	1.011	1.020	1.028	1.035	1.041
			API gravity	9.7	8.5	7.2	6.1	5.2	4.4
	65	149	Density, g/cc	0.990	1.000	1.009	1.017	1.025	1.032
			API gravity	11.4	10.0	8.7	7.6	6.6	5.6
Venezuela	25	77	Density, g/cc	1.024	1.032	1.040	1.048	1.054	1.061
			API gravity	6.7	5.6	4.6	3.5	2.7	1.9
	45	113	Density, g/cc	1.012	1.020	1.029	1.037	1.044	1.051
			API gravity	8.3	7.2	6.0	5.0	4.0	3.1
	65	149	Density, g/cc	1.000	1.009	1.018	1.027	1.034	1.041
			API gravity	10.0	8.7	7.5	6.3	5.3	4.4

The API gravity of a feedstock (ASTM D287) is calculated directly from the specific gravity. The specific gravity of bitumen shows a fairly wide range of variation. The largest degree of variation is usually due to local conditions that affect material close to the faces, or exposures, occurring in surface oil sand beds. There are also variations in the specific gravity of the bitumen found in beds that have not been exposed to weathering or other external factors. The range of specific gravity usually varies over the range of the order of 0.995–1.04.

A very important property of the Athabasca bitumen (which also accounts for the success of the hot water separation process) is the variation in density (specific gravity) of the bitumen with temperature. Over the temperature range 30°C–130°C (85°F–265°F) the bitumen is lighter than water. Flotation of the bitumen (with aeration) on the water is facilitated, hence the logic of the hot water separation process (Speight, 2005, 2009).

2.3.4 METAL CONTENT

Heteroatoms (nitrogen, oxygen, sulfur, and *metals)* are found in every crude oil, and the concentrations have to be reduced to convert the oil to transportation fuel. The reason is that if nitrogen and sulfur are present in the final fuel during combustion, nitrogen oxides (NO_x) and sulfur oxides (SO_x) form, respectively. In addition, metals affect many upgrading processes adversely, poisoning catalysts in refining and causing deposits in combustion.

Heteroatoms do affect every aspect of refining. Sulfur is usually the most concentrated and is fairly easy to remove; many commercial catalysts are available, which routinely remove 90% of the sulfur. Nitrogen is more difficult to remove than sulfur, and there are fewer catalysts that are specific for nitrogen. Metals cause particular problems because they poison catalysts used for sulfur and nitrogen removal, as well as other processes such as catalytic cracking.

Metals cause problems during petroleum refining because they poison the catalysts used for sulfur and nitrogen removal, as well as catalysts other processes such as catalytic cracking. Heavy oils and residua contain relatively high proportions of metals either in the form of salts or as organometallic constituents (such as the metalloporphyrins), which are extremely difficult to remove from the feedstock. Indeed, the nature of the process by which residua are produced virtually dictates that all the metals in the original crude oil are concentrated in the residuum (Speight, 2000). Those metallic constituents that may actually *volatilize* under the distillation conditions and appear in the higher-boiling distillates are the exceptions here. The deleterious effect of metallic constituents on the catalyst is known, and serious attempts have been made to develop catalysts that can tolerate a high concentration of metals without serious loss of catalyst activity or catalyst life.

A variety of tests have been designated for the determination of metals on petroleum products (ASTM D1318, ASTM D3340, ASTM D3341, ASTM D3605). At the time of this writing, the specific test for the determination of metals in whole feeds has not been designated. However, this task can be accomplished by the combustion of the sample so that only inorganic ash remains. The ash can then be digested with an acid and the solution examined for metal species by atomic absorption (AA) spectroscopy or by inductively coupled argon plasma (ICP) spectrometry.

2.3.5 Surface and Interfacial Tension

Surface tension is a measure of the force acting at a boundary between two phases. If the boundary is between a liquid and a solid or between a liquid and a gas (air) the attractive forces are referred to as surface tension, but the attractive forces between two immiscible liquids are referred to as *interfacial tension.*

Temperature and molecular weight have a significant effect on surface tension (Tables 2.5 and 2.6). For example, in the normal hydrocarbon series, a rise in temperature leads to a decrease in the surface tension, but an increase in molecular weight increases the surface tension. A similar trend, that is, an increase in molecular weight causing an increase in surface tension, also occurs in the acrylic series and, to a lesser extent, in the alkylbenzene series.

The surface tension of petroleum and petroleum products has been studied for many years. The narrow range of values (approximately 24–38 dynes/cm) for such widely diverse materials as gasoline (26 dynes/cm), kerosene (30 dynes/cm), and the lubricating fractions (34 dynes/cm) has rendered the surface tension of little value for any attempted characterization. However, it is generally acknowledged that nonhydrocarbon materials dissolved in an oil reduce the surface tension: polar compounds, such as soaps and fatty acids, are particularly active. The effect is marked at low concentrations up to a critical value beyond which further additions cause little change; the critical value corresponds closely with that required for a monomolecular layer on the exposed surface, where it is adsorbed and accounts for the lowering.

A high proportion of the complex phenomena shown by emulsions and foams can be traced to these induced surface tension effects. Dissolved gases, even hydrocarbon gases, lower the surface tension of oils, but the effects are less dramatic and the changes probably result from dilution. The matter is presumably of some importance in petroleum production engineering in which the viscosity and surface tension of the reservoir fluid may govern the amount of oil recovered under certain conditions.

On the other hand, although petroleum products show little variation in surface tension, within a narrow range, the *interfacial tension* of petroleum, especially of petroleum products, against aqueous solutions provides valuable information (ASTM D971). Thus, the interfacial tension of petroleum is subject to the same constraints as surface tension, that is, differences in composition, molecular weight, and so on. When oil–water systems are involved, the pH of the aqueous phase influences the tension at the interface; the change is small for highly refined oils, but increasing pH causes a rapid decrease for poorly refined, contaminated, or slightly oxidized oils.

A change in interfacial tension between oil and alkaline water has been proposed as an index for following the refining or deterioration of certain products, such as turbine and insulating oils. When surface or interfacial tensions are lowered by the presence of solutes, which tend to concentrate on the surface, time is required to obtain the final concentration and hence the final value of

TABLE 2.5
Surface Tension of Selected Hydrocarbons

Hydrocarbon		Surface Tension			
	°C	20	38	93	
	°F	68	100	200	
n-Pentane		16.0	14.0	8.0	dyne/cm
		16.0	14.0	8.0	mN/m
n-Hexane		18.4	16.5	10.9	dyne/cm
		18.4	16.5	10.9	mN/m
n-Heptane		20.3	18.6	13.1	dyne/cm
		20.3	18.6	13.1	mN/m
n-Octane		21.8	20.2	14.9	dyne/cm
		21.8	20.2	14.9	mN/m
Cyclopentane		22.4			dyne/cm
		22.4			mN/m
Cyclohexane		25.0			dyne/cm
		25.0			mN/m
Tetralin		35.2			dyne/cm
		35.2			mN/m
Decalin		29.9			dyne/cm
		29.9			mN/m
Benzene		28.8			dyne/cm
		28.8			mN/m
Toluene		28.5			dyne/cm
		28.5			mN/m
Ethylbenzene		29.0			dyne/cm
		29.0			mN/m
n-Butylbenzene		29.2			dyne/cm
		29.2			mN/m

TABLE 2.6
**Effect of Temperature on the Surface
Tension of Athabasca Bitumen**

Temperature		Surface Tension	
°C	°F	dyne/cm	mN/m
21.1	70.0	35.3	35.3
23.3	74.0	34.7	34.7
30.0	86.0	30.1	30.1
43.3	110.0	27.3	27.3
51.7	125.0	28.0	28.0
65.6	150.0	25.4	25.4
73.9	165.0	22.5	22.5
82.2	180.0	21.0	21.0
87.8	190.0	18.9	18.9
95.6	204.0	20.0	20.0
104.0	219.0	19.2	19.2
123.9	255.0	18.2	18.2

the tension. In such systems, dynamic and static tension must be distinguished; the first concerns the freshly exposed surface having nearly the same composition as the body of the liquid; it usually has a value only slightly less than that of the pure solvent. The static tension is that existing after equilibrium concentrations have been reached at the surface.

The interfacial tension between oil and distilled water provides an indication of compounds in the oil that have an affinity for water. The measurement of interfacial tension has received special attention because of its possible use in predicting when an oil in constant use will reach the limit of its serviceability. This interest is based on the fact that oxidation decreases the interfacial tension of the oil. Furthermore, the interfacial tension of turbine oil against water is lowered by the presence of oxidation products, impurities from the air or rust particles, and certain antirust compounds intentionally blended in the oil. Thus, a depletion of the antirust additive may cause an increase in interfacial tension, whereas the formation of oxidation products or contamination with dust and rust lowers the interfacial tension.

In following the performance of oil in service, a decrease in interfacial tension indicates oxidation, if it is known that antirust additives and contamination with dust and rust are absent. In the absence of contamination and oxidation products, an increase in interfacial tension indicates a depletion trend in the antirust additive. Very minor changes over appreciable periods of time signify satisfactory operating conditions. The addition of makeup oil to a system introduces further complications in following the effects of service on the interfacial tension of a particular charge of oil.

2.3.6 VISCOSITY

Viscosity is the force in dynes required to move a plane of 1 cm² area at a distance of 1 cm from another plane of 1 cm² area through a distance of 1 cm in 1 s. In the centimeter–gram–second (cgs) system, the unit of viscosity is the poise or centipoise (0.01 P). Two other terms in common use are *kinematic viscosity* and *fluidity*. The kinematic viscosity is the viscosity in centipoises divided by the specific gravity, and the unit is the stoke (cm²/s), although centistokes (0.01 cSt) is in more common usage; fluidity is simply the reciprocal of viscosity. The viscosity (ASTM D445, ASTM D88, ASTM D2161, ASTM D341, ASTM D2270) of crude oils varies markedly over a very wide range. Values vary from less than 10 cP at room temperature to many thousands of centipoises at the same temperature.

In the early days of the petroleum industry, viscosity was regarded as the *body* of petroleum, a significant number for lubricants or for any liquid pumped or handled in quantity. The changes in viscosity with temperature, pressure, and rate of shear are pertinent not only in lubrication but also for such engineering concepts as heat transfer. The viscosity and relative viscosity of different phases, such as gas, liquid oil, and water, are determining influences in producing the flow of reservoir fluids through porous oil-bearing formations. The rate and amount of oil production from a reservoir are often governed by these properties.

Many types of instruments have been proposed for the determination of viscosity. The simplest and most widely used are capillary types (ASTM D445), and the viscosity is derived from the following equation:

$$\mu = \frac{\pi r^4 P}{8nl}$$

where
   r is the tube radius
   l is the tube length
   P is the pressure difference between the ends of a capillary
   n is the *coefficient of viscosity*
   µ is the quantity discharged in unit time

Not only are such capillary instruments the most simple, but when designed in accordance with known principle and used with known necessary correction factors, they are probably the most accurate viscometers available. It is usually more convenient, however, to use relative measurements, and for this purpose, the instrument is calibrated with an appropriate standard liquid of known viscosity.

Batch flow times are generally used; in other words, the time required for a fixed amount of sample to flow from a reservoir through a capillary is the datum actually observed. Any features of technique that contribute to longer flow times are usually desirable. Some of the principal capillary viscometers in use are those of Cannon-Fenske, Ubbelohde, Fitzsimmons, and Zeitfuchs.

The Saybolt Universal Viscosity (SUS) (ASTM D88) is the time in seconds required for the flow of 60 mL of petroleum from a container, at constant temperature, through a calibrated orifice. The Saybolt Furol Viscosity (SFS) (ASTM D88) is determined in a similar manner except that a larger orifice is employed.

As a result of the various methods for viscosity determination, it is not surprising that much effort has been spent on interconversion of the several scales, especially converting Saybolt to Kinematic Viscosity (ASTM D2161),

$$\text{Kinematic Viscosity} = a \times \text{Saybolt s} + b/\text{Saybolt s}$$

In this equation, a and b are constants.

The Saybolt Universal Viscosity equivalent to a given kinematic viscosity varies slightly with the temperature at which the determination is made because the temperature of the calibrated receiving flask used in the Saybolt method is not the same as that of the oil. Conversion factors are used to convert kinematic viscosity from 2 to 70 cSt at 38°C (100°F) and 99°C (210°F) to equivalent Saybolt Universal Viscosity in seconds. Appropriate multipliers are listed to convert kinematic viscosity over 70 cSt. For a Kinematic Viscosity determined at any other temperature, the equivalent Saybolt Universal Value is calculated by the use of the Saybolt equivalent at 38°C (100°F) and a multiplier that varies with the temperature:

$$\text{Saybolt s at } 100°F \ (38°C) = \text{cSt} \times 4.635$$

$$\text{Saybolt s at } 210°F \ (99°C) = \text{cSt} \times 4.667$$

Various studies have also been conducted on the effect of temperature on viscosity since the viscosity of petroleum, or a petroleum product, decreases as the temperature increases. The rate of change appears to depend primarily on the nature or composition of the petroleum, but other factors, such as volatility, may also have a minor effect. The effect of temperature on viscosity is generally represented by the following equation:

$$\log \log (n + c) = A + B \log T$$

where
  n is the absolute viscosity
  T is the temperature
  A and B are the constants

This equation has been sufficient for most purposes and has come into very general use. The constants A and B vary widely with different oils, but c remains fixed at 0.6 for all oils having a viscosity over 1.5 cSt; it increases only slightly at lower viscosity (0.75 at 0.5 cSt). The viscosity–temperature characteristics of any oil, so plotted, thus create a straight line, and the parameters A and B are equivalent to the intercept and slope of the line. To express the viscosity and viscosity–temperature characteristics of an oil, the slope and the viscosity at one temperature must be known; the usual practice is to select 38°C (100°F) and 99°C (210°F) as the observation temperatures.

Suitable conversion tables are available (ASTM D341), and each table or chart is constructed in such a way that for any given petroleum or petroleum product, the viscosity–temperature points result in a straight line over the applicable temperature range. Thus, only two viscosity measurements need to be made at temperatures far enough apart to determine a line on the appropriate chart from which the approximate viscosity at any other temperature can be read. The charts can be applicable only to measurements made in the temperature range in which a given petroleum oil is a Newtonian liquid. The oil may cease to be a simple liquid near the cloud point because of the formation of wax particles or, near the boiling point, because of vaporization. However, the charts do not give accurate results when either the cloud point or boiling point is approached but they are useful over the Newtonian range for estimating the temperature at which oil attains a desired viscosity.

Since the viscosity–temperature coefficient of lubricating oil is an important expression of its suitability, a convenient number to express this property is very useful, and hence, a viscosity index (ASTM D2270) was derived. It is established that naphthenic oils have higher viscosity–temperature coefficients than do paraffinic oils at equal viscosity and temperatures. The Dean and Davis scale was based on the assignment of a zero value to a typical naphthenic crude oil and that of 100 to a typical paraffinic crude oil; intermediate oils were rated by the following formula:

$$\text{Viscosity index} = \frac{L-U}{L-H \times 100}$$

where
   L and H are the viscosities of the zero and 100 index reference oils, both having the same viscosity at 99°C (210°F)
   U is that of the unknown, all at 38°C (100°F)

Originally, the viscosity index was calculated from Saybolt Viscosity data, but subsequently data were provided for kinematic viscosity.

The viscosity of petroleum fractions increases on the application of pressure, and this increase may be very large. The pressure coefficient of viscosity correlates with the temperature coefficient, even when oils of widely different types are compared. A plot of the logarithm of the kinematic viscosity against pressure for several oils has given reasonably linear results up to approximately 20,000 psi, and the slopes of the isotherms are such that extrapolated values for a given oil intersect. At higher pressures, the viscosity decreases with increasing temperature, as at atmospheric pressure; in fact, viscosity changes of small magnitude are usually proportional to density changes, whether these are caused by pressure or by temperature.

The classification of lubricating oil by viscosity is a matter of some importance. A useful system is that of the Society of Automotive Engineers. Each oil class carries an index designation. For those classes designated by letter and number, maximum viscosity and minimum viscosity are specified at −18°C (0°F); those designated by number are only specified in viscosity at 99°C (210°F). Viscosity is also used in specifying several grades of fuel oils and in setting the requirement for kerosene and insulating oil.

2.4 THERMAL PROPERTIES

The thermal properties of petroleum are those properties (or characteristics) that determine how petroleum will behave (or react) when it is subjected to excessive heat or heat fluctuations over time.

As with all properties of petroleum, a collection of standard test methods is instrumental in the evaluation and assessment of the thermal properties (Speight, 2015). These standards allow petroleum refineries and other geological and chemical processing plants to appropriately examine and process petroleum in a safe and efficient manner.

2.4.1 ANILINE POINT

The *aniline point* of a liquid was originally defined as the *consolute* or *critical solution temperature* of the two liquids, that is, the minimum temperature at which they are miscible in all proportions. The term is now most generally applied to the temperature at which exactly equal parts of the two are miscible. This value is more conveniently measured than the original value and is only a few tenths of a degree lower for most substances.

Although it is an arbitrary index (ASTM D611), the aniline point is of considerable value in the characterization of petroleum products. For oils of a given type, it increases slightly with the molecular weight; for those of a given molecular weight, it increases rapidly with increasing paraffinic character. As a consequence, it was one of the first properties proposed for the group analysis of petroleum products with respect to aromatic and naphthene content. It is used, alternately, even in one of the more recent methods. The simplicity of the determination makes it attractive for the rough estimation of aromatic content when that value is important for functional requirements, as in the case of the solvent power of naphtha and the combustion characteristics of gasoline and diesel fuel.

2.4.2 CARBON RESIDUE

Petroleum products are mixtures of many compounds that differ widely in their physical and chemical properties. Some of them may be vaporized in the absence of air at atmospheric pressure without leaving an appreciable residue. Other nonvolatile compounds leave a *carbonaceous residue* when destructively distilled under such conditions. This residue is known as carbon residue when determined in accordance with prescribed procedure.

The carbon residue is a property that can be correlated with several other properties of petroleum (Speight, 2000; Figure 2.2); hence, it also presents indications of the volatility of the crude oil and

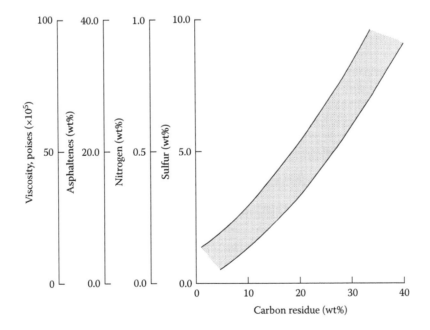

FIGURE 2.2 Relationship of carbon residue to other physical properties.

the coke-forming (or gasoline-producing) propensity. However, tests for carbon residue are sometimes used to evaluate the carbonaceous depositing characteristics of fuels used in certain types of oil-burning equipment and internal combustion engines.

The mechanical design and operating conditions of such equipment have such a profound influence on carbon deposition during service that comparison of carbon residues between oils should be considered as giving only a rough approximation of relative deposit-forming tendencies. A more precise relationship between carbon residue and hydrogen content, H/C atomic ratio, nitrogen content, and sulfur content has been shown to exist. These data can provide more precise information about the anticipated behavior of a variety of feedstocks in thermal processes.

Because of the extremely small values of carbon residue obtained by the Conradson and Ramsbottom methods when applied to the lighter distillate fuel oils, it is customary to distill such products to 10% residual oil and determine the carbon residue thereof. Such values may be used directly in comparing fuel oils, as long as it is kept in mind that the value is that for a residuum oil and are not to be compared with the carbon residue of the whole feedstock.

There are two older methods for determining the carbon residue of a petroleum or petroleum product: the *Conradson method* (ASTM D189) and the *Ramsbottom method* (ASTM D524). Both are applicable to the relatively nonvolatile portion of petroleum and petroleum products, which partially decompose when distilled at a pressure of 1 atmosphere. However, crude oil that contains ash-forming constituents will have an erroneously high carbon residue by either method unless the ash is first removed from the oil; the degree of error is proportional to the amount of ash.

A third method, involving micropyrolysis of the sample, is also available as a standard test method (ASTM D4530). The method requires smaller sample amounts and was originally developed as a *thermogravimetric method*. The carbon residue produced by this method is often referred to as the *microcarbon residue* (*MCR*). Agreements between the data from the three methods are good, making it possible to interrelate all of the data from carbon residue tests (Long and Speight, 1989).

Even though the three methods have their relative merits, there is a tendency to advocate the use of the more expedient microcarbon method to the exclusion of the Conradson and Ramsbottom methods because of the lesser amounts required in the *microcarbon* method, which is somewhat less precise in practical technique.

The mechanical design and operating conditions of such equipment have such a profound influence on carbon deposition during service that comparison of carbon residues between oils should be considered as giving only a rough approximation of relative deposit-forming tendencies. Recent work has focused on the carbon residue of the different fractions of crude oils, especially the asphaltene constituents (Figure 2.3). A more precise relationship between carbon residue and hydrogen content, H/C atomic ratio, nitrogen content, and sulfur content has been shown to exist. These data can provide more precise information about the anticipated behavior of a variety of feedstocks in thermal processes. Thus, there is a fairly universal linear correlation between the carbon residue (Conradson) and the H/C ratio:

$$H/C = 171 - 0.0115CR \text{ (Conradson)}$$

This equation holds within two limits; at H/C values = 171, where the carbon residue is zero (no coke formation) and H/C = 0.5, where the carbon residue is 100 (all the material converts to coke under test conditions). There is a relationship between the carbon residue (Conradson) and the nitrogen content.

Because of the extremely small values of carbon residue obtained by the Conradson and Ramsbottom methods when applied to the lighter distillate fuel oils, it is customary to distill such products to 10% residual oil and determine the carbon residue thereof. Such values may be used directly in comparing fuel oils, as long as it is kept in mind that the values are carbon residues on 10% residual oil and are not to be compared with straight carbon residues.

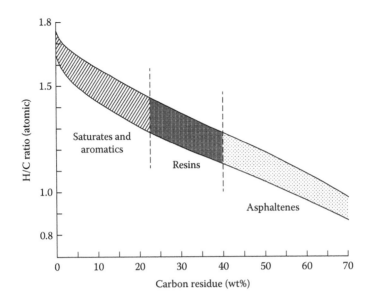

FIGURE 2.3 Relationship of thermal coke yields to feedstock fractions.

2.4.3 CRITICAL PROPERTIES

A study of the pressure, volume, and temperature relationships of a pure component reveals a particular unique state where the properties of a liquid and vapor become indistinguishable from each other. At that state, the latent heat of vaporization becomes zero and no volume change occurs when the liquid is vaporized. This state is called the critical state, and the appropriate parameters of the state are termed the critical pressure (P_C), critical volume (V_C), and critical temperature (T_C). It is an important characteristic of the critical state for a pure component with values of P or T greater than either P_C or T_C, the vapor and liquid states cannot coexist at equilibrium, and thus, P_C and T_C represent the maximum values of P and T at which phase separation can occur.

Since the critical state of a component is unique, it is perhaps not surprising that knowledge of P_C, T_C, and V_C allows many predictions to be made concerning the physical properties of substances. These predictions are based on the law of corresponding states that states that substances behave in the same way when they are in the same state with reference to the critical state. The particular corresponding state is characterized by its reduced properties, that is, $T_r = T/T_C$, $Pr = P/P_C$, $Vr = V/V_C$.

The use of this concept permits generalized plots in terms of reduced properties to be drawn that are then applicable to all substances (which obey the law) and can be of great value in determining thermodynamic relationships. It is rare in petroleum engineering to have to deal with pure substances, and unfortunately, the application of the law of corresponding states to mixtures is complicated by the fact that the use of the true critical point for a mixture does not yield correct values of reduced properties for accurate prediction from generalized charts. For a mixture, the critical state no longer represents the maximum temperature and pressure at which a liquid and vapor phase can coexist and phase separation can occur under retrograde conditions.

For engineering purposes, this difficulty is resolved by the use of pseudocritical conditions, which are based on the molal average critical temperatures and pressures of the compounds of the mixture. Although the use of pseudo-reduced conditions for mixtures of hydrocarbons is generally satisfactory, this is not true for states near the true critical, nor, in general, for mixtures of vapor and liquid.

The temperature, pressure, and volume at the critical state are of considerable interest in petroleum physics, particularly in connection with modern high-pressure, high-temperature refinery operations and in correlating pressure–temperature–volume relationships for other states. Critical data are known for most of the lower-molecular-weight pure hydrocarbons, and standard methods are generally used for such determinations.

The *critical point* of a pure compound is the equilibrium state in which its gaseous and liquid phases are indistinguishable and coexistent; they have the same intensive properties. However, localized variations in these phase properties may be evident experimentally. The definition of the critical point of a mixture is the same. However, mixtures generally have a maximum temperature or pressure at other than the true critical point; *maximum* here denotes the greatest value at which two phases can coexist in equilibrium.

Thus, when a pure compound is heated at atmospheric pressure, it eventually reaches its boiling point and is completely vaporized at a constant temperature unless the pressure is increased. If the pressure is increased, the compound is completely condensed and cannot be vaporized again unless the temperature is also increased. This mechanism, alternately increasing the pressure and temperature, functions until at some high temperature and pressure it is found that the material cannot be condensed regardless of the amount of pressure applied. This point is called the "critical point," and the temperature and pressure at the critical point are called the "critical temperature" and "critical pressure," respectively.

The liquid phase and vapor phase merge at the critical point so that one phase cannot be distinguished from the other. No volume change occurs when a liquid is vaporized at the critical point, and no heat is required for vaporization but the coefficient of expansion has become large.

Limited information concerning the behavior of complex mixtures is required that the pseudocritical temperature and pseudocritical pressure be used for many petroleum fractions and products. The *pseudocritical point* is defined as the molal average critical temperature and pressure of the several constituents that make up a mixture. It may be used as the critical point of a mixture in computing reduced temperatures and pressures. However, in computing the pressure–volume–temperature relations of mixtures by the use of the pseudocritical point, it must be recognized that the values are not accurate in the region of the critical point and that it cannot be applied to mixtures of gas and liquid.

In the correlation of many properties, reduced properties are useful. *Reduced properties* are defined as the ratio of the actual value of the property to its critical value. Thus, for volume, temperature, or pressure, the relationships are as follows:

$$\text{Reduced volume, } V_R = \frac{V}{V_c}$$

where
  V is the volume at specified conditions
  V_c is the volume at the critical point

Similarly,

$$\text{Reduced temperature } T_R = T/T_c$$

$$\text{Reduced volume } P_R = P/P_c$$

where
  T and P are the temperature and volume, respectively, at specified conditions
  T_c and V_c are the temperature and volume, respectively, at the critical point

2.4.4 ENTHALPY

Enthalpy (heat content) is the heat energy necessary to bring a system from a reference state to a given state. Enthalpy is a function only of the end states and is the integral of the specific heats with respect to temperature between the limit states, plus any latent heats of transition that occur within the interval. The usual reference temperature is 0°C (32°F). Enthalpy data are easily obtained from specific heat data by graphic integration, or if the empirical equation given for specific heat is sufficiently accurate, from the following equation:

$$H = 1/d \, (0.388 + 0.000225t^2 - 12.65)$$

Generally, only differences in enthalpy are required in engineering design, that is, the quantity of heat necessary to heat (or cool) a unit amount of material from one temperature to another. Such calculations are very simple since the quantities are arithmetically additive, and the enthalpy for such a change of state is merely the difference between the enthalpies of the end states.

2.4.5 HEAT OF COMBUSTION

The heat of combustion (ASTM D240) is a direct measure of fuel energy content and is determined as the quantity of heat liberated by the combustion of a quantity of fuel with oxygen in a standard bomb calorimeter.

Chemically, the heat of combustion is the energy (heat) released when an organic compound is burned to produce water (H_2O_{liquid}), carbon dioxide (CO_{2gas}), sulfuric acid ($H_2SO_{4liquid}$), and nitric acid ($HNO_{3liquid}$). The value can be calculated using a theoretical equation based upon the elemental composition of the feedstock:

$$H_g/4.187 = 8{,}400C + 27{,}765H + 1{,}500N + 2{,}500S - 2{,}650O$$

where
H_g is given in kilojoules per kilogram (1.0 kJ/kg = 0.43 Btu/lb)
C, H, N, S, and O are the normalized weight fractions for these elements in the sample

The gross heats of combustion of crude oil and its products are given with fair accuracy by the following equation:

$$Q = 12{,}400 - 2{,}100d^2$$

where d is the 60/60°F specific gravity. Deviation is generally less than 1% although many highly aromatic crude oils show considerably higher values; the ranges for crude oil are 10,000–11,600 calories/g and the heat of combustion of heavy oil and tar sand bitumen is considerably higher (Table 2.7).

For gasoline, the heat of combustion is 11,000–11,500 calories/g and for kerosene (and diesel fuel) it falls in the range of 10,500–11,200 calories/g. Finally, the heat of combustion for fuel oil is on the order of 9,500–11,200 calories/g. Heats of combustion of petroleum gases may be calculated from the analysis and data for the pure compounds. Experimental values for gaseous fuels may be obtained by measurement in a water flow calorimeter, and heats of combustion of liquids are usually measured in a bomb calorimeter.

For thermodynamic calculation of equilibria useful in hydrocarbon research, combustion data of extreme accuracy are required because the heats of formation of water and carbon dioxide are large in comparison with those in the hydrocarbons. Great accuracy is also required of the specific heat data for the calculation of free energy or entropy. Much care must be exercised in selecting values

TABLE 2.7
Heat of Combustion of Canadian Heavy Oil and
Tar Sand Bitumen

Heavy Oil or Bitumen	Heat of Combustion		
	Btu/lb	Calories/g	kJ/kg
Athabasca			
Mildred Lake	18,030	10,025	41,940
Carbonate			
Grosmont	17,570–17,650	9,765–9,810	40,865–41,050
Cold Lake			
Clearwater	17,975–18,300	9,990–10,170	41,810–42,530
Lloydminster	17,975–18,285	9,990–10,165	41,810–42,530
Peace River	17,750–18,020	9,880–10,020	41,350–42,530
Wabasca	17,875–18,400	9,935–10,230	41,580–42,800

from the literature for these purposes, since many of those available were determined before the development of modern calorimetric techniques.

An alternative criterion of energy content is the aniline gravity product (AGP) (ASTM D1405) that is in reasonable agreement with the calorific value. It is the product of the API gravity and the aniline point (ASTM D611) of the sample.

2.4.6 LATENT HEAT

There are two properties that represent phase transformations: the latent heat of fusion and the latent heat of vaporization. The latent heat of fusion, defined as the quantity of heat necessary to change a unit weight of solid to a liquid without any temperature change, has received only intermittent attention but, nevertheless, some general rules have been formulated. For hydrocarbons, latent heats of fusion commerce at approximately 15 calories/g for methane, rising to 40 calories/g for octane and then gradually approaching a limiting value of 55 calories/g. Branched paraffins usually have a lower latent heat of fusion than the normal isomers; paraffin wax has a latent heat of fusion in the range 50–60 calories/g.

The *latent heat of vaporization*, defined as the amount of heat required to vaporize a unit weight of a liquid at its atmospheric boiling point, is perhaps the most important property of the two and has received considerably more attention because of its connection with equipment design. The latent heat of vaporization at the atmospheric boiling point generally increases with increasing molecular weight and, for the normal paraffins, generally decreases with increasing temperature and pressure.

2.4.7 LIQUEFACTION AND SOLIDIFICATION

Petroleum and the majority of petroleum products are liquids at ambient temperature, and problems that may arise from solidification during normal use are not common. Nevertheless, there are feedstocks and products that are semisolid to solid.

The *melting point* is a test (ASTM D87, ASTM D127) that is widely used by suppliers of wax and by the wax consumers; it is particularly applied to the highly paraffinic or crystalline waxes. Quantitative prediction of the melting point of pure hydrocarbons is difficult, but the melting point tends to increase qualitatively with the molecular weight and with symmetry of the molecule.

Unsubstituted and symmetrically substituted compounds (e.g., benzene, cyclohexane, p-xylene, and naphthalene) melt at higher temperatures relative to the paraffin compounds of similar molecular weight: the unsymmetrical isomers generally melt at lower temperatures than the aliphatic hydrocarbons of the same molecular weight.

Unsaturation affects the melting point principally by its alteration of symmetry; thus, ethane (−172°C, −278°F) and ethylene (−169.5°C, −273°F) differ only slightly, but the melting points of cyclohexane (6.2°C, 21°F) and cyclohexane (−104°C, −155°F) contrast strongly. All types of highly unsymmetrical hydrocarbons are difficult to crystallize; asymmetrically branched aliphatic hydrocarbons as low as octane and most substituted cyclic hydrocarbons comprise the greater part of the lubricating fractions of petroleum, crystallize slowly, if at all, and on cooling merely take the form of glass-like solids.

Although the melting points of petroleum and petroleum products are of limited usefulness, except to estimate the purity or perhaps the composition of waxes, the reverse process, *solidification*, has received attention in petroleum chemistry. In fact, solidification of petroleum and petroleum products has been differentiated into four categories, namely, *freezing point*, *congealing point*, *cloud point*, and *pour point.*

Petroleum becomes more or less a plastic solid when cooled to sufficiently low temperatures. This is due to the congealing of the various hydrocarbons that constitute the oil. The cloud point of a petroleum oil is the temperature at which paraffin wax or other solidifiable compounds present in the oil appear as a haze when the oil is chilled under definitely prescribed conditions (ASTM D2500, ASTM D3117). As cooling is continued, all petroleum oils become more and more viscous and flow becomes slower and slower. The pour point of a petroleum oil is the lowest temperature at which the oil pours or flows under definitely prescribed conditions when it is chilled without disturbance at a standard rate (ASTM D97).

The solidification characteristics of a petroleum product depend on its grade or kind. For grease, the temperature of interest is that at which fluidity occurs, commonly known as the *dropping point*. The dropping point of grease is the temperature at which the grease passes from a plastic solid to a liquid state and begins to flow under the conditions of the test (ASTM D566, ASTM D2265). For another type of plastic solid, including petrolatum and microcrystalline wax, both *melting point* and *congealing point* are of interest.

The *melting point* of wax is the temperature at which the wax becomes sufficiently fluid to drop from the thermometer; the *congealing point* is the temperature at which melted petrolatum ceases to flow when allowed to cool under definitely prescribed conditions (ASTM D938).

For paraffin wax, the *solidification temperature* is of interest. For such purposes, the *melting point* is the temperature at which the melted paraffin wax begins to solidify, as shown by the minimum rate of temperature change, when cooled under prescribed conditions. For pure or essentially pure hydrocarbons, the solidification temperature is the freezing point, the temperature at which a hydrocarbon passes from a liquid to a solid state (ASTM D1015, ASTM D1016).

The relationship of *cloud point*, *pour point*, *melting point*, and *freezing point* to one another varies widely from one petroleum product to another. Hence, their significance for different types of product also varies. In general, cloud, melting, and freezing points are of more limited value, and each is of narrower range of application than the pour point.

The *cloud point* of petroleum or a petroleum product is the temperature at which paraffin wax or other solidifiable compounds present in the oil appear as a haze when the sample is chilled under definitely prescribed conditions (ASTM D2500, ASTM D3117).

To determine the *cloud point* and the *pour point* (ASTM D97, ASTM D5853, ASTM D5949, ASTM D5950, ASTM D5985), the oil is contained in a glass test tube fitted with a thermometer and immersed in one of three baths containing coolants. The sample is dehydrated and filtered at a temperature 25°C (45°F) higher than the anticipated cloud point mentioned earlier. It is then placed in a test tube and cooled progressively in coolants held at −1°C to +2°C (30°F to 35°F), −18°C to −20°C (−4°F to 0°F), and −32°C to −35°C (−26°F to −31°F), respectively. The sample is inspected for cloudiness at temperature intervals of 1°C (2°F). If conditions or oil properties are such that reduced temperatures are required to determine the pour point, alternate tests are available, which accommodate the various types of samples. Related to the *cloud point*, the wax appearance temperature or *wax appearance point* is also determined (ASTM D3117).

The *pour point* of petroleum or a petroleum product is determined using this same technique (ASTM D97), and it is the lowest temperature at which the oil pours or flows. It is actually 2°C (3°F) above the temperature at which the oil ceases to flow under these definitely prescribed conditions when it is chilled without disturbance at a standard rate. To determine the pour point, the sample is first heated to 46°C (115°F) and cooled in air to 32°C (90°F) before the tube is immersed in the same series of coolants as used for the determination of the *cloud point*. The sample is inspected at temperature intervals of 2°C (3°F) by withdrawal and holding horizontal for 5 seconds until no flow is observed during this time interval.

Cloud and pour points are useful for predicting the temperature at which the observed viscosity of oil deviates from the true (Newtonian) viscosity in the low-temperature range. They are also useful for the identification of oils or when planning the storage of oil supplies, as low temperatures may cause handling difficulties with some oils.

The pour point of a crude oil was originally applied to crude oil that had a high wax content. More recently, the pour point, like the viscosity, is determined principally for use in pumping arid pipeline design calculations. Difficulty occurs in these determinations with waxy crude oils that begin to exhibit irregular flow behavior when wax begins to separate. These crude oils possess viscosity relationships that are difficult to predict in pipeline operation. In addition, some waxy crude oils are sensitive to heat treatment that can also affect their viscosity characteristics. This complex behavior limits the value of viscosity and pour point tests on waxy crude oils. At the present time, long crude oil pipelines and the increasing production of waxy crude oils make an assessment of the pumpability of a wax-containing crude oil through a given system a matter of some difficulty that can often only be resolved after field trials. Consequently, considerable work is in progress to develop a suitable laboratory pumpability test (such as described in IP 230) that gives an estimate of minimum handling temperature and minimum line or storage temperature.

2.4.8 Pressure–Volume–Temperature Relationships

Hydrocarbon vapors, like other gases, follow the ideal gas law (i.e., PV = RT) only at relatively low pressures and high temperatures, that is, far from the critical state. Several more empirical equations have been proposed to represent the gas laws more accurately, such as the well-known van der Waals equation, but they are either inconvenient for calculation or require the experimental determination of several constants. A more useful device is to use the simple gas law and to induce a correction, termed the *compressibility factor*, μ, so that the equation takes the following form:

$$PV = \mu RT$$

For hydrocarbons, the compressibility factor is very nearly a function only of the reduced variables of state, that is, a function of the pressure and temperature divided by the respective critical values. The compressibility factor method functions excellently for pure compounds but may become ambiguous for mixtures because the critical constants have a slightly different significance. However, the use of pseudocritical temperature and pressure values is generally lower than the true values, permitting the compressibility factor to be employed in such cases.

2.4.9 Specific Heat

Specific heat is defined as the quantity of heat required to raise a unit mass of material through one degree of temperature (ASTM D2766).

Specific heats are extremely important engineering quantities in refinery practice because they are used in all calculations on heating and cooling petroleum products. Many measurements have been made on various hydrocarbon materials, but the data for most purposes may be summarized by the following general equation:

$$C = 1/d(0.388 + 0.00045t)$$

C is the specific heat at a specified temperature (°F) of an oil whose specific gravity 60/60°F is d; thus, specific heat increases with temperature and decreases with specific gravity.

2.4.10 THERMAL CONDUCTIVITY

The thermal conductivity K of hydrocarbons (in cgs units) is given by the following equation:

$$K = 0.28/d(1 - 0.00054) \times 10^{-3}$$

where d is the specific gravity.

The value for solid paraffin wax is approximately 0.00056, nearly independent of temperature and wax type; the oil equation holds satisfactorily for waxes above the melting point.

2.4.11 VOLATILITY

The volatility of a liquid or liquefied gas may be defined as its tendency to vaporize, that is, to change from the liquid to the vapor or gaseous state. Because one of the three essentials for combustion in a flame is that the fuel be in the gaseous state, volatility is a primary characteristic of liquid fuels.

The vaporizing tendencies of petroleum and petroleum products are the basis for the general characterization of liquid petroleum fuels, such as liquefied petroleum gas, natural gasoline, motor and aviation gasoline, naphtha, kerosene, gas oil, diesel fuel, and fuel oil (ASTM D2715). A test (ASTM D6) also exists for determining the loss of material when crude oil and asphaltic compounds are heated. Another test (ASTM D20) is a method for the distillation of road tars that might also be applied to estimating the volatility of high-molecular-weight residues.

For some purposes, it is necessary to have information on the initial stage of vaporization. To supply this need, flash and fire, vapor pressure, and evaporation methods are available. The data from the early stages of the several distillation methods are also useful. For other uses, it is important to know the tendency of a product to partially vaporize or to completely vaporize, and in some cases to know if small quantities of high-boiling components are present. For such purposes, chief reliance is placed on the distillation methods.

The *flash point* of petroleum or a petroleum product is the temperature to which the product must be heated under specified conditions to give off sufficient vapor to form a mixture with air that can be ignited momentarily by a specified flame (ASTM D56, ASTM D92, ASTM D93). The *fire point* is the temperature to which the product must be heated under the prescribed conditions of the method to burn continuously when the mixture of vapor and air is ignited by a specified flame (ASTM D92).

From the viewpoint of safety, information about the flash point is of most significance at or slightly above the maximum temperatures (30°C–60°C, 86°F–140°F) that may be encountered in storage, transportation, and use of liquid petroleum products, in either closed or open containers. In this temperature range, the relative fire and explosion hazard can be estimated from the flash point. For products with flash point below 40°C (104°F), special precautions are necessary for safe handling. Flash points above 60°C (140°F) gradually lose their safety significance until they become indirect measures of some other quality.

The flash point of a petroleum product is also used to detect contamination. A substantially lower flash point than expected for a product is a reliable indicator that a product has become contaminated with a more volatile product, such as gasoline. The flash point is also an aid in establishing the identity of a particular petroleum product.

A further aspect of volatility that receives considerable attention is the vapor pressure of petroleum and its constituent fractions. The *vapor pressure* is the force exerted on the walls of a closed container by the vaporized portion of a liquid. Conversely, it is the force that must be exerted on the liquid to prevent it from vaporizing further (ASTM D323). The vapor pressure increases with temperature for any given gasoline, liquefied petroleum gas, or other product. The temperature at which the vapor pressure of a liquid, either a pure compound of a mixture of many compounds, equals 1 atmosphere (14.7 psi, absolute) is designated as the boiling point of the liquid.

In each homologous series of hydrocarbons, the boiling points increase with molecular weight and structure also has a marked influence since it is a general rule that branched paraffin isomers have lower boiling points than the corresponding *n*-alkane. In any given series, steric effects notwithstanding, there is an increase in boiling point with an increase in carbon number of the alkyl side chain. This particularly applies to alkyl aromatic compounds where alkyl-substituted aromatic compounds can have higher boiling points than polycondensed aromatic systems. And this fact is very meaningful when attempts are made to develop hypothetical structures for asphaltene constituents (Speight, 1994, 2014).

The boiling points of petroleum fractions are rarely, if ever, distinct temperatures; it is, in fact, more correct to refer to the boiling ranges of the various fractions. To determine these ranges, the petroleum is tested in various methods of distillation, either at atmospheric pressure or at reduced pressure. In general, the limiting molecular weight range for distillation at atmospheric pressure without thermal degradation is 200–250, whereas the limiting molecular weight range for conventional vacuum distillation is 500–600.

As an early part of characterization studies, a correlation was observed between the quality of petroleum products and their hydrogen content since gasoline, kerosene, diesel fuel, and lubricating oil are made up of hydrocarbon constituents containing high proportions of hydrogen. Thus, it is not surprising that test to determine the volatility of petroleum and petroleum products was among the first to be defined. Indeed, volatility is one of the major tests for petroleum products, and it is inevitable that all products will, at some stage of their history, be tested for volatility characteristics.

Distillation involves the general procedure of vaporizing the petroleum liquid in a suitable flask either at *atmospheric pressure* (ASTM D86, ASTM D2892) or at *reduced pressure* (ASTM D1160), and the data are reported in terms of one or more of the following seven items:

1. *Initial boiling point* is the thermometer reading in the neck of the distillation flask when the first drop of distillate leaves the tip of the condenser tube. This reading is materially affected by a number of test conditions, namely, room temperature, rate of heating, and condenser temperature.
2. *Distillation temperatures* are usually observed when the level of the distillate reaches each 10% mark on the graduated receiver, with the temperatures for the 5% and 95% marks often included. Conversely, the volume of the distillate in the receiver, that is, the percentage recovered, is often observed at specified thermometer readings.
3. *End point* or *maximum temperature* is the highest thermometer reading observed during distillation. In most cases, it is reached when the entire sample has been vaporized. If a liquid residue remains in the flask after the maximum permissible adjustments are made in heating rate, this is recorded as indicative of the presence of very high-boiling compounds.
4. *Dry point* is the thermometer reading at the instant the flask becomes dry and is for special purposes, such as for solvents and for relatively pure hydrocarbons. For these purposes, dry point is considered more indicative of the final boiling point than end point or maximum temperature.

5. *Recovery* is the total volume of distillate recovered in the graduated receiver and *residue* is the liquid material, mostly condensed vapors, left in the flask after it has been allowed to cool at the end of distillation. The residue is measured by transferring it to an appropriate small graduated cylinder. Low or abnormally high residues indicate the absence or presence, respectively, of high-boiling components.

6. *Total recovery* is the sum of the liquid recovery and residue; *distillation loss* is determined by subtracting the total recovery from 100%. It is, of course, the measure of the portion of the vaporized sample that does not condense under the conditions of the test. Like the initial boiling point, distillation loss is affected materially by a number of test conditions, namely, condenser temperature, sampling and receiving temperatures, barometric pressure, heating rate in the early part of the distillation, and others. Provisions are made for correcting high distillation losses for the effect of low barometric pressure because of the practice of including distillation loss as one of the items in some specifications for motor gasoline.

7. *Percentage evaporated* is the percentage recovered at a specific thermometer reading or other distillation temperatures, or the converse. The amounts that have been evaporated are usually obtained by plotting observed thermometer readings against the corresponding observed recoveries plus, in each case, the distillation loss. The initial boiling point is plotted with the distillation loss as the percentage evaporated. Distillation data are considerably reproducible, particularly for the more volatile products.

One of the main properties of petroleum that serves to indicate the comparative ease with which the material can be refined is the volatility (Speight, 2015). Investigation of the volatility of petroleum is usually carried out under standard conditions, thereby allowing comparisons to be made between data obtained from various laboratories. Thus, nondestructive distillation data (U.S. Bureau of Mines method) show that, not surprisingly, bitumen is a higher-boiling material than the more conventional crude oils (Tables 2.8 and 2.9). There is usually little, or no, gasoline (naphtha) fraction in bitumen, and the majority of the distillate falls in the gas oil–lubrication distillate range (>260°C, >500°F). In excess of 50% v/v of the bitumen is nondistillable under the conditions of the test, and the yield of the nonvolatile material corresponds very closely to the asphaltic (asphaltene constituents plus resin constituents) content of each feedstock. In fact, detailed fractionation of the sample might be of secondary importance. Thus, it must be recognized that the general shape of a one-plate distillation curve is often adequate for making engineering calculations, correlating with other physical properties, and predicting the product slate.

There is also another method that is increasing in popularity for application to a variety of feedstocks and that is the method commonly known as "simulated distillation" (ASTM D2887) (Carbognani et al., 2012). The method has been well researched in terms of method development and application (Romanowski and Thomas, 1985; Schwartz et al., 1987; Neer and Deo, 1995). The benefits of the technique include good comparisons with other ASTM distillation data as well as the application to higher-boiling fractions of petroleum. In fact, data output includes the provision of the corresponding Engler profile (ASTM D86) as well as the prediction of other properties such as vapor pressure and flash point. When it is necessary to monitor product properties, as is often the case during refining operations, such data provide a valuable aid to process control and online product testing.

For a more detailed distillation analysis of feedstocks and products, a low-resolution, temperature-programmed gas chromatographic analysis has been developed to simulate the time-consuming true boiling point distillation. The method relies on the general observation that hydrocarbons are eluted from a nonpolar adsorbent in the order of their boiling points. The regularity of the elution order of the hydrocarbon components allows the retention times to be equated to distillation temperatures, and the term "simulated distillation by gas chromatography" (or *simdis*) is used throughout the industry to refer to this technique.

TABLE 2.8
Distillation Profile of Conventional Crude Oil (Leduc, Woodbend, Upper Devonian, Alberta, Canada) and Selected Properties of the Fractions

| | Boiling Range | | | Wt% | Specific | API | Sulfur | Carbon Residue |
	°C	°F	Wt%	Cumulative	Gravity	Gravity	Wt%	(Conradson)
Whole crude oil				100.0	0.828	39.4	0.4	1.5
Fraction[a]								
1	0–50	0–122	2.6	2.6	0.650	86.2		
2	50–75	122–167	3.0	5.6	0.674	78.4		
3	75–100	167–212	5.2	10.8	0.716	66.1		
4	100–125	212–257	6.6	17.4	0.744	58.7		
5	125–150	257–302	6.3	23.7	0.763	54.0		
6	150–175	302–347	5.5	29.2	0.783	49.2		
7	175–200	347–392	5.3	34.5	0.797	46.0		
8	200–225	392–437	5.0	39.5	0.812	42.8		
9	225–250	437–482	4.7	44.2	0.823	40.4		
10	250–275	482–527	6.6	50.8	0.837	37.6		
11	<200	<392	5.4	56.2	0.852	34.6		
12	200–225	392–437	4.9	61.1	0.861	32.8		
13	225–250	437–482	5.2	66.3	0.875	30.2		
14	250–275	482–527	2.8	69.1	0.883	28.8		
15	275–300	527–572	6.7	75.4	0.892	27.0		
Residuum	>300	>572	22.6	98.4	0.929	20.8		6.6
Distillation loss				1.6				

[a] Distillation at 765 mm Hg then at 40 mm Hg for fractions 11–15.

Simulated distillation by gas chromatography is often applied in the petroleum industry to obtain true boiling point data for distillates and crude oils (Speight, 2001). Two standardized methods (ASTM D2887 and ASTM D3710) are available for the boiling point determination of petroleum fractions and gasoline, respectively. The ASTM D2887 method utilizes nonpolar, packed gas chromatographic columns in conjunction with flame ionization detection. The upper limit of the boiling range covered by this method is approximately 540°C (1000°F) atmospheric equivalent boiling point. Recent efforts in which high-temperature gas chromatography was used have focused on extending the scope of the ASTM D2887 method for higher-boiling petroleum materials to 800°C (1470°F) atmospheric equivalent boiling point.

2.5   ELECTRICAL PROPERTIES

Understanding of how petroleum behaves and why different crude oils differ in properties is also possible with an atomistic understanding allowed by quantum mechanics. The combination of physics, chemistry, and the focus on the relationship between the properties of a material and its electrical properties allows uses to be designed and also provides a knowledge base for a variety of chemical and engineering applications.

2.5.1   CONDUCTIVITY

From the fragmentary evidence available, the electrical conductivity of petroleum fractions is small but measurable (Penzes and Speight, 1974; Fotland et al., 1993; Fotland and Anfindsen, 1996). For

TABLE 2.9
Distillation Profile of Bitumen (Athabasca, McMurray Formation, Upper Cretaceous, Alberta, Canada) and Selected Properties of the Fractions

Feedstock	Boiling Range °C	Boiling Range °F	Wt%	Wt% Cumulative	Specific Gravity	API Gravity	Sulfur Wt%	Carbon Residue (Conradson)
Whole bitumen				100.0	1.030	5.9	5.8	19.6
Fraction[a]								
1	0–50	0–122	0.0	0.0				
2	50–75	122–167	0.0	0.0				
3	75–100	167–212	0.0	0.0				
4	100–125	212–257	0.0	0.0				
5	125–150	257–302	0.9	0.9				
6	150–175	302–347	0.8	1.7	0.809	43.4		
7	175–200	347–392	1.1	2.8	0.823	40.4		
8	200–225	392–437	1.1	3.9	0.848	35.4		
9	225–250	437–482	4.1	8.0	0.866	31.8		
10	250–275	482–527	11.9	19.9	0.867	31.7		
11	<200	<392	1.6	21.5	0.878	29.7		
12	200–225	392–437	3.2	24.7	0.929	20.8		
13	225–250	437–482	6.1	30.8	0.947	17.9		
14	250–275	482–527	6.4	37.2	0.958	16.2		
15	275–300	527–572	10.6	47.8	0.972	14.1		
Residuum	>300	>572	49.5	97.3				39.6

[a] Distillation at 762 mm Hg and then at 40 mm Hg for fractions 11–15.

example, the normal hydrocarbons (from hexane up) have an electrical conductivity smaller than 10^{-16} Ω/cm; benzene itself has an electrical conductivity of 4.4×10^{-17} Ω/cm, and cyclohexane has an electrical conductivity of 7×10^{-18} Ω/cm. It is generally recognized that hydrocarbons do not usually have an electrical conductivity larger than 10^{-18} Ω/cm. Thus, it is not surprising that the electrical conductivity of hydrocarbon oils is also exceedingly small—on the order of 10^{-19} to 10^{-12} Ω/cm.

Available data indicate that the observed conductivity is frequently more dependent on the method of measurement and the presence of trace impurities than on the chemical type of the oil. Conduction through oils is not ohmic insofar as the current is not proportional to field strength: in some regions, it is observed to increase exponentially with the latter. Time effects are also observed, the current being at first relatively large and decreasing to a smaller steady value. This is partly because of electrode polarization and partly because of ions removed from the solution. Most oils increase in conductivity with rising temperatures.

2.5.2 DIELECTRIC CONSTANT

The *dielectric constant* (ε) of a substance may be defined as the ratio of the capacity of a condenser with the material between the condenser plates C to that with the condenser empty and under vacuum C_0:

$$\varepsilon = \frac{C}{C_0}$$

The dielectric constant of petroleum and petroleum products may be used to indicate the presence of various constituents, such as asphaltene constituents, resin constituents, or oxidized materials.

Furthermore, the dielectric constant of petroleum products that are used in equipment, such as condensers, may actually affect the electrical properties and performance of that equipment (ASTM D877).

The dielectric constant of hydrocarbons and hence most crude oils and their products is usually low and decreases with an increase in temperature. It is also noteworthy that for hydrocarbons, hydrocarbon fractions, and products the dielectric constant is approximately equal to the square of the refractive index. Polar materials have dielectric constants greater than the square of the refractive index.

2.5.3 Dielectric Strength

The dielectric strength, or breakdown voltage (ASTM D877), is the greatest potential gradient or potential that an insulator can withstand without permitting an electric discharge. The property is, in the case of oils as well as other dielectric materials, somewhat dependent on the method of measurement, that is, on the length of path through which the breakdown occurs, the composition, shape, and condition of the electrode surfaces, and the duration of the applied potential difference.

The standard test used in North America is applied to oils of petroleum origin for use in cables, transformers, oil circuit breakers, and similar apparatus. Oils of high purity and cleanliness show nearly the same value under standard conditions, generally ranging from 30 to 35 kV. For alkanes, dielectric strength has been shown to increase linearly with liquid density, and the value for a mineral oil fits the data well. For n-heptane, a correlation was found between the dielectric strength and the density changes with temperature. There are many reasons that the dielectric strength of an insulator may fail. The most important appears to be the presence of some type of impurity, produced by corrosion, oxidation, thermal or electrical cracking, or gaseous discharge; invasion by water is a common trouble.

2.5.4 Dielectric Loss and Power Factors

A condenser insulated with an ideal dielectric shows no dissipation of energy when an alternating potential is applied. The charging current, technically termed the *circulating current*, lags exactly 90° in phase angle behind the applied potential, and the energy stored in the condenser during each half-cycle is completely recovered in the next. No real dielectric material exhibits this ideal behavior; that is, some energy is dissipated under alternating stress and appears as heat. Such a lack of efficiency is broadly termed *dielectric loss*.

Ordinary conduction comprises one component of dielectric loss. Here, the capacitance-held charge is partly lost by short circuit through the medium. Other effects in the presence of an alternating field occur, and a dielectric of zero conductivity may still exhibit losses. Suspended droplets of another phase undergo spheroidal oscillation by electrostatic induction effects and dissipate energy as heat as a consequence of the viscosity of the medium. Polar molecules oscillate as electrets and dissipate energy on collision with others. All such losses are of practical importance when insulation is used in connection with alternating current equipment.

The measure of the dielectric loss is the power factor. This is defined as the factor k in the following relation:

$$k = \frac{W}{EI}$$

where W is the power in watts dissipated by a circuit portion under voltage E and passing current, I.

From ac theory, the power factor is recognized as the cosine of the phase angle between the voltage and current where a pure sine wave form exists for both; it increases with a use in temperature. When an insulating material serves as the dielectric of a condenser the power factor is an intrinsic property of the dielectric. For practical electrical equipment, low-power factors for the insulation are of course always desirable; petroleum oils are generally excellent in this respect, having values of the order of 0.0005, comparable with fused quartz and poly-styrene resin constituents. The power factor of pure hydrocarbons is extremely small. Traces of polar impurities, however, cause a striking increase. All electrical oils, therefore, are drastically refined and handled with care to avoid contamination; insoluble oxidation products are particu-larly undesirable.

2.5.5 STATIC ELECTRIFICATION

Dielectric liquids, particularly light naphtha, may acquire high static charges on flowing through or being sprayed from metal pipes. The effect seems to be associated with colloidally dispersed contaminants, such as oxidation products, which can be removed by drastic filtration or adsorption. Since a considerable fire hazard is involved that a variety of methods have been studied for mini-mizing the danger.

For large-scale storage, avoidance of surface agitation and the use of floating metal roofs on tanks are beneficial. High humidity in the surrounding atmosphere is helpful in lowering the static charge, and radioactive materials have been used to try to induce discharge to ground. A variety of additives have been found, which increase the conductivity of petroleum liquids, thus lowering the degree of electrification; chromium salts of alkylated salicylic acids and other salts of alkylated sulfosuccinic acids are employed in low concentrations, say 0.005%.

2.6 OPTICAL PROPERTIES

Optical property means the response of petroleum to exposure to electromagnetic radiation and, in particular, to visible light. Such properties, while not often used in the past, are not finding use in environmental cases.

Among their many impacts, oil pollutants modify light fields above and below the water surface. These modifications are manifested by the attenuation of the light passing through an oiled water surface by changes in light absorption in the seawater column due to the formation of an emulsion and by the scattering of light by particles of such an emulsion. The optical properties of an emulsion can be described by the attenuation-specific cross section and the absorption-specific cross section, which depend on the optical characteristics of the oil and the seawater, on the size distribution of oil droplets, and on their concentration.

In the case of pollution with fresh crude, the oil undergoes weathering, which alters its prop-erties and, hence, changes in the optical properties of a petroleum–seawater emulsion due to oil weathering.

2.6.1 OPTICAL ACTIVITY

The occurrence of optical activity in petroleum is universal and is a general phenomenon not restricted to a particular type of crude oil, such as the paraffinic or naphthenic crude oils. Petroleum is usually *dextrorotatory*, that is, the plane of polarized light is rotated to the right, but there are known *laevorotatory* crude oils, that is, the plane of polarized light is rotated to the left, and some crude oils have been reported to be optically inactive.

Examination of the individual fractions of optically active crude oils shows that the rota-tory power increases with molecular weight (or boiling point) to pronounced maxima and then decreases again. The rotatory power appears to be concentrated in certain fractions, the maximum lying at a molecular weight of approximately 350–400; this maximum is approximately the same for all crude oils. The occurrence of optically active compounds in unaltered natural petroleum has been a strong argument in favor of a rather low-temperature origin of petroleum from organic raw materials.

A magnetic field causes all liquids to exhibit optical rotation, usually in the same direction as that of the magnetizing current; this phenomenon is known as the Faraday effect (θ) and may be expressed by the following relation:

$$\theta = pth$$

where
θ is the total angle of rotation
t is the thickness of substance through which the light passes
h is the magnetic field
p, which is constant, is an intrinsic property of the substance, usually termed the Verdet constant (minutes of arc/cm per G)

There have been some attempts to use the Verdet constant in studying the constitution of hydrocarbons by physical property correlation.

2.6.2 REFRACTIVE INDEX

The *refractive index* is the ratio of the velocity of light in a vacuum to the velocity of light in the substance. The measurement of the refractive index is very simple (ASTM D1218) and requires small quantities of material and, consequently, has found wide use in the characterization of hydrocarbons and petroleum samples.

For closely separated fractions of similar molecular weight, the values increase in the order paraffin, naphthene, and aromatic. For polycyclic naphthenes and polycyclic aromatics, the refractive index is usually higher than that of the corresponding monocyclic compounds. For a series of hydrocarbons of essentially the same type, the refractive index increases with molecular weight, especially in the paraffin series. Thus, the refractive index can be used to provide valuable information about the composition of hydrocarbon (petroleum) mixtures; as with density, low values indicate paraffinic materials and higher values indicate the presence of aromatic compounds. However, the combination of refractive index and density may be used to provide even more definite information about the nature of a hydrocarbon mixture and, hence, the use of the refractivity intercept (n – d/2).

The refractive and specific dispersion as well as the molecular and specific refraction have all been advocated for use in the characterization of petroleum and petroleum products.

The *refractive dispersion* of a substance is defined as the difference between its refractive indices at two specified wavelengths of light. Two lines commonly used to calculate dispersions are the C (6563 Å, red) and F (4861 Å, blue) lines of the hydrogen spectrum. The *specific dispersion* is the refractive dispersion divided by the density at the same temperature:

$$\text{Specific dispersion} = \frac{nF - nC}{d}$$

This equation is of particular significance in petroleum chemistry because all the saturated hydrocarbons, naphthene and paraffin, have nearly the same value irrespective of molecular weight, whereas aromatics are much higher and unsaturated aliphatic hydrocarbons are intermediate.

Specific refraction is the term applied to the quantity defined by the following expression:

$$\frac{n-1}{(n^2 + 2)d} = C$$

where
   n is the refractive index
   d is the density
   C is a constant independent of temperature

Molecular refraction is the specific refraction multiplied by molecular weight; its particular usefulness lies in the fact that it is very nearly additive for the components of a molecule; that is, numerical values can be assigned to atoms and structural features, such as double bonds and rings. The value for any pure compound is then approximately the sum of such component constants for the molecule.

2.7   SPECTROSCOPIC PROPERTIES

Spectroscopic studies have played an important role in the evaluation of petroleum and of petroleum products for the last three decades, and many of the methods are now used as standard methods of analysis for refinery feedstocks and products. The application of these methods to feedstocks and products is a natural consequence for the refiner.

The methods include the use of *mass spectrometry* to determine the (1) hydrocarbon types in middle distillates (ASTM D2425), (2) hydrocarbon types of gas oil saturate fractions (ASTM D2786), (3) hydrocarbon types in low-olefin gasoline (ASTM D2789), and (4) aromatic types of gas oil aromatic fractions (ASTM D3239). *Nuclear magnetic resonance spectroscopy* has been developed as a standard method for the determination of hydrogen types in aviation turbine fuels (ASTM D3701). *X-ray fluorescence spectrometry* has been applied to the determination of sulfur in various petroleum products (ASTM D2622, ASTM D4294).

Infrared spectroscopy is used for the determination of benzene in motor and/or aviation gasoline (ASTM D4053), while ultraviolet spectroscopy is employed for the evaluation of mineral oils (ASTM D2269) and for determining the naphthalene content of aviation turbine fuels (ASTM D1840).

Other techniques include the use of *flame emission spectroscopy* for determining trace metals in gas turbine fuels (ASTM D3605) and the use of *absorption spectrophotometry* for the determination of the alkyl nitrate content of diesel fuel (ASTM D4046). *Atomic absorption* has been employed as a means of measuring the lead content of gasoline (ASTM D3237) and also for the manganese content of gasoline (ASTM D3831) as well as for determining the barium, calcium, magnesium, and zinc contents of lubricating oils (ASTM D4628). *Flame photometry* has been employed as a means of measuring the lithium/sodium content of lubricating greases (ASTM D3340) and the sodium content of residual fuel oil (ASTM D1318).

Nowhere is the contribution of spectroscopic studies more emphatic than in application to the delineation of structural types in the heavier feedstocks. This has been necessary because of the unknown nature of these feedstocks by refiners. One particular example is the *ndM. method* (ASTM D3238), which is designed for the carbon distribution and structural group analysis of petroleum oils. Later investigators have taken structural group analysis several steps further than the ndM. method.

It is also appropriate at this point to give a brief description of other methods that are used for the identification of the constituents of petroleum (Yen, 1984).

It is not intended to convey here that any one of these methods can be used for identification purposes. However, although these methods may fall short of complete acceptability as methods for the characterization of individual constituents of feedstocks, they can be used as methods by which an overall evaluation of the feedstock may be obtained in terms of molecular types.

2.7.1 INFRARED SPECTROSCOPY

Conventional infrared spectroscopy yields information about the functional features of various petroleum constituents. For example, infrared spectroscopy will aid in the identification of N–H and O–H functions, the nature of polymethylene chains, the C–H out-of-place bending frequencies, and the nature of any polynuclear aromatic systems.

With the recent progress of *Fourier transform infrared spectroscopy (FTIR)*, quantitative estimates of the various functional groups can also be made. This is particularly important for application to the higher-molecular-weight solid constituents of petroleum (i.e., the asphaltene fraction). It is also possible to derive structural parameters from infrared spectroscopic data, and these are (1) saturated hydrogen to saturated carbon ratio, (2) paraffinic character, (3) naphthenic character, (4) methyl group content, and (5) paraffin chain length.

In conjunction with proton magnetic resonance (see the next section), structural parameters such as the fraction of paraffinic methyl groups to aromatic methyl groups can be obtained.

2.7.2 MASS SPECTROMETRY

Mass spectrometry can play a key role in the identification of the constituents of feedstocks and products. The principal advantages of mass spectrometric methods are (1) high reproducibility of quantitative analyses, (2) the potential for obtaining detailed data on the individual components and/or carbon number homologues in complex mixtures, and (3) a minimal sample size required for analysis. The ability of mass spectrometry to identify individual components in complex mixtures is unmatched by any modern analytical technique. Perhaps the exception is gas chromatography.

However, there are disadvantages arising from the use of mass spectrometry, and these are as follows: (1) the limitation of the method to organic materials that are volatile and stable at temperatures up to 300°C (570°F) and (2) the difficulty of separating isomers for absolute identification. The sample is usually destroyed, but this is seldom a disadvantage.

Nevertheless, in spite of these limitations, mass spectrometry does furnish useful information about the composition of feedstocks and products even if this information is not as exhaustive as might be required. There are structural similarities that might hinder the identification of individual components. Consequently, identification by type or by homologue will be more meaningful since similar structural types may be presumed to behave similarly in processing situations. Knowledge of the individual isomeric distribution may add only a little to an understanding of the relationships between composition and processing parameters.

Mass spectrometry should be used discriminately where a maximum amount of information can be expected. The heavier nonvolatile feedstocks are for practical purposes, beyond the useful range of routine mass spectrometry. At the elevated temperatures necessary to encourage volatility, thermal decomposition will occur in the inlet and any subsequent analysis would be biased to the low-molecular-weight end and to the lower-molecular-weight products produced by the thermal decomposition.

On the other hand, the occurrence of high-molecular-weight hydrocarbons in ozokerite and a waxy yellow crude oil from the Uinta Basin (Utah) has been studied successfully by field ionization mass spectrometry (FIMS) (Del Río and Philp, 1999). The spectra consisted predominantly of molecular ions ranging up to near mass 2000 and correspond to several series of hydrocarbons ranging up to C_{110}. The use of method permitted the range of hydrocarbons identified in geological materials to be extended far beyond that identified by the usual chromatographic techniques. Moreover, from the spectra it was possible to extract the molecular ions corresponding to series of hydrocarbons with different degrees of unsaturation or ring closures. The ozokerite solid bitumen consisted mainly of series of branched alkanes (C_nH_{2n+2}) and cyclic alkanes (C_nH_{2n} and C_nH_{2n-2}) up to C_{110} with a predominance of monocyclic alkanes in the high-molecular-weight region (above C_{40}). The waxy yellow crude oil, on the other hand, contained only acyclic compounds, mainly *n*-alkanes, ranging up to C_{100}.

2.7.3 NUCLEAR MAGNETIC RESONANCE SPECTROSCOPY

Nuclear magnetic resonance spectroscopy has frequently been employed for general studies and for the structural studies of petroleum constituents (Bouquet and Bailleul, 1982; Hasan et al., 1989). In fact, proton magnetic resonance (PMR) studies (along with infrared spectroscopic studies) were, perhaps, the first studies of the modern era that allowed structural inferences to be made about the polynuclear aromatic systems that occur in the high-molecular-weight constituents of petroleum.

In general, the proton (hydrogen) types in petroleum fractions can be subdivided into five types, which subdivide the hydrogen distribution into (1) aromatic hydrogen, (2) substituted hydrogen next to an aromatic ring, (3) naphthenic hydrogen, (4) methylene hydrogen, and (5) terminal methyl hydrogen remote from an aromatic ring. Other ratios are also derived from which a series of structural parameters can be calculated. However, it must be remembered that the structural details of structural entities obtained by the use of physical techniques are, in many cases, derived by inference and it must be recognized that some signals can be obscured by intermolecular interactions. This, of course, can cause errors in deduction reasoning that can have a substantial influence on the outcome of the calculations (Ebert et al., 1984, 1987; Ebert, 1990; Speight, 1994, 2014, 2015).

It is in this regard that *carbon-13 magnetic resonance* (CMR) can play a useful role. Since carbon magnetic resonance deals with analyzing the carbon distribution types, the obvious structural parameter to be determined is the aromaticity, f_a. A direct determination from the various carbon-type environments is one of the better methods for the determination of aromaticity (Snape et al., 1979). Thus, through a combination of proton and carbon magnetic resonance techniques, refinements can be made on the structural parameters and for the solid-state high-resolution CMR technique additional structural parameters can be obtained (Weinberg et al., 1981).

2.8 CHROMATOGRAPHIC PROPERTIES

Chromatography is the collective term for a set of laboratory techniques for the separation of mixtures. Typically, the mixture is dissolved in a fluid (*mobile phase*) that carries it through a structure holding another material (*stationary phase*). The various constituents of the mixture travel at different speeds, causing them to separate. The separation is based on differential partitioning between the mobile and stationary phases. Subtle differences in the partition coefficient of different compounds result in differential retention on the stationary phase and thus changing the separation.

A chromatographic technique may be preparative or analytical. The purpose of preparative chromatography is to separate the components of a mixture for more advanced use (and is thus a form of purification). Analytical chromatography generally requires smaller amounts of material and is for measuring the relative proportions of analytes in a mixture. The two are not mutually exclusive.

2.8.1 ADSORPTION CHROMATOGRAPHY

Adsorption chromatography has helped to characterize the group composition of crude oils and hydrocarbon products since the beginning of this century. The type and relative amount of certain hydrocarbon classes in the matrix can have a profound effect on the quality and performance of the hydrocarbon product, and two standard test methods have been used predominantly over the years (ASTM D2007, ASTM D4124). The fluorescent indicator adsorption (FIA) method (ASTM D1319) has served for more than 30 years as the official method of the petroleum industry for measuring the paraffinic, olefinic, and aromatic content of gasoline and jet fuel. The technique consists of displacing a sample under *iso*-propanol pressure through a column packed with silica gel in the presence of fluorescent indicators specific to each hydrocarbon family. Despite its widespread use, fluorescent indicator adsorption has numerous limitations (Suatoni and Garber, 1975; Miller et al., 1983; Norris and Rawdon, 1984).

The segregation of individual components from a mixture can be achieved by the application of adsorption chromatography in which the adsorbent is either packed in an open tube (column chromatography) or shaped in the form of a sheet (thin-layer chromatography, TLC). A suitable solvent is used to elute from the bed of the adsorbent. Chromatographic separations are usually performed to determine the composition of a sample. Even with such complex samples as petroleum, some information about the chemical structure of a fraction can be gained from the separation data.

In the present context, the challenge is the nature of the heteroatomic species in the heavier feedstocks. It is these constituents that are largely responsible for coke formation and catalyst deactivation during refining operations. Therefore, it is these constituents that are the focus of much of the study. An ideal integrated separation scheme for the analysis of the heteroatomic constituents should therefore meet several criteria:

1. The various compound types should be concentrated into a reasonable number of discrete fractions, and each fraction should contain specific types of the heteroatomic compounds. It is also necessary that most of the heterocompounds be separated from the hydrocarbons and sulfur compounds that may constitute the bulk of the sample.
2. Perhaps most important, the separation should be reproducible insofar as the yields of the various fractions and the distribution of the compound types among the fractions should be constant within the limits of experimental error.
3. The separation scheme should be applicable to high-boiling distillates and heavy feedstocks such as residua since heteroatomic compounds often predominate in these feedstocks.
4. The separation procedures should be relatively simple to perform and free of complexity.
5. Finally, the overall separation procedure should yield quantitative or, at worst, near quantitative recovery of the various heteroatomic species present in the feedstock. There should be no significant loss of these species to the adsorbent or, perhaps more important, any chemical alteration of these compounds. Should chemical alteration occur, it will give misleading data that could have serious effects on refining predictions or on geochemical observations.

Group-type analysis by means of chromatography has been applied to a wide variety of petroleum types and products (Speight, 2014, 2015). These types of analysis are often abbreviated by the names PONA (paraffins, olefins, naphthenes, and aromatics), PIONA (paraffins, *iso*-paraffins, olefins, naphthenes, and aromatics), PNA (paraffins, naphthenes, and aromatics), PINA (paraffins, *iso*-paraffins, naphthenes, and aromatics), or SARA (saturates, aromatics, resin constituents, and asphaltene constituents).

The USBM-API (U.S. Bureau of Mines–American Petroleum Institute) method allows fractionation of petroleum samples into acids, bases, neutral nitrogen compounds, saturates, and mono-, di-, and polyaromatic compounds. Multidimensional techniques, that is, the combination of two or more chromatographic techniques, can be very useful to gain further information about the individual components of chemical groups. Compounds can be isolated and identified from complex matrices and detailed *fingerprinting* of petroleum constituents is feasible (Altgelt and Gouw, 1979).

2.8.2 Gas Chromatography

Gas–liquid chromatography (GLC) is a method for separating the volatile components of various mixtures. It is, in fact, a highly efficient fractionating technique, and it is ideally suited to the quantitative analysis of mixtures when the possible components are known and the interest lies only in determining the amounts of each present. In this type of application, gas chromatography has taken over much of the work previously done by the other techniques; it is now the preferred technique for the analysis of hydrocarbon gases, and gas chromatographic in-line monitors are having increasing application in refinery plant control.

Thus, it is not surprising that gas chromatography has been used extensively for individual component identification, as well as percentage composition, in the gaseous boiling ranges (ASTM D2163, ASTM D2504, ASTM D2505, ASTM D2593, ASTM D2597, ASTM D2712, ASTM D4424, ASTM D4864, ASTM D5303, ASTM D6159), in the gasoline boiling ranges (ASTM D2427, ASTM D3525, ASTM D3606, ASTM D3710, ASTM D4815, ASTM D5134, ASTM D5441, ASTM D5443, ASTM D5501, ASTM D5580, ASTM D5599, ASTM D5623, ASTM D5845, ASTM D5986), in higher boiling ranges such as diesel fuel (ASTM D3524), aviation gasoline (ASTM D3606), engine oil, motor oil, and wax (ASTM D5442), as well as for the boiling range distribution of petroleum fractions (ASTM D2887, ASTM D5307) or the purity of solvents using capillary gas chromatography (ASTM D2268).

The evolution of gas–liquid chromatography has been a major factor in the successful identification of petroleum constituents. It is, however, almost impossible to apply this technique to the higher-boiling petroleum constituents because of the comparatively low volatility. It is this comparative lack of volatility in the higher-molecular-weight, asphaltic constituents of petroleum that brought about another type of identification procedure, namely, carbon-type analysis.

The technique has proved to be an exceptional and versatile instrumental tool for analyzing compounds that are of low molecular weight and that can be volatilized without decomposition. However, these constraints limit the principal applicability in petroleum science to feedstock identification when the composition is known to be in the low to medium boiling range. The use of this technique for direct component analysis in the heavy fractions of petroleum is subject to many limitations (Speight, 2001).

For example, the number of possible components of a certain molecular weight range increases markedly with increasing molecular weight. Furthermore, there is a corresponding sharp decrease in physical property differences between isometric structures as the molecular weight increases. Thus, it is very difficult, and on occasion almost impossible, to separate and identify single components in the heavier fractions of petroleum by gas chromatography. Indeed, the molecular weights of the constituents dictate that long residence times are necessary. This is inevitably accompanied by the requirement of increased column temperature, which decreases the residence time on the column but, at the same time, increases the possibility of thermal decomposition.

The operation of the instrumentation for gas–liquid chromatography is fairly straightforward and involves passing a carrier gas passes through a controller to the column (packed with an adsorbent) at the opening of which is a sample injector. The carrier gas then elutes the components of the mixture through the column to the detector at the end of which may be another gas flow monitor. Any gas, such as helium, argon, nitrogen, or hydrogen that is easily distinguishable from the components in the mixture may be used as the carrier gas.

Column dimensions vary, but for analytic purposes a packed column may be 6 ft (2 m) long by 3 inch (6 mm) in diameter. It is also necessary to use a *dissolving* liquid as part of the column substance. This remains stationary on the adsorbent and effects partition of the components of the mixture. The solid support is usually a porous material that allows passage of the gas. For example, kieselguhr (diatomaceous earth), which can absorb up to 40% by weight of a liquid without appearing to be overly moist, is commonly used. The supporting material should not adsorb any of the components of the mixture and must therefore be inert.

Individual components of mixtures are usually identified by their respective retention times, that is, the time required for the component to traverse through the column under the specified conditions. Although tables for retention time data are available, it is more common in practice to determine the retention times of the pure compounds. The retention time of any component is itself a function of the many variables of column operation, such as the flow rate of the carrier gas and column temperature, and exact duplication of other operator's conditions may be difficult, if not impossible.

The sample size used in gas chromatography may vary upward from a microliter, and there is no theoretical upper limit to the size of the sample that may be handled if the equipment is built to accommodate it. The technique can be used for the analysis of mixtures of volatile vaporizable compounds boiling at any temperature between absolute zero (−273°C, 459°F) and 450°C (840°F). Identification of any substance that can be heated sufficiently without decomposing to give a vapor pressure of a few millimeters mercury is also possible.

The use of gas–liquid chromatography for direct component analysis in the *higher-boiling* fractions of petroleum, such as residua, is beset by many problems, not the least of which is the low volatility and tendency for adsorption on solids by the higher-molecular-weight constituents. The number of possible components in any given molecular weight range increases markedly with the molecular weight (Speight, 2015), and there is a *significant drop* in the differences in physical properties among similar structural entities. This limits the ability of gas–liquid chromatography, and unless the sample has been fractionated by other techniques to reduce the complexity, complete component analysis is difficult, if not impossible.

The mass spectrometer identifies chemical compounds principally in terms of molecular type and molecular weight, and for many problems, therefore, it becomes necessary to use additional means of identification; the integrated gas–liquid chromatography infrared system is a very valuable complement to the mass spectrometer technique.

Considerable attention has also been given to trapping devices to collect gas chromatographic fractions for examination by one or more of the spectroscopic techniques. At the same time, developments in preparative gas–liquid chromatography have contributed even more to the compositional studies of petroleum and its products. With column size of 4–6 inches in diameter and capable of dealing with sample sizes of 200 mL or more, there is every possibility that gas–liquid chromatography will replace distillation in such areas as standard crude oil assay work.

Gas–liquid chromatography also provides a simple and convenient method for determining *n*-paraffin distribution throughout the petroleum distillate range. In this method, the *n*-paraffins are first separated by activated chemical destruction of the sieve with hydrofluoric acid, and the identity of the individual paraffins is determined chromatographically. This allows *n*-paraffin distribution throughout the boiling range 170°C–500°C (340°F–930°F) to be determined.

Gas chromatographic process analyzers have become very important in petroleum refineries. In some refineries, more samples are analyzed automatically by process chromatographs than are analyzed with laboratory instruments. These chromatographs are usually fully automatic. In some cases, after an analysis the instrument even makes automatic adjustments to the refinery unit. The chromatographs usually determine from 1 to 10 components, and the analyses are repeated at short intervals (15–20 minutes) over 24 hours.

A more recent, very important development in gas chromatography is its combination with a mass spectrometer as the detector. The technique in which gas chromatography is combined with spectrometry (GC/MS) has proved to be a powerful tool for identifying many compounds at very low levels in a wide range of boiling matrix (Poirier and Das, 1984; Lin et al., 1989). By the combination of the two techniques in one instrument, the onerous trapping of fractions from the gas chromatographic column is avoided and higher sensitivities can be attained. In passing through the gas chromatographic column, the sample is separated more or less according to its boiling point.

In view of the molecular characterizing nature of spectrometric techniques, it is not surprising that considerable attention has been given to the combined use of gas–liquid chromatography and these techniques. In recent years, the use of the *mass spectrometer* to monitor continuously the effluent of a chromatographic column has been reported, and considerable progress has been made in the development of rapid scan infrared spectrometers for this purpose. The mass spectrometer, however, has the advantage that the quantity of material required for the production of a spectrum is considerably less than that necessary to produce an infrared spectrum.

Although insufficient component resolution is observed in most cases, the eluting compounds at any time are usually closely related to each other in boiling point and molecular weight or both and are free from interfering lower and higher-molecular-weight species. Because of the reduced complexity of the gas chromatographic fractions, mass spectrometric scans carried out at regular intervals yield simpler spectra from which compound classes can more easily be determined.

Pyrolysis gas chromatography can be used for information on the gross composition of heavy petroleum fractions. In this technique, the sample under investigation is pyrolyzed, and the products are introduced into a gas chromatography system for analysis. There has also been extensive use of pyrolysis gas chromatography by geochemists to correlate crude oil with source rock and to derive geochemical characterization parameters from oil-bearing strata.

In the technique of inverse gas–liquid chromatography, the sample under study is used as the stationary phase, and a number of volatile test compounds are chromatographed on this column. The interaction coefficient determined for these compounds is a measure of certain qualities of the liquid phase. The coefficient is therefore indicative of the chemical interaction of the solute with the stationary phase. The technique has been used largely for the studies of asphalt.

2.8.3 Gel Permeation Chromatography

There are two additional techniques that have evolved from the more recent development of chromatographic methods.

The first technique, *gel filtration chromatography* (*GFC*), has been successfully employed for application to aqueous systems by biochemists for more than five decades. The technique was developed using soft, cross-linked dextran beads. The second technique, *gel permeation chromatography* (*GPC*), employs semirigid, cross-linked polystyrene beads. In either technique, the packing particles swell in the chromatographic solvent forming a porous gel structure.

The distinction between the methods is based on the degree of swelling of the packing; the dextran swells to a much greater extent than the polystyrene. Subsequent developments of rigid porous packings of glass, silica, and silica gel have led to their use and classification as packings for gel permeation chromatography.

Gel permeation chromatography, also called "size exclusion chromatography" (*SEC*), which in its simplest representation, consists of employing column(s) packed with gels of varying pore sizes in a liquid chromatograph (Carbognani, 1997). Under conditions of constant flow, the solutes are injected onto the top of the column, whereupon they appear at the detector in order of decreasing molecular weight. The separation is based on the fact that the larger solute molecules cannot be accommodated within the pore systems of the gel beads and thus are eluted first. On the other hand, the smaller solute molecules have increasing volume within the beads, depending upon their relative size, and require more time to elute.

Thus, it is possible, with careful flow control, calibration, injection, and detection (usually by refractive index or UV absorption), to obtain an accurate chromatographic representation of the molecular weight distribution of the solute (Carbognani, 1997). This must of course assume that there is no chemical or physical interaction between the solute and the gel that negates the concept of solute size and pore size. For example, highly polar, small molecules that could associate in solution and are difficult to dissociate could conceivably appear in the *incorrect* molecular weight range.

In theory, gel permeation chromatography is an attractive technique for the determination of the number of average molecular weight (M_n) distribution of petroleum fractions. However, it is imperative to recognize that petroleum contains constituents of widely differing polarity, including nonpolar paraffins and naphthenes (alicyclic compounds), moderately polar aromatics (mononuclear and condensed), and polar nitrogen, oxygen, and sulfur species. Each particular compound type interacts with the gel surface to a different degree. The strength of the interaction increases with increasing polarity of the constituents and with decreasing polarity of the solvent. It must therefore be anticipated that the ideal linear relationship of log M_n against elution volume V_e that may be operative for nonpolar hydrocarbon species cannot be expected to remain in operation. It must also be recognized that the lack of realistic standards of known number average molecular weight distribution and of chemical nature similar to that of the constituents of petroleum for calibration purposes may also be an issue. However, gel permeation chromatography has been employed in the study of petroleum constituents, especially the heavier constituents, and has yielded valuable data (Baltus and Anderson, 1984; Reynolds and Biggs, 1988; Speight, 2001, 2015).

The adoption of gel permeation chromatography represents a novel approach to the identification of the constituents since the method is not limited by the vapor pressure of the constituents. However, the situation is different with heavy petroleum samples. These are not homologous mixtures differing only in molecular weight. In any particular crude oil, a large variety of molecular species, varying from paraffinic molecules to the polynuclear aromatic ring systems, may not follow the assumed physical relationships that the method dictates from use with polymers.

Gel permeation chromatography is the separation method that comes closest to differentiating by molecular weight only and is almost unaffected by chemical composition (hence the alternate name size exclusion chromatography). The method actually achieves the separation of analytes by molecular size and has been used to measure molecular weights (Altgelt and Guow, 1979), although there is some question about the value of the data when the method is applied to asphaltene constituents (Speight et al., 1985).

Size exclusion chromatography is usually practiced with refractive index detection and yields a mass profile (concentration versus time or elution volume) that can be converted to a mass versus molecular weight plot by means of a calibration curve. The combination of size exclusion chromatography with element specific detection has widened this concept to provide the distribution of heterocompounds in the sample as a function of elution volume and molecular weight.

The use of size exclusion chromatography with reverse phase high-performance liquid chromatography (HPLC) with a graphite furnace atomic absorption (GFAA) detector has been described for measuring the distribution of vanadium and nickel in high-molecular-weight petroleum fractions, including the asphaltene fraction. Using variants of this technique, inductively coupled and direct current plasma atomic emission spectroscopy (ICP and DCP), the method was extended and improved the former size exclusion chromatography-graphite furnace atomic absorption (SEC-GFAA) method allowing the separation to be continuously monitored.

The combination of gel permeation chromatography with another separation technique also allows the fractionation of a sample separately by molecular weight and by chemical structure. This is particularly advantageous for the characterization of the heavier fractions of petroleum materials because there are limitations to the use of other methods. Thus, it is possible to obtain a matrix of fractions differing in molecular weight and in chemical structure. It is also considered advisable to first fractionate a feedstock by gel permeation chromatography to avoid overlap of the functionality that might occur in different molecular weight species in the separation by other chromatographic methods.

In short, the gel permeation chromatographic technique concentrates all of a specific functional type into one fraction, recognizing that there will be a wide range of molecular weight species in that fraction. This is especially true when the chromatographic feedstock is a whole feed rather than a distillate fraction.

2.8.4 High-Performance Liquid Chromatography

High-performance liquid chromatography (*HPLC*), particularly in the normal phase mode, has found great utility in separating different hydrocarbon group types and identifying specific constituent types (Colin and Vion, 1983; Miller et al., 1983). However, a severe shortcoming of most high-performance liquid chromatographic approaches to a hydrocarbon group type of analysis is the difficulty in obtaining accurate response factors applicable to different distillate products. Unfortunately, accuracy can be compromised when these response factors are used to analyze hydrotreated and hydrocracked materials having the same boiling range. In fact, significant changes in the hydrocarbon distribution within a certain group type cause the analytic results to be misleading for such samples because of the variation in response with carbon number exhibited by most routinely used HPLC detectors.

Of particular interest is the application of the HPLC technique to the identification of the molecular types in nonvolatile feedstocks such as residua. The molecular species in the asphaltene fraction have been of particular interest (Colin and Vion, 1983; Felix et al., 1985), leading to the identification of the size of polynuclear aromatic systems in the asphaltene constituents.

Several recent high-performance liquid chromatographic separation schemes are particularly interesting since they also incorporate detectors not usually associated with conventional hydrocarbon group types of analyses (Matsushita et al., 1981; Miller et al., 1983; Norris and Rawdon, 1984; Rawdon, 1984; Schwartz and Brownlee, 1986). The ideal detector for a truly versatile and accurate hydrocarbon group type of analysis is one that is sensitive to hydrocarbons but demonstrates a response independent of carbon number.

In general, the amount of information that can be derived from any chromatographic separation, however effective, depends on the detectors (Pearson and Gharfeh, 1986). As the field of application for high-performance liquid chromatographic has increased, the limitations of commercially available conventional detectors, such as ultraviolet/visible absorption (UV/VIS) and refractive index (RI), have become increasingly restrictive to the growth of the technique. This has led to a search for detectors capable of producing even more information. The so-called hyphenated techniques are the outcome of this search.

The general advantages of high-performance liquid chromatography method are as follows: (1) each sample is analyzed *as received*, (2) the boiling range of the sample is generally immaterial, (3) the total time per analysis is usually of the order of minutes, and (4) the method can be adapted for onstream analysis.

2.8.5 Ion-Exchange Chromatography

Ion-exchange chromatography is widely used in the analyses of petroleum fractions for the isolation and preliminary separation of acid and basic components (Speight, 2001). This technique has the advantage of greatly improving the quality of a complex operation, but it can be a very time-consuming separation.

Ion-exchange resin constituents are prepared from aluminum silicates, synthetic resin constituents, and polysaccharides. The most widely used resin constituents have a skeletal structure of polystyrene cross-linked with varying amounts of divinylbenzene derivatives. They have a loose gel structure of cross-linked polymer chains through which the sample ions must diffuse to reach most of the exchange sites. Since ion-exchange resin constituents are usually prepared as beads that are several hundred micrometers in diameter, most of the exchange sites are located at points quite distant from the surface. Because of the polyelectrolyte nature of these organic resin constituents, they can absorb large amounts of water or solvents and swell to volumes considerably larger than the dried gel. The size of the particle is determined by the intermolecular spacing between the polymeric chains of the 3D polyelectrolyte resin.

Cation-exchange chromatography is now used primarily to isolate the nitrogen constituents in a petroleum fraction. The relative importance of these compounds in petroleum has arisen because of their deleterious effects in many petroleum refining processes. They reduce the activity of cracking and hydrocracking catalysts and contribute to gum formation, color, odor, and poor storage properties of the fuel. However, not all basic compounds isolated by cation-exchange chromatography are nitrogen compounds. Anion-exchange chromatography is used to isolate the acid components (such as carboxylic acids and phenols) from petroleum fractions.

2.8.6 SIMULATED DISTILLATION

Gas–liquid chromatography has also been found useful for the preparation of simulated distillation curves. By integrating increments of the total area of the chromatogram and relating these to the boiling points of the components within each increment, which are calculated from the known boiling points of the easily recognizable *n*-paraffins, simulated boiling point data are produced.

Distillation is the most widely used separation process in the petroleum industry (Speight and Ozum, 2002; Parkash, 2003; Gary et al., 2007; Speight, 2014). In fact, knowledge of the boiling range of crude feedstocks and finished products has been an essential part of the determination of feedstock quality since the start of the refining industry. The technique has been used for control of plant and refinery processes as well as for predicting product slates. Thus, it is not surprising that routine laboratory-scale distillation tests have been widely used for determining the boiling ranges of crude feedstocks and a whole slate of refinery products (Speight, 2015).

There are some limitations to the routine distillation tests. For example, although heavy crude oils contain volatile constituents, it is not always advisable to use distillation for the identification of these volatile constituents. Thermal decomposition of the constituents of petroleum is known to occur at approximately 350°C (660°F). Thermal decomposition of the constituents of the heavier, but immature, crude oil has been known to commence at temperatures as low as 200°C (390°F), however. Thus, thermal alteration of the constituents and erroneous identification of the decomposition products as *natural* constituents is always a possibility.

On the other hand, the limitations to the use of distillation as an identification technique may be economic, and detailed fractionation of the sample may also be of secondary importance. There have been attempts to combat these limitations, but it must be recognized that the general shape of a one-plate distillation curve is often adequate for making engineering calculations, correlating with other physical properties, and predicting the product slate.

However, a low-resolution, temperature-programmed gas chromatographic analysis has been developed to simulate the time-consuming true boiling point distillation (ASTM D2887). The method relies on the general observation that hydrocarbons are eluted from a nonpolar adsorbent in the order of their boiling points. The method has been well researched in terms of method development and application (MacAllister and DeRuiter, 1985; Romanowski and Thomas, 1985; Schwartz et al., 1987). The benefits of the technique include good comparisons with other ASTM distillation data as well as application to higher-boiling fractions of petroleum (Speight, 2001, 2015).

The full development of simulated distillation as a routine procedure has been made possible by the massive expansion in gas chromatographic instrumentation (such as the introduction of automatic temperature programming) since the 1960s. In fact, a fully automated simulated distillation system, under computer control, can operate continuously to provide finished reports in a choice of formats that agree well with true boiling point data. For example, data output includes the provision of the corresponding Engler profile (ASTM D86) as well as the prediction of other properties, such as vapor pressure and flash point.

Simulated distillation by gas chromatography is applied in the petrochemical industry to obtain true boiling point distributions of distillates and crude oils. Two standardized methods, ASTM D2887 and ASTM D3710, are available for the boiling point determination of petroleum fractions and gasoline, respectively. The ASTM D2887 method utilizes nonpolar, packed gas chromatographic columns in conjunction with flame ionization detection. The upper limit of the boiling range covered by this method is to approximately 540°C (1000°F) atmospheric equivalent boiling point. Recent efforts in which high-temperature gas chromatography was used have focused on extending the scope of the ASTM D2887 methods for higher-boiling petroleum materials to 800°C (1470°F) atmospheric equivalent boiling point.

2.8.7 SUPERCRITICAL FLUID CHROMATOGRAPHY

A supercritical fluid is defined as a substance above its critical temperature that has properties not usually found at ambient temperatures and pressures. The use of a fluid under supercritical conditions conveys upon the fluid extraction capabilities that allows the opportunity to improve the recovery of a solute (Taylor, 1996).

In supercritical fluid chromatography, the mobile phase is a substance maintained at a temperature a few degrees above its critical point. The physical properties of this substance are intermediate to those of a liquid and of a gas at ambient conditions. Hence, it is preferable to designate this condition as the supercritical phase.

In a chromatographic column, the supercritical fluid usually has a density approximately one-third to one-fourth of that of the corresponding liquid when used as the mobile phase; the diffusivity is approximately 1/100 that of a gas and approximately 200 times that of the liquid. The viscosity is of the same order of magnitude as that of the gas. Thus, for chromatographic purposes, such a fluid has more desirable transport properties than a liquid. In addition, the high density of the fluid results in a 1000-fold better solvency than that of a gas. This is especially valuable for analyzing high-molecular-weight compounds.

A primary advantage of chromatography using supercritical mobile phases results from the mass transfer characteristics of the solute. The increased diffusion coefficients of supercritical fluids compared with liquids can lead to greater speed in separations or greater resolution in complex mixture analyses. Another advantage of supercritical fluids compared with gases is that they can dissolve thermally labile and nonvolatile solutes and, upon expansion (decompression) of this solution, introduce the solute into the vapor phase for detection. Although supercritical fluids are sometimes considered to have superior solvating power, they usually do not provide any advantages in solvating power over liquids given a similar temperature constraint. In fact, many unique capabilities of supercritical fluids can be attributed to the poor solvent properties obtained at lower fluid densities. This dissolution phenomenon is increased by the variability of the solvent power of the fluid with density as the pressure or temperature changes.

The solvent properties that are most relevant for supercritical fluid chromatography are the critical temperature, polarity, and any specific solute–solvent intermolecular interactions (such as hydrogen bonding) that can enhance solubility and selectivity in a separation. Nonpolar or low-polarity solvents with moderate critical temperatures (e.g., nitrous oxide, carbon dioxide, ethane, propane, pentane, xenon, sulfur hexafluoride, and various Freon derivatives) have been well explored for use in supercritical fluid chromatography. Carbon dioxide has been the fluid of choice in many supercritical fluid chromatography applications because of its low critical temperature (31°C, 88°F), nontoxic nature, and lack of interference with most detection methods (Smith et al., 1988).

2.9 MOLECULAR WEIGHT

The molecular weight (formula weight) of a compound is the sum of the atomic weights of all the atoms in a molecule and can be determined by a variety of methods (Cooper, 1989). Petroleum, being a complex mixture of (at least) several thousand constituents, requires qualification of the molecular weight as either (1) number average molecular weight or (2) weight average molecular weight.

The *number average molecular weight* is the ordinary arithmetic mean or average of the molecular weights of the individual constituents. It is determined by measuring the molecular weight of n molecules, summing the weights, and dividing by n.

The *weight average molecular weight* is a way of describing the molecular weight of a complex mixture such as petroleum even if the molecular constituents are not of the same type and exist in different sizes.

Even though refining produces, in general, lower-molecular-weight species than those originally in the feedstock, there is still the need to determine the molecular weight of the original constituents as well as the molecular weights of the products as a means of understanding the process. For those original constituents and products, for example, resin constituents and asphaltene constituents, that have little or no volatility, *vapor pressure osmometry* has been proven to be of considerable value (Blondel-Telouk et al., 1995).

A particularly appropriate method involves the use of different solvents (at least two), and the data are then extrapolated to infinite dilution. There has also been the use of different temperatures for a particular solvent after which the data are extrapolated to room temperature (Speight et al., 1985; Speight, 1987). In this manner, different solvents are employed, and the molecular weight of a petroleum fraction (particularly the asphaltene constituents) can be determined for which it can be assumed that there is little or no influence from any intermolecular forces. In summary, the molecular weight may be as close to the real value as possible.

In fact, it is strongly recommended that to negate concentration effects and temperature effects, the molecular weight determination be carried out at three different concentrations at three different temperatures. The data for each temperature are then extrapolated to zero concentration and the zero concentration data at each temperature are then extrapolated to room temperature (Speight, 1987).

2.10   USE OF THE DATA

The data derived from the evaluation techniques described here can be employed to give the refiner an indication of the means by which the crude feedstock should be processed as well as for the prediction of product properties (Dolbear et al., 1987; Wallace and Carrigy, 1988; Speight, 2014, 2015). Other properties (Table 2.1) may also be required for further feedstock evaluation, or, more likely, for comparison between feedstocks even though they may not play any role in dictating which refinery operations are necessary. An example of such an application is the calculation of product yields for delayed coking operations by using the carbon residue and the API gravity (Table 2.10) of the feedstock.

Nevertheless, it must be emphasized that to proceed from the raw evaluation data to full-scale production is not the preferred step; further evaluation of the processability of the feedstock is

---

TABLE 2.10

Use of Carbon Residue Test Data to Estimate Product Yield from Delayed Coking

Wilmington

Coke, wt%	$39.68 - 1.60 \times °API$
Gas ($\leq C_4$), wt%	$11.27 - 0.14 \times °API$
Gasoline, wt%	$20.5 - 0.36 \times °API$
Gas oil, wt%	$28.55 + 2.10 \times °API$
Gasoline, vol%	$\dfrac{186.5}{131.5 + °API} \times wt\% \text{ gasoline}$
Gas oil, vol%	$\dfrac{155.5}{131.5 + °API} \times wt\% \text{ gas oil}$

East Texas

Coke, wt%	$45.76 - 1.78 \times °API$
Gas ($\leq C_4$), wt%	$11.92 - 0.16 \times °API$
Gasoline, wt%	$20.5 - 0.36 \times °API$
Gasoline, vol%	$\dfrac{186.5}{131.5 + °API} \times wt\% \text{ gasoline}$
Gas oil, vol%	$\dfrac{155.5}{131.5 + °API} \times wt\% \text{ gas oil}$

usually necessary through the use of a pilot-scale operation. To take the evaluation of a feedstock one step further, it may then be possible to develop correlations between the data obtained from the actual plant operations (as well as the pilot plant data) with one or more of the physical properties determined as part of the initial feedstock evaluation.

However, it is essential that when such data are derived, the parameters employed should be carefully specified. For example, the data presented in the tables were derived on the basis of straight-run residua having API gravity less than 18°. The gas oil end point was of the order of 470°C–495°C (875°F–925°F), the gasoline end point was 205°C (400°F), and the pressure in the coke drum was standardized at 35–45 psi. Obviously, there are benefits to the derivation of such specific data, but the numerical values, although representing only an approximation, may vary substantially when applied to different feedstocks (Speight, 1987).

Evaluation of petroleum from known physical properties may also be achieved by the use of the refractivity intercept. Thus, if refractive indices of hydrocarbons are plotted against the respective densities, straight lines of constant slope are obtained, one for each homologous series; the intercepts of these lines with the ordinate of the plot are characteristic, and the refractivity intercept is derived from the following formula:

$$\text{Refractivity intercept} = \frac{n - d}{2}$$

The intercept cannot differentiate accurately among all series, which restricts the number of different types of compounds that can be recognized in a sample. The technique has been applied to nonaromatic olefin-free materials in the gasoline range by assuming the additivity of the constant on a volume basis.

Following from this, an equation has been devised that is applicable to straight-run lubricating distillates if the material contains between 25% and 75% of the carbon present in naphthenic rings:

$$\text{Refractivity intercept} = 1.0502 - 0.00020\%C_N$$

Although not specifically addressed in this chapter, the fractionation of petroleum (Speight, 2014, 2015) also plays a role, along with the physical testing methods, of evaluating petroleum as a refinery feedstock. For example, by careful selection of an appropriate technique, it is possible to obtain a detailed overview of feedstock or product composition that can be used for process predictions. Using the adsorbent separation as an example, it becomes possible to develop one or more petroleum *maps* and determine how a crude oil might behave under specified process conditions.

This concept has been developed to the point where various physical parameters can be represented graphically as the ordinates and abscissa. However, it must be recognized that such *maps* do not give any indication of the complex interactions that occur between, for example, such fractions as the asphaltene constituents and resin constituents (Koots and Speight, 1975; Speight, 1994), but it does allow predictions of feedstock behavior. It must also be recognized that such a representation varies for different feedstocks.

In summary, evaluation of feedstock behavior from test data is not only possible but has been practiced for decades. And such evaluations will continue for decades to come. However, it is essential to recognize that the derivation of an equation for the predictability of behavior will not suffice (with a reasonable degree of accuracy) for all feedstocks. Many of the data are feedstock dependent because they incorporate the complex reactions of the feedstock constituents with each other. Careful testing and evaluation of the behavior of each feedstock and blend of feedstocks is recommended. If this is not done, incompatibility or instability (Speight, 2014, 2015) can result, leading to higher-than-predicted yields of thermal or catalytic coke.

REFERENCES

Altgelt, K.H. and Gouw, T.H. 1979. *Chromatography in Petroleum Analysis.* Dekker, New York.

ASTM C1109. 2015. *Standard Practice for Analysis of Aqueous Leachates from Nuclear Waste Materials Using Inductively Coupled Plasma-Atomic Emission Spectroscopy.* Annual Book of Standards. ASTM International, West Conshohocken, PA.

ASTM C1111. 2015. *Standard Test Method for Determining Elements in Waste Streams by Inductively Coupled Plasma-Atomic Emission Spectroscopy.* Annual Book of Standards. ASTM International, West Conshohocken, PA.

ASTM D6. 2015. *Standard Test Method for Loss on Heating of Oil and Asphaltic Compounds.* Annual Book of Standards. ASTM International, West Conshohocken, PA.

ASTM D20. 2015. *Standard Test Method for Distillation of Road Tars.* Annual Book of Standards. ASTM International, West Conshohocken, PA.

ASTM D56. 2015. *Standard Test Method for Flash Point by Tag Closed Cup Tester.* Annual Book of Standards. ASTM International, West Conshohocken, PA.

ASTM D70. 2015. *Standard Test Method for Density of Semi-Solid Bituminous Materials (Pycnometer Method).* Annual Book of Standards. ASTM International, West Conshohocken, PA.

ASTM D71. 2015. *Standard Test Method for Relative Density of Solid Pitch and Asphalt (Displacement Method).* Annual Book of Standards. ASTM International, West Conshohocken, PA.

ASTM D86. 2015. *Standard Test Method for Distillation of Petroleum Products at Atmospheric Pressure.* Annual Book of Standards. ASTM International, West Conshohocken, PA.

ASTM D87. 2015. *Standard Test Method for Melting Point of Petroleum Wax (Cooling Curve).* Annual Book of Standards. ASTM International, West Conshohocken, PA.

ASTM D88. 2015. *Standard Test Method for Saybolt Viscosity.* Annual Book of Standards. ASTM International, West Conshohocken, PA.

ASTM D92. 2015. *Standard Test Method for Flash and Fire Points by Cleveland Open Cup Tester.* Annual Book of Standards. ASTM International, West Conshohocken, PA.

ASTM D93. 2015. *Standard Test Methods for Flash Point by Pensky-Martens Closed Cup Tester.* Annual Book of Standards. ASTM International, West Conshohocken, PA.

ASTM D97. 2015. *Standard Test Method for Pour Point of Petroleum Products.* Annual Book of Standards. ASTM International, West Conshohocken, PA.

ASTM D127. 2015. *Standard Test Method for Drop Melting Point of Petroleum Wax, Including Petrolatum.* Annual Book of Standards. ASTM International, West Conshohocken, PA.

ASTM D129. 2015. *Standard Test Method for Sulfur in Petroleum Products (General High Pressure Decomposition Device Method).* Annual Book of Standards. ASTM International, West Conshohocken, PA.

ASTM D139. 2015. *Standard Test Method for Float Test for Bituminous Materials.* Annual Book of Standards. ASTM International, West Conshohocken, PA.

ASTM D189. 2015. *Standard Test Method for Conradson Carbon Residue of Petroleum Products.* Annual Book of Standards. ASTM International, West Conshohocken, PA.

ASTM D240. 2015. *Standard Test Method for Heat of Combustion of Liquid Hydrocarbon Fuels by Bomb Calorimeter.* Annual Book of Standards. ASTM International, West Conshohocken, PA.

ASTM D287. 2015. *Standard Test Method for API Gravity of Crude Petroleum and Petroleum Products (Hydrometer Method).* Annual Book of Standards. ASTM International, West Conshohocken, PA.

ASTM D323. 2015. *Standard Test Method for Vapor Pressure of Petroleum Products (Reid Method).* Annual Book of Standards. ASTM International, West Conshohocken, PA.

ASTM D341. 2015. *Standard Practice for Viscosity-Temperature Charts for Liquid Petroleum Products.* Annual Book of Standards. ASTM International, West Conshohocken, PA.

ASTM D445. 2015. *Standard Test Method for Kinematic Viscosity of Transparent and Opaque Liquids (and Calculation of Dynamic Viscosity).* Annual Book of Standards. ASTM International, West Conshohocken, PA.

ASTM D482. 2015. *Standard Test Method for Ash from Petroleum Products.* Annual Book of Standards. ASTM International, West Conshohocken, PA.

ASTM D524. 2015. *Standard Test Method for Ramsbottom Carbon Residue of Petroleum Products.* Annual Book of Standards. ASTM International, West Conshohocken, PA.

ASTM D566. 2015. *Standard Test Method for Dropping Point of Lubricating Grease.* Annual Book of Standards. ASTM International, West Conshohocken, PA.

ASTM D611. 2015. *Standard Test Methods for Aniline Point and Mixed Aniline Point of Petroleum Products and Hydrocarbon Solvents*. Annual Book of Standards. ASTM International, West Conshohocken, PA.

ASTM D664. 2015. *Standard Test Method for Acid Number of Petroleum Products by Potentiometric Titration*. Annual Book of Standards. ASTM International, West Conshohocken, PA.

ASTM D877. 2015. *Standard Test Method for Dielectric Breakdown Voltage of Insulating Liquids Using Disk Electrodes*. Annual Book of Standards. ASTM International, West Conshohocken, PA.

ASTM D938. 2015. *Standard Test Method for Congealing Point of Petroleum Waxes, Including Petrolatum*. Annual Book of Standards. ASTM International, West Conshohocken, PA.

ASTM D971. 2015. *Standard Test Method for Interfacial Tension of Oil against Water by the Ring Method*. Annual Book of Standards. ASTM International, West Conshohocken, PA.

ASTM D974. 2015. *Standard Test Method for Acid and Base Number by Color-Indicator Titration*. Annual Book of Standards. ASTM International, West Conshohocken, PA.

ASTM D1015. 2015. *Standard Test Method for Freezing Points of High-Purity Hydrocarbons*. Annual Book of Standards. ASTM International, West Conshohocken, PA.

ASTM D1016. 2015. *Standard Test Method for Purity of Hydrocarbons from Freezing Points*. Annual Book of Standards. ASTM International, West Conshohocken, PA.

ASTM D1018. 2015. *Standard Test Method for Hydrogen in Petroleum Fractions*. Annual Book of Standards. ASTM International, West Conshohocken, PA.

ASTM D1160. 2015. *Standard Test Method for Distillation of Petroleum Products at Reduced Pressure*. Annual Book of Standards. ASTM International, West Conshohocken, PA.

ASTM D1217. 2015. *Standard Test Method for Density and Relative Density (Specific Gravity) of Liquids by Bingham Pycnometer*. Annual Book of Standards. ASTM International, West Conshohocken, PA.

ASTM D1218. 2015. *Standard Test Method for Refractive Index and Refractive Dispersion of Hydrocarbon Liquids*. Annual Book of Standards. ASTM International, West Conshohocken, PA.

ASTM D1266. 2015. *Standard Test Method for Sulfur in Petroleum Products (Lamp Method)*. Annual Book of Standards. ASTM International, West Conshohocken, PA.

ASTM D1298. 2015. *Standard Test Method for Density, Relative Density, or API Gravity of Crude Petroleum and Liquid Petroleum Products by Hydrometer Method*. Annual Book of Standards. ASTM International, West Conshohocken, PA.

ASTM D1318. 2015. *Standard Test Method for Sodium in Residual Fuel Oil (Flame Photometric Method)*. Annual Book of Standards. ASTM International, West Conshohocken, PA.

ASTM D1319. 2015. *Standard Test Method for Hydrocarbon Types in Liquid Petroleum Products by Fluorescent Indicator Adsorption*. Annual Book of Standards. ASTM International, West Conshohocken, PA.

ASTM D1405. 2015. *Standard Test Method for Estimation of Net Heat of Combustion of Aviation Fuels*. Annual Book of Standards. ASTM International, West Conshohocken, PA.

ASTM D1480. 2015. *Standard Test Method for Density and Relative Density (Specific Gravity) of Viscous Materials by Bingham Pycnometer*. Annual Book of Standards. ASTM International, West Conshohocken, PA.

ASTM D1481. 2015. *Standard Test Method for Density and Relative Density (Specific Gravity) of Viscous Materials by Lipkin Bicapillary Pycnometer*. Annual Book of Standards. ASTM International, West Conshohocken, PA.

ASTM D1534. 2015. *Standard Test Method for Approximate Acidity in Electrical Insulating Liquids by Color-Indicator Titration*. Annual Book of Standards. ASTM International, West Conshohocken, PA.

ASTM D1552. 2015. *Standard Test Method for Sulfur in Petroleum Products (High-Temperature Method)*. Annual Book of Standards. ASTM International, West Conshohocken, PA.

ASTM D1555. 2015. *Standard Test Method for Calculation of Volume and Weight of Industrial Aromatic Hydrocarbons and Cyclohexane*. Annual Book of Standards. ASTM International, West Conshohocken, PA.

ASTM D1657. 2015. *Standard Test Method for Density or Relative Density of Light Hydrocarbons by Pressure Hydrometer*. Annual Book of Standards. ASTM International, West Conshohocken, PA.

ASTM D1757. 2015. *Standard Test Method for Sulfur in Ash from Coal and Coke*. Annual Book of Standards. ASTM International, West Conshohocken, PA.

ASTM D1840. 2015. *Standard Test Method for Naphthalene Hydrocarbons in Aviation Turbine Fuels by Ultraviolet Spectrophotometry*. Annual Book of Standards. ASTM International, West Conshohocken, PA.

ASTM D2007. 2015. *Standard Test Method for Characteristic Groups in Rubber Extender and Processing Oils and Other Petroleum-Derived Oils by the Clay-Gel Absorption Chromatographic Method.* Annual Book of Standards. ASTM International, West Conshohocken, PA.

ASTM D2161. 2015. *Standard Practice for Conversion of Kinematic Viscosity to Saybolt Universal Viscosity or to Saybolt Furol Viscosity.* Annual Book of Standards. ASTM International, West Conshohocken, PA.

ASTM D2163. 2015. *Standard Test Method for Determination of Hydrocarbons in Liquefied Petroleum (LP) Gases and Propane/Propene Mixtures by Gas Chromatography.* Annual Book of Standards. ASTM International, West Conshohocken, PA.

ASTM D2265. 2015. *Standard Test Method for Dropping Point of Lubricating Grease Over Wide Temperature Range.* Annual Book of Standards. ASTM International, West Conshohocken, PA.

ASTM D2268. 2015. *Standard Test Method for Analysis of High-Purity n-Heptane and Isooctane by Capillary Gas Chromatography.* Annual Book of Standards. ASTM International, West Conshohocken, PA.

ASTM D2269. 2015. *Standard Test Method for Evaluation of White Mineral Oils by Ultraviolet Absorption.* Annual Book of Standards. ASTM International, West Conshohocken, PA.

ASTM D2270. 2015. *Standard Practice for Calculating Viscosity Index from Kinematic Viscosity at 40 and 100°C.* Annual Book of Standards. ASTM International, West Conshohocken, PA.

ASTM D2425. 2015. *Standard Test Method for Hydrocarbon Types in Middle Distillates by Mass Spectrometry.* Annual Book of Standards. ASTM International, West Conshohocken, PA.

ASTM D2427. 2015. *Standard Test Method for Determination of C_2 through C_5 Hydrocarbons in Gasolines by Gas Chromatography.* Annual Book of Standards. ASTM International, West Conshohocken, PA.

ASTM D2500. 2015. *Standard Test Method for Cloud Point of Petroleum Products.* Annual Book of Standards. ASTM International, West Conshohocken, PA.

ASTM D2504. 2015. *Standard Test Method for Non-Condensable Gases in C2 and Lighter Hydrocarbon Products by Gas Chromatography.* Annual Book of Standards. ASTM International, West Conshohocken, PA.

ASTM D2505. 2015. *Standard Test Method for Ethylene, Other Hydrocarbons, and Carbon Dioxide in High-Purity Ethylene by Gas Chromatography.* Annual Book of Standards. ASTM International, West Conshohocken, PA.

ASTM D2593. 2015. *Standard Test Method for Butadiene Purity and Hydrocarbon Impurities by Gas Chromatography.* Annual Book of Standards. ASTM International, West Conshohocken, PA.

ASTM D2597. 2015. *Standard Test Method for Analysis of Demethanized Hydrocarbon Liquid Mixtures Containing Nitrogen and Carbon Dioxide by Gas Chromatography.* Annual Book of Standards. ASTM International, West Conshohocken, PA.

ASTM D2622. 2015. *Standard Test Method for Sulfur in Petroleum Products by Wavelength Dispersive X-Ray Fluorescence Spectrometry.* Annual Book of Standards. ASTM International, West Conshohocken, PA.

ASTM D2712. 2015. *Standard Test Method for Hydrocarbon Traces in Propylene Concentrates by Gas Chromatography.* Annual Book of Standards. ASTM International, West Conshohocken, PA.

ASTM D2715. 2015. *Standard Test Method for Volatilization Rates of Lubricants in Vacuum.* Annual Book of Standards. ASTM International, West Conshohocken, PA.

ASTM D2766. 2015. *Standard Test Method for Specific Heat of Liquids and Solids.* Annual Book of Standards. ASTM International, West Conshohocken, PA.

ASTM D2786. 2015. *Standard Test Method for Hydrocarbon Types Analysis of Gas-Oil Saturates Fractions by High Ionizing Voltage Mass Spectrometry.* Annual Book of Standards. ASTM International, West Conshohocken, PA.

ASTM D2887. 2015. *Standard Test Method for Boiling Range Distribution of Petroleum Fractions by Gas Chromatography.* Annual Book of Standards. ASTM International, West Conshohocken, PA.

ASTM D2892. 2015. *Standard Test Method for Distillation of Crude Petroleum (15-Theoretical Plate Column).* Annual Book of Standards. ASTM International, West Conshohocken, PA.

ASTM D3117. 2015. *Standard Test Method for Wax Appearance Point of Distillate Fuels.* Annual Book of Standards. ASTM International, West Conshohocken, PA.

ASTM D3120. 2015. *Standard Test Method for Trace Quantities of Sulfur in Light Liquid Petroleum Hydrocarbons by Oxidative Microcoulometry.* Annual Book of Standards. ASTM International, West Conshohocken, PA.

ASTM D3177. 2015. *Standard Test Methods for Total Sulfur in the Analysis Sample of Coal and Coke.* Annual Book of Standards. ASTM International, West Conshohocken, PA.

ASTM D3178. 2015. *Standard Test Methods for Carbon and Hydrogen in the Analysis Sample of Coal and Coke.* Annual Book of Standards. ASTM International, West Conshohocken, PA.

ASTM D3179. 2015. *Standard Test Methods for Nitrogen in the Analysis Sample of Coal and Coke.* Annual Book of Standards. ASTM International, West Conshohocken, PA.

ASTM D3228. 2015. *Standard Test Method for Total Nitrogen in Lubricating Oils and Fuel Oils by Modified Kjeldahl Method.* Annual Book of Standards. ASTM International, West Conshohocken, PA.

ASTM D3237. 2015. *Standard Test Method for Lead in Gasoline by Atomic Absorption Spectroscopy.* Annual Book of Standards. ASTM International, West Conshohocken, PA.

ASTM D3238. 2015. *Standard Test Method for Calculation of Carbon Distribution and Structural Group Analysis of Petroleum Oils by the n-d-M Method.* Annual Book of Standards. ASTM International, West Conshohocken, PA.

ASTM D3239. 2015. *Standard Test Method for Aromatic Types Analysis of Gas-Oil Aromatic Fractions by High Ionizing Voltage Mass Spectrometry.* Annual Book of Standards. ASTM International, West Conshohocken, PA.

ASTM D3339. 2015. *Standard Test Method for Acid Number of Petroleum Products by Semi-Micro Color Indicator Titration.* Annual Book of Standards. ASTM International, West Conshohocken, PA.

ASTM D3340. 2015. *Standard Test Method for Lithium and Sodium in Lubricating Greases by Flame Photometer.* Annual Book of Standards. ASTM International, West Conshohocken, PA.

ASTM D3341. 2015. *Standard Test Method for Lead in Gasoline—Iodine Monochloride Method.* Annual Book of Standards. ASTM International, West Conshohocken, PA.

ASTM D3343. 2015. *Standard Test Method for Estimation of Hydrogen Content of Aviation Fuels.* Annual Book of Standards. ASTM International, West Conshohocken, PA.

ASTM D3525. 2015. *Standard Test Method for Gasoline Diluent in Used Gasoline Engine Oils by Gas Chromatography.* Annual Book of Standards. ASTM International, West Conshohocken, PA.

ASTM D3605. 2015. *Standard Test Method for Trace Metals in Gas Turbine Fuels by Atomic Absorption and Flame Emission Spectroscopy.* Annual Book of Standards. ASTM International, West Conshohocken, PA.

ASTM D3606. 2015. *Standard Test Method for Determination of Benzene and Toluene in Finished Motor and Aviation Gasoline by Gas Chromatography.* Annual Book of Standards. ASTM International, West Conshohocken, PA.

ASTM D3701. 2015. *Standard Test Method for Hydrogen Content of Aviation Turbine Fuels by Low Resolution Nuclear Magnetic Resonance Spectrometry.* Annual Book of Standards. ASTM International, West Conshohocken, PA.

ASTM D3710. 2015. *Standard Test Method for Boiling Range Distribution of Gasoline and Gasoline Fractions by Gas Chromatography.* Annual Book of Standards. ASTM International, West Conshohocken, PA.

ASTM D3831. 2015. *Standard Test Method for Manganese in Gasoline by Atomic Absorption Spectroscopy.* Annual Book of Standards. ASTM International, West Conshohocken, PA.

ASTM D4045. 2015. *Standard Test Method for Sulfur in Petroleum Products by Hydrogenolysis and Rateometric Colorimetry.* Annual Book of Standards. ASTM International, West Conshohocken, PA.

ASTM D4046. 2015. *Standard Test Method for Alkyl Nitrate in Diesel Fuels by Spectrophotometry.* Annual Book of Standards. ASTM International, West Conshohocken, PA.

ASTM D4052. 2015. *Standard Test Method for Density, Relative Density, and API Gravity of Liquids by Digital Density Meter.* Annual Book of Standards. ASTM International, West Conshohocken, PA.

ASTM D4053. 2015. *Standard Test Method for Benzene in Motor and Aviation Gasoline by Infrared Spectroscopy.* Annual Book of Standards. ASTM International, West Conshohocken, PA.

ASTM D4057. 2015. *Standard Practice for Manual Sampling of Petroleum and Petroleum Products.* Annual Book of Standards. ASTM International, West Conshohocken, PA.

ASTM D4124. 2015. *Standard Test Method for Separation of Asphalt into Four Fractions.* Annual Book of Standards. ASTM International, West Conshohocken, PA.

ASTM D4294. 2015. *Standard Test Method for Sulfur in Petroleum and Petroleum Products by Energy Dispersive X-Ray Fluorescence Spectrometry.* Annual Book of Standards. ASTM International, West Conshohocken, PA.

ASTM D4424. 2015. *Standard Test Method for Butylene Analysis by Gas Chromatography.* Annual Book of Standards. ASTM International, West Conshohocken, PA.

ASTM D4470. 2015. *Standard Test Method for Static Electrification.* Annual Book of Standards. ASTM International, West Conshohocken, PA.

ASTM D4530. 2015. *Standard Test Method for Determination of Carbon Residue (Micro Method).* Annual Book of Standards. ASTM International, West Conshohocken, PA.

ASTM D4628. 2015. *Standard Test Method for Analysis of Barium, Calcium, Magnesium, and Zinc in Unused Lubricating Oils by Atomic Absorption Spectrometry.* Annual Book of Standards. ASTM International, West Conshohocken, PA.

ASTM D4815. 2015. *Standard Test Method for Determination of MTBE, ETBE, TAME, DIPE, Tertiary-Amyl Alcohol and C1 to C4 Alcohols in Gasoline by Gas Chromatography.* Annual Book of Standards. ASTM International, West Conshohocken, PA.

ASTM D4864. 2015. *Standard Test Method for Determination of Traces of Methanol in Propylene Concentrates by Gas Chromatography.* Annual Book of Standards. ASTM International, West Conshohocken, PA.

ASTM D5002. 2015. *Standard Test Method for Density and Relative Density of Crude Oils by Digital Density Analyzer.* Annual Book of Standards. ASTM International, West Conshohocken, PA.

ASTM D5134. 2015. *Standard Test Method for Detailed Analysis of Petroleum Naphtha through n-Nonane by Capillary Gas Chromatography.* Annual Book of Standards. ASTM International, West Conshohocken, PA.

ASTM D5291. 2015. *Standard Test Methods for Instrumental Determination of Carbon, Hydrogen, and Nitrogen in Petroleum Products and Lubricants.* Annual Book of Standards. ASTM International, West Conshohocken, PA.

ASTM D5303. 2015. *Standard Test Method for Trace Carbonyl Sulfide in Propylene by Gas Chromatography.* Annual Book of Standards. ASTM International, West Conshohocken, PA.

ASTM D5307. 2015. *Standard Test Method for Determination of Boiling Range Distribution of Crude Petroleum by Gas Chromatography.* Annual Book of Standards. ASTM International, West Conshohocken, PA.

ASTM D5441. 2015. *Standard Test Method for Analysis of Methyl Tert-Butyl Ether (MTBE) by Gas Chromatography.* Annual Book of Standards. ASTM International, West Conshohocken, PA.

ASTM D5442. 2015. *Standard Test Method for Analysis of Petroleum Waxes by Gas Chromatography.* Annual Book of Standards. ASTM International, West Conshohocken, PA.

ASTM D5443. 2015. *Standard Test Method for Paraffin, Naphthene, and Aromatic Hydrocarbon Type Analysis in Petroleum Distillates through 200°C by Multi-Dimensional Gas Chromatography.* Annual Book of Standards. ASTM International, West Conshohocken, PA.

ASTM D5501. 2015. *Standard Test Method for Determination of Ethanol and Methanol Content in Fuels Containing Greater than 20% Ethanol by Gas Chromatography.* Annual Book of Standards. ASTM International, West Conshohocken, PA.

ASTM D5580. 2015. *Standard Test Method for Determination of Benzene, Toluene, Ethylbenzene, p/m-Xylene, o-Xylene, C9 and Heavier Aromatics, and Total Aromatics in Finished Gasoline by Gas Chromatography.* Annual Book of Standards. ASTM International, West Conshohocken, PA.

ASTM D5599. 2015. *Standard Test Method for Determination of Oxygenates in Gasoline by Gas Chromatography and Oxygen Selective Flame Ionization Detection.* Annual Book of Standards. ASTM International, West Conshohocken, PA.

ASTM D5623. 2015. *Standard Test Method for Sulfur Compounds in Light Petroleum Liquids by Gas Chromatography and Sulfur Selective Detection.* Annual Book of Standards. ASTM International, West Conshohocken, PA.

ASTM D5845. 2015. *Standard Test Method for Determination of MTBE, ETBE, TAME, DIPE, Methanol, Ethanol and t-Butanol in Gasoline by Infrared Spectroscopy.* Annual Book of Standards. ASTM International, West Conshohocken, PA.

ASTM D5853. 2015. *Standard Test Method for Pour Point of Crude Oils.* Annual Book of Standards. ASTM International, West Conshohocken, PA.

ASTM D5949. 2015. *Standard Test Method for Pour Point of Petroleum Products (Automatic Pressure Pulsing Method).* Annual Book of Standards. ASTM International, West Conshohocken, PA.

ASTM D5950. 2015. *Standard Test Method for Pour Point of Petroleum Products (Automatic Tilt Method).* Annual Book of Standards. ASTM International, West Conshohocken, PA.

ASTM D5986. 2015. *Standard Test Method for Determination of Oxygenates, Benzene, Toluene, C8–C12 Aromatics and Total Aromatics in Finished Gasoline by Gas Chromatography/Fourier Transform Infrared Spectroscopy.* Annual Book of Standards. ASTM International, West Conshohocken, PA.

ASTM D6159. 2015. *Standard Test Method for Determination of Hydrocarbon Impurities in Ethylene by Gas Chromatography.* Annual Book of Standards. ASTM International, West Conshohocken, PA.

ASTM E258. 2015. *Standard Test Method for Total Nitrogen in Organic Materials by Modified Kjeldahl Method.* Annual Book of Standards. ASTM International, West Conshohocken, PA.

ASTM E385. 2015. *Standard Test Method for Oxygen Content Using a 14-MeV Neutron Activation and Direct-Counting Technique*. Annual Book of Standards. ASTM International, West Conshohocken, PA.

ASTM E777. 2015. *Standard Test Method for Carbon and Hydrogen in the Analysis Sample of Refuse-Derived Fuel*. Annual Book of Standards. ASTM International, West Conshohocken, PA.

Baltus, R.E. and Anderson, J.L. 1984. Comparison of GPC elution and diffusion coefficients of asphaltenes. *Fuel*, 63: 530.

Blondel-Telouk, A., Loiseleur, H., Barreau, A., Béhar, E., and Jose, J. 1995. Determination of the average molecular weight of petroleum cuts by vapor pressure depression. *Fluid Phase Equilibria*, 110: 315–339.

Bouquet, M. and Bailleul, A. 1982. Nuclear magnetic resonance in the petroleum industry. In: *Petroanalysis '81. Advances in Analytical Chemistry in the Petroleum Industry 1975–1982*. G.B. Crump (Editor). John Wiley & Sons, Chichester, UK.

Carbognani, L. 1997. Fast monitoring of C_{20}-C_{160} crude oil alkanes by size-exclusion chromatography-evaporative light scattering detection performed with silica columns. *Journal of Chromatography A*, 788: 63–73.

Carbognani, L., Díaz-Gómez, L., Oldenburg, T.B.P., and Pereira-Almao, P. 2012. Determination of molecular masses for petroleum distillates by simulated distillation. *CT&F—Ciencia, Tecnología y Futuro*, 4(5): 43–55.

Colin, J.M. and Vion, G. 1983. Routine hydrocarbon group-type analysis in refinery laboratories by high-performance liquid chromatography. *Journal of Chromatography*, 280: 152–158.

Cooper, A.R. 1989. *Determination of Molecular Weight*. John Wiley & Sons, Hoboken, NJ.

Del Río, J.C. and Philp, R.P. 1999. Field ionization mass spectrometric study of high molecular weight hydrocarbons in a crude oil and a solid bitumen. *Organic Geochemistry*, 30: 279–286.

Dolbear, G.E., Tang, A., and Moorehead, E.L. 1987. Upgrading studies with California, Mexican, and Middle Eastern heavy oils. In: *Metal Complexes in Fossil Fuels*. R.H. Filby and J.F. Branthaver (Editors). Symposium Series No. 344. American Chemical Society, Washington, DC, p. 220.

Ebert, L.B. 1990. Comment on the study of asphaltenes by X-Ray diffraction. *Fuel Science and Technology International*, 8: 563–569.

Ebert, L.B., Mills, D.R., and Scanlon, J.C. 1987. Preprints. *American Chemical Society, Division of Petroleum Chemistry*, 32(2): 419.

Ebert, L.B., Scanlon, J.C., and Mills, D.R. 1984. X-ray diffraction of *n*-paraffins and stacked aromatic molecules: insights into the structure of petroleum asphaltenes. *Liquid Fuels Technology*, 2(3): 257–286.

Felix, G., Bertrand, C., and Van Gastel, F. 1985. Hydroprocessing of heavy oils and residua. *Chromatographia*, 20(3): 155–160.

Fotland, P. and Anfindsen, H. 1996. Electrical conductivity of asphaltenes in organic solvents. *Fuel Science and Technology International*, 14: 101–115.

Fotland, P., Anfindsen, H., and Fadnes, F.H. 1993. Detection of asphaltene precipitation and amounts precipitated by measurement of electrical conductivity. *Fluid Phase Equilibria*, 82: 157–164.

Gary, J.G., Handwerk, G.E., and Kaiser, M.J. 2007. *Petroleum Refining: Technology and Economics*, 5th Edition. CRC Press/Taylor & Francis Group, Boca Raton, FL.

Ghashghaee, M. 2015. Predictive correlations for thermal upgrading of petroleum residues. *Journal of Analytical and Applied Pyrolysis*, 115: 326–336.

Hasan, M., Ali, M.F., and Arab, M. 1989. Structural Characterization of Saudi Arabian extra light and light crudes by 1-H and 13-C NMR spectroscopy. *Fuel*, 68: 801–803.

Koots, J.A. and Speight, J.G. 1975. The relation of petroleum resins to asphaltenes. *Fuel*, 54: 179.

Long, R.B. and Speight, J.G. 1989. Studies in petroleum composition. I: Development of a compositional map for various feedstocks. *Revue de l'Institut Français du Pétrole*, 44: 205.

MacAllister, D.J. and DeRuiter, R.A. 1985. Further development and application of simulated distillation for enhanced oil recovery. Paper No. SPE 14335. *60th Annual Technical Conference*. Society of Petroleum Engineers, Las Vegas, NV, September 22–25.

Matsushita, S., Tada, Y., and Ikushige. 1981. Rapid hydrocarbon group analysis of gasoline by high-performance liquid chromatography. *Journal of Chromatography*, 208: 429–432.

Miller, R.L., Ettre, L.S., and Johansen, N.G. 1983. Quantitative analysis of hydrocarbons by structural group type in gasoline and distillates. Part II. *Journal of Chromatography*, 259: 393.

Neer, L.A. and Deo, M.D. 1995. Simulated distillation of oils with a wide carbon number distribution. *Journal of Chromatographic Science*, 33: 133–138.

Norris, T.A. and Rawdon, M.G. 1984. Determination of hydrocarbon types in petroleum liquids by supercritical fluid chromatography with flame ionization detection. *Analytical Chemistry*, 56: 1767–1769.

Parkash, S. 2003. *Refining Processes Handbook*. Gulf Professional Publishing, Elsevier, Amsterdam, the Netherlands.

Penzes, S. and Speight, J.G. 1974. Electrical conductivities of bitumen fractions in non-aqueous solvents. *Fuel*, 53: 192.

Rawdon, M. 1984. Modified flame ionization detector for supercritical fluid chromatography. *Analytical Chemistry*, 56: 831–832.

Reynolds, J.G. and Biggs, W.R. 1988. Analysis of residuum demetallation by size exclusion chromatography with element specific detection. *Fuel Science and Technology International*, 6: 329.

Romanowski, L.J. and Thomas, K.P. 1985. Steamflooding of preheated tar sand. Report No. DOE/FE/60177-2326. U.S. Department of Energy, Washington, DC.

Schwartz, H.E. and Brownlee, R.G. 1986. Use of reversed-phase chromatography in carbohydrate analysis. *Journal of Chromatography*, 353: 77.

Schwartz, H.E., Brownlee, R.G., Boduszynski, M.M., and Su, F. 1987. Simulated distillation of high-boiling petroleum fractions by capillary supercritical chromatography and vacuum thermal gravimetric analysis. *Analytical Chemistry*, 59: 1393–1401.

Shafizadeh, A., McAteer, G., and Sigmon, J. 2003. High-acid crudes. *Proceedings. Crude Oil Quality Group Meeting*. New Orleans, LA, January 30.

Sheridan, M. 2006. California crude oil production and imports. Staff Paper, Report No. CERC-600-2006-006. Fossil Fuels Office, Fuels and Transportation Division, California Energy Commission. Sacramento, CA, April.

Snape, C.E., Ladner, W.R., and Bartle, K.D. 1979. Survey of carbon-13 chemical shifts—Application to coal-derived materials. *Analytical Chemistry*, 51: 2189–2198.

Speight, J.G. 1987. Initial reactions in the coking of residua. Preprints. *American Chemical Society, Division of Petroleum Chemistry*, 32(2): 413.

Speight, J.G. 1994. Chemical and physical studies of petroleum asphaltenes. In: *Asphaltenes and Asphalts, I. Developments in Petroleum Science, 40*. T.F. Yen and G.V. Chilingarian (Editors). Elsevier, Amsterdam, the Netherlands, Chapter 2.

Speight, J.G. 2000. *The Desulfurization of Heavy Oils and Residua*, 2nd Edition. Marcel Dekker, New York.

Speight, J.G. 2001. *Handbook of Petroleum Analysis*. John Wiley & Sons, Hoboken, NJ.

Speight, J.G. 2005. Natural bitumen (tar Sands) and heavy oil. In: *Coal, Oil Shale, Natural Bitumen, Heavy Oil and Peat. Encyclopedia of Life Support Systems (EOLSS)*. Developed under the Auspices of the UNESCO. EOLSS Publishers, Oxford, UK.

Speight, J.G. 2009. *Enhanced Recovery Methods for Heavy Oil and Tar Sands*. Gulf Publishing Company, Houston, TX.

Speight, J.G. 2014. *The Chemistry and Technology of Petroleum*, 5th Edition. CRC Press/Taylor & Francis Group, Boca Raton, FL.

Speight, J.G. 2015. *Handbook of Petroleum Product Analysis*, 2nd Edition. John Wiley & Sons, Hoboken, NJ.

Speight, J.G. and Ozum, B. 2002. *Petroleum Refining Processes*. Marcel Dekker, New York.

Speight, J.G., Wernick, D.L., Gould, K.A., Overfield, R.E., Rao, B.M.L., and Savage, D.W. 1985. Molecular weights and association of asphaltenes: A critical review. *Revue de l'Institut Français du Pétrole*, 40: 27.

Suatoni, J.C. and Garber, H.R. 1975. HPLC preparative group-type separation of olefins from synfuels. *Journal of Chromatographic Science*, 13: 367.

Wallace, D. (Editor). 1988. *A Review of Analytical Methods for Bitumens and Heavy Oils*. Alberta Oil Sands Technology and Research Authority, Edmonton, Alberta, Canada.

Wallace, D. and Carrigy, M.A. 1988. New analytical results on oil sands from deposits throughout the world. *Proceedings of the Third UNITAR/UNDP International Conference on Heavy Crude and Tar Sands*. R.F. Meyer (Editor). Alberta Oil Sands Technology and Research Authority, Edmonton, Alberta, Canada.

Wallace, D., Starr, J., Thomas, K.P., and Dorrence S.M. 1988. *Characterization of Oil Sand Resources*. Alberta Oil Sands Technology and Research Authority, Edmonton, Alberta, Canada.

Weinberg, V.L., Yen, T.F., Gerstein, B.C., and Murphy, P.D. 1981. Characterization of pyrolyzed asphaltenes by diffuse reflectance-Fourier transform infrared and dipolar dephasing-solid state ^{13}C nuclear magnetic resonance spectroscopy. Preprints. *American Chemical Society, Division of Petroleum Chemistry*, 26(4): 816–824.

Yen, T.F. 1984. Characterization of heavy oil. In: *The Future of Heavy Crude and Tar Sands*. R.F. Meyer, J.C. Wynn, and J.C. Olson (Editors). McGraw-Hill, New York.

3 Feedstock Composition

3.1 INTRODUCTION

Petroleum, heavy oil, and tar sand bitumen are not usually found where the precursors were laid down, but in reservoirs where accumulation occurs after it has migrated from the source rocks through geologic strata. In addition, the theory that the petroleum precursors form a mix that is often referred to as "protopetroleum" (also referred to as "primordial precursor soup" or "petroleum porridge") is an acceptable generalization (Speight, 2014). And the molecular types in any specified fraction are limited by the nature of the precursors of petroleum, their chemical structures, and the physical conditions that are prevalent during the maturation (conversion of the precursors) processes.

This concept has resulted in the consideration of petroleum as a continuum of molecular types, and the nature of continuum is dictated by the proportions of the precursors that form the *protopetroleum* after which the prevalent conditions become operational in the formation of the final crude oil product. With this in mind, it might be anticipated that similar molecular types occur in heavy oil and bitumen as occurs in conventional petroleum. It then becomes a question of degree as well as molecular weight.

In the natural state, crude oil, heavy oil, extra heavy oil, and tar sand bitumen are not homogeneous materials, and the physical characteristics differ depending on where the material was produced. This is due to the fact that any of these feedstocks from different geographical locations will naturally have unique properties. In its natural, unrefined state, crude oil, heavy oil, extra heavy oil, and tar sand bitumen range in density and consistency from very thin, lightweight, and volatile fluidity to an extremely thick, semisolid (Speight, 2014). There is also a tremendous gradation in the color that ranges from a light, golden yellow (conventional crude oil) to black tar sand bitumen.

Thus, petroleum, heavy oil, and tar sand bitumen are not (within the individual categories) uniform materials, and the chemical and physical (fractional) composition of each of these refinery feedstocks can vary not only with the location and age of the reservoir or deposit but also with the depth of the individual well within the reservoir or deposit. On a molecular basis, the three feedstocks are complex mixtures containing (depending upon the feedstock) hydrocarbons with varying amounts of hydrocarbonaceous constituents that contain sulfur, oxygen, and nitrogen, as well as constituents containing vanadium, nickel, iron, and copper. The hydrocarbon content may be as high as 97% w/w, for example, in a light crude oil or less than 50% w/w in heavy crude oil and bitumen (Speight, 2014).

Heavy oil, bitumen, and residua (as a result of the concentration effect of distillation) contain more heteroatomic species and less hydrocarbon constituents than conventional crude oil. Thus, to obtain more gasoline and other liquid fuels, there have been different approaches to refining the heavier feedstocks as well as the recognition that knowledge of the constituents of these higher-boiling feedstocks is also of some importance. The problems encountered in processing the heavier feedstocks can be equated to the *chemical character* and the *amount* of complex, higher-boiling constituents in the feedstock. Refining these materials is not just a matter of applying know-how derived from refining *conventional* crude oils but requires knowledge of the *chemical structure* and *chemical behavior* of these more complex constituents.

The purpose of this chapter is to present a brief overview of the types of constituents that are found in petroleum, heavy oil, and bitumen and to also provide brief descriptions of the chemistry and physics of thermal decomposition.

3.2 ELEMENTAL COMPOSITION

The analysis of feedstocks for the percentages by weight of carbon, hydrogen, nitrogen, oxygen, and sulfur (elemental composition, ultimate composition) is perhaps the first method used to examine the general nature, and perform an evaluation, of a feedstock. The atomic ratios of the various elements to carbon (i.e., H/C, N/C, O/C, and S/C) are frequently used for indications of the overall character of the feedstock. It is also of value to determine the amounts of trace elements, such as vanadium and nickel, in a feedstock since these materials can have serious deleterious effects on catalyst performance during refining by catalytic processes.

For example, *carbon content* can be determined by the method designated for coal and coke (ASTM D3178, 2015) or by the method designated for municipal solid waste (ASTM E777, 2015). There are also methods designated for (1) *hydrogen content*, ASTM D1018 (2015), ASTM D3178 (2015), ASTM D3343 (2015), ASTM D3701 (2015), and ASTM E777 (2015); (2) *nitrogen content*, ASTM D3179 (2015), ASTM D3228 (2015), ASTM E258 (2015), and ASTM E778 (2015); (3) *oxygen content*, ASTM E385 (2015); and (4) *sulfur content*, ASTM D1266 (2015), ASTM D1552 (2015), ASTM D1757 (2015), ASTM D2622 (2015), ASTM D3177 (2015), ASTM D4045 (2015), and ASTM D4294 (2015). For all feedstocks, the higher the atomic hydrogen–carbon ratio, the higher is its value as refinery feedstock because of the lower hydrogen requirements for upgrading. Similarly, the lower the heteroatom content, the lower the hydrogen requirements for upgrading. Thus, inspection of the elemental composition of feedstocks is an initial indication of the quality of the feedstock and, with the molecular weight, indicates the molar hydrogen requirements for upgrading (Figure 3.1).

However, it has become apparent, with the introduction of the heavier feedstocks into refinery operations, that these ratios are not the only requirement for predicting feedstock character before refining. The use of more complex feedstocks (in terms of chemical composition) has added a new dimension to refining operations. Thus, although atomic ratios, as determined by elemental analyses, may be used on a comparative basis between feedstocks, there is now no guarantee that a particular feedstock will behave as predicted from these data. Product slates cannot be predicted accurately, if at all, from these ratios. Additional knowledge such as defining the various chemical reactions of the constituents as well as the reactions of these constituents with each other also plays a role in determining the processability of a feedstock.

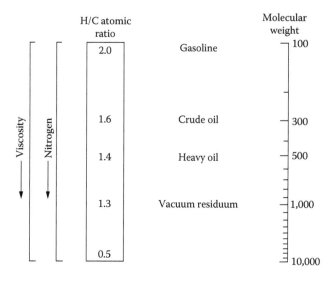

FIGURE 3.1   Atomic hydrogen/carbon ratio and molecular weight of feedstocks.

In summary, petroleum contains carbon, hydrogen, nitrogen, oxygen, sulfur, and metals (particularly nickel and vanadium), and the amounts of these elements in a whole series of crude oils vary over fairly narrow limits:

Carbon: 83.0%–87.0%
Hydrogen: 10.0%–14.0%
Nitrogen: 0.1%–2.0%
Oxygen: 0.05%–1.5%
Sulfur: 0.05%–6.0%
Metals: (Ni and V) <1000 ppm

This narrow range is contradictory to the wide variation in physical properties from the lighter, more mobile crude oils at one extreme to the heavier asphaltic crude oils at the other extreme (see also Charbonnier et al., 1969; Draper et al., 1977). And because of the narrow range of carbon and hydrogen content, it is not possible to classify petroleum on the basis of carbon content as coal is classified; carbon contents of coal can vary from as low as 75% w/w in lignite to 95% w/w in anthracite (Speight, 2013). Of course, other subdivisions are possible within the various carbon ranges of the coals, but petroleum is restricted to a much narrower range of elemental composition.

The elemental analysis of oil sand bitumen has also been widely reported (Speight, 1990, 2009), but the data suffer from the disadvantage that identification of the source is too general (i.e., Athabasca bitumen that covers several deposits) and is often not site specific. In addition, the analysis is quoted for separated bitumen, which may have been obtained by any one of several procedures and may therefore not be representative of the total bitumen on the sand. However, recent efforts have focused on a program to produce sound, reproducible data from samples for which the origin is carefully identified (Wallace et al., 1988).

Like conventional petroleum of the data that are available, the elemental composition of oil sand bitumen is generally constant and, like the data for petroleum, falls into a narrow range:

Carbon: 83.4%–0.5%
Hydrogen: 10.4%–0.2%
Nitrogen: 0.4%–0.2%
Oxygen: 1.0%–0.2%
Sulfur: 5.0%–0.5%
Metals: (Ni and V) >1000 ppm

The major exception to these narrow limits is the oxygen content that can vary from as little as 0.2% to as high as 4.5%. This is not surprising, since when oxygen is estimated by difference the analysis is subject to the accumulation of all of the errors in the other elemental data. In addition, bitumen is susceptible to aerial oxygen and the oxygen content is very dependent on sample history.

Although several generalities can be noted from the ultimate composition, these can only give indications of how the material might behave during processing.

The viscosity of tar sand bitumen is related to its hydrogen-to-carbon atomic ratio and hence the required supplementary heat energy for thermal extraction processes. It also affects the bitumen's distillation curve or thermodynamic characteristics, its gravity, and its pour point. Atomic hydrogen-to-carbon ratios as low as 1.3 have been observed for tar sand bitumen, although an atomic hydrogen-to-carbon ratio of 1.5 is more typical. The higher the hydrogen–carbon ratio of bitumen, the higher is its value as refinery feedstock because of the lower hydrogen requirements. Elements related to the hydrogen–carbon ratio are distillation curve, bitumen gravity, pour point, and bitumen viscosity.

The occurrence of sulfur in bitumen as organic or elemental sulfur or in produced gas as compounds of oxygen and hydrogen is an expensive nuisance. It must be removed from the bitumen at some point in the upgrading and refining process. Sulfur contents of some tar sand bitumen can exceed 10% w/w. Elements related to sulfur content are hydrogen content, hydrogen–carbon ratio, nitrogen content, distillation curve, and viscosity.

The nitrogen content of tar sand bitumen can be as high as 1.3% by weight and nitrogen-containing constituents complicate the refining process by poisoning the catalysts employed in the refining process. Elements related to nitrogen content are sulfur content, hydrogen content, hydrogen–carbon ratio, bitumen viscosity, distillation profile, and viscosity.

Furthermore, heteroatoms in feedstocks affect every aspect of refining. The occurrence of *sulfur* in feedstocks as organic or elemental sulfur or in produced gas as compounds of oxygen (SO_x) and hydrogen (H_2S) is an expensive aspect of refining. It must be removed at some point in the upgrading and refining process. Sulfur contents of many crude oil are on the order of 1% by weight, whereas the sulfur content of tar sand bitumen can exceed 5% or even 10% by weight. Of all of the hetero-elements, sulfur is usually the easiest to remove and many commercial catalysts are available that routinely remove 90% of the sulfur from a feedstock (Speight, 2000).

The *nitrogen* content of petroleum is usually less than 1% by weight, but the nitrogen content of tar sand bitumen can be as high as 1.5% by weight. The presence of nitrogen complicates refining by poisoning the catalysts employed in the various processes. Nitrogen is more difficult to remove than sulfur, and there are fewer catalysts that are specific for nitrogen. If the nitrogen is not removed, the potential for the production of nitrogen oxides (NO_x) during processing and use becomes real.

Metals (particularly *vanadium* and *nickel*) are found in every most crude oils. Heavy oils and residua contain relatively high proportions of metals either in the form of salts or as organometallic constituents (such as the metalloporphyrins), which are extremely difficult to remove from the feedstock. Indeed, the nature of the process by which residua are produced virtually dictates that all the metals in the original crude oil are concentrated in the residuum (Speight, 2000). The metallic constituents that may actually *volatilize* under the distillation conditions and appear in the higher-boiling distillates are the exceptions here.

Metal constituents of feedstocks cause problems by poisoning the catalysts used for sulfur and nitrogen removal as well as the catalysts used in other processes such as catalytic cracking. Thus, serious attempts are being made to develop catalysts that can tolerate a high concentration of metals without serious loss of catalyst activity or catalyst life.

A variety of tests have been designated for the determination of metals in petroleum products (ASTM D1318, 2015; ASTM D3340, 2015; ASTM D3341, 2015; ASTM D3605, 2015). Determination of metals in whole feeds can be accomplished by combustion of the sample so that only inorganic ash remains. The ash can then be digested with an acid and the solution examined for metal species by atomic absorption spectroscopy or by inductively coupled argon plasma spectrometry.

3.3 CHEMICAL COMPOSITION

Crude oil, heavy oil, and tar sand bitumen contain an extreme range of organic functionality and molecular size. In fact, the variety is so great that it is unlikely that a complete compound-by-compound description for even a single crude oil would not be possible. As already noted, the composition of petroleum can vary with the location and age of the field in addition to any variations that occur with the depth of the individual well. Two adjacent wells are more than likely to produce petroleum with very different characteristics.

In very general terms (and as observed from elemental analyses), petroleum, heavy oil, bitumen, and residua are a complex composition of (1) hydrocarbons, (2) nitrogen compounds, (3) oxygen compounds, (4) sulfur compounds, and (5) metallic constituents. However, this general definition is not adequate to describe the composition of petroleum as it relates to the behavior of these feedstocks. Indeed, the consideration of hydrogen-to-carbon atomic ratio, sulfur content, and API gravity is no longer adequate for the task of determining refining behavior.

Furthermore, the molecular composition of petroleum can be described in terms of three classes of compounds: saturates, aromatics, and compounds bearing heteroatoms (sulfur, oxygen, or nitrogen). Within each class, there are several families of related compounds: (1) saturated constituents include normal alkanes, branched alkanes, and cycloalkanes (paraffins, *iso*-paraffins, and naphthenes, in petroleum terms), (2) alkene constituents (olefins) are rare to the extent of being considered an oddity, (3) monoaromatic constituents range from benzene to multiple fused ring analogs (naphthalene, phenanthrene, etc.), (4) thiol constituents (mercaptan constituents) contain sulfur as do thioether derivatives and thiophenederivatives, and (5) nitrogen- and oxygen-containing constituents are more likely to be found in polar forms (pyridines, pyrroles, phenols, carboxylic acids, amides, etc.) than in nonpolar forms (such as ethers). The distribution and characteristics of these molecular species account for the rich variety of crude oils.

Feedstock behavior during refining is better addressed through the consideration of the molecular makeup of the feedstock (perhaps, by analogy, just as genetic makeup dictates human behavior). The occurrence of amphoteric species (i.e., compounds having a mixed acid/base nature) is rarely addressed, as well as the phenomenon of molecular size or the occurrence of specific functional types (Figure 3.4), which can play a major role in the interactions between the constituents of a feedstock. All of these items are important in determining the feedstock behavior during refining operations.

An understanding of the chemical types (or composition) of any feedstock can lead to an understanding of the chemical aspects of processing the feedstock. Processability is not only a matter of knowing the elemental composition of a feedstock; it is also a matter of understanding the bulk properties as they relate to the chemical or physical composition of the material. For example, it is difficult to understand, *a priori*, the process chemistry of various feedstocks from the elemental composition alone. From such data, it might be surmised that the major difference between a heavy crude oil and a more conventional material is the H/C atomic ratio alone. This property indicates that a heavy crude oil (having a lower H/C atomic ratio and being more aromatic in character) would require more hydrogen for upgrading to liquid fuels. This is indeed true, but much more information is necessary to understand the *processability* of the feedstock.

With the necessity of processing crude oil residua, heavy oil, and tar sand bitumen, to obtain more gasoline and other liquid fuels, there has been the recognition that knowledge of the constituents of these higher-boiling feedstocks is also of some importance. Indeed, the problems encountered in processing the heavier feedstocks can be equated to the *chemical character* and the *amount* of complex, higher-boiling constituents in the feedstock. Refining these materials is not just a matter of applying know-how derived from refining *conventional* crude oils but requires knowledge of the *chemical structure* and *chemical behavior* of these more complex constituents.

However, heavy crude oil and bitumen are extremely complex, and very little direct information can be obtained by distillation. It is not possible to isolate and identify the constituents of the heavier feedstocks (using analytical techniques that rely upon volatility). Other methods of identifying the chemical constituents must be employed. Such techniques include a myriad of fractionation procedures as well as methods designed to draw inferences related to the hydrocarbon skeletal structures and the nature of the heteroatomic functions.

The hydrocarbon content of petroleum may be as high as 97% by weight (e.g., in the lighter paraffinic crude oils) or as low as 50% by weight or less as illustrated by the heavy asphaltic crude oils. Nevertheless, crude oils with as little as 50% hydrocarbon components are still assumed to retain most of the essential characteristics of the hydrocarbons. It is, nevertheless, the nonhydrocarbon (sulfur, oxygen, nitrogen, and metal) constituents that play a large part in determining the processability of the crude oil and will determine the processability of crude oil, heavy oil, and tar sand bitumen in the future (Speight, 2011). But there is more to the composition of petroleum than the hydrocarbon content. The inclusion of organic compounds of sulfur, nitrogen, and oxygen serves only to present crude oils as even more complex mixtures, and the appearance of appreciable amounts of these nonhydrocarbon compounds causes some concern in the refining of crude oils. Even though the concentration of nonhydrocarbon constituents (i.e., those organic compounds containing one or more sulfur, oxygen, or nitrogen atoms) in certain fractions may be quite small, they tend to concentrate in the higher-boiling fractions of petroleum. Indeed, their influence on the processability of the petroleum is important irrespective of their molecular size and the fraction in which they occur. It is, nevertheless, the nonhydrocarbon (sulfur, oxygen, nitrogen, and metal) constituents that play a large part in determining the processability of the crude oil and their influence on the processability of the petroleum is important irrespective of their molecular size (Green et al., 1989; Speight, 2000, 2014, 2015). The occurrence of organic compounds of sulfur, nitrogen, and oxygen serves only to present crude oil as even more complex mixture, and the appearance of appreciable amounts of these nonhydrocarbon compounds causes some concern in crude oil refining. The nonhydrocarbon constituents (i.e., those organic compounds containing one or more sulfur, oxygen, or nitrogen atoms) tend to concentrate in the higher-boiling fractions of petroleum (Speight, 2000). In addition, as the feedstock series progresses to higher-molecular-weight feedstocks from crude oil to heavy crude oil to tar sand bitumen, not only does the number of the constituents increase but the molecular complexity of the constituents also increases (Figure 3.2).

The presence of traces of nonhydrocarbon compounds can impart objectionable characteristics to finished products, leading to discoloration and/or lack of stability during storage. On the other

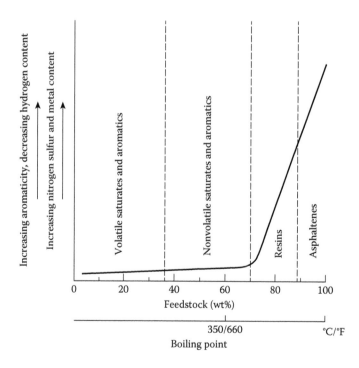

FIGURE 3.2 Variation of compound types in petroleum with the boiling point—for heavy oil and tar sand bitumen the line from 0% to 70% is steeper.

hand, catalyst poisoning and corrosion are the most noticeable effects during refining sequences when these compounds are present. It is therefore not surprising that considerable attention must be given to the nonhydrocarbon constituents of petroleum as the trend in the refining industry, of late, has been to process more heavy crude oil as well as residua that contain substantial proportions of these nonhydrocarbon materials.

3.3.1 HYDROCARBON CONSTITUENTS

The isolation of pure compounds from petroleum is an exceedingly difficult task, and the overwhelming complexity of the hydrocarbon constituents of the higher-molecular-weight fractions and the presence of compounds of sulfur, oxygen, and nitrogen are the main causes for the difficulties encountered. It is difficult on the basis of the data obtained from synthesized hydrocarbons to determine the identity or even the similarity of the synthetic hydrocarbons to those that constitute many of the higher-boiling fractions of petroleum. Nevertheless, it has been well established that the hydrocarbon components of petroleum are composed of paraffinic, naphthenic, and aromatic groups (Table 3.1). Olefin groups are not usually found in crude oils, and acetylene-type hydrocarbons are very rare indeed. It is therefore convenient to divide the hydrocarbon components of petroleum into the following three classes:

1. *Paraffins*, which are saturated hydrocarbons with straight or branched chains, but without any ring structure
2. *Naphthenes*, which are saturated hydrocarbons containing one or more rings, each of which may have one or more paraffinic side chains (more correctly known as "alicyclic hydrocarbons")

TABLE 3.1
Hydrocarbon and Heteroatom Types in Crude Oil, Heavy Oil, and Tar Sand Bitumen

Class	Compound Types
Saturated hydrocarbons	*n*-Paraffins
	Iso-Paraffins and other branched paraffins
	Cycloparaffins (naphthenes)
	Condensed cycloparaffins (including steranes, hopanes)
	Alkyl side chains on ring systems
Unsaturated hydrocarbons	Olefins, nonindigenous; present in products of thermal reactions
Aromatic hydrocarbons	Benzene systems
	Condensed aromatic systems
	Condensed naphthene–aromatic systems
	Alkyl side chains on ring systems
Saturated heteroatomic systems	Alkyl sulfides
	Cycloalkyl sulfides
Sulfides	Alkyl side chains on ring systems
Aromatic heteroatomic systems	Furans (single-ring and multi-ring systems)
	Thiophenes (single-ring and multi-ring systems)
	Pyrroles (single-ring and multi-ring systems)
	Pyridines (single-ring and multi-ring systems)
	Mixed heteroatomic systems
	Amphoteric (acid–base systems)
	Alkyl side chains on ring systems

3. *Aromatics*, which are hydrocarbons containing one or more aromatic nuclei, such as benzene, naphthalene, and phenanthrene ring systems, which may be linked up with (substituted) naphthene rings and/or paraffinic side chains

3.3.1.1 Paraffin Hydrocarbons

The proportion of paraffins in crude oil varies with the type of crude, but within any one crude oil, the proportion of paraffinic hydrocarbons usually decreases with increasing molecular weight and there is a concomitant increase in aromaticity and the relative proportion of heteroatoms (nitrogen, oxygen, and sulfur) (Figure 3.2).

The relationship between the various hydrocarbon constituents of crude oils is one of hydrogen addition or hydrogen loss (Figure 3.3). This is an extremely important aspect of petroleum composition, and there is no reason to deny the occurrence of these interconversion schemes during the formation, maturation, and *in situ* alteration of petroleum. Indeed, a scheme of this type lends even more credence to the complexity of petroleum within the hydrocarbon series alone and also supports current contentions that the high-molecular-weight constituents (resin constituents and asphaltene constituents) are structurally related to the lower-boiling constituents rather than proposals that invoke the existence of highly condensed polynuclear aromatic systems.

The abundance of the different members of the same homologous series varies considerably in absolute and relative values. However, in any particular crude oil or crude oil fraction, there may be a small number of constituents forming the greater part of the fraction, and these have been referred to as the "predominant constituents" (Bestougeff, 1961). This generality may also apply to other constituents and is very dependent upon the nature of the source material as well as the relative amounts of the individual source materials prevailing during maturation conditions (Speight, 2014).

Normal paraffin hydrocarbons (n-*paraffins, straight-chain paraffins*) occur in varying proportions in most crude oils. In fact, paraffinic petroleum may contain up to 20%–50% by weight *n*-paraffins in the gas oil fraction. However, naphthenic or asphaltic crude oils sometimes contain only very small amounts of normal paraffins.

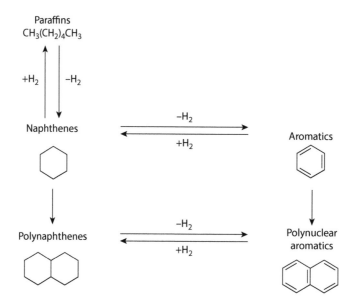

FIGURE 3.3 Interrelationship of the hydrocarbon types in petroleum.

Considerable quantities of *iso-paraffins* have been noted to be present in the straight-run gasoline fraction of petroleum. The 2- and 3-methyl derivatives are the most abundant, and the 4-methyl derivative is present in small amounts, if at all, and it is generally accepted that the slightly branched paraffins predominate over the highly branched materials. It seems that the *iso*-paraffins occur throughout the boiling range of petroleum fractions. The proportion tends to decrease with increasing boiling point; it appears that if the *iso*-paraffins are present in lubricating oils their amount is too small to have any significant influence on the physical properties of the lubricating oils.

As the molecular weight (or boiling point) of the petroleum fraction increases, there is a concomitant decrease in the amount of free paraffins in the fraction. In certain types of crude oil, there may be no paraffins at all in the vacuum gas oil (VGO) fraction. For example, in the paraffinic crude oils, free paraffins will separate as a part of the asphaltene fraction but in the naphthenic crude oils, free paraffins are not expected in the gas oil and asphaltene fractions. The vestiges of paraffins in the asphaltenes fractions occur as alkyl side chains on aromatic and naphthenic systems. And, these alkyl chains can contain 20 or more carbon atoms (Speight, 1994, 2014).

3.3.1.2 Cycloparaffin Hydrocarbons

Although only a small number of representatives have been isolated so far, cyclohexane derivatives, cyclopentane derivatives, and decahydronaphthalene (decalin) derivatives (naphthenes) are largely represented in oil fractions. Petroleum also contains polycyclic naphthenes, such as terpenes, and such molecules (often designated bridge-ring hydrocarbons) occur even in the heavy gasoline fractions (boiling point 150°C–200°C, 300°F–390°F). Naphthene rings may be built up of a varying number of carbon atoms, and among the synthesized hydrocarbons there are individual constituents with rings of the three-, four-, five-, six-, seven-, and eight-carbon atoms. It is now generally believed that crude oil fractions contain chiefly five- and six-carbon rings. Only naphthenes with five- and six-membered rings have been isolated from the lower-boiling fractions. Thermodynamic studies show that naphthene rings with five- and six-carbon atoms are the most stable. The naphthenic acids contain chiefly cycle pentane as well as cyclohexane rings.

Cycloparaffin derivatives (naphthene derivatives) are represented in all fractions in which the constituent molecules contain more than five-carbon atoms. Several series of cycloparaffin derivatives, usually containing five- or six-membered rings or their combinations, occur as polycyclic structures. The content of cycloparaffin derivatives in petroleum varies up to 60% of the total hydrocarbons. However, the cycloparaffin content of different boiling range fractions of a crude oil may not vary considerably and generally remains within rather close limits. Nevertheless, the structure of these constituents may change within the same crude oil as a function of the molecular weight or boiling range of the individual fractions as well as from one crude oil to another.

The principal structural variation of naphthenes is the number of rings present in the molecule. The mono- and bicyclic naphthenes are generally the major types of cycloparaffin derivatives in the lower-boiling fractions of petroleum, with boiling point or molecular weight increased by the presence of alkyl chains. The higher-boiling fractions, such as the lubricating oils, may contain two to six rings per molecule. As the molecular weight (or boiling point) of the petroleum fraction increases, there is a concomitant increase in the amount of cycloparaffin (naphthene) species in the fraction. In the asphaltic (naphthenic) crude oils, the gas oil fraction can contain considerable amounts of naphthenic ring systems that increase even more in consideration of the molecular types in the asphaltenes. However, as the molecular weight of the fraction increases, the occurrence of condensed naphthene ring systems and alkyl-substituted naphthene ring systems increases.

There is also the premise that the naphthene ring systems carry alkyl chains that are generally shorter than the alkyl substituents carried by aromatic systems. There are indications from spectroscopic studies that the short chains (methyl and ethyl) appear to be characteristic substituents of the aromatic portion of the molecule, whereas a limited number (one or two) of longer chains may be attached to the cycloparaffin rings. The total number of chains, which is in general four to six, as well as their length, increases according to the molecular weight of the naphthene–aromatic compounds.

In the asphaltene constituent, free condensed naphthenic ring systems may occur, but general observations favor the occurrence of combined aromatic–naphthenic systems that are variously substituted by alkyl systems. There is also general evidence that the aromatic systems are responsible for the polarity of the asphaltene constituents. The heteroatoms are favored to occur on or within the aromatic (pseudo-aromatic) systems (Speight, 1994, 2014).

3.3.1.3 Aromatic Hydrocarbons

The concept of the occurrence of identifiable aromatic systems in nature is a reality and the occurrence of monocyclic and polycyclic aromatic systems in natural product chemicals is well documented (Sakarnen and Ludwig, 1971; Durand, 1980; Weiss and Edwards, 1980). However, one source of aromatic systems that is often ignored is petroleum (Eglinton and Murphy, 1969; Tissot and Welte, 1978; Brooks and Welte, 1984). Therefore, attempts to identify such systems in the nonvolatile constituents of petroleum should be an integral part of the repertoire of the petroleum chemist as well as the domain of the natural product chemist.

Crude oil is a mixture of compounds and aromatic compounds are common to all petroleum, and it is the difference in extent that becomes evident upon examination of a series of petroleum. By far, the majority of these aromatics contain paraffinic chains, naphthene rings, and aromatic rings side by side.

There is a general increase in the proportion of aromatic hydrocarbons with increasing molecular weight. However, aromatic hydrocarbons without the accompanying naphthene rings or alkyl-substituted derivatives seem to be present in appreciable amounts only in the lower petroleum fractions. Thus, the limitation of instrumentation notwithstanding, it is not surprising that spectrographic identification of such compounds has been concerned with these low-boiling aromatics.

All known aromatics are present in gasoline fractions, but the benzene content is usually low compared to the benzene homologues, such as toluene and the xylene isomer. In addition to the 1- and 2-methylnaphthalenes, other simple alkylnaphthalene derivatives have also been isolated from crude oil. Aromatics without naphthene rings appear to be relatively rare in the heavier fractions of petroleum (e.g., lubricating oils). In the higher-molecular-weight fractions, the rings are usually condensed together. Thus, components with two aromatic rings are presumed to be naphthalene derivatives and those with three aromatic rings may be phenanthrene derivatives. Currently, and because of the consideration of the natural product origins of petroleum, phenanthrene derivatives are favored over anthracene derivatives.

In summation, all hydrocarbon compounds that have aromatic rings, in addition to the presence of alkyl chains and naphthenic rings within the same molecule, are classified as aromatic compounds. Many separation procedures that have been applied to petroleum (Speight, 2014, 2015) result in the isolation of a compound as an *aromatic* even if there is only one such ring (i.e., six-carbon atoms) that is substituted by many more than six-carbon nonaromatic atoms.

It should also be emphasized that in the higher-boiling petroleum fractions, many polycyclic structures occur in naphthene–aromatic systems. The naphthene–aromatic hydrocarbons, together with the naphthenic hydrocarbon series, form the major content of higher-boiling petroleum fractions. Usually, the different naphthene–aromatic components are classified according to the number of aromatic rings in their molecules. The first to be distinguished is the series with an equal number of aromatic and naphthenic rings. The first members of the bicyclic series C_9–C_{11} are the simplest, such as the 1-methyl-, 2-methyl, and 4-methylindanes and 2-methyl- and 7-methyltetralin. Tetralin and methyl-, dimethyl-, methyl ethyl-, and tetramethyltetralin have been found in several crude oils, particularly in the heavier, naphthenic, crude oils and there are valid reasons to believe that this increase in the number of rings and side-chain complexity continues into the heavy oil and bitumen feedstocks.

Of special interest in the present context are the aromatic systems that occur in the nonvolatile asphaltene fraction (Speight, 1994). These polycyclic aromatic systems are complex molecules that fall into a molecular weight and boiling range where very little is known about model compounds (Speight, 1994, 2014). There has not been much success in determining the nature of such systems in the higher-boiling constituents of petroleum, that is, the residua or nonvolatile constituents. In fact, it has been generally assumed that as the boiling point of a petroleum fraction increases, so does the number of condensed rings in a polycyclic aromatic system. To an extent, this is true but the simplicities of such assumptions cause an omission of other important structural constituents of the petroleum matrix, the alkyl substituents, the heteroatoms, and any polycyclic systems that are linked by alkyl chains or by heteroatoms.

The active principle is that petroleum is a continuum (Chapters 12 and 13) and has natural product origins (Long, 1979, 1981; Speight, 1981, 1994, 2014). As such, it might be anticipated that there is a continuum of aromatic systems throughout petroleum that might differ from volatile to nonvolatile fractions but which, in fact, are based on natural product systems. It might also be argued that substitution patterns of the aromatic nucleus that are identified in the volatile fractions, or in any natural product counterparts, also apply to the nonvolatile fractions.

The application of thermal techniques to study the nature of the volatile thermal fragments from petroleum asphaltenes has produced some interesting data relating to the nature of the aromatic systems and the alkyl side chins in crude oil, heavy oil, and bitumen (Speight, 1971; Ritchie et al., 1979a,b; Schucker and Keweshan, 1980; Gallegos, 1981). These thermal techniques have produced strong evidence for the presence of small (1–4 rings) aromatic systems (Speight and Pancirov, 1984; Speight, 1987). There was a preponderance of single-ring (cycloparaffin and alkylbenzene) species as well as the domination of saturated material over aromatic material.

Further studies using pyrolysis/gas chromatography/mass spectrometry (py/gc/ms) showed that different constituents of the asphaltene fraction produce the same type of polycyclic aromatic systems in the volatile matter but the distribution was not constant (Speight and Pancirov, 1984). It was also possible to compute the hydrocarbon distribution from which a noteworthy point here is preponderance of single-ring (cycloparaffin and alkylbenzene) species as well as the domination of saturated material over aromatic material. The emphasis on low-molecular-weight material in the volatile products is to be anticipated on the basis that more complex systems remain as nonvolatile material and, in fact, are converted to coke.

One other point worthy of note is that the py/gc/ms program does not accommodate nitrogen and oxygen species whether or not they be associated with aromatic systems. This matter is resolved, in part, not only by the concentration of nitrogen and oxygen in the nonvolatile material (coke) but also by the overall low proportions of these heteroatoms originally present in the asphaltenes (Speight and Pancirov, 1984). The major drawback to the use of the py/gc/ms technique to the study of the aromatic systems in asphaltenes is the amount of material that remains as a nonvolatile residue.

Of all of the methods applied to determining the types of aromatic systems in petroleum asphaltenes, one with considerable potential, but given the least attention, is ultraviolet spectroscopy (Lee et al., 1981; Bjorseth, 1983). Typically, the ultraviolet spectrum of an asphaltene shows two major regions with very little fine structure. Interpretation of such a spectrum can only be made in general terms. However, the technique can add valuable information about the degree of condensation of polycyclic aromatic ring systems through the auspices of high-performance liquid chromatography (HPLC) (Lee et al., 1981; Bjorseth, 1983; Monin and Pelet, 1983; Felix et al., 1985; Killops and Readman, 1985; Speight, 1986). Indeed, when this approach is taken, the technique not only confirms the complex nature of the asphaltene fraction but also allows further detailed identifications to be made of the individual functional constituents of asphaltenes. The constituents of the fraction produce a multicomponent chromatogram, but subfractions produce a less complex and much narrower chromatograph that may even approximate a single peak, which may prove much more difficult to separate by a detector.

These data provide strong indications of the ring-size distribution of the polycyclic aromatic systems in petroleum asphaltenes. For example, from an examination of various functional subfractions (Figures 3.4 and 3.5), it was shown that amphoteric species and basic nitrogen species contain polycyclic aromatic systems having two to six rings per system. On the other hand, acid subfractions (phenolic/carboxylic functions) and neutral polar subfractions (amides/imino functions) contain few if any polycyclic aromatic systems having more than three rings per system. Moreover, the differences in the functionality of the subfractions result in substantial differences in thermal and catalytic reactivity that can lead to unanticipated phase separation and, subsequently, coke formation in a thermal reactor as well as structural orientation on, and blocking of, the active sites on a catalyst. This is especially the case when the behavior of the functional types that occur in the various high-boiling fractions of heavy oil and tar sand bitumen is considered (Figure 3.6) (Speight, 2014).

In all cases, the evidence favored the preponderance of the smaller (one-to-four) ring systems (Speight, 1986). But perhaps what is more important about these investigations is that the data show that the asphaltene fraction is a complex mixture of compound types that confirms fractionation studies and *cannot be adequately represented* by any particular formula that is construed to be

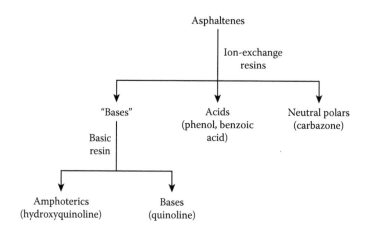

FIGURE 3.4   Separation of asphaltene constituents based on functionality (polarity).

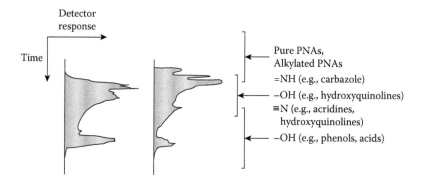

FIGURE 3.5   The complexity of the asphaltene fraction as shown by HPLC—the figure also refutes the concept of using an average structure to determine the behavior and properties of the asphaltene fraction.

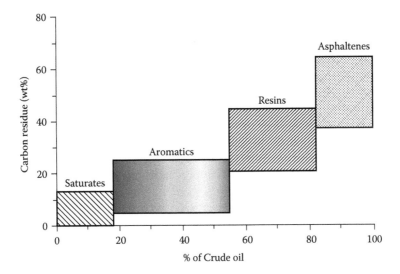

FIGURE 3.6 Subfractionation of the various fractions of crude oil yield subfractions with different carbon residues—particularly worthy of note in this instance is the spread of carbon residue yield for the subfractions of the asphaltene fraction.

average. Therefore, the concept of a large polycyclic aromatic ring system as the central feature of the asphaltene fraction must be abandoned for lack of evidence.

In summary, the premise is that petroleum is a natural product and that the aromatic systems are based on identifiable structural systems that are derived from natural product precursors.

3.3.1.4 Unsaturated Hydrocarbons

The presence of olefins ($RCH=CHR^1$) in petroleum has been under dispute for many years because there are investigators who claim that olefins are actually present. In fact, these claims usually refer to distilled fractions, and it is very difficult to entirely avoid cracking during the distillation process. Nevertheless, evidence for the presence of considerable proportions of olefins in Pennsylvanian crude oils has been obtained; spectroscopic and chemical methods showed that the crude oils, as well as all distillate fractions, contained up to 3% w/w olefins. Hence, although the opinion that petroleum does not contain olefins requires some revision, it is perhaps reasonable to assume that the Pennsylvania crude oils may hold an exceptional position and that olefins are present in crude oil in only a few special cases. The presence of diene derivatives ($RCH=CH=CHR'$) and acetylene derivatives ($RC\equiv CR'$) is considered to be extremely unlikely.

In summary, a variety of hydrocarbon compounds occur throughout petroleum. Although the amount of any particular hydrocarbon varies from one crude oil to another, the family from which that hydrocarbon arises is well represented.

3.3.2 Nonhydrocarbon Constituents

The previous sections present some indication of the types and nomenclature of the organic hydrocarbons that occur in various crude oils. Thus, it is not surprising that petroleum, which contains only hydrocarbons, is, in fact, an extremely complex mixture. The phenomenal increase in the number of possible isomers for the higher hydrocarbons makes it very difficult, if not impossible in most cases, to isolate individual members of any one series having more than, say, 12-carbon atoms.

Inclusion of organic compounds of nitrogen, oxygen, and sulfur serves only to present crude oil as an even more complex mixture than was originally conceived. Nevertheless, considerable progress has been made in the isolation and/or identification of the lower-molecular-weight hydrocarbons, as well as accurate estimations of the overall proportions of the hydrocarbon types present in petroleum. Indeed, it has been established that as the boiling point of the petroleum fraction increases, not only the number of the constituents but the molecular complexity of the constituents also increases (Figure 3.2) (Speight, 2000, 2014).

Crude oils contain appreciable amounts of organic nonhydrocarbon constituents, mainly sulfur-, nitrogen-, and oxygen-containing compounds and, in smaller amounts, organometallic compounds in solution and inorganic salts in colloidal suspension. These constituents appear throughout the entire boiling range of the crude oil but tend to concentrate mainly in the heavier fractions and in the nonvolatile residues.

Although their concentration in certain fractions may be quite small, their influence is important. For example, the decomposition of inorganic salts suspended in the crude can cause serious breakdowns in refinery operations; the thermal decomposition of deposited inorganic chlorides with evolution of free hydrochloric acid can give rise to serious corrosion problems in the distillation equipment. The presence of organic acid components, such as mercaptan derivatives and acid derivatives, can also promote metallic corrosion. In catalytic operations, passivation and/or poisoning of the catalyst can be caused by the deposition of traces of metals (vanadium and nickel) or by chemisorption of nitrogen-containing compounds on the catalyst, thus necessitating the frequent regeneration of the catalyst or its expensive replacement.

The presence of traces of nonhydrocarbons may impart objectionable characteristics in finished products, such as discoloration, lack of stability on storage, or a reduction in the effectiveness of organic lead antiknock additives. It is thus obvious that a more extensive knowledge of these compounds and of their characteristics could result in improved refining methods and even in finished products of better quality. The nonhydrocarbon compounds, particularly the porphyrins and related compounds, are also of fundamental interest in the elucidation of the origin and nature of crude oils.

Although sulfur is the most important (certainly the most abundant) heteroatom (i.e., nonhydrocarbon) present in petroleum with respect to the current context, other nonhydrocarbon atoms can exert a substantial influence not only on the nature and properties of the products but also on the nature and efficiency of the process. Such atoms are nitrogen, oxygen, and metals, and because of their influence on the process, some discussion of each is warranted here. Furthermore, knowledge of their surface-active characteristics is of help in understanding problems related to the migration of oil from the source rocks to the actual reservoirs.

3.3.2.1 Sulfur Compounds

Although the concentration of heteroatom constituents in certain fractions may be quite small, their influence is important. For example, the decomposition of inorganic salts suspended in the crude can cause serious breakdowns in refinery operations; the thermal decomposition of deposited inorganic chlorides with evolution of free hydrochloric acid can give rise to serious corrosion problems in the distillation equipment. The presence of organic acid components, such as mercaptan derivatives and acid derivatives, can also promote metallic corrosion. In catalytic operations, passivation and/or poisoning of the catalyst can be caused by the deposition of traces of metals (vanadium and nickel) or by the chemisorption of nitrogen-containing compounds on the catalyst, thus necessitating the frequent regeneration of the catalyst or its expensive replacement.

Sulfur compounds are among the most important heteroatomic constituents of petroleum, and although there are many varieties of sulfur compounds (Table 3.2) (Speight, 2000, 2014), the prevailing conditions during the formation, maturation, and even *in situ* alteration may dictate that

TABLE 3.2
Nomenclature and Types of Organic Sulfur Compounds

RSH	Thiols (mercaptans)
RSR′	Sulfides
	Cyclic sulfides
RSSR′	Disulfides
	Thiophene
	Benzothiophene
	Dibenzothiophene
	Naphthobenzothiophene

only preferred types exist in any particular crude oil. Nevertheless, sulfur compounds of one type or another are present in all crude oils (Thompson et al., 1976). In general, the higher the density of the crude oil (or the lower the API gravity of the crude oil) the higher the sulfur content and the total sulfur in the crude oil can vary from approximately 0.04% w/w for light crude oil to approximately 5.0% for heavy crude oil and tar sand bitumen. However, the sulfur content of crude oils produced from broad geographic regions varies with time, depending on the composition of newly discovered fields, particularly those in different geological environments.

The presence of sulfur compounds in finished petroleum products often produces harmful effects. For example, in gasoline, sulfur compounds are believed to promote the corrosion of engine parts, especially under winter conditions, when water-containing sulfur dioxide from the combustion may accumulate in the crankcase. In addition, mercaptan derivatives in hydrocarbon solution cause the corrosion of copper and brass in the presence of air and also affect lead susceptibility and color stability. Free sulfur is also corrosive, as are sulfide derivatives, disulfide derivatives, and thiophene derivatives, which are detrimental to the octane number response to tetraethyllead. However, gasoline with a sulfur content between 0.2% and 0.5% has been used without obvious harmful effect. In diesel fuels, sulfur compounds increase wear and can contribute to the formation of engine deposits. Although high sulfur content can sometimes be tolerated in industrial fuel oils, the situation for lubricating oils is that a high content of sulfur compounds in lubricating oils seems to lower resistance to oxidation and increases the deposition of solids.

Although it is generally true that the proportion of sulfur increases with the boiling point during distillation (Speight, 2000, 2014), the middle fractions may actually contain more sulfur than higher-boiling fractions as a result of decomposition of the higher-molecular-weight compounds during the distillation. High sulfur content is generally considered harmful in most petroleum products, and the removal of sulfur compounds or their conversion to less deleterious types is an important part of refinery practice. The distribution of the various types of sulfur compounds varies markedly among crude oils of diverse origin as well as between the various heavy feedstocks, but fortunately some of the sulfur compounds in petroleum undergo thermal reactions at relatively low temperatures. If elemental sulfur is present in the oil, a reaction, with the evolution of hydrogen sulfide, begins at approximately 150°C (300°F) and is very rapid at 220°C (430°F), but organically bound sulfur compounds do not yield hydrogen sulfide until higher temperatures are reached. Hydrogen sulfide is, however, a common constituent of many crude oils, and some crude oils with >1% w/w sulfur are often accompanied by a gas having substantial properties of hydrogen sulfide.

Various thiophene derivatives have also been isolated from a variety of crude oils; benzothiophene derivatives are usually present in the higher-boiling petroleum fractions. On the other hand, disulfides are not regarded as true constituents of crude oil but are generally formed by the oxidation of thiols during processing:

$$2R-SH + [O] \rightarrow R-S-S-R + H_2O$$

3.3.2.2 Nitrogen Compounds

Nitrogen in petroleum may be classified arbitrarily as basic and nonbasic. The basic nitrogen compounds (Table 3.3), which are composed mainly of pyridine homologues and occur throughout the boiling ranges, have a decided tendency to exist in the higher-boiling fractions and residua. The nonbasic nitrogen compounds, which are usually of the pyrrole, indole, and carbazole types, also occur in the higher-boiling fractions and residua.

In general, the nitrogen content of crude oil is low and generally falls within the range 0.1%–0.9% w/w, although early work indicates that some crude oil may contain up to 2% nitrogen. However, crude oils with no detectable nitrogen or even trace amounts are not uncommon, but in general the more asphaltic the oil, the higher its nitrogen content. Insofar as an approximate correlation exists between the sulfur content and the API gravity of crude oils (Speight, 2000, 2014), there also exists a correlation between the nitrogen content and the API gravity of crude oil. It also follows that there is an approximate correlation between the nitrogen content and the carbon residue: the higher the nitrogen content, the higher the carbon residue. The presence of nitrogen in petroleum is of much greater significance in refinery operations than might be expected from the small amounts present. Nitrogen compounds can be responsible for the poisoning of cracking catalysts, and they also contribute to gum formation in such products as domestic fuel oil. The trend in recent years toward cutting deeper into the crude to obtain stocks for catalytic cracking has accentuated the harmful effects of the nitrogen compounds, which are concentrated largely in the higher-boiling portions.

Basic nitrogen compounds with a relatively low molecular weight can be extracted with dilute mineral acids; equally strong bases of higher molecular weight remain unextracted because of unfavorable partitioning between the oil and aqueous phases. A method has been developed in which the nitrogen compounds are classified as basic or nonbasic, depending on whether they can be titrated with perchloric acid in a 50:50 solution of glacial acetic acid and benzene. The application of this method has shown that the ratio of basic to total nitrogen is approximately constant (0 to 30 ± 0.05) irrespective of the source of the crude. Indeed, the ratio of basic to total nitrogen was found to be approximately constant throughout the entire range of distillate and residual fractions.

TABLE 3.3
Nomenclature and Types of Organic Nitrogen Compounds

Nonbasic

Pyrrole	C_4H_5N	
Indole	C_8H_7N	
Carbazole	$C_{12}H_9N$	
Benzo(a)carbazole	$C_{16}H_{11}N$	

Basic

Pyridine	C_5H_5N	
Quinoline	C_9H_7N	
Indoline	C_8H_9N	
Benzo(f)quinoline	$C_{13}H_9N$	

Nitrogen compounds extractable with dilute mineral acids from petroleum distillates were found to consist of alkyl pyridine derivatives, alkyl quinoline derivatives, and alkyl *iso*-quinoline derivatives carrying alkyl substituents, as well as pyridine derivatives in which the substituent was a cyclopentyl or cyclohexyl group. The compounds that cannot be extracted with dilute mineral acids contain the greater part of the nitrogen in petroleum and are generally of the carbazole, indole, and pyrrole types.

3.3.2.3 Oxygen Compounds

Oxygen in organic compounds can occur in a variety of forms (ROH, ArOH, ROR', RCO_2H, $ArCO_2H$, RCO_2R, $ArCO_2R$, $R_2C{=}O$ as well as the cyclic furan derivatives, where R and R' are alkyl groups and Ar is an aromatic group) in nature, so it is not surprising that the more common oxygen-containing compounds occur in petroleum (Speight, 2014). The total oxygen content of crude oil is usually less than 2% w/w, although larger amounts have been reported, but when the oxygen content is phenomenally high it may be that the oil has suffered prolonged exposure to the atmosphere either during or after production. However, the oxygen content of petroleum increases with the boiling point of the fractions examined; in fact, the nonvolatile residua may have oxygen contents up to 8% w/w. Although these high-molecular-weight compounds contain most of the oxygen in petroleum, little is known concerning their structure, but those of lower molecular weight have been investigated with considerably more success and have been shown to contain carboxylic acids and phenols.

It has generally been concluded that the carboxylic acids in petroleum with fewer than eight-carbon atoms per molecule are almost entirely aliphatic in nature; monocyclic acids begin at C_6 and predominate above C_{14}. This indicates that the structures of the carboxylic acids correspond with those of the hydrocarbons with which they are associated in the crude oil. In the range in which paraffins are the prevailing type of hydrocarbon, the aliphatic acids may be expected to predominate. Similarly, in the ranges in which monocycloparaffin derivatives and dicycloparaffin derivatives prevail, one may expect to find principally monocyclic and dicyclic acid derivatives, respectively.

Although comparisons are frequently made between the sulfur and nitrogen contents and such physical properties as the API gravity, it is not the same with the oxygen contents of crude oils. It is possible to postulate, and show, that such relationships exist. However, the ease with which some of the crude oil constituents can react with oxygen (aerial or dissolved) to incorporate oxygen functions into their molecular structure often renders the exercise somewhat futile if meaningful deductions are to be made.

Carboxylic acid derivatives (RCO_2H) may be less detrimental than other heteroatom constituents because there is the high potential for decarboxylation to a hydrocarbon and carbon dioxide at the temperatures (>340°C, >645°F) used during distillation of flashing (Speight and Francisco, 1990):

$$RCO_2H \rightarrow RH + CO_2$$

In addition to the carboxylic acids and phenolic compounds (ArOH, where Ar is an aromatic moiety), the presence of ketones (>C=O), esters [>C(=O)–OR], ethers (R–O–R), and anhydrides >C(=O)–O–(O=)C<] has been claimed for a variety of crude oils. However, the precise identification of these compounds is difficult because most of them occur in the higher-molecular-weight nonvolatile residua. They are claimed to be products of the air blowing of the residua, and their existence in virgin crude oil, heavy oil, or bitumen may yet need to be substantiated.

3.3.2.4 Metallic Constituents

Metallic constituents are found in every crude oil, and the concentrations have to be reduced to convert the oil to transportation fuel. Metals affect many upgrading processes and cause particular problems because they poison catalysts used for sulfur and nitrogen removal as well as other processes such as catalytic cracking. The trace metals Ni and V are generally orders of magnitude higher than other metals in petroleum, except when contaminated with coproduced brine salts (Na, Mg, Ca, Cl) or corrosion products gathered in transportation (Fe).

The occurrence of metallic constituents in crude oil is of considerably greater interest to the petroleum industry than might be expected from the very small amounts present. Even minute amounts of iron, copper, and particularly nickel and vanadium in the charging stocks for catalytic cracking affect the activity of the catalyst and result in increased gas and coke formation and reduced yields of gasoline. In high-temperature power generators, such as oil-fired gas turbines, the presence of metallic constituents, particularly vanadium in the fuel, may lead to ash deposits on the turbine rotors, thus reducing clearances and disturbing their balance. More particularly, damage by corrosion may be very severe. The ash resulting from the combustion of fuels containing sodium and especially vanadium reacts with refractory furnace linings to lower their fusion points and so cause their deterioration.

Thus, the ash residue left after burning of a crude oil is due to the presence of these metallic constituents, part of which occur as inorganic water-soluble salts (mainly chlorides and sulfates of sodium, potassium, magnesium, and calcium) in the water phase of crude oil emulsions (Abdel-Aal et al., 2016). These are removed in the desalting operations, either by evaporation of the water and subsequent water washing or by breaking the emulsion, thereby causing the original mineral content of the crude to be substantially reduced. Other metals are present in the form of oil-soluble organo-metallic compounds as complexes, metallic soaps, or in the form of colloidal suspensions, and the total ash from desalted crude oils is of the order of 0.1–100 mg/L. Metals are generally found only in the nonvolatile portion of crude oil (Altgelt and Boduszynski, 1994; Reynolds, 1998).

Two groups of elements appear in significant concentrations in the original crude oil associated with well-defined types of compounds. Zinc, titanium, calcium, and magnesium appear in the form of organometallic soaps with surface-active properties adsorbed in the water/oil interfaces and act as emulsion stabilizers. However, vanadium, copper, nickel, and part of the iron found in crude oils seem to be in a different class and are present as oil-soluble compounds. These metals are capable of complexing with pyrrole pigment compounds derived from chlorophyll and hemoglobin and are almost certain to have been present in plant and animal source materials. It is easy to surmise that the metals in question are present in such form, ending in the ash content. Evidence for the presence of several other metals in oil-soluble form has been produced, and thus, zinc, titanium, calcium, and magnesium compounds have been identified in addition to vanadium, nickel, iron, and copper. Examination of the analyses of a number of crude oil for iron, nickel, vanadium, and copper indicates a relatively high vanadium content, which usually exceeds that of nickel, although the reverse can also occur.

Distillation concentrates the metallic constituents in the residues (Reynolds, 1998), although some can appear in the higher-boiling distillates, but the latter may be due in part to entrainment. Nevertheless, there is evidence that a portion of the metallic constituents may occur in the distillates by volatilization of the organometallic compounds present in the petroleum. In fact, as the percentage of overhead obtained by vacuum distillation of a reduced crude is increased, the amount of metallic constituents in the overhead oil is also increased. The majority of the vanadium, nickel, iron, and copper in residual stocks may be precipitated along with the asphaltenes by hydrocarbon solvents. Thus, removal of the asphaltenes with *n*-pentane reduces the vanadium content of the oil by up to 95% with substantial reductions in the amounts of iron and nickel.

3.3.2.5 Porphyrins

Porphyrins are a naturally occurring chemical species that exist in petroleum and usually occur in the nonbasic portion of the nitrogen-containing concentrate (Bonnett, 1978; Reynolds, 1998). They are not usually considered among the usual nitrogen-containing constituents of petroleum, nor are they considered a metallo-containing organic material that also occurs in some crude oils. As a result of these early investigations, there arose the concept of porphyrins as biomarkers that could establish a link between compounds found in the geosphere and their corresponding biological precursors.

Porphyrins are derivatives of porphine that consists of four pyrrole molecules joined by methine (–CH=) bridges (Figure 3.7). The methine bridges establish conjugated linkages between the component pyrrole nuclei, forming a more extended resonance system. Although the resulting structure retains much of the inherent character of the pyrrole components, the larger conjugated system gives increased aromatic character to the porphine molecule. Furthermore, the imine functions (–NH–) in the porphine system allow metals such as nickel to be included into the molecule through chelation (Figure 3.8).

A large number of different porphyrin compounds exist in nature or have been synthesized. Most of these compounds have substituents other than hydrogen on many of the ring carbons. The nature of the substituents on porphyrin rings determines the classification of a specific porphyrin compound into one of various types according to one common system of nomenclature (Bonnet, 1978). Porphyrins also have well-known trivial names or acronyms that are often in more common usage than the formal system of nomenclature. When one or two double bonds of a porphyrin are hydrogenated, a chlorin or a phlorin is the result. Chlorin derivatives are components of

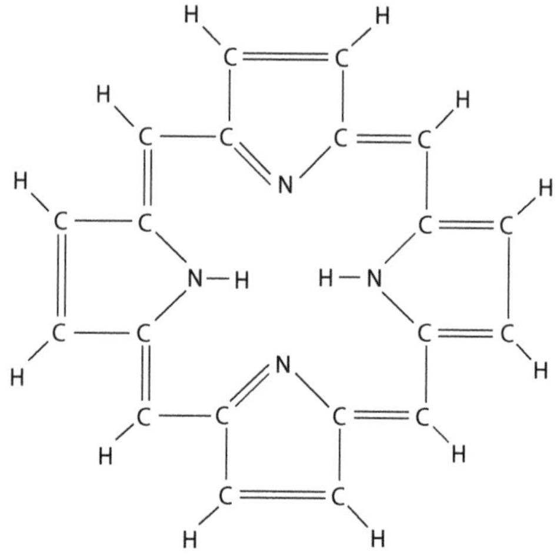

FIGURE 3.7 Porphine—the basic structure of porphyrins.

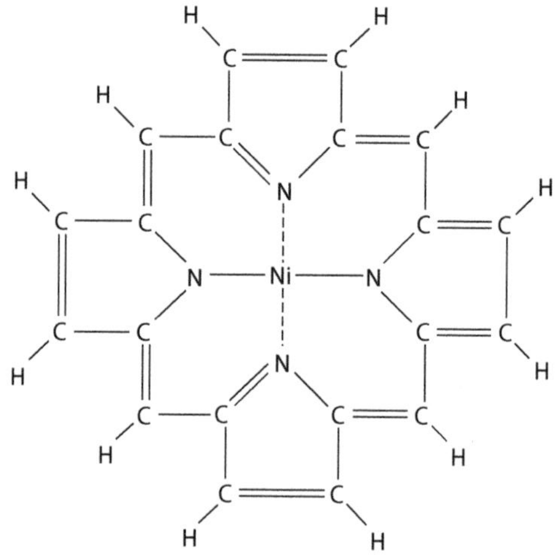

FIGURE 3.8 Nickel chelate of porphine.

chlorophyll and possess an *iso*-cyclic ring formed by two methylene groups bridging a pyrrole-type carbon to a methine carbon. Geological porphyrins that contain this structural feature are assumed to be derived from chlorophylls. Etioporphyrin derivatives are also commonly found in geological materials and have no substituents (other than hydrogen) on the methine carbons. Benzoporphyrin derivatives and tetrahydrobenzoporphyrin derivatives have also been identified in geological materials. These compounds have either a benzene ring or a hydrogenated benzene ring fused onto a pyrrole unit.

Almost all crude oil, heavy oil, and bitumen contain detectable amounts of vanadium and nickel porphyrin derivatives. More mature, lighter crude oils usually contain only small amounts of these compounds. Heavy oils may contain large amounts of vanadium and nickel porphyrin derivatives. Vanadium concentrations of over 1000 ppm are known for some crude oil, and a substantial amount of the vanadium in these crude oils is chelated with porphyrins. In high-sulfur crude oil of marine origin, vanadium porphyrin derivatives are more abundant than nickel porphyrin derivatives. Low-sulfur crude oils of lacustrine origin usually contain more nickel porphyrin derivatives than vanadium porphyrin derivatives.

Of all the metals in the periodic table, only vanadium and nickel have been proven definitely to exist as chelates in significant amounts in a large number of crude oils and tar sand bitumen. Geochemical reasons for the absence of substantial quantities of porphyrins chelated with metals other than nickel and vanadium in crude oils, heavy oils, and tar sand bitumen have been advanced (Hodgson et al., 1967; Baker, 1969; Baker and Palmer, 1978; Baker and Louda, 1986; Filby and Van Berkel, 1987; Quirke, 1987).

If the vanadium and nickel contents of crude oils are measured and compared with porphyrin concentrations, it is usually found that not all the metal content can be accounted for as porphyrin constituents (Reynolds, 1998). In some crude oils, as little as 10% w/w of total metals appears to be chelated with porphyrins. Only rarely can all measured nickel and vanadium in a crude oil be accounted for as porphyrin type. Currently, some investigators believe that part of the vanadium and nickel in crude oils is chelated with ligands that are not porphyrins. These metal chelates are referred to as nonporphyrin metal chelates or complexes (Reynolds et al., 1987).

Finally, during the fractionation of petroleum, the metallic constituents (metalloporphyrins and nonporphyrin metal chelates) are concentrated in the asphaltene fraction. The deasphaltened oil contains smaller concentrations of porphyrins than the parent materials and usually very small concentrations of nonporphyrin metals.

3.4   CHEMICAL COMPOSITION BY DISTILLATION

Although distillation is presented in more detail elsewhere (Chapter 7), it is appropriate to mention distillation here insofar as it is a method by which the constituents of petroleum can be separated and identified. Distillation is a means of separating chemical compounds (usually liquids) through differences in the respective vapor pressures. In the mixture, the components evaporate such that the vapor has a composition determined by the chemical properties of the mixture. Distillation of a given component is possible, if the vapor has a higher proportion of the given component than the mixture. This is caused by the given component having a higher vapor pressure and a lower boiling point than the other components.

By the nature of the process, it is theoretically impossible to completely separate and purify the individual components of petroleum when the possible number of isomers is considered for the individual carbon numbers that occur within the paraffin family (Table 3.4). When other types of compounds are included, such as the aromatic derivatives and heteroatom derivatives, even though the maturation process might limit the possible number of isomeric permutations (Tissot and Welte, 1978), the potential number of compounds in petroleum is still (in a sense) astronomical.

However, petroleum can be separated into a variety of fractions on the basis of the boiling points of the petroleum constituents. Such fractions are primarily identified by their respective boiling

TABLE 3.4

Boiling Point of the *n*-Isomers of the Various Paraffins and the Number of Possible Isomers Associated with Each Carbon Number

Number of Carbon Atoms	Boiling Point *n*-Isomer (°C)	Boiling Point *n*-Isomer (°F)	Number of Isomers
5	36	97	3
10	174	345	75
15	271	519	4,347
20	344	651	366,319
25	402	755	36,797,588
30	450	841	4,111,846,763
40	525	977	62,491,178,805,831

ranges and, to a lesser extent, by chemical composition. However, it is often obvious that as the boiling ranges increase, the nature of the constituents remains closely similar and it is the number of the substituents that caused the increase in boiling point. It is through the recognition of such phenomena that molecular design of the higher-boiling constituents can be achieved (Figure 3.9). Invoking the existence of structurally different constituents in the nonvolatile fractions from those identifiable constituents in the lower-boiling fractions is unnecessary (considering the nature of the precursors and maturation paths) and irrational (Speight, 1994, 2014). For example, the predominant types of condensed aromatic systems in petroleum are derivatives of phenanthrene and there it is to be anticipated that the higher peri-condensed homologues (Figure 3.10) will be present in resin constituents and asphaltene constituents rather than the derivatives of kata-condensed polynuclear aromatic system (Figure 3.11).

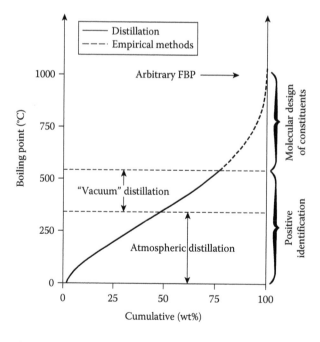

FIGURE 3.9 Separation of petroleum constituents by distillation.

Chrysene

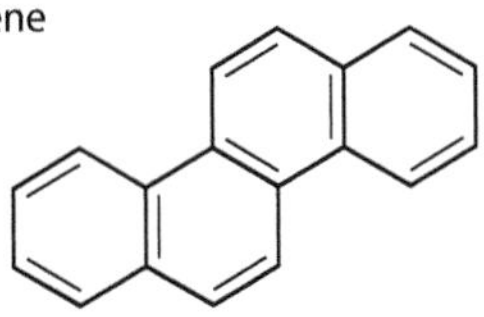

Picene

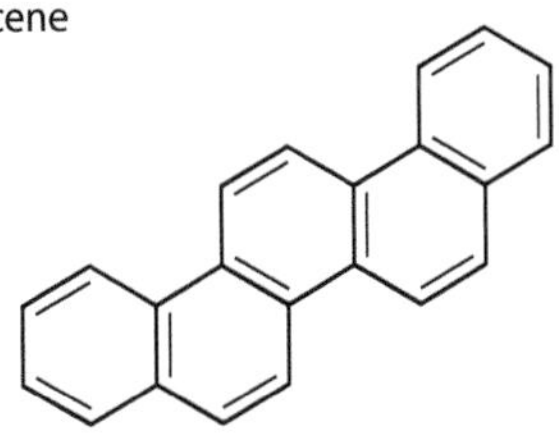

1,2,5,6-Dibenzanthracene

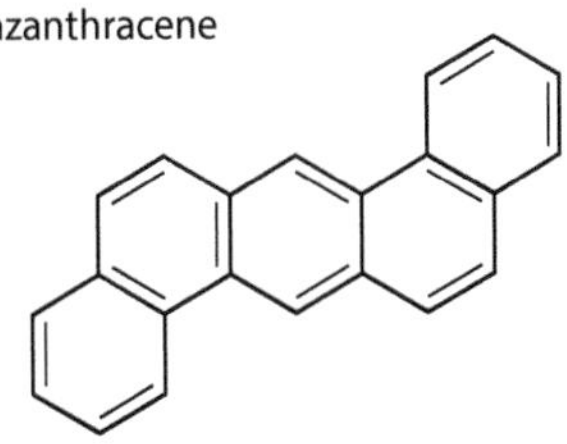

FIGURE 3.10   Examples of peri-condensed aromatic systems.

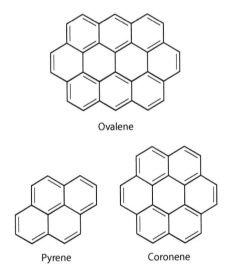

Ovalene

Pyrene   Coronene

FIGURE 3.11   Examples of kata-condensed (cata-condensed) aromatic systems.

3.4.1   Gases and Naphtha

Methane is the main hydrocarbon component of petroleum gases with lesser amounts of ethane, propane, butane, *iso*-butane, and some C_4^+ hydrocarbons. Other gases, such as hydrogen, carbon dioxide, hydrogen sulfide, and carbonyl sulfide, are also present.

Saturated constituents with lesser amounts of mono- and diaromatics dominate the naphtha fraction. While naphtha covers the boiling range of gasoline, most of the raw petroleum naphtha molecules have low octane number. However, most raw naphtha is processed further and combined with other process naphtha and additives to formulate commercial gasoline.

Within the saturated constituents in petroleum gases and naphtha, every possible paraffin from methane (CH_4) to n-decane (normal decane, n-$C_{10}H_{22}$) is present. Depending upon the source, one of these low-boiling paraffins may be the most abundant compound in a crude oil reaching several percent. The *iso*-paraffins begin at C_4 with *iso*-butane as the only isomer of n-butane. The number of isomers grows rapidly with carbon number, and there may be increased difficulty in dealing with multiple isomers during analysis.

In addition to aliphatic molecules, the saturated constituents consist of cycloalkanes (naphthenes) with predominantly five- or six-carbon rings. Methyl derivatives of cyclopentane and cyclohexane, which are commonly found at higher levels than the parent unsubstituted structures, may be present (Tissot and Welte, 1978). Fused ring dicycloalkane derivatives such as cis-decahydronaphthalene (cis-decalin) and trans-decahydronaphthalene (trans-decalin) and hexahydro-indane are also common, but bicyclic naphthene derivatives separated by a single bond, such as cyclohexyl cyclohexane, are not.

The numerous aromatic constituents in petroleum naphtha begin with benzene, but its C_1–C_3 alkylated derivatives are also present (Tissot and Welte, 1978). Each of the alkyl benzene homologues through the 20 isomeric C_4 alkyl benzenes has been isolated from crude oil along with various C_5 derivatives. Benzene derivatives having fused cycloparaffin rings (naphthene aromatics) such as indane and tetralin have been isolated along with a number of their methyl derivatives. Naphthalene is included in this fraction, while the 1- and 2-methyl naphthalene derivatives and higher homologues of fused two ring aromatics appear in the mid-distillate fraction.

Sulfur-containing compounds are the only heteroatom compounds to be found in this fraction (Mair, 1964; Rall et al., 1972). Usually, the total amount of the sulfur in this fraction accounts for less than 1% of the total sulfur in the crude oil. In naphtha from high-sulfur (sour) petroleum, 50%–70% of the sulfur may be in the form of mercaptan derivatives (thiol derivatives). Over 40 individual thiols have been identified, including all the isomeric C_1 to C_6 compounds plus some C_7 and C_8 isomers plus thiophenol (Rall et al., 1972). In naphtha from low-sulfur (sweet) crude oil, the sulfur is distributed between sulfides (thioether derivatives) and thiophene derivatives. In these cases, the sulfides may be in the form of both linear (alkyl sulfides) and five- or six-ring cyclic (thiacyclane) structures. The sulfur structure distribution tends to follow the distribution hydrocarbons; that is, naphthenic oils with a high cycloalkane content tend to have a high thiacyclane content. Typical alkyl thiophene derivatives in naphtha have multiple short side chains or exist as naphthene–thiophene derivatives (Rall et al., 1972). Methyl and ethyl disulfides have been confirmed to be present in some crude oils in analyses that minimized their possible formation by oxidative coupling of thiols (Aksenova and Kayanov, 1980; Freidlina and Skorova, 1980).

3.4.2 MIDDLE DISTILLATES

Saturated species are the major component in the mid-distillate fraction of petroleum but aromatics, which now include simple compounds with up to three aromatic rings, and heterocyclic compounds are present and represent a larger portion of the total. Kerosene, jet fuel, and diesel fuel are all derived from raw middle distillate, which can also be obtained from cracked and hydroprocessed refinery streams.

Within the saturated constituents, the concentration of n-paraffins decreases regularly from C_{11} to C_{20}. Two isoprenoid species (pristane = 2,6,10,14-tetramethylpentadecane and phytane = 2,6,10,14-tetramethylhexadecane) are generally present in crude oils in sufficient concentration to be seen as irregular peaks alongside the n-C_{17} and n-C_{18} peaks in a gas chromatogram. These isoprene derivatives, believed to arise as fragments of ancient precursors, have relevance as simple biomarkers, to the genesis of petroleum. The distribution of pristane and phytane relative to their neighboring n-C_{17} and n-C_{18} peaks has been used to aid the identification of crude oils and to detect the onset of biodegradation. The ratio of pristane to phytane has also been used for the assessment of the oxidation and reduction environment in which ancient organisms were converted into petroleum (Hunt, 1979).

Mono- and dicycloparaffin derivatives with five or six carbons per ring constitute the bulk of the naphthenes in the middle distillate boiling range, decreasing in concentration as the carbon number increases (Tissot and Welte, 1978), and the alkylated naphthenes may have a single long side chain as well as one or more methyl or ethyl groups. Similarly, substituted three-ring naphthenes have been detected by gas chromatography and adamantane has been found in crude oil (Hunt, 1979; Lee et al., 1981).

The most abundant aromatics in the mid-distillate boiling fractions are di- and tri-methyl naphthalene derivatives. Other one and two ring aromatics are undoubtedly present in small quantities as either naphthene or alkyl homologues in the C_{11}–C_{20} range. In addition to these homologues of alkylbenzenes, tetralin, and naphthalene derivatives, the mid-distillate contains some fluorene derivatives and phenanthrene derivatives. The phenanthrene structure appears to be favored over that of anthracene structure (Tissot and Welte, 1978), and this appears to continue through the higher-boiling fractions of petroleum (Speight, 1994).

The five-membered heterocyclic constituents in the mid-distillate range are primarily the thiacyclane derivatives, benzothiophene derivatives, and dibenzothiophene derivatives with lesser amounts of dialkyl-, diaryl, and aryl–alkyl sulfide derivatives (Aksenova and Kayanov, 1980; Freidlina and Skorova, 1980). Alkylthiophenes are also present. As with the naphtha fractions, these sulfur species account for a minimal fraction of the total sulfur in the crude.

Although only trace amounts (usually ppm levels) of nitrogen are found in the middle distillate fractions, both neutral and basic nitrogen compounds have been isolated and identified in fractions boiling below 343°C (650°F) (Hirsch et al., 1974). Pyrrole derivatives and indole derivatives account for approximately two-thirds of the nitrogen, while the remainder is found in the basic alkylated pyridine and alkylated quinoline compounds.

The saturate constituents contribute less to the vacuum gas oil (VGO) than the aromatic constituents but more than the polar constituents that are now present at percentage rather than trace levels. Vacuum gas oil is occasionally used as a heating oil, but most commonly it is processed by catalytic cracking to produce naphtha or extraction to yield lubricating oil.

Within the vacuum gas oil, saturates, distribution of paraffins, *iso*-paraffins, and naphthenes are highly dependent upon the petroleum source. Generally, the naphthene constituents account for approximately two-thirds 60% of the saturate constituents, but the overall range of variation is from <20% to >80%. In most samples, the *n*-paraffins from C_{20} to C_{44} are still present in sufficient quantity to be detected as distinct peaks in gas chromatographic analysis.

The bulk of the saturated constituents in VGO consists of *iso*-paraffins and especially naphthene species, although isoprenoid compounds, such as squalane (C_{30}) and lycopane (C_{40}), have been detected. Analytical techniques show that the naphthenes contain from one to more than six fused rings accompanied by alkyl substitution. For mono- and diaromatics, the alkyl substitution typically involves several methyl and ethyl substituents. Hopanes and steranes have also been identified and are also used as internal markers for estimating the biodegradation of crude oils during bioremediation processes (Prince et al., 1994).

The aromatic constituents in vacuum gas oil may contain one to six fused aromatic rings that may bear additional naphthene rings and alkyl substituents in keeping with their boiling range. Mono- and diaromatics account for approximately 50% of the aromatics in petroleum vacuum gas oil samples. Analytical data show the presence of up to four fused naphthenic rings on some aromatic compounds. This is consistent with the suggestion that these species originate from the aromatization of steroids. Although present at lower concentration, alkyl benzenes and naphthalene derivatives show one long side chain and multiple short side chains.

The fused ring aromatic compounds (having three or more rings) in petroleum include phenanthrene, chrysene, and picene, as well as fluoranthene, pyrene, benzo(a)pyrene, and benzo(ghi) perylene. The most abundant reported individual phenanthrene compounds appear to be the three derivatives. In addition, phenanthrene derivatives outnumber anthracene derivatives by as much as 100:1. In addition, chrysene derivatives are favored over pyrene derivative.

Heterocyclic constituents are significant contributors to the vacuum gas oil fraction. In terms of sulfur compounds, thiophene and thiacyclane sulfur predominate over sulfide sulfur. Some molecules even contain more than one sulfur atom. The benzothiophene derivatives and dibenzothiophene derivatives are the prevalent thiophene forms of sulfur. In the vacuum gas oil range, the nitrogen-containing compounds include higher-molecular-weight pyridine derivatives, quinoline derivatives, benzoquinoline derivatives, amide derivatives, indole derivatives, carbazole derivative, and molecules with two nitrogen atoms (diaza compounds) with three and four aromatic rings that are especially prevalent (Green et al., 1989). Typically, approximately one-third of the compounds are basic, that is, pyridine and its benzo derivatives, while the remainder is present as neutral species (amide derivatives and carbazole derivatives). Although benzo- and dibenzo-quinoline derivatives found in petroleum are rich in sterically hindered structures, hindered and unhindered structures have been found to be present at equivalent concentrations in source rocks. This has been rationalized as geochromatography in which the less polar (hindered) structures moved more readily to the reservoir and are not adsorbed on any intervening rocks structures.

Oxygen levels in the vacuum gas oil parallel the nitrogen content. Thus, the most commonly identified oxygen compounds are the carboxylic acids and phenols, collectively called naphthenic acids (Seifert and Teeter, 1970).

3.4.3 Vacuum Residua

The vacuum residuum (*vacuum bottoms*, typically 950°F$^+$ or 1050°F$^+$) is the most complex of petroleum and, in many cases, may even resemble heavy oil or extra heavy oil or tar sand bitumen in composition. Vacuum residua contain the majority of the heteroatoms originally in the petroleum and molecular weight of the constituents range up to several thousand (as near as can be determined but subject to method dependence). The fraction is so complex that the characterization of individual species is virtually impossible, no matter what claims have been made or will be made. Separation of vacuum residua by group type becomes difficult and confused because of the multi-substitution of aromatic and naphthenic species as well as of the presence of multiple functionalities in single molecules.

Classically, *n*-pentane or *n*-heptane precipitation is used as the initial step for the characterization of vacuum residuum. The insoluble fraction, the pentane or heptane asphaltenes, may be as much as 50% by weight of a vacuum residuum. The pentane- or heptane-soluble portion (maltene) constituents of the residuum are then fractionated chromatographically into several solubility or adsorption classes for characterization. However, in spite of claims to the contrary, the method is not a separation by chemical type. Kit is a separation by solubility and adsorption. However, the separation of the asphaltene constituents does, however, provide a simple way to remove some of the highest molecular weight and most polar components, but the asphaltene fraction is so complex that compositional detail based on average parameters is of questionable value.

For the 565°C$^+$ (1050°F$^+$) fractions of petroleum, the levels of nitrogen and oxygen may begin to approach the concentration of sulfur. These elements consistently concentrate in the most polar fractions to the extent that every molecule contains more than one heteroatom. At this point, structural identification is somewhat fruitless and characterization techniques are used to confirm the presence of the functionalities found in lower-boiling fractions such as acids, phenols, nonbasic (carbazole-type) nitrogen, and basic (quinoline-type) nitrogen.

The nickel and vanadium that are concentrated into the vacuum residuum appear to occur in two forms: (1) porphyrins and (2) nonporphyrins (Reynolds, 1998). Because the metalloporphyrins can provide insights into petroleum maturation processes, they have been studied extensively and several families of related structures have been identified. On the other hand, the nonporphyrin metals remain not clearly identified although some studies suggest that some of the metals in these compounds still exist in a tetra-pyrrole (porphyrin-type) environment (Pearson and Green, 1993).

It is more than likely that, in a specific residuum molecule, the heteroatoms are arranged in different functionalities, making an incredibly complex molecule. Considering how many different combinations are possible, the chances of determining every structure in a residuum are very low. Because of this seemingly insurmountable task, it may be better to determine ways of utilizing the residuum rather attempting to determine (at best questionable) molecular structures.

3.5 FRACTIONAL COMPOSITION

Refining petroleum involves subjecting the feedstock to a series of integrated physical and chemical unit processes (Figure 3.12) as a result of which a variety of products are generated. In some of the processes, for example, distillation, the constituents of the feedstock are isolated unchanged, whereas in other processes (such as cracking) considerable changes occur in the constituents. Feedstocks can be defined (on a *relative* or *standard* basis) in terms of three or four general fractions: asphaltenes, resins, saturates, and aromatics (Figure 3.13). Thus, it is possible to compare interlaboratory investigations and thence to apply the concept of predictability to refining sequences and potential products. Recognition that refinery behavior is related to the composition of the feedstock has led to a multiplicity of attempts to establish petroleum and its fractions as compositions of matter. As a result, various analytical techniques have been developed for the identification and quantification of *every molecule* in the lower-boiling fractions of petroleum. It is now generally recognized that the name *petroleum* does not describe a composition of matter but rather a mixture of various organic compounds that includes a wide range of molecular weights and molecular types that exist in balance with each other (Speight, 1994; Long and Speight, 1998). There must also be some questions of the advisability (perhaps *futility* is a better word) of attempting to describe *every molecule* in petroleum. The true focus should be to what ends these molecules can be used.

The fractionation methods available to the petroleum industry allow a reasonably effective degree of separation of hydrocarbon mixtures (Speight, 2014, 2015). However, the problems are separating the petroleum constituents without alteration of their molecular structure and obtaining these constituents in a substantially pure state. Thus, the general procedure is to employ techniques that segregate the constituents according to molecular size and molecular type.

It is more generally true, however, that the success of any attempted fractionation procedure involves not only the application of one particular technique but also the utilization of several integrated techniques, especially those techniques involving the use of chemical and physical properties to differentiate among the various constituents. For example, the standard processes of physical fractionation used in the petroleum industry are those of distillation and solvent treatment, as well as adsorption by surface-active materials. Chemical procedures depend on specific reactions, such as the interaction of olefins with sulfuric acid or the various classes of adduct formation. Chemical fractionation is often but not always successful because of the complex nature of crude oil. This may result in unprovoked chemical reactions that have an adverse effect on the fractionation and the resulting data. Indeed, caution is advised when using methods that involve chemical separation of the constituents.

The order in which the several fractionation methods are used is determined not only by the nature and/or composition of the crude oil but also by the effectiveness of a particular process and its compatibility with the other separation procedures to be employed. Thus, although there are wide variations in the nature of refinery feedstocks, there have been many attempts to devise standard methods of petroleum fractionation. However, the various laboratories are inclined to adhere firmly to, and promote, their own particular methods. Recognition that no one particular method may satisfy all the requirements of petroleum fractionation is the first step in any fractionation study. This is due, in the main part, to the complexity of petroleum not only from the

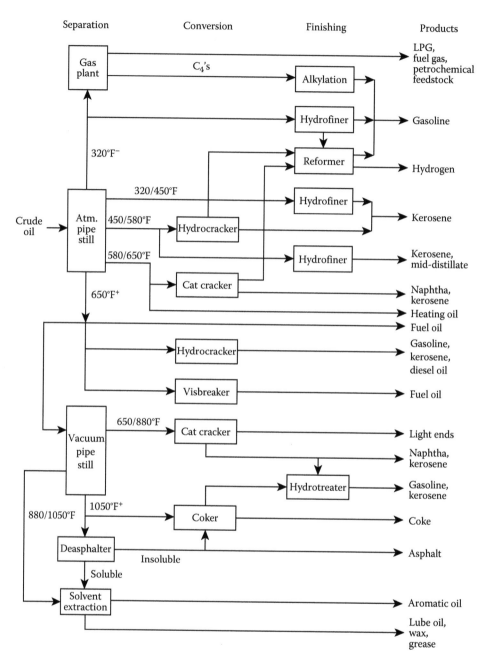

FIGURE 3.12 Representation of a refinery.

distribution of the hydrocarbon species but also from the distribution of the heteroatom (nitrogen, oxygen, and sulfur) species.

3.5.1 SOLVENT METHODS

Fractionation of petroleum by distillation is an excellent means by which the volatile constituents can be isolated and studied. However, the nonvolatile residuum, which may actually constitute from

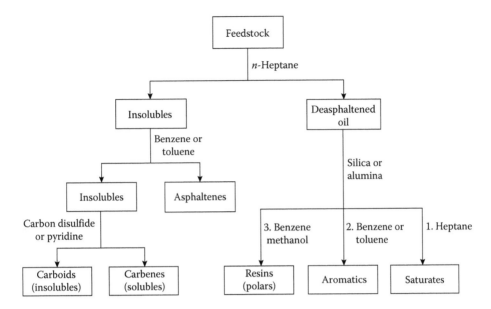

FIGURE 3.13 Separation scheme for various feedstocks.

1% to 60% of the petroleum, cannot be fractionated by distillation without the possibility of thermal decomposition, and as a result, alternative methods of fractionation have been developed.

The distillation process separates *light* (lower-molecular-weight) and *heavy* (higher-molecular-weight) constituents by virtue of their volatility and involves the participation of a vapor phase and a liquid phase. These are, however, physical processes that involve the use of two liquid phases, usually a solvent phase and an oil phase.

Solvent methods have also been applied to petroleum fractionation on the basis of molecular weight. The major molecular weight separation process used in the laboratory as well as in the refinery is solvent precipitation. Solvent precipitation occurs in a refinery in a deasphalting unit (Figure 3.14) and is essentially an extension of the procedure for separation by molecular weight, although some separation by polarity might also be operative. The deasphalting process is usually applied to the higher-molecular-weight fractions of petroleum such as atmospheric and vacuum residua.

These fractionation techniques can also be applied to cracked residua, asphalt, bitumen, and even to virgin petroleum, but in the last case the possibility of losses of the lower-boiling constituents is apparent; hence, the recommended procedure for virgin petroleum is, first, distillation followed by fractionation of the residua.

The simplest application of solvent extraction consists of mixing petroleum with another liquid, which results in the formation of two phases. This causes distribution of the petroleum constituents over the two phases; the dissolved portion (the extract) and the nondissolved portion (the raffinate).

The ratio of the concentration of any particular component in the two phases is known as the distribution coefficient K:

$$K = \frac{C_1}{C_2}$$

where C_1 and C_2 are the concentrations of the components in the various phases. The distribution coefficient is usually constant and may vary only slightly, if at all, with the concentration of the other components. In fact, the distribution coefficients may differ for the various components of the mixture to such an extent that the ratio of the concentrations of the various components in the

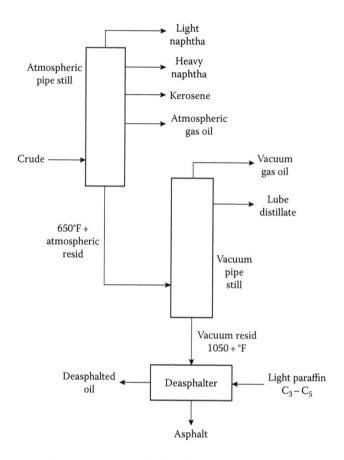

FIGURE 3.14 Placement of a deasphalting unit in the refinery (see also Figure 3.12).

solvent phase differs from that in the original petroleum; this is the basis for solvent extraction procedures.

It is generally molecular type, not molecular size, which is responsible for the solubility of species in various solvents. Thus, solvent extraction separates petroleum fractions according to type, although within any particular series there is a separation according to molecular size. Lower-molecular-weight hydrocarbons of a series (the light fraction) may well be separated from their higher-molecular-weight homologues (the heavy fraction) by solvent extraction procedures.

In general, it is advisable that selective extraction be employed with fairly narrow boiling range fractions. However, the separation achieved after one treatment with the solvent is rarely complete, and several repetitions of the treatment are required. Such repetitious treatments are normally carried out by the movement of the liquids countercurrently through the extraction equipment (*countercurrent extraction*), which affords better yields of the extractable materials.

The list of compounds that have been suggested as selective solvents for the preferential extraction fractionation of petroleum contains a large selection of different functional types (Table 3.5). However, before any extraction process is attempted, it is necessary to consider the following criteria: (1) the differences in the solubility of the petroleum constituents in the solvent should be substantial, (2) the solvent should be significantly less dense or denser than the petroleum (product) to be separated to allow easier countercurrent flow of the two phases, and (3) separation of the solvent from the extracted material should be relatively easy.

It may also be advantageous to consider other properties, such as viscosity, surface tension, and the like, as well as the optimal temperature for the extraction process. Thus, aromatics can be

TABLE 3.5
Compound Types Used for the
Selective Extraction of Petroleum

Esters	$R-C\overset{\displaystyle O}{\underset{OR'}{\big\langle}}$
Alcohols	$R-OH$
Aldehydes	$R-C\overset{\displaystyle O}{\underset{H}{\big\langle}}$
Acids	$R-C\overset{\displaystyle O}{\underset{OH}{\big\langle}}$
Ketones	$R-\overset{\displaystyle O}{\underset{\displaystyle \|}{C}}-R'$
Amines	$R-NH_2$
Amides	$R-C\overset{\displaystyle O}{\underset{NH_2}{\big\langle}}$
Nitrocompounds	$R-NO_2$
Nitriles	$R-CN$

Note: R and R′ are alkyl aromatic radicals, and if both are alkyl or both are aromatic they may or may not be the same.

separated from naphthene and paraffinic hydrocarbons by the use of selective solvents. Furthermore, aromatics with differing numbers of aromatic rings that may exist in various narrow boiling fractions can also be effectively separated by solvent treatment.

The separation of crude oil into two fractions—(1) the asphaltene fraction and (2) the maltene fraction—is conveniently brought about by the use of low-molecular-weight paraffin hydrocarbons, which were recognized to have selective solvency for hydrocarbons, and simply relatively low-molecular-weight hydrocarbon derivatives. The more complex, the higher the molecular weight value of compounds that are precipitated particularly well by the addition of 40 volumes of *n*-pentane or *n*-heptane in the methods generally preferred at present (Speight et al., 1984; Speight, 1994) although hexane is used occasionally (Yan et al., 1997). It is no doubt that there is a separation of the chemical components with the most complex structures from the mixture, and this fraction, which should correctly be called "*n*-pentane asphaltenes" or "*n*-heptane asphaltenes," is qualitatively and quantitatively reproducible (Figure 3.13).

Variation in the solvent type also causes significant changes in asphaltene yield. For example, in the ease of a Western Canadian bitumen and, indeed, for conventional crude oils, branched-chain paraffins or terminal olefins do not precipitate the same amount of asphaltenes as do the corresponding normal paraffins (Mitchell and Speight, 1973). Cycloparaffin derivatives (naphthene derivatives) have a remarkable effect on asphaltene yield and give results totally unrelated to those from any other nonaromatic solvent. For example, when cyclopentane, cyclohexane, or their methyl derivatives are employed as precipitating media, only approximately 1% w/w of the material remains insoluble.

To explain those differences, it was necessary to consider the solvent power of the precipitating liquid, which can be related to molecular properties (Hildebrand et al., 1970). The solvent power of nonpolar solvents has been expressed as a solubility parameter (δ) and equated to the internal pressure of the solvent, that is, the ratio between the surface tension (γ) and the cubic root of the molar volume (V):

$$\delta_1 = {}^3\sqrt{V}$$

Alternatively, the solubility parameter of nonpolar solvents can be related to the energy or vaporization ΔR^v and the molar volume:

$$\delta_2 = \left(\frac{\Delta E^v}{V}\right)^{1/2v}$$

Also

$$\delta_2 = \left(\Delta H^v - \frac{RT}{V}\right)^{1/2}$$

where
    ΔH^v is the heat of vaporization
    R is the gas constant
    T is the absolute temperature

The introduction of a polar group (heteroatom function) into the molecule of the solvent has significant effects on the quantity of precipitate. For example, treatment of a residuum with a variety of ethers or treatment of asphaltenes with a variety of solvents illustrates this point (Speight, 1979). In the latter instance, it was not possible to obtain data from addition of the solvent to the whole feedstock *per se* since the majority of the nonhydrocarbon materials were not miscible with the feedstock. It is nevertheless interesting that, as with the hydrocarbons, the amount of precipitate, or asphaltene solubility, can be related to the solubility parameter.

The solubility parameter allows an explanation of certain apparent anomalies, for example, the insolubility of asphaltenes in pentane and the near complete solubility of the materials in cyclopentane. Moreover, the solvent power of various solvents is in agreement with the derivation of the solubility parameter; for any one series of solvents, the relationship between the amount of precipitate (or asphaltene solubility) and the solubility parameter δ is quite regular.

In any method used to isolate asphaltenes as a separate fraction, standardization of the technique is essential. For many years, the method of asphaltenes separation was not standardized, and even now it remains subject to the preferences of the standard organizations of different countries. The use of both *n*-pentane and *n*-heptane has been widely advocated, and although *n*-heptane is becoming the deasphalting liquid of choice, this is by no means a hard-and-fast rule. And it must be recognized that large volumes of solvent may be required to effect a reproducible separation, similar to amounts required for consistent asphaltene separation. It is also preferable that the solvents be of sufficiently low boiling point that complete removal of the solvent from the fraction can be effected and, most important, the solvent must not react with the feedstock. Hence, the preference for the use of hydrocarbon liquids in standard methods, although the several standard methods that have been used are not unanimous in the ratio of hydrocarbon liquid to feedstock.

Method	Deasphalting Liquid	Volume (mL/g)
ASTM D893 (2015)	*n*-Pentane	10
ASTM D2007 (2015)	*n*-Pentane	10
ASTM D3279 (2015)	*n*-Heptane	100
ASTM D4124 (2015)	*n*-Heptane	100

However, it must be recognized that some of these methods were developed for use with feedstocks other than heavy oil and adjustments are necessary.

Although *n*-pentane and *n*-heptane are the solvents of choice in the laboratory, other solvents can be used (Speight, 1979) and cause the separation of asphaltenes as brown-to-black powdery materials. In the refinery, supercritical low-molecular-weight hydrocarbons (e.g., liquid propane, liquid butane, or mixtures of both) are the solvents of choice and the product is a semisolid (tacky) to solid asphalt. The amount of asphalt that settles out of the paraffin/residuum mixture depends on the size of the paraffin, the temperature, and the paraffin-to-feedstock ratio (Corbett and Petrossi, 1978; Speight et al., 1984; Speight, 2014, 2015).

Insofar as industrial solvents are very rarely one compound, it was also of interest to note that the physical characteristics of two different solvent types, in this case benzene and *n*-pentane, are additive on a mole-fraction basis (Mitchell and Speight, 1973) and also explain the variation of solubility with temperature. The data also show the effects of blending a solvent with the bitumen itself and allowing the resulting solvent–heavy oil blend to control the degree of bitumen solubility. Varying proportions of the hydrocarbon alter the physical characteristics of the oil to such an extent that the amount of precipitate (asphaltenes) can be varied accordingly within a certain range.

At constant temperature, the quantity of precipitate first increases with the increasing ratio of solvent to feedstock and then reaches a maximum (Speight et al., 1984). In fact, there are indications that when the proportion of solvent in the mix is <35% little or no asphaltenes are precipitated. In addition, when pentane and the lower-molecular-weight hydrocarbon solvents are used in large excess, the quantity of precipitate and the composition of the precipitate change with increasing temperature (Mitchell and Speight, 1973).

Contact time between the hydrocarbon and the feedstock also plays an important role in asphaltene separation (Speight et al., 1984). Yields of the asphaltenes reach a maximum after approximately 8 hours, which may be ascribed to the time required for the asphaltene particles to agglomerate into particles of a *filterable size* as well as the diffusion-controlled nature of the process. Heavier feedstocks also need time for the hydrocarbon to penetrate their mass.

After removal of the asphaltene fraction, further fractionation of petroleum is also possible by variation of the hydrocarbon solvent. For example, liquefied gases, such as propane and butane, precipitate as much as 50% by weight of the residuum or bitumen. The precipitate is a black, tacky, semisolid material, in contrast to the pentane-precipitated asphaltenes, which are usually brown, amorphous solids. Treatment of the propane precipitate with pentane then yields the insoluble brown, amorphous asphaltenes and soluble, near-black, semisolid resins, which are, as near as can be determined, equivalent to the resins isolated by adsorption techniques (Speight, 2014, 2015).

3.5.2 Adsorption Methods

Separation by adsorption chromatography essentially commences with the preparation of a porous bed of finely divided solid, the adsorbent. The adsorbent is usually contained in an open tube (column chromatography); the sample is introduced at one end of the adsorbent bed and induced to flow through the bed by means of a suitable solvent. As the sample moves through the bed, the various components are held (adsorbed) to a greater or lesser extent depending on the chemical nature of the component. Thus, those molecules that are strongly adsorbed spend considerable time on the adsorbent surface rather than in the moving (solvent) phase, but components that are slightly adsorbed move through the bed comparatively rapidly.

It is essential that, before the application of the adsorption technique to the petroleum, the asphaltenes first be completely removed, for example, by any of the methods outlined in the previous section. The prior removal of the asphaltenes is essential insofar as they are usually difficult to remove from the earth or clay and may actually be irreversibly adsorbed on the adsorbent.

By definition, the *saturate fraction* consists of paraffins and cycloparaffin derivatives (naphthene derivatives). The single-ring *naphthene derivatives*, or *cycloparaffin derivatives*, present in petroleum are primarily alkyl-substituted cyclopentane and cyclohexane rings. The alkyl groups are usually quite short, with methyl, ethyl, and isopropyl groups the predominant substituents. As the molecular weight of the naphthenes increases, the naphthene fraction contains more condensed rings with six-membered rings predominating. However, five-membered rings are still present in the complex higher-molecular-weight molecules.

The *aromatic fraction* consists of those compounds containing an aromatic ring and varies from *monoaromatics* (containing one benzene ring in a molecule) to *diaromatics* (substituted naphthalene) to *triaromatics* (substituted phenanthrene). Higher condensed ring systems (*tetra-aromatics*, *penta-aromatics*) are also known but are somewhat less prevalent than the lower ring systems, and each aromatic type will have increasing amounts of condensed ring naphthene attached to the aromatic ring as molecular weight is increased.

However, depending upon the adsorbent employed for the separation, a compound having an aromatic ring (i.e., six-carbon aromatic atoms) carrying side chains consisting *in toto* of more than six-carbon atoms (i.e., more than six-carbon nonaromatic atoms) will appear in the aromatic fraction.

Careful monitoring of the experimental procedures and the nature of the adsorbent has been responsible for the successes achieved with this particular technique. Early procedures consisted of warming solutions of the petroleum fraction with the adsorbent and subsequent filtration. This procedure has continued to the present day, and separation by adsorption is used commercially in plant operations in the form of clay treatment of crude oil fractions and products. In addition, the proportions of each fraction are subject to the ratio of adsorbent to deasphalted oil.

It is also advisable, once a procedure using a specific adsorbent has been established, that the same type of adsorbent be employed for future fractionation since the ratio of the product fractions varies from adsorbent to adsorbent. It is also very necessary that the procedure be used with caution and that the method not only be reproducible but quantitative recoveries be guaranteed; reproducibility with only, say, 85% of the material recoverable is not a criterion of success.

There are two procedures that have received considerable attention over the years and these are (1) the method by the U.S. Bureau of Mines—American Petroleum Institute (USBM-API) method and (2) the saturates–aromatics–resin constituents–asphaltene constituents (SARA) method. This latter method is often also called the saturates–aromatics–polar constituents–asphaltene constituents (SAPA) method. These two methods are used as representing the standard methods of petroleum fractionation. Other methods are also noted, especially when the method has added further meaningful knowledge to compositional studies (Speight, 2014, 2015).

However, there are precautions that must be taken when attempting to separate heavy feedstocks (heavy oil, tar sand bitumen) or polar feedstocks into constituent fractions. The disadvantages in using ill-defined adsorbents are that adsorbent performance differs with the same feed and, in certain instances, may even cause chemical and physical modification of the feed constituents. The use of a chemical reactant like sulfuric acid should only be advocated with caution since feedstocks react differently and may even cause irreversible chemical changes and/or emulsion formation (Abdel-Aal et al., 2016). These advantages may be of little consequence when it is not, for various reasons, the intention to recover the various product fractions *in toto* or in the original state, but in terms of the compositional evaluation of different feedstocks the disadvantages are very real.

In summary, the terminology used for the identification of the various methods might differ. However, in general terms, group-type analysis of petroleum is often identified by the acronyms for the names: PONA (paraffins, olefins, naphthenes, and aromatics), PIONA (paraffins, *iso*-paraffins, olefins, naphthenes, and aromatics), PNA (paraffins, naphthenes, and aromatics), PINA (paraffins, *iso*-paraffins, naphthenes, and aromatics), or SARA (saturates, aromatics, resins, and asphaltenes). However, it must be recognized that the fractions produced by the use of different adsorbents will differ in content and will also be different from fractions produced by solvent separation techniques.

The variety of fractions isolated by these methods and the potential for the differences in composition of the fractions makes it even more essential that the method is described accurately and that it is reproducible not only in any one laboratory but also between various laboratories.

3.5.3 CHEMICAL METHODS

Methods of fractionation using chemical reactants are entirely different in nature from the methods described in the preceding sections. Although several methods using chemical reactants have been applied to fractionation, methods such as adsorption, solvent treatment, and treatment with alkali (Speight, 2014, 2015) are often applied to product purification as well as separation.

The method of chemical separation commonly applied to separate crude oil into various fractions is treatment with sulfuric acid and since this method has also been applied in the refinery but with limited success in the fractionation of heavy oil and/or bitumen due to the formation of complex sulfates and difficult-to-break emulsions (Speight, 2014, 2015). Obviously, the success of this fractionation method is feedstock dependent and, in conclusion, it would appear that the test be left more as a method of product cleaning for which it was originally designed rather than a method of separation of the various fractions.

3.6 USE OF THE DATA

In the simplest sense, crude oil is a composite of four major fractions that are defined by the method of separation (Figure 3.13) (Speight, 2014, 2015), but, more important, the behavior and properties of any feedstock are dictated by composition (Speight, 2014, 2015). Although early studies were primarily focused on the composition and behavior of asphalt, the techniques developed for those investigations have provided an excellent means of studying heavy feedstocks (Tissot, 1984). Later studies have focused not only on the composition of petroleum and its major operational fractions but on further fractionation that allows different feedstocks to be compared on a relative basis and to provide a very simple but convenient feedstock *map*.

Such a map does not give any indication of the complex interrelationships of the various fractions (Koots and Speight, 1975), although predictions of feedstock behavior are possible using such data. It is necessary to take the composition studies one step further using subfractionation of the major fractions to obtain a more representative indication of petroleum composition.

Thus, by careful selection of an appropriate technique, it is possible to obtain an overview of petroleum composition that can be used for behavioral predictions. By taking the approach one step further and by assiduous collection of various subfractions from the original bulk fractions (Figure 3.13), it becomes possible to develop the petroleum map and add an extra dimension to compositional studies (Figure 3.15). Petroleum and heavy feedstocks then appear more as a continuum than as four specific fractions. Such a concept has also been applied to the asphaltene fraction of petroleum in which asphaltenes are considered a complex state of matter based on molecular weight and polarity (Long, 1979, 1981; Speight, 1994).

Furthermore, petroleum can be viewed as consisting of two continuous distributions, one of molecular weight and the other of molecular type. Using data from molecular weight studies and elemental analyses, the number of nitrogen and sulfur atoms in the aromatic fraction and in the polar aromatic fraction can also be exhibited. These data showed that not only can every molecule in the resins and asphaltenes has more than one sulfur atom or more than one nitrogen atom but also some molecules probably contain both sulfur and nitrogen. As the molecular weight of the aromatic fraction decreases, the sulfur and nitrogen contents of the fractions also decrease. In contrast to the sulfur-containing molecules, which appear in both the naphthene aromatics and the polar aromatic fractions, the oxygen compounds present in the heavy fractions of petroleum are normally found in the polar aromatics fraction.

More recent work (Long and Speight, 1989) involved the development of a different type of compositional map using the molecular weight distribution and the molecular type distribution as coordinates. The separation involved the use of an adsorbent such as clay, and the fractions were characterized by *solubility parameter* as a measure of the polarity of the molecular types. The molecular weight distribution can be determined by gel permeation chromatography. Using these

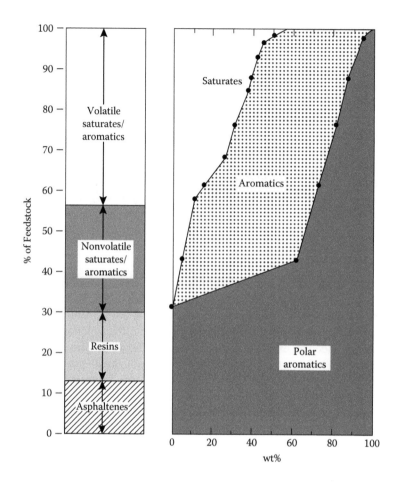

FIGURE 3.15 A petroleum *map* based on simple fractionation.

two distributions, a map of composition can be prepared using molecular weight and solubility parameter as the coordinates for plotting the two distributions. Such a composition map can provide insights into many separation and conversion processes used in petroleum refining.

The molecular type was characterized by the polarity of the molecules, as measured by the increasing adsorption strength on an adsorbent. At the time of the original concept, it was unclear how to characterize the continuum in molecular type or polarity. For this reason, the molecular type coordinate of their first maps was the yield of the molecular types ranked in order of increasing polarity. However, this type of map can be somewhat misleading because the areas are not related to the amounts of material in a given type. The horizontal distance on the plot is a measure of the yield, and there is not a continuous variation in polarity for the horizontal coordinate. It was suggested that the solubility parameter of the different fractions could be used to characterize both polarity and adsorption strength.

In order to attempt to remove some of these potential ambiguities, more recent developments of this concept have focused on the solubility parameter, estimated by the values for the eluting solvents that remove the fractions from the adsorbent. The simplest maps that can be derived using the solubility parameter are produced with the solubility parameters of the solvents used in solvent separation procedures and equating these parameters to the various fractions (Figure 3.16).

Thus, a composition map can be used to show where a particular physical or chemical property tends to concentrate on the map. For example, the *coke-forming propensity*, that is, the amount of the carbon residue, is shown for various regions on the map for a sample of atmospheric residuum (Figure 3.17)

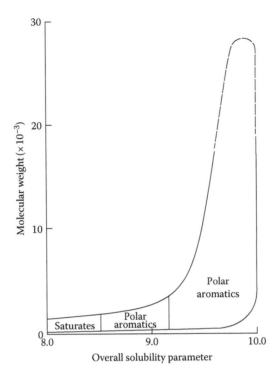

FIGURE 3.16 A petroleum *map* based on molecular weight and solubility parameter.

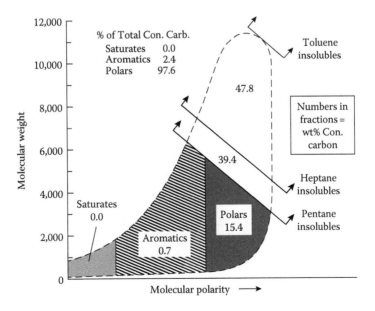

FIGURE 3.17 Property prediction using a petroleum *map*.

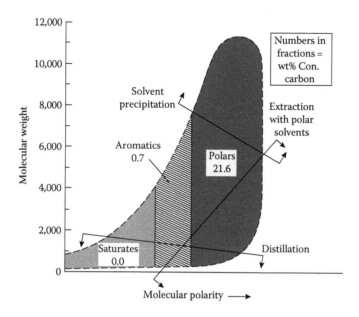

FIGURE 3.18 Separation efficiency using a petroleum *map*.

(Long and Speight, 1998). The plot shows molecular weight plotted against weight percent yield in order of increasing polarity. The dashed line is the envelope of composition of the total sample. The slanted lines show the boundaries of solvent-precipitated fractions, and the vertical lines show the boundaries of the fractions obtained by clay adsorption of the pentane-deasphalted oil.

A composition map can be very useful for predicting the effectiveness of various types of separations or conversions of petroleum (Figure 3.18) (Long and Speight, 1998). These processes are adsorption, distillation, solvent precipitation with relatively nonpolar solvents, and solvent extraction

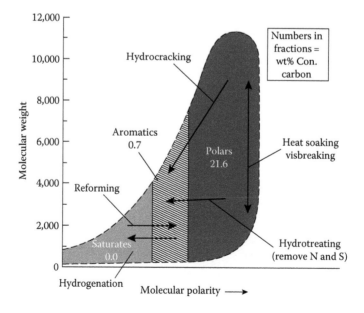

FIGURE 3.19 Mapping feedstock processability.

with polar solvents. The vertical lines show the cut points between saturates aromatics and polar aromatics as determined by clay chromatography. The slanted lines show how distillation, extraction, and solvent precipitation can divide the composition map. The line for distillation divides the map into distillate, which lies below the dividing line, and bottoms, which lies above the line. As the boiling point of the distillate is raised, the line moves upward, including higher-molecular-weight materials and more of the polar species in the distillate and rejecting lower-molecular-weight materials from the bottoms. As more of the polar species are included in the distillate, the carbon residue of the distillate rises. In contrast to the cut lines generated by separation processes, conversion processes move materials in the composition from one molecular type to another (Figure 3.19) (Long and Speight, 1998).

The ultimate decision in the choice of any particular fractionation technique must be influenced by the need for the data. For example, there are those needs that require only that the crude oil be separated into four bulk fractions. On the other hand, there may be the need to separate the crude oil into many subfractions in order to define specific compound types (Green et al., 1988; Vogh and Reynolds, 1988). Neither method is incorrect; each method is merely being used to answer the relevant questions about the character of the crude oil.

REFERENCES

Abdel-Aal, H.K., Aggour, M.A., and Fahim, M.A. 2016. *Petroleum and Gas Filed Processing.* CRC Press/ Taylor & Francis Group, Boca Raton, FL.

Aksenov, V.S. and Kanayanov, V.F. 1980. Regularities in composition and structures of native sulfur compounds from petroleum. *Proceedings of Ninth International Symposium on Organic Sulfur Chemistry.* Riga, USSR, June 9–14.

Altgelt, K.H. and Boduszynski, M.M. 1994. *Compositional Analysis of Heavy Petroleum Fractions.* Marcel Dekker, New York.

ASTM D893. 2015. *Standard Test Method for Insolubles in Used Lubricating Oils.* Annual Book of Standards. ASTM International, West Conshohocken, PA.

ASTM D1018. 2015. *Standard Test Method for Hydrogen in Petroleum Fractions.* Annual Book of Standards. ASTM International, West Conshohocken, PA.

ASTM D1266. 2015. *Standard Test Method for Sulfur in Petroleum Products (Lamp Method).* Annual Book of Standards. ASTM International, West Conshohocken, PA.

ASTM D1318. 2015. *Standard Test Method for Sodium in Residual Fuel Oil (Flame Photometric Method).* Annual Book of Standards. ASTM International, West Conshohocken, PA.

ASTM D1552. 2015. *Standard Test Method for Sulfur in Petroleum Products (High-Temperature Method).* Annual Book of Standards. ASTM International, West Conshohocken, PA.

ASTM D1757. 2015. *Standard Test Method for Sulfur in Ash from Coal and Coke.* Annual Book of Standards. ASTM International, West Conshohocken, PA.

ASTM D2007. 2015. *Standard Test Method for Characteristic Groups in Rubber Extender and Processing Oils and Other Petroleum-Derived Oils by the Clay-Gel Absorption Chromatographic Method.* Annual Book of Standards. ASTM International, West Conshohocken, PA.

ASTM D2622. 2015. *Standard Test Method for Sulfur in Petroleum Products by Wavelength Dispersive X-ray Fluorescence Spectrometry.* Annual Book of Standards. ASTM International, West Conshohocken, PA.

ASTM D3177. 2015. *Standard Test Methods for Total Sulfur in the Analysis Sample of Coal and Coke.* Annual Book of Standards. ASTM International, West Conshohocken, PA.

ASTM D3178. 2015. *Standard Test Methods for Carbon and Hydrogen in the Analysis Sample of Coal and Coke.* Annual Book of Standards. ASTM International, West Conshohocken, PA.

ASTM D3179. 2015. *Standard Test Methods for Nitrogen in the Analysis Sample of Coal and Coke.* Annual Book of Standards. ASTM International, West Conshohocken, PA.

ASTM D3228. 2015. *Standard Test Method for Total Nitrogen in Lubricating Oils and Fuel Oils by Modified Kjeldahl Method.* Annual Book of Standards. ASTM International, West Conshohocken, PA.

ASTM D3279. 2015. *Standard Test Method for n-Heptane Insolubles.* Annual Book of Standards. ASTM International, West Conshohocken, PA.

ASTM D3340. 2015. *Standard Test Method for Lithium and Sodium in Lubricating Greases by Flame Photometer.* Annual Book of Standards. ASTM International, West Conshohocken, PA.

ASTM D3341. 2015. *Standard Test Method for Lead in Gasoline—Iodine Monochloride Method*. Annual Book of Standards. ASTM International, West Conshohocken, PA.

ASTM D3343. 2015. *Standard Test Method for Estimation of Hydrogen Content of Aviation Fuels*. Annual Book of Standards. ASTM International, West Conshohocken, PA.

ASTM D3605. 2015. *Standard Test Method for Trace Metals in Gas Turbine Fuels by Atomic Absorption and Flame Emission Spectroscopy*. Annual Book of Standards. ASTM International, West Conshohocken, PA.

ASTM D3701. 2015. *Standard Test Method for Hydrogen Content of Aviation Turbine Fuels by Low Resolution Nuclear Magnetic Resonance Spectrometry*. Annual Book of Standards. ASTM International, West Conshohocken, PA.

ASTM D4045. 2015. *Standard Test Method for Sulfur in Petroleum Products by Hydrogenolysis and Rateometric Colorimetry*. Annual Book of Standards. ASTM International, West Conshohocken, PA.

ASTM D4124. 2015. *Standard Test Method for Separation of Asphalt into Four Fractions*. Annual Book of Standards. ASTM International, West Conshohocken, PA.

ASTM D4294. 2015. *Standard Test Method for Sulfur in Petroleum and Petroleum Products by Energy Dispersive X-ray Fluorescence Spectrometry*. Annual Book of Standards. ASTM International, West Conshohocken, PA.

ASTM E258. 2015. *Standard Test Method for Total Nitrogen in Organic Materials by Modified Kjeldahl Method*. Annual Book of Standards. ASTM International, West Conshohocken, PA.

ASTM E385. 2015. *Standard Test Method for Oxygen Content Using a 14-MeV Neutron Activation and Direct-Counting Technique*. Annual Book of Standards. ASTM International, West Conshohocken, PA.

ASTM E777. 2015. *Standard Test Method for Carbon and Hydrogen in the Analysis Sample of Refuse-Derived Fuel*. Annual Book of Standards. ASTM International, West Conshohocken, PA.

ASTM E778. 2015. *Standard Test Methods for Nitrogen in Refuse-Derived Fuel Analysis Samples*. Annual Book of Standards. ASTM International, West Conshohocken, PA.

Baker, E.W. 1969. In: *Organic Geochemistry*. G. Eglinton and M.T.J. Murphy (Editors). Springer-Verlag, New York.

Baker, E.W. and Louda, J.W. 1986. In: *Biological Markers in the Sedimentary Record*. R. B. Johns (Editor). Elsevier, Amsterdam, the Netherlands.

Baker, E.W. and Palmer, S.E. 1978. In: *The Porphyrins. Volume I. Structure and Synthesis. Part A*. D. Dolphin (Editor). Academic Press, New York.

Bestougeff, M.A. 1961. J. Études Methodes Separation Immediate Chromatogr. *Comptes Réndus (Paris)*, 55.

Bjorseth, A. 1983. *Handbook of Polycyclic Aromatic Hydrocarbons*. Marcel Dekker, New York.

Bonnett, R. 1978. In: *The Porphyrins. Volume I. Structure and Synthesis. Part A*. D. Dolphin (Editor). Academic Press, New York.

Brooks, J. and Welte, D.H. 1984. *Advances in Petroleum Geochemistry*. Academic Press, New York.

Charbonnier, R.P., Draper, R.G., Harper, W.H., and Yates, A. 1969. Analyses and Characteristics of Oil Samples from Alberta. Information Circular IC 232. Department of Energy Mines and Resources, Mines Branch, Ottawa, Ontario, Canada.

Corbett, L.W. and Petrossi, U. 1978. Differences in distillation and solvent separated asphalt residua. *Industrial & Engineering Chemistry Product Research and Development*, 17: 342.

Draper, R.G., Kowalchuk, E., and Noel, G. 1977. Analyses and characteristics of crude oil samples performed between 1969 and 1976. Report ERP/ERL 77-59 (TR). Energy, Mines, and Resources, Ottawa, Ontario, Canada.

Durand, B. 1980. *Kerogen: Insoluble Organic Matter from Sedimentary Rocks*. Editions Technip, Paris, France.

Eglinton, G. and Murphy, B. 1969. *Organic Geochemistry: Methods and Results*. Springer-Verlag, New York.

Felix, G., Bertrand, C., and Van Gastel, F. 1985. Hydroprocessing of heavy oils and residua. *Chromatographia*, 20: 155–160.

Filby, R.H. and Van Berkel, G.J. 1987. In: *Metal Complexes in Fossil Fuels*. R.H. Filby and J.F. Branthaver (Editors). Symposium Series No. 344. American Chemical Society, Washington, DC, p. 2.

Freidlina, I.K. and Skorova, A.E. (Editors). 1980. *Organic Sulfur Chemistry*. Pergamon Press, New York.

Gallegos, E.J.J. 1981. Alkylbenzenes derived from carotenes in coals by GC/MS. *Chromatographic Science*, 19: 177–182.

Green, J.A., Green, J.B., Grigsby, R.D., Pearson, C.D., Reynolds, J.W., Sbay, I.Y., Sturm, O.P. Jr. et al. 1989. *Analysis of Heavy Oils: Method Development and Application to Cerro Negro Heavy Petroleum*, Volumes I and II. NIPER-452 (DE90000200. Research Institute, National Institute for Petroleum and Energy Research (NIPER), Bartlesville, OK.

Green, J.B., Grizzle, P.L., Thomson, P.S., Shay, J.Y., Diehl, B.H., Hornung, K.W., and Sanchez, V. 1988. Report No. DE88 001235. Contract FC22-83F460149. U.S. Department of Energy, Washington, DC.

Hildebrand, J.H., Prausnitz, J.M., and Scott, R.L. 1970. *Regular Solutions*. Van Nostrand-Reinhold, New York.

Hirsch, D.E., Cooley, J.E., Coleman, H.J., and Thompson, C.J. 1974. Qualitative characterization of aromatic concentrates of crude oils from GPC analysis. Report 7974. Bureau of Mines, U.S. Department of the Interior, Washington, DC.

Hodgson, G.W., Baker, B.L., and Peake, E. 1967. In: *Fundamental Aspects of Petroleum Geochemistry*. B. Nagy and U. Columbo (Editors). Elsevier, Amsterdam, the Netherlands, Chapter 5.

Killops, S.D. and Readman, J.W. 1985. HPLC fractionation and GC-MS determination of aromatic hydrocarbons for oils and sediments. *Organic Geochemistry*, 8: 247–257.

Koots, J.A. and Speight, J.G. 1975. The relationship of petroleum resins to asphaltenes. *Fuel*, 54: 179.

Lee, M.L., Novotny, M.S., and Bartle, K.D. 1981. *Analytical Chemistry of Polycyclic Aromatic Compounds*. Academic Press Inc., New York.

Long, R.B. 1979. The concept of asphaltenes. Preprints. *American Chemical Society, Division of Petroleum Chemistry*, 24(4): 891.

Long, R.B. 1981. The concept of asphaltenes. In: *The Chemistry of Asphaltenes*. J.W. Bunger and N. Li (Editors). Advances in Chemistry Series No. 195. American Chemical Society, Washington, DC.

Long, R.B. and Speight, J.G. 1989. Studies in petroleum composition. I: Development of a compositional map for various feedstocks. *Revue de l'Institut Français du Pétrole*, 44: 205.

Long, R.B. and Speight, J.G. 1998. The composition of petroleum. In: *Petroleum Chemistry and Refining*. Taylor & Francis Group, Washington, DC, Chapter 1.

Mitchell, D.L. and Speight, J.G. 1973. The solubility of asphaltenes in hydrocarbon solvents. *Fuel*, 52: 149.

Monin, J.C. and Pelet, R. 1983. In: *Advances in Organic Geochemistry*. M. Bjorev (Editor). John Wiley & Sons, New York.

Pearson, C.D. and Green, J.B. 1993. Vanadium and nickel complexes in petroleum resid acid, base, and neutral fractions. *Energy & Fuels*, 7: 338–346.

Prince, R.O., Elmendoff, D.L., Lute, B.R., Hsu, C.S., Hath, C., Sunnis, B.P., Decherd, G., Douglas, D., and Butler, E. 1994. *Environmental Science and Technology*, 28: 142.

Quirke, J.M.E. 1987. In: *Metal Complexes in Fossil Fuels*. R.H. Filby and J.F. Branthaver (Editors). Symposium Series No. 344. American Chemical Society, Washington, DC, p. 74.

Rall, H.T., Thompson, C.J., Coleman H.J., and Hopkins, R.L. 1972. Sulfur compounds in crude oil. Bulletin 659. Bureau of Mines, U.S. Department of the Interior, Washington, DC.

Reynolds, J.G. 1998. Metals and heteroatoms in heavy crude oils. In: *Petroleum Chemistry and Refining*. J.G. Speight (Editor). Taylor & Francis Group, Washington, DC, Chapter 3.

Reynolds, J.G., Biggs, W.E., and Bezman, S.A. 1987. In: *Metal Complexes in Fossil Fuels*. R.H. Filby and J.F. Branthaver (Editors). Symposium Series No. 344. American Chemical Society, Washington, DC, p. 205.

Ritchie, R.G.S., Roche, R.S., and Steedman, W. 1979a. *Fuel*, 58: 523.

Ritchie, R.G.S., Roche, R.S., and Steedman, W. 1979b. *Chemistry and Industry*, 25.

Sakarnen, K.V. and Ludwig, C.H. 1971. *Lignins: Occurrence, Formation, Structure and Reactions*. John Wiley & Sons, New York.

Schucker, R.C. and Keweshan, C.F. 1980. Reactivity of Cold Lake asphaltenes. Preprints. *American Chemical Society, Division of Fuel Chemistry*, 25: 155.

Seifert, W.K. and Teeter, R.M. 1970. Identification of polycyclic aromatic and heterocyclic crude oil carboxylic acids. *Analytical Chemistry*, 42: 750–758.

Speight, J.G. 1971. Thermal cracking of athabasca bitumen, athabasca asphaltenes, and athabasca deasphalted heavy oil. *Fuel*, 49: 134.

Speight, J.G. 1979. *Studies on Bitumen Fractionation: (a) Fractionation by a Cryoscopic Method (b) Effect of Solvent Type on Asphaltene Solubility*. Information Series No. 84. Alberta Research Council, Edmonton, Alberta, Canada.

Speight, J.G. 1981. Asphaltenes as an organic natural product and their influence on crude oil properties. *Proceedings of Division of Geochemistry, American Chemical Society*. New York.

Speight, J.G. 1986. Polynuclear aromatic systems in petroleum. Preprints. *American Chemical Society, Division of Petroleum Chemistry*, 31(4): 818.

Speight, J.G. 1987. Initial reactions in the coking of residua. Preprints. *American Chemical Society, Division of Petroleum Chemistry*, 32(2): 413.

Speight, J.G. 1990. Tar sands. In: *Fuel Science and Technology Handbook*. Marcel Dekker, New York, Chapter 12.

Speight, J.G. 1994. Chemical and physical studies of petroleum asphaltenes. In: *Asphaltenes and Asphalts. I. Developments in Petroleum Science*, Volume 40. T.F. Yen and G.V. Chilingarian (Editors). Elsevier, Amsterdam, the Netherlands, Chapter 2.

Speight, J.G. 2000. *The Desulfurization of Heavy Oils and Residua*, 2nd Edition. Marcel Dekker, New York.

Speight, J.G. 2009. *Enhanced Recovery Methods for Heavy Oil and Tar Sands*. Gulf Publishing Company, Houston, TX.

Speight, J.G. 2011. *The Refinery of the Future*. Gulf Professional Publishing, Elsevier, Oxford, UK.

Speight, J.G. 2013. *The Chemistry and Technology of Coal*, 3rd Edition. CRC Press/Taylor & Francis Group, Boca Raton, FL.

Speight, J.G. 2014. *The Chemistry and Technology of Petroleum*, 5th Edition. CRC Press/Taylor & Francis Group, Boca Raton, FL.

Speight, J.G. 2015. *Handbook of Petroleum Product Analysis*, 2nd Edition. John Wiley & Sons, Hoboken, NJ.

Speight, J.G. 1990. *Handbook of Petroleum Analysis*. John Wiley & Sons, Hoboken, NJ.

Speight, J.G. and Francisco, M.A. 1990. Studies in petroleum composition IV: Changes in the nature of chemical constituents during crude oil distillation. *Revue de l'Institut Français du Pétrole*, 45: 733.

Speight, J.G., Long, R.B., and Trowbridge, T.D. 1984. Factors influencing the separation of asphaltenes from heavy petroleum feedstocks. *Fuel*, 63: 616.

Speight, J.G. and Pancirov, R.J. 1984. Structural types in asphaltenes as deduced from pyrolysis-gas chromatography-mass spectrometry. *Liquid Fuels Technology*, 2: 287.

Thompson, C.J., Ward, C.C., and Ball, J.S. 1976. Characteristics of World's Crude oils and Results of API Research Project 60. Report BERC/RI-76/8. Bartlesville Energy Technology Center, Bartlesville, OK.

Tissot, B.P. (Editor). 1984. *Characterization of Heavy Crude Oils and Petroleum Residues*. Editions Technip, Paris, France.

Tissot, B.P. and Welte, D.H. 1978. *Petroleum Formation and Occurrence*. Springer-Verlag, New York.

Vogh, J.W. and Reynolds, J.W. 1988. Report No. DE88 001242. Contract FC22-83FE60149. U.S. Department of Energy, Washington, DC.

Wallace, D., Starr, J., Thomas, K.P., and Dorrence, S.M. 1988. *Characterization of Oil Sands Resources*. Alberta Oil Sands Technology and Research Authority, Edmonton, Alberta, Canada.

Weiss, V. and Edwards, J.M. 1980. *The Biosynthesis of Aromatic Compounds*. John Wiley & Sons, New York.

Yan, J., Plancher, H., and Morrow, N.R. 1997. Wettability changes induced by adsorption of asphaltenes. Paper No. SPE 37232. *Proceedings of SPE International Symposium on Oilfield Chemistry*. Houston, TX. Society of Petroleum Engineers, Richardson, TX.

4 Introduction to Refining Processes

4.1 INTRODUCTION

In the crude state, petroleum has minimal value, but when refined, it provides high-value liquid fuels, solvents, lubricants, and many other products (Speight and Ozum, 2002; Parkash, 2003; Hsu and Robinson, 2006; Gary et al., 2007; Speight, 2011a,b, 2014). The fuels derived from petroleum contribute approximately one-third to one-half of the total world energy supply and are used not only for transportation fuels (i.e., gasoline, diesel fuel, and aviation fuel, among others) but also to heat buildings. Petroleum products have a wide variety of uses that vary from gaseous and liquid fuels to near-solid machinery lubricants. In addition, the residue of many refinery processes, asphalt—a once-maligned by-product—is now a premium value product for highway surfaces, roofing materials, and miscellaneous waterproofing uses.

In a very general sense, petroleum refining can be traced back over 5000 years to the times when asphalt materials and oils were isolated from areas where natural seepage occurred (Abraham, 1945; Forbes, 1958; Hoiberg, 1960). Any treatment of the asphalt (such as hardening in the air prior to use) or of the oil (such as allowing for more volatile components to escape prior to use in lamps) may be considered to be refining under the general definition of refining. However, petroleum refining as we know it is a very recent science and many innovations evolved during the twentieth century.

Briefly, petroleum refining is the separation of petroleum into fractions and the subsequent treating of these fractions to yield marketable products (Speight and Ozum, 2002; Parkash, 2003; Hsu and Robinson, 2006; Gary et al., 2007; Speight, 2011a,b, 2014). In fact, a refinery is essentially a group of manufacturing plants that vary in number with the variety of products produced (Figure 4.1). Refinery processes must be selected and products are manufactured to give a balanced operation in which petroleum is converted into a variety of products in amounts that are in accord with the demand for each. For example, the manufacture of products from the lower-boiling portion of petroleum automatically produces a certain amount of higher-boiling components. If the latter cannot be sold as, say, heavy fuel oil, these products will accumulate until refinery storage facilities are full. To prevent the occurrence of such a situation, the refinery must be flexible and be able to change operations as needed. This usually means more processes: thermal processes to change an excess of heavy fuel oil into more gasoline with coke as the residual product, or a vacuum distillation process to separate the heavy oil into lubricating oil blend stocks and asphalt.

As the basic elements of crude oil, hydrogen and carbon form the main input into a refinery, combining into thousands of individual constituents, and the economic recovery of these constituents varies with the individual petroleum according to its particular individual qualities and the processing facilities of a particular refinery. In general, crude oil, once refined, yields three basic groupings of products that are produced when it is broken down into cuts or fractions (Table 4.1). The gas and gasoline cuts form the lower-boiling products and are usually more valuable than the higher-boiling fractions and provide gas (liquefied petroleum gas), naphtha, aviation fuel, motor fuel, and feedstocks for the petrochemical industry (Table 4.2). Naphtha, a precursor to gasoline and solvents, is produced from the light and middle range of distillate cuts (sometimes referred to collectively as light gas oil) and is also used as a feedstock for the petrochemical industry (Table 4.3). The middle distillates refer to products from the middle boiling range of petroleum and include kerosene, diesel

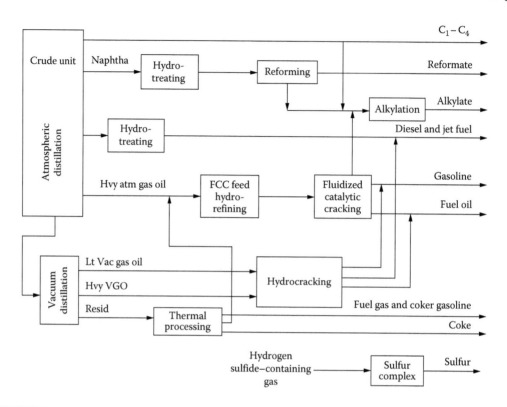

FIGURE 4.1   Schematic overview of a refinery.

TABLE 4.1
Boiling Fractions of Crude Petroleum

	Boiling Range[a]	
Fraction	(°C)	(°F)
Light naphtha	−1–150	30–300
Gasoline	−1–180	30–355
Heavy naphtha	150–205	300–400
Kerosene	205–260	400–500
Light gas oil	260–315	400–600
Heavy gas oil	315–425	600–800
Lubricating oil	>400	>750
Vacuum gas oil	425–600	800–1100
Residuum	>510	>950

[a] For convenience, boiling ranges are converted to the nearest 5°.

fuel, distillate fuel oil, and light gas oil. Waxy distillate and lower-boiling lubricating oils are some-times included in the middle distillates. The remainder of the crude oil includes the higher-boiling lubricating oils, gas oil, and residuum (the nonvolatile fraction of the crude oil). The residuum can also produce heavy lubricating oils and waxes but is more often sued for asphalt production. The complexity of petroleum is emphasized insofar as the actual proportions of light, medium, and heavy fractions vary significantly from one crude oil to another.

TABLE 4.2
Production of Starting Materials for Petrochemical Processing

Starting Feedstock	Process	Product
Petroleum	Distillation	Light ends
		Methane
		Ethane
		Propane
		Butane
	Catalytic cracking	Ethylene
		Propylene
		Butylenes
		Higher olefins
	Coking	Ethylene
		Propylene
		Butylenes
		Higher olefins
Natural gas	Refining/gas processing	Methane
		Ethane
		Propane
		Butane

TABLE 4.3
Sources of Naphtha

Process	Primary Product	Secondary Process	Secondary Product
Atmospheric distillation	Light naphtha	Cracking	Petrochemical feedstocks
	Heavy naphtha	Catalytic cracking	Light naphtha
	Gas oil	Catalytic cracking	Light naphtha
	Gas oil	Hydrocracking	Light naphtha
Vacuum distillation	Gas oil	Catalytic cracking	Light naphtha
		Hydrocracking	Light naphtha
	Residuum	Coking	Light naphtha
		Hydrocracking	Light naphtha

The refining industry has been the subject of the four major forces that affect most industries and that have hastened the development of new petroleum refining processes: (1) the demand for products such as gasoline, diesel, fuel oil, and jet fuel; (2) feedstock supply, specifically the changing quality of crude oil and geopolitics between different countries and the emergence of alternate feed supplies such as bitumen from tar sand, natural gas, and coal; (3) environmental regulations that include more stringent regulations in relation to sulfur in gasoline and diesel; and (4) technology development such as new catalysts and processes.

In the early days of the twentieth century, refining processes were developed to extract kerosene for lamps. Any other products were considered to be unusable and were usually discarded. Thus, the first refining processes were developed to purify, stabilize, and improve the quality of kerosene. However, the invention of the internal combustion engine led (at approximately the time of World War I) to a demand for gasoline for use in increasing quantities as a motor fuel for cars and trucks.

This demand on the lower-boiling products increased, particularly when the market for aviation fuel developed. Thereafter, refining methods had to be constantly adapted and improved to meet the quality requirements and needs of car and aircraft engines.

Since then, the general trend throughout refining has been to produce more products from each barrel of petroleum and to process those products in different ways to meet the product specifications for use in modern engines. Overall, the demand for gasoline has rapidly expanded, and demand has also developed for gas oils and fuels for domestic central heating and fuel oil for power generation, as well as for light distillates and other inputs, derived from crude oil, for the petrochemical industries.

As the need for the lower-boiling products developed, petroleum yielding the desired quantities of the lower-boiling products became less available, and refineries had to introduce conversion processes to produce greater quantities of lighter products from the higher-boiling fractions. The means by which a refinery operates in terms of producing the relevant products depends not only on the nature of the petroleum feedstock but also on its configuration (i.e., the number of types of the processes that are employed to produce the desired product slate), and the refinery configuration is, therefore, influenced by the specific demands of a market. Therefore, refineries need to be constantly adapted and upgraded to remain viable and responsive to ever-changing patterns of crude supply and product market demands. As a result, refineries have been introducing increasingly complex and expensive processes to gain higher yields of lower-boiling products from the higher-boiling fractions and residua.

To convert crude oil into desired products in an economically feasible and environmentally acceptable manner, refinery processes for crude oil are generally divided into three categories: (1) separation processes, of which distillation is the prime example; (2) conversion processes, of which coking and catalytic cracking are prime example; and (3) finishing processes, of which hydrotreating to remove sulfur is a prime example.

The simplest refinery configuration is the *topping refinery*, which is designed to prepare feedstocks for petrochemical manufacture or for the production of industrial fuels. The topping refinery consists of tankage, a distillation unit, recovery facilities for gases and light hydrocarbons, and the necessary utility systems (steam, power, and water-treatment plants). Topping refineries produce large quantities of unfinished oils and are highly dependent on local markets, but the addition of hydrotreating and reforming units to this basic configuration results in a more flexible *hydroskimming refinery*, which can also produce desulfurized distillate fuels and high-octane gasoline. These refineries may produce up to half of their output as residual fuel oil, and they face increasing market loss as the demand for low-sulfur (even no-sulfur) high-sulfur fuel oil increases.

The most versatile refinery configuration is the *conversion refinery*. A conversion refinery incorporates all the basic units found in both the topping and hydroskimming refineries, but it also features gas oil conversion plants such as catalytic cracking and hydrocracking units, olefin conversion plants such as alkylation or polymerization units, and, frequently, coking units for sharply reducing or eliminating the production of residual fuels. Modern conversion refineries may produce two-thirds of their output as unleaded gasoline, with the balance distributed between liquefied petroleum gas, jet fuel, diesel fuel, and a small quantity of coke. Many such refineries also incorporate solvent extraction processes for manufacturing lubricants and petrochemical units with which to recover propylene, benzene, toluene, and xylenes for further processing into polymers.

Finally, the yields and quality of refined petroleum products produced by any given oil refinery depend on the mixture of crude oil used as feedstock and the configuration of the refinery facilities. Light/sweet crude oil is generally more expensive and has inherent great yields of higher-value low-boiling products such as naphtha, gasoline, jet fuel, kerosene, and diesel fuel. Heavy sour crude oil is generally less expensive and produces greater yields of lower-value higher-boiling products that must be converted into lower-boiling products (Speight, 2013, 2014).

Since a refinery is a group of integrated manufacturing plants (Figure 4.1) that are selected to give a balanced production of saleable products in amounts that are in accord with the demand for each, it is necessary to prevent the accumulation of nonsaleable products, and the refinery must be flexible and be able to change operations as needed. The complexity of petroleum is emphasized insofar as the actual amounts of the products vary significantly from one crude oil to another (Speight, 2014, 2016). In addition, the configuration of refineries may vary from refinery to refinery. Some refineries may be more oriented toward the production of gasoline (large reforming and/or catalytic cracking), whereas the configuration of other refineries may be more oriented toward the production of middle distillates such as jet fuel and gas oil.

This chapter presents an introduction to petroleum refining in order for the reader to place each process in the correct context of the refinery.

4.2 DEWATERING AND DESALTING

Petroleum is recovered from the reservoir mixed with a variety of substances: gases, water, and dirt (minerals). Thus, refining actually commences with the production of fluids from the well or reservoir and is followed by pretreatment operations that are applied to the crude oil either at the refinery or prior to transportation. Pipeline operators, for instance, are insistent upon the quality of the fluids put into the pipelines; therefore, any crude oil to be shipped by pipeline or, for that matter, by any other form of transportation must meet rigid specifications in regard to water and salt content. In some instances, sulfur content, nitrogen content, and viscosity may also be specified. Field separation, which occurs at a field site near the recovery operation, is the first attempt to remove the gases, water, and dirt that accompany crude oil coming from the ground. The separator may be no more than a large vessel that gives a quieting zone for gravity separation into three phases: gases, crude oil, and water containing entrained dirt.

Thus, the first step in petroleum processing, even before the crude oil enters the refinery, occurs at the wellhead (Abdel-Aal et al., 2016). It is at this stage that fluids from the well are separated into crude oil, natural gas, and water phases using a gas–oil separator. The separators can be horizontal, vertical, or spherical and are generally classified into two types based on the number of phases to separate: (1) two-phase separators, which are used to separate gas from oil in oil fields or gas from water for gas fields, and (2) three-phase separators, which are used to separate the gas from the liquid phase and water from oil. The liquid (oil, emulsion) leaves at the bottom through a level control or an exit valve. The gas leaves the vessel at the top, passing through a mist extractor to remove the small liquid droplets in the gas. Separators can also be categorized according to their operating pressure: (1) low-pressure units can tolerate pressures on the order of 10–180 psi, (2) medium-pressure separators operate from 230 to 700 psi, and (3) high-pressure units can tolerate pressures of 975–1500 psi. Even after this type of separation and before separation of petroleum into its various constituents can proceed, there is the need to clean the petroleum. This is often referred to as desalting and dewatering in which the goal is to remove water and the constituents of the brine that accompany the crude oil from the reservoir to the wellhead during recovery operations.

Desalting is a water-washing operation performed at the production field and at the refinery site for additional crude oil cleanup (Figure 4.2). If the petroleum from the separators contains water and dirt, water washing can remove much of the water-soluble minerals and entrained solids. If these crude oil contaminants are not removed, they can cause operating problems during refinery processing, such as equipment plugging and corrosion as well as catalyst deactivation.

The usual practice is to blend crude oils of similar characteristics, although fluctuations in the properties of the individual crude oils may cause significant variations in the properties of the blend over a period of time. Blending several crude oils prior to refining can eliminate the frequent need to change the processing conditions that may be required to process each of the crude oils individually.

However, simplification of the refining procedure is not always the end result. Incompatibility of different crude oils, which can occur if, for example, a paraffinic crude oil is blended with heavy asphaltic oil, can cause sediment formation in the unrefined feedstock or in the products, thereby complicating the refinery process.

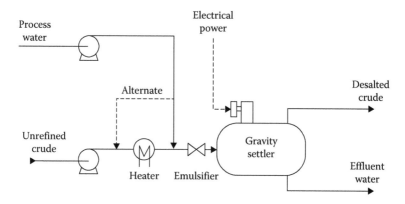

FIGURE 4.2 An electrostatic desalting unit. (From OSHA Technical Manual, Section IV, Chapter 2, Petroleum refining processes, 1999, http://www.osha.gov/dts/osta/otm/otm_iv/otm_iv_2.html.)

4.3 DISTILLATION

In the early stages of refinery development, when illuminating and lubricating oils were the main products, distillation was the major, and often only, refinery process. At that time, gasoline was a minor product but, as the demand for gasoline increased, conversion processes were developed, because distillation could no longer supply the necessary quantities.

It is possible to obtain products ranging from gaseous materials taken off at the top of the distillation column to a nonvolatile residue or reduced crude (*bottoms*), with correspondingly lighter materials at intermediate points. The reduced crude may then be processed by vacuum, or steam, distillation in order to separate the high-boiling lubricating oil fractions without the danger of decomposition, which occurs at high (>350°C, >660°F) temperatures. Atmospheric distillation may be terminated with a lower-boiling fraction (*cut*) if it is felt that vacuum or steam distillation will yield a better-quality product or if the process appears to be economically more favorable. Not all crude oils yield the same distillation products (Table 4.1)—although there may be variations by several degrees in the boiling ranges of the fractions as specified by different companies—and the nature of the crude oil dictates the processes that may be required for refining.

Distillation was the first method by which petroleum was refined. The original technique involved a batch operation in which the still was a cast-iron vessel mounted on brickwork over a fire and the volatile materials were passed through a pipe or gooseneck that led from the top of the still to a condenser. The latter was a coil of pipe (*worm*) immersed in a tank of running water.

Heating a batch of crude petroleum caused the more volatile, lower-boiling components to vaporize and then condense in the worm to form naphtha. As the distillation progressed, the higher-boiling components became vaporized and were condensed to produce kerosene: the major petroleum product of the time. When all of the possible kerosene had been obtained, the material remaining in the still was discarded. The still was then refilled with petroleum and the operation repeated.

The capacity of the stills at that time was usually several barrels of petroleum and if often required 3 or more days to distill (*run*) a batch of crude oil. The simple distillation as practiced in the 1860s and 1870s was notoriously inefficient. The kerosene was more often than not contaminated by naphtha, which distilled during the early stages, or by heavy oil, which distilled from the residue during the final stages of the process. The naphtha generally rendered the kerosene so flammable explosions accompanied that ignition. On the other hand, the presence of heavier oil adversely affected the excellent burning properties of the kerosene and created a great deal of smoke. This condition could be corrected by redistilling (*rerunning*) the kerosene, during which process the more volatile fraction (*front end*) was recovered as additional naphtha, while the kerosene residue (*tail*) remaining in the still was discarded.

The 1880s saw the introduction of the continuous distillation of petroleum. The method employed a number of stills coupled together in a row, and each still was heated separately and was hotter than the preceding one. The stills were arranged so that oil flowed by gravity from the first to the last. Crude petroleum in the first still was heated so that a light naphtha fraction distilled from it before the crude petroleum flowed into the second still, where a higher temperature caused the distillation of a heavier naphtha fraction. The residue then flowed to the third still where an even higher temperature caused kerosene to distill. The oil thus progressed through the battery to the last still, where destructive distillation (thermal decomposition; cracking) was carried out to produce more kerosene. The residue from the last still was removed continuously for processing into lubricating oils or for use as fuel oil.

In the early 1900s, a method of partial (or selective) condensation was developed to allow a more exact separation of petroleum fractions. A partial condenser was inserted between the still and the conventional water-cooled condenser. The lower section of the tower was packed with stones and insulated with brick so that the heavier less volatile material entering the tower condensed and drained back into the still. Noncondensed material passed into another section where more of the less volatile material was condensed on air-cooled tubes, and the condensate was withdrawn as a petroleum fraction. The noncondensable (overhead) material from the air-cooled section entered a second tower that also contained air-cooled tubes and often produced a second fraction. The volatile material remaining at this stage was then condensed in a water-cooled condenser to yield a third fraction. The van Dyke tower is essentially one of the first stages in a series of improvements that ultimately led to the distillation units found in modern refineries, which separate petroleum fractions by fractional distillation.

4.3.1 Distillation at Atmospheric Pressure

The present-day petroleum distillation unit is, like the battery of the 1800s, a collection of distillation units but, in contrast to the early battery units, a tower is used in the typical modern refinery (Figure 4.3) and brings about a fairly efficient degree of fractionation (separation).

The feed to a distillation tower is heated by flow through pipes arranged within a large furnace. The heating unit is known as a pipe still heater or pipe still furnace, and the heating unit and the fractional distillation tower make up the essential parts of a distillation unit or pipe still. The pipe still furnace heats the feed to a predetermined temperature—usually a temperature at which a predetermined portion of the feed will change into vapor. The vapor is held under pressure in the pipe in the furnace until it discharges as a foaming stream into the fractional distillation tower. The unvaporized or liquid portion of the feed descends to the bottom of the tower to be pumped away as a bottom nonvolatile product, while the vaporized material passes up the tower to be fractionated into gas oils, kerosene, and naphtha.

Pipe still furnaces vary greatly and, in contrast to the early units where capacity was usually 200–500 bbl per day, can accommodate 25,000 bbl or more of crude petroleum per day. The walls and ceiling are insulated with firebrick, and the interior of the furnace is partially divided into two sections: a smaller convection section where the oil first enters the furnace and a larger section (fitted with heaters) and where the oil reaches its highest temperature.

Another twentieth-century innovation in distillation is the use of heat exchangers that are also used to preheat the feed to the furnace. These exchangers are bundles of tubes arranged within a shell so that a feedstock passes through the tubes in the opposite direction a heated feedstock passing through the shell. By this means, cold crude oil is passed through a series of heat exchangers where hot products from the distillation tower are cooled before entering the furnace and as a heated feedstock. This results in a saving of heater fuel and is a major factor in the economical operation of modern distillation units.

All of the primary fractions from a distillation unit are equilibrium mixtures and contain some of the lower-boiling constituents that are characteristic of a lower-boiling fraction. The primary fractions are *stripped* of these constituents (*stabilized*) before storage or further processing.

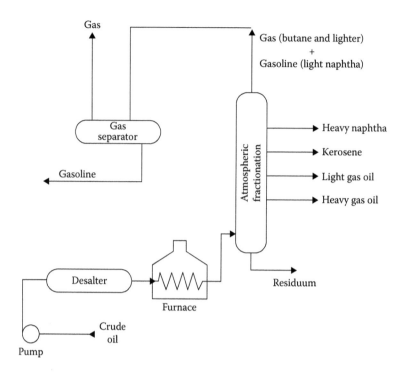

FIGURE 4.3 An atmospheric distillation unit. (From OSHA Technical Manual, Section IV, Chapter 2, Petroleum refining processes, 1999, http://www.osha.gov/dts/osta/otm/otm_iv/otm_iv_2.html.)

4.3.2 DISTILLATION UNDER REDUCED PRESSURE

Distillation under reduced pressure (vacuum distillation) as applied to the petroleum refining industry is truly a technique of the twentieth century and has since wide use in petroleum refining. Vacuum distillation evolved because of the need to separate the less volatile products, such as lubricating oils, from the petroleum without subjecting these high-boiling products to cracking conditions. The boiling point of the heaviest cut obtainable at atmospheric pressure is limited by the temperature (ca. 350°C; ca. 660°F) at which the residue starts to decompose (*crack*). When the feedstock is required for the manufacture of lubricating oils, further fractionation without cracking is desirable and this can be achieved by distillation under vacuum conditions.

Operating conditions for vacuum distillation (Figure 4.4) are usually 50–100 mm of mercury (atmospheric pressure = 760 mm of mercury). In order to minimize large fluctuations in pressure in the vacuum tower, the units are necessarily of a larger diameter than the atmospheric units. Some vacuum distillation units have diameters on the order of 45 ft (14 m). By this means, a heavy gas oil may be obtained as an overhead product at temperatures of approximately 150°C (300°F), and lubricating oil cuts may be obtained at temperatures of 250°C–350°C (480°F–660°F), feed and residue temperatures being kept below the temperature of 350°C (660°F), above which cracking will occur. The partial pressure of the hydrocarbons is effectively reduced still further by the injection of steam. The steam added to the column, principally for the stripping of asphalt in the base of the column, is superheated in the convection section of the heater.

The fractions obtained by vacuum distillation of the reduced crude (atmospheric residuum) from an atmospheric distillation unit depend on whether or not the unit is designed to produce lubricating or vacuum gas oils. In the former case, the fractions include (1) heavy gas oil, which is an overhead product and is used as catalytic cracking stock or, after suitable treatment, a light lubricating oil; (2) lubricating oil (usually three fractions—light, intermediate, and heavy), which

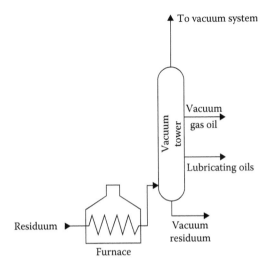

To vacuum system

Vacuum tower

Vacuum gas oil

Lubricating oils

Residuum

Vacuum residuum

Furnace

FIGURE 4.4   A vacuum distillation unit. (From OSHA Technical Manual, Section IV, Chapter 2, Petroleum refining processes, 1999, http://www.osha.gov/dts/osta/otm/otm_iv/otm_iv_2.html.)

is obtained as a sidestream product; and (3) asphalt (or residuum), which is the bottom product and may be used directly as, or to produce, asphalt and which may also be blended with gas oils to produce a heavy fuel oil.

In the early refineries, distillation was the prime means by which products were separated from crude petroleum. As the technologies for refining evolved into the twenty-first century, refineries became much more complex (Figure 4.1), but distillation remained the prime means by which petroleum is refined. Indeed, the distillation section of a modern refinery (Figures 4.3 and 4.4) is the most flexible section in the refinery since conditions can be adjusted to process a wide range of refinery feedstocks from the lighter crude oils to the heavier more viscous crude oils. However, the maximum permissible temperature (in the vaporizing furnace or heater) to which the feedstock can be subjected is 350°C (660°F). Thermal decomposition occurs above this temperature that, if it occurs within a distillation unit, can lead to coke deposition in the heater pipes or in the tower itself with the resulting failure of the unit.

The contained use of atmospheric and vacuum distillation has been a major part of refinery operations during this century, and no doubt will continue to be employed throughout the remainder of the century as the primary refining operation.

4.3.3   AZEOTROPIC AND EXTRACTIVE DISTILLATION

As the twentieth century evolved, distillation techniques in refineries became more sophisticated to handle a wider variety of crude oils to produce marketable products or feedstocks for other refinery units. However, it became apparent that the distillation units in the refineries were incapable of producing specific product fractions. In order to accommodate this type of product demand, refineries have, in the latter half of this century, incorporated azeotropic distillation and extractive distillation in their operations.

All compounds have definite boiling temperatures, but a mixture of chemically dissimilar compounds will sometimes cause one or both of the components to boil at a temperature other than that expected. A mixture that boils at a temperature lower than the boiling point of any of the components is an azeotropic mixture. When it is desired to separate close-boiling components, the addition of a nonindigenous component will form an azeotropic mixture with one of the components of the mixture thereby lowering the boiling point by the formation of an azeotrope and facilitate separation by distillation.

The separation of these components of similar volatility may become economic if an *entrainer* can be found, which effectively changes the relative volatility. It is also desirable that the entrainer be reasonably cheap, stable, nontoxic, and readily recoverable from the components. In practice, it is probably this last-named criterion that limits severely the application of extractive and azeotropic distillation. The majority of successful processes are those in which the entrainer and one of the components separate into two liquid phases on cooling if direct recovery by distillation is not feasible. A further restriction in the selection of an azeotropic entrainer is that the boiling point of the entrainer be in the range of 10°C–40°C (18°F–72°F) below that of the components.

4.4 THERMAL PROCESSES

Cracking was used commercially in the production of oils from coal and shales before the petroleum industry began, and the discovery that the heavier products could be decomposed to lighter oils was used to increase the production of kerosene and was called cracking distillation.

The precise origins of cracking distillation are unknown. It is rumored that, in 1861, the attending stillman had to leave his charge for a longer time than he intended (the reason is not known) during which time the still overheated. When he returned he noticed that the distillate in the collector was much more volatile than anticipated at that particular stage of the distillation. Further investigation led to the development of cracking distillation (i.e., thermal degradation with the simultaneous production of distillate).

Cracking distillation (thermal decomposition with simultaneous removal of distillate) was recognized as a means of producing the valuable lighter product (kerosene) from heavier nonvolatile materials. In the early days of the process (1870–1900) the technique was very simple—a batch of crude oil was heated until most of the kerosene had been distilled from it and the overhead material had become dark in color. At this point, distillation was discontinued and the heavy oils were held in the hot zone, during which time some of the high-molecular-weight components were decomposed to produce lower-molecular-weight products. After a suitable time, distillation was continued to yield light oil (kerosene) instead of the heavy oil that would otherwise have been produced.

The yields of kerosene products were usually markedly increased by means of cracking distillation, but the technique was not suitable for gasoline production. As the need for gasoline arose in the early 1900s, the necessity of prolonging the cracking process became apparent and a process known as pressure cracking evolved.

Pressure cracking was a batch operation in which, as an example, gas oil (200 bbl) was heated to approximately 425°C (800°F) in stills that had been reinforced to operate at pressures as high as 95 psi. The gas oil was held under maximum pressure for 24 hours, while fires maintained the temperature. Distillation was then started and during the next 48 hours produces a lighter distillate (100 bbl) that contained the gasoline components. This distillate was treated with sulfuric acid to remove unstable gum-forming components and then redistilled to produce a cracked gasoline (boiling range).

The large-scale production of cracked gasoline was first developed by Burton in 1912. The process employed batch distillation in horizontal shell stills and operated at approximately 400°C (ca. 750°F) and 75–95 psi. It was the first successful method of converting heavier oils into gasoline. Nevertheless, heating a bulk volume of oil was soon considered cumbersome, and during the years 1914–1922, a number of successful continuous cracking processes were developed. By these processes, gas oil was continuously pumped through a unit that heated the gas oil to the required temperature, held it for a time under pressure, and then discharged the cracked material into distillation equipment where it was separated into gases, gasoline, gas oil, and tar.

The tube-and-tank cracking process is not only typical of the early (post-1900) cracking units but also is one of the first units on record in which the concept of reactors (soakers) being onstream/offstream is realized. Such a concept departs from the true batch concept and allowed a greater degree of continuity. In fact, the tube-and-tank cracking unit may be looked upon as a forerunner of the delayed coking operation.

In the tube-and-tank process, a feedstock (at that time a gas oil) was preheated by exchange with the hot products from the unit pumped into the cracking coil, which consisted of several hundred feet of very strong pipe that lined the inner walls of a furnace where oil or gas burners raised the temperature of the gas oil to 425°C (800°F). The hot gas oil passed from the cracking coil into a large reaction chamber (soaker) where the gas oil was held under the temperature and pressure conditions long enough for the cracking reactions to be completed. The cracking reactions formed coke that, in the course of several days, filled the soaker. The gas oil stream was then switched to a second soaker, and the first soaker was cleaned out by drilling operations similar to those used in drilling an oil well.

The cracked material (other than coke) left the onstream soaker to enter an evaporator (tar separator) maintained under a much lower pressure than the soaker where, because of the lower pressure, all of the cracked material, except the tar, became vaporized. The vapor left the top of the separator where it was distilled into separate fractions—gases, gasoline, and gas oil. The tar that was deposited in the separator was pumped out for use as asphalt or as a heavy fuel oil.

Early in the development of tube-and-tank thermal cracking, it was found that adequate yields of gasoline could not be obtained by one passage of the stock through the heating coil; attempts to increase the conversion in one pass brought about undesirable high yields of gas and coke. It was better to crack to a limited extent, remove the products, and recycle the rest of the oil (or a distilled fraction free of tar) for repeated partial conversion. The high-boiling constituents once exposed to cracking were so changed in composition as to be more refractory than the original feedstock.

With the onset of the development of the automobile, the most important part of any refinery became the gasoline-manufacturing facilities. Among the processes that have evolved for gasoline production are thermal cracking, catalytic cracking, thermal reforming, catalytic reforming, polymerization, alkylation, coking, and distillation of fractions directly from crude petroleum.

When kerosene was the major product, gasoline was the portion of crude petroleum too volatile to be included in kerosene. The refiners of the 1890s and early 1900s had no use for it and often dumped an accumulation of gasoline into the creek or river that was usually nearby. As the demand for gasoline increased with the onset of World War I and the ensuing 1920s, more crude oil had to be distilled not only to meet the demand for gasoline but also to reduce the overproduction of the heavier petroleum fractions, including kerosene.

The problem of how to produce more gasoline from less crude oil was solved in 1913 by the incorporation of cracking units into refinery operations in which fractions heavier than gasoline were converted into gasoline by thermal decomposition. The early (pre-1940) processes employed for gasoline manufacture were processes in which the major variables involved were feedstock type, time, temperature, and pressure, which need to be considered to achieve the cracking of the feedstock to lighter products with minimal coke formation.

As refining technology evolved throughout this century, the feedstocks for cracking processes became the residuum or heavy distillate from a distillation unit. In addition, the residual oils produced as the end products of distillation processes and even some of the heavier virgin oils often contain substantial amounts of asphaltic materials, which preclude the use of the residuum as fuel oils or lubricating stocks. However, subjecting these residua directly to thermal processes has become economically advantageous, since, on the one hand, the end result is the production of lower-boiling salable materials; on the other hand, the asphaltic materials in the residua are regarded as the unwanted coke-forming constituents.

As the thermal processes evolved and catalysts were employed with more frequency, poisoning of the catalyst with a concurrent reduction in the lifetime of the catalyst became a major issue for refiners. To avoid catalyst poisoning, it became essential that as much of the nitrogen and metals (such as vanadium and nickel) as possible should be removed from the feedstock. The majority of the heteroatoms (nitrogen, oxygen, and sulfur) and the metals are contained in, or associated with, the asphaltic fraction (residuum). It became necessary that this fraction be removed from cracking feedstocks.

With this as the goal, a number of thermal processes, such as tar separation (flash distillation), vacuum flashing, visbreaking, and coking, came into wide usage by refiners and were directed at upgrading feedstocks by the removal of the asphaltic fraction. The method of deasphalting with liquid hydrocarbon, gases such as propane, butane, or *iso*-butane, became a widely used refinery operation in the 1950s and was very effective for the preparation of residua for cracking feedstocks. In this process, the desirable oil in the feedstock is dissolved in the liquid hydrocarbon and asphaltic materials remain insoluble.

Operating conditions in the deasphalting tower depend on the boiling range of the feedstock and the required properties of the product. Generally, extraction temperatures can range from 55°C to 120°C (130°F to 250°F) with a pressure of 400–600 psi. Hydrocarbon/oil ratios on the order of 6:1 to 10:1 by volume are typically used.

4.4.1 THERMAL CRACKING

One of the earliest conversion processes used in the petroleum industry is the thermal decomposition of higher-boiling materials into lower-boiling products. This process is known as thermal cracking, and the exact origins of the process are unknown. The process was developed in the early 1900s to produce gasoline from the "unwanted" higher-boiling products of the distillation process. However, it was soon learned that the thermal cracking process also produced a wide slate of products varying from highly volatile gases to nonvolatile coke.

The heavier oils produced by cracking are light and heavy gas oils as well as a residual oil that could also be used as heavy fuel oil. Gas oils from catalytic cracking were suitable for domestic and industrial fuel oils or as diesel fuels when blended with straight-run gas oils. The gas oils produced by cracking were also a further important source of gasoline. In a once-through cracking operation, all of the cracked material is separated into products and may be used as such. However, the gas oils produced by cracking (cracked gas oils) are more resistant to cracking (more refractory) than gas oils produced by distillation (straight-run gas oils) but could still be cracked to produce more gasoline. This was achieved using a later innovation (post-1940) involving a recycle operation in which the cracked gas oil was combined with fresh feed for another trip through the cracking unit. The extent to which recycling was carried out affected the yield of gasoline from the process.

The majority of the thermal cracking processes use temperatures of 455°C–540°C (850°F–1005°F) and pressures of 100–1000 psi; the Dubbs process may be taken as a typical application of an early thermal cracking operation. The feedstock (reduced crude) is preheated by direct exchange with the cracking products in the fractionating columns. Cracked gasoline and heating oil are removed from the upper section of the column. Light and heavy distillate fractions are removed from the lower section and are pumped to separate heaters. Higher temperatures are used to crack the more refractory light distillate fraction. The streams from the heaters are combined and sent to a soaking chamber where additional time is provided to complete the cracking reactions. The cracked products are then separated in a low-pressure flash chamber where a heavy fuel oil is removed as bottoms. The remaining cracked products are sent to the fractionating columns.

Mild cracking conditions, with a low conversion per cycle, favor a high yield of gasoline components, with low gas and coke production, but the gasoline quality is not high, whereas more severe conditions give increased gas and coke production and reduced gasoline yield (but of higher quality). With limited conversion per cycle, the heavier residues must be recycled, but these recycle oils become increasingly refractory upon repeated cracking, and if they are not required as a fuel oil stock they may be coked to increase gasoline yield or refined by means of a hydrogen process.

The thermal cracking of higher-boiling petroleum fractions to produce gasoline is now virtually obsolete. The antiknock requirements of modern automobile engines together with the different nature of crude oils (compared to those of 50 or more years ago) have reduced the ability of the thermal cracking process to produce gasoline on an economic basis. Very few new units have been installed since the 1960s, and some refineries may still operate the older cracking units.

4.4.2 Visbreaking

Visbreaking (viscosity breaking) is essentially a process of the post-1940 era and was initially introduced as a mild thermal cracking operation that could be used to reduce the viscosity of residua to allow the products to meet fuel oil specifications. Alternatively, the visbroken residua could be blended with lighter product oils to produce fuel oils of acceptable viscosity. By reducing the viscosity of the residuum, visbreaking reduces the amount of light heating oil that is required for blending to meet the fuel oil specifications. In addition to the major product, fuel oil, material in the gas oil and gasoline boiling range is produced. The gas oil may be used as additional feed for catalytic cracking units or as heating oil.

In a typical visbreaking operation (Figure 4.5), a crude oil residuum is passed through a furnace where it is heated to a temperature of 480°C (895°F) under an outlet pressure of approximately 100 psi. The heating coils in the furnace are arranged to provide a soaking section of low heat density, where the charge remains until the visbreaking reactions are completed and the cracked products are then passed into a flash distillation chamber. The overhead material from this chamber is then fractionated to produce a low-quality gasoline as an overhead product and light gas oil as bottom. The liquid products from the flash chamber are cooled with a gas oil flux and then sent to a vacuum fractionator. This yields a heavy gas oil distillate and a residual tar of reduced viscosity.

4.4.3 Coking

Coking is a thermal process for the continuous conversion of heavy, low-grade oils into lighter products. Unlike visbreaking, coking involved compete thermal conversion of the feedstock into volatile

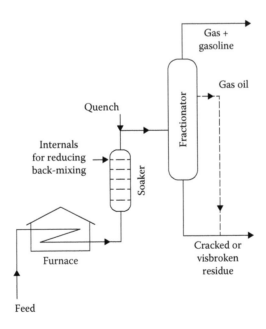

FIGURE 4.5 A soaker visbreaker. (From OSHA Technical Manual, Section IV, Chapter 2, Petroleum refining processes, 1999, http://www.osha.gov/dts/osta/otm/otm_iv/otm_iv_2.html.)

TABLE 4.4
Comparison of Visbreaking with Delayed Coking and Fluid Coking

Visbreaking

Purpose: to reduce the viscosity of fuel oil to acceptable levels
Conversion is not a prime purpose
Mild (470°C–495°C; 880°F–920°F) heating at pressures of 50–200 psi
Reactions quenched before going to completion
Low conversion (10%) to products boiling less than 220°C (430°F)
Heated coil or drum (soaker)

Delayed Coking

Purpose: to produce maximum yields of distillate products
Moderate (480°C–515°C; 900°F–960°F) heating at pressures of 90 psi
Reactions allowed to proceed to completion
Complete conversion of the feedstock
Soak drums (845°F–900°F) used in pairs (one onstream and one offstream being decoked)
Coked until drum solid
Coke removed hydraulically from offstream drum
Coke yield: 20%–40% by weight (dependent upon feedstock)
Yield of distillate boiling below 220°C (430°F): ca. 30% (but feedstock dependent)

Fluid Coking

Purpose: to produce maximum yields of distillate products
Severe (480°C–565°C; 900°F–1050°F) heating at pressures of 10 psi
Reactions allowed to proceed to completion
Complete conversion of the feedstock
Oil contacts refractory coke
Bed fluidized with steam; heat dissipated throughout the fluid bed
Higher yields of light ends (<C_5) than delayed coking
Less coke make than delayed coking (for one particular feedstock)

products and coke (Table 4.4). The feedstock is typically a residuum, and the products are gases, naphtha, fuel oil, gas oil, and coke. The gas oil may be the major product of a coking operation and serves primarily as a feedstock for catalytic cracking units. The coke obtained is usually used as fuel but specialty uses, such as electrode manufacture, production of chemicals, and metallurgical coke are also possible and increases the value of the coke. For these uses, the coke may require treatment to remove sulfur and metal impurities.

After a gap of several years, the recovery of heavy oils either through secondary recovery techniques from oil sand formations caused a renewal of interest in these feedstocks in the 1960s and, henceforth, for coking operations. Furthermore, the increasing attention paid to reducing atmospheric pollution has also served to direct some attention to coking, since the process not only concentrates pollutants such as feedstock sulfur in the coke but also can usually yield volatile products that can be conveniently desulfurized.

Investigations of technologies that result in the production of coke are almost as old as the refining industry itself, but the development of the modern coking processes can be traced in the 1930s with many units being added to refineries in the 1940–1970 era. Coking processes generally utilize longer reaction times than the older thermal cracking processes and, in fact, may be considered to be descendants of the thermal cracking processes.

4.4.3.1   Delayed Coking

Delayed coking is a semicontinuous process (Figure 4.6) in which the heated charge is transferred to large soaking (or coking) drums, which provide the long residence time needed to allow the cracking reactions to proceed to completion. The feed to these units is normally an atmospheric residuum, although cracked residua are also used.

The feedstock is introduced into the product fractionator where it is heated and lighter fractions are removed as sidestreams. The fractionator bottoms, including a recycle stream of heavy product, are then heated in a furnace whose outlet temperature varies from 480°C to 515°C (895°F to 960°F). The heated feedstock enters one of a pair of coking drums where the cracking reactions continue. The cracked products leave as overheads, and coke deposits form on the inner surface of the drum. To give continuous operation, two drums are used; while one is onstream, the other is being cleaned. The temperature in the coke drum ranges from 415°C to 450°C (780°F to 840°F) with pressures from 15 to 90 psi.

Overhead products go to the fractionator, where naphtha and heating oil fractions are recovered. The nonvolatile material is combined with preheated fresh feed and returned to the reactor. The coke drum is usually onstream for approximately 24 hours before becoming filled with porous coke after which the coke is removed hydraulically. Normally, 24 hours is required to complete the cleaning operation and to prepare the coke drum for subsequent use onstream.

4.4.3.2   Fluid Coking

Fluid coking is a continuous process (Figure 4.7) that uses the fluidized-solid technique to convert atmospheric and vacuum residua to more valuable products. The residuum is coked by being sprayed into a fluidized bed of hot, fine coke particles, which permits the coking reactions to be conducted at higher temperatures and shorter contact times than can be employed in delayed coking. Moreover, these conditions result in decreased yields of coke; greater quantities of more valuable liquid product are recovered in the fluid coking process.

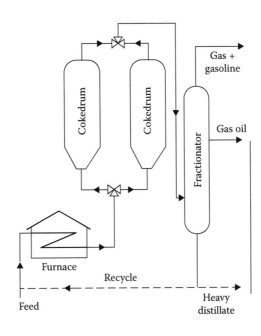

FIGURE 4.6   A delayed coker. (From OSHA Technical Manual, Section IV, Chapter 2, Petroleum refining processes, 1999, http://www.osha.gov/dts/osta/otm/otm_iv/otm_iv_2.html.)

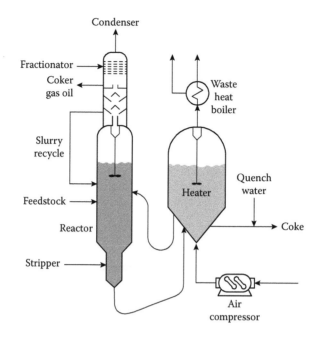

FIGURE 4.7 A fluid coker.

Fluid coking uses two vessels, a reactor and a burner; coke particles are circulated between these to transfer heat (generated by burning a portion of the coke) to the reactor. The reactor holds a bed of fluidized coke particles, and steam is introduced at the bottom of the reactor to fluidize the bed.

Flexicoking (Figure 4.8) is also a continuous process that is a direct descendent of fluid coking. The unit uses the same configuration as the fluid coker but has a gasification section in which excess coke can be gasified to produce refinery fuel gas. The flexicoking process was designed during the

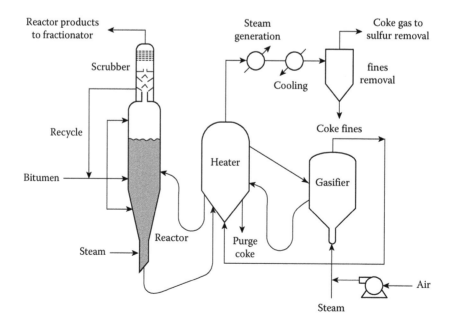

FIGURE 4.8 Flexicoking process.

late 1960s and the 1970s as a means by which excess coke-make could be reduced in view of the gradual incursion of the heavier feedstocks in refinery operations. Such feedstocks are notorious for producing high yields of coke (>15% by weight) in thermal and catalytic operations.

4.5 CATALYTIC PROCESSES

There are many processes in a refinery that employ a catalyst to improve process efficiency (Table 4.5). The original incentive arose from the need to increase gasoline supplies in the 1930s and 1940s. Since cracking could virtually double the volume of gasoline from a barrel of crude oil, cracking was justifiable on this basis alone.

In the 1930s, thermal cracking units produced approximately 50% of the total gasoline. The octane number of this gasoline was approximately 70 compared to approximately 60 for straight-run (distilled) gasoline. The thermal reforming and polymerization processes that were developed during the 1930s could be expected to further increase the octane number of gasoline to some extent, but an additional innovation was needed to increase the octane number of gasoline to enhance the development of more powerful automobile engines.

4.5.1 CATALYTIC CRACKING

In 1936, a new cracking process opened the way to higher-octane gasoline—this process was catalytic cracking. This process is basically the same as thermal cracking but differs by the use of a catalyst, which is not (in theory) consumed in the process, and directs the course of the cracking reactions to produce more of the desired higher-octane hydrocarbon products.

Catalytic cracking has a number of advantages over thermal cracking: (1) the gasoline produced has a higher octane number; (2) the catalytically cracked gasoline consists largely of *iso*-paraffins

---

TABLE 4.5
Summary of Catalytic Cracking Processes

Conditions
Solid acidic catalyst (such as silica–alumina and zeolite).
Temperature: 480°C–540°C (900°F–1000°F (solid/vapor contact).
Pressure: 10–20 psi.
Provisions needed for continuous catalyst replacement with heavier feedstocks (residua).
Catalyst may be regenerated or replaced.

Feedstocks
Gas oils and residua
Residua pretreated to remove salts (metals)
Residua pretreated to remove high molecular weight (asphaltic constituents)

Products
Lower molecular weight than feedstock
Some gases (feedstock and process parameters dependent)
Iso-paraffins in product
Coke deposited on catalyst

Variations
Fixed bed
Moving bed
Fluidized bed

---

and aromatics, which have high octane numbers and greater chemical stability than monoolefins and diolefins that are present in much greater quantities in thermally cracked gasoline. Substantial quantities of olefinic gases suitable for polymer gasoline manufacture and smaller quantities of methane, ethane, and ethylene are produced by catalytic cracking. Sulfur compounds are changed in such a way that the sulfur content of catalytically cracked gasoline is lower than in thermally cracked gasoline. Catalytic cracking produces less heavy residual or tar and more of the useful gas oils than does thermal cracking. The process has considerable flexibility, permitting the manufacture of both motor and aviation gasoline and a variation in the gas oil yield to meet changes in the fuel oil market.

The last 40 years have seen substantial advances in the development of catalytic processes. This has involved not only rapid advances in the chemistry and physics of the catalysts themselves but also major engineering advances in reactor design. For example, the evolution of the design of the catalyst beds from fixed beds to moving beds to fluidized beds. Catalyst chemistry/physics and bed design have allowed major improvements in process efficiency and product yields.

Catalytic cracking is another innovation that truly belongs to the twentieth century and is regarded as the modern method for converting high-boiling petroleum fractions, such as gas oil, into gasoline and other low-boiling fractions. Thus, catalytic cracking in the usual commercial process involves contacting a gas oil fraction with an active catalyst under suitable conditions of temperature, pressure, and residence time so that a substantial part (>50%) of the gas oil is converted into gasoline and lower-boiling products, usually in a single-pass operation.

However, during the cracking reaction, carbonaceous material is deposited on the catalyst, which markedly reduces its activity, and removal of the deposit is very necessary. This is usually accomplished by burning the catalyst in the presence of air until catalyst activity is reestablished.

The several processes currently employed in catalytic cracking differ mainly in the method of catalyst handling, although there is overlap with regard to catalyst type and the nature of the products.

The catalyst, which may be an activated natural or synthetic material, is employed in bead, pellet, or microspherical form and can be used as a fixed bed, moving bed, or fluid bed. The fixed-bed process was the first process to be used commercially and uses a static bed of catalyst in several reactors, which allows a continuous flow of feedstock to be maintained. Thus, the cycle of operations consists of (1) flow of feedstock through the catalyst bed, (2) discontinuance of feedstock flow and removal of coke from the catalyst by burning, and (3) insertion of the reactor onstream. The moving-bed process uses a reaction vessel (in which cracking takes place) and a kiln (in which the spent catalyst is regenerated) and catalyst movement between the vessels is provided by various means.

The *fluid-bed catalytic cracking process* (Figure 4.9) differs from the fixed-bed and moving-bed processes, insofar as the powdered catalyst is circulated essentially as a fluid with the feedstock. The several fluid catalytic cracking processes in use differ primarily in mechanical design. Side-by-side reactor–regenerator construction along with unitary vessel construction (the reactor either above or below the regenerator) is the two main mechanical variations.

4.5.2 CATALYSTS

Natural clays have long been known to exert a catalytic influence on the cracking of oils, but it was not until about 1936 that the process using silica–alumina catalysts was developed sufficiently for commercial use. Since then, catalytic cracking has progressively supplanted thermal cracking as the most advantageous means of converting distillate oils into gasoline. The main reason for the wide adoption of catalytic cracking is the fact that a better yield of higher-octane gasoline can be obtained than by any known thermal operation. At the same time, the gas

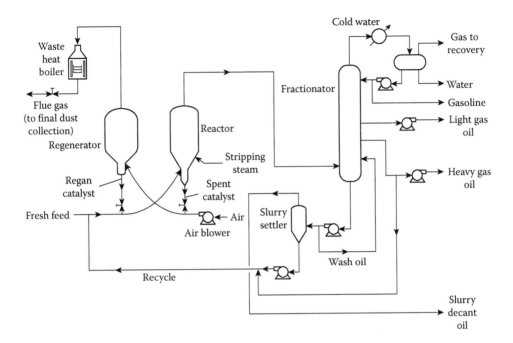

FIGURE 4.9 A fluid catalytic cracking unit.

produced consists mostly of propane and butane with less methane and ethane. The production of heavy oils and tars, higher in molecular weight than the charge material, is also minimized, and both the gasoline and the uncracked "cycle oil" are more saturated than the products of thermal cracking.

The major innovations of the twentieth century lie not only in reactor configuration and efficiency but also in catalyst development. There is probably not an oil company in the United States that does not have some research and development activity related to catalyst development. Much of the work is proprietary and, therefore, can only be addressed here in generalities.

The cracking of crude oil fractions occurs over many types of catalytic materials, but high yields of desirable products are obtained with hydrated aluminum silicates. These may be either activated (acid-treated) natural clays of the bentonite type of synthesized silica–alumina or silica–magnesia preparations. Their activity to yield essentially the same products may be enhanced to some extent by the incorporation of small amounts of other materials such as the oxides of zirconium, boron (which has a tendency to volatilize away on use), and thorium. Natural and synthetic catalysts can be used as pellets or beads and also in the form of powder; in either case, replacements are necessary because of attrition and gradual loss of efficiency. It is essential that they be stable to withstand the physical impact of loading and thermal shocks and that they withstand the action of carbon dioxide, air, nitrogen compounds, and steam. They also should be resistant to sulfur and nitrogen compounds and synthetic catalysts, or certain selected clays, appear to be better in this regard than average untreated natural catalysts.

The catalysts are porous and highly adsorptive, and their performance is affected markedly by the method of preparation. Two chemically identical catalysts having pores of different size and distribution may have different activity, selectivity, temperature coefficients of reaction rates, and responses to poisons. The intrinsic chemistry and catalytic action of a surface may be independent of pore size, but small pores produce different effects because of the manner in which hydrocarbon vapors are transported into and out of the pore systems.

4.6 HYDROPROCESSES

The use of hydrogen in thermal processes is perhaps the single most significant advance in refining technology during the twentieth century. The process uses the principle that the presence of hydrogen during a thermal reaction of a petroleum feedstock will terminate many of the coke-forming reactions and enhance the yields of the lower-boiling components such as gasoline, kerosene, and jet fuel (Table 4.6).

Hydrogenation processes for the conversion of petroleum fractions and petroleum products may be classified as destructive and nondestructive. Destructive hydrogenation (hydrogenolysis or hydrocracking) is characterized by the conversion of the higher-molecular-weight constituents in a feedstock to lower-boiling products. Such treatment requires severe processing conditions and the use of high hydrogen pressures to minimize polymerization and condensation reactions that lead to coke formation.

Nondestructive or simple hydrogenation is generally used for the purpose of improving product quality without appreciable alteration of the boiling range. Mild processing conditions are employed so that only the more unstable materials are attacked. Nitrogen, sulfur, and oxygen compounds undergo reaction with the hydrogen to remove ammonia, hydrogen sulfide, and water, respectively. Unstable compounds that might lead to the formation of gums, or insoluble materials, are converted to more stable compounds.

TABLE 4.6
Summary of Hydrocracking Processes

Conditions

Solid acid catalyst (silica–alumina with rare earth metals, various other options)

Temperature: 260°C–450°C (500°F–845°F (solid/liquid contact)

Pressure: 1000–6000 psi hydrogen

Frequent catalysts renewal for heavier feedstocks

Gas oil: catalyst life up to 3 years

Heavy oil/tar sand bitumen: catalyst life less than 1 year

Feedstocks

Refractory (aromatic) streams

Coker oils

Cycle oils

Gas oils

Residua (as a full hydrocracking or hydrotreating option)

In some cases, asphaltic constituents (S, N, and metals) removed by deasphalting

Products

Lower-molecular-weight paraffins

Some methane, ethane, propane, and butane

Hydrocarbon distillates (full range depending on the feedstock)

Residual tar (recycle)

Contaminants (asphaltic constituents) deposited on the catalyst as coke or metals

Variations

Fixed bed (suitable for liquid feedstocks)

Ebullating bed (suitable for heavy feedstocks)

4.6.1 HYDROTREATING

Distillate hydrotreating (Figure 4.10) is carried out by charging the feed to the reactor, together with hydrogen in the presence of catalysts such as tungsten–nickel sulfide, cobalt–molybdenum–alumina, nickel oxide–silica–alumina, and platinum–alumina. Most processes employ cobalt–molybdena catalysts that generally contain approximately 10% of molybdenum oxide and less than 1% of cobalt oxide supported on alumina. The temperatures employed are in the range of 260°C–345°C (500°F–655°F), while the hydrogen pressures are approximately 500–1000 psi.

The reaction generally takes place in the vapor phase but, depending on the application, may be a mixed-phase reaction. Generally, it is more economical to hydrotreat high-sulfur feedstocks prior to catalytic cracking than to hydrotreat the products from catalytic cracking. The advantages are that (1) sulfur is removed from the catalytic cracking feedstock, and corrosion is reduced in the cracking unit; (2) carbon formation during cracking is reduced so that higher conversions result; and (3) the cracking quality of the gas oil fraction is improved.

4.6.1.1 Hydrofining

Hydrofining is a process that first went onstream in the 1950s and is one example of the many hydroprocesses available. It can be applied to lubricating oils, naphtha, and gas oils. The feedstock is heated in a furnace and passed with hydrogen through a reactor containing a suitable metal oxide catalyst, such as cobalt and molybdenum oxides on alumina. Reactor operating conditions range from 205°C to 425°C (400°F to 800°F) and from 50 to 800 psi and depend on the kind of feedstock and the degree of treating required. Higher-boiling feedstocks, high sulfur content, and maximum sulfur removal require higher temperatures and pressures.

After passing through the reactor, the treated oil is cooled and separated from the excess hydrogen that is recycled through the reactor. The treated oil is pumped to a stripper tower where hydrogen sulfide, formed by the hydrogenation reaction, is removed by steam, vacuum, or flue gas, and the finished product leaves the bottom of the stripper tower. The catalyst is not usually regenerated; it is replaced after use for approximately 1 year.

4.6.2 HYDROCRACKING

Hydrocracking is similar to catalytic cracking, with hydrogenation superimposed and with the reactions taking place either simultaneously or sequentially. Hydrocracking was initially used to upgrade low-value distillate feedstocks, such as cycle oils (high aromatic products from a catalytic

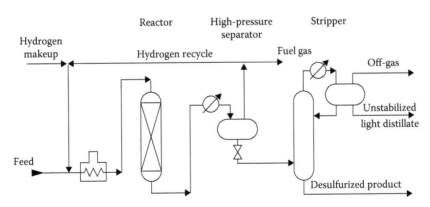

FIGURE 4.10 A distillate hydrotreater for hydrodesulfurization. (From OSHA Technical Manual, Section IV, Chapter 2, Petroleum refining processes, 1999, http://www.osha.gov/dts/osta/otm/otm_iv/otm_iv_2.html.)

cracker that usually are not recycled to extinction for economic reasons), thermal and coker gas oils, and heavy-cracked and straight-run naphtha. These feedstocks are difficult to process by either catalytic cracking or reforming, since they are characterized usually by a high polycyclic aromatic content and/or by high concentrations of the two principal catalyst poisons—sulfur and nitrogen compounds.

The older hydrogenolysis type of hydrocracking practiced in Europe during, and after, World War II used tungsten or molybdenum sulfides as catalysts and required high reaction temperatures and operating pressures, sometimes in excess of approximately 3000 psi for continuous operation. The modern hydrocracking processes (Figure 4.11) were initially developed for converting refractory feedstocks (such as gas oils) to gasoline and jet fuel, but process and catalyst improvements and modifications have made it possible to yield products from gases and naphtha to furnace oils and catalytic cracking feedstocks.

A comparison of hydrocracking with hydrotreating is useful in assessing the parts played by these two processes in refinery operations. Hydrotreating of distillates may be defined simply as the removal of nitrogen–sulfur and oxygen-containing compounds by selective hydrogenation. The hydrotreating catalysts are usually cobalt plus molybdenum or nickel plus molybdenum (in the sulfide) form impregnated on an alumina base. The hydrotreated operating conditions are such that appreciable hydrogenation of aromatics will not occur: 1000–2000 psi hydrogen and 370°C (700°F). The desulfurization reactions are usually accompanied by small amounts of hydrogenation and hydrocracking.

Hydrocracking is an extremely versatile process that can be utilized in many different ways such as conversion of the high-boiling aromatic streams that are produced by catalytic cracking or by

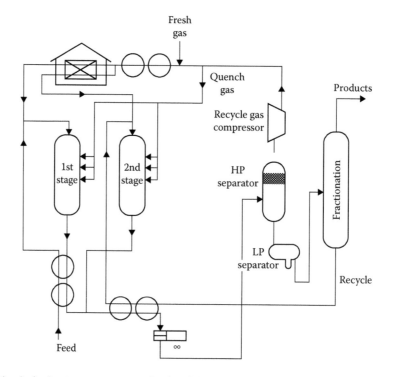

FIGURE 4.11 A single-stage or two-stage (optional) hydrocracking unit. (From OSHA Technical Manual, Section IV, Chapter 2, Petroleum refining processes, 1999, http://www.osha.gov/dts/osta/otm/otm_iv/otm_iv_2.html.)

coking processes. To take full advantage of hydrocracking, the process must be integrated in the refinery with other process units.

The commercial processes for treating, or finishing, petroleum fractions with hydrogen all operate in essentially the same manner. The feedstock is heated and passed with hydrogen gas through a tower or reactor filled with catalyst pellets. The reactor is maintained at a temperature of 260°C–425°C (500°F–800°F) at pressures from 100 to 1000 psi, depending on the particular process, the nature of the feedstock, and the degree of hydrogenation required. After leaving the reactor, excess hydrogen is separated from the treated product and recycled through the reactor after the removal of hydrogen sulfide. The liquid product is passed into a stripping tower where steam removes dissolved hydrogen and hydrogen sulfide and, after cooling, the product is taken to product storage or, in the case of feedstock preparation, pumped to the next processing unit.

The manufacture of base stocks for lubricating oil production is an essential part of modern refining. In the past four decades, the majority of the expansion of lubricating oil production is being achieved by production using catalytic hydroprocessing (hydrocracking and hydroisomerization) because of the demand for higher-quality lube base oils. Base oils are subdivided into a number of categories: groups I, II, II, and IV. Group I base oils are typically conventional solvent refined products. Groups II and III were added to lubricant classifications in the early 1990s to represent low sulfur, low aromatic, and high viscosity index (VI) lubricants with good oxidative stability and soot handling. The reduction of wax content in the lubricants also improves the operating range and engine, low-temperature performance via improved pour and cloud point. The first catalytic-based plants were introduced in the 1980s, but at that time, the catalytic route only produced conventional base oil (group I). In the 1990s, hydroisomerization was introduced to produce base oils with higher stability. Hydroisomerization has propagated such that a considerable amount of lube base oils is produced in this manner.

4.7 REFORMING PROCESSES

When the demand for higher-octane gasoline developed during the early 1930s, attention was directed to ways and means of improving the octane number of fractions within the boiling range of gasoline. Straight-run (distilled) gasoline frequently had very low octane numbers, and any process that would improve the octane numbers would aid in meeting the demand for gasoline with higher octane number. Such a process (called thermal reforming) was developed and used widely, but to a much lesser extent than thermal cracking. Thermal reforming was a natural development from older thermal cracking processes; cracking converts heavier oils into gasoline whereas reforming converts (reforms) gasoline into higher-octane gasoline. The equipment for thermal reforming is essentially the same as for thermal cracking, but higher temperatures are used.

4.7.1 THERMAL REFORMING

In carrying out thermal reforming, a feedstock such as 205°C (400°F) end-point naphtha or a straight-run gasoline is heated to 510°C–595°C (950°F–1100°F) in a furnace, much the same as a cracking furnace, with pressures from 400 to 1000 psi. As the heated naphtha leaves the furnace, it is cooled or quenched by the addition of cold naphtha. The material then enters a fractional distillation tower where any heavy products are separated. The remainder of the reformed material leaves the top of the tower to be separated into gases and reformate. The higher octane number of the reformate is due primarily to the cracking of longer-chain paraffins into higher-octane olefins.

The products of thermal reforming are gases, gasoline, and residual oil or tar, the latter being formed in very small amounts (approximately 1%). The amount and quality of the gasoline, known as reformate, are very dependent on the temperature. A general rule is the higher the reforming temperature, the higher the octane number, but the lower the yield of reformate.

Thermal reforming is less effective and less economical than catalytic processes and has been largely supplanted. As it used to be practiced, a single-pass operation was employed at temperatures in the range of 540°C–760°C (1000°F–1140°F) and pressures of approximately 500–1000 psi. The degree of improvement of the octane number depended on the extent of conversion but was not directly proportional to the extent of crack per pass. However, at very high conversions, the production of coke and gas became prohibitively high. The gases produced were generally olefinic, and the process required either a separate gas polymerization operation or one in which C_3–C_4 gases were added back to the reforming system.

More recent modifications of the thermal reforming process due to the inclusion of hydrocarbon gases with the feedstock are known as gas reversion and polyforming. Thus, olefinic gases produced by cracking and reforming can be converted into liquids boiling in the gasoline range by heating them under high pressure. Since the resulting liquids (polymers) have high octane numbers, they increase the overall quantity and quality of gasoline produced in a refinery.

4.7.2 Catalytic Reforming

The catalytic reforming process was commercially nonexistent in the United States prior to 1940. The process is really a process of the 1950s and showed phenomenal growth in 1953–1959 time period.

Like thermal reforming, catalytic reforming converts low-octane gasoline into high-octane gasoline (reformate). When thermal reforming could produce reformate with research octane numbers of 65–80 depending on the yield, catalytic reforming produces reformate with octane numbers on the order of 90–95. Catalytic reforming is conducted in the presence of hydrogen over hydrogenation–dehydrogenation catalysts, which may be supported on alumina or silica–alumina. Depending on the catalyst, a definite sequence of reactions takes place, involving structural changes in the feedstock. This more modern concept actually rendered thermal reforming somewhat obsolescent.

The commercial processes available for use can be broadly classified as the moving-bed, fluid-bed, and fixed-bed types. The fluid-bed and moving-bed processes used mixed nonprecious metal oxide catalysts in units equipped with separate regeneration facilities. Fixed-bed processes use predominantly platinum-containing catalysts in units equipped for cycle, occasional, or no regeneration.

Catalytic reformer feeds are saturated (i.e., not olefinic) materials; in the majority of cases that feed may be a straight-run naphtha but other by-product low-octane naphtha (e.g., coker naphtha) can be processed after treatment to remove olefins and other contaminants. Hydrocracker naphtha that contains substantial quantities of naphthenes is also a suitable feed.

Dehydrogenation is a main chemical reaction in catalytic reforming, and hydrogen gas is consequently produced in large quantities. The hydrogen is recycled through the reactors where the reforming takes place to provide the atmosphere necessary for the chemical reactions and also prevents the carbon from being deposited on the catalyst, thus extending its operating life. An excess of hydrogen above whatever is consumed in the process is produced, and, as a result, catalytic reforming processes are unique in that they are the only petroleum refinery processes to produce hydrogen as a by-product.

Catalytic reforming is usually carried out by feeding a naphtha (after pretreating with hydrogen if necessary) and hydrogen mixture to a furnace where the mixture is heated to the desired temperature, 450°C–520°C (840°F–965°F) and then passed through fixed-bed catalytic reactors at hydrogen pressures of 100–1000 psi (Figure 4.12). Normally, pairs of reactors are used in series with heaters located between adjoining reactors in order to compensate for the endothermic reactions taking place. Sometimes, as many as four or five reactors are kept onstream in series, while one or more is being regenerated.

The onstream cycle of any one reactor may vary from several hours to many days, depending on the feedstock and reaction conditions.

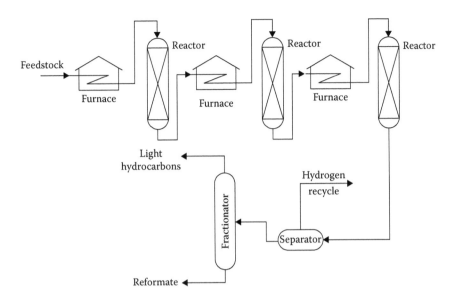

FIGURE 4.12 Catalytic reforming. (From OSHA Technical Manual, Section IV, Chapter 2, Petroleum refining processes, 1999, http://www.osha.gov/dts/osta/otm/otm_iv/otm_iv_2.html.)

4.7.3 Catalysts

The composition of a reforming catalyst is dictated by the composition of the feedstock and the desired reformate. The catalysts used are principally molybdena–alumina, chromia–alumina, or platinum on a silica–alumina or alumina base. The nonplatinum catalysts are widely used in regenerative process for feeds containing, for example, sulfur, which poisons platinum catalysts, although pretreatment processes (e.g., hydrodesulfurization) may permit platinum catalysts to be employed.

The purpose of platinum on the catalyst is to promote dehydrogenation and hydrogenation reactions, that is, the production of aromatics, participation in hydrocracking, and rapid hydrogenation of carbon-forming precursors. For the catalyst to have an activity for the isomerization of both paraffins and naphthenes—the initial cracking step of hydrocracking—and to participate in paraffin dehydrocyclization, it must have an acid activity. The balance between these two activities is most important in a reforming catalyst. In fact, in the production of aromatics from cyclic saturated materials (naphthenes), it is important that hydrocracking be minimized to avoid loss of the desired product and, thus, the catalytic activity must be moderated relative to the case of gasoline production from a paraffinic feed, where dehydrocyclization and hydrocracking play an important part.

4.8 ISOMERIZATION PROCESSES

Catalytic reforming processes provide high-octane constituents in the heavier gasoline fraction, but the normal paraffin components of the lighter gasoline fraction, especially butanes, pentanes, and hexanes, have poor octane ratings. The conversion of these normal paraffins to their isomers (isomerization) yields gasoline components of high octane rating in this lower-boiling range. Conversion is obtained in the presence of a catalyst (aluminum chloride activated with hydrochloric acid), and it is essential to inhibit side reactions such as cracking and olefin formation.

Isomerization—another innovation of the twentieth century—found initial commercial applications during World War II for making high-octane aviation gasoline components and additional feed for alkylation units. The lowered alkylate demands in the post-World War II period led to the majority of the butane isomerization units being shut down. In recent years, the greater demand for high-octane motor fuel has resulted in new butane isomerization units being installed.

4.8.1 Processes

The earliest process of note was the production of *iso*-butane, which is required as an alkylation feed. The isomerization may take place in the vapor phase, with the activated catalyst supported on a solid phase, or in the liquid phase with a dissolved catalyst. In the process, pure butane or a mixture of isomeric butanes (Figure 4.13) is mixed with hydrogen (to inhibit olefin formation) and passed to the reactor at 110°C–170°C (230°F–340°F) and 200–300 psi. The product is cooled, the hydrogen separated, and the cracked gases are then removed in a stabilizer column. The stabilizer bottom product is passed to a superfractionator where the normal butane is separated from the *iso*-butane.

Present isomerization applications in petroleum refining are used with the objective of providing additional feedstock for alkylation units or high-octane fractions for gasoline blending (Table 4.7). Straight-chain paraffins (*n*-butane, *n*-pentane, *n*-hexane) are converted to respective *iso*-compounds by continuous catalytic (aluminum chloride, noble metals) processes. Natural gasoline or light straight-run gasoline can provide feed by first fractionating as a preparatory step. High volumetric yields (>95%) and 40%–60% conversion per pass are characteristic of the isomerization reaction.

4.8.2 Catalysts

During World War II, aluminum chloride was the catalyst used to isomerize butane, pentane, and hexane. Since then, supported metal catalysts have been developed for use in high-temperature processes that operate in the range of 370°C–480°C (700°F–900°F) and 300–750 psi, while aluminum chloride plus hydrogen chloride are universally used for the low-temperature processes.

Nonregenerable aluminum chloride catalyst is employed with various carriers in a fixed-bed or liquid contactor. Platinum or other metal catalyst processes utilized fixed-bed operation and can

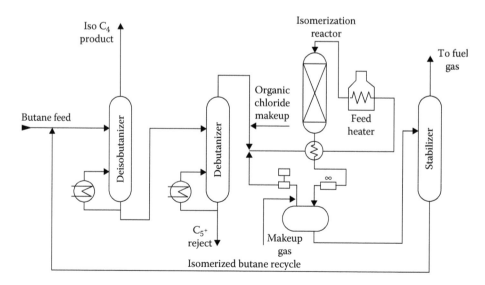

FIGURE 4.13 A butane isomerization unit. (From OSHA Technical Manual, Section IV, Chapter 2, Petroleum refining processes, 1999, http://www.osha.gov/dts/osta/otm/otm_iv/otm_iv_2.html.)

TABLE 4.7
Component Streams for Gasoline

Stream	Producing Process	Boiling Range (°C)	(°F)
Paraffinic			
Butane	Distillation	0	32
	Conversion		
Iso-pentane	Distillation	27	81
	Conversion		
	Isomerization		
Alkylate	Alkylation	40–150	105–300
Isomerate	Isomerization	40–70	105–160
Naphtha	Distillation	30–100	85–212
Hydrocrackate	Hydrocracking	40–200	105–390
Olefinic			
Catalytic naphtha	Catalytic cracking	40–200	105–390
Cracked naphtha	Steam cracking	40–200	105–390
Polymer	Polymerization	60–200	140–390
Aromatic			
Catalytic reformate	Catalytic reforming	40–200	105–390

be regenerable or nonregenerable. The reaction conditions vary widely depending on the particular process and feedstock: 40°C–480°C (100°F–900°F) and 150–1000 psi.

4.9 ALKYLATION PROCESSES

The combination of olefins with paraffins to form higher *iso*-paraffins is termed *alkylation*. Since olefins are reactive (unstable) and are responsible for exhaust pollutants, their conversion to high-octane *iso*-paraffins is desirable when possible. In refinery practice, only *iso*-butane is alkylated by reaction with *iso*-butene or normal butene and *iso*-octane is the product. Although alkylation is possible without catalysts, commercial processes use aluminum chloride, sulfuric acid, or hydrogen fluoride as catalysts, when the reactions can take place at low temperatures, minimizing undesirable side reactions, such as polymerization of olefins.

Alkylate is composed of a mixture of *iso*-paraffins that have octane numbers that vary with the olefins from which they were made. Butylenes produce the highest octane numbers, propylene the lowest, and pentylene derivatives the intermediate values. All alkylates, however, have high octane numbers (>87) that make them particularly valuable. The alkylation process is another twentieth-century refinery innovation and developments in petroleum processing in the late 1930s and during World War II was directed toward the production of high-octane liquids for aviation gasoline. The sulfuric acid process was introduced in 1938, and hydrogen fluoride alkylation was introduced in 1942. Rapid commercialization took place during the war to supply military needs, but many of these plants were shut down at the end of the war.

In the mid-1950s, aviation gasoline demand started to decline, but motor gasoline quality requirements rose sharply. Wherever practical, refiners shifted the use of alkylate to premium motor fuel. To aid in the improvement of the economics of the alkylation process and also the sensitivity of the premium gasoline pool, additional olefins were gradually added to alkylation feed. New plants were built to alkylate propylene and the butylenes (butanes) produced in the refinery rather than the butane–butylene stream formerly used.

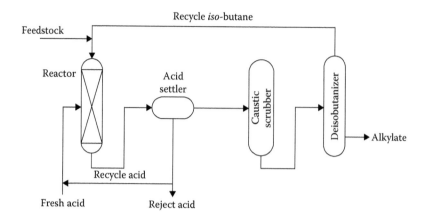

FIGURE 4.14 An alkylation unit (sulfuric acid catalyst). (From OSHA Technical Manual, Section IV, Chapter 2, Petroleum refining processes, 1999, http://www.osha.gov/dts/osta/otm/otm_iv/otm_iv_2.html.)

4.9.1 PROCESSES

The alkylation reaction as now practiced in petroleum refining is the union, through the agency of a catalyst, of an olefin (ethylene, propylene, butylene, and amylene) with *iso*-butane to yield high-octane branched-chain hydrocarbons in the gasoline boiling range. Olefin feedstock is derived from the gas produced in a catalytic cracker, while *iso*-butane is recovered by refinery gases or produced by catalytic butane isomerization.

To accomplish this, either ethylene or propylene is combined with *iso*-butane at 50°C–280°C (125°F–450°F) and 300–1000 psi in the presence of metal halide catalysts such as aluminum chloride. Conditions are less stringent in catalytic alkylation; olefins (propylene, butylene derivatives, or pentylene derivatives) are combined with *iso*-butane in the presence of an acid catalyst (sulfuric acid or hydrofluoric acid) at low temperatures and pressures (1°C–40°C, 30°F–105°F, and 14.8–150 psi) (Figure 4.14).

4.9.2 CATALYSTS

Sulfuric acid, hydrogen fluoride, and aluminum chloride are the general catalysts used commercially. Sulfuric acid is used with propylene and higher-boiling feeds, but not with ethylene, because it reacts to form ethyl hydrogen sulfate. The acid is pumped through the reactor and forms an air emulsion with reactants, and the emulsion is maintained at 50% acid. The rate of deactivation varies with the feed and *iso*-butane charge rate. Butene feeds cause less acid consumption than the propylene feeds.

Aluminum chloride is not widely used as an alkylation catalyst but when employed, hydrogen chloride is used as a promoter and water is injected to activate the catalyst as an aluminum chloride/hydrocarbon complex. Hydrogen fluoride is used for the alkylation of higher-boiling olefins, and the advantage of hydrogen fluoride is that it is more readily separated and recovered from the resulting product.

4.10 POLYMERIZATION PROCESSES

In the petroleum industry, polymerization is the process by which olefin gases are converted to liquid products that may be suitable for gasoline (polymer gasoline) or other liquid fuels. The feedstock usually consists of propylene and butylenes from cracking processes or may even be selective olefins for dimer, trimer, or tetramer production.

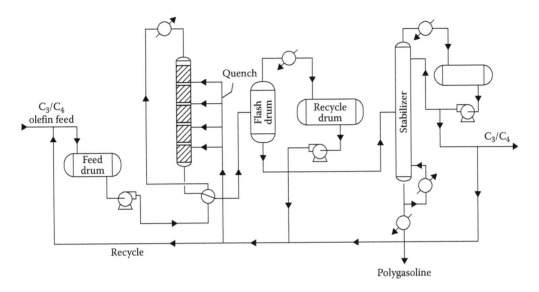

FIGURE 4.15 A polymerization unit. (From OSHA Technical Manual, Section IV, Chapter 2, Petroleum refining processes, 1999, http://www.osha.gov/dts/osta/otm/otm_iv/otm_iv_2.html.)

Polymerization is a process that can claim to be the earliest process to employ catalysts on a commercial scale. Catalytic polymerization came into use in the 1930s and was one of the first catalytic processes to be used in the petroleum industry.

4.10.1 PROCESSES

Polymerization may be accomplished thermally or in the presence of a catalyst at lower temperatures. Thermal polymerization is regarded as not being as effective as catalytic polymerization but has the advantage that it can be used to "polymerize" saturated materials that cannot be induced to react by catalysts. The process consists of vapor-phase cracking of, for example, propane and butane followed by prolonged periods at the high temperature (510°C–595°C, 950°F–1100°F) for the reactions to proceed to near completion.

Olefins can also be conveniently polymerized by means of an acid catalyst (Figure 4.15). Thus, the treated, olefin-rich feed stream is contacted with a catalyst (sulfuric acid, copper pyrophosphate, phosphoric acid) at 150°C–220°C (300°F–425°F) and 150–1200 psi, depending on feedstock and product requirement.

4.10.2 CATALYSTS

Phosphates are the principal catalysts used in polymerization units; the commercially used catalysts are liquid phosphoric acid, phosphoric acid on kieselguhr, copper pyrophosphate pellets, and phosphoric acid film on quartz. The latter is the least active, but the most used and easiest one to regenerate simply by washing and recoating; the serious disadvantage is that tar must occasionally be burned off the support. The process using liquid phosphoric acid catalyst is far more responsible to attempts to raise production by increasing temperature than the other processes.

4.11 SOLVENT PROCESSES

Many refineries also incorporate solvent extraction processes (also called *solvent refining processes*) for manufacturing lubricants and petrochemical units that are used to recover propylene, benzene, toluene, and xylenes for further processing into polymers (Speight and Ozum, 2002; Parkash, 2003; Hsu and Robinson, 2006; Gary et al., 2007; Speight, 2011a,b, 2014). While all solvent processes serve a purpose, the processes included here are (1) solvent deasphalting and (2) solvent dewaxing. In these processes, the feedstock is contacted directly with the solvents in order to disrupt the molecular forces within the feedstock and extract a specific fraction as the desired soluble product or as the raffinate leaving an insoluble product.

4.11.1 DEASPHALTING

Solvent deasphalting processes are a major part of refinery operations (Speight and Ozum, 2002; Parkash, 2003; Hsu and Robinson, 2006; Gary et al., 2007; Speight, 2011a,b) and are not often appreciated for the tasks for which they are used. In the solvent deasphalting processes, an alkane is injected into the feedstock to disrupt the dispersion of components and causes the polar constituents to precipitate. Propane (or sometimes propane/butane mixtures) is extensively used for deasphalting and produces a deasphalted oil (DAO) and propane deasphalter asphalt (PDA or PD tar) (Dunning and Moore, 1957). Propane has unique solvent properties: at lower temperatures (38°C–60°C; 100°C–140°C), paraffins are very soluble in propane, and at higher temperatures (approximately 93°C; 200°F), all hydrocarbons are almost insoluble in propane.

A *solvent deasphalting* unit (Figure 4.16) processes the residuum from the vacuum distillation unit and produces DAO used as feedstock for a fluid catalytic cracking unit (FCCU) and the asphaltic residue (deasphalter tar, deasphalter bottoms) that, as a residual fraction, can only be used to produce asphalt or as a blend stock or visbreaker feedstock for low-grade fuel oil. Solvent deasphalting processes have not realized their maximum potential. With ongoing improvements in energy efficiency, such processes would display its effects in a combination with other processes. Solvent deasphalting allows removal of sulfur and nitrogen compounds as well as metallic constituents by balancing yield with the desired feedstock properties (Ditman, 1973).

The propane deasphalting process is similar to solvent extraction in that a packed or baffled extraction tower or rotating disc contactor is used to mix the feedstock with the solvent. In the tower method, four to eight volumes of propane are fed to the bottom of the tower for every volume of feed flowing down from the top of the tower. The oil, which is more soluble in the propane, dissolves and flows to the top. The asphaltene and resins flow to the bottom of the tower where they are removed

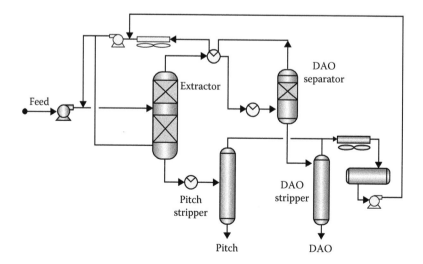

FIGURE 4.16 Propane deasphalting. (From OSHA Technical Manual, Section IV, Chapter 2, Petroleum refining processes, 1999, http://www.osha.gov/dts/osta/otm/otm_iv/otm_iv_2.html.)

in a propane mix. Propane is recovered from the two streams through two-stage flash systems followed by steam stripping in which propane is condensed and removed by cooling at high pressure in the first stage and at low pressure in the second stage. The asphalt recovered can be blended with other asphalts or heavy fuels or can be used as feed to the coker.

The major process variables are temperature, pressure, solvent-to-oil ratio, and solvent type. Pressure and temperature are both variables because the solvent power of light hydrocarbon is approximately proportional to the density of the solvent. Higher temperature always results in decreased yield of deasphalted oil. On the other hand, increasing solvent-to-oil ratio increases the recovery of deasphalted oil with an increase in viscosity. However, for the given product quality that can be maintained with change in temperature, solvent-to-oil ratio increases the yield of deasphalted oil. It has been shown that the solvent power of paraffin solvent increases with an increase in solvent molecular weight.

4.11.2 DEWAXING

Paraffinic crude oils often contain microcrystalline or paraffin waxes. The crude oil may be treated with a solvent such as methyl–ethyl–ketone methyl ethyl ketone (MEK) to remove this wax before it is processed. This is not common practice, however, and *solvent dewaxing processes* are designed to remove wax from lubricating oils to give the product good fluidity characteristics at low temperatures (e.g., low pour points) rather than from the whole crude oil. The mechanism of solvent dewaxing involves either the separation of wax as a solid that crystallizes from the oil solution at low temperature or the separation of wax as a liquid that is extracted at temperatures above the melting point of the wax through preferential selectivity of the solvent. However, the former mechanism is the usual basis for commercial dewaxing processes.

In the 1930s, two types of stocks, naphthenic and paraffinic, were used to make motor oils. Both types were solvent extracted to improve their quality, but in the high-temperature conditions encountered in service the naphthenic type could not stand up as well as the paraffinic type. Nevertheless, the naphthenic type was the preferred oil, particularly in cold weather, because of its fluidity at low temperatures. Prior to 1938, the highest-quality lubricating oils were of the naphthenic type and were phenol treated to pour points of −40°C to −7°C (−40°F to 20°F), depending on the viscosity of the oil. Paraffinic oils were also available and could be phenol treated to higher-quality oil, but their wax content was so high that the oils were solid at room temperature.

Dewaxing of lubricating oil base stocks is necessary to ensure that the oil will have the proper viscosity at lower ambient temperatures. Two types of dewaxing processes are used: selective hydrocracking and solvent dewaxing. In selective hydrocracking, one or two zeolite catalysts are used to selectively crack the wax paraffins. Solvent dewaxing is more prevalent. In solvent dewaxing, the oil feed is diluted with solvent to lower the viscosity, chilled until the wax is crystallized, and then filtered to remove the wax. Solvents used for the process include propane and mixtures of MEK with methyl *iso*-butyl ketone (MIBK) or MEK with toluene.

The lowest viscosity paraffinic oils were dewaxed by the cold press method to produce oils with a pour point of 2°C (35°F). The light paraffin distillate oils contained a paraffin wax that crystallized into large crystals when chilled and could thus readily be separated from the oil by the cold press filtration method. The more viscous paraffinic oils (intermediate and heavy paraffin distillates) contained amorphous or microcrystalline waxes, which formed small crystals that plugged the filter cloths in the cold press and prevented filtration. Because the wax could not be removed from intermediate and heavy paraffin distillates, the high-quality, high-viscosity lubricating oils in them could not be used except as cracking stock.

Methods were therefore developed to dewax these high-viscosity paraffinic oils. The methods were essentially alike in that the waxy oil was dissolved in a solvent that would keep the oil in solution; the wax separated as crystals when the temperature was lowered. The processes differed chiefly in the use of the solvent. Commercially used solvents were naphtha, propane, sulfur dioxide,

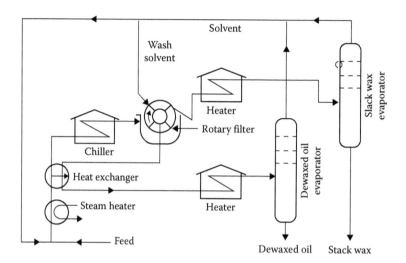

FIGURE 4.17 A solvent dewaxing unit. (From OSHA Technical Manual, Section IV, Chapter 2, Petroleum refining processes, 1999, http://www.osha.gov/dts/osta/otm/otm_iv/otm_iv_2.html.)

acetone–benzene, trichloroethylene, ethylene dichloride–benzene (*Barisol*), methyl ethyl ketone–benzene (*benzol*), methyl-*n*-butyl ketone, and methyl-*n*-propyl ketone.

The process as now practiced involves mixing the feedstock with one to four times its volume of the ketone (Figure 4.17). The mixture is then heated until the oil is in solution and the solution is chilled at a slow, controlled rate in double-pipe, scraped-surface exchangers. Cold solvent, such as filtrate from the filters, passes through the 2-inch annular space between the inner and outer pipes and chills the waxy oil solution flowing through the inner 6-inch pipe.

4.12 PETROLEUM PRODUCTS

Petroleum products, in contrast to *petrochemicals*, are those bulk fractions that are derived from petroleum and have commercial value as a bulk product (Speight and Ozum, 2002; Parkash, 2003; Hsu and Robinson, 2006; Gary et al., 2007; Speight, 2014). In the strictest sense, petrochemicals are also petroleum products, but they are individual chemicals that are used as the basic building blocks of the chemical industry.

The use of petroleum and its products was established in pre-Christian times and is known largely through documentation by many of the older civilizations (Abraham, 1945; Forbes, 1958; Hoiberg, 1960; Speight, 2014), and thus, the use of petroleum and the development of related technology are not such a modern subject as we are inclined to believe. However, there have been many changes in emphasis on product demand since petroleum first came into use some 5–6 millennia before the present time (Speight and Ozum, 2002; Parkash, 2003; Hsu and Robinson, 2006; Gary et al., 2007; Speight, 2014). It is these changes in product demand that have been largely responsible for the evolution of the industry from the asphalt used in ancient times to the gasoline and other liquid fuels of today.

Petroleum is an extremely complex mixture of hydrocarbon compounds, usually with minor amounts of nitrogen-containing, oxygen-containing, and sulfur-containing compounds as well as trace amounts of metal-containing compounds (Chapter 3). In addition, the properties of petroleum vary widely (Chapter 3). Thus, petroleum is not used in its raw state. A variety of processing steps are required to convert petroleum from its raw state to products that have well-defined properties.

The constant demand for products, such as liquid fuels, is the main driving force behind the petroleum industry. Other products, such as lubricating oils, waxes, and asphalt, have also added to the popularity of petroleum as a national resource. Indeed, fuel products that are derived from petroleum supply more than half of the world's total supply of energy. Gasoline, kerosene, and diesel oil provide fuel for automobiles, tractors, trucks, aircraft, and ships. Fuel oil and natural gas are used to heat homes and commercial buildings as well as to generate electricity. Petroleum products are the basic materials used for the manufacture of synthetic fibers for clothing and in plastics, paints, fertilizers, insecticides, soaps, and synthetic rubber. The uses of petroleum as a source of raw material in manufacturing are central to the functioning of modern industry.

Product complexity has made the industry unique among industries. Indeed, current analytical techniques that are accepted as standard methods for, as an example, the aromatics content of fuels (ASTM D1319, 2015; ASTM D2425, 2015; ASTM D2549, 2015; ASTM D2786, 2015; ASTM D2789, 2015), as well as proton and carbon nuclear magnetic resonance methods, yield different information. Each method will yield the "% aromatics" in the sample, but the data must be evaluated within the context of the method.

Product complexity becomes even more meaningful when various fractions from different types of crude oil as well as fractions from synthetic crude oil are blended with the corresponding petroleum stock. The implications for refining the fractions to salable products increase. However, for the main part, the petroleum industry was inspired by the development of the automobile and the continued demand for gasoline and other fuels. Such a demand has been accompanied by the demand for other products: diesel fuel for engines, lubricants for engine and machinery parts, fuel oil to provide power for the industrial complex, and asphalt for roadways.

Unlike processes, products are more difficult to place on an individual evolutionary scale. Processes changed and evolved to accommodate the demand for, say, higher-octane fuels, longer-lasting asphalt, or lower sulfur coke. In this section, a general overview of some petroleum products is presented to show the raison d'être of the industry. Another consideration that must be acknowledged is the change in character and composition of the original petroleum feedstock. In the early days of the petroleum industry, several products were obtained by distillation and could be used without any further treatment. Nowadays, the different character and composition of the petroleum dictate that any liquids obtained by distillation must go through one or more of the several available product improvement processes. Such changes in feedstock character and composition have caused the refining industry to evolve in a direction so that such changes in the input into the refinery can be accommodated.

It must also be recognized that adequate storage facilities for the gases, liquids, and solids that are produced during the refining operations are also an essential part of a refinery. Without such facilities, refineries would be incapable of operating efficiently.

The customary processing of petroleum does not usually involve the separation and handling of pure hydrocarbons. Indeed, petroleum-derived products are always mixtures: occasionally simple but more often very complex. Thus, for the purposes of this chapter, such materials as the gross fractions of petroleum (e.g., gasoline, naphtha, kerosene, and the like) that are usually obtained by distillation and/or refining are classed as *petroleum products*; asphalt and other solid products (e.g., wax) are also included in this division.

4.13 PETROCHEMICALS

The petrochemical industry began in the 1920s as suitable by-products became available through improvements in the refining processes. It developed parallel with the oil industry and has rapidly expanded since the 1940s, with the oil refining industry providing plentiful cheap raw materials.

A *petrochemical* is any chemical (as distinct from fuels and petroleum products) manufactured from petroleum (and natural gas) and used for a variety of commercial purposes. The definition, however, has been broadened to include the whole range of aliphatic, aromatic, and naphthenic organic chemicals, as well as carbon black and such inorganic materials as sulfur and ammonia. Petroleum and natural gas are made up of hydrocarbon molecules, which comprise one or more

TABLE 4.8

Hydrocarbon Intermediates Used in the Petrochemical Industry

Carbon Number	Hydrocarbon Type		
	Saturated	Unsaturated	Aromatic
1	Methane		
2	Ethane	Ethylene	
		Acetylene	
3	Propane	Propylene	
4	Butanes	*n*-Butenes	
		Iso-Butene	
		Butadiene	
5	Pentanes	*Iso*-Pentenes	
		(*Iso*-Amylenes)	
		Isoprene	
6	Hexanes	Methylpentenes	Benzene
	Cyclohexane		
7		Mixed heptenes	Toluene
8		Di-isobutylene	Xylenes
			Ethylbenzene
			Styrene
9			Cumene
12		Propylene tetramer	
		Tri-isobutylene	
18			Dodecylbenzene
6–18		*n*-Olefins	
11–18	*n*-Paraffins		

carbon atoms, to which hydrogen atoms are attached. Currently, through a variety of intermediates (Table 4.8) oil and gas are the main sources of the raw materials (Table 4.9) because they are the least expensive, most readily available, and can be processed most easily into the primary petrochemicals. Primary petrochemicals include olefins (ethylene, propylene, and butadiene), aromatics (benzene, toluene, and the isomers of xylene), and methanol. Thus, petrochemical feedstocks can be classified into three general groups: olefins, aromatics, and methanol; a fourth group includes inorganic compounds and synthesis gas (mixtures of carbon monoxide and hydrogen). In many instances, a specific chemical included among the petrochemicals may also be obtained from other sources, such as coal, coke, or vegetable products. For example, materials such as benzene and naphthalene can be made from either petroleum or coal, while ethyl alcohol may be of petrochemical or vegetable origin.

As stated earlier, some of the chemicals and compounds produced in a refinery are destined for further processing and as raw material feedstocks for the fast-growing petrochemical industry. Such nonfuel uses of crude oil products are sometimes referred to as its nonenergy uses. Petroleum products and natural gas provide two of the basic starting points for this industry: methane from natural gas and naphtha and refinery gases.

Petrochemical intermediates are generally produced by chemical conversion of primary petrochemicals to form more complicated derivative products. Petrochemical derivative products can be made in a variety of ways: directly from primary petrochemicals, through intermediate products that still contain only carbon and hydrogen, and through intermediates that incorporate chlorine, nitrogen, or oxygen in the finished derivative. In some cases, they are finished products; in others, more steps are needed to arrive at the desired composition.

TABLE 4.9

Sources of Petrochemical Intermediates

Hydrocarbon	Source
Methane	Natural gas
Ethane	Natural gas
Ethylene	Cracking processes
Propane	Natural gas, catalytic reforming, cracking processes
Propylene	Cracking processes
Butane	Natural gas, reforming and cracking processes
Butene(s)	Cracking processes
Cyclohexane	Distillation
Benzene	Catalytic reforming
Toluene	Catalytic reforming
Xylene(s)	Catalytic reforming
Ethylbenzene	Catalytic reforming
Alkylbenzenes	Alkylation
>C_9	Polymerization

Of all the processes used, one of the most important is polymerization. It is used in the production of plastics, fibers, and synthetic rubber, the main finished petrochemical derivatives. Some typical petrochemical intermediates are as follows: vinyl acetate for paint, paper, and textile coatings, vinyl chloride for polyvinyl chloride (PVC), and resin manufacture, ethylene glycol for polyester textile fibers, and styrene that is important in rubber and plastic manufacturing. The end products number in the thousands, some going on as inputs into the chemical industry for further processing. The more common products made from petrochemicals include adhesives, plastics, soaps, detergents, solvents, paints, drugs, fertilizer, pesticides, insecticides, explosives, synthetic fibers, synthetic rubber, and flooring and insulating materials.

4.14 ANCILLARY OPERATIONS

Refineries typically utilize primary and secondary wastewater treatment. Primary wastewater treatment consists of the separation of oil, water, and solids. After primary treatment, the wastewater can be discharged to a publicly owned treatment works (POTW) or undergo secondary treatment before being discharged directly to surface waters under a National Pollution Discharge Elimination System (NPDES) permit. In secondary treatment, dissolved oil and other organic pollutants may be consumed biologically by microorganisms.

Sulfur is removed from a number of refinery process off-gas streams (sour gas) in order to meet the SO_x emissions limits of the Clean Air Act (CAA) and to recover saleable elemental sulfur. Process off-gas streams, or sour gas, from the coker, catalytic cracking unit, hydrotreating units, and hydroprocessing units can contain high concentrations of hydrogen sulfide mixed with light refinery fuel gases. Before elemental sulfur can be recovered, the fuel gases (primarily methane and ethane) need to be separated from the hydrogen sulfide. This is typically accomplished by dissolving the hydrogen sulfide in a chemical solvent. Solvents most commonly used are amines, such as diethanolamine (DEA) (Chapter 16).

A number of chemicals (mostly alcohols and ethers) are added to motor fuels to either improve performance or meet federal and state environmental requirements. Since the 1970s, alcohols (methanol and ethanol) and ethers have been added to gasoline to increase octane levels and reduce carbon monoxide generation in place of the lead additives that were being phased out as required by the 1970 Clean Air Act (CAA). In 1990, the more stringent Clean Air Act amendments established minimum and maximum amounts of chemically combined oxygen in motor fuels as well as an upper limit on vapor pressure. As a result, alcohol additives have been increasingly supplemented or replaced with a number of different ethers that are better able to meet both the new oxygen requirements and the vapor pressure limits.

The most common ethers being used as additives are methyl tertiary butyl ether (MTBE) and tertiary amyl methyl ether (TAME). Many of the larger refineries manufacture their own supplies of MTBE and TAME by reacting *iso*-butylene and/or *iso*-amylene with methanol. Smaller refineries usually buy their supplies from chemical manufacturers or the larger refineries. *Iso*-butylene is obtained from a number of refinery sources including the light naphtha from the FCCU and coking units, the by-product from steam cracking of naphtha or light hydrocarbons during the production of ethylene and propylene, catalytic dehydrogenation of *iso*-butane, and conversion of tertiary butyl alcohol recovered as a by-product in the manufacture of propylene oxides.

Heat exchangers are used throughout petroleum refineries to heat or cool petroleum process streams. The heat exchangers consist of bundles of pipes, tubes, plate coils, or steam coils enclosing heating or cooling water, steam, or oil to transfer heat indirectly to or from the oil process stream. The bundles are cleaned periodically to remove accumulations of scales, sludge, and any oily residues.

Blowdown systems provide for the safe handling and disposal of liquid and gases that are either automatically vented from the process units through pressure relief valves or that are manually drawn from units. Recirculated process streams and cooling water streams are often manually purged to prevent the continued buildup of contaminants in the stream. Part or all of the contents of equipment can also be purged to the blowdown system prior to shutdown before normal or emergency shutdowns. Blowdown systems utilize a series of flash drums and condensers to separate the blowdown into its vapor and liquid components.

There is no one refinery process that will produce (for example) sales gasoline, sales diesel fuel, sales fuel oil, or sales lubricating oil. Each product is a composite mixture of stream from various units and blending (with the necessary additives so that the product can meet specifications) is the final operation before sales (Speight, 2014). The blending operation consists of mixing the products in various proportions to meet specifications such as vapor pressure, specific gravity, sulfur content, viscosity, octane number, cetane index, initial boiling point, and pour point. Blending can be carried out in-line or in batch blending tanks.

Storage tanks are used throughout the refining process to store crude oil and intermediate process feeds for cooling and further processing. Finished petroleum products are also kept in storage tanks before transport off-site. Storage tank bottoms are mixtures of iron rust from corrosion, sand, water, and emulsified oil and wax, which accumulate at the bottom of tanks. Liquid tank bottoms (primarily water and oil emulsions) are periodically drawn off to prevent their continued buildup. Tank bottom liquids and sludge are also removed during periodic cleaning of tanks for inspection.

REFERENCES

Abdel-Aal, H.K., Aggour, M.A., and Fahim, M.A. 2016. *Petroleum and Gas Filed Processing.* CRC Press/ Taylor & Francis Group, Boca Raton, FL.

Abraham, H. 1945. *Asphalts and Allied Substances*, Volume I. Van Nostrand Scientific Publishers, New York.

ASTM D1319. 2015. *Standard Test Method for Hydrocarbon Types in Liquid Petroleum Products by Fluorescent Indicator Adsorption.* Annual Book of Standards. ASTM International, West Conshohocken, PA.

ASTM D2425. 2015. *Standard Test Method for Hydrocarbon Types in Middle Distillates by Mass Spectrometry.* Annual Book of Standards. ASTM International, West Conshohocken, PA.

ASTM D2549. 2015. *Standard Test Method for Separation of Representative Aromatics and Non-aromatics Fractions of High-Boiling Oils by Elution Chromatography.* Annual Book of Standards. ASTM International, West Conshohocken, PA.

ASTM D2786. 2015. *Standard Test Method for Hydrocarbon Types Analysis of Gas-Oil Saturates Fractions by High Ionizing Voltage Mass Spectrometry*. Annual Book of Standards. ASTM International, West Conshohocken, PA.

ASTM D2789. 2015. *Standard Test Method for Hydrocarbon Types in Low Olefinic Gasoline by Mass Spectrometry*. Annual Book of Standards. ASTM International, West Conshohocken, PA.

Ditman, J.G. 1973. Solvent deasphalting. *Hydrocarbon Processing*, 52(5): 110.

Dunning, H.N. and Moore, J.W. 1957. Propane removes asphalts from crudes. *Petroleum Refiner*, 36(5): 247–250.

Forbes, R.J. 1958. *A History of Technology, V.* Oxford University Press, Oxford, UK.

Gary, J.H., Handwerk, G.E., and Kaiser, M.J. 2007. *Petroleum Refining: Technology and Economics*, 5th Edition. CRC Press/Taylor & Francis Group, Boca Raton, FL.

Hoiberg, A.J. 1960. *Bituminous Materials: Asphalts, Tars and Pitches, I & II*. Interscience, New York.

Hsu, C.S. and Robinson, P.R. (Editors). 2006. *Practical Advances in Petroleum Processing*, Volume 1 and Volume 2. Springer Science, New York.

OSHA Technical Manual. 1999. Section IV, Chapter 2: Petroleum refining processes. http://www.osha.gov/dts/osta/otm/otm_iv/otm_iv_2.html.

Parkash, S. 2003. *Refining Processes Handbook*. Gulf Professional Publishing, Elsevier, Amsterdam, the Netherlands.

Speight, J.G. 2011a. *An Introduction to Petroleum Technology, Economics, and Politics*. Scrivener Publishing, Salem, MA.

Speight, J.G. 2011b. *The Refinery of the Future*. Gulf Professional Publishing, Elsevier, Oxford, UK.

Speight, J.G. 2013. *Heavy and Extra Heavy Oil Upgrading Technologies*. Gulf Professional Publishing, Elsevier, Oxford, UK.

Speight, J.G. 2014. *The Chemistry and Technology of Petroleum*, 5th Edition. CRC Press/Taylor & Francis Group, Boca Raton, FL.

Speight, J.G. 2016. *Introduction to Enhanced Recovery Methods for Heavy Oil and Tar Sands*, 2nd Edition. Gulf Publishing Company, Elsevier, Waltham, MA.

Speight, J.G. and Ozum, B. 2002. *Petroleum Refining Processes*. Marcel Dekker, New York.

5 Refining Chemistry

5.1 INTRODUCTION

Crude oil is rarely used in its raw form but must instead be processed into its various products, generally as a means of forming products with hydrogen content different from that of the original feedstock. Thus, the chemistry of the refining process is concerned primarily with the production not only of better products but also of salable materials.

Crude oil contains many thousands of different compounds that vary in molecular weight from methane (CH_4, molecular weight 16) to more than 2000 (Boduszynski, 1987, 1988; Speight, 1994). This broad range in molecular weights results in boiling points that range from $-160°C$ ($-288°F$) to temperatures on the order of nearly $1100°C$ ($2000°F$). Many of the constituents of crude oil are paraffins. Remembering that the word *paraffin* was derived from the Latin *parum-maffinis*, which means *little affinity* or *little reactivity*, it must have come as a great surprise that hydrocarbons, including paraffins, can undergo a diversity of reactions, thereby influencing the chemistry of refining depending upon the source of the crude oil (Smith, 1994; Laszlo, 1995; Yen, 1998).

The major refinery products are liquefied petroleum gas (*LPG*), gasoline, jet fuel, solvents, kerosene, middle distillates (known as *gas oil* outside the United States), residual fuel oil, and asphalt. In the United States, with its high demand for gasoline, refineries typically upgrade their products much more than in other areas of the world, where the heavy end products, like residual fuel oil, are used in industry and power generation.

Understanding refining chemistry not only allows for an explanation of the means by which these products can be formed from crude oil but also offers a chance of predictability. This is very necessary when the different types of crude oil accepted by refineries are considered. And the major processes by which these products are produced from crude oil constituents involve thermal decomposition.

There are various theories relating to the thermal decomposition of organic molecules and this area of petroleum technology has been the subject of study for several decades (Hurd, 1929; Fabuss et al., 1964; Fitzer et al., 1971). The relative reactivity of petroleum constituents can be assessed on the basis of bond energies, but the thermal stability of an organic molecule is dependent upon the bond strength of the weakest bond. And even though the use of bond energy data is a method for predicting the reactivity or the stability of specific bonds under designed conditions, the reactivity of a particular bond is also subject to its environment. Thus, it is not only the reactivity of the constituents of petroleum that are important in processing behavior but also the stereochemistry of the constituents as they relate to one another, which is also of some importance (Speight, 2014). It must be appreciated that the stereochemistry of organic compounds is often a major factor in determining reactivity and properties (Eliel and Wilen, 1994).

In the present context, it is necessary to recognize that (*parumaffinis* or not) most hydrocarbons decompose thermally at temperatures above approximately $650°F$ ($340°C$), so the high boiling points of many petroleum constituents cannot be measured directly and must be estimated from other measurements. And in the present context, it is as well that hydrocarbons decompose at elevated temperates and thereby lies the route to many modern products. For example, in a petroleum refinery, the highest-value products are transportation fuels: (1) gasoline (boiling range $35°C–220°C$, $95°F–425°F$), (2) jet fuel (boiling range $175°C–290°C$, $350°F–550°F$), and (3) diesel fuel (boiling range $175°C–370°C$, $350°F–700°F$). These boiling ranges are not always precise to the degree and are subject to variation and depend upon the process used for their production. In winter, gasoline will typically (inn cold regions) have butane added to the mix (to facilitate cold starting), thereby changing the boiling range to 0°C–220°C (32°F–425°F). The fuels are produced by thermal decomposition of a variety of hydrocarbons, including high-molecular-weight paraffins. Less than one-third of a typical crude oil distills in these ranges, and thus, the goal of refining chemistry might be stated simply as the methods by which crude oil is converted to these fuels. It must be recognized that refining involves a wide variety of chemical reactions, but the production of liquid fuels is the focus of a refinery.

Refining processes involve the use of various thermal and catalytic processes to higher-molecular-weight constituents to lower-boiling products (Speight and Ozum, 2002; Parkash, 2003; Hsu and Robinson, 2006; Gary et al., 2007; Speight, 2014). This efficiency translates into a strong economic advantage, leading to the widespread use of conversion processes in refineries today. However, in order to understand the principles of catalytic cracking, understanding the principles of adsorption and reaction on solid surfaces is valuable (Samorjai, 1994; Masel, 1995).

A refinery is a complex network of integrated unit processes for the purpose of producing a variety of products from crude oil (Chapter 4) (Speight and Ozum, 2002; Parkash, 2003; Hsu and Robinson, 2006; Gary et al., 2007; Speight, 2014). Refined products establish the order in which the individual refining units will be introduced, and the choice from among several types of units and the size of these units is dependent upon economic factors. The trade-off among product types, quantity, and quality influences the choice of one kind of processing option over another.

Each refinery has its own range of preferred crude oil feedstock from which a desired distribution of products is obtained. Nevertheless, refinery processes can be divided into three major types:

1. *Separation*: Division of crude oil into various streams (or fractions) depending on the nature of the crude material
2. *Conversion*: Production of salable materials from crude oil, usually by skeletal alteration, or even by alteration of the chemical type, of the crude oil constituents
3. *Finishing*: Purification of various product streams by a variety of processes that essentially remove impurities from the product; for convenience, processes that accomplish molecular alteration, such as *reforming*, are also included in this category

The *separation* and *finishing processes* may involve distillation or even treatment with a *wash* solution, either to remove impurities or, in the case of distillation, to produce a material boiling over a narrower range and the chemistry of these processes can be represented by simple equations, even to the disadvantage of oversimplification of the process (Speight, 2014). The inclusion of reforming processes in this category is purely for descriptive purposes rather than being representative of the chemistry involved. Reforming processes produce streams that allow the product to be *finished* as the term applies to product behavior and utility.

Conversion processes are, in essence, processes that change the number of carbon atoms per molecule, alter the molecular hydrogen-to-carbon ratio, or change the molecular structure of the material without affecting the number of carbon atoms per molecule. These latter processes (*isomerization processes*) essentially change the shape of the molecule(s) and are used to improve the quality of the product (Speight, 2014).

Nevertheless, the chemistry of conversion process may be quite complex (King et al., 1973), and an understanding of the chemistry involved in the conversion of a crude oil to a variety of products is essential to an understanding of refinery operations. Therefore, the purpose of this chapter is to serve as an introduction to the chemistry involved in these conversion processes so that the subsequent chapters dealing with refining are easier to visualize and understand. However, understanding refining chemistry from the behavior of model compounds under refining conditions is not as straightforward as it may appear (Ebert et al., 1987).

The complexity of the individual reactions occurring in an extremely complex mixture and the *interference* of the products with those from other components of the mixture is unpredictable. Or the *interference* of secondary and tertiary products with the course of a reaction and, hence, with the formation of primary products may also be cause for concern. Hence, caution is advised when applying the data from model compound studies to the behavior of petroleum, especially the molecularly complex heavy oils. These have few, if any, parallels in organic chemistry.

5.2 CRACKING CHEMISTRY

The term *cracking* applies to the decomposition of petroleum constituents that is induced by elevated temperatures (>350°C, >660°F) whereby the higher-molecular-weight constituents of petroleum are converted to lower-molecular-weight products. Cracking reactions involve carbon–carbon bond rupture and are thermodynamically favored at high temperature.

5.2.1 THERMAL CRACKING

With the dramatic increases in the number of gasoline-powered vehicles, distillation processes (Chapters 4 and 7) were not able to completely fill the increased demand for gasoline. In 1913, the *thermal cracking process* was developed and is the phenomenon by which higher-boiling (higher-molecular-weight) constituents in petroleum are converted into lower-boiling (lower-molecular-weight) products' application of elevated temperatures (usually on the order of >350°C, >660°F).

Thermal cracking is the oldest and in principle the simplest refinery conversion process. The temperature and pressure depend on the type of feedstock and the product requirements as well as the residence time. Thermal cracking processes allow the production of lower-molecular-weight products such as the constituents of liquefied petroleum gas (LPG) and naphtha/gasoline constituents from higher-molecular-weight fraction such as gas oils and residua. The simplest thermal cracking process—the visbreaking process (Chapters 4 and 8)—is used to upgrade fractions such as distillation residua and other heavy feedstocks (Chapters 4 and 7) to produce fuel oil that meets specifications or feedstocks for other refinery processes.

Thus, cracking is a phenomenon by which higher-boiling constituents (higher-molecular-weight constituents) in petroleum are converted into lower-boiling (lower-molecular-weight) products. However, certain products may interact with one another to yield products having higher molecular weights than the constituents of the original feedstock. Some of the products are expelled from the system as, say, gases, gasoline-range materials, kerosene-range materials, and the various intermediates that produce other products such as coke. Materials that have boiling ranges higher than gasoline and kerosene may (depending upon the refining options) be referred to as *recycle* stock, which is recycled in the cracking equipment until conversion is complete.

In thermal cracking processes, some of the lower-molecular-weight products are expelled from the system as gases, gasoline-range materials, kerosene-range materials, and the various intermediates that produce other products such as coke. Materials that have boiling ranges higher than gasoline and kerosene may (depending upon the refining options) be referred to as *recycle* stock, which is recycled in the cracking equipment until conversion is complete.

5.2.1.1 General Chemistry

Thermal cracking is a *free radical* chain reaction. A free radical (in which an atom or a group of atoms possessing an unpaired electron) is very reactive (often difficult to control), and it is the mode of reaction of free radicals that determines the product distribution during thermal cracking (i.e., noncatalytic thermal decomposition). In addition, a significant feature of hydrocarbon free radicals is the resistance to isomerization during the existence of the radical. For example, thermal cracking does not produce any degree of branching in the products (by migration of an alkyl group) other than that already present in the feedstock. Nevertheless, the classical chemistry of free radical formation and behavior involves the following chemical reactions—it can only be presumed that the formation of free radicals during thermal (noncatalytic) cracking follows similar paths:

1. *Initiation reaction*, where a single molecule breaks apart into two free radicals. Only a small fraction of the feedstock constituents may actually undergo initiation, which involves breaking the bond between two carbon atoms rather than the thermodynamically stronger bond between a carbon atom and a hydrogen atom.

$$CH_3CH_3 \rightarrow 2CH_3\cdot$$

2. *Hydrogen abstraction reaction* in which the free radical abstracts a hydrogen atom from another molecule:

$$CH_3\cdot + CH_3CH_3 \rightarrow CH_4 + CH_3CH_2\cdot$$

3. *Radical decomposition reaction* in which a free radical decomposes into an alkene:

$$CH_3CH_2\cdot \rightarrow CH_2=CH_2 + H\cdot$$

4. *Radical addition reaction* in which a radical reacts with an alkene to form a single, larger free radical:

$$CH_3CH_2\cdot + CH_2=CH_2 \rightarrow CH_3CH_2CH_2CH_2\cdot$$

5. *Termination reaction* in which two free radicals react with each other to produce the products—two common forms of termination reactions are *recombination reactions* (in which two radicals combine to form one molecule) and *disproportionation reactions* (in which one free radical transfers a hydrogen atom to the other to produce an alkene and an alkane):

$$CH_3\cdot + CH_3CH_2\cdot \rightarrow CH_3CH_2CH_3$$

$$CH_3CH_2\cdot + CH_3CH_2\cdot \rightarrow CH_2=CH_2 + CH_3CH_3$$

The smaller free radicals, hydrogen, methyl, and ethyl are more stable than the larger radicals. They will tend to capture a hydrogen atom from another hydrocarbon, thereby forming a saturated hydrocarbon and a new radical. In addition, many thermal cracking processes and many different chemical reactions occur simultaneously. Thus, an accurate explanation of the mechanism of the thermal cracking reactions is difficult. The primary reactions are the decomposition of higher-molecular-weight species into lower-molecular-weight products.

As the molecular weight of the hydrocarbon feedstock increases, the reactions become much more complex, leading to a wider variety of products. For example, using a more complex hydrocarbon (dodecane, $C_{12}H_{26}$) as the example, two general types of reaction occur during cracking:

1. The decomposition of high-molecular-weight constituents into lower-molecular-weight constituents (*primary reactions*):

$$CH_3(CH_2)_{10}CH_3 \rightarrow CH_3(CH_2)_8CH_3 + CH_2=CH_2$$

$$CH_3(CH_2)_{10}CH_3 \rightarrow CH_3(CH_2)_7CH_3 + CH_2=CHCH_3$$

$$CH_3(CH_2)_{10}CH_3 \rightarrow CH_3(CH_2)_6CH_3 + CH_2=CHCH_2CH_3$$

$$CH_3(CH_2)_{10}CH_3 \rightarrow CH_3(CH_2)_5CH_3 + CH_2\!\!=\!\!CH(CH_2)_2CH_3$$

$$CH_3(CH_2)_{10}CH_3 \rightarrow CH_3(CH_2)_4CH_3 + CH_2\!\!=\!\!CH(CH_2)_3CH_3$$

$$CH_3(CH_2)_{10}CH_3 \rightarrow CH_3(CH_2)_3CH_3 + CH_2\!\!=\!\!CH(CH_2)_4CH_3$$

$$CH_3(CH_2)_{10}CH_3 \rightarrow CH_3(CH_2)_2CH_3 + CH_2\!\!=\!\!CH(CH_2)_5CH_3$$

$$CH_3(CH_2)_{10}CH_3 \rightarrow CH_3CH_2CH_3 + CH_2\!\!=\!\!CH(CH_2)_6CH_3$$

$$CH_3(CH_2)_{10}CH_3 \rightarrow CH_3CH_3 + CH_2\!\!=\!\!CH(CH_2)_7CH_3$$

$$CH_3(CH_2)_{10}CH_3 \rightarrow CH_4 + CH_2\!\!=\!\!CH(CH_2)_8CH_3$$

2. Reactions by which some of the primary products interact to form higher-molecular-weight materials (secondary reactions):

$$CH_2\!\!=\!\!CH_2 + CH_2\!\!=\!\!CH_2 \rightarrow CH_3CH_2CH\!\!=\!\!CH_2$$

$$RCH\!\!=\!\!CH_2 + R^1CH\!\!=\!\!CH_2 \rightarrow \text{cracked residuum} + \text{coke} + \text{other products}$$

Thus, from the chemistry of the thermal decomposing of pure compounds (and assuming little interference from other molecular species in the reaction mixture), it is difficult but not impossible to predict the product types that arise from the thermal cracking of various feedstocks. However, during thermal cracking, all of the reactions illustrated earlier can and do occur simultaneously and to some extent are uncontrollable. However, one of the significant features of hydrocarbon free radicals is their resistance to isomerization, for example, migration of an alkyl group and, as a result, thermal cracking does not produce any degree of branching in the products other than that already present in the feedstock.

Data obtained from the thermal decomposition of pure compounds indicate certain decomposition characteristics that permit predictions to be made of the product types that arise from the thermal cracking of various feedstocks. For example, normal paraffins are believed to form, initially, higher-molecular-weight material, which subsequently decomposes as the reaction progresses. Other paraffinic materials and (terminal) olefins are produced. An increase in pressure inhibits the formation of low-molecular-weight gaseous products and therefore promotes the formation of higher-molecular-weight materials.

Furthermore, for saturated hydrocarbons, the connecting link between gas-phase pyrolysis and liquid-phase thermal degradation is the concentration of alkyl radicals. In the gas phase, alkyl radicals are present in low concentration and undergo unimolecular radical decomposition reactions to form α-olefins and smaller alkyl radicals. In the liquid phase, alkyl radicals are in much higher concentration and prefer hydrogen abstraction reactions to radical decomposition reactions. It is this preference for hydrogen abstraction reactions that gives liquid-phase thermal degradation a broad product distribution.

Branched paraffins react somewhat differently to the normal paraffins during cracking processes and produce substantially higher yields of olefins having one fewer carbon atom than the parent hydrocarbon. Cycloparaffins (naphthenes) react differently to their noncyclic counterparts and are somewhat more stable. For example, cyclohexane produces hydrogen, ethylene, butadiene, and benzene: Alkyl-substituted cycloparaffins decompose by means of scission of the alkyl chain to produce an olefin and a methyl or ethyl cyclohexane.

The aromatic ring is considered fairly stable at moderate cracking temperatures (350°C–500°C, 660°F–930°F). Alkylated aromatics, like the alkylated naphthenes, are more prone to dealkylation than to ring destruction. However, ring destruction of the benzene derivatives occurs above

500°C (930°F), but condensed aromatics may undergo ring destruction at somewhat lower temperatures (450°C, 840°F).

Generally, the relative ease of cracking of the various types of hydrocarbons *of the same molecular weight* is given in the following descending order: (1) paraffins, (2) olefins, (3) naphthenes, and (4) aromatics. To remove any potential confusion, paraffins are the least stable and aromatics are the most stable.

Within any type of hydrocarbon, the higher-molecular-weight hydrocarbons tend to crack easier than the lighter ones. Paraffins are by far the easiest hydrocarbons to crack with the rupture most likely to occur between the first and second carbon bonds in the lighter paraffins. However, as the molecular weight of the paraffin molecule increases, rupture tends to occur near the middle of the molecule. The main secondary reactions that occur in thermal cracking are polymerization and condensation.

Two extremes of the thermal cracking in terms of product range are represented by high-temperature processes: (1) steam cracking and (2) pyrolysis. Steam cracking is a process in which feedstock is decomposed into lower-molecular-weight (often unsaturated) saturated hydrocarbons. In the process, a gaseous or liquid hydrocarbon feed like such as ethane or naphtha is diluted with steam and briefly heated in a furnace (at approximately 850°C, 1560°F) in the absence of oxygen at a short residence time (often on the order of milliseconds). After the cracking temperature has been reached, the products are rapidly quenched in a heat exchanger. The products produced in the reaction depend on the composition of the feedstock, the feedstock/steam ratio, the cracking temperature, and the residence time. Pyrolysis processes require temperatures on the order of 750°C–900°C (1380°F–1650°F) to produce high yields of low-molecular-weight products, such as ethylene, for petrochemical use. Delayed coking, which uses temperature on the order of 500°C (930°F), is used to produce distillates from nonvolatile residua as well as coke for fuel and other uses—such as the production of electrodes for the steel and aluminum industries.

5.2.1.2 Asphaltene Chemistry

Petroleum, heavy oils, and residua contain heptane-insoluble asphaltene constituents and resin constituents that, because of the content of polynuclear aromatic compounds and polar functionalities, provide hurdles to conversion. The high thermal stability of polynuclear aromatic systems prevents thermal decomposition to lower-boiling-point products and usually results in the production of substantial yields of thermal coke. Furthermore, the high concentrations of heteroatom compounds (nitrogen, oxygen, sulfur) and metals (vanadium and nickel) in heavy oils and residua have an adverse effect on catalysts. Therefore, process choice often favors thermal process, but catalytic processes can be used as long as catalyst replacement and catalyst regeneration is practiced.

Asphaltene constituents and, to a lesser extent, resin constituents can cause major problems in refineries through unanticipated coke formation and/or through excessive coke formation. Recognition of this is a step in the direction of mitigating the problem, and improvement in heavy feedstock conversion may be sought through the use of specific chemical additives. However, to improve the conversion of heavy feedstocks, it is necessary to understand the chemistry of conversion.

The thermal decomposition of the more complex asphaltene constituents has received some attention (Magaril and Aksenova, 1967, 1968, 1970a,b, 1972; Magaril and Ramazaeva, 1969; Magaril et al., 1970, 1971; Schucker and Keweshan, 1980; Speight, 2014). The thermal reaction is believed to be first order although there is the potential that it is, in fact, a multiorder reaction process, but because of the multiplicity of the reactions that occur, it appears as a pseudo-first-order process. However, it is definite that there is an induction period before coke begins to form that seems to be triggered by phase separation of reacted asphaltene product (Magaril and Aksenova, 1967, 1968, 1970a,b, 1972; Magaril and Ramazaeva, 1969; Magaril et al., 1970, 1971; Speight, 1987, 2014; Wiehe, 1993). The organic nitrogen originally in the asphaltene constituents invariably undergoes thermal reaction to concentrate in the nonvolatile coke (Speight, 1970, 1989; Vercier, 1981). In scheme, the chemistry of asphaltene coking has been suggested to involve the thermolysis of thermally labile bonds to form reactive species that react with each other (*condensation*) to form coke. However, not all the original aromatic carbon in the asphaltene constituents forms coke. Volatile aromatic species are eliminated during thermal decomposition, and it must be assumed that some of the original aliphatic carbon plays a role in coke formation.

It is more likely that the initial reactions of asphaltene constituents involve thermolysis of pendant alkyl chains to form lower-molecular-weight, higher-polarity species that are often referred to as carbenes and carboids (Chapters 2 and 3), which then react to form coke. The reactions involve unimolecular thermolysis of aromatic–alkyl systems of the asphaltene constituents to produce volatile species (paraffins and olefins) and nonvolatile species (aromatics) (Speight, 1987, 1994; Wiehe, 1994; Schabron and Speight, 1997).

It is also interesting to note that although the aromaticity of the resin and asphaltene constituents is approximately equivalent to the yield of thermal coke, not all the original aromatic carbon in the asphaltene constituents forms coke. Volatile aromatic species are eliminated during thermal decomposition, and it must be assumed that some of the original aliphatic carbon plays a role in coke formation. Its precise nature has yet to be determined, but the process can be represented as involving a multireaction process involving series and parallel reactions (Speight, 2014).

As examples of thermal cracking, in the delayed coking process, the feedstock is heated to high temperatures (480°C–500°C; 895°F–930°F) in a furnace and then reaction is allowed to continue in a cylindrical, insulated drum. The volatile products pass overhead into a fractionator and coke accumulates in the drum. Any high-boiling liquid product from the fractionator is recycled to the coker furnace. When the drum fills up with coke, the reacting feedstock is directed to a second drum. The coke is removed from the first drum by hydraulic drilling and cutting after which the drum is ready for the next 16–24-hour reaction cycle. During this process, the asphaltene and resin constituents in the feedstock are converted to coke in accordance with their respective carbon residue values (ca. 50% w/w for asphaltene constituents and ca. 35% w/w for resin constituents) (Speight, 2014).

Nitrogen species also appear to contribute to the pattern of the thermolysis insofar as the carbon–carbon bonds adjacent to ring nitrogen undergo thermolysis quite readily (Fitzer et al., 1971; Speight, 1998). Thus, the initial reactions of asphaltene decomposition involve the thermolysis of aromatic–alkyl bonds that are enhanced by the presence of heterocyclic nitrogen (Speight, 1987). Thus, the molecular species within the asphaltene fraction, which contain nitrogen and other heteroatoms (and have lower volatility than the pure hydrocarbons), are the prime movers in the production of coke (Speight, 1987). Such species, containing various polynuclear aromatic systems, can be denuded of the attendant hydrocarbon moieties and are undoubtedly insoluble (Bjorseth, 1983; Dias, 1987, 1988) in the surrounding hydrocarbon medium. The next step is gradual carbonization of these heteroatom-rich entities to form coke.

Thus, coke formation is a complex thermal process involving both chemical reactions and thermodynamic behavior. The challenges facing process chemistry and physics are determining (1) the means by which petroleum constituents thermally decompose, (2) the nature of the products of thermal decomposition, (3) the subsequent decomposition of the *primary* thermal products, (4) the interaction of the products with each other, (5) the interaction of the products with the original constituents, and (6) the influence of the products on the composition of the liquids.

The goal is to mitigate coke formation by elimination or modification of the prime chemical reactions in the formation of incompatible products during the processing of feedstocks containing asphaltene constituents, particularly those reactions in which the insoluble lower-molecular-weight products (*carbenes* and *carboids*) (Chapter 3) are formed (Speight, 1987, 1992, 2014; Wiehe, 1992; Wiehe, 1993).

5.2.1.3 Biomass Chemistry

The utilization of biomass to produce valuable products by thermal processes is an important aspect of biomass technology. Biomass pyrolysis gives usually rise to three phases: (1) gases, (2) condensable liquids, and (3) char/coke. However, there are various types of related kinetic pathways ranging from very simple paths to more complex paths, and all usually include several elementary processes occurring in series or in competition. As anticipated, the kinetic paths are different for cellulose, lignin, and hemicelluloses (biomass main basic components) and also for usual biomasses according to their origin, composition, and inorganic contents.

The main biomass constituents—hemicellulose, cellulose, and lignin—can be selectively devolatilized into value-added chemicals. This thermal breakdown is guided by the order of thermochemical stability of the biomass constituents that ranges from hemicellulose (as the least stable constituent) to the more stable—lignin exhibits an intermediate thermal degradation behavior. Thus, wood constituents are decomposed in the order of hemicellulose–cellulose–lignin, with a restricted decomposition of the lignin at relatively low temperatures. With prolonged heating, condensation of the lignin takes place, whereby thermally largely stable macromolecules develop. Whereas both hemicellulose and cellulose exhibit a relatively high devolatilization rate over a relatively narrow temperature range, thermal degradation of lignin is a slow-rate process that commences at a lower temperature when compared to cellulose.

Since the thermal stabilities of the main biomass constituents partially overlap and the thermal treatment is not specific, a careful selection of temperatures, heating rates, and gas and solid residence times is required to make a discrete degasification possible when applying a stepwise increase in temperature. Depending on these process conditions and parameters such as composition of the biomass and the presence of catalytically active materials, the product mixture is expected to contain degradation products from hemicellulose, cellulose, or lignin.

5.2.2 Catalytic Cracking

Catalytic cracking is the thermal decomposition of petroleum constituents in the presence of a catalyst (Pines, 1981). Thermal cracking has been superseded by catalytic cracking as the process for gasoline manufacture. Indeed, gasoline produced by catalytic cracking is richer in branched paraffins, cycloparaffins, and aromatics, which all serve to increase the quality of the gasoline. Catalytic cracking also results in the production of the maximum amount of butene derivatives and butane derivatives (C_4H_8 and C_4H_{10}) rather than the production of ethylene and ethane (C_2H_4 and C_2H_6).

Catalytic cracking processes evolved in the 1930s from research on petroleum and coal liquids. The petroleum work came to fruition with the invention of acid cracking. The work to produce liquid fuels from coal, most notably in Germany, resulted in metal sulfide hydrogenation catalysts. In the 1930s, a catalytic cracking catalyst for petroleum that used solid acids as catalysts was developed using acid-treated clay minerals. Clay minerals are a family of crystalline aluminosilicate solids, and the acid treatment develops acidic sites by removing aluminum from the structure. The acid sites also catalyze the formation of coke, and Houdry developed a moving-bed process that continuously removed the cooked beads from the reactor for regeneration by oxidation with air.

Although thermal cracking is a free radical (neutral) process, catalytic cracking is an ionic process involving carbonium ions, which are hydrocarbon ions having a positive charge on a carbon atom. The formation of carbonium ions during catalytic cracking can occur by (1) addition of a proton from an acid catalyst to an olefin and/or (2) abstraction of a hydride ion (H^-) from a hydrocarbon by the acid catalyst or by another carbonium ion. However, carbonium ions are not formed by cleavage of a carbon–carbon bond.

In essence, the use of a catalyst permits alternate routes for cracking reactions, usually by lowering the free energy of activation for the reaction. The acid catalysts first used in catalytic cracking were amorphous solids composed of approximately 87% silica (SiO_2) and 13% alumina (Al_2O_3)

and were designated low-alumina catalysts. However, this type of catalyst is now being replaced by crystalline aluminosilicates (zeolites) or molecular sieves.

The first catalysts used for catalytic cracking were acid-treated clay minerals, formed into beads. In fact, clay minerals are still employed as catalyst in some cracking processes (Speight, 2014). Clay minerals are a family of crystalline aluminosilicate solids, and the acid treatment develops acidic sites by removing aluminum from the structure. The acid sites also catalyze the formation of coke, and the development of a moving-bed process that continuously removed the cooked beads from the reactor reduced the yield of coke; clay regeneration was achieved by oxidation with air.

Clays are natural compounds of silica and alumina, containing major amounts of the oxides of sodium, potassium, magnesium, calcium, and other alkali and alkaline earth metals. Iron and other transition metals are often found in natural clays, substituted for the aluminum cations. Oxides of virtually every metal are found as impurity deposits in clay minerals.

Clay minerals are layered crystalline materials. They contain large amounts of water within and between the layers (Keller, 1985). Heating the clays above 100°C can drive out some or all of this water; at higher temperatures, the clay structures themselves can undergo complex solid-state reactions. Such behavior makes the chemistry of clays a fascinating field of study in its own right. Typical clays include kaolinite, montmorillonite, and illite (Keller, 1985). They are found in most natural soils and in large, relatively pure deposits, from which they are mined for applications ranging from adsorbents to paper making.

Once the carbonium ions are formed, the modes of interaction constitute an important means by which product formation occurs during catalytic cracking. For example, isomerization either by hydride ion shift or by methyl group shift, both of which occur readily. The trend is for stabilization of the carbonium ion by the *movement* of the charged carbon atom toward the center of the molecule, which accounts for the isomerization of α-olefins to internal olefins when carbonium ions are produced. Cyclization can occur by internal addition of a carbonium ion to a double bond, which, by continuation of the sequence, can result in the aromatization of the cyclic carbonium ion.

Like the paraffins, naphthenes do not appear to isomerize before cracking. However, the naphthenic hydrocarbons (from C_9 upward) produce considerable amounts of aromatic hydrocarbons during catalytic cracking. Reaction schemes similar to that outlined here provide possible routes for the conversion of naphthenes to aromatics. Alkylated benzenes undergo nearly quantitative dealkylation to benzene without apparent ring degradation below 500°C (930°F). However, polymethlyl benzene derivatives undergo disproportionation and isomerization with very little benzene formation.

Catalytic cracking can be represented by simple reaction schemes. However, questions have arisen as to how the cracking of paraffins is initiated. Several hypotheses for the initiation step in catalytic cracking of paraffins have been proposed (Cumming and Wojciechowski, 1996). The Lewis site mechanism is the most obvious, as it proposes that a carbenium ion is formed by the abstraction of a hydride ion from a saturated hydrocarbon by a strong Lewis acid site: a tricoordinated aluminum species (Figure 5.1). On Brønsted sites, a carbenium ion may be readily formed from an olefin by the addition of a proton to the double bond or, more rarely, via the abstraction of a hydride ion from a paraffin by a strong Brønsted proton. This latter process requires the formation of hydrogen as an initial product. This concept was, for various reasons that are of uncertain foundation, often neglected.

It is therefore not surprising that the earliest cracking mechanisms postulated that the initial carbenium ions are formed only by the protonation of olefins generated either by thermal cracking or present in the feed as an impurity. For a number of reasons, this proposal was not convincing, and in the continuing search for initiating reactions it was even proposed that electrical fields associated with the cations in the zeolite are responsible for the polarization of reactant paraffins, thereby activating them for cracking. More recently, however, it has been convincingly shown that a pentacoordinated carbonium ion can be formed on the alkane itself by protonation, if a sufficiently strong Brønsted proton is available (Cumming and Wojciechowski, 1996).

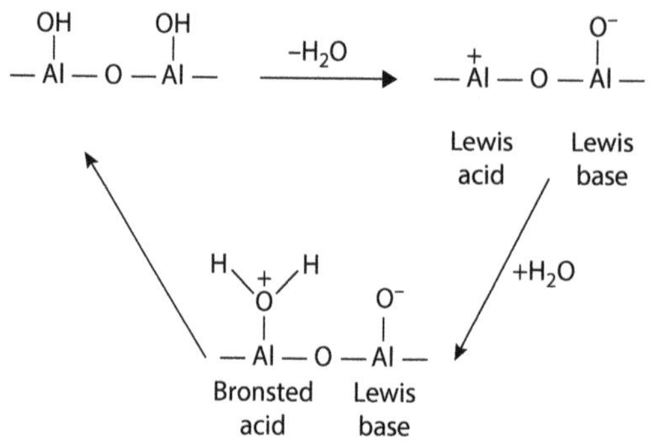

FIGURE 5.1 Acidic and basic sites.

Coke formation is considered, with just cause to be a malignant side reaction of normal carbenium ions. However, while chain reactions dominate events occurring on the surface, and produce the majority of products, certain less desirable bimolecular events have a finite chance of involving the same carbenium ions in a bimolecular interaction with one another. Of these reactions, most will produce a paraffin and leave carbene/carboid-type species on the surface. This carbene/carboid-type species can produce other products, but the most damaging product will be the one that remains on the catalyst surface and cannot be desorbed and results in the formation of coke or remains in a noncoke form but effectively blocks the active sites of the catalyst.

A general reaction sequence for coke formation from paraffins involves oligomerization, cyclization, and dehydrogenation of small molecules at active sites within zeolite pores:

Alkanes   → alkenes
Alkenes   → oligomers
Oligomers   → naphthenes
Naphthenes → aromatics
Aromatics   → coke

Whether or not these are the true steps to coke formation can only be surmised. The problem with this reaction sequence is that it ignores sequential reactions in favor of consecutive reactions. And it must be accepted that the chemistry leading up to coke formation is a complex process, consisting of many sequential and parallel reactions.

There is a complex and little understood relationship between coke content, catalyst activity, and the chemical nature of the coke. For instance, the atomic hydrogen/carbon ratio of coke depends on how the coke was formed; its exact value will vary from system to system (Cumming and Wojciechowski, 1996). And it seems that catalyst decay is not related in any simple way to the hydrogen-to-carbon atomic ratio of the coke, or to the total coke content of the catalyst, or any simple measure of coke properties. Moreover, despite many and varied attempts, there is currently no consensus as to the detailed chemistry of coke formation. There is, however, much evidence and good reason to believe that catalytic coke is formed from carbenium ions that undergo addition, dehydrogenation and cyclization, and elimination side reactions in addition to the mainline chain propagation processes (Cumming and Wojciechowski, 1996).

5.2.3 Dehydrogenation

Dehydrogenation is a class of chemical reactions by means of which less saturated and more reactive compounds can be produced. There are many important conversion processes in which hydrogen is directly or indirectly removed. In the current context, the largest-scale dehydrogenations are those of hydrocarbons such as the conversion of paraffin derivatives to olefin derivatives and olefin derivatives to diolefin derivatives:

$$-CH_2CH_2CH_2CH_3 \rightarrow -CH_2CH_2CH=CH_2$$

$$-CH_2CH_2CH=CH_2 \rightarrow -CH=CHCH=CH_2$$

Another example is the conversion of cycloparaffin derivatives to aromatic derivatives—the simplest example of which is the conversion of cyclohexane to benzene:

$$C_6H_{12} \rightarrow C_6H_6 + 3H_2$$

Dehydrogenation reactions of less specific character occur frequently in the refining and petrochemical industries, where many of the processes have names of their own. Some in which dehydrogenation plays a large part are pyrolysis, cracking, gasification by partial combustion, carbonization, and reforming.

The common primary reactions of pyrolysis are dehydrogenation and carbon bond scission. The extent of one or the other varies with the starting material and operating conditions, but because of its practical importance, methods have been found to increase the extent of dehydrogenation and, in some cases, to render it almost the only reaction.

Dehydrogenation is essentially the removal of hydrogen from the parent molecule. For example, at 550°C (1025°F) n-butane loses hydrogen to produce butene-1 and butene-2. The development of selective catalysts, such as chromic oxide (chromia, Cr_2O_3) on alumina (Al_2O_3), has rendered the dehydrogenation of paraffins to olefins particularly effective, and the formation of higher-molecular-weight material is minimized. The extent of dehydrogenation (vis-à-vis carbon–carbon bond scission) during the thermal cracking of petroleum varies with the starting material and operating conditions, but because of its practical importance, methods have been found to increase the extent of dehydrogenation and, in some cases, to render it almost the only reaction.

Naphthenes are somewhat more difficult to dehydrogenate, and cyclopentane derivatives form only aromatics if a preliminary step to form the cyclohexane structure can occur. Alkyl derivatives of cyclohexane usually dehydrogenate at 480°C–500°C (895°F–930°F), and polycyclic naphthenes are also quite easy to dehydrogenate thermally. In the presence of catalysts, cyclohexane and its derivatives are readily converted into aromatics; reactions of this type are prevalent in catalytic cracking and reforming. Benzene and toluene are prepared by the catalytic dehydrogenation of cyclohexane and methyl cyclohexane, respectively.

Polycyclic naphthenes can also be converted to the corresponding aromatics by heating at 450°C (840°F) in the presence of a chromia–alumina (Cr_2O_3–Al_2O_3) catalyst. Alkyl aromatic derivatives also dehydrogenate to various products. For example, styrene is prepared by the catalytic dehydrogenation of ethylbenzene. Other alkylbenzenes can be dehydrogenated similarly; *iso*-propyl benzene yields α-methyl styrene.

In general, dehydrogenation reactions are difficult reactions that require high temperatures for favorable equilibria as well as for adequate reaction velocities. Dehydrogenation reactions—using reforming reactions as the example (Figure 5.2)—are endothermic and, hence, have high heat requirements and active catalysts are usually necessary. Furthermore, since permissible hydrogen partial pressures are inadequate to prevent coke deposition, periodic regenerations are often

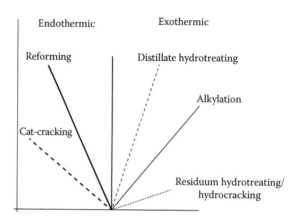

FIGURE 5.2 Endothermic and exothermic reactions in a refinery.

necessary. Because of these problems with pure dehydrogenations, many efforts have been made to use oxidative dehydrogenations in which oxygen or another oxidizing agent combines with the hydrogen removed. This expedient has been successful with some reactions where it has served to overcome thermodynamic limitations and coke formation problems.

The endothermic heat of pure dehydrogenation may be supplied through the walls of tubes (2–6 inches id), by preheating the feeds, by adding hot diluents, by reheaters between stages, or by heat stored in periodically regenerated fixed or fluidized-solid catalyst beds. Usually, fairly large temperature gradients will have to be tolerated, either from wall to center of tube, from inlet to outlet of bed, or from start to finish of a processing cycle between regenerations. The ideal profile of a constant temperature (or even a rising temperature) is seldom achieved in practice. In oxidative dehydrogenation reactions, the complementary problem of temperature rise because of exothermic nature of the reaction is encountered. Other characteristic problems met in dehydrogenations are the needs for rapid heating and quenching to prevent side reactions, the need for low pressure drops through catalyst beds, and the selection of reactor materials that can withstand the operating conditions.

Selection of operating conditions for a straight dehydrogenation reaction often requires a compromise. The temperature must be high enough for a favorable equilibrium and for a good reaction rate, but not as high as to cause excessive cracking or catalyst deactivation. The rate of the dehydrogenation reaction diminishes as conversion increases, not only because equilibrium is approached more closely, but also because in many cases reaction products act as inhibitors. The ideal temperature profile in a reactor would probably show an increase with distance, but practically attainable profiles normally are either flat or show a decline. Large adiabatic beds in which the decline is steep are often used. The reactor pressure should be as low as possible without excessive recycle costs or equipment size. Usually, the pressure is close to near-atmospheric pressure, but reduced pressures have been used in the Houdry butane dehydrogenation process. In any case, the catalyst bed must be designed for a low pressure drop.

Rapid preheating of the feed is desirable to minimize cracking. Usually, this is done by mixing preheated feedstock with superheated diluent just as the two streams enter the reactor. Rapid cooling or quenching at the exit of the reactor is usually necessary to prevent condensation reactions of the olefinic products. Materials of construction must be resistant to attack by hydrogen, capable of prolonged operation at high temperature, and not be unduly active for the conversion of hydrocarbons to carbon. Alloy steels containing chromium are usually favored although steel alloys containing nickel are also used, but these latter alloys can cause problems arising from carbon formation. If steam is not present, traces of sulfur compounds may be needed to avoid carbonization. Both steam and sulfur compounds act to keep metal walls in a passive condition.

5.2.4 Dehydrocyclization

Catalytic aromatization involving the loss of 1 mol of hydrogen followed by ring formation and further loss of hydrogen has been demonstrated for a variety of paraffins (typically *n*-hexane and *n*-heptane). Thus, *n*-hexane can be converted to benzene, heptane is converted to toluene, and octane is converted to ethyl benzene and o-xylene. Conversion takes place at low pressures, even atmospheric, and at temperatures above 300°C (570°F), although 450°C–550°C (840°F–1020°F) is the preferred temperature range.

The catalysts are metals (or their oxides) of the titanium, vanadium, and tungsten groups and are generally supported on alumina; the mechanism is believed to be dehydrogenation of the paraffin to an olefin, which in turn is cyclized and dehydrogenated to the aromatic hydrocarbon. In support of this, olefins can be converted to aromatics much more easily than the corresponding paraffins.

5.3 HYDROGENATION

The purpose of hydrogenating petroleum constituents is (1) to improve existing petroleum products or develop new products or even new uses, (2) to convert inferior or low-grade materials into valuable products, and (3) to transform higher-molecular-weight constituents into liquid fuels. The distinguishing feature of the hydrogenating processes is that, although the composition of the feedstock is relatively unknown and a variety of reactions may occur simultaneously, the final product may actually meet all the required specifications for its particular use (Furimsky, 1983; Speight, 2000).

Hydrogenation processes for the conversion of petroleum and petroleum products may be classified as *nondestructive* and *destructive* (Speight and Ozum, 2002; Parkash, 2003; Hsu and Robinson, 2006; Gary et al., 2007; Speight, 2014). Nondestructive, or simple, hydrogenation is generally used for the purpose of improving product (or even feedstock) quality without appreciable alteration of the boiling range. Treatment under such mild conditions is often referred to as *hydrotreating* or *hydrofining* and is essentially a means of eliminating nitrogen, oxygen, and sulfur as ammonia, water, and hydrogen sulfide, respectively. On the other hand, the latter process (*hydrogenolysis* or *hydrocracking*) is characterized by the rupture of carbon–carbon bonds and is accompanied by hydrogen saturation of the fragments to produce lower-boiling products. Such treatment requires rather high temperatures and high hydrogen pressures, the latter to minimize coke formation. Many other reactions, such as isomerization, dehydrogenation, and cyclization, can occur under these conditions (Dolbear et al., 1987).

The chemistry of the hydroprocesses can be defined by the type of catalysis employed, either (1) metal catalysis or acid (2) catalysis:

1. *Metal catalysis*

2. Acid catalysis

One advantage with catalytic processing is the flexibility it offers in feedstock processing. Heavier feedstocks require more extensive processing, which can be accomplished by higher conversion severity in the hydrocracking unit. This may result in lower yields of some products, but the converted products from the hydrocracking unit are primarily higher-value transportation fuel such as high-quality diesel and jet fuel.

5.3.1 HYDROTREATING

It is generally recognized that the higher the hydrogen content of a petroleum product, especially the fuel products, the better is the quality of the product. This knowledge has stimulated the use of hydrogen-adding processes in the refinery.

5.3.1.1 General Chemistry

Hydrotreating (i.e., hydrogenation without simultaneous occurrence of thermal decomposition) (Chapters 4 and 10) is used for saturating olefins or for converting aromatics to naphthenes as well as for heteroatom removal. Under atmospheric pressure, olefins can be hydrogenated up to approximately 500°C (930°F), but beyond this temperature dehydrogenation commences. The presence of hydrogen changes the nature of the products (especially the decreasing coke yield) by preventing the buildup of precursors that are incompatible in the liquid medium and form coke (Magaril and Aksenova, 1967, 1968, 1970a,b, 1972; Magaril and Ramazaeva, 1969; Magaril et al., 1970, 1971; Speight and Moschopedis, 1979; Schucker and Keweshan, 1980; Speight, 2014).

In contrast to the visbreaking process, in which the general principle is the production of products for use as fuel oil, the hydroprocessing is employed to produce a slate of products for use as liquid fuels. Nevertheless, the decomposition of asphaltene constituents is, again, an issue, and just as models consisting of large polynuclear aromatic systems are inadequate to explain the chemistry of visbreaking, they are also of little value for explaining the chemistry of hydrocracking.

5.3.1.2 Asphaltene Chemistry

The asphaltene constituents present complex modes of thermal cracking (Speight, 2014) and, in fact, it is at this point that the thermal chemistry of model compounds decreases in use in terms of understanding the thermal cracking of petroleum. As stated earlier, the thermal behavior of model compounds may not (does not) reflect the true thermal behavior of a complex mixture such as petroleum and the thermal cracking of petroleum residua cannot be described by a single activation energy model. The complexity of the individual reactions occurring in a residuum of similar mixture and the *interference* of the products with those from other components of the mixture are unpredictable. Or the *interference* of secondary and tertiary products with the course of a reaction and, hence, with the formation of primary products may also be cause for concern. Hence, caution is advised when applying the data from model compound studies to the behavior of petroleum, especially the molecularly complex heavy oils. These have few, if any, parallels in organic chemistry.

Recognition that the thermal behavior of petroleum is related to composition has led to a multiplicity of attempts to establish petroleum and its fractions as compositions of matter. As a result, various analytical techniques have been developed for the identification and quantification of *every molecule* in the lower-boiling fractions of petroleum. However, the name *petroleum* does not describe a composition of matter but rather a mixture of various organic compounds that includes a wide range of molecular weights and molecular types that exist in balance with each other (Speight, 2014).

In a mixture as complex as petroleum, the reaction processes can only be generalized because of difficulties in analyzing not only the products but also the feedstock as well as the intricate and complex nature of the molecules that make up the feedstock. The formation of coke from the higher-molecular-weight and high-polarity constituents (resin fraction and asphaltene fraction) of petroleum is detrimental to process efficiency and to catalyst performance. Although little has been acknowledged here of the role of low-molecular-weight polar species (resin constituents) in coke formation, the resin constituents are presumed to be lower-molecular-weight analogs of the asphaltene constituents. This being the case, similar reaction pathways may apply.

Deposition of solids or incompatibility is still possible when asphaltene constituents interact with catalysts, especially acidic support catalysts, through the functional groups, for example, the basic nitrogen species just as they interact with adsorbents. And there is a possibility for interaction of the asphaltene with the catalyst through the agency of a single functional group in which the remainder of the asphaltene molecule remains in the liquid phase. There is also a less desirable option in which the asphaltene reacts with the catalyst at several points of contact, causing immediate incompatibility on the catalyst surface.

During the early stages of the hydrotreating process, the chemistry of the asphaltene constituents follows the same routes as the thermal chemistry (Ancheyta et al., 2005b). Thus, initially there is an increase in the amount of asphaltene constituents followed by a decrease indicating that, in the early stages of the process, resin constituents are being converted to asphaltene material by aromatization and by some dealkylation.

In addition, aromatization and dealkylation of the original asphaltene constituents yields of asphaltene products that are of higher polarity and lower molecular weight than that of original asphaltene constituents. Analogous to the thermal processes, this produces an overall asphaltene fraction that is more polar material and also of lower molecular weight. As the hydrotreating process proceeds, the amount of asphaltene constituents precipitated decreases due to conversion of the asphaltene constituents to products. At more prolonged onstream times, there is a steady increase in the yield of the asphaltene constituents. This is accompanied by a general increase in the molecular weight of the precipitated material.

As predicted from the chemistry of the thermal reactions of the asphaltene constituents, there is a steady increase in aromaticity (reflected as a decrease in the hydrogen/carbon atomic ratio) with onstream time. This is due to (1) aromatization of naphthene ring system that is present in asphaltene constituents, (2) cyclodehydrogenation of alkyl chains to form other naphthene ring systems, (3) dehydrogenation of the new naphthene ring systems to form more aromatic rings, and (4) dealkylation of aromatic ring systems.

As the reaction progresses, the aromatic carbon atoms in the asphaltene constituents show a general increase and the degree of substitution of the aromatic rings decreases. Again this is in keeping with the formation of products from the original asphaltene constituents (carbenes, carboids, and eventually coke) that have an increased aromaticity and a decreased number of alkyl chains as well as a decrease in the alkyl chain length. Thus, as the reaction progresses with increased onstream time, *new* asphaltene constituents are formed, in which, relative to the original asphaltene constituents, the *new* species have increased aromaticity coupled with a lesser number of alkyl chains that are shorter than the original alkyl chains.

5.3.1.3 Catalysts

A wide variety of metals are active hydrogenation catalysts; those of most interest are nickel, palladium, platinum, cobalt, iron, nickel-promoted copper, and copper chromite. Special preparations of the first three are active at room temperature and atmospheric pressure. The metallic catalysts are easily poisoned by sulfur-containing and arsenic-containing compounds, and even by other metals. To avoid such poisoning, less effective but more resistant metal oxides or sulfides are frequently employed, generally those of tungsten, cobalt, chromium, or molybdenum.

Alternatively, catalysts poisoning can be minimized by mild hydrogenation to remove nitrogen, oxygen, and sulfur from feedstocks in the presence of more resistant catalysts, such as cobalt–molybdenum–alumina ($Co-Mo-Al_2O_3$). The reactions involved in nitrogen removal are somewhat analogous to those of the sulfur compounds and follow a stepwise mechanism to produce ammonia and the relevant substituted aromatic compound.

Hydrotreating catalysts consist of metals impregnated on a porous alumina support. Almost all of the surface area is found in the pores of the alumina (200–300 m²/g), and the metals are dispersed in a thin layer over the entire alumina surface within the pores. This type of catalyst does display a huge catalytic surface for a small weight of catalyst. Cobalt (Co), molybdenum (Mo), and nickel (Ni) are the most commonly used metals for desulfurization catalysts. The catalysts are manufactured with the metals in an oxide state. In the active form, they are in the sulfide state, which is obtained by sulfiding the catalyst either prior to use or with the feed during actual use. Any catalyst that exhibits hydrogenation activity will catalyze hydrodesulfurization to some extent. However, group VIB metals (chromium, molybdenum, and tungsten) are particularly active for desulfurization, especially when promoted with metals from the iron group (iron, cobalt, nickel).

Hydrodesulfurization and demetallization occur simultaneously on the active sites within the catalyst pore structure. Sulfur and nitrogen occurring in residua are converted to hydrogen sulfide and ammonia in the catalytic reactor, and these gases are scrubbed out of the reactor effluent gas stream. The metals in the feedstock are deposited on the catalyst in the form of metal sulfides and cracking of the feedstock to distillate produces a laydown of carbonaceous material on the catalyst; both events poison the catalyst and activity or selectivity suffers. The deposition of carbonaceous material is a fast reaction that soon equilibrates to a particular carbon level and is controlled by hydrogen partial pressure within the reactors. On the other hand, metal deposition is a slow reaction that is directly proportional to the amount of feedstock passed over the catalyst.

Removal of sulfur from the feedstock results in a gradual increase in catalyst activity, returning almost to the original activity level. As with ammonia, the concentration of the hydrogen sulfide can be used to control precisely the activity of the catalyst. Nonnoble metal–loaded zeolite catalysts have an inherently different response to sulfur impurities since a minimum level of hydrogen sulfide is required to maintain the nickel–molybdenum and nickel–tungsten in the sulfide state.

Alternatively, catalysts poisoning can be minimized by mild hydrogenation to remove nitrogen, oxygen, and sulfur from feedstocks in the presence of more resistant catalysts, such as cobalt–molybdenum–alumina ($Co-Mo-Al_2O_3$). The reactions involved in nitrogen removal are somewhat analogous to those of the sulfur compounds and follow a stepwise mechanism to produce ammonia and the relevant substituted aromatic compound.

5.3.2 Hydrocracking

Hydrocracking (Chapters 4 and 11) is a thermal process (>350°C, >660°F) in which hydrogenation accompanies cracking. Relatively high pressure (100–2000 psi) is employed, and the overall result is usually a change in the character or quality of the products. The wide range of products possible from hydrocracking is the result of combining catalytic cracking reactions with hydrogenation

(Dolbear, 1998; Hajji et al., 2010). The reactions are catalyzed by dual-function catalysts in which the cracking function is provided by silica–alumina (or zeolite) catalysts, and platinum, tungsten oxide, or nickel provides the hydrogenation function.

5.3.2.1 General Chemistry

Essentially all the initial reactions of catalytic cracking occur, but some of the secondary reactions are inhibited or stopped by the presence of hydrogen. For example, the yields of olefins and the secondary reactions that result from the presence of these materials are substantially diminished and branched-chain paraffins undergo demethanation. The methyl groups attached to secondary carbons are more easily removed than those attached to tertiary carbon atoms, whereas methyl groups attached to quaternary carbons are the most resistant to hydrocracking.

The effect of hydrogen on naphthenic hydrocarbons is mainly that of ring scission followed by immediate saturation of each end of the fragment produced. The ring is preferentially broken at favored positions, although generally all the carbon–carbon bond positions are attacked to some extent. For example, methylcyclopentane is converted (over a platinum–carbon catalyst) to 2-methylpentane, 3-methylpentane, and n-hexane.

Aromatic hydrocarbons are resistant to hydrogenation under mild conditions, but under more severe conditions the main reactions are conversion of the aromatic to naphthenic rings and scissions within the alkyl side chains. The naphthenes may also be converted to paraffins. However, polynuclear aromatics are more readily attacked than the single-ring compounds, the reaction proceeding by a stepwise process in which one ring at a time is saturated and then opened. For example, naphthalene is hydrocracked over a molybdenum oxide molecular catalyst to produce a variety of low-molecular-weight paraffins ($\leq C_6$).

There have been many attempts to analyze the chemical kinetics of the hydrocracking reaction (Ancheyta et al., 2005a; Bahmani et al., 2007). Lump models have been used for several years for kinetic modeling of complex reactions. Catalyst screening, process control, basic process studies, and dynamic modeling, among others, are areas in which lump kinetic models are extensively applied. The main disadvantages of lump models are their simplicity in predicting product yields, the dependency of kinetic parameters on feed properties, and the use of an invariant distillation range of products, which, if changed, necessitates further experiments and parameter estimation. Structure-oriented lumping models are more detailed approaches that express the chemical transformations in terms of typical molecule structures. These models describe reaction kinetics in terms of a relatively large number of pseudo-components, and hence, they do not completely eliminate lumps. In addition, the dependency of rate parameters on feedstock properties is present.

Models based on continuous mixtures (continuous theory of lumping) overcome some of these deficiencies by considering the properties of the reaction mixture, the underlying pathways, and the associated selectivity of the reactions. The common parameter of characterization is the true boiling point temperature, since during reaction it changes continuously inside the reactor as the residence time increases. However, the dependency of model parameters on feedstock properties is still present. Distillation curves, either chromatographic or physical, also present some difficulties when analyzing heavy oils since initial and final boiling points are not accurate during experimentation. In fact, for many purposes, 10% and 90% boiling points are commonly utilized instead of initial boiling point and final boiling point, respectively.

The single event concept uses elementary steps of cation chemistry, which consists of a limited number of types of steps involving series of homologous species. The number of rate coefficients to be determined from experimental information can be reduced and are modeled based upon transition state theory and statistical thermodynamics. With this approach, parameter values are not dependent on feed properties. However, even though the number of parameters can be diminished, detailed and sufficient experimental data are necessary.

However, the complexity of the feedstocks—especially heavy feedstocks such as heavy oil, extra heavy oil tar sand bitumen, and residua—suggests (rightly or wrongly) that models based on lumping theory will continue to be used for the study of the reaction kinetics of the hydrocracking process. However, more accurate approaches are still required for a better understanding and representation of heavy oil hydrocracking kinetics. Because of the changing nature of heavy feedstocks and even refinery gas oils, whether or not the issues of reaction kinetics will ever be solved remains in question. In fact, dealing with one fraction of constituents—the asphaltene fraction—is itself a problem that can only be solved in general terms, being subject to the chemical and physical character of the asphaltene constituents (recalling that the asphaltene fraction is not a homogeneous chemical and physical fraction) as well as the react configuration and the process parameters. The same rationale can be applied with justification to the resin fraction, which is also a heterogeneous chemical and physical fraction (Speight, 2014).

5.3.2.2 Asphaltene Chemistry

In terms of hydroprocessing, the means by which asphaltene constituents are desulfurized, as one step of a hydrocracking operation, is also suggested as part of the process. This concept can then be taken one step further to show the dealkylation of the aromatic systems as a definitive step in the hydrocracking process (Speight, 1987). If catalytic processes are employed, complex molecules (such as those occurring in the asphaltene fraction) or those formed during the process are not sufficiently mobile (or are too strongly adsorbed by the catalyst) to be saturated by the hydrogenation components. Such molecular species eventually degrade to coke, which deactivates the catalyst sites and eventually interfere with the process.

It may be that the chemistry of hydrocracking has to be given serious reconsideration insofar as the data show that the initial reactions of the asphaltene constituents appear to be the same as the reactions under thermal conditions where hydrogen is not present. Rethinking of the process conditions and the potential destruction of the catalyst by the deposition of carbenes and carboids require further investigation of the chemistry of asphaltene hydrocracking.

If these effects are prevalent during hydrocracking high-asphaltene-content feedstocks, the option may be to hydrotreat the feedstock first and then to hydrocrack the hydrotreated feedstock. There are indications that such hydrotreatment can (at some obvious cost) act beneficially in the overall conversion of the feedstocks to liquid products.

The resin fraction has received somewhat less attention than the asphaltene fractions but the chemical and physical heterogeneity of the fraction remains unresolved and is believed to match the chemical and physical heterogeneity of the asphaltene fraction (Koots and Speight, 1975; Andersen and Speight, 2001).

5.3.2.3 Catalysts

The reactions of hydrocracking require a dual-function catalyst with high cracking and hydrogenation activities. The catalyst base, such as acid-treated clay, usually supplies the cracking function or alumina or silica–alumina that is used to support the hydrogenation function supplied by metals, such as nickel, tungsten, platinum, and palladium. These highly acid catalysts are very sensitive to nitrogen compounds in the feed, which break down the conditions of reaction to give ammonia and neutralize the acid sites.

Hydrocracking catalysts typically contain separate hydrogenation and cracking functions. Palladium sulfide and promoted group VI sulfides (nickel molybdenum or nickel tungsten) provide the hydrogenation function. These active compositions saturate aromatics in the feed, saturate olefins formed in the cracking, and protect the catalysts from poisoning by coke. Zeolites or amorphous silica–alumina provide the cracking functions. The zeolites are usually type Y (faujasite), ion exchanged to replace sodium with hydrogen and make up 25%–50% of the catalysts. Pentasils (silicalite or ZSM-5) may be included in dewaxing catalysts.

Hydrocracking catalysts, such as nickel (5% by weight) on silica–alumina, work best on feedstocks that have been hydrotreated to low nitrogen and sulfur levels. The nickel catalyst then operates well at 350°C–370°C (660°F–700°F) and a pressure of approximately 1500 psi to give good conversion of feed to lower-boiling liquid fractions with minimum saturation of single-ring aromatics and a high *iso*-paraffin to *n*-paraffin ratio in the lower-molecular-weight paraffins.

Catalysts containing platinum or palladium (approximately 0.5% wet) on a zeolite base appear to be somewhat less sensitive to nitrogen than are nickel catalysts, and successful operation has been achieved with feedstocks containing 40 ppm nitrogen. This catalyst is also more tolerant of sulfur in the feed, which acts as a temporary poison, the catalyst recovering its activity when the sulfur content of the feed is reduced. With catalysts of higher hydrogenation activity, such as platinum on silica–alumina, direct isomerization occurs. The product distribution is also different, and the ratio of low- to intermediate-molecular-weight paraffins in the breakdown product is reduced.

Catalyst poisoning can be minimized by mild hydrogenation to remove nitrogen, oxygen and sulfur from feedstocks in the presence of more resistant catalysts, such as cobalt–molybdenum–alumina ($Co-Mo-Al_2O_3$).

5.3.3 Solvent Deasphalting

Heavy feedstocks are mixtures of various hydrocarbon classes including saturates, aromatics, and asphaltenes as well as hydrocarbonaceous materials in which carbon and hydrogen have been replaced by heteroatoms (Speight, 2014). The chemistry of the deasphalting process does not, to the purist, involve the organic chemistry of the system but is more related to the physical chemistry of the system insofar as the outcome of the process is dictated more by physical solvent–feedstock relationship rather than by organic chemical reactions.

It is postulated that the asphaltene constituents are effectively *peptized* by association with the resin constituents and exist at the center of a colloidal particle or micelle. The actual structure of the micelle is not known with any degree of certainty (Speight, 1994, 2014). When the entire micelle system contains sufficient constituents for the formation of the semicontinuous outer region, the asphaltenes are fully peptized by the outer ring that is more compatible (less prone to a separate phase). Low-boiling liquid paraffin solvents partially or completely dissolve the peptizing agents and the asphaltene constituents, being incompatible with the liquid phase, separate and precipitate. As the molecular weight of the liquid paraffin solvent decreases, the solubility of the resins and of some of the heavier and more aromatic hydrocarbon in the paraffin solvent decreases (as evidenced by the propane deasphalting process). Thus, lower-molecular-weight paraffin solvents (such as propane) precipitate a tacky material consisting of asphaltene constituents, resin constituents, and other aromatic constituents (or naphthene–aromatic constituents with some long-chain paraffins' derivative) as may be insoluble in the paraffin-dominated liquid phase.

5.3.3.1 Effects of Temperature and Pressure

As evidenced from the laboratory separation of asphaltene constituents and the fractionation of feedstocks (Speight, 1994, 2014, 2015), temperature and pressure are both variables because the solvent power of low-boiling hydrocarbon is approximately proportional to the density of the solvent. That is, decreasing temperature or increasing pressure will increase the average molecular weight of hydrocarbons soluble in solvent-rich phase. For propane, at temperatures below 80°C (176°F) and a pressure above the vapor pressure of propane, temperature is the most important factor in determining the solubility of the feedstock constituents. At temperatures near the critical region, pressure is also an important factor, as properties of the liquid propane approach those of gaseous propane. Thus, higher temperatures typically result in decreased yields of deasphalted oil, although the converse has also been observed (Speight, 1994, 2014). This is accompanied with the decrease in viscosity and molecular weight range of the deasphalted oil.

Thus, during normal operation, when both the solvent composition and the extraction pressure are fixed, the yields and qualities of the various products recovered in the solvent deasphalting unit are controlled by adjusting its operating temperature. Increasing the extraction temperature reduces the solubility of the heavier components of the feedstock, which results in improved deasphalted oil quality but reduced yield of deasphalted oil yield. Subsequent increases in the extractor temperature can further improve the quality of the deasphalted oil by causing further rejection of asphaltene constituents.

Generally, the control of the process may become difficult when rapid changes in temperature occur especially near the critical region because at conditions close to the critical point, the rate of change of solubility is very large. For practical applications, the lower operating temperature is set by the viscosity of the asphaltene phase. The upper limit is to stay below the critical temperature while maintaining the desired yield of deasphalted oil and stable operation. In some cases, a temperature gradient may be maintained along the length of the column with the higher temperature at the top of the column to generate an internal reflux by precipitation of dissolved heavier material—which improves the quality of the deasphalted oil—but a high rate of internal reflux can limit the capacity of the extraction column.

The operating pressure of the extractor is based on the composition of the solvent, which is being used. In the process, sufficient operating pressure must be maintained to ensure the solvent/feedstock mixture in the extractor is in the liquid state. Although the unit may be designed for a range of operating pressure, once it is in operation the extractor pressure may not be typically considered a control variable.

5.3.3.2 Effects of the Solvent-to-Oil Ratio

In general, increasing solvent-to-oil ratio increases the recovery of deasphalted oil with increase in viscosity. The yield of deasphalted oil can be further adjusted with other variables, such as solvent type or the temperature. At higher ratios of solvent to oil, the quality of the deasphalted can be improved by increasing the extraction temperature but with variable decreases in the yield of deasphalted oil. The solvent-to-oil ratio is important from the standpoint of solvent selectivity and the yield advantage at a given product quality at higher solvent ratio varies from feedstock to feedstock and needs to be estimated.

5.3.3.3 Effects of Solvent Types

The yield and quality of the products, which are recovered in a solvent deasphalting unit, are directly related to the solvent composition. Low-boiling hydrocarbons such as propane, butane, and pentane can be used for the process. The solvent power (dissolving ability) of light hydrocarbons increases with increased molecular weight, and when the lowest practical process temperature is reached with a particular solvent, it is necessary to use higher-molecular-weight solvents in order to recover the maximum yield of deasphalted oil. As the molecular weight of the solvent increases, the yield of deasphalted oil also increases but, concurrently, the quality of the deasphalted oil declines, which is reflected in higher viscosity, higher specific gravity, and higher propensity of coke formation (as determined by the Conradson carbon residue text method (Speight, 2014, 2015). Since the deasphalted oil is usually processed in a conversion unit designed to utilize highly active, metal-sensitive catalysts that are incapable of economically processing feedstocks containing more than a few parts per million of organometallics, solvent selection must consider both the desired quantity and quality of the recovered products.

With the continually changing nature of refinery feedstocks, it is preferable that solvent deasphalting units have the flexibility to operate with different solvents. Factors such as market supply and technology of downstream processes can change during the operating life of a solvent deasphalting unit in a refinery. The continued development and improvement of catalyst performance (in catalytic processes) may allow feedstocks with higher metal content and a higher propensity of form coke (measured as the Conradson carbon residue) to produce even higher yields of distillate products (Chapters 9 through 11). In addition, the option for solvent flexibility in a solvent deasphalting unit needs to be a consideration during the design stage of a solvent process (Chapter 12).

5.4   ISOMERIZATION

The importance of *isomerization* in petroleum refining operations is twofold. First, the process is valuable in converting *n*-butane into *iso*-butane, which can be alkylated to liquid hydrocarbons in the gasoline boiling range. Second, the process can be used to increase the octane number of the paraffins boiling in the gasoline boiling range by converting some of the *n*-paraffins present into *iso*-paraffins.

The process involves contact of the hydrocarbon feedstock and a catalyst under conditions favorable to good product recovery. The catalyst may be aluminum chloride promoted with hydrochloric acid or a platinum-containing catalyst. Both are very reactive and can lead to undesirable side reactions along with isomerization. These side reactions include disproportionation and cracking, which decrease the yield and produce olefinic fragments that may combine with the catalyst and shorten its life. These undesired reactions are controlled by such techniques as the addition of inhibitors to the hydrocarbon feed or by carrying out the reaction in the presence of hydrogen.

Paraffins are readily isomerized at room temperature, and the reaction is believed to occur by means of the formation and rearrangement of carbonium ions. The chain-initiating ion R^+ is formed by the addition of a proton from the acid catalyst to an olefin molecule, which may be added, present as an impurity, or formed by dehydrogenation of the paraffin. Except for butane, the isomerization of paraffins is generally accompanied by side reactions involving carbon–carbon bond scissions when catalysts of the aluminum halide type are used. Products boiling both higher and lower than the starting material are formed, and the disproportionation reactions occur with the pentanes and higher paraffins ($>C_5$) are caused by nonpromoted aluminum halide. A substantial pressure of hydrogen tends to minimize these side reactions.

The ease of paraffin isomerization increases with molecular weight, but the extent of disproportionation reactions also increases. Conditions can be established under which isomerization takes place only with the butanes, but this is difficult for the pentanes and higher hydrocarbons. At 27°C (81°F) over aluminum bromide ($AlBr_3$), the equilibrium mixture of *n*-pentane and *iso*-pentane contains over 70% of the branched isomer; at 0°C (32°F) approximately 90% of the branched isomer is present. Higher- and lower-boiling hydrocarbon products, hexane derivatives, heptane derivatives, and *iso*-butane are also formed in side reactions—even at 0°C (32°F)—and in increased amounts when the temperature is raised. Although the thermodynamic conditions are favorable, neo-pentane $[C(CH_3)_4]$ does not appear to isomerize under these conditions.

Olefins are readily isomerized; the reaction involves either movement of the position of the double bond (hydrogen atom shift) or skeletal alteration (methyl group shift). The double-bond shift may also include a reorientation of the groups around the double bond to bring about a cis-trans isomerization. Thus, 1-butene is isomerized to a mixture of cis- and trans-2-butene. Cis (same side) and trans (opposite side) refer to the spatial arrangement of the methyl groups with respect to the double bond. Olefin derivatives having a terminal double bond are the least stable and isomerize more rapidly than olefin derivatives in which the double bond carries the maximum number of alkyl groups.

Naphthenes can isomerize in various ways; for example, in the case of cyclopropane (C_3H_6) and cyclobutane (C_4H_8), ring scission can occur to produce an olefin. Carbon–carbon rupture may also occur in any side chains to produce polymethyl derivatives, whereas cyclopentane (C_5H_{10}) and cyclohexane (C_6H_{12}) rings may expand and contract, respectively.

The isomerization of alkyl aromatic derivatives may involve changes in the side-chain configuration, disproportionation of the substituent groups, or their migration about the nucleus. The conditions needed for isomerization within attached long side chains of alkylbenzenes and alkyl naphthalene derivatives are also those for the scission of such groups from the ring. Such isomerization, therefore, does not take place unless the side chains are relatively short. The isomerization of ethylbenzene to xylenes, and the reverse reaction, occurs readily.

Disproportionation of attached side chains is also a common occurrence; higher and lower alkyl substitution products are formed. For example, xylenes disproportionate in the presence of hydrogen fluoride–boron trifluoride or aluminum chloride to form benzene, toluene, and higher alkylated products; ethylbenzene in the presence of boron trifluoride forms a mixture of benzene and 1,3-diethylbenzene.

5.5 ALKYLATION

Alkylation in the petroleum industry refers to a process for the production of high-octane motor fuel components by the combination of olefins and paraffins. The reaction of *iso*-butane with olefins, using an aluminum chloride catalyst, is a typical alkylation reaction.

In acid-catalyzed alkylation reactions, only paraffins with tertiary carbon atoms, such as *iso*-butane and *iso*-pentane react with the olefin. Ethylene is slower to react than the higher olefins. Olefins higher than propene may complicate the products by engaging in hydrogen exchange reactions.

Cycloparaffins, especially those containing tertiary carbon atoms, are alkylated with olefins in a manner similar to the *iso*-paraffins; the reaction is not as clean, and the yields are low because of the several side reactions that take place. Aromatic hydrocarbons are more easily alkylated than the *iso*-paraffins by olefins. Cumene (*iso*-propylbenzene) is prepared by alkylating benzene with propene over an acid catalyst. The alkylating agent is usually an olefin, although cyclopropane, alkyl halides, aliphatic alcohols, ethers, and esters may also be used. The alkylation of aromatic hydrocarbons is presumed to occur through the agency of the carbonium ion.

Thermal alkylation is also used in some plants, but like thermal cracking, it is presumed to involve the transient formation of neutral free radicals and therefore tends to be less specific in production distribution.

5.6 POLYMERIZATION

Polymerization is a process in which a substance of low molecular weight is transformed into one of the same composition but of higher molecular weight while maintaining the atomic arrangement present in the basic molecules. It has also been described as the successive addition of one molecule to another by means of a functional group, such as that present in an aliphatic olefin.

In the petroleum industry, polymerization is used to indicate the production of, say, gasoline components that fall into a specific (and controlled) molecular weight range, hence the term *polymer gasoline*. Furthermore, it is not essential that only one type of monomer be involved:

$$CH_3CH=CH_2 + CH_2=CH_2 \rightarrow CH_3CH_2CH_2CH=CH_2$$

This type of reaction is correctly called *copolymerization*, but polymerization in the true sense of the word is usually prevented, and all attempts are made to terminate the reaction at the dimer or trimer (three monomers joined together) stage. It is the 4- to 12-carbon compounds that are required as the constituents of liquid fuels. However, in the petrochemical section of the refinery, polymerization, which results in the production of, say, polyethylene, is allowed to proceed until materials of the required high molecular weight have been produced.

5.7 PROCESS CHEMISTRY

In a mixture as complex as petroleum, the reaction processes can only be generalized because of difficulties in analyzing not only the products but also the feedstock as well as the intricate and complex nature of the molecules that make up the feedstock. The formation of coke from the higher-molecular-weight and higher-polarity constituents of a given feedstock is detrimental to process efficiency and to catalyst performance (Speight, 1987; Dolbear, 1998).

Refining the constituents of heavy oil and bitumen has become a major issue in modern refinery practice. The limitations of processing heavy oils and residua depend to a large extent on the amount of higher-molecular-weight constituents (i.e., asphaltene constituents) present in the feedstock (Speight, 1984, 2000, 2004; Schabron and Speight, 1997) that are responsible for high yields of coke in thermal and catalytic processes (Chapters 8 and 9).

5.7.1 THERMAL CHEMISTRY

When petroleum is heated to temperatures in excess of 350°C (660°F), the rate of thermal decomposition of the constituents increases significantly. The higher the temperature, the shorter the time to achieve a given conversion and the *severity* of the process conditions is a combination of residence time of the crude oil constituents in the reactor and the temperature needed to achieve a given conversion.

Thermal conversion does not require the addition of a catalyst. This approach is the oldest technology available for residue conversion, and the severity of thermal processing determines the conversion and the product characteristics. As the temperature and residence time are increased, the primary products undergo further reaction to produce various secondary products, and so on, with the ultimate products (coke and methane) being formed at extreme temperatures of approximately 1000°C (1830°F).

The thermal decomposition of petroleum asphaltene constituents has received some attention (Magaril and Aksenova, 1968; Magaril and Ramazaeva, 1969; Magaril et al., 1970, 1971; Schucker and Keweshan, 1980; Speight, 1990, 1998). Special attention has been given to the nature of the volatile products of asphaltene decomposition mainly because of the difficulty of characterizing the nonvolatile coke.

The organic nitrogen originally in the asphaltene constituents invariably undergoes thermal reaction to concentrate in the nonvolatile coke (Speight, 1970, 1989, 2014; Vercier, 1981). Thus, although asphaltene constituents produce high yields of thermal coke, little is known of the actual chemistry of coke formation. In a more general scheme, the chemistry of asphaltene coking has been suggested to involve the thermolysis of thermally labile bonds to form reactive species that then react with each other to form coke. As part of the coke-forming process, the highly aromatic and highly polar (refractory) products separate from the surrounding oil medium as an insoluble phase and proceed to form coke.

It is also interesting to note that although the aromaticity of the asphaltene constituents is approximately equivalent to the yield of thermal coke (Figure 5.3), not all the original aromatic carbon in the asphaltene constituents forms coke. Volatile aromatic species are eliminated during thermal decomposition, and it must be assumed that some of the original aliphatic carbon plays a role in coke formation.

Various patterns of thermal behavior have been observed for the constituents of petroleum feedstocks (Table 5.1). Since the chemistry of thermal and catalytic cracking has been studied and well resolved, there has been a tendency to focus on the refractory (nonvolatile) constituents. These constituents of petroleum generally produce coke in yields varying from almost zero to more than 60% w/w (Figure 5.4). As an aside, it should also be noted that the differences in thermal behavior of the different subfractions of the asphaltene fraction detract from the concept of average structure. However, the focus of thermal studies has been, for obvious reasons, on the asphaltene constituents that produce thermal coke in amounts varying from approximately 35% by weight to approximately 65% by weight. Petroleum mapping techniques often show the nonvolatile constituents, specifically the asphaltene constituents and the resin constituents, producing coke while the volatile constituents produce distillates. It is often ignored that the asphaltene

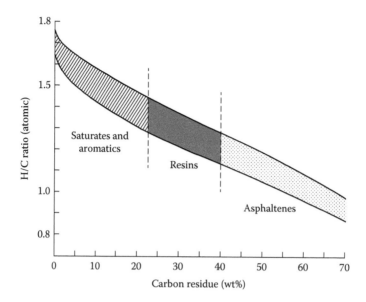

FIGURE 5.3 Yields of thermal coke for various petroleum fractions as determined by the Conradson carbon residue test.

TABLE 5.1
General Indications of Feedstock Cracking

Feedstock Type	Characterization Factor (K)	Naphtha Yield (% v/v)	Coke Yield (% w/w)	Relative Reactivity (Relative Decomposition)
Aromatic	11.0(1)	35.0	13.5	Refractory
Aromatic	11.2(2)	49.6	12.5	Refractory
Aromatic	11.2(1)	37.0	11.5	Refractory
Aromatic–naphthenic	11.4(2)	47.0	9.1	Intermediate
Aromatic–naphthenic	11.4(1)	39.0	9.0	Intermediate
Naphthenic	11.6(2)	45.0	7.1	Intermediate
Naphthenic	11.6(1)	40.0	7.2	Intermediate
Naphthenic–paraffinic	11.8(2)	43.0	5.3	High
Naphthenic–paraffinic	11.8(1)	41.0	6.0	High
Naphthenic–paraffinic	12.0(2)	41.5	4.0	High
Naphthenic–paraffinic	12.0(1)	41.5	5.3	High
Paraffinic	12.2(2)	40.0	3.0	High

Note: 1, cycle oil/cracked feedstocks, 60% conversion; 2, straight-run/uncracked feedstocks, 60% conversion.

constituents also produce high yields (35%–65% by weight) of volatile thermal products that vary from condensable liquids to gases.

It has been generally thought that the chemistry of coke formation involves immediate condensation reactions to produce higher-molecular-weight, condensed aromatic species. And there is the claim that coking is a bimolecular process. However, more recent approaches to the chemistry of coking render the bimolecular process debatable. The rate of decomposition will vary with the nature of the individual constituents, thereby giving rise to the perception of second-order or even multiorder kinetics. The initial reactions of asphaltene constituents involve thermolysis of pendant

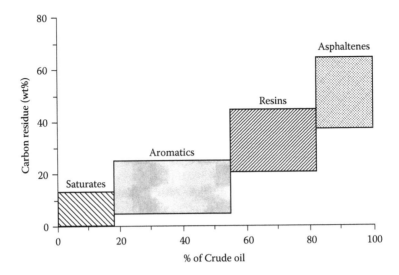

FIGURE 5.4 Illustration of the yields of thermal coke from fractions and subfractions of a specific crude oil as determined by the Conradson carbon residue test.

alkyl chains to form lower-molecular-weight, higher-polarity species (carbenes and carboids), which then react to form coke. Indeed, as opposed to the bimolecular approach, the initial reactions in the coking of petroleum feedstocks that contain asphaltene constituents appear to involve unimolecular thermolysis of asphaltene aromatic–alkyl systems to produce volatile species (paraffins and olefins) and nonvolatile species (aromatics) (Figure 5.5) (Speight, 1987; Schabron and Speight, 1997).

Thermal studies using model compounds confirm that the volatility of the fragments is a major influence in carbon residue formation and a pendant-core model for the high-molecular-weight constituents of petroleum has been proposed (Wiehe, 1994). In such a model, the scission of alkyl side chains occurs, thereby leaving a polar core of reduced volatility that commences to produce a carbon residue (Speight, 1994; Wiehe, 1994). In addition, the pendant-core model also suggests that even one-ring aromatic cores can produce a carbon residue if multiple bonds need to be broken before a core can volatilize (Wiehe, 1994).

In support of the participation of asphaltene constituents in sediment or coke formation, it has been reported that the formation of a coke-like substance during heavy oil upgrading is dependent upon several factors (Storm et al., 1997): (1) the degree of polynuclear condensation in the

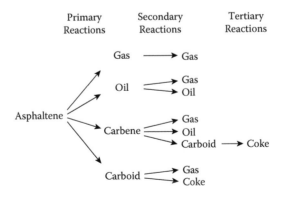

FIGURE 5.5 Multireaction sequence for the thermal decomposition of asphaltene constituents.

feedstock, (2) the average number of alkyl groups on the polynuclear aromatic systems, (3) the ratio of heptane-insoluble material to the pentane-insoluble/heptane-soluble fraction, and (4) the hydrogen-to-carbon atomic ratio of the pentane-insoluble/heptane-soluble fraction. These findings correlate quite well with the proposed chemistry of coke or sediment formation during the processing of heavy feedstocks and even offer some predictability since the characteristics of the whole feedstocks are evaluated.

Nitrogen species also appear to contribute to the pattern of the thermolysis. For example, the hydrogen or carbon–carbon bonds adjacent to ring nitrogen undergo thermolysis quite readily, as if promoted by the presence of the nitrogen atom (Fitzer et al., 1971; Speight, 1998). If it can be assumed that heterocyclic nitrogen plays a similar role in the thermolysis of asphaltene constituents, the initial reactions therefore involve thermolysis of aromatic–alkyl bonds that are enhanced by the presence of heterocyclic nitrogen. An ensuing series of secondary reactions, such as aromatization of naphthenic species and condensation of the aromatic ring systems, then leads to the production of coke. Thus, the initial step in the formation of coke from asphaltene constituents is the formation of volatile hydrocarbon fragments and nonvolatile heteroatom-containing systems.

It has been reported that as the temperature of a 1-methylnaphthalene is raised from 100°C (212°F) to 400°C (750°F), there is a progressive decrease in the particle size of the asphaltene constituents (Thiyagarajan et al., 1995). Furthermore, there is also the inference that the structural integrity of the asphaltene particle is compromised and that irreversible thermochemistry has occurred. Indeed, that is precisely what is predicted and expected from the thermal chemistry of asphaltene fraction and molecular weight studies of asphaltene fraction.

An additional corollary to this work is that conventional models of petroleum asphaltene constituents (which, despite evidence to the contrary, invoked the concept of a large polynuclear aromatic system) offer little, if any, explanation of the intimate events involved in the chemistry of coking. Models that invoke the concept of the asphaltene fractions as a complex solubility class with molecular entities composed of smaller polynuclear aromatic systems are more in keeping with the present data. But the concept of an average structure is not in keeping with the complexity of the fraction and the chemical or thermal reactions of the constituents (Speight, 1994, 2014). Little has been acknowledged here of the role of low-molecular-weight, low-polarity species (resin constituents) in coke formation. However, it is worthy of note that the resin constituents are presumed to be lower-molecular-weight analogs of the asphaltene constituents. This being the case, similar reaction pathways may apply (Koots and Speight, 1975; Speight, 1994, 2014).

Thus, it is now considered more likely that molecular species within the asphaltene fraction, which contains nitrogen and other heteroatoms (and have lower volatility than the pure hydrocarbons), are the prime movers in the production of coke (Speight, 1987). Such species, containing various polynuclear aromatic systems, can be denuded of the attendant hydrocarbon moieties and are undoubtedly insoluble (Bjorseth, 1983; Dias, 1987, 1988) in the surrounding hydrocarbon medium. The next step is gradual carbonization of such entities to form coke (Cooper and Ballard, 1962; Magaril and Aksenova, 1968; Magaril and Ramzaeva, 1969; Magaril et al., 1970).

Thermal processes (such as visbreaking and coking) are the oldest methods for crude oil conversion and are still used in modern refineries. The thermal chemistry of petroleum constituents has been investigated for more than five decades, and the precise chemistry of the lower-molecular-weight constituents has been well defined because of the bountiful supply of pure compounds. The major issue in determining the thermal chemistry of the nonvolatile constituents is, of course, their largely unknown chemical nature and therefore the inability to define their thermal chemistry with any degree of accuracy. Indeed, it is only recently that some light has been cast on the thermal chemistry of the nonvolatile constituents.

Thus, the challenges facing process chemistry and physics are determining (1) the means by which petroleum constituents thermally decompose, (2) the nature of the products of thermal decomposition, (3) the subsequent decomposition of the *primary* thermal products, (4) the interaction of the products with each other, (5) the interaction of the products with the original constituents, and (6) the influence of the products on the composition of the liquids.

When petroleum is heated to temperatures over approximately 410°C (770°F), the thermal or free radical reactions start to crack the mixture at significant rates. Thermal conversion does not require the addition of a catalyst; therefore, this approach is the oldest technology available for residue conversion. The severity of thermal processing determines the conversion and the product characteristics.

Asphaltene constituents are major components of residua and heavy oils and their thermal decomposition has been the focus of much attention (Wiehe, 1993; Gray, 1994; Speight, 1994). The thermal decomposition not only produces high yields (40 wt%) of coke but also, optimistically and realistically, produces equally high yields of volatile products (Speight, 1970). Thus, the challenge in studying the thermal decomposition of asphaltene constituents is to decrease the yields of coke and increase the yields of volatile products.

Several chemical models describe the thermal decomposition of asphaltene constituents (Wiehe, 1993; Gray, 1994; Speight, 1994). Using these available asphaltene models as a guide, the prevalent thinking is that the asphaltene nuclear fragments become progressively more polar as the paraffinic fragments are stripped from the ring systems by scission of the bonds (preferentially) between the carbon atoms alpha and beta to the aromatic rings.

The higher-polarity polynuclear aromatic systems that have been denuded of the attendant hydrocarbon moieties are somewhat less soluble in the surrounding hydrocarbon medium than their *parent* systems (Bjorseth, 1983; Dias, 1987, 1988). Two factors are operative in determining the solubility of the polynuclear aromatic systems in the liquid product. The alkyl moieties that have a solubilizing effect have been removed, and there is also enrichment of the liquid medium in paraffinic constituents. Again, there is an analogy with the deasphalting process (Chapter 12), except that the paraffinic material is a product of the thermal decomposition of the asphaltene molecules and is formed *in situ* rather than being added separately.

The coke has a lower hydrogen-to-carbon atomic ratio than the hydrogen-to-carbon ratio of any of the constituents present in the original crude oil. The hydrocarbon products *may* have a higher hydrogen-to-carbon atomic ratio than the hydrogen-to-carbon ratio of any of the constituents present in the original crude oil or hydrogen-to-carbon atomic ratios at least equal to those of many of the original constituents. It must also be recognized that the production of coke and volatile hydrocarbon products is accompanied by a shift in the hydrogen distribution.

Mild-severity and high-severity processes are frequently used for the processing of residue fractions, whereas conditions similar to those of *ultrapyrolysis* (high temperature and very short residence time) are used commercially only for cracking ethane, propane, butane, and light distillate feeds to produce ethylene and higher olefins.

The formation of solid sediments, or coke, during thermal processes is a major limitation on processing. Furthermore, the presence of different types of solids shows that solubility controls the formation of solids. And the tendency for solid formation changes in response to the relative amounts of the light ends, middle distillates, and residues and to their changing chemical composition during the process (Gray, 1994). In fact, the prime mover in the formation of incompatible products during the processing of feedstocks containing asphaltene constituents is the nature of the primary thermal decomposition products, particularly those fractions that are designated as *carbenes* and *carboids* (Chapters 2 and 3) (Speight, 1987, 1992, 2014; Wiehe, 1992, 1993).

Coke formation during the thermal treatment of petroleum residua is postulated to occur by a mechanism that involves the liquid–liquid-phase separation of reacted asphaltene constituents (which may be *carbenes*) to form a phase that is of low hydrogen content that is substantially nonreactive. The unreacted asphaltene constituents were found to be the fraction with the highest rate of thermal reaction but with the least extent of reaction. This not only described the appearance and disappearance of asphaltene constituents but also quantitatively described the variation in molecular weight and hydrogen content of the asphaltene constituents with reaction time. Thus, the main features in coke formation are as follows: (1) an *induction period* prior to coke formation, (2) a maximum concentration of asphaltene constituents in the reacting liquid, (3) a decrease in the asphaltene concentration that parallels the decrease in heptane-soluble material, and (4) high reactivity of the unconverted asphaltene constituents.

The induction period has been observed experimentally by many previous investigators (Levinter et al., 1966, 1967; Magaril and Aksenova, 1968; Valyavin et al., 1979; Takatsuka et al., 1989a) and makes visbreaking and the Eureka processes possible. The postulation that coke formation is triggered by the phase separation of asphaltene constituents (Magaril et al., 1971) led to the use of linear variations of the concentration of each fraction with reaction time, resulting in the assumption of zero-order kinetics rather than first-order kinetics. More recently (Yan, 1987), coke formation in visbreaking was described as resulting from a phase-separation step, but the phase-separation step was not included in the resulting kinetic model for coke formation.

This model represents the conversion of asphaltene constituents over the entire temperature range and of heptane-soluble materials in the coke induction period as first-order reactions. The data also show that the four reactions give simultaneously lower aromatic and higher aromatic products, on the basis of other evidence (Wiehe, 1992). Also, the previous work showed that residua fractions can be converted without completely changing solubility classes (Magaril et al., 1971) and that coke formation is triggered by the phase separation of converted asphaltene constituents.

The maximum solubility of these product asphaltene constituents is proportional to the total heptane-soluble materials, as suggested by the observation that the decrease in asphaltene constituents parallels the decrease of heptane-soluble materials. Finally, the conversion of the insoluble product asphaltene constituents into toluene-insoluble coke is pictured as producing a heptane-soluble by-product, which provides a mechanism for the heptane-soluble conversion to deviate from first-order behavior once coke begins to form. In support of this assumption, it is known (Langer et al., 1961) that partially hydrogenated refinery process streams provide reactive hydrogen and as a result, inhibit coke formation during residuum thermal conversion. Thus, the heptane-soluble fraction of a residuum that contains naturally occurring partially hydrogenated aromatics can provide reactive hydrogen during thermal reactions. As the conversion proceeds, the concentration of asphaltene cores continues to increase and the heptane-soluble fraction continues to decrease until the solubility limit, S_L, is reached. Beyond the solubility limit, the excess asphaltene core, A^*_{ex}, phase separate to form a second liquid phase that contains little reactive hydrogen. In this new phase, asphaltene radical–asphaltene radical recombination is quite frequent, causing a rapid reaction to form solid coke and a by-product of a heptane-soluble core.

The asphaltene concentration varies little in the coke induction period (Wiehe, 1993) but then decreases once coke begins to form. Observing this, it might be concluded that asphaltene constituents are unreactive, but it is the high reactivity of the asphaltene constituents down to the asphaltene core that offsets the generation of asphaltene cores from the heptane-soluble materials to keep the overall asphaltene concentration nearly constant.

Previously, it was demonstrated (Schucker and Keweshan, 1980; Savage et al., 1988) that the hydrogen-to-carbon atomic ratio of the asphaltene constituents decreases rapidly with reaction time for asphaltene thermolysis and then approaches an asymptotic limit at long reaction times, which provides qualitative evidence for asphaltene cracking down to a core.

The measurement of the molecular weight of the asphaltene fraction and the various sub-fractions is known to give different values depending on the technique, the solvent, and the (Dickie and Yen, 1967; Moschopedis et al., 1976; Speight et. al., 1985; Speight, 2014, 2015). As shown by small-angle x-ray (Kim and Long, 1979) and neutron (Overfield et al., 1989) scattering, asphaltene constituents tend to self-associate and form aggregates.

Thus, coke formation is a complex process involving both chemical reactions and thermodynamic behavior. Reactions that contribute to this process are (1) cracking of side chains from aromatic groups, (2) dehydrogenation of naphthenes to form aromatics, (3) condensation of aliphatic structures to form aromatics, (4) condensation of aromatics to form higher fused-ring aromatics, and (5) dimerization or oligomerization reactions. Loss of side chains always accompanies thermal cracking, and dehydrogenation and condensation reactions are favored by hydrogen-deficient conditions.

The importance of solvents in coking has been recognized for many years (e.g., Langer et al., 1961), but their effects have often been ascribed to hydrogen donor reactions rather than phase behavior. The separation of the phases depends on the solvent characteristics of the liquid. Addition of aromatic solvents suppresses phase separation, whereas paraffins enhance separation. Microscopic examination of coke particles often shows evidence for the presence of mesophase, spherical domains that exhibit the anisotropic optical characteristics of liquid crystals.

This phenomenon is consistent with the formation of a second liquid phase; the mesophase liquid is denser than the rest of the hydrocarbon, has a higher surface tension, and probably wets metal surfaces better than the rest of the liquid phase. The mesophase characteristic of coke diminishes as the liquid phase becomes more compatible with the aromatic material.

The phase-separation phenomenon that is the prelude to coke formation can also be explained by the use of the solubility parameter, δ, for petroleum fractions and for the solvents (Yen, 1984; Speight, 1994, 2014). As an extension of this concept, there is sufficient data to draw a correlation between the atomic hydrogen/carbon ratio and the solubility parameter for hydrocarbons and the constituents of the lower-boiling fractions of petroleum (Speight, 1994). Recognition that hydrocarbon liquids can dissolve polynuclear hydrocarbons, a case in which there is usually less than a three-point difference between the lower solubility parameter of the solvent and the higher solubility parameter of the solute. Thus, a parallel, or near-parallel, line can be assumed, which allows the solubility parameter of the asphaltene constituents and resin constituents to be estimated.

By this means, the solubility parameter of asphaltene constituents can be estimated to fall in the range 9–12, which is in keeping with the asphaltene fraction being composed of a mixture of different compound types with an accompanying variation in polarity. Removal of alkyl side chains from the asphaltene constituents decreases the hydrogen-to-carbon atomic ratio (Wiehe, 1993; Gray, 1994) and increases the solubility parameter, thereby bringing about a concurrent decrease of the asphaltene product in the hydrocarbon solvent.

In fact, on the molecular weight–polarity diagram for asphaltene constituents, carbenes and carboids can be shown as lower-molecular-weight, highly polar entities in keeping with molecular fragmentation models (Speight, 1994). If this increase in polarity and solubility parameter (Mitchell and Speight, 1973) is too drastic relative to the surrounding medium (Figure 5.6), phase separation will occur. Furthermore, the available evidence favors a multistep mechanism rather than a stepwise mechanism (Figure 5.7) as the means by which the thermal decomposition of petroleum constituents occurs.

Any chemical or physical interactions (especially thermal effects) that cause a change in the solubility parameter of the solute relative to that of the solvent will also cause *incompatibility* which can be called *instability*, *phase separation*, *sediment formation*, or *sludge formation*.

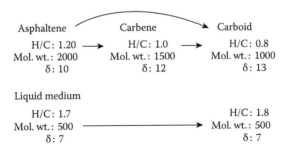

FIGURE 5.6 Illustration of the changes in the solubility parameter of the various fractions of petroleum during thermal treatment.

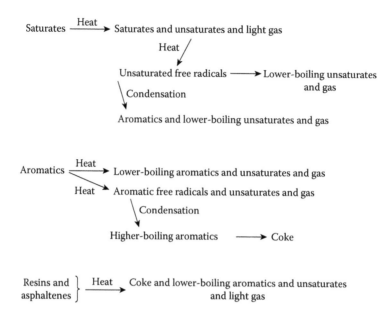

FIGURE 5.7 Simplified schematic of the thermal decomposition of petroleum constituents.

Instability or *incompatibility* resulting in the separation of solids during refining can occur during a variety of process, either by intent (such as in the deasphalting process) or inadvertently when the separation is detrimental to the process (Table 5.2) (Mushrush and Speight, 1995; Speight, 2014). Thus, separation of solids occurs whenever the solvent characteristics of the liquid phase are no longer adequate to maintain polar and/or high-molecular-weight material in solution. Examples of such occurrences are as follows: (1) asphaltene separation, which occurs when the paraffin content or character of the liquid medium increases; (2) wax separation, which occurs when there is a drop in temperature or the aromatic content or character of the liquid medium increases; (3) sludge or sediment formation in a reactor, which occurs when the solvent characteristics of the liquid medium change so that asphalt or wax materials separate; (4) coke formation, which occurs at high temperatures and commences when the solvent power of the liquid phase is not sufficient to maintain the coke precursors in solution; and (5) sludge or sediment formation in fuel products, which occurs because of the interplay of several chemical and physical factors.

TABLE 5.2
Properties Related to Instability and Incompatibility

Property	Comments
Asphaltene constituents	Interact readily with catalyst
	Thermal alteration leading to phase separation
	Phase separation in paraffinic medium
Heteroatom constituents	Thermally labile
	React readily with oxygen
	Provide polarity to feedstock (or products)
Aromatic constituents	May be incompatible with paraffin medium
	Phase separation of paraffin constituents
Nonasphaltene constituents	Thermal alteration that causes changes in polarity
	Phase separation of polar species in products

This mechanism also appears to be operable during residua hydroconversion, which has included a phase-separation step (the formation of *dry sludge*) in a kinetic model but was not included as a preliminary step to coke formation in a thermal cracking model (Takatsuka et al., 1989a,b; Andersen and Speight, 2001; Speight, 2004a,b,c; Ancheyta et al., 2005a).

5.7.2 HYDROCONVERSION CHEMISTRY

There have also been many attempts to focus attention on the asphaltene constituents during hydrocracking studies. The focus has been on the macromolecular changes that occur by investigation of the changes to the generic fractions, that is, the asphaltene constituents, the resin constituents, and the other fractions that make up such a feedstock (Ancheyta and Speight, 2007). In terms of hydroprocessing, the means by which asphaltene constituents are desulfurized, as one step of a hydrocracking operation, is also suggested as part of the process. This concept can then be taken one step further to show the dealkylation of the aromatic systems as a definitive step in the hydrocracking process (Speight, 1987).

When catalytic processes are employed, complex molecules (such as those that may be found in the original asphaltene fraction or those formed during the process) are not sufficiently mobile (or are too strongly adsorbed by the catalyst) to be saturated by the hydrogenation components. Hence, these molecular species continue to condense and eventually degrade to coke. These deposits deactivate the catalyst sites and eventually interfere with the process.

Several noteworthy attempts have been made to focus attention on the asphaltene constituents during hydroprocessing studies. The focus has been on the macromolecular changes that occur by investigation of the changes in the generic fractions, that is, the asphaltene constituents, the resin constituents, and the other fractions that make up such a feedstock. This option suggests that the overall pathway by which hydrotreating and hydrocracking of heavy oils and residua occur involves a stepwise mechanism:

Asphaltene constituents → Resin-type constituents (polar aromatics)

Resin-type constituents → Aromatic constituents

Aromatic constituents → Saturate constituents

A direct step from either the asphaltene constituents or the resin constituents to the saturates is not considered a predominant pathway for hydroprocessing.

The means by which asphaltene constituents are desulfurized, as one step of a hydrocracking operation, is also suggested as part of this process. This concept can then be taken one step further to show the dealkylation of the aromatic systems as a definitive step in the hydrocracking process (Speight, 1987). It is also likely that molecular species (within the asphaltene fraction) that contain nitrogen and other heteroatoms, and have lower volatility than their hydrocarbon analogs, are the prime movers in the production of coke (Speight, 1987).

When catalytic processes are employed, complex molecules such as those that may be found in the original asphaltene fraction or those formed during the process are not sufficiently mobile (or are too strongly adsorbed by the catalyst) to be saturated by the hydrogenation components and, hence, continue to condense and eventually degrade to coke. These deposits deactivate the catalyst sites and eventually interfere with the hydroprocess.

A convenient means of understanding the influence of feedstock on the hydrocracking process is through a study of the hydrogen content (hydrogen-to-carbon atomic ratio) and molecular weight (carbon number) of the feedstocks and products. Such data show the extent to which the carbon number must be reduced and/or the relative amount of hydrogen must be added to generate the desired lower-molecular-weight, hydrogenated products. In addition, it is possible to use data for hydrogen usage in residuum processing, where the relative amount of hydrogen consumed in the process can be shown to be dependent upon the sulfur content of the feedstock.

5.7.3 Chemistry in the Refinery

Thermal cracking processes are commonly used to convert petroleum residua into distillable liquid products, although thermal cracking processes as used in the early refineries are no longer in use. Examples of modern thermal cracking processes are *visbreaking* and *coking* (*delayed coking*, *fluid coking*, and *flexicoking*) (Chapter 8). In all of these processes, the simultaneous formation of sediment or coke limits the conversion to usable liquid products. However, for the purposes of this section, the focus will be on the visbreaking and hydrocracking processes. The coking processes in which the reactions are taken to completion with the maximum yields of products are not a part of this discussion.

5.7.3.1 Visbreaking

To study the thermal chemistry of petroleum constituents, it is appropriate to select the visbreaking process (a *carbon rejection process*) (Chapters 4 and 8) and the hydrocracking process (a *hydrogen addition process*) (Chapters 4 and 11) as used in a modern refinery. The processes operate under different conditions (Figure 5.8) and have different levels of conversion (Figure 5.9), and although they do offer different avenues for conversion, these processes are illustrative of the thermal chemistry that occurs in refineries.

The visbreaking process (Chapters 4 and 8) is primarily a means of reducing the viscosity of heavy feedstocks by *controlled thermal decomposition* insofar as the hot products are quenched before complete conversion can occur (Speight and Ozum, 2002; Parkash, 2003; Hsu and Robinson, 2006; Gary et al., 2007; Speight, 2014). However, the process is often plagued by sediment formation in the products. This sediment, or sludge, must be removed if the products are to meet fuel oil specifications. The process (Figure 5.10) uses the mild thermal cracking (*partial conversion*) as a relatively low-cost and low-severity approach to improving the viscosity characteristics of the residue without attempting significant conversion to distillates. Low residence times are required to avoid coking reactions, although additives can help to suppress coke deposits on the tubes of the furnace (Allan et al., 1983).

A visbreaking unit consists of a reaction furnace, followed by quenching with a recycled oil, and fractionation of the product mixture. All of the reaction in this process occurs as the oil flows through the tubes of the reaction furnace. The severity is controlled by the flow rate through the

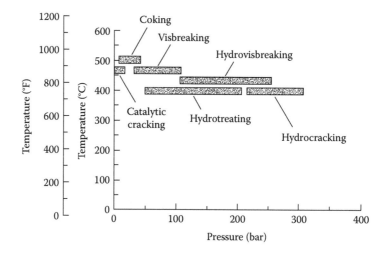

FIGURE 5.8 Temperature and pressure ranges for various processes.

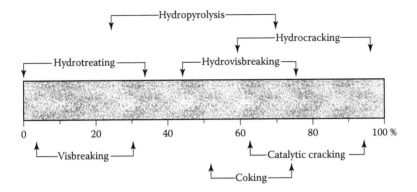

FIGURE 5.9 Feedstock conversion in various processes.

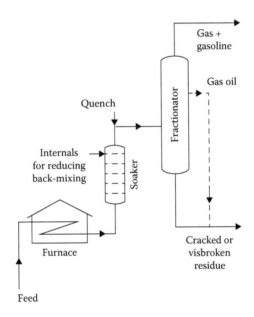

FIGURE 5.10 The visbreaking process using a soaker. (From OSHA Technical Manual, Section IV, Chapter 2, Petroleum refining processes, 1999, http://www.osha.gov/dts/osta/otm/otm_iv/otm_iv_2.html.)

furnace and the temperature; typical conditions are 475°C–500°C (885°F–930°F) at the furnace exit with a residence time of 1–3 minutes. Operation for 3–6 months onstream (continuous use) is possible before the furnace tubes must be cleaned and the coke removed. The operating pressure in the furnace tubes varies over a considerable range depending on the degree of vaporization and the residence time desired. For a given furnace tube volume, a lower operating pressure will reduce the actual residence time of the liquid phase.

The reduction in viscosity of the unconverted residue tends to reach a limiting value with conversion, although the total product viscosity can continue to decrease. Conversion of residue in visbreaking follows first-order reaction kinetics (Henderson and Weber, 1965). The minimum viscosity of the unconverted residue can lie outside the range of allowable conversion if sediment begins to form (Rhoe and de Blignieres, 1979). When pipelining of the visbreaker product is a process objective, a diluent such as gas condensate can be added to achieve a further reduction in viscosity.

The high viscosity of the heavier feedstocks and residua is thought to be due to entanglement of the high-molecular-weight components of the oil and the formation of ordered structures in the liquid phase. Thermal cracking at low conversion can remove side chains from the asphaltene constituents and break bridging aliphatic linkages. A 5%–10% conversion of atmospheric residue to naphtha is sufficient to reduce the entanglements and structures in the liquid phase and give at least a fivefold reduction in viscosity.

The stability of visbroken products is also an issue that might be addressed at this time. Using this simplified model, visbroken products might contain polar species that have been denuded of some of the alkyl chains and that, on the basis of solubility, might be more rightly called *carbenes* and *carboids*, but an induction period is required for phase separation or agglomeration to occur. Such products might initially be *soluble* in the liquid phase but after the induction period, cooling, and/or diffusion of the products, incompatibility (phase separation, sludge formation, agglomeration) occurs.

On occasion higher temperatures are employed in various reactors as it is often assumed that, if no side reactions occur, longer residence times at a lower temperature are equivalent to shorter residence times at a higher temperature. However, this assumption does not acknowledge the change in thermal chemistry that can occur at the higher temperatures, irrespective of the residence time. Thermal conditions can, indeed, induce a variety of different reactions in crude oil constituents so that selectivity for a given product may change considerably with temperature. The onset of secondary, tertiary, and even quaternary reactions under the more extreme high-temperature conditions can convert higher-molecular-weight constituents of petroleum to low-boiling distillates, butane, propane, ethane, and (ultimately) methane. Caution is advised in the use of extreme temperatures.

Obviously, the temperature and residence time of the asphaltene constituents in the reactor are key to the successful operation of a visbreaker. A visbreaking unit must operate in temperature and residence time regimes that do not promote the formation of sediment (often referred to as coke). However, as already noted, there is a *break point* above which considering might be increased but the possibility of sediment deposition increases. At the temperatures and residence times outside of the most beneficial temperature and residence time regimes, thermal changes to the asphaltene constituents cause phase separation of a solid product that then progresses to coke. Furthermore, it is in such operations that models derived from *average parameters* can be ineffective and misleading. For example, the amphoteric constituents of the asphaltene fraction are more reactive than the less polar constituents (Speight, 2014). The thermal products from the amphoteric constituents form first and will separate out from the reaction matrix before other products (Figure 5.11). Under such conditions, models based on average structural parameters or on average properties will not predict early phase separation to the detriment of the product and the process as a whole.

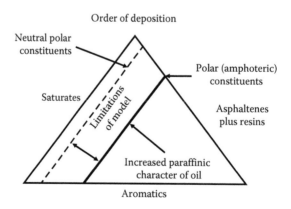

FIGURE 5.11 The limitations of the visbreaking process.

Knowing the actual nature of the subtypes of the asphaltene constituents is obviously beneficial and will allow steps to be taken to correct any such unpredictable occurrence. Indeed, the concept of hydrovisbreaking (visbreaking in the presence of hydrogen) could be of valuable assistance when high-asphaltene-content feedstocks are used.

5.7.3.2 Hydroprocessing

Hydrotreating is the (relatively) low-temperature removal of heteroatomic species by the treatment of a feedstock or product in the presence of hydrogen (Chapters 4 and 10). On the other hand, *hydrocracking* (Figure 5.12) is the thermal decomposition of a feedstock in which carbon–carbon bonds are cleaved in addition to the removal of heteroatomic species (Speight, 2014). The presence of hydrogen changes the nature of the products (especially the decreasing coke yield) by preventing the buildup of precursors that are incompatible in the liquid medium and form coke (Magaril and Aksenova, 1968; Magaril and Ramazaeva, 1969; Magaril et al., 1970; Speight and Moschopedis, 1979). In fact, the chemistry involved in the reduction of asphaltene constituents to liquids using models in which where the polynuclear aromatic system borders on graphitic is difficult to visualize let alone justify. However, the *paper chemistry* derived from the use of a molecularly designed model composed of smaller polynuclear aromatic systems is much easier to visualize (Speight, 1994, 2014). But precisely how asphaltene constituents react with the catalysts is open to much more speculation.

In contrast to the visbreaking process, in which the general principle is the production of products for use as fuel oil, the hydroprocessing is employed to produce a slate of products for use as liquid fuels. Nevertheless, the decomposition of asphaltene constituents is, again, an

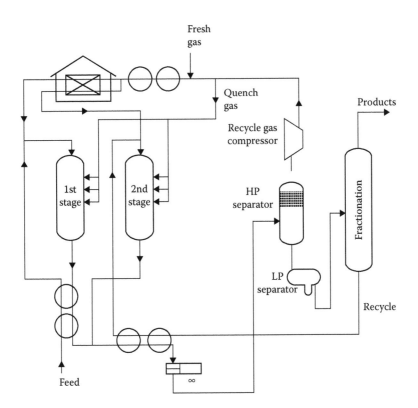

FIGURE 5.12 A two-stage hydrocracking unit. (From OSHA Technical Manual, Section IV, Chapter 2, Petroleum refining processes, 1999, http://www.osha.gov/dts/osta/otm/otm_iv/otm_iv_2.html.)

issue, and just as models consisting of large polynuclear aromatic systems are inadequate to explain the chemistry of visbreaking, they are also of little value for explaining the chemistry of hydrocracking.

Deposition of solids or incompatibility is still possible when asphaltene constituents interact with catalysts, especially acidic support catalysts, through the functional groups, for example, the basic nitrogen species just as they interact with adsorbents. And there is a possibility for interaction of the asphaltene with the catalyst through the agency of a single functional group in which the remainder of the asphaltene molecule remains in the liquid phase. There is also a less desirable option in which the asphaltene reacts with the catalyst at several points of contact, causing immediate incompatibility on the catalyst surface.

There is evidence to show that during the early stages of the hydrotreating process, the chemistry of the asphaltene constituents follows the same routes as the thermal chemistry (Ancheyta et al., 2005a). Thus, initially there is an increase in the amount of asphaltene constituents followed by a decrease indicating that, in the early stages of the process, resin constituents are being converted to asphaltene material by aromatization and by some dealkylation. In addition, aromatization and dealkylation of the original asphaltene constituents yield asphaltene products that are of higher polarity and lower molecular weight than that of original asphaltene constituents. Analogous to the thermal processes, this produces an overall asphaltene fraction that is more polar material and also of lower molecular weight. As the hydrotreating process proceeds, the amount of asphaltene constituents precipitated decreases due to conversion of the asphaltene constituents to products. At more prolonged onstream times, there is a steady increase in the yield of the asphaltene constituents. This is accompanied by a general increase in the molecular weight of the precipitated material.

These observations are in keeping with observations for the thermal reactions of asphaltene constituents in the absence in hydrogen where the initial events are a reduction in the molecular weight of the asphaltene constituents leading to lower molecular weight by more polar products that are derived from the asphaltene constituents but are often referred to as *carbenes* and *carboids*. As the reaction progresses, these derived products increase in molecular weight and eventually become insoluble in the reaction medium, deposit on the catalyst, and form coke.

REFERENCES

Allan, D.E., Martinez, C.H., Eng, C.C., and Barton, W.J. 1983. Visbreaking gains renewed interest. *Chemical Engineering Progress*, 79(1): 85–89.

Ancheyta, J., Centeno, G., Trejo, F., Betancourt, G., and Speight, J.G. 2005b. Asphaltene characterization as a function of time on-stream during hydroprocessing of maya crude. *Catalysis Today*, 109: 162.

Ancheyta, J., Sánchez, S., and Rodrıguez, M.A. 2005a. Kinetic modeling of hydrocracking of heavy oil fractions: A review. *Catalysis Today*, 109: 76–92.

Ancheyta, J. and Speight, J.G. 2007. Feedstock evaluation and composition. In: *Hydroprocessing of Heavy Oils and Residua*. J. Ancheyta and J.G. Speight (Editors). CRC Press/Taylor & Francis Group, Boca Raton, FL, Chapter 2.

Andersen, S.I. and Speight, J.G. 2001. Petroleum reins: Separation, character, and role in petroleum. *Petroleum Science and Technology*, 19: 1.

Bahmani, M., Sadighi, S., Mashayekhi, M., SeifMohaddecy, S.R., and Vakili, D. 2007. Maximizing naphtha and diesel yields of an industrial hydrocracking unit with minimal changes. *Petroleum & Coal*, 49(1): 16–20.

Bjorseth, A. 1983. *Handbook of Polycyclic Aromatic Hydrocarbons*. Marcel Dekker, New York.

Boduszynski, M.M. 1987. Composition of heavy petroleum. 1. Molecular weight, hydrogen deficiency, and heteroatom concentration as a function of atmospheric equivalent boiling point up to 1400°F (760°C). *Energy & Fuels*, 1: 2–11.

Boduszynski, M.M. 1988. Composition of heavy petroleum. 2. Molecular characterization. *Energy & Fuels*, 2: 597–613.

Cooper, T.A. and Ballard, W.P. 1962. In: *Advances in Petroleum Chemistry and Refining*, Volume 6. K.A. Kobe and J.J. McKetta (Editors). Interscience, New York, Chapter 4.

Cumming, K.A. and Wojciechowski, B.W. 1996. *Catalysis Reviews—Science and Engineering*, 38: 101–157.

Dias, J.R. 1987. *Handbook of Polycyclic Hydrocarbons. Part A. Benzenoid Hydrocarbons*. Elsevier, New York.

Dias, J.R. 1988. *Handbook of Polycyclic Hydrocarbons. Part B. Polycyclic Isomers and Heteroatom Analogs of Benzenoid Hydrocarbons*. Elsevier, New York.

Dickie, J.P. and Yen, T.F. 1967. Macrostructures of the asphaltic fractions by various instrumental methods. *Analytical Chemistry*, 39: 1847–1852.

Dolbear, G.E. 1998. Hydrocracking: Reactions, catalysts, and processes. In: *Petroleum Chemistry and Refining*. J.G. Speight (Editor). Taylor & Francis Group, Washington, DC, Chapter 7.

Dolbear, G.E., Tang, A., and Moorehead, E.L. 1987. Upgrading studies with California, Mexican, and Middle Eastern heavy oils. In: *Metal Complexes in Fossil Fuels*. R.H. Filby and J.F. Branthaver (Editors). American Chemical Society, Washington, DC, pp. 220–232.

Ebert, L.B., Mills, D.R., and Scanlon, J.C. 1987. Preprints. *American Chemical Society, Division of Petroleum Chemistry*, 32(2): 419.

Eliel, E. and Wilen, S. 1994. *Stereochemistry of Organic Compounds*. John Wiley & Sons, New York.

Fabuss, B.M., Smith, J.O., and Satterfield, C.N., 1964. Thermal cracking of pure saturated hydrocarbons. In: *Advances in Petroleum Chemistry and Refining*, Volume 3. J.J. McKetta (Editor). John Wiley & Sons, New York, pp. 156–201.

Fitzer, E., Mueller, K., and Schaefer, W. 1971. The chemistry of the pyrolytic conversion of organic compounds to carbon. *Chemistry and Physics of Carbon*, 7: 237–383.

Furimsky, E. 1983. Thermochemical and mechanistic aspects of removal of sulfur, nitrogen and oxygen from petroleum. *Erdöl und Kohle*, 36: 518.

Gary, J.G., Handwerk, G.E., and Kaiser, M.J. 2007. *Petroleum Refining: Technology and Economics*, 5th Edition. CRC Press/Taylor & Francis Group, Boca Raton, FL.

Gray, M.R. 1994. *Upgrading Petroleum Residues and Heavy Oils*. Marcel Dekker, New York.

Hajji, A.A., Muller, H., and Koseoglu, O.R. 2010. Molecular details of hydrocracking feedstocks. *Saudi Aramco Journal of Technology Spring*, 1–12.

Henderson, J.H. and Weber, L. 1965. Physical upgrading of heavy oils by the application of heat. *Journal of Canadian Petroleum Technology*, 4: 206–212.

Hsu, C.S. and Robinson, P.R. 2006. *Practical Advances in Petroleum Processing*, Volumes 1 and 2. Springer, New York.

Hurd, C.D. 1929. *The Pyrolysis of Carbon Compounds*. The Chemical Catalog Company Inc., New York.

Keller, W.D. 1985. Clays. In: *Kirk Othmer Concise Encyclopedia of Chemical Technology*. M. Grayson (Editor). Wiley Interscience, New York, p. 283.

Kim, H. and Long, R.B. 1979. *Journal of Industrial and Engineering Chemistry Fundamentals*, 18: 60.

King, P.J., Morton, F., and Sagarra, A. 1973. In: *Modern Petroleum Technology*. G.D. Hobson and W. Pohl (Editors). Applied Science Publishers, Essex, UK.

Koots, J.A. and Speight J.G. 1975. The relation of petroleum resins to asphaltenes. *Fuel*, 54: 179–184.

Langer, A.W., Stewart, J., Thompson, C.E., White, H.T., and Hill, R.M. 1961. *Journal of Industrial and Engineering Chemistry*, 53: 27.

Laszlo, P. 1995. *Organic Reactions: Logic and Simplicity*. John Wiley & Sons, New York.

Levinter, M.E., Medvedeva, M.I., Panchenkov, G.M., Agapov, G.I., Galiakbarov, M.F., and Galikeev, R.K. 1967. *Khimiya i Tekhnologiya Topliv i Masel*, 4: 20.

Levinter, M.E., Medvedeva, M.I., Panchenkov, G.M., Aseev, Y.G., Nedoshivin, Y.N., Finkelshtein, G.B., and Galiakbarov, M.F. 1966. *Khimiya i Tekhnologiya Topliv i Masel*, 9: 31.

Magaril, R.Z. and Akensova, E.I. 1967. Mechanism of coke formation during the cracking of petroleum tars. *IzvestiaVysshUcheb. Zaved., Neft Gas*, 10(11): 134–136.

Magaril, R.Z. and Akensova, E.I. 1968. Study of the mechanism of coke formation in the cracking of petroleum resins. *International Journal of Chemical Engineering*, 8(4): 727–729.

Magaril, R.Z. and Aksenova, E.I. 1970a. Mechanism of coke formation in the thermal decompositon of asphaltenes. *Khimiya i Tekhnologiya Topliv i Masel*, 15(7): 22–24.

Magaril, R.Z. and Aksenova, E.I. 1970b. Kinetics and mechanism of coking asphaltenes. *Khim. IzvestiaVyssh. Ucheb. Zaved. Neft Gaz.*, 13(5): 47–53.

Magaril, R.Z. and Aksenova, E.I. 1972. Coking kinetics and mechanism of asphaltenes. *Khim. Kim Tekhnol., Tr. Tyumen Ind. Inst.*, 169–172.

Magaril, R.Z. and Ramazaeva, L.F. 1969. Study of carbon formation in the thermal decomposition of asphaltenes in solution. *IzvestiaVyssh. Ucheb. Zaved. Neft Gaz.*, 12(1): 61–64.

Magaril, R.Z., Ramazaeva, L.F., and Aksenova, E.I. 1970. Kinetics of coke formation in the thermal processing of petroleum. *Khimiya i Tekhnologiya Topliv i Masel*, 15(3): 15–16.

Magaril, R.Z., Ramazaeva, L.F., and Aksenova, E.I. 1971. Kinetics of the formation of coke in the thermal processing of crude oil. *International Journal of Chemical Engineering*, 11(2): 250–251.

Masel, R.I. 1995. *Principles of Adsorption and Reaction on Solid Surfaces*. John Wiley & Sons, New York.

Mitchell, D.L. and Speight, J.G. 1973. The solubility of asphaltenes in hydrocarbon solvents. *Fuel*, 52: 149.

Moschopedis, S.E., Fryer, J.F., and Speight, J.G. 1976. An investigation of asphaltene molecular weights. *Fuel*, 55: 227.

Mushrush, G.W. and Speight, J.G. 1995. *Petroleum Products: Instability and Incompatibility*. Taylor & Francis Group, Philadelphia, PA.

OSHA Technical Manual. 1999. Section IV, Chapter 2: Petroleum refining processes. http://www.osha.gov/dts/osta/otm/otm_iv/otm_iv_2.html.

Overfield, R.E., Sheu, E.Y., Sinha, S.K., and Liang, K.S. 1989. SANS study of asphaltene aggregation. *Fuel Science and Technology International*, 7: 611.

Parkash, S. 2003. *Refining Processes Handbook*. Gulf Professional Publishing, Elsevier, Amsterdam, the Netherlands.

Pines, H. 1981. *The Chemistry of Catalytic Hydrocarbon Conversions*. Academic Press, New York.

Rhoe, A. and de Blignieres, C. 1979. *Hydrocarbon Processing*, 58(1): 131–136.

Samorjai, G.A. 1994. *Introduction to Surface Chemistry and Catalysis*. John Wiley & Sons, New York.

Savage, P.E., Klein, M.T., and Kukes, S.G. 1988. Asphaltene reaction pathways 3. Effect of reaction environment. *Energy & Fuels*, 2: 619–628.

Schabron, J.F. and Speight, J.G. 1997. An evaluation of the delayed coking product yield of heavy feedstocks using asphaltene content and carbon residue. *Revue de l'Institut Français du Pétrole*, 52(1): 73–85.

Schucker, R.C. and Keweshan, C.F. 1980. Reactivity of cold lake asphaltenes. Preprints. *American Chemical Society, Division of Fuel Chemistry.*, 25: 155.

Smith, M.B. 1994. *Organic Synthesis*. McGraw-Hill Inc., New York.

Speight, J.G. 1970. Thermal cracking of athabasca bitumen, athabasca asphaltenes, and athabasca deasphalted heavy oil. *Fuel*, 49: 134.

Speight, J.G. 1984. Upgrading heavy oils and residua: The nature of the problem. In: *Catalysis on the Energy Scene*. S. Kaliaguine and A. Mahay (Editors). Elsevier, Amsterdam, the Netherlands, p. 515.

Speight, J.G. 1987. Initial reactions in the coking of residua. Preprints. *American Chemical Society, Division of Petroleum Chemistry*, 32(2): 413.

Speight, J.G. 1989. Thermal decomposition of asphaltenes. *Neftekhimiya*, 29: 732.

Speight, J.G. 1990. The chemistry of the thermal degradation of petroleum asphaltenes. *ActaPetroleiSinica*, 6(1): 29.

Speight, J.G. 1992. A chemical and physical explanation of incompatibility during refining operations. *Proceedings of Fourth International Conference on the Stability and Handling of Liquid Fuels*. US. Department of Energy (DOE/CONF-911102), Washington, DC, p. 169.

Speight, J.G. 1994. Chemical and physical studies of petroleum asphaltenes. In: *Asphalts and Asphaltenes*, Volume 1. T.F. Yen and G.V. Chilingarian (Editors). Elsevier, Amsterdam, the Netherlands, Chapter 2.

Speight, J.G. 1998. Thermal chemistry of petroleum constituents. In: *Petroleum Chemistry and Refining*. J.G. Speight (Editor). Taylor & Francis Group, Washington, DC. Chapter 5.

Speight, J.G. 2000. *The Desulfurization of Heavy Oils and Residua*, 2nd Edition. Marcel Dekker, New York.

Speight, J.G. 2004a. New approaches to hydroprocessing. *Catalysis Today*, 98(1–2): 55–60.

Speight, J.G. 2004b. Petroleum asphaltenes. Part 1: Asphaltenes, Resins, and the structure of petroleum. *Revue de l'Institut Français du Pétrole*, 59: 467.

Speight, J.G. 2004c. Petroleum asphaltenes. Part 2: The effect of asphaltene and resin constituents on recovery and refining processes. *Revue de l'Institut Français du Pétrole—Oil & Gas Science and Technology*, 59(5): 479–488.

Speight, J.G. 2014. *The Chemistry and Technology of Petroleum*, 5th Edition. CRC Press/Taylor & Francis Group, Boca Raton, FL.

Speight, J.G. 2015. *Handbook of Petroleum Product Analysis*, 2nd Edition. John Wiley & Sons, Hoboken, NJ.

Speight, J.G. and Moschopedis, S.E. 1979. The production of low-sulfur liquids and coke from athabasca bitumen. *Fuel Processing Technology*, 2: 295.

Speight, J.G. and Ozum, B. 2002. *Petroleum Refining Processes*. Marcel Dekker, New York.

Speight, J.G., Wernick, D.L., Gould, K.A., Overfield, R.E., Rao, B.M.L., and Savage, D.W. 1985. Molecular weights and association of asphaltenes: A critical review. *Revue de l'Institut Français du Pétrole*, 40: 51.

Storm, D.A., Decanio, S.J., Edwards, J.C., and Sheu, E.Y. 1997. Sediment formation during heavy oil upgrading. *Petroleum Science and Technology*, 15: 77.

Takatsuka, T., Kajiyama, R., Hashimoto, H., Matsuo, I., and Miwa, S.A. 1989a. *Journal of Chemical Engineering of Japan*, 22: 304.

Takatuska, T., Wada, Y., Hirohama, S., and Fukui, Y.A. 1989b. *Journal of Chemical Engineering of Japan*, 22: 298.

Thiyagarajan, P., Hunt, J.E., Winans, R.E., Anderson, K.B., and Miller, J.T. 1995. Temperature dependent structural changes of asphaltenes in 1-methylnaphthalene. *Energy & Fuels*, 9: 629.

Valyavin, G.G., Fryazinov, V.V., Gimaev, R.H., Syunyaev, Z.I., Vyatkin, Y.L., and Mulyukov, S.F. 1979. *Khimiya i Tekhnologiya Topliv i Masel*, 8: 8.

Vercier, P. 1981. Programmed pyrolysis, programmed combustion, and specific nitrogen and sulfur detection. In: *The Chemistry of Asphaltenes*. J.W. Bunger and N.C. Li (Editors). Advances in Chemistry Series No. 195. American Chemical Society, Washington, DC, pp. 203–217.

Wiehe, I.A. 1992. A solvent-resid phase diagram for tracking resid conversion. *Industrial & Engineering Chemistry Research*, 31: 530–536.

Wiehe, I.A. 1993. A phase-separation kinetic model for coke formation. *Industrial & Engineering Chemistry Research*, 32: 2447–2454.

Wiehe, I.A. 1994. The pendant-core building block model of petroleum residua. *Energy & Fuels*, 8: 536–544.

Yan, T.Y. 1987. Coker formation in the visbreaking process. Preprints. *American Chemical Society, Division of Petroleum Chemistry*, 32: 490.

Yen, T.F. 1984. In: *The Future of Heavy Crude Oil and Tar Sands*. R.F. Meyer, J.C. Wynn, and J.C. Olson (Editors). McGraw-Hill, New York.

Yen, T.F. 1998. Correlation between heavy crude sources and types and their refining and upgrading methods. *Proceedings of Seventh UNITAR International Conference on Heavy Crude and Tar Sand*. Beijing, China, Volume 2, pp. 2137–2144.

6 Refinery Reactors

6.1 INTRODUCTION

Petroleum refining has evolved continuously in response to changing demand for better and different products. The original requirement was to produce kerosene as a cheaper and better source of light than whale oil. The development of the internal combustion engine led to the production of gasoline and diesel fuels. In addition, the evolution of the airplane created a need for high-octane aviation gasoline and then for jet fuel, which required further processing of the kerosene fraction. This modern refinery produces a variety of products including many required feedstocks for the petrochemical industry, and each product is the result of application of one or more reactor configurations during the refining process.

The subject of initiated chemical reaction engineering as it applies to the petroleum refining industry evolved primarily to choose, size, and determine the optimal operating conditions for a reactor with the purpose of producing a specific set of products for petroleum-related (such as liquid fuels) and petrochemical application. This involves (1) the chemical reaction, (2) the chemical changes to the feedstock that are expected to occur, (3) the chemical nature of the products vis-à-vis the chemical nature of the feedstock, (4) the physical nature of the products vis-à-vis the physical nature of the feedstock, and (5) the rate of the reaction. Thus, the initial task in approaching the description of a chemically reacting system is to understand the answers to these five criteria from which reactor design can eventually occur. Furthermore, each reaction is often represented by stoichiometrically simple equations (Chapter 5) that often are not truly representative of the refinery process. Furthermore, the term *simple reaction* should be avoided since a stoichiometrically simple reaction does not occur in a simple manner. In fact, most refinery processes proceed through complicated sequences of *chemical steps* involving reactive intermediates that do not appear in the stoichiometric representations of the processes. The identification of these intermediates and the role that they play in the process are a necessity for the design and design of refinery reactors.

In discussions of the chemistry of refinery processes, in chemical kinetics, the terms *mechanism* and *model* receive frequent use and are used to indicate a plausible but *assumed* sequence of steps for a given reaction. However, the various levels of detail in investigating reaction mechanisms, sequences, and steps are so different; the terms *mechanism* and *model* (with the associated descriptors) are often associated with much speculation. An example is the ongoing attempts to assign molecular parameters to the higher-molecular-weight species in petroleum feedstocks (Speight, 2014) and thence basing process designs and reactors design on these assumptions. As worthy is such efforts may seem, the assumptions employed can lead to erroneous design and development of refinery processes. As any chemically reacting system proceeds from reactants to products, a number of species (*reactive intermediates*) are produced, reach a reaction-specific concentration, and then move to produce the products.

The complexity of the refinery-related chemistry and refinery-related processes dictates that reactor engineering and reactor design play an extremely important role in petroleum refining—the basis of economical and safe operation dictates that reactor selection and design must be suitable for the feedstock and for adapting to the changing nature of refinery feedstocks (Furimsky, 1998; Davis and Davis, 2003; Robinson, 2006; Salmi et al., 2011). Chemical reactions in petroleum refining include a huge spectrum of unique properties and include (1) contact between the reactants, (2) the presence or absence of a catalyst is used, (3) whether heat is evolved or absorbed, and (4) the rate of the reaction. In addition, the reactors are used for the conversion of feedstocks into products, which can be a batch-type reactor or a continuously operating reactor (Fogler, 2006; Salmi et al., 2011).

If the desired product purity cannot be achieved in the reactor—as is often the case—one or several separation units are installed after the actual reactor. Common separation units include distillation, absorption, extraction, or crystallization equipment (Versteeg et al., 1997). A chemical reactor coupled with a separation unit constitutes a unit process that is a part of the refinery system. The role of the reactor is crucial for the whole process: product quality from the chemical reactor determines the following process steps, such as type, structure, and operation principles of separation units.

Most processes that are relevant to petroleum refining are carried out in the presence of a catalyst. Homogeneous or homogeneously catalyzed reactions can be facilitated in simple tube or tank reactors. For heterogeneous catalytic reactions (Jones and Pujado, 2006), the reactor typically has a solid catalyst phase, which is not consumed as the reaction takes place. The catalyst is placed in the reactor to enhance reaction velocity. Heterogeneous catalytic reactions are commonly carried out in packed-bed reactors, in which the reacting gas or liquid flows through a stagnant catalyst layer. If catalyst particles are very small, they can be set in motion and a fluidized bed can be considered. In case the catalytic reactor contains both a gas phase and a liquid phase, it is referred to as a three-phase reactor. If catalyst particles are immobile, the reactor is typically referred to as a *trickle-bed reactor*, which can operate under pressure (Ng and Chu, 1987; Haure et al., 1989; Gianetto and Specchia, 1992; Saroha and Nigam, 1996; Al-Dahhan et al., 1997).

In catalytic three-phase reactors, a gas phase, a liquid phase, and a solid catalyst phase coexist. Some of the reactants and/or products are in the gas phase under the prevailing conditions (temperature and pressure). The gas components diffuse through the gas–liquid interface, dissolve in the liquid, diffuse through the liquid film to the liquid bulk phase, and diffuse through the liquid film around the catalyst particle to the catalyst surface, where the chemical reaction takes place. If catalyst particles are porous, chemical reaction and diffusion take place simultaneously in the catalyst pores. The product molecules are transported in the opposite direction.

The size of the catalyst particle is of considerable importance for catalytic three-phase reactors. Catalyst particles can be very small and are suspended in the liquid phase. Catalyst particles of a size similar to those used in two-phase packed-bed reactors can also be used in three-phase reactors. Small catalyst particles are mainly used in bubble columns, stirred-tank reactors, and fluidized beds. Slurry phase reactors—especially reactors used in hydrocracking processes (Chapter 11)—offer a thermal cracking process in the presence of hydrogen and dispersed catalyst. The dispersed catalyst is typically in powder form and may be a natural ore (particularly iron ore, powdered coal) or an oil-soluble salt that might contain metals such as cobalt, molybdenum, nickel, tungsten, and manganese. On the other hand, packed-bed reactors typically contain large catalyst particles.

Catalytic three-phase processes are of enormous industrial importance. Catalytic three-phase processes exist in the petroleum industry and are used in hydrodesulfurization and hydrodemetallization processes, for the removal of oxygen and nitrogen from oil fractions (hydrodeoxygenation and hydrodenitrogenation), and in the hydrogenation of aromatic compounds (dearomatization) (Table 6.1). The production of synthetic fuels (Fischer–Tropsch synthesis) is a three-phase system. For some catalytic two-phase processes, competing three-phase processes have been developed. Oxidation of sulfur dioxide (SO_2) to sulfur trioxide (SO_3) over an active carbon catalyst and methanol (CH_3OH) synthesis can be carried out in three-phase slurry reactors.

This type of reactor is considered to be the simplest reactor type for use in catalytic processes where a gas and liquid (normally both reagents) are present in the reactor, and accordingly, it is extensively used in processing plants. Typical examples of the use of the trickle-bed reactor (downflow fixed-bed reactor) in refineries are (1) liquid-phase hydrogenation, (2) hydrodesulfurization, and (3) hydrodenitrogenation. Most commercial trickle-bed reactors operate adiabatically at high temperatures and high pressures and generally involve hydrogen and organic liquids. Kinetics and/or thermodynamics of reactions conducted in trickle-bed reactors often necessitate high temperatures. A variant of the downflow fixed-bed reactor (Figure 6.1) is the radial-flow fixed-bed

TABLE 6.1

Types of Reactors Used in a Refinery

Reactor(s)	Type	Purpose
Naphtha hydrotreater	Vapor-phase catalytic	Remove S and N from catalytic and virgin naphtha before catalytic reforming
Catalytic reformer	Vapor-phase catalytic	Convert paraffins to higher-octane aromatics and *iso*-paraffins
Alkylation unit	Liquid-phase catalytic (H_2SO_4 or HF)	Combine *iso*-paraffins with olefins to gasoline
Distillate hydrotreating cat-feed hydrotreater	Trickle-phase catalytic	Remove S and N and saturate aromatics
Hydrocracker	Trickle-phase catalytic	Convert gas oils, coker gas oil, and light catalytic cycle oil to lighter products
Fluid catalytic cracker unit	Vapor-phase catalytic	Convert vacuum gas oil to catalytic naphtha
Coker	Semibatch thermal	Convert residuum to gas oil and coke
Residuum hydrotreater	Trickle-phase catalytic/thermal	Convert heavy residuum to lighter distillates, removing metals (Ni, V), S, and N

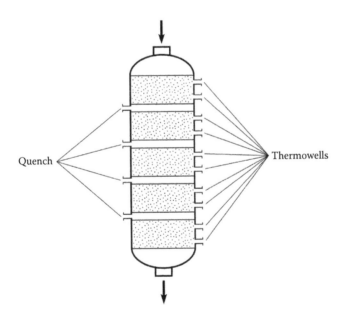

FIGURE 6.1 A downflow fixed-bed reactor.

reactor in which the feed enters the top of the reactor and flows through the bed in a radial direction and then flows out through the base of the reactor instead of flowing downward through the catalyst bed (Figure 6.2).

Commercial hydrocracking processes mainly use two types of reactors: fixed trickle-bed reactor (TBR) and ebullated-bed reactor (EBR), and in both cases when processing heavy feedstocks, three phases are present (Ancheyta et al., 2005). The advantages of using fixed-bed reactors are the relative simplicity of scale-up and operation; the reactors operate in downflow mode, with liquid and gas (mainly hydrogen) flowing down over the catalytic bed, as in the distillate hydrotreating process (Figure 6.3). The major issue that arises with this type of reactor is the accumulation of metals and coke in the mouth of the catalytic pores, blocking the access of reactants to the internal surface.

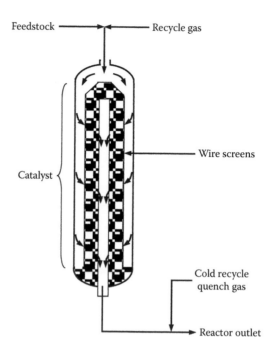

FIGURE 6.2 The radial-flow fixed-bed reactor (RF-FBR).

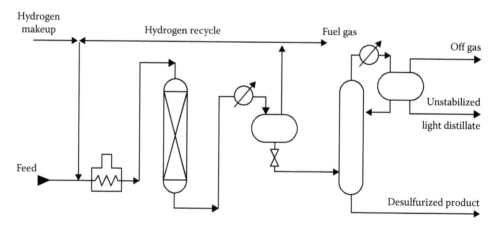

FIGURE 6.3 Distillate hydrotreating. (From OSHA Technical Manual, Section IV, Chapter 2: Petroleum refining processes, 1999, http://www.osha.gov/dts/osta/otm/otm_iv/otm_iv_2.html.)

The ebullated-bed reactors eliminate this difficulty by fluidizing the catalyst. Metals are deposited in the catalyst inventory allowing for uniform deactivation. The catalyst is continuously added and removed in order to keep the catalytic activity at a certain constant level. In general, ebullated-bed technology is most applicable for highly exothermic reactions and for feedstocks that are difficult to process in a fixed-bed reactor due to high levels of contaminants.

Ebullated-bed reactors are designed to hydroprocess heavy feedstocks that contain high amounts of metal constituents and asphaltene constituents (Chapter 11). The function of these reactors is to overcome some of the deficiencies of the fixed-bed reactors. In the ebullated-bed reactor, the feedstock and hydrogen are fed in an upflow mode through the catalyst bed, expanding and back-mixing the bed, minimizing bed plugging, and consequently reducing the effects of pressure drop. The mixture of the gas (makeup and recycle hydrogen) and feedstock (plus any recycled stream) enter the reactor where mixing through the gas–liquid mixer, spargers, and catalyst support grid plate occurs. The product quality is constantly maintained at a high level by intermittent catalyst addition and withdrawal, which is one of the features included to eliminate the need to shut down for catalyst replacement.

The hydroprocessing ebullated-bed reactor is a three-phase system (i.e., gas, liquid, and solid catalysts) in which oil is separated from the catalyst at the top of the reactor and recirculated to the bottom of the bed to mix with the new feed. The large liquid recycle causes the reactor to behave as a continuously stirred-tank reactor. The reactor is provided with an ebullating pump to maintain liquid circulation within the reactor and maintain the reactor at isothermal conditions so that there is no need for quenches within the reactor. Any unconverted feedstock is recirculated back to the reactor with a small amount of diluent to improve fluidity and thus overall conversion. Fresh catalyst is added to the top of the reactor, and spent catalyst is withdrawn from the bottom of the reactor. The inventory of catalyst in the reactor is maintained at the desired level by adjusting the catalyst addition rate equal to the withdrawal rate plus any losses. The catalyst replacement rate can be adjusted to suit feedstock properties, process parameter, product slate, and product quality requirements.

Homogeneous and homogeneously catalyzed gas–liquid reactions take place in the liquid phase in which gaseous reactants dissolve and react with other reactants that are primarily present in the liquid phase. Typical constructions to be used as gas–liquid reactors are column and tank reactors. Liquid–liquid reactors principally resemble gas–liquid reactors, but the gas phase is replaced by another liquid phase. The reactions can principally take place in either of, or even both, the liquid phases.

The most complicated systems are represented by reactors in which the solid phase is consumed—or solid particles are generated—while a reaction takes place. For this kind of reaction, similar types of reactors are utilized as for heterogeneous catalytic reactions: (1) packed-bed reactors and (2) fluidized-bed reactors (Table 6.1). However, it is not the configuration of the reactor itself but the chemistry involved in the industrial process that plays a central role in the production of new substances. The process chemistry decides, to a large extent, the choice of the reactor.

In a refinery, reactor selection and design are the basis of economical and safe operation. Chemical reactions in petroleum refining include a huge spectrum of unique properties (Froment and Bischoff, 1990; Zhukova et al., 1990). This includes how the reactants are contacted, whether a catalyst is used, how much heat is evolved or absorbed, and how fast the reaction takes place. This article guides the reader in selecting and designing reactors that will best carry out the reactions of interest. The reactor types discussed focus on those in a petroleum refinery, but many can be used in chemical processing as well. Reactor and reaction engineering play a vital role in petroleum processing.

Any errors (even small errors) in equipment sizing or yield translate to unnecessary expense, and it is extremely important that the refinery engineer correctly size and specify the process parameters for a refinery reactor. The process chemist and the process engineer must have a clear understanding of a reactor at each of the three stages of development, which are (1) the laboratory reactor, which explores new reaction conditions, catalyst formulations, feedstock types, and reaction kinetics, (2) development of the reactor to simulate the commercial operation by employing recycle streams to achieve a realistic assessment of the reactor behavior, and (3) sizing of the reactor for operation at the commercial scale and adiabatic operation. In the second stage, isothermal conditions are usually maintained in the reactor, but if heat release is a concern, such as residuum hydrotreating, then the reactor should be tested under adiabatic conditions to establish the adiabatic reaction temperature and to determine the amount of heat that must be removed in the final commercial design. At this stage, examining the potential for catalyst deactivation, product yield variations, and the manner in which changing feedstock quality can influence the process. Finally, as part of the third-stage investigations, the commercial-size reactor must be investigated for safe startup and shutdown as well as operation under steady-state conditions.

By way of definition, an adiabatic process is a thermodynamic process during which no energy is transferred as heat across the boundaries of the system. This does not exclude energy transfer as work. The adiabatic process provides a rigorous conceptual basis for the theory used to expound the first law of thermodynamics, and as such it is a key concept in thermodynamics. Some chemical and physical processes occur so rapidly that they may be conveniently described by the adiabatic approximation, meaning that there is not enough time for the transfer of energy as heat to take place to or from the system. An adiabatic reactor is designed to enhance and promote such reactions.

Thus, the purpose of this chapter is to present an introduction to design and selection and the characteristics of the chemical reaction of interest. The reactor types discussed focus on those in a petroleum refinery and whether the reaction occurs in the vapor, liquid, or mixed vapor–liquid phase. More specifically, a refinery must select reactors according to the process, such as (1) naphtha-processing reactors where reaction may be in the gaseous phase, (2) reactors for processing kerosene and middle distillate fractions that react partially in the gas phase and the liquid phase, and (3) reactors that are used for processing residua, which react completely in the liquid phase. The chapter is especially to the activities of the modern refinery that may well evolve into a refinery that accepted other nonpetroleum feedstocks for refining in the future (Chapter 17). In such cases, reactor technology and motivation of reactors for processing such feedstocks will become a major event in many refineries.

Thus, the chapter will serve as an introduction to reactor technology with emphasis on the types of reactors used in refineries and presents an introduction to the fundamentals of reactor technology as it applies to the various refinery processes.

6.2 REACTOR TYPES

The reactors used during petroleum refining are among the most complex and difficult to model and design. The composition and properties of the various petroleum feedstocks (varying from distillates such as naphtha to nonvolatile residua and tar sand bitumen) that are converted in refinery reactors have properties that are such that the reaction system can involve various phases, catalysts, reactor configuration, and continuous catalyst addition that serves to make the development and design of the reactor a serious challenge. In addition, the presence of an unknown (petrochemical processes being the general exception) number (but at least a number in the hundreds) of constituents undergoing different chemical reactions leading to a multitude of reaction paths and competing for the active sites of catalysts contributes to increasing the complexity of the reactor design.

The choice of a suitable reactor type and, hence design of the reactors, is dictated by the process parameters as well as the nature and boiling range (i.e., physical and chemical properties) of the feedstock (Fogler, 2006; Salmi et al., 2011). Typically, the higher the boiling point of the feed, the higher the reaction severity, especially in the high-pressure hydroprocesses. Hence, the various reactors used in petroleum refining must be designed for noncatalytic process and for catalytic processes, which establishes reactor size. However, one of the issues that often arise when defining reactor types is that the names of the various reactors are often arbitrary and difficult to define. However, as a word of caution, reactor definitions based on reactor use do not always signify the correct or standard name for the reactor—the name may be based on the reactor type, bed type, or process type and the names may be intermingled—and there may be more than one name for a particular reactor type. Where possible, simple names are used in this section and are often based on the use for which the reactor was designed and constructed.

In the design of a reactor for a refinery process, the first issue is whether the process should be operated in the batch mode (discontinuous mode) or continuous mode. In the refining industry, the tendency is toward continuous operation other than planned shutdowns for reactor maintenance. There is no general rule for the selection of the operation mode but economic balance, scale of production, long reaction times, flexibility of production, and nature of the process, and the product may dictate the selection of batch or semibatch operation. However, batch reactors are therefore often used for small production rates such as fine chemicals and specialties where reaction condition can be adjusted to product specification or quality.

In terms of production flexibility, the same reactor is often used not only for different products but also for different process operations such as heating, reacting, solvent evaporation, cooling, blending with additives, besides standard cyclic operations such as reactor initial conditioning, gas evacuation, reactants charge, product discharge, and reactor washing (Donati and Paludetto, 1999).

6.2.1 BATCH REACTORS

In the batch mode of operation, the reactants and any additional components of the reaction mixture such as catalysts are loaded into the reactor where they remain for a well-defined set of reaction parameters (under fixed conditions). In the course of this process, the composition of the content of the reactor changes continuously, that is, the reactor operates in an *unsteady mode*. The advantages of the batch reactor lie with its versatility. A single vessel can carry out a sequence of different operations without the need to break containment. This is particularly useful when processing toxic or highly potent compounds.

The batch reactor may be as simple as a pipe that is operated batchwise and then shut down for emptying/cleaning. The continuously stirred-tank reactor is typically used in kinetic studies because the reaction rate is derived directly from the inlet and out concentrations, and it may simulate operation in a larger commercial reactor such as an ebullated bed where the high recycle rate approximates complete mixing. For continuous processing, almost any petroleum fraction may be fed over a fixed bed of catalyst in a plug flow reactor, with vapor-phase operation for naphtha and trickle phase for distillates and residuum.

A microbatch reactor, such as the tubing bomb reactor is a common, inexpensive device to develop data. The reactants and, optionally, the catalyst are changed in the small reactor, sealed, and then pressured. To start the reaction, the tubing bomb is typically immersed in a heated fluidized sand bath for a specified length of time with agitation. Shortly after immersion in the heated sand bath, the reactor pressure is increased to a final level, close to commercial conditions. To stop the reaction, the microreactor is pulled out of the heated bath and rapidly quenched in a cooling fluid.

The *batch reactor* is the generic term for a type of vessel widely used in the many process industries as well as in the petroleum industry—the analog is the laboratory-scale flask in which chemicals are mixed and reacted. Vessels of this type are used for a variety of process operations such as solids dissolution, product mixing, chemical reactions, batch distillation, crystallization, liquid–liquid extraction, and polymerization. In some cases, they are not referred to as reactors but have a name that reflects the role they perform such as a crystallizer (which might be used in the petrochemical industry) or a bioreactor (which might be used in the biomass-to-biofuel industry). To a point the delayed coking drum can be considered to be a batch reactor or a semibatch reactor insofar as the reactions that produce coke and distillate products are allowed to proceed to completion in the drum before the drum is taken batchwise and cleaned. The drums are used in pairs and when one of the pairs is offstream, the other is onstream (Chapter 8).

The batch reactor has high flexibility and allows high conversion of reactants to products as the reaction time may be arbitrarily long. This flexibility of batch reactors allows to adjust the reaction condition in various reaction phases and therefore to tailor the process variables to process specifications. This is an additional reason to prefer batch operation, in some cases even for large-scale productions, as in the plastic industry. Additional reasons, in favor of batch and semibatch operations, belong more to the R&D practice and attitude. Many reactions are first investigated in batch lab equipment and scaling up by enlarging vessels, without kinetic experiments or other engineering evaluations may appear as the easiest way. This theorem is however far to be proven especially when transport phenomena and mixing effects are relevant to the examined process.

A typical batch reactor consists of a tank with an agitator and an integral heating/cooling system. These vessels may vary in size from less than 1 L to more than 15,000 L. They are usually fabricated in stainless steel, glass-lined steel, or suitable alloy. The liquids and/or solid feedstocks are usually charged via connections in the top cover of the reactor. Vapors and gases also discharge through connections in the top and liquids are usually discharged out of the bottom. The usual agitator arrangement is a centrally mounted driveshaft with an overhead drive unit. Impeller blades are mounted on the shaft. A wide variety of blade designs are used, and typically, the blades cover approximately two-thirds of the diameter of the reactor. Where viscous products are handled, anchor-shaped paddles are often used, which have a close clearance between the blade and the vessel walls. Most batch reactors also use baffles. These are stationary blades that break up the flow caused by the rotating agitator. These may be fixed to the vessel cover or mounted on the interior of the side walls.

Despite significant improvements in agitator blade and baffle design, mixing in large batch reactors is ultimately constrained by the amount of energy that can be applied. On large vessels, mixing energies of more than 5 W/L can put an unacceptable burden on the cooling system. High agitator loads can also create shaft stability problems. Where mixing is a critical parameter, the batch reactor is not the ideal solution. Much higher mixing rates can be achieved by using smaller flowing systems with high-speed agitators, ultrasonic mixing, or static mixers.

6.2.1.1 Heating and Cooling Systems

Products within batch reactors usually liberate or absorb heat during processing. Even the action of stirring stored liquids generates heat. In order to hold the reactor contents at the desired temperature, heat has to be added or removed by a cooling jacket or cooling pipe. Heating/cooling coils or external jackets are used for heating and cooling batch reactors. Heat transfer fluid passes through the jacket or coils to add or remove heat.

Temperature control is one of the key functions of a reactor. Poor temperature control can severely affect both yield and product quality. It can also lead to boiling or freezing within the reactor, which may stop the reactor from working altogether. In extreme cases, poor temperature control can lead to severe overpressure, which can be destructive on the equipment and potentially dangerous. Within the petrochemical and pharmaceutical industries, external cooling jackets are generally preferred as they make the vessel easier to clean. The performance of these jackets can be defined by three parameters: (1) response time to modify the jacket temperature, (2) uniformity of jacket temperature, and (3) stability of the jacket temperature. It can be argued that the heat transfer coefficient is also an important parameter. It has to be recognized however that large batch reactors with external cooling jackets have severe heat transfer constraints by virtue of design. It is difficult to achieve better than 100 W/L even with ideal heat transfer conditions. By contrast, continuous reactors can deliver cooling capacities in excess of 10,000 W/L. For processes with very high heat loads, there are better solutions than batch reactors.

Fast temperature control response and uniform jacket heating and cooling are particularly important for crystallization processes or operations where the product or process is very temperature sensitive. There are several types of batch reactor cooling jackets: (1) single external jacket, (2) half coil jacket, and (3) constant flux cooling jacket.

The *single external jacket* design consists of an outer jacket that surrounds the vessel. Heat transfer fluid flows around the jacket and is injected at high *velocity* via nozzles. The temperature in the jacket is *regulated* to control heating or cooling. The single jacket is probably the oldest design of external cooling jacket. Despite being a tried and tested solution, it has some limitations. On large vessels, it can take many minutes to adjust the temperature of the fluid in the cooling jacket. This results in sluggish temperature control. The distribution of *heat transfer* fluid is also far from ideal, and the heating or cooling tends to vary between the side walls and bottom dish. Another issue to consider is the inlet temperature of the heat transfer fluid that can oscillate (in response to the temperature control valve) over a wide temperature range to cause hot or cold spots at the jacket inlet points.

The *half coil jacket* is made by welding a half pipe around the outside of the vessel to create a semicircular flow channel. The heat transfer fluid passes through the channel in a plug flow fashion. A large reactor may use several coils to deliver the heat transfer fluid. Like the single jacket, the temperature in the jacket is regulated to control heating or cooling. The plug flow characteristics of a half coil jacket permit faster displacement of the heat transfer fluid in the jacket (typically less than 60 seconds). This is desirable for good temperature control. It also provides good distribution of heat transfer fluid that avoids the problems of nonuniform heating or cooling between the side walls and bottom dish. Like the single jacket design, however, the inlet heat transfer fluid is also vulnerable to large oscillations (in response to the temperature control valve) in temperature.

The *constant flux cooling jacket* (*coflux jacket*) is a relatively recent development. It is not a single jacket but has a series of 20 or more small jacket elements. The temperature control valve operates by opening and closing these channels as required. By varying the heat transfer area in this way, the process temperature can be regulated without altering the jacket temperature. The constant flux jacket has very fast temperature control response (typically less than 5 seconds) due to the short length of the flow channels and high velocity of the heat transfer fluid. Like the half coil jacket the heating/cooling *flux* is uniform. Because the jacket operates at substantially constant temperature, however, the inlet temperature oscillations seen in other jackets are absent. An unusual feature of this type jacket is that process heat can be measured very sensitively. This allows the user to perform tasks such as (1) monitoring the rate of reaction for detecting end points, (2) controlling addition rates, and (3) controlling crystallization.

Batch reactors are often used in the process industry. Batch reactors also have many laboratory applications, such as small-scale production and inducing fermentation for beverage products. They also have many uses in medical production. Batch reactors are generally considered expensive to run, as well as variable product reliability. They are also used for experiments of reaction kinetics, volatiles, and thermodynamics. Batch reactors are also highly used in wastewater treatment and are effective in reducing the biological oxygen demand (BODF) of influent untreated water.

Batch reactors are very versatile and are used for a variety of different unit operations (such as batch distillation, storage, crystallization, and liquid–liquid extraction) in addition to chemical reactions. Batch reactors represent an effective and economic solution for many types of slow reactions. In a batch reactor, good temperature control is achieved when the heat added or removed by the heat exchange surface is equal to the heat generated or absorbed by the process material. For flowing reactors made up of tubes or plates, satisfying the *heat added to the heat generated* relationship does not deliver good temperature control since the rate of process heat liberation/absorption varies at different points within the reactor. Controlling the outlet temperature does not prevent hot/cold spots within the reactor. Hot or cold spots caused by exothermic or endothermic activity can be eliminated by relocating the temperature sensor (T) to the point where the hot/cold spots exist. This however leads to overheating or overcooling downstream of the temperature sensor.

Hot/cold spots are created when the reactor is treated as a single stage for temperature control. Hot/cold spots can be eliminated by moving the temperature sensor. This however causes overcooling or overheating downstream of the temperature sensor. Many different types of plate or tube reactors use simple feedback control of the product temperature. However, this approach is only suitable for processes where the effects of hot/cold spots do not compromise safety, quality, or yield. The disadvantages of the batch reactor are that there are idle periods for loading, unloading, heating, and control, and regulation of an unsteady process requires considerable instrumentation and efforts.

In the refining industry, the batch-type reactor is not commonly used but in the petrochemical industry the reactor may be used when small quantities of a product are manufactured or if the reactor will be used for the production of a variety of different petrochemical products.

In summary, the batch operation mode involves loading the reaction vessel with the reactants and the chemical reaction is allowed to proceed until the desired conversion of reactants into products has taken place. A more common approach is the semibatch mode of operation or preferentially the continuous operation mode of operations of the reactor in which the reactants are fed continuously into the reaction vessel and a product flow is continuously removed from the vessel.

6.2.2 SEMIBATCH REACTORS

Semibatch reactors lie between batch and continuous reactors in terms of operation. Semibatch reactors occupy a middle ground between batch and continuous reactors. They are open systems (similar to the continuously stirred-tank reactors) and run on an unsteady-state basis like batch reactors. They usually consist of a single stirred tank, similar to a batch reactor. The *half-pipe coil jacketed* reactor can be used in semibatch operations. In a semibatch process, an initial amount of reactants is charged into the reactor. The reactor is then started, and additional reactants are added continuously to the reaction mix in the vessel, which is then allowed to run until the desired conversion is achieved. At this point, the products and remaining reactants are removed from the tank and the process can be started once more.

Semibatch reactors are not used as often as other reactor types. However, they can be used for many two-phase (i.e., solid–liquid) reactions. Also, semibatch reactors are used when a reaction has many unwanted side reactions or has a high heat of reaction, and by limiting the introduction of fresh reactants, the potential problems can be mitigated.

6.2.3 CONTINUOUS REACTORS

Continuous reactors (also referred to as *flow reactors*) carry material as a flowing stream. In use, reactants are continuously fed into the reactor and emerge as continuous stream of product. Continuous reactors are used for a wide variety of processes within the refining industry. A survey of the continuous reactor market will throw up a daunting variety of shapes and types of machine. Beneath this variation however lies a relatively small number of key design features that determine the capabilities of the reactor. When classifying continuous reactors, it can be more helpful to look at these design features rather than the whole system.

In the semicontinuous mode of operation, individual reactants can be entered discontinuously, while others are entered continuously—the products can be withdrawn continuously or discontinuously (batchwise). The output from a continuous reactor can be altered by varying the run time. However, conditions within a continuous reactor change as the product passes along the flow channel. In an ideal reactor, the design of the flow channel is optimized to adapt the any changing conditions, and this is achieved by operating the reactor as a series of stages. Within each stage, the ideal heat transfer conditions can be achieved by varying the surface-to-volume ratio or the cooling/heating flux. Thus, stages where process heat output is very high either use extreme heat transfer fluid temperatures or have high surface-to-volume ratios (or both). By using a series of stages, extreme cooling/heating conditions can be employed at the hot/cold spots without suffering overheating or overcooling as the reactants and product pass through the various stages. Thus, larger flow channels can be used, which permit (1) a higher flow rate, (2) a lower pressure drop, and (3) a reduced tendency for channel blockage.

Mixing is another important feature for continuous reactors since efficient mixing improves the efficiency of heat and mass transfer. In terms of flow through the reactor, the ideal condition for a continuous reactor is plug flow (since this delivers uniform residence time within the reactor) but there is the potential (or reality) for flow conflict between efficient mixing and plug flow since mixing generates axial as well as radial movement of the fluid. In tube-type reactors (with or without static mixing), adequate mixing can be achieved without seriously compromising plug flow, and accordingly, these types of reactor are sometimes referred to as plug flow reactors.

Continuous reactors can be classified in terms of the mixing mechanism as follows: (1) mixing by diffusion, (2) mixing by pumping, and (3) mixing by agitation. Diffusion mixing relies on concentration or temperature gradients within the product. This approach is common with micro reactors where the channel thicknesses are very small and heat can be transmitted to and from the heat transfer surface by conduction. In larger channels and for some types of reaction mixture (especially immiscible fluids), mixing by diffusion is not always practical. However, if the product is continuously pumped through the reactor, the pump can be used to promote mixing—if the fluid velocity is sufficiently high, the induced turbulent flow conditions will promote mixing. The disadvantage with this approach is that it leads to long reactors with high pressure drops and high minimum flow rates and is especially true where the reaction is slow or the product has high viscosity. This problem can be reduced with the use of static mixers that are typically represented as baffles in the flow channel that promote mixing. Static mixers can be effective but still tend to require relatively long flow channels and generate a relatively high pressure drop. The oscillatory baffled reactor is a specialized form of static mixer where the direction of process flow is cycled to allow static mixing of the reactants with a low net flow through the reactor that allows the reactor to be kept comparatively short.

On the other hand, some continuous reactors use mechanical agitation for mixing (rather than the product transfer pump). While this option adds complexity to the reactor design, it offers significant advantages insofar as (1) efficient mixing can be maintained irrespective of product throughput or product viscosity and (2) the method eliminates the need for long flow channels and high pressure drops. However, the use of mechanical agitators does create strong axial mixing, but a nonbeneficial effect can be negated by subdividing the reactor into a series of mixed stages that are separated by small plug flow channels.

The most familiar form of continuous reactor of this type is the continuously stirred-tank reactor (CSTR) (Figure 6.4), which is essentially a batch reactor used in a continuous flow. The disadvantage with a single-stage continuously stirred-tank reactor is that it can be relatively wasteful on product during startup and shutdown. The reactants are also added to a mixture that is rich in product. For some types of process, this can have an impact on quality and yield. These problems are managed by using multistage continuously stirred-tank reactors. At the large scale, conventional batch reactors can be used for the continuously stirred-tank reactor stages.

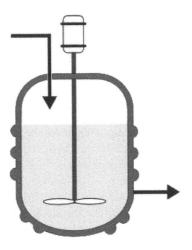

FIGURE 6.4   Schematic of a continuously stirred-tank reactor.

A continuously stirred-tank reactor often refers to a model used to estimate the key unit operation variables when using a continuous agitated-tank reactor to reach a specified output. The behavior of a continuously stirred-tank reactor is often approximated or modeled by that of a continuous ideally stirred-tank reactor (CISTR). All calculations performed with continuous ideally stirred-tank reactors assume perfect mixing. In a perfectly mixed reactor, the output composition is identical to composition of the material inside the reactor, which is a function of residence time and rate of reaction. If the residence time is 5–10 times the mixing time, this approximation is valid for engineering purposes. The continuous ideally stirred-tank reactor model is often used to simplify engineering calculations and can be used to describe research reactors. In practice, it can only be approached, in particular, in industrial-size reactors.

6.2.4 PLUG FLOW REACTORS

The plug flow reactor is probably the most commonly used reactor in catalyst evaluation because it is simply a tube filled with catalyst that reactants are fed into. However, for catalyst evaluation, it is difficult to measure the reaction rate because concentration changes along the axis, and there are frequently temperature gradients, too. Furthermore, because the fluid velocity next to the catalyst is low, the chance for mass transfer limitations through the film around the catalyst is high.

The design of larger commercial reactors provides a significant challenge because heat effects are typically substantial and vary with the endothermic cat cracking or reforming reactions to the highly exothermic hydrotreating and hydrocracking reactions; the flow regime deviates from the ideals of plug flow and perfect mixing.

6.2.5 FLASH REACTORS

As an extension of the fluidized-bed family of separation processes, the flash reactor (FR) (or transport reactor) employs turbulent fluid introduced at high velocities to encourage chemical reaction with feedstocks and subsequently achieve separation through the chemical conversion of desired substances to different phases and streams. A flash reactor consists of a main reaction chamber and an outlet for separated products to enter downstream processes. Flash reactor vessels facilitate a low gas and solid retention (and hence reactant contact time) for industrial applications that give rise to a high-throughput, pure product and less than ideal thermal distribution when compared to other fluidized-bed reactors. Due to these properties as well as its relative simplicity, flash reactors have the potential for use for pretreatment and posttreatment processes where these strengths of the flash reactor are prioritized the most. While a variety of applications are available for a flash reactor, a general set of operating parameters are used and the important parameters to consider when designing a flash reactor are (1) fluid velocity and flow configuration, (2) solid retention time, (3) refractory lining material, and (4) feed and fluid type.

In the flash reactor, gas is introduced from the bottom at an elevated temperature and high velocity, with a slight drop in velocity experienced at the central part of the vessel. The design of the vessel can vary in shapes and sizes (i.e., from pipeline to an egg-like shape) to promote the vertical circulation of the gases and particulate matter. Whatever the choice of shape, the configuration should be designed to increase the velocity of the fluid at the bottom of the chamber, thereby allowing for higher-molecular-weight (denser) feedstock constituents to be in a continuous circulation that promotes a reaction site for separation processes. The method of feed delivery varies depending on the phase—solid feedstocks may be delivered using a conveyor while fluid feedstocks are vaporized and sprayed directly into the flash reactor. The feedstock is then contacted with a continuously circulating hot gas that interacts throughout the chamber with the incoming feed. The product mixture is sent to a separator where an exhaust vent emits gaseous products.

A relatively fast fluid velocity is usually required in flash reactor operations to encourage a continuous particle distribution throughout the reactor's vessel. This minimizes the slip velocity (average velocity difference of different fluids in a pipe) of the column and provides a positive impact on heat and mass transfer rates, thereby allowing for the use of smaller diameter vessels that can lower operating costs. Also, the use of a vertical fluid flow configuration results in inefficient mixing in the horizontal and vertical direction and can cause low product quality. The use of a fast fluid velocity also ensures a short solid feed retention time that is beneficial for reactions that require a purer product and higher throughput. However, if the operating conditions for a reaction require an extended reaction time, this can be implemented by introducing a recycle mode in which the fluid in the flash reactor can be recirculated with the feedstock to allow for additional contact time. Due to the high-temperature requirements for flash reactor operations, a refractory lining is required to reinforce and maintain vessel integrity and operability and such a lining also serves to isolate the high temperature of the flash chamber from the ambient temperature.

In the *centrifugal flash reactor*, unlike other flash reactor designs, the powdered feed is contacted on a solid heat carrier rather than a gaseous carrier. The design involves the use of a heated rotating plate that disperses the feed powder particles for a short duration that is achieved by the use of centrifugal forces that compress the powder onto the surface of the plate, thereby allowing for direct contact between the particles and hot metal, and that enables a higher heat transfer rate. In the *pipeline flash reactor*, which, as the name implies, is in the form of a pipe, allows it to be easily integrated into new process systems and modifications and extensions be easily added to the pipeline flash reactor to accommodate the requirements of certain processes. In the reactor, the reactants come into contact with each other in the pipe rather than a mixing vessel, and this eliminates the need for extra mixing tanks. Due to the nature of the device, the reactants processed in a pipeline flash reactor will have short retention times, although adding backflow options into the system can be used to increase retention time if required.

The versatility of flash/transport reactors makes them suitable for a wide range of quality-sensitive separation processes. In addition, flash reactor applications do not typically require any posttreatment or pretreatment systems due to a lack of waste generated. For example, in some aspects of gas processing, especially in the glycol dehydration of gas streams (Chapter 16), a flash tank separator (which consists of a device that reduces the pressure of the glycol solution stream) is employed to enable methane and other hydrocarbons to vaporize (*flash*) and separate from the glycol.

6.2.6 Slurry Reactors

Slurry reactors are three-phase reactors (solid–liquid–gas), which can be used to react solids, liquids, and gases simultaneously. They usually consist of a catalyst (solid) suspended in a liquid, through which a gas is bubbled and can operate in either semibatch or continuous mode. The slurry phase Fischer–Tropsch reactor is an example of such a reactor that is used in a gas-to-liquid (GTL) process plant to convert natural gas into diesel fuel (Chadeesingh, 2011).

Inside the reactor are catalyst pellets suspended in a liquid. Gas reactant is bubbled into the reactor, and the gas is absorbed into the liquid from the bubble surface. The absorbed gas then diffuses through the liquid to the catalyst surface, at which point it diffuses into the catalyst pellet and the reaction takes place. Thus, using the production of diesel fuel as the example, synthesis gas (syngas) enters at the bottom of the reactor into heated feedstock. The gas then reacts with the assistance of suspended catalyst to form the methanol (methyl alcohol) product. Unreacted gas and methanol vapor exit though the top of the reactor. Once out of the reactor, the methanol is condensed to a liquid.

The advantage of a slurry reactor with small and finely dispersed catalyst particles is that the diffusion resistance inside the catalyst particles seldom limits the reaction, whereas the diffusion resistance can be a limiting factor in packed-bed reactors. The temperature in the slurry reactor is rather constant, and no hot spot phenomena occur. In slurry reactors, it is also possible to regenerate the catalyst (fluidized bed). However, the separation of small catalyst particles from the suspension may introduce problems. The high degree of back-mixing is usually less efficient for the reaction kinetics, which results in a lower conversion of the reactants than under plug flow conditions. For autocatalytic reactions, there is the opposite effect since some degree of back-mixing can enhance the reaction rate. The flow pattern in a three-phase fluidized bed is usually much closer to complete back-mixing than to a plug flow. Because of the higher liquid flow velocities, larger particles can be used than in bubble columns.

Slurry reactors are most frequently used when a liquid reactant must be contacted with a solid catalyst, and when a reaction has a high heat of reaction. They may be used in such refinery applications as *hydrogenation* and in petrochemical operations such as oxidation, *hydroformylation*, and *ethynylation*.

6.3 BED TYPES

6.3.1 PACKED BEDS

A packed-bed reactor (or fixed-bed reactor) is a hollow tube, pipe, or other vessel that is filled with a packing material. The packing can be randomly filled with small objects such as Raschig rings or the reactor can be specifically designed with structured packing (Ellenberger and Krishna, 1999; Stockfleth and Brunner, 1999). These reactors are filled with solid catalyst particles (such as zeolite pellets or granular activated carbon), most often used to catalyze gas reactions (Fogler, 2006). The chemical reaction takes place on the surface of the catalyst. The advantage of using a packed-bed reactor is the higher conversion per weight of catalyst than other catalytic reactors. The conversion is based on the amount of the solid catalyst rather than the volume of the reactor.

The purpose of a packed bed is typically to improve contact between two phases in a process and can be used in a distillation process, a chemical process, or a scrubber as well as a heat exchanger. On the other hand, a packed column is a type of packed bed used in separation processes, such as absorption, stripping, and distillation (Fogler, 2006). The column can be filled with random dumped packing (*random packed column*) or with structured packing sections, which are arranged or stacked (*stacked packed column*). In the column, liquids tend to wet the surface of the packing and the vapors pass across this wetted surface, where mass transfer takes place. Packing material can be used instead of trays to improve separation in distillation columns. Packing offers the advantage of a lower pressure drop across the column (when compared to plates or trays), which is beneficial while operating under vacuum. Differently shaped packing materials have different surface areas and void space between the packing. Both of these factors affect packing performance.

Another factor in performance, in addition to the packing shape and surface area, is the liquid and vapor distribution that enters the packed bed. The number of theoretical stages required to make a given separation is calculated using a specific vapor-to-liquid ratio. If the liquid and vapor are not evenly distributed across the superficial tower area as it enters the packed bed, the liquid-to-vapor ratio will not be correct and the required separation will not be achieved. The packing will appear to not be working properly. The *height equivalent to a theoretical plate* (HETP) will be greater than expected. The problem is not the packing itself but the maldistribution of the fluids entering the packed bed. These columns can contain liquid distributors and redistributors, which help to distribute the liquid evenly over a section of packing, increasing the efficiency of the mass transfer. The design of the liquid distributors used to introduce the feed and reflux to a packed bed is critical to making the packing perform at maximum efficiency.

Packed columns have a continuous vapor-equilibrium curve, unlike conventional tray distillation in which every tray represents a separate point of vapor–liquid equilibrium. However, when modeling packed columns it is useful to compute a number of theoretical plates to denote the separation efficiency of the packed column with respect to more traditional trays. In design, the number of necessary theoretical equilibrium stages is first determined and then the packing height equivalent to a theoretical equilibrium stage (the *height equivalent to a theoretical plate,* [HETP]) is also determined. The total packing height required is the number of theoretical stages multiplied by the height equivalent to a theoretical plate.

Processes based upon multiphase reactions occur in a broad range of application areas and form the basis for the manufacture of a large variety of intermediate and consumer end products (Salmi et al., 2011). Some examples of multiphase reactor technology uses include (1) the upgrading and conversion of petroleum feedstocks and intermediates; (2) the conversion of coal-derived chemicals or synthesis gas into fuels, hydrocarbons, and oxygenates; (3) the manufacture of bulk commodity chemicals that serve as monomers and other basic building blocks for higher chemicals and polymers; (4) the manufacture of pharmaceuticals or chemicals that are used in "the specialty chemical markets as drugs or pharmaceuticals; and (5) the conversion of undesired chemical or petroleum processing by-products into environmentally acceptable or recyclable products (Duduković et al., 1999).

Multiphase catalytic packed-bed reactors (PBRs) operate in two modes: (1) trickle operation, with a continuous gas phase and a distributed liquid phase and the main mass transfer resistance located in the gas, and (2) bubble operation, with a distributed gas and a continuous liquid phase and the main mass transfer resistance located in the liquid phase (Krishna and Sie, 1994; Duduković et al., 1999; Carbonell, 2000). For three-phase reactions (gas and liquid phases in contact with a solid catalyst), the common modes of operation are trickle- or packed-bed reactors, in which the catalyst is stationary, and slurry reactors, in which the catalyst is suspended in the liquid phase. In these reactors, gas and liquid move cocurrently down flow or gas is fed countercurrently upflow. Commercially, the former is the most used reactor, in which the liquid phase flows mainly through the catalyst particles in the form of films, rivulets, and droplets.

A trickle-bed reactor (TBR) is a reactor that uses the downward movement of a liquid and gas over a packed bed of catalyst particles. Trickle-bed reactors are the most widely used type of three-phase reactors. The gas and liquid cocurrently flow downward over a fixed bed of catalyst particles. Liquid trickles down, while gas phase is continuous. In a trickle-bed reactor, various flow regimes are distinguished, depending on gas and liquid flow rates, fluid properties, and packing characteristics. Approximate dimensions of commercial trickle-bed reactors vary but can be on the order of 30 ft high and 7 ft in diameter (Boelhouwer, 2001).

In the earlier decades of the refining industry, fixed-bed hydrotreating reactors (trickle-bed reactors) were used predominantly for the processing of the lower-boiling feedstocks (such as naphtha and middle distillates) but at present they are also used for the hydroprocessing of heavier feeds, such as petroleum residua.

In fixed-bed reactors, the liquid hydrocarbon trickles down through the fixed catalyst bed from top to bottom of the reactor. Hydrogen gas passes cocurrently through the bed. A single tailor-made optimum catalyst having high hydrodesulfurization activity and low metal tolerance can handle feedstocks having less than 25 ppm w/w of metal for cycle length of 1 year. For feedstocks containing metals in the range of 25–50 ppm w/w, a dual catalysts system is more effective. In such a system, one catalyst having a higher metal tolerance is placed in the front of the reactor, whereas the second catalyst located in the tail end of the reactor is generally of higher desulfurization activity (Kressmann et al., 1998; Scheffer et al., 1998; Liu et al., 2009). A triple catalyst system consisting of a hydrodemetallization catalyst, a hydrodemetallization/hydrodesulfurization catalyst, and a refining catalyst is generally used for feedstocks where the metal content is in the range of 100–150 ppm w/w for a typical cycle length of 1 year. For feeds with metal content higher than this range, the onstream life of the hydrodemetallization catalyst is short (usually a matter of months) and for achieving a cycle length of 1 year, a swing reactor fixed-bed concept has been introduced. In the swing reactor fixed-bed system, the two reactors operate in a switchable mode and the catalyst can be unloaded/reloaded without disturbing the continuous operation of the system. The swing reactors are generally followed by various fixed-bed reactors in series containing hydrodesulfurization and other hydrofining catalysts.

However, when the feedstock contains a high amount of metals and other impurities, for example, asphaltenes, the use of trickle-bed reactors has to be carefully examined according to the catalyst cycle life. Alternatively, moving-bed reactors and ebullated-bed reactors have demonstrated reliable operation with the heavy feedstocks. Depending on the feedstock, the catalyst life may vary in the order of months or a year. It is then clear that the timescale of deactivation influences the choice of reactor (Moulijn et al., 2001).

Trickle-bed reactors are multifunctional reactive systems that allow a unique way of achieving reaction goals in the refining industry (Nigam and Larachi, 2005) and are extensively used in the petroleum industry for the hydrotreatment of petroleum distillates, the removal of impurities such as sulfur and nitrogen, and hydrocracking. These reactors are also being used for HDT of residue, where hydrodesulfurization (HDS) and hydrodemetallization (HDM) are the key reactions. On the other hand, in place of the trickle-bed concept, the catalyst may be suspended in the liquid feedstock in which case the reactor is a slurry reactor. Furthermore, these types of reactors are essential to meet the ultralow specification of the transportation fuel. Process intensification by operating at elevated pressures and temperatures (as well as a better understanding of flow behavior) helps in better design and scale-up (Nigam and Larachi, 2005).

Although the physical aspects of the trickle-bed reactor are relatively simple, the hydrodynamics in the reactor are extremely complex. It is for this reason that trickle-bed reactors have been extensively studied over the past several decades; however, the understanding of the hydrodynamics still leaves much to be desired. In the trickle-bed reactor, the reactions take place in the liquid phase, since much of the residual feed and product constituents do not vaporize at reactor pressure and temperature. The feedstock in the reactor is saturated with hydrogen gas because the partial pressure of hydrogen is very high and hydrogen is available in great excess. The oil and hydrogen reactant molecules diffuse through the liquid oil filling the catalyst pores and adsorb onto the catalyst surface where the hydrotreating reactions take place. Large molecular constituents tend to adsorb more strongly onto the catalyst surface than smaller molecules, and thus, the high-molecular-weight constituents tend to dominate the reactions on the catalyst when they can successfully diffuse into the catalyst pores. The product molecules must then desorb from the catalyst surface and diffuse out through the liquid that fills the catalyst pores.

Based on the direction of the fluid flow, packed-bed reactors can then be classified as trickle-bed reactors with cocurrent gas–liquid downflow, trickle-bed reactors with countercurrent gas–liquid flow, and packed-bubble reactors, where gas and liquid are contacted in cocurrent upflow. To carry out the catalyst and reactor selection and process design properly, knowledge of what each reactor type can and cannot do is very important. When a fixed-bed reactor is chosen, the issue is whether to use an upflow mode of operations or a downflow mode of operation.

Bubble column reactors are often operated in a semibatch mode with the gas phase as the continuous phase and the liquid with the suspended catalyst particles in batch. This is a typical way of producing chemicals in smaller amounts. Good mixing of the gas–solid–liquid mixture is important in bubble columns. Mixing can be enhanced by the use of a gas lift or a circulation pump with an ejector. The back-mixing of the suspension of liquid and catalyst particles is more intensive than that of the gas phase. Because of back-mixing, bubble columns are mostly rather isothermal.

The flow profile in a bubble column is determined by the gas flow velocity and the cross-sectional area of the column. At low gas velocities, all gas bubbles are assumed to have the same size. In this regime, we have a *homogeneous bubble flow*. If the gas velocity is increased in a narrow column, a *slug flow* is developed. In a slug flow, the bubbles fill the entire cross section of the reactor. Small bubbles exist in the liquid between the slugs, but the "majority of the gas is in the form of large bubbles. In wider bubble column vessels, a bubble size distribution is developed; this is called *a heterogeneous flow*. The flow properties in a bubble column are of considerable importance for the performance of three-phase reactors. The flow properties determine the gas volume fraction and the size of the interfacial area in the column. The flow profiles have a crucial impact on the reactor performance.

In the case of catalytic packed beds with two-phase flow (Trambouze, 1990), such as those used for straight-run naphtha hydrodesulfurization, from a reaction engineering perspective, a large catalyst-to-liquid volume ratio and plug flow of both phases are preferred, and catalyst deactivation is very slow or negligible, which facilitates reactor modeling and design. However, for three-phase catalytic reactors such as those employed for the hydrotreating of middle distillates and heavy petroleum fractions, the reaction occurs between the dissolved gas and the liquid-phase reactant at the surface of the catalyst, and the choice of upflow versus downflow operation can be based on rational considerations regarding the limiting reactant at the operating conditions of interest (Topsøe et al., 1996; Duduković et al., 2002).

The flow properties are of utmost importance for packed beds used in three-phase reactions. The most common operation policy is to allow the liquid to flow downward in the reactor. The gas phase can flow upward or downward in a concurrent or a countercurrent flow. The name of this reactor— the *trickle-bed* reactor—is indicative of flow conditions in the reactor, as the liquid flows downward in a laminar flow wetting the catalyst particles efficiently (*trickling flow*). It is also possible to allow both the gas and the liquid to flow upward in the reactor. In this case, no trickling flow can develop, and the reactor is called a *packed-bed* or a *fixed-bed* reactor.

At low gas and liquid flows, a *trickle flow* dominates; if the flow rates are higher, a *pulsed flow* develops in the reactor. At low gas and high liquid flows, the liquid phase is continuous and gas bubbles flow through the liquid phase. At high gas velocities, the gas phase is continuous and the liquid droplets are dispersed in the gas flow (*spray flow*). Trickle-bed reactors are usually gas upflow fixed bed, operated under *trickle* or *pulse* flow conditions. Both the gas and liquid phases approach plug flow conditions in a *trickle-bed* reactor. Different flow patterns also develop in these reactors, depending on the gas and the liquid flow rates. At low gas and high liquid flow rates, a bubble flow prevails, the bubbles flowing through the continuous liquid phase. At higher gas and low liquid velocities, the liquid is dispersed in the gas and the flow type is called a *spray flow*. At higher gas and low liquid flow rates, a *slug flow* develops in the reactor, and the bubble size distribution becomes very uneven. In this kind of packed-bed reactor, the gas phase is close to a plug flow, but the liquid phase is partially back-mixed.

The main advantage with packed beds is the flow pattern. Conditions approaching a plug flow are advantageous for most reaction kinetics. Diffusion resistance in catalyst particles may sometimes reduce the reaction rates, but for strongly exothermic reactions, effectiveness factors higher than unity (1) can be obtained. Hot spots appear in highly exothermic reactions, and these can have negative effects on the chemical stability and physical sustainability of the catalyst. If the catalyst in a packed bed is poisoned, it must be replaced, which is a cumbersome procedure. A packed bed is sometimes favorable, because the catalyst poison is accumulated in the first part of the bed and deactivation can be predicted in advance. In the hydrogenation of sulfur-containing aromatic compounds over nickel catalysts in a packed bed, the sulfur is adsorbed as a multimolecular layer on the catalyst at the inlet of the reactor. However, this layer works as a catalyst poison trap.

For *bubble flow*, the solid particles are evenly distributed in the reactor. This flow pattern resembles fluidized beds where only a liquid phase and a solid catalyst phase exist. At high gas velocities, a flow pattern called *aggregative fluidization* develops. In aggregative fluidization, the solid particles are unevenly distributed, and the conditions resemble those of a fluidized bed with a gas phase and a solid catalyst phase. Between these extreme flow areas, there exists a *slug flow* domain, which has the characteristics typical of both extreme cases. An uneven distribution of gas bubbles is characteristic for a *slug flow*.

Recently, a novel technology for three-phase processes has been developed: the monolith catalyst, sometimes also called the *frozen slurry reactor*. Similar to catalytic gas-phase processes, the active catalyst material and the catalyst carrier are fixed to the monolith structure. The gas and liquid flow through fluidized beds working in a countercurrent mode also exist. Because of gravity, the particles rise only to a certain level in the reactor. The liquid and gas phases are transported out of the reactor and can be separated by decanting.

6.3.2 Plug Flow Reactors

Plug flow, or tubular, reactors consist of a hollow pipe or tube through which reactants flow. These reactors are usually operated at steady state, and the reactants are continually consumed as they flow down the length of the reactor.

Plug flow reactors may be configured as one long tube or a number of shorter tubes. They range in diameter from a few centimeters to several meters. The choice of diameter is based on construction cost, pumping cost, the desired residence time, and heat transfer needs. Typically, long small diameter tubes are used with high reaction rates and larger-diameter tubes are used with slow reaction rates. Plug flow reactors have a wide variety of applications in either gas- or liquid-phase systems. Common industrial uses of tubular reactors are in gasoline production, oil cracking, synthesis of ammonia from its elements, and the oxidation of sulfur dioxide to sulfur trioxide.

6.3.3 Fluidized-Bed Reactors

A fluidized-bed reactor (FBR) is a type of reactor that can be used to carry out a variety of multiphase reactions. In this type of reactor, a fluid (gas or liquid) is passed through a granular solid material (usually a catalyst possibly configured as small spheres) at a sufficiently high velocity to suspend the solid and cause it to behave as though it were a fluid (Figure 6.5). This process (fluidization) imparts many advantages to the fluidized-bed reactor and, as a result, the fluidized-bed reactor is now used in many industrial applications.

In the reactor, the solid substrate (the catalytic material upon which chemical species react) material in the fluidized-bed reactor is typically supported by a *porous* plate, known as a distributor. The fluid is then forced through the distributor up through the solid material. At lower fluid velocities, the solids remain in place as the fluid passes through the voids in the material. As the

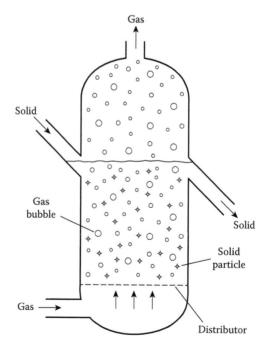

FIGURE 6.5   Schematic of a fluidized-bed reactor.

fluid velocity is increased, the reactor will reach a stage where the force of the fluid on the solids is enough to balance the weight of the solid material. This stage is known as incipient fluidization and occurs at this minimum fluidization velocity. Once this minimum velocity is surpassed, the contents of the reactor bed begin to expand and swirl around and the reactor is, at that point, a fluidized-bed reactor. Depending on the operating conditions and properties of the solid phase, various flow regimes can be observed in this reactor.

Due to the intrinsic fluidlike behavior of the solid material, fluidized beds do not experience poor mixing as in packed beds. This complete mixing allows for a uniform product that can often be hard to achieve in other reactor designs. The elimination of radial and axial concentration gradients allows for better fluid–solid contact, which is essential for reaction efficiency and quality. Also, many chemical reactions require the addition or removal of heat. Local hot or cold spots within the reaction bed, often a problem in packed beds, are avoided in a fluidized situation such as a fluidized-bed reactor. In other reactor types, these local temperature differences, especially hot spots, can result in product degradation. Thus, fluidized-bed reactors are well suited to exothermic reactions. Moreover, the fluidized-bed nature of these reactors allows for the ability to continuously withdraw product and introduce new reactants into the reaction vessel. Operating at a continuous process state allows refiners to produce the products more efficiently due to the removal of startup conditions in batch processes.

However, as in any design, the fluidized-bed reactor does have its drawbacks, which any reactor designer must take into consideration. Because of the expansion of the bed materials in the reactor, a larger vessel is often required than that for a packed-bed reactor. This larger vessel means that more must be spent on initial capital costs. Also, the requirement for the fluid to suspend the solid material necessitates that a higher fluid velocity is attained in the reactor. In order to achieve this, more pumping power and thus higher energy costs are needed. In addition, the pressure drop associated with deep beds also requires additional pumping power. Furthermore, the high gas velocities present in this style of reactor often result in fine particles becoming entrained in the fluid. These captured particles are then carried out of the reactor with the fluid, where they must be separated. This can be a very difficult and expensive problem to address depending on the design and function of the reactor. This may often continue to be a problem even with other entrainment reducing technologies. In addition, the fluidlike behavior of the fine solid particles within the bed eventually results in the wear of the reactor vessel that can require expensive maintenance and upkeep for the reaction vessel and pipes.

In a fluidized bed, the finely crushed catalyst particles are fluidized because of the movement of the liquid. Three-phase fluidized beds usually operate in a concurrent mode with gas and liquid flowing upward. However, bubbles fill the cross section of the channels. This flow type is called a *slug flow*. At higher gas flow rates, the smaller bubbles in a *slug flow* merge, and the resulting flow is called a *Taylor flow* or a *churn flow*. At even higher gas flow rates, the gas phase becomes continuous, and a gas–liquid dispersion develops. The flow is called an *annular flow,* and it is very inefficient and undesirable in three-phase systems. A monolith catalyst must always work under *bubble flow* or *slug flow* conditions; a *slug flow* gives the best mass transfer rates. Monolith catalysts are used in, for example, hydrogenation and dehydrogenation reactions.

Three different flow patterns can be observed in a fluidized-bed reactor. For *bubble flow,* the solid particles are evenly distributed in the reactor. This flow pattern resembles fluidized beds where only a liquid phase and a solid catalyst phase exist. At high gas velocities, a flow pattern called *aggregative fluidization* develops. In aggregative fluidization, the solid particles are unevenly distributed, and the conditions resemble those of a fluidized bed with a gas phase and a solid catalyst phase. Between these extreme flow areas, there exists a *slug flow* domain, which has the characteristics typical of both extreme cases. An uneven distribution of gas bubbles is characteristic for a *slug flow*. The flow pattern in a three-phase fluidized bed is usually much closer to complete back-mixing than to a plug flow. Because of the higher liquid flow velocities, larger particles can be used than in bubble columns.

6.3.4 Downflow Fixed-Bed Reactors

The reactor design commonly used in hydrodesulfurization of distillates is the fixed-bed reactor design in which the feedstock enters at the top of the reactor and the product leaves at the bottom of the reactor (Figure 6.1). The catalyst remains in a stationary position (fixed bed) with hydrogen and petroleum feedstock passing in a downflow direction through the catalyst bed. The hydrodesulfurization reaction is exothermic, and the temperature rises from the inlet to the outlet of each catalyst bed. With a high hydrogen consumption and subsequent large temperature rise, the reaction mixture can be quenched with cold recycled gas at intermediate points in the reactor system. This is achieved by dividing the catalyst charge into a series of catalyst beds, and the effluent from each catalyst bed is quenched to the inlet temperature of the next catalyst bed.

In a fixed-bed reactor, the flow pattern is much closer to plug flow, and the ratio of liquid to solid catalyst is small. If there are highly exothermic reactions such as those occurring in hydrotreating of unsaturated feeds (light cycle oil from fluid catalytic cracking units) (Sadeghbeigi, 2000), the reactions can be controlled by recycling of the liquid product stream, although this may not be practical if the product is not relatively stable under reaction conditions or if very high conversion is desired, as in hydrodesulfurization, since recycling causes the system to approach the behavior of a continuously stirred-tank reactor (CSTR). For such high-temperature increases, the preferred solution is quenching with hydrogen.

The extent of desulfurization is controlled by raising the inlet temperature to each catalyst bed to maintain constant catalyst activity over the course of the process. Fixed-bed reactors are mathematically modeled as plug flow reactors with very little back-mixing in the catalyst beds. The first catalyst bed is poisoned with vanadium and nickel at the inlet to the bed and may be a cheaper catalyst (*guard bed*). As the catalyst is poisoned in front of the bed, the temperature exotherm moves down the bed and the activity of the entire catalyst charge declines, thus requiring a raise in the reactor temperature over the course of the process sequence. After catalyst regeneration, the reactors are opened and inspected, and the high-metal-content catalyst layer at the inlet to the first bed may be discarded and replaced with fresh catalyst. The catalyst loses activity after a series of regenerations and, consequently, after a series of regenerations it is necessary to replace the complete catalyst charge. In the case of very high-metal-content feedstocks (such as residua), it is often necessary to replace the entire catalyst charge rather than to regenerate it. This is due to the fact that the metal contaminants cannot be removed by economical means during rapid regeneration, and the metals have been reported to interfere with the combustion of carbon and sulfur, catalyzing the conversion of sulfur dioxide (SO_2) to sulfate (SO_4^{2-}) that has a permanent poisoning effect on the catalyst.

Fixed-bed hydrodesulfurization units are generally used for distillate hydrodesulfurization and may also be used for residuum hydrodesulfurization but require special precautions in processing. The residuum must undergo two-stage electrostatic desalting so that salt deposits do not plug the inlet to the first catalyst bed and must be low in vanadium and nickel content to avoid plugging the beds with metal deposits. Hence the need for a guard bed in residuum hydrodesulfurization reactors.

During the operation of a fixed-bed reactor, contaminants entering with fresh feed are filtered out and fill the voids between catalyst particles in the bed. The buildup of contaminants in the bed can result in the channeling of reactants through the bed and reducing the hydrodesulfurization efficiency. As the flow pattern becomes distorted or restricted the pressure drop throughout the catalyst bed increases. If the pressure drop becomes high enough physical damage to the reactor internals can result. When high pressure drops are observed throughout any portion of the reactor, the unit is shut down and the catalyst bed is skimmed and refilled.

With fixed-bed reactors, a balance must be reached between reaction rate and pressure drop across the catalyst bed. As catalyst particle size is decreased, the desulfurization reaction rate increases but so does the pressure drop across the catalyst bed. Expanded-bed reactors do not have this limitation, and small 1/32 inch (0.8 mm) extrudate catalysts or fine catalysts may be used without increasing the pressure drop.

6.3.5 UPFLOW EXPANDED-BED REACTORS

Expanded-bed reactors are applicable to distillates but are commercially used for very heavy, high metals, and/or dirty feedstocks having extraneous fine solids material. They operate in such a way that the catalyst is in an expanded state so that the extraneous solids pass through the catalyst bed without plugging. They are isothermal, which conveniently handles the high-temperature exotherms associated with high hydrogen consumptions. Since the catalyst is in an expanded state of motion, it is possible to treat the catalyst as a fluid and to withdraw and add catalyst during operation.

Expanded beds of catalyst are referred to as particulate fluidized insofar as the feedstock and hydrogen flow upward through an expanded bed of catalyst with each catalyst particle in independent motion. Thus, the catalyst migrates throughout the entire reactor bed. Expanded-bed reactors are mathematically modeled as back-mix reactors with the entire catalyst bed at one uniform temperature. Spent catalyst may be withdrawn and replaced with fresh catalyst on a daily basis. Daily catalyst addition and withdrawal eliminate the need for costly shutdowns to change out catalyst and also result in a constant equilibrium catalyst activity and product quality. The catalyst is withdrawn daily and has a vanadium, nickel, and carbon content that is representative on a macroscale of what is found throughout the entire reactor. On a microscale, individual catalyst particles have ages from that of fresh catalyst to as old as the initial catalyst charge to the unit, but the catalyst particles of each age group are so well dispersed in the reactor that the reactor contents appear uniform.

In the unit, the feedstock and hydrogen recycle gas enter the bottom of the reactor, pass up through the expanded catalyst bed, and leave from the top of the reactor. Commercial expanded-bed reactors normally operate with 1/32 inch (0.8 mm) extrudate catalysts that provide a higher rate of desulfurization than the larger catalyst particles used in fixed-bed reactors. With extrudate catalysts of this size, the upward liquid velocity based on fresh feedstock is not sufficient to keep the catalyst particles in an expanded state. Therefore, for each part of the fresh feed, several parts of product oil are taken from the top of the reactor, recycled internally through a large vertical pipe to the bottom of the reactor, and pumped back up through the expanded catalyst bed. The amount of catalyst bed expansion is controlled by the recycle of product oil back up through the catalyst bed.

The expansion and turbulence of gas and oil passing upward through the expanded catalyst bed are sufficient to cause almost complete random motion in the bed (particulate fluidized). This effect produces the isothermal operation. It also causes almost complete back-mixing. Consequently, in order to effect near complete sulfur removal (over 75%), it is necessary to operate with two or more reactors in series. The ability to operate at a single temperature throughout the reactor or reactors, and to operate at a selected optimum temperature rather than an increasing temperature from the start to the end of the run, results in more effective use of the reactor and catalyst contents. When all these factors are put together, that is, the use of a smaller catalyst particle size, isothermal, fixed temperature throughout run, back-mixing, daily catalyst addition, and constant product quality, the reactor size required for an expanded bed is often smaller than that required for a fixed bed to achieve the same product goals. This is generally true when the feeds have high initial boiling points and/or the hydrogen consumption is very high.

6.3.6 EBULLATING BED REACTORS

In fixed-bed hydrocracking reactors, such as those used to process vacuum gas oil, heavy feedstocks (such as residua, heavy oil, extra heavy oil, and tar sand bitumen) will reduce the catalyst cycle life because of the content of asphaltene constituents (and resin constituents), trace metals (nickel, vanadium, and iron), or particulate matter. Recalling that fixed-bed units designed to process residue remove metals and other contaminants with upstream guard beds or onstream catalyst replacement technology, ebullated-bed units in which the catalyst within the reactor bed is not fixed can and do process significant amounts of heavy feedstocks (Speight, 2013, 2014). In the ebullated-bed unit, fresh catalyst is added and spent catalyst is removed continuously. Consequently, catalyst life does not impose limitations on feed selection or conversion.

In such a process, the hydrocarbon feed stream enters the bottom of the reactor and flows upward through the catalyst—the catalyst is kept in suspension by the pressure of the fluid feed. Ebullating bed reactors are capable of converting the most problematic feeds, such as atmospheric residua, vacuum residua, heavy oil, and tar sand bitumen (all of which have a high content of asphaltene constituents as well as metal constituents, sulfur constituents, and constituents ready to form sediment), to lower-boiling, more valuable products while simultaneously removing the contaminants.

In the unit, the hydrogen-rich recycle gas is bubbled up through a mixture of oil and catalyst particles that provides three-phase turbulent mixing, which is needed to ensure a uniform temperature distribution. At the top of the reactor, the catalyst is disengaged from the process fluids, which are separated in downstream flash drums. Some of the catalyst is withdrawn and replaced with fresh catalyst while the majority of the catalyst is returned to the reactor.

The function of the catalyst is to remove contaminants such as sulfur and nitrogen heteroatoms, which accelerate the deactivation of the catalyst, while cracking (converting) the feed to lighter products. Because ebullating bed reactors perform both hydrotreating and hydrocracking functions, they are considered to be dual-purpose reactors. Ebullating bed catalysts are made of pellets that are less than 1 mm in size to facilitate suspension by the liquid phase in the reactor.

In contrast to fixed-bed hydrocracking units for vacuum gas oil, ebullating bed units are suitable for processing residua and other heavy feedstocks, such as tar sand bitumen. The main advantages are as follows: (1) high conversion of atmospheric residue, up to 90% v/v; (2) better product quality than many other residue conversion processes, especially delayed coking; and (3) long run length—catalyst life does not limit these units since fresh catalyst is added and spent catalyst is removed continuously and (subject only to mechanical issues) the units can typically run for a much longer time than fixed-bed units for the same heavy feedstock.

Examples of processes that use ebullated-bed reactors are (1) H-Oil process, (2) LC-Fining process, and (3) T-Star process.

6.3.7 Demetallization Reactors

The demetallization reactor (*guard reactor*, *guard bed reactor*) is a reactor that is placed in front of hydrocracking reactors to remove contaminants, particularly metals, prior to hydrocracking. Such reactors may employ an inexpensive catalyst to remove metals from expanded-bed feed. Spent demetallization catalyst can be loaded to more than 30% vanadium. A catalyst support having large pores is preferentially demetallized with a low degree of desulfurization.

Feedstocks that have relatively high metal contents (>300 ppm) substantially increase catalyst consumption because the metals poison the catalyst, thereby requiring frequent catalyst replacement. The usual desulfurization catalysts are relatively expensive for these consumption rates, but there are catalysts that are relatively inexpensive and can be used in the first reactor to remove a large percentage of the metals. Subsequent reactors downstream of the first reactor would use normal hydrodesulfurization catalysts. Since the catalyst materials are proprietary, it is not possible to identify them here. However, it is understood that such catalysts contain little or no metal promoters, that is, nickel, cobalt, and molybdenum. Metals' removal on the order of 90% has been observed with these materials.

Thus, one method of controlling demetallization is to employ separate smaller *guard reactors* just ahead of the fixed-bed hydrodesulfurization reactor section. The preheated feed and hydrogen pass through the guard reactors that are filled with an appropriate catalyst for demetallization that is often the same as the catalyst used in the hydrodesulfurization section. The advantage of this system is that it enables replacement of the most contaminated catalyst (*guard bed*), where pressure drop is highest, without having to replace the entire inventory or shut down the unit. The feedstock is alternated between guard reactors while catalyst in the idle guard reactor is being replaced.

When the expanded-bed design is used, the first reactor could employ a low-cost catalyst (5% of the cost of Co/Mo catalyst) to remove the metals, and the subsequent reactors can use the more selective hydrodesulfurization catalyst. The demetallization catalyst can be added continuously without taking the reactor out of service and the spent demetallization catalyst can be loaded to more than 30% vanadium, which makes it a valuable source of vanadium.

6.4   PROCESS PARAMETERS

Reactor configurations within petroleum hydroprocessing units may include catalysts beds that are fixed or moving (Kundu et al., 2003; Ancheyta and Speight, 2007). Most hydroprocessing reactors are fixed-bed reactors and units with fixed-bed reactors must be shut down to remove the spent catalyst when catalyst activity declines below an acceptable level (due to the accumulation of coke, metals, and other contaminants)—there are also hydroprocessing reactors with moving or ebullating catalyst beds. A further example of a fixed-bed reactor is in catalytic reforming of naphtha to produce branched chain alkanes, cycloalkanes, and aromatic hydrocarbons usually using platinum or a platinum–rhenium alloy on an alumina support.

Using the hydrodesulfurization process as an example, all hydrodesulfurization processes react a feedstock with hydrogen to produce hydrogen sulfide and a desulfurized hydrocarbon product (Figure 6.3). The feedstock is preheated and mixed with hot recycle gas containing hydrogen and the mixture is passed over the catalyst in the reactor section at temperatures between 290°C and 445°C (550°F and 850°F) and pressures between 150 and 3000 psi (Table 6.2). The reactor effluent is then cooked by heat exchange, and desulfurized liquid hydrocarbon product and recycle gas are separated at essentially the same pressure as used in the reactor. The recycle gas is then scrubbed and/or purged of the hydrogen sulfide and light hydrocarbon gases, mixed with fresh hydrogen makeup, and preheated prior to mixing with hot hydrocarbon feedstock.

The recycle gas scheme is used in the hydrodesulfurization process to minimize physical losses of expensive hydrogen. Hydrodesulfurization reactions require a high hydrogen partial pressure in the gas phase to maintain high desulfurization reaction rates and to suppress carbon laydown (catalyst deactivation). The high hydrogen partial pressure is maintained by supplying hydrogen to the reactors at several times the chemical hydrogen consumption rate. The majority of the unreacted hydrogen is cooled to remove hydrocarbons, recovered in the separator, and recycled for

TABLE 6.2
Process Parameters for Hydrodesulfurization

Parameter	Naphtha	Residuum
Temperature, °C	300–400	340–425
Temperature, °F	570–750	645–800
Pressure, atm	515–1000	800–2500
LHSV	4.0–10.0	0.2–1.0
H_2 recycle rate, scf/bbl	400–1000	3000–5000
Catalyst life, months	36–120	6–12
Sulfur removal, % w/w	99.9	85.0
Nitrogen removal, % w/w	99.5	40.0

further utilization. Hydrogen is physically lost in the process by solubility in the desulfurized liquid hydrocarbon product and from losses during the scrubbing or purging of hydrogen sulfide and light hydrocarbon gases from the recycle gas.

The operating conditions in distillate hydrodesulfurization are dependent upon the feedstock as well as the desired degree of desulfurization or quality improvement. Kerosene and light gas oils are generally processed at mild severity and high throughput, whereas light catalytic cycle oils and thermal distillates require slightly more severe conditions. Higher-boiling distillates, such as vacuum gas oils and lube oil extracts, require the most severe conditions. The principal variables affecting the required severity in distillate desulfurization are (1) hydrogen partial pressure, (2) space velocity, (3) reaction temperature, and (4) feedstock properties.

6.4.1 PARTIAL PRESSURE

The important effect of *hydrogen partial pressure* is the minimization of coking reactions. If the hydrogen pressure is too low for the required duty at any position within the reaction system, premature aging of the remaining portion of catalyst will be encountered. In addition, the effect of hydrogen pressure on desulfurization varies with feed boiling range. For a given feed, there exists a threshold level above which hydrogen pressure is beneficial to the desired desulfurization reaction. Below this level, desulfurization drops off rapidly as hydrogen pressure is reduced.

6.4.2 SPACE VELOCITY

As the *space velocity* is increased, desulfurization is decreased but increasing the hydrogen partial pressure and/or the reactor temperature can offset the detrimental effect of increasing space velocity.

Reactors for *endothermic processes* (catalytic cracking, reforming, coking) require heat input to maintain the reaction temperature in the cracking zone and is shown on the far right in the endothermic region. Burning coke off the catalyst in the regenerator provides this heat and the recirculating catalyst transfers that energy to the cracking reaction in the riser of the fluid catalytic cracking unit. In the reforming reactor, the dehydrogenation reaction is highly endothermic and requires a reactor system of three to four reactors in series, with interstage heating between the reactors because the reaction temperature drop in each stage must be increased so that the reaction rate does not slow down too much. Reactors for *thermally neutral processes* (such as isomerization processes) that involve skeletal rearrangement of the molecular constituents of the feedstock but no change in the molecular weight does not cause any cooling or heating of the feed stream. *Exothermic processes* include hydrotreating, hydrocracking, and alkylation processes. Hydrocracking is highly exothermic owing to aromatic saturation reactions. Although the molecular weight is reduced by the cracking reaction, this is preceded by hydrogenation reactions, for example, aromatic ring saturation, which is necessary before the ring opening can occur. The alkylation process is also quite exothermic because higher-molecular-weight compounds are formed from *iso*-butane and olefins. Distillate and naphtha hydrotreating also release heat when organosulfur and nitrogen compounds (i.e., dibenzothiophene and pyridine) are converted to hydrogen sulfide and ammonia, respectively.

Many types of reactors have entered the field of petroleum refining, but they can be roughly divided into three types: (1) batch, (2) continuous stirred-tank reactor, CSTR, and (3) continuous plug flow reactor (Butt, 2000; Fogler, 2006).

6.4.3 TEMPERATURE

A higher *reaction temperature* increases the rate of desulfurization at constant feed rate, and the start-of-run temperature is set by the design desulfurization level, space velocity, and hydrogen partial pressure. The capability to increase temperature as the catalyst deactivates is built into the most process or unit designs. Temperatures of 415°C (780°F) and above result in excessive coking reactions and higher than normal catalyst aging rates. Therefore, units are designed to avoid the use of such temperatures for any significant part of the cycle life.

6.4.4 CATALYST LIFE

Catalyst life depends on the charge stock properties and the degree of desulfurization desired. The only permanent poisons to the catalyst are metals in the feedstock that deposit on the catalyst, usually quantitatively, causing permanent deactivation as they accumulate. However, this is usually of little concern except when deasphalted oils are used as feedstocks since most distillate feedstocks contain low amounts of metals. Nitrogen compounds are a temporary poison to the catalyst, but there is essentially no effect on catalyst aging except that caused by a higher-temperature requirement to achieve the desired desulfurization. Hydrogen sulfide can be a temporary poison in the reactor gas and recycle gas scrubbing is employed to counteract this condition.

Providing that pressure drop buildup is avoided, cycles of 1 year or more and ultimate catalyst life of 3 years or more can be expected. The catalyst employed can be regenerated by normal steam–air or recycle combustion gas–air procedures. The catalyst is restored to near fresh activity by regeneration during the early part of its ultimate life. However, permanent deactivation of the catalyst occurs slowly during usage and repeated regenerations, so replacement becomes necessary.

6.4.5 FEEDSTOCK

The different types of streams that can undergo hydroprocessing range from heavy feedstocks of resid and vacuum gas oil to lighter feedstocks of naphtha and distillate. Naphtha, or gasoline, is hydroprocessed to remove contaminants such as sulfur, which is harmful to downstream operations (such as precious metal reforming catalyst). Diesel hydroprocessing removes sulfur to meet fuel requirements and saturates aromatics. The purpose of resid and vacuum gas oil hydroprocessing is to remove metals, sulfur, and nitrogen (e.g., *hydrotreating*), as well as to convert high-molecular-weight hydrocarbons into lower-molecular-weight hydrocarbons (e.g., *hydrocracking*). Thus, it is not surprising that the character of the *feedstock properties*, especially the feed boiling range, has a definite effect on the ultimate design of the desulfurization unit and process flow. In addition, there is a definite relationship between the percent by weight sulfur in the feedstock and the hydrogen requirements. Also, the reaction rate constant in the kinetic relationships decreases rapidly with increasing average boiling point in the kerosene and light gas oil range but much more slowly in the heavy gas oil range. This is attributed to the difficulty in removing sulfur from ring structures present in the entire heavy gas oil boding range.

The hydrodesulfurization of light (low-boiling) distillate (naphtha or kerosene) is one of the more common catalytic processes since it is usually used as a pretreatment of such feedstocks prior to deep hydrodesulfurization or prior to catalytic reforming. Hydrodesulfurization of such feedstocks is required because sulfur compounds poison the precious metal catalysts used in reforming, and desulfurization can be achieved under relatively mild conditions and is near quantitative (Table 6.3). If the feedstock arises from a cracking operation, hydrodesulfurization will be accompanied by some degree of saturation resulting in increased hydrogen consumption. The hydrodesulfurization of low-boiling (naphtha) feedstocks is usually a gas-phase reaction and may employ the catalyst in fixed beds and (with all of the reactants in the gaseous phase) only minimal diffusion problems are encountered within the catalyst pore system. It is, however, important that the feedstock be completely volatile before entering the reactor as there may be the possibility of pressure variations (leading to less satisfactory results) if some of the feedstock enters the reactor in the liquid phase and is vaporized within the reactor.

TABLE 6.3

Hydrodesulfurization of Various Naphtha Fractions

Feedstock	Boiling Range		Sulfur	Desulfurization[a]
	°C	°F	wt%	%
Visbreaker naphtha	65–220	150–430	1.00	90
Visbreaker–coker naphtha	65–220	150–430	1.03	85
Straight-run naphtha	85–170	185–340	0.04	99
Catalytic naphtha (light)	95–175	200–350	0.18	89
Catalytic naphtha (heavy)	120–225	250–440	0.24	71
Thermal naphtha (heavy)	150–230	300–450	0.28	57

[a] Process conditions, Co–Mo on alumina, 260°C–370°C/500°F–700°F, 200–500 psi hydrogen.

In applications of this type, the sulfur content of the feedstock may vary from 100 ppm to 1%, and the necessary degree of desulfurization to be effected by the treatment may vary from as little as 50% to more than 99%. If the sulfur content of the feedstock is particularly low, it will be necessary to presulfide the catalyst. For example, if the feedstock only has 100–200 ppm sulfur, several days may be required to sulfide the catalyst as an integral part of the desulfurization process even with complete reaction of all of the feedstock sulfur to, say, cobalt and molybdenum (catalyst) sulfides. In such a case, presulfiding can be conveniently achieved by the addition of sulfur compounds to the feedstock or by the addition of hydrogen sulfide to the hydrogen.

Generally, hydrodesulfurization of naphtha feedstocks to produce catalytic reforming feedstocks is carried to the point where the desulfurized feedstock contains less than 20 ppm sulfur. The net hydrogen produced by the reforming operation may actually be sufficient to provide the hydrogen consumed in the desulfurization process. The hydrodesulfurization of middle distillates is also an efficient process and applications include predominantly the desulfurization of kerosene, diesel fuel, jet fuel, and heating oils that boil over the general range 250°C–400°C (480°F–750°F). However, with this type of feedstock, hydrogenation of the higher-boiling catalytic cracking feedstocks has become increasingly important where hydrodesulfurization is accomplished alongside the saturation of condensed-ring aromatic compounds as an aid to subsequent processing.

Under the relatively mild processing conditions used for the hydrodesulfurization of these particular feedstocks, it is difficult to achieve complete vaporization of the feed. Process conditions may dictate that only part of the feedstock is actually in the vapor phase and that sufficient liquid phase is maintained in the catalyst bed to carry the larger molecular constituents of the feedstock through the bed. If the amount of liquid phase is insufficient for this purpose, molecular stagnation (leading to carbon deposition on the catalyst) will occur.

Hydrodesulfurization of middle distillates causes a more marked change in the specific gravity of the feedstock, and the amount of low-boiling material is much more significant when compared with the naphtha-type feedstock. In addition, the somewhat more severe reaction conditions (leading to a designated degree of hydrocracking) also cause an overall increase in hydrogen consumption when middle distillates are employed as feedstocks in place of the naphtha.

High-boiling distillates, such as the atmospheric and vacuum gas oils, are not usually produced as a refinery product but merely serve as feedstocks to other processes for conversion to lower-boiling materials. For example, gas oils can be desulfurized to remove more than 80% of the sulfur originally in the gas oil with some conversion of the gas oil to lower-boiling materials. The treated gas oil (which has a reduced carbon residue as well as lower sulfur and nitrogen contents relative to the untreated material) can then be converted to lower-boiling products in, say, a catalytic cracker where an improved catalyst life and volumetric yield may be noted. The conditions used for the hydrodesulfurization of gas oil may be somewhat more severe than the conditions employed for the hydrodesulfurization of middle distillates with, of course, the feedstock in the liquid phase.

In summary, the hydrodesulfurization of the low-, middle-, and high-boiling distillates can be achieved quite conveniently using a variety of processes. One major advantage of this type of feedstock is that the catalyst does not become poisoned by metal contaminants in the feedstock since only negligible amounts of these contaminants will be present. Thus, the catalyst may be regenerated several times and onstream times between catalyst regenerations (while varying with the process conditions and application) may be of the order of 3–4 years.

Residuum hydroconversion requires a substantially different catalyst to the catalyst used for distillate hydroconversion. The catalyst used is frequently a low metal loading with molybdenum (Mo) with a special pore size distribution that is less subject to pore plugging by coke—more specifically, a bimodal pore size distribution is frequently used. Because residuum frequently has high levels of vanadium and nickel in it, the catalyst activity may actually increase initially and then slowly decrease as the promotional effects of the metals adsorbed on the hydrocracking catalyst are counterbalanced by deactivation because of coking. The residuum hydroconversion reactors may be either fluid bed or (more likely) ebullated bed. Because of the high rates of catalyst coking, most moving-bed residuum conversion reactors have a capability for fresh catalyst addition and spent catalyst withdrawal while the reactor is operating.

Processes that employ the ebullated-bed concept are the H-Oil process and the LC-Fining process (Speight, 2013, 2014). In the H-Oil reactor, a recycle pump, located either internally or externally, circulates the reactor fluids down through a central downcomer and then upward through a distributor plate and into the ebullated catalyst bed. The reactor is usually well insulated and operated adiabatically. Although the H-Oil reactor is loaded with catalyst, not all of the reactions are catalyzed; some are thermal reactions, like thermal cracking, which depend on liquid holdup and not on how much catalyst is present.

In the LC-Fining reactor, hydrogen reacted within an expanded catalyst bed that is maintained in turbulence by liquid upflow so as to achieve efficient isothermal operation. Product quality is constantly maintained at a high level by intermittent catalyst addition and withdrawal. Reactor products flow to the high-pressure separator, to the low-pressure separator, and then to product fractionation; recycled hydrogen is separated and purified. Process features include onstream catalyst addition and withdrawal, thereby eliminating the need to shut down for catalyst replacement. The expanded-bed reactors operate at near isothermal conditions without the need for quenches within the reactor.

REFERENCES

Al-Dahhan, M.H., Larachi, F., Duduković, M.P., and Laurent, A. 1997. High pressure trickle-bed reactors: A review. *Industrial and Engineering Chemistry Research*, 36: 3292.

Ancheyta, J., Sánchez, S., and Rodríguez, M.A. 2005. Kinetic modeling of hydrocracking of heavy oil fractions: A review. *Today*, 109: 76–92.

Ancheyta, J. and Speight, J.G. 2007. Feedstock evaluation and composition. In: *Hydroprocessing of Heavy Oils and Residua*. J. Ancheyta and J.G. Speight (Editors). CRC Press/Taylor & Francis Group, Boca Raton, FL, Chapter 2.

Boelhouwer, J.G. 2001. Non-steady operation of trickle-bed reactors: Hydrodynamics, mass and heat transfer. PhD thesis. Technische Universiteit Eindhoven, Eindhoven, the Netherlands.

Butt, J.B. 2000. *Reaction Kinetics and Reactor Design*, 2nd Edition. Marcel Dekker, New York.

Carbonell, R.G. 2000. Multiphase flow models in packed beds. *Revue de l'Institut Français du Pétrole, Oil and Gas Science and Technology*, 55(4): 417.

Chadeesingh, R. 2011. The Fischer-Tropsch process. In: *The Biofuels Handbook*. J.G. Speight (Editor). The Royal Society of Chemistry, London, UK, Part 3, Chapter 5, pp. 476–517.

Davis, M.R.E. and Davis, R.J. 2003. *Fundamentals of Chemical Reaction Engineering*. McGraw-Hill, New York.

Donati, G. and Paludetto, R. 1999. Batch and semi-batch catalytic reactors (from theory to practice). *Catalysis Today*, 52: 183–195.

Duduković, M.P., Larachi, F., and Mills, P.L. 1999. Multiphase reactors—Revisited. *Chemical Engineering Science*, 54: 1975–1995.

Duduković, M.P., Larachi, F., and Mills, P.L. 2002. Multiphase catalytic reactors: A perspective on current knowledge and future trends. *Catalysis Reviews Science and Engineering*, 44: 123.

Ellenberger, J. and Krishna, R. 1999. Counter-current operation of structured catalytically packed distillation columns: Pressure drop, holdup and mixing. *Chemical Engineering Science*, 54: 1339.

Fogler, S.H. 2006. *Elements of Chemical Reaction Engineering*, 4th Edition. Prentice Hall, Englewood Cliffs, NJ.

Froment, G.F. and Bischoff, K.B. 1990. *Chemical Reactor Analysis and Design*, 2nd Edition. John Wiley & Sons, New York.

Furimsky, E. 1998. Selection of catalysts and reactors for hydroprocessing. *Applied Catalysis A*, 171: 177–206.

Gianetto, A. and Specchia, V. 1992. Trickle-bed reactors: State of the art and perspectives. *Chemical Engineering Science*, 47: 3197.

Haure, P.M., Hudgins, R.R., and Silveston, P.L. 1989. Periodic operation of a trickle-bed reactor. *AIChE Journal*, 35: 1437.

Jones, D.S.J. and Pujado, P.P. (Editors). 2006. *Handbook of Petroleum Processing*. Springer Science, Heidelberg, Germany.

Kressmann, S., Morel, F., Harlé, V., and Kasztelan, S. 1998. Recent developments in fixed-bed catalytic residue upgrading. *Catalysis Today*, 43: 203–215.

Krishna, R. and Sie, S.T. 1994. Strategies for multiphase reactor selection. *Chemical Engineering Science*, 49: 4029.

Kundu, A., Nigam, K.D.P., Duquenne, A.M., and Delmas, H. 2003. Recent developments on hydroprocessing reactors. *Reviews in Chemical Engineering*, 19: 531–605.

Liu, Y., Gao, L., Wen, L., and Zong, B. 2009. Recent advances in heavy oil hydroprocessing technologies. *Recent Patents in Chemical Engineering*, 2: 22–36.

Moulijn, J.A., van Diepen, A.E., and Kapteijn, F. 2001. Catalyst deactivation: Is it predictable? What to do? *Applied Catalysis A*, 212: 3–16.

Ng, K.M. and Chu, C.F. 1987. Trickle-bed reactors. *Chemical Engineering Progress*, 83: 55.

Nigam, K.D.P. and Larachi, F. 2005. Process intensification in trickle-bed reactors. *Chemical Engineering Science*, 60: 5880–5894.

OSHA Technical Manual. 1999. Section IV, Chapter 2: Petroleum refining processes. http://www.osha.gov/dts/osta/otm/otm_iv/otm_iv_2.html.

Robinson, K.K. 2006. *Reactor Engineering. Encyclopedia of Chemical Processing*. CRC Press/Taylor & Francis Group, Boca Raton, FL.

Sadeghbeigi, R. 2000. *Fluid Catalytic Cracking Handbook*, 2nd Edition. Gulf Professional Publishing, Elsevier, Oxford, UK.

Salmi, T.O., Mikkola, J.P., and Wärnå, J.P. 2011. *Chemical Reaction Engineering and Reactor Technology*. CRC Press/Taylor & Francis Group, Boca Raton, FL.

Saroha, A.K. and Nigam, K.D.P. 1996. Trickle-bed reactors. *Reviews in Chemical Engineering*, 12: 207.

Scheffer, B., van Koten, M.A., Robschlager, K.W., and de Boks, F.C. 1998. The shell residue hydroconversion process: Development and achievements. *Catalysis Today*, 43: 217–224.

Speight, J.G. 2013. *Heavy and Extra Heavy Oil Upgrading Technologies*. Gulf Professional Publishing, Elsevier, Oxford, UK.

Speight, J.G. 2014. *The Chemistry and Technology of Petroleum*, 5th Edition. CRC Press/Taylor & Francis Group, Boca Raton, FL.

Stockfleth, R. and Brunner, G. 1999. Hydrodynamics of a packed countercurrent column for gas extraction. *Industrial and Engineering Chemistry Research*, 38: 4000.

Topsøe, H., Clausen, B.S., and Massoth, F.E. 1996. *Hydrotreating Catalysis: Science and Technology*. Springer-Verlag, Berlin, Germany.

Trambouze, P. 1990. Countercurrent two-phase flow fixed bed catalytic reactors. *Chemical Engineering Science*, 45: 2269.

Versteeg, G.F., Visser, J.B.M., van Dierendonck, L.L., and Kuipers, J.A.M. 1997. Absorption accompanied with chemical reaction in trickle-bed reactors. *Chemical Engineering Science*, 52: 4057.

Zhukova, T.B., Paisarenko, V.N., and Kafarov, V.V. 1990. Modelling and design of industrial reactors with a stationery bed of catalyst and two-phase gas-liquid flow: A review. *International Chemical Engineering*, 30: 57.

Section II

Refining

7 Pretreatment and Distillation

7.1 INTRODUCTION

Petroleum in the unrefined state is of limited value and of limited use. Refining is required to produce products that are attractive to the marketplace. Thus, petroleum refining is a series of integrated steps by which the crude oil is converted into salable products with the desired qualities and in the amounts dictated by the market (Speight and Ozum, 2002; Parkash, 2003; Hsu and Robinson, 2006; Gary et al., 2007; Speight, 2014a). In fact, a refinery is essentially a group of manufacturing plants that vary in number with the variety of products produced. Refinery processes must be selected and products manufactured to give a balanced operation. *Crude oil must be converted into products according to the demand for each.*

The petroleum refinery of the current century is a much more complex operation (Figure 7.1) than those refineries of the early 1900s and even of the immediate years following World War II (Jones, 1995; Speight, 2011, 2014a). Early refineries were predominantly distillation units, perhaps with ancillary units to remove objectionable odors from the various product streams. The refinery of the 1930s was somewhat more complex but was essentially a distillation unit (Speight, 2014a), but at this time cracking and coking units were starting to appear in the scheme of refinery operations. These units were not what we imagine today as a cracking and coking unit but were the forerunners of today's units. Also at this time, asphalt was becoming a recognized petroleum product. Finally, current refineries (Figure 7.1) are a result of major evolutionary trends and are highly complex operations. Most of the evolutionary adjustments to refineries have occurred during the decades since the commencement of World War II. In the petroleum industry, as in many other industries, *supply* and *demand* are key factors in efficient and economic operation. Innovation is also a key.

A refinery is an integrated group of manufacturing plants (Figure 7.1) that vary in number with the variety of products produced. Refinery processes must be selected and products manufactured to give a balanced operation: that is, crude oil must be converted into products according to the rate of sale of each. For example, the manufacture of products from the lower-boiling portion of petroleum automatically produces a certain amount of higher-boiling components. If the latter cannot be sold as, say, heavy fuel oil, they accumulate until refinery storage facilities are full. To prevent the occurrence of such a situation, the refinery must be flexible and able to change operations as needed. This usually means more processes to accommodate the ever-changing demands of the market (Speight and Ozum, 2002; Parkash, 2003; Hsu and Robinson, 2006; Gary et al., 2007; Speight, 2014a). This could be reflected in the inclusion of a cracking process to change an excess of heavy fuel oil into more gasoline with coke as the residual product or inclusion of a vacuum distillation process to separate the heavy oil into lubricating oil stocks and asphalt.

Thus, to accommodate the sudden changes in market demand, a refinery must include the following: (1) all necessary nonprocessing facilities; (2) adequate tank capacity for storing crude oil, intermediate, and finished products; (3) a dependable source of electrical power; (4) material-handling equipment; (5) workshops and supplies for maintaining a continuous 24 hours/day, 7 day/week operation; (6) waste disposal and water-treating equipment; and (7) and product-blending facilities (Speight and Ozum, 2002; Parkash, 2003; Hsu and Robinson, 2006; Gary et al., 2007; Speight, 2014a).

Petroleum refining is a very recent science and many innovations have evolved during the twentieth century. The purpose of this chapter is to illustrate the initial processes that are applied

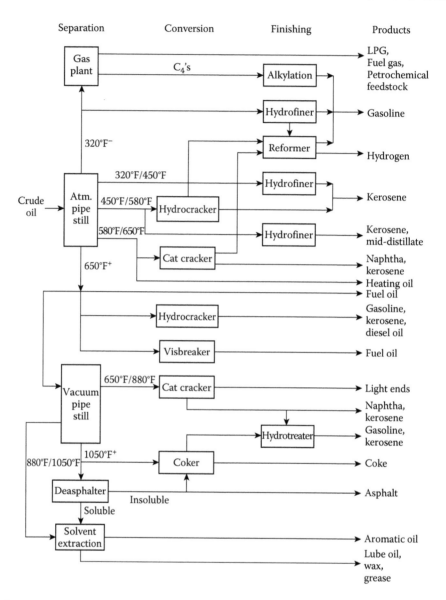

FIGURE 7.1 General schematic of a refinery.

to a feedstock in a refinery. The first processes (desalting and dewatering) focus on the cleanup of the feedstock, particularly the removal of the troublesome brine constituents (Burris, 1992). This is followed by distillation to remove the volatile constituents with the concurrent production of a residuum that can be used as a cracking (coking) feedstock or as a precursor to asphalt. Current methods of bitumen processing (Speight, 2000, 2014a) involve direct use of the bitumen as feedstock for delayed or fluid coking (Chapter 8). Other methods of feedstock treatment that involve the concept of volatility are also included here even though some of the methods (such as *stripping, rerunning,* and the like) might also be used for product purification. However, the distillation step is a viable step that produces additional valuable high-boiling fractions from the bitumen.

7.2 DEWATERING AND DESALTING

Even though distillation is, to all appearances, the first step in crude oil refining, it should be recognized that crude oil that is contaminated by salt water either from the well or during transportation to the refinery must be treated to remove the emulsion (Burris, 1992; Abdel-Aal et al., 2016). If salt water is not removed, the materials of construction of the heater tubes and column intervals are exposed to chloride ion attack and the corrosive action of hydrogen chloride, which may be formed at the temperature of the column feed.

Most crude oils contain traces of salt through the salt water produced with the crude feedstock (Speight, 2000, 2014a; Speight and Ozum, 2002; Parkash, 2003; Hsu and Robinson, 2006; Gary et al., 2007). Desalting operations are necessary to remove salt from the brines that are present with the crude oil after recovery (Abel-Aal et al., 2016). The salt can decompose in the heater to form hydrochloric acid and cause corrosion of the fractionator overhead equipment. In order to remove the salt, water is injected into the partially preheated crude, and the stream is thoroughly mixed so that the water extracts practically all the salt from the oil. The mixture of oil and water is separated in a desalter, which is a large vessel in which the water settles out of the oil, a process that may be accelerated by the addition of chemicals or by electrical devices. The salt-laden water is automatically drained from the bottom of the desalter. Failure to remove the brine to acceptable levels can result in unacceptable levels of hydrogen chloride produced during refining. The hydrogen chloride will cause corrosion to equipment even to the point of weakening the equipment and causing fires and explosions.

Generally, removal of this unwanted water has been fairly straightforward, involving wash tanks or a heater. Removal of this same water along with salt concentration reduction presents a completely different set of problems. These can be overcome with relative ease if the operator is willing to spend many thousands of dollars each year for water, fuel, power, and chemical additives. Conversely, if some time is spent in the initial design stages to determine the best methods for water removal to achieve lower bottom sediment and water (BS&W) remnants, mixing, and injection, then a system can be designed that will operate at a greatly reduced annual cost. In most cases, yearly savings on operational expenses will pay for the complete installation in 1–3 years.

The practice of desalting crude oil is an old process and can occur at the wellhead or (depending on the level of salt in the crude oil) at the refinery (Speight, 2000, 2014a; Speight and Ozum, 2002; Parkash, 2003; Hsu and Robinson, 2006; Gary et al., 2007). Indeed, refineries have been successfully desalting crude oil to less than 5 lb per thousand bbl for many years, and mechanical and electrostatic desalting have been improved greatly. However, very little attention has been given to the use of dilution water, probably due to the general availability of both freshwater and wastewater in and near refineries. Three general approaches have been taken to the desalting of crude petroleum (Figure 7.2). Numerous variations of each type have been devised, but the selection of a particular process depends on the type of salt dispersion and the properties of the crude oil.

The salt or brine suspensions may be removed from crude oil by heating (90°C–150°C, 200°F–300°F) under pressure (50–250 psi) that is sufficient to prevent vapor loss and then allowing the material to settle in a large vessel. Alternatively, coalescence is aided by passage through a tower packed with sand, gravel, and the like.

The common removal technique is to dilute the original brine with fresher water so that the salt content of water that remains after separation treatment is acceptable, perhaps 10 lb per thousand barrels of crude oil, or less. In areas where freshwater supplies are limited, the economics of this process can be critical. However, crude oil desalting techniques in the field have improved with the introduction of the electrostatic coalescing process (Figure 7.2). Even when adequate supplies of freshwater are available for desalting operations, preparation of the water for dilution purposes may still be expensive.

Requirements for dilution water ratios based on water salinity calculations (Table 7.1; Figure 7.3) can be calculated as a material balance and by combining the arithmetic mean of material balance

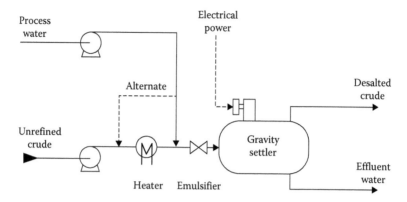

FIGURE 7.2   General methods for desalting crude oil.

TABLE 7.1
Pounds of Salt per 1000 bbl of Crude Oil

Salt Content of Water (ppm)	Volume Percent Water Content in Oil (bbl Water)				
	1.00 (10)	0.50 (5)	0.20 (2)	0.10 (1)	0.05 (0.5)
10,000	35.00	17.50	7.00	3.50	1.75
20,000	70.00	35.00	14.00	7.00	3.50
30,000	105.00	52.50	21.00	10.00	5.25
40,000	140.00	70.00	28.00	14.00	7.00
50,000	175.00	87.50	35.00	17.50	8.75
100,000	350.00	175.00	70.00	35.00	17.50
150,000	525.00	262.50	105.00	52.50	26.25
200,000	700.00	300.00	140.00	70.00	35.00

and water injection and dispersion for contact efficiency, very low dilution water use rates can be achieved. This can be highly significant in an area where production rates of 100,000 bbl/day are common and freshwater supply is limited.

Desalter units generally will produce a dehydrated stream containing like amounts of bottom sediment and water (BS&W) from each stage. Therefore, bottom sediment and water can be considered as *pass through volume*, and the dilution water added is the amount of water to be recycled. The recycle pump, however, generally is oversized to compensate for difficult emulsion conditions and upsets in the system. *Dilution water* calculations for a two-stage system using recycle are slightly more complicated than for a single-stage process (Table 7.2; Figure 7.4) or a two-stage process without recycle (Table 7.3; Figure 7.5).

The common approach to desalting crude oil involves use of the two-stage desalting system (Figure 7.5) in which dilution water is injected between stages after the stream water content has been reduced to a very low level by the first stage. Further reduction is achieved by adding the second-stage recycle pump. The *second-stage water* is much lower in sodium chloride (NaCl) than the produced stream inlet water due to addition of dilution water. By recycling this water to the first stage, both salt reduction and dehydration are achieved in the first stage. The water volume to be recycled is assumed to be the same as dilution water injection volume.

A very low bottom sediment and water content at the first-stage desalter exit requires a high percentage of dilution water to properly contact dispersed, produced water droplets and achieve desired

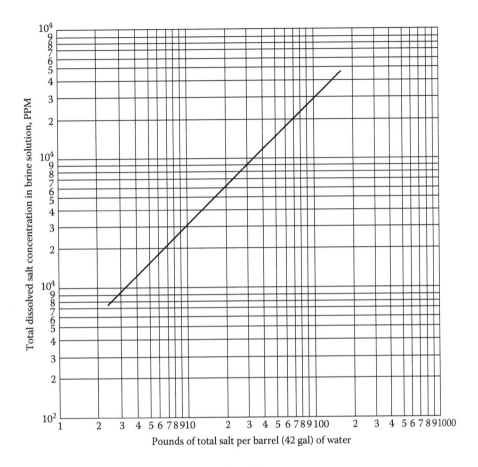

FIGURE 7.3   Pounds of salt per barrel (42 U.S. gallons) of water.

salt concentration reduction. This percentage dilution water varies with the strength of the water/ oil emulsion and oil viscosity. Empirical data show that the range is from 4.0% to as high as 10%. Obviously, this indicates that the mixing efficiency of 80% is not valid when low water contents are present. Additional field data show that the use rates for the use of low dilution water can be maintained and still meet the required mixing efficiencies. When 99.9% of the produced water has been removed, the remaining 0.1% consists of thousands of very small droplets more or less evenly distributed throughout the oil. To contact them would require either a large amount of dilution water dispersed in the oil or a somewhat smaller amount with better droplet dispersion. Whatever the required amount, it can be attained without exceeding the dilution water rates shown in the earlier example.

Water contained in each desalter unit is an excellent source of volume ratio makeup through the use of a recycle option (Figure 7.6). The amount of water recycled to the first stage must be the same as the dilution water injection rate to maintain the water level in the second-stage unit. The volume of water recycled to the second-stage inlet can be any amount, since it immediately rejoins the controlled water volume in the lower portion of the desalter unit (*internal recycle*). An additional pump may be used (Figure 7.6) for recycling first-stage water to first-stage inlet. The first-stage internal recycle is not necessarily required for each installation and is dependent upon amount of produced water present in the inlet stream. In terms of water requirements, *internal recycle* may be ignored since it does not add salt or water volume to the stream process.

Emulsions may also be broken by addition of treating agents, such as soaps, fatty acids, sulfonates, and long-chain alcohols. When a chemical is used for emulsion breaking during desalting, it

TABLE 7.2

Calculation for Required Amount of Dilution Water for a Single-Stage Process[a]

$Z = BK_3$

$$K_3 = \frac{AK_1 + YK_2E}{A + YE}$$

where

$$A = \frac{1000X_1}{1 - X_1} = \text{water in inlet stream, bbl}$$

$$B = \frac{1000x_2}{1 - X_2} = \text{water in clean oil, bbl}$$

$C = A + Y = $ water to desalter inlet, bbl

Y is the injection water (varies with each problem), bbl

$V = A + Y - B = $ water to disposal, bbl

E is the mixing efficiency of Y with A (as a function); assume 80%

K_1 is the salt per barrel of water in produced oil stream, lb

K_2 is the salt per barrel of dilution water, lb

$K_3 = \dfrac{AK_1 + YK_2E}{A + YE}$, which is the salt per barrel of water to desalter inlet, lb

X_1 is the fraction of water in produced oil stream

X_2 is the fraction of water in clean oil outlet

Z is the salt in outlet clean oil per 1000 bbl of net oil, lb

[a] See Figure 7.4.

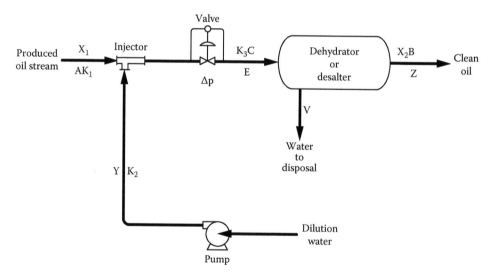

FIGURE 7.4   A single-stage desalting system.

may be added at one or more of three points in the system. First, it may be added to the crude oil before it is mixed with freshwater. Second, it may be added to the freshwater before mixing with the crude oil. Third, it may be added to the mixture of crude oil and water. A high-potential field across the settling vessel also aids coalescence and breaks emulsions, in which case dissolved salts and impurities are removed with the water.

TABLE 7.3

Calculation for Required Amount of Dilution Water for a Two-Stage Process[a]

$$VK_5 = \frac{BAK_1(R-C) + SYK_3(R-E_2C)}{SR + BE_1(C-R)}$$

$$V = B + Y - C$$

$$Z = BK_2 + YK_3 - VK_5$$

where

$A = \dfrac{1000X_1}{1-X_1} = $ water in inlet stream to facility, bbl

$B = \dfrac{1000X_2}{1-X_2} = $ water in effluent of first stage, bbl

$C = \dfrac{1000X_3}{1-X_3} = $ water in effluent of second-stage desalter, bbl

Y is the injection water of lower salinity than inlet water (A), bbl

V is the recycle water to injection in first-stage inlet line, bbl

E_1 is the mixing efficiency of V with A (as a fraction), assumed 80%

E_2 is the mixing efficiency of Y with B (as a fraction), assumed 80%

X_1 is the fraction of water in inlet stream to facility

X_2 is the fraction of water in first-stage outlet oil

X_3 is the fraction of water in second-stage outlet oil

K_1 is the salt per barrel of water to facility, lb

$K_2 = \dfrac{AK_1 + VE_1K_5}{A + VE_1}$, which is the salt per barrel of water to first-stage desalter, lb

K_3 is the salt per barrel of dilution water, lb

$K_4 = \dfrac{BK_2 + YE_2K_3}{B + E_2Y}$, which is the salt per barrel of water to second-stage desalter, lb

K_5 is the salt per barrel of water to recycle injection into inlet line to first stage, lb

$S = A + E_1 V$, which is the water to first-stage desalter, bbl

$R = B + E_2Y$, which is the water to second-stage desalter, bbl

Z is the salt in outlet per 1000 bbl of net oil, lb

Note: All water volumes are per 1000 bbl net oil, all salt contents are pounds of total dissolved salts per barrel of water, and Y varies with each individual problem.

[a] See Figure 7.5.

If the oil entering the desalter is not hot enough, it may be too viscous to permit proper mixing and complete separation of the water and the oil, and some of the water may be carried into the fractionator. If, on the other hand, the oil is too hot, some vaporization may occur, and the resulting turbulence can result in improper separation of oil and water. The desalter temperature is therefore quite critical, and normally a bypass is provided around at least one of the exchangers so that the temperature can be controlled. The optimum temperature depends upon the desalter pressure and the quantity of light material in the crude, but is normally approximately 120°C (250°F), 100°C (212°F), being lower for low pressures and light crude oils. The average water injection rate is 5% of the charge.

Regular laboratory analyses will monitor the desalter performance, and the desalted crude should normally not contain more than one kilogram of salt per 1000 barrels of feed.

Good desalter control is indicated by the chloride content of the overhead receiver water and should be on the order of 10–30 ppm chlorides. If the desalter operation appears to be satisfactory but the

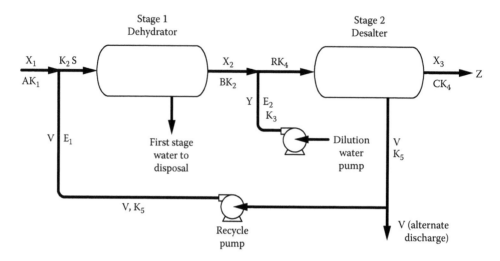

FIGURE 7.5   A two-stage desalting system.

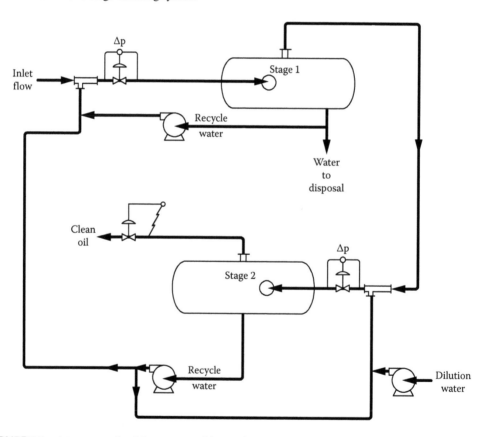

FIGURE 7.6   A two-stage desalting system with recycle.

chloride content in the overhead receiver water is greater than 30 ppm, then caustic should be injected at the rate of 1–3 lb per 1000 barrels of charge to reduce the chloride content to the range of 10–30 ppm. Salting out will occur below 10 and severe corrosion above 30 ppm. Another controlling factor on the overhead receiver water is pH. This should be controlled between pH 5.5 and 6.5. Ammonia injection into the tower top section can be used as a control for this. In addition to electrical methods for desalting, desalting may also be achieved by using the concept of a packed column (Figure 7.2) that facilitates the separation of the crude oil and brine through the agency of an adsorbent.

Finally, *flashing* the crude oil feed can frequently reduce corrosion in the principal distillation column. In the flashing operation, desalted crude is heat exchanged against other heat sources that are available to recover maximum heat before crude is charged to the heater, which ultimately supplies all the heat required for operation of the atmospheric distillation unit. Having the heater transfer temperature offset, the flow of fuel to the burners allows control of the heat input. The heater transfer temperature is merely a convenient control, and the actual temperature, which has no great significance, will vary from 320°C (610°F) to as high as 430°C (805°F), depending on the type of crude oil and the pressure at the bottom of the fractionating tower.

7.3 DISTILLATION

Distillation has remained a major refinery process and a process to which crude oil that enters the refinery is subjected. In early refineries, distillation was the primary means by which products were separated from crude petroleum. As the technologies for refining evolved into the twentieth century, refineries became much more complex but distillation remained the prime means by which petroleum is refined. Indeed, the distillation section of a modern refinery is the most flexible unit in the refinery since conditions can be adjusted to process a wide range of refinery feedstocks from the lighter crude oils to the heavier, more viscous crude oils. Generally, the maximum permissible temperature (in the vaporizing furnace or heater) to which the feedstock can be subjected is 350°C (660°F). The rate of thermal decomposition increases markedly above this temperature, although higher temperatures (up to approximately 395°C, 745°F) are part of the specifications for some distillation units—serious cracking does not occur at these higher temperatures but is subject to the properties of the crude oil feedstock and the residence time of the feedstock in the hot zone. If unplanned cracking occurs within a distillation unit, coke deposition can occur in the heater pipes or in the tower itself, resulting in failure of the distillation unit.

Generally, the maximum permissible temperature of the feedstock in the vaporizing furnace is the factor limiting the range of products in a single-stage (atmospheric) column. Thermal decomposition or *cracking* of the constituents begins as the temperature of the oil approaches 350°C (660°F), and the rate increases markedly above this temperature. However, the decomposition is time dependent, and temperatures on the order of 395°C (745°F) may be employed provided the residence time of the feedstocks in the hot zone does not cause thermal decomposition of the constituents. Thermal decomposition is generally regarded as being undesirable because the coke-like material produced tends to be deposited on the tubes with consequent formation of hot spots and eventual failure of the affected tubes. In the processing of lubricating oil stocks, an equally important consideration in the avoidance of these high temperatures is the deleterious effect on the lubricating properties. However, there are occasions when cracking distillation might be regarded as beneficial and the still temperature will be adjusted accordingly. In such a case, the products will be named accordingly using the prefix *cracked*, for example, *cracked residuum* in which case the term *pitch* (Chapter 1) is applied.

Based upon chemical characteristics, a very approximate estimation of the potential for thermal decomposition of various feedstocks can be made using the Watson characterization factor (Speight, 2014a), K_w:

$$K_w = \frac{T_b^{1/3}}{sp\,gr}$$

where
T_b is the mean average boiling point in degrees Rankine ([°R] = [°F] + 459.67)
sp gr is the specific gravity

The characterization factor ranges from approximately 10 for paraffinic crude oil to approximately 15 for highly aromatic crude oil. On the assumption that the components of paraffinic crude oil are more thermally labile than the components of aromatic crude oil, it might be supposed that a relationship between the characterization factor and temperature is viable. However, the relationship is so broad that it may not be sufficiently accurate to help the refiner. There are occasions when cracking distillation might be regarded as beneficial and the still temperature will be adjusted accordingly. In such a case, the products will be named accordingly using the prefix *cracked*, for example, *cracked residuum* in which case the arbitrary term *pitch* is applied.

In the modern sense, distillation was the first method by which petroleum was refined (Speight, 2014a). As petroleum refining evolved, distillation became a formidable means by which various products were separated. Further evolution saw the development of topping or skimming or hydroskimming refineries (Figure 7.7) and conversion refineries (Figure 7.8) named for the manner in which petroleum was treated in each case. And many of these configurations exist in the world of modern refining. However, of all the units in a refinery, the distillation section comprising the atmospheric unit and the vacuum unit is required to have the greatest flexibility in terms of variable quality of feedstock and range of product yields. This flexibility is somewhat reduced because of the tendency to omit the distillation section when heavy oil, extra heavy oils, and tar sand bitumen enter the refinery. Thus, refinery configurations can be adapted to the properties of the feedstocks that may dictate no distillation or a simple removal of any volatile constituents.

The simplest refinery configuration, called a *topping refinery*, is designed to remove volatile constituents from a feedstock under simple conditions. It consists of tankage, a distillation unit, recovery facilities for gases and light hydrocarbons, and the necessary utility systems (steam, power, and water-treatment plants). Topping refineries produce large quantities of unfinished oils and are highly dependent on local markets, but the addition of hydrotreating and reforming units to this basic configuration results in a more flexible hydroskimming refinery (Figure 7.7), which can also produce desulfurized distillate fuels and high octane gasoline.

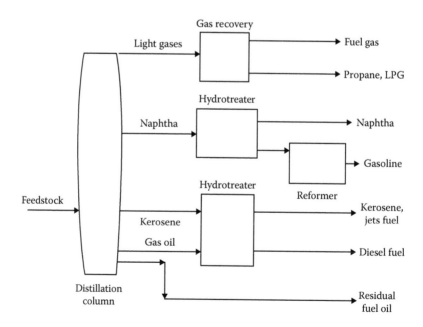

FIGURE 7.7 A hydroskimming refinery (see also Figure 7.8).

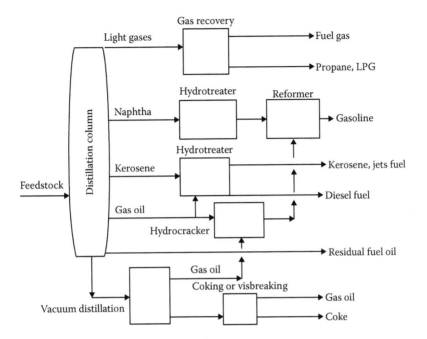

FIGURE 7.8   A conversion refinery (see also Figure 7.7).

The most versatile refinery configuration today is known as the conversion refinery (Figure 7.8). A conversion refinery incorporates all the basic building blocks found in both the topping and hydroskimming refineries, but it also features gas oil conversion plants such as catalytic cracking and hydrocracking units, olefin conversion plants such as alkylation or polymerization units, and, frequently, coking units for sharply reducing or eliminating the production of residual fuels. Modern conversion refineries may produce two-thirds of their output as unleaded gasoline, with the balance distributed between high-quality jet fuel, liquefied petroleum gas (LPG), low-sulfur diesel fuel, and a small quantity of petroleum coke. Many such refineries also incorporate solvent extraction processes for manufacturing lubricants and petrochemical units with which to recover high-purity propylene, benzene, toluene, and xylenes for further processing into polymers.

A multitude of separations are accomplished by distillation, but its most important and primary function in the refinery is its use of the distillation tower and the temperature gradients therein (Figure 7.9) for the separation of crude oil into fractions that consists of varying amounts of different components (Table 7.4; Figure 7.10) (Speight, 2000, 2014a). Thus, it is possible to obtain products ranging from gaseous materials taken off the top of the distillation column to a nonvolatile atmospheric residuum (*atmospheric bottoms, reduced crude*) with correspondingly lower-boiling materials (gas, gasoline, naphtha, kerosene, and gas oil) taken off at intermediate points with each crude oil providing different amounts of the various fractions (Figure 7.11) (Diwekar, 1995; Jones, 1995; Speight, 2000, 2014a; Speight and Ozum, 2002; Parkash, 2003; Hsu and Robinson, 2006; Gary et al., 2007).

The reduced crude may then be processed by vacuum or steam distillation to separate the high-boiling lubricating oil fractions without the danger of decomposition, which occurs at high (>350°C, 660°F) temperatures (Speight, 2000, 2014a). Indeed, atmospheric distillation may be terminated with a lower-boiling fraction (*boiling cut*) if it is thought that vacuum or steam distillation will yield a better-quality product or if the process appears to be economically more favorable.

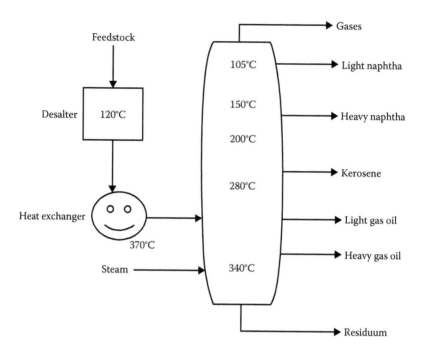

FIGURE 7.9   Temperature variation in an atmospheric distillation tower.

It should be noted at this point that not all crude oils yield the same distillation products because of the differences in composition (Charbonnier et al., 1969; Coleman et al., 1978; Speight, 2012). In fact, the nature of the crude oil dictates the processes that may be required for refining. Petroleum can be classified according to the nature of the distillation residue, which in turn depends on the relative content of hydrocarbon types: paraffins, naphthenes, and aromatics. For example, a *paraffin-base crude oil* produces distillation cuts with higher proportions of paraffins than asphalt

TABLE 7.4
Boiling Fractions of Petroleum

	Boiling Range		
Fraction	**0°C**	**°F**	**Uses**
Fuel gas	−160 to −40	−260 to −40	Refinery fuel
Propane	−40	−40	Liquefied petroleum gas (LPG)
Butane(s)	−12 to −1	11 to 30	Increases volatility of gasoline, advantageous in cold climates
Light naphtha	−1 to 150	30 to 300	Gasoline components, may be (with heavy naphtha) reformer feedstock
Heavy naphtha	150 to 205	300 to 400	Reformer feedstock; with light gas oil, jet fuels
Gasoline	−1 to 180	30 to 355	Motor fuel
Kerosene	205 to 260	400 to 500	Fuel oil
Stove oil	205 to 290	400 to 550	Fuel oil
Light gas oil	260 to 315	500 to 600	Furnace and diesel fuel components
Heavy gas oil	315 to 425	600 to 800	Feedstock for catalytic cracker
Lubricating oil	>400	>750	Lubrication
Vacuum gas oil	425 to 600	800 to 1100	Feedstock for catalytic cracker
Residuum	>600	>1100	Heavy fuel oil, asphalts

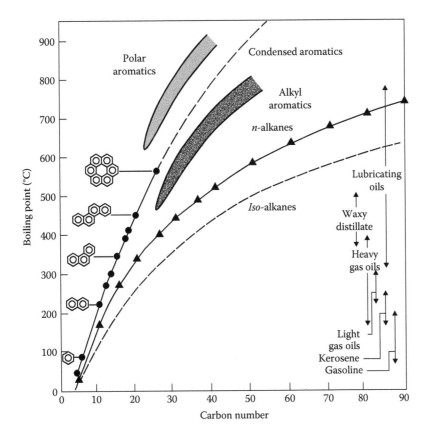

FIGURE 7.10 Variation of distillate composition with boiling range.

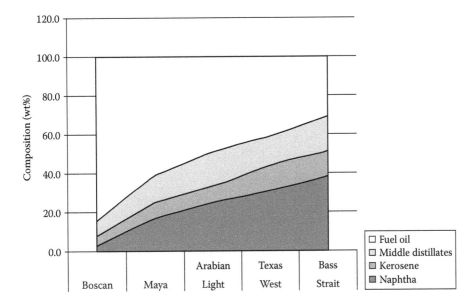

FIGURE 7.11 Distillation fractionation of different feedstocks.

base crude. The converse is also true; that is, an *asphalt-base crude oil* produces materials with higher proportions of cyclic compounds. A *paraffin-base crude oil* yields wax distillates rather than the lubricating distillates produced by the naphthenic-base crude oils. The residuum from paraffin-base petroleum is referred to as "cylinder stock" rather than "asphaltic bottoms," which is the name often given to the residuum from distillation of *naphthenic crude oil*. It is emphasized that, in these cases, it is not a matter of the use of archaic terminology but a reflection of the nature of the product and the petroleum from which it is derived.

7.3.1 Distillation at Atmospheric Pressure

Distillation columns are the most commonly used separation units in a refinery. Operation is based on the difference in boiling temperatures of the liquid mixture components, and on recycling countercurrent gas–liquid flow. The properly organized temperature distribution up the column results in different mixture compositions at different heights. While multicomponent interphase mass transfer is a common phenomenon for all column types, the flow regimes are very different depending on the internal elements used. The two main types are a tray column and a packed column, the latter equipped with either random or structured packing. Different types of distillation columns are used for different processes, depending on the desired liquid holdup, capacity (flow rates), and pressure drop but each column is a complex unit, combining many structural elements.

The present-day petroleum distillation unit is, in fact, a collection of distillation units that enable a fairly efficient degree of fractionation to be achieved. In contrast to the early units, which consisted of separate stills, a tower is used in the modern-day refinery. In fact, of all the units in a refinery, the distillation unit (Figure 7.12) is required to have the greatest flexibility in terms of variable

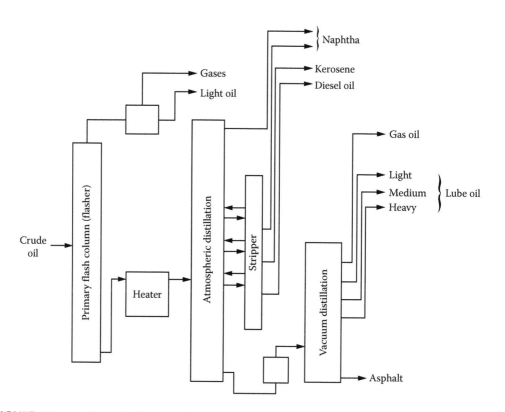

FIGURE 7.12 A refinery distillation section.

quality of feedstock and range of product yields. Thus, crude oil can be separated into gasoline, kerosene, diesel oil, gas oil, and other products, by distillation at atmospheric pressure. Distillation is an operation in which vapors rising through fractionating decks in a tower are intimately contacted with liquid descending across the decks so that higher-boiling components are condensed, and concentrate at the bottom of the tower while the lighter ones are concentrated at the top or pass overhead.

It is common practice to use furnaces to heat the feedstock only when distillation temperatures above 205°C (400°F) are required. Lower temperatures (such as that used in the redistillation of naphtha and similar low-boiling products) are provided by heat exchangers and/or steam reboilers. Thus, the desalted feedstock is generally pumped to the unit directly from a storage tank, and it is important that charge tanks be drained completely free from water before charging to the unit. The crude feedstock is heat exchanged against whatever other heat sources are available to recover maximum heat before crude is charged to the heater, which ultimately supplies all the heat required for operation of the distillation unit.

The feed to a fractional distillation tower is heated by flow through pipe arranged within a large furnace. The heating unit is known as a *pipe still heater* or *pipe still furnace*, and the heating unit and the fractional distillation tower make up the essential parts of a distillation unit or pipe still. The pipe still furnace heats the feed to a predetermined temperature, usually a temperature at which a calculated portion of the feed changes into vapor. The vapor is held under pressure in the pipe still furnace until it discharges as a foaming stream into the fractional distillation tower. Here the vapors pass up the tower to be fractionated into gas oil, kerosene, and naphtha while the nonvolatile or liquid portion of the feed descends to the bottom of the tower to be pumped away as a bottom product.

Pipe still furnaces vary greatly in size, shape, and interior arrangement and can accommodate 25,000 bbl or more of crude petroleum per day. The walls and ceiling are insulated with firebrick, and gas or oil burners are inserted through one or more walls. The interior of the furnace is partially divided into two sections: a smaller convection section where the oil first enters the furnace and a larger section into which the burners discharge and where the oil reaches its highest temperature.

Heat exchangers are also used to preheat the feedstock before it enters the furnace. These exchangers are bundles of tubes arranged within a shell so that a stream passes through the tubes in the opposite direction of a stream passing through the shell. Thus, cold crude oil, by passing through a series of heat exchangers where hot products from the distillation tower are cooled, before entering the furnace and saving of heat in this manner, may be a major factor in the economical operation of refineries.

Steam reboilers may take the form of a steam coil in the bottom of the fractional distillation tower or in a separate vessel. In the latter case, the bottom product from the tower enters the reboiler where part is vaporized by heat from the steam coil. The hot vapor is directed back to the bottom of the tower and provides part of the heat needed to operate the tower. The nonvolatile product leaves the reboiler and passes through a heat exchanger, where its heat is transferred to the feed to the tower. Steam may also be injected into a fractional distillation tower not only to provide heat but also to induce boiling to take place at lower temperatures. Reboilers generally increase the efficiency of fractionation, but a satisfactory degree of separation can usually be achieved more conveniently by the use of a *stripping* section. The *stripping operation* occurs in that part of the tower below the point at which the feed is introduced. The more volatile components are stripped from the descending liquid. Above the feed point (the rectifying section), the concentration of the less volatile component in the vapor is reduced.

If water is entrained in the charge, it will vaporize in the exchangers and in the heater, and cause a high-pressure drop through that equipment. If a slug of water should be charged to the unit, the quantity of steam generated by its vaporization is so much greater than the quantity of vapor obtained from the same volume of oil, that the decks in the fractionating column could be damaged. Water expands in volume 1600 times upon vaporization at 100°C (212°F) at atmospheric pressure.

The feed to a fractional distillation tower is heated by flow through pipe arranged within a large furnace. The heating unit is known as a *pipe still heater* or *pipe still furnace*, and the heating unit and the fractional distillation tower make up the essential parts of a distillation unit or pipe still (Figure 7.12). The pipe still furnace heats the feed to a predetermined temperature, usually a temperature at which a calculated portion of the feed changes into vapor. The vapor is held under pressure in the pipe still furnace until it discharges as a foaming stream into the fractional distillation tower. Here the vapors pass up the tower to be fractionated into gas oil, kerosene, and naphtha, while the nonvolatile or liquid portion of the feed descends to the bottom of the tower to be pumped away as a bottom product.

Heat exchangers are used to preheat the crude oil feedstock before entry into the distillation unit. In order to reduce the cost of operating a crude unit, as much heat as possible is recovered from the hot streams by heat exchanging them with the cold crude charge. The number of heat exchangers within the crude unit and cross heat exchange with other units will vary with unit design. A record should be kept of heat exchanger outlet temperatures so that fouling can be detected and possibly corrected before the capacity of the unit is affected.

Crude entering the flash zone of the fractionating column flashes into the vapor that rises up the column and the liquid residue that drops downward. This flash is a very rough separation; the vapors contain appreciable quantities of heavy ends, which must be rejected downward into reduced crude, while the liquid contains lighter products, which must be stripped out. In the distillation of crude petroleum, light naphtha and gases are removed as vapor from the top of the tower, heavy naphtha, kerosene, and gas oil are removed as sidestream products and reduced crude is taken from the bottom of the tower.

Having the heater transfer temperature reset, flow of fuel to the burners controls the heat input. The heater transfer temperature is merely a convenient control, and the actual temperature, which has no great significance, will vary from 320°C (610°F) to as high as 430°C (805°F), depending on the type of crude and the pressure at the bottom of the fractionating tower. It is noteworthy that if the quantity of gasoline and kerosene in a crude is reduced, the transfer temperature required for the same operation will be increased, even though the *lift* is less. However, at such temperatures, the residence time of the crude oil and its fractions exerts considerable influence on the potential for racking reactions to occur. This is particularly important in determining the properties of the atmospheric residuum.

External reflux that is returned to the top of the fractionator passes downward against the rising vapors. Lighter components of the reflux are vaporized and return to the top of the column while the heavier components in the rising vapors are condensed and return down the column. Thus, there is an internal reflux stream flowing from the top of the fractionator all the way back to the flash zone and becoming progressively heavier as it descends.

The products heavier than the net overhead are obtained by withdrawing portions of the internal reflux stream. The end point of a sidestream fraction (*side cut*) will depend on the quantity withdrawn. If the sidestream fraction withdrawal rate is increased, the extra product is material that was formerly flowing down the fractionator as internal reflux. Since the internal reflux below the *drawoff* is reduced, heavier vapors can now rise to that point and result in a heavier product. Changing the drawoff rate is the manner in which sidestream fractions are kept on end point specifications. The temperature of the drawoff decks is an indication of the end point of the product drawn at that point and the drawoff rate can be controlled to hold a constant deck temperature and therefore a specification product.

The degree of fractionation is generally judged by measuring the number of degrees centigrade between the 95% point of the lighter product and the 5% point of the heavier product. The initial boiling point (IBP) and the final boiling point (FBP) can be used but the initial boiling point varies with the intensity of efficiency of the stripping operation. Fractionation can be improved by increasing the reflux in the fractionator, which is done by raising the transfer temperature. There may be occasions when the internal reflux necessary to achieve satisfactory fractionation between the heavier products is so great that if it was supplied from the top of the fractionator the upper decks will flood. An *intermediate circulating reflux* (ICR) solves this problem. Some internal reflux is withdrawn, pumped through a cooler, or exchanger, and returned colder a few decks higher in the column. This cold oil return condenses extra vapors to liquid and increases the internal reflux below that point. Improvement in the fractionation between the light and heavy gas oil can be achieved by increasing the heater transfer temperature, which would cause the top reflux to increase and then restore the top reflux to its former rate by increasing the circulating reflux rate. Even though the heater transfer temperature is increased, the extra heat is recovered by exchange with crude oil feedstock and, as a result, the heater duty will only increase slightly.

Sometimes a fractionator will be *pulled dry* insofar as the rate at which a product is being withdrawn is greater than the quantity of internal reflux in the fractionator. All the internal reflux then flows to the stripper, the decks below the drawoff run dry, and therefore no fractionation takes place, while at the same time there is insufficient material to maintain the level in the stripper, and the product pump will tend to lose suction. It is necessary then to either lower the product withdrawal rate or to increase the internal reflux in the tower by raising the transfer temperature or by reducing the rate at which the next lightest product is being withdrawn.

Pipe still furnaces vary greatly in size, shape, and interior arrangement and can accommodate 25,000 bbl or more of crude petroleum per day. The walls and ceiling are insulated with firebrick, and gas or oil burners are inserted through one or more walls. The interior of the furnace is partially divided into two sections: a smaller convection section where the oil first enters the furnace and a larger section into which the burners discharge and where the oil reaches its highest temperature.

Steam reboilers may take the form of a steam coil in the bottom of the fractional distillation tower or in a separate vessel. In the latter case, the bottom product from the tower enters the reboiler where part is vaporized by heat from the steam coil. The hot vapor is directed back to the bottom of the tower and provides part of the heat needed to operate the tower. The nonvolatile product leaves the reboiler and passes through a heat exchanger, where its heat is transferred to the feed to the tower. Steam may also be injected into a fractional distillation tower not only to provide heat but also to induce boiling to take place at lower temperatures. Reboilers generally increase the efficiency of fractionation, but a satisfactory degree of separation can usually be achieved more conveniently by the use of a *stripping* section.

The *stripping section* is the part of the tower below the point at which the feed is introduced; the more volatile components are stripped from the descending liquid. Above the feed point (the rectifying section), the concentration of the less volatile component in the vapor is reduced. The stripping section is necessary because the flashed residue in the bottom of the fractionator and the sidestream products have been in contact with lighter boiling vapors. These vapors must be removed to meet flash point specifications and to drive the light ends into lighter and (usually) more valuable products.

Steam, usually superheated steam, is used to strip these light ends. Generally, sufficient steam is used to meet a flash point specification and, while a further increase in the quantity of steam may raise the initial boiling point of the product slightly, the only way to substantially increase the initial boiling point of a specific product is to increase the yield of the next lighter product. Provided, of course, the fractionator has enough internal reflux to accomplish an efficient separation of the feedstock constituents.

All of the stripping steam that is condensed in the overhead receiver and must be drained off because refluxing water will upset the balance of activities in the fractionator. If the end point of the overhead product is very low, water may not pass overhead and will accumulate on the upper decks and cause the tower to flood thereby reducing efficiency and perhaps even shut down the tower. If the latter occurs, and if distillation is the first (other than desalting) process to which a crude oil is subjected in a refinery, the economic consequences for the refinery operation can be substantial.

In simple refineries, cut points can be changed slightly to vary yields and balance products, but the more common practice is to produce relatively narrow fractions and then process (or blend) to meet product demand. Since all these primary fractions are equilibrium mixtures, they all contain some proportion of the lighter constituents characteristic of a lower-boiling fraction and so are stripped of these constituents, or stabilized, before further processing or storage. Thus, gasoline is stabilized to a controlled butanes–pentanes content, and the overhead may be passed to superfractionators, towers with a large number of plates that can produce nearly pure C_1–C_4 hydrocarbons (methane to butanes, CH_4 to C_4H_{10})—the successive columns termed deethanizers, depropanizers, debutanizers, and whichever separation columns are still necessary.

Kerosene and gas oil fractions are obtained as sidestream products from the atmospheric tower (*primary tower*), and these are treated in stripping columns (i.e., vessels of a few bubble trays) into which steam is injected and the volatile overhead from the stripper is returned to the primary tower. Steam is usually introduced by the stripping section of the primary column to lower the temperature at which fractionation of the heavier ends of the crude can occur.

The specifications for most petroleum products make it extremely difficult to obtain marketable material by distillation only. In fact, the purpose of atmospheric distillation is considered the provision of fractions that serve as feedstocks for intermediate refining operations and for blending. Generally this is carried out at atmospheric pressure, although light crude oils may be *topped* at an elevated pressure and the residue then distilled at atmospheric pressure.

The *topping* operation differs from normal distillation procedures insofar as the majority of the heat is directed to the feed stream rather than by reboiling the material in the base of the tower. In addition, products of volatility intermediate between that of the overhead fractions and bottoms (residua) are withdrawn as sidestream products. Furthermore, steam is injected into the base of the column and the sidestream strippers to adjust and control the initial boiling range (or point) of the fractions. Topped crude oil must always be *stripped* with steam to elevate the flash point or to recover the final portions of gas oil. The composition of the topped crude oil is a function of the temperature of the vaporizer (or *flasher*).

All products are cooled before being sent to storage. Low-boiling products should be restrained to temperatures below 60°C (140°F) in order to reduce vapor losses in storage, but the need to store higher-boiling products below such temperatures is not as acute, unless facile oxidation of the product at higher temperatures is possible. If a product is being charged to another unit as feedstock, there may be an advantage in transmitting the hot product to the unit. However, caution is advised if a product is leaving a unit at temperatures in excess of 100°C (212°F) if there is any possibility of it entering a tank with water bottoms. The hot oil could readily boil the water and cause the roof to detach from the tank, perhaps violently!

7.3.2 Distillation at a Reduced Pressure

The boiling range of the highest boiling fraction that can be produced at atmospheric pressure is limited by the temperature at which the residue starts to decompose or *crack*. If the atmospheric residuum is required for the manufacture of lubricating oils further fractionation without cracking may be desirable, and this may be achieved by distillation under vacuum. The residua produced by distillation under reduced pressure have properties markedly different from the residua produced by distillation at atmospheric pressure (Table 7.5).

Vacuum distillation as applied to the petroleum refining industry is a technique that has seen wide use in petroleum refining. Vacuum distillation evolved because of the need to separate the

TABLE 7.5
Properties of Atmospheric (b.p. >650°F) and Vacuum Residua (b.p. >1050°F)

Feedstock	Gravity API	Sulfur wt%	Nitrogen wt%	Nickel ppm	Vanadium ppm	Asphaltenes (Heptane) wt%	Carbon Residue (Conradson) wt%
Arabian light >650°F	17.7	3.0	0.2	10.0	26.0	1.8	7.5
Arabian light >1050°F	8.5	4.4	0.5	24.0	66.0	4.3	14.2
Arabian heavy >650°F	11.9	4.4	0.3	27.0	103.0	8.0	14.0
Arabian heavy >1050°F	7.3	5.1	0.3	40.0	174.0	10.0	19.0
Alaska North Slope >650°F	15.2	1.6	0.4	18.0	30.0	2.0	8.5
Alaska North Slope >1050°F	8.2	2.2	0.6	47.0	82.0	4.0	18.0
Lloydminster (Canada) >650°F	10.3	4.1	0.3	65.0	141.0	14.0	12.1
Lloydminster (Canada) >1050°F	8.5	4.4	0.6	115.0	252.0	18.0	21.4
Kuwait >650°F	13.9	4.4	0.3	14.0	50.0	2.4	12.2
Kuwait >1050°F	5.5	5.5	0.4	32.0	102.0	7.1	23.1
Tia Juana >650°F	17.3	1.8	0.3	25.0	185.0		9.3
Tia Juana >1050°F	7.1	2.6	0.6	64.0	450.0		21.6
Taching >650°F	27.3	0.2	0.2	5.0	1.0	4.4	3.8
Taching >1050°F	21.5	0.3	0.4	9.0	2.0	7.6	7.9

less volatile products, such as lubricating oils, from the petroleum without subjecting these high-boiling products to cracking conditions. The boiling point of the heaviest cut obtainable at atmospheric pressure is limited by the temperature (ca. 350°C; ca. 660°F) at which the residue starts to decompose or *crack*, unless *cracking distillation* is preferred. When the feedstock is required for the manufacture of lubricating oils, further fractionation without cracking is desirable, and this can be achieved by distillation under vacuum conditions.

In order to maximize the production of gas oil and lighter components from the residuum of an atmospheric distillation unit (reduced crude), the residuum can be further distilled in a vacuum distillation unit (Figure 7.12). Residuum distillation is conducted at a low pressure in order to avoid thermal decomposition or cracking at high temperatures. A stock that boils at 400°C (750°F) at 0.1 psi (50 mm) would not boil until approximately 500°C (930°F) at atmospheric pressure and petroleum constituents commence thermal decomposition (cracking) at approximately 350°C (660°F) (Speight, 2000, 2014a). In the vacuum unit, almost no attempt is made to fractionate the products. It is only desired to vaporize the gas oil, remove the entrained residuum, and condense the liquid product as efficiently as possible. Vacuum distillation units that produce lubricating oil fractions are completely different in both design and operation.

In the vacuum tower, the reduced crude is charged through a heater into the vacuum column in the same manner as whole crude is charged to an atmospheric distillation unit. However, whereas the flash zone of an atmospheric column may be at 14–18.5 psi (760–957 mm), the pressure in a vacuum column is very much lower, generally less than 0.8 psi (less than 40 mm Hg) in the flash zone to less 0.2 psi (less than 10 mm Hg) at the top of the vacuum tower. The vacuum heater transfer temperature is generally used for control, even though the pressure drop along the transfer line makes the temperature at that point somewhat meaningless. The flash zone temperature has much greater significance.

The heater transfer and flash zone temperatures are generally varied to meet the vacuum bottoms specification, which is probably either a gravity (or viscosity) specification for fuel oil or a *penetration specification* for asphalt. The penetration of an asphalt is the depth in 1/100 cm to which a needle carrying a 100 g weight sinks into a sample at 25°C (77°F) in 5 seconds (ASTM D5), so that the lower the penetration, the heavier the residuum or asphalt. If the flash zone temperature is too high the crude can start to crack and produce gases that overload the ejectors and break the vacuum. It is then necessary to lower the temperature and if a heavier residuum product is still required, an attempt should be made to obtain a better vacuum.

Slight cracking may occur without seriously affecting the vacuum, and the occurrence of cracking can be established by a positive result from the Oliensis Spot Test (Speight, 2014a, 2015). This test is a convenient laboratory test that indicates the presence of cracked components as sediment by the separation of the sediment when a 20% solution of asphalt in naphtha is dropped on a filter paper. However, some crude oils yield a residuum that exhibit a positive test for the presence of sediment (solid phase) in the residuum. If a negative test result is required, operation at the highest vacuum and lowest temperature should be attempted. Since the degree of cracking depends on both the temperature and the time (*residence time* in the hot zone) during which the oil is exposed to that temperature, the level of the residuum in the bottom of the tower should be held at a minimum, and its temperature reduced by recirculating some of the residuum from the outlet of the residuum/crude oil heat exchanger to the bottom of the column. Quite often, when the level of the residuum rises the column vacuum falls because of cracking due to increased residence time.

The flash zone temperature will vary widely and is dependent on the source of the crude oil, residuum specifications, the quantity of product taken overhead, and the flash zone pressure, and temperatures from below 315°C (600°F) to more than 425°C (800°F) have been used in commercial operations. Some vacuum distillation units are provided with facilities to strip the residuum with steam, and this will tend to lower the temperature necessary to meet an asphalt specification, but an excessive quantity of steam will overload the jets.

The distillation of high-boiling lubricating oil stocks may require pressures as low as 0.29–0.58 psi (15–30 mm Hg), but operating conditions are more usually 0.97–1.93 psi (50–100 mm Hg). Volumes of vapor at these pressures are large and pressure drops must be small to maintain control, so vacuum columns are necessarily of large diameter. Differences in vapor pressure of different fractions are relatively larger than for lower-boiling fractions, and relatively few plates are required. Under these conditions, a heavy gas oil may be obtained as an overhead product at temperatures of approximately 150°C (300°F). Lubricating oil fractions may be obtained as sidestream products at temperatures of 250°C–350°C (480°F–660°F). The feedstock and residue temperatures being kept below the temperature of 350°C (660°F), above which the rate of thermal decomposition (cracking) increases (Speight, 2000, 2014a). The partial pressure of the hydrocarbons is effectively reduced yet further by the injection of steam. The steam added to the column, principally for the stripping of bitumen in the base of the column, is superheated in the convection section of the heater.

When trays similar to those used in the atmospheric column are used in vacuum distillation, the column diameter may be extremely high, up to 45 ft (14 m). To maintain low-pressure drops across the trays, the liquid seal must be minimal. The low holdup and the relatively high viscosity of the liquid limit the tray efficiency, which tends to be much lower than in the atmospheric column. The vacuum is maintained in the column by removing the noncondensable gas that enters the column by way of the feed to the column or by leakage of air.

The fractions obtained by vacuum distillation of reduced crude depend on whether the run is designed to produce lubricating or vacuum gas oils. In the former case, the fractions include (1) *heavy gas oil*, an overhead product and is used as catalytic cracking stock or, after suitable treatment, a light lubricating oil; (2) *lubricating oil* (usually three fractions: light, intermediate, and heavy) obtained as a sidestream product; and (3) *residuum*, the nonvolatile product that may be used directly as asphalt or to asphalt. The residuum may also be used as a feedstock for a coking operation or blended with gas oils to produce a heavy fuel oil. However, if the *reduced* crude is not required as a source of lubricating oils, the lubricating and heavy gas oil fractions are combined or, more likely, removed from the *residuum* as one fraction and used as a catalytic cracking feedstock.

Three types of high-vacuum units for long residue upgrading have been developed for commercial application: (1) feedstock preparation units, (2) lube oil high-vacuum units, and (3) high-vacuum units for asphalt production.

The *feedstock preparation units* make a major contribution to deep conversion upgrading and produce distillate feedstocks for further upgrading in catalytic crackers, hydrocracking units, and coking units. To obtain an optimum waxy distillate quality a wash oil section is installed between feed flash zone and waxy distillate drawoff. The wash oil produced is used as fuel component or recycled to feed. The flashed residue (short residue) is cooled by heat exchange against long residue feed. A slipstream of this cooled short residue is returned to the bottom of the high-vacuum column as quench to minimize cracking (maintain low bottom temperature).

Lube oil high-vacuum units are specifically designed to produce high-quality distillate fractions for lube oil manufacturing. Special precautions are therefore taken to prevent thermal degradation of the distillates produced. The units are of the *wet* type. Normally, three sharply fractionated distillates are produced (spindle oil, light machine oil, and medium machine oil). Cut points between those fractions are typically controlled on their viscosity quality. Spindle oil and light machine oil are subsequently steam stripped in dedicated strippers. The distillates are further processed to produce lubricating base oil. The short residue is normally used as feedstock for the solvent deasphalting process to produce deasphalted oil, an intermediate for bright stock manufacturing. *High-vacuum units for asphalt production* are designed to produce straight-run asphalt and/or feedstocks for residuum blowing to produce *blown asphalt* that meets specifications. In principle, these units are designed on the same basis as feed preparation units, which may also be used to provide feedstocks for asphalt manufacturing.

Deep cut vacuum distillation involving a revamp of the vacuum distillation unit to cut deeper into the residue is one of the first options available to the refiner. In addition to the limits of the major equipment, other constraints include (1) the vacuum gas oil quality specification required by downstream conversion units, (2) the minimum flash zone pressure achievable, and (3) the maximum heater outlet temperature achievable without excessive cracking. These constraints typically limit the cut point (true boiling point) to 560°C–590°C (1040°F–1100°F) although units are designed for cut points (true boiling point) as high as 627°C (1160°F).

Prior to 1960, most of the trays in a vacuum tower were conventional designed to provide as low a pressure drop as possible. Many of these standard trays have been replaced by grid packing that provides very low-pressure drops as well as a high tray efficiency. Up to this time, flash zone temperature reduction was enhanced by steam stripping of the residuum but with the new grid packing the use of steam to enhance flash temperature has been eliminated and most modern units are *dry* vacuum units.

If the *reduced* crude is not required as a source of lubricating oils, the lubricating and heavy gas oil fractions are combined or, more likely, removed from the *residuum* as one fraction and used as a catalytic cracking feedstock. The continued use of atmospheric and vacuum distillation has been a major part of refinery operations during this century and no doubt will continue to be employed, at least into the beginning decades of the twenty-first century, as the primary refining operation.

The vacuum residuum (vacuum bottoms) must be handled more carefully than most refinery products since the pumps that handle hot heavy material have a tendency to lose suction. Recycling cooled residuum to the column bottom thereby reducing the tendency of vapor to form in the suction line can minimize this potential situation. It is also important that the residuum pump be sealed in such a manner so as to prevent the entry of air. In addition, and since most vacuum residua are solid at ambient temperature, all vacuum residua handling equipment must either be kept active, or flushed out with gas oil, when it is shut down. Steam tracing alone may be inadequate to keep the residuum fluid, but where this is done, the high-pressure steam should be used.

The vacuum residuum from a vacuum tower is sometimes cooled in open box units, as shell-and-tube units are not efficient in this service. It is often desirable to send residuum to storage at high temperature to facilitate blending. If it is desired to increase the temperature of the residuum, it is better to do so by lowering the level of water in the open box and not by lowering the water temperature. If the water in the box is too cold, the residuum can solidify on the inside wall of the tube and insulate the hot residuum in the central core from the cooling water. Lowering the water temperature can actually result in a hotter product. When the residuum is sent to storage at over 100°C (212°F), care should be taken to insure that the tank is absolutely free from water. Residuum coolers should always be flushed out with gas oil immediately once the residuum flow stops, since melting the contents of a cooler is a slow process.

The vapor rising above the flash zone will entrain some residuum that cannot be tolerated in cracking unit charge. The vapor is generally washed with gas oil product, sprayed into the *slop wax* section. The mixture of gas oil and entrained residuum is known as slop wax, and it is often circulated over the decks to improve contact, although the circulation rate may not be critical. The final stage of entrainment removal is obtained by passing the rising vapors through a metallic mesh demister blanket through which the fresh gas oil is sprayed.

Most of the gas oil spray is vaporized by the hot rising vapors and returned up the column. Some slop wax must be yielded in order to reject the captured entrainment. The amount of spray to the demister blanket is generally varied so that the yield of slop wax necessary to maintain the level in the slop wax pan is approximately 5% of the charge. If the carbon residue or the metals content of the heavy vacuum gas oil is high, a greater percentage of slop wax must be withdrawn or circulated. Variation in the color of the gas oil product is a valuable indication of the effectiveness of entrainment control.

Slop wax is a mixture of gas oil and residuum, and it can be recirculated through the heater to the flash zone and reflashed. If, however, a crude contains volatile metal compounds, these will be recycled with the slop wax and can finally rise into the gas oil. Where volatile metals are a problem, it is necessary either to yield slop wax as a product, or to make lighter asphalt, which will contain the metal compounds returned with the slop wax.

The scrubbed vapor rising above the demister blanket is the product, and no further fractionation is required. It is only desired to condense these vapors as efficiently as possible. This could be done in a shell-and-tube condenser, but these are inefficient at low pressures, and the high-pressure drop through such a condenser would raise the flash zone pressure. The most efficient method is to contact the hot vapors with liquid product that has been cooled by pumping through heat exchangers.

Finally, confusion often arises because of the different scales used to measure the vacuum. Positive pressures are commonly measured as kilograms per square-centimeter gauge, which are kilograms per square centimeter above atmospheric pressure, which is 1.035 kg/cm^2 or 14.7 psi. Another means of measurement is to measure in millimeters of mercury in which atmospheric pressure (sea level) is 760 mm of mercury absolute while a perfect vacuum is 0 mm absolute.

7.3.3 DISTILLATION TOWERS

Distillation towers (distillation columns) are made up of several components, each of which is used either to transfer heat energy or enhance material transfer. A typical distillation column consists of several major parts: (1) a vertical shell where is separation of the components is carried out; (2) column internals such as trays, or plates, or packings that are used to enhance component separation; (3) a reboiler to provide the necessary vaporization for the distillation process; (4) a condenser to cool and condense the vapor leaving the top of the column; and (5) a reflux drum to hold the condensed vapor from the top of the column so that liquid (reflux) can be recycled back to the column. The vertical shell houses the column internals and together with the condenser and reboiler constitutes a distillation column.

In a petroleum distillation unit, the feedstock liquid mixture is introduced usually near the middle of the column to a tray known as the feed tray. The feed tray divides the column into a top (enriching, rectification) section and a bottom (stripping) section. The feed flows down the column where it is collected at the bottom in the reboiler. Heat is supplied to the reboiler to generate vapor. The source of heat input can be any suitable fluid, although in most chemical plants this is normally steam. In refineries, the heating source may be the output streams of other columns. The vapor raised in the reboiler is reintroduced into the unit at the bottom of the column. The liquid removed from the reboiler is known as the bottoms.

The vapor moves up the column, and as it exits the top of the unit, it is cooled by a condenser. The condensed liquid is stored in a holding vessel known as the reflux drum. Some of this liquid is recycled back to the top of the column and this is called the reflux. The condensed liquid that is removed from the system is known as the distillate or top product. Thus, there are internal flows of vapor and liquid within the column as well as external flows of feeds and product streams, into and out of the column.

The tower is divided into a number of horizontal sections by metal trays or plates, and each is the equivalent of a still. The more trays, the more redistillation, and hence the better is the fractionation or separation of the mixture fed into the tower. A tower for fractionating crude petroleum may be 13 ft in diameter and 85 ft high according to a general formula:

$$c = 220d^2r$$

where
    c is the capacity in bbl/day
    d is the diameter in feet
    r is the amount of residuum expressed as a fraction of the feedstock

A tower stripping unwanted volatile material from gas oil may be only 3 or 4 ft in diameter and 10 ft high with less than 20 trays (Speight, 2014a). Towers concerned with the distillation of liquefied gases are only a few feet in diameter but may be up to 200 ft in height. A tower used in the fractionation of crude petroleum may have from 16 to 28 trays, but one used in the fractionation (superfractionation) of liquefied gases may have 30–100 trays. The feed to a typical tower enters the vaporizing or flash zone, an area without trays. The majority of the trays are usually located above this area. The feed to a bubble tower, however, may be at any point from top to bottom with trays above and below the entry point, depending on the kind of feedstock and the characteristics desired in the products.

7.3.3.1 Tray Towers

The tray tower typically combines the open flow channel with weirs, down comers, and heat exchangers. Free surface flow over the tray is disturbed by gas bubbles coming through the perforated tray, and possible leakage of liquid dropping through the upper tray.

Liquid collects on each tray to a depth of, say, several inches and the depth controlled by a dam or weir. As the liquid level rises, excess liquid spills over the weir into a channel (downspout), which carries the liquid to the tray below. The temperature of the trays is progressively cooler from bottom to top. The bottom tray is heated by the incoming heated feedstock, although in some instances a steam coil (reboiler) is used to supply additional heat. As the hot vapors pass upward in the tower, condensation occurs onto the trays until refluxing (simultaneous boiling of a liquid and condensing of the vapor) occurs on the trays. Vapors continue to pass upward through the tower, whereas the liquid on any particular trays spills onto the tray below, and so on until the heat at a particular point is too intense for the material to remain liquid. It then becomes vapor and joins the other vapors passing upward through the tower. The whole tower thus simulates a collection of several (or many) stills, with the composition of the liquid at any one point or on any one tray remaining fairly consistent. This allows part of the refluxing liquid to be tapped off at various points as *sidestream* products.

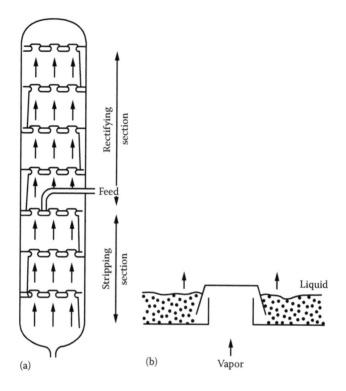

FIGURE 7.13   Cross section of (a) a distillation tower and (b) a bubble cap.

The efficient operation of the distillation, or fractionating, tower requires the rising vapors to mix with the liquid on each tray. This is usually achieved by installing a short chimney on each hole in the plate and a cap with a serrated edge (*bubble cap*, hence *bubble-cap tower*) over each chimney (Figure 7.13). The cap forces the vapors to go below the surface of the liquid and to bubble up through it. Since the vapors may pass up the tower at substantial velocities, the caps are held in place by bolted steel bars.

Perforated trays are also used in fractionating towers. This tray is similar to the bubble-cap tray but has smaller holes (~3 inches, 6 mm, vs. 2 inches, 50 mm). The liquid spills back to the tray below through weirs and is actually prevented from returning to the tray below through the holes by the velocity of the rising vapors. Needless to say, a minimum vapor velocity is required to prevent return of the liquid through the perforations.

As a result, flashed vapors rise up the fractionating column through the trays and countercurrent to the internal reflux flowing down the column. The lightest product, which is generally gasoline, passes overhead and is condensed in the overhead receiver. If the crude oil contains any noncondensable gas, it will leave the receiver as a gas, and can be recovered by other equipment, which should be operated to obtain the minimum flash zone pressure. The temperature at the top of the fractionator is a good measure of the end point of the gasoline and is controlled by returning some of the condensed gasoline (as reflux) to the top of the column. Increasing the reflux rate lowers the top temperature and results in the net overhead product having a lower end point. The loss in net overhead product must be removed on the next lower draw tray. This decreases the initial boiling point of material from this tray. Increasing the heater transfer temperature increases the heat input and demands more reflux to maintain the same top temperature.

Usually, trays are horizontal, flat, specially prefabricated metal sheets, which are placed at a regular distance in a vertical cylindrical column. Trays have two main parts: (1) the part where vapor (gas) and liquid are being contacted (the *contacting area*) and (2) the part where vapor and liquid are separated, after having been contacted (the *downcomer area*). Classification of trays is based on (1) the type of plate used in the contacting area, (2) the type and number of downcomers making up the downcomer area, (3) the direction and path of the liquid flowing across the contacting area of the tray, (4) the vapor (gas) flow direction through the (orifices in) the plate, and (5) the presence of baffles, packing, or other additions to the contacting area to improve the separation performance of the tray.

Common plate types, for use in the *contacting area* are as follows: (1) *bubble cap* tray in which caps are mounted over risers fixed on the plate (the caps come in a wide variety of sizes and shapes, round, square, and rectangular [tunnel]); (2) *sieve* trays that come with different hole shapes (round, square, triangular, rectangular [slots], star), various hole sizes (from approximately 2 mm to approximately 25 mm), and several punch patterns (triangular, square, rectangular); and (3) the *valve* tray that also is available in a variety of valve shapes (round, square, rectangular, triangular), valve sizes, valve weights (light and heavy), and orifice sizes and is either fixed or floating valves.

Trays usually have one or more downcomers. The type and number of downcomers used mainly depends on the amount of *downcomer area* required to handle the liquid flow. Single-pass trays are trays with one downcomer delivering the liquid from the next higher tray, a single bubbling area across which the liquid passes to contact the vapor and one downcomer for the liquid to the next lower tray. Trays with multiple downcomers and hence multiple liquid passes can have a number of layout geometries. The downcomers may extend, in parallel, from wall to wall, as in. The downcomers may be rotated 90° (or 180°) on successive trays. The downcomer layout pattern determines the liquid *flow path* arrangement and liquid *flow direction* in the *contacting area* of the trays. Giving a preferential direction to the vapor flowing through the orifices in the plate will induce the liquid to flow in the same direction. In this way, liquid flow rate and flow direction, as well as liquid height, can be manipulated. The presence of *baffles*, *screen mesh* or *demister mats*, loose or restrained dumped *packing*, and/or the addition of other devices in the *contacting area* can be beneficial for improving the contacting performance of the tray, namely, its separation efficiency.

The most important parameter of a tray is its separation performance and four parameters are of importance in the design and operation of a tray column: (1) the level of the tray efficiency, in the normal operating range; (2) the vapor rate at the "upper limit," that is, the maximum vapor load; (3) the vapor rate at the "lower limit," that is, the minimum vapor load; and (4) the tray pressure drop.

The separation performance of a tray is the basis of the performance of the column as a whole. The primary function of, for instance, a distillation column is the separation of a feed stream in (at least) one top product stream and one bottom product stream. The quality of the separation performed by a column can be judged from the purity of the top and bottom product streams. The specification of the impurity levels in the top and bottom streams and the degree of recovery of pure products set the targets for a successful operation of a distillation column. It is evident that tray efficiency is influenced by (1) the specific component under consideration (this holds specially for multicomponent systems in which the efficiency can be different for each component, because of different diffusivities, diffusion interactions, and different stripping factors) and (2) the vapor flow rate; usually, increasing the flow rate increases the effective mass transfer rate, while it decreases the contact time at the same time. These counteracting effects lead to a roughly constant efficiency value, for a tray in its normal operating range. Upon approaching the lower operating limit a tray starts weeping and loses efficiency.

7.3.3.2 Packed Towers

A packed tower reactor (packed column reactor) is similar to a trickle-bed reactor (Chapter 6), where liquid film flows down over the packing surface in contact with the upward gas flow. A small fragment of packing geometry can be accurately analyzed assuming the periodic boundary conditions, which allows calibration of the porous media model for a big packing segment. The packing in a distillation column creates a surface for the liquid to spread on thereby providing a high surface area for mass transfer between the liquid and the vapor.

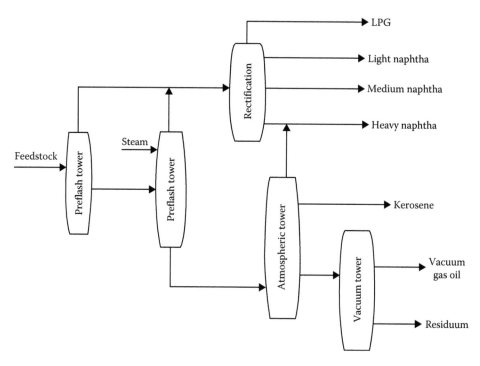

FIGURE 7.14 The D2000 distillation process.

Variations in both the atmospheric and vacuum distillation protocols, including the tower internals, are claimed to improve process efficiency and economics. For example, the D2000 process (Figure 7.14) uses progressive distillation to minimize the total energy consumption required for separation. The process is normally applied for new topping units or new integrated topping/ vacuum units. Incorporation of a vacuum flasher into the distillation circuit (Figure 7.15) is claimed

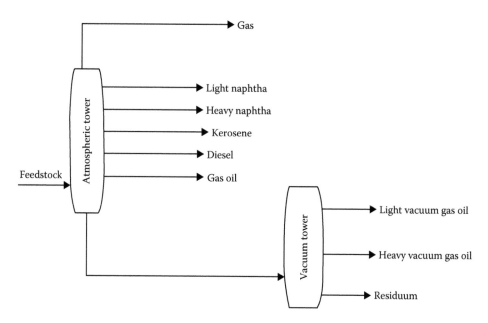

FIGURE 7.15 Distillation with the inclusion of a vacuum flasher.

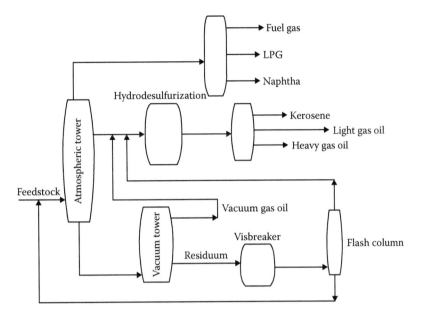

FIGURE 7.16 Distillation followed by high vacuum distillation, hydrodesulfurization, and visbreaking.

to produce an increased yield of distillate materials as well as the usual vacuum residuum. Finally, integration of a crude distillation unit, a hydrodesulfurization unit, a high vacuum unit, and a visbreaker (Figure 7.16) (Chapter 8) also improves efficiency.

In summary, the continued use of atmospheric and vacuum distillation has been a major part of refinery operations during this century and no doubt will continue to be employed, at least into the midpoint and latter decades of the twenty-first century, as the primary refining operation (Speight, 2011).

7.4 OTHER DISTILLATION PROCESSES

Atmospheric distillation and vacuum distillation provide the primary fractions from crude oil to use as feedstocks for other refinery processes for conversion into products. Many of these subsequent processes involve fractional distillation, and some of the procedures are so specialized and used with such frequency that they are identified by name.

7.4.1 STRIPPING

Stripping is a fractional distillation operation carried out on each sidestream product immediately after it leaves the main distillation tower. Since perfect separation is not accomplished in the main tower, unwanted components are mixed with those of the sidestream product. The purpose of stripping is to remove the more volatile components and thus reduce the flash point of the sidestream product. Thus, a sidestream product enters at the top tray of a stripper, and as it spills down the four to six trays, steam injected into the bottom of the stripper removes the volatile components. The steam and volatile components leave the top of the stripper to return to the main tower. The stripped sidestream product leaves at the bottom and, after being cooled in a heat exchanger, goes to storage. Since strippers are short, they are arranged one above another in a single tower; each stripper, however, operates as a separate unit.

A tower stripping unwanted volatile material from gas oil may be only 3 or 4 ft in diameter and 10 ft high with less than 20 trays (Table 7.6). Towers concerned with the distillation of liquefied gases are only a few feet in diameter but may be up to 200 ft in height. A tower used in the

TABLE 7.6
Number of Required Trays According to Tower Function

Degree of Rectification	No. of Trays	Ratio of Vapor to Feed
1. Stripping still	10–20	Vapor = 20% of feed
2. Primary fractionator	20–40	Vapor = 35%–40% of feed
3. Secondary fractionator	40–50	Feed = 50% of vapor
4. Splitter	50–70	Feed = 25% of vapor
5. Superfractionator	70–100	Feed = 10% of vapor

fractionation of crude petroleum may have from 16 to 28 trays, but one used in the fractionation (superfractionation) of liquefied gases may have 30–100 trays. The feed to a typical tower enters the vaporizing or flash zone, an area without trays. The majority of the trays are usually located above this area. The feed to a bubble tower, however, may be at any point from top to bottom with trays above and below the entry point, depending on the kind of feedstock and the characteristics desired in the products.

7.4.2 RERUNNING

Rerunning is a general term covering the *redistillation* of any material and indicating, usually, that a large part of the material is distilled overhead. Stripping, in contrast, removes only a relatively small amount of material as an overhead product. A rerun tower may be associated with a crude distillation unit that produces wide boiling range naphtha as an overhead product. By separating the wide-cut fraction into a light and heavy naphtha, the rerun tower acts in effect as an extension of the crude distillation tower.

The product from chemical treating process of various fractions may be rerun to remove the treating chemical or its reaction products. If the volume of material being processed is small, a *shell still* may be used instead of a continuous fractional distillation unit. The same applies to gas oils and other fractions from which the *front end* or *tail* must be removed for special purposes.

7.4.3 STABILIZATION

The gaseous and more volatile liquid hydrocarbons produced in a refinery are collectively known as *light hydrocarbons* or *light ends* (Table 7.7). Light ends are produced in relatively small quantities from crude petroleum and in large quantities when gasoline is manufactured by cracking and reforming. When a naphtha or gasoline component at the time of its manufacture is passed through a condenser, most of the light ends do not condense and are withdrawn and handled as a gas. A considerable part of the light ends, however, can remain dissolved in the condensate, thus forming a liquid with a high vapor pressure, which may be categorized as unstable and stabilization is required (Abdel-Aal et al., 2016).

Liquids with high vapor pressures may be stored in refrigerated tanks or in tanks capable of withstanding the pressures developed by the gases dissolved in the liquid. The more usual procedure, however, is to separate the light ends from the liquid by a distillation process generally known as *stabilization*. Enough of the light ends are removed to make a stabilized liquid, that is, a liquid with a low enough vapor pressure to permit its storage in ordinary tanks without loss of vapor. The simplest stabilization process is a stripping process. Light naphtha from a crude tower, for example, may be pumped into the top of a tall, small-diameter fractional distillation tower operated under a pressure of 50–80 psi. Heat is introduced at the bottom of the tower by a steam reboiler. As the naphtha cascades down the tower, the light ends separate and pass up the tower to leave as

TABLE 7.7

Hydrocarbon Constituents of the *Light Ends* Fraction

Hydrocarbon	Carbon Atoms	Mol. Wt.	Boiling Range		Uses
			°C	°F	
Methane	1	16	−182	−296	Fuel gas
Ethane	2	30	−89	−128	Fuel gas
Ethylene	2	28	−104	−155	Fuel gas, petrochemicals
Propane	3	44	−42	−44	Fuel gas, LPG
Propylene	3	42	−48	−54	Fuel gas, petrochemicals, polymer gasoline
Iso-Butane	4	58	−12	11	Alkylate, motor gasoline
n-Butane	4	58	−1	31	Automotive gasoline
Iso-Butylene	4	56	−7	20	Synthetic rubber and chemicals, polymer gasoline, alkylate, motor gasoline
Butylene-1[a]	4	56	−6	21	Synthetic rubber and chemicals,
Butylene-2[a]	4	56	1	34	alkylate, polymer gasoline, motor gasoline
Iso-Pentane	5	72	28	82	Automotive and aviation gasolines
n-Pentane	5	72	36	97	Automotive and aviation gasolines
Pentylenes	5	70	30	86	Automotive and aviation gasolines
Iso-Hexane	6	86	61	141	Automotive and aviation gasolines
n-Hexane	6	86	69	156	Automotive and aviation gasolines

[a] Numbers refer to the positions of the double bond. For example, butylene-1 (or butene-1 or but-1-ene) is $CH_3CH_2CH=CH_2$, and butylene-2 (or butene-2 or but-2-ene) is $CH_3CH=CHCH_3$.

an overhead product. Since reflux is not used, considerable amounts of liquid hydrocarbons pass overhead with the light ends.

Stabilization is usually a more precise operation than that just described. An example of more precise stabilization can be seen in the handling of the mixture of hydrocarbons produced by cracking. The overhead from the atmospheric distillation tower that fractionates the cracked mixture consists of light ends and cracked gasoline with light ends dissolved in it. If the latter is pumped to the usual type of tank storage, the dissolved gases cause the gasoline to boil, with consequent loss of the gases and some of the liquid components. To prevent this, the gasoline and the gases dissolved in it are pumped to a stabilizer maintained under a pressure of approximately 100 psi and operated with reflux. This fractionating tower makes a cut between the highest boiling gaseous component (butane) and the lowest boiling liquid component (pentane). The bottom product is thus a liquid free of all gaseous components, including butane; hence the fractionating tower is known as a *debutanizer* (Figure 7.17). The debutanizer bottoms (gasoline constituents) can be safely stored, whereas the overhead from the debutanizer contains the butane, propane ethane, and methane fractions. The butane fraction, which consists of all the hydrocarbons containing four carbon atoms, is particularly needed to give easy starting characteristics to gasoline. It must be separated from the other gases and blended with gasoline in amounts that vary with the season: more in the winter and less in the summer. Separation of the butane fraction is effected by another distillation in a fractional distillation tower called a *depropanizer*, since its purpose is to separate propane and the lighter gases from the butane fraction.

The depropanizer is very similar to the debutanizer, except that it is smaller in diameter because of the smaller volume being distilled and is taller because of the larger number of trays required to make a sharp cut between the butane and propane fractions. Since the normally gaseous propane must exist as a liquid in the tower, a pressure of 200 psi is maintained. The bottom product, known as the butane fraction, stabilizer bottoms, or refinery casinghead, is a high-vapor-pressure material that must be stored in refrigerated tanks or pressure tanks. The depropanizer overhead, consisting of

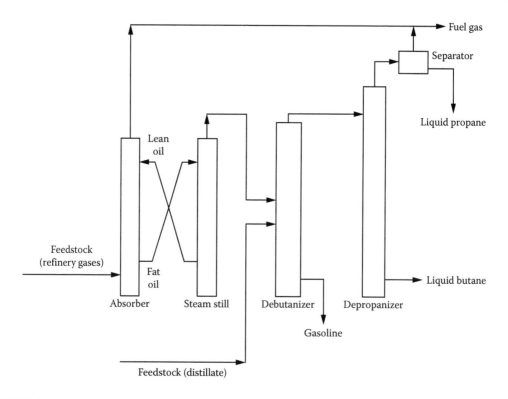

FIGURE 7.17   A light ends plant.

propane and lighter gases, is used as a petrochemical feedstock or as a refinery fuel gas, depending on the composition.

A depentanizer is a fractional distillation tower that removes the pentane fraction from a debutanized (butane-free) fraction. Depentanizers are similar to debutanizers and have been introduced recently to segregate the pentane fractions from cracked gasoline and reformate. The pentane fraction when added to a premium gasoline makes this gasoline extraordinarily responsive to the demands of an engine accelerator.

The gases produced as overhead products from crude distillation, stabilization, and depropanization units may be delivered to a gas absorption plant for the recovery of small amounts of butane and higher-boiling hydrocarbons. The gas absorption plant consists essentially of two towers. One tower is the absorber where the butane and higher-boiling hydrocarbons are removed from the lighter gases.

The gas mixture enters at the bottom of the tower and rises to the top. As it does this, it contacts the lean oil, which absorbs the butane and higher-boiling hydrocarbons but not the lower-boiling hydrocarbons. The latter leave the top of the absorber as dry gas. The lean oil that has become enriched with butane and higher-boiling hydrocarbons is now termed *fat oil*. This is pumped from the bottom of the absorber into the second tower, where fractional distillation separates the butane and higher-boiling hydrocarbons as an overhead fraction and the oil, once again lean oil, as the bottom product.

The condensed butane and higher-boiling hydrocarbons are included with the refinery *casing-head bottoms* or *stabilizer bottoms*. The dry gas is frequently used as fuel gas for refinery furnaces. It contains propane and propylene, however, which may be required for liquefied petroleum gas for the manufacture of polymer gasoline or petrochemicals. Separation of the propane fraction (propane and propylene) from the lighter gases is accomplished by further distillation in a fractional distillation tower similar to those previously described and particularly designed to handle liquefied gases. Further separation of hydrocarbon gases is required for petrochemical manufacture.

7.4.4 SUPERFRACTIONATION

The term *superfractionation* is sometimes applied to a highly efficient fractionating tower used to separate ordinary petroleum products. For example, to increase the yield of furnace fuel oil, heavy naphtha may be redistilled in a tower that is capable of making a better separation of the naphtha and the fuel oil components. The latter, obtained as a bottom product, is diverted to furnace fuel oil.

Fractional distillation as normally carried out in a refinery does not completely separate one petroleum fraction from another. One product overlaps another, depending on the efficiency of the fractionation, which in turn depends on the number of trays in the tower, the amount of reflux used, and the rate of distillation. Kerosene, for example, normally contains a small percentage of hydrocarbons that (according to their boiling points) belong in the naphtha fraction and a small percentage that should be in the gas oil fraction. Complete separation is not required for the ordinary uses of these materials, but certain materials, such as solvents for particular purposes (hexane, heptane, and aromatics), are required as essentially pure compounds. Since they occur in mixtures of hydrocarbons they must be separated by distillation, with no overlap of one hydrocarbon with another. This requires highly efficient fractional distillation towers that are especially designed for this purpose and referred to as superfractionators. Several towers with 50–100 trays operating at high reflux ratios may be required to separate a single compound with the necessary purity.

7.4.5 AZEOTROPIC DISTILLATION

Azeotropic distillation is the use of a third component to separate two close-boiling components by means of the formation of an azeotropic mixture between one of the original components and the third component to increase the difference in the boiling points and facilitates separation by distillation.

Sometimes the separation of a desired compound calls for azeotropic distillation. All compounds have definite boiling temperatures, but a mixture of chemically dissimilar compounds sometimes causes one or both of the components to boil at a temperature other than that expected. The separation of these components of similar volatility may become economical if an *entrainer* can be found that effectively changes the relative volatility. It is also desirable that the entrainer be reasonably cheap, stable, nontoxic, and readily recoverable from the components. In practice, it is probably this last criterion that severely limits the application of extractive and azeotropic distillation. The majority of successful processes, in fact, are those in which the entrainer and one of the components separate into two liquid phases on cooling if direct recovery by distillation is not feasible.

A further restriction in the selection of an azeotropic entrainer is that the boiling point of the entrainer be in the range 10°C–40°C (18°F–72°F) below that of the components. Thus, although the entrainer is more volatile than the components and distills off in the overhead product, it is present in a sufficiently high concentration in the rectification section of the column.

The five methods for separating azeotropic mixtures are as follows: (1) *homogeneous azeotropic distillation* where the liquid separating agent is completely miscible; (2) *heterogeneous azeotropic distillation*, or more commonly, *azeotropic distillation* where the liquid separating agent (the *entrainer*) forms one or more azeotropes with the other components in the mixture and causes two liquid phases to exist over a wide range of compositions, which is the key to making the distillation sequence work; (3) *distillation using ionic salts*, where the salts dissociate in the liquid mixture and alters the relative volatilities sufficiently that the separation become possible; (4) *pressure-swing distillation* where a series of column operating at different pressures are used to separate binary azeotropes that change appreciably in composition over a moderate pressure range or where a separating agent that forms a pressure-sensitive azeotrope is added to separate a pressure-insensitive azeotrope; and (5) *reactive distillation* where the separating agent reacts preferentially and reversibly with one of the azeotropic constitutes after which the reaction product is distilled from the nonreacting components and the reaction is reversed to recover the initial component.

In *simple distillation* a multicomponent liquid mixture is slowly boiled in a heated zone and the vapors are continuously removed as they form and, at any instant in time, the vapor is in equilibrium with the liquid remaining on the still. Because the vapor is always richer in the more volatile components than the liquid, the liquid composition changes continuously with time, becoming more and more concentrated in the least volatile species. A simple *distillation residue curve* is a means by which the changes in the composition of the liquid residue curves on the pot changes with time (Speight, 2014a). A *residue curve map* is a collection of the liquid residue curves originating from different initial compositions. Residue curve maps contain the same information as phase diagrams, but represent this information in a way that is more useful for understanding how to synthesize a distillation sequence to separate a mixture.

All of the residue curves originate at the light (lowest boiling) pure component in a region, move toward the intermediate boiling component, and end at the heavy (highest boiling) pure component in the same region. The lowest temperature nodes are termed as *unstable nodes*, as all trajectories leave from them; while the highest temperature points in the region are termed *stable nodes*, as all trajectories ultimately reach them. The point that the trajectories approach from one direction and end in a different direction (as always is the point of intermediate boiling component) is termed "saddle point." Residue curve that divide the composition space into different distillation regions is called distillation boundaries.

Many different residue curve maps are possible when azeotropes are present. Ternary mixtures containing only one azeotrope may exhibit six possible residue curve maps that differ by the binary pair forming the azeotrope and by whether the azeotrope is minimum or maximum boiling. By identifying the limiting separation achievable by distillation, residue curve maps are also useful in synthesizing separation sequences combining distillation with other methods.

However, the separation of components of similar volatility may become economical if an *entrainer* can be found that effectively changes the relative volatility. It is also desirable that the entrainer be reasonably cheap, stable, nontoxic, and readily recoverable from the components. In practice, it is probably this last criterion that severely limits the application of extractive and azeotropic distillation. The majority of successful processes, in fact, are those in which the entrainer and one of the components separate into two liquid phases on cooling if direct recovery by distillation is not feasible. A further restriction in the selection of an azeotropic entrainer is that the boiling point of the entrainer be in the range 10°C–40°C (18°F–72°F) below that of the components. Thus, although the entrainer is more volatile than the components and distills off in the overhead product, it is present in a sufficiently high concentration in the rectification section of the column.

7.4.6 EXTRACTIVE DISTILLATION

Extractive distillation is the use of a third component to separate two close-boiling components in which one of the original components in the mixture is extracted by the third component and retained in the liquid phase to facilitate separation by distillation. In the process, the difference in volatility of the components to be separated is enhanced by the addition of a solvent or an entrainer.

Using acetone–water as an extractive solvent for butanes and butenes, butane is removed as overhead from the extractive distillation column with acetone–water charged at a point close to the top of the column. The bottom product of butenes and the extractive solvent are fed to a second column where the butenes are removed as overhead. The acetone–water solvent from the base of this column is recycled to the first column.

Extractive distillation (Figure 7.18) may also be used for the continuous recovery of individual aromatics, such as benzene, toluene, or xylene(s), from the appropriate petroleum fractions. *Prefractionation* concentrates a single aromatic cut into a close boiling cut, after which the aromatic concentrate is *extractively distilled* with a solvent (usually phenol) for benzene or toluene recovery. Mixed cresylic acids (cresol derivatives and methyl phenol derivatives) are used as the solvent for xylene recovery.

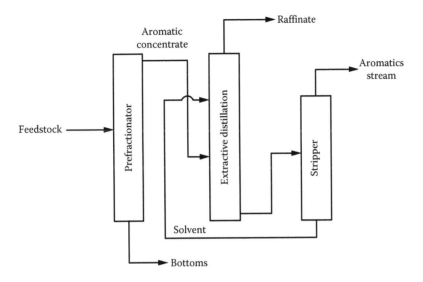

FIGURE 7.18   Extractive distillation for aromatics recovery.

In general, none of the fractions or combinations of fractions separated from crude petroleum are suitable for immediate use as petroleum products. Each must be separately refined by treatments and processes that vary with the impurities in the fraction and the properties required in the finished product. The simplest treatment is the washing of a fraction with a lye solution to remove sulfur compounds. The most complex is the series of treatments—solvent treating, dewaxing, clay treating or hydrorefining, and blending—required to produce lubricating oils. On rare occasions no treatment of any kind is required. Some crude oils yield a light gas oil fraction that is suitable as furnace fuel oil or as a diesel fuel.

Extractive distillation is successful because the solvent is specially chosen to interact differently with the components of the original mixture, thereby altering their relative volatilities. Because these interactions occur predominantly in the liquid phase, the solvent is continuously added near the top of the extractive distillation column so that an appreciable amount is present in the liquid phase on all of the trays below. The mixture to be separated is added through second feed point further down the column. In the extractive column, the component having the greater volatility, not necessarily the component having the lowest boiling point, is taken overhead as a relatively pure distillate. The other component leaves with the solvent via the column bottoms. The solvent is separated from the remaining components in a second distillation column and then recycled back to the first column.

One of the most important steps in developing a successful (economical) extractive distillation sequence is selecting a good solvent. In general, selection criteria for the solvent include the following; the process (1) should enhance significantly the natural relative volatility of the key component, (2) should not require an excessive ratio of solvent to nonsolvent (because of cost of handling in the column and auxiliary equipment, (3) should not lead to the formation of two phases, and (4) should allow the desired product to be easily separable from the bottom product.

No single solvent or solvent mixture satisfies all of the criteria for use in extractive distillation. In general, none of the fractions or combinations of fractions separated from crude petroleum is suitable for immediate use as petroleum products. Each fraction must be separately refined by processes that vary with the impurities in the fraction and the properties required in the finished product (Speight and Ozum, 2002; Parkash, 2003; Hsu and Robinson, 2006; Gary et al., 2007; Speight, 2014a). The simplest treatment is the washing of a fraction with a lye solution to remove sulfur compounds.

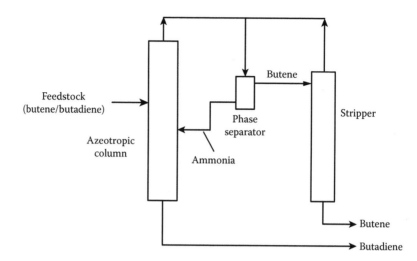

FIGURE 7.19   Separation of butene and butadiene by azeotropic distillation.

Two processes illustrate the similarities and differences between azeotropic distillation and extractive distillation. Both have been used for the separation of C_4 hydrocarbons (Figures 7.19 and 7.20). Thus, butadiene and butene may be separated by the use of liquid ammonia, which forms an azeotrope with butene. The ammonia–butene azeotrope overhead from the azeotropic distillation is condensed, cooled, and allowed to separate into a butene layer and a heavier ammonia layer. The butene layer is fed to a second column, where the ammonia is removed as a butene–ammonia azeotrope, and the remaining butene is recovered as bottom product. The ammonia layer is returned to the lower section of the first azeotropic distillation column. Butadiene is recovered as bottom product from this column.

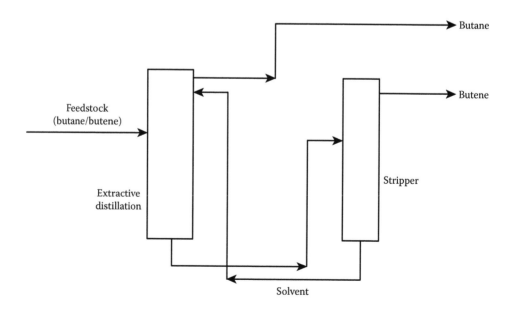

FIGURE 7.20   Separation of butane and butene by extractive distillation.

7.5   OPTIONS FOR HEAVY FEEDSTOCKS

In order to further distill heavy oil, tar sand bitumen, or residuum, or topped crude, reduced pressure is required to prevent thermal cracking and the process takes place in one or more vacuum distillation towers. Since the heavy feedstocks are expected to contribute a growing fraction of hydrocarbon fuels production, changes can be expected in terms of the actual unit internals, unit operation, and prevention of corrosion. Innovations to the distillation units will most likely be more subtle than a complete restructuring of the distillation section of the refinery and will focus on (1) changes to the internal packing to prevent fouling within the distillation system, and (2) the use of metal alloy systems to mitigate corrosion.

7.5.1   DISTILLATION OPERATIONS

The principles of vacuum distillation resemble those of fractional distillation, and, except that larger-diameter columns are used to maintain comparable vapor velocities at the reduced pressures, the equipment is also similar. The internal designs of some vacuum towers are different from atmospheric towers in that random packing and demister pads are used instead of trays. A typical first-phase vacuum tower may produce gas oil, lubricating oil base stock, and a heavy residuum for propane deasphalting. A second-phase tower operating at lower vacuum may distill surplus residuum from the atmospheric tower, which is not used for lube-stock processing, and surplus residuum from the first vacuum tower not used for deasphalting.

Vacuum towers are typically used to separate catalytic cracking feedstock from surplus residuum and heavy oil and tar sand bitumen have fewer components distilling at atmospheric pressure and under vacuum than conventional petroleum. Nevertheless, some heavy oil still pass through the distillation stage of a refinery before further processing is undertaken. In addition, a vacuum tower has recently been installed at the Syncrude Canada plant to offer an additional process option for upgrading tar sand bitumen (Speight, 2005, 2014a). The installation of such a tower as a means of refining heavy feedstocks (with the possible exception of the residua that are usually produced through a vacuum tower) is a question of economics and the ultimate goal of the refinery in terms of product slate. After distillation, the residuum from the heavy oil might pass to a cracking unit such as visbreaking or coking to produce salable products. Catalytic cracking of the residuum or the whole heavy oil is also an option but is very dependent on the constituents of the feedstock and their interaction with the catalyst.

Thus, there is the potential for applying a related concept to the deep distillation of the heavy feedstocks. The continued and projected increased influx of heavy oil, extra heavy oil, and tar sand bitumen into refineries will require reassessment of the need for refinery distillation. Nevertheless, vacuum distillation is an option for tar sand bitumen processes in which the distillation unit is employed to collect as much valuable high-vacuum gas oil as possible from the bitumen before the residuum is sent to a conversion unit. This option can assist in balancing the overall technical efficiency and economic efficiency of the bitumen refinery. Furthermore, if partial conversion (such as the use of visbreaking) is an option for processing heavy feedstocks or partial upgrading during recovery of the feedstock, distillation will still find a use in refineries.

Moreover, as feedstocks change in composition, the distillation unit will be required to achieve higher degrees of efficiency to produce the precursors to hydrocarbon fuels as well as feedstocks for other units that will eventually produce hydrocarbon fuels through cracking. This will more likely be achieved by changes in the internals of the distillation units as well as changes to the overlay use of the units. The overall effects will be for refineries to create the option to take deeper cuts into the crude oil feedstock leaving a harder redid to be used as feedstocks for the cracking units.

Catalytic distillation (reactive distillation) is a branch of reactive distillation that combines the processes of distillation and catalysis (Ng and Rempel, 2002; Harmsen, 2007). Catalytic distillation is a reactor technology that combines a heterogeneous catalytic reaction and the separation of reactants and products via distillation in a single reactor/distillation column. The heterogeneous catalyst provides the sites for catalytic reactions and also the interfacial surface for liquid/vapor separation. The distinct difference between the catalytic distillation column and the conventional distillation column lies in the placement of solid catalysts usually incorporated in a packing within the distillation column to provide a reaction section in addition to the traditional trays or random packings used for separations in the stripping the rectifying sections of the distillation column. Mass transfer characteristics of the catalytic distillation column packing in the reaction zone have significant influence on the product yield and selectivity. The benefits of catalytic distillation include energy and capital savings, enhanced conversion and product selectivity, longer catalyst lifetime, and reduction of waste streams. The first commercial application of CD was for the production of methyl tertiary butyl ether (MTBE). There are many other possible applications of CD such as the hydration of olefins, alkylation reactions, esterification reactions, hydrolysis, aldol condensation, hydrogenation, desulfurization, and oligomerization of olefins.

For example, catalytic distillation finds application for reversible reactions, such as methyl tetrabutyl ether (MTBE) and ethyl tributyl ether (ETBE) synthesis, so as to shift an unfavorable equilibrium by continuous reaction product withdrawal (DeCroocq, 1997). Catalytic distillation can provide also several advantages in selective hydrogenation of C_3, C_4, and C_5 cuts for petrochemical use. Inserting the catalyst in the fractionation column improves mercaptan derivatives removal, catalyst fouling resistance, and selective hydrogenation performances by modifying the reaction mixture composition along the column.

Fouling and *foaming* are frequent occurrence in distillation towers. Chemical reactions and surface phenomena in fouling and foaming systems can further complicate their predictability. Several techniques for dealing with such unpredictable problems include monitoring of tower conditions, selection of tower internals, and pretreatment of recycle streams. These methods will be improved and developed to the point where they are operative in all distillation units. Furthermore, distillation efficiency is limited by the undesired coke deposition, resulting in a significant loss of distillation efficiency. When a residuum is heated to pyrolysis temperatures (>350°C, 650°F), there is typically an induction period before coke formation begins. To avoid fouling, refiners often stop heating well before coke forms, using arbitrary criteria, but cessation of the heating can result in less than maximum distillate yield.

Over the past three decades, a better understanding of the chemistry and physics of coking has evolved (Chapter 5) (Speight, 2014a) and improved designs based on primary internals have allowed an increase in the amount of gas oil produced with increases in cut point from approximately 520°C (970°F) to 590°C (1095°F). As continuing inroads are made into the chemistry of coking, future distillation units will show improvements in the design of the internals leading to process equivalents of the laboratory spinning band distillation units. Thus, with the potential for an increase in the influx of heavy oil, tar sand bitumen, and biomass to refineries, there may be a resurgence of interest in the application of *reactive distillation* in refineries, which is a process where the still is also a chemical reactor. Separation of the product from the reaction mixture does not need a separate distillation step, which saves energy (for heating) and materials. This technique is especially useful for equilibrium-limited reactions and conversion can be increased far beyond what is expected by the equilibrium due to the continuous removal of reaction products from the reactive zone. This helps reduce capital and investment costs and may be important for sustainable development due to a lower consumption of resources. The suitability of reactive distillation for a particular reaction depends on various factors such as volatility of the reactants and products along with the feasible reaction and distillation temperature. Hence, the use of reactive distillation for every reaction may not be feasible. Exploring the candidate reactions for reactive distillation is an area that needs considerable attention to expand the domain of reactive distillation processes.

However, the conditions in the reactive column are suboptimal both as a distillation column and a chemical reactor, since the reactive column combines these. In addition, the introduction of an *in situ* separation process in the reaction zone or vice versa leads to complex interactions between vapor–liquid equilibrium, mass transfer rates, diffusion, and chemical kinetics, which poses a great challenge for design and synthesis of these systems. Side reactors, where a separate column feeds a reactor and vice versa, are better for some reactions if the optimal conditions of distillation and reaction differ too much.

Membranes may offer future alternatives to distillation. Membranes have started to enter the refinery for hydrogen recovery (see earlier) but are also being developed for other separations. Current membrane systems will probably be most effective in hybrid distillation processes to perform a first, crude, low-energy, low-cost separation, leaving the polishing operation for distillation. If high selectivity could be achieved with membranes, there is the potential to replace distillation in many separation processes.

7.5.2 CORROSION

Refinery distillation units run as efficiently as possible to reduce costs. One of the major issues that occurs in distillation units and decreases efficiency is corrosion of the metal components found throughout the process line of the hydrocarbon refining process (Speight, 2014b). Corrosion causes the failure of parts in addition to dictating the shutdown schedule of the unit, which can cause shutdown of the refinery. Attempts to block such corrosive influences will be a major issue of future refineries.

Furthermore, in addition to the corrosive properties of high-acid crude oils, sulfur may be present in crude oil as hydrogen sulfide (H_2S), as compounds (such as mercaptan derivatives, sulfide derivatives, disulfide derivatives, and thiophene derivatives), or as elemental sulfur. Each crude oil has different amounts and types of sulfur compounds but, generally, the proportion, stability, and complexity of the compounds are greater in heavier crude oils. Hydrogen sulfide is a primary contributor to corrosion in refinery processing units. Other corrosive substances are elemental sulfur and mercaptan derivatives.

Heavy feedstocks (of the types that are relevant to this chapter, i.e., heavy oil, extra heavy oil, and tar sand bitumen) contain inorganic salts such as sodium chloride, magnesium chloride, and calcium chloride in suspension or dissolved in entrained water (brine). These salts must be removed or neutralized before processing to prevent catalyst poisoning, equipment corrosion, and fouling. Salt corrosion is caused by the hydrolysis of some metal chlorides to hydrogen chloride (HCl) and the subsequent formation of hydrochloric acid when crude is heated. Hydrogen chloride may also combine with ammonia to form ammonium chloride (NH_4Cl), which causes fouling and corrosion.

The sections of the process susceptible to corrosion include (but may not be limited to) preheat exchanger (HCl and H_2S), preheat furnace and bottoms exchanger (H_2S and sulfur compounds), atmospheric tower and vacuum furnace (H_2S, sulfur compounds, and organic acids), vacuum tower (H_2S and organic acids), and overhead (H_2S, HCl, and water). Where sour crudes are processed, severe corrosion can occur in furnace tubing and in both atmospheric and vacuum towers where metal temperatures exceed 450°F. Wet hydrogen sulfide also will cause cracks in steel. When processing high-nitrogen crudes, nitrogen oxides can form in the flue gases of furnaces, and these oxides are corrosive to steel when cooled to low temperatures (nitric and nitrous acids are formed) in the presence of water.

As a first step in the refining process, to reduce corrosion, plugging, and fouling of equipment and to prevent poisoning the catalysts in processing units, these contaminants must be removed by desalting (dehydration). However, the desalting operation does not always remove all of the corrosive elements and hydrogen chloride may be a product of the thermal treatment that occurs as part of the distillation process. Inadequate desalting can cause fouling of heater tubes and heat exchangers throughout the refinery. Fouling restricts product flow and heat transfer and leads to failures due to increased pressures and temperatures. Corrosion, which occurs due to the presence of hydrogen sulfide, hydrogen chloride, naphthenic (organic) acids, and other contaminants in the crude oil, also causes equipment failure. Neutralized salts (ammonium chlorides and sulfides), when moistened by condensed water, can cause corrosion.

Corrosion occurs in various forms in the distillation section of the refinery and is manifested by events such as pitting corrosion from water droplets, embrittlement from chemical attack if the dewatering and desalting unit has not operated efficiently, and stress corrosion cracking from sulfide attack.

High-temperature corrosion of distillation units will continue to be a major concern to the refining industry. The presence of naphthenic acid and sulfur compounds considerably increases corrosion in the high-temperature parts of the distillation units and equipment failures have become a critical safety and reliability issue. The difference in process conditions, materials of construction, and blend processed in each refinery and especially the frequent variation in crude diet increases the problem of correlating corrosion of a unit to a certain type of crude oil. In addition, a large number of interdependent parameters influence the high-temperature crude corrosion process.

Naphthenic acid corrosion is differentiated from sulfidic corrosion by the nature of the corrosion (pitting and impingement) and by its severe attack at high velocities in crude distillation units. Crude feedstock heaters, furnaces, transfer lines, feed and reflux sections of columns, atmospheric and vacuum columns, heat exchangers, and condensers are among the type of equipment subject to this type of corrosion.

From a materials standpoint, carbon steel can be used for refinery components. Carbon steel is resistant to the most common forms of corrosion, particularly from hydrocarbon impurities at temperatures below 205°C (400°F), but other corrosive chemicals and high-temperature environments prevent its use everywhere. Common replacement materials are low alloy steel containing chromium and molybdenum, with stainless steel containing more chromium dealing with more corrosive environments. More expensive materials commonly used are nickel titanium and copper alloys. These are primarily saved for the most problematic areas where extremely high temperatures or very corrosive chemicals are present.

Attempts to mitigate corrosion will continue to use a complex system of monitoring, preventative repairs, and careful use of materials. *Monitoring methods* include both offline checks taken during maintenance and online monitoring. Offline checks measure corrosion after it has occurred, telling the engineer when equipment must be replaced based on the historical information he has collected.

Blending of refinery feedstocks (with the inherent danger of phase separation and incompatibility) will continue to be used to mitigate the effects of corrosion as will blending, inhibition, materials upgrading, and process control. Blending will be used to reduce the naphthenic acid content of the feed, thereby reducing corrosion to an acceptable level. However, while blending of heavy and light crude oils can change shear stress parameters and might also help reduce corrosion, there is also the potential for incompatibility of heavy and light crude oils.

In summary, refinery distillation may appear to be waning in terms of under the processing of such heavy feedstocks but it is definitely not out.

REFERENCES

Abdel-Aal, H.K., Aggour, M.A., and Fahim, M.A. 2016. *Petroleum and Gas Field Processing.* CRC Press/ Taylor & Francis Group, Boca Raton FL.

ASTM D5. 2015. *Standard Test Method for Penetration of Bituminous Materials.* Annual Book of Standards, ASTM International, West Conshohocken, PA.

Burris, D.R. 1992. Desalting crude oil. In: *Petroleum Processing Handbook.* J.J. McKetta (Editor). Marcel Dekker, New York, pp. 666.

Charbonnier, R.P., Draper, R.G., Harper, W.H., and Yates, Y. 1969. Analyses and characteristics of oil samples from Alberta. Information Circular No. IC 232. Department of Energy Mines and Resources, Mines Branch, Ottawa, Ontario, Canada.

Coleman, H.J., Shelton, E.M., Nicholls, D.T., and Thompson, C.J. 1978. Analysis of 800 crude oils from United States oilfields. Report No. BETC/RI-78/14. Technical Information Center, Department of Energy, Washington, DC.

DeCroocq, D. 1997. Major scientific and technical challenges about development of new processes in refining and petrochemistry. *Revue Institut Français de Pétrole*, 52(5): 469–489.

Diwekar, U.M. 1995. *Batch Distillation: Simulation, Optimal Design, and Control*. Taylor & Francis Group, Philadelphia, PA.

Gary, J.G., Handwerk, G.E., and Kaiser, M.J. 2007. *Petroleum Refining: Technology and Economics*, 5th Edition. CRC Press/Taylor & Francis Group, Boca Raton, FL.

Harmsen, G.J. 2007. Reactive distillation: The front-runner of industrial process intensification. A full review of commercial applications, research, scale-up, design and operation. *Chemical Engineering and Processing*, 46: 774–780.

Hsu, C.S. and Robinson, P.R. 2006. *Practical Advances in Petroleum Processing*, Volumes 1 and 2. Springer, New York.

Jones, D.S.J. 1995. *Elements of Petroleum Processing*. John Wiley & Sons, Chichester, UK.

Ng, F.T.T. and Rempel, G.L. 2002. *Catalytic Distillation. Encyclopedia of Catalysis*. John Wiley & Sons, Hoboken, NJ.

Parkash, S. 2003. *Refining Processes Handbook*. Gulf Professional Publishing, Elsevier, Amsterdam, the Netherlands.

Speight, J.G. 2000. *The Desulfurization of Heavy Oils and Residua*, 2nd Edition. Marcel Dekker, New York.

Speight, J.G. 2005. Natural bitumen (tar sands) and heavy oil. In: *Coal, Oil Shale, Natural Bitumen, Heavy Oil and Peat. Encyclopedia of Life Support Systems (EOLSS)*. Developed under the Auspices of the UNESCO. EOLSS Publishers, Oxford, UK.

Speight, J.G. 2011. *The Refinery of the Future*. Gulf Professional Publishing, Elsevier, Oxford, UK.

Speight, J.G. 2012. *Crude Oil Assay Database*. Knovel, Elsevier, New York.

Speight, J.G. 2014a. *The Chemistry and Technology of Petroleum*, 5th Edition. CRC Press/Taylor & Francis Group, Boca Raton, FL.

Speight, J.G. 2014b. *Oil and Gas Corrosion Prevention*. Gulf Professional Publishing, Elsevier, Oxford, UK.

Speight, J.G. 2015. *Handbook of Petroleum Product Analysis*, 2nd Edition. John Wiley, Hoboken, NJ.

Speight, J.G. and Ozum, B. 2002. *Petroleum Refining Processes*. Marcel Dekker, New York.

8 Thermal Cracking Processes

8.1 INTRODUCTION

Distillation (Chapter 7) has remained a major refinery process, and almost every crude oil that enters a refinery is subjected to this process. However, not all crude oils yield the same distillation products. In fact, the nature of the crude oil dictates the processes that may be required for refining. And balancing product yield with demand is a necessary part of refinery operations (Speight and Ozum, 2002; Parkash, 2003; Hsu and Robinson, 2006; Gary et al., 2007; Speight, 2014). However, the balancing of product yield and market demand, without the manufacture of large quantities of fractions having low commercial value, has long required processes for the conversion of hydrocarbons of one molecular weight range and/or structure into some other molecular weight range and/or structure. Basic processes for this are still the so-called cracking processes in which relatively high-boiling constituents carbons are cracked, that is, thermally decomposed into lower-molecular-weight, smaller, lower-boiling molecules, although reforming alkylation, polymerization, and hydrogen-refining processes have wide applications in making premium-quality products (Speight and Ozum, 2002; Parkash, 2003; Hsu and Robinson, 2006; Gary et al., 2007; Speight, 2014).

After 1910 and the conclusion of World War I, the demand for automotive (and other) fuels began to outstrip the market requirements for kerosene and refiners, needing to stay abreast of the market pull, were pressed to develop new technologies to increase gasoline yields. There being finite amounts of straight-run distillate fuels in crude oil, refiners had, of necessity, the urgency to develop processes to produce additional amounts of these fuels. The conversion of coal and oil shale to liquid through the agency of cracking had been known for centuries, and the production of various spirits from petroleum though thermal methods had been known since at least the inception of Greek fire in earlier centuries.

The discovery that higher-molecular-weight (higher-boiling) materials could be decomposed to lower-molecular-weight (lower-boiling) products was used to increase the production of kerosene and was called *cracking distillation*. In the process, a batch of crude oil was heated until most of the kerosene was distilled from it and the overhead material became dark in color. At this point the still fires were lowered, the rate of distillation decreased, and the heavy oils were held in the hot zone, during which time some of the large hydrocarbons were decomposed and rearranged into lower-molecular-weight products. After a suitable time, the still fires were increased and distillation continued in the normal way. The overhead product, however, was light oil suitable for kerosene instead of the heavy oil that would otherwise have been produced. Thus, it was not surprising that such technologies were adapted for the fledgling petroleum industry.

The earliest processes, which involved thermal cracking, consisted of heating heavier oils (for which there was a low market requirement) in pressurized reactors and thereby cracking, or splitting, their large molecules into the smaller ones that form the lighter, more valuable fractions such as gasoline, kerosene, and light industrial fuels. Gasoline manufactured by thermal cracking processes performed better in automobile engines than gasoline derived from straight distillation of crude petroleum. The development of more powerful aircraft engines in the late 1930s gave rise to a need to increase the combustion characteristics of gasoline to improve engine performance. Thus, during World War II and the late 1940s, improved refining processes involving the use of catalysts led to further improvements in the quality of transportation fuels and further increased their supply. These improved processes, including catalytic cracking of residua and other heavy feedstocks (Chapter 9), alkylation (Chapter 13), polymerization (Chapter 13), and isomerization (Chapter 13), enabled the petroleum industry to meet the demands of high-performance combat aircraft and, after the war, to supply increasing quantities of transportation fuels.

The 1950s and 1960s brought a large-scale demand for jet fuel and high-quality lubricating oils. The continuing increase in demand for petroleum products also heightened the need to process a wider variety of crude oils into high-quality products. Catalytic reforming of naphtha (Chapter 13) replaced the earlier thermal reforming process and became the leading process for upgrading fuel qualities to meet the needs of higher-compression engines. Hydrocracking, a catalytic cracking process conducted in the presence of hydrogen (Chapter 11), was developed to be a versatile manufacturing process for increasing the yields of either gasoline or jet fuels.

In the early stages of thermal cracking process development, processes were generally classified as either liquid phase, high pressure (350–1500 psi), low temperature (400°C–510°C, 750°F–950°F) or vapor phase, low pressure (less than 200 psi), high temperature (540°C–650°C, 1000°F–1200°F). In reality, the processes were mixed phase with no process really being entirely liquid or vapor phase, but the classification (like many classifications of crude oil and related areas) was still used as a matter of convenience (Speight and Ozum, 2002; Parkash, 2003; Hsu and Robinson, 2006; Gary et al., 2007; Speight, 2014).

The early were classified as liquid-phase processes and had the following advantages over vapor-phase processes: (1) large yields of gasoline of moderate octane number, (2) low gas yields, (3) ability to use a wide variety of charge stocks, (4) long cycle time due to low coke formation, and (5) flexibility and ease of control. However, the vapor-phase processes had the advantages of operation at lower pressures and the production of a higher-octane gasoline due to the increased production of olefins and light aromatics. However, there were many disadvantages that curtailed the development of vapor-phase processes: (1) temperatures were required which the steel alloys available at the time could not tolerate, (2) there were high gas yields and resulting losses since the gases were normally not recovered, and (3) there was a high production of olefinic compounds that created naphtha with poor stability (increased tendency to form undesirable gum) (Mushrush and Speight, 1995; Speight, 2014) that required subsequent treating of the gasoline to stabilize it against gum formation. The vapor-phase processes were not considered suitable for the production of large quantities of gasoline but did find application in petrochemical manufacture due to the high concentration of olefins produced.

It is generally recognized that the most important part of any refinery, after the distillation units, is the gasoline (and liquid fuels) manufacturing facilities; other facilities are added to manufacture additional products as indicated by technical feasibility and economic gain. More equipment is used in the manufacture of gasoline, the equipment is more elaborate, and the processes more complex than for any other product. Among the processes that have been used for liquid fuels production are thermal cracking, catalytic cracking, thermal reforming, catalytic reforming, polymerization, alkylation, coking, and distillation of fractions directly from crude petroleum (Figure 8.1). Each of these processes may be carried out in a number of ways, which differ in details of operation, or essential equipment, or both (Speight and Ozum, 2002; Parkash, 2003; Hsu and Robinson, 2006; Gary et al., 2007; Speight, 2014).

Thermal processes are essentially processes that decompose, rearrange, or combine hydrocarbon molecules by the application of heat. The major variables involved are feedstock type, time, temperature, and pressure and, as such, are usually considered in promoting cracking (thermal decomposition) of the heavier molecules to lighter products and in minimizing coke formation. Thus, one of the earliest processes used in the petroleum industry is the noncatalytic conversion of higher-boiling petroleum stocks into lower-boiling products, known as *thermal cracking*.

The thermal decomposition (cracking) of high-molecular-weight hydrocarbons to lower-molecular-weight and normally more valuable hydrocarbons has long been practiced in the petroleum refining industry. Although catalytic cracking has generally replaced thermal cracking, noncatalytic cracking processes using high temperature to achieve the decomposition are still in operation. In several cases, thermal cracking processes to produce specific desired products or

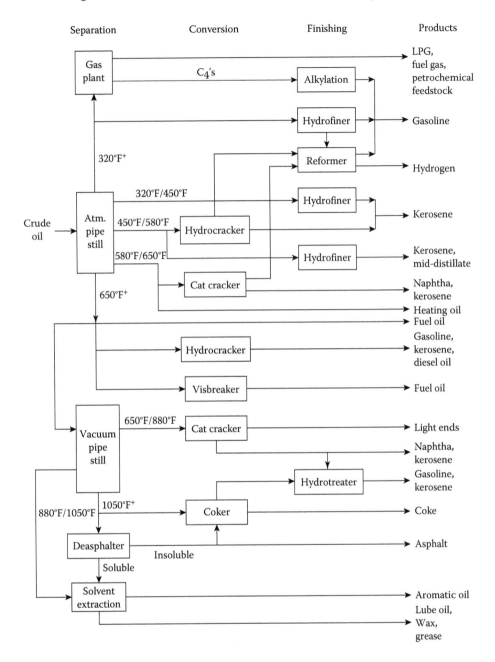

FIGURE 8.1 Schematic representation of a refinery showing placement of the various conversion units.

to dispose of specific undesirable charge streams are being operated or installed. The purpose of this chapter is to provide basic information to assist the practicing engineer/petroleum refiner to (1) determine if a particular thermal cracking process would be suitable for a specific application and could fit into the overall operation, (2) develop a basic design for a thermal cracking process, and (3) operate an existing or proposed process.

Conventional *thermal cracking* is the thermal decomposition, under pressure, of high-molecular-weight constituents (higher molecular weight and higher boiling than gasoline constituents) to form lower-molecular-weight (and lower-boiling) species. Thus, the thermal cracking process is designed to produce gasoline from higher-boiling charge stocks, and any unconverted or mildly cracked charge components (compounds which have been partially decomposed but are still higher boiling than gasoline) are usually recycled to extinction to maximize gasoline production. A moderate quantity of light hydrocarbon gases is also formed. As thermal cracking proceeds, reactive unsaturated molecules are formed that continue to react and can ultimately create higher-molecular-weight species that are relatively hydrogen deficient and readily form coke. Thus, they cannot be recycled without excessive coke formation and are therefore removed from the system as cycle fuel oil.

When petroleum fractions are heated to temperatures over 350°C (660°F), the rates of the thermal cause thermal decomposition to proceed at significant rates (Speight, 2000, 2014). Thermal decomposition does not require the addition of catalyst; therefore, this approach is the oldest technology available for residue conversion. The severity of thermal processing determines the conversion and the product characteristics. Thermal treatment of residues ranges from mild treatment for reduction of viscosity to *ultrapyrolysis* (high-temperature cracking at very short residence time) for complete conversion to olefins and light ends. The higher the temperature, the shorter the time required to achieve a given conversion but, in many cases, with a change in the chemistry of the reaction. The severity of the process conditions is the combination of reaction time and temperature to achieve a given conversion.

Thermal reactions, however, can give rise to a variety of different reactions so that selectivity for a given product changes with temperature and pressure. The mild- and high-severity processes are frequently used for processing of residua, while conditions similar to *ultrapyrolysis* (high temperature and very short residence time) are only used commercially for cracking ethane, propane, butane, and light distillate feeds to produce ethylene and higher olefins. Sufficiently high temperatures convert oils entirely to gases and coke; cracking conditions are controlled to produce as much as possible of the desired product, which is usually gasoline but may be cracked gases for petrochemicals or a lower viscosity oil for use as a fuel oil. The feedstock, or cracking stock, may be almost any fraction obtained from crude petroleum, but the greatest amount of cracking is carried out on gas oils, a term that refers to the portion of crude petroleum boiling between the fuel oils (kerosene and/or stove oil) and the residuum. Residua are also cracked, but the processes are somewhat different from those used for gas oils.

Thus, thermal conversion processes are designed to increase the yield of lower-boiling products obtainable from petroleum either directly (by means of the production of gasoline components from higher-boiling feedstocks) or indirectly (by production of olefins and the like, which are precursors of the gasoline components). These processes may also be characterized by the physical state (liquid and/or vapor phase) in which the decomposition occurs. The state depends on the nature of the feedstock as well as conditions of pressure and temperature.

From the chemical viewpoint the products of cracking are very different from those obtained directly from crude petroleum—the products are nonindigenous to petroleum because they are created from petroleum by application of an external force (heat). When a 12-carbon atom hydrocarbon typical of straight-run gas oil is cracked or broken into two parts, one may be a 6-carbon paraffin hydrocarbon and the other a 6-carbon olefin hydrocarbon:

$$CH_3(CH_2)_{10}CH_3 \rightarrow CH_3(CH_2)_4CH_3 + CH_2{=}CH(CH_2)_3CH_3$$

The paraffin may be the same as is found in straight-run (distilled) gasoline, but the olefin is new. Furthermore, the paraffin has an octane number approaching 0 but the olefin has an octane number approaching 100. Hence naphtha formed by cracking (*cracked gasoline*) has a higher octane number than straight-run gasoline. In addition to a large variety of olefins, cracking produces high-octane aromatic and branched-chain hydrocarbons in higher proportions than are found in straight-run gasoline. Diolefins are produced but in relatively small amounts; they are undesirable in gasoline because they readily combine to form gum. The overall complexity of such a reaction is illustrated by the following equations in which the products are subject to the position of bond scission within the starting molecule:

$$CH_3(CH_2)_{10}CH_3 \rightarrow CH_3(CH_2)_8CH_3 + CH_2=CH_2$$

$$CH_3(CH_2)_{10}CH_3 \rightarrow CH_3(CH_2)_7CH_3 + CH_2=CHCH_3$$

$$CH_3(CH_2)_{10}CH_3 \rightarrow CH_3(CH_2)_6CH_3 + CH_2=CHCH_2CH_3$$

$$CH_3(CH_2)_{10}CH_3 \rightarrow CH_3(CH_2)_5CH_3 + CH_2=CH(CH_2)_2CH_3$$

$$CH_3(CH_2)_{10}CH_3 \rightarrow CH_3(CH_2)_4CH_3 + CH_2=CH(CH_2)_3CH_3$$

$$CH_3(CH_2)_{10}CH_3 \rightarrow CH_3(CH_2)_3CH_3 + CH_2=CH(CH_2)_4CH_3$$

$$CH_3(CH_2)_{10}CH_3 \rightarrow CH_3(CH_2)_2CH_3 + CH_2=CH(CH_2)_5CH_3$$

$$CH_3(CH_2)_{10}CH_3 \rightarrow CH_3CH_2CH_3 + CH_2=CH(CH_2)_6CH_3$$

$$CH_3(CH_2)_{10}CH_3 \rightarrow CH_3CH_3 + CH_2=CH(CH_2)_7CH_3$$

$$CH_3(CH_2)_{10}CH_3 \rightarrow CH_4 + CH_2=CH(CH_2)_8CH_3$$

Furthermore, the primary products (unless the reaction conditions are monitored carefully) will react further to yield secondary, tertiary, and even quaternary products.

The hydrocarbons with the least thermal stability are the paraffins, and the olefins produced by the cracking of paraffins are also reactive. Cycloparaffin derivatives (naphthene derivatives) are less easily cracked, their stability depending mainly on any side chains present, but ring splitting may occur, and dehydrogenation can lead to the formation of unsaturated naphthenes and aromatics. Aromatics are the most stable (*refractory*) hydrocarbons, the stability depending on the length and stability of side chains. Very severe thermal cracking of high-molecular-weight constituents can result in the production of excessive amounts of coke.

The higher-boiling oils produced by cracking are light and heavy gas oils as well as a residual oil, which in the case of thermal cracking is usually (erroneously) called tar and in the case of catalytic cracking is called *cracked fractionator bottoms*. The residual oil may be used as heavy fuel oil, and gas oils from catalytic cracking are suitable as domestic and industrial fuel oils or as diesel fuels if blended with straight-run gas oils. Gas oils from thermal cracking must be mixed with straight-run (distilled) gas oils before they become suitable for domestic fuel oils and diesel fuels.

The gas oils produced by cracking are an important source of gasoline and, in a once-through cracking operation, all of the cracked material is separated into products and may be used as such. However, the cracked gas oils are more resistant to cracking (more refractory) than straight-run gas oils but can still be cracked to produce more gasoline. This is done in a recycling operation in which the cracked gas oil is combined with fresh feed for another trip through the cracking unit. The operation may be repeated until the cracked gas oil is almost completely decomposed (*cracking to extinction*) by recycling (*recycling to extinction*) the higher-boiling product, but it is more usual to withdraw part of the cracked gas oil from the system according to the need for fuel oils. The extent to which recycling is carried out affects the amount or yield of cracked gasoline resulting from the process.

The gases formed by cracking are particularly important because of their chemical properties and their quantity. Only relatively small amounts of paraffinic gases are obtained from crude oil, and these are chemically inactive. Cracking produces both paraffinic gases (e.g., propane, C_3H_8) and olefinic gases (e.g., propene, C_3H_6); the latter are used in the refinery as the feed for polymerization plants where high-octane polymer gasoline is made. In some refineries, the gases are used to make alkylate, a high-octane component for aviation gasoline and for motor gasoline. In particular, the cracked gases are the starting points for many petrochemicals (Speight, 2014).

The importance of solvents in coking has been recognized for many years (Langer et al., 1961, 1962), but their effects have often been ascribed to hydrogen-donor reactions rather than phase behavior. The separation of the phases depends on the solvent characteristics of the liquid. Addition of aromatic solvents will suppress phase separation while paraffins will enhance separation. Microscopic examination of coke particles often shows evidence for the presence of a *mesophase*, which are spherical domains that exhibit the anisotropic optical characteristics of liquid crystal. This phenomenon is consistent with the formation of a second liquid phase; the mesophase liquid is denser than the rest of the hydrocarbon, has a higher surface tension, and likely wets metal surfaces better than the rest of the liquid phase. The mesophase characteristic of coke diminishes as the liquid phase becomes more compatible with the aromatic material (Speight, 1990, 2000).

Thermal cracking of higher-boiling materials to produce motor gasoline is now becoming an obsolete process, since the antiknock requirement of modern automobile engines has outstripped the ability of the thermal cracking process to supply an economical source of high-quality fuel. New units are rarely installed, but a few refineries still operate thermal cracking units built in previous years.

In summary, the cracking of petroleum constituents can be visualized as a series of simple thermal conversions (Chapter 5). The reactions involve the formation of transient highly reactive species that may react further in several ways to produce the observed product slate (Germain, 1969; Speight, 2000, 2014). Thus, even though chemistry and physics can be used to explain feedstock reactivity the main objective of feedstock evaluation (Chapter 2) is to allow a degree of predictability of feedstock behavior in thermal processes (Speight, 2000, 2014). And in such instances, chemical principles must be combined with engineering principles to understand feedstock processability and predictability of feedstock behavior. In the simplest sense, process planning can be built on an understanding of the following three parameter groups (Speight, 2015a): (1) feedstock properties, (2) process parameters, and (3) equipment parameters.

Feedstock properties such as carbon residue (potential coke formation), sulfur content (hydrogen needs for desulfurization), metallic constituents (catalyst rejuvenation), nitrogen content (catalyst rejuvenation), naphthenic or paraffinic character through the use of a characterization factor or similar indicator (potential for cracking in different ways to give different products), and, to a lesser extent, asphaltene content (coke formation) since this last parameter is related to several of the previous parameters.

Process parameters such as time–temperature–pressure relationships (distillate and coke yields), feedstock recycle ratio (distillate and coke yields plus overall conversion), and coke formation (lack of liquid production when liquids are the preferred products).

Equipment parameters such as batch operation, semicontinuous operation, or continuous operation (residence time and contact with the catalyst, if any), coke removal, and unit capacity that also dictates residence time.

However, it is not the purpose of this text to present the details of these three categories but they should be kept in mind when considering and deciding upon the potential utility of any process presented throughout this and subsequent chapters.

8.2   THERMAL CRACKING

As the demand for gasoline increased with the onset of automobile sales, the issue of how to produce more gasoline from less crude oil was solved in 1913 by the incorporation of cracking units into refinery operations in which fractions higher boiling than gasoline were converted into gasoline by thermal decomposition.

The origins of cracking are unknown. There are records that illustrate the use of naphtha in *Greek fire* almost 2000 years ago (Speight, 2014), but whether the naphtha was produced naturally by distillation or by cracking distillation is not clear. Cracking was used commercially in the production of oils from coal and oil shale before the beginning of the modern petroleum industry. From this, the discovery that the higher-boiling materials from petroleum could be decomposed to lower-molecular-weight products was used to increase the production of kerosene and was called *cracking distillation* (Kobe and McKetta, 1958).

The precise origins of the modern version of cracking distillation, as applied in the modern petroleum industry, are also unknown. However, it is essential to recognize that the production of volatile product by the destructive distillation of wood and coal was known for many years, if not decades or centuries, before the birth of the modern petroleum industry. Indeed, the production of *spirits of fire* (i.e., naphtha, the flammable constituent of *Greek fire*) was known from early times. The occurrence of bitumen at Hit (Mesopotamia) that was used as mastic by the Assyrians was further developed for use in warfare though the production of naphtha by destructive distillation.

At the beginning of the twentieth century, the yields of gasoline and kerosene fractions were usually markedly increased by means of *cracking distillation*, but the technique was not entirely suitable for gasoline production. As the need for gasoline arose, the necessity of prolonging the cracking process became apparent and led to a process known as *pressure cracking* which is a batch operation in which feedstock was heated to approximately 425°C (800°F) in stills (shell stills) especially reinforced to operate at pressures as high as 95 psi for 24 hours. Distillation was then started and, during the next 48–72 hours, a low-boiling distillate was obtained, which was treated with sulfuric acid to remove unstable gum-forming components (olefins and diolefins) and then redistilled to produce a naphtha (cracked gasoline, boiling range <205°C, <400°F) and residual fuel oil (Stephens and Spencer, 1956).

The Burton cracking process for the large-scale production of naphtha (cracked gasoline) was first used in 1912. The process employed batch distillation in horizontal shell stills and operated at approximately 400°C (ca. 750°F) and 75–95 psi and was the first successful method of converting higher-boiling feedstocks to gasoline. However, batch heating gas oil was considered inefficient, and during the years 1914–1922 a number of successful continuous cracking processes were developed. In these processes gas oil was continuously pumped through a unit that heated the gas oil to the required temperature, held it for a time under pressure, and then the cracked product was discharged into a distillation unit for separation into gases, gasoline, gas oil, and cracked residuum (often called *tar*).

The *tube and tank cracking process* is typical of the early continuous cracking processes. Gas oil, preheated by exchange with the hot products of cracking, was pumped into the cracking coil (up to several hundred feet long) that lined the inner walls of a furnace where oil or gas burners raised the temperature of the gas oil to 425°C (800°F). The hot gas oil passed from the cracking coil into a reaction chamber (soaker) where the gas oil was held under these temperature and pressure conditions until the cracking reactions to be completed. The cracking reactions formed coke, which over the course of several days filled the soaker. The gas oil stream was then switched to a second soaker, and drilling operations similar to those used in drilling an oil well cleaned out the first soaker. The cracked material (other than coke) left the onstream soaker to enter an evaporator (tar separator) maintained under a much lower pressure than the soaker, where, because of the lower pressure, all the cracked material except the tar became vaporized. The vapor left the top of the separator, where it was distilled into separate fractions: gases, gasoline, and gas oil. The tar that was deposited in the separator was pumped out for use as asphalt or as a heavy fuel oil.

Shortly thereafter, in 1921, a more advanced thermal cracking process which operated at 750°F–860°F (400°C–460°C) was developed (*Dubbs process*). In the process, a reduced crude (such as an atmospheric residuum or a topped crude oil) was the feedstock and the process also employed the concept of recycling in which the gas oil was combined with fresh (heavy) feedstock for further cracking. In a typical application of conventional thermal cracking (Figure 8.2),

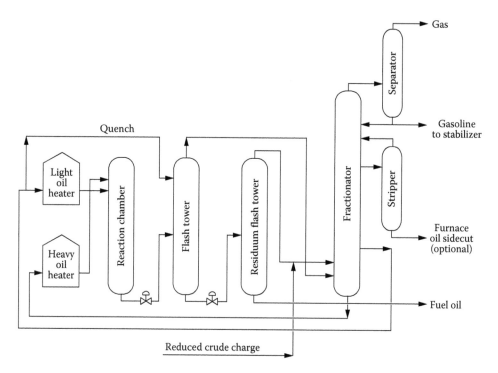

FIGURE 8.2   A thermal cracking unit.

the feedstock (reduced crude, i.e., residuum or flashed crude oil) is preheated by direct exchange with the cracked products in the fractionating columns. Cracked gasoline and middle distillate fractions were removed from the upper section of the column. Light and heavy distillate fractions were removed from the lower section and are pumped to separate heaters. Higher temperatures were used to crack the more refractory light distillate fraction. The streams from the heaters were combined and sent to a soaking chamber, where additional time is provided to complete the cracking reactions. The cracked products were then separated in a low-pressure flash chamber where a heavy fuel oil is removed as bottoms. The remaining cracked products were sent to the fractionating columns.

As refining technology evolved throughout the twentieth century, the feedstocks for cracking processes became the residuum or gas oil from a distillation unit. In addition, the residual oil produced as the end product of distillation processes, and even some of the higher-boiling crude oil constituents, often contains substantial amounts of asphaltic materials, which preclude use of the residuum as fuel oils or lubricating stocks (Speight, 2000, 2014; Speight and Ozum, 2002; Parkash, 2003; Hsu and Robinson, 2006; Gary et al., 2007).

However, subjecting these residua directly to thermal processes has become economically advantageous since, on the one hand, the end result is the production of lower-boiling products but, on the other hand, the asphaltene constituents and the resin constituents that are concentrated in residua are precursors to high yields of thermal coke (i.e., coke formed in noncatalytic processes) (Chapter 3). Although new thermal cracking units are now under development for heavy oil and tar sand bitumen (Speight, 2008, 2014), processes that can be regarded as having evolved from the original concept of thermal cracking are visbreaking and the various coking processes (Speight, 2014).

Low pressures (<100 psi) and temperatures in excess of 500°C (930°F) tend to produce lower-molecular-weight hydrocarbons than those produced at higher pressures (400–1000 psi) and at temperatures below 500°C (930°F). The reaction time is also important; light feeds (gas oils) and recycle oils require longer reaction times than the readily cracked heavy residues. Recycle of the light oil (middle distillate or fuel oil) fraction also affects the product slate of the thermal cracker

TABLE 8.1

Example of Thermal Cracking in Reduced Crude (Flasher Residuum)

Feedstock	
API gravity	25.0
IBP	227.0°C; 440.0°F
Cracking parameters	
Temperature	500.0°; 930.0°F
Soaker pressure	225.0 psig; 1550.0 kPa
Product yields, vol%	
Gasoline[a]	
Naphtha	57.5
Heating oil (light oil)	0.0
Residuum	37.5
API gravity	7.4
Heating oil[b]	
Gas	1.0
Naphtha	42.0
Heating oil (light oil)	23.0
Residuum	34.0
API gravity	8.0

[a] Light oil recycle.
[b] No recycle of light oil.

(Table 8.1). Mild cracking conditions (defined here as a low conversion per cycle) favor a high yield of gasoline components with low gas and coke production, but the gasoline quality is not high, whereas more severe conditions give increased gas and coke production and reduced gasoline yield (but of higher quality). With limited conversion per cycle, the heavier residues must be recycled. However, the recycled oils become increasingly refractory upon repeated cracking, and if they are not required as a fuel oil stock they may be subjected to a coking operation to increase gasoline yield or refined by means of a hydrogen process.

8.3   VISBREAKING

Visbreaking (*viscosity reduction, viscosity breaking*), a mild form of thermal cracking, was developed in the late 1930s to produce more desirable and valuable products (Speight and Ozum, 2002; Hsu and Robinson, 2006; Gary et al., 2007; Stell et al., 2009a,b; Carrillo and Corredor, 2013; Speight, 2014). The processes regarded as having evolved from the original concept of thermal cracking are visbreaking and the various coking processes (Table 8.2). The process is a relatively mild, liquid-phase thermal cracking process used to convert heavy, high viscosity feedstocks to lower viscosity fractions suitable for use in heavy fuel oil. This ultimately results in less production of fuel oil since less cutter stock (low viscosity diluent) is required for blending to meet fuel oil viscosity specifications. The cutter stock no longer required in fuel oil may then be used in more valuable products. A secondary benefit from the visbreaking operation is the production of gas oil and gasoline streams that usually have higher product values than the visbreaker charge. Visbreaking produces a small quantity of light hydrocarbon gases and a larger amount of gasoline and remains a process of promise for heavy feedstocks (Stark and Falkler, 2008; Stark et al., 2008).

TABLE 8.2

Summary of the Various Thermal Cracking Processes

Thermal Cracking

Prime purpose: conversion: prime purpose

Semicontinuous process

High conversion

Process configuration: various, depending on feedstock

Visbreaking

Prime purpose: viscosity reduction

Low conversion (10%) to products boiling less than 220°C (430°F)

Mild (470°C–495°C; 880°F–920°F) heating at pressures of 50–200 psi

Thermal reactions quenched before going to completion

Heated coil or drum (soaker)

Delayed Coking

Prime purpose: conversion to distillates

Complete conversion of the feedstock

Moderate–short residence time in hot zone (480°C–515°C; 900°F–960°F)

Pressures on the order of 90 psi

Thermal reactions allowed to proceed to completion out of hot zone

Soak drums (845°F–900°F) used in pairs (one onstream and one offstream for decoking)

Coke yield: 20%–40% w/w (dependent upon feedstock)

Fluid Coking

Prime purpose: conversion to distillates

Complete conversion of the feedstock

Severe–longer residence time in hot zone (480°C–565°C; 900°F–1050°F)

Pressure on the order of 10 psi

Thermal reactions allowed to proceed to completion in hot zone

Coke bed fluidized with steam; heat dissipated throughout the fluid bed

Higher yields of hydrocarbon gases ($<C_5$) than delayed coking

Less coke yield than delayed coking (for one particular feedstock)

The process can also be used as the first step in upgrading heavy feedstocks (Schucker, 2003). In such a process, the heavy feedstock is first thermally cracked using visbreaking or hydrovis-breaking technology to produce a product that is lower in molecular weight and boiling point than the feed. The product is then deasphalted using an alkane solvent at a solvent-to-feed ratio of less than 2 wherein separation of solvent and deasphalted oil from the asphaltenes is achieved through the use of a two-stage membrane separation system in which the second stage is a centrifugal membrane.

Visbreaking, unlike conventional thermal cracking, typically does not employ a recycle stream. Conditions are too mild to crack a gas oil recycle stream and the unconverted residual stream, if recycled, would cause excessive heater coking. The boiling range of the product residual stream is extended by visbreaking so that light and heavy gas oils can be fractionated from the product residual stream, if desired. In some present applications, the heavy gas oil stream is recycled and cracked to extinction in a separate higher-temperature heater with the production of products that are lower boiling than the original feedstock (Table 8.3) (Ballard et al., 1992; Speight and Ozum, 2002; Parkash, 2003; Negin and Van Tine, 2004; Hsu and Robinson, 2006; Gary et al., 2007; Speight, 2014). Low residence times are required to avoid polymerization and coking reactions, although additives can help to suppress coke deposits on the tubes of the furnace.

TABLE 8.3

Examples of Product Yields and Product Properties for Visbreaking Various Feedstocks

	Feedstock									
	Louisiana Vacuum Residuum	Arabian Light Atmospheric Residuum	Arabian Light Atmospheric Residuum	Arabian Light Atmospheric Residuum	Arabian Light Vacuum Residuum	Arabian Light Vacuum Residuum	Arabian Light Vacuum Residuum	Kuwait Atmospheric Residue	Iranian Light Vacuum Residue	Athabasca Bitumen
Feedstock										
Gravity, API	11.9	15.9	16.9	16.9	7.1	6.9	6.9	14.4	8.2	8.6
Carbon residue[a]	10.6	8.5			20.3			9.4	22.0	13.5
Sulfur, wt%	0.6	3.0	3.0	3.0	4.0	4.0	4.0	4.1	3.5	4.8
Product yields,[b] vol%										
Naphtha (<425°F; <220°C)	6.2	8.0	7.8	7.8	6.0	8.1	8.1	4.4	3.5	4.8
Light gas oil (425°F–645°F; 220°C–340°C)	6.3	15.0	11.9		16.0	10.5		16.9	13.1	21.0
Heavy gas oil (645°F–1000°F; 340°C–545°C)	15.0			70.8		20.8				35.0
Residuum	88.4	74.0	79.7	20.9	76.0	60.5	91.8	76.6	79.9	34.0
Gravity, API	11.4	13.5	14.7	1.3	3.5	0.8	7.2	11.0	5.5	
Carbon residue[a]	15.0									
Sulfur, wt%	0.6	3.5	3.2	5.0	4.7	4.6	4.0	4.4	3.8	

[a] Conradson.

[b] A blank product yield line indicates that the yield of the lower-boiling product has been included in the yield of the higher-boiling product.

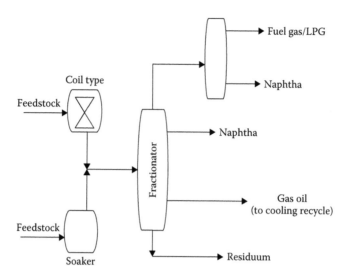

FIGURE 8.3 General representation of a visbreaking unit.

Visbreaking conditions range from 455°C to 510°C (850°F to 950°F) at a short residence time and from 50 to 300 psi at the heating coil outlet. It is the short residence time that brings to visbreaking the concept of being a mild thermal reaction in contrast to, for example, the delayed coking process where residence times are much longer and the thermal reactions are allowed to proceed to completion. The visbreaking process uses a quench operation to terminate the thermal reactions. Liquid-phase cracking takes place under these low-severity conditions to produce some naphtha, as well as material in the kerosene and gas oil boiling range. The gas oil may be used as additional feed for catalytic cracking units, or as heating oil.

There are two visbreaking processes, which are available commercially (Figure 8.3), the first process is the coil, or furnace, type and the second is the soaker type: (1) the coal visbreaking process and (2) the soaker visbreaking process. In the coil visbreaking unit (sometimes referred to as the *furnace visbreaking unit*), the cracking process occurs in the furnace tubes (or coils). Material exiting the furnace is quenched to halt the cracking reactions: frequently this is achieved by heat exchange with the virgin material being fed to the furnace, which in turn is a good energy efficiency step—a stream of cold oil (such as gas oil) is used to the same effect after which the gas oil is recovered and reused. The extent of the cracking reaction is controlled by regulation of the speed of flow of the oil through the furnace tubes. The quenched oil then passes to a fractionator where the products of the cracking (gas, LPG, gasoline, gas oil, and tar) are separated and recovered.

The *coil visbreaking process* (Figure 8.4) achieves conversion by high-temperature cracking within a dedicated soaking coil in the furnace. With conversion primarily achieved as a result of temperature and residence time, coil visbreaking is described as a high-temperature, short-residence-time route. The main advantage of the coil-type design is the two-zone fired heater that provides better control of the material being heated and, with the coil-type design, decoking of the heater tubes is accomplished more easily by the use of steam–air decoking.

In the alternative *soaker visbreaking process* (Figure 8.5), the bulk of the cracking reaction occurs not in the furnace but in a drum located after the furnace (the *soaker*) in which the heated feedstock is held at an elevated temperature for a predetermined period of time to allow cracking to occur before being quenched and then passed to a fractionator. In soaker visbreaking, lower temperatures are used than in coil visbreaking. Consequently, the soaker visbreaking process is described as a low-temperature, high-residence-time route. Product quality and yields from the coil and soaker drum design are essentially the same at a specified severity being independent of visbreaker configuration. By providing the residence time required to achieve the desired reaction,

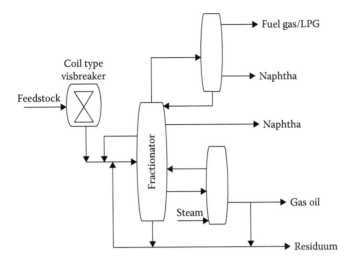

FIGURE 8.4 A coil-type visbreaker.

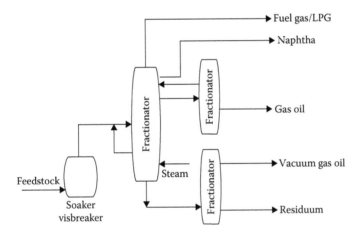

FIGURE 8.5 A soaker-type visbreaker.

the soaker drum design allows the heater to operate at a lower outlet temperature (thereby saving fuel) but there are disadvantages.

In the process, the heavy feedstock is passed through a furnace where it is heated to a temperature of 480°C (895°F) under an outlet pressure of approximately 100 psi. The cracked products are then passed into a flash-distillation chamber. The overhead material from this chamber is then fractionated to produce naphtha and light gas oil. The liquid products from the flash chamber are cooled with a gas oil flux and then sent to a vacuum fractionator. This yields a heavy gas oil distillate and a residuum of reduced viscosity—a 5%–10% v/v conversion of residuum to naphtha is usually sufficient to afford at least an approximate fivefold reduction in viscosity. Reduction in viscosity is also accompanied by a reduction in the pour point. An alternative option is to use lower furnace temperatures and longer times, achieved by installing a soaking drum between the furnace and the fractionator. The disadvantage of this approach is the need to remove coke from the soaking drum.

The higher heater outlet temperature specified for a coil visbreaker is an important advantage of coil visbreaking. The higher heater outlet temperature is used to recover significantly higher quantities of heavy visbroken gas oil. This capability cannot be achieved with a soaker visbreaker without the addition of a vacuum flasher. In terms of product yield, there is little difference between the two options (soaker visbreaker compared to coil visbreaker) approaches. However, each offers significant advantages in particular situations. For example, the cracking reaction forms coke as a by-product. In coil visbreaking, this lays down in the tubes of the furnace and will eventually lead to fouling or blocking of the tubes. The lower temperatures used in the soaker approach mean that these units use less fuel. In cases where a refinery buys fuel to support process operations, any savings in fuel consumption could be extremely valuable. In such cases, soaker visbreaking may be advantageous. In fact, most of the existing visbreaker are the soaker type, which utilize a soaker drum in conjunction with a fired heater to achieve conversion and which reduces the temperature required to achieve conversion while producing a stable residue product, thereby increasing the heater run length and reducing the frequency of unit shutdown for heater decoking.

Decoking is accomplished by a high-pressure water jet. First, the top and bottom heads of the coke drum are removed after which a hole is drilled in the coke from the top to the bottom of the vessel and a rotating stem is lowered through the hole, spraying a water jet sideways. The high-pressure jet cuts the coke into lumps, which fall out the bottom of the drum for subsequent loading into trucks or railcars for shipment to customers. Typically, coke drums operate on fixed cycles that depend upon the feedstock and the coking parameters. Cokers produce no liquid residue but yield up to 30% coke by weight. Much of the low-sulfur product is employed to produce electrodes for the electrolytic smelting of aluminum.

The main disadvantage is the decoking operation of the heater and soaker drum and, although decoking requirements of the soaker drum design are not as frequent as those of the coil-type design, the soaker design requires more equipment for coke removal and handling. The customary practice of removing coke from a drum is to cut it out with high-pressure water thereby producing a significant amount of coke-laden water that needs to be handled, filtered, and then recycled for use again.

The Shell soaker visbreaking process is suitable for the production of fuel oil by residuum (atmospheric residuum, vacuum residuum, or solvent deasphalter bottoms) and viscosity reduction with a maximum production of distillates. The basic configuration of the process includes the heater, soaker, and fractionator and more recently a vacuum flasher to recover more distillate products (Speight and Ozum, 2002; Hsu and Robinson, 2006; Gary et al., 2007; Speight, 2014). The cut point of the HGO stream taken from the vacuum flasher is approximately 520°C (970°F). In the process, the feedstock is preheated before entering the visbreaker heater, where the residue is heated to the required cracking temperature. Heater effluent is sent to the soaker drum where most of the thermal cracking and viscosity reduction takes place under controlled conditions. Soaker drum effluent is flashed and then quenched in the fractionator and the flashed vapors are fractionated into gas, naphtha, gas oil, and visbreaker residue. The visbreaker residue is steam stripped in the bottom of the fractionator and pumped through the cooling circuit for further processing. Visbreaker gas oil, which is recovered as a side stream, is steam stripped, cooled, and sent for further processing. As expected, product yields are dependent on feed type and product specifications. The heavy gas oil stream for the visbreaker can be used as feedstock for a thermal distillate cracking unit or for a catalytic cracker for the production of lower-boiling distillate products.

Other variations of visbreaking technology include the Tervahl-T and Tervahl-H processes. The Tervahl-T alternative (Figure 8.6) includes only the thermal section to produce a synthetic crude oil with better transportability by having reduced viscosity and greater stability. The Tervahl-H alternative adds hydrogen that also increases the extent of the desulfurization and decreases the carbon residua. The Aquaconversion process (Figure 8.7) is a hydrovisbreaking process that uses catalyst-activated transfer of hydrogen from water added to the feedstock. Reactions that lead to coke formation are suppressed and there is no separation of asphaltene-type material (Marzin et al., 1998).

Visbreaking conditions range from 455°C to 510°C (850°F to 950°F) at a short residence time and from 50 to 300 psi at the heating coil outlet. It is the short residence time that brings to visbreaking the concept of being a mild thermal reaction. This is in contrast to, for example, the delayed coking process where residence times are much longer and the thermal reactions are allowed to

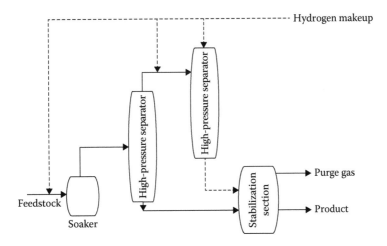

FIGURE 8.6   The Tervahl-T and Tervahl-H process configurations.

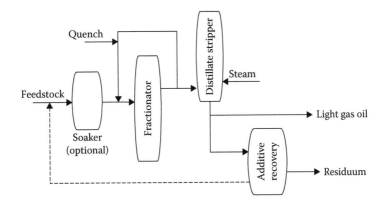

FIGURE 8.7   The aquaconversion process.

proceed to completion. The visbreaking process uses a quench operation to terminate the thermal reactions. Liquid-phase cracking takes place under these low-severity conditions to produce some naphtha, as well as material in the kerosene and gas oil boiling range. The gas oil may be used as additional feed for catalytic cracking units or as heating oil.

Atmospheric and vacuum residua are the usual feedstocks to a visbreaker although extra heavy oil and tar sand bitumen are also likely feedstocks. The heavy feedstocks will typically achieve a conversion to gas, naphtha, and gas oil in the order of 10%–50% w/w, depending on the severity and feedstock characteristics. The conversion of the residua to distillate (low-boiling products) is commonly used as a measurement of the severity of the visbreaking operation and the conversion is determined as the amount of 345°C$^+$ (650°F$^+$) material present in the atmospheric residuum or the 482°C$^+$ (900°F$^+$) material present in the vacuum residuum that is converted (*visbroken*) into lower-boiling components.

The extent of feedstock conversion is limited by a number of feedstock characteristics, such as asphaltene content (Figure 8.8) that varies with the type of feedstock and, hence the type of residuum, and, more particularly, carbon residue (Figure 8.9). In very general terms, paraffinic feedstocks will have a low heptane–asphaltene content (0%–8% by weight) whereas naphthenic feedstock will have a much higher heptane–asphaltene content (10%–20% by weight) with the mixed crude oils having intermediate values. Of course, when the heptane–asphaltenes are

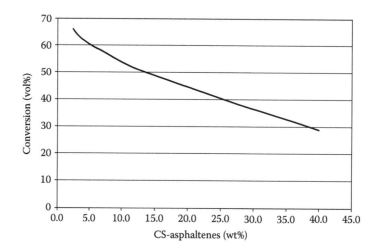

FIGURE 8.8   Relationship of visbreaker conversion to asphaltene content.

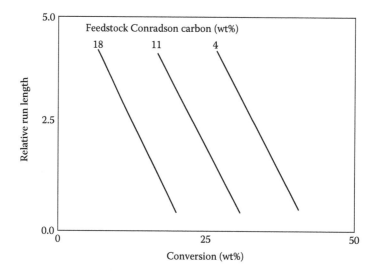

FIGURE 8.9   Relationship of visbreaker conversion time to feedstock carbon residue.

concentrated in the residua (through distillation), the proportions of the asphaltenes will be much higher. Thus, feedstocks with a high heptane–asphaltene content will result in an overall lower conversion than feedstocks with a lower heptane–asphaltene content while maintaining production of a stable fuel oil from the visbreaker bottoms. Minimizing the sodium content to almost a negligible amount and minimizing the Conradson carbon weight percent will result in longer cycle run lengths.

In addition, variations in feedstock quality will impact the level of conversion obtained at a specific severity. For example, for a given feedstock, as the severity is increased, the viscosity of the 205°C$^+$ (400°F$^+$) visbroken residue (often referred to as *visbroken tar* or *visbreaker tar*) initially decreases and then, at higher severity levels, increases dramatically, indicating the formation of coke precursors and their initial phase separation as sediment. The point at which this viscosity reversal occurs differs from feedstock to feedstock but can be estimated from the amount of low-molecular-weight hydrocarbon gases ($\leq C_3$) (Negin and Van Tine, 2004).

Thus, a crude oil residuum is passed through a furnace where it is heated to a temperature of 480°C (895°F) under an outlet pressure of approximately 100 psi. The heating coils in the furnace are arranged to provide a soaking section of low heat density, where the charge remains until the visbreaking reactions are completed. The cracked products are then passed into a flash-distillation chamber. It is advisable to maintain the flash zone temperature as low as possible to minimize the potential for coking. Under fixed flashing conditions, increasing the yield of the residuum will reduce this temperature.

The overhead material from this chamber is then fractionated to produce a low-quality gasoline as an overhead product and light gas oil as bottoms. The liquid products from the flash chamber are cooled with a gas oil flux and then sent to a vacuum fractionator. This yields a heavy gas oil distillate and a residuum of reduced viscosity. A quench oil may also be used to terminate the reactions and will also influence the temperature of the flash zone.

A 5%–10% v/v conversion of atmospheric residua to naphtha is usually sufficient to afford at least an approximate fivefold reduction in viscosity. Reduction in viscosity is also accompanied by a reduction in the pour point. However, the reduction in viscosity of distillation residua tends to reach a limiting value with conversion, although the total product viscosity can continue to decrease. The minimum viscosity of the unconverted residue can lie outside the range of allowable conversion if sediment begins to form. When shipment of the visbreaker product by pipeline is the process objective, addition of a diluent such as gas condensate can be used to achieve a further reduction in viscosity.

Conversion of residua in visbreaking follows first-order reaction kinetics. The minimum viscosity of the unconverted residue can lie outside the range of allowable conversion if sediment begins to form. When pipelining of the visbreaker product is the process objective, addition of a diluent such as gas condensate can be used to achieve a further reduction in viscosity.

Briefly, fouling (a deposit buildup in refinery processes that impedes heat transfer and/or reduces throughput) is the leading cause of diminished efficiency and productivity in refineries. The energy lost due to this inefficiency must be supplied by burning additional fuel or reducing feed. While most fouling is caused by the deposition of heavier hydrocarbon species coming directly from the crude oil, a small undetermined percentage is related to corrosion and scale deposits, either actively participating as loose corrosion products or by scale acting as a substrate for hydrocarbon deposition. Fouling will also occur in the drum of a soaker visbreaker, though the lower temperatures used in the soaker drum lead to fouling at a much slower rate. Coil visbreaking units therefore require frequent decoking. Soaker drums require far less frequent attention but their being taken out of service normally requires a complete halt to the operation.

Thus, the severity of the visbreaking operation is generally limited by the stability of the visbroken product—generally a prelude to the onset of fouling. If over-cracking occurs, the resulting fuel oil may form excessive deposits in storage or when used as a fuel in a furnace. The visbreaking correlations presented are based on operating to levels where the fuel oil quality will be limited by this test. This severity level is well within the operating limits that would be imposed by excessive coke formation in properly designed visbreaking furnaces.

The main limitation of the visbreaking process, and for that matter all thermal processes is that the products can be unstable. Thermal cracking at low pressure gives olefins, particularly in the naphtha fraction. These olefins give a very unstable product, which tends to undergo secondary reactions to form gum and intractable residua. Product stability of the visbreaker residue is a main concern in selecting the severity of the visbreaker operating conditions. Severity, or the degree of conversion, can cause phase separation of the fuel oil even after cutter stock blending. Increasing visbreaking severity and percent conversion will initially lead to a reduction in the visbroken fuel oil viscosity. However, visbroken fuel oil stability will decrease as the level of severity—and hence conversion—is increased beyond a certain point, dependent on feedstock characteristics.

The instability of the visbroken fuel oil is related to the asphaltene constituents and their thermal present in the residuum. Asphaltenes are heavy nonvolatile compounds that can be classified according to their solubility in various solvents. The asphaltene constituents can be thermally altered during visbreaking operations. In addition, during visbreaking, some of the high-molecular-weight constituents, including some of the asphaltene constituents, are converted to lower-boiling and medium-boiling paraffinic components, some of which are removed from the residuum. The asphaltenes and thermally altered, being unchanged, are thus concentrated in the product residuum (that may contain new paraffinic material) and if the extent of the visbreaking reaction is too high, the asphaltene constituents or altered asphaltene constituents will phase tend to precipitate in the product fuel oil, creating an unstable fuel oil.

A common method of measuring the amount of asphaltenes in petroleum is by addition of a low-boiling liquid hydrocarbon such as *n*-pentane (ASTM D 893, ASTM D2007) or *n*-heptane (ASTM D3279, ASTM D4124) by which treatment the asphaltenes separate as a solid (Speight, 2000, 2014). Since the amount of asphaltenes in the visbreaking unit charge residuum may limit the severity of the visbreaking operations, the *n*-pentane insoluble content or *n*-heptane insoluble content of the feedstock is used as the correlating parameter in various visbreaking correlations. However, these correlations can be visbreaker and feedstock dependent and application from one unit to another and one feedstock to another may be misleading.

Sulfur in the visbroken residuum can also be an issue since the sulfur content of the visbreaker residuum is often higher (approximately 0.5% w/w or greater) than the sulfur in the feedstock. Therefore, it can be difficult to meet the commercial sulfur specifications of the refinery product residual fuel oil, and blending with low-sulfur cutter stocks may be required.

Visbreaking, like thermal cracking, is a *first-order reaction*. However, due to the visbreaking severity limits imposed by fuel oil instability, operating conditions do not approach the level where secondary reactions, polymerization and condensation, occur to any significant extent. The first-order reaction rate equation altered to fit the visbreaking reaction is

$$K = \left(\frac{1}{t}\right)\left(\frac{\ln 100}{X_1}\right)$$

where
   K is the first-order reaction velocity constant, 1/s
   t is the time at thermal conversion conditions, s
   $X_1 = 900°F^+$ visbroken residuum yield, vol%

A simplified graphical representation of the yields of the various product with conversion can be constructed (Figure 8.10) with the understanding that different feedstocks will require different graphical presentations.

The visbreaking reaction is first order and velocity constant data as a function of visbreaking furnace outlet temperature can be presented graphically (Figure 8.11). The thermal conversion reactions are generally assumed to start at 425°C (800°F) although some visbreaking occurs below this temperature. Therefore, the residence time in the 425°C–450°C (800°F–865°F) reaction zone should be 613 seconds.

The central piece of *equipment* in any thermal process is, and visbreaking is with no exception, the heater. The heater must be adequate to efficiently supply the heat required to accomplish the desired degree of thermal conversion. A continually increasing temperature gradient designed to give most of the temperature increase in the front part of the heater tubes with only a slow rate of increase near the outlet is preferred. Precision control of time and temperature is usually not critical in the processes covered in this article. Usually, all that is required is to design to some target temperature range and then adjust actual operations to achieve the desired cracking. In the higher-temperature processes (e.g., ethylene manufacture), temperature control does become of prime importance due to equilibrium considerations.

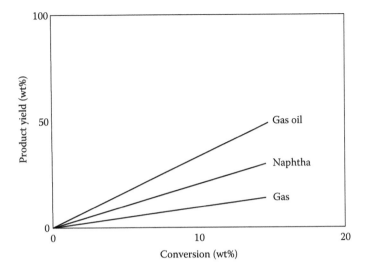

FIGURE 8.10 Trends for visbreaker product yields with feedstock conversion.

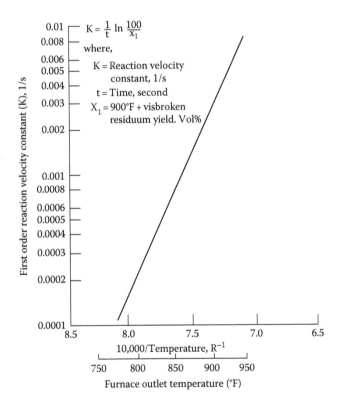

FIGURE 8.11 First-order reaction velocity constants for visbreaking.

Equipment design to minimize coke formation is of importance. The excessive production of coke adversely affects the thermal cracking process in the following ways: (1) reduces heat transfer rates, (2) increases pressure drops, (3) creates overheating, (4) reduces run time, and (5) requires the expense of removing the coke from the equipment. In addition, the metallurgy of the equipment, specifically the heater tubes and pumps, in the high temperature, corrosive environments must be

TABLE 8.4

Relationship of the Feedstock Flow Rate to the Tube Diameter

Total Charge Rate (Fresh Feed + Recycle) (bbl/Day)	Internal Diameter of Tube (inch)
3,000	2–3
6,000	3–4
12,000	4–41/2[a]

[a] Parallel tubes of smaller diameters would be preferable to one large tube. In this case, two 3 inch diameter parallel tubes may be preferred.

adequate to prevent expensive destruction and replacement of equipment. In the early days of thermal cracking, the metallurgy of the heater tubes was not of sufficient quality to permit extended periods of high temperatures. Modern improvements in the quality of steel have extended the durability of the thermal cracking equipment.

The advances in heater design have reached a point where very efficient furnace and heating tube arrangements can be built that give the refiner the desired thermal cracking operation. The practice of the refiner is generally to set the specifications the heater is expected to meet for the specific application and have a heater manufacturer prepare a suitable design. Proper *tube size* selection depends upon minimizing pressure drop while obtaining good turbulence for proper heat transfer that is also dependent upon the charge rate (Table 8.4) that ultimately affects the residence time and, therefore, the extent of the conversion.

The charge stock *liquid velocity* should be sufficient to provide enough turbulence to ensure a good rate of heat transfer and to minimize coking. A minimum linear cold 15.6°C (60°F) velocity on the order of 5 ft/s for a 100% liquid charge rate is considered to be sufficient. The maximum velocity would be limited to approximately 10 ft/s due to excessive pressure drop. The velocities at the higher cracking temperatures would, of course, be greater due to the partial vaporization of the charge.

Most of the *heat supplied* to the charge stock is radiant heat. The convection section of the heater is used primarily to supply preheat to the charge prior to the main heating in the radiant section. The *heat transfer rate* in the convection section will range from 3,000 to 10,000 Btu/ft^2 of tube outside area per hour with an average rate of 5,000 Btu/ft^2/h. The heating rates in the radiant section will range from 8,000 to 20,000 Btu/ft^2/h depending upon the charge stock, with heavier oil generally requiring the lower heating rate.

The heating tube *outlet temperature* will depend upon the charge stock being processed and the degree of thermal conversion required. The outlet temperature will vary from a minimum of 425°C (800°F) for visbreaking to a maximum of 595°C (1100°F) for thermal reforming. The combustion chamber temperature will range from 650°C to 870°C (1200°F to 1600°F) at a point approximately 1 ft below the radiant tubes. Flue gas temperatures are usually high (425°C–595°C, 800°F–1100°F) particularly since the heavy charge stock is usually entering the heater at a high temperature from a fractionating tower. An exception to the charge entering at a high temperature would be when charging gasoline to a thermal reformer. However, since thermal reforming requires high temperatures, flue gas temperatures will also be high.

Since it is desirable to maintain different temperature increase rates throughout the charge heating, that is, rapid increase at the beginning of the heating coil and a lower rate near the outlet, zone temperature control within the furnace is desired. A three-zone furnace is preferred with the first zone giving the greatest rate of temperature increase and the last zone the least.

Coke formation limits the operation of the heater, and techniques should be employed to minimize coke formation in the heater tubes. Coking occurs on the walls of the tubes, particularly where turbulence is low and temperature is high. Maintaining sufficient turbulence assists in limiting coke formation. Baffles within the tubes are sometimes used but water injection into the charge stream is the preferred method. Water is usually injected at the inlet although water also may be injected at additional points along the heater tubes. The water, in addition to providing turbulence in the heater tubes as it is vaporized to steam, also provides a means to control temperature. The optimum initial point of water injection into the heater tubes is at the point of incipient cracking where coke would start to form. An advantage to this injection point is the elimination of the additional pressure drop that would have been created by the presence of water between the heater inlet and the point of incipient cracking.

The preferred method to remove coke (decoke) from the heater tubes is to burn off the coke using a steam–air mixture. The heater tubes, therefore, should be capable of withstanding temperatures up to 760°C (1400°F) at low pressures for limited time periods. The heater tubes along with the tube supports should be designed to handle the thermal expansion extremes that would be encountered. Mechanical means, such as drills, can also be used to remove coke, but most modern heaters use the steam–air combustion technique. Parallel heaters may be employed so that one can be decoked while permitting cracking to proceed in the other heater(s).

The metallurgy of thermal cracking units is variable although alloy steel tubes of 7%–9% chromium are usually satisfactory to resist sulfur corrosion in thermal cracking heaters. If the hydrogen sulfide content of the cracked products exceeds 0.1 mol% in the cracking zone, a higher alloy steel may be required. Stabilized stainless steel, such as Type 321 or 347, would be suitable in this case. Other alloys, such as the Inconel or Incoloy alloys, could also be used. Seamless tubes with welded return bends are now normally used in heaters. Flanged return bends were used in earlier thermal cracking units to facilitate cleaning. However, use of steam air to burn out the coke essentially eliminates the need for flanged fittings that, in turn, reduces the possibility of dangerous leaks.

A useful tool to aid in the design and operation of thermal cracking units is the soaking volume factor (SVF). This factor combines time, temperature, and pressure of thermal cracking operations into a single numerical value. The SVF is defined as the *equivalent* coil volume in cubic feet per daily barrel of charge (fresh plus recycle) if the cracking reaction had occurred at 800°F and 750 psi:

$$SVF_{750psi/800F} = \frac{1}{F} RK_p dV$$

where
  $SVF_{750psi/800F}$ is the SVF at base reaction conditions of 750 psi gauge pressure and 800°F, cubic feet of coil volume per total charge throughput in barrels per day
  F is the charge (fresh plus recycle) throughput rate, barrels per day
  R is the ratio of reaction velocity constant at temperature Y and reaction velocity constant at 800°F
  K_p is the pressure correction factor for pressures other than 750 psi gauge
  dV is the incremental coil volume, cubic feet

The ratio of reaction velocity constants may be obtained from graphical formats (Figure 8.12) and should not be obtained from such plots of reaction velocity constants since there is a correction for the effect of temperature on the volume of the reacting material that needs to be corrected and incorporated into ion; this correction has been incorporated into the data (Figure 8.13). The pressure correction factor, K_p, may be obtained graphically (Figure 8.14).

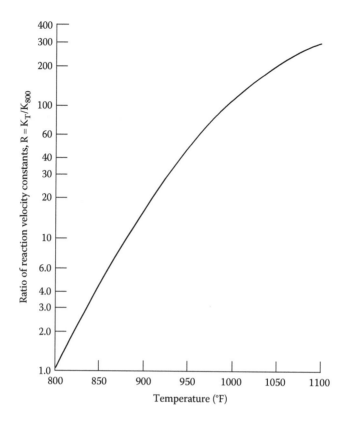

FIGURE 8.12   Change of visbreaker reaction velocity constant with temperature in excess of 425°C (800°F).

When an additional soaking drum is used, the SYF for the soaking drum should be added to the coil SVF. The SVF for the drum may be determined from:

$$SVF_D = \frac{DV}{F(K_{TD})(K_p)}$$

where
   SVF_D is the SVF of the drum
   DV is the volume of drum, ft^3
   F is the charge (fresh plus recycle) throughput rate, bbl/day
   K_{TD} is the reaction velocity constant for the mean drum temperature
   K_p is the pressure correction factor for the mean drum pressure

The SVF will range from 0.03 for visbreaking of heavy residual stocks to approximately 1.2 for light gas oil cracking. The SVF is a numerical expression of cracking rate and thus can be correlated with product yield and quality SVF may also be translated into cracking coils and still volumes of known dimensions under design conditions of temperature and pressure.

A cracking unit seldom operates very long at design conditions. Charge stock quality changes, desired product yields and qualities change, or additional capacity is required. These changes require a SVF that is different than the design SVF. The SVF may be varied by (1) varying pressure at constant temperature and feed rate; (2) varying temperature at constant feed rate, the pressure gradient varying with the effect upon cracking rate and fluid density in the cracking coil; and

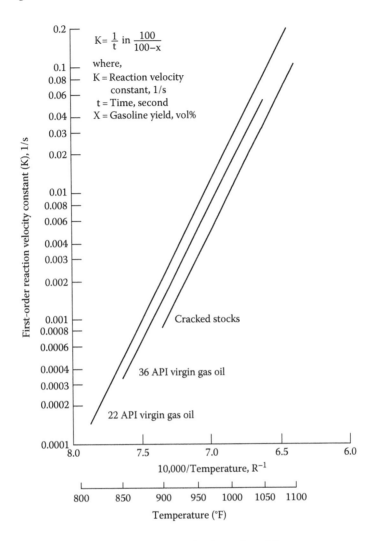

FIGURE 8.13   First-order reaction velocity constants for thermal cracking.

(3) varying the soaking volume at constant temperature and pressure by varying heater feed rate and/or varying the number of tubes in the section above 425°C (800°F).

With the advent of higher firing rate, better efficiency heaters, the use of external *soaking drums* to provide additional reaction times is of less importance in thermal cracking operations. In modern units the coil in the heater is usually sufficient to provide the temperature–time relationships required. A possible exception would be the case where it is desirable to crack a considerable amount of heavy residual stock. The temperature required probably could not be successfully obtained in a heater coil without excessive coking. A reaction chamber (soaking drum) is employed where the hotter, cleaner light gas oil is used to supply heat to the heavier dirty oil stream in a soaking drum. A low-temperature light gas oil stream is also frequently used to wet the walls of the soaking drum to minimize coking. Parallel soakers could be used to allow one to be decoked while the other is used for the cracking operations.

The *pumps* used in thermal cracking operations must be capable of operation for extended periods handling a high temperature (above 230°C, 450°F, and up to 345°C, 650°F) corrosive liquid. In addition, since coke particles are formed in thermal cracking, the pumps must be able to withstand the potential erosion of the metal parts by the coke particles. In the early days of thermal

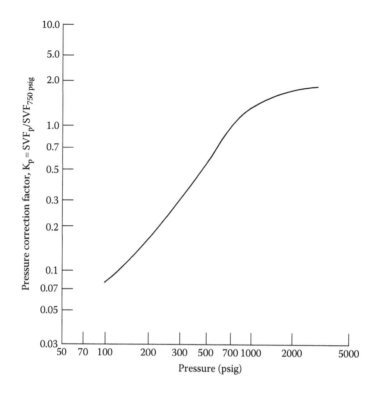

FIGURE 8.14 Pressure correction for the SVF with pressure in excess of 750 psig (5170 kPa).

cracking, reciprocating pumps were commonly used. In later units, centrifugal pumps have been used. A preferred centrifugal pump would be of the coke-crushing type or may have open impellors with case wear plates substituted for the front rings. The metal should be 12% chromium steel alloy or a higher alloy if serious corrosion is potential.

Heat exchangers should be constructed to provide easy cleaning since the high temperatures and coke particles can create extensive fouling of the exchangers. The downstream processing equipment (*flash drums, separators, fractionating towers*) are typically of standard design, and no special design specifications are required other than minimizing potential coke buildup. This can be accomplished by designing the equipment so there would be no significant holdup or dormant spots in the process equipment where coke could accumulate.

In thermal cracking operations there is a considerable amount of *excess heat that* cannot be economically utilized within the cracking unit itself. When a thermal cracking unit is being considered, it is desirable to construct the unit in conjunction with some other unit, such as a crude still, which could utilize the excess heat to preheat the crude oil charge. Alternately, the excess heat could be used in steam generation facilities.

Visbreaking may be the most underestimated and/or undervalued process in a refinery. The process may find rejuvenated use not only for heavy feedstock (including tar sand bitumen) but also for biofeedstocks. These visbreaking processes posses sufficient hardware flexibility to accommodate feedstock blending (petroleum feedstocks and biofeedstocks) and of the unit as well as a high measure of reliability and predictive operations/maintenance, thereby minimizing unplanned shutdowns.

The severity of visbreaker operation is generally limited by the stability requirement of the product as well as the extent of fouling and coke laydown in the visbreaker heater (Speight, 2015a). The former requirement means that the stability of the residue must be sufficient to ensure that the finished fuel resulting from blending with diluents (that are less aromatic than the residue) is stable and that asphaltene flocculation does not occur. Where the residue is converted to an emulsion, blend stability is improved and severity/conversion can be increased, subject to acceptable levels of heater fouling and coke deposition (Miles, 2009). Operational modifications, such as increasing steam injection or recycling heavy distillates from the visbreaker fractionator, may help mitigate coking tendency and enhance yield, while some relatively low-cost options to increase heater capacity might be implemented in certain instances.

In terms of processing biofeedstocks, many biofeedstocks have a high oxygen content and high mineral content which could (even when blended) disqualify the use of the biomaterial as a feedstock to a hydroprocessing unit. Refiners are very wary of high-oxygen and high-mineral feedstocks because of the increased hydrogen requirements (hydrogen is an expensive refinery commodity) to remove the oxygen from the hydrocarbon products with the appearance of the additional hydrogen as water. However, blending a biofeedstock with a resid as feedstock to a visbreaking unit to produce additional fuel products is a concept that could pay dividends and provide refineries with a source of fuels to supplement petroleum feedstocks. In the visbreaker, the feedstock is converted to overhead (volatile products) and coke (if the unit is operated beyond the typical operating point or coke-forming threshold) (Speight, 2014). The majority of the nitrogen, sulfur, and minerals appear in the coke. Oxygen often appears in the volatile product as water and carbon dioxide, unfortunately removing valuable hydrogen from the internal hydrogen management system.

Alternatively, another option is the preparation of a feedstock that is acceptable to a refinery. In particular, any process that reduces the mineral matter in the biofeedstock and reduces the oxygen content in the biofeed would be a benefit.

This can be accomplished by one or two preliminary treatment steps (such as the visbreaking process) in which the feedstock is demineralized and the oxygen constituents are removed as overhead (volatile) material giving the potential for the production of a fraction rich in oxygen functions that may be of some use to the chemical industry. Such a process might have to be established at a biofeedstock production site, unless the refinery has the means by which to accommodate the feedstock in an already existing unit.

In a manner similar to the visbreaking process where the biofeedstock is blended with a residuum, the biofeedstock alone would be heated in a visbreaker-type reactor (at a lower temperature than the conventional visbreaking temperature) to the point where hydrocarbons (or alcohols) are evolved and coke starts to form. As the coke forms, the mineral matter is deposited with the coke and the oxygen constituents are deoxygenized leaving a (predominantly) hydrocarbon product, which as a liquid that will ensure easy separation from the coke and mineral matter.

8.4 COKING

Coking is a thermal process for the continuous conversion of residua into lower-boiling products. The feedstock can be atmospheric residuum, vacuum residuum, or cracked residuum and the products are gases, naphtha, fuel oil, gas oil, and coke. Coking processes generally utilize longer reaction times than thermal cracking processes. To accomplish this, drums or chambers (reaction vessels) are employed, but it is necessary to use two or more such vessels so that coke removal can be accomplished in those vessels not onstream without interrupting the semicontinuous nature of the process. Gas oil may be the major product of a coking operation and serves primarily as a feedstock for catalytic cracking units. The coke can be used as fuel but processing for specialty uses, such as electrode manufacture, production of chemicals, and metallurgical coke, is also possible. For these latter uses, the coke may require treatment to remove sulfur and metal impurities—calcined petroleum coke can be used for making anodes for aluminum manufacture and a variety of carbon or graphite products such as brushes for electrical equipment.

Thus, coking is a thermal cracking-type operation used to convert low-grade feedstocks such as straight-run and cracked residua to coke, gas, and distillates. Two types of petroleum coking processes are presently operating: (1) delayed coking, which uses multiple coking chambers to permit continuous feed processing wherein one drum is making coke and one drum is being decoked; and (2) fluid coking, which is a fully continuous process where product coke can be withdrawn as a fluidized solid.

Crude oil residua obtained from the vacuum distillation tower as a nonvolatile (bottoms) fraction, heavy oil, and tar sand bitumen are the usual feedstocks to coking units. Atmospheric tower bottoms (long residua) may be charged to coking units but it is generally not attractive to thermally degrade the gas oil fraction contained in the longer residua. Other feedstocks to coking units are deasphalter bottoms (often referred to as *deasphalter pitch*) and tar sand bitumen, and cracked residua (thermal tars). The products are gases, naphtha, fuel oil, gas oil, and coke. The gas oil may be the major product of a coking operation and serves primarily as a feedstock for catalytic cracking units. The coke obtained is usually used as fuel, but processing for specialty uses, such as electrode manufacture, production of chemicals, and metallurgical coke, is also possible and increases the value of the coke. For these uses, the coke may require treatment to remove sulfur and metal impurities. Furthermore, the increasing attention paid to reducing atmospheric pollution has also served to direct some attention to coking, since the process not only concentrates such pollutants as feedstock sulfur in the coke but also usually yields products that can be conveniently subjected to desulfurization processes.

Coking processes have the virtue of eliminating the residue fraction of the feed, at the cost of forming a solid carbonaceous product. The yield of coke in a given coking process tends to be proportional to the carbon residue content of the feed (measured as the Conradson carbon residue) (Speight, 2014, 2015b). The data (Table 8.5) illustrate how the yield of coke from delayed and fluid coking varies with Conradson carbon residue of the feed. The formation of large quantities of coke is a severe drawback unless the coke can be put to use. Calcined petroleum coke can be used for making anodes for aluminum manufacture and a variety of carbon or graphite products such as brushes for electrical equipment. These applications, however, require a coke that is low in mineral matter and sulfur.

If the feedstock produces a high-sulfur, high-ash, high-vanadium coke, one option for use of the coke is combustion of the coke to produce process steam (and large quantities of sulfur dioxide unless the coke is first gasified or the combustion gases are scrubbed). Another option is stockpiling. For some feedstocks, particularly from heavy oil, the combination of poor coke properties for anode use, limits on sulfur dioxide emissions, and loss of liquid product volume have tended to relegate coking processes to a strictly secondary role in any new upgrading facilities.

8.4.1 Delayed Coking

Delayed coking is the oldest, most widely used process and has changed very little since the process was first brought onstream approximately 80 years ago. It is a semicontinuous (semibatch) process in which the heated charge is transferred to large coking (or soaking) drums that provide the long

TABLE 8.5
Relationship of Coke Yield to API Gravity

Carbon Residue (wt%)	°API	Coke Yield, Delayed Coker	Weight % Fluid Coker
1		0	
5	26	8.5	3
10	16	18	11.5
15	10	27.5	17
20	6	35.5	23
25	3.5	42	29
30	2		34.5
40	−2.5		46

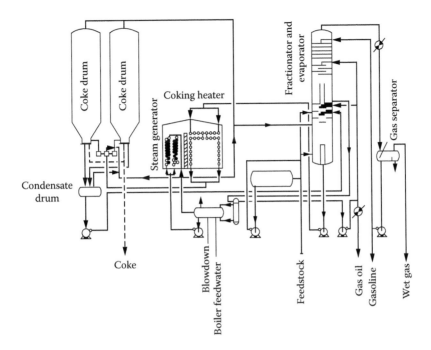

FIGURE 8.15 The delayed coking process.

residence time needed to allow the cracking reactions to proceed to completion (McKinney, 1992; Feintuch and Negin, 2004). The process (Figure 8.15) is widely used for treating residua and is particularly attractive when the green coke produced can be sold for anode or graphitic carbon manufacture or when there is no market for fuel oils. The process uses long reaction times in the liquid phase to convert the residue fraction of the feed to gases, distillates, and coke. The condensation reactions that give rise to the highly aromatic coke product also tend to retain sulfur, nitrogen, and metals, so that the coke is enriched in these elements relative to the feed.

In the process (Figure 8.15), the feedstock is charged to the fractionator and subsequently charged with an amount of recycle material (usually approximately 10%, but as much as 25%, of the total feedstock) from the coker fractionator through a preheater and then to one of a pair of coke drums; the heater outlet temperature varies from 480°C to 515°C (895°F to 960°F) to produce the various products (Table 8.6). The cracked products leave the drum as overheads to the fractionator and coke deposits form on the inner surface of the drum. The majority of the sulfur originally in the feedstock remains in the coke (Table 8.7). A pair of coke drums is used so that while one drum is on stream, the other is being cleaned allowing continuous processing and the drum operation cycle is typically 48 hours (Table 8.8). The temperature in the coke drum ranges from 415°C to 450°C (780°F to 840°F) at pressures from 15 to 90 psi (103 to 620 kPa).

Fractionators separate the overhead products from the coke drum into fuel gas (low-molecular-weight gases up to and including ethane), propane and propylene ($CH_3CH_2CH_3$–$CH_3CH=CH_2$), butane–butene ($CH_3CH_2CH_2CH_3$–$CH_3CH_2CH=CH_2$), naphtha, light gas oil, and heavy gas oil. Yields and product quality vary widely due to the broad range of feedstock used for coking units (Table 8.6).

Coker naphtha typically has a boiling range of up to 220°C (430°F), is olefinic, and must be upgraded by hydrogen processing for removal of olefins and sulfur. They are then used conventionally for reforming to gasoline or as chemical feedstocks. Middle distillates, boiling in the range of 220°C–360°C (430°F–680°F), are also hydrogen treated for improved storage stability, sulfur removal, and nitrogen reduction. They can then be used for either diesel or burner fuels or thermally processed to lower-boiling naphtha. The gas oil boiling up to approximately 510°C (950°F) end

TABLE 8.6
Examples of Product Yields and Product Properties for Delayed Coking Various Feedstocks

Feedstock	Louisiana Residuum	Kuwait Residuum	Kuwait Residuum[b]	West Texas Residuum	West Texas Residuum[b]	Oklahoma Residuum	Oklahoma Residuum	Alaska NS Residuum	California Residuum	California Residuum	Midcontinent Residuum	Middle East Residuum	Middle East Residuum	Venezuela Residuum
API gravity	12.3	6.7	16.1	8.9	15.2	13.0	16.8	7.4	12.0	12.0	12.3	7.4	8.2	2.6
Carbon residue[c]	13.0	19.8	9.1	17.8	9.3	14.1	8.0	18.1	9.4	9.6	11.3	20.0	15.6	23.3
Sulfur, wt%	0.7	5.2	0.7	3.0	0.6	1.2		2.0	1.6	1.6	0.4	4.2	3.4	4.4
Product yields, vol%														
Naphtha (95°F–925°F; 35°C–220°C)	22.8	26.7	22.0	28.9	20.1		10.7		22.5	15.7	16.0	12.6	17.4	10.0
Light gas oil[a] (425°F–645°F; 220°C–340°C)	18.4	28.0	41.9	16.5	31.7	20.4	36.5	36.5						
Heavy gas oil (645°F–1000°F; 340°C–540°C)	37.6	18.4	19.1	26.4	27.5	57.2	16.7	16.7	72.3		56.5	50.8	48.5	50.3
Coke	23.7	30.2	18.5	28.4	20.7	23.6	20.8	19.1	21.6	21.6	21.0	28.7	24.9	31.0
Sulfur, wt%	1.3	7.5	1.7	4.5	1.6									

Feedstock	North Africa Residuum	North Africa Residuum	SE Asia Residuum	Arkansas Residuum	Tia Juana Residuum	Alaska NS Residuum[d]	Alaska NS Residuum	Arabian Light Residuum	Arabian Light Residuum	Mexican Residuum	Santa Maria	Athabasca Bitumen
API gravity	15.2	12.8	17.1	15.3	8.5	7.4	7.4	6.9	16.9	4.0	7.2	7.3
Carbon residue[c]	16.7	5.2	11.1	11.5	22.0	18.1	18.1	22.0	3.0	22.0	14.8	17.9
Sulfur, wt%	0.7	0.6	0.5	2.8	2.9	2.0	2.0	2.9	4.0	5.3	6.7	5.3
Product yields, vol%												
Naphtha (95°F–925°F; 35°C–220°C)	19.9	18.5	20.4	13.5	25.6	15.0	12.5	19.1	14.2	21.4	22.4	20.3
Light gas oil[a] (425°F–645°F; 220°C–340°C)					26.4						16.2	
Heavy gas oil (645°F–1000°F; 340°C–540°C)	46.0	65.3	54.5	63.0	13.8	44.9	51.2	48.4	79.6	33.0	36.8	58.8
Coke	26.4	10.0	17.7	22.6	33.0	30.2	27.2	32.8	15.4	35.1	19.8	21.0
Sulfur, wt%						2.6	2.6	5.6	4.8			8.0

[a] A blank product line indicates that the yield of the lower-boiling product has been included in the yield of the higher-boiling product.
[b] Hydrodesulfurized.
[c] Conradson.
[d] 35 psig compared to ~14–18 psig for the oilier delayed cokers.

TABLE 8.7
Relationship of Feedstock Sulfur to Coke Sulfur

Feedstock	API Gravity	Sulfur in Feed (wt%)	Sulfur in Coke (wt%)	%S in Coke/%S in Feedstock
Elk Basin, WY, residuum	2.5	3.5	6.5	1.83
Hawkins, TX, residuum	4.5	4.5	7.0	1.55
Kuwait, residuum	6.0	5.37	10.8	2.01
Athabasca (Canada), bitumen	7.3	5.3	5.65	1.06
West Texas, residuum	—	3.5	3.06	0.875
Boscan (Venezuela), crude oil	10.0	5.0	5.0	1.0
East Texas, residuum	10.5	1.26	2.57	2.04
Texas Panhandle, residuum	18.9	0.6	0.6	1.00

TABLE 8.8
Typical Time Cycle of Coke Drums in a Delayed Coker

Operation	Time (Hours)
Coking	24
Decoking	24
Switch drums	0.5
Steam, cool	6.0
Drain, unhead decoke	7.0
Reheat, warm-up	9.0
Spare time, contingency	1.5

point may be charged to a fluid catalytic cracking unit immediately or after hydrogen upgrading when low sulfur is a requirement.

As noted earlier (Table 8.8), the coke drums are on a 48-hour cycle. The coke drum is usually onstream for approximately 24 hours before becoming filled with porous coke after which time the coke is removed by the following procedure: (1) the coke deposit is cooled with water; (2) one of the heads of the coking drum is removed to permit the drilling of a hole through the center of the deposit; and (3) a hydraulic cutting device, which uses multiple high-pressure water jets, is inserted into the hole and the wet coke is removed from the drum. Typically, 24 hours is required to complete the cleaning operation and to prepare the coke drum for subsequent use onstream (Table 8.8).

A well-designed delayed coker will have an operating efficiency of better than 95%, although delayed coking units are generally scheduled for shutdown for cleaning and repairs on a 12- to 18-month schedule, depending on the most economical cycle for the refinery. In terms of process efficiency, the feedstock *heater* and the *coke drums* are the most critical parts of the delayed coking process. The function of the heater or furnace is to preheat the charge quickly, to avoid preliminary decomposition, to the required temperature. Since coking is endothermic, the furnace outlet temperature must be approximately 55°C (100°F) higher than the coke drum temperature to provide the necessary process heat. The heater run length is a function of coke laydown in heater tubes, and careful design is necessary to avoid premature shutdown with cycle lengths preferably at least I year. When the charge stock is derived from crude distillation, double desalting is desirable since salt deposits will shorten heater cycles.

The heater for a delayed coking unit does not require as broad and operating range as a thermal cracking or visbreaking heater where both contact time and temperature can be varied to achieve the desired level of conversion. The coker heater must reach a fixed outlet temperature for the required coke drum temperatures. Thus, the coker heater requires a short residence time, high radiant heat flux, and good control of heat distribution.

The function of the coke drum is to provide the residence time required for the coking reactions to proceed to completion and to accumulate the coke. In sizing coke drums, a superficial vapor velocity in the range of 0.3–0.5 ft/s is used and coke drums with heights of 97 ft (30 m) have been constructed and approach a practical limit for hydraulic coke cutting. Drum diameters up to 26 ft (8 m) have been commonly used, and larger drums are feasible for efficient processing. Various types of level detectors are used to permit drum filling to within 7–8 ft (2–2.5 m) of the upper tangent line of the drum monitor coke height in the drum during onstream service.

Hydraulic cutters are used to remove coke from the drum and the first step is to bore a vertical pilot hole through the coke after which cutting heads with horizontally directed nozzles then undercut the coke and drop it out of the bottom of the drum. Hydraulic pressures in the range of 3000–3600 psi are used in the 26 ft diameter coking drums.

In regard to the process parameters and product yields, an increase in the coking temperature (1) decreases coke production, (2) increases liquid yield, and (3) increases gas oil end point. On the other hand, increasing pressure and/or recycle ratio (1) increases gas yield, (2) increases coke yield, (3) decreases liquid yield, and (4) decreases gas oil end point. As an example, increasing the pressure from the currently designed 15–35 psi (Table 8.6) causes the higher-boiling products to remain in the hot zone longer causing further decomposition and an increase in yield of the naphtha fraction, a decrease in the yield of the middle distillate–gas oil fraction, and an increase in the yield of coke.

In the past many delayed coking units were designed to provide complete conversion of atmospheric residue to naphtha, kerosene, and other low-boiling products. However, some units have been designed to minimize coke and produce heavy coker gas oil (HCGO) that is catalytically upgraded. The yield slate for a delayed coker can be varied to meet a refiner's objectives through the selection of operating parameters. Furthermore, delayed coking has an increasingly important role to play in the integration of modem petroleum refineries because of the inherent flexibility of the process to handle even the heaviest of residues. The flexibility of operation inherent in delayed coking permits refiners to process a wide variety of crude oils including those containing heavy, high-sulfur residua.

Low-pressure coking is a process designed for a once-through, low-pressure operation. The process is similar to delayed coking except that recycling is not usually practiced and the coke chamber operating conditions are 435°C (815°F), 25 psi. Excessive coking is inhibited by the addition of water to the feedstock in order to quench and restrict further reactions of the reactive intermediates.

High-temperature coking is a semicontinuous process designed to convert asphaltic residua to gas oil and coke as the primary products. In the process, the feedstock is transported to the heater (370°C, 700°F, 30 psi) and finally to the coking unit, where temperatures may be as high as 980°C–1095°C (1800°F–2000°F). Volatile materials are fractionated, and after the cycle is complete, coke is collected for sulfur removal before storage.

Delayed coking is likely to remain the workhorse of thermal cracking processes for the foreseeable future. Online spalling, decoking techniques have been developed, based on successful, similar application on delayed coker heaters. Heater operation (in delayed coking units and in visbreaker units) is improved by online spalling of the heater pipes. When an online pipe is to be spalled, flow is diverted to the offline pipe allowing for full operation of the coker heater. In another embodiment, a thermal transfer resistant zone plate is movably mounted in the radiant section of the coker heater. By moving the zone plate from an operating position to a spalling position and adjusting the temperature of the plurality of burners, the temperature of the pipes in the zone of the heater radiant section to be spalled can be lowered, while the temperature in the remaining zones of the heater radiant section is fully operational.

The delayed coking process will remain a preferred residue upgrading option because of its ability to handle the heaviest, contaminated crudes. Globally, approximately one-third of installed residue upgrading plant is by delayed coking. Although a mature process, in recent years many developments have taken place, including (1) development of automated coke drum unheeding devices, allowing the operator to carry out the decoking procedure safely from a remote location; (2) understanding of process parameters affecting yields, coker product qualities, and coke qualities (e.g., shot coke); and (3) design and operation of major equipment items, in particular coke drums (allowing shorter coking cycles) and the delayed coker heater (online spalling/decoking and minimization of coking in furnace tubes).

Considering the need for expanding heavy feedstock processing (Speight and Ozum, 2002; Parkash, 2003; Hsu and Robinson, 2006; Gary et al., 2007; Speight, 2014), there will be a need to incorporate operational flexibility into the original design of the delayed coker and make a few key equipment choices with long-term goals in mind. In recent projects and licensing proposals, refineries are also incorporating unique long-term rationales into their designs (Wodnik and Hughes, 2005). For example, some locations intentionally leave plot space and specify design criteria around the coker to allow for easy placement of an additional pair of coke drums with minimal debottlenecking of existing assets. This decision is based upon future plans to construct either another sour-crude train or other projects to make more coker feedstock available from existing refinery units. These units utilize the benefit of being online and generating earnings to help pay for the future expansion projects. Incorporating distillate recycle in the processing scheme or at least designing the coker so that recycle technology can be added later at minimal cost is another design strategy possibility.

8.4.2 Fluid Coking

Throughout the history of the refining industry, with only short-term exceptions, there has been a considerable economic driving force for upgrading residua. This has led to the development of processes to reduce residua yields such as thermal cracking, visbreaking, delayed coking, vacuum distillation, and deasphalting. The process is also valuable for conversion of heavy oil, extra heavy oil, and tar sand bitumen to distillates (Speight and Ozum, 2002; Parkash, 2003; Hsu and Robinson, 2006; Gary et al., 2007; Speight, 2014). In the process, the hot feedstock is sprayed on to a fluidized bed of hot, fine coke particles, which permits the coking reactions to be conducted at higher temperatures and shorter contact times that can be employed in delayed coking. These conditions result in decreased yields of coke; greater quantities of more valuable liquid product are recovered in the fluid coking process.

As a brief history, in the late 1940s and early 1950s, there was a large incentive to develop a continuous process to convert heavy vacuum residua into lighter, more valuable products. During this period, fluid coking using the principle of fluidized solids was developed and contact coking, using the principle of a moving solid bed, was also developed and the first commercial fluid coker went onstream in late 1954. During the late 1960s, environmental considerations indicated that, in many areas, it would no longer be possible to utilize high-sulfur coke as a boiler fuel. This and other environmental considerations resulted in the development of Flexicoking to convert the coke product from a fluid coker into clean fuel. The first commercial Flexicoking unit went onstream in 1976.

Fluid coking (Figure 8.16) is a continuous process that uses the fluidized solids technique to convert residua, including vacuum residua and cracked residua, to more valuable products (Table 8.9) (Roundtree, 1997). This coking process allows improvement in the yield of distillates by reducing the residence time of the cracked vapors and also allows simplified handling of the coke product. Heat for the process is supplied by partial combustion of the coke with the remaining coke being drawn as product. The new coke is deposited in a thin fresh layer (ca. 0.005 mm, 5 μm) on the outside surface of the circulating coke particle, giving an onion skin effect.

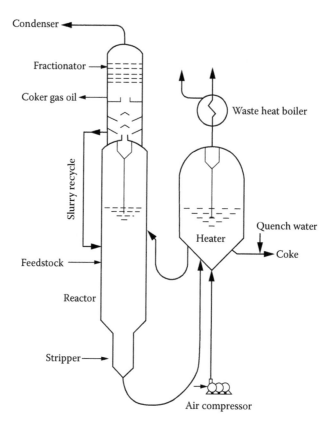

FIGURE 8.16 The fluid coking process.

The equipment for the fluid coking process is similar to that used in fluid catalytic cracking (Chapter 9) and follows comparable design concepts except that the fluidized coke solids replace catalyst. Small particles of coke made in the process circulate in a fluidized state between the vessels and are the heat transfer medium and, thus, the process requires no high-temperature preheat furnace.

Fluid coking uses two vessels, a reactor and a burner; coke particles are circulated between these to transfer heat (generated by burning a portion of the coke) to the reactor (Figure 8.16) (Blaser, 1992). The reactor holds a bed of fluidized coke particles, and steam is introduced at the bottom of the reactor to fluidize the bed. The feed coming from the bottom of a vacuum tower at, for example, 260°C–370°C (500°F–700°F) is injected directly into the reactor. The temperature in the coking vessel ranges from 480°C to 565°C (900°F to 1050°F), with short residence times of the order of 15–30 seconds, and the pressure is substantially atmospheric so the incoming feed is partly vaporized and partly deposited on the fluidized coke particles. The material on the particle surface then cracks and vaporizes, leaving a residue that dries to form coke. The vapor products pass through cyclones that remove most of the entrained coke.

Vapor products leave the bed and pass through cyclones that are necessary for removal of the entrained coke. The cyclones discharge the vapor into the bottom of a scrubber and any coke dust remaining after passage through the cyclones is scrubbed out with a pump-around stream and the products are cooled to condense the heavy tar. The resulting slurry is recycled to the reactor. The scrubber overhead vapors are sent to a fractionator where they are separated into wet gas, naphtha, and various gas oil fractions. The wet gas is compressed and further fractionated into the desired components.

TABLE 8.9
Examples of Product Yields and Product Properties for Fluid Coking Various Feedstocks

Feedstock	LA Basin Vacuum Residuum	LA Basin Vishbreaker Residuum	LA Basin Deasphalter Bottoms	Texas Vacuum Residuum	Kuwait Vacuum Residuum	Louisiana Atmospheric Residuum	Louisiana Vacuum Residuum	Hawkins Vacuum Residuum	Middle East Vacuum Residuum	Tia Juana Vacuum Residuum
API gravity	6.7	−5.0	−1.0	17.3	5.6	17.8	11.6	4.2	5.1	8.5
Carbon residue[a]	17.0	41.0	33.0	11.0	21.8	5.0	13.0	24.5	21.4	22.0
Sulfur, wt%	2.1	2.1	2.3	0.7	5.5	0.5	0.6	4.3	3.4	2.9
Product yields,[b] vol%										
Naphtha (95°F–425°F; 35°C–220°C)	17.0	14.0	18.0	21.0	21.0	17.0	21.0	19.5	15.4	20.7
Gas oil (425°F–1000°F; 220°C–540°C)	62.0	32.0	45.0	69.0	48.0	74.0	61.0	53.0	55.1	48.3
Coke, wt%	21.0	48.0	36.0	12.0	28.0	8.0	17.0	27.5	26.4	20.0

Feedstock	Tia Juana Vacuum Residuum	Arabian Light Vacuum Residuum	Arabian Heavy Vacuum Residuum	Arabian Heavy Vacuum Residuum	Arabian Heavy Vacuum Residuum	Iranian Heavy Vacuum Residuum	Bachaquero Vacuum Residuum	Bachaquero Vacuum Residuum	Zaca Vacuum Residuum	West Texas Bitumen
API gravity	7.9	6.5	3.2	4.4	3.3	5.1	2.6	2.6	4.7	−0.2
Carbon residue[a]	23.3	19.2	28.5	24.4	27.8	21.4	21.4	26.5	19.0	34.0
Sulfur, wt%	3.0	4.3	5.6	5.3	6.0	3.4	3.7	3.7	7.8	4.6
Product yields,[b] vol%										
Naphtha (95°F–425°F; 35°C–220°C)	15.1	15.5	14.4	15.0	17.0	15.4	14.7	14.8	20.5	13.4
Gas oil (425°F–1000°F; 220°C–540°C)	52.2	58.0	37.5	47.7	45.0	55.1	48.3	47.7	61.0	36.9
Coke, wt%	29.2	23.4	35.2	30.4	34.0	26.4	32.9	33.2	17.5	43.9

[a] Conradson.
[b] A blank product yield line indicates that the yield of the lower-boiling product has been included in the yield of the higher-boiling product.

In the reactor, the coke particles flow down through the vessel into the stripping zone. The stripped coke then flows down a standpipe and through a slide valve that controls the reactor bed level. A riser carries the *cold coke* to the burner. Air is introduced to the burner to burn part of the coke to provide reactor heat. The hot coke from the burner flows down a standpipe through a slide valve that controls coke flow and thus the reactor bed temperature. A riser carries the hot coke to the top of the reactor bed. Combustion products from the burner bed pass through two stages of cyclones to recover coke fines and return them to the burner bed.

Coke is withdrawn from the burner to keep the solids inventory constant. To aid in keeping the coke from becoming too coarse, large particles are selectively removed as product in a quench elutriator drum and coke fines are returned to the burner. The product coke is quenched with water in the quench elutriator drum and pneumatically transported to storage. A simple jet attrition system in the reactor provides additional seed coke to maintain a constant particle size within the system.

Due to the higher thermal cracking severity used in fluid coking compared to delayed coking, the products are somewhat more olefinic than the products from delayed coking. In general, products are handled for upgrading in a comparable manner from both coking processes. Currently, delayed coking and fluid coking have been the processes of choice for conversion of Athabasca bitumen to liquid products for more than four decades (Figure 8.17) (Speight, 1990, 2000, 2014). Both processes are termed the *primary conversion processes* for the tar sand plants in Ft. McMurray, Alberta, Canada. The unstable liquid product streams are hydrotreated before recombining to the synthetic crude oil.

Coke, being a product of the process, must be withdrawn from the system to keep the solids inventory from increasing. The net coke produced is removed from the burner bed through a quench elutriator drum, where water is added for cooling and cooled coke is withdrawn and sent to storage. During the course of the coking reaction, the particles tend to grow in size. The size of the coke particles remaining in the system is controlled by a grinding system within the reactor.

The coke product from the fluidized process is a laminated sphere with an average particle size of 0.17–0.22 mm (170–220 µm), readily handled by fluid transport techniques. It is much harder and denser than delayed coke, and in general is not as desirable for manufacturing formed products.

The yields of products are determined by the feed properties, the temperature of the fluid bed, and the residence time in the bed. The lower limit on operating temperature is set by the behavior of the fluidized coke particles. If the conversion to coke and light ends is too slow, the coke particles agglomerate in the reactor, a condition known as *bogging*. The use of a fluidized bed reduces

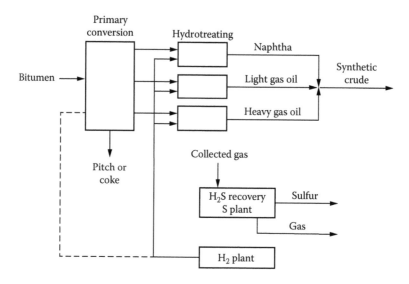

FIGURE 8.17 Processing sequence for tar sand bitumen.

the residence time of the vapor-phase products in comparison to delayed coking, which in turn reduces cracking reactions. The yield of coke is thereby reduced, and the yield of gas oil and olefins increased. An increase of 5°C (9°F) in the operating temperature of the fluid-bed reactor typically increases gas yield by 1% w/w and naphtha by approximately 1% w/w.

The disadvantage of burning the coke to generate process heat is that sulfur from the coke is liberated as sulfur dioxide (SO_2). The gas stream from the coke burner also contains carbon monoxide (CO), carbon dioxide (CO_2), and nitrogen (N_2). An alternate approach is to use a coke gasifier to convert the carbonaceous solids to a mixture of carbon monoxide (CO), carbon dioxide (CO_2), and hydrogen (H_2).

The liquid products from the coker can, following cleanup via commercially available gas oil hydrodesulfurization technology (Chapter 10), provide large quantities of low-sulfur fuel (<0.2% by weight sulfur). The incentive for fluid coking or Flexicoking increases relative to alternate processing, such as direct hydroprocessing, as feedstock quality (Conradson carbon, metals, sulfur, nitrogen, etc.) decreases. Changes in yields and product quality result from a change from a low cut point, high reactor temperature operation, to a high cut point operation with a lower reactor temperature (Table 8.10).

Fluid coke is used in electrodes for aluminum manufacture, in silicon carbide manufacture, in ore sintering operations, and as fuel. The coke from a feedstock containing a large amount of contaminants may not be suitable for these uses, either from a product contamination or environmental standpoint. The Flexicoking process overcomes this problem by converting part of the gross coke to a gas that can be burned in process furnaces and boilers. The coke fines from a Flexicoking unit contain most of the metals in the feedstock and may be suitable for metals recovery.

The fluid coking processes can be used to produce a high yield of low-sulfur fuel oil as well as to completely eliminate residual fuel and asphalt from the refinery product slate (Table 8.11). The different distributions are obtained by varying the fluid coker/Flexicoker operating conditions and changing the downstream processing of the coker reactor products. In fact, there are many *process variations* that can be used to adapt the process to particular refining situations. Once-through or partial recycle coking can be used where there is a small market for heavy fuel oil or where a quantity of high-sulfur material can be blended into the fuel oil pool.

In reference to the process parameters, the *reactor temperature* is normally set at 510°C–540°C (950°F–1000°F). Low temperature favors high liquid yields and reduces the unsaturation of the gas but increases the reactor holdup requirements. The burner temperature is normally 55°C–110°C (100°F–200°F) above the reactor temperature. Regulating the amount of coke sent to the reactor from the burner controls the reactor temperature. Burner temperature is controlled by the air rate to the burner.

Low pressure provides maximum gas oil recycle cut point, minimizes steam requirements, and reduces air blower horsepower. Reactor pressure normally adjusts to the gas compressor suction pressure but is higher due to the pressure drop through the piping, the condenser, the fractionation tower, and the reactor cyclone. The unit pressure balance required for coke circulation is normally controlled at a fixed differential pressure relative to the reactor sets the burner pressure. *Reactor coke level* is controlled by the cold coke slide valve on the transfer line from the reactor to the burner and *burner coke level* is controlled by the coke withdrawal rate.

In all coking processes, *product yields* are a function of feed properties, the severity of the operation, and recycle cut point (Table 8.12). Severity is a function of time and temperature since low severity and high gas oil cut point favor high liquid yields whereas high severity and low gas oil cut point increase coke and gas yields. Data from these sources indicate that the gross coke yield is directly related to feedstock Conradson carbon residue. Coke quality (Table 8.13) and gas quality (Table 8.14) are also important.

In most cases, high liquid yield and minimum coke and gas yields are required and, in theory, two cracking rates should be considered. The first is the rate at which the liquid cracks and vaporizes after initially laying down on the coke particles that determines the reactor holdup. The second is that vapor-phase cracking determines the distribution of the products between gas, naphtha, and

TABLE 8.10

Flexibility of Operations in Fluid Coking Units Allows Changes to Product Quality

Feed characteristics

Conradson carbon, wt%	15.5
Gravity, °API	6.4
LV below 1000°F, %	8.0
Sulfur, wt%	2.6
Nitrogen, wt%	1.0
Nickel, ppm	283
Vanadium, ppm	126

Yields	Low Reactor Temperature, High Cut Point Maximum Gas Oil		High Reactor Temperature, Low Cut Point Low Metals Gas Oil	
	wt%	LV%	wt%	LV%
Hydrogen sulfide	0.5		0.7	
Hydrogen	0.1		0.2	
C_1–C_3	8.0		9.6	
C_4	1.6	2.8	2.0	3.5
C_5 to 215°F (ASTM)	4.2	6.2	5.1	7.5
215°F–400°F	8.6	11.4	10.4	13.9
400°F to end point	58.4	62.5	51.8	56.3
Gross coke	18.5		20.2	
Net coke	10.0		10.6	
Product qualities				
C_5 to 215 naphtha				
Gravity, °API	71.6		71.6	
Sulfur, wt%	0.41		0.43	
Nitrogen, wt%	0.009		0.009	
215°F–400°F				
Gravity	52.3		52.2	
Sulfur, wt%	0.66		0.69	
Nitrogen, wt%	0.036		0.025	
Gas oil				
Gravity, °API	16.9		18.4	
Sulfur, wt%	2.36		2.28	
Nitrogen, wt%	0.73		0.63	
Nickel, ppm	2.7		0.43	
Vanadium, ppm	0.5		0.05	
Conradson carbon, wt%	2.3		1.0	
Aniline point, °F	140		110	
Coke				
Sulfur, wt%	3.4		3.4	
Nickel, ppm	1520		1400	
Vanadium, ppm	680		620	

gas oil. The vapor residence time can be determined from the reactor volume and the volume flow of hydrocarbon vapor and steam, and can be divided into time in the fluid bed and time in the disperse phase. The former is a function of the coke holdup or weight space velocity (W/H/W) that is normally expressed as reciprocal hours. For maximum liquid yield, the secondary cracking time should be kept at a minimum and, thus, it is normally desirable to design the unit for the maximum operable W/H/W.

TABLE 8.11
Effect of Flexicoking on Product Yields

	Crude Composition	Flexicoking Gas Oil Hydrodesulfurization	Flexicoking Gas Oil Hydrodesulfurization and Cat. Cracking
Yields, LV% crude			
Gas, FOEB	3	7	12
Naphtha	16	23	47
Middle distillate	26	31	35
Gas oil (LSFO)	25	34	
Residuum	30		
Total	100	95	94

TABLE 8.12
Examples of Product Yields and Product Properties for Flexicoking Various Feedstocks

Vacuum Residuum Properties	Arabian Light	Iranian Heavy	Arabian Heavy	Bachaquero	West Texas Sour Asphalt
Gravity, ° API	6.5	5.1	4.4	2.6	−0.2
Conradson carbon, wt%	19.2	21.4	24.4	26.5	34.0
Sulfur, wt%	4.29	3.43	5.34	3.66	4.6
Nitrogen, wt%	0.34	0.77	0.41	0.81	0.65
V + Ni, ppm	90	525	225	1040	137
Flexicoking yields on vacuum residuum					
C_3 gas, wt%	9.8	9.9	10.7	10.6	11.3
C_4 saturates, wt%	0.6	0.6	0.6	0.7	0.7
C_4 unsaturates, wt%	1.3	1.3	1.4	1.4	1.5
C_5–360°F naphtha					
wt%	11.2	11.0	10.6	10.4	9.2
LV%	15.5	15.4	15.0	14.8	13.4
360°F–975°F gas oil,					
wt%	53.7	50.8	46.3	43.7	33.4
LV%	58.0	55.1	50.4	47.7	36.9
Gross coke, wt%	23.4	26.4	30.4	33.2	43.9
Purge coke, wt%	1.1	1.2	1.4	1.5	2.0
Coke gas, FOE vol%	13.1	15.5	18.3	21.3	30.0

The maximum rate at which feed can be injected into a fluid coker is limited by a condition known as *bogging*. The conditions required to avoid a bogging are as follows: (1) the feedstock must be uniformly distributed over the entire surface of the heat transfer medium; (2) the layer of feed material on the particles should not be too great; the thickness of the sticky plastic layer depends on the specific flow rate of feedstock, its coking factor, and the recirculation rate of the heat transfer medium; (3) the bed temperature and the initial temperature of the heat transfer medium should be sufficiently high that the first stage of the process is completed in a short time; and (4) the heat transfer medium should not consist of particles that are too fine. The heat reserve of the granules should be sufficient to cover the entire energy requirements in connection with heating the feedstock, supplying the energy for the endothermic cracking reaction, and evaporating the decomposition products. If the

TABLE 8.13

Representative Properties of Fluid Coke and Flexicoking Coke

	Flexicoke	Fluid Coke
Bulk density, lb/ft^3	50	60
Particle density, lb/ft^3	85	95
Surface area, m^2/g	70	<12
Average particle size, μm	120	170–40
Sulfur, wt%	2.0	6.0

TABLE 8.14

Representative Gas Composition from Flexicoking Operations

	After Particulate Removal	After Sulfur Removal
H$_2$S, vppm	7100	<10
COS, vppm	150	<15
NH$_3$, vppm[a]	<3	<3
HCN, vppm	<3	Nil
Solids, lb/Mscf	0.0042	Nil
Sulfur, wt% FOE basis	9.7	<0.04

[a] Below detectable limits of 3 vppm.

feed injection rate exceeds the vaporization rate for an extended period of time, the thickness of the tacky oil film on the particles will increase until the particles rapidly agglomerate, causing the bed to lose fluidity. When fluidization is lost, the heat transfer rate is greatly reduced, further aggravating the condition. Coke circulation cannot be maintained due to the loss of reactor fluidization.

For comparative purposes because of the similarity of the processes, there are some notable differences between the operation of a fluid coker and a fluid catalytic cracking unit (FCCU) and some of these differences tend to make the fluid coker easier to operate. The fluid coker heat balance is very easy to maintain, as there is always an excess of carbon to burn whereas a fluid catalytic cracking unit has a sensitive interaction between heat balance and intensity balance, and therefore between carbon burned and carbon produced, which complicates control, especially during operating changes, start-up, and shutdown.

In addition, recovery from upsets caused by loss of utilities such as steam and air is normally easier and faster with a fluid coker than with a fluid catalytic cracking unit. The fluid coker normally operates well at low feed rates and turndown to low rate is normally limited by the ability of the tower to maintain fractionation of the products. The fluid coker proper can operate at any feed rate that will provide enough coke to heat balance.

However, the fluid coker has some inherent features that can create problems if proper precautions are not followed. The heavy residuum can set up if the lines are not properly heat traced and insulated. Low reactor temperature results in reactor bogging. If the particle size of the circulating coke is not properly controlled, the size can grow to the point that coke circulation problems are encountered. The feed nozzles must be maintained and occasionally cleaned to prevent poor feed distribution followed by excessive agglomerate formation. Control of the reactor bed level is critical since an excessively high bed level will flood the reactor cyclone and allow coke to be carried to the scrubber where it will plug the heavy oil circuits.

Along similar lines to the fluid coking process, the *rapid thermal processing* (RTP process, now the HTL process or heavy-to-light upgrading technology) was developed by Ivanhoe Energy Inc. in the 1980s. The process uses a circulating transport bed of hot sand to rapidly induce thermal cracking of the heavy feedstock in the absence of air to produce a light synthetic crude oil (Veith, 2007; Koshka et al., 2008; Silverman et al., 2011).

8.4.3 Flexicoking

Flexicoking is a direct descendent of fluid coking (Figure 8.16) and uses the same configuration as the fluid coker but includes a gasification section in which excess coke can be gasified to produce refinery fuel gas (Figure 8.18) (Roundtree, 1997; Marano, 2003). The Flexicoking process was designed during the late 1960s and the 1970s as a means by which excess coke could be reduced in view of the gradual incursion of the heavier feedstocks in refinery operations. Such feedstocks are notorious for producing high yields of coke (>15% by weight) in thermal and catalytic operations.

In the process, the heavy feedstock enters the scrubber for direct contact heat exchange with the overhead product vapors from the reactor. The higher-boiling products (>525°C/>975°F) present in the overhead condense in the scrubber and return to the reactor as a recycle stream with fresh feedstock. Lower-boiling overhead constituents in the scrubber go to a conventional fractionator and also to light-ends recovery. The feedstock is thermally cracked in the reactor fluidized bed to a range of gas and liquid products, and coke. The coke inventory is maintained by circulating the bed coke from the reactor to the heater via the cold coke transfer line. In the heater, the coke is heated by the gasifier products and circulated back to the reactor via the hot coke transfer line to supply the heat that sustains the thermal cracking process.

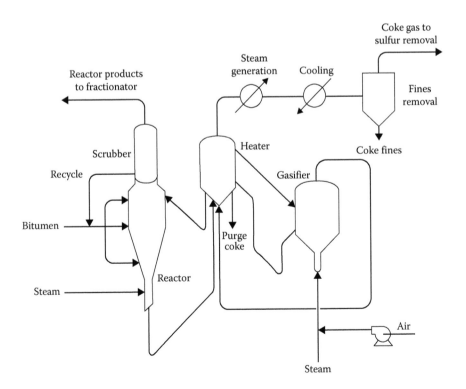

FIGURE 8.18 The Flexicoking process.

Excess coke is converted to a low-heating value gas in a fluid-bed gasifier with steam and air. The air is supplied to the gasifier to maintain temperatures of 830°C–1000°C (1525°F–1830°F) but is insufficient to burn all of the coke. The gasifier products, consisting of a gas and coke mixture, return to the heater to heat up the coke. The gas exits the heater overhead and goes to steam generation, to dry/wet particulate removal, and to desulfurization. The clean flexi-gas is then ready for use as fuel in refinery boilers and furnaces and/or for steam and power generation. Approximately 95% w/w/ of the coke generated in the reactor is converted in the process. Only a small amount of product coke is collected as fines from the flexi-gas and purged from the heater to extract feed metals.

A typical gas product, after removal of hydrogen sulfide, contains carbon monoxide (CO, 18%), carbon dioxide (CO_2, 10%), hydrogen (H_2, 15%), nitrogen (N_2, 51%), water (H_2O, 5%), and methane (CH_4, 1%). The heater is located between the reactor and the gasifier, and it serves to transfer heat between the two vessels. The heater temperature is controlled by the rate of coke circulation between the heater and the gasifier. Adjusting the air rate to the gasifier controls the unit inventory of coke and the gasifier temperature is controlled by steam injection to the gasifier.

Yields of liquid products from Flexicoking are the same as from fluid coking, because the coking reactor is unaltered. The main drawback of gasification is the requirement for a large additional reactor, especially if high conversion of the coke is required. Units are designed to gasify 60%–97% w/w of the coke from the reactor. Even with the gasifier, the product coke will contain more sulfur than the feed, which limits the attractiveness of even the most advance of coking processes.

The Flexicoking process produces a clean fuel gas with a heating value of approximately 90 Btu/ft³ or higher. The coke gasification can be controlled to burn approximately 95% of the coke to maximize production of coke gas or at a reduced level to produce both gas and a coke that has been desulfurized by approximately 65%. This flexibility permits adjustment for coke market conditions over a considerable range of feedstock properties. Fluid coke is currently being used in power plant boilers.

Fluid coking and Flexicoking are versatile processes that are applicable to a wide range of heavy feedstocks and provide a variety of products. The feedstock should have a carbon residue in excess of 5% w/w and there is no upper limit on the carbon residue. Suitable feedstocks include vacuum residua, asphalt, tar sand bitumen, and visbreaker residuum.

8.5   OTHER PROCESSES

Typically, in terms of upgrading tar sand bitumen, the bitumen was processed (on the surface) by delayed coking or fluid coking to produce a synthetic crude oil (Figure 8.17) (Speight, 2013, 2014). In a variation of the *bitumen upgrading process*, bitumen from the separation plant is first sent to an atmospheric distillation tower where the resulting products are naphtha, light gas oil, heavy gas oil, and residue (NRCAN, 2015). Each of the naphtha, light gas oil, and heavy gas oil streams is then sent to its own separate hydrotreater to remove sulfur and nitrogen by adding hydrogen. The resulting products from each of the separate hydrotreaters are then combined to produce synthetic crude oil. The residuum from the atmospheric distillation tower is further separated into two streams, heavy vacuum gas oil and vacuum residue. The heavy vacuum gas oil is mixed with the heavy gas oil from the atmospheric distillation tower to undergo hydrotreating. The vacuum residuum is further upgraded either in a hydroconverter or in a coker to produce lighter liquid product and by-product coke (in the coker). The overhead product from the hydroconverter or coker is further separated into naphtha, light gas oil, and heavy gas oil by distillation. Each of these three streams is mixed with the corresponding liquid product from the atmospheric distillation tower and then sent to the separate hydrotreaters to remove sulfur and nitrogen by adding hydrogen. The final products from the upgrading facility are synthetic crude oil and coke.

The *decarbonizing* thermal process is designed to minimize coke and gasoline yields but, at the same time, to produce maximum yields of gas oil. The process is essentially the same as the delayed coking process, but lower temperatures and pressures are employed. For example, pressures range from 10 to 25 psi, heater outlet temperatures may be 485°C (905°F), and coke drum temperatures may be of the order of 415°C (780°F). Decarbonizing in this sense of the term should not be confused with *propane decarbonizing*, which is essentially a solvent deasphalting process (Chapter 12).

Low-pressure coking is a process designed for a once-through, low-pressure operation. The process is similar to delayed coking except that recycling is not usually practiced and the coke chamber operating conditions are 435°C (815°F), 25 psi. Excessive coking is inhibited by the addition of water to the feedstock in order to quench and restrict the reactions of the reactive intermediates.

High-temperature coking is a semicontinuous thermal conversion process designed for high-melting asphaltic residua that yields coke and gas oil as the primary products. The coke may be treated to remove sulfur to produce a low-sulfur coke (≤5%), even though the feedstock contained as much as 5% w/w sulfur. In the process, the feedstock is transported to the pitch accumulator, then to the heater (370°C, 700°F, 30 psi), and finally to the coke oven, where temperatures may be as high as 980°C–1095°C (1800°F–2000°F). Volatile materials are fractionated, and after the cycle is complete, coke is collected for sulfur removal and quenching before storage.

Mixed-phase cracking (also called *liquid-phase cracking*) is a continuous thermal decomposition process for the conversion of heavy feedstocks to products boiling in the gasoline range. The process generally employs rapid heating of the feedstock (kerosene, gas oil, reduced crude, or even whole crude), after which it is passed to a reaction chamber and then to a separator where the vapors are cooled. Overhead products from the flash chamber are fractionated to gasoline components and recycle stock, and flash chamber bottoms are withdrawn as a heavy fuel oil. Coke formation, which may be considerable at the process temperatures (400°C–480°C, 750°F–900°F), is minimized by use of pressures in excess of 350 psi.

Vapor-phase cracking is a high-temperature (545°C–595°C, 1000°F–1100°F), low-pressure (<50 psi) thermal conversion process that favors dehydrogenation of feedstock (gaseous hydrocarbons to gas oils) components to olefins and aromatics. Coke is often deposited in heater tubes, causing shutdowns. Relatively large reactors are required for these units.

Selective thermal cracking is a thermal conversion process that utilizes different conditions depending on the nature of the feedstock. For example, a heavy oil may be cracked at 494°C–515°C (920°F–960°F) and 300–500 psi; a lighter gas oil may be cracked at 510°C–530°C (950°F–990°F) and 500–700 psi (Figure 8.19). Each feedstock has its own particular characteristics that dictate the

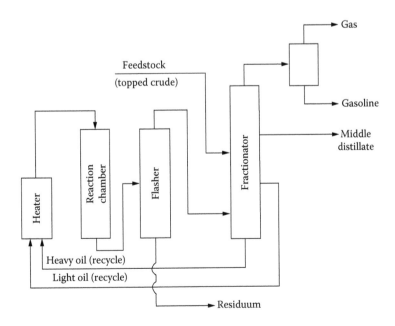

FIGURE 8.19   Selective thermal cracking.

optimum conditions of temperature and pressure for maximum yields of the products. These factors are utilized in selective combination of cracking units in which the more refractory feedstocks are cracked for longer periods of time or at higher temperatures than the less stable feedstocks, which are cracked at lower temperatures.

The process eliminates the accumulation of stable low-boiling material in the recycle stock and also minimizes coke formation from high-temperature cracking of the higher-boiling material. The end result is the production of fairly high yields of gasoline, middle distillates, and olefin gases.

The thermal cracking of naphtha involves the upgrading of low-octane fractions of catalytic naphtha to higher quality material. The process is designed, in fact, to upgrade the heavier portions of naphtha, which contain virgin feedstock, and to remove naphthenes, as well as paraffins. Some heavy aromatics are produced by condensation reactions, and substantial quantities of olefins occur in the product streams.

8.6   OPTIONS FOR HEAVY FEEDSTOCKS

The limitations of processing the more complex difficult-to-convert heavy oil, residua, and tar bitumen depend to a large extent on the amount of nonvolatile higher-molecular-weight constituents, which also contain the majority of the heteroatoms (i.e., nitrogen, oxygen, sulfur, and metals such as nickel and vanadium) (Chapter 1). The chemistry of the thermal reactions of some of these constituents dictates that certain reactions, once initiated, cannot be reversed and proceed to completion and coke is the eventual product (Chapter 3) (Speight and Ozum, 2002; Hsu and Robinson, 2006; Gary et al., 2007; Speight, 2014).

Upgrading residua, which are similar in character to some heavy oils and tar sand bitumen, began with the introduction of desulfurization processes that were designed to reduce the sulfur content of residua as well as some heavy crude oils and products therefrom. In the early days, the goal was desulfurization but, in later years, the processes were adapted to a 10%–30% partial conversion operation, as intended to achieve desulfurization and obtain low-boiling fractions simultaneously, by increasing severity in operating conditions.

Refinery evolution has seen the introduction of a variety of heavy feedstock cracking processes (some use catalysts and are, of necessity, included here). These processes are different from one another in cracking method, cracked product patterns, and product properties and will be employed in refineries according to their respective features.

8.6.1   Aquaconversion Process

The *Aquaconversion process* is a hydrovisbreaking technology that uses catalyst-activated transfer of hydrogen from water added to the feedstock. Reactions that lead to coke formation are suppressed and there is no separation of asphaltene-type material (Marzin et al., 1998; Speight and Ozum, 2002; Gary et al., 2007; Speight, 2014). The important aspect of the Aquaconversion technology is that it does not produce any solid by-product such as coke, nor requires any hydrogen source or high-pressure equipment. In addition, the Aquaconversion process can be implanted in the production area, and thus the need for external diluent and its transport over large distances is eliminated.

8.6.2   Asphalt Coking Technology (ASCOT) Process

The *ASCOT process* (asphalt coking technology process) is a residual oil upgrading process that integrates the delayed coking process and the deep solvent deasphalting process (low energy deasphalting [LEDA]) (Bonilla and Elliot, 1987). Removing the deasphalted oil fraction prior to application of the delayed coking process has two benefits: (1) in the coking process this fraction is thermally cracked to extinction, degrading this material as an FCC feedstock, and (2) thermally cracking this material to extinction results in conversion of a significant portion to coke.

In the process, the vacuum residuum is brought to the desired extraction temperature and then sent to the extractor where the solvent (straight-run naphtha, coker naphtha) flows upward and extracts soluble material from the down-flowing feedstock. The solvent-deasphalted phase leaves the top of the extractor and flows to the solvent recovery system where the solvent is separated from the deasphalted oil and recycled to the extractor. The deasphalted oil is sent to the delayed coker where it is combined with the heavy coker gas oil from the coker fractionator and sent to the heavy coker gas oil stripper where low-boiling hydrocarbons are stripped off and returned to the fractionator. The stripped deasphalted oil/heavy coker gas oil mixture is removed from the bottom of the stripper and used to provide heat to the naphtha stabilizer–reboiler before being sent to battery limits as a cracking stock. The raffinate phase containing the asphalt and some solvent flows at a controlled rate from the bottom of the extractor and is charged directly to the coking section.

The solvent contained in the asphalt and deasphalted oil is condensed in the fractionator overhead condensers, where it can be recovered and used as lean oil for a propane/butane recovery in the absorber, eliminating the need to recirculate lean oil from the naphtha stabilizer. The solvent introduced in the coker heater and coke drums results in a significant reduction in the partial pressure of asphalt feed, compared with a regular delayed coking unit. The low asphalt partial pressure results in low coke and high liquid yields in the coking reaction.

With the ASCOT process, there is a significant reduction in by-product fuel as compared to either solvent deasphalting or delayed coking and the process can be tailored to process a specific quantity or process to a specific quality of cracking stock (Speight and Ozum, 2002; Hsu and Robinson, 2006; Gary et al., 2007; Speight, 2014).

8.6.3 CHERRY-P PROCESS

The Cherry-P process (comprehensive heavy ends reforming refinery process) is a process for the conversion of heavy crude oil or residuum into distillate and a cracked residuum. In this process, the principal aim is to upgrade heavy petroleum residues at conditions between those of conventional visbreaking and delayed coking. Although coal is added to the feedstock, it is not intended to be a coprocessing feedstock but the coal is intended to act as a scavenger to prevent the buildup of coke on the reactor wall. The use of scavengers in the process is projected to increase (Stark and Falkler, 2008; Stark et al., 2008; Speight, 2014).

In the process, the feedstock is mixed with coal powder in a slurry mixing vessel (without a catalyst or hydrogen), heated in the furnace, and fed to the reactor where the feedstock undergoes thermal cracking reactions for several hours at a temperature higher than 400°C–450°C (750°F–840°F) and under pressure (70–290 psi) with a residence time on the order of 1–5 hours. Gas and distillate from the reactor are sent to a fractionator and the cracked residuum is extracted out of the system after distilling low-boiling fractions by the flash drum and vacuum flasher to adjust its softening point. Distillable product yields of 44% by weight on total feed are reported (Speight and Ozum, 2002; Hsu and Robinson, 2006; Gary et al., 2007; Speight, 2014). Since this yield is obtained when using anthracite, the proportion that is derived from the coal is likely to be very low and unlikely to cause compatibility reactions in downstream reactors due to the presence of phenols and other polar species (Speight, 1990, 2014).

8.6.4 CONTINUOUS COKING PROCESS

A new coking process that can accept heavy feedstock and continuously discharge vapor and dry petroleum coke particles has been developed (Sullivan, 2011). The process promotes a rapid recovery of volatiles from the resid enabling recovery of more volatiles. It also causes the carbonization reactions to proceed more rapidly, and it produces uniform composition and uniform size of coke particles that have a low volatiles content.

The new process uses a kneading and mixing action to continuously expose new resid surface to the vapor space and causes a more complete removal of volatiles from the produced petroleum coke. Not only are more valuable volatiles recovered, the volatiles are likely to be richer in middle distillates. As a result of kneading/mixing action by the reactor/devolatilizer, new surfaces of the residuum mass are continuously exposed to the gas phase, enhancing the rapid mass transfer of volatiles into the gas phase. The volatiles are then rapidly cooled to retard degradation. With the rapid reduction of volatiles content in the resid mass, the carbonization reaction rates are accelerated, enabling continuous and rapid production of solid petroleum coke particles. The short contact time of the volatiles with the hot residuum minimizes thermal degradation of volatiles.

Concurrently with the carbonization reactions and the formation of coke, some cracking of side chains off the larger molecules likely occurs. These smaller, low-boiling molecules produced from cracking reactions join the population of the indigenous volatiles. Some volatiles may be generated even after the solid coke is formed. In the delayed coking process, many of these late forming volatiles remain trapped in the coke. The process promotes the release of these late forming cracked volatiles, allowing them to escape into the gas phase by breaking the solid coke into small particles.

In addition to the recovery of additional and more valuable volatiles, there are other benefits of the new process compared to the delayed coking process. The consumption of utilities is less because no steam or water is required. Since there is no quenching, energy from the hot coke is recovered. The process is continuous so is never opened to the atmosphere. There is no cutting procedure as in the delayed coking process where high-pressure water is used to cut the coke out of the drums. Therefore, no volatiles and no coke particles are released into the atmosphere.

8.6.5 Decarbonizing Process

The thermal *decarbonizing process* (not to be confused with the propane decarbonizing process, which is a deasphalting process) is designed to minimize coke and gasoline yields but, at the same time, to produce maximum yields of gas oil. Decarbonizing in this sense of the term should not be confused with *propane decarbonizing*, which is essentially a solvent deasphalting process (Speight and Ozum, 2002; Hsu and Robinson, 2006; Speight, 2014). Thermal decarbonizing is, in many respects, similar to the delayed coking process, but lower temperatures and pressures are employed. For example, heater outlet temperatures may be 485°C (905°F) and coke drum temperatures may be of the order of 415°C (780°F) while pressures range from 10 to 25 psi.

8.6.6 ET-II Process

The ET-II process is a thermal cracking process for the production of distillates and cracked residuum for use as metallurgical coke and is designed to accommodate feedstocks such as heavy oils, atmospheric residua, and vacuum residua (Kuwahara, 1987). The distillate (*cracked oil*) is suitable as a feedstock to hydrocracker and fluid catalytic cracking. The basic technology of the ET-II process is derived from that of the original Eureka process.

In the process, the feedstock is heated up to 350°C (660°F) by passage through the preheater and fed into the bottom of the fractionator, where it is mixed with recycle oil, and the high-boiling fraction of the cracked oil. The ratio of recycle oil to feedstock is within the range of 0.1%–0.3% by weight. The feedstock mixed with recycle oil is then pumped out and fed into the cracking heater, where the temperature is raised to approximately 490°C–495°C (915°F–925°F), and the outflow is fed to the stirred-tank reactor where it is subjected to further thermal cracking. Both cracking and condensation reactions take place in the reactor.

The cracked oil and gas products, together with steam from the top of the reactor, are introduced into the fractionator where the oil is separated into two fractions, *cracked light oil* and *vacuum gas oil, and pitch* (Speight and Ozum, 2002; Hsu and Robinson, 2006; Gary et al., 2007; Speight, 2014).

8.6.7 EUREKA PROCESS

The Eureka process is a thermal cracking process to produce a cracked oil and aromatic residuum from heavy residual materials (Aiba et al., 1981; Speight and Ozum, 2002; Hsu and Robinson, 2006; Gary et al., 2007; Ohba et al., 2008; AlHumaidan et al., 2013a,b; Speight, 2014). The cracking reactions occur under lower cracked oil partial pressure by introducing steam into the reactor. The unconverted cracked residuum (pitch) in the reactor behaves as a homogeneous system that provides stable and trouble-free operating conditions. The cracked oil is further hydrotreated, cracked, and/or hydrocracked to produce marketable fuels, and the cracked residuum is utilized as a boiler fuel or as a gasification (partial oxidation) feedstock for hydrogen production or synthesis gas production (Chapter 14).

In this process (Figure 8.20), the heavy feedstock is fed to the preheater and then enters the bottom of the fractionator, where it is mixed with the recycle oil. The mixture is then fed to the reactor system that consists of a pair of reactors operating alternately. In the reactor, thermal cracking reaction occurs in the presence of superheated steam which is injected to strip the cracked products out of the reactor and supply a part of heat required for cracking reaction. At the end of the reaction, the bottom product is quenched. The oil and gas products (and steam) pass from the top of the reactor to the lower section of the fractionator, where a small amount of entrained material is removed by a wash operation. The upper section is an ordinary fractionator, where the heavier fraction of cracked oil is drawn as a side stream. The process bottoms (pitch) can be used as boiler fuel, as partial oxidation feedstock for producing hydrogen and carbon monoxide, and as binder pitch for manufacturing metallurgical coke (Speight and Ozum, 2002; Hsu and Robinson, 2006; Gary et al., 2007; Speight, 2014).

The process reactions proceed at lower cracked oil partial pressure by injecting steam into the reactor, keeping petroleum pitch in a homogeneous liquid state and, unlike a conventional delayed coker, a higher cracked oil yield can be obtained. A wide range of residua can be used as feedstock, such as atmospheric and vacuum residue of petroleum crude oils, various cracked residues, asphalt products from solvent deasphalting and native asphalt. After hydrotreating, the cracked oil is used as feedstock for a fluid catalytic cracker or hydrocracker.

The original Eureka process uses two batch reactors, while the newer ET-II and the HSC process both employ continuous reactors.

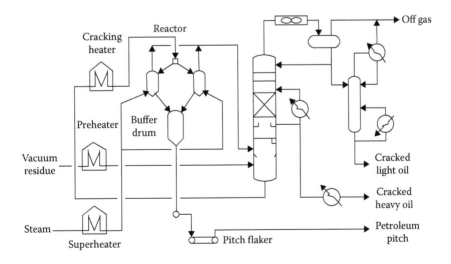

FIGURE 8.20 The Eureka process.

8.6.8 FTC Process

The FTC process (fluid thermal cracking process) is a heavy oil and residuum upgrading process in which the feedstock is thermally cracked to produce distillate and coke (Miyauchi et al., 1981, 1987; Speight and Ozum, 2002; Hsu and Robinson, 2006; Gary et al., 2007; Speight, 2014).

The feedstock, mixed with recycle stock from the fractionator, is injected into the cracker and is immediately absorbed into the pores of the particles by capillary force where it is subjected to thermal cracking. In consequence, the surface of the noncatalytic particles is kept dry and good fluidity is maintained allowing a good yield of, and selectivity for, middle distillate products. Hydrogen-containing gas from the fractionator is used for the fluidization in the cracker. Excessive coke caused by the metals accumulated on the particle is suppressed under the presence of hydrogen. The particles with deposited coke from the cracker are sent to the gasifier, where the coke is gasified and converted into carbon monoxide (CO), hydrogen (H_2), carbon dioxide (CO_2), and hydrogen sulfide (H_2S) with steam and air. Regenerated hot particles are returned to the cracker.

8.6.9 HSC Process

The HSC process (high conversion soaker cracking process) is a cracking process designed for moderate conversion, higher than visbreaking but lower than coking (Watari et al., 1987; Speight and Ozum, 2002; Hsu and Robinson, 2006; Gary et al., 2007; Speight, 2014). The process is an advanced continuous thermal cracking technology with a proprietary soaking drum, featuring a wide range of conversion levels between visbreaking and coking while producing pumpable liquid residue at process temperature. A broad range of heavy feedstocks such as heavy crude, long and short residue with high contents of sulfur and heavy metals and even visbroken residue can be charged to the HSC process. The cracked distillates from the HSC process are mostly light and heavy gas oils with fewer unsaturated compounds than coker distillates. The heavy gas oil fraction serves as the feedstock to the fluid catalytic cracking unit and the cracked residue can be used as the fuel for boiler at the power station. The process uses no hydrogen, no catalyst, and no high-pressure equipment. The process economics is benefited by low investment cost and low utilities consumptions due to its simple process scheme as visbreaking process. The process features less gas make and a higher yield of distillate compared to other thermal cracking processes. The process can be used to convert a wide range of feedstocks with high sulfur and metal content, including heavy oils, oil sand bitumen, residua, and visbroken residua. As a note of interest, the HSC process employs continuous reactors whereas the original Eureka process (q.v.) uses two batch reactors.

In the process, the preheated feedstock enters the bottom of the fractionator, where it is mixed with the recycle oil. The mixture is pumped up to the charge heater and fed to the soaking drum (ca. atmospheric pressure, steam injection at the top and bottom), where sufficient residence time is provided to complete the thermal cracking. In the soaking drum, the feedstock and some product flows downward passing through a number of perforated plates while steam with cracked gas and distillate vapors flow through the perforated plates countercurrently.

The volatile products from the soaking drum enter the fractionator where the distillates are fractionated into desired product oil streams, including a heavy gas oil fraction. The cracked gas product is compressed and used as refinery fuel gas after sweetening. The cracked oil product after hydrotreating is used as fluid catalytic cracking or hydrocracker feedstock. The residuum is suitable for use as boiler fuel, road asphalt, binder for the coking industry, and as a feedstock for partial oxidation.

8.6.10 Mixed-Phase Cracking Process

The mixed-phase cracking process (also called *liquid-phase cracking*) is a continuous thermal decomposition process for the conversion of heavy feedstocks to products boiling in the gasoline range. The process generally employs rapid heating of the feedstock (kerosene, gas oil, reduced crude, or even whole crude), after which it is passed to a reaction chamber and then to a separator where the vapors are cooled. Overhead products from the flash chamber are fractionated to gasoline components and recycle stock, and flash chamber bottoms are withdrawn as a heavy fuel oil. Coke formation, which may be considerable at the process temperatures (400°C–480°C, 750°F–900°F), is minimized by use of pressures in excess of 350 psi.

8.6.11 SELECTIVE CRACKING PROCESS

Selective cracking is a thermal conversion process that utilizes different conditions depending on feedstock composition. For example, a heavy oil may be cracked at 494°C–515°C (920°F–960°F) and 300–500 psi; a lighter gas oil may be cracked at 510°C–530°C (950°F–990°F) and 500–700 psi (Moschopedis et al., 1998; Speight and Ozum, 2002; Hsu and Robinson, 2006; Gary et al., 2007; Speight, 2014).

Each feedstock has its own particular characteristics that dictate the optimum conditions of temperature and pressure for maximum yields of the products. These factors are utilized in selective combination of cracking units in which the more refractory feedstocks are cracked for longer periods of time or at higher temperatures than the less stable feedstocks, which are cracked at lower temperatures.

The process eliminates the accumulation of stable low-boiling material in the recycle stock and also minimizes coke formation from high-temperature cracking of the higher-boiling material. The end result is the production of fairly high yields of gasoline, middle distillates, and olefin gases.

8.6.12 SHELL THERMAL CRACKING PROCESS

The Shell thermal distillate cracking unit is based on the principle of converting high-boiling feedstocks to lower-boiling products (Speight and Ozum, 2002; Hsu and Robinson, 2006; Gary et al., 2007; Speight, 2014). Thermal cracking of heavy oil takes place in the liquid phase in a furnace, at elevated pressure and temperature and the products are residuum and distillate products.

In the process, the feedstock (e.g., heavy gas oils from the atmospheric distillation unit or vacuum gas oil from the vacuum distillation unit) is sent to a surge drum. Liquid from this drum is pumped to the distillate heater, which typically operates at a pressure of approximately 490°C (915°C) and 290 psi. Under these conditions, the cracking reactions take place in the liquid phase.

Fluid from the distillate heater is then routed to the combi tower where separation is achieved between residue, gas oil, and lighter products. In addition, a heavy gas oil fraction is taken from the combi tower, returned to the surge drum, and then recycled through the distillate heater. The bottom product of the combi tower is routed to a vacuum flasher where heavy gas oil is recovered from the residuum stream and routed back to the distillate heater. The vacuum flashed residuum from the vacuum flasher can be routed to fuel oil blending or can be used internally as refinery fuel. The recycling of heavy gas oil from both the vacuum flasher and the combi tower to the distillate heater means that all of the heavy gas oil is converted. Light gas oil from the combi tower is first stripped and is then routed to a hydrotreater. Alternatively, the light gas oil can be used as cutter stock.

The feedstock and product requirements of the thermal distillate cracking process are flexible, and the process has the capability to optimize conversion through adjustment of the heavy gas oil recycle rate.

The *Shell deep thermal gasoil process* is a combination of the *Shell deep thermal conversion process* and the *Shell thermal gasoil process*. In this alternative high conversion scheme, the heavy gas oil (HGO) from the atmospheric distillation unit and the vacuum gas oil (VGO) from the vacuum flasher or vacuum distillation unit are cracked in a distillate thermal cracking heater into lower-boiling range gasoil.

A related process, the *deep thermal conversion process* (DTC process) (Figure 8.21) offers a bridge between visbreaking and coking and provides maximum distillate yields by applying deep

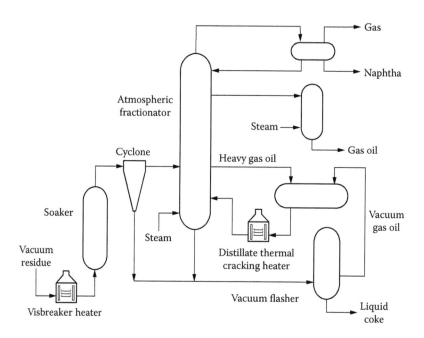

FIGURE 8.21 The DTC process.

thermal conversion to vacuum residua followed by vacuum flashing of the products. In the process, the heated vacuum residuum is charged to the heater and from there to the soaker where conversion occurs. The products are then led to an atmospheric fractionator to produce gases, naphtha, kerosene, and gas oil. The fractionator residuum is sent to a vacuum flasher that recovers additional gas oil and distillate. The next steps for the coke are dependent on its potential use and it may be isolated as *liquid* coke (pitch, cracked residuum) or solid coke. The process yields a maximum of distillates from heavy feedstocks (such as vacuum residua) and by vacuum flashing the cracked residue. The liquid coke, not suitable for blending to commercial fuel, is used for specialty products, gasification and/or combustion, for example, to generate power and/or hydrogen.

8.6.13 Tervahl-T Process

The Tervahl-T process offers options that allow the process to accommodate differences in the feedstock as well as the desired sale of products. In the process (Peries et al., 1988; Speight and Ozum, 2002; Hsu and Robinson, 2006; Gary et al., 2007; Speight, 2014), the feedstock is heated to the desired temperature using the coil heater and heat recovered in the stabilization section and held for a specified residence time in the soaking drum. The soaking drum effluent is quenched and sent to a conventional stabilizer or fractionator where the products are separated into the desired streams (Speight and Ozum, 2002; Hsu and Robinson, 2006; Gary et al., 2007; Speight, 2014). The gas produced from the process is used for fuel.

In the related Tervahl-H process (a hydrogenation process but covered here for convenient comparison with the Tervahl-T process), the feedstock and hydrogen-rich stream are heated using heat recovery techniques and fired heater and held in the soak drum as in the Tervahl-T process. The gas and oil from the soaking drum effluent are mixed with recycle hydrogen and separated in the hot separator where the gas is cooled and passed through a separator and recycled to the heater and soaking drum effluent. The liquids from the hot and cold separator are sent to the stabilizer section where purge gas and synthetic crude are separated. The gas is used as fuel and the synthetic crude can now be transported or stored.

REFERENCES

Aiba, T., Kaji, H., Suzuki, T., and Wakamatsu, T. 1981. The Eureka process. *Chemical Engineering Progress*, February: 37.

AlHumaidan, F., Hauser, A., Al-Rabiah, H., Lababidi, H., and Bouresli, R. 2013a. Studies on thermal cracking behavior of vacuum residues in Eureka process. *Fuel*, 109: 635–646.

AlHumaidan, F., Haitham, M.S., Lababidi, H., and Al-Rabiah, H. 2013b. Thermal cracking kinetics of Kuwaiti vacuum residues in Eureka process. *Fuel*, 109: 923–931.

ASTM D893. 2015. *Standard Test Method for Insolubles in Used Lubricating Oils*. Annual Book of Standards. ASTM International, West Conshohocken, PA.

ASTM D2007. 2015. *Standard Test Method for Characteristic Groups in Rubber Extender and Processing Oils and Other Petroleum-Derived Oils by the Clay-Gel Absorption Chromatographic Method*. Annual Book of Standards. ASTM International, West Conshohocken, PA.

ASTM D3279. 2015. *Standard Test Method for n-Heptane Insolubles*. Annual Book of Standards. ASTM International, West Conshohocken, PA.

ASTM D4124. 2015. *Standard Test Method for Separation of Asphalt into Four Fractions*. Annual Book of Standards. ASTM International, West Conshohocken, PA.

Ballard, W.P., Cottington, G.I., and Cooper, T.A. 1992. Cracking, thermal. In: *Petroleum Processing Handbook*. J.J. McKetta (Editor). Marcel Dekker, New York, p. 309.

Blaser, D.E. 1992. Coking, petroleum (fluid). In: *Petroleum Processing Handbook*. J.J. McKetta (Editor). Marcel Dekker, New York, p. 255.

Bonilla, J. and Elliott, J.D. 1987. Asphalt coking method. U.S. Patent 4,686,027. August 11.

Carrillo, J.A. and Corredor, L.M. 2013. Heavy crude oil upgrading: Jazmin crude. *Advances in Chemical Engineering and Science*, 3: 46–55.

Feintuch, H.M. and Negin, K.M. 2004. FW delayed coking process. In: *Handbook of Petroleum Refining Processes*, 2nd Edition. R.A. Meyers (Editor). McGraw-Hill, New York, Chapter 12.33.

Gary, J.G., Handwerk, G.E., and Kaiser, M.J. 2007. *Petroleum Refining: Technology and Economics*, 5th Edition. CRC Press/Taylor & Francis Group, Boca Raton, FL.

Germain, J.E. 1969. *Catalytic Conversion of Hydrocarbons*. Academic Press Inc., New York.

Hsu, C.S. and Robinson, P.R. 2006. *Practical Advances in Petroleum Processing*, Volumes 1 and 2. Springer, New York.

Kobe, K.A. and McKetta, J.J. 1958. *Advances in Petroleum Chemistry and Refining*. Interscience, New York.

Koshka, E., Kuhach, J., and Veith, E. 2008. Improving Athabasca bitumen development economics through integration with HTL upgrading. *Proceedings of the World Heavy Oil Congress*. Alberta Department of Energy, Edmonton, Alberta, Canada, March.

Kuwahara, I. 1987. The ET-II process. *Koagaku Kogaku*, 51: 1.

Langer, A.W., Stewart, J., Thompson, C.E., White, H.T., and Hill, R.M. 1961. Thermal hydrogenation of crude residua. *Industrial & Engineering Chemistry*, 53: 27–30.

Langer, A.W., Stewart, J., Thompson, C.E., White, H.T., and Hill, R.M. 1962. Hydrogen donor diluent visbreaking of residua. *Industrial & Engineering Chemical Process Design and Development*, 1: 309–312.

Marano, J.J. 2003. Refinery technology profiles: Gasification and supporting technologies. Report prepared for the U.S. Department of Energy, National Energy Technology Laboratory, United States Energy Information Administration, Washington, DC, June.

Marzin, R., Pereira, P., McGrath, M.J., Feintuch, H.M., and Thompson, G. 1998. A new option for residue conversion and heavy oil upgrading. *Oil & Gas Journal*, 97(44): 79.

McKinney, J.D. 1992. Coking, petroleum (delayed and fluid). In: *Petroleum Processing Handbook*. J.J. McKetta (Editor). Marcel Dekker, New York, p. 245.

Miles, J. 2009. Maximizing distillate yields and refinery economics—An alternative solution to conventional fuel oil production or residue conversion. *Proceedings Session A. 14th Annual Meeting—European Refining Technology Conference*. November 11.

Miyauchi, T., Furusaki, S., and Morooka, Y. 1981. Upgrading resid. In: *Advances in Chemical Engineering*. Academic Press Inc., New York, Chapter 11.

Miyauchi, T., Tsutsui, T., and Nozaki, Y. 1987. A new fluid thermal cracking process for upgrading resid. Paper 65B. *Proceedings of the Spring National Meeting*. American Institute of Chemical Engineers, Houston, TX, March 29.

Moschopedis, S.E., Ozum, B., and Speight, J.G. 1998. Upgrading heavy oils. *Reviews in Process Chemistry and Engineering*, 1(3): 201–259.

Mushrush, G.W. and Speight, J.G. 1995. *Petroleum Products: Instability and Incompatibility.* Taylor & Francis Group, Philadelphia, PA.

Negin, K.M. and Van Tine, F.M. 2004. FW/UOP visbreaking process. In: *Handbook of Petroleum Refining Processes*, 2nd Edition. R.A. Meyers (Editor). McGraw-Hill, New York, Chapter 12.3.

NRCAN. 2015. Upgrading and refining process development. National Research Council of Canada, Ottawa, Ontario, Canada. http://www.nrcan.gc.ca/energy/oil-sands/upgrading/5879; accessed August 15, 2015.

Ohba, T., Shibutani, I., Watari, R., Inomata, J., and Nagata, H. 2008. The advanced EUREKA process: Environment friendly thermal cracking process. Paper No. WPC-19-2856. *Proceedings of the 19th World Petroleum Congress.* Madrid, Spain, June 29–July 3.

Parkash, S. 2003. *Refining Processes Handbook.* Gulf Professional Publishing, Elsevier, Amsterdam, the Netherlands.

Peries, J.P., Quignard, A., Farjon, C., and Laborde, M. 1988. Thermal and catalytic ASVAHL processes under hydrogen pressure for converting heavy crudes and conventional residues. *Revue Institut Français Du Pétrole*, 43(6): 847–853.

Roundtree, E.M. 1997. Fluid coking. In: *Handbook of Petroleum Refining Processes*, 2nd Edition. R.A. Meyers (Editor). McGraw-Hill, New York, Chapter 12.1.

Schucker, R.C. 2003. Heavy oil upgrading process. U.S. Patent 6,524,469. February 25.

Silverman, M.A., Pavel, S.K., and Hillerman, M.D. 2011. HTL heavy oil upgrading: A key solution for heavy oil upstream and midstream operations. Paper No. WHOC11-419. *Proceedings of the World Heavy Oil Congress.* Edmonton, Alberta, Canada, March 14–17.

Speight, J.G. 1990. Tar sands. In: *Fuel Science and Technology Handbook.* J.G. Speight (Editor). Marcel Dekker, New York, Chapters 12–16.

Speight, J.G. 2000. *The Desulfurization of Heavy Oils and Residua*, 2nd Edition. Marcel Dekker, New York.

Speight, J.G. 2008. *Synthetic Fuels Handbook: Properties, Processes and Performance.* McGraw-Hill, New York.

Speight, J.G. 2013. *Heavy and Extra Heavy Oil Upgrading Technologies.* Gulf Professional Publishing, Elsevier, Oxford, UK.

Speight, J.G. 2014. *The Chemistry and Technology of Petroleum*, 5th Edition. CRC Press/Taylor & Francis Group, Boca Raton, FL.

Speight, J.G. 2015a. *Fouling in Refineries.* Gulf Professional Publishing, Elsevier, Oxford, UK.

Speight, J.G. 2015b. *Handbook of Petroleum Product Analysis*, 2nd Edition. John Wiley & Sons, Hoboken, NJ.

Speight, J.G. and Ozum, B. 2002. *Petroleum Refining Processes.* Marcel Dekker, New York.

Stark, J.L. and Falkler, T. 2008. Method for improving liquid yield during thermal cracking of hydrocarbons. U.S. Patent 7,425,259. September 16.

Stark, J.L., Falkler, T., Weers, J.J., and Zetlmeisl, M.J. 2008. Method for improving liquid yield during thermal cracking of hydrocarbons. U.S. Patent 7,416,654. August 26.

Stell, R.C., Balinsky, G.J., McCoy, J.N., and Keusenkothen, P.F. 2009a. Process and apparatus for cracking hydrocarbon feedstock containing resid. U.S. Patent 7,588,737. September 15.

Stell, R.C., Dinicolantonio, A.R., Frye, J.M., Spicer, D.B., McCoy, J.N., and Strack, R.D. 2009b. Process for steam cracking heavy hydrocarbon feedstocks. U.S. Patent 7,578,929. August 25.

Stephens, M.M. and Spencer, O.F. 1956. *Petroleum Refining Processes.* Pennsylvania State University Press, University Park, PA.

Sullivan, D.W. 2011. New continuous coking process. *Proceedings of the 14th Topical Symposium on Refinery Processing. AIChE Spring Meeting and Global Congress on Process Safety.* Chicago, IL, March 13–17.

Veith, E.J. 2007. Performance of heavy-to-light-crude-oil upgrading process. *Proceedings of the SPE International Oil Conference and Exhibition.* Veracruz, Mexico, June 27–30. Society of Petroleum Engineers, Richardson, TX.

Watari, R., Shoji, Y., Ishikawa, T., Hirotani, H., and Takeuchi, T. 1987. Paper AM-87-43. *Annual Meeting.* National Petroleum Refiners Association, San Antonio, TX.

Wodnik, R. and Hughes, G.C. 2005. Delayed coking advances. *Petroleum Technology Quarterly*, Q4: 1–6.

9 Catalytic Cracking Processes

9.1 INTRODUCTION

Catalytic cracking is different to thermal cracking insofar as a catalyst is used for the catalytic process. The mechanism of catalytic cracking is also different insofar as the mechanism of the thermal process involves free radical intermediates whereas the mechanism of the catalytic process involves ionic intermediates (Table 9.1). However, there has been the claim that, in some cases, the feedstock constituents decompose by thermolysis before the species may come into contact with the catalyst.

Catalytic cracking is widely used to convert heavy oils into more valuable naphtha (a blend stock for gasoline manufacture) and other low-boiling products. As the demand for gasoline increased, catalytic cracking replaced thermal cracking with the evolution of catalytic cracking. Fluid catalytic cracking (FCC) refers to the behavior of the catalyst during this process insofar as the fine, powdered catalyst (typically zeolites, which have a particle size on the order of 70 µm) takes on the properties of a fluid when it is mixed with the vaporized feed. Fluidized catalyst particles circulate continuously between the reaction zone and the regeneration zone.

In terms of process parameters, catalytic cracking is typically performed at temperatures ranging from 485°C to 540°C (900°F to 1000°F) and pressures up to 100 psi. Feedstocks for the process have typically been gas oil fractions (Table 9.2), but the focus is shifting to gas oil–residua blends, gas oil–heavy oil blends, and gas oil–bitumen (Shidhaye et al., 2015). In some cases, heavy oils have been blended with the minimum amount of gas oil (added as a flux) as the feedstock to catalytic cracking units. In the process, the feedstock enters the unit at temperatures on the order of 485°C–540°C (900°F–1000°F), and the circulating catalyst provides heat from the regeneration zone to the oil feed. Carbon (coke) is burned off the catalyst in the regenerator, raising the catalyst temperature to 620°C–735°C (1150°F–1350°F, before the catalyst returns to the reactor).

The preferred feedstock to a fluid catalytic cracking unit has been and continues to be the portion of the crude oil that has an initial boiling point (at atmospheric pressure) of approximately 275°C (525°F) up to the initial boiling point of the atmospheric residuum (345°C, 655°F) (Table 9.2). On occasion, the vacuum gas oil (boiling range: 345°C–510°C or 345°C–565°C; 655°F–950°F or 655°F–1050°F) may also be used as feedstock to the fluid catalytic cracking unit. However, the changing slate of refinery feedstocks has caused this to change. Currently, the feedstocks for catalytic cracking can be any one (or blends) of the following: (1) straight-run gas oil, (2) vacuum gas oil, (3) atmospheric residuum, and (4) vacuum residuum, with special emphasis on the heavier feedstocks (Lifschultz, 2005; Ross et al., 2005; Speight, 2014). If blends of the earlier feedstocks are employed, compatibility of the constituents of the blends (i.e., no phase separation) must be assured, or excessive coke (and metals) will be laid down onto the catalyst.

In addition, there are several *pretreatment options* for the feedstocks that offer process benefits and these are (1) deasphalting to prevent excessive coking on catalyst surfaces; (2) demetallization, that is, removal of nickel, vanadium, and iron to prevent catalyst deactivation; (3) use of a short residence time as a means of preparing the feedstock; (4) hydrotreating or mild hydrocracking to prevent excessive coking in the fluid catalytic cracking unit; and (5) blending with an aromatic gas oil type to prevent phase separation; and (6) staged partial conversion (Birch and Ulivieri, 2000; Speight, 2000; Patel et al., 2002, 2004; Speight and Ozum, 2002; Hsu and Robinson, 2006; Gary et al., 2007; Dziabala et al., 2011). Hydrotreating the feedstock to the fluid catalytic cracker improves the yield and quality of naphtha (Table 9.3) and reduces the sulfur oxide (SO_x) emissions from the

TABLE 9.1

Comparison of Thermal Cracking and Catalytic Cracking

Thermal Cracking

- No catalyst
- Free radical reaction mechanism
- Moderate yields of naphtha and other distillates
- Feedstock-dependent gas yields
- Low-to-moderate product selectivity
- Low octane naphtha
- Low-to-moderate yields of C_4 olefins
- Low-to-moderate yields of aromatics

Catalytic Cracking

- Uses a catalyst
- Ionic reaction mechanism
- High yields of naphtha and other distillates
- Lower gas yields than the thermal process
- High product selectivity
- Low yields of n-alkanes—high-octane naphtha
- Chain branching and high yields of C_4 olefins
- High yields of aromatics

catalytic cracker unit (Sayles and Bailor, 2005). Refineries wishing to process heavier crude oil may only have the option to desulfurize the resulting high-sulfur naphtha produced in the process.

On a global basis, the effect of declining crude quality (Speight, 2011a, 2014) may be looked upon as influencing FCC feedstock quality and amount. However, this will be a secondary factor compared with the changes required in the refined products slate. In addition to the heavier viscous crude oils—as a blend or as a hydrotreated feedstock—the production of synthetic crude oil

TABLE 9.2

Preferred Composition of the FCC Feedstock

Test	Data Range
Gravity, API	19.5–23.0
Density at 15°C, kg/L	0.9153–0.9366
Distillation (D 1160) °C (°F)	
Initial boiling point	275 (525)
End point	345 (655)
Flash point	116–143 (240–290)
Pour point	17–38 (60–100)
Viscosity at 50°C (122°F), cSt	20–50
Sulfur, % w/w	1.1–1.4
Carbon residue, % w/w	0.1–0.5
Aniline point	73–79 (163–174)
Asphaltene content, % w/w	<2
Nitrogen, ppm	1200–1700
Basic nitrogen, ppm	400–600
Vanadium, ppm	<0.05
Nickel, ppm	<0.10

TABLE 9.3

**Feedstock and Product Data for the Fluid Catalytic Process
with and without Feedstock Hydrotreating**

	Without Feedstock Hydrotreating	With Feedstock Hydrotreating
Feedstock (>370°C, >700°F)		
API	15.1	20.1
Sulfur, % w/w	3.3	0.5
Nitrogen, % w/w	0.2	0.1
Carbon residue, % w/w	8.9	4.9
Nickel + vanadium, ppm	51.0	7.0
Products		
Naphtha (C5–221°C, C5–430°F), % v/v	50.6	58.0
Light cycle oil (221°C–360°C, 430°F–680°F), % v/v	21.4	18.2
Residuum (>360°C, > 680°F), % w/w	9.7	7.2
Coke, % w/w	10.3	7.0

from tar sand bitumen will increase dramatically in the next decade (Patel, 2007; Speight, 2008, 2009, 2011a). For example, the synthetic crude oil from Canadian tar sand sources is projected to increase to 3.0 million bpd by 2015. With Canadian reserves in excess of 170 billion barrels (170×10^9 barrels) of viable oil, economic forecasts predict that tar sand deposits will continue to be a significant crude source (and, hence, feedstock to the catalytic cracking unit) for the foreseeable future (Schiller, 2011). With the increasing focus to reduce sulfur content in fuels, the role of *desulfurization* in the refinery becomes more and more important. Currently, the process of choice is the hydrotreater in which the unfinished fuel is hydrotreated to remove sulfur from the fuel. Hydrotreating of the feedstock to the catalytic cracking unit can increase conversion by 8%–12% v/v and with most feedstocks (Salazar-Sotelo et al., 2004). Thus, it will be possible to reduce the sulfur content of the naphtha/gasoline product to levels low enough to meet the future low-sulfur gasoline pool specifications.

Finally, the use of bio-feedstocks (such as animal fats, vegetable oils, cellulosic materials, and lignin) in the fluid catalytic cracking unit will be used to increase the yield of light cycle oil and will also provide high-quality products in terms of cetane number (Speight, 2008). Practical implementation in a refinery will be accompanied by blending with vacuum gas oil or resid (Speight, 2011a). From a strategic point of view, refiners should not try to compete with biofuels producers, but rather try to use renewable feedstocks in traditional petroleum refining processes and make products that are compatible with conventional hydrocarbon fuels (Speight, 2008, 2011a).

Furthermore, fluid catalytic cracking technology represents one of the most expanded processes producing the precursors to liquid fuels (naphtha and kerosene) and automobile fuels from gas oil distillates and from heavy feedstocks. A key factor is the use of active, stable, and selective (*tailor-made*) catalysts to convert specific feedstocks (especially heavy feedstocks) into desired products. Thus, the refinery process (Figure 9.1) can be applied to a variety of feedstocks ranging from gas oil to heavy oil. It is one of several practical applications used in a refinery that employ a catalyst to improve process efficiency (Table 9.4). The original incentive to develop cracking processes arose from the need to increase gasoline supplies and, since cracking could virtually double the volume of naphtha from a barrel of crude oil, the purpose of cracking was wholly justified.

In the 1930s, thermal cracking units produced approximately half the total naphtha manufactured, the octane number of which was approximately 70 compared to approximately 60 for

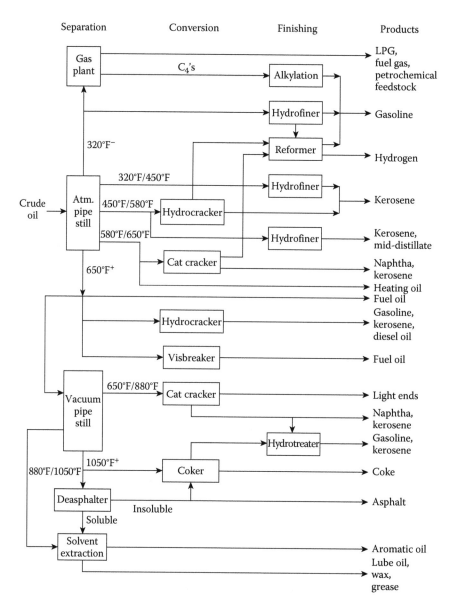

FIGURE 9.1 Generalized refinery layout showing relative placement of the catalytic cracking units.

straight-run naphtha. These were usually blended together with light ends and sometimes with polymer gasoline and reformatted to form a gasoline base stock with an octane number of approximately 65. The addition of *tetraethyllead* (*ethyl fluid*) increased the octane number to approximately 70 for the *regular grade* gasoline and 80 for *premium grade* gasoline. The thermal reforming and polymerization processes that were developed during the 1930s could be expected to further increase the octane number of gasoline to some extent, but something new was needed to break the octane barrier that threatened to stop the development of more powerful automobile engines. In 1936, a new cracking process opened the way to higher-octane gasoline; this process was catalytic cracking. Since that time the use of catalysts in the petroleum industry has spread to other processes (Table 9.4) (see also Bradley et al., 1989).

   Catalytic cracking is basically the same as thermal cracking, but it differs by the use of a catalyst that is not (in theory) consumed in the process (Table 9.5). The catalyst directs the course of the

TABLE 9.4

Refinery Processes That Employ Catalysts to Enhance Reactivity

Process	Materials Charged	Products Recovered	Temperature of Reaction	Type of Reaction
Cracking	Gas oil, fuel oil, heavy feedstocks	Gasoline, gas, and fuel oil	875°F–975°F (470°C–525°C)	Dissociation or splitting of molecules
Hydrogenation	Gasoline to heavy feedstocks	Low-boiling products	400°F–850°F (205°C–455°C)	Mild hydrogenation: cracking; removal of sulfur, nitrogen, oxygen, and metallic compounds
Reforming	Gasolines, naphthas	High-octane gasolines, aromatics	850°F–1000°F (455°C–535°C)	Dehydrogenation, dehydroisomerization, isomerization, hydrocracking, dehydrocyclization
Isomerization	Butane, C_4H_{10}	Iso-Butane, C_4H_{10}		Rearrangement
Alkylation	Butylene and iso-butane, C_4H_8 and C_4H_{10}	Alkylate, C_8H_{18}	32°F–50°F (0°C–10°C)	Combination
Polymerization	Butylene, C_4H_8	Octene, C_8H_{16}	300°F–350°F (150°C–175°C)	Combination

TABLE 9.5

Summary of Catalytic Cracking Processing Conditions

Condition
  Solid acid catalyst (silica–alumina, zeolite, others)
  900°F–1000°F (solid/vapor contact)
  10–20 psig
Feeds
  Virgin naphthas to atmospheric residua
  Pretreated to remove salts (metals)
  Pretreated to remove asphalts
Products
  Lower-molecular-weight components
  C_3–C_4 gases > C_2 gases
  Iso-Paraffins
  Coke (fuel)
Variations
  Fixed bed
  Moving bed
  Fluidized bed

cracking reactions to produce more of the desired products that can be used for the production of better quality gasoline and other liquid fuels (Table 9.6) (Avidan and Krambeck, 1990).

Catalytic cracking has a number of advantages over thermal cracking. The naphtha produced by catalytic cracking has a higher octane number and consists largely of iso-paraffins and aromatics. The iso-paraffins and aromatic hydrocarbons have high octane numbers and greater chemical stability than mono-olefins and diolefins. The olefins and diolefins are present in much greater quantities in thermally cracked naphtha. Furthermore, olefins (e.g., $RCH=CH_2$ where R=H or an alkyl group) and smaller quantities of methane (CH_4) and ethane ($CH_3 \cdot CH_3$) are produced by catalytic

TABLE 9.6

Comparison of Thermal Cracking and Catalytic Cracking

Hydrocarbon	Catalytic Cracking	Thermal Cracking
n-Paraffins	Extensive breakdown to C_2 and larger fragments. Product largely in C_2–C_6 range contains many branched aliphatics. Few normal α-olefins above C_4.	Extensive breakdown to C_2 fragments, with much C_1 and C_2. Prominent amounts of C_4–C_{n-1} normal α olefins. Aliphatics largely unbranched.
Iso-Paraffins	Cracking rate relative to n-paraffins increased considerably by the presence of tertiary carbon atoms.	Cracking rate increased to a relatively small degree by the presence of tertiary carbon atoms.
Naphthenes	Crack at about same rate as paraffins with similar numbers of tertiary carbon atoms. Aromatics produced, with much hydrogen transfer to unsaturates.	Crack at lower rate than normal paraffins. Aromatics produced with little hydrogen transfer to unsaturates.
Unsubstituted aromatics	Little reaction; some condensation to biaryls.	Little reaction; some condensation to biaryls.
Alkyl aromatics (substituents C_3 or larger)	Entire alkyl group cracked next to ring and removed as olelin. Crack at much higher rate than paraffins.	Alkyl group cracked to leave one or two carbon atoms attached to ring. Crack at lower rate than paraffins.
n-Olefins	Product similar to that from n-paraffins but more olefinic.	Product similar to that from n-paraffins but more olefinic.
All olefins	Hydrogen transfer is an important reaction, especially with tertiary olefins. Crack at much higher rate than corresponding paraffins.	Hydrogen transfer is a minor reaction, with little preference for tertiary olefins. Crack at about same rate as corresponding paraffins.

cracking and are suitable for petrochemical use (Speight and Ozum, 2002; Parkash, 2003; Hsu and Robinson, 2006; Gary et al., 2007; Speight, 2014). Sulfur compounds are changed in such a way that the sulfur content of naphtha produced by catalytic cracking is lower than the sulfur content of naphtha produced by thermal cracking. Catalytic cracking produces less residuum and more of the useful gas oil constituents than thermal cracking. Finally, the process has considerable flexibility, permitting the manufacture of both automobile gasoline and aviation gasoline and a variation in the gas oil production to meet changes in the fuel oil market.

Catalytic cracking in the usual commercial process involves contacting a feedstock (usually a gas oil fraction) with a catalyst under suitable conditions of temperature, pressure, and residence time. By this means, a substantial part (>50%) of the feedstock is converted into naphtha and lower-boiling products, usually in a single-pass operation. However, during the cracking reaction, carbonaceous material is deposited on the catalyst, which markedly reduces its activity, and removal of the deposit is very necessary. The carbonaceous deposit arises from the presence of high-molecular-weight polar species (Chapter 3) in the feedstock. Removal of the deposit from the catalyst is usually accomplished by burning in the presence of air until catalyst activity is reestablished.

The reactions that occur during catalytic cracking are complex (Germain, 1969), but there is a measure of predictability now that catalyst activity is better understood. The major catalytic cracking reaction exhibited by paraffins is carbon–carbon bond scission into a lighter paraffin and olefin. Bond rupture occurs at certain definite locations on the paraffin molecule, rather than randomly as in thermal cracking. For example, paraffins tend to crack toward the center of the molecule, the long chains cracking in several places simultaneously. Normal paraffins usually crack at carbon–carbon bonds or still nearer the center of the molecule. On the other hand, iso-paraffins tend to rupture between carbon atoms that are, respectively, β and γ to a tertiary carbon. In either case, catalytic cracking tends to yield products containing three or four carbon atoms rather than the one- or two-carbon-atom molecules produced in thermal cracking.

As in thermal cracking (Chapter 8), high-molecular-weight constituents usually crack more readily than small molecules, unless there has been some recycle, and the constituents of the recycle stream have become more refractory and are less liable to decompose. Paraffins having more than six carbon atoms may also undergo rearrangement of their carbon skeletons before cracking, and a minor amount of dehydrocyclization also occurs, yielding aromatics and hydrogen.

Olefins are the most reactive class of hydrocarbons in catalytic cracking and tend to crack from 1,000 to 10,000 times faster than in thermal processes. Severe cracking conditions destroy olefins almost completely, except for those in the low-boiling naphtha and gaseous hydrocarbon range and, as in the catalytic cracking of paraffins, *iso*-olefins crack more readily than *n*-olefins. The olefins tend to undergo rapid isomerization and yield mixtures with an equilibrium distribution of double-bond positions. In addition, the chain-branching isomerization of olefins is fairly rapid and often reaches equilibrium. These branched-chain olefins can then undergo hydrogen transfer reactions with naphthenes and other hydrocarbons. Other olefin reactions include polymerization and condensation to yield aromatic molecules, which in turn may be the precursors of coke formation.

In catalytic cracking, the cycloparaffin (naphthene) species crack more readily than paraffins but not as readily as olefins. Naphthene cracking occurs by ring rupture and by rupture of alkyl chains to yield olefins and paraffins, but formation of methane and the C_2 hydrocarbons (ethane, $CH_3 \cdot CH_3$, ethylene, $CH_2 = CH_2$, and acetylene, $CH \equiv CH$) is relatively minor.

Aromatic hydrocarbons exhibit wide variations in their susceptibility to catalytic cracking. The benzene ring is relatively inert, and condensed-ring compounds, such as naphthalene, anthracene, and phenanthrene, crack very slowly. When these aromatics crack, a substantial part of their *conversion* is reflected in the amount of coke deposited on the catalyst. Alkylbenzenes with attached groups of C_2 or larger primarily form benzene and the corresponding olefins, and heat sensitivity increases as the size of the alkyl group increases.

The several processes currently employed commercially for in catalytic cracking differ mainly in the method of catalyst handling, although there is an overlap with regard to catalyst type and the nature of the products. The catalyst, which may be an activated natural or synthetic material, is employed in bead, pellet, or microspherical form and can be used as a *fixed-bed*, *moving-bed*, or *fluid-bed* configurations.

The *fixed-bed process* was the first to be used commercially and uses a static bed of catalyst in several reactors, which allows a continuous flow of feedstock to be maintained. Thus, the cycle of operations consists of (1) flow of feedstock through the catalyst bed, (2) discontinuance of feedstock flow and removal of coke from the catalyst by burning, and (3) insertion of the reactor onstream.

The *moving-bed process* uses a reaction vessel in which cracking takes place and a kiln in which the spent catalyst is regenerated, and catalyst movement between the vessels is provided by various means.

The *fluid-bed process* differs from the fixed-bed and moving-bed processes insofar as the powdered catalyst is circulated essentially as a fluid with the feedstock (Sadeghbeigi, 1995). The several fluid catalytic cracking processes in use differ primarily in mechanical design. Side-by-side reactor–regenerator configuration or a configuration where the reactor is either above or below the regenerator is the main mechanical variation. From a flow standpoint, all fluid catalytic cracking processes contact the feedstock and any recycle streams with the finely divided catalyst in the reactor.

Feedstocks may range from naphtha fractions (included in normal heavier feedstocks for upgrading) to an atmospheric residuum (*reduced crude*). Feed preparation (to remove *metallic constituents* and *high-molecular-weight nonvolatile materials*) is usually carried out through the application of any one of several other processes: coking, propane deasphalting, furfural extraction, vacuum distillation, viscosity breaking, thermal cracking, and hydrodesulfurization (Speight, 2000).

The major process variables are temperature, pressure, catalyst/oil ratio (ratio of the weight of catalyst entering the reactor per hour to the weight of oil charged per hour), and space velocity (weight or volume of the oil charged per hour per weight or volume of catalyst in the reaction zone). Wide flexibility in product distribution and quality is possible through control of these variables along with the extent of internal recycling is necessary. Increased conversion can be obtained by applying higher temperature or higher pressure. Alternatively, lower space velocity and higher catalyst/oil ratio will also contribute to an increased conversion.

When cracking is conducted in a single stage, the more reactive hydrocarbons may be cracked, with a high conversion to gas and coke, in the reaction time necessary for reasonable conversion of the more refractory hydrocarbons. However, in a two-stage process, gas and naphtha from a short-reaction-time, high-temperature cracking operation are separated before the main cracking reactions take place in a second-stage reactor. For the short time of the first stage, a flow line or vertical riser may act as the reactor, and some conversion is effected with minimal coke formation. Cracked gases are separated and fractionated; the catalyst and residue, together with recycle oil from a second-stage fractionator, pass to the main reactor for further cracking. The products of this second-stage reaction are gas, naphtha and gas oil streams, and recycle oil.

Most fluid catalytic cracking units (FCCs or FCCUs) are operated to maximize conversion to naphtha and LPG (Speight, 2011a, 2014). In the current context, the catalyst, which may be an activated natural or a synthetic material, is employed in bead, pellet, or microspherical form in any one (or all) of the several available or *fluidized-bed* (*fluid-bed*) configurations, which differ primarily in mechanical design (Sadeghbeigi, 1995; Hunt, 1997; Johnson and Niccum, 1997; Ladwig, 1997; Hemler and Smith, 2004; Speight, 2011a, 2014). In addition, as the worldwide consumption of fuels has increased, product demand pattern has continued to shift toward distillate fuels such as gasoline, diesel, and kerosene–jet fuel with varying demand for heavy fuels (Figure 9.2) (Ross et al., 2005). On the other hand, the octane number of the naphtha is also enhanced by overcracking the middle boiling point fraction with low octane number. This technique is more effective in the case of that octane number enhancement in FCC naphtha and if an increase in propylene yield has a priority over naphtha production (Buchanan et al., 1996; Imhof et al., 2005).

The last 60 years have seen substantial advances in the development of catalytic processes. This has involved not only rapid advances in the chemistry and physics of the catalysts themselves but also major engineering advances in reactor design; for example, the evolution of the design of the catalyst beds from *fixed beds* to *moving beds* to *fluidized beds*. Catalyst chemistry and physics and bed design have allowed major improvements in process efficiency and product yields (Sadeghbeigi, 1995). In terms of catalyst use, the most important concerns of the petroleum refining industry in the near future are (1) meeting the growing market of cleaner fuels, (2) gradual substitution of

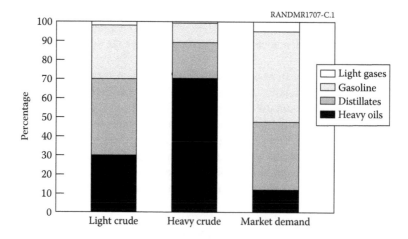

FIGURE 9.2 Range of products in light and heavy crude oils prior to refining and the market demand for refined products.

scarce light low-sulfur refinery feedstocks by the heavier high-sulfur feedstocks, (3) the decreasing demand for heavy fuel oil, and (4) the need to update processing operations (Speight and Ozum, 2002; Swaty, 2005; Hsu and Robinson, 2006; Gary et al., 2007; Gembicki et al., 2007; Speight, 2011a; Bridjanian and Khadem Samimi, 2011; Letzsch, 2011).

As the trend toward processing heavy crude oil and tar sand bitumen increases (Patel, 2007; Speight, 2011a), evolving environmental mandates require lower levels of sulfur in the final fuel product and a reduction in emissions of sulfur dioxide (EPA, 2010). Reducing the sulfur concentration requires not only more efficient process options and specialized catalysts, especially options required to process heavy oil and tar sand bitumen (Gembicki et al., 2007; Patel, 2007; Runyan, 2007). Furthermore, stricter environmental regulations are on the horizon. That venerable, almost revered, Bunker fuel oil was, in the past, released to markets that served as an outlet for a large percentage of the organic sulfur in the refinery feedstock. The International Convention on the Reduction of Pollution from Ships (MARPOL, 2005) has mandated a staged reduction in the allowable sulfur content of maritime fuels, which will drop to less than 0.5% w/w by 2020. In short, the continued emphasis for refinery operations in the foreseeable future 2020 will focus on (1) the production of clean fuels and (2) heavy feedstock upgrading to produce transportation fuels.

9.2 FIXED-BED PROCESSES

Although fixed-bed catalytic cracking units have been phased out of existence, they represented an outstanding chemical engineering commercial development by incorporating a fully automatic instrumentation system that provided a short-time reactor/purge/regeneration cycle, a novel molten salt heat transfer system, and a flue gas expander for recovering power to drive the regeneration air compressor. Historically, the Houdry fixed-bed process (Figure 9.3), which went onstream in June 1936, was the first of the modern catalytic cracking processes. Only the McAfee batch process that employed a metal halide catalyst but that has long since lost any commercial significance preceded it.

In a *fixed-bed process*, the catalyst in the form of small lumps or pellets is made up in layers or beds in several (four or more) catalyst-containing drums called converters (Figure 9.3).

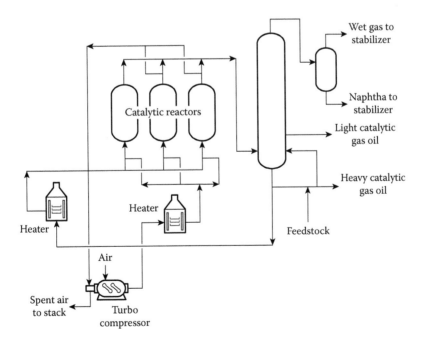

FIGURE 9.3 The Houdry fixed-bed catalytic cracking process.

Feedstock vaporized at approximately 450°C (840°F) and less than 7–15 psi pressure is passed through one of the converters where the cracking reactions take place. After a short time, deposition of coke on the catalyst renders the catalyst ineffective, and using a synchronized valve system, the feed stream was passed into a converter while the catalyst in the first converter was regenerated by carefully burning the coke deposits with air. After approximately 10 minutes, the catalyst is ready to go onstream again.

The requirement of complete vaporization necessarily limited feeds to those with a low boiling range, and higher-boiling feedstock constituents are retained in a separator before the feed is passed into the bottom of the upflow fixed bed reactors. The catalyst consisted of a pelletized natural silica alumina catalyst and was held in reactors or *cases* approximately 11 ft (3.4 m) in diameter and 38 ft (11.6 m) length for a 15,000 bbl/day unit. Cracked products are passed through the preheat exchanger, condensed, and fractionated in a conventional manner. The reactors operated at approximately 30 psi and 480°C (900°F).

The heat of reaction and some of the required feed circulating a molten salt through vertical tubes distributed through the reactor beds. The reaction cycle of an individual reactor was approximately 10 minutes, after which the feed was automatically switched to a new reactor that had been regenerated. The reactor was purged with steam for approximately 5 minutes and then isolated by an automatic cycle timer. Regeneration air was introduced under close control and carbon was burned off at a rate at which the recirculating salt stream could control the bed temperature. This stream comprised a mixture of potassium nitrate (KNO_3) and sodium nitrate ($NaNO_3$), which melts at 140°C (284°F), and was cooled in the reactors through which feed was being processed. The regeneration cycle lasted approximately 10 minutes. The regenerated bed was then purged of oxygen and automatically cut back into cracking service. There were three to six reactors in a unit. Naphtha yields diminished over the life of the catalysts (18 months) from 52% by volume to 42% by volume, based on fresh feedstock.

Equilibrium was never reached in this cyclic process. The gas oil conversion, that is, the amount of feed converted to lighter components, was high at the start of a reaction cycle and progressively diminished as the carbon deposit accumulated on the catalyst until regeneration was required. Multiple parallel reactors were used to approach a steady-state process. However, the resulting process flows were still far from steady state. The reaction bed temperature varied widely during reaction and regeneration periods, and the temperature differential within the bed during each cycle was considerable.

Fixed-bed catalytic cracking units have now generally been replaced by moving-bed or fluid-bed processes.

9.3 MOVING-BED PROCESSES

The fixed-bed process had obvious capacity and mechanical limitations that needed improvement and, thus, was replaced by a moving-bed process in which the hot salt systems were eliminated. The catalyst was lifted to the top of the reactor system and flowed by gravity down through the process vessels. The plants were generally limited in size to units processing up to approximately 30,000; these units have been essentially replaced by larger fluid solids units.

In the moving-bed processes, the catalyst is a pelletized form (approximately 0.125 inch, 3 mm) diameter beads that flow by gravity from the top of the unit through a seal zone to the reactor that operates at approximately 10 psi and at temperatures on the order of 455°C–495°C (850°F–925°F). The catalyst then flows down through another seal and countercurrent through a stripping zone to the regenerator or kiln that operates at a pressure that is close to atmospheric. In early moving-bed units, built around 1943, bucket elevators were used to lift the catalyst to the top of the structure. In later units, built approximately in 1949, a pneumatic lift was used. This pneumatic lift permitted higher catalyst circulation rates, which in turn permitted injection of all liquid feedstocks, as well as feedstocks that had a higher boiling range. A primary air stream was used to convey the

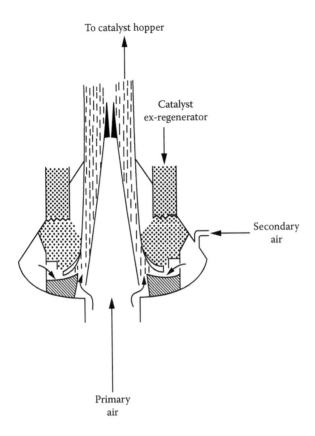

FIGURE 9.4   Catalyst pickup system.

catalyst (Figure 9.4). A secondary air stream was injected through an annulus into which the cata-lyst could flow. Varying the secondary air rate varied the circulation rate.

The lift pipe is tapered to a larger diameter at the top and minimizes erosion and catalyst attrition at the top. This taper is also designed so that total collapse of circulation will not occur instanta-neously when a specific concentration or velocity of solids, below which particles tend to drop out of the flowing gas stream, is experienced. The taper can be designed so that this potential separation of solids is preceded by a pressure instability that can alert the operators to take corrective action.

The *Airlift Thermofor Catalytic Cracking (Socony Airlift TCC) process* (Figure 9.5) is a moving-bed, reactor-over-generator continuous process for conversion of heavy gas oils into lighter high-quality naphtha and middle distillate fuel oils. Feed preparation may consist of flashing in a tar separator to obtain vapor feed, and the tar separator bottoms may be sent to a vacuum tower from which the liquid feed is produced.

The gas–oil vapor–liquid flows downward through the reactor concurrently with the regenerated synthetic bead catalyst. The catalyst is purged by steam at the base of the reactor and gravitates into the kiln, or regeneration is accomplished by the use of air injected into the kiln. Approximately 70% of the carbon on the catalyst is burned in the upper kiln burning zone and the remainder in the bottom-burning zone. Regenerated, cooled catalyst enters the lift pot, where low-pressure air transports it to the surge hopper above the reactor for reuse.

The *Houdriflow catalytic cracking process* (Figure 9.6) is a continuous, moving-bed process employing an integrated single vessel for the reactor and regenerator kiln. The charge stock, sweet or sour, can be any fraction of the crude boiling between naphtha and soft asphalt. The catalyst is trans-ported from the bottom of the unit to the top in a gas lift employing compressed flue gas and steam.

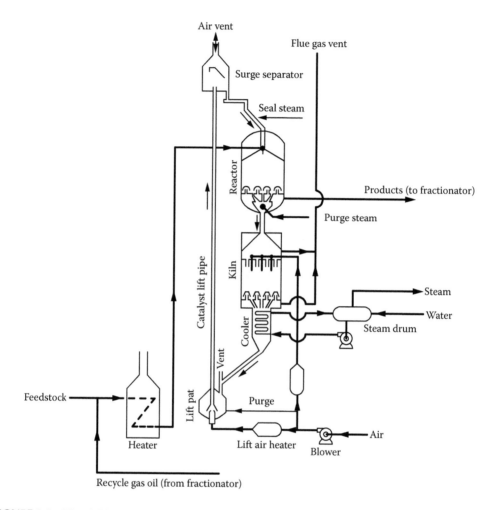

FIGURE 9.5 The airlift Thermofor catalytic cracking process.

The reactor feed and catalyst pass concurrently through the reactor zone to a disengager section, in which vapors are separated and directed to a conventional fractionation system. The spent catalyst, which has been steam purged of residual oil, flows to the kiln for regeneration, after which steam and flue gas are used to transport the catalyst to the reactor.

Houdresid catalytic cracking process (Figure 9.7) is a process that uses a variation of the continuously moving catalyst bed designed to obtain high yields of high-octane naphtha and light distillate from reduced crude charge. Residuum cuts ranging from crude tower bottoms to vacuum bottoms, including residua high in sulfur or nitrogen, can be employed as the feedstock, and the catalyst is synthetic or natural (Alvarenga Baptista et al., 2010a,b). Although the equipment employed is similar in many respects to that used in Houdriflow units, novel process features modify or eliminate the adverse effects and catalyst and product selectivity usually resulting when heavy metals—iron, nickel, copper, and vanadium—are present in the fuel. The Houdresid catalytic reactor and catalyst-regenerating kiln are contained in a single vessel. Fresh feed plus recycled gas oil is charged to the top of the unit in a partially vaporized state and mixed with steam.

Suspensoid catalytic cracking process (Figure 9.8) was developed from the thermal cracking process carried out in tube and tank units. Small amounts of powdered catalyst or a mixture with the feedstock are pumped through a cracking coil furnace. Cracking temperatures are 550°C–610°C (1025°F–1130°F), with pressures of 200–500 psi. After leaving the furnace, the cracked material

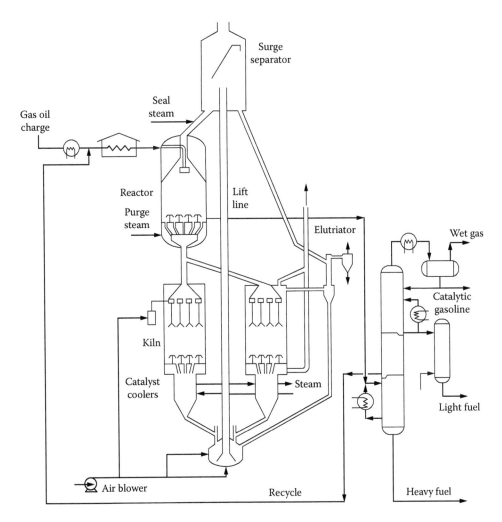

FIGURE 9.6 The Houdriflow moving-bed catalytic cracking process.

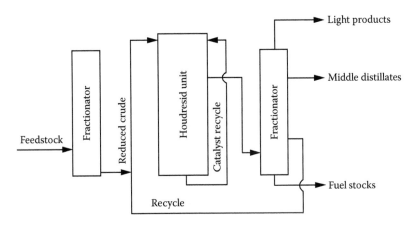

FIGURE 9.7 The Houdresid catalytic cracking process.

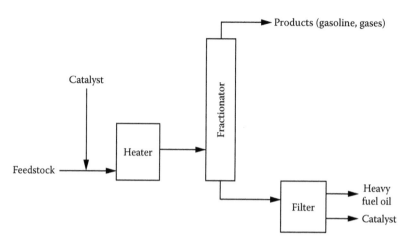

FIGURE 9.8 The suspensoid catalytic cracking process.

enters a tar separator where the catalyst and tar are left behind. The cracked vapors enter a bubble tower where they are separated into two parts, gas oil and pressure distillate. The latter is separated into naphtha and gases. The spent catalyst is filtered from the tar, which is used as a heavy industrial fuel oil. The process is actually a compromise between catalytic and thermal cracking. The main effect of the catalyst is to allow a higher cracking temperature and to assist mechanically in keeping coke from accumulating on the walls of the tubes. The normal catalyst employed is spent clay obtained from the contact filtration of lubricating oils (2–10 lb per barrel of feed).

9.4 FLUID-BED PROCESSES

The application of fluidized solids techniques to catalytic cracking resulted in a major process breakthrough. It was possible to transfer all of the regeneration heat to the reaction zone. Much larger units could be built and higher-boiling feedstocks could be processed. In fact, the improvement in catalysts and unit configurations has permitted the catalytic cracking of higher-boiling (poorer quality) feedstocks such as residua. Presently, there are a number of processes that allow catalytic cracking of heavy oils and residua (Speight and Ozum, 2002; Parkash, 2003; Hsu and Robinson, 2006; Gary et al., 2007; Speight, 2014).

The first fluid catalytic cracking units were Model I upflow units in which the catalyst flowed up through the reaction and regeneration zones in a riser type of flow regime (Figure 9.9). Originally, the Model I unit was designed to feed a reduced crude to a vaporizer furnace where all of the gas oil was vaporized and fed, as vapor, to the reactor. The nonvolatile residuum (bottoms) bypassed the cracking section. The original Model I upflow design (1941) was superseded by the Model II downflow design (1944) (Figure 9.10) followed by the Model III (1947) balanced-pressure design with the later introduction of the Model IV low-elevation design (Figure 9.11).

Of the catalytic cracking process concepts, the fluid catalytic cracking process is the most widely used process and is characterized by the use of a finely powdered catalyst that is moved through the processing unit (Figure 9.12). The catalyst particles are of such a size that when *aerated* with air or hydrocarbon vapor, the catalyst behaves like a liquid and can be moved through pipes. Thus, vaporized feedstock and fluidized catalyst flow together into a reaction chamber where the catalyst, still dispersed in the hydrocarbon vapors, forms beds in the reaction chamber and the cracking reactions take place. The cracked vapors pass through cyclones located in the top of the reaction chamber, and the catalyst powder is thrown out of the vapors by centrifugal force. The cracked vapors then enter the bubble towers where fractionation into light- and heavy-cracked gas oils, cracked naphtha, and cracked gases takes place.

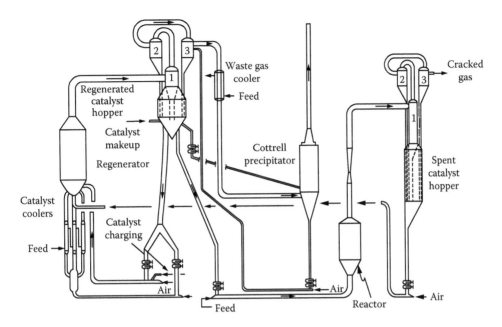

FIGURE 9.9   The Model I catalytic cracking unit.

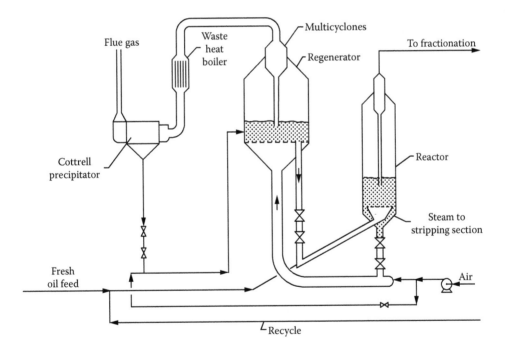

FIGURE 9.10   The Model II catalytic cracking unit.

Since the catalyst in the reactor becomes contaminated with coke, the catalyst is continuously withdrawn from the bottom of the reactor and lifted by means of a stream of air into a regenerator where the coke is removed by controlled burning. The regenerated catalyst then flows to the fresh feed line, where the heat in the catalyst is sufficient to vaporize the fresh feed before it reaches the reactor, where the temperature is approximately 510°C (950°F).

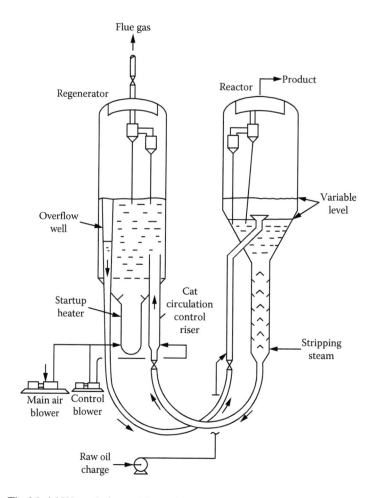

FIGURE 9.11   The Model IV catalytic cracking unit.

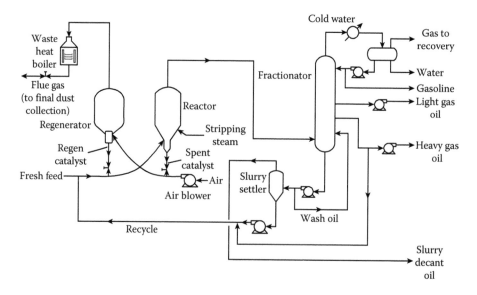

FIGURE 9.12   The fluid-bed catalytic cracking process.

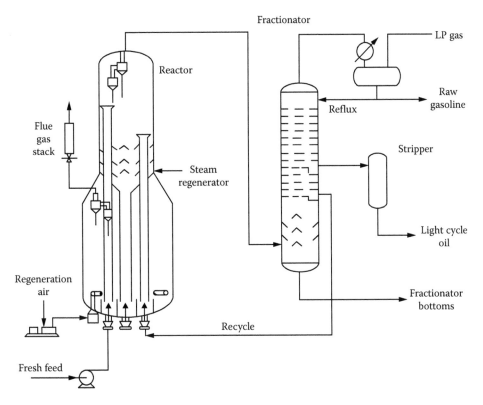

FIGURE 9.13 The orthoflow catalytic cracking unit.

The *Model IV fluid-bed catalytic cracking unit* (Figure 9.11) involves a process in which the catalyst is transferred between the reactor and regenerator by means of U bends and the catalyst flow rate can be varied in relation to the amount of air injected into the spent-catalyst U bend. Regeneration air, other than that used to control circulation, enters the regenerator through a grid, and the reactor and regenerator are mounted side by side.

The *orthoflow fluid-bed catalytic cracking process* (Figure 9.13) uses the unitary vessel design, which provides a straight-line flow of catalyst and thereby minimizes the erosion encountered in pipe bends. Commercial Orthoflow designs are of three types: models A and C, with the regenerator beneath the reactor, and model B, with the regenerator above the reactor. In all cases, the catalyst-stripping section is located between the reactor and the regenerator. All designs employ the heat-balanced principle incorporating fresh feed–recycle feed cracking.

The *Universal Oil Products (UOP) fluid-bed catalytic cracking process* (Figure 9.14) is adaptable to the needs of both large and small refineries. The major distinguishing features of the process are (1) elimination of the air riser with its attendant large expansion joints, (2) elimination of considerable structural steel supports, and (3) reduction in regenerator and in air-line size through use of 15–18 psi pressure operation. The UOP process is also designed to produce low-molecular-weight olefins (for alkylation, polymerization, etherification, or petrochemicals), liquefied petroleum gas, high-octane naphtha, distillates, and fuel oils.

In the process, a side-by-side reactor/regenerator configuration and a preacceleration zone is used to condition the regenerated catalyst before feed injection—the riser terminates in a vortex separation system. A high-efficiency stripper then separates the remaining hydrocarbons from the catalyst, which is then reactivated in a combustorstyle regenerator. The reactor zone features a short-contact-time riser, state-of-the-art riser termination device for quick separation of catalyst and vapor (with high hydrocarbon containment technology), and a portion of the stripped (carbonized) catalyst from the

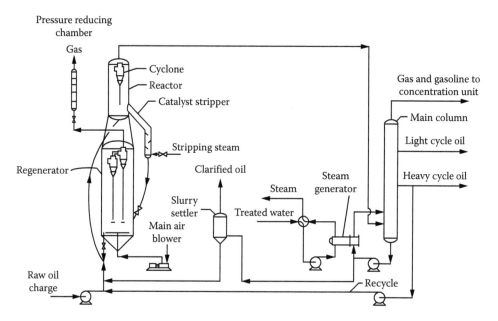

FIGURE 9.14 A *stacked* catalytic cracking unit.

reactor is blended with the hot regenerated catalyst in a proprietary mixing chamber for delivery to the riser. Additionally, the recycling of cooler partially spent catalyst back to the base of the riser lowers the reactor inlet temperature, which results in a reduction of undesirable thermally produced products, including dry gas. The ability to vary the carbonized/regenerated catalyst ratio provides considerable flexibility to handle changes in feedstock quality and enables a real-time switch between gasoline, olefin, and distillate operating modes. For heavier feedstocks, a two-stage regenerator is used—in the first stage, the bulk of the carbon is burned from the catalyst, forming a mixture of carbon monoxide and carbon dioxide. In the second stage, where the remaining coke is burned from the catalyst resulting in only low levels of carbon on the regenerated catalyst, a catalyst cooler is located between the stages.

The Shell two-stage fluid-bed catalytic cracking process (Figure 9.15) was devised to permit greater flexibility in shifting product distribution when dictated by demand. Thus, feedstock is first contacted with cracking catalyst in a riser reactor, that is, a pipe in which fluidized catalyst and vaporized oil flow concurrently upward, and the total contact time in this first stage is of the order of seconds. High temperatures 470°C–565°C (875°F–1050°F) are employed to reduce undesirable coke deposits on catalyst without destruction of naphtha by secondary cracking. Other operating conditions in the first stage are a pressure of 16 psi and a catalyst/oil ratio of 3:1 to 50:1, and volume conversion ranges between 20% and 70% have been recorded.

All or part of the unconverted or partially converted gas oil product from the first stage is then cracked further in the second-stage fluid-bed reactor. Operating conditions are 480°C–540°C (900°F–1000°F) and 16 psi with a catalyst/oil ratio of 2:1 to 12:1. Conversion in the second stage varies between 15% and 70%, with an overall conversion range of 50%–80%.

The residuum cracking unit M.W. Kellogg Company/Phillips Petroleum Company (Figure 9.16) offers conversions up to 85% w/w of atmospheric residua or equivalent feedstocks (Table 9.7). The unit is similar to the Orthoflow C unit (Figure 9.13), but there are some differences that enhance performance on residua. The catalyst flows from the regenerator through a plug valve that controls the flow to hold the reactor temperature. Steam is injected upstream of the feed point to accelerate the catalyst and disperse it so as to avoid high rates of coke formation at the feed point. The feedstock, atomized with steam, is then injected into this stream through a multiple nozzle arrangement. The flow rates are adjusted to control the contact time in the riser since the effects of metals poisoning on

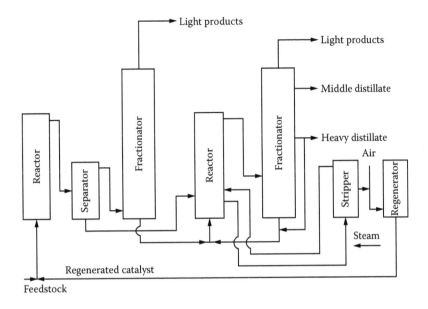

FIGURE 9.15 A two-stage fluid catalytic cracking process.

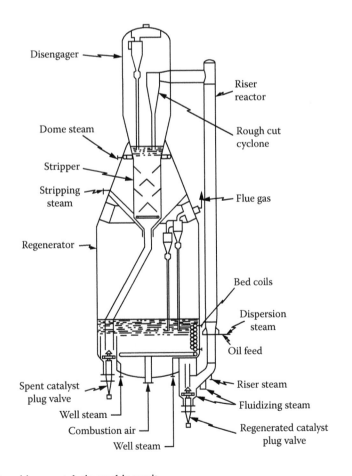

FIGURE 9.16 A residuum catalytic cracking unit.

TABLE 9.7
Representative Product Slate for Residuum Catalytic Cracking

	Vol%		Wt%
Product yield			
C$_5$/400	59.7		48.2
Light cycle oil	13.3		13.3
Decant oil	5.0		5.7
Coke		13.3	
Product quality			
Gasoline, API		−56	
Sulfur, wt%		0.25	
Octane			
RON		90	
MON		79	
Light cycle oil, API			
Sulfur, wt%		2.2	
Decant oil, API			
Sulfur, wt%		5.5	

Abbreviations: RON, research octane number; MON, motor octane number.

yields are claimed to be largely a function of the time that the catalyst and oil are in contact. Passing the mix through a rough cut cyclone stops the reaction.

The *Gulf residuum process* consists of cracking a residuum that has been previously hydrotreated to low sulfur and metal levels. In this case, high conversions are obtained but coke yield and hydrogen yield are kept at conventional levels by keeping metals on catalyst low.

Other processes include the deep catalytic cracking (DCC) process (Figure 9.17) that is a fluidized catalytic process for selectively cracking a variety of feedstocks to low-molecular-weight

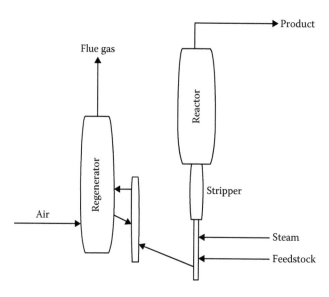

FIGURE 9.17   The *deep* catalytic cracking process.

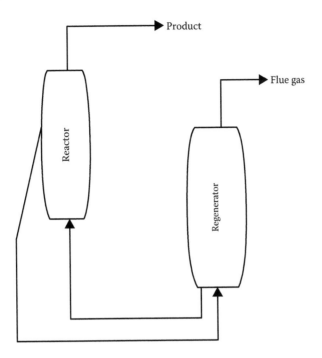

FIGURE 9.18   The Flexicracking process.

olefins (Chapin and Letzsch, 1994). A traditional reactor/regenerator unit design is employed with a catalyst having physical properties much like those of a catalyst for a fluid catalytic cracking unit. The unit may be operated in two operational modes: maximum propylene (Type I) or maximum iso-olefins (Type II) and each mode utilizes a unique catalyst as well as reaction conditions. The Type I unit uses both riser and bed cracking at relatively severe reactor conditions while the Type II unit uses only riser cracking like a modern fluid catalytic cracking (FCC) unit at milder conditions. The overall flow scheme of the process is similar to that of a conventional fluid catalytic cracking unit, but changes in the areas of catalyst development, process variable selection and severity, and gas plant design enable the production of higher yields of olefins than the conventional fluid catalytic cracking processes. The products are light olefins, high-octane naphtha, light cycle oil, dry gas, and coke. Propylene yields over 24% w/w are achievable with paraffin feedstocks.

There is also the Flexicracking process (Figure 9.18) that is designed for conversion of gas oils, residua, and deasphalted oils to distillates (Draemel, 1992).

9.5   COKE FORMATION AND ADDITIVES

9.5.1   COKE FORMATION

The formation of coke deposits has been observed in virtually every unit in operation, and the deposits that can be very thick with thicknesses up to 4 ft have been reported (McPherson, 1984). Coke has been observed to form where condensation of hydrocarbon vapors occurs. The reactor walls and plenum offer a colder surface where hydrocarbons can condense. Higher-boiling constituents in the feedstock may be very close to their dew point, and they will readily condense and form coke nucleation sites on even slightly cooler surfaces.

Nonvaporized feedstock droplets readily collect to form coke precursors on any available surface since the high-boiling feedstock constituents do not vaporize at the mixing zone of the riser. Thus, it is not surprising that residuum processing makes this problem even worse. Low residence time cracking also contributes to coke deposits since there is less time for heat to transfer to feed droplets and vaporize them. This is an observation in line with the increase in coking when short-contact-time *riser crackers* (*q.v.*) were replacing the longer residence time fluid bed reactors.

Higher-boiling feedstocks that have high aromaticity result in higher yields of coke. Furthermore, polynuclear aromatics and aromatics containing heteroatoms (i.e., nitrogen, oxygen, and sulfur) are more facile coke makers than simpler aromatics (Hsu and Robinson, 2006). Feedstock quality alone is not a foolproof method of predicting where coking will occur. However, it is known that feedstock hydrotreaters rarely have coking problems. The hydrotreating step mitigates the effect of the coke formers and coke formation is diminished.

The recognition that significant postriser cracking occurs in commercial catalytic cracking units resulting in substantial production of dry gas and other low-value products (Avidan and Krambeck, 1990). There are two mechanisms by which this postriser cracking occurs, thermal and dilute phase catalytic cracking.

Thermal cracking results from extended residence times of hydrocarbon vapors in the reactor disengaging area and leads to high dry gas yields via nonselective free radical cracking mechanisms. On the other hand, dilute phase catalytic cracking results from extended contact between catalyst and hydrocarbon vapors downstream of the riser. While much of this undesirable cracking was eliminated in the transition from bed to riser cracking, there is still a substantial amount of nonselective cracking occurring in the dilute phase due to the significant catalyst holdup.

Many catalytic cracking units are equipped with advanced riser termination systems to minimize postriser cracking (Long et al., 1993). However, due to the complexity and diversity of catalytic cracking units, there are many variations of these systems and many such as closed cyclones and many designs are specific to the unit configuration but all serve the same fundamental purpose of reducing the undesirable postriser reactions. Furthermore, there are many options for taking advantage of reduced postriser cracking to improve yields. A combination of higher reactor temperature, higher cat/oil ratio, higher feed rate, and/or poorer quality feed is typically employed. Catalyst modification is also appropriate and typical catalyst objectives such as low coke and dry gas selectivity are reduced in importance due to the process changes, while other features such as activity stability and bottoms cracking selectivity become more important for the new unit constraints.

Certain catalyst types seem to increase coke deposit formation. For example, these catalysts (some rare earth zeolites) that tend to form aromatics from naphthenes as a result of secondary hydrogen transfer reactions and the catalysts contribute to coke formation indirectly because the products that they produce have a greater tendency to be coke precursors. In addition, high-zeolite-content, low-surface-area cracking catalysts are less efficient at heavy oil cracking than many amorphous catalysts because the nonzeolite catalysts contained a matrix that was better able to crack heavy oils and convert the coke precursors. The active matrix of some modern catalysts serves the same function.

Once coke is formed, it is matter of where it will appear. Coke deposits are most often found in the reactor (or disengager), transfer line, and slurry circuit and cause major problems in some units such as increased pressure drops, when a layer of coke reduces the flow through a pipe, or plugging, when chunks of coke spall off and block the flow completely. Deposited coke is commonly observed in the reactor as a black deposit on the surface of the cyclone barrels, reactor dome, and walls. Coke is also often deposited on the cyclone barrels 180° away from the inlet. Coking within the cyclones can be potentially very troublesome since any coke spalls going down into the dip leg could restrict catalyst flow or jam the flapper valve. Either situation reduces cyclone efficiency and can increase catalyst losses from the reactor. Coke formation also occurs at nozzles, which can increase the nozzle pressure drop. It is possible for steam or instrument nozzles to be plugged completely, a serious problem in the case of unit instrumentation.

Coking in the transfer line between the reactor and main fractionator is also common, especially at the elbow where it enters the fractionator. Transfer line coking causes pressure drop and spalling and can lead to reduced throughput. Furthermore, any coke in the transfer line that spalls off can pass through the fractionator into the circulating slurry system where it is likely to plug up exchangers, resulting in lower slurry circulation rates and reduced heat removal. Pressure balance is obviously affected if the reactor has to be run at higher pressures to compensate for transfer line coking. On units where circulation is limited by low slide valve differentials, coke laydown may then indirectly reduce catalyst circulation. The risk of a flow reversal is also increased. In units with reactor grids, coking increases grid pressure drop, which can directly affect the catalyst circulation rate.

Shutdowns and startups can aggravate problems due to coking. The thermal cycling leads to differential expansion and contraction between the coke and the metal wall that will often cause the coke to spall in large pieces. Another hazard during shutdowns is the possibility of an internal fire when the unit is opened up to the atmosphere. Proper shutdown procedures that ensure that the internals have sufficiently cooled before air enters the reactor will eliminate this problem. In fact, the only defense against having coke plugging problems during start up is to thoroughly clean the unit during the turnaround and remove all the coke. If strainers on the line(s), they will have to be cleaned frequently.

The two basic principles to minimize coking are to avoid dead spots and prevent heat losses. An example of minimizing *dead spots* is using purge steam to sweep out stagnant areas in the disengager system. The steam prevents collection of high-boiling condensable products in the cooler regions. Steam also provides a reduced partial pressure or steam distillation effect on the high-boiling constituents and cause enhanced vaporization at lower temperatures. Steam for purging should preferably be superheated since medium-pressure low-velocity steam in small pipes with high heat losses is likely to be very wet at the point of injection and will cause more problems. *Cold spots* are often caused by heat loss through the walls in which case increased thermal resistance might help reduce coking. The transfer line, being a common source of coke deposits, should be as heavily insulated as possible, provided that stress-related problems have been taken into consideration.

In some cases, changing catalyst type or the use of an additive (*q.v.*) can alleviate coking problems. The catalyst types that appear to result in the least coke formation (not delta coke or catalytic coke) contain low or zero earth zeolites with moderate matrix activities. Eliminating heavy recycle streams can lead to reduced coke formation. Since clarified oil is a desirable feedstock to make needle coke in a coker, then it must also be a potential coke maker in the disengager.

One of the trends in recent years has been to improve product yields by means of better feed atomization. The ultimate objective is to produce an oil droplet small enough so that a single particle of catalyst will have sufficient energy to vaporize it. This has the double benefit of improving cracking selectivity and reducing the number of liquid droplets that can collect to form coke nucleation sites.

9.5.2 ADDITIVES

In addition to cracking catalyst described earlier, a series of *additives* has been developed that catalyze or otherwise alter the primary catalyst's activity/selectivity or act as pollution control agents. Additives are most often prepared in microspherical form to be compatible with the primary catalysts and are available separately in compositions that (1) enhance naphtha octane and light olefin formation, (2) selectively crack heavy cycle oil, (3) passivate vanadium and nickel present in many heavy feedstocks, (4) oxidize coke to carbon dioxide, and (5) reduce sulfur dioxide emissions.

Both vanadium and nickel deposit on the cracking catalyst and are extremely deleterious when present in excess of 3000 ppm on the catalyst. Formulation changes to the catalyst can improve tolerance to vanadium and nickel but the use of additives that specifically passivate either metal is often preferred.

9.6 PROCESS VARIABLES

9.6.1 FEEDSTOCK QUALITY

Generally, the ability of any single unit to accommodate wide variations in feedstock is an issue related to the flexibility of the process (Navarro et al., 2015). Initially, catalytic cracking units were designed to process gas oil feedstocks but many units have been modified successfully, and new units designed, to handle more complex feedstocks, and feedstock blends, containing residua.

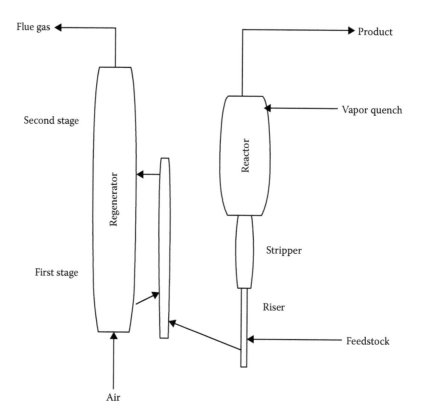

FIGURE 9.19 A residuum catalytic cracking unit.

Vacuum gas oil (ibp 315°C–345°C, 600°F–650°F; fbp 510°C–565°C, 950°F–1050°F), as produced by vacuum flashing or vacuum distillation, is the usual feedstock with final boiling point being limited by the carbon forming constituents (measured by the Conradson carbon residue) or metals content since both properties have adverse effects on cracking characteristics. The vacuum residua (565°C+, 1050°F+) are occasionally included in cat cracker feed when the units (residuum cat crackers) are capable of handling such materials (Figure 9.19). In such cases, if the residua are either relatively low in terms of carbon-forming constituents and metals (as, e.g., a residuum from a waxy crude) so that the effects of these properties are relatively small. Many units also recycle slurry oil (455°C+, 850°F+) and a heavy cycle oil stream. Gas oils from thermal cracking or coking processes (Chapter 8), gas oils from hydrotreating processes (Chapter 10), and gas oils from deasphalting processes (Chapter 12) are often included in feedstocks for catalytic cracking units (Speight and Ozum, 2002; Parkash, 2003; Hsu and Robinson, 2006; Gary et al., 2007; Speight, 2014).

The general feedstock quality effects can be indicated by characterization factor, K:

$$K = \frac{\left(MABP\right)^{1/3}}{\text{Specific gravity at } 60°F/60°F}$$

MABP is the mean average boiling point expressed in degrees Rankin (°R = °F + 460). However, it is a single parameter such that this can only reflect general trends (Table 9.8) and even then the accuracy and meaningful nature of the data may be very questionable. Generally speaking, coke yield increases as the characterization factor (K) decreases or as the feed becomes less paraffinic

TABLE 9.8
General Indications of Feedstock Cracking

Characterization Factor, K	Naphtha Yield (vol%)	Coke Yield (wt%)	Feedstock Type	Relative Reactivity (Relative Crackability)
11.0[a]	35.0	13.5	Aromatic	Refractory; estimated coke yield
11.2[b]	49.6	12.5	Aromatic	Refractory
11.2[a]	37.0	11.5	Aromatic	Refractory
11.4[b]	47.0	9.1	Aromatic–naphthenic	Intermediate
11.4[a]	39.0	9.0	Aromatic–naphthenic	Intermediate
11.6[b]	45.0	7.1	Naphthenic	Intermediate
11.6[a]	40.0	7.2	Naphthenic	Intermediate
11.8[b]	43.0	5.3	Naphthenic–paraffinic	High
11.8[a]	41.0	6.0	Naphthenic–paraffinic	High
12.0[b]	41.5	4.0	Naphthenic–paraffinic	High
12.0[a]	41.5	5.3	Naphthenic–paraffinic	High
12.2[b]	40.0	3.0	Paraffinic	High

[a] Cycle oil/cracked feedstocks; 60% conversion.
[b] Straight-run/uncracked feedstocks; 60% conversion.

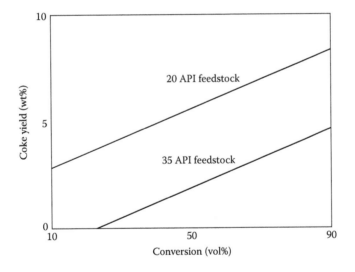

FIGURE 9.20 Relationship of coke yield to conversion and feedstock API gravity.

and the API gravity decreases as the conversion increases (Figure 9.20) (Maples, 2000). With straight-run gas oils, naphtha yield increases as the characterization factor decreases (i.e., the paraffin character of the oil decreases), but the opposite effect is obtained with cracked stocks or cycle oils.

Either molecular weight, or average boiling point, or feed boiling range is an important feedstock characteristic in determining cat cracking yields and product quality. In general, for straight-run fractions, thermal sensitivity (increased thermal decomposition or cracking) increases as molecular weight increases; coke and naphtha production (at constant processing conditions) also increase with the heavier feedstocks but, not to become too enthusiastic approximately the word *increase*, coke yield also increases (Figure 9.21).

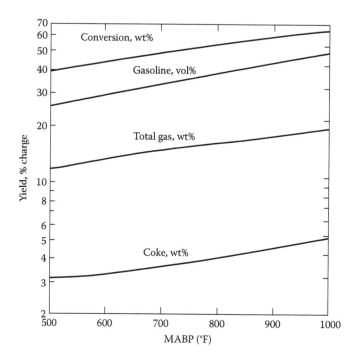

FIGURE 9.21   Relationship of product slate to feedstock boiling point.

However, the characterization factor and feed boiling range are generally insufficient to characterize a feedstock for any purpose other than approximate comparisons. A more detailed description of the feedstock is needed to reflect and predict the variations in feedstock composition and cracking behavior (Speight, 2000).

Irrespective of the source of the high-boiling feedstocks, a number of issues typically arise when these materials are processed in a fluid catalytic cracking unit, although the magnitude of the problem can vary substantially.

Heavy feedstocks have high levels of contaminants (Table 9.9) that will affect the process and must be removed. Examples are the carbon-forming constituents that yield high levels of (Conradson) carbon residue, and the overall coke production (as carbon on the catalyst) is high. Burning this coke requires additional regeneration air that might be constraint that limits the capacity of the unit. Metals in the heavy feedstocks also deposit (almost quantitatively) on the catalyst where two significant effects are caused. First, the deposited metals can accelerate certain metal-catalyzed dehydrogenation reactions, thereby contributing to light-gas (hydrogen) production and to the formation of additional coke. A second, and more damaging, effect is the situation in which the deposition of the metals causes a decline in catalyst activity because of the limited access to the active catalytic sites. This latter effect is normally controlled by catalyst makeup practices (adding and withdrawing catalyst).

The amount of sulfur and nitrogen in the products, waste streams, and flue gas generally increases when high-boiling feedstocks are processed because these feed components typically have higher sulfur and nitrogen contents than gas oil. However, in the case of nitrogen, the issue is not only one of higher nitrogen levels in the products but also (because of the feedstock nitrogen is basic in character) catalyst poisoning that reduces the useful activity of the catalyst.

Heat balance control may be the most immediate and troublesome aspect of processing high-boiling feedstocks. As the contaminant carbon increases, the first response is usually to increase regenerator temperature. Adjustments in operating parameters can be made to assist in this control, but eventually, a point will be reached for heavier feedstocks when the regenerator temperature is

TABLE 9.9

Feedstock Contaminants That Affect Catalytic Cracking Processes[a]

Contaminants	Effect on Catalyst	Mitigation	Process
Sulfur	Catalyst fouling Deactivation of active sites	Hydrodesulfurization	Hydroprocessing
Nitrogen	Adsorption of basic nitrogen Destruction of active sites	Hydrodemetallization	Hydroprocessing
Metals	Fouling of active sites Fouling of pores	Demetallization	Demet, Met-X
Particulate matter	Deactivation of active sites Pore plugging	Filter/pretreatment	Clay filtration/guard bed
Coke precursors	Formation of coke Catalyst fouling Deactivation of active sites Pore plugging	Remove asphaltene constituents Remove resin constituents	Mild hydrocracking/ hydroprocessing

[a] Also applicable to hydrocracking processes.

too high for good catalytic performance. At this point, some external heat removal from the regenerator is required and would necessitate a mechanical modification like a catalyst cooler.

For the last two decades, demetallized oil (produced by the extraction of a vacuum-tower bottoms stream using a light paraffinic solvent) has been included as a component of the feedstock in fluid catalytic cracking units. Modern solvent extraction processes, such as the Demex process, provide a higher demetallized oil yield than is possible in the propane-deasphalting process that has been used to prepare fluid catalytic cracker feedstock, as well as demetallized feedstocks for other processes. Consequently, the demetallized oil is more heavily contaminated. In general, demetallized oils are still good cracking stocks, but most feedstocks can be further improved by hydrotreating to reduce contaminant levels and to increase their hydrogen content thereby becoming a more presentable and process-friendly feedstock.

In many cases, atmospheric residua have been added as a blended component to feedstocks for existing fluid catalytic cracking units as a means of converting the highest boiling constituents of crude oil. In fact, in some cases the atmospheric residuum has ranged from a relatively low proportion of the total feed all the way to a situation in which it represents the entire feed to the unit. To improve the handling of these high-boiling feedstocks, several units have been revamped to upgrade them from their original gas oil designs whereas other units have taken a stepwise approach to residuum processing whereby modifications to the operating conditions and processing techniques are made as more experience is gained in the processing of residua.

9.6.2 FEEDSTOCK PREHEATING

In a heat-balanced commercial operation, increasing the temperature of the feed to a cracking reactor reduces the heat that must be supplied by combustion of the coked catalyst in the regenerator. Feedstock preheating is usually supplied by heat exchange with hot product streams, a fired preheater, or both. When feed rate, recycle rate, and reactor temperature are held constant as feed preheat is increased, the following changes in operation result: (1) The catalyst/oil ratio (catalyst circulation rate) is decreased to hold the reactor temperature constant. (2) Conversion and all conversion-related yields, including coke, decline due to the decrease in catalyst/oil ratio and severity. (3) The regenerator temperature will usually increase. Although the total heat released in the regenerator and the air required by the regenerator are reduced by the lower coke yield, the lower catalyst

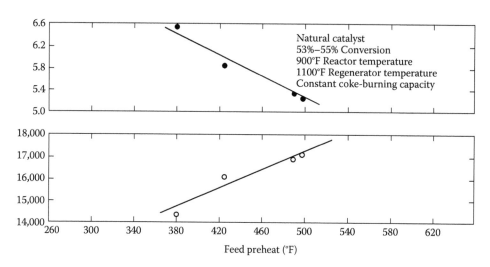

FIGURE 9.22   The effect of feedstock preheating on coke yield.

circulation usually overrides this effect and results in an increase in regenerator temperature, and (4) as a result of the lower catalyst circulation rate, residence time in the stripper and overall stripper efficiency is increased and liquid recovery is increased and a corresponding decrease in coke usually results. Advantage is usually taken of these feed preheat effects, including the reduced air requirement, by increasing the total feed rate until coke production again requires all of the available air (Figure 9.22).

9.6.3   FEEDSTOCK PRESSURE

The pressure in a catalytic cracking unit is generally set slightly above atmospheric by balancing the yield and quality debits of high pressure plus increased regeneration air compression costs against improved era king and regeneration kinetics, the lower cost of smaller vessels, plus, in some cases, power recovery from the regenerator stack gases. Representative yield and product effects (Table 9.10) show, at the same conversion level, that coke and naphtha yields are increased marginally and light gas yields are reduced at the higher pressure. The sulfur content of the naphtha fraction is reduced.

Pressure levels in commercial units are generally in the range 15–35 psi. Lowering the partial pressure of the reacting gases with steam will improve yields somewhat, but the major beneficial effect of feed injection steam is that it atomizes the feed to small droplets that will vaporize and react quickly. If feed is not atomized, it will soak into the catalyst and possibly crack to a higher coke make.

Both pressure and partial pressure of the feedstock, or steam/feedstock ratio, are generally established in the design of a commercial unit and thus are usually not available as independent variables over any significant range. However, in some units, injector steam is varied over a narrow range to balance carbon make with regeneration carbon burn off.

9.6.4   FEEDSTOCK CONVERSION

All of the independent variables in catalytic cracking have a significant effect on conversion that is truly a dependent variable but can be shown as a function of API gravity as well as a variety of other functions (Maples, 2000). The detailed effects of changing conversion depend on which the conversion is changed, that is, by temperature, space velocity, or catalyst/oil ratio, and catalyst activity.

TABLE 9.10
Yield Variations with Pressure

Total pressure, psi gauge	12	50
Conversion, vol%	65–67	65–66
Yields		
H_2, wt%	0.2	0.1
C_1, wt%	1.9	1.2
C_2^{2-}, wt%	1.4	0.9
C_2, wt%	1.2	1.5
C_3^{2-}, wt%	5.3	3.9
C_3, wt%	3.9	4.4
iC_4^{2-}, vol%	11.8	10.7
nC_4^{2-}, vol%	2.0	1.2
iC_4, vol%	4.4	3.5
nC_4, vol%	2.4	3.2
iC_5, vol%	6.3	5.6
C_5^{2-}, vol%	3.1	3.0
nC_5, vol%	1.4	1.6
C_6, 284°F, 90% pt, vol%	16.7	19.6
Heavy naphtha, vol%	8.2	7.2
C_6, 410°F, EP gasoline, vol%	24.9	26.8
Cycle oil, vol%	32.9	33.3
Carbon, wt%	6.1	6.8
Quality, C_6, 410°F, gasoline		
API	39	44
Sulfur, wt%	0.14	0.09
RON clear	98	98
ASTM clear	86	84

Increasing conversion increases yields of naphtha and all light products up to a conversion level of 60%–80% by volume in most cases. At this high conversion level, secondary reactions become sufficient to cause a decrease in the yields of olefins and naphtha. However, the feedstock, operating conditions, catalyst activity, and other parameters are the points at which this occurs.

9.6.5 REACTOR TEMPERATURE

The principal effects of increasing reactor temperature at constant conversion are to decrease naphtha yield and coke yield and to increase yields of methane (CH_4), ethane (C_2H_6), propane ($CH_3CH_2CH_3$), and total butane (C_4H_{10}) yield; yields of the pentanes (C_5H_{12}) and higher-molecular-weight paraffin decrease while olefin yields are increased.

The effect of reactor temperature on a commercial unit is, of course, considerably more complicated as variables other than temperature must be changed to maintain heat balance. For example, in order to increase the temperature of a reactor at constant fresh feed rate, the interrelated changes of recycle rate, space velocity, and feedstock preheat are required to maintain heat balance on the unit by increasing circulation rate and coke yield. The combined effects of higher reactor temperature and higher conversion resulted in the following additional changes: (1) increased yields of butanes and propane, (2) naphtha yield is increased, (3) the yield of light catalytic cycle oil is decreased. Thus, the effects of an increase in reactor temperature on an operating unit reflect not only the effects of temperature *per se* but also the effects of several concomitant changes such as increased conversion and increased catalyst/oil ratio.

9.6.6 Recycle Rate

With most feedstocks and catalyst, naphtha yield increases with increasing conversion up to a point, passes through a maximum, and then decreases. This phenomenon (*overcracking*) is due to the increased thermal stability (*refractory character*) of the unconverted feed as conversion increases and the destruction of naphtha through secondary reactions, primarily cracking of olefins. The onset of secondary reactions and the subsequent leveling off or decrease in naphtha yield can be avoided by recycling a portion of the reactor product, usually a fractionator product with a boiling points on the order of 345°C–455°C (650°F–850°F). Other tests have shown the following effects of increasing recycle rate when space velocity was simultaneously adjusted to maintain conversion constant: (1) the naphtha yield increased significantly; (2) the coke yield decreases appreciably; (3) there is a decrease in the yield of dry gas components, propylene, and propane; (4) the yield of butane decreased while the yield of butylene increased; and (5) the yields of light catalytic cycle oil and clarified oil increased but heavy catalytic cycle oil yield decreased.

With the introduction of high activity zeolite catalysts, it was found that in once-through cracking operations with no recycle the maximum in naphtha yield was located at much higher conversions. In effect, the higher activity catalysts were allowing higher conversions to be obtained at severity levels that significantly reduced the extent of secondary reactions (*overcracking*). Thus, on many units employing zeolite catalysts, recycle has been eliminated or reduced to less than 15% of the fresh feed rate.

9.6.7 Space Velocity

The role of space velocity as an independent variable arises from its relation to *catalyst contact time* or *catalyst residence time*. Thus,

$$\Theta = \frac{60}{(\text{WHSV} \times \text{C/O})}$$

where
    Θ is catalyst residence time in minutes
    WHSV is the weight hourly space velocity on a total weight basis
    C/O is the catalyst/oil weight ratio

The catalyst/oil ratio is a dependent variable so that catalyst time becomes directly related to the weight hourly space velocity. When catalyst contact time is low, secondary reactions are minimized, thus naphtha yield is improved and light as and coke yields are decreased. In dense-bed units, the holdup of catalyst in the reactor can be controlled within limits, usually by a slide valve in the spent catalyst standpipe, and feed rate can also be varied within limits. Thus, there is usually some freedom to increase space velocity and reduce catalyst residence time. In a riser-type reactor, holdup and feed rate are not independent and space velocity is not a meaningful term. Nevertheless, with both dense-be and riser-type reactors, contact times are usually minimized to improve selectivity. An important step in this direction was the introduction of high activity zeolite catalyst. These catalysts require short contact times for optimum performance and have generally moved cracking operations in the direction of minimum holdup in dense-bed reactors or replacement of dense-bed reactors with short-contact-time riser reactors.

Strict comparisons of short-contact-time riser cracking versus the longer contact time dense-bed mode of operations are generally not available due to differences in cat activity, carbon content of the regenerated catalyst, or factors other than contact time but inherent in the two modes of operation. However, in general, improvements in catalyst activity have resulted in the need for less catalyst in the reaction zone.

Many units are designed with only riser cracking; that is, no dense-bed catalyst reactor cracking occurs and all cracking is done in the catalyst/oil transfer lines leading into the reactor cyclone vessel. However, in some of these instances the reactor temperature must be increased to the 550°C–565°C (1020°F–1050°F) range in order to increase the intensity of cracking conditions to achieve the desired conversion level. This is because not enough catalyst can be held in the riser zone, since the length of the riser is determined by the configuration and elevation of the major vessels in the unit. Alternatively, super-active catalysts can be used to achieve the desired conversion in the riser.

Significant selectivity disadvantages have not been shown if a dispersed catalyst phase or even a very small dense bed is provided downstream of the transfer line riser cracking zone. In this case the cracking reaction can be run at a lower temperature, say 510°C (950°F), which will reduce light gas make and increase naphtha yield when compared to the higher-temperature (550°C–565°C, 1020°F–1050°F) operation.

9.6.8 Catalyst Activity

Catalyst activity as an independent variable is governed by the capability of the unit to control the carbon content of the spent catalyst and the quantity and quality of fresh catalyst that can be continuously added to the unit. The carbon content of the regenerated catalyst is generally maintained at the lowest practical level to obtain the selectivity benefits of low carbon on the catalyst. Thus, catalyst addition is, in effect, the principal determinant of catalyst activity.

The deliberate withdrawal of catalyst over and above the inherent loss rate through regenerator stack losses and decant or clarified oil, if fly, and a corresponding increase in fresh catalyst addition rate is generally not practiced as a means of increasing the activity level of the circulating catalyst. If a higher activity is needed, the addition of a higher activity fresh catalyst to the minimum makeup rate to maintain inventory is usually the more economical route. The general effects of increasing activity are to permit a reduction in severity and thus reduce the extent of secondary cracking reactions. Higher activity typically results in more naphtha and less coke. In other cases, higher activity catalysts are employed to increase the feed rate at essentially constant conversion and constant coke production so that the coke burning or regenerator air compression capacities are fully utilized.

9.6.9 Catalyst/Oil Ratio

The dependent variable catalyst/oil ratio is established by the unit heat balance and coke make that in turn are influenced by almost every independent variable. Since catalyst/oil ratio changes are accompanied by one or more shifts in other variables, the effects of catalyst/oil ratio are generally associated with other effects. A basic relation, however, in all catalyst/oil ratio shifts is the effect on conversion and carbon yield. At constant space velocity and temperature, increasing catalyst/oil ratio increases conversion (Figure 9.23). In addition to increasing conversion, higher catalyst/oil ratios generally increase coke yield at a constant conversion. This increase in coke is related to the hydrocarbons entrapped in the pores of the catalyst and carried through the stripper to the regenerator. Thus, this portion of the catalyst/oil ratio effect is highly variable and depends not only on the catalyst/oil ratio change but also on the catalyst porosity and stripper conditions (Figure 9.24). The following changes, in addition to increased coke yield, accompany an increase in the catalyst/oil ratio in the range of 5–20 at constant conversion, reactor temperature, and catalyst activity: (1) decreased hydrogen yield, (2) decreased methane to butane(s) yields, and (3) little effect on the naphtha yield or octane number.

The catalyst/oil ratio (v/v) ranges from 5:1 to 30:1 for the different processes, although most processes are operated to 10:1. By comparison, the catalyst/oil volume ratio for moving-bed processes may be substantially lower than 10:1. In a traditional FCC unit, increasing the catalyst/oil ratio to increase conversion also increases the coke yield and catalyst circulation to the regenerator.

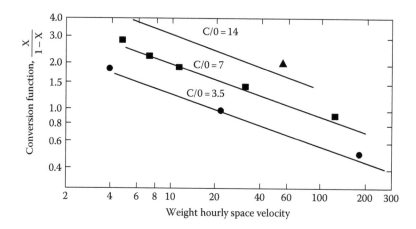

FIGURE 9.23   Effect of catalyst/oil ratio and space velocity on feedstock conversion.

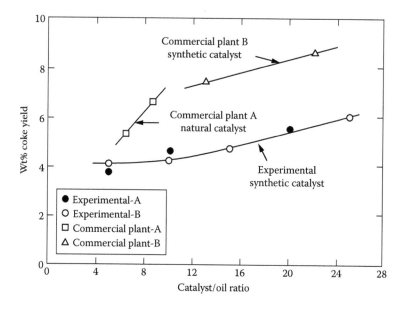

FIGURE 9.24   Effect of catalyst/oil ratio on coke yield.

A high catalyst oil ratio is necessary to maintain high reaction temperature by transferring enough heat from regenerator to reactor in commercial units. It is possible, using a specific type of catalyst (containing low acid density zeolite) to suppress hydrogen transfer and maximize olefin production (Soni et al., 2009; Fujiyama et al., 2010).

The UOP RxCat technology provides the ability to increase both conversion and selectivity by recycling a portion of the carbonized catalyst back to the base of the reactor riser. The carbonized catalyst circulated from the stripper back to the base of the riser is effectively at the same temperature as the reactor. Since the recycle catalyst adds no heat to the system, the recycle is heat balance neutral. For the first time, the catalyst circulation up the riser can be varied independently from the catalyst circulation rate to the regenerator and is decoupled from the unit heat balance (Wolschlag and Couch, 2010b; Wolschlag, 2011). When process conditions are changed so that an increase in the catalyst/oil ratio occurs, an increase in conversion is also typically observed.

By increasing the catalyst/oil ratio, the effects of operating at high reaction temperature (thermal cracking) are minimized. High catalyst/oil ratio maintains heat balance thereby achieving high reaction temperature. It also increases conversion and maximizes light olefins production (Maadhah et al., 2008).

9.6.10 Regenerator Temperature

Catalyst circulation, coke yield, and feedstock preheating are the principal determinants of regenerator temperature that is generally allowed to respond as a dependent variable within limits. Mechanical or structural specifications in the regenerator section generally limit regenerator temperature to a maximum value specific to each unit. However, in some cases the maximum temperature may be set by catalyst stability. In either case, if regenerator temperature is too high it can be reduced by decreasing feedstock preheat; catalyst circulation is then increased to hold a constant reactor temperature and this increased catalyst circulation will carry more heat from the regenerator and lower the regenerator temperature. The sequence of events is actually more complicated as the shift in catalyst/oil ratio and, to a lesser extent, the shift in carbon content of the regenerated catalyst will change coke make and the heat release in the regenerator.

9.6.11 Regenerator Air Rate

The amount of air required for regeneration depends primarily on coke production. Regenerators have been operated with only a slight excess of air leaving the dense phase. With less air, carbon content of the spent catalyst increases and a reduction in coke yield is required of air, burning of carbon monoxide (CO) to carbon dioxide (CO_2) above the dense bed will occur. This *after burning* must be controlled as the extremely high temperature because of the absence of the heat sink provided by the catalyst.

In terms of catalyst regeneration, higher stability catalysts are available and regenerator temperatures can be increased by 38°C–65°C (100°F–150°F) up to the 720°C–745°C (1325°F–1370°F) range without significant thermal damage to the catalyst. At these higher temperatures, the oxidation of carbon monoxide to carbon dioxide is greatly accelerated, and the regenerator can be designed to absorb the heat of combustion in the catalyst under controlled conditions. The carbon burning rate is improved at the higher temperatures, and there usually are selectivity benefits associated with the lower carbon on regenerated catalyst; this high-temperature technique usually results in carbon on regenerated catalyst levels of 0.05% by weight or less. In addition, the use of catalysts containing promoters for the oxidation of carbon monoxide to carbon dioxide produces a major effect of high-temperature regeneration, that is, low-carbon-monoxide-content regenerator stack gases and the resultant regenerator conditions may result in a lower-carbon-content regenerated catalyst.

9.6.12 Process Design

Process design improvement will continue to focus on (1) modification of existing units, (2) commercialized technology changes, and (3) new directions in processing technology to allow processing of a variety of feedstocks (Chen, 2006; Freel and Graham, 2011).

9.6.12.1 Modifications for Existing Units

The feed injection system is by far the most critical breakthrough of FCC reactor design. In addition, a higher regenerator temperature is used to achieve more complete catalyst regeneration. The typical riser top temperature is on the order of 510°C–565°C (950°F–1050°F), but typical regenerated catalyst temperature is much higher—on the order of 675°C–760°C (1250°F–1400°F). Feed injection reduces thermal cracking reactions by cooling off the lower riser quickly through rapid mixing and vaporization of the feed.

As the FCC feedstock moves to heavy crude oil, feed vaporization is more difficult. However, the newest generation of side-entry FCC feed nozzles generates more uniform feed distribution (and more rapid mixing) as a result of better control of homogeneity of two-phase flow and atomization at the nozzle exit using two-phase choke flow. Some older FCC units still retain the original feed injection system located at the bottom of the riser (bottom-entry nozzles). A new generation of feed injection technology uses a similar side-entry atomization mechanism. For catalyst circulation, the bottom-entry nozzles have the advantage of reducing pressure drop through the riser. This system also enables longer riser residence time if riser height is limited.

The newest generation feed nozzles optimize the temperature profile in the riser and substantially reduces dry gas yield, thereby increasing naphtha yield. These results are in line with the expectation that better feed injection design reduces thermal cracking reactions, which are the primary source for dry gas. Process/hardware technologies to improve light cycle oil yield include proper feed injection systems and risers/reaction zone designs as claimed by Petrobras, Shell, and Sinopec in their latest commercial processes.

The FCC riser is known for its shortcomings of density and velocity variations but the newest riser internal technology minimizes these shortcomings and promotes ideal plug flow. Improved riser reaction termination technology sharpens the termination of reactions by the combination of the unique design of primary stripper cyclones and close coupled secondary cyclones as well as designs to reduce coke formation (Hedrick and Palmas, 2011). Due to the development of highly active zeolite FCC catalyst, the reaction time has been shortened significantly to a few seconds in the modern riser reactor. Since catalytic cracking reactions can only occur after the vaporization of the liquid hydrocarbon feedstock, mixing and feed vaporization must take place in the riser as quickly as possible; otherwise, thermal cracking reactions will dominate (Chen, 2004). An efficient product separator suppresses side reactions (oligomerization and hydrogenation of light olefins) and coke formation accelerated by condensation.

The high yields of products generated in the highly selective reactor riser environment must be preserved in the rest of the reaction system. Improvements in the design of riser separation or termination systems focus on the rapid disengagement of catalyst from the cracked products in a highly contained system. Product vapors are quickly directed to the fractionation system for thermal quench and recovery (Ross et al., 2005; Fujiyama et al., 2011). The design improves both the separation system to reduce dry gas and the stripping system to reduce coke. In addition, the pressure drop is extremely low in order to limit dip-leg immersion requirements to seal the positive pressure separator and so that the capacity of the unit will not be limited. The dip-leg size and flux are optimized to minimize gas entrainment with the catalyst and even allow for stripping within the dip-legs.

Prospective techniques for fluid catalytic cracking control technology fall into four general categories: (1) hydrodesulfurization, (2) catalyst additives, (3) scrubbing, and (4) chemical reaction such as selective noncatalytic reduction (SNCR) and selective catalytic reduction (SCR) (Bouziden et al., 2002; Couch et al., 2004). Also, hydrotreating of the FCC feedstock will decrease feedstock sulfur and nitrogen thereby decreasing sulfur emissions from the unit as well as nitrogen in the coke on catalyst, and consequently nitrogen oxide (NOx) emissions from the regenerator.

9.6.12.2 Commercial Technology Changes

Dramatic yield improvements have been demonstrated commercially using the latest fluid catalytic cracking process technology advances. Riser termination designs, in particular, have received much attention, for new units as well as for revamped units. Significant postriser cracking occurs in commercial FCC units resulting in substantial production of dry gas and other lower-value products (Avidan and Krambeck, 1990). Furthermore, the major process licensors have developed advanced riser termination systems to minimize postriser cracking (Draemel, 1992; Khouw, 1992; Wrench and Glascow, 1992; Long et al., 1993; Upson and Wegerer, 1993), and many units have implemented these in both new unit and revamp applications. In addition, some refiners have implemented their own "in-house" designs for the same purpose.

Due to the complexity and diversity of existing FCC units as well as new unit design differences, there are many variations of these systems such as (1) closed cyclones, (2) close-coupled cyclones, (3) direct connected cyclones, (4) coupled cyclones, (5) high containment systems, and (6) short-contact-time systems. There are differences in the specific designs, and some may be more appropriate for specific unit configurations than others, but all serve the same fundamental purpose of reducing the undesirable postriser reactions.

Proper catalyst selection is essential to realizing the maximum potential benefit from these hardware improvements. Both fresh catalyst properties and catalyst management policy are important variables. These include the use of (1) high activity catalyst for maximum conversion or throughput, (2) high matrix catalyst for reducing slurry yield, and (3) a metals tolerant catalyst for processing resid feed. In all cases, improved yields are achieved when incorporating catalyst change effects, which may not have been possible without the advanced hardware, due to unit constraints.

The downer is a gas–solid cocurrent downflow reactor that has the potential to overcome the drawback of a conventional upflow reactor (or a riser) caused by back-mixing of catalyst. In the downer, gas and solid catalysts move downward together with the assist of the gravity; this can avoid the back-mixing of catalyst in the reactor.

The operation of the downer is affected by various key parameters including (1) recycled catalyst flow rate, (2) superficial gas velocity, (3) spent catalyst flow rate, and (4) carbon content on the spent catalyst. The parameters that affect the temperature of the downer regenerator should be carefully selected as they have the most significant effect to a regeneration process. High regeneration temperature could deactivate the catalyst permanently, but low-temperature operation lowers the regeneration performance (Chuachuensuk et al., 2010).

A downflow reactor system has been adopted for the high-severity fluid catalytic cracking (HS-FCC) process. The downer permits higher catalyst/oil ratios because the lifting of catalyst by vaporized feed is not required. As with most reactor designs involving competing reactions and secondary product degradation, there is a concern over catalyst-feed contacting, back-mixing, and control of the reaction time and temperature. The downflow reactor would ensure plug flow without back-mixing (Maadhah et al., 2008).

The development of a highly active zeolite catalyst has led to the reaction time being decreased to a few seconds in the riser reactor. The short contact time (short residence time, <0.5 second) of feed and product hydrocarbons in the downer should be favorable for minimizing thermal cracking. Undesirable secondary reactions such as polymerization reactions and hydrogen transfer reactions, which consume olefins, are suppressed. In order to attain the short residence time, the catalyst and the products have to be mixed and dispersed at the reactor inlet and separated immediately at the reactor outlet. For this purpose, a high-efficiency product separator has been developed capable of suppressing side reactions (oligomerization and hydrogenation of light olefins) and coke formation accelerated by condensation (Nishida and Fujiyama, 2000). The short-contact-time reactor affords (1) minimal back-mixing and erosion, (2) efficient catalyst/oil contacting, (3) reduced hydrogen transfer, and (4) high yield selectivity.

The overcracking of naphtha to gases is minimized by reducing the contact time between catalyst and hydrocarbon products. Addition of ZSM-5 additive enhances the octane number in FCC naphtha by overcracking of naphtha fraction. On the other hand, the octane number enhancement is achieved by overcracking of the middle boiling point fraction with low octane number (Buchanan et al., 1996). Short-contact-time riser cracking is inherently more flexible than a typical fluid catalytic cracking unit because the product slate can be easily adjusted to maximize propylene, maximize naphtha, or produce combinations such as propylene plus ethylene or propylene plus naphtha (Jakkula et al., 1997). This process flexibility is a key variable in maximizing profitability in a given market scenario.

The UOP millisecond catalytic cracking process (MSCC process) involves injection of the feedstock perpendicular to a downflowing steam of catalyst (Schnaith et al., 1998; Harding et al., 2001). The basic MSCC reactor configuration consists of an injection zone, a central dilute phase disengaging zone, the lower dense phase collection zone, and an upper inertial separation zone. The short contact time combined with the low-volume reaction zone reduces secondary cracking reactions and produces more naphtha and less coke compared with conventional fluid catalytic cracking. Another benefit is that the low coke yield allows heavier feedstocks (conventionally high coke-make feedstocks) to be processed.

Improvements in the design of riser separation or termination systems focus on the rapid disengagement of catalyst from the cracked products in a highly contained system. Product vapors are quickly directed to the fractionation system for thermal quench and recovery (Ross et al., 2005). Using a two-riser system, a heavy feedstock can be treated by use of a two riser system (two-stage riser fluid catalytic cracking). The spent catalyst from the other of the two risers is fed to the inlet of the first riser to produce relatively mild cracking conditions. Improved total naphtha plus distillate yields are achieved and the novel two riser system facilitates heat balancing of the system (Krambeck and Pereira, 1986; Shan et al., 2003).

9.6.12.3 New Directions

Continued research and technology developments for fluid catalytic cracking will focus on (1) widening the boiling range of the feed that can be processed in the unit, (2) maximizing diesel yield and light olefin yield, and (3) providing operational flexibility to allow the unit to take advantage of favorable market opportunities.

A new hydrocarbon conversion process, *high-severity fluid catalytic cracking (HS-FCC)*, has been developed to maximize propylene production in oil refineries (Fujiyama et al., 2005; Maadhah et al., 2008). The yield of propylene was maximized using a combination of three factors: (1) catalyst properties, (2) reaction conditions, and (3) reactor design. Optimization of reaction conditions and catalyst development found that high reaction temperature accelerated catalytic cracking rather than hydrogen transfer. As a result, the olefin/paraffin ratio of the product was higher at high reaction temperatures.

The special features of this process include (1) rapid feed vaporization, (2) downflow reactor, (3) high severity, (4) short contact time, and (5) high catalyst/oil ratio. Since the FCC process involves successive reactions, the desired products such as olefins and naphtha are considered intermediate products. A suppression of back-mixing by using the downer reactor is the key to achieving maximum yield of these intermediates. Compared to conventional FCC processes, the HS-FCC has modifications in the reactor/regenerator and stripper sections (Fujiyama, 1999; Ino and Ikeda, 1999; Fujiyama et al., 2000; Nishida and Fujiyama, 2000; Ino et al., 2003).

Recent enhancements made to resid fluid catalytic cracking units permit feeding significant amounts of heavy crudes, while simultaneously improving yields and service factors. Traditional technology has been modified in key areas including: (1) catalyst design to accommodate higher metals feed and to minimize the amount of coke formed on the catalyst, (2) feed injection, (3) riser pipe design and catalyst/oil product separation to minimize overcracking, (4) regenerator design improvements to handle high coke output and avoid damage to catalyst structure, and (5) overall reactor/regenerator design concepts.

These developments have allowed fluid catalytic cracking units to substantially increase residue processing capabilities and substantial portions of refinery residua are processes (as blends with gas oils) in fluidized bed units thereby increasing naphtha and diesel production. This will not only continue but also increase in the future.

9.7 CATALYSTS

Commercial synthetic catalysts are amorphous and contain more silica than is called for by the preceding formulae; they are generally composed of 10%–15% alumina (Al_2O_3) and 85%–90% silica (SiO_2). The natural materials, montmorillonite, a nonswelling bentonite, and halloysite, are hydrosilicates of aluminum, with a well-defined crystal structure and approximate composition of

Al_2O_3 $4Si_2O·xH_2O$. Some of the newer catalysts contain up to 25% alumina and are reputed to have a longer active life.

However, cracking occurs over many types of catalytic materials, and cracking catalysts can differ markedly in both activity to promote the cracking reaction and in the quality of the products obtained from cracking the feedstocks (Gates et al., 1979; Wojciechowski and Corma, 1986; Stiles and Koch, 1995; Cybulski and Moulijn, 1998; Occelli and O'Connor, 1998; Domokos et al., 2010). Activity can be related directly to the total number of active (acid) sites per unit weight of catalyst and also to the acidic strength of these sites. Differences in activity and acidity regulate the extent of various secondary reactions occurring and thus the product quality differences. The acidic sites are considered to be Lewis- or Brønsted-type acid sites, but there is much controversy as to which type of site predominates.

Briefly, and by way of a historical introduction, the first acid catalyst, tested for cracking of heavy petroleum fraction, was aluminum chloride ($AlCl_3$) but the problems with catalyst manipulation, corrosion, and waste treatment or disposal put the use of this catalyst at a serious disadvantage. In the 1940s, silica–alumina catalysts were created and showed great improvement over the use of catalytic clay minerals natural clay catalysts. After natural aluminosilicate minerals were found to be adequate to the task, synthetic aluminosilicates were prepared and showed enhanced cracking properties. Both types (natural and synthetic) of aluminosilicates were known for the presence of Lewis acid sites. The early synthetic amorphous aluminosilicate catalysts contained approximately 13% w/w alumina (Al_2O_3), which was boosted to 25% w/w alumina in the mid-1950s. In the original fixed-bed process, activated bentonite was used (probably in the form of pellets). For the Thermofor catalytic cracking unit, the catalysts were of spherical shape (diameter: approximately 1–2 mm). In 1948, the first spray-dried catalyst was introduced and the microspherical particles (50–100 μm) were produced with the similar particle size distribution as the ground catalysts. However, the spherical particles showed both improved fluidization properties as well as significant reduction of attrition losses.

The most significant advance came in 1962 when zeolite catalysts (particularly zeolite-Y at that time) were incorporated into the silica–alumina structures after which advances in catalysts have produced the greatest overall performance of fluid catalytic cracking units over the last 50 years. The presence of the zeolites has resulted in the presence of strong Brønsted acid sites with very easily accessible Lewis acid sites also being present. These new types of catalysts possess the properties required of a successful catalyst: activity, stability, selectivity, correct pore size, resistance to fouling, and low cost. The first commercial zeolite catalysts were introduced in 1964, and zeolite catalysts remain in use in modern refineries.

From this point on, catalyst development has proceeded at a rapid rate and the development of active and stable catalysts (typically acid catalysts) has paralleled equipment design and development. The ultimate goal is the development of catalysts that are resistant to the obnoxious constituents of heavy feedstocks. Generally, the philosophy of the catalyst preparation for fluid catalytic cracking units is to have weak acid centers in macro-porous part of catalyst particles to insure precracking of heavy feedstock constituents to lower-molecular-weight producers that enter to the mesopores with stronger acidity. Cracking in mesopores leads to even low-molecular-weight products that can enter the zeolite micropores and crack over strongest zeolite acid centers into the desired products, typically naphtha constituents.

9.7.1 Catalyst Types

The first cracking catalysts were acid-leached *montmorillonite clays*. The acid leach was to remove various metal impurities, principally iron, copper, and nickel that could exert adverse effects on the cracking performance of the catalyst. The catalysts first used in fixed-bed and moving-bed reactor systems in the form of shaped pellets. Later, with the development of the fluid catalytic cracking process, the clay catalysts were made in the form of a ground, sized powder. Clay catalysts are relatively inexpensive and have been used extensively for many years.

The desire to have catalysts that were uniform in composition and catalytic performance led to the development of *synthetic catalysts*. The first synthetic cracking catalyst consisting of 87% silica (SiO_2) and 13% alumina (Al_2O_3) was used in pellet form and used in fixed-bed units in 1940. Catalysts of this composition were ground and sized for use in fluid catalytic cracking units. In 1944 catalysts in the form of beads approximately 2.5–5.0 mm diameter were introduced and comprised approximately 90% silica and 10% alumina and were extremely durable. One version of these catalysts contained a minor amount of chromia (Cr_2O_3) to act as an oxidation promoter.

Neither silica (SiO_2) nor alumina (Al_2O_3) alone is effective in promoting catalytic cracking reactions. In fact, they (and also activated carbon) promote hydrocarbon decompositions of the thermal type. Mixtures of anhydrous silica and alumina ($SiO_2 \cdot Al_2O_3$) or anhydrous silica with hydrated alumina ($2SiO_2 \cdot 2Al_2O_3 \cdot 6H_2O$) are also essentially noneffective. A catalyst having appreciable cracking activity is obtained only when prepared from hydrous oxides followed by partial dehydration (*calcining*). The small amount of water remaining is necessary for proper functioning.

Commercial synthetic catalysts are amorphous and contain more silica than is called for by the preceding formulae; they are generally composed of 10%–15% alumina (Al_2O_3) and 85%–90% silica (SiO_2). The natural materials, montmorillonite, a nonswelling bentonite, and halloysite, are hydrosilicates of aluminum, with a well-defined crystal structure and approximate composition of $Al_2O_3\ 4Si_2O \cdot xH_2O$. Some of the newer catalysts contain up to 25% alumina and are reputed to have a longer active life.

The catalysts are porous and highly adsorptive, and their performance is affected markedly by the method of preparation. Two catalysts that are chemically identical but have pores of different size and distribution may have different activity, selectivity, temperature coefficient of reaction rate, and response to poisons. The intrinsic chemistry and catalytic action of a surface may be independent of pore size, but small pores appear to produce different effects because of the manner and time in which hydrocarbon vapors are transported into and out of the interstices.

In addition to synthetic catalysts comprising silica–alumina, other combinations of *mixed oxides* were found to be catalytically active and were developed during the 1940s. These systems included silica (SiO_2), magnesia (MgO), silica–zirconia (SiO_2–ZrO), silica–alumina–magnesia, silica–alumina–zirconia, and alumina–boria (Al_2O_3–B_2O_3). Of these, only silica–magnesia was used in commercial units, but operating difficulties developed with the regeneration of the catalyst, which at the time demanded a switch to another catalyst. Further improvements in silica–magnesia catalysts have since been made. High yields of desirable products are obtained with hydrated aluminum silicates. These may be either activated (acid-treated natural clays of the bentonite type) or synthesized silica–alumina or silica–magnesia preparations. Both the natural and the synthetic catalysts can be used as pellets or beads, and also in the form of powder; in either case replacements are necessary because of attrition and gradual loss of efficiency (DeCroocq, 1984; Le Page et al., 1987).

During the 1940–1962 period, the cracking catalysts used most widely commercially were the aforementioned acid-leached clays and silica–alumina. The latter was made in two versions: *low alumina* (approximately 13% Al_2O_3) and *high alumina* (approximately 25% Al_2O_3) contents. High alumina content catalysts showed a higher equilibrium activity level and surface area.

During the 1958–1960 period, *semisynthetic catalysts* of silica–alumina catalyst were used in which approximately 25%–35% kaolin was dispersed throughout the silica–alumina gel. These catalysts could be offered at a lower price and therefore were disposable but they were marked by a lower catalytic activity and greater stack losses because of increased attrition rates. One virtue of the semisynthetic catalysts was that a lesser amount of adsorbed, unconverted, high-molecular-weight products on the catalyst were carried over to the stripper zone and regenerator. This resulted in a higher yield of more valuable products and also smoother operation of the regenerator as local hot spots were minimized.

The catalysts must be stable to physical impact loading and thermal shocks and must withstand the action of carbon dioxide, air, nitrogen compounds, and steam. They should also be resistant to sulfur compounds; the synthetic catalysts and certain selected clays appear to be better in this regard than average untreated natural catalysts.

Commercially used cracking catalysts are *insulator catalysts* possessing strong protonic (acidic) properties. They function as catalyst by altering the cracking process mechanisms through an alternative mechanism involving *chemisorption* by *proton donation* and *desorption*, resulting in cracked oil and theoretically restored catalyst. Thus, it is not surprising that all cracking catalysts are poisoned by proton-accepting vanadium.

The catalyst/oil volume ratios range from 5:1 to 30:1 for the different processes, although most processes are operated to 10:1. However, for moving-bed processes the catalyst/oil volume ratios may be substantially lower than 10:1.

Crystalline *zeolite catalysts* having molecular sieve properties were introduced as selective adsorbents in the 1955–1959 period. In a relatively short time period, all the cracking catalyst manufacturers were offering their versions of zeolite catalysts to refiners. The intrinsically higher activity of the crystalline zeolites vis-à-vis conventional amorphous silica–alumina catalysts coupled with the much higher yields of naphtha and decreased coke and light ends yields served to revitalize research and development in the mature refinery process of catalytic cracking.

A number of *zeolite catalysts* have been mentioned as having catalytic cracking properties, such as synthetic faujasite (X- and Y-types), offretite, mordenite, and erionite. Of these, the faujasites have been most widely used commercially. While faujasite is synthesized in the sodium form, base exchange removes the sodium with other metal ions that, for cracking catalysts, include magnesium, calcium, rare earths (mixed or individual), and ammonium. In particular, mixed rare earths alone or in combination with ammonium ions have been the most commonly used forms of faujasite in cracking catalyst formulations. Empirically, X-type faujasite has a stoichiometric formula of Na_2O Al_2O_3 $2.5SiO_2$ and Y-type faujasite has Na_2O Al_2O_3 $4.8SiO_2$. Slight variations in the silica/alumina (SiO_2/Al_2O_3) ratio exist for each of the types. Rare earth exchanged Y-type faujasite retains much of its crystallinity after steaming at 825°C (1520°F) with steam for 12 hours. The rare earth form X-faujasite, while thermally stable in dry air, will lose its crystallinity at these temperatures in the presence of steam.

9.7.2 CATALYST MANUFACTURE

While each manufacturer has developed proprietary procedures for making silica–alumina catalyst, the general procedure consists of (1) the gelling of dilute sodium silicate solution ($Na_2O \cdot 3.25SiO_2 \cdot xH_2O$) by addition of an acid (H_2SO_4, CO_2) or an acid salt such as aluminum sulfate, (2) aging the hydrogel under controlled conditions, (3) adding the prescribed amount of alumina as aluminum sulfate and/or sodium aluminate, (4) adjusting the pH of the mixture, and (5) filtering the composite mixture. After filtering, the filter cake can either be (1) washed free of extraneous soluble salts by a succession of slurrying and filtration steps and spray dried or (2) spray dried and then washed free of extraneous soluble salts before flash drying the finished catalyst.

There are a number of critical areas in the preparative processes that affect the physical and catalytic properties of the finished catalyst. Principal among them is the concentration and temperature of the initial sodium silicate solution. The amount of acid added to effect gelation, the length of time of aging the gel, the method and conditions of adding the aluminum salt to the gel, and its incorporation therein. Under a given set of conditions the product catalyst is quite reproducible in both physical properties and catalytic performance.

During the 1940–1962 period, a number of improvements were made in silica–alumina catalyst manufacture. These included continuous production lines versus batch-type operation, introduction of spray drying to eliminate grinding and sizing of the catalyst while reducing catalyst losses as fines, improved catalyst stability by controlling pore volume, and improved wash procedures to remove extraneous salts from high alumina content catalysts to improve equilibrium catalyst performance.

Zeolite cracking catalysts are made by dispersing or imbedding the crystals in a matrix. The matrix is generally amorphous silica–alumina gel and may also contain finely divided clay.

The zeolite content of the composite catalyst is generally in the range of 5%–16% by weight. If clay (e.g., kaolin) is used in the matrix, it is present in an amount of 25%–45% by weight, the remainder being the silica–alumina hydrogel *glue* that binds the composite together. The zeolite may be preexchanged to the desired metal form and calcined to lock the exchangeable metal ions into position before compositing with the other ingredients. In an alternate scheme, sodium-form zeolite is composited with the other components, washed, and then treated with a dilute salt solution of the desired metal ions before the final drying step.

As stated earlier, the matrix generally consists of silica–alumina, but several catalysts have been commercialized and contain (1) silica–magnesia and kaolin, and (2) synthetic montmorillonite-mica and/or kaolin as the matrix for faujasite.

9.7.3 Catalyst Selectivity

In the catalytic cracking process, the most abundant products are those having 3, 4, and 5 carbon atoms. On a weight basis the 4-carbon-atom fraction is the largest. The differences between the catalysts of the mixed oxide type lie in the relative action toward promoting the individual reaction types included in the overall cracking operation. For example, silica–magnesia catalyst under a given set of cracking conditions will give a higher conversion to cracked products than silica–alumina catalyst. However, the products from a silica–magnesia (SiO_2–MgO) catalyst have a higher average molecular weight, hence a lower volatility, lesser amounts of highly branched/acyclic isomers, but more olefins among the naphtha boiling range products (C_4—220°C, 430°F) than the products from a silica–alumina catalyst. With these changes in composition, the naphtha from cracking with a silica magnesia catalyst is of lower octane number.

These differences between catalysts may also be described as differences in the intensity of the action at the individual active catalytic centers. That is, a catalyst such as silica–alumina would give greater intensity of reaction than silica–magnesia as observed from the nature and yields of the individual cracked products and the automobile gasoline octane number. Titration of these two catalysts shows silica–alumina to have a lower acid titer than silica–magnesia but the acid strength of the sites is higher.

While each of the individual component parts in these catalysts is essentially nonacidic, when mixed together properly, they give rise to a titratable acidity as described earlier. Many of the secondary reactions occurring in the cracking process may also be promoted with strong mineral acids, such as concentrated sulfuric and phosphoric acids, aluminum halides, hydrogen fluoride, and hydrogen fluoride–boron trifluoride (BF_3) mixtures. This parallelism lends support to the concept of the active catalytic site as being acidic. Zeolites have a much higher active site density (titer) than the amorphous mixed oxides, which may account in large part for their extremely high cracking propensity. In addition, these materials strongly promote complex hydrogen transfer reactions among the primary products so that the recovered cracked products have a much lower olefin and higher paraffin content than are obtained with the amorphous mixed oxide catalysts. This hydrogen transfer propensity of zeolites to saturate primary cracked product olefins to paraffins minimizes the reaction of polymerizing the olefins to form a coke deposit, thus accounting in part for the much lower coke yields with zeolite catalysts than with amorphous catalysts.

Activity of the catalyst varies with faujasite content as does the selectivity of the catalyst to coke and naphtha. As the faujasite content drops below 5% by weight the catalyst starts to show some of the cracking properties of the matrix, while for zeolite contents of 10% by weight or higher, very little change in selectivity patterns is noted. The various ion exchanged forms of the faujasite can result in slightly different cracking properties; for example, using high-cerium-content mixed rare earths improves carbon burning rates in the regenerator, use of H-form of faujasite improves selectivity to propane–pentane fractions, use of a minor amount of copper form faujasite increases light olefin yield and naphtha octanes.

9.7.4 Catalyst Deactivation

A cracking catalyst should maintain its cracking activity with little change in product selectivity as it ages in a unit. A number of factors contribute to degrade the catalyst: (1) the combination of high temperature, steam partial pressure, and time, (2) impurities present in the fresh catalyst, and (3) impurities picked up by the catalyst from the feedstock while in use. Under normal operating conditions, the catalyst experiences temperatures of 480°C–515°C (900°F–960°F) in the reactor and steam stripper zones and temperatures of 620°C–720°C (1150°F–1325°F) and higher in the regenerator accompanied by a substantial partial pressure of steam. With mixed oxide amorphous gel catalysts, the plastic nature of the gel is such that the surface area and pore volume decrease rather sharply in the first few days of use and then at a slow inexorable rate thereafter. This plastic flow also results in a loss in the number and strength of the active catalytic sites.

Zeolite catalysts comprising both amorphous gel and crystalline zeolite degrade from instability of the gel, as stated earlier, and also from loss in crystallinity. The latter also results from the combined effects of time, temperature, and steam partial pressure. When crystallinity is lost, the amorphous residue is relatively low in activity, approximating that of the amorphous gel matrix. The rate of degradation of the amorphous gel component may not be the same as that of the zeolite crystals; for example, the gel may degrade rapidly and through thermoplastic flow effectively coat the crystals and interfere with the diffusion of hydrocarbons to the catalytic sites in the zeolite. Catalyst manufacturers try to combine high stability in the matrix with high stability zeolite crystals in making zeolite catalysts.

Residual impurities in freshly manufactured catalysts are principally sodium and sulfate. These result from the use of sodium silicate and aluminum sulfate in making the silica–alumina gel matrix and subsequent washing of the composite catalyst with ammonium sulfate to remove sodium. Generally, the sodium content of the amorphous gel is <01% w/w (as Na_2O) and sulfate is <0.5% w/w.

With zeolite catalysts, the residual sodium may be primarily associated with the zeolite, so that sodium levels may range from approximately 0.2%–0.80% for the composite catalyst. Sulfate levels in zeolite catalysts are still <0.5%. An excessive amount of sodium reacts with the silica in the matrix under regenerator operating conditions and serves as a flux to increase the rate of surface area and pore volume loss. Sodium faujasite is not as hydrothermally stable as other metal-exchanged (e.g., mixed rare earths) forms of faujasite. It is most desirable to reduce the sodium content of the faujasite component to <5.0% by weight (as Na_2O) with rare earths or with mixtures of rare earths and ammonium ions.

Finally, catalysts can degrade as a result of impurities picked up from the feed being processed. These impurities are sodium, nickel, vanadium, iron, and copper. Sodium as laid down on the catalyst not only acts to neutralize active acid sites, reducing catalyst activity, but also acts as a flux to accelerate matrix degradation. Freshly deposited metals are effective *poisons* to cracking catalysts because of the loss of active surface area by metal deposition (Figure 9.25) (Otterstedt et al., 1986). Zeolite catalysts are less responsive to metal contaminants than amorphous gel catalysts. Hence equilibrium catalysts can tolerate low levels of these metals so long as they have enough time to become buried. A sudden deposition of fresh metals can cause adverse effects on unit performance. Metals levels on equilibrium catalysts reflect the metals content of the feeds being processed typical ranges are 200–1200 ppm V, 150–500 ppm Ni, and 5–45 ppm Cu. Sodium levels are in the range of 0.25%–0.8% by weight (as Na_2O).

9.7.5 Catalyst Stripping

Catalyst leaving the reaction zone is fluidized with reactor product vapors that must be removed and recovered with the reactor product. In order to accomplish this, the catalyst is passed into a stripping zone where most of the hydrocarbon is displaced with steam.

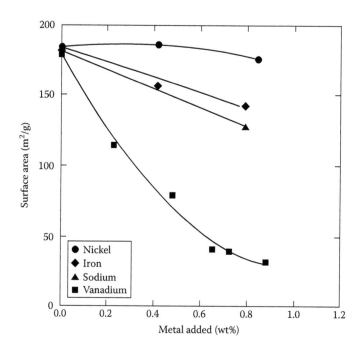

FIGURE 9.25 Effect of metals deposition on the catalyst.

Stripping is generally done in a countercurrent contact zone where shed baffles or contactors are provided to insure equal vapor flow up through the stripper and efficient contacting. Stripping can be accomplished in a dilute catalyst phase. Generally, a dense phase is used, but with lighter feeds or higher reactor temperature and high conversion operations a significant portion of the contacting can be done in a dilute phase.

The amount of hydrocarbon carried to the regenerator is dependent upon the amount of stripping steam used per pound of catalyst and the pressure and temperature at which the stripper operates.

Probes at the stripper outlet have been used to measure the composition of the hydrocarbon vapors leaving the stripper. When expressed as percent of coke burned in the regenerator, the strippable hydrocarbon is only 2%–5%. Very poor stripping is shown when the hydrogen content of the regenerator coke is on the order of 10% w/w, or higher. Good stripping is shown by 6%–9% by weight hydrogen levels.

The proper level of stripping is found in many operating units by reducing the stripping steam until there is a noticeable effect or rise in regenerator temperature. Steam is then marginally increased above this rate. In some units, stripping steam is used as a control variable to control the carbon burning rate or differential temperature between the regenerator bed and cyclone inlets.

In summary, the catalytic cracking unit is an extremely dynamic unit, primarily because there are three major process flow streams (the catalyst, hydrocarbon, and regeneration air), all of which interact with each other. Problems can arise in the equipment and flowing streams that are sometimes difficult to diagnose because of the complex effects they can create (Table 9.11).

9.7.6 Catalyst Treatment

The latest technique developed by the refining industry to increase naphtha yield and quality is to treat the catalysts from the cracking units to remove metal poisons that accumulate on the catalyst. Nickel, vanadium, iron, and copper compounds contained in catalytic cracking feedstocks are deposited on the catalyst during the cracking operation, thereby adversely affecting both catalyst

TABLE 9.11
General Commentary on the Fluid Catalytic Cracking Process

Problem	Symptoms	Causes	Data Required	Corrective Action
Fresh Feed Quality Problems				
(a) Meals in feed	(1) High H_2 make (2) General increase in light ends (3) Excessive coke production (4) Higher reactor velocities (5) Overloaded gas compressor	(1) Pitch entrainment at atmosphere and vacuum feed preparation unit (A&V) (2) Feed type change (3) Abnormal A&V operation	(1) Feed nickel equivalent (2) Feed CCR (3) H_2 make SCF/bbl fresh feed (4) Actual versus predicted yields (5) A&V operating conditions (6) Fresh feed color	(1) Lower metals in feed. (2) Feed segregation. (3) Catalyst replacement program.
(b) Feed contamination with heavy hydrocarbons	(1) Excessive coke make (2) Unexplained increase in air requirements at same conversion (3) Poor weight balance	(1) Leak in exchanger train (2) Partly open valves (3) Fractionator bottoms entrainment in heart cut recycle	(1) Feed CCR (2) Feed RI and °API (3) Recycle stream inspections	(1) Isolate leaking exchangers. (2) Minimize possibility of leaks by pressure balance.
(c) Feed contamination with light hydrocarbons	(1) Unsteady preheat header pressures and flows (2) Poor weight balance (3) Shift in yields distribution	(1) Leak in exchanger train (2) Partly open valves	(1) Feed °API (2) Feed front end distillation	(1) As earlier.
(d) Water in feed	(1) Vibration in preheat system (2) Unsteady flows and temperatures (3) Severe upset if large amounts of water in feed	(1) Water in feed tankage (2) Leaks from steam-out connections (3) Trapped water from idle equipment	(1) Water in feed	(1) Check for steam leaks on tank heaters. (2) Isolate and swing to high suction on catalyst feed tank. (3) Ensure that idle equipment is well drained and hot before being brought onstream.
Catalyst Problems				
(a) Catalyst contamination	(1) See metals in feed	(1) Low catalyst replacement rate (2) Metals in feed	(1) Metals on catalyst (2) H_2 production (3) Product yield distribution	(1) Consider lowering metals in feed; A&V opera Lion. (2) Catalyst replacement program. (3) Feed segregation.

(Continued)

TABLE 9.11 (*Continued*)

General Commentary on the Fluid Catalytic Cracking Process

Problem	Symptoms	Causes	Data Required	Corrective Action
(b) Sodium on catalyst	(1) See sintering	(1) Salt in feed (2) Treated boiler feed water used in regenerator sprays	(1) Historical Na content of catalyst	(1) Minimize sources of Na input to system.
(c) Sintering of catalyst	(1) Apparent decrease in catalyst activity (2) Increase in carbon on regenerated catalyst (3) Change in product yield distribution	(1) High Na and V on catalyst (2) Excessive regenerator temperatures (3) Low bed stability (4) Excessive or prolonged use of torch oil (5) Localized high temperatures	(1) Sintering index (2) EASC total and burnable carbon (3) Metals on catalyst (4) Yield distribution (5) Predicted and measured activity (6) Regenerator operating conditions	(1) Catalyst replacement. (2) Minimize Na input with seawater or salt in feed or with regenerator sprays. (3) Consider lower V content in feed. (4) Use high stability catalysts. (5) Review regenerator operations.
(d) Coarse catalyst	(1) Poor circulation (2) Poor regeneration (3) Change in yield distribution (4) Poorer stripping	(1) Loss of lines	(1) Roller analysis (2) H/C ratio (3) Yield distribution (4) Operating conditions in general	(1) Minimize catalyst losses by lowering regenerator velocity. (2) Change to fiber catalyst. (3) Consider use of attriter.
(e) Attrition	(1) Higher catalyst losses with increasing fines content	(1) High velocity stream into dense phase (2) Fragile catalyst	(1) Check unit for high velocity streams exceeding 200 ft/s into catalyst. (2) Missing ROs, blast steam, partly open valves, etc. (3) Check fresh catalyst properties. (4) Study catalyst loss pattern.	(1) Eliminate or reduce high velocity fluid injection. (2) Follow up fresh catalyst supplies.

(*Continued*)

TABLE 9.11 (*Continued*)
General Commentary on the Fluid Catalytic Cracking Process

Problem	Symptoms	Causes	Data Required	Corrective Action
Fresh Feed Quality Problem				
(a) Reactor cyclone failure	(1) High catalyst in fractionator bottoms (2) Frequent fractionator bottoms pump plugging (3) Loss of catalyst fines (4) Catalyst losses become progressively higher	(1) Erosion and/or corrosion (2) Pressure surges	(1) Catalyst content in fractionator bottoms (2) Cyclone pressure drop	(1) Minimize reactor velocity. (2) Review new methods in cyclone design and repair. (3) Review operating procedures and history of past occurrences.
(b) Eroded or plugged grid holes	(1) Change in grid ΔP (2) Decline in reactor efficiency (3) Change in yield distribution (4) Low overflow well level if grid eroded, high if plugged	(1) Lumps of coke or refractory in catalyst (2) Failure of grid hole inserts (3) Hole velocity too high (4) Feed injector velocity too low	(1) Grid pressure drop (2) Product yield pattern (3) Declining catalyst activity as determined by unit tracking despite adequate catalyst replacement rate	(1) Review methods for grid design and repair (2) Change reactor–regenerator differential pressure controller (DPRC) setting and/or control air to maintain circulation and normal overflow well level. (3) Check feed injector operation.
(c) Unsteady bed temperature and reactor pressure				See water in feed.
(d) Coking in overhead line	(1) Rise in ΔP between reactor outlet and fractionator inlet	(1) Condensation of reactor products in overhead line	(1) Pressure survey	(1) Ensure that overhead line insulation is in good condition.
(e) Poor catalyst stripping	(1) Unexplained increase in coke (2) Higher H/C ratio	(1) Insufficient stripping steam (2) Poor catalyst steam contacting (3) Catalyst properties (4) Low reactor temp. (5) Inaccurate steam flow controller (FRC)	(1) H/C ratios (2) Stripper tests	(1) Check stripping steam rate. (2) Consider higher reactor temperature. (3) Consider lower circulation rate. (4) Review stripper design.

(Continued)

TABLE 9.11 (*Continued*)

General Commentary on the Fluid Catalytic Cracking Process

Problem	Symptoms	Causes	Data Required	Corrective Action
Regenerator Problems				
(a) Cyclone failure	(1) Increase in catalyst losses (2) Catalyst losses become progressively higher	(1) Excessive temperatures and allowable stresses exceeded (2) Erosion	(1) Pressure differential between dilute phase and plenum (2) Measure of catalyst losses (3) Regenerator velocity	(1) Minimize catalyst feed. (2) Minimize regenerator velocity. (3) Review new methods in cyclone design and repair. (4) Review operating procedures and history.
(b) Plenum chamber failure	(1) Increase in catalyst losses above normal level	(1) Excessive temperatures and stress cracking (2) Impingement of plenum sprays	(1) As earlier (2) Check conditions of spray nozzles	(1) As earlier.
(c) Failure of internal seals	(1) Uneven bed and cyclone inlet temperatures (2) Uneven O_2 distribution in dilute phase (3) Salt and pepper catalyst (4) Surging of catalyst bed (5) Drop in grid Δp (6) Increase in catalyst losses (7) Afterburning	(1) Erosion (2) Pressure bump (3) Stresses too high (4) Abnormal conditions with auxiliary burner on start-up	(1) Grid Δp (2) Catalyst appearance (3) O_2 analysis of gas at dilute phase sprays (4) Historical operating data	(1) Avoid pressure surges on start-up. (2) Maximize air to grid (minimize control air). (3) Low regenerator pressure to increase velocity through grid. (4) Review operating history and seal design.
(d) Hole in overflow well	(1) Unstable overthrow well level and catalyst circulation (2) Unsteady overflow well density (3) High regenerator holdup (4) Uneven temperatures (5) Uneven O_2 distribution	(1) Abnormal stresses (2) Erosion	(1) Overflow well level and densities (2) Historical regenerator operator conditions, particularly holdup and temperatures	(1) Alter standpipe aeration. (2) Adjust DPRC and circulation. (3) Maximize control air (minimize air to grid). (4) Review design.

(*Continued*)

TABLE 9.11 (*Continued*)
General Commentary on the Fluid Catalytic Cracking Process

Problem	Symptoms	Causes	Data Required	Corrective Action
(e) Grid hole plugging and or erosion	(1) Uneven bed and cyclone temperatures (2) Uneven O_2 distribution (3) Increased catalyst losses (4) Surging of catalyst bed (5) Increased sintering	(1) Lumps of catalyst or refractory in catalyst bed (2) Low bed stability	(1) Catalyst sintering index (2) Bed stability (3) O_2 analyses of dilute phase	(1) Maintain grid pressure drop at 30% bed pressure drop. (2) Maximize air through grid. (3) Review grid design. (4) Check condition of catalyst hopper to ensure that debris does not enter regenerator.
(f) Stuck or failed trickle valves	(1) High catalyst losses but no increase with time (2) Loss of catalyst fines: increase in coarseness index	(1) Binding of hinge rings by (a) Clearances not large enough (b) Oxidation scale or pieces of lining (c) Nonuniform wear of moving parts (2) Installation angles not correct (3) Improper material	(1) Careful examination of trickle valves on shutdown	(1) Revise or replace as required to meet specifications.
(g) Plugged dip legs	(1) As earlier	(1) Spalled refractory forming partial and eventually final plug (2) Air-out periods with a lot of water/steam in vessel	(1) Careful examination of dip legs or use of probes, lights, balls, etc., to ensure dip legs are free before start-up	(1) Lower bed level to allow catalyst to flow out of dip leg.
(h) Excessive input of steam or air	(1) Higher level of catalyst losses (2) High cyclone pressure drop (3) Indication of attrition (higher catalyst fines content despite higher losses)	(1) Missing restriction orifices (2) Large restriction orifices (3) Partially open valves on steam, water, or air lines (4) Malfunctioning steam traps (5) Metering errors	(1) Equipment survey to make sure all restriction orifices (ROs) are in place, etc.	(1) Make careful detailed check of unit for and correct equipment conditions listed under causes.
(i) Bed stability too low	(1) Increase in temperature below the grid	(1) Insufficient air through grid to support catalyst bed (2) Eroded grid holes	(1) Grid ΔP (2) Historical records of temperature below grid	(1) Maximize air through grid. (2) If problem expected to persist, redesign grid for lower air rates.

(Continued)

TABLE 9.11 (*Continued*)

General Commentary on the Fluid Catalytic Cracking Process

Problem	Symptoms	Causes	Data Required	Corrective Action
(j) Surging of catalyst bed	(1) Erratic or cycling instrument records on holdup, density, and overflow well (2) Unsteady circulation and heat balance (3) Catalyst sintering (4) Uneven catalyst regeneration	(1) Seal failures (2) Grid hole erosion (3) Hole in overflow well (4) Poor bed stability (5) Poor DPRC control	(1) Pressure survey (2) Operating condition survey (3) Bed stability calculation	(1) See action required for seal failures, grid hole erosion, poor bed stability, poor DPRC control.

(k) Unsteady pressure differential control	(1) Fluctuating regenerator pressure
	(2) Unsteady circulation and overflow well level
	(3) Unsteady rector temperature
	(4) Catalyst shifts between reactor and regenerator
		(1) Poor stack slide valve performance
		  (a) Sticky slide valves
		  (b) Poor slide valve instrument performance
		(2) Unsteady circulation
			(1) Unit performance records	(1) Check out slide valves and associated instrumentation.
				(2) Check adequacy of U-bend or standpipe aeration.

(l) Regenerator hot spots	(1) High temperatures on vessel shells or U-bends	(1) Damaged refractory	(1) Change in paint color (2) Glow at night (3) Surface temperature measurements	(1) Cool the spot with steam or water, avoiding abrupt changes in temperature due to interpretation of cooling.
(m) Rough catalyst circulation	(1) Unsteady regenerator and reactor temperatures and holdups (2) Unsteady control air (3) Fluctuating overflow well level and U-bend densities (4) U-bend vibration	(1) Improper aeration (2) Coarse catalyst (3) Fluctuating DPRC (4) Water in aeration medium (5) Poor performance of control air system	(1) U-bend aeration pattern and pressure survey (2) Standpipe and U-bend densities (3) Circulation rate (4) Catalyst roller analysis	(1) Change aeration. (2) Put control air on manual to determine if it is the cause. (3) Check DPRC system. (4) Make sure aeration medium is water-free.

(*Continued*)

TABLE 9.11 (*Continued*)
General Commentary on the Fluid Catalytic Cracking Process

Problem	Symptoms	Causes	Data Required	Corrective Action
(n) High carbon on catalyst. Carbon buildup	(1) Dark catalyst (2) Dilute phase temperature decreases relative to dense-bed temperature (3) Low excess O_2 (4) Apparent loss in catalyst activity	(1) Excessive coke from (a) Increase in operating intensity (b) Poorer feedstock (c) Poor catalyst stripping (d) Heavier recycle (e) Leakage of fraction bottoms into feed (2) Low excess O_2 (3) Poor air distribution (4) False O_2 recorder readings (5) Feed and recycle meter Errors	(1) Carbon on catalyst analysis (2) Feed quality (3) Recycle boiling range and CCR (4) Check O_2 levels in regenerator (5) Check O_2 recorder for accuracy (6) Check stripper performance	(1) Lower intensity of operation. (2) Increase air to regenerator. (3) Inject lighter feed. (4) Improve normal feed and or recycle if possible. (5) Improve catalyst stripping. (6) Check meter accuracy.
(o) Afterburning	(1) Excessive dilute phase and cyclone temperatures (2) Trend to lighter catalyst (3) Increase in dilute phase temperature relative to dense bed (4) High excess O_2 (5) Higher CO_2/CO ratios	(1) Insufficient coke production: (a) Decrease of intensity (b) Swing to better feed (c) Lighter recycle (2) Too much excess air (3) False O_2 recorder readings (4) Feed and recycle meter errors (5) Cyclone steam meter errors	(1) Same as in (n) (1) above except (6)	(1) Decrease air. (2) Marginal use of torch oil. (3) Increase operating intensity. (4) Check cyclone stream rate.
(p) Low catalyst circulation rate	(1) Inability to lower regenerator temperature (2) Excessive feed preheat requirements	(1) Partial blockage of U-bends (2) Too much stripping steam (3) Improper aeration (4) Control air too low (5) DPRC setting inadequate	(1) As in (o) above	(1) Change DPRC setting. (2) Check control air system and increase control air rate. (3) Ensure adequate aeration. (4) Check unit design.

(*Continued*)

TABLE 9.11 (*Continued*)

General Commentary on the Fluid Catalytic Cracking Process

Problem	Symptoms	Causes	Data Required	Corrective Action
Fractionator Problems				
(a) Poor split between LOGO and recycle	(1) High overlaps between heating oil and recycle stream distillation (2) Shift in yield pattern (3) Increase in recycle volume at constant operating conditions	(1) Inadequate steam to recycle stripper (2) Improper tray loading in tower (3) Stripper malfunction	(1) Heating oil and recycle distillations (2) Fractionator pressure survey (3) Pumparound heat removal (4) Detailed fractionator and stripper analysis (5) Consider equipment x-rays	(1) Ensure adequate stripping steam rates. (2) Adjust pumparound heat duties. (3) Consider equipment changes.
(b) Fractionator bottoms too light	(1) High bottoms API (2) Excessive bottoms yield (3) Dark color of recycle stream	(1) Too much heat removal in bottoms pumparound (2) Tower bottom liquid too cold (3) Poor liquid–vapor contact in shed section (4) Lower operating intensity (5) Heart cut recycle rate too low (6) Leaks into bottoms system	(1) Fractionator bottoms inspections (2) Tower operating conditions in the lower half (3) Check on cracking intensity (4) Check recycle flow rate; meter accuracy	(1) Lower heat removal from tower bottoms consistent with coking considerations. (2) Maximize heart cut recycle. (3) Consider equipment changes.
(c) Coking	(1) Frequent plugging of exchangers and pumps with coke (2) Poor heat transfer in exchangers (3) High insolubles in fractionator bottoms	(1) Excessive temperatures in tower bottom (2) Low bottoms API (3) High liquid residence time in tower bottoms	(1) History of bottoms API and bottoms sediment and water content (2) Tower bottoms operating conditions (3) Calculations of bottom liquid residence time (4) History of equipment fouling	(1) Review recommended limitations of tower bottoms operating conditions and adjust operation accordingly. (2) Consider changing pump screen size.
(d) Salt fouling of fractionator top	(1) High pressure drop in top trays (2) Flooding in the top section	(1) Salts in fresh feed (2) TPA return too cold	(1) Pressure and temperature survey of tower top	(1) Cautiously inject water to TPA.

(*Continued*)

TABLE 9.11 (*Continued*)

General Commentary on the Fluid Catalytic Cracking Process

Problem	Symptoms	Causes	Data Required	Corrective Action
	(3) Poor split between overhead and first sidestream		(2) TPA operating conditions	(2) Adjust tower temperature.
	(4) Salt in TPA pumps		(3) Chlorides in fresh feed	(3) Adjust TPA return temperature.
	(5) Chlorides in overhead water		(4) Chlorides in distillate drum water	(4) Consider installing screens in TPA pump suctions.
			(5) Split between overhead product and first sidestream	
Miscellaneous Problems				
(a) Air blower low turbine efficiency	(1) Excessive steam usage	(1) Turbine fouling	(1) Steam supply and exhaust conditions	(1) Improve steam quality.
	(2) Unusual exhaust steam conditions	(2) Turbine blade wear	(2) Calculation of turbine efficiency	(2) Consider turbine wash or shutdown for cleaning.
	(3) Decrease in rpm at maximum steam	(3) Quality of steam supply	(3) Blower operation conditions	
(b) Surplus heat in unit	(1) High regeneration temperature	(1) Too much feed preheat	(1) Unit temperatures and heat balances	(1) Back off on feed injection temperature.
	(2) Low control air blower rate (see also low circulation rate)	(2) Reactor temperature set too low	(2) Circulation rate	(2) Increase circulation rate.
			(3) Coke make	(3) Increase reactor temperature and lower holdup.
				(4) Adjust DPRC.
				(5) Use dilute phase sprays
(c) Insufficient heat in unit	(1) Low regenerator temperature	(1) Insufficient feed preheat	(1) Unit temperatures and heat balances	(1) Increase preheat if possible.
	(2) High control air if on automatic setting	(2) Reactor temperature set too high	(2) Operating intensity level	(2) Increase fresh catalyst additions.
		(3) Not enough carbon produced:	(3) Circulation rate	(3) Take out sprays if any.
		(a) Catalyst activity low	(4) Coke production	(4) Use torch oil.
		(b) Lighter recycle	(5) Feed and recycle inspections	
		(c) Lighter feed		
		(d) Lower reactor holdup		

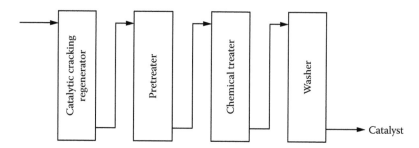

FIGURE 9.26 The Demet process.

activity and selectivity. Increased catalyst metal contents affect catalytic cracking yields by increasing coke formation, decreasing naphtha and butane and butylene production, and increasing hydrogen production.

The recent commercial development and adoption of cracking catalyst-treating processes definitely improve the overall catalytic cracking process economics.

9.7.6.1 Demet

A cracking catalyst is subjected to two pretreatment steps (Figure 9.26). The first step effects vanadium removal; the second, nickel removal, to prepare the metals on the catalyst for chemical conversion to compounds (chemical treatment step) that can be readily removed through water washing (catalyst wash step). The treatment steps include use of a sulfurous compound followed by chlorination with an anhydrous chlorinating agent (e.g., chlorine gas) and washing with an aqueous solution of a chelating agent such as citric acid ($HO_2CCH_2C(OH)(CO_2H)CH_2CO_2H$, 2-hydroxy-1,2,3-propanetricarboxylic acid). The catalyst is then dried and further treated before returning to the cracking unit.

9.7.6.2 Met-X

This process consists of cooling, mixing, and ion-exchange separation, filtration, and resin regeneration. Moist catalyst from the filter is dispersed in oil and returned to the cracking reactor in a slurry (Figure 9.27). On a continuous basis, the catalyst from a cracking unit is cooled and then transported

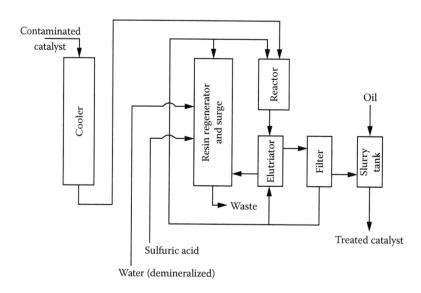

FIGURE 9.27 The Met-X process.

to a stirred reactor and mixed with an ion-exchange resin (introduced as slurry). The catalyst-resin slurry then flows to an elutriator for separation. The catalyst slurry is taken overhead to a filter, and the wet filter cake is slurried with oil and pumped into the catalytic cracker feed system. The resin leaves the bottom of the elutriator and is regenerated before returning to the reactor.

9.7.7　Recent Advances

Recent advances in FCC catalysts have concentrated on modifying zeolite Y for improved coke selectivity, higher cracking activity, and greater stability through manipulation of extra-framework aluminum or through the generation of mesoporous nature of the zeolite crystals. Extra-framework aluminum is introduced either by steaming or via ion exchange. The development of improved FCC catalysts includes modifying a single crystal structure to achieve multiple catalytic objectives (Degnan, 2000).

In order to meet demands for a higher-octane product, an increasing number of refiners has switched to using ultra-stable Y (USY) zeolite type catalysts. These catalysts, which feature zeolites with a reduced unit cell size, are a proven means of increasing the octane number of FCC naphtha (which is used for gasoline production in the blending operation). However, there are trade-offs associated with using a reduced unit cell size zeolite—one of the more important properties being affected is catalyst activity. In addition, the use of ZSM-5 (a Pentasil zeolite concentrate that typically incorporates up to 25%–40% Pentasil) enhances the octane number in FCC naphtha by overcracking of naphtha/gasoline fraction. This reaction mainly occurs on the stronger acid site of ZSM-5 than that of the USY catalyst. Comparing to the result with typical FCC catalyst, the yield of LPG is increased by overcracking and the octane number is enhanced for the mixed catalyst, although the yield of FCC naphtha is decreased. The increase of LPG yield is due to production of propylene in the case of mixed catalyst (Buchanan et al., 1996; Imhof et al., 2005; Li et al., 2007).

9.7.7.1　Matrix, Binder, and Zeolite

The *matrix* in the fluid catalytic cracking catalyst is often considered to be that part of the catalyst other than the zeolites—it may or may not have catalytic activity. Very often it does possess sufficient catalytic activity toward some components of the feedstock, and in this case, they are described by the term *active matrix*. The matrix consists of two main components—first a synthetic component like amorphous silica, silica–alumina gel, or silica–magnesia gel, which serve as the binder and also exhibit catalytic properties. The other component is natural or chemically modified clay, such as kaolinite, halloysite, or montmorillonite. The clays provide mechanical stability.

The functions of the *fillers* and *binders* incorporated into the FCC catalysts are often similar to those performed by the matrix. Sometimes additional fillers like kaolin may be provided for physical integrity and as a more efficient fluidizing medium. The binder performs the all-important function of holding the catalyst, the matrix, and the filler glued together. This is especially important when the catalyst contains a higher amount of zeolites. The filler and binder minimize the production of catalyst fines in the reactor–regenerator system and help to control or mitigate catalyst losses.

9.7.7.2　Additives

In addition to what to cracking catalyst described earlier, a series of *additives* has been developed that catalyze or otherwise alter the primary catalyst's activity/selectivity or act as pollution control agents. Additives are most often prepared in microspherical form to be compatible with the primary catalysts and are available separately in compositions that (1) enhance naphtha octane and light olefin formation, (2) selectively crack heavy cycle oil, (3) passivate vanadium and nickel present in many heavy feedstocks, (4) oxidize coke to carbon dioxide, and (5) reduce sulfur dioxide emissions. Both vanadium and nickel deposit on the cracking catalyst are extremely deleterious when present in excess of 3000 ppm on the catalyst. Formulation changes to the catalyst can improve tolerance to vanadium and nickel but the use of additives that specifically passivate either metal is often preferred.

The ability of small amounts of ZSM-5 added to the FCC unit to improve naphtha octane number while producing more light olefins has prompted a substantial amount of process and catalyst research into zeolite-based FCC additives. Significant advances have been made in stabilizing ZSM-5 to harsh FCC regenerator conditions, which, in turn, have led to reductions in the level of ZSM-5 needed to achieve desired uplifts and wider use of the less expensive additives (Degnan, 2000).

9.7.7.3 Metal Traps

The performance of fluid catalytic cracking catalysts for processing heavy feedstocks is often determined by the tolerance of the catalyst to metal contaminants. Different approaches are used to manage contaminant metals such as riser pipe design (Hedrick and Palmas, 2011). Metals traps have been introduced into catalysts in order to protect the zeolite from poisonous metals that have accumulated in large amounts as a result of lowering the amount of catalyst use, and in order to give the catalyst metal tolerance so that the formation of hydrogen and coke due to the dehydrogenation over poisonous metals would be reduced.

As a result, catalyst manufacturers have developed more stable zeolites and a series of vanadium traps to increase the ability of the zeolite to handle vanadium. These traps are based on barium (Ba), titanium (Ti), rare earth elements, and other elements. Some are more effective than others, but the basic idea is the same, that is, to keep the vanadium away from the zeolite by binding to the surface of an inactive particle (Harding et al., 2001).

9.7.7.4 Low Rare Earths

Rare earth elements inhibit alumina removal from a zeolite, a higher concentration of acid sites will be found in a rare earth exchanged catalyst. This improves both the activity and the hydrothermal stability of the catalyst. On average, these sites are weaker and in closer proximity to each other than those found in a more highly dealuminated catalyst characterized by lower unit cell size measurements.

The addition of rare earth into the zeolite inhibits the degree of unit cell size shrinkage during equilibration in the regenerator. Steam in the FCC regenerator removes active acidic alumina from the zeolite. Rare earth inhibits the extraction of aluminum from the zeolite's structure (alumina removal), which in turn increases the equilibrium unit cell size for fluid catalytic cracking catalysts. Since reducing the equilibrium unit cell size of a fluid catalytic cracking catalyst has the effect of improving octane, adding rare earth decreases the octane.

Because rare earth oxides promote hydrogen transfer, the yield of C_3 and C_4 olefins in the LPG fraction will be lower. The resulting reduction in the total LPG yield results in a reduction in the wet gas yield, which can have a major effect on plant operations, as compressor capacity is often the limiting factor for fluid catalytic cracking unit throughput.

9.7.7.5 Catalysts for Olefin Production

The market for propene derivatives and butene derivatives is cyclical and depends on the local demand for plastics precursors and alkylate feed. There are several key determinants of light olefin selectivity (Harding et al., 2001; Soni et al., 2009). In addition to increasing reactor temperature and using ZSM-5, catalysts can be designed with lower hydrogen transfer (to reduce the conversion of isobutylene to *iso*-butane) and moderate matrix activity (to increase the C4-olefin content of the product).

The addition of ZSM-5 is the single most important method to increase the yields of light olefins at the expense of naphtha. Catalyst producers have developed methods to increase the concentration of the ZSM-5 in their additives in order to avoid dilution effects at high levels of additive. The breakthrough technologies in this area involve the stabilization of the ZSM-5 to hydrothermal deactivation at higher concentrations in the additive.

Current commercial catalysts range from 10% to 25% ZSM-5, although evolving techniques are expected to allow much higher levels in an additive particle. Several recent efforts have also been made to change the selectivity of ZSM-5 by increasing the Si/Al ratio to increase the ratio of butene to propene. However, this approach reduces the overall activity quite significantly.

9.7.7.6 Catalysts for Jet and Diesel Production

In markets dominated by fluid catalytic cracking–based refineries, the need to increase distillate production has taken on a new dimension, posing interesting challenges, while presenting some unique opportunities (Yung and Pouwels, 2008). Much attention on the possibility of tighter specifications for aromatics, density, boiling range, and cetane number can be expected. New technologies are being developed to address these challenges. For example, a commercialized moderate-pressure hydrocracking suite of technologies can provide an economic solution to desulfurize heavy gas oil (HGO) with options to convert part of the feed into valuable higher-quality distillate products and to maximize production of Jet A1 and diesel products (Degnan, 2000; Hilbert et al., 2008).

The catalysts are large pore, highly dealuminated Y-zeolites as well as improved base metal combinations that have been tailored to operate well at low hydrogen partial pressures. Processes have been developed around USY-based catalysts for partial conversion of vacuum gas oils, cracked gas oils, deasphalted oil, and FCC light cycle oil. The processes comprise a dual catalyst system consisting of an amorphous hydrotreating catalyst (normally $NiMo/Al_2O_3$) and a metal-containing USY-based hydrocracking catalyst. Single-stage, single-pass conversions are typically in the range of 30%–60% w/w. The process requirements are similar to those used in vacuum gas oil hydrodesulfurization, which has led to several catalytic hydrodesulfurization revamps (Degnan, 2000).

For cracking catalysts, there is usually a trade-off between high activity and high middle distillate selectivity. The objective for new cracking catalyst development is often to increase activity while maintaining selectivity or vice versa. This can be achieved by altering the catalyst's acidic function and/or the hydrogenation function, as is illustrated by the UOP Unicracking process (Ackelson, 2004; Abdo et al., 2008).

In addition to control and optimization of process conditions in the Unicracking process, catalyst design principles when applied to this problem take advantage of improved molecular-scale definition of feed compositions, appropriate selection of catalytic materials, and enhanced knowledge of the impact of specific process conditions on catalyst performance to deliver the needed level of activity and selectivity to meet cycle length and product yields (Motaghi et al., 2010).

In the early 1990s, Zeolyst introduced Z-603, a zeolite catalyst with high activity and also good middle distillate selectivity. The improved performance of the catalyst versus other zeolite and amorphous silica–alumina catalysts was the result of two major developments—a new dealuminated Y-zeolite used in conjunction with an amorphous silica–alumina catalysts powder and an improved hydrogenation function. This catalyst has been used successfully at Shell's Godorf (Germany) refinery and at Alliance Refining Company's Rayong (Thailand) refinery. Because of its higher activity and stability, Z-603 has extended the cycle length beyond the capability of an amorphous silica–alumina catalysts with only a small reduction in middle distillate yield relative to the amorphous silica–alumina catalysts (Huve et al., 2004).

Continuing development efforts have led to a new range of cracking catalysts (for example, Z-623 and Z-613) with improved middle distillate yields/activities. Z-623 is significantly more active than the Z-603 catalyst with very little middle distillate yield loss. The benefits of the Z-623 catalyst have been demonstrated in a number of commercial cases. At Alliance Refining Company, the catalyst has made it possible to increase the hydrocracker's cycle length from 2 years to more than 3 years. At Shell's Godorf refinery, the catalyst was a key component of the hydrocracker revamp that has produced a 15% v/v increase in throughput of feedstock without any cycle length reduction (Huve et al., 2004).

9.7.7.7 New Directions

With the increasing focus to reduce sulfur content in fuels, the role of *desulfurization* in the refinery becomes more and more important. Currently, the process of choice is the hydrotreater to remove the sulfur from the fuel. Because of the increased attention for fuel desulfurization, various new process concepts are being developed with various claims of efficiency and effectiveness.

Resid fluid catalytic cracking is an important component in the upgrading of heavy oils, with unit profitability depending upon the extent to which heavy hydrocarbons in the feed are cracked into valuable products. The product slate, in turn, depends upon the feed characteristics, the catalyst, the hardware, and the operating conditions.

Product recycling and multiple reaction sections are the prevalent technology trends. Improving feed injectors, riser termination and catalyst separation devices, strippers, and regenerator components are revamp options for existing units. As feedstocks get heavier, the trend toward a higher stripper residence time and, consequently, increased mass transfer between entrained hydrocarbons and steam will continue. Moreover, the role of the regenerator continues to evolve because of reduction requirements in the carbon dioxide emissions. The need for development of flue gas treating systems points to the need for continued innovation in that area.

The use of bio-feedstocks (such as animal fats and vegetable oils) in the fluid catalytic cracking will be used to increase the yield of light cycle oil and will also provide high-quality products. Practical implementation in a refinery will, more than likely, be accompanied by blending with vacuum gas oil or residuum (Speight, 2011a).

Oils derived from bio-feedstocks are generally best upgraded by HZSM-5 or ZSM-5, as these zeolite catalysts promote high yields of liquid products and propylene. Unfortunately, these feeds tend to coke easily, and high acid numbers and undesirable by-products such as water and carbon are, and will continue to be, additional challenges. Waxy feeds obtained from biomass gasification followed by Fischer–Tropsch synthesis to hydrocarbon liquids and waxes (biomass to liquids) are especially suited for increasing light cycle oil production in the catalytic cracking unit, due to the high paraffinic character, low-sulfur content, and low-aromatic yield of the feed. A major disadvantage for biomass-to-liquid products is the intrinsically low coke yield that can disrupt the heat balance of the catalytic cracking unit.

The major developmental opportunity area in the next decade may well be at the refinery–petrochemical interface. Processes that maximize olefins and aromatics while integrating clean fuels production will continue to look attractive to refiners who are seeing their fuel product margins come under increasing competitive pressure.

Furthermore, in order to alleviate growing concerns over energy security and global climate change, the fluid catalytic cracking unit will take on two additional roles: (1) acceptance of biomass feedstocks and (2) reduction in carbon dioxide emissions. With committed efforts by refiners and technology developers, the process will continue to demonstrate adaptability amid changing market requirements through the installation of dual-radius feedstock distributors and spent catalyst distributors (Wolschlag and Couch, 2010a).

Methods to reduce emissions from the FCC unit and other refinery operations lend themselves to two conflicting categories: those that hinder the formation of carbon dioxide and those that aim to produce a pure carbon dioxide stream for capture. Improving unit energy efficiency, reducing coke yields, and shifting vacuum gas oil to the hydrocracking unit belong to the former category, while carbon capture methods typically comprise the latter (Spoor, 2008). As refiners continue to develop a comprehensive strategy to reduce carbon dioxide emissions, a balance between these two categories must be found. This happy medium will be influenced greatly by the FCC unit heat balance, which determines the amount of coke burned in the regenerator in present unit designs.

9.8 OPTIONS FOR HEAVY FEEDSTOCKS

The processes described later are the evolutionary offspring of the fluid catalytic cracking and the residuum catalytic cracking processes. Some of these newer processes use catalysts with different silica/alumina ratios as acid support of metals such as Mo, Co, Ni, and W. In general, the first catalyst used to remove metals from oils was the conventional hydrodesulfurization (HDS) catalyst. Diverse natural minerals are also used as raw material for elaborating catalysts addressed to the upgrading of heavy fractions. Among these minerals are clays; manganese nodules; bauxite activated with vanadium (V), nickel (Ni), chromium (Cr), iron (Fe), and cobalt (Co), as well as the high-iron-oxide-content iron laterites and clay minerals such as sepiolite minerals; and mineral nickel and transition metal sulfides supported on silica and alumina. Other kinds of catalysts, such as vanadium sulfide, are generated *in situ*, possibly in colloidal states.

In the past decades, in spite of the difficulty of handling heavy feedstocks, residuum fluidized catalytic cracking (RFCC) has evolved to become a well-established approach for converting a significant portion of the heavier fractions of the crude barrel into a high-octane gasoline blending component. Residuum fluidized catalytic cracking, which is an extension of conventional fluid catalytic cracking technology for applications involving the conversion of highly contaminated residua, has been commercially proven on feedstocks ranging from gas oil–residuum blends to atmospheric residua, as well as blends of atmospheric and vacuum residua blends. In addition to high naphtha yields, the residuum fluidized catalytic cracking unit also produces gaseous, distillate, and fuel oil-range products.

The product quality from the residuum fluidized catalytic cracker is directly affected by its feedstock quality. In particular, and unlike hydrotreating, the residuum fluidized catalytic cracking redistributes sulfur among the various products, but does not remove sulfur from the products unless, of course, one discounts the sulfur that is retained by any coke formed on the catalyst. Consequently, tightening product specifications have forced refiners to hydrotreat some, or all, of the products from the resid cat cracking unit. Similarly, in the future the emissions of sulfur oxides (SO_x) from a resid cat cracker may become more of an obstacle for residue conversion projects. For these reasons, a point can be reached where the economic operability of the unit can be sufficient to justify hydrotreating the feedstock to the cat cracker.

As an integrated conversion block, residue hydrotreating and residuum fluidized catalytic cracking complement each other and can offset many of the inherent deficiencies related to residue conversion.

9.8.1 ASPHALT RESIDUAL TREATING (ART) PROCESS

The ART process is a process for increasing the production of transportation fuels and reduces heavy fuel oil production, without hydrocracking (Bartholic, 1981a,b, 1989; Bartholic and Haseltine, 1981; Speight and Ozum, 2002; Parkash, 2003; Hsu and Robinson, 2006; Gary et al., 2007; Speight, 2014). This is a flexible selective vaporization process that can be used for removing essentially all of the metals and substantial proportion of the carbon residue, nitrogen, and sulfur compounds from the heavy feedstock (Bartholic and Haseltine, 1981). In addition, the process can be considered as an efficient carbon rejection process followed by catalytic hydrogenation.

In the process, the preheated feedstock (which may be whole crude, atmospheric residuum, vacuum residuum, or tar sand bitumen) is injected into a stream of fluidized, hot catalyst (trade name: ArtCat). Complete mixing of the feedstock with the catalyst is achieved in the contactor, which is operated within a pressure–temperature envelope to ensure selective vaporization. The vapor and the contactor effluent are quickly and efficiently separated from each other and entrained hydrocarbons are stripped from the contaminant (containing spent solid) in the stripping section. The contactor vapor effluent and vapor from the stripping section are combined and rapidly quenched in a quench drum to minimize product degradation. The cooled products are then transported to a conventional fractionator that is similar to that found in a fluid catalytic cracking unit. Spent solid from the stripping section is transported to the combustor bottom zone for carbon burn-off.

The process is claimed to require lower hydrogen consumption in comparison to direct hydrocracking processes (Suchanek and Moore, 1986).

In the combustor, coke is burned from the spent solid that is then separated from combustion gas in the surge vessel. The surge vessel circulates regenerated catalyst streams to the contactor inlet for feed vaporization and to the combustor bottom zone for premixing. The components of the combustion gases include carbon dioxide (CO_2), nitrogen (N_2), oxygen (O_2), sulfur oxides (SO_x), and nitrogen oxides (NO_x) that are released from the catalyst with the combustion of the coke in the combustor. The concentration of sulfur oxides in the combustion gas requires treatment for their removal.

9.8.2 RESIDUE FLUID CATALYTIC CRACKING PROCESS

The residue fluid catalytic cracking process (HOC process) is a version of the fluid catalytic cracking process that has been adapted to conversion of residua that contain high amounts of metal and asphaltene constituents. Depending on quality and product objectives, feedstocks with vanadium-plus-nickel content of 5–30 ppm and a carbon residue on the order of 5%–10% w/w can be processed without feed pretreatment. The process, when coupled with hydrodesulfurization, provides a particularly effective means for meeting demands for naphtha or diesel fuel blending stock, and low-molecular-weight olefins. The asphaltene constituents are converted to coke (which deposits on the catalyst) and distillate from the cracked fragments. The catalyst is regenerated by combusting its carbon deposits, and the heat generated from the combustion of carbon deposits is used to produce high pressure steam.

In the process, a residuum is desulfurized, and the nonvolatile fraction from the hydrodesulfurization unit is charged to the residuum fluid catalytic cracking unit. The reaction system is an external vertical riser terminating in a closed cyclone system. Dispersion steam in amounts higher than that used for gas oils is used to assist in the vaporization of any volatile constituents of heavy feedstocks. A two-stage stripper is utilized to remove hydrocarbons from the catalyst. Hot catalyst flows at low velocity in dense phase through the catalyst cooler and returns to the regenerator. Regenerated catalyst flows to the bottom of the riser to meet the feed. The coke deposited on the catalyst is burned off in the regenerator along with the coke formed during the cracking of the gas oil fraction. If the feedstock contains high proportions of metals, control of the metals on the catalyst requires excessive amounts of catalyst withdrawal and fresh catalyst addition. This problem can be addressed by feedstock pretreatment.

The feedstocks for the process are rated on the basis of carbon residue and content of metals. Thus, *good-quality feedstocks* have less than 5% by weight carbon residue and less than 10 ppm metals. *Medium-quality feedstocks* have greater than 5% but less than 10% by weight carbon residue and greater than 10 but less than 30 ppm metals. *Poor-quality feedstocks* have greater than 10 but less than 20% by weight carbon residue and greater than 30 but less than 150 ppm metals. Finally, *bad-quality feedstocks* have greater than 20% by weight carbon residue and greater than 150 ppm metals. One might question the value of this rating of the feedstocks for the HOC process since these feedstock ratings can apply to virtually many fluid catalytic cracking processes.

The process is similar to the fluid catalytic cracking process in configuration, catalysts, and product handling but differs from gas oil cracking in that the heat release, due to burning the coke produced from the asphaltenes in the charge, is considerably greater. In addition, the much higher content of feedstock metals—particularly nickel and vanadium—requires special consideration in catalyst development and in operation. With the need to convert residual fuels to naphtha and middle distillates, the installation of new heavy oil cracking units as well as the conversion of fluid catalytic cracking units to handle residua has become a refining necessity.

Special catalysts are required for heavy oil cracking units because of the required specifications for activity and selectivity. Some of the catalysts have a high zeolite content or they have pore structures that avoid trapping large molecules and cause coke production. Poisons such as sodium and vanadium accelerate the deactivation rate of catalyst and high amounts of sodium are usually avoided by double desalting of the crude oil. While customized catalysts can improve heavy oil cracking operations, optimization can also take place by closer process control. For example, high cracking activity, if used correctly, can override the adverse dehydrogenation activity of the metals. Thus, low contact time in risers, along with rapid and efficient separation of catalyst and oil vapors, is preferred in reduced crude cracking. Also, heavy catalytic gas oil or slurry oil recycle produces higher yields of coke and higher heat release than it removes through vaporization, and therefore should be minimized or eliminated. In addition to reducing coke yield, lower regenerator temperatures can be realized by direct heat removal or lower heat generation.

9.8.3 Heavy Oil Treating Process

The heavy oil treating (HOT) process is a catalytic cracking process for upgrading heavy feedstocks such as topped crude oils, vacuum residua, and solvent deasphalted bottoms using a fluidized bed of iron ore particles (Speight and Ozum, 2002; Speight, 2014).

The main section of the process consists of three fluidized reactors and separate reactions take place in each reactor (*cracker*, *regenerator*, and *desulfurizer*):

$$Fe_3O_4 + asphaltene\ constituents \rightarrow coke/Fe_3O_4 + Oil + Gas\ (in\ the\ cracker)$$

$$3FeO + H_2O \rightarrow Fe_3O_4 + H_2\ (in\ the\ cracker)$$

$$Coke/Fe_3O_4 + O_2 \rightarrow 3FeO + CO + CO_2\ (in\ the\ regenerator)$$

$$FeO + SO_2 + 3CO \rightarrow FeS + 3CO_2\ (in\ the\ regenerator)$$

$$3FeS + 5O_2 \rightarrow Fe_3O_4 + 3SO_2\ (in\ the\ desulfurizer)$$

In the *cracker*, heavy oil cracking and the steam–iron reaction take place simultaneously under the conditions similar usual to thermal cracking. Any unconverted feedstock is recycled to the cracker from the bottom of the scrubber. The scrubber effluent is separated into hydrogen gas, liquefied petroleum gas (LPG), and liquid products that can be upgraded by conventional technologies to priority products.

In the *regenerator*, coke deposited on the catalyst is partially burned to form carbon monoxide in order to reduce iron tetroxide and to act as a heat supply. In the *desulfurizer*, sulfur in the solid catalyst is removed and recovered as molten sulfur in the final recovery stage.

9.8.4 R2R Process

The R2R process is a fluid catalytic cracking process for conversion of heavy feedstocks (Heinrich and Mauleon, 1994; Inai, 1994; Speight and Ozum, 2002; Speight, 2014).

In the process, the feedstock is vaporized upon contacting hot regenerated catalyst at the base of the riser and lifts the catalyst into the reactor vessel separation chamber where rapid disengagement of the hydrocarbon vapors from the catalyst is accomplished by both a special solids separator and cyclones. The bulk of the cracking reactions takes place at the moment of contact and continues as the catalyst and hydrocarbons travel up the riser. The reaction products, along with a minute amount of entrained catalyst, then flow to the fractionation column. The stripped spent catalyst, deactivated with coke, flows into the Number 1 regenerator.

Partially regenerated catalyst is pneumatically transferred via an air riser to the Number 2 regenerator, where the remaining carbon is completely burned in a dryer atmosphere. This regenerator is designed to minimize catalyst inventory and residence time at high temperature while optimizing the coke-burning rate. Flue gases pass through external cyclones to a waste heat recovery system. Regenerated catalyst flows into a withdrawal well and after stabilization is charged back to the oil riser.

9.8.5 REDUCED CRUDE OIL CONVERSION PROCESS

In recent years, because of a trend for low-boiling products, most refineries perform the operation by partially blending residua into vacuum gas oil. However, conventional fluid catalytic cracking processes have limits in residue processing, so residue fluid catalytic cracking processes have lately been employed one after another. Because the residue fluid catalytic cracking process enables efficient naphtha production directly from residues, it will play the most important role as a residue cracking process, along with the residue hydroconversion process. Another role of the *residuum fluid catalytic cracking process* is to generate high-quality gasoline blending stock and petrochemical feedstock. Olefins (propene, the isomeric butene derivatives, and the isomeric pentene derivatives) serve as feed for alkylating processes, for polymer gasoline, as well as for additives for reformulated gasoline.

In the reduced crude oil conversion process (RCC process), the clean regenerated catalyst enters the bottom of the reactor riser where it contacts low-boiling hydrocarbon *lift gas* that accelerates the catalyst up the riser prior to feed injection (Speight and Ozum, 2002; Speight, 2014). At the top of the lift gas zone the feed is injected through a series of nozzles located around the circumference of the reactor riser.

The catalyst/oil disengaging system is designed to separate the catalyst from the reaction products and then rapidly remove the reaction products from the reactor vessel. Spent catalyst from reaction zone is first steam stripped to remove adsorbed hydrocarbon and then routed to the regenerator. In the regenerator all of the carbonaceous deposits are removed from the catalyst by combustion, restoring the catalyst to an active state with a very low carbon content. The catalyst is then returned to the bottom of the reactor riser at a controlled rate to achieve the desired conversion and selectivity to the primary products.

9.8.6 SHELL FCC PROCESS

The Shell FCC process is designed to maximize the production of distillates from residua (Table 9.12) (Speight and Ozum, 2002; Speight, 2014). In the process, the preheated feedstock (vacuum gas oil, atmospheric residuum) is mixed with the hot regenerated catalyst. After reaction in a riser, volatile materials and catalyst are separated after which the spent catalyst is immediately stripped of entrained and adsorbed hydrocarbons in a very effective multistage stripper. The stripped catalyst

TABLE 9.12
Feedstock and Product Data for the Shell FCC Process

	Residuum[a]	Residuum[a]
Feedstock		
API	18.2	13.4
Sulfur, wt%	1.1	1.3
Carbon residue, % w/w	1.2	4.7
Products		
Gasoline C5–221°C, C5–430°F), wt%	49.5	46.2
Light cycle oil (221°C–370°C, 430°F–700°F), wt%	20.1	19.1
Heavy cycle oil (>370°C, >700°F), wt%	5.9	10.8
Coke, wt%	5.9	7.6

[a] Unspecified.

gravitates through a short standpipe into a single vessel, simple, reliable, and yet efficient catalyst regenerator. Regenerative flue gas passes via a cyclone/swirl tube combination to a power recovery turbine. From the expander turbine the heat in the flue gas is further recovered in a waste heat boiler. Depending on the environmental conservation requirements, a de-NO_xing, de-SO_xing, and particulate emission control device can be included in the flue gas train.

There is a claim that feedstock pretreatment of bitumen (by hydrogenation) prior to fluid catalytic cracking (or for that matter any catalytic cracking process) can result in enhanced yield of naphtha. It is suggested that mild hydrotreating be carried out upstream of a fluid catalytic cracking unit to provide an increase in yield and quality of distillate products (Long et al., 1993; Speight and Ozum, 2002; Parkash, 2003; Hsu and Robinson, 2006; Gary et al., 2007; Speight, 2014). This is in keeping with earlier work (Speight and Moschopedis, 1979) where mild hydrotreating of bitumen was reported to produce low-sulfur liquids that would be amenable to further catalytic processing.

9.8.7 S&W FLUID CATALYTIC CRACKING PROCESS

The S&W FCC process is also designed to maximize the production of distillates from residua (Table 9.13) (Speight and Ozum, 2002; Speight, 2014). In the process, the heavy feedstock is injected into a stabilized, upward flowing catalyst stream whereupon the feedstock–steam–catalyst mixture travels up the riser and is separated by a high-efficiency inertial separator. The product vapor goes overhead to the main fractionator.

The spent catalyst is immediately stripped in a staged, baffled stripper to minimize hydrocarbon carryover to the regenerator system. The first regenerator (650°C–700°C, 1200°F–1290°F) burns 50%–70% of the coke in an incomplete carbon monoxide combustion mode running countercurrently. This relatively mild, partial regeneration step minimizes the significant contribution of hydrothermal catalyst deactivation. The remaining coke is burned in the second regenerator (ca. 775°C, 1425°F) with an extremely low steam content. Hot clean catalyst enters a withdrawal well that stabilizes its fluid qualities prior to being returned to the reaction system.

9.9 OTHER OPTIONS

Catalytic cracking is widely used in the petroleum refining industry to convert heavy oils into more valuable naphtha and lighter products. As the demand for higher-octane gasoline has increased,

TABLE 9.13
Feedstock and Product Data for the S&W FCC Process

	Residuum[a]	Residuum[a]
Feedstock		
API	24.1	22.3
Sulfur, wt%	0.8	1.0
Carbon residue, % w/w	4.4	6.5
Products		
Naphtha, vol%	61.5	60.2
Light cycle oil, vol%	16.6	17.5
Heavy cycle oil, vol%	5.6	6.6
Coke, wt%	7.1	7.8
Conversion, vol%	77.7	75.9

[a] Unspecified.

catalytic cracking has replaced thermal cracking. Two of the most intensive and commonly used catalytic cracking processes in petroleum refining are fluid catalytic cracking and hydrocracking. In the fluid catalytic cracking process, the fine, powdery catalyst (typically zeolites, which have an average particle size of approximately 70 μm) takes on the properties of a fluid when it is mixed with the vaporized feed. Fluidized catalyst circulates continuously between the reaction zone and the regeneration zone.

Several process innovations have been introduced in the form of varying process options, some using piggyback techniques (where one process works in close conjunction with another process; please see earlier), there are other options that have not yet been introduced or even invented but may well fit into the refinery of the future.

In the fluid catalytic cracker (FCC), the major developments are in integration with sulfur removal to produce low-sulfur naphtha without octane loss (Babich and Moulijn, 2003; Rama Rao et al., 2011). This development will build on the development of new catalysts (see earlier). Furthermore, recent enhancements made to RFCC units to permit feeding significant amounts of heavy crudes, while simultaneously improving yields and service factors, have focused on improved feed injection and dispersion, reduced contact time of products and catalyst, improved separation of products and catalyst, and regenerator heat removal. Traditional technology has been modified in key areas including (1) catalyst design to accommodate higher metals feed and to minimize the amount of coke formed on the catalyst, (2) feed injection, (3) riser pipe design and catalyst/oil product separation to minimize overcracking, (4) regenerator design improvements to handle high coke output and avoid damage to catalyst structure, and (5) overall reactor/regenerator design concepts.

These developments have allowed fluid catalytic cracking units to substantially increase residue processing and substantial portions of refinery residua are processes (as blends with gas oils) in fluidized bed units thereby increasing naphtha and diesel production. This will not only continue but also increase in the future. Furthermore, power recovery turbines can be installed on catalytic cracking units to produce power from the pressure of the off-gases of the catalytic cracking. The technology is commercially proven but is not in current general use.

Hydrotreating of the feedstock to the catalytic cracking unit can increase conversion by 8%–12% v/v and with most feeds, it will be possible to reduce the sulfur content of the gasoline/naphtha product to levels low enough to meet the future low-sulfur gasoline pool specifications. With the increasing focus to reduce sulfur content in fuels, the role of *desulfurization* in the refinery becomes more and more important. Currently, the process of choice is the hydrotreater, in which hydrogen is added to the fuel to remove the sulfur from the fuel. Some hydrogen may be lost to reduce the octane number of the fuel, which is undesirable. Because of the increased attention for fuel desulfurization various new process concepts are being developed with various claims of efficiency and effectiveness.

The major developments in desulfurization of three main routes are advanced hydrotreating (new catalysts, catalytic distillation) and reactive adsorption (such as the use of metal oxides that will chemically abstract sulfur) Babich and Moulijn (2003). Using such concepts that are already onstream in some refineries, the number of units performing chemical desulfurization will increase in the future.

There are also proposals (some of which have been put into practice) to bypass the atmospheric and vacuum distillation units by feeding crude oil directly into a thermal cracking process, which would provide sufficient flexibility to supply a varying need of products with a net energy savings.

Light cycle oil from catalytic cracking units will be increased by modifying feedstock composition, introducing improved catalysts and additives, and modifying operating conditions (such as recycle ratio, temperature, catalyst/oil ratio). The addition of an active alumina matrix is a common feature to help refiners increase light cycle oil production when cracking heavy feeds. A comprehensive survey of patent literature in the report found the use of an inorganic additive to occur more than once in catalyst formulations, and metal-doped anionic clays and amorphous silicoaluminophosphates (SAPO) are cited, among other inventions. There is some overlap in catalysts tailored for the production of light cycle oil and catalysts tailored for residuum feedstocks.

Refineries equipped to process heavy crudes have, so far, reported high refining margins because they can take advantage of less expensive heavy oils. Resid fluid catalytic cracking is an important component in the upgrading of such crudes, with unit profitability depending upon the extent to which heavy hydrocarbons in the feed are cracked into valuable products. The product slate, in turn, depends upon the feed characteristics, the catalyst, the hardware, and the operating conditions. Exemplifying a trend toward heavier feedstocks, the majority of the fluid bed catalytic cracking units scheduled to start up before 2015 are expected to process heavy vacuum gas oil and/or residuum feedstocks.

Complete process technologies for residuum catalytic cracking units are the most comprehensive approach to improve resid processing operations, but present, as might be anticipated, the most expensive. Product recycle and multiple reaction sections will be the prevalent technology trends. Improving feed injectors, riser termination and catalyst separation devices, strippers, and regenerator components are revamp options for existing units. As feedstocks get heavier, the trend toward a higher stripper residence time and, consequently, increased mass transfer between entrained hydrocarbons and steam will continue. Moreover, the role of the regenerator continues to evolve because of reduction requirements in the carbon dioxide emissions. The need for development of flue gas treating systems points to the need for continued innovation in that area.

Process/hardware technologies to improve light cycle oil yield from the fluid catalytic cracking unit will include improved feed injection systems and riser pipe and reaction zone designs. Finally, the use of bio-feedstocks (such as namely animal fats and vegetable oils) in the fluid catalytic cracking will be used to increase the yield of light cycle oil and will also provide high-quality products in terms of cetane number. In fact, over the next two decades, fluid catalytic cracking units will take on two additional roles—a user of biomass feedstocks and to reduce carbon dioxide emissions—to alleviate growing concerns over energy security and global warming.

The implementation of bio-feedstock processing techniques in petroleum refineries can result in a competitive advantage for both refiners and society at large. First, the processes provide refineries with alternative feeds that are renewable and could be lower in cost than petroleum. Second, they can reduce the costs of producing fuels and chemicals from bio-feedstocks by utilizing the existing production and distribution systems for petroleum-based products and avoiding the establishment of parallel systems. Finally, the use of bio-feedstocks provides a production base for fuels and chemicals that is less threatened by changes in government policies toward fossil fuel feeds and renewable energies.

Bio-feedstocks that are able to be processed in the fluid catalytic cracking unit can be categorized as biomass-derived oils (both lignocellulosic materials and free carbohydrates) or triglycerides and their free fatty acids. The operating conditions and catalysts used for each type of feed to achieve a desired product slate vary, and each feed comes with inherent advantages and disadvantages. Most of the research work completed to date has been performed on relatively pure bio-feedstocks as opposed to blends of bio-based materials with traditional catalytic cracker feedstocks. Practical implementation in a refinery will, more than likely, be accompanied by blending with vacuum gas oil or residuum. In fact, biomass constituents can be blended with the feedstock and fed to a fluid catalytic cracking unit (Speight, 2011a,b). The acidity of the oil (caused by the presence of oxygen functions) acidity can be reduced by means of a moderate thermal treatment at temperatures in the range of 320°C–420°C (610°F–790°F).

Oils derived from bio-feedstocks oils are generally best upgraded by HZSM-5 or ZSM-5, as these zeolite catalysts promote high yields of liquid products and propylene. Unfortunately, these feeds tend to coke easily, and high acid numbers and undesirable by-products such as water and carbon are, and will continue to be, additional challenges. Waxy feeds obtained from biomass gasification followed by Fischer–Tropsch synthesis to hydrocarbon liquids and waxes (biomass to liquids) are especially suited for increasing light cycle oil production in the catalytic cracking unit, due to the high paraffinic character, low sulfur content, and low aromatics yield of the feed. A major disadvantage for biomass-to-liquids products is the intrinsically low coke yield that can disrupt the heat balance of the catalytic cracking unit. In terms of processability, triglycerides are the best suited bio-feedstock for the catalytic cracker. These materials generally produce high-quality diesel, high-octane naphtha, and are low in sulfur.

Finally, the use of microwaves to generate heat in a uniform and controlled fashion is well known, and microwave technology with catalysts is another technology that has been applied recently in upgrading of heavy oil (Mutyala et al., 2010; Lam et al., 2012). For example, catalytic hydroconversion of residue from coal liquefaction by using microwave irradiation with a Ni catalyst (Wang et al., 2008), microwave-assisted desulfurization of heavy and sour crude oil using iron powder as catalyst (Leadbeater and Khan, 2008). Also upgrading of Athabasca bitumen with microwave-assisted catalytic pyrolysis was carried out in one study. Silicon carbide is used for pyrolysis and nickel and molybdenum nanoparticles are used to enhance catalyst performance. The results of the work suggest that microwave heating with nanoparticles catalyst can be a useful tool for the upgrading of heavy crudes such as bitumen because of rapid heating and energy efficiency (Jeon et al., 2012).

REFERENCES

Abdo, S.F., Thakkar, V., Ackelson, D.B., Wang, L., and Rossi, R.J. 2008. Maximize diesel with UOP enhanced two-stage Unicracking™ technology. *Proceedings of the 18th Annual Saudi-Japan Symposium.* Dhahran, Saudi Arabia, November 16–17.

Ackelson, D. 2004. UOP unicracking process for hydrocracking. In: *Handbook of Petroleum Refining Processes.* R.A. Meyers (Editor). McGraw-Hill, New York, Chapter 7.2.

Alvarenga Baptista, C.M. de L., Cerqueira, H.S., and Sandes, E.F. 2010a. Process for fluid catalytic cracking of hydrocarbon feedstocks with high levels of basic nitrogen. U.S. Patent 7,736,491. June 15.

Alvarenga Baptista, C.M. de L., Moreira, E.M., and Cerqueira, H.S. 2010b. Process for fluid catalytic cracking of hydrocarbon feedstocks with high levels of basic nitrogen. U.S. Patent 7,744,745. June 29.

Avidan, A.A. and Krambeck, F.J. 1990. FCC closed cyclone system eliminates post riser cracking. *Proceedings of the Annual Meeting of National Petrochemical and Refiners Association.*

Babich, I.V. and Moulijn, J.A. 2003. Science and technology of novel processes for deep desulfurization of oil refinery streams: A review. *Fuel*, 82: 607–631.

Bartholic, D.B. 1981a. Preparation of FCC charge from residual fractions. U.S. Patent 4,243,514. January 6.

Bartholic, D.B. 1981b. Upgrading petroleum and residual fractions thereof. U.S. Patent 4,263,128. April 21.

Bartholic, D.B. 1989. Process for upgrading tar sand bitumen. U.S. Patent 4,804,459. February 14.

Bartholic, D.B. and Haseltine, R.P. 1981. *Oil & Gas Journal*, 79(45): 242.

Birch, C.H. and Ulivieri, R. 2000. *ULS Gasoline and Diesel Refining Study.* Pervin and Gertz Inc., Houston, TX.

Bouziden, G., Gentile, K., and Kunz, R.G. 2002. Selective catalytic reduction of NO$_x$ from fluid catalytic cracking case study: BP whiting refinery. Paper ENV-03-128. *NPRA Meeting. National Environmental and Safety Conference.* New Orleans, LA, April 23–24.

Bradley, S.A., Gattuso, M.J., and Bertolacini, R.J. 1989. Characterization and catalyst development. Symposium Series No. 411. American Chemical Society, Washington, DC.

Bridjanian, H. and Khadem Samimi, A. 2011. Bottom of the barrel, an important challenge of the petroleum refining industry. *Petroleum & Coal*, 53(1): 13–21.

Buchanan, J.S., Santiesteban, J.G., and Haag, W.O. 1996. Mechanistic considerations in acid-catalyzed cracking of olefins. *Journal of Catalysis*, 158: 279–287.

Chapin, L. and Letzsch, W. 1994. Deep catalytic cracking maximize olefin production. Paper No. AM-94-43. *Proceedings of the NPRA Annual Meeting.* San Antonio, TX, March 20–22.

Chen, Y.-M. 2004. Recent advances in FCC technology. *Proceedings of the 2004 AIChE Annual Meeting.* Austin, TX, November 7–12.

Chen, Y.-M. 2006. Recent advances in FCC technology. *Powder Technology*, 163: 2–8.

Chuachuensuk, A., Paengjuntuek, W., Kheawhom, S., and Arpornwichanopa, A. 2010. *Proceedings of the 20th European Symposium on Computer Aided Process Engineering—ESCAPE20.* S. Pierucci and G. Buzzi Ferraris (Editors). Elsevier, Amsterdam, the Netherlands.

Couch, K.A., Siebert, K.D., and Van Opdorp, P.J. 2004. Controlling FCC yield and emissions. *Proceedings of the NPRA Annual Meeting*. March.

Cybulski, A. and Moulijn, J.A. (Editors). 1998. *Structured Catalysts and Reactors*. Marcel Dekker, New York.

DeCroocq, D. 1984. *Catalytic Cracking of Heavy Petroleum Hydrocarbons*. Editions Technip, Paris, France.

Degnan, T.F. 2000. Applications of zeolites in petroleum refining. *Topics in Catalysis*, 13: 349–356.

Domokos, L., Jongkind, H., Stork, W.H.J., and Van Den Tol-Kershof, J.M.H. 2010. Catalyst composition, its preparation and use. U.S. Patent 7,749,937. July 6.

Draemel, D.C. 1992. Flexicracking IIIR—ER&E's latest cat cracking design. *Proceedings of the JPI Petroleum Refining Conference*. Japanese Petroleum Institute, Tokyo, Japan.

Dziabala, B., Thakkar, V.P., and Abdo, S.F. 2011. Combination of mild hydrotreating and hydrocracking for making low sulfur diesel and high octane naphtha. U.S. Patent 8,066,867. November 29.

EPA. 2010. Available and emerging technologies for reducing greenhouse gas emissions from the petroleum refining industry. Sector Policies and Programs Division, Office of Air Quality Planning and Standards, United States Environmental Protection Agency, Research Triangle Park, NC.

Freel, B. and Graham, R.G. 2011. Products produced from rapid thermal processing of heavy hydrocarbon feedstocks. U.S. Patent 8,062,503. November 22.

Fujiyama, Y. 1999. Process for fluid catalytic cracking of oils. U.S. Patent 5,904,837. May 18.

Fujiyama, Y., Adachi, M., Okuhara, T., and Yamamoto, S. 2000. Process for fluid catalytic cracking of heavy fraction oils. U.S. Patent 6,045,690. April 4.

Fujiyama, Y., Al-Tayyar, M.H., Dean, C.F., Aitani, A., Redhwi, H.H., Tsutsui, T., and Mizuta, K. 2010. Development of high severity FCC process for maximizing propylene production—Catalyst development and optimization of reaction conditions. *Journal of the Japan Petroleum Institute*, 53(6): 336–341.

Fujiyama, Y., Okuhara, T., and Uchiura, A. 2011. Method of designing gas-solid separator. U.S. Patent 8,070,846. December 6.

Fujiyama, Y., Redhwi, H., Aitani, A., Saeed, R., and Dean, C. 2005. Demonstration plant for new FCC technology yields increased propylene. *Oil & Gas Journal*, 26: 62–67.

Gary, J.G., Handwerk, G.E., and Kaiser, M.J. 2007. *Petroleum Refining: Technology and Economics*, 5th Edition. CRC Press/Taylor & Francis Group, Boca Raton, FL.

Gates, B.C., Katzer, J.R., and Schuit, G.C.A. 1979. *Chemistry of Catalytic Processes*. McGraw-Hill Inc., New York.

Gembicki, V.A., Cowan, T.M., and Brierley, G.R. 2007. Update processing operations to handle heavy feedstocks. *Hydrocarbon Processing*, 86(2): 41–53.

Germain, G.E. 1969. *Catalytic Conversion of Hydrocarbons*. Academic Press Inc., New York.

Harding, R.H., Peters, A.W., and Nee, J.R.D. 2001. New developments in FCC catalyst technology. *Applied Catalysis A: General*, 221: 389–396.

Hedrick, B.W. and Palmas, P. 2011. Process for contacting high contaminated feedstocks with catalyst in an FCC unit. U.S. Patent 8,062,506. November 22.

Heinrich, G. and Mauleon, J.-L. 1994. The R2R process: 21st century FCC technology. *Révue Institut Français du Pétrole*, 49(5): 509–520.

Hemler C.L. and Smith, L.F. 2004. UOP fluid catalytic cracking process. In: *Handbook of Petroleum Refining Processes*. R.A. Meyers (Editor). McGraw-Hill, New York, Chapter 3.3.

Hilbert, T.L., Chitnis, G.K., Umansky, B.S., Kamienski, P.W., Patel, V., and Subramanian, A. 2008. Consider new technology to produce clean diesel. *Hydrocarbon Processing*, 87(2): 47–56.

Hsu, C.S. and Robinson, P.R. 2006. *Practical Advances in Petroleum Processing*, Vols. 1 and 2. Springer, New York.

Hunt, D.A. 1997. In: *Handbook of Petroleum Refining Processes*. R.A. Meyers (Editor). McGraw-Hill, New York, Chapter 3.5.

Huve, L.G., Creyghton, E.J., Ouwehand, C., van Veen, J.A.R., and Hanna, A. 2004. New catalyst technologies expand hydrocrackers' flexibility and contributions. Report No. CRI424/0704. Criterion Catalysts and Technologies, Shell Global Solutions International BV, Amsterdam, the Netherlands.

Imhof, P., Rautiainen, E.P.H., and Gonzalez, J.A. 2005. Maximize propylene yields. *Hydrocarbon Processing*, 84(9): 109–114.

Inai, K. 1994. Operation results of the R2R process. *Revue Institut Français du Pétrole*, 49(5): 521–527.

Ino, T. and Ikeda, S. 1999. Process for fluid catalytic cracking of heavy fraction oil. Unites States Patent 5,951,850. September 14.

Ino, T., Okuhahra, T., Abul-Hamayel, M., Aitani, A., and Maghrabi, A. 2003. Fluid catalytic cracking process for heavy oil. U.S. Patent No. 6,656,346. December 2.

Jakkula, J., Hiltunen, J., and Niemi, V.M. 1997. Short contact time catalytic cracking process: Results from bench scale unit. Paper 29241. *Proceedings of the 15th World Petroleum Congress*. Beijing, China, October 12–17.

Jeon, S.G., Kwak, N.S., Rho, N.S., Ko, C.H., Na, J.-G., Yi, K.B., and Park, S.B. 2012. Catalytic pyrolysis of Athabasca bitumen in H2 atmosphere using microwave irradiation. *Chemical Engineering Research and Design*, 90: 1292–1296.

Johnson, T.E. and Niccum, P.K. 1997. In: *Handbook of Petroleum Refining Processes*. R.A. Meyers (Editor). McGraw-Hill, New York, Chapter 3.2.

Khouw, F.H.H. 1992. Shell residue FCC technology: Challenges and opportunities in a changing environment. *JPI Petroleum Refining Conference*.

Krambeck, F.J. and Pereira, C.J. 1986. FCC processing scheme with multiple risers. U.S. Patent 4,606,810. August 19.

Ladwig, P.K. 1997. In: *Handbook of Petroleum Refining Processes*. R.A. Meyers (Editor). McGraw-Hill, New York, Chapter 3.1.

Lam, S.S., Russell, A.D., Lee, C.L., and Chase, H.A. 2012. Microwave-heated pyrolysis of waste automotive engine oil: Influence of operation parameters on the yield, composition, and fuel properties of pyrolysis oil. *Fuel*, 92: 327–339.

Le Page, J.F., Cosyns, J., Courty, P., Freund, E., Franck, J.P., Jacquin, Y., Juguin, B. et al. 1987. *Applied Heterogeneous Catalysis*. Editions Technip, Paris, France.

Leadbeater, N.E. and Khan, M.R. 2008. Microwave-promoted desulfurization of heavy oil and a review of recent advances on process technologies for upgrading of heavy and sulfur containing crude oil. *Energy Fuel*, 22: 1836–1839.

Letzsch, W. 2011. Innovation drives new catalyst development. *Catalyst of Hydrocarbon Processing*, C93–C94.

Li, X., Li, C., Zhang, J., Yang, C., and Shan, H. 2007. Effects of temperature and catalyst to oil weight ratio on the catalytic conversion of heavy oil to propylene using ZSM-5 and USY catalysts. *Journal of Natural Gas Chemistry*, 16(1): 92–99.

Lifschultz, D.K. 2005. Oil refiner's gathering storm: Help is on the way. *Hydrocarbon Processing*, 84(9): 59–62.

Long, S.L., Johnson, A.R., and Dharia, D. 1993. Advances in residual oil FCC. Paper No. AM-93-50. *Proceedings of the Annual Meeting of National Petrochemical and Refiners Association*.

Maadhah, A.G., Fujiyama, Y., Redhwi, H., Abul-Hamayel, M., Aitani, A., Saeed, M., and Dean, C. 2008. A new catalytic cracking process to maximize refinery propylene. *The Arabian Journal for Science and Engineering*, 33(1B): 17–28.

Maples, R.E. 2000. *Petroleum Refinery Process Economics*, 2nd Edition. PennWell Corporation, Tulsa, OK.

MARPOL. 2005. International convention for the prevention of pollution from ships (MARPOL).
Annex I: Regulations for the prevention of pollution by oil (entered into force October 2, 1983).
Annex II: Regulations for the control of pollution by noxious liquid substances in bulk (entered into force October 2, 1983).
Annex III: Prevention of pollution by harmful substances carried by sea in packaged form (entered into force July 1, 1992).
Annex IV: Prevention of pollution by sewage from ships (entered into force September 27, 2003).
Annex V: Prevention of pollution by garbage from ships (entered into force December 31, 1988).
Annex VI: Prevention of air pollution from ships (entered into force May 19, 2005).

McPherson, L.J. 1984. Causes of FCC reactor coke deposits identified. *Oil & Gas Journal*, 10: 139.

Motaghi, M., Shree, K., and Krishnamurthy, S. 2010. Consider new methods for bottom of the barrel processing—Part 1. *Hydrocarbon Processing*, 89(2): 35–40.

Mutyala, S., Fairbridge, C., Jocelyn Paré, J.R., Bélanger, J.M.R., Ng, S., and Hawkins, R. 2010. Microwave applications to oil sands and petroleum: A review. *Fuel Processing Technology*, 91: 127–135.

Navarro, U., Ni, M., and Orlicki, D.F. 2015. FCC 101: How to estimate product yields cost-effectively and improve operations. *Hydrocarbon Processing*, 94(2): 41–52.

Nishida, S. and Fujiyama, Y. 2000. Separation device. U.S. Patent 6,146,597. November 14.

Occelli, M.L. and O'Connor, P. 1998. *Fluid Cracking Catalysts*. Marcel Dekker, New York.

Otterstedt, J.E., Gevert, S.B., Jaras, S.G., and Menon, P.G. 1986. Processing heavy oils. *Applied Catalysis*, 22: 159–179.

Parkash, S. 2003. *Refining Processes Handbook*. Gulf Professional Publishing, Elsevier, Amsterdam, the Netherlands.

Patel, R., Moore, H., and Hamari, B. 2004. FCC hydrotreater revamp for low-sulfur gasoline. *Proceedings of the NPRA Annual Meeting*. San Antonio, TX, March. National Petrochemical and Refiners Association, Washington, DC.

Patel, R., Zeuthen, P., and Schaldemose, M. 2002. Advanced FCC feed pretreatment technology and catalysts improves FCC profitability. *Proceedings of the NPRA Annual Meeting*. San Antonio, TX, March 17–19. National Petrochemical and Refiners Association, Washington, DC.

Patel, S. 2007. Canadian oil sands—Opportunities, technologies, and challenges. *Hydrocarbon Processing*, 86(2): 65–74.

Rama Rao, M., Soni, D., Siele, G.M., and Bhattacharyya, D. 2011. Convert bottom-of-the-barrel and diesel into light olefins. *Hydrocarbon Processing*, 90(2): 46–49.

Ross, J., Roux, R., Gauthier, T., and Anderson, L.R. 2005. Fine-tune FCC operations for changing fuels market. *Hydrocarbon Processing*, 84(9): 65–73.

Runyan, J. 2007. Is bottomless-barrel refining possible. *Hydrocarbon Processing*, 86(9): 81–92.

Sadeghbeigi, R. 1995. *Fluid Catalytic Cracking: Design, Operation, and Troubleshooting of FCC Facilities*. Gulf Publishing Company, Houston, TX.

Salazar-Sotelo, D., Maya-Yescas, R., Mariaca-Domínguez, E., Rodríguez-Salomón, S., and Aguilera-López, M. 2004. Effect of hydrotreating FCC feedstock on product distribution. *Catalysis Today*, 98(1–2): 273–280.

Sayles, S. and Bailor, J. 2005. Upgrade FCC hydrotreating. *Hydrocarbon Processing*, 84(9): 87–90.

Schiller, R. 2011. Effect of synthetic crude feedstocks on FCC yield. *Refinery Operations*, 2(4): 1–2.

Schnaith, M.W., Sexson, A., Tru, D., Bartholic, D.B., Lee, Y.K., Yoo, I.S., and Kang, H.S. 1998. *Oil & Gas Journal*, 96(25): 53.

Shan, H.H., Zhao, W., He, C.Z., Zhang, J.F., and Yang, C.H. 2003. Maximum FCC diesel yield with TSRFCC technology. Preprints. *Division of Fuel Chemistry, American Chemical Society*, 48(2): 710–711.

Shidhaye, H., Kukade, S., Kumar, P., Rao, P.V.C., and Choudary, N.V. 2015. Improve FCC margins by processing more vacuum resid in feed. *Hydrocarbon Processing*, 94(12): 35–38.

Soni, D., Rama Rao, M., Saidulu, G., Bhattacharyya, D., and Satheesh, V.K. 2009. Catalytic cracking process enhances production of olefins. *Petroleum Technology Quarterly*, 14(Q4): 95–100.

Speight, J.G. 2000. *The Desulfurization of Heavy Oils and Residua*, 2nd Edition. Marcel Dekker, New York.

Speight, J.G. 2008. *Synthetic Fuels Handbook: Properties, Processes, and Performance*. McGraw-Hill, New York.

Speight, J.G. 2009. *Enhanced Recovery Methods for Heavy Oil and Tar Sands*. Gulf Publishing Company, Houston, TX.

Speight, J.G. 2011a. *The Refinery of the Future*. Gulf Professional Publishing, Elsevier, Oxford, UK.

Speight, J.G. (Editor). 2011b. *Biofuels Handbook*. Royal Society of Chemistry, London, UK.

Speight, J.G. 2014. *The Chemistry and Technology of Petroleum*, 5th Edition. CRC Press/Taylor & Francis Group, Boca Raton, FL.

Speight, J.G. and Moschopedis, S.E. 1979. The production of low-sulfur liquids and coke from Athabasca bitumen. *Fuel Processing Technology*, 2: 295.

Speight, J.G. and Ozum, B. 2002. *Petroleum Refining Processes*. Marcel Dekker, New York.

Spoor, R.M. 2008. Low-carbon refinery: Dream or reality. *Hydrocarbon Processing*, 87(11): 113–117.

Stiles A.B. and Koch, T.A. 1995. *Catalyst Manufacture*. Marcel Dekker, New York.

Suchanek, A.J. and Moore, A.S. 1986. Modern residue upgrading by ART. *Proceedings of the NPRA National Meeting*. San Antonio, TX, March 23–25. National Petrochemical and Refiners Association, Washington, DC.

Swaty, T.E. 2005. Global refinery industry trends: The present and the future. *Hydrocarbon Processing*, 84(9): 35–46.

Upson, L.L. and Wegerer, D.E. 1993. Rapid disengager techniques in riser design. *Proceedings of the Third International Symposium on Advances in Fluid Catalytic Cracking*. August. American Chemical Society.

Wang, T.X., Zong, Z.M., Zhang, V.W., Wei, Y.B., Zhao, W., Li, B.M., and Wei, X.Y. 2008. Microwave-assisted hydroconversion of demineralized coal liquefaction residues from Shenfu and Shengli coals. *Fuel*, 87: 498–507.

Wojciechowski, B.W. and Corma, A. 1986. *Catalytic Cracking: Catalysts, Chemistry, and Kinetics*. Marcel Dekker, New York.

Wolschlag, L.M. 2011. Innovations developed using sophisticated engineering tools. *Proceedings of the AIChE 2011 Regional Process Technology Conference*. October 6–7.

Wolschlag, L.M. and Couch, K.A. 2010a. Upgrade FCC performance. *Hydrocarbon Processing*, 89(9): 57–65.

Wolschlag, L.M. and Couch, K.A. 2010b. UOP FCC innovations developed using sophisticated engineering tools. Report No. AM-10-109. UOP LLC, Des Plaines, IL.

Wrench, R.E. and Glascow, P.E. 1992. FCC hardware options for the modern cat cracker. AIChE Symposium Series #29, 1992.

Yung, K.Y. and Pouwels, A.C. 2008. Fluid catalytic cracking—A diesel producing machine. *Hydrocarbon Processing*, 87(2): 79–83.

10 Hydrotreating Processes

10.1 INTRODUCTION

Hydroprocessing (covers the process terms "hydrotreating" and "hydrocracking") is a refining technology in which a feedstock is treated with hydrogen at temperature and under pressure that can affect the product slate in refineries by strategic placement (Figure 10.1). In fact, the use of hydrogen in thermal processes is perhaps the single most significant advance in refining technology during the twentieth century (Scherzer and Gruia, 1996; Dolbear, 1998).

The hydrotreating process uses the principle that the presence of hydrogen during *mild* thermal treatment reaction of a petroleum feedstock removes the heteroatoms and metals (Hunter et al., 2010). On the other hand, the presence of hydrogen during cracking (*hydrocracking*) terminates many of the coke-forming reactions and enhances the yields of the lower-boiling components, such as gasoline, kerosene, and jet fuel. However, hydrocracking (Chapter 11) should not be regarded as a competitor for catalytic cracking (Chapter 9). Catalytic cracking units normally use gas oil distillates as feedstocks, whereas hydrocracking feedstock usually consists of refractive gas oils derived from cracking and coking operations. Hydrocracking is a supplement to, rather than a replacement for, catalytic cracking.

In hydrotreating, the feedstock is reacted with hydrogen at elevated temperatures, in the range of 300°C–450°C (570°F–840°F), and elevated pressures, in the range of 120–2200 psi under the presence of a hydrogenation catalyst typically cobalt–molybdenum (Co–Mo) or nickel–molybdenum (Ni–Mo) on gamma-alumina (γ-Al_2O_3). The catalysts are produced in the oxide forms and are sulfided before their use in the process. During hydrotreating, the heteroatoms are removed in the form of hydrogen sulfide (H_2S), ammonia (NH_3), and water (H_2O); simultaneously, metals species such as vanadium (V) and nickel (Ni) are also removed by hydrodemetallization reactions. By hydrotreating, hydrogen is also added to the feedstock, which in the case of the heavy feedstocks reduces the tendency for coke formation during subsequent thermal or catalytic cracking processes. Fouling of hydrotreating catalysts by metal deposits as well as by coke depositions becomes unavoidable, which may cause expensive plant shut downs for the replacement of the catalyst. Catalyst reactivation and replacement of poisoned catalyst by the fresh catalyst are important elements of reactor design.

The distinguishing feature of the hydroprocesses is that, although the composition of the feedstock is relatively unknown and a variety of reactions may occur simultaneously, the final product may actually meet all the required specifications for its particular use. Thus, the purposes of refinery hydroprocesses are (1) to improve existing petroleum products, (2) to enable petroleum products to meet market specifications, (3) to develop new products or even new uses for existing products, (4) to convert inferior or low-grade materials into valuable products, and (5) to transform near-solid residua into liquid fuels.

There is a rough correlation between the quality of petroleum products and their hydrogen content (Dolbear, 1998). It so happens that desirable aviation gasoline, kerosene, diesel fuel, and lubricating oil are made up of hydrocarbons containing high proportions of hydrogen (Figure 10.2). In addition, it is usually possible to convert olefins and higher-molecular-weight constituents to paraffins and monocyclic hydrocarbons by hydrogen-addition processes. These facts have for many years encouraged attempts to employ hydrogenation for refining operations; despite considerable technical success, such processes were not economically possible until low-priced hydrogen became available as a result of the rise of reforming, which converts naphthenes to aromatics with the release of hydrogen.

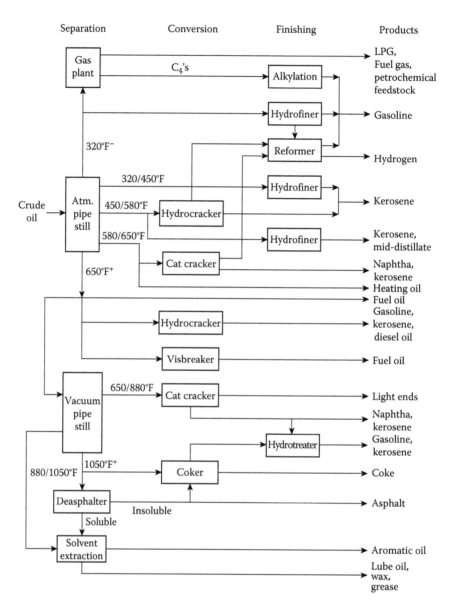

FIGURE 10.1 Generalized refinery layout showing relative placement of the catalytic cracking units.

As already noted, hydroprocesses for the conversion of petroleum fractions and petroleum products may be classified as *nondestructive* and *destructive*. Nondestructive hydrogenation is characterized by the removal of heteroatom constituents as the hydrogenated analogs:

$$RSR^1 + H_2 \rightarrow RH + R^1H + H_2S$$

$$2RN(R_2)R^1 + 3H_2 \rightarrow 2RH + 2R^1H + 2R^2H + 2NH_3$$

$$ROR^1 + H_2 \rightarrow RH + R^1H + H_2O$$

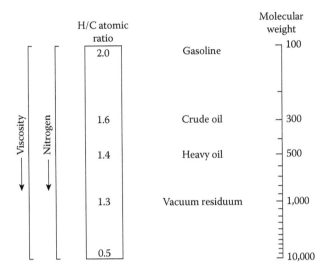

FIGURE 10.2 Atomic hydrogen–carbon ratios of various feedstocks.

In addition, the saturation of olefin derivatives in the products from thermal processes also occurs:

$$RCH=CHR^1 + H_2 \rightarrow R\text{-}CH_2CH_2\text{-}R^1$$

Aromatic constituents may also be saturated to produce cycloaliphatic derivatives (naphthene derivatives) by this treatment:

$$C_6H_6 + 3H_2 \rightarrow C_6H_{12}$$

10.1.1 BENZENE CYCLOHEXANE

Any metals (usually nickel and vanadium) present in the feedstock are usually removed during hydrogen processing as a cost to the hydrogen consumption and not, of course, as their hydrogen analogs but by deposition on the catalyst through changes in the chemical properties of the metal-containing constituents by the high temperatures and the presence of hydrogen. These two process parameters progressively affect the ability of the organic structures or retain the metals within the organic matrix and deposition ensues.

Thus, hydrotreating of distillates may be defined simply as the removal of sulfur, nitrogen, and oxygen compounds as well as olefinic compounds by selective hydrogenation. The hydrotreating catalysts are usually cobalt plus molybdenum (CoS–MoS) or nickel plus molybdenum (NiS–MoS) sulfides impregnated on an alumina (Al_2O_3) base. The hydrotreating operating conditions 1000–2000 psi hydrogen and approximately 370°C (700°F) are such that appreciable hydrogenation of aromatics does not occur. The desulfurization reactions are invariably accompanied by small amounts of hydrogenation and hydrocracking, the extent of which depends on the nature of the feedstock and the severity of desulfurization. In summary, hydrotreating (nondestructive hydrogenation) is generally used for the purpose of improving product quality without appreciable alteration of the boiling range. Mild processing conditions are employed so that only the more unstable materials are attacked.

Destructive hydrogenation (*hydrogenolysis* or *hydrocracking*) is characterized by the cleavage of carbon–carbon linkages accompanied by hydrogen saturation of the fragments to produce lower-boiling products:

$$RCH_2CH_2R^1 + H_2 \rightarrow RCH_3 + CH_3R^1$$

Such treatment requires processing thermal regimes that are similar to those used in catalytic cracking (Chapter 9) and the use of high hydrogen pressures to minimize the reactions that lead to coke formation.

The major differences between *hydrotreating* and *hydrocracking* are the time at which the feed-stock remains at reaction temperature and the extent of the decomposition of the nonheteroatom constituents and products. The upper limits of hydrotreating conditions may overlap with the lower limits of hydrocracking conditions (Figure 10.3) with a lower overall conversion for hydrotreating (Figure 10.4). And where the reaction conditions overlap, feedstocks to be hydrotreated will generally be exposed to the reactor temperature for shorter periods; hence the reason why hydrotreating conditions may be referred to as *mild*. All is relative.

Unsaturated compounds, such as olefins, are not indigenous to petroleum and are produced during cracking processes and need to be removed from product streams because of the tendency of unsaturated compounds and heteroatomic polar compounds to form gum and sediment (Speight, 2014). On the other hand, aromatic compounds are indigenous to petroleum and some may be formed during cracking reactions. The most likely explanation is that the aromatic compounds present in product streams are related to the aromatic compounds originally present in petroleum but now having shorter alkyl side chains. Thus, in addition to olefins, product streams will contain

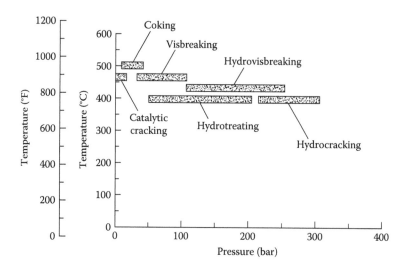

FIGURE 10.3   Temperature and pressure ranges for various processes.

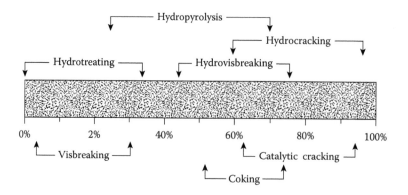

FIGURE 10.4   Feedstock conversion in various processes.

a range of aromatic compounds that have to be removed to enable many of the product streams to meet product specifications.

Of the aromatic constituents, the polycyclic aromatics are first partially hydrogenated before cracking of the aromatic nucleus takes place. The sulfur and nitrogen atoms are converted to hydrogen sulfide and ammonia, but a more important role of the hydrogenation is probably to hydrogenate the coke precursors rapidly and prevent their conversion to coke.

One of the problems in the processing of high-sulfur and high-nitrogen feeds is the large quantity of hydrogen sulfide and ammonia that are produced. Substantial removal of both compounds from the recycle gas can be achieved by the injection of water in which, under the high-pressure conditions employed, both hydrogen sulfide and ammonia are very soluble compared with hydrogen and hydrocarbon gases. The solution is processed in a separate unit for the recovery of anhydrous ammonia and hydrogen sulfide.

Hydrotreating is carried out by charging the feed to the reactor, together with a portion of all the hydrogen produced in the catalytic reformer. Suitable catalysts are tungsten–nickel sulfide, cobalt–molybdenum–alumina, nickel oxide–silica–alumina, and platinum–alumina. Most processes employ Co–Mo catalysts, which generally contain approximately 10% molybdenum oxide and less than 1% cobalt oxide supported on alumina. The temperatures employed are in the range of 300°C–345°C (570°F–850°F), and the hydrogen pressures are approximately 500–1000 psi.

The reaction generally takes place in the vapor phase but, depending on the application, may be a mixed-phase reaction. The reaction products are cooled in a heat exchanger and led to a high-pressure separator where hydrogen gas is separated for recycling. Liquid products from the high-pressure separator flow to a low-pressure separator (stabilizer) where dissolved light gases are removed. The product may then be fed to a reforming or cracking unit if desired.

With the influx of heavy feedstocks in refineries, hydrotreating has taken on a new role insofar as hydrotreating units are usually placed upstream of units where catalyst deactivation may occur from feed impurities, or to lower impurities in finished products, like jet fuel or diesel. A large refinery may have five or more hydrotreaters, and three types of hydrotreaters are typically found in all refineries: (1) the naphtha hydrotreater, which pretreats feed to the reformer, (2) the kerosene hydrotreater, also referred to as the *middle distillate hydrotreater*, which is used to treat middle distillates from the atmospheric crude tower, and (3) the gas oil hydrotreater, also referred to as the *diesel hydrotreater*, which treats gas oil from the atmospheric crude tower or pretreats vacuum gas oil entering a cracking unit. It is this hydrotreater that may also be used to treat blends of gas oil with heavier feedstocks.

In the process, the feedstock to the hydrotreater is mixed with hydrogen-rich gas before entering a fixed-bed reactor. In the presence of a catalyst, hydrogen reacts with the oil feed to produce hydrogen sulfide, ammonia, saturated hydrocarbons, and other free metals. The metals remain on the surface of the catalyst and other products leave the reactor with the oil-hydrogen stream. Oil is separated from the hydrogen-rich gas stream, and any remaining light ends (C_4 and lighter) are removed in the stripper. The gas stream is treated to remove hydrogen sulfide and then it is recycled to the reactor (Speight and Ozum, 2002; Parkash, 2003; Hsu and Robinson, 2006; Gary et al., 2007; Speight, 2014).

Most hydrotreating reactions are carried out below 425°C (800°F) to minimize cracking. Product streams vary considerably depending on (1) the characteristics of the feedstock, (2) the catalyst, and (3) the process parameters. The predominant reaction type is hydrodesulfurization, although many reactions take place in hydrotreating including denitrogenation, deoxygenation, hydrogenation, and hydrocracking. Almost all hydrotreating reactions are exothermic and, depending on the specific conditions, a temperature rise through the reactor of 3°C–11°C (5°F–20°F) is usually observed.

Generally, it is more economical to hydrotreat high-sulfur feedstocks before catalytic cracking than to hydrotreat the products from catalytic cracking. The advantages are as follows: (1) the products require less finishing; (2) sulfur is removed from the catalytic cracking feedstock, and corrosion is reduced in the cracking unit; (3) carbon formation during cracking is reduced and higher conversions result; and (4) the catalytic cracking quality of the gas oil fraction is improved.

One of the chief problems with the processing of residua is the deposition of metals, in particular vanadium, on the catalyst. It is not possible to remove vanadium from the catalyst, which must therefore be replaced when deactivated, and the time taken for catalyst replacement can significantly reduce the unit time efficiency. Fixed-bed catalysts tend to plug owing to solids in the feed or carbon deposits when processing residual feeds. As mentioned previously, the highly exothermic reaction at high conversion gives difficult reactor design problems in heat removal and temperature control.

It is the physical and chemical composition of a feedstock that plays a large part not only in determining the nature of the products that arise from refining operations but also in determining the precise manner by which a particular feedstock should be processed. Furthermore, it is apparent that the conversion of heavy oils and residua requires new lines of thought to develop suitable processing scenarios. Indeed, the use of thermal (*carbon rejection*) processes and of hydrothermal (*hydrogen addition*) processes, which were inherent in the refineries designed to process lighter feedstocks, have been a particular cause for concern. This has brought about the evolution of processing schemes that accommodate the heavier feedstocks. As a point of reference, an example of the former option is the delayed coking process in which the feedstock is converted to overhead with the concurrent deposition of coke, for example, that used by Suncor, Inc., at their oil sands plant (Speight, 2013, 2014).

The hydrogen addition concept is illustrated, in part, by the hydrotreating processes in which hydrogen is used in an attempt to remove the heteroatoms and metals as well as to *stabilize* the reactive fragments produced by the low degree of hydrocracking, thereby decreasing their potential for recombination to heavier products and ultimately to coke. The choice of processing schemes for a given hydrotreating application depends upon the nature of the feedstock as well as the product requirements. The process can be simply illustrated as a single-stage or as a two-stage operation (Figure 10.5).

The petroleum industry often employs two-stage processes in which the feedstock undergoes both hydrotreating and hydrocracking. In the first, or pretreating, stage, the main purpose is conversion of nitrogen compounds in the feed to hydrocarbons and to ammonia by hydrogenation and mild hydrocracking. Typical conditions are 340°C–390°C (650°F–740°F), 150–2500 psi, and a catalyst contact time of 0.5–1.5 hours. Up to 1.5% w/w hydrogen is absorbed, partly by conversion of the nitrogen compounds but chiefly by aromatic compounds that are hydrogenated. Thus, the single-stage process can be used to facilitate hydrotreating and the two-stage process may then be used for hydrocracking primarily to produce distillates from high-boiling feedstocks. Both processes use an extinction-recycling technique to maximize the yields of the desired product. Significant conversion

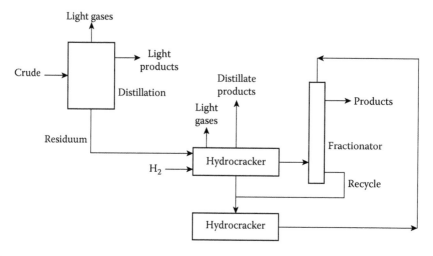

FIGURE 10.5   A single- and two-stage hydroprocessing configuration.

of heavy feedstocks can be accomplished by including the second stage and hydrocracking at high severity. For some applications, the products boiling up to 340°C (650°F) can be blended to give the desired final product.

Product distribution and quality vary considerably depending upon the nature of the feedstock constituents as well as on the process. Different process configurations will produce variations in the product slate from any one particular feedstock and the *feedstock recycle* option adds another dimension to variations in the product slate (Table 10.1).

In modern refineries, hydrotreating is one of the several process options that can be applied to the production of liquid fuels from the heavier feedstocks (Speight, 2000). A most important aspect of the modern refinery operation is the desired product slate, which dictates the matching of a process with any particular feedstock to overcome differences in feedstock composition. Hydrogen consumption is also a parameter that varies with feedstock composition (Tables 10.2 and 10.3), thereby indicating the need for a thorough understanding of the feedstock constituents if the process is to be employed to maximum efficiency.

A convenient means of understanding the influence of feedstock on the hydrotreating process is through a study of the hydrogen content (H/C atomic ratio) and molecular weight (carbon number)

TABLE 10.1
Effect of Recycle on Product Distribution

	Once-Through	Recycle
Conversion, vol%	90	95
Conversion per pass, vol%	90	70
LHSV, based on fresh feed	Base	Base
Hydrogen partial pressure	Base	−20%
Catalyst concentration, wt%	Base	+60%
Hydrogen consumption, wt%	Base	−13%
C_1–C_4 yield, wt%	Base	−10%
Liquid yield, wt%	Base	+3%–6%
vol%	Base	+3%/6%

TABLE 10.2
Hydrogen Consumption during Hydrotreating Various Feedstocks

	°API	Sulfur (wt%)	Carbon Residue (Conradson) (wt%)	Nitrogen (wt%)	Hydrogen (scf/bbl)
Venezuela, atmospheric	15.3–17.2	2.1–2.2	9.9–10.4	—	425–730
Venezuela, vacuum	4.5–7.5	2.9–3.2	20.5–21.4	—	825–950
Boscan (whole crude)	10.4	5.6	—	0.52	1100
Tia Juana, vacuum	7.8	2.5	21.4	0.52	490–770
Bachaquero, vacuum	5.8	3.7	23.1	0.56	1080–1260
West Texas, atmospheric	17.7–17.9	2.2–2.5	8.4	—	520–670
West Texas, vacuum	10.0–13.8	2.3–3.2	12.2–14.8	—	675–1200
Khafji, atmospheric	15.1–15.7	4.0–4.1	11.0–12.2	—	725–800
Khafji, vacuum	5.0	5.4	21.0	—	1000–1100
Arabian light, vacuum	8.5	3.8	—	—	435–1180
Kuwait, atmospheric	15.7–17.2	3.7–4.0	8.6–9.5	0.20–0.23	470–815
Kuwait, vacuum	5.5–8.0	5.1–5.5	16.0	—	290–1200

TABLE 10.3
Additional Hydrogen Consumption Caused by Nitrogen and Metals during Hydrodesulfurization

Metals	
V + Ni (ppm)	Corrections (%) to Hydrogen Consumption
0–100	−2
200	1
300	2.5
400	4
500	6.5
600	9
700	12
800	16
900	21
1000	28
1100	38
1200	50

Nitrogen Compounds	Additional Hydrogen Required	
	(mol H_2/Compound)	scf/bbl Feed
Saturated amines	1	83
Pyrrolidine	2	167
Nitriles, pyrroline, alkyl cyanides	3	250
Pyrrole, nitroparaffins	4	334
Analine, pyridine	5	417
Indole	7	584

of the various feedstocks or products (Figure 10.6). These data show the carbon number and/or the relative amount of hydrogen must be added to generate the desired heteroatom removal. In addition, it is also possible to use data for hydrogen usage in residuum processing (Figure 10.7), where the relative amount of hydrogen consumed in the process can be shown to be dependent upon the sulfur content of the feedstock.

The commercial processes for treating or finishing petroleum fractions with hydrogen operate in essentially the same manner with similar parameters. The feedstock is heated and passed with hydrogen gas through a tower or reactor filled with catalyst pellets. The reactor is maintained at a temperature of 260°C–425°C (500°F–800°F) at pressures from 100 to 1000 psi, depending on the particular process, the nature of the feedstock, and the degree of hydrogenation required. After leaving the reactor, excess hydrogen is separated from the treated product and recycled through the reactor after removal of hydrogen sulfide. The liquid product is passed into a stripping tower, where steam removes dissolved hydrogen and hydrogen sulfide, and after cooling the product is run to finished product storage or, in the case of feedstock preparation, pumped to the next processing unit.

It is most important to reduce the nitrogen content of the product oil to less than 0.001% w/w (10 ppm). This stage is usually carried out with a bifunctional catalyst containing hydrogenation promoters, for example, nickel and tungsten or molybdenum sulfides, on an acid support, such

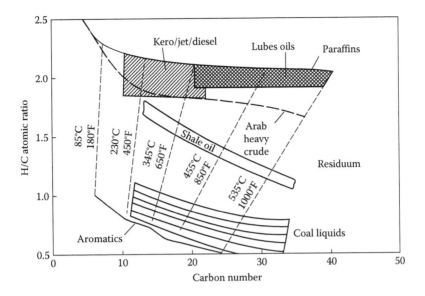

FIGURE 10.6 Representation of process chemistry and hydrogen requirements.

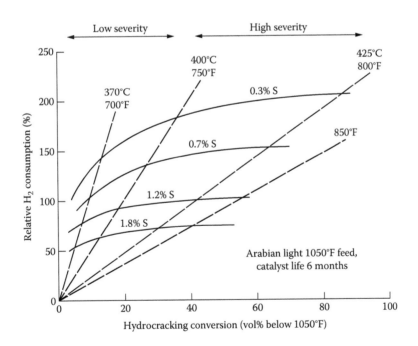

FIGURE 10.7 Hydrogen requirements in hydrotreating and hydrocracking processes.

as silica-alumina. The metal sulfides hydrogenate aromatics and nitrogen compounds and prevent deposition of carbonaceous deposits; the acid support accelerates nitrogen removal as ammonia by breaking carbon–nitrogen bonds. The catalyst is generally used as 1/8 by 1/8 inch (0.32 by 0.32 cm) or 1/16 by 1/8 inch (0.16 by 0.32 cm) pellets, formed by extrusion.

Most of the hydrocracking is accomplished in the second stage, which resembles the first but uses a different catalyst. Ammonia and some gasoline are usually removed from the first-stage product, and then the remaining oil, which is low in nitrogen compounds, is passed over the second-stage catalyst. Again, typical conditions are 300°C–370°C (600°F–700°F), 1500–2500 psi hydrogen pressure, and 0.5–1.5-hour contact time; 1%–1.5% by weight hydrogen may be absorbed. Conversion to naphtha and kerosene is seldom complete in one contact with the catalyst, so the lighter oils are removed by distillation of the products, and the heavier, higher-boiling product is combined with fresh feed and recycled over the catalyst until it is completely converted.

10.2 HYDRODESULFURIZATION

At this point, and in the context of this chapter, it is pertinent that there be a discussion of the techniques for desulfurization and concurrent demetallization of various feedstocks. Thus, desulfurization is the removal of sulfur or sulfur compounds from crude oil and crude oil products and demetallization is the removal of metals or metal-containing constituents (such as porphyrins) from crude oil and crude oil products.

By way of introduction, sulfur content and metals content vary with crude oil type (Figure 10.8) and may or may not present a problem to the refiner, depending on the amount of metals present and the downstream processing required. Vanadium (V) and nickel (Ni), the primary metals found in petroleum, can range from less than 1 part per million (ppm by weight) in some crude oils to as high

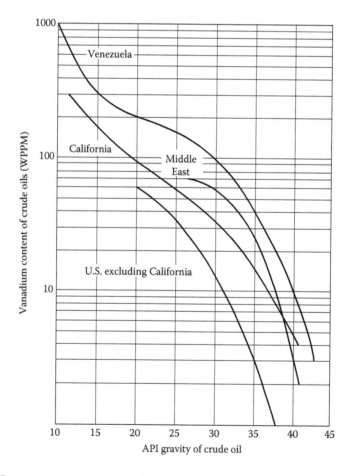

FIGURE 10.8 General representation of metal content of various feedstocks.

as 1100 ppm vanadium and 85 ppm nickel for Boscan (Venezuela) crude oil. Vanadium is usually present in higher concentrations than nickel for Middle East crude oils and Venezuelan crude oils. However, for many U.S. crude oils, particularly those from California, the nickel content is higher than the vanadium content. Other metals, such as sodium and iron, are also found in quantities up to 100 and 60 ppm, respectively, though usually much lower.

A number of methods are available for segregating metals from crude oil. For example, deasphalting removes metals insofar as they appear in the separated asphalt (Speight, 2000) and coking processes cause the metals to concentrate in the coke (Speight, 2000); there is a similar segregation of sulfur but it is not as dramatic as the metal segregation (Speight, 2000). Hydrodemetallization and hydrodesulfurization processes are the most effective and will be discussed here.

Hydrotreating processes, in particular the hydrodesulfurization of petroleum residua, are catalytic processes. Hydrocarbon feedstock and hydrogen are passed through a catalyst bed at elevated temperatures and pressures. Some of the sulfur atoms attached to hydrocarbon molecules react with hydrogen on the surface of the catalyst to form hydrogen sulfide (H_2S), and thermodynamic equilibrium calculations show that these reactions can be driven to almost 100% completion (Speight, 2000).

Hydrodesulfurization and demetallization occur simultaneously on the active sites within the catalyst pore structure. Sulfur and nitrogen occurring in residua are converted to hydrogen sulfide and ammonia in the catalytic reactor and these gases are scrubbed out of the reactor effluent gas stream. The metals in the feedstock are deposited on the catalyst in the form of metal sulfides and cracking of the feedstock to distillate produces a laydown of carbonaceous material on the catalyst; both events poison the catalyst and activity or selectivity suffers. The deposition of carbonaceous material is a fast reaction that soon equilibrates to a particular carbon level and is controlled by hydrogen partial pressure within the reactors. On the other hand, metal deposition is a slow reaction that is directly proportional to the amount of feedstock passed over the catalyst.

The life of a catalyst used to hydrotreat petroleum residua is dependent on the rate of carbon deposition and the rate at which organometallic compounds decompose and form metal sulfides on the surface. Several different metal complexes exist in the asphaltene fraction of the residuum and an explicit reaction mechanism of decomposition that would be a perfect fit for all of the compounds is not possible. However, in general terms, the reaction can be described as hydrogen (A) dissolved in the feedstock contacting an organometallic compound (B) at the surface of the hydrotreating catalyst and producing a metal sulfide (C) and a hydrocarbon (D):

$$A + B \rightarrow C + D$$

Different rates of reaction may occur with various types and concentrations of metallic compounds. For example, a medium-metal-content feedstock will generally have a lower rate of demetallization compared to high-metal-content feedstock. And, although individual organometallic compounds decompose according to both first- and second-order rate expressions, for reactor design, a second-order rate expression is applicable to the decomposition of residuum as a whole.

Finally, it has been recognized over the past three decades that desulfurization of gas oil or heavy feedstock that is subsequently fed to fluid catalytic cracking units allows the feedstock to be catalytically cracked without excessive sulfur oxide emissions in the regenerator flue gas. There are additional advantages such as (1) higher conversions of the feedstock, (2) higher yields of naphtha, (3) less sulfur in the naphtha and other distillates, (4) lower yields of coke, and (5) reduced consumption of the cracking catalyst consumption.

Most of the metals in the catalytic cracking feedstock are (1) removed during desulfurization, eliminating a principal source of cracking catalyst deactivation or (2) kore preferably for the heavy feedstocks, removed by use of a guard bed (demetallization reactor) or mild thermal treatment in which the metals are separated as part of the initially formed coke.

10.2.1 PROCESS CONFIGURATION

All hydrodesulfurization processes react hydrogen with a hydrocarbon feedstock to produce hydrogen sulfide and a desulfurized hydrocarbon product (Figure 10.5). The feedstock is preheated and mixed with hot recycle gas containing hydrogen, and the mixture is passed over the catalyst in the reactor section at temperatures between 290°C and 445°C (550°F–850°F) and pressures between 150 and 3000 psig. The reactor effluent is then cooked by heat exchange, and desulfurized liquid hydrocarbon product and recycle gas are separated at essentially the same pressure as used in the reactor. The recycle gas is then scrubbed and/or purged of the hydrogen sulfide and light hydrocarbon gases, mixed with fresh hydrogen makeup, and preheated prior to mixing with hot hydrocarbon feedstock.

The recycle gas scheme is used in the hydrodesulfurization process to minimize physical losses of expensive hydrogen. Hydrodesulfurization reactions require a high hydrogen partial pressure in the gas phase to maintain high desulfurization reaction rates and to suppress carbon laydown (catalyst deactivation). The high hydrogen partial pressure is maintained by supplying hydrogen to the reactors at several times the chemical hydrogen consumption rate. The majority of the unreacted hydrogen is cooled to remove hydrocarbons, recovered in the separator, and recycled for further utilization. Hydrogen is physically lost in the process by solubility in the desulfurized liquid hydrocarbon product, and from losses during the scrubbing or purging of hydrogen sulfide and light hydrocarbon gases from the recycle gas.

10.2.2 DOWNFLOW FIXED-BED REACTORS

The reactor design commonly used in hydrodesulfurization of distillates is the fixed-bed reactor design in which the feedstock enters at the top of the reactor and the product leaves at the bottom of the reactor (Figure 10.9). The catalyst remains in a stationary position (fixed-bed) with hydrogen and petroleum feedstock passing in a downflow direction through the catalyst bed. The hydrodesulfurization reaction is exothermic, and the temperature rises from the inlet to the outlet of each catalyst bed. With a high hydrogen consumption and subsequent large temperature rise, the reaction

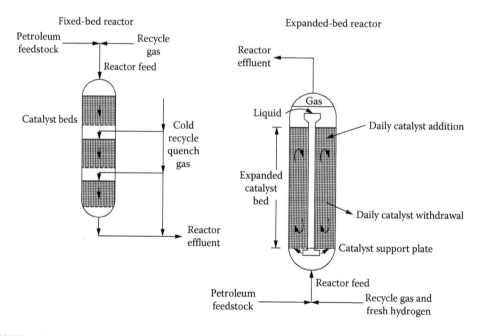

FIGURE 10.9 Reactor design for hydrodesulfurization processes.

mixture can be quenched with cold recycled gas at intermediate points in the reactor system. This is achieved by dividing the catalyst charge into a series of catalyst beds, and the effluent from each catalyst bed is quenched to the inlet temperature of the next catalyst bed.

The extent of desulfurization is controlled by raising the inlet temperature to each catalyst bed to maintain constant catalyst activity over the course of the process. Fixed-bed reactors are mathematically modeled as plug-flow reactors with very little back mixing in the catalyst beds. The first catalyst bed is poisoned with vanadium and nickel at the inlet to the bed and may be a cheaper catalyst (*guard bed*). As the catalyst is poisoned in the front of the bed, the temperature exotherm moves down the bed and the activity of the entire catalyst charge declines thus requiring a raise in the reactor temperature over the course of the process sequence. After catalyst regeneration, the reactors are opened and inspected, and the high-metal-content catalyst layer at the inlet to the first bed may be discarded and replaced with fresh catalyst. The catalyst loses activity after a series of regenerations and, consequently, after a series of regenerations it is necessary to replace the complete catalyst charge. In the case of very-high-metal-content feedstocks (such as residua), it is often necessary to replace the entire catalyst charge rather than to regenerate it. This is due to the fact that the metal contaminants cannot be removed by economical means during rapid regeneration, and the metals have been reported to interfere with the combustion of carbon and sulfur, catalyzing the conversion of sulfur dioxide (SO_2) to sulfate (SO_4^{2-}) that has a permanent poisoning effect on the catalyst.

Fixed-bed hydrodesulfurization units are generally used for distillate hydrodesulfurization and may also be used for residuum hydrodesulfurization but require special precautions in processing. The residuum must undergo two-stage electrostatic desalting so that salt deposits do not plug the inlet to the first catalyst bed and the residuum must be low in vanadium and nickel content to avoid plugging the beds with metal deposits. Hence, the need for a guard bed in residuum hydrodesulfurization reactors.

During the operation of a fixed-bed reactor, contaminants entering with fresh feed are filtered out and fill the voids between catalyst particles in the bed. The buildup of contaminants in the bed can result in the channeling of reactants through the bed and reducing the hydrodesulfurization efficiency. As the flow pattern becomes distorted or restricted, the pressure drop throughout the catalyst bed increases. If the pressure drop becomes high enough, physical damage to the reactor internals can result. When high-pressure drops are observed throughout any portion of the reactor, the unit is shut down and the catalyst bed is skimmed and refilled.

With fixed-bed reactors, a balance must be reached between reaction rate and pressure drop across the catalyst bed. As catalyst particle size is decreased, the desulfurization reaction rate increases but so does the pressure drop across the catalyst bed. Expanded-bed reactors do not have this limitation and small 1/32 inch (0.8 mm) extrudate catalysts or fine catalysts may be used without increasing the pressure drop.

10.2.3 UPFLOW EXPANDED-BED REACTORS

Expanded-bed reactors are applicable to distillates, but are commercially used for very heavy, high metals, and/or dirty feedstocks having extraneous fine solids material. They operate in such a way that the catalyst is in an expanded state so that the extraneous solids pass through the catalyst bed without plugging. They are isothermal, which conveniently handles the high-temperature exotherms associated with high hydrogen consumptions. Since the catalyst is in an expanded state of motion, it is possible to treat the catalyst as a fluid and to withdraw and add catalyst during operation.

Expanded beds of catalyst (Figure 10.10) are referred to as particulate fluidized insofar as the feedstock and hydrogen flow upward through an expanded bed of catalyst with each catalyst particle in independent motion. Thus, the catalyst migrates throughout the entire reactor bed. Expanded-bed reactors are mathematically modeled as back-mix reactors with the entire catalyst bed at one uniform temperature. Spent catalyst may be withdrawn and replaced with fresh catalyst on a daily

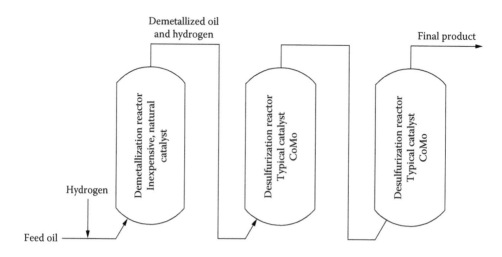

FIGURE 10.10   Representation of guard bed placement for hydrodemetallization and hydrodesulfurization.

basis. Daily catalyst addition and withdrawal eliminate the need for costly shutdowns to change out catalyst and also result in a constant equilibrium catalyst activity and product quality. The catalyst is withdrawn daily with a vanadium, nickel, and carbon content, which is representative on a macroscale of what is found throughout the entire reactor. On a microscale, individual catalyst particles have ages from that of fresh catalyst to as old as the initial catalyst charge to the unit but the catalyst particles of each age group are so well dispersed in the reactor that the reactor contents appear uniform.

In the unit (Figure 10.9), the feedstock and hydrogen recycle gas enter the bottom of the reactor, pass up through the expanded catalyst bed, and leave from the top of the reactor. Commercial expanded-bed reactors normally operate with 1/32 inch (0.8 mm) extrudate catalysts that provide a higher rate of desulfurization than the larger catalyst particles used in fixed-bed reactors. With extrudate catalysts of this size, the upward liquid velocity based on fresh feedstock is not sufficient to keep the catalyst particles in an expanded state. Therefore, for each part of the fresh feed, several parts of product oil are taken from the top of the reactor, recycled internally through a large vertical pipe to the bottom of the reactor, and pumped back up through the expanded catalyst bed. The amount of catalyst-bed expansion is controlled by the recycle of product oil backup through the catalyst bed.

The expansion and turbulence of gas and oil passing upward through the expanded catalyst bed are sufficient to cause almost complete random motion in the bed (particulate fluidized). This effect produces the isothermal operation. It also causes almost complete back-mixing. Consequently, in order to effect near complete sulfur removal (over 75%), it is necessary to operate with two or more reactors in series. The ability to operate at a single temperature throughout the reactor or reactors, and to operate at a selected optimum temperature rather than an increasing temperature from the start to the end of the run, results in more effective use of the reactor and catalyst contents. When all these factors are put together, that is, use of a smaller catalyst particle size, isothermal, fixed temperature throughout run, back-mixing, daily catalyst addition, and constant product quality, the reactor size required for an expanded bed is often smaller than that required for a fixed bed to achieve the same product goals. This is generally true when the feeds have high initial boiling points and/or the hydrogen consumption is very high.

10.2.4   DEMETALLIZATION REACTORS

The demetallization reactor (guard-bed reactor) is used when feedstocks that have relatively high-metal contents (>300 ppm) substantially increase catalyst consumption because the metals poison the catalyst, thereby requiring frequent catalyst replacement. The usual desulfurization catalysts are relatively expensive for these consumption rates, but there are catalysts that are relatively inexpensive and can be used in the first reactor to remove a large percentage of the metals. Subsequent reactors downstream of the first reactor would use normal hydrodesulfurization catalysts. Since the catalyst materials are proprietary, it is not possible to identify them here. However, it is understood that such catalysts contain little or no metal promoters, that is, nickel, cobalt, and molybdenum. Metals removal on the order of 90% has been observed with these materials.

Thus, one method of controlling demetallization is to employ separate smaller *guard reactors* just ahead of the fixed-bed hydrodesulfurization reactor section. The preheated feed and hydrogen pass through the guard reactors that are filled with an appropriate catalyst for demetallization that is often the same as the catalyst used in the hydrodesulfurization section. The advantage of this system is that it enables replacement of the most contaminated catalyst (*guard bed*), where pressure drop is highest, without having to replace the entire inventory or shut down the unit. The feedstock is alternated between guard reactors while catalyst in the idle guard reactor is being replaced.

When the expanded-bed design is used, the first reactor could employ a low-cost catalyst (5% of the cost of Co–Mo catalyst) to remove the metals, and subsequent reactors can use the more selective hydrodesulfurization catalyst (Figure 10.10). The demetallization catalyst can be added continuously without taking the reactor out of service, and the spent demetallization catalyst can be loaded to more than 30% vanadium, which makes it a valuable source of vanadium.

10.2.5 Catalysts

The selection of the catalysts must take into consideration the properties of the feed to be hydroprocessed. Differences in feedstock composition influence the choice of processing schemes. The heteroatom content and molecular weight distribution are important aspects of feedstock properties and, in addition, for heavy feedstocks the amount of asphaltene constituents and metals in the feed should be determined (Chapters 2 and 3). In fact, based on the variation in feedstock properties, a catalyst or catalytic system suitable for hydroprocessing the various feedstocks does not exist and the catalyst must be designed according to the feedstock properties and process parameters.

Hydrodesulfurization catalysts typically consist of metals impregnated on a porous alumina support (Furimsky, 1998; Ancheyta and Speight, 2007). Almost all of the surface area is found in the pores of the alumina (200–300 m²/g), and the metals are dispersed in a thin layer over the entire alumina surface within the pores. This type of catalyst does display a huge catalytic surface for a small weight of catalyst. Co, Mo, and Ni are the most commonly used metals for desulfurization catalysts. The catalysts are manufactured with the metals in an oxide state. In the active form they are in the sulfide state, which is obtained by sulfiding the catalyst either prior to use or with the feed during actual use. Any catalyst that exhibits hydrogenation activity will catalyze hydrodesulfurization to some extent. However, the Group VIB metals (chromium, molybdenum, and tungsten) are particularly active for desulfurization, especially when promoted with metals from the iron group (iron, cobalt, and nickel).

Co–Mo catalysts are by far the most popular choice for desulfurization, particularly for straight-run petroleum fractions. Ni–Mo catalysts are often chosen instead of Co–Mo catalysts when higher activity for saturation of polynuclear aromatic compounds saturation or nitrogen removal is required, or when more refractory sulfur compounds such as those in cracked feedstocks must be desulfurized. In some applications, Ni–Co–Mo catalysts appear to offer a useful balance of hydrotreating activity. Nickel–tungsten (Ni–W) is usually chosen only when very high activity for aromatics saturation is required along with activity for sulfur and nitrogen removal. There are several different compositions for available catalysts (Table 10.4).

Co–Mo and Ni–Mo catalysts resist poisoning and are the most universally applied catalysts for hydrodesulfurization of everything from naphtha to residua. In addition, Co–Mo and Ni–Mo catalysts promote both demetallization and desulfurization. The vanadium deposition rate at a given

TABLE 10.4
Representative Metal Content of Catalysts

	Co–Mo	Ni–Mo	Ni–Co–Mo	Ni–W
Metal (wt%)				
Cobalt	2.5		1.5	
Nickel		2.5	2.3	4.0
Molybdenum	10.0	10.0	11.0	
Tungsten				16.0

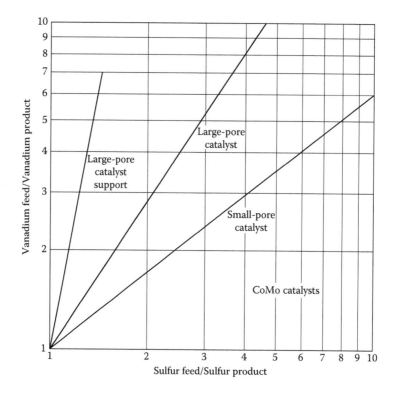

FIGURE 10.11 General relationship between vanadium and sulfur removal for different catalysts.

desulfurization level is a function of the pore structure of the alumina support and the types of metals on the support (Figure 10.11), whereas a catalyst support having small pores preferentially removes sulfur with a low degree of demetallization (Figure 10.12).

10.3 DISTILLATE HYDRODESULFURIZATION

10.3.1 Processes

Hydrotreating (catalytic hydrodesulfurization) of naphtha is widely applied to prepare charge for catalytic reforming and isomerization processes. It is accomplished by passing a feedstock together with hydrogen over a fixed catalyst bed at elevated temperature and pressure (Figure 10.13). Although the main purpose is to remove sulfur, denitrogenation, deoxygenation, and olefin saturation reactions occur simultaneously with desulfurization. These reactions are also beneficial since

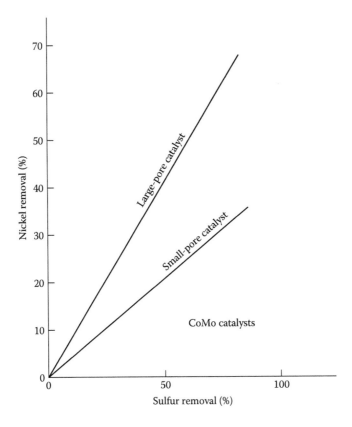

FIGURE 10.12   General relationship between nickel and sulfur removal for different catalysts.

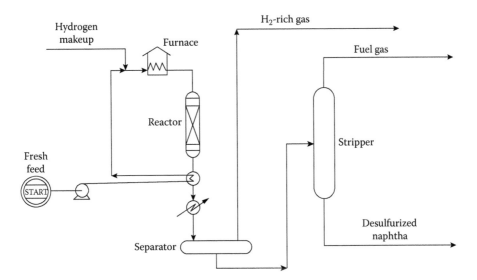

FIGURE 10.13   Representation of a naphtha hydrotreater.

the noble metal catalysts used in reforming and isomerization can be poisoned by olefins, oxygen, and nitrogen compounds as well as by sulfur in the feedstocks. Sulfur and nitrogen limitations can in some cases be 0.5 ppm or less. Failure to adequately remove these contaminants can lead to poor yields, low catalyst activity, and brief catalyst life.

In terms of specific processes, *hydrofining* may be used to upgrade low-quality, high-sulfur naphtha, thus increasing the supply of catalytic reformer feedstock, solvent naphtha, and other naphtha-type materials. The sulfur content of kerosene can be reduced with improved color, odor, and wick-char characteristics. The tendency of kerosene to form smoke is not affected since aromatics, which cause smoke, are not affected by the mild hydrofining conditions. Cracked gas oil having a high sulfur content can be converted to excellent fuel oil and diesel fuel by reduction in sulfur content and by the elimination of components that form gum and carbon residues.

This process can be applied to lubricating oil, naphtha, and gas oil. The feedstock is heated in a furnace and passed with hydrogen through a reactor containing a suitable metal oxide catalyst, such as cobalt and molybdenum oxides or alumina (Figure 10.14). Hydrogen is obtained from catalytic reforming units. Reactor operating conditions range from 205°C to 425°C (400°F to 800°F) and from 50 to 800 psi, depending on the kind of feedstock and the degree of treatment required. Higher-boiling feedstocks, high sulfur content, and maximum sulfur removal require higher temperatures and pressures.

After passing through the reactor, the treated oil is cooled and separated from the excess hydrogen, which is recycled through the reactor. The treated oil is pumped to a stripper tower, where hydrogen sulfide formed by the hydrogenation reaction is removed by steam, vacuum, or flue gas, and the finished product leaves the bottom of the stripper tower. The catalyst is not usually regenerated; it is replaced after use for approximately 1 year.

The *autofining process* differs from other hydrorefining processes in that an external source of hydrogen is not required. Sufficient hydrogen to convert sulfur to hydrogen sulfide is obtained by dehydrogenation of naphthenes in the feedstock. The processing equipment is similar to that used in hydrofining (Figure 10.15). The catalyst is cobalt and molybdenum oxides on alumina, and operating conditions are usually 340°C–425°C (650°F–800°F) at pressures of 100–200 psi. Hydrogen formed by dehydrogenation of naphthenes in the reactor is separated from the treated oil and is then recycled through the reactor. The catalyst is regenerated with steam and air at 200–1000-hour intervals, depending on whether light or heavy feedstocks have been processed. The process is used for the same purpose as hydrofining but is limited to fractions with end points no higher than 370°C (700°F).

Early desulfurization processes, such as caustic, amine, and clay treating, were not as successful with middle distillates as they were with lower-boiling feedstocks such as naphtha. More sophisticated extraction processes utilizing hydrogen fluoride or sulfur dioxide contacting were

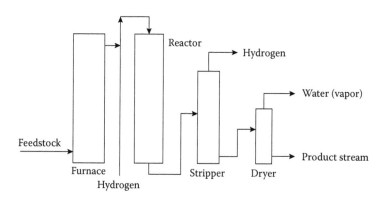

FIGURE 10.14 The hydrofining process.

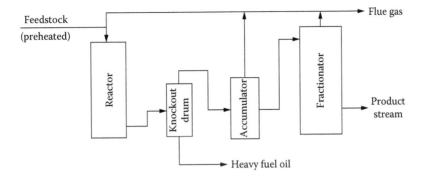

FIGURE 10.15 The autofining process.

somewhat more applicable to the removal of sulfur from diesel, gas oils, and cycle oils. Such processes did suffice in producing satisfactory distillate fuels although substantial yield losses were inherent.

The availability of cheaper hydrogen from catalytic reforming led to the development of many middle distillate hydrotreating (hydrodesulfurization) processes in the 1950s. These processes result in superior treatment of distillates to improve color, odor, corrosion properties, thermal stability, and burning characteristics in addition to accomplishing essentially complete sulfur removal.

In a typical middle distillate hydrodesulfurization process (Figure 10.16), the feedstock is mixed with fresh and recycled hydrogen and heated under pressure to the proper reactor temperature. The feedstock–hydrogen mixture is charged to the reactor, passing downflow through the catalyst. In the reactor, fresh feed is hydrotreated and a limited amount of hydrogenation, isomerization, and cracking occurs to produce a small amount of C_1 through C_5 paraffins. In addition, sulfur compounds are converted to hydrogen sulfide and nitrogen compounds are converted to ammonia. Olefins are

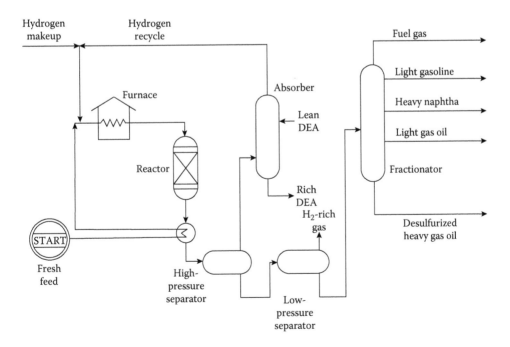

FIGURE 10.16 Representation of a gas oil hydrotreater.

TABLE 10.5

Yield and Product Properties for Desulfurization of Kuwaiti Crude Oil Vacuum Gas Oil

Yield, % of Charge (Average for Cycle)

H_2S, wt%	2.64
NH_3, wt%	0.03
C_1–C_4, wt%	0.70
C_5–375°F, vol%	1.08
375–650°F, vol%	12.05
650°F + gas oil, vol%	87.74

Property	Feed	C_5–375°F	375°F–650°F	650°F[a]
Gravity, °API	24.1	51.0	35.0	29.0
Sulfur, wt%	2.60	<0.01	0.02	0.13
Nitrogen, wt%	0.07	—	0.005	0.05
Aniline point, °F	174.4	—	—	186.3
Distillation, °F				
10%	682	—	358	659
50%	810	—	464	791
90%	961	—	600	954

[a] Hydrogen consumption, 380 scf/bbl.

also saturated. These reactions are exothermic and, in the cases of vacuum gas oils or unsaturated feedstocks, reactor temperature rise is regulated by the use of cold recycle gas quench.

Reactor effluent is cooled and enters the high-pressure separator where the oil is separated from the hydrogen sulfide and hydrogen-rich gas. Hydrogen sulfide is scrubbed from the gas (optional for light distillate units) and the hydrogen-rich gas is recycled. The liquid is passed through a low-pressure separator and stripper to remove the remaining light ends and dissolved hydrogen sulfide. Fractionation of the liquid product is sometimes employed, especially on higher-boiling feedstocks and cracked stocks would show similar upgrading and product slate to that from gas oil (Table 10.5) but would entail much greater hydrogen consumption.

The *Unifining process* is a regenerative, fixed-bed, catalytic process to desulfurize and hydrogenate refinery distillates of any boiling range. Contaminating metals, nitrogen compounds, and oxygen compounds are eliminated, along with sulfur. The catalyst is a cobalt–molybdenum–alumina type that may be regenerated *in situ* with steam and air.

Ultrafining is a regenerative, fixed-bed, catalytic process to desulfurize and hydrogenate refinery stocks from naphtha up to and including lubricating oil. The catalyst is Co–Mo on alumina and may be regenerated *in situ* using an air-stream mixture. Regeneration requires 10–20 hours and may be repeated 50–100 times for a given batch of catalyst; catalyst life is 2–5 years depending on feedstock.

The *Isomax process* is a two-stage, fixed-bed catalyst system (Figure 10.17) that operates under hydrogen pressures from 500 to 1500 psi in a temperature range of 205°C–370°C (400°F–700°F), for example, with middle distillate feedstocks. Exact conditions depend on the feedstock and product requirements, and hydrogen consumption is of the order of 1000–1600 ft³/bbl of feed processed. Each stage has a separate hydrogen recycling system. Conversion may be balanced to provide products for variable requirements, and recycling can be taken to extinction if necessary. Fractionation can also be handled in a number of different ways to yield desired products.

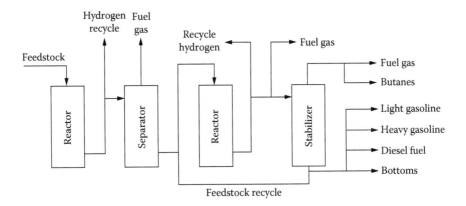

FIGURE 10.17   The Isomax process.

10.3.2   PROCESS PARAMETERS

The operating conditions in distillate hydrodesulfurization are dependent upon the stock to be charged as well as the desired degree of desulfurization or quality improvement. Kerosene and light gas oils are generally processed at mild severity and high throughput whereas light catalytic cycle oils and thermal distillates require slightly more severe conditions. Higher-boiling distillates, such as vacuum gas oils and lube oil extracts, require the most severe conditions (Table 10.6).

The principal variables affecting the required severity in distillate desulfurization are (1) hydrogen partial pressure, (2) space velocity, (3) reaction temperature, (4) catalyst life, and (5) feedstock properties.

10.3.2.1   Hydrogen Partial Pressure

The important effect of *hydrogen partial pressure* is the minimization of coking reactions. If the hydrogen pressure is too low for the required duty at any position within the reaction system, premature aging of the remaining portion of catalyst will be encountered. In addition, the effect of hydrogen pressure on desulfurization varies with feed boiling range. For a given feed, there exists a threshold level above which hydrogen pressure is beneficial to the desired desulfurization reaction. Below this level, desulfurization drops off rapidly as hydrogen pressure is reduced.

10.3.2.2   Space Velocity

As the *space velocity* is increased, desulfurization is decreased but increasing the hydrogen partial pressure and/or the reactor temperature can offset the detrimental effect of increasing space velocity.

TABLE 10.6
Representative Process Parameters for
Hydrotreating Distillates

	Kerosene and Light Gas Oils	Heavy Gas Oils and Lube Oil Extracts
Total pressure, psi gauge	100–1000	500–1500
Reactor temperature, °F	450–800	650–800
Space velocity, V/h/V	2–10	1–3

In terms of the kinetics, the presence of a complex mixture of sulfur-bearing compounds results in an apparent reaction order between first and second. First-order behavior has been shown if either liquid holdup or effective catalyst wetting is accounted for.

10.3.2.3   Reaction Temperature

A higher *reaction temperature* increases the rate of desulfurization at constant feed rate, and the start-of-run temperature is set by the design desulfurization level, space velocity, and hydrogen partial pressure. The capability to increase temperature as the catalyst deactivates is built into the most process or unit designs. Temperatures of 415°C (780°F) and above result in excessive coking reactions and higher than normal catalyst aging rates. Therefore, units are designed to avoid the use of such temperatures for any significant part of the cycle life.

10.3.2.4   Catalyst Life

Catalyst life depends on the charge stock properties and the degree of desulfurization desired. The only permanent poisons to the catalyst are metals in the feedstock that deposit on the catalyst, usually quantitatively, causing permanent deactivation as they accumulate. However, this is usually of little concern except when deasphalted oils are used as feedstocks since most distillate feedstocks contain low amounts of metals. Nitrogen compounds are a temporary poison to the catalyst, but there is essentially no effect on catalyst aging except that caused by a higher temperature requirement to achieve the desired desulfurization. Hydrogen sulfide can be a temporary poison in the reactor gas and recycle gas scrubbing is employed to counteract this condition.

Providing that pressure drop buildup is avoided, cycles of 1 year or more and ultimate catalyst life of 3 years or more can be expected. The catalyst employed can be regenerated by normal steam-air or recycle combustion gas-air procedures. The catalyst is restored to near fresh activity by regeneration during the early part of its ultimate life. However, permanent deactivation of the catalyst occurs slowly during usage and repeated regenerations, so replacement becomes necessary.

10.3.2.5   Feedstock Effects

The character of the *feedstock properties*, especially the feed boiling range, has a definite effect on the ultimate design of the desulfurization unit and process flow. In agreement, there is a definite relationship between the percent by weight sulfur in the feedstock and the hydrogen requirements (Figure 10.18) (Maples, 2000; Speight, 2000; Ancheyta and Speight, 2007).

In addition, the reaction rate constant in the kinetic relationships decreases rapidly with increasing average boiling point in the kerosene and light gas oil range but much more slowly in the heavy gas oil range. This is attributed to the difficulty in removing sulfur from ring structures present in the entire heavy gas oil boding range.

The hydrodesulfurization of light (low-boiling) distillate (naphtha) is one of the more common catalytic hydrodesulfurization processes since it is usually used as a pretreatment of such feedstocks prior to catalytic reforming. Hydrodesulfurization of such feedstocks is required because sulfur compounds poison the precious-metal catalysts used in reforming and desulfurization can be achieved under relatively mild conditions and is near quantitative. If the feedstock arises from a cracking operation, hydrodesulfurization will be accompanied by some degree of saturation resulting in increased hydrogen consumption.

The hydrodesulfurization of low-boiling (naphtha) feedstocks is usually a gas-phase reaction and may employ the catalyst in fixed beds and (with all of the reactants in the gaseous phase) only minimal diffusion problems are encountered within the catalyst pore system. It is, however, important that the feedstock be completely volatile before entering the reactor as there may be the possibility of pressure variations (leading to less satisfactory results) if some of the feedstock enters the reactor in the liquid phase and is vaporized within the reactor.

In applications of this type, the sulfur content of the feedstock may vary from 100 ppm to 1% and the necessary degree of desulfurization to be effected by the treatment may vary from as little as

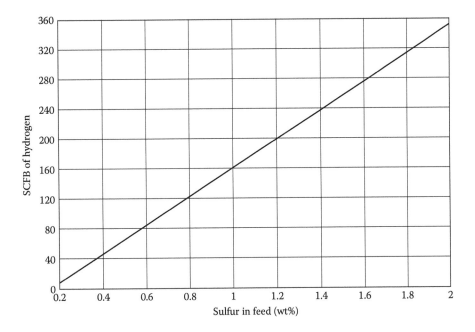

FIGURE 10.18   General relationship of feedstock sulfur to hydrogen requirements.

50% to more than 99%. If the sulfur content of the feedstock is particularly low, it will be necessary to pre-sulfide the catalyst. For example, if the feedstock only has 100–200 ppm sulfur, several days may be required to sulfide the catalyst as an integral part of the desulfurization process even with complete reaction of all of the feedstock sulfur to, say, cobalt and molybdenum (catalyst) sulfides. In such a case, pre-sulfiding can be conveniently achieved by addition of sulfur compounds to the feedstock or by addition of hydrogen sulfide to the hydrogen.

Generally, hydrodesulfurization of naphtha feedstocks (Table 10.7) to produce catalytic reforming feedstocks is carried to the point where the desulfurized feedstock contains less than 20 ppm sulfur. The net hydrogen produced by the reforming operation may actually be sufficient to provide the hydrogen consumed in the desulfurization process.

TABLE 10.7
Hydrodesulfurization[a] of Various Naphtha Fractions

Feedstock	Boiling Range		Sulfur (wt%)	Desulfurization (%)
	°C	°F		
Visbreaker naphtha	65–220	150–430	1.00	90
Visbreaker–coker naphtha	65–220	150–430	1.03	85
Straight-run naphtha	85–170	185–340	0.04	99
Catalytic naphtha (light)	95–175	200–350	0.18	89
Catalytic naphtha (heavy)	120–225	250–440	0.24	71
Thermal naphtha (heavy)	150–230	300–450	0.28	57

[a] Process conditions: Co–Mo on alumina, 260°C–370°C (500°F–700°F), 200–500 psi (1380–3440 kPa) hydrogen.

TABLE 10.8
Hydrodesulfurization of Middle Distillates

	Straight-Run Middle Distillate	Cracked Middle Distillate
Feedstock		
Specific gravity	0.844	0.901
°API	36.2	25.5
ASTM distillation, °C		
IBP	238	227
10 vol%	265	242
50 vol%	288	256
90 vol%	332	277
EP	368	293
Sulfur, wt%	1.20	2.34
Aniline point, °C	76.0	32.4
Pour point, °C	−7	−28
Conradson carbon (10% residuum)	0.024	0.34
Viscosity, cSt (38°C)	4.14	2.38
Process conditions		
Catalyst	**Cobalt Molybdate on Alumina**	**Cobalt Molybdate on Alumina Type**
Temperature, °C	315–430	315–430
Pressure, psi	250–1000	250–1000
Product		
Specific gravity	0.832	0.876
°API	38.6	30.0
ASTM distillation, °C		
IBP	222	203
10 vol%	259	227
50 vol%	287	250
90 vol%	330	275
EP	336	293
Sulfur, wt%	0.09	0.35
Aniline point, °C	79.5	36.0
Pour point, °C	−6	−28
Conradson carbon (10% residuum)	0.022	0.05
Viscosity, cSt (38°C)	3.77	2.22

The hydrodesulfurization of middle distillates is also an efficient process (Table 10.8), and applications include predominantly the desulfurization of kerosene, diesel fuel, jet fuel, and heating oils that boil over the general range 250°C–400°C (480°F–750°F). However, with this type of feedstock, hydrogenation of the higher-boiling catalytic cracking feedstocks has become increasingly important where hydrodesulfurization is accomplished alongside the saturation of condensed-ring aromatic compounds as an aid to subsequent processing.

Under the relatively mild processing conditions used for the hydrodesulfurization of these particular feedstocks, it is difficult to achieve complete vaporization of the feed. Process conditions may dictate that only part of the feedstock is actually in the vapor phase and that sufficient liquid phase is maintained in the catalyst bed to carry the larger molecular constituents of the feedstock through the bed. If the amount of liquid phase is insufficient for this purpose, molecular stagnation (leading to carbon deposition on the catalyst) will occur.

Hydrodesulfurization of middle distillates causes a more marked change in the specific gravity of the feedstock, and the amount of low-boiling material is much more significant when compared with the naphtha-type feedstock. In addition, the somewhat more severe reaction conditions (leading to a designated degree of hydrocracking) also leads to an overall increase in hydrogen consumption when middle distillates are employed as feedstocks in place of the naphtha.

High-boiling distillates, such as the atmospheric and vacuum gas oils, are not usually produced as a refinery product but merely serve as feedstocks to other processes for conversion to lower-boiling materials. For example, gas oils can be desulfurized to remove more than 80% of the sulfur originally in the gas oil with some conversion of the gas oil to lower-boiling materials (Table 10.9). The treated gas oil (which has a reduced carbon residue as well as lower sulfur and nitrogen contents relative to the untreated material) can then be converted to lower-boiling products in, say, a catalytic cracker where an improved catalyst life and volumetric yield may be noted.

The conditions used for the hydrodesulfurization of a gas oil may be somewhat more severe than the conditions employed for the hydrodesulfurization of middle distillates with, of course, the feedstock in the liquid phase.

In summary, the hydrodesulfurization of the low-, middle-, and high-boiling distillates can be achieved quite conveniently using a variety of processes. One major advantage of this type of feedstock is that the catalyst does not become poisoned by metal contaminants in the feedstock since only negligible amounts of these contaminants will be present. Thus, the catalyst may be regenerated several times and onstream times between catalyst regeneration (while varying with the process conditions and application) may be of the order of 3–4 years (Table 10.10).

TABLE 10.9
Hydrodesulfurization of Gas Oil

	Source	
	Kuwait	Khafji
Feedstock		
Boiling range, °C	370–595	370–595
Specific gravity	0.935	0.929
°API	19.8	20.2
Sulfur, wt%	3.25	3.05
Process conditions		
Temperature, °C		370–430
Pressure, psi		100–500
Hydrogen consumption, scf/bbl	420	400
Product		
Naphtha, C_5–205°C		
Yield, vol%	1.7	1.7
Specific gravity	0.802	0.802
°API	45.0	45.0
Sulfur, wt%	0.02	0.02
High boilers (>205°C)		
Yield, vol%	99.6	99.5
Specific gravity	0.897	0.893
°API	26.3	27.0
Sulfur, wt%	0.50	0.48

TABLE 10.10
Process Parameters for Hydrodesulfurization of Different Feedstocks

| Feedstock | Boiling Range | | Process Condition | | | | | | | Catalyst Life | |
| | | | Temperature | | Hydrogen Pressure | | | | | | |
	°C	°F	°C	°F	psi	kg/cm²	Hydrogen Rate (scf/bbl)	LHSV		Months	bbl/lb
Naphtha	70–170	160–340	300–370	570–700	100–450	7–31.5	250–1,500	5–8		36–48	500–1200
Kerosene	160–240	320–465	330–370	625–700	150–500	10.5–35	500–1,500	4–6		36–48	300–600
Gas oil	240–350	465–660	340–400	645–750	150–700	10.5–49	1,000–2,000	2–6		36–48	200–400
Vacuum gas oil	350–650	660–1200	360–400	680–750	450–800	31.5–56	1,000–4,000	1–3		36–48	50–350
Residua	>650	>1200	370–450	700–840	750–2250	52.5–157.5	1,500–10,000	0.5–2		12–24	2–50

10.4 HEAVY FEEDSTOCK HYDRODESULFURIZATION

The objectives of upgrading heavy feedstocks (such as residua, heavy oil, extra heavy oil, and tar sand bitumen) are (1) reduction of metals (such as nickel, vanadium, and iron), (2) reduction in the sulfur content, (3) reduction in the amount of coke formers in the feedstock, (4) reduction in nitrogen, and, the last but certainly not the least, (5) conversion of the asphaltene and resin constituents into lower-molecular-weight easier-to-refine molecular species (such as naphtha, middle distillate, and gas oil) in order to produce higher-value feedstocks for heavy oil conversion units.

Advances made in hydrotreating processes have made the utilization of heavier feedstocks almost a common practice for many refineries. Upgrading processes can be used for the upgrading of atmospheric ($650°F^+$, $345°C^+$) and vacuum ($1050°F^+$, $565°C^+$) residua, heavy oil, and tar sand bitumen. However, there is no hydroprocessing process that can universally be applicable to upgrade all heavy feedstocks. As a result, several hydroprocessing processes are developed for different commercial applications, and many other processes are in their development stages.

Two routes exist for residue upgrading: (1) carbon rejection, such as coking processes, and (2) hydrogen addition, such as hydrotreating processes that use a fixed-bed unit or an ebullated-bed unit (Speight and Ozum, 2002; Parkash, 2003; Hsu and Robinson, 2006; Gary et al., 2007; Speight, 2014). The hydrogen addition route is a more expensive option relative to a carbon rejection option but results in a significantly higher yield of liquid products. The two major process designs for the hydrogen addition approach are (1) the ebullated-bed process and (2) the fixed-bed process. Both the fixed-bed and the ebullated-bed processes require a catalyst system with a pore size distribution to match the changing molecular structure of the feedstock constituents. The catalyst can be designed for high-metal uptake capacity and moderate sulfur conversion to be applied in the front-end reactor when processing high-metal-containing feedstocks (>70 ppm vanadium). On the other hand, the catalyst may be designed for moderate metals removal capacity but higher activity for sulfur and conversion of the coke precursors, which is applied in front-end reactors when processing feedstocks with a lower metal content (<70 ppm vanadium) or in middle reactors when processing high-metal-containing feedstocks. Catalysts with a high propensity for sulfur, removal of coke precursors, and removal of nitrogen are applied in the middle and/or tail-end reactors.

10.4.1 Processes

In refining heavy feedstocks, hydrodesulfurization (HDS) processes and hydrodemetallization (HDM) processes are used to reduce or eliminate poisoning of sophisticated and expensive catalysts that are used in the downstream refining steps (i.e., fluid catalytic cracking [FCC], reforming, and hydrotreating). The hydrodemetallization process is a pretreatment process for the heavy feedstock by which metals and part of heteroatom contaminates are removed along with conversion of the residue to a lighter fraction. The hydrodemetallization process uses low-cost catalysts either in a fixed bed or moving bed reactor operating at moderate temperatures and pressures (580–2900 psi) and at relatively high liquid hourly space velocity (LHSV). Guard-bed catalysts are often used also.

Processes for the direct desulfurization of residua (Figure 10.19) have a similar flow to distillate hydrodesulfurization but with distinguishing features such as the catalyst compositions and shapes employed. Examples of such processes are the RDS/VRDS process for hydrotreating atmospheric residua and vacuum residua (Figure 10.20) and the Residfining process (Figure 10.21) (a derivative of the hydrofining process) (Speight, 2000).

10.4.1.1 Resid Desulfurization and Vacuum Resid Desulfurization Process

The resid desulfurization (RDS) process and the vacuum resid desulfurization (VRDS) process are designed to remove sulfur, nitrogen, asphaltene, and metal contaminants from residua and are also capable of accepting whole crude oils or topped crude oils as feedstocks. The major product of the processes is a low-sulfur fuel oil and the amount of gasoline and middle distillates is maintained

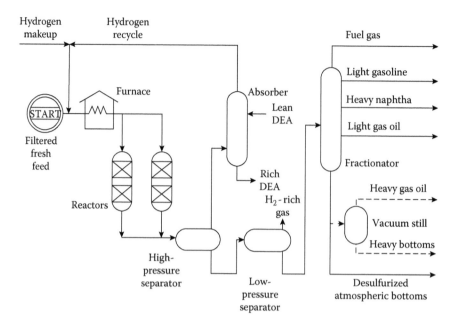

FIGURE 10.19   Representation of a residuum hydrotreater.

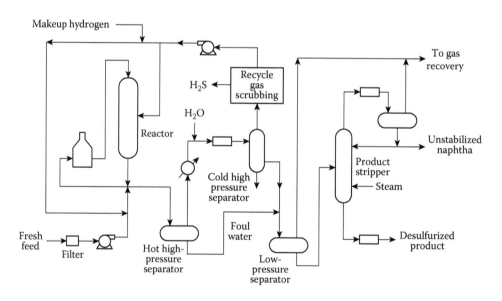

FIGURE 10.20   Representation of the RDS and VRDS processes.

at a minimum to conserve hydrogen. The basic elements of each process are similar (Figure 10.20) and consist of a once-through operation of the feedstock coming into contact with hydrogen and the catalyst in a downflow reactor that is designed to maintain activity and selectivity in the presence of deposited metals. Moderate temperatures and pressures are employed to reduce the incidence of hydrocracking and, hence, minimize production of low-boiling distillates (Table 10.11). The combination of a desulfurization step and a vacuum residuum desulfurizer (VRDS) is often seen as an attractive alternate to the atmospheric residuum desulfurizer (RDS) because the combination route uses less hydrogen for a similar investment cost. Both the RDS and the VRDS processes can be

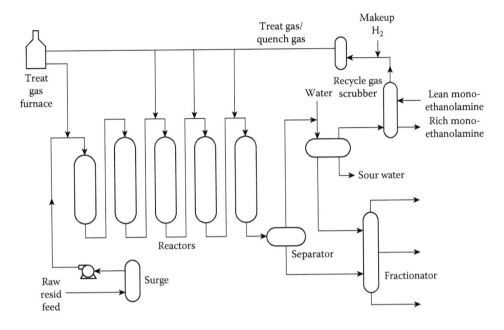

FIGURE 10.21 Representation of the Residfining process.

TABLE 10.11

Process Data for the Resid Desulfurization and Vacuum Resid Desulfurization Processes

	Arabian Light		Arabian Heavy	
	Atmospheric	Vacuum	Atmospheric	Vacuum
Feed properties				
Gravity, °API	17.7	6.5	16.8	6.1
Sulfur, wt%	3.2	4.1	3.9	5.1
Distillation, vol%				
Boiling below 650°F	5	—	16	—
Boiling below 1000°F	67	1	60	10
H_2 consumption (chemical), scf/bbl	560	750	780	960
Product yields				
$\leq C_4$, wt%	3.5	—	4.7	—
C_5–350°F, vol%	1.4	5.7	1.4	6.4
350°F–650°F, vol%	11.1		25.0	
650°F–1000°F, vol%	64.4	18.1	43.4	26.6
1000°F+, vol%	24.3	78.4	31.4	69.9
Sulfur content, wt%				
650°F+	0.46	1.0	0.55	1.0
1000°F+	0.97	1.2	0.88	1.25

coupled with other processes (such as delayed coking, fluid catalytic cracking, and solvent deasphalting) to achieve the most optimum refining performance.

10.4.1.2 Residfining Process

The Residfining process is a catalytic fixed-bed process for the desulfurization and demetallization of residua (Table 10.12). The process can also be used to pretreat residua to suitably low contaminant levels prior to catalytic cracking. In the process (Figure 10.21), liquid feed to the unit is filtered, pumped to pressure, preheated and combined with treat gas prior to entering the reactors. A guard reactor would typically be employed to prevent plugging/fouling of the main reactors with provisions to periodically remove the guard while keeping the main reactors online. The temperature rise associated with the exothermic reactions is controlled utilizing either a gas or liquid quench. A train of separators is employed to separate the gas and liquid products. The recycle gas is scrubbed to remove ammonia (NH_3) and hydrogen sulfide (H_2S). It is then combined with fresh makeup hydrogen before being reheated and recombined with fresh feed. The liquid product is sent to a fractionator where the product is fractionated.

The different catalysts allow other minor differences in operating conditions and peripheral equipment. Primary differences include the use of higher purity hydrogen makeup gas (usually 95% or greater), inclusion of filtration equipment in most cases, and facilities to upgrade the off-gases to maintain higher concentration of hydrogen in the recycle gas. Most of the processes utilize downflow operation over fixed-bed catalyst systems but exceptions to this are the H-Oil and LC-Fining processes (which are predominantly conversion processes) that employ upflow designs and ebullating catalyst systems with continuous catalyst removal capability, and the Shell Process (a conversion process) that may involve the use of a *bunker flow* reactor (Figure 10.22) ahead of the main reactors to allow periodic changeover of catalyst.

10.4.1.3 Other Options

The primary objective in most of the residue desulfurization processes is to remove sulfur with minimum consumption of hydrogen. Substantial percentages of nitrogen, oxygen, and metals are also removed from the feedstock. However, complete elimination of other reactions is not feasible and, in addition, hydrocracking, thermal cracking, and aromatic saturation reactions occur to some extent. Certain processes, that is, H-Oil (Figure 10.23) using a single-stage or a two-stage reactor (Figure 10.24) and LC-Fining (Figure 10.25) using an expanded-bed reactor can be designed to accomplish greater amounts of hydrocracking to yield larger quantities of lighter distillates at the expense of desulfurization.

TABLE 10.12
Residfining Data for Two Atmospheric Residua

	Gach Saran (650°F⁺)	Arabian Heavy (650°F⁺)
Feedstock		
°API	15.0	12.3
Sulfur, wt%	2.50	4.19
Products		
°API	19.6	20.7
Sulfur, wt%	0.3	0.3
C5/400°F, vol%	3.4	6.0
400°F⁺, vol%	98.1	96.4
Chemical H_2 consumption, scf/bbl		

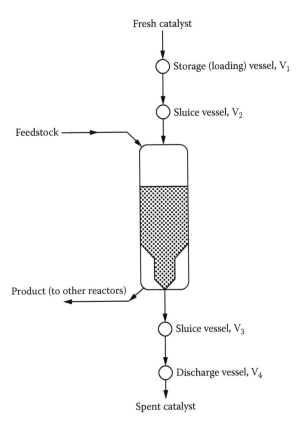

FIGURE 10.22 Representation of the bunker reactor for a residuum hydrotreater.

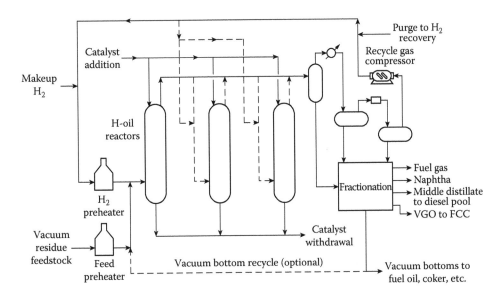

FIGURE 10.23 The H-Oil process.

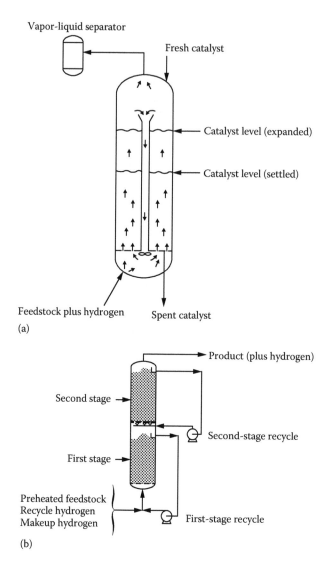

Vapor-liquid separator

Fresh catalyst

Catalyst level (expanded)

Catalyst level (settled)

Feedstock plus hydrogen

Spent catalyst

(a)

Product (plus hydrogen)

Second stage

Second-stage recycle

First stage

Preheated feedstock
Recycle hydrogen
Makeup hydrogen

First-stage recycle

(b)

FIGURE 10.24 H-Oil process: (a) single-stage reactor and (b) two-stage reactor.

Removal of nitrogen is much more difficult than removal of sulfur. For example, nitrogen removal may be only approximately 25%–30% when sulfur removal is at a 75%–80% level. Metals are removed from the feedstock in substantial quantities and are mainly deposited on the catalyst surface and exist as metal sulfides at processing conditions. As these deposits accumulate, the catalyst pores eventually become blocked and inaccessible, thus catalyst activity is lost.

Desulfurization of residua is considerably more difficult than desulfurization of distillates (including vacuum gas oil) because many more contaminants are present and very large, complex molecules are involved. The most difficult portion of feed in residue desulfurization is the asphaltene fraction that forms coke readily, and it is essential that these large molecules be prevented from condensing with each other to form coke, which deactivates the catalyst. This is accomplished by selection of proper catalysts, use of adequate hydrogen partial pressure, and assuring intimate contact of the hydrogen-rich gases and oil molecules in the process design.

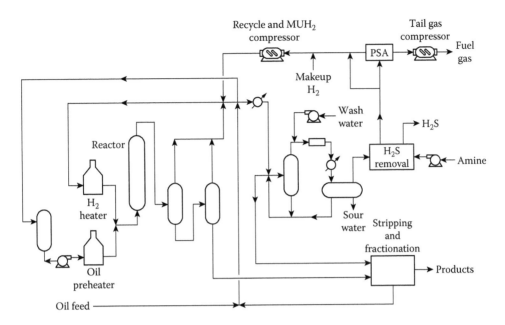

FIGURE 10.25 The LC-fining process.

10.4.2 PROCESS PARAMETERS

10.4.2.1 Catalyst Types

The general *catalyst types* used in residuum desulfurization are combinations of metal oxides on alumina (Al_2O_3) or silica-stabilized alumina (SiO_2–Al_2O_3) supports. Molybdenum always seems to be one of the metals, with cobalt and/or nickel being used in combination with the molybdenum in many cases. The supports are usually tailored to the process objectives since different support can be made to accomplish particular goals of a specific unit. For example, smaller pored catalysts will tend to remove less metals than larger pored catalysts and be more active for desulfurization reactions. However, the metal-holding capacity of the small-pored catalyst will also be less than the large-pored catalyst, which results in a sacrifice of catalyst life. The better catalyst selection would depend on the combination of these two characteristics that will allow the best life activity relationship for a given application. In some cases, critical combinations of large and small pore sizes are used to arrive at the best catalyst for a given feedstock and operating conditions. The objectives of the process unit are also important since metals removal would be more critical in an application to produce residual feedstock for catalytic cracking versus an application to produce residual fuel.

Catalyst size and shape are also important factors in residue desulfurization processes. Smaller size contributes to improved desulfurization and demetallization but pressure drop considerations increase in importance.

10.4.2.2 Metals Accumulation

While many of the conventional design criteria in distillate desulfurization (hydrogen partial pressure, degree of desulfurization, gas circulation rates) must be considered in residuum desulfurization, an additional important criterion is the effect of *metals accumulation* on the catalyst. The effective life of a particular catalyst will vary depending on its pore structure and total pore volume. It is also dependent upon the particular feedstock being processed and the operating conditions employed. As mentioned previously, many of the competitive processes use different catalyst characteristics that have been tailored to achieve the objectives of most concern to the individual licensor. Thus some processes will remove more metals from the feedstock while others will reject metals to a greater extent.

In general terms, catalysts that show better selectivity for metals removal will also hold more total metals before they become inoperable for the required desulfurization duty. This holding capacity for metals has been defined as a *saturation level* that increases with decreasing size for a given catalyst. However, the selectivity for demetallization over desulfurization reactions also increases with decreasing size. The combination of these effects results in an optimum particle size to maximize the cycle life.

10.4.2.3 Catalyst Activity

A gradual loss of *catalyst activity* occurs during normal operation of the residue process. Therefore, a gradual increase in catalyst temperature is required through the cycle to maintain the desired product sulfur content. This loss in activity is caused by deposition of coke and metals (from nickel and vanadium in the feedstock) on the catalyst surface and in the catalyst pores. Ultimate *catalyst life* is directly related to the total metals tolerance of the catalyst, which is a function of particle size, shape, and pore size and volume. Metals deposited cause permanent deactivation of the catalyst and preclude the restoration of catalyst activity by normal regeneration procedures. Spent catalysts are either discarded or returned to reclaimers for recovery of the various metals of value. The amount of coke deposited on the catalyst depends primarily on hydrogen partial pressure but is also influenced by the asphaltene content of the feedstock. Higher hydrogen pressure decreases coking while higher asphaltene content increases coking.

Besides the effect of hydrogen partial pressure on *catalyst aging*, maintenance of adequate amounts of hydrogen within the system is required. Circulation rates and purity requirements are set to avoid a shortage of hydrogen anywhere in the reaction system in order to prevent undesirable side reactions.

10.4.2.4 Temperature and Space Velocity

Temperature and *space velocity* are very important variables that influence the operation of the process. In a given system, a reduction in space velocity without an appropriate reduction in temperature will result in feedstock over-treating. This will lead to irreversible premature aging of the catalyst by virtue of increased coking and incremental metals deposition.

Since *catalyst temperature* is increased over the length of an operating cycle, both yields of lighter materials and properties of the remaining feedstock are affected. The magnitude of the variations will depend on the catalyst selected for the operation and the operating conditions employed (Table 10.13). The product properties affected to the greatest extent are viscosity and pour point and, with changes in distillate yields, indicate that cracking reactions are increasing as the run progresses (due to temperature increases that lead to a greater degree of thermal cracking).

10.4.2.5 Feedstock Effects

The problems encountered in hydrotreating heavy feedstocks can be equated to the amount of higher-boiling constituents (Speight, 2000; Ancheyta and Speight, 2007). Processing these feedstocks is not just a matter of applying know-how derived from refining *conventional* crude oils that often used hydrogen-to-carbon (H/C) atomic ratios as the main criterion for determining process options, but requires knowledge of several other properties (Table 10.14). The materials are not only complex in terms of the carbon number and boiling point ranges but also because a large part of this *envelope* falls into a range about which very little is known about model compounds (Figure 10.26). It is also established that the majority of the higher-molecular-weight materials produce coke (with some liquids), while the majority of the lower-molecular-weight constituents produce liquids (with some coke). It is to the latter trend that hydrotreating is aimed.

TABLE 10.13
Yields and Properties for Desulfurization of Kuwaiti Crude Oil Atmospheric Residuum

Yield (% of Hydrodesulfurization Charge)	Feed	1% Sulfur 650°F+ Fuel[a]		0.3% Sulfur 375°F+ Fuel[a]		0.1% Sulfur 375°F+ Fuel[a]	
H_2S, wt%		3.07–3.14	(3.10)	3.73–3.74	(3.73)	3.93–3.94	(3.93)
NH_3, wt%		0.08–0.07	(0.08)	0.13–0.12	(0.12)	0.17–0.17	(0.17)
C_1–C_4, wt%		0.27–1.10	(0.62)	0.33–1.67	(0.89)	0.40–2.07	(1.14)
C_5–375°F naphtha, vol%		1.4–4.1	(2.6)	2.0–6.4	(3.8)	2.5–7.6	(4.6)
375°F–650°F distillate, vol%		6.6–11.2	(8.5)	8.9–17.4	(12.5)	9.1–20.8	(14.0)
650°F+ residue, vol%		92.7–86.5	(90.1)	90.4–78.9	(85.6)	89.9–74.6	(83.5)
Chemical hydrogen consumption, scf/bbl			(497)		(650)		(725)
Residue product properties							
Gravity, °API	16.6	20.5–22.1	(21.2)	22.0–23.9	(22.8)	22.5–24.4	(23.3)
Sulfur, wt%	3.8	1.0		0.32–0.35	(0.33)	0.10–0.11	(0.11)
Carbon residue, wt%	9.0	5.6–6.1	(5.8)	3.6–4.1	(3.8)	3.0–3.6	(3.2)
Nitrogen, wt%	0.22	0.17–0.19	(0.18)	0.13–0.15	(0.14)	0.09–0.11	(0.10)
Pour point, °F	+60	+65–+55	(+60)	+70–+20	(+35)	+60–0	(+15)
Nickel, ppm	15	4.5–4.8	(4.6)	1.3–1.6	(1.5)	0.4–0.5	(0.5)
Vanadium, ppm	45	7.9–8.6	(8.2)	2.4–2.8	(2.6)	1.0–1.3	(1.2)
Viscosity, SUS at 210°F	250	122–88	(108)	104–70	(90)	94–64	(81)

[a] Values in parentheses indicate average values for cycle.

TABLE 10.14
Properties of Different Residua

Feedstock	Gravity (°API)	Sulfur (wt%)	Nitrogen (wt%)	Nickel (ppm)	Vanadium (ppm)	Asphaltenes (Heptane) (wt%)	Carbon Residue (Conradson) (wt%)
Arabian light >650°F	17.7	3.0	0.2	10.0	26.0	1.8	7.5
Arabian light >1050°F	8.5	4.4	0.5	24.0	66.0	4.3	14.2
Arabian heavy >650°F	11.9	4.4	0.3	27.0	103.0	8.0	14.0
Arabian heavy >1050°F	7.3	5.1	0.3	40.0	174.0	10.0	19.0
Alaska North Slope >650°F	15.2	1.6	0.4	18.0	30.0	2.0	8.5
Alaska North Slope >1050°F	8.2	2.2	0.6	47.0	82.0	4.0	18.0
Lloydminster (Canada) >650°F	10.3	4.1	0.3	65.0	141.0	14.0	12.1
Lloydminster (Canada) >1050°F	8.5	4.4	0.6	115.0	252.0	18.0	21.4
Kuwait >650°F	13.9	4.4	0.3	14.0	50.0	2.4	12.2
Kuwait >1050°F	5.5	5.5	0.4	32.0	102.0	7.1	23.1
Tia Juana >650°F	17.3	1.8	0.3	25.0	185.0		9.3
Tia Juana >1050°F	7.1	2.6	0.6	64.0	450.0		21.6
Taching >650°F	27.3	0.2	0.2	5.0	1.0	4.4	3.8
Taching >1050°F	21.5	0.3	0.4	9.0	2.0	7.6	7.9
Maya >650°F	10.5	4.4	0.5	70.0	370.0	16.0	15.0

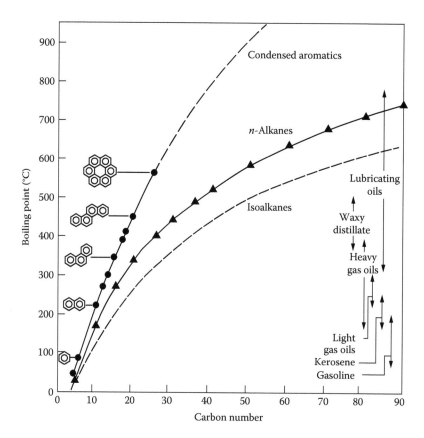

FIGURE 10.26   Relationship of carbon number and boiling range for various organic compounds.

It is the physical and chemical composition of a feedstock that plays a large part not only in determining the nature of the products that arise from refining operations but also in determining the precise manner by which a particular feedstock should be processed. Furthermore, it is apparent that the conversion of residua requires new lines of thought to develop suitable processing scenarios. Indeed, the use of thermal (*carbon rejection*) and hydrothermal (*hydrogen addition*) processes that were inherent in the refineries designed to process lighter feedstocks has been a particular cause for concern and has brought about the evolution of processing schemes that will accommodate the heavier feedstocks. However, processes based upon carbon rejection are not chemically efficient since they degrade usable portions of the feedstock to coke.

Thus, there is the potential for the application of more efficient conversion processes to heavy feedstock refining. Hydroprocessing, that is, hydrotreating (in the present context) is probably the most versatile of petroleum refining processes because of its applicability to a wide range of feedstocks. In fact, hydrotreating can be applied to the removal of heteroatoms from heavier feedstocks as a first step in preparing the feedstock for other process options.

In actual practice, the reactions that are used to chemically define the processes, that is, hydrodesulfurization, hydrodenitrogenation, and hydrocracking reactions (Chapters 5 and 11), can occur (or be encouraged to occur) in any one particular process. Thus, hydrodesulfurization may be accompanied, in all likelihood, a degree of hydrocracking as determined by the refiner thereby producing not only products that are low in sulfur but also low-boiling products. Thus, the choice of processing schemes for a given hydroprocess depends upon the nature of the feedstock as well as the product requirements. The process can be simply illustrated as a single-stage or as a two-stage operation (Figure 10.5).

The single-stage process can be used to produce gasoline but is more often used to produce middle distillate from heavy vacuum gas oils and may be used to remove the heteroatoms from residua with a specified degree of hydrocracking. The two-stage process was developed primarily to produce high yields of gasoline from straight-run gas oils and the first stage may actually be a purification step to remove sulfur-containing and nitrogen-containing organic materials. Both processes use an extinction/recycle technique to maximize the yields of the desired product. Significant conversion of heavy feedstocks can be accomplished by hydrocracking at high severity. For some applications, the products boiling up to 340°C (650°F) can be blended to give the desired final product.

In reality, no single bottom-of-the-barrel processing scheme is always the best choice. Refiners must consider the potential of proven processes, evaluate the promise of newer ones and choose based on the situation. The best selection will always depend on the kind of crude oil, the market for products, financial, and environmental consideration. Although there are no simple solutions, the available established processes and the growing number of new ones under development offer some reasonable choices. The issue then becomes how to most effectively handle the asphaltene fraction of the feedstock at the most reasonable cost. Solutions to this processing issue can be separated into two broad categories: (1) conversion of asphaltenes into another, salable product and (2) use of the asphaltenes by concentration into a marketable, or useable, product such as asphalt.

The hydrodesulfurization process variables (Speight, 2000) usually require some modification to accommodate the various feedstocks that are submitted for this particular aspect of refinery processing (Table 10.10). The main point of this section is to outline the hydrotreating process with particular reference to the heavier oils and residua. However, some reference to the lighter feedstocks is warranted. This will serve as a base point to indicate the necessary requirements for heavy oil and residuum hydrodesulfurization.

One particular aspect of the hydrotreating process that needs careful monitoring, with respect to feedstock type, is the exothermic nature of the reaction. The heat of the reaction is proportional to the hydrogen consumption and with the more saturated lower-boiling feedstocks where hydrocracking may be virtually eliminated, the overall heat production during the reaction may be small, leading to a more controllable temperature profile. However, with heavy feedstocks where hydrogen consumption is appreciable (either by virtue of the hydrocracking that is necessary to produce a usable product or by virtue of the extensive hydrodesulfurization that must occur), it may be desirable to provide internal cooling of the reactor. This can be accomplished by introducing cold recycle gas to the catalyst bed to compensate for excessive heat. One other generalization may apply to the lower-boiling feedstocks in the hydrodesulfurization process. The process may actually have very little effect on the properties of the feedstock (assuming that hydrocracking reactions are negligible)—removal of sulfur will cause some drop in specific gravity that could give rise to volume recoveries approaching (or even above) 100%. Furthermore, with the assumption that cracking reactions are minimal, there may be a slight lowering of the boiling range due to sulfur removal from the feedstock constituents. However, the production of lighter fractions is usually small and may only amount to some 1%–5% by weight of the products boiling below the initial boiling point of the feedstock.

One consideration for heavy feedstocks is that it may be more economical to hydrotreat and desulfurize high-sulfur feedstocks before catalytic cracking than to hydrotreat the products from catalytic cracking. This approach (DeCroocq, 1984; Speight, 2000, 2014) has the potential for several advantages, such as: (1) the products require less finishing, (2) sulfur is removed from the catalytic cracking feedstock, and corrosion is reduced in the cracking unit, (3) coke formation is reduced, (4) higher feedstock conversions, and (5) the potential for better-quality products. The downside is that many of the heavier feedstocks act as hydrogen sinks in terms of their ability to interact with the expensive hydrogen. A balance of the economic advantages/disadvantages must be struck on an individual feedstock basis.

In terms of the feedstock composition, it must be recognized that when catalytic processes are employed for heavy feedstocks, complex molecules (such as those that may be found in the original

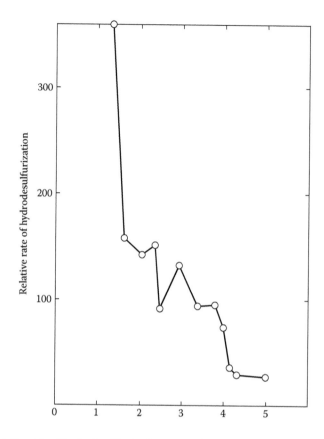

FIGURE 10.27   Relationship of hydrodesulfurization rate to feedstock sulfur.

asphaltene fraction) or those formed during the process are not sufficiently mobile. They are also too strongly adsorbed by the catalyst to be saturated by the hydrogenation component and, hence, continue to react and eventually degrade to coke. These deposits deactivate the catalyst sites and eventually interfere with the hydroprocess by causing a decrease in the relative rate of hydrodesulfurization (Figure 10.27).

Heavy feedstocks, such as residua, require more severe hydrodesulfurization conditions to produce low-sulfur liquid product streams that can then, as is often now desired, be employed as feedstocks for other refining operations. Hydrodesulfurization of the heavier feedstocks is normally accompanied by a high degree of hydrocracking, and thus the process conditions required to achieve 70%–90% desulfurization will also effect substantial conversion of the feedstock to lower-boiling products. In addition, the extent of hydrodesulfurization reaction of heavy feedstocks is dependent upon the temperature, and the reaction rate increases with increase in temperature.

In contrast to the lighter feedstocks that may be subjected to the hydrodesulfurization operation, the process catalysts are usually susceptible to poisoning by nitrogen (and oxygen) compounds and metallic salts (in addition to the various sulfur-compound types) that tend to be concentrated in residua or exist as an integral part of the heavy oil matrix. Thus, any processing sequence devised for hydrodesulfurization of residua must be capable of accommodating the constituents that adversely affect the ability of the catalyst to function in the most efficient manner possible.

The conditions employed for the hydrodesulfurization of the heavier feedstocks may be similar to those applied to the hydrodesulfurization of gas oil fractions but with the tendency to increased pressures. However, carbon deposition on, and metal contamination of, the catalyst is much greater when residua and heavy oils are employed as feedstocks and, unless a low level of desulfurization is acceptable, frequent catalyst regeneration is necessary.

A wide choice of commercial processes is available for the catalytic hydrodesulfurization of residua. The suitability of any particular process depends not only upon the nature of the feedstock but also on the degree of desulfurization that is required. There is also a dependence on the relative amounts of the lower-boiling products that are to be produced as feedstocks for further refining and generation of liquid fuels.

There is, however, one aspect of feedstock properties that has not yet been discussed fully and that is feedstock composition. This particular aspect of the nature of the feedstock is, in fact, related to the previous section where the influence of various feedstock types on the hydrodesulfurization process was noted, but it is especially relevant when residua and heavy oils from various sources are to be desulfurized.

One of the major drawbacks to defining the influence of the feedstock on the process is that the research with respect to feedstocks has been fragmented. In every case, a conventional catalyst has been used, and the results obtained are only valid for the operating conditions, reactor system, and catalyst used.

More rigorous correlation is required and there is a need to determine the optimum temperature for each type of sulfur compound. In order to obtain a useful model, the intrinsic kinetics of the reaction for a given catalyst should also be known. In addition, other factors that influence the desulfurization process such as: (1) catalyst inhibition or deactivation by hydrogen sulfide, (2) effect of nitrogen compounds, and (3) the effects of various solvents should also be included in order to obtain a comprehensive model that is independent of the feedstock.

Residua and other heavy feedstocks contain impurities other than sulfur, nitrogen, and oxygen, and the most troublesome of these impurities are the organometallic compounds of nickel and vanadium. The metal content of a residuum can vary from several parts per million to more than 1000 per million (Table 10.15), and there does seem to be a more than chance relationship between the metals content of a feedstock and its physical properties (Reynolds, 1998; Speight, 2000, 2014). In the hydrodesulfurization of the heavier feedstocks, the metals (nickel plus vanadium) are an important factor since large amounts (over 150 ppm) will cause rapid deterioration of the catalyst. The free metals, or the sulfides, deposit on the surface of the catalyst and within the pores of the catalyst, thereby poisoning the catalyst by making the more active catalyst sites inaccessible to the feedstock and the hydrogen. This results in frequent replacement of an expensive process commodity unless there are adequate means by which the catalyst can be regenerated.

The problem of metal deposition on the hydrodesulfurization catalysts has generally been addressed using any one of three methods: (1) suppressing deposition of the metals on the catalyst, (2) development of a catalyst that will accept the metals and can tolerate high levels of metals without marked reduction in the hydrodesulfurization capabilities of the catalyst, and (3) removal of the metal contaminants before the hydrodesulfurization step. The first two methods involve a careful and deliberate choice of the process catalyst as well as the operating conditions. However, these methods may only be viable for feedstocks with less than 150 ppm total metals since the decrease in catalyst activity is directly proportional to the metals content of the feedstock. There are, however, catalysts that can tolerate substantial proportions of metals within their porous structure before the desulfurizing capability drops to an unsatisfactory level. Unfortunately, data on such catalysts are extremely limited because of their proprietary nature, and details are not always available but tolerance levels for metals that are equivalent to 15%–65% by weight of the catalyst have been quoted.

The third method may be especially applicable to feedstocks with a high-metal content and requires a separate demetallization step just prior to the hydrodesulfurization reactor by use of a guard-bed reactor. Such a step might involve passage of the feedstock through a demetallization chamber that contains a catalyst with a high selectivity for metals but whose activity for sulfur removal is low. Nevertheless, demetallization applied as a separate process can be used to generate

TABLE 10.15
Sulfur Content and Metal Content of Various Feedstocks

Feedstock	°API	V + Ni (ppm)	% S
Alaska, Simpson	~4	100–212	2.4–3.1
Alaska, Hurl State	~6	72–164	1.0–1.9
Arabian light, atmospheric	16.1–17.3	34–42	2.8–3.0
Arabian light, vacuum	6.1–8.2	101–112	4.0–4.2
Arabian light, vacuum	11.2	71	3.71
Arabian medium, atmospheric	~15.0	51–140	4.0–4.3
Arabian heavy, atmospheric	11.0–17.2	102–140	4.2–4.6
Safaniya	11.0–12.7	102–130	4.3
Khursaniyah	15.1	41	4.0
Khafji, atmospheric	14.7–15.7	95–125	4.0–43
Khafji, vacuum	5.0	252	5.4
Wafra, Eocene	9.7	125	5.6
Wafra, Ratawi	14.2	100	4.7–5.2
Agha Jari	16.6	92–110	2.3–2.5
Agha Jari, vacuum	7.5	274	3.8
Iranian light, vacuum	9.5	150–316	3.1–3.2
Iranian heavy, atmospheric	15.0–19.8	179–221	2.5–2.8
Alaska, Kuparup	~11	64–153	1.5–2.4
Alaska, Put River	~10	34–83	1.5–2.1
Alaska, Sag River	~8	52–129	1.6–2.2
California, coastal	4.3	300	2.3
California, atmospheric	13.7	134	1.73
California, deasphalted	13.8	24	0.95
Delta, Louisiana	11.3	25	0.93
Mid-continent, vacuum	14.5	25	1.2
Wyoming, vacuum	6.3	292	4.56
Salamanca, Mexico	5.4–10.7	182–250	3.2–3.8
Gachsaran	15.8–17.0	210–220	2.4–2.6
Gachsaran, vacuum	9.2–10.0	329–359	3.1–3.3
Darius, vacuum	12.8	61	4.7
Sassan, atmospheric	18.3	33	3.3
Kuwait, atmospheric	10.2–16.7	60–65	3.8–4.1
Kuwait, vacuum	5.5–8.3	110–199	5.1–5.8
Zubair, Iraq	18.1	43	3.3
Khurais	14.6	31	3.3
Murban	24.0–25.0	2–10	1.5–1.6
Qatar marine	25.2	35	2.77
Middle East, vacuum	8.3–17.6	40–110	3.0–5.8
Pilon (tar belt)	9.7	608	3.92
Monogus (tar belt)	12.0	254	2.5
Morichal (tar belt)	9.6–12.4	233–468	2.1–4.1
Melones II (tar belt)	9.9	424	3.3
Tucupita	16.0	129	1.03
Bachaquero (lake)	5.8–14.6	415–970	2.2–3.5
Boscan (lake)	9.5–10.4	1185–1750	5.2–5.9
Lagunillas (lake)	~16.5	271–400	2.2–2.6
Taparita (lake)	16.9	490	2.4

(*Continued*)

TABLE 10.15 (*Continued*)
Sulfur Content and Metal Content of Various Feedstocks

Feedstock	°API	V + Ni (ppm)	% S
Tia Juana (lake)	6.5–7.8	380–705	2.4–3.5
Tia Juana (lake)	~8.0	327–534	2.6–4.0
Urdaneta	11.7	480	2.68
Venezuela, atmospheric	11.8–17.2	200–460	2.1–2.8
Venezuela, vacuum	4.5–7.5	690–760	2.9–3.2
West Texas, atmospheric	15.4–17.9	39–40	2.5–3.7
West Texas, vacuum	7.3–10.5	65–88	3.2–4.6

TABLE 10.16
Process Data for Desulfurization and Demetallization of a Residuum

Feedstock Data	Type of Operation	
	Desulfurization	Demetallization/Desulfurization
Gravity, °API	11.8	12.6
Sulfur, wt%	2.8	2.8
Vanadium, ppm	375	398
Nickel, ppm	55	57
975°F+, vol%	55	55

	Desulfurization			Demetallization/Desulfurization		
Product Yield	Yield (vol%)	Sulfur (wt%)	Vanadium (ppm)	Yield (vol%)	Sulfur (wt%)	Vanadium (ppm)
C_4 end point	103.5	0.64	170	103.3	0.64	88
Chemical H_2 consumption, scf/bbl	—	720	—	—	680	—
Desulfurization, %[a]	—	77	—	—	77	—

[a] For two Co–Mo reactors in series.

low-metal feedstocks and will allow a more active, and stable, desulfurization system so that a high degree of desulfurization can be achieved on high-metal feedstocks with an acceptable duration of operation (Table 10.16; Figure 10.10).

10.5 OTHER OPTIONS

Environmental concerns and newly enacted rules and regulations mean the feedstock that petroleum products are expected to meet lower and lower limits on contaminates, such as sulfur and nitrogen. New regulations require the removal of sulfur compositions from liquid hydrocarbons, such as those used in gasoline, diesel fuel, and other transportation fuels.

In the hydrodesulfurization process, high temperatures and pressures may be required to obtain the desired low levels of sulfur. High-temperature processing of olefinic naphtha, however, may result in a lower grade fuel due to saturation of olefins leading to an octane loss. Low octane gasoline may require additional refining, isomerization, blending, and the like to produce higher quality fuels suitable for use in gasoline products. Such extra processing adds additional cost, expense, and complexity to the process, and may result in other undesirable changes in the products. As a result, the future will see processes that are more chemically precise in hydrotreating and that offer higher efficiency and conversion to selected products on a basis that is not practiced currently.

Five of the most common approaches to upgrading hydrotreaters for clean-fuel production (in order of increasing capital cost) are currently and will continue (at least for the next two decades) to be (1) developing higher activity and more resilient catalysts, (2) replacing reactor internals for increased efficiency, (3) adding reactor capacity to accommodate heavy feedstocks and increase gasoline–diesel production, (4) increasing hydrogen partial pressure, (5) process design and hardware that are more specialized and focused on process schemes that effectively reduce hydrogen consumption.

Finally, hydrotreating of residua requires considerably different catalysts and process flows, depending on the specific operation so that efficient hydroconversion through uniform distribution of liquid, hydrogen-rich gas, and catalyst across the reactor is assured. There will also be automated demetallization of fixed-bed systems as well as more units that operate as ebullating-bed hydrocrackers (Chapter 11).

Finally, severe hydrotreating requires high-purity (>99% v/v) hydrogen, while less severe hydrotreating can employ low-purity (<90% v/v) hydrogen. Refiners will continue to optimize hydrogen use by cascading hydrogen through the refinery. High-purity hydrogen will continue to be used only where required, and the low-purity hydrogen purged from these applications will be used for services that do not require the high-purity hydrogen.

10.5.1 Catalyst Technology

Conventional hydroprocessing catalysts are generally in the form of a carrier of a refractory oxide material on which hydrogenation metals are deposited, the choice and amount of each component being determined by the end use. Refractory oxide materials usual in the art are amorphous or crystalline forms of alumina, silica and combinations thereof. These oxide materials can have some intrinsic catalytic activity but often only provide the support on which active metal compounds are held. Generally, the thermal stability, low surface area, and poor mechanical strength have all hindered the commercial exploitation of certain metal oxide–supported catalyst systems. The intrinsic activity of hydrogenation metals-on-catalyst is superior to alumina-based catalysts. Catalyst synthesis will attempt to harness the intrinsic activity of various metals and remedy the deficiencies that currently plague low-metal loading and thermal instability by using mixed oxides.

Use of ultra-deep desulfurization of liquid hydrocarbon fuels such as gasoline, diesel, and jet fuel to satisfy new environmental regulations and fuel cell applications is receiving increased attention and will continue to receive attention. Conventional hydrodesulfurization technology (HDS) is difficult and costly to use to remove sulfur compounds from liquid hydrocarbon fuels to levels suitable to match environmental regulations. Several nonhydrodesulfurization-based desulfurization technologies for use with liquid fuels have been initiated. These technologies include (1) *biodesulfurization* (Ranson and Rivas, 2008) and (2) adsorptive desulfurization (Song et al., 2010). Both of these technologies will continue to receive attention and have a high potential for incorporation into future refineries. In fact, the biodesulfurization technology is likely to see application during microbial enhanced oil recovery processes.

The current trend in hydroprocessing is the treatment of heavy sour feeds that contain compounds such as sulfur, nitrogen, aromatics, iron, and other undesirable components. These compounds pose significant problems with catalyst poisoning; however, developments are keeping pace with increased demand. In light of growing demand for ultralow-sulfur diesel (ULSD), light cycle oil hydrotreating is receiving much attention. Feeds such as these are typically high in heavy metals, which will require additional unit modifications and/or the installation of guard beds/reactors.

In fact, the Topsøe ultra-low sulfur diesel process yields a maximum of distillates by applying deep thermal conversion of the vacuum residue feed and by vacuum flashing the cracked residue (Egebjeng et al., 2011). The process is a hydrotreating process that combines a high-activity catalyst as well as state-of-the-art reactor internals and can be applied over a very wide range of reactor pressures.

In addition, hydrotreating feedstocks prior to sending the feedstock to the fluid catalytic cracking units is another important focus and will continue to be important or even increase in importance. Many fluid catalytic cracking units incorporate pretreaters (in the form of hydrotreating the feedstocks or guard beds/reactors) to meet their naphtha-gasoline sulfur requirements. Installation of such reactor units will necessarily increase as heavy feedstocks are incorporated into gas oils (fed to the fluid catalytic cracking unit) or become the sole feedstock for the catalytic cracking unit. Proven technology is available to remove sulfur, metals, and asphaltene content while converting an important part of the feed to lighter quality products. This technology will improve and operations will be varied to upgrade the following typical feedstocks: atmospheric and vacuum residua, tar sand bitumen, deasphalter bottoms, and bio-feedstocks.

Thus, with the increasing focus to reduce sulfur content in fuels, the role of *desulfurization* in the refinery becomes more and more important. Currently, the process of choice is the hydrotreater, in which hydrogen is added to the fuel to remove the sulfur from the fuel. Some hydrogen may be lost to reduce the octane number of the fuel, which is undesirable. Because of the increased attention for fuel desulfurization, various new process concepts are being developed with various claims of efficiency and effectiveness.

The major developments in desulfurization have three main routes: advanced hydrotreating (new catalysts, catalytic distillation, processing at mild conditions), reactive adsorption (type of adsorbent used, process design), and oxidative desulfurization (catalyst, process design).

The demand for low sulfur transportation fuels requires that refiners evaluate the many different options for reaching the target. Selection of catalyst types is one of the important decisions and depends on (1) feedstock, (2) operation conditions, (3) hydrogen availability, and product properties may play a role (Tippet et al., 1999).

Besides the issues related to the legislative drive for removing sulfur the refiners will be faced with a growing demand for diesel fuel, which may be met by producing less low-value products such as heating oil. This can be done by converting heavy fractions by hydrocracking or mild hydrocracking (hydrotreating) processes, or one may adopt upgrading processes, for example, for light cycle oil. The latter alternative will require innovative technology that not only removes refractory sulfur species in the presence of high amounts of nitrogen but also performs a certain degree of ring opening to reach a reasonable product cetane number. The innovative might well involve (in part) the use of guard beds or reactor or the use of scavengers (such as calcium oxide, alone or supported on the catalyst) that will remove sulfur and coke formers during the hydrotreating process.

Catalyst development will accelerate, including catalysts for pretreating feedstocks to the fluid catalytic cracking unit that will eliminate much of the need for naphtha–gasoline posttreating (Topsøe et al., 2004). The catalysts will have multiple functions, such as (1) optimized hydrodesulfurization and (2) minimized hydrogen consumption. The addition of metals such as iron, tungsten, niobium, boron, and phosphorus to catalyst compositions; and the use of unsupported nanoparticles will increase and begin (if not already beginning) commercial utilization.

The challenges that the refining industry is facing in regard to hydrodesulfurization processes in the next two decades requite for major developments within hydroprocessing catalyst technology (Lautenschlager Moro, 2003). Areas such as (1) catalyst supports, (2) catalyst morphology, and (3) reaction pathways will continue to provide new opportunities for the development of improved commercial hydrotreating catalysts. Breakthrough technology is essential if catalysts are to be developed that are able to exhibit high activity to produce the desire products.

10.5.2 GASOLINE AND DIESEL FUEL POLISHING

The DuPont Isotherming process provides a means to upgrade gasoil, heavy gas oil, coker gas oil, deasphalted oil, and FCC cycle oils. The products are low-sulfur, low-nitrogen FCC feed, low-sulfur gasoline-kerosene type products, and low-nitrogen fuels and/or downstream feedstocks. In the process, fresh feedstock, after heat exchange, is combined with recycle product and hydrogen in a mixer internal to the reactor. The liquid feed with soluble hydrogen is fed to the Isotherming reactor/bed one where partial desulfurization, removal of metals, or even mild hydrocracking occurs. The stream is resaturated with additional hydrogen in a second mixer and fed to the second Isotherming reactor/bed where further mild hydrocracking takes place. The treated oil from the second Isotherming bed may then be fed to additional Isotherming beds to achieve the desired level of conversion. Treated oil from the last bed is recycled back to the inlet of bed one. This recycle stream delivers recycled hydrogen to the reactors and also acts as a heat sink; thus, a nearly isothermal reactor operation is achieved.

This technology provides a solution to the daunting challenges refiners face to reduce sulfur in finished products to ultra-low levels such as (1) kerosene hydrotreating, (2) diesel hydrotreating, (3) vacuum gas oil hydrotreating, (4) dewaxing, as well as an option for mild hydrocracking (Chapter 11). The technology provides the hydrogen necessary for reactions through a saturated liquid recycle stream. The hydrotreated liquid recycle stream also acts as a heat sink, resulting in few hot spots, reduced light-ends make, and decreased catalyst deactivation.

Biodesulfurization is only one of several concepts by which gasoline and diesel fuel might be polished, that is, sulfur removed to an extremely low, if not to a zero, level. At this point, it is pertinent that a brief review of the potential methods for fuel polishing should be introduced. Briefly, the two desulfurization processes used for fuels purification (desulfurization) are (1) sweetening and (2) hydrotreating. Sweetening is effective only against mercaptans, which are the predominant species in light gasoline. Hydrotreating is effective against all sulfur species and is more widely used.

In the sweetening process, a light naphtha stream is washed with amine to remove hydrogen sulfide and then reacted with caustic, which promotes the conversion of mercaptans to disulfides.

$$R\text{-}SH \rightarrow RSSR$$

The disulfides can subsequently be extracted and removed in what is referred to as extractive sweetening.

In the hydrotreating process, the feed is reacted with hydrogen, in the presence of a solid catalyst. The hydrogen removes sulfur by conversion to hydrogen sulfide, which is subsequently separated and removed from the reacted stream. As the reaction is favored by both temperature and pressure, hydrotreaters are typically designed and operated at approximately 370°C (700°F) and 1000–2000 psi hydrogen. The lower ends of the ranges typically apply to gasoline desulfurization, while gas oil desulfurization requires a more severe operation.

Hydrogen is provided in the form of treating gas at a purity that is typically around on the order of 90% by volume although gas with as little as 60% by volume hydrogen is reputed to be used. Hydrogen is produced by catalytic reformers or hydrogen generation units (Speight and Ozum, 2002; Parkash, 2003; Hsu and Robinson, 2006; Gary et al., 2007; Speight, 2014) and distributed to the hydrotreaters through a refinery-wide network.

In a hydrotreating unit, feed and treating gas are combined and brought to the reaction temperature and pressure, prior to entering the reactor. The reactor is a vessel preloaded with solid catalyst, which promotes the reaction. The catalyst is slowly deactivated by the continuous exposure to high temperatures and by the formation of a coke layer on its surface. Refineries have to shut down the units periodically and regenerate or replace the catalyst.

The severity of operation of an existing unit can be increased by increasing the reaction temperature but there is a negative impact on the catalyst life. The severity of operation can also be increased by increasing the catalyst volume of the unit. In this case the typical solution is to add a second reactor identical to the existing one, doubling the reactor volume. The pressure of an existing unit cannot be changed to increase its severity, because the pressure is related to material of construction and thickness of metal surfaces. If higher pressure is required, the typical solution is to install a new unit and use the existing one for a less severe service.

One new technology is the use of *adsorption by metal oxides* in which the oxides react either by physical adsorption or by chemical adsorption insofar as adsorption followed by chemical reaction is promoted. The major distinction of this type of process from conventional hydrotreating is that the sulfur in the sulfur-containing compounds adsorbs to the catalyst after the feedstock–hydrogen mixture interacts with the catalyst. The catalyst does need to be regenerated constantly.

Another option involves *sulfur oxidization* in which a petroleum and water emulsion is reacted with hydrogen peroxide (or another oxidizer) to convert the sulfur in sulfur-containing compounds to sulfones. The sulfones are separated from the hydrocarbons for post-processing. The major advantages of this new technology include low reactor temperatures and pressures, short residence time, no emissions, and no hydrogen requirement. The technology preferentially treats dibenzothiophene derivatives, one of streams that are most difficult to desulfurize.

One way to add to the supply of ultra-low sulfur fuels is to turn to a non-oil-based diesel. The *Fischer–Tropsch process*, for example, can be used to convert natural gas to a synthetic, sulfur-free diesel fuel. Commercial viability of gas-to-liquid projects depends (in addition to capital costs) on the market for petroleum products and possible price premiums for gas-to-liquid fuels as well as the value of any by-products.

A second way to avoid desulfurization is with *biodiesel* made from vegetable oil or animal fats. Although other processes are available, most biodiesel is made with a base-catalyzed reaction. In the process, fat or oil is reacted with an alcohol, such as methanol, in the presence of a catalyst to produce glycerin and methyl esters or biodiesel. The methanol is charged in excess to assist in quick conversion and recovered for reuse. The catalyst, usually sodium or potassium hydroxide, is mixed with the methanol. Biodiesel is a strong solvent and can dissolve paint as well as deposits left in fuel lines by petroleum-based diesel, sometimes leading to engine problems. Biodiesel also freezes at a higher temperature than petroleum-based diesel.

10.5.3 BIODESULFURIZATION

Refiners are being continually challenged to produce products with ever-decreasing levels of sulfur. At the same time, the supplies of light, low sulfur crude oil that favor distillate production are limited and even decreasing. Generally, the sulfur content of petroleum continues to rise with the accompanying decrease in API gravity and an increase in the proportion of residua in the crude oil. These factors require the crude oil to be processed more severely to produce gasoline and other transportation fuels. Thus, many refineries are now configured for maximum gasoline production that also includes increasingly processing highly aromatic distillate by-products, such as light cycle oil, for the additional feedstock to produce more distillate.

Biocatalyst desulfurization of petroleum constituents is one of a number of possible modes of applying biologically based processing to the needs of the petroleum industry in terms of processing and spill cleanup (McFarland et al., 1998; Setti et al., 1999). In addition, *mycobacterium goodie* has been found to desulfurize benzothiophene (Li et al., 2005). The desulfurization product was identified as α-hydroxystyrene. This strain appeared to have the ability to remove organic sulfur from a broad range of sulfur species in gasoline. When straight-run gasoline containing various organic sulfur compounds was treated with immobilized cells of *mycobacterium goodie* for 24 hours at 40°C (104°F), the total sulfur content significantly decreased, from 227 to 71 ppm at 40°C. Furthermore, when immobilized cells were incubated at 40°C (104°F) with *mycobacterium goodie*, the sulfur content of the gasoline decreased from 275 to 54 ppm in two consecutive reactions.

A dibenzothiophene-degrading bacterial strain, *Nocardia* sp., was able to convert dibenzothiophene to 2-hydroxybiphenyl as the end metabolite through a sulfur-specific pathway (Chang et al., 1998). Other organic sulfur compounds, such as thiophene derivatives, thiazole derivatives, sulfides, and disulfides, were also desulfurized by *Nocardia* sp. When a sample in which dibenzothiophene was dissolved in hexadecane and treated with growing cells, the dibenzothiophene was desulfurized in approximately 80 hours.

The soil-isolated strain microcobe identified as *Rhodococcus erythropolis* can efficiently desulfurize benzonaphthothiophene (Yu et al., 2006). The desulfurization product was α-hydroxy-β-phenyl-naphthalene. Resting cells were able to desulfurize diesel oil (total organic sulfur, 259 ppm) after hydrodesulfurization and the sulfur content of diesel oil was reduced by 94.5% after 24 hours at 30°C (86°F). Biodesulfurization of crude oils was also investigated, and after 72 hours at 30°C (86°F), 62.3% of the total sulfur content in Fushun crude oil (initial total sulfur content, 3210 ppm) and 47.2% of the sulfur in Sudanese crude oil (initial total sulfur, 1237 ppm) were removed (see also Abbad-Andaloussi et al., 2003).

Heavy crude oil recovery, facilitated by microorganisms, was suggested in the 1920s and received growing interest in the 1980s as microbial enhanced oil recovery. However, such projects have been slow to get under way although *in situ* biosurfactant and biopolymer applications continue to garner interest (Van Hamme et al., 2003). In fact, studies have been carried out on biological methods of removing heavy metals such as nickel and vanadium from petroleum distillate fractions, coal-derived liquid shale, bitumen, and synthetic fuels. However, further characterization on the biochemical mechanisms and bioprocessing issues involved in petroleum upgrading are required in order to develop reliable biological processes.

For upgrading options, the use of microbes has to show a competitive advantage of enzyme over the tried-and-true chemical methods prevalent in the industry. Currently, the range of reactions using microbes is large but is usually related to production of bioactive compounds or precursors. But the door is not closed and the issues of biodesulfurization and bioupgrading remain open for the challenge of bulk petroleum processing. These drawbacks limit the applicability of this technology to specialty chemicals and steer it away from bulk petroleum processing.

Biodesulfurization is, therefore, another technology to remove sulfur from the feedstock. However, several factors may limit the application of this technology, however. Many ancillary processes novel to petroleum refining would be needed, including a biocatalyst fermenter to regenerate the bacteria. The process is also sensitive to environmental conditions such as sterilization, temperature, and residence time of the biocatalyst. Finally, the process requires the existing hydrotreater to continue in operation to provide a lower sulfur feedstock to the unit and is more costly than conventional hydrotreating. Nevertheless, the limiting factors should not stop the investigations of the concept and work should be continued with success in mind.

Once the concept has been proven on the scale that a refiner would require, the successful microbial technology will most probably involve a genetically modified bacterial strain for (1) upgrading distillates and other petroleum fractions in refineries, (2) upgrading crude petroleum upstream, and (3) dealing with environmental problems that ace industry, especially in areas related to spillage of crude oil and products. These developments are part of a wider trend to use bioprocessing to make products and do many of the tasks that are accomplished currently by conventional chemical processing. If commercialized for refineries, however, biologically based approaches will be at scales and with economic impacts beyond anything previously seen in industry.

In addition, the successful biodesulfurization process will, most likely, be based on naturally occurring aerobic bacteria that can remove organically bound sulfur in heterocyclic compounds without degrading the fuel value of the hydrocarbon matrix. Because of the susceptibility of bacteria to heat, the process will need to operate at temperatures and pressures close to ambient and also use air to promote sulfur removal from the feedstock.

10.5.4 Bio-Feedstocks

There are also bio-feedstock issues that will become relevant in the refinery of the future, not the least of which will be the incorporation of bio-feedstocks into refinery hydrotreaters. For example, there are already reports of (1) refining extracted bio-oils being combined with refinery streams for hydroprocessing, (2) hydrogenation of animal fats to produce a high-cetane diesel-range product, and (3) hydrogenation of palm oil. However, there are issues related to quality of hydrotreated vegetable oils in terms of high paraffin content, low filter plugging points, and low density that must and will be resolved. Resolution of such issues will lead to recommendations on the means by which bio-feedstocks can be (or will be) incorporated into existing hydrotreating units based on process efficiency as well as economic considerations.

REFERENCES

Abbad-Andaloussi, S., Warzywoda, M., and Monot, F. 2003. *Revue Institut Français Du Pétrole*, 58(4): 505–513.

Ancheyta, J. and Speight, J.G. (Editors). 2007. Feedstock evaluation and composition. In: *Hydroprocessing of Heavy Oils and Residua*. CRC Press/Taylor & Francis Group, Boca Raton, FL.

Chang, J.H., Rhee, S.K., Chang Y.K., and Chang H.N. 1998. *Biotechnology Progress*, 14(6): 851–855.

DeCroocq, D. 1984. *Catalytic Cracking of Heavy Petroleum Hydrocarbons*. Editions Technip, Paris, France.

Dolbear, G.E. 1998. Hydrocracking: Reactions, catalysts, and processes. In: *Petroleum Chemistry and Refining*. J.G. Speight (Editor). Taylor & Francis Group, Washington, DC, Chapter 7.

Egebjeng, R., Knudsen, K., and Grennfelt, E. 2011. Bigger is better: Industrial-scale production of renewable diesel. *Proceedings. NPRA Annual Meeting*. San Antonio, TX, March. National Petrochemical & Refiners Association, Washington, DC.

Furimsky, E. 1998. Selection of catalysts and reactors for hydroprocessing. *Applied Catalysis A: General*, 171: 177–206.

Gary, J.G., Handwerk, G.E., and Kaiser, M.J. 2007. *Petroleum Refining: Technology and Economics*, 5th Edition. CRC Press/Taylor & Francis Group, Boca Raton, FL.

Hsu, C.S. and Robinson, P.R. 2006. *Practical Advances in Petroleum Processing*, Volumes 1 and 2. Springer, New York.

Hunter, M.G., Vivas, A.H., Jensen, L.S., and Low, G.G. 2010. Partial conversion hydrocracking process and apparatus. U.S. Patent 7,763,218. July 27.

Khan, M.R. and Patmore, D.J. 1998. Heavy oil upgrading process. In: *Petroleum Chemistry and Refining*. J.G. Speight (Editor). Taylor & Francis Group, Washington, DC, Chapter 6.

Lautenschlager Moro, L.F. 2003. Process technology in the petroleum refining industry/current situation and future trends. *Computers and Chemical Engineering*, 27: 1303–1305.

Li, F., Xu, P., Feng, J., Meng, L., Zheng, Y., Luo, L., and Ma, C. 2005. *Applied and Environmental Microbiology*, 71(1): 276–281.

Maples, R.E. 2000. *Petroleum Refinery Process Economics*, 2nd Edition. PennWell Corporation, Tulsa, OK.

McFarland, B.L., Boron, D.J., Deever, W., Meyer, J.A., Johnson, A.R., and Atlas, R.M. 1998. *Critical Reviews in Microbiology*, 24: 99–147.

Parkash, S. 2003. *Refining Processes Handbook*. Gulf Professional Publishing, Elsevier, Amsterdam, the Netherlands.

Ranson, I. and Rivas, C.M. 2008. Biodesulfurization of hydrocarbons. U.S. Patent 7,338,795. March 4.

Reynolds, J.G. 1998. Metals and heteroatoms in heavy crude oils. In: *Petroleum Chemistry and Refining*. J.G. Speight (Editor). Taylor & Francis Group, Washington, DC, Chapter 3.

Scherzer, J. and Gruia, A.J. 1996. *Hydrocracking Science and Technology*. Marcel Dekker, New York.

Setti, L., Farinelli, P., Di Martino, S., Frassinetti, S., Lanzarini, G., and Pifferia, P.G. 1999. *Applied Microbiology and Biotechnology*, 52: 111–117.

Song, C., Ma, X., Watanabe, S., and Sun, F. 2010. Oxidatively regenerable adsorbents for sulfur removal. U.S. Patent 7,731,837. June 8.

Speight, J.G. 2000. *The Desulfurization of Heavy Oils and Residua*, 2nd Edition. Marcel Dekker, New York.

Speight, J.G. 2013. *Heavy and Extra Heavy Oil Upgrading Technologies*. Gulf Professional Publishing, Elsevier, Oxford, UK.

Speight, J.G. 2014. *The Chemistry and Technology of Petroleum*, 5th Edition. CRC Press/Taylor & Francs Group, Boca Raton, FL.

Speight, J.G. and Ozum, B. 2002. *Petroleum Refining Processes*. Marcel Dekker, New York.

Tippet, T., Knudsen, K.G., and Cooper, B.C. 1999. *Proceedings. NPRA Annual Meeting*. Paper AM-99-06.

Topsøe, H., Egeberg, R.G., and Knudsen, K.G. 2004. Preprints. *American Chemical Society, Division of Fuel Chemistry*, 49(2): 568.

Van Hamme, J.D., Singh, A., and Ward, O.P. 2003. *Microbiology and Molecular Biology Reviews*, 67(4): 503–549.

Yu, B., Xu, P., Shi, Q., and Ma, C. 2006. *Applied and Environmental Microbiology*, 72: 54–58.

11 Hydrocracking

11.1 INTRODUCTION

Hydrocracking is a refining technology that, like hydrotreating (Chapter 10), also falls under the general process umbrella of *hydroprocessing*. The outcome is the conversion of a variety of feedstocks to a range of products (Table 11.1), and units to accomplish this goal can be found at various points in a refinery (Figure 11.1).

Hydrocracking is a more recent process development compared to the older thermal cracking, visbreaking, and coking. In fact, the use of hydrogen in thermal processes is perhaps the single most significant advance in refining technology during the twentieth century (Bridge et al., 1981; Scherzer and Gruia, 1996; Dolbear, 1998; Rana et al., 2007). The ability of refiners to cope with the renewed trend toward distillate production from heavier feedstocks with low atomic hydrogen–carbon ratios (Figure 11.2) has created a renewed interest in hydrocracking. Without the required conversion units, heavier crude oils produce in lower yields of naphtha and middle distillate. To maintain current naphtha and middle distillate production levels, additional conversion capacity is required because of the differential in the amount of distillates produced from light crude oil and the distillate products produced from heavier crude oil (Figure 11.3).

The concept of hydrocracking allows the refiner to produce products having a lower molecular weight with a higher hydrogen content and a lower yield of coke. In summary, hydrocracking facilities add flexibility to refinery processing and to the product. The hydrocracking process is more severe than the hydrotreating process (Chapters 5 and 10) there being the intent, in hydrocracking processes, to convert the feedstock to lower-boiling products rather than to treat the feedstock for heteroatom and metals removal only. Process parameters (Figures 11.4 and 11.5) emphasize the relatively severe nature of the hydrocracking process.

The older hydrogenolysis type of hydrocracking practiced in Europe during and after World War II was tungsten sulfide (WS_2) or molybdenum sulfide (MoS) as catalysts. These processes required high reaction temperatures and operating pressures, sometimes in excess of approximately 3000 psi for continuous operation. The modern hydrocracking processes were initially developed for converting refractory feedstocks to naphtha and jet fuel; process and catalyst improvements and modifications have made it possible to yield products from gases and naphtha to furnace oils and catalytic cracking feedstocks.

Hydrocracking is an extremely versatile process that can be utilized in many different ways, and one of the advantages of hydrocracking is its ability to break down high-boiling aromatic stocks produced by catalytic cracking or coking. To take full advantage of hydrocracking, the process must be integrated in the refinery with other process units (Figure 11.1). In naphtha production, for example, the hydrocracker product must be further processed in a catalytic reformer as it has a high naphthene content and relatively low octane number. The high naphthene content makes the hydrocracker naphtha an excellent feed for catalytic reforming, and good yields of high-octane-number gasoline can be obtained.

If high-molecular-weight petroleum fractions are *pyrolyzed*, that is, if no hydrogenation occurs, progressive cracking and condensation reactions generally lead to the final products. These products are usually (1) gaseous and low-boiling liquid compounds of high hydrogen content, (2) liquid material of intermediate molecular weight with a hydrogen–carbon atomic ratio differing more or less from that of the original feedstock, depending on the method of operation, and (3) material of high molecular weight, such as coke, possessing a lower hydrogen–carbon atomic ratio than the starting material. Highly aromatic or refractory recycle stocks or gas oils that contain varying proportions

TABLE 11.1

Refinery Processes That Employ Hydrocracking

	Process Characteristics						
Feedstock	Hydro-cracking	Aromatics Removal	Sulfur Removal	Nitrogen Removal	Metals Removal	Olefins Removal	Products
Naphtha			✓	✓			Reformer feedstock
	✓					✓	Liquefield petroleum gas (LPG)
Gas oil							
Atmospheric	✓						Naphtha
Vacuum			✓	✓	✓		Catalytic cracker feedstock
	✓	✓	✓				Diesel fuel
	✓	✓	✓				Kerosene
	✓	✓	✓				Jet fuel
	✓						Naphtha
	✓						LPG
	✓	✓					Lubricating oil
Residuum	✓						Diesel fuel (others)

of highly condensed aromatic structures (e.g., naphthalene and phenanthrene) usually crack, in the absence of hydrogen, to yield intractable residues and coke.

An essential difference between *pyrolysis* (thermal decomposition, usually in the absence of any added agent) and *hydrogenolysis* (thermal decomposition in the presence of hydrogen and a catalyst or a hydrogen-donating solvent) of petroleum is that in pyrolysis a certain amount of polymerized heavier products, like cracked residuum and coke, is always formed along with the light products, such as gas and naphtha. During hydrogenolysis (*destructive hydrogenation*), polymerization may be partly or even entirely prevented so that only light products are formed. The prevention of coke formation usually results in an increased distillate (e.g., naphtha) yield. The condensed type of molecule, such as naphthalene or phenanthrene, is one that is closely associated with the formation of coke, but in an atmosphere of hydrogen and in contact with catalysts, these condensed molecules are converted into lower-molecular-weight saturated compounds that boil within the gasoline range.

The mechanism of hydrocracking is basically similar to that of catalytic cracking, but with concurrent hydrogenation. The catalyst assists in the production of carbonium ions via olefin intermediates, and these intermediates are quickly hydrogenated under the high-hydrogen partial pressures employed in hydrocracking. The rapid hydrogenation prevents adsorption of olefins on the catalyst and, hence, prevents their subsequent dehydrogenation, which ultimately leads to coke formation so that long onstream times can be obtained without the necessity of catalyst regeneration.

One of the most important reactions in hydrocracking is the partial hydrogenation of polycyclic aromatics followed by rupture of the saturated rings to form substituted monocyclic aromatics. The side chains may then be split off to give *iso*-paraffins. It is desirable to avoid excessive hydrogenation activity of the catalyst so that the monocyclic aromatics become hydrogenated to naphthenes; furthermore, repeated hydrogenation leads to loss in octane number, which increases the catalytic reforming required to process the hydrocracked naphtha.

Side chains of three- or four-carbon atoms are easily removed from an aromatic ring during catalytic cracking, but the reaction of aromatic rings with shorter side chains appears to be quite different. For example, hydrocracking single-ring aromatics containing four or more methyl groups

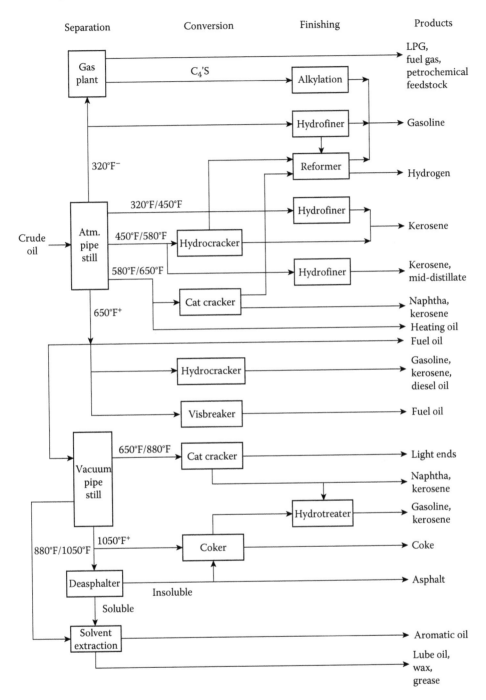

FIGURE 11.1 Generalized refinery layout showing relative placement of the catalytic cracking units.

produces largely *iso*-butane and benzene. It may be that successive isomerization of the feed molecule adsorbed on the catalyst occurs until a four-carbon side chain is formed, which then breaks off to yield *iso*-butane and benzene. Overall, coke formation is very low in hydrocracking since the secondary reactions and the formation of the precursors to coke are suppressed as the hydrogen pressure is increased.

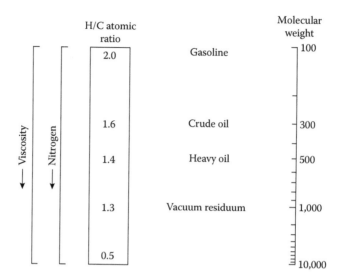

FIGURE 11.2 Atomic hydrogen–carbon ratios for various feedstocks.

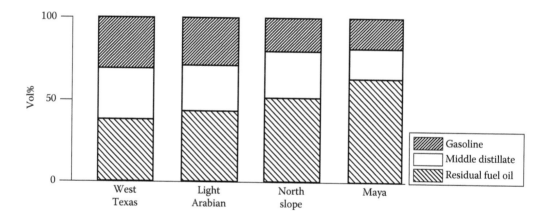

FIGURE 11.3 Hydrocracking product yields from various feedstocks.

When applied to heavy feedstocks, the hydrocracking process can be used for processes such as (1) fuel oil desulfurization and (2) feedstock conversion to lower-boiling distillates. The products from hydrocracking are composed of either saturated or aromatic compounds; no olefins are found. In making gasoline, the lower paraffins formed have high octane numbers; for example, the five- and six-carbon number fractions have leaded research octane numbers on the order of 99–100. The remaining naphtha has excellent properties as a feed to catalytic reforming, producing a highly aromatic naphtha that is capable of a high octane number. Both types of naphtha are suitable for premium-grade automobile gasoline. Another attractive feature of hydrocracking is the low yield of gaseous components, such as methane, ethane, and propane, which are less desirable than naphtha. When making jet fuel, more hydrogenation activity of the catalysts is used, since jet fuel contains more saturates than gasoline.

Like many refinery processes, the problems encountered in hydrocracking heavy feedstocks can be directly equated to the amount of complex, higher-boiling constituents that may require pretreatment (Speight, 2000, 2014; Moulton and Erwin, 2005; Stratiev and Petkov, 2009; Bridjanian and

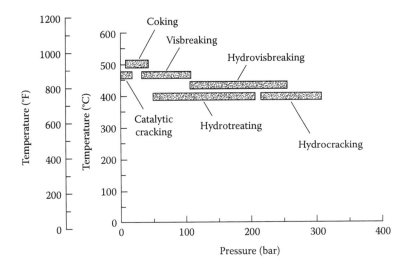

FIGURE 11.4 Temperature and pressure ranges for various processes.

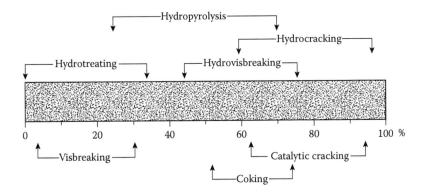

FIGURE 11.5 Feedstock conversion in various processes.

Khadem Samimi, 2011). Processing these feedstocks is not merely a matter of applying know-how derived from refining *conventional* crude oils but requires a knowledge of composition (Chapters 2 and 3). The materials are not only complex in terms of the carbon number and boiling point ranges (Figure 11.6) but also because a large part of this *envelope* falls into a range of model compounds (Figure 11.7) and very little is known about the properties. It is also established that the majority of the higher-molecular-weight materials produce coke (with some liquids), but the majority of the lower-molecular-weight constituents produce liquids (with some coke). It is to both of these trends that hydrocracking is aimed (Figure 11.8).

It is the physical and chemical composition of a feedstock that plays a large part not only in determining the nature of the products that arise from refining operations but also in determining the precise manner by which a particular feedstock should be processed. Furthermore, it is apparent that the conversion of heavy feedstocks requires new lines of thought to develop suitable processing scenarios (Babich and Moulijn, 2003). Indeed, the use of thermal (*carbon rejection*) processes and of hydrothermal (*hydrogen addition*) processes, which were inherent in the refineries designed to process lighter feedstocks, has been a particular cause for concern. This has brought about the evolution of processing schemes that accommodate the heavier feedstocks (Khan and Patmore, 1998;

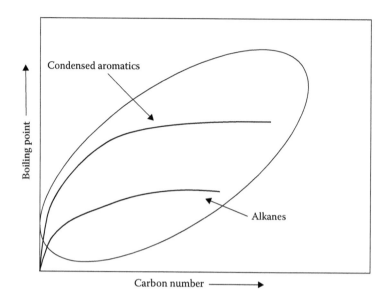

FIGURE 11.6   General boiling point/carbon number envelope for crude oil.

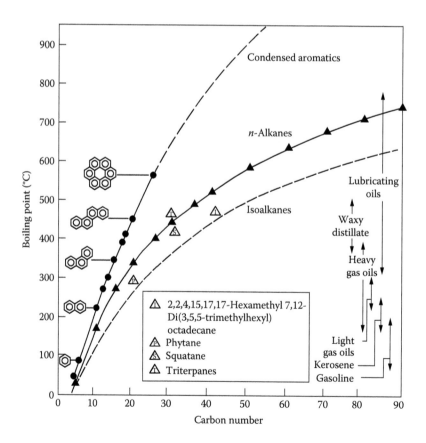

FIGURE 11.7   General boiling point/carbon number envelope for pure compounds showing crude oil distillation fractions.

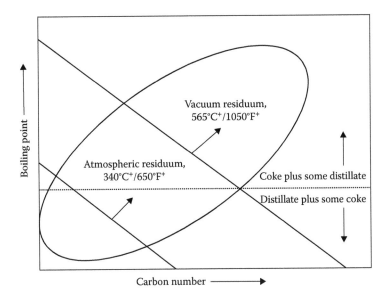

FIGURE 11.8   Thermal processing trends for crude oil fractions.

Speight, 2014). As a point of reference, an example of the former option is the delayed coking process in which the feedstock is converted to overhead with the concurrent deposition of coke, for example, that used by Suncor, Inc., at their oil sands plant (Speight, 1990, 2014).

The hydrogen addition concept is illustrated by the hydrocracking process in which hydrogen is used in an attempt to *stabilize* the reactive fragments produced during the cracking, thereby decreasing their potential for recombination to heavier products and ultimately to coke. The choice of processing schemes for a given hydrocracking application depends upon the nature of the feedstock as well as the product requirements. The process can be simply illustrated as a single-stage or as a two-stage operation (Figure 11.9) (Hansen et al., 2010).

The single-stage process can be used to produce naphtha but is more often used to produce middle distillate from heavy vacuum gas oils. The single-stage process may contain two reactors but without separation. On the other hand, in the two-stage processes, the undesirable products (such as hydrogen sulfide and ammonia) of the first stage are eliminated before the second stage.

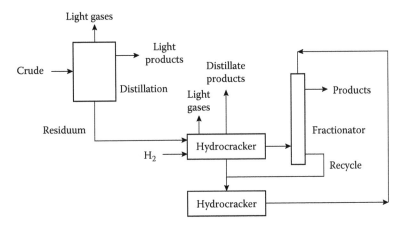

FIGURE 11.9   A single-stage and two-stage hydrocracking unit.

The most common reactor design for heavy feedstock hydroprocessing is the downflow and fixed-bed reactor. The single stage with recycle is a commonly used configuration. In this process option, the uncracked residual hydrocarbon oil from the bottom of reaction product fractionation tower is recycled back into the single reactor for further cracking. However, for single-stage hydrocracking of heavy feedstocks, it is advisable that the feedstock should first be hydrotreated to remove ammonia and hydrogen sulfide or the catalyst used in the single reactor must be capable of both hydrotreating and hydrocracking.

The two-stage process was developed primarily to produce high yields of naphtha from straight-run gas oil, and the first stage may actually be a purification step to remove sulfur-containing (as well as nitrogen-containing) organic materials. In terms of sulfur removal, it appears that nonasphaltene sulfur may be removed before the more refractory asphaltene sulfur (Figure 11.10), thereby requiring thorough desulfurization. This is a good reason for processes to use an extinction-recycling technique to maximize desulfurization and the yields of the desired product. Significant conversion of heavy feedstocks can be accomplished by hydrocracking at high severity. For some applications, the products boiling up to 340°C (650°F) can be blended to give the desired final product.

The two-stage hydrocracker process configuration uses two reactors, and the residual product from the bottom of reaction product fractionation tower is recycled back into the second reactor for further cracking. Since the first-stage reactor accomplishes both hydrotreating and hydrocracking, the second-stage reactor feed is virtually free of ammonia and hydrogen sulfide. This permits the use of high-performance noble metal (palladium, platinum) catalysts that are susceptible to poisoning by sulfur or nitrogen compounds. The process is best suited for large units and for processing difficult, high-nitrogen feedstocks. Almost all of the unconverted bottoms are recycled and conversion levels of 95%–99% can be achieved. The two-stage hydrocracker is typically installed as a stand-alone unit and does not involve integration with any other units. In any multibed hydrocracking reactor, particulates can accumulate and affect the top-bed catalyst performance, even with sophisticated automatic backwash feed filters, which are considered necessary for all vacuum gas oil. Furthermore, the two-stage hydrocracker process configuration is best suited for large units and for processing difficult, high-nitrogen feedstocks, and offers higher conversion of feedstock through two distinct reactor stages with intermediate fractionation. This helps to maximize the yield of high-quality products (kerosene and heavy diesel). Other configurations, such as the single-stage design (once-through and recycle mode) (Figure 11.11), can be used to produce base oils for

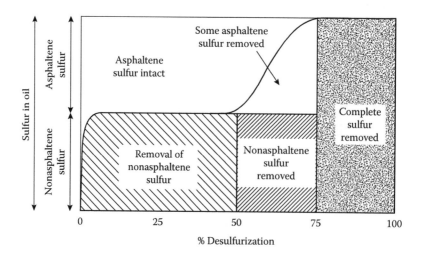

FIGURE 11.10 Trends for sulfur removal from crude oil.

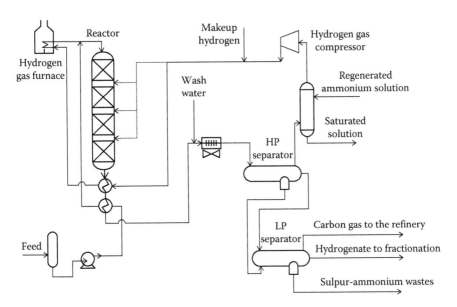

FIGURE 11.11 The once-through configuration.

lubricating oil blending or as feedstocks for fluid catalytic cracking units as well as feedstocks for ethylene cracker feedstocks.

Hydrocracking is similar to catalytic cracking, with hydrogenation superimposed and with the reactions taking place either simultaneously or sequentially. Hydrocracking was initially used to upgrade low-value distillate feedstocks, such as cycle oils (highly aromatic products from a catalytic cracker that usually are not recycled to extinction for economic reasons), thermal and coker gas oils, and heavy-cracked and straight-run naphtha. These feedstocks are difficult to process by either catalytic cracking or reforming, since they are usually characterized by a high polycyclic aromatic content and/or by high concentrations of the two principal catalyst poisons, sulfur and nitrogen compounds.

While whole families of catalysts are required depending on feed available and the desired product slate or product character, the number of process stages is also important to catalysts choice. Generally, one of the three options is utilized by the refinery. Thus, depending on the feedstock being processed and the type of plant design employed (*single-stage* or *two-stage*), flexibility can be provided to vary product distribution among the following principal end products.

Fundamentally, the trend toward lower API gravity feedstocks is related to an increase in the hydrogen–carbon atomic ratio of crude oils (Figure 11.2) because of the higher content of residuum. This can be overcome by upgrading methods that lower this ratio by adding hydrogen, rejecting carbon, or using a combination of both methods. Though several technologies exist to upgrade heavy feedstocks (Speight, 2014), selection of the optimum process units is very much dependent on each refiner's needs and goals, with the market pull being the prime motivator. Furthermore, processing options systems to *dig deeper into the barrel* by converting more of the higher-boiling materials to distillable products should not only be cost-effective and reliable but also flexible. Hydrocracking adds that flexibility and offers the refiner a process that can handle varying feeds and operate under diverse process conditions. Utilizing different types of catalysts can modify the product slate produced but reactor design and number of processing stages play a role in this flexibility.

Finally, a word about conversion measures for upgrading processes. Such measures are necessary for *any conversion process* but more particularly for hydrocracking processes where hydrogen management is an integral, and essential, part of process design.

The objective of any upgrading process is to convert heavy feedstock into marketable products by reducing their heteroatom (nitrogen, oxygen, and sulfur) and metal contents, modifying their asphaltenic structures (reducing coke precursors), and converting the high-molecular-weight polar species large molecules into lower-molecular-weight and lower-boiling hydrocarbon products. Upgrading processes are evaluated on the basis of liquid yield (i.e., naphtha, distillate, and gas oil), heteroatom removal efficiency, feedstock conversion (FC), carbon mobilization (CM), and hydrogen utilization (HU), along with other process characteristics. Definitions of FC, CM, and HU are

$$FC = (Feedstock_{IN} - Feedstock_{OUT})/Feedstock_{IN} \times 100$$

$$CM = Carbon_{LIQUIDS}/Carbon_{FEEDSTOCK} \times 100$$

$$HU = Hydrogen_{LIQUIDS}/Hydrogen_{FEEDSTOCK} \times 100$$

High carbon mobilization (CM < 100%) and high hydrogen utilization (HU) correspond to high feedstock conversion (FC) processes involving hydrogen addition such as hydrocracking. Since hydrogen is added, hydrogen utilization can be greater than 100%. These tasks can be achieved by using thermal and/or catalytic processes. Low carbon mobilization and low hydrogen utilization correspond to low feedstock conversion such as coking (carbon rejection) processes (Chapter 8). Maximum efficiency from an upgrading process can be obtained by maximizing the liquid yield and its quality by minimizing the gas (C_1–C_4) yield simultaneously. Under these operating conditions the hydrogen consumption would be the most efficient, that is, hydrogen is consumed to increase the liquid yield and its quality. Several process optimization models can be formulated if the reaction kinetics is known.

11.2 PROCESSES AND PROCESS DESIGN

Hydrocracking has become an indispensable processing technology to modern petroleum refining and petrochemical industry due to its flexibility to feedstocks and product scheme and high-quality products. Particularly, high-quality naphtha, jet fuel, diesel, and lube base oil can be produced through this technology. The hydrocracker provides a better balance of naphtha and distillates, improves gasoline yield, octane quality, and can supplement the fluid catalytic cracker to upgrade heavy feedstocks. In the hydrocracker, light fuel oil is converted into lighter products under a high hydrogen pressure and over a hot catalyst bed—the main products are naphtha, jet fuel, and diesel oil.

The objective of the hydrocracking process is to convert higher-molecular-weight feedstocks into high-quality, lower-molecular-weight products such as naphtha and kerosene from which liquid fuels can be produced. The choice of processing schemes for a given hydrocracking application depends upon the nature of the feedstock as well as the product requirements (Rashid, 2007; Speight, 2011a, 2014). A two-stage process is typically employed. In the first (pretreat) step, polyaromatic compounds are saturated and organic nitrogen and sulfur are converted to ammonia and hydrogen sulfide. The organic nitrogen contained in the feedstock would otherwise inhibit the activity of the cracking catalyst. In the second (cracking) step, higher-molecular-weight hydrocarbon molecules are preferentially cracked over an acidic metal-containing hydrocracking catalyst. The product yields and product properties are determined by the feedstock, the cracking catalyst selectivity, and the process conditions.

A particular feature of the hydrocracking process, as compared with its alternatives, is its flexibility with respect to product production and the relatively high quality of the products. On the whole, hydrocracking can handle a wider range of feedstock than catalytic cracking, although the latter process has seen some recent catalyst developments that narrowed the gap. There are also examples where hydrocracking is complementary rather than alternative to the other conversion process; an example, cycle oils, which cannot be recycled to extinction in the catalytic cracker, can be processed in the hydrocracker. In addition, since biomass gasification and Fischer–Tropsch conversion are considered promising routes in next-generation biofuels developments (Speight, 2008, 2011a,b, 2014), refiners should closely monitor the latest refinery-related advances as well as future directions.

11.2.1 Process Design

The simplest form of the hydrocracking process is the *single-stage process* in which the layout of the reactor section generally resembles that of a hydrotreating unit. This configuration finds application in cases where only a moderate degree of conversion (say 60% or less) is required. It may well apply to process where the feedstock is pretreated prior to introduction to a fluid catalytic cracking unit. Another form of hydrocracking process for heavier feedstocks is a *two-stage* operation (Figure 11.9). Generally, the first stage of the two-stage unit resembles a *single-stage once-through* (SSOT) unit. This configuration uses only one reactor and any uncracked residual hydrocarbon oil from the bottom of the reaction product fractionation (distillation) tower is not recycled for further cracking. For single-stage hydrocracking, either the feedstock must first be hydrotreated to remove ammonia and hydrogen sulfide or the catalyst used in the single reactor must be capable of both hydrotreating and hydrocracking. This flow scheme has been very popular since it can be used to maximize the yield of transportation fuels and is an attempt to combat the adverse effect of ammonia and nitrogen compounds on catalyst activity. Similarly, the *series-flow* version of the multistage hydrocracker has also been developed. The two-stage flow scheme has been very popular since it maximizes the yield of transportation fuels and has the flexibility to produce naphtha and kerosene to meet seasonal swings in demand for fuels.

Thus, the two-stage hydrocracker consists of two reactor stages together with a product distillation section. The choice of catalyst in each reaction stage depends on the product slate required and the character of the feedstock. In general, however, the first-stage catalyst is designed to remove nitrogen and heavy aromatics from raw petroleum stocks. The second-stage catalyst carries out a selective hydrocracking reaction on the cleaner oil produced in the first stage. Both reactor stages have similar process flow schemes. The oil feed is combined with a preheated mixture of makeup hydrogen and hydrogen-rich recycle gas and heated to reactor inlet temperature via a feed-effluent exchanger and a reactor charge heater. The reactor charge heater design philosophy is based on many years of safe operation with such two-phase furnaces. The feed-effluent exchangers take advantage of special high-pressure exchanger design features developed by Chevron engineers to give leak-free end closures. From the charge heater, the partially vaporized feed enters to the top of the reactor. The catalyst is loaded in separate beds in the reactor with facilities between the beds for quenching the reaction mix and ensuring good flow distribution through the catalyst.

The reactor effluent is cooled through a variety of heat exchangers including the feed-effluent exchanger and one or more air coolers. Deaerated condensate is injected into the first-stage reactor effluent before the final air cooler in order to remove ammonia and some of the hydrogen sulfide. This prevents solid ammonium bisulfide from depositing in the system. A body of expertise in the field of materials selection for hydrocracker cooling trains is quite important for proper design.

The reactor effluent leaving the air cooler is separated into hydrogen-rich recycle gas, a sour water stream, and a hydrocarbon liquid stream in the high-pressure separator. The sour water effluent stream is often then sent to a plant for ammonia recovery and for purification so water can be recycled back to the hydrocracker. The hydrocarbon-rich stream is pressure reduced and fed to the distillation section after light products are flashed off in a low-pressure separator. The hydrogen-rich gas stream from the high-pressure separator is recycled back to the reactor feed by using a recycle compressor. Sometimes with sour feeds, the first-stage recycle gas is scrubbed with an amine system to remove hydrogen. If the feed sulfur level is high, this option can improve the performance of the catalyst and result in less costly materials of construction.

The distillation section consists of a hydrogen sulfide (H_2S) stripper and a recycle splitter. This latter column separates the product into the desired cuts. The column bottoms stream is recycled back to the second-stage feed. The recycle cut point is changed depending on the light products needed. It can be as low as 160°C (320°F) if naphtha production is maximized (for aromatics) or as high as 380°C (720°F) if a low pour point diesel is needed. Between these two extremes a recycle cut point of 260°C–285°C (500°F–550°F) results in high yields of high smoke point low freeze point jet fuel.

In the *series-flow* configuration, the principal difference is the elimination of first-stage cooling and gas/liquid separation and the ammonia removal step. The effluent from the first stage is mixed with more recycle gas and routed direct to the inlet of the second reactor. In contrast with the amorphous catalyst of the two-stage process, the second reactor in series flow generally has a zeolite catalyst, based on crystalline silica–alumina. As in the two-stage process, any material that is not converted to the product boiling range is recycled from the fractionation section.

An *single-stage once-through* (SSOT) unit resembles the first stage of the two-stage configuration. This type of hydrocracker usually requires the least capital investment. The feedstock is not completely converted to lighter products. For this application, the refiner must have a demand for a highly refined heavy oil. In many refining situations, such an oil product can be used as lube oil plant feed or as feedstock to the fluid catalytic cracking unit or in low-sulfur oil blends or as ethylene plant feed. It also lends itself to stepwise construction of a future two-stage hydrocracker for full feed conversion.

A *single-stage recycle* (SSREC) unit converts a heavy oil completely into light products with a flow scheme resembling the second stage of the two-stage plant. Such a unit maximizes the yield of naphtha, jet fuel, or diesel depending on the recycle cut point used in the distillation section. This type of unit is more economical than the more complex two-stage unit when plant design capacity is less than approximately 10,000–15,000 bbl/day. Commercial SSREC plants have operated to produce low pour point diesel fuel from waxy Middle Fast vacuum gas oils. Recent emphasis has been placed on the upgrading of lighter gas oils into jet fuels.

Building on the theme of *one-* or *two-stage* hydrocracking, the *once-through partial conversion* (OTPC) concept evolved. This concept offers the means to convert heavy vacuum gas oil feed into high-quality gasoline, jet fuel, and diesel products by a partial conversion operation. The advantage is lower initial capital investment and also lower utilities consumption than a plant designed for total conversion. Because total conversion of the higher-molecular-weight compounds in the feedstock is not required, once-through hydrocracking can be carried out at lower temperatures and in most cases at lower hydrogen partial pressures than in recycle hydrocracking, where total conversion of the feedstock is normally an objective.

Proper selection of the types of catalysts employed can even permit partial conversion of heavy gas oil feeds to diesel and lighter products at the low hydrogen partial pressures for which gas oil hydrotreaters are normally designed. This so-called mild hydrocracking has been attracting a great deal of interest from refiners who have existing hydrotreaters and wish to increase their refinery's conversion of fuel oil into lower-boiling higher-value products (Table 11.2). Mild hydrocracking is characterized by relatively low conversion (20%–40% v/v of the feedstock) compared to conventional hydrocracking, which gives 70%–100% v/v conversion of heavy distillate at high pressure. The mild hydrocracking route produces low-sulfur products through conversion of vacuum gas oil.

The *recycle hydrocracking unit* is designed to operate at hydrogen partial pressures from approximately 1200 to 2300 psi depending on the type of feed being processed. Hydrogen partial pressure is set in the design in part depending on required catalyst cycle length but also to enable the catalyst to convert high-molecular-weight polynuclear aromatic and naphthene compounds that must be hydrogenated before they can be cracked. Hydrogen partial pressure also affects properties of the hydrocracked products that depend on hydrogen uptake, such as jet fuel aromatics content and smoke point and diesel cetane number. In general, the higher the feed endpoint, the higher the required hydrogen partial pressure necessary to achieve satisfactory performance of the plant.

TABLE 11.2
Feedstocks and Respective Products

Feed	Products
Straight-run gas oil	Liquefied petroleum gas (LPG)
Vacuum gas oil	Motor gasolines
Fluid catalytic cracking oils and decant oil	Reformer feeds
Coker gas oil	Aviation turbine fuels
Thermally cracked stock	Diesel fuels
Solvent deasphalted residual oil	Heating oils
Straight-run naphtha	Solvents and thinners
Cracked naphtha	Lube oils
	Petrochemical feedstocks
	Ethylene feed pretreatment process (FPP)

Once-through partial conversion hydrocracking of a given feedstock may be carried out at hydrogen partial pressures significantly lower than required for recycle total conversion hydrocracking. The potential higher catalyst deactivation rates experienced at lower hydrogen partial pressures can be offset by using higher activity catalysts and designing the plant for lower catalyst space velocities. Catalyst deactivation is also reduced by the elimination of the recycle stream. The lower capital cost resulting from the reduction in plant operating pressure is much more significant than the increase resulting from the possible additional catalyst requirement and larger volume reactors. Additional capital cost savings from once-through hydrocracking result from the reduced overall required hydraulic capacity of the plant for a given fresh feed rate as a result of the elimination of a recycle oil stream. Hydraulic capacity at the same fresh feed rate is 30%–40% lower for a once-through plant compared to one designed for recycle.

Utilities savings for a once-through versus recycle operation arise from lower pumping and compression costs as a result of the lower design pressure possible and also lower hydrogen consumption. Additional savings are realized as a result of the lower oil and gas circulation rates required, since recycle of oil from the fractionator bottoms is not necessary. Lower capital investment and operating costs are obvious advantages of once-through hydrocracking compared to a recycle design. This type of operation may be adaptable for use in an existing gas oil hydrotreater or in a heavy feedstock desulfurization unit. The change from hydrotreating to hydrocracking service will require some modifications and capital expenditure, but in most cases these changes will be minimal.

One disadvantage of once-through hydrocracking compared to a recycle operation is a somewhat reduced flexibility for varying the ratio of naphtha to middle distillate that is produced. A greater quantity of naphtha can be produced by increasing conversion, and production of jet fuel plus diesel can also be increased. But selectivity for higher-boiling products is also a function of conversion. Selectivity decreases as once-through conversion increases. If conversion is increased too much, the yield of desired product will decrease, accompanied by an increase in light ends and gas production. Higher yields of naphtha or jet fuel plus diesel are possible from a recycle than from a once-through operation. However, the fact that unconverted oil is produced by the plant is not necessarily a disadvantage. The unconverted oil produced by once-through hydrocracking is a high-quality, low-sulfur, and low-nitrogen material that is an excellent feedstock for an FCC unit or ethylene pyrolysis furnace or a source of high viscosity index lube oil base stock. The properties of the oil are a function of the degree of conversion and other plant operating conditions.

Middle distillate products made by once-through hydrocracking are generally higher in aromatics content of poorer burning quality than those produced by recycle hydrocracking. However, the quality is generally better than produced by catalytic cracking or from straight run. Middle distillate product quality improves as the degree of conversion increases and as the hydrogen partial pressure is increased.

The hydrocracking process employs high-activity catalysts that produce a significant yield of light products. Catalyst selectivity to middle distillate is a function of both the conversion level and operating temperature, with values in excess of 90% being reported in commercial operation. In addition to the increased hydrocracking activity of the catalyst, percentage desulfurization and denitrogenation at start-of-run conditions are also substantially increased. End of cycle is reached when product sulfur has risen to the level achieved in conventional vacuum gas oil (VGO) hydrodesulfurization process.

An important consideration, however, is that commercial hydrocracking units are often limited by design constraints of an existing vacuum gas oil hydrotreating units. Thus, the proper choice of catalyst(s) is critical when searching for optimum performance. Typical commercial distillate hydrocracking (DHC) catalysts contain both the hydrogenation (metal) and cracking (acid sites) functions required for service in existing desulfurization units.

11.2.2 FEEDSTOCKS AND HYDROGEN REQUIREMENTS

Processing heavier feedstocks poses many challenges to hydrocracking operations in terms of producing the desired products from various feedstocks (Figures 11.2 and 11.3). It is also important to consider various other aspects of the hydrocracking unit design to maximize catalyst utilization as well as pretreatment of the feedstock using a hydrotreating unit (Morel et al., 2009; Dziabala et al., 2011).

The hydrocracking process operating parameters (400°C–815°C; 750°F–1500°F and 1000–2000 psi) depend on the nature of the feedstock and the relative rates of the two competing reactions: hydrogenation and cracking. Under these conditions, heavy aromatic feedstock is converted into lighter products under a wide range of very high pressures and high temperatures. When the feedstock has a high paraffinic content, the primary function of hydrogen is to prevent the formation of polycyclic aromatic compounds. Hydrogenation also serves to convert sulfur and nitrogen compounds present in the feedstock to hydrogen sulfide and ammonia.

It is estimated that the refining industry will require more than 14 trillion ft^3 of hydrogen to meet processing requirements between 2010 and 2030 with the Asia Pacific region and the Middle East region representing approximately 40% of global requirements (Vauk et al., 2008). In any such scenario, hydrogen management is key (Patel et al., 2005; Luckwal and Mandal, 2009). The chemical hydrogen consumption for the production of ultra-low-sulfur diesel (ULSD) will be significantly higher than the consumption for low-sulfur diesel production (Morel et al., 2009). Several factors influence the amount of the increased hydrogen consumption, including (1) catalyst selection for low-sulfur diesel versus ultra-low sulfur diesel, (2) the amount of cracked material boiling over 325°C (620°F) in the feedstock, (3) unit pressure, and (4) the flow rate (LHSV).

Many refineries chose to use Co/Mo catalyst in the production of low-sulfur diesel. The Co/Mo catalyst provided an effective route to sulfur removal in lower pressure units with the added benefit of lower hydrogen consumption since the Co/Mo catalysts have lower hydrogenation activity compared to the Ni/Mo catalyst. Moreover, the Co/Mo catalysts actually have higher activity for direct hydrodesulfurization and less organic nitrogen inhibition. Refiners with feed that contains a high amount of high-boiling cracked material currently using Co/Mo catalyst will experience a large increase in hydrogen consumption when switching to Ni/Mo catalysts for ULSD production.

Hydrogen consumption increases significantly when producing a 10 ppm sulfur product compared to production of a 500 ppm sulfur product. Increases in chemical hydrogen consumption from 40% to 100% can be expected. In addition, the higher consumption results in higher heat release. The increased heat release requires more quench gas (as much as 950 ft^3/bbl) to limit the temperature rise.

Because of the operating temperatures and presence of hydrogen, the equipment must be capable of handling high-sulfur feedstocks and must be able to withstand the possibility of severe corrosion. When processing high-nitrogen feedstock, ammonia and hydrogen sulfide form ammonium hydrosulfide (bisulfide), which causes serious corrosion at temperatures below the water dew point.

When hydrogen availability is depleted by new hydroprocessing demands, the refinery needs to be flexible and alternative sources of hydrogen are needed (Vauk et al., 2008). Each refinery's configuration, crude slate, and new production requirements are different, so various hydrogen network optimization considerations must be considered (Patel et al., 2005; Luckwal and Mandal, 2009). Hydrogen network optimization is increasingly more important to maximizing higher-value transportation fuels production, to minimizing clean fuels investments, and to overall refinery profitability. Refiners must seek to unlock the hydrogen contained in refinery off gases, improve hydroprocessing network purities, and consider new hydrogen production options. The optimum solution for an overall refinery hydrogen network will continue to be complex and involve the screening of numerous options.

11.2.3 DESIGN IMPROVEMENTS

Operating severity increases with heavier feedstocks so catalyst loading and high-performance reactor internals are becoming even more important to get the most out of the catalyst. Furthermore, the importance of feed filtering, particulate, and metal trapping must be included within the design of the technology. Guard reactors are now (and will continue to be) recognized as being essential in the hydrocracking processes to protect catalysts in subsequent reactors from contaminants in feedstocks that are not previously hydrotreated—the purpose of the guard reactor is to convert organic sulfur and nitrogen compounds to hydrogen sulfide and ammonia and to reduce the metal content in the feed to the hydrocracking units (Table 11.3).

Reactor internals are a critical part of any hydroprocessing technology package. Maximum catalytic performance can now be achieved through good vapor/liquid mixing and distribution across the catalyst bed. Good distribution results in maximum catalyst utilization from start to end of run, maximizing product selectivity. The reactor internals that are used in a hydrocracker

TABLE 11.3
Feedstock Contaminants That Affect Hydrocracking Processes[a]

Contaminants	Effect on Catalyst	Mitigation	Process
Sulfur	Catalyst fouling Deactivation of active sites	Hydrodesulfurization	Hydroprocessing
Nitrogen	Adsorption of basic nitrogen Destruction of active sites	Hydrodemetallization	Hydroprocessing
Metals	Fouling of active sites Fouling of pores	Demetallization	Demet, Met-X
Particulate matter	Deactivation of active sites Pore plugging	Filter/pretreatment	Clay filtration/guard bed
Coke precursors	Formation of coke	Remove asphaltene constituents	Mild hydrocracking/ hydroprocessing
	Catalyst fouling Deactivation of active sites Pore plugging	Remove resin constituents	

[a] Also applicable to catalytic cracking processes.

can play an important role in determining its capital cost, as they can improve the volume available for catalyst and therefore can help to reduce the reactor size. Reactor internals also have a major influence on its performance as they can drive its onstream factor, utilization, cycle length, product yields, and even product quality. In fact, reactor internals should be custom designed and fabricated for each application to ensure maximum performance over the desired range of operating conditions—off-the-shelf units are not as popular as they were in the past. Consequently, reactor internals are being designed to (1) distribute gas and liquid uniformly, (2) minimize thermal instabilities, and (3) maximize reactor catalyst inventory and catalyst utilization. This is now recognized as being critical for applications with more stringent product specifications, such as ultra-low-sulfur diesel production.

Reactor temperature controls and safety instrumented systems are now used in the hydrocracking process found in many refineries. A significant change in feed flow rate can result in a temperature runaway due to rapid change of the hydrogen-to-hydrocarbon ratio. Significant changes in the feed flow rate are the result of failures in feed flow controllers and feed pumps. The temperature rise in this scenario is moderately fast, but recovery is possible through automatic and manual adjustments of quench rates and readjustment of feed flow rates. Excessive temperature of the reactor feed can also result in temperature runaway. Excessive temperature of the reactor feedstock is possible as the result of a failure of temperature control in the charge heater such that maximum firing occurs.

Due to the highly exothermic reactions, coupled with the high apparent activation energy (a measure of the sensitivity for the activity to temperature changes) in the second stage of hydrocracking, advanced control for the hydrocracker reactor has become essential for maintaining a reliable hydrocracking operation. Poor control in the reactor is still often reported—even with the implementation of the advanced control mechanism—due to poor reactor internals and the problematic overall control scheme implemented for the hydrocracking reactor.

The conventional control scheme using heater firing control has been found to be insufficient for the hydrocracker—attempting to maintain a constant reactor inlet temperature (a precursor for temperature runaway) with poor internals and deficient overall reactor control scheme leads to further operating problems. With a properly designed quench box and an improved reactor control scheme, the bed inlet temperature spread can be reduced to less than 3°C (5°F), and the temperature variation at the reactor inlet can be reduced to less than 1.5°C (3°F) with a low capital investment.

Reactor internals are a critical part of hydrocracking technology. Maximum catalytic performance can only be achieved through efficient vapor/liquid mixing and distribution across the catalyst bed. Good distribution results in maximum catalyst utilization from start to end of run, maximizing product selectivity and catalyst run length. Included in this area for improvement are (1) the inlet diffuser, (2) the liquid distribution tray, (3) the vapor/liquid distribution tray, (4) the catalyst support grid, (5) the quench zone, and (6) the outlet collector.

Options have also arisen for taking advantage of reduced post-riser cracking to improve product yields. A combination of higher reactor temperature, higher cat/oil ratio, higher feed rate, and/or poorer quality feed is typically employed. The types of modifications to the catalyst are also necessary (on a unit and feedstock basis) to complement these designs, particularly for revamp applications.

The concept of single-stage and double-stage hydroprocessing operations has long been recognized (Speight, 2011a, 2014) and, with the advent of heavier feedstocks as part of the reactor feed, is becoming a necessity. However, the performance of a hydroprocessing operation is not only determined by the number of stages and the catalyst loaded but also by the design of the reactor internals. Antifouling trays are employed to reduce pressure drop buildup and maximize unit run length. Increased efficiency and technologies to counteract fouling are particularly important when operators are processing increasingly difficult feedstocks, such as heavy feedstocks (Kunnas et al., 2010).

Most designs in commercial operation process a variety of feedstocks in multiple flow schemes. The heavier feed components in the blends cover heavy coker gas oil, heavy vacuum gas oil, light cycle oil, and deasphalted oil. These feedstock blends are processed in hydrocracking units with various objectives and flow schemes including single-stage once-through and recycle as well as two-stage and separate hydrotreating flow schemes.

In a single-stage unit, the feed is first hydrotreated and the reactor effluent goes through gas/liquid separation. The hydrocracking reactor effluent goes through gas/liquid separation and then to the fractionator. This configuration uses recycle for optimization of yield and processing severity. The flow scheme is designed to assure that high-quality product is produced in terms of ultra-low sulfur diesel with a cetane index over 55.

In hydrocrackers that process vacuum gas oils or other feeds with similar boiling ranges, the typical once-through conversion exceeds 60% w/w. If the unconverted oil is recycled, the overall conversion can exceed 95% w/w. As with mild hydrocracking, the unconverted bottoms are high-value oils, which usually are sent to fluid catalytic cracking units, lube plants, or olefin plants. For heavier feedstocks, conversions are much lower, especially in fixed-bed units.

The limitation in fixed-bed reactor is the catalyst bed poisoning with time. Catalyst life depends on the rate of deactivation by coke and metal deposits and sintering of the active phases. Information regarding the activity, selectivity, and deactivation of the individual catalyst is, therefore, highly desirable for optimizing reactor loading in the multicatalyst systems. These parameters have to be optimum for hydroprocessing operation, which can be achieved by properly matching the type of reactor and catalyst, along with properties of heavy feeds. However, in ebullated-bed units, the conversion of 565°C+ (1050°F+) residue can exceed 60% w/w.

Many hydrocrackers in the refineries operate in mild hydrocracking mode. For these units, the main objectives are to obtain a certain minimum conversion and to meet specific product properties such as sulfur content, density, and cetane number. Typical pressures are in the 850–1560 psig range. Typical conversion is 10%–20% for lower pressure units and 30%–50% for higher pressure units. The demand for refined products has increased to the extent that refiners desire larger hydrocracking reactors that can operate at higher pressures with design conditions that are even more severe. New feedstocks for refineries utilize more difficult to "crack" crudes; demand for reactors that can withstand higher temperatures (>450°C, >840°F); and higher hydrogen partial pressures likewise is increasing. Under such severe processing conditions, reactor vessels are constructed from low alloy chromium (Cr)–molybdenum (Mo) steel of various grades.

As an example of mild hydrocracking (and hydrotreating), the DOW Isotherming process provides a means to upgrade gas oil, heavy gas oil, coker gas oil, deasphalted oil, and FCC cycle oils. The products are low-sulfur, low-nitrogen FCC feed, low-sulfur gasoline–kerosene-type products, and low-nitrogen fuels and/or downstream feedstocks. In the process, fresh feedstock, after heat exchange, is combined with recycle product and hydrogen in a mixer internal to the reactor. The liquid feed with soluble hydrogen is fed to the Isotherming reactor/bed one where partial desulfurization, removal of metals, or even mild hydrocracking occurs. The stream is resaturated with additional hydrogen in a second mixer and fed to the second Isotherming reactor/bed where further mild hydrocracking takes place. The treated oil from the second Isotherming bed may then be fed to additional Isotherming beds to achieve the desired level of conversion. Treated oil from the last bed is recycled back to the inlet of bed one. This recycle stream delivers recycled hydrogen to the reactors and also acts as a heat sink; thus, a nearly isothermal reactor operation is achieved.

Removal of sulfur and nitrogen from feedstocks is affected by the chemical composition of supported molybdate catalysts. Cobalt and nickel, when added to these catalysts, have a promoting effect on these reactions. However, the relative rates always follow the same trend; that is, the hydrodesulfurization is the fastest reaction followed by hydrodenitrogenation. The hydrodenitrogenation (HDN) reaction rates are dependent on the catalyst, the concentration of the reactive species (organic nitrogen and hydrogen), and the reaction temperature. Ni/Mo catalysts generally have higher denitrogenation activity than the Co/Mo catalysts. The affinity of the organic nitrogen compounds for the catalytic sites has been reported as extremely high, and if the organic nitrogen concentration exceeds 20 ppm as much as 90% of the catalyst sites may be occupied.

Ammonia inhibits the hydrocracking catalyst activity, requiring higher operating temperatures to achieve target conversion, but this has been found to result in better liquid yields than would be the case if no ammonia were present. There is no interstage product separation in single-stage or series-flow operation. Two-stage hydrocrackers employ interstage product separation that removes hydrogen sulfide and ammonia and the second-stage hydrocracking catalyst is exposed to lower levels of these gases. However, caution is needed as some two-stage hydrocracker designs do result in high levels of hydrogen sulfide in the second stage.

Aromatics saturation can be achieved by the use of a pretreatment step. The main objectives for the pretreated catalyst are (1) the removal of organic nitrogen and sulfur from the feedstock to levels which allow the second-stage catalysts to better perform the hydrocracking function and (2) the initiation of the sequence of hydrocracking reactions by saturation of the aromatic compounds in the feedstock. Pretreated catalysts must have adequate activity to achieve both objectives within the operating limits of the unit (hydrogen partial pressure, temperature, and LHSV).

A hydrocracking step is necessary for obtaining high-quality fuels from Fischer–Tropsch wax. Isomerization is an important reaction that takes place during the hydroconversion process. The amount and the type of the isomers in the produced fuels heavily influence both cold flow properties and cetane number (Gamba et al., 2010). The reaction temperature and the space velocity exhibit a considerable impact on the conversion as well as on the *iso*-paraffin to *n*-paraffin ratio and on the boiling range of the hydrocracking product. At severe reactor conditions, increased hydrogenolysis activity of the base metal catalyst will be observed, resulting in an increased methane formation.

The quality of the diesel fuel product shows a strong dependence on the conversion achieved from hydrocracking. Cetane numbers of up to 80 can be recoded under low-conversion conditions due to lower *iso*-paraffin to *n*-paraffin ratios of the hydrocracking products, whereas high-conversion conditions lead to an increase in the *iso*-paraffin to *n*-paraffin ratio which leads to improved cold flow properties with cold filter plugging points down to or below −27°C (−17°F) (Olschar et al., 2007).

The original ebullated-bed process—the H-Oil process—has evolved into various configurations that have the potential to play major refinery roles up to and beyond the year 2020. The H-Oil$_{DC}$ process (H-Oil *distillate cracking process*, previously known as the T-Star process) is a specially engineered, ebullated-bed process for the treatment of vacuum gas oil, deasphalted oils, or coal-derived oils that typically contain metals, coke precursors, nitrogen, or suspended solids that cause rapid catalyst fouling and contamination. The ebullated-bed reactor system overcomes problems encountered with fixed-bed reactors when processing these difficult feedstocks or when high processing severity is required. The bed of catalyst is fluidized (ebullated) by the lift of hydrogen and feed oil and recycled reactor liquid. The ebullated catalyst is well mixed and can be added and withdrawn from the reactor while at operating conditions. Because of the ability to replace the catalyst bed incrementally, the reactor can operate reliably for 4–5 years between turnarounds, with high yields of middle distillate and constant product properties over time. Process conversion levels range from low conversion to above 80% w/w depending on the refinery scheme. The catalysts that have been developed for this process provide high refining activity, metal tolerance, and selective conversion to the diesel fraction. In addition, the catalysts have excellent mechanical properties that allow high-temperature operation and minimal fines production.

Because of the ability to replace the catalyst bed incrementally, the H-Oil$_{DC}$ reactor can operate indefinitely—typically, 4–5 years between turnarounds to coincide with the inspection and maintenance schedule for a fluid catalytic cracking unit. The difficult processing requirements which result from stricter environmental regulations and the processing of heavy feedstocks makes H-Oil$_{DC}$ a preferred choice for pretreatment of fluid catalytic cracker feedstocks. Axens has already eight commercially operating ebullated-bed reactor units.

Two additional advanced generation H-Oil processes are also available: H-Oil$_{HCC}$ and H-Oil$_{RC.}$ The H-Oil$_{HCC}$ is a heavy feedstock crude conversion process that produces synthetic crude oil. The objective of the unit is to enable just enough conversion to reduce viscosity and increase stability so that the product can be readily transported to an upgrading center.

In the H-Oil$_{RC}$ process, the classical H-Oil process has been significantly upgraded to increase conversion levels, increase product stability, and reduce costs. Among the improvements made to the traditional technology are the integration of an interstage separator between reactors in series and the application of cascade catalyst utilization. The process is suitable for feedstocks having (1) high metal content, (2) a high coke-forming propensity, and (3) high asphaltene contents, which automatically coordinates with a high coke-forming propensity. The process can have two different objectives: (1) at high conversion, to produce stable products or (2) at moderate conversion, to produce a synthetic crude oil. Different catalysts are available as a function of the feedstock and the required objectives. A unit can operate for 3-year run lengths at constant catalyst activity with conversion in the 50%–80% v/v range with hydrodesulfurization as high as 85%.

The shortage of good source of lubricant base oil is creating a high demand for feedstocks and units that produce such materials. Generally, very high viscosity index base oils are produced by severe hydrocracking of vacuum distillate fractions in a fuels hydrocracker. Another route for very high viscosity index base oil production is hydroisomerization, in which slack wax, produced by solvent dewaxing, is converted to branched-chain paraffins under hydroisomerization conditions with an appropriate catalyst. The conversion rate of the process is approximately 80%–85%, and the residual wax is typically removed by solvent dewaxing to improve the low-temperature fluidity of the finished product.

Mild hydrocracking technology and catalysts enable a medium conversion of heavier feedstocks to lighter and more valuable products. The Topsøe mild hydrocracking technology portfolio includes staged partial conversion and back-end shift. Staged partial conversion is a new pretreatment technology designed to produce low-sulfur FCC feed to allow ultra-low sulfur diesel production without gasoline posttreatment (Section 2.2). Back-end shift technology significantly reduces the distillation temperature by selective hydrocracking of the heavy hydrocarbons present in the back-end distillation with high diesel yields and moderate hydrogen consumption.

For the heavy feedstocks, which will increase in amounts in terms of hydrocracking feedstocks, reactor designs will continue to focus on online catalyst addition and withdrawal. Fixed-bed designs have suffered from (1) mechanical inadequacy when used for the heavier feedstocks and (2) short catalyst lives—6 months or less—even though large catalyst volumes are used (LHSV typically of 0.5–1.5). Refiners will attempt to overcome these shortcomings by innovative designs, allowing better feedstock flow and catalyst utilization or online catalyst removal. For example, the onstream catalyst replacement (OCR) process, in which a lead, moving-bed reactor is used to demetallize heavy feed ahead of the fixed-bed hydrocracking reactors, has seen some success. The OCR process enables refiners to process heavy feedstocks with more than 400 ppm metals (Ni + V) or to achieve deeper desulfurization. The life of the downstream catalyst is improved substantially and problems from pressure drop buildup are reduced.

11.3 CATALYSTS

Several types of catalysts are used in hydrocracking process, and the catalysts combine cracking activity and hydrogenation activity to achieve the conversion of specific feedstocks into desirable products. Hydrocracking reactions require a dual-function catalyst with high cracking and hydrogenation activities (Katzer and Sivasubramanian, 1979). The former activity (high cracking activity) is provided by an acidic support or by a clay support whereas the hydrogenation activity is provided by metals on the support. The catalyst base, such as acid-treated clay, usually supplies the cracking function or alumina or silica–alumina that is used to support the hydrogenation function supplied by metals, such as nickel, tungsten, platinum, and palladium. These highly acid catalysts are very sensitive to nitrogen compounds in the feed, which break down the conditions of reaction to give ammonia and neutralize the acid sites. As many heavy gas oils contain substantial amounts of nitrogen (up to approximately 2500 ppm), a purification stage is frequently required. Denitrogenation and desulfurization can be carried out using cobalt–molybdenum or nickel–cobalt–molybdenum on alumina or silica–alumina.

Acid sites (crystalline zeolite, amorphous silica–alumina, mixture of crystalline zeolite and amorphous oxides) provide cracking activity. Metals (noble metal such as palladium and platinum or nonnoble metal sulfides such as molybdenum, tungsten, cobalt, or nickel) provide hydrogenation–dehydrogenation activity. These metals catalyze the hydrogenation of feedstocks making them more reactive for cracking and heteroatom removal as well as reducing the coke yield. Zeolite-based hydrocracking catalysts have the following advantages: (1) greater acidity resulting in greater cracking activity, (2) better thermal/hydrothermal stability, (3) better naphtha selectivity, (4) better resistance to nitrogen and sulfur compounds, (5) low coke-forming tendency, and (6) easy regenerability.

Palladium sulfide and promoted group VI sulfides (nickel molybdenum or nickel tungsten) provide the hydrogenation function. These active compositions saturate aromatics in the feed, saturate olefins formed in the cracking, and protect the catalysts from poisoning by coke. Zeolites or amorphous silica–alumina provide the cracking functions. The zeolites are usually type Y (faujasite) ion exchanged to replace sodium with hydrogen and make up 25%–50% of the catalysts. Pentasils (silicalite or ZSM-5) may be included in dewaxing catalysts.

Hydrocracking catalysts, such as nickel (5% by weight) on silica–alumina, work best on feedstocks that have been hydrofined to low nitrogen and sulfur levels. The nickel catalyst then operates well at 350°C–370°C (660°F–700°F) and a pressure of approximately 1500 psi to give good conversion of feed to lower-boiling liquid fractions with minimum saturation of single-ring aromatics and a high *iso*-paraffin to *n*-paraffin ratio in the lower-molecular-weight paraffins.

The catalyst used in a single-stage process comprises a hydrogenation function in combination with a strong cracking function. Sulfided metals such as cobalt, molybdenum, and nickel provide the hydrogenation function. An acidic support, usually alumina, attends to the cracking function. Nitrogen compounds and ammonia produced by hydrogenation interfere with acidic activity of the catalyst. In the cases where high/full conversion is required, the reaction temperatures and run lengths of interest in commercial operation can no longer be adhered to. Moreover, conversion asymptotes out with increasing hydrogen pressure (Speight, 2014) so more hydrogen in the reactor is not the answer. In fact, it becomes necessary to switch to a different reactor bed system or to a multistage process, in which the cracking reaction mainly takes place in an additional reactor. The operating temperature influences reaction selectivity since the activation energy for hydrotreating reactions is much lower than the activating energy for the hydrocracking reaction. Therefore, raising the temperature in a heavy feedstock hydrotreater increases the extent of hydrocracking relative to hydrotreating, which also increases the hydrogen consumption.

Multistage reaction catalysts take advantage of staged reactions with different catalytic attributes much the same way that staged hydrotreating catalyst loading permits different reaction zones in a fixed-bed reactor vessel. The concept of staged reactions is not new to the refining industry, but its application to a circulating system such as the fluid catalytic cracking system is a step forward in catalyst technology. The multistage reaction catalyst platform uses existing catalyst technologies and combines two or more existing functionalities within the same catalyst particle. The location of the various stages can be specifically engineered to achieve maximum value for the refiner. This staging approach can be applied to allow processing of heavier feedstocks or to maximize specific product yields.

The transition metal sulfides such as molybdenum (Mo), cobalt (Co), and nickel (Ni) are still the industry favorites, because of their excellent hydrogenation, hydrodesulfurization (HDS), and hydrodenitrogenation (HDN) activities, as well as their availability and cost. However, greater attention will be paid to the size of the particles, pore volume and size distribution, pore diameter and the shape of the particles to maximize utilization of the catalyst. Thus, the future challenge for refiners will be to obtain higher conversion of heavier and more refractive feedstocks, to make cleaner transportation fuels in larger quantities, while reducing the emissions of greenhouse gases. One of the main constraints to meeting these objectives is that deeper conversion of heavier oil tends to result in reduced stability of the unconverted product, leading to higher fouling rates in the reactors and downstream equipment and thus reduced reliability of the process.

Moreover, the increasing importance of hydrodesulfurization (HDS) and hydrodenitrogenation (HDN) in petroleum processing in order to produce clean-burning fuels has led to a surge of research on the chemistry and engineering of heteroatom removal, with sulfur removal being the most prominent focus. Most of the earlier works are focused on (1) catalyst characterization by physical methods, (2) on low-pressure reaction studies of model compounds having relatively high reactivity, (3) on process development, or (4) on cobalt–molybdenum (Co–Mo) catalysts, nickel–molybdenum (Ni–Mo) catalysts, or nickel–tungsten (Ni–W) catalysts supported on alumina, often doped by fluorine or phosphorus.

The need to develop catalysts that can carry out deep hydrodesulfurization and deep hydrodenitrogenation has become even more pressing in view of recent environmental regulations limiting the amount of sulfur and nitrogen emissions. The development of a new generation of catalysts to achieve this objective of low nitrogen and sulfur levels in the processing of different feedstocks presents an interesting challenge for catalyst development. Recent developments (from the catalyst and process aspects) have focused on improvement of (1) the desulfurization level, (2) the denitrogenation level, (3) the run length, and (4) the conversion level of fixed-bed hydroprocessing units. These improvements include the development of a multiple swing-reactor system as well as of complex associations of guard-bed materials and catalysts with particle size, activity, pore size, and shape grading.

The deposition of coke and metals onto the catalyst diminishes the cracking activity of hydrocracking catalysts. Basic nitrogen plays a major role because of the susceptibility of such compounds for the catalyst and their predisposition to form coke (Speight, 2000, 2014). However, zeolite catalysts can operate in the presence of substantial concentrations of ammonia, in marked contrast to silica–alumina catalysts, which are strongly poisoned by ammonia. Similarly, sulfur-containing compounds in a feedstock adversely affect the noble metal hydrogenation component of hydrocracking catalysts. These compounds are hydrocracked to hydrogen sulfide, which converts the noble metal to the sulfide form. The extent of this conversion is a function of the hydrogen and hydrogen sulfide partial pressures.

Removal of sulfur from the feedstock results in a gradual increase in catalyst activity, returning almost to the original activity level. As with ammonia, the concentration of the hydrogen sulfide can be used to control precisely the activity of the catalyst. Nonnoble metal-loaded zeolite catalysts have an inherently different response to sulfur impurities since a minimum level of hydrogen sulfide is required to maintain the nickel–molybdenum and nickel–tungsten in the sulfide state.

Hydrodenitrogenation is more difficult to accomplish than hydrodesulfurization, but the relatively smaller amounts of nitrogen-containing compounds in conventional crude oil made this of little concern to refiners. However, the trend to heavier feedstocks in refinery operations, which are richer in nitrogen than the conventional feedstocks, has increased the awareness of refiners to the presence of nitrogen compounds in crude feedstocks. For the most part, however, hydrodesulfurization catalyst technology has been used to accomplish hydrodenitrogenation although such catalysts are not ideally suited to nitrogen removal. However, in recent years, the limitations of hydrodesulfurization catalysts when applied to hydrodenitrogenation have been recognized and there has been the need to manufacture catalysts more specific to nitrogen removal.

The character of the hydrotreating processes is chemically very simple since they essentially involve removal of sulfur and nitrogen as hydrogen sulfide and ammonia, respectively:

$$S_{feedstock} + H_2 \rightarrow H_2S$$

$$2N_{feedstock} + 3H_2 \rightarrow 2NH_3$$

However, nitrogen is the most difficult contaminant to remove from feedstocks, and processing conditions are usually dictated by the requirements for nitrogen removal.

In general, any catalyst capable of participating in hydrogenation reactions may be used for hydrodesulfurization. The sulfides of hydrogenating metals are particularly used for hydrodesulfurization, and catalysts containing cobalt, molybdenum, nickel, and tungsten are widely used on a commercial basis. *Hydrotreating catalysts* are usually cobalt–molybdenum catalysts and under the conditions whereby nitrogen removal is accomplished, desulfurization usually occurs as well as oxygen removal. Indeed, it is generally recognized that fullest activity of the hydrotreating catalyst is not reached until some interaction with the sulfur (from the feedstock) has occurred, with part of the catalyst metals converted to the sulfides. Too much interaction may of course lead to catalyst deactivation.

The reactions of hydrocracking require a dual-function catalyst with high cracking and hydrogenation activities. The catalyst base, such as acid-treated clay, usually supplies the cracking function or alumina or silica–alumina that is used to support the hydrogenation function supplied by metals, such as nickel, tungsten, platinum, and palladium. These highly acid catalysts are very sensitive to nitrogen compounds in the feed, which break down the conditions of reaction to give ammonia and neutralize the acid sites. As many heavy gas oils contain substantial amounts of nitrogen (up to approximately 2500 ppm), a purification stage is frequently required. Denitrogenation and desulfurization can be carried out using cobalt–molybdenum or nickel–cobalt–molybdenum on alumina or silica–alumina.

Hydrocracking catalysts typically contain separate hydrogenation and cracking functions. Palladium sulfide and promoted group VI sulfides (nickel molybdenum or nickel tungsten) provide the hydrogenation function. These active compositions saturate aromatics in the feed, saturate olefins formed in the cracking, and protect the catalysts from poisoning by coke. Zeolites or amorphous silica–alumina provide the cracking functions. The zeolites are usually type Y (faujasite) ion exchanged to replace sodium with hydrogen and make up 25%–50% of the catalysts. Pentasils (silicalite or ZSM-5) may be included in dewaxing catalysts.

Hydrocracking catalysts, such as nickel (5% by weight) on silica–alumina, work best on feedstocks that have been hydrofined to low nitrogen and sulfur levels. The nickel catalyst then operates well at 350°C–370°C (660°F–700°F) and a pressure of approximately 1500 psi to give good conversion of feed to lower-boiling liquid fractions with minimum saturation of single-ring aromatics and a high *iso*-paraffin to *n*-paraffin ratio in the lower-molecular-weight paraffins.

The poisoning effect of nitrogen can be offset to a certain degree by operation at a higher temperature. However, the higher temperature tends to increase the production of material in the methane (CH_4) to butane (C_4H_{10}) range and decrease the operating stability of the catalyst so that it requires more frequent regeneration. Catalysts containing platinum or palladium (approximately 0.5% wet) on a zeolite base appear to be somewhat less sensitive to nitrogen than are nickel catalysts, and successful operation has been achieved with feedstocks containing 40 ppm nitrogen. This catalyst is also more tolerant of sulfur in the feed, which acts as a temporary poison, the catalyst recovering its activity when the sulfur content of the feed is reduced.

On such catalysts as nickel or tungsten sulfide on silica–alumina, isomerization does not appear to play any part in the reaction, as uncracked normal paraffins from the feedstock tend to retain their normal structure. Extensive splitting produces large amounts of low-molecular-weight (C_3–C_6) paraffins, and it appears that a primary reaction of paraffins is catalytic cracking followed by hydrogenation to form *iso*-paraffins. With catalysts of higher hydrogenation activity, such as platinum on silica–alumina, direct isomerization occurs. The product distribution is also different, and the ratio of low- to intermediate-molecular-weight paraffins in the breakdown product is reduced.

In addition to the chemical nature of the catalyst, the physical structure of the catalyst is also important in determining the hydrogenation and cracking capabilities, particularly for heavy feedstocks. When gas oils and heavy feedstocks are used, the feedstock is present as liquids under the conditions of the reaction. Additional feedstock and the hydrogen must diffuse through this liquid before reaction can take place at the interior surfaces of the catalyst particle.

At high temperatures, reaction rates can be much higher than diffusion rates and concentration gradients can develop within the catalyst particle. Therefore, the choice of catalyst porosity is an important parameter. When feedstocks are to be hydrocracked to liquefied petroleum gas and naphtha, pore diffusion effects are usually absent. High surface area (approximately 300 m$_2$/g) and low to moderate porosity (from 12 Å pore diameter with crystalline acidic components to 50 Å or more with amorphous materials) catalysts are used. With reactions involving high-molecular-weight feedstocks, pore diffusion can exert a large influence, and catalysts with large pore diameters are necessary for more efficient conversion.

Aromatic hydrogenation in petroleum refining may be carried out over supported metal or metal sulfide catalysts depending on the sulfur and nitrogen levels in the feedstock. For hydrorefining of feedstocks that contain appreciable concentrations of sulfur and nitrogen, sulfided nickel–molybdenum (Ni–Mo), nickel–tungsten (Ni–W), or cobalt–molybdenum (Co–Mo) on alumina (γ-A$_2$-O$_3$) catalysts are generally used, whereas supported noble metal catalysts have been used for sulfur- and nitrogen-free feedstocks. Catalysts containing noble metals on Y-zeolites have been reported to be more sulfur tolerant than those on other supports. Within the series of cobalt- or nickel-promoted group VI metal (Mo or W) sulfides supported on γ-Al$_2$O$_3$, the ranking for hydrogenation is

$$Ni–W > Ni–Mo > Co–Mo > Co–W$$

Nickel–tungsten (Ni–W) and nickel–molybdenum (Ni–Mo) on Al$_2$O$_3$ catalysts are widely used to reduce sulfur, nitrogen, and aromatics levels in petroleum fractions by hydrotreating.

Molybdenum sulfide (MoS$_2$), usually supported on alumina, is widely used in petroleum processes for hydrogenation reactions. It is a layered structure that can be made much more active by addition of cobalt or nickel. When promoted with cobalt sulfide (CoS), making what is called *cobalt–moly* catalysts, it is widely used in hydrodesulfurization (HDS) processes. The nickel sulfide (NiS)-promoted version is used for hydrodenitrogenation (HDN) as well as hydrodesulfurization (HDS). The closely related tungsten compound (WS$_2$) is used in commercial hydrocracking catalysts. Other sulfides (iron sulfide, FeS, chromium sulfide, Cr$_2$S$_3$, and vanadium sulfide, V$_2$S$_5$) are also effective and used in some catalysts. A valuable alternative to the base metal sulfides is palladium sulfide (PdS). Although it is expensive, palladium sulfide forms the basis for several very active catalysts while clay minerals which are also used as cracking catalysts for heavy feedstocks, especially for demetallization of heavy crude oil are much cheaper.

The choice of hydrogenation catalyst depends on what the catalyst designer wishes to accomplish. In catalysts to make naphtha, for instance, vigorous cracking is needed to convert a large fraction of the feed to the kinds of molecules that will make a good gasoline blending stock. For this vigorous cracking, a vigorous hydrogenation component is needed. Since palladium is the most active catalyst for this, the extra expense is warranted. On the other hand, many refiners wish only to make acceptable diesel, a less demanding application. For this, the less expensive molybdenum sulfides are adequate.

The cracking reaction results from attack of a strong acid on a paraffinic chain to form a carbonium ion (carbocation, e.g., R$^+$) (Dolbear, 1998). Strong acids come in two fundamental types, Brønsted and Lewis acids. *Brønsted acids* are the familiar proton-containing acids; *Lewis acids* are a broader class including inorganic and organic species formed by positively charged centers. Both kinds have been identified on the surfaces of catalysts; sometimes both kinds of sites occur on the same catalyst. The mixture of Brønsted and Lewis acids sometimes depends on the level of water in the system.

Examples of Brønsted acids are the familiar proton-containing species such as sulfuric acid (H$_2$SO$_4$). Acidity is provided by the very active hydrogen ion (H$^+$), which has a very high positive charge density. It seeks out centers of negative charge such as the pi electrons in aromatic centers. Such reactions are familiar to organic chemistry students, who are taught that bromination of aromatics takes place by attack of the bromonium ion (Br$^+$) on such a ring system. The proton in strong acid systems behaves in much the same way, adding to the pi electrons and then migrating to a site of high electron density on one of the carbon atoms. These acids all have high positive charge densities. Examples are aluminum chloride ($AlCl_3$) and bromonium ion (Br^+). Such strong positive species have become known as Lewis acids. This class obviously includes proton acids, but the latter are usually designated Brønsted acids in honor of the Danish chemist J.N. Brønsted, who contributed greatly to the understanding of the thermodynamics of aqueous solutions.

In reactions with hydrocarbons, both Lewis and Brønsted acids can catalyze cracking reactions. For example, the proton in Brønsted acids can add to an olefinic double bond to form a carbocation. Similarly, a Lewis acid can abstract a hydride from the corresponding paraffin to generate the same intermediate (Dolbear, 1998). Although these reactions are written to show identical intermediates in the two reactions, in real catalytic systems the intermediates would be different. This is because the carbocations would probably be adsorbed on surface sites that would be different in the two kinds of catalysts.

Zeolites and amorphous silica–alumina provide the cracking function in hydrocracking catalysts. Both of these have similar chemistry at the molecular level, but the crystalline structure of the zeolites provides higher activities and controlled selectivity not found in the amorphous materials.

Chemists outside the catalyst field are often surprised that a solid can have strong acid properties. In fact, many solid materials have acid strength matching that of concentrated sulfuric acid. Some specific examples are (1) amorphous silica–alumina, SiO_2/Al_2O_3, (2) zeolites, (3) activated acid-leached clay minerals, (4) aluminum chloride ($AlCl_3$) and many related metal chlorides, (5) amorphous silica magnesia compounds (SiO_2/MgO), (6) chloride-promoted alumina ($Al_2O_3 \cdot Cl$), and (7) phosphoric acid supported on silica gel, H_3PO_4/SiO_2. Each of these is applied in one or more commercial catalysts in the petroleum refining industry. For commercial hydrocracking catalysts, only zeolites and amorphous silica–alumina are used commercially.

In 1756, Baron Axel F. Cronstedt, a Swedish mineralogist, made the observation that certain minerals, when they were heated sufficiently, bubbled as if they were boiling. He called the substances *zeolites* (from the Greek *zeo*, to boil, and *lithos*, stone), which are now known to consist primarily of silicon, aluminum, and oxygen and to host an assortment of other elements. In addition, zeolites are highly porous crystals veined with submicroscopic channels. The channels contain water (hence the bubbling at high temperatures), which can be eliminated by heating (combined with other treatments) without altering the crystal structure (Occelli and Robson, 1989).

Typical naturally occurring zeolites include *analcite* (also called *analcime*) $Na(AlSi_2O_6)$ and *faujasite* $Na_2Ca(AlO_2)2(SiO_2)4 \cdot H_2O$ that is the structural analog of the synthetic *zeolite X* and *zeolite Y*. *Sodalite* ($Na_8[(Al_2O_2)_6(SiO_2)_6]Cl_2$) contains the truncated octahedral structural unit known as the *sodalite cage* that is found in several zeolites. The corners of the faces of the cage are defined by either four or six Al/Si atoms, which are joined together through oxygen atoms. The zeolite structure is generated by joining *sodalite* cages through the four-Si/Al rings, so enclosing a cavity or *super cage* bounded by a cube of eight *sodalite* cages and readily accessible through the faces of that cube (*channels* or *pores*). Joining sodalite cages together through the six-Si/Al faces generates the structural frameworks of faujasite, zeolite X, and zeolite Y. In zeolites, the effective width of the pores is usually controlled by the nature of the cation (M^+ or M^{2+}).

Natural zeolites form hydrothermally (e.g., by the action of hot water on volcanic ash or lava), and synthetic zeolites can be made by mixing solutions of aluminates and silicates and maintaining the resulting gel at temperatures of 100°C (212°F) or higher for appropriate periods. *Zeolite-A* can form at temperatures below 100°C (212°F), but most zeolite syntheses require hydrothermal conditions (typically 150°C/300°F at the appropriate pressure). The reaction mechanism appears to involve dissolution of the gel and precipitation as the crystalline zeolite and the identity of the zeolite produced depends on the composition of the solution. Aqueous alkali metal hydroxide solutions favor zeolites with relatively high aluminum contents, while the presence of organic molecules such as amines or alcohols favors highly siliceous zeolites such as *silicalite* or ZSM-5. Various tetra-alkyl ammonium cations favor the formation of certain specific zeolite structures and are known as *template* ions, although it should not be supposed that the channels and cages form simply by the wrapping of alumina–silica fragments around suitably shaped cations.

Zeolite catalysts have also found use in the refining industry during the last two decades. Like the silica–alumina catalysts, zeolites also consist of a framework of tetrahedrons usually with a silicon atom or an aluminum atom at the center. The geometric characteristics of the zeolites are responsible for their special properties, which are particularly attractive to the refining industry (DeCroocq, 1984). Specific zeolite catalysts have shown up to 10,000 times more activity than the so-called conventional catalysts in specific cracking tests. The mordenite-type catalysts are particularly worthy of mention since they have shown up to 200 times greater activity for hexane cracking in the temperature range of 360°C–400°C (680°F–750°F).

Other zeolite catalysts have also shown remarkable adaptability to the refining industry. For example, the resistance to deactivation of the type Y zeolite catalysts containing either noble or nonnoble metals is remarkable, and catalyst life of up to 7 years has been obtained commercially in processing heavy gas oils in the Unicracking-JHC processes. Operating life depends on the nature of the feedstock, the severity of the operation, and the nature and extent of operational upsets. Gradual catalyst deactivation in commercial use is counteracted by incrementally raising the operating temperature to maintain the required conversion per pass. The more active a catalyst, the lower is the temperature required. When processing for naphtha, lower operating temperatures have the additional advantage that less of the feedstock is converted to *iso*-butane.

Any given zeolite is distinguished from other zeolites by structural differences in its unit cell, which is a tetrahedral structure arranged in various combinations. Oxygen atoms establish the four vertices of each tetrahedron, which are bound to, and enclose, either a silicon (Si) or an aluminum (Al) atom. The vertex oxygen atoms are each shared by two tetrahedrons, so that every silicon atom or aluminum atom within the tetrahedral cage is bound to four neighboring caged atoms through an intervening oxygen. The number of aluminum atoms in a unit cell is always smaller than, or at most equal to, the number of silicon atoms because two aluminum atoms never share the same oxygen.

The aluminum is actually in the ionic form and can readily accommodate electrons donated from three of the bound oxygen atoms. The electron donated by the fourth oxygen imparts a negative, or anionic, charge to the aluminum atom. This negative charge is balanced by a cation from the alkali metal or the alkaline earth groups of the periodic table. Such cations are commonly sodium, potassium, calcium, or magnesium. These cations play a major role in many zeolite functions and help to attract polar molecules, such as water. However, the cations are not part of the zeolite framework and can be exchanged for other cations without any effect on crystal structure.

Zeolites provide the cracking function in many hydrocracking catalysts, as they do in fluid catalytic cracking catalysts. The zeolites are crystalline aluminosilicates, and in almost all commercial catalysts today, the zeolite used is faujasite. Pentasil zeolites, including silicalite and ZSM-5, are also used in some catalysts for their ability to crack long-chain paraffins selectively.

Typical levels are 25%–50% by wt. zeolite in the catalysts, with the remainder being the hydrogenation component and a silica (SiO_2) or alumina (Al_2O_3) binder. Exact recipes are guarded as trade secrets.

Crystalline zeolite compounds provide a broad family of solid acid catalysts. The chemistry and structures of these solids are beyond the scope of this book. What is important here is that the zeolites are not acidic as crystallized. They must be converted to acidic forms by ion exchange processes. In the process of doing this conversion, the chemistry of the crystalline structure is often changed. This complication provides tools for controlling the catalytic properties, and much work has been done on understanding and applying these reactions as a way to make catalysts with higher activities and more desirable selectivity.

As an example, the zeolite faujasite crystallizes with the composition $SiO_2(NaAlO_2)_x(H_2O)_y$. The ratio of silicon to aluminum, expressed here by the subscript x, can be varied in the crystallization from 1 to greater than 10. What does not vary is the total number of silicon and aluminum atoms per unit cell, 192. For legal purposes to define certain composition of matter patents, zeolites with a ratio of 1–1.5 are called typex; those with ratio greater than 1.5 are type y.

Both silicon and aluminum in zeolites are found in tetrahedral oxide sites. The four oxides are shared with another silicon or aluminum (except that two aluminum ions are never found in adjacent, linked tetrahedral). Silicon with a plus four charge balances exactly half of the charge of the oxide ions it is linked to; since all of the oxygens are shared, silicon balances all of the charge around it and is electrically neutral. Aluminum, with three positive charges, leaves one charge unsatisfied. Sodium neutralizes this charge.

The sodium, as expected from its chemistry, is not linked to the oxides by covalent bonds as the silicon and aluminum are. The attraction is simply *ionic*, and sodium can be replaced by other cations by ion exchange processes. In extensive but rarely published experiments, virtually every metallic and organic cation has been exchanged into zeolites in studies by catalyst designers.

The most important ion exchanged for sodium is the proton. In the hydrogen ion form, faujasite zeolites are very strong acids, with strengths approaching that of oleum. Unfortunately, direct exchange using mineral acids such as hydrochloric acid is not practical. The acid tends to attack the silica–alumina network, in the same way that strong acids attack clays in the activation processes developed by Houdry. The technique adopted to avoid this problem is indirect exchange, beginning with exchange of ammonium ion for the sodium. When heated to a few hundred degrees, the ammonium decomposes, forming gaseous ammonia and leaving behind a proton:

$$R^- NH_4^+ \rightarrow R^- H^+ + NH_3 \uparrow$$

The step is accompanied by a variety of solid-state reactions that can change the zeolite structure in subtle but in important ways. This chemistry and the related structural alterations have been described in many articles.

While zeolites provided a breakthrough that allowed catalytic hydrocracking to become commercially important, continued advances in the manufacture of amorphous silica–alumina made these materials competitive in certain kinds of applications. This was important, because patents controlled by Unocal and Exxon dominated the application of zeolites in this area. Developments in amorphous catalysts by Chevron and UOP allowed them to compete actively in this area.

Typical catalysts of this type contain 60–80 wt% of the silica–alumina, with the remainder being the hydrogenation component. The compositions of these catalysts are closely held secrets. Over the years, broad ranges of silica/alumina molar ratios have been used in various cracking applications, but silica is almost always in excess for high acidity and stability. A typical level might be 25 wt% alumina (Al_2O_3).

Amorphous silica–alumina is made by a variety of precipitation techniques. The whole class of materials traces its beginnings to silica gel technology, in which sodium silicate is acidified to precipitate the hydrous silica–alumina sulfate; sulfuric acid is used as some or all of the acid for this precipitation, and a mixed gel is formed. The properties of this gel, including acidity and porosity, can be varied by changing the recipe—concentrations, order of addition, pH, temperature, aging time, and the like. The gels are isolated by filtration and washed to remove sodium and other ions.

Careful control of the precipitation allows the pore size distributions of amorphous materials to be controlled, but the distributions are still much broader than those in the zeolites. This limits the activity and selectivity. One effect of the reduced activity has been that these materials have been applied only in making middle distillates: diesel and turbine fuels. At higher process severities, the poor selectivity results in production of unacceptable amounts of methane (CH_4) to butane (C_4H_{10}) hydrocarbons.

Hydrocarbons, especially aromatic hydrocarbons, can react in the presence of strong acids to form coke. This coke is a complex polynuclear aromatic material that is low in hydrogen. Coke can deposit on the surface of a catalyst, blocking access to the active sites and reducing the activity of the catalyst. Coke poisoning is a major problem in fluid catalytic cracking catalysts, where coked catalysts are circulated to a fluidized bed combustor to be regenerated. In hydrocracking, coke deposition is virtually eliminated by the catalyst's hydrogenation function.

However, the product referred to as *coke* is not a single material. The first products deposited are tarry deposits that can, with time and temperature, continue to become more complex. In a well-designed hydrocracking system, the hydrogenation function adds hydrogen to the tarry deposits. This reduces the concentration of coke precursors on the surface. There is, however, a slow accumulation of coke that reduces activity over a 1- to 2-year period. Refiners respond to this slow reduction in activity by raising the average temperature of the catalyst bed to maintain conversions. Eventually, however, an upper limit to the allowable temperature is reached and the catalyst must be removed and regenerated.

Catalysts carrying coke deposits can be regenerated by burning off the accumulated coke. This is done by service in rotary or similar kilns rather than leaving catalysts in the hydrocracking reactor, where the reactions could damage the metals in the walls. Removing the catalysts also allows inspection and repair of the complex and expensive reactor internals, discussed in the following. Regeneration of a large catalyst charge can take weeks or months, so refiners may own two catalyst loads, one in the reactor, one regenerated and ready for reload. The thermal reactions also convert the metal sulfide hydrogenation functions to oxides and may result in agglomeration. Excellent progress has been made since the 1970s in regenerating hydrocracking catalysts; similar regeneration of hydrotreating catalysts is widely practiced.

After combustion to remove the carbonaceous deposits, the catalysts are treated to disperse active metals. Vendor documents claim more than 95% recovery of activity and selectivity in these regenerations. Catalysts can undergo successive cycles of use and regeneration, providing long functional life with these expensive materials.

Hydrocracking allows refiners the potential to balance fuel oil supply and demand by adding vacuum gas oil cracking capacity. Situations where this is the case include (1) refineries with no existing vacuum gas oil cracking capacity, (2) refineries with more vacuum gas oil available than vacuum gas oil conversion capacity, (3) refineries where addition of heavy feedstock conversion capacity has resulted in production of additional cracking feedstocks boiling in the vacuum gas oil range (e.g., coker gas oil), and (4) refineries that have one of the two types of vacuum gas oil conversion units but could benefit from adding the second type. In some cases, a refiner might add both gas oil cracking and heavy feedstock conversion capacity simultaneously.

Those refiners who do choose gas oil cracking as part of their strategy for balancing residual fuel oil supply and demand must decide whether to select a hydrocracker or a fluid catalytic cracking unit. Although the two processes have been compared vigorously over the years, neither process has evolved to be the universal choice for gas oil cracking. Both processes have their advantages and disadvantages, and process selection can be properly made only after careful consideration of many case-specific factors. Among the most important factors are (1) product slate required, (2) amount of flexibility required to vary the product slate, (3) product quality (specifications) required, and (4) the need to integrate the new facilities in a logical and cost-effective way with any existing facilities.

The type of catalyst used can influence the product slate obtained. For example, for a mild hydrocracking operation at constant temperature, the selectivity of the catalyst varies from approximately 65% to approximately 90% by volume. Indeed, several catalytic systems have now been developed with a group of catalysts specifically for mild hydrocracking operations. Depending on the type of catalyst, they may be run as a single catalyst or in conjunction with a hydrotreating catalyst. Insight into catalyst nanostructures is leading to the development of high-activity catalysts, which provide solutions and designs to meet many product specifications. In addition, such insights are leading to optimization of hydrocracker units with respect to yield structure, product properties, throughput, and onstream efficiency, resulting in improved refinery margins.

The development of new hydrocracking catalysts is very dependent on new or modified materials. Topsøe has found unique methods of preparing hydroprocessing catalysts, and through an extensive understanding of the chemistry has demonstrated a high level of expertise in making catalyst carriers with a uniform distribution of acidic sites and hydrogenation metal sites. Many zeolite hydrocracking catalysts are now offered in the tri-lobe shape which reduces the diffusion path and decrease the pressure drop. The design significantly enhances the accessibility of the active catalyst sites and thus provides a substantial enhancement of the catalyst activity.

In order to optimize overall unit performance, catalysts with pore size distribution to match the changing molecular structure of the oil as it processes through the reactor system are necessary. They are applied in the front-end reactor when processing high-metal-containing feedstocks (>70 ppm vanadium). Catalysts which exhibit the high activity for sulfur, coke precursors, and nitrogen conversion are applied in the middle and/or tail end reactors.

11.4   OPTIONS FOR HEAVY FEEDSTOCKS

The goal of *heavy feedstock hydroconversion* is to convert feedstocks to low-sulfur liquid product oils or, in some cases, to pretreat feedstocks for fluid catalytic cracking processes. Some of the processes available for hydroprocessing heavy feedstocks are presented in the following. However, when applied to heavy feedstocks (heavy oil, tar sand bitumen, and residua), the problems encountered can be directly equated to the amount of complex, higher-boiling constituents that may require pretreatment (Speight and Moschopedis, 1979; Speight, 2000, 2014; Speight and Ozum, 2002; Parkash, 2003; Hsu and Robinson, 2006; Gary et al., 2007). Furthermore, the majority of the higher-molecular-weight materials produce high yields (35%–60% by weight) of coke. It is this trend of coke formation that hydrocracking offers some relief. Thus, the major goal of *heavy feedstock hydroconversion* is cracking of heavy feedstocks with desulfurization, metal removal, denitrogenation, and asphaltene conversion. However, asphaltene constituents and metal-containing constituents exert a strong deactivating influence on the catalyst, which markedly decreases the hydrogenolysis rate of sulfur compounds, practically without having impact on coke formation. In addition, nitrogen-containing compounds are adsorbed on acid sites, blocking the sites and thereby lowering catalyst activity. Thus, during the hydrocracking of heavy feedstocks, preliminary feedstock hydrodesulfurization and demetallization over special catalyst are advantageous.

The processes that follow are available for conversion of heavy feedstocks to a variety of product slates and are listed in alphabetical order with no other preference in mind.

11.4.1   Aquaconversion

The Aquaconversion process is a hydrovisbreaking technology that uses catalyst-activated transfer of hydrogen from water added to the feedstock. Reactions that lead to coke formation are suppressed, and there is no separation of asphaltene-type material (Marzin et al., 1998; Pereira et al., 1998, 2001).

Typically, visbreaking technology is limited in conversion level because of the stability of the resulting product. The Aquaconversion process extends the maximum conversion level within the stability specification by adding a homogeneous catalyst in the presence of steam. This allows hydrogen from the water to be transferred to the resid when contacted in a coil-soaker system normally used for the visbreaking process. The hydrogen incorporation is much lower than that obtained when using a deep hydroconversion process under high hydrogen partial pressure. Nevertheless, it is high enough to saturate the free radicals, formed within the thermal process, which would normally lead to coke formation. With hydrogen incorporation, a higher conversion level can be reached and thus higher API and viscosity improvements while maintaining syncrude stability.

The important aspect of the Aquaconversion technology is that it does not produce coke, nor does it require any hydrogen source or high-pressure equipment. In addition, the Aquaconversion process can be implanted in the production area, and thus the need for external diluent and its transport over large distances is eliminated. Light distillates from the raw crude can be used as diluent for both the production and desalting processes. Also some catalyst processes have been developed such as catalytic aquathermolysis, which is used widely for upgrading heavy oil (Fan et al., 2004, 2009; Li et al., 2007; Wen et al., 2007; Chen et al., 2009). In this process to maximize the upgrading effect, the suitable catalysts should be chosen.

11.4.2 Asphaltenic Bottom Cracking Process

The asphaltenic bottom cracking (ABC) process can be used for distillate production (Table 11.4), hydrodemetallization, asphaltene cracking, and moderate hydrodesulfurization as well as sufficient resistance to coke fouling and metal deposition using such feedstocks as heavy oil, vacuum residua, thermally cracked residua, solvent deasphalted bottoms, and bitumen with fixed catalyst beds (Kressman et al., 1998).

The process can be combined with (1) solvent deasphalting for complete or partial conversion of the residuum or (2) hydrodesulfurization to promote the conversion of residue, to treat feedstock

TABLE 11.4
Feedstock and Product Data for the ABC Process

	Arabian Light Vacuum Residuum	Arabian Heavy Vacuum Residuum	Cerro Negro Vacuum Residuum
Feedstock			
API	7.0	5.1	1.7
Sulfur, wt%	4.0	5.3	4.3
Carbon residue, % w/w	20.8	23.3	23.6
C_7-asphaltenes	7.0	13.1	19,819.8
Nickel, ppm	223.0	52.0	150.0
Vanadium, ppm	76.0	150.0	640.0
Products			
Naphtha, vol%	6.5	7.7	15.1
API	57.2	57.2	54.7
Distillate, vol%	16.0	19.8	21.3
API	34.2[a]	34.2[a]	32.5[a]
Vacuum gas oil, vol%	34.3	38.1	32.8
API	24.7	21.6	15.4
Sulfur, % w/w	0.2	1.7	0.5
Vacuum residuum, vol%	46.2	37.9	34.7
API	10.6	7.8	<0.0
Sulfur, wt%	0.6	1.7	2.2
Carbon residue, wt%	13.6	26.5	13.6
C_7-asphaltenes, wt%			
Nickel, ppm	9.0	45.0	117.0
Vanadium, ppm	11.0	75.0	371.0
Conversion	55.0	60.0	60.0

Source: Speight, J.G., *The Chemistry and Technology of Petroleum*, 4th edn., CRC Press/Taylor & Francis Group, Boca Raton, FL, 2007.

[a] Estimated.

with high metals, and to increase catalyst life or (3) hydrovisbreaking to attain high conversion of the heavy feedstock with favorable product stability.

In the process, the feedstock is pumped up to the reaction pressure and mixed with hydrogen. The mixture is heated to the reaction temperature in the charge heater after a heat exchange and fed to the reactor.

In the reactor, hydrodemetallization and subsequent asphaltene cracking with moderate hydrodesulfurization take place simultaneously under conditions similar to residuum hydrodesulfurization. The reactor effluent gas is cooled, cleaned up and recycled to the reactor section, while the separated liquid is distilled into distillate fractions and vacuum residue which is further separated by deasphalting into deasphalted oil and asphalt using butane or pentane (Chapter 12).

In the case of the ABC–hydrodesulfurization catalyst combination, the ABC catalyst is placed upstream of the hydrodesulfurization catalyst and can be operated at a higher temperature than the hydrodesulfurization catalyst under conventional heavy feedstock hydrodesulfurization conditions. In the VisABC process, a soaking drum is provided after heater, when necessary. Hydrovisbroken oil is first stabilized by the ABC catalyst through hydrogenation of coke precursors and then desulfurized by the HDS catalyst.

11.4.3 CANMET Process

The CANMET hydrocracking process was designed to process heavy feedstocks. The process is now no longer available from the licensors (the Government of Canada) and has been incorporated into the UOP Uniflex process. Nevertheless, the CANMET process is worthy of description here—if only for historical purposes—because of the novel aspect in the successful use of a scavenger to enhance the hydrocracking process and decrease coke formation. The process was specially developed for use with Athabasca bitumen but can accommodate a variety of heavy feedstocks, including atmospheric residua and vacuum residua (Table 11.5) (Pruden, 1978; Pruden et al., 1993). The process was a high-conversion, high-demetallization, heavy feedstock hydrocracking process which, using an additive to inhibit coke formation, achieves conversion of high-boiling-point hydrocarbons into lighter products. Initially developed to upgrade tar sand bitumen and heavy oils of Alberta, an ongoing program of development has broadened the technology to processing offshore heavy oils and the bottom of the barrel from so-called conventional crude oils. The process did not use a catalyst but employs a low-cost additive to inhibit coke formation and allow high conversion of heavy feedstocks (such as heavy oil and bitumen) into lower-boiling products using a single reactor. The process was unaffected by high levels of feed contaminants such as sulfur, nitrogen, and metals. Conversion of over 90% of the $525°C^+$ ($975°F^+$) fraction into distillates was attained.

In the process (Figure 11.12), the feedstock and recycle hydrogen gas were heated to reactor temperature in separate heaters. A small portion of the recycle gas stream and the required amount of additive are routed through the oil heater to prevent coking in the heater tubes. The outlet streams from both heaters are fed to the bottom of the reactor. The vertical reactor vessel is free of internal equipment and operates in a three-phase mode. The solid additive particles are suspended in the primary liquid hydrocarbon phase through which the hydrogen and product gases flow rapidly in bubble form. The reactor exit stream was quenched with cold recycle hydrogen prior to the high-pressure separator. The heavy liquids are further reduced in pressure to a hot medium pressure separator and from there to fractionation. The spent additive leaves with the heavy fraction and remains in the unconverted heavy feedstock. The vapor stream from the hot high-pressure separator is cooled stepwise to produce middle distillate and naphtha that are sent to fractionation. High-pressure purge of low-boiling hydrocarbon gases is minimized by a sponge oil circulation system. Product naphtha will be hydrotreated and reformed, light gas oil will be hydrotreated and sent to the distillate pool, the heavy gas oil will be used as a feedstock for the FCC unit, and the pitch will be sold.

The additive, prepared from iron sulfate $[Fe_2(SO_4)_3]$, is used to promote hydrogenation and effectively eliminate coke formation. The effectiveness of the dual-role additive permits the use

TABLE 11.5
CANMET Process Feedstock and Product Data

Feedstock[a]

API gravity	4.4
Sulfur, wt%	5.1
Nitrogen, wt%	0.6
Asphaltenes, wt%	15.5
Carbon residue, wt%	20.6
Metals, ppm	
Ni	80
V	170
Residuum (>525°C; >975°F), wt%	
Products,[b] wt%	
Naphtha (C_5–204°C; 400°F)	19.8
Nitrogen, wt%	0.1
Sulfur, wt%	0.6
Distillate (204°C–343°C; 400°F–650°F)	33.5
Nitrogen, wt%	0.4
Sulfur, wt%	1.8
Vacuum gas oil (343°C–534°C; 650°F–975°F)	28.5
Nitrogen, wt%	0.6
Sulfur, wt%	2.3
Residuum (>534°C; >975°F)	4.5
Nitrogen, wt%	1.6
Sulfur, wt%	3.1

[a] Cold Lake (Canada) heavy oil vacuum residuum.
[b] Residuum: 93.5% by weight.

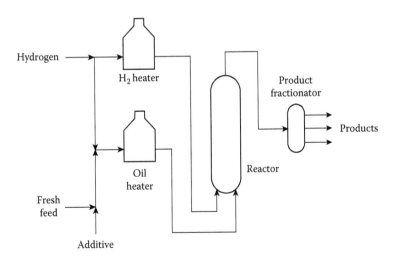

FIGURE 11.12 CANMET process.

of operating temperatures that give high conversion in a single-stage reactor. The process maximizes the use of reactor volume and provides a thermally stable operation with no possibility of temperature runaway. In terms of the additive, the use of iron sulfate is reminiscent of the older *red mud process* which used an iron-containing mud (an iron-containing slurry) to convert coal to liquids.

11.4.4 Chevron RDS Isomax and VRDS Process

The RDS/VRDS process (like the Residfining process) is designed to hydrotreat vacuum gas oil, atmospheric residuum, vacuum residuum, and other heavy feedstocks to remove sulfur metallic constituents while part of the feedstock is converted to lower-boiling products. In the case of heavy feedstocks, the asphaltene content is reduced. The process consists of a once-through operation and is ideally suited to produce feedstocks for residuum fluid catalytic crackers or delayed coking units to achieve minimal production of residual products in a refinery.

The basic elements of each process are similar and consist of a once-through operation of the feedstock coming into contact with hydrogen and the catalyst in a downflow reactor that is designed to maintain activity and selectivity in the presence of deposited metals. Moderate temperatures and pressures are employed to reduce the incidence of hydrocracking and, hence, minimize production of low-boiling distillates. The combination of a desulfurization step and a heavy feedstock desulfurizer (VRDS) is often seen as an attractive alternate to the atmospheric residuum desulfurizer (RDS). In addition, either RDS option or the VRDS option can be coupled with other processes (such as delayed coking, fluid catalytic cracking, and solvent deasphalting) to achieve the most optimum refining performance.

The *Chevron deasphalted oil hydrotreating process* is designed to desulfurize heavy feedstocks that have had the asphaltene fraction removed by prior application of a deasphalting process. The principal product is a low-sulfur fuel oil that can be used as a blending stock or as a feedstock for a fluid catalytic unit. The process employs a downflow, fixed-bed reactor containing a highly selective catalyst that provides extensive desulfurization at low pressures with minimal cracking and, therefore, low consumption of hydrogen.

In the process, which is designed for heavy feedstocks including a wide range of residua and other heavy feedstocks, the feedstock and hydrogen are charged to the reactors in a once-through operation. The catalyst combination can be varied significantly according to feedstock properties to meet the required product qualities. Product separation is done by the hot separator, cold separator, and fractionator. Recycle hydrogen passes through a hydrogen sulfide absorber. The *onstream catalyst replacement* (OCR) reactor technology improves catalyst utilization and increases run length with high-metal, heavy feedstocks. This technology allows spent catalyst to be removed from one or more reactors and replaced with fresh while the reactors continue to operate normally. The novel use of upflow reactors in the onstream catalyst replacement technology provides increased tolerance of feed solids while maintaining low-pressure drop. A related technology (upflow reactor technology) uses a multibed upflow reactor for minimum pressure drop in cases where onstream catalyst replacement is not necessary. Onstream catalyst replacement technology and upflow reactor technology are particularly well suited to revamp existing RDS/VRDS units for additional throughput or heavier feedstock. The products (residuum FCC feedstock, coker feedstock, solvent deasphalter feedstock, or low-sulfur fuel oil, and vacuum gas oil) are suitable for further upgrading by fluid catalytic cracking units or hydrocrackers for naphtha/mid-distillate manufacture. Mid-distillate material can be directly blended into low-sulfur diesel or further hydrotreated into ultra-low-sulfur diesel (ULSD). Thus, the process can be integrated with residuum FCC units to minimize catalyst consumption, improve yields, and reduce sulfur content of FCC products. RDS/VRDS also can be used to substantially improve the yields of downstream coking units and solvent deasphalting units.

11.4.5 ENI SLURRY-PHASE TECHNOLOGY

The advent of slurry-phase hydrocracking into the refineries has caused much interest. This technology adopts high operating pressures and can achieve near complete (if not, complete) conversion of the residuum while producing finished saleable products (Motaghi et al., 2010). Slurry-phase hydrocracking can be used to convert heavy feedstock in the presence of hydrogen under severe process conditions—on the order of 450°C (840°F) and 2000–3000 psi. To prevent excessive coking, finely powdered additives are added to the feedstock. Inside the reactor, the feedstock/powder mixture behaves as a single phase due to the small size of the additive particles.

For example, in the ENI slurry-phase process, which is based on an organic oleo-soluble molybdenum compound and the catalyst precursor, the catalyst is added to the feedstock before it enters the reactor (Bellussi et al., 2013). In the process, fresh feedstock is sent to the fresh feed heater and then mixed with the proprietary catalyst makeup and sent to the upflow slurry bubble column reactor. Hot hydrogen is also sent to the slurry reactor providing the thermocatalytic hydroconversion of the feedstock. The reactor effluent is collected in a hot high-pressure separator where a gas-vapor stream and a vacuum gas oil stream are separated. The stream is subsequently cooled and sent to a cold high-pressure separator to separate the gas stream, rich in hydrogen, and the hydrocarbon liquid stream. The liquid stream from the cold high-pressure separator is sent to the light distillates stabilizer from which low-boiling distillates are separated and sent to battery limit. The reaction occurs at 400°C–450°C (750°F–840°F) and at approximately 2200 psi with hydrogen fed from the bottom of the reactor. Under the reaction conditions, the catalyst precursor forms highly dispersed molybdenum sulfide (MoS_2) nanoparticles. The *in situ* formation of the catalyst preparation method enables the dispersion of molybdenum sulfide (MoS_2) mainly as single layers within the slurry reactor.

The unconverted nonvolatile fraction at the bottom of the vacuum distillation column, containing all of the catalyst, is recycled back to the reactor, and only a small part of the heavy fraction is purged (1%–3% w/w of the fresh feedstock) to avoid the accumulation of coke precursors and of Ni and V sulfides from the organometallic compounds contained in the feedstock. With the purge, a limited amount of molybdenum is also removed; therefore, an equivalent amount is fed continuously to the reactor to maintain concentration constant. The purge can be used as a fuel in the cement or steel industries. In order to facilitate the handling of the purge and its blending with other streams, the viscosity can be adjusted by adding a small amount of a low-value flow improver (such as vacuum gas oil). The purge can also be treated in a centrifugal decanter to recover the liquid fraction, which is recycled back to the reactor, and a solid product (cake) containing heavy hydrocarbons, coke, and concentrated metal sulfides. The cake can be processed further to recover the metals (molybdenum, vanadium, and nickel). Since the largest part of the catalyst is not lost, but is recycled to the reaction section, the process can operate at a higher catalyst concentration than in the case with other slurry technologies.

The process is very flexible with regard to the feedstock and can accept feedstocks such as vacuum residua from different heavy crudes, tar sand bitumen, and refinery visbroken tar. The typical overall performance of the process is (1) metal removal, >99%; (2) Conradson carbon residue reduction, >97%; (3) sulfur reduction, >85%; and (4) nitrogen reduction, >40%. Furthermore, because of the recycling of unconverted products and the dispersed catalyst, the process has the ability to reach total conversion of the feedstock.

11.4.6 GULF RESID HYDRODESULFURIZATION PROCESS

The Gulf Resid HDS process is a regenerative fixed-bed process to upgrade heavy feedstocks by catalytic hydrogenation to refined heavy fuel oils or to high-quality catalytic charge stocks (Figure 11.13). Desulfurization and quality improvement are the primary purposes of the process,

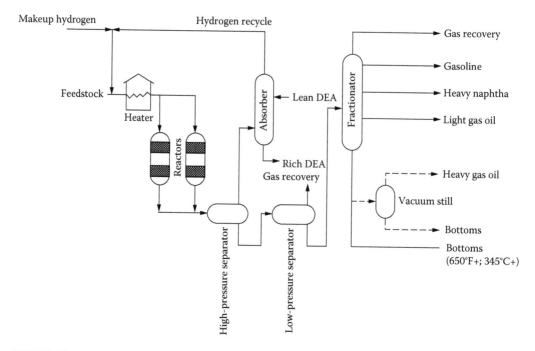

FIGURE 11.13 Gulf residua hydrodesulfurization process.

but if the operating conditions and catalysts are varied, light distillates can be produced and the viscosity of heavy material can be lowered. Long onstream cycles are maintained by reducing random hydrocracking reactions to a minimum, and whole crude oils, virgin, or cracked residua may serve as feedstock. This process is suitable for the desulfurization of high-sulfur residua (atmospheric and vacuum) to produce low-sulfur fuel oils or catalytic cracking feedstocks. In addition, the process can be used, through alternate design types, to upgrade high-sulfur crude oils or bitumen that are unsuited for the more conventional refining techniques.

The process has three basic variations—the Type II unit, the Type III unit, and the Type IV unit—with the degree of desulfurization and process severity increasing from Type I to Type IV. Thus, liquid products from Types III and IV units can be used directly as catalytic cracker feedstocks and perform similarly to virgin gas oil fractions, whereas liquid products from the Type II unit usually need to be vacuum-flashed to provide a feedstock suitable for a catalytic cracker.

In the process, fresh, filtered feedstock is heated together with hydrogen and recycle gas and charged to the downflow reactor from which the liquid product goes to fractionation after flashing to produce the various product streams. Each process type is basically similar to its predecessor but will differ in the number of reactors. For example, modifications necessary to convert the Type II to the Type III process consist of the addition of a reactor and related equipment, while the Type III process can be modified to a Type IV process by the addition of a third reactor section. Types III and IV are especially pertinent to the problem of desulfurizing heavy oils and residua since they have the capability of producing extremely low-sulfur liquids from high-sulfur residua (Table 11.6).

The catalyst is a metallic compound supported on pelleted alumina and may be regenerated *in situ* with air and steam or flue gas through a temperature cycle of 400°C–650°C (750°F–1200°F). Onstream cycles of 4–5 months can be obtained at desulfurization levels of 65%–75%, and catalyst life may be as long as 2 years.

TABLE 11.6
Gulf Residua HDS Process Feedstock and Product Data

Properties	South Louisiana	West Texas	Light Arabian	Kuwait
°API	22.4	19.3	18.5	16.6
Sulfur, wt%	0.46	2.2	2.93	3.8
Nitrogen, wt%	0.16	0.20	0.16	0.21
Carbon residue (Ramsbottom), wt%	1.76	5.5	6.79	8.3
Nickel, ppm	3.2	7.1	7.3	15
Vanadium, ppm	1.7	14.0	27.0	45
Yield on crude, vol%	48	44.1	41	53
Hydrodesulfurization type	II	II	IV	III
Desulfurization, %	87	85	95.9	92
°API	24.0	23.8	25.1	25.7
Sulfur, wt%	0.05	0.33	0.12	0.28
Nitrogen, wt%	0.10	0.14	0.09	0.14
Carbon residue (Ramsbottom), wt%	1.4	2.21	2.28	2.83
Nickel, ppm	0.1	1.1	0.5	1.1
Vanadium, ppm	0.1	0.7	0.3	0.8
Demetallization, %	97.1	87.5	95.3	94.6

11.4.7   H-G Hydrocracking Process

The H-G hydrocracking process may be designed with either a single- or a two-stage reactor system for conversion of light and heavy gas oils to lower-boiling fractions (Figure 11.14). The feedstock is mixed with recycle gas oil, makeup hydrogen, and hydrogen-rich recycle gas, and then heated and charged to the reactor. The reactor effluent is cooled and sent to a high-pressure separator, where hydrogen-rich gas is flashed off, scrubbed, and then recycled to the reactor. Separator liquid passes

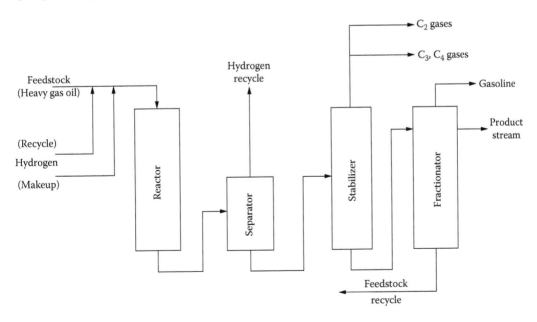

FIGURE 11.14   The H-G hydrocracking process.

to a stabilizer for removal of butanes and lighter products, and the bottoms are taken to a fractionator for separation; any unconverted material is recycled to the reactor.

11.4.8 H-Oil Process

The H-Oil process (Speight and Ozum, 2002; Speight, 2014) is a catalytic process (Table 11.7) that uses a single-stage, two-stage, or three-stage ebullated-bed reactor in which, during the reaction, considerable hydrocracking takes place (Figure 11.15). The process is used to upgrade sulfur-containing heavy feedstock crude oils and residual stocks to low-sulfur distillates, thereby reducing fuel oil yield (Table 11.8). A modification of H-Oil called Hy-C cracking converts heavy distillates to middle distillates and kerosene.

The process is designed for hydrogenation of heavy feedstocks in an ebullated-bed reactor to produce upgraded petroleum products (Speight and Ozum, 2002; Speight, 2014). The process is able to convert all types of feedstocks to either distillate products as well as to desulfurize and

TABLE 11.7

General Process Parameters for Ebullated-Bed (H-Oil and LC-Fining) Processes

Parameter	H-Oil	LC-Fining
Temperature, °C	415–440	385–450
Temperature, °F	780–825	725–840
Pressure, psi	2440–3000	1015–2740
Hydrogen/bbl	1410	1350
Conversion	45–90	40–95
HDS	55–90	60–90
HDM	65–90	50–95

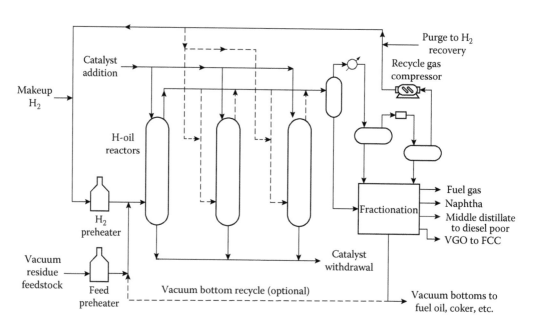

FIGURE 11.15 The H-Oil process.

TABLE 11.8
H-Oil Process Feedstock and Product Data

	Unspecified	Arabian Medium Vacuum Residuum[a] 65% Conv	90% Conv	Unspecified[b] 40% Desulf	80% Desulf	Athabasca Bitumen
Feedstock						
Gravity, °API	8.8	4.9	4.9	14.0	14.0	8.3
Sulfur, wt%	2.0	5.4	5.4	4.0	4.0	4.9
Nitrogen, wt%	0.9					0.5
Carbon residue, wt%	14.6			9.2	9.2	
Metals, ppm		128.0	128.0			
Ni	80.0					
V	170.0					
Residuum (>525°C, >975°F), wt%	71.7			50.0	50.0	50.3
Products, wt%						
Naphtha (C$_5$–204°C; 400°F)	33.0	17.6	23.8	1.8	5.0	16.0
Sulfur, wt%	<0.1				0.1	1.0
Distillate (204°C–343°C; 400°F–650°F)	40.0	22.1	36.5			43.0
Sulfur, wt%	0.4					2.0
Vacuum gas oil (343°C–534°C; 650°F–975°F)	24.0	34.0	37.1	12.8	15.0	26.4
Sulfur, wt%	1.0			0.7	0.3	3.5
Residuum (>534°C; >975°F)	11.0	33.2	9.5	86.6	81.0	16.0
Sulfur, wt%	2.0			2.5	1.0	5.7

[a] Conv, conversion.
[b] Desulf, desulfurization.

demetallize heavy feedstocks for feed to coking units or fluid catalytic cracking units, for production of low-sulfur fuel oil, or for production to asphalt blending. A modification of the H-Oil process (Hy-C cracking process) converts high-boiling distillates to middle distillates and kerosene (Table 11.9).

A wide variety of process options can be used with the H-Oil process depending on the specific needs. In all cases, a catalytic ebullated-bed reactor system is used to provide an efficient hydroconversion. The system ensures uniform distribution of liquid, hydrogen-rich gas, and catalyst across the reactor. The ebullated-bed system operates under essentially isothermal conditions, exhibiting little temperature gradient across the bed (Kressmann et al., 2000). The heat of reaction is used to bring the feed oil and hydrogen up to the reactor temperature.

In the process, feedstock (which may be combined with recycled residuum) and hydrogen are fed upward through the reactors as a liquid–gas mixture at a velocity such that catalyst is in continuous motion. A catalyst of small particle size can be used, giving efficient contact among gas, liquid, and solid with good mass and heat transfer. Part of the reactor effluent is recycled back through the reactors for temperature control and to maintain the requisite liquid velocity. The entire bed is held within a narrow temperature range, which provides essentially an isothermal operation with an exothermic process. Because of the movement of catalyst particles in the liquid–gas medium,

TABLE 11.9
Feedstock and Product Data for the H-Oil Process

Feedstock	Arabian Medium Vacuum Residuum 65%[a]	Arabian Medium Vacuum Residuum 90%[a]	Athabasca Bitumen
API gravity	4.9	4.9	8.3
Sulfur, % w/w	5.4	5.4	4.9
Nitrogen, % w/w			0.5
Carbon residue, % w/w			
Metals, ppm	128.0	128.0	
Ni			
V			
Residuum (>525°C, >975°F), % w/w			50.3
Products, % w/w [b]			
Naphtha (C$_5$–204°C, 400°F)	17.6	23.8	16.0
Sulfur, % w/w			1.0
Distillate (204°C–343°C, 400°F–650°F)	22.1	36.5	43.0
Sulfur, wt%			2.0
Vacuum gas oil (343°C–534°C, 650°F–975°F)	34.0	37.1	26.4
Sulfur, wt%			3.5
Residuum (>534°C, >975°F)	33.2	9.5	16.0
Sulfur, wt%			5.7

Source: Speight, J.G., *The Chemistry and Technology of Petroleum*, 4th edn., CRC Press/Taylor & Francis Group, Boca Raton, FL, 2007.

[a] % conversion.

[b] % desulfurization.

deposition of tar and coke is minimized and fine solids entrained in the feed do not lead to reactor plugging. The catalyst can also be added and withdrawn from the reactor without destroying the continuity of the process. The reactor effluent is cooled by exchange and separates into vapor and liquid. After scrubbing in a lean oil absorber, hydrogen is recycled and the liquid product is either stored directly or fractionated before storage and blending.

A catalyst of small particle size can be used, giving efficient contact among gas, liquid, and solid with good mass and heat transfer. Part of the reactor effluent is recycled back through the reactors for temperature control and to maintain the requisite liquid velocity. The entire bed is held within a narrow temperature range, which provides essentially an isothermal operation with an exothermic process. Because of the movement of catalyst particles in the liquid–gas medium, deposition of tar and coke is minimized and fine solids entrained in the feed do not lead to reactor plugging. The catalyst can also be added and withdrawn from the reactor without destroying the continuity of the process. The reactor effluent is cooled by exchange and separates into vapor and liquid. After scrubbing in a lean oil absorber, hydrogen is recycled and the liquid product is either stored directly or fractionated before storage and blending.

A variation of this process (the HDH resid hydrocracking process) was originally developed for the upgrading of heavy oils from Orinoco Oil Belt, Venezuela. In the process, the heavy feedstock is slurried with a low-cost catalyst and fed into a series of upflow bubbling (slurry) reactors operating at 420°C–480°C temperature under hydrogen partial pressure. The reaction products are fractionated using a high-pressure hot separator.

While ebullated-bed processes are continuous and produce higher levels of liquid fuels (no coke), it is not always possible to achieve complete heavy feedstock conversion and the unit may still produce 20%–30% v/v of heavy-resid product (Motaghi et al., 2010). Ebullated beds have also been prone to high operating costs, and have sometimes been plagued with poor operability. The quality of liquid products, although improved over coking, still requires secondary processing to produce clean fuels. The inability to achieve near complete conversion requires further processing of unconverted resid. As a result, ebullated-bed technologies have not achieved huge deployment, which, when coupled with the high capital cost, makes them the least robust at low-oil price scenarios.

In terms of advancement of ebullated-bed technology, the original H-Oil process has evolved, during the past decade, into various configurations that have the potential to play major roles in heavy feedstock upgrading up to and beyond the year 2020. A new development in H-Oil process technology is interstage separation for a two-stage unit design (Kressman et al., 2004). In this configuration, an additional vessel is fed to the first-stage reactor effluent (mixed phase) and separates it into vapor and liquid products. The interstage liquid is fed to the second-stage reactor and the vapor to the overhead of the hot high-pressure separator located after the second-stage reactor. With interstage separation, off-loading of the first-stage reactor gas results in improved reaction kinetics in the second-stage reactor since the amount of gas holdup in the reactor is greatly reduced and increasing liquid holdup enables greater conversion of the feedstock.

The H-Oil$_{DC}$ process (previously known as the T-Star process) is a specially engineered, ebullated-bed process for the treatment of vacuum gas oils. Because of the ability to replace the catalyst bed incrementally, the H-Oil$_{DC}$ reactor can operate indefinitely—typically, 4–5 years between turnarounds to coincide with the inspection and maintenance schedule for a fluid catalytic cracking unit. The difficult processing requirements which result from stricter environmental regulations and the processing of heavy feedstocks makes H-Oil$_{DC}$ a preferred choice for pretreatment of fluid catalytic cracker feedstocks.

The H-Oil$_{HCC}$ is a heavy crude conversion process that produces synthetic crude oil. The objective of the unit is to enable just enough conversion to reduce viscosity and increase stability so that the product can be readily transported to an upgrading center. Among the improvements made to the traditional H-Oil technology are the integration of an interstage separator between reactors in series and the application of cascade catalyst utilization. The result is (1) an increase in conversion levels, (2) an increase in product stability, and (3) reduced processing costs.

11.4.9 HYCAR PROCESS

Briefly, *hydrovisbreaking*, a noncatalytic process, is conducted under similar conditions to visbreaking and involves treatment with hydrogen under mild conditions. The presence of hydrogen leads to more stable products (lower *flocculation threshold* of the asphaltene-type constituents) than can be obtained with straight visbreaking, which means that higher conversions can be achieved, producing a lower viscosity product.

The HYCAR process is composed fundamentally of three parts: (1) visbreaking, (2) hydrodemetallization, and (3) hydrocracking. In the visbreaking section, the heavy feedstock (e.g., vacuum residuum or bitumen) is subjected to moderate thermal cracking while no coke formation is induced. The visbreaker oil is fed to the demetallization reactor in the presence of catalysts, which provides sufficient pore for diffusion and adsorption of high-molecular-weight constituents. The product from this second stage proceeds to the hydrocracking reactor, where desulfurization and denitrogenation take place along with hydrocracking.

11.4.10 HYVAHL-F PROCESS

The process is used to hydrotreat atmospheric and vacuum residua to convert the feedstock to naphtha and middle distillates (Peries et al., 1988; Billon et al., 1994; Speight and Ozum, 2002; Speight, 2014).

The main features of this process are its dual-catalyst system and its fixed-bed swing-reactor concept. The first catalyst has a high capacity for metals (to 100% by w/w new catalyst) and is used for both hydrodemetallization (HDM) and most of the conversion. This catalyst is resistant to fouling, coking, and plugging by asphaltene constituents (as well as by reacted asphaltene constituents) and shields the second catalyst from the same. Protected from metal poisons and deposition of coke-like products, the highly active second catalyst can carry out its deep hydrodesulfurization (HDS) and refining functions. Both catalyst systems use fixed beds that are more efficient than moving beds and are not subject to attrition problems. The swing-reactor design reserves two of the hydrodemetallization reactors as guard reactors: one of them can be removed from service for catalyst reconditioning and put on standby while the rest of the unit continues to operate. More than 50% of the metals are removed from the feed in the guard reactors.

In the process, the preheated feedstock enters one of the two guard reactors where a large proportion of the nickel and vanadium are adsorbed and hydroconversion of the high-molecular-weight constituents commences. Meanwhile, the second guard reactor catalyst undergoes a reconditioning process and then is put on standby. From the guard reactors, the feedstock flows through a series of hydrodemetallization reactors that continue the metals removal and the conversion of heavy ends. The next processing stage, hydrodesulfurization (HDS), is where most of sulfur, some of the nitrogen, and the residual metals are removed. A limited amount of conversion also takes place. From the final reactor, the gas phase is separated, hydrogen is recirculated to the reaction section, and the liquid products are sent to a conventional fractionation section for separation into naphtha, middle distillates, and heavier streams.

A related process—the Hyvahl-M process—employs countercurrent moving-bed reactors and is recommended for feedstocks containing substantial amounts of metals and asphaltene constituents.

11.4.11   IFP Hydrocracking Process

The process features a dual-catalyst system: the first catalyst is a promoted nickel–molybdenum amorphous catalyst. It acts to remove sulfur and nitrogen and hydrogenate aromatic rings. The second catalyst is a zeolite that finishes the hydrogenation and promotes the hydrocracking reaction.

In the single-stage process, the first reactor effluent is sent directly to the second reactor, followed by the separation and fractionation steps (Figure 11.16). The fractionator bottoms are recycled to the second reactor or sold. In the two-stage process (Figure 11.16), feedstock and hydrogen are heated and sent to the first reaction stage where conversion to products occurs (RAROP, 1991, p. 85). The reactor effluent phases are cooled and separated and the hydrogen-rich gas is compressed and recycled. The liquid leaving the separator is fractionated, the middle distillates and lower-boiling streams (Table 11.10) (Speight and Ozum, 2002; Speight, 2014) are sent to storage, and the high-boiling stream is transferred to the second reactor section and then recycled back to the separator section.

11.4.12   Isocracking Process

The hydrocracker is a high-pressure, moderate-temperature conversion unit and the designs include single-stage once-through, single-stage recycle and two-stage recycle processes. A two-stage hydrocracker with intermediate distillation represents the most common process configuration for maximizing middle distillates. The process has been applied commercially in the full range of process flow schemes: single-stage, once-through liquid; single-stage, partial recycle of heavy oil; single-stage recycle to extinction of the feedstock (100% conversion); and two-stage recycle to extinction of the feedstock (Figure 11.17) (Bridge, 1997; Speight and Ozum, 2002; Speight, 2014). The preferred flow scheme will depend on the feed properties, the processing objectives, and, to some extent, the

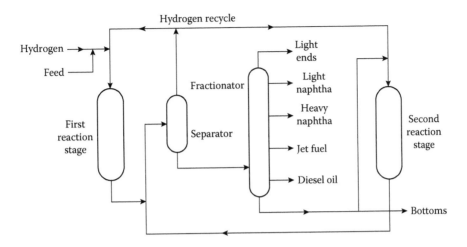

FIGURE 11.16 The IFP process.

TABLE 11.10
IFP Hydrocracking Process Feedstock and Product Data

Feedstock: Vacuum gas oil, 350°C–570°C (660°F–1060°F)

API gravity, °API	20.6
Sulfur, wt%	2.3
Nitrogen, wt%	1.0
Carbon residue, wt%	0.5
Products	
Light naphtha, wt%	6.0
Sulfur, ppm	<1
Distillate (heavy naphtha, jet fuel, diesel fuel), wt%	91.0
Sulfur, ppm	<50

specified feed rate. The process uses multibed reactors and, in most applications, a number of catalysts are used in a reactor. The catalysts are dual function being a mixture of hydrous oxides (for cracking) and heavy metal sulfides (for hydrogenation) (Bridge, 1997). The catalysts are used in a layered system to optimize the processing of the feedstock that undergoes changes in its properties along the reaction pathway (Speight and Ozum, 2002; Speight, 2014).

In the process, the feedstock (typically a blend of heavy gas oils) is sent to the first stage of the hydrocracker and is severely hydrotreated. Most of the sulfur and nitrogen compounds are removed from the feedstock and many of the aromatics are saturated. In addition, significant conversion to light products occurs in the first stage. The liquid product from the first stage is then sent to a common fractionation section where, to prevent overcracking, lower-boiling products are removed by distillation. The unconverted material from the bottom of the fractionator is routed to the second-stage reactor section. The second reaction stage saturates almost all of the aromatics and cracks the oil feed to light products. Due to the aromatics saturation, the second stage produces excellent quality products. The liquid product from the second stage is then sent to the common fractionator, where light products are distilled. The second stage operates in a recycle-to-extinction mode with per pass conversions ranging from 50% to 80% v/v. The overhead liquid and vapor from the hydrocracker fractionator is further processed in a light ends recovery unit where fuel gas and liquefied petroleum gas (LPG) and naphtha are separated. The hydrogen supplied to the reactor sections of

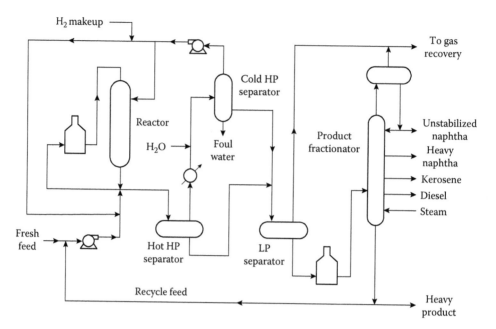

FIGURE 11.17   The Isocracking process.

the hydrocracker comes from reformers or steam reformers. The hydrogen is compressed in stages until it reaches system pressure of the reactor sections.

In most commercial Isocracking units, the entire fractionator bottoms fraction is recycled or all of it is drawn as heavy product, depending on whether the low-boiling or high-boiling products are of greater value. If the low-boiling distillate products (naphtha or naphtha/kerosene) are the most valuable products, the higher-boiling-point distillates (like diesel) can be recycled to the reactor for conversion rather than drawn as a product (Tables 11.11 and 11.12) (RAROP, 1991, p. 83; Khan and Patmore, 1998). Product distribution depends upon the feedstock and (as anticipated) the product yield is very much dependent upon the catalyst and the process parameters.

TABLE 11.11
Isocracking Process Feedstock and Product Data

Feedstock: Vacuum gas oil, 360°C–530°C (680°F–985°F)

API gravity, °API	22.6
Sulfur, wt%	2.2
Nitrogen, wt%	0.6
Carbon residue, wt%	0.3
Metals	
Ni	0.1
V	0.3
Products	
Naphtha (C$_5$–124°C; C$_5$–255°F), wt%	15.9
Sulfur, ppm	<2
Distillate (124°C–295°C; 255°F–565°F), wt%	51.6
Sulfur, ppm	<5
Heavy distillate (295°C–375°C; 565°F–705°F), wt%	42.3
Sulfur, ppm	<5

TABLE 11.12
Isocracking Process Desulfurization Data

Operation	Conventional Desulfurization	Severe Desulfurization	Mild Isocracking
% HDS	90.0	99.8	99.6
Yield, LV (%)			
Naphtha	0.2	1.5	3.5
Light isomate	17.2	30.8	37.1
Heavy isomate	84.0	70.0	62.5
Feed			
Gravity, °API	22.6	22.6	23.0
Sulfur, wt%	2.67	2.67	2.57
Nitrogen, ppm	720	720	617
Ni + V, ppm	0.2	0.2	—
Distillation, ASTM, °F	579–993	579–993	552–1031
Light isomate			
Gravity, °API	30.9	37.8	34.0
Sulfur, wt%	0.07	0.002	0.005
Nitrogen, ppm	18	20	20
Pour point, °F	18	14	18
Cetane index	51.5	53.0	53.5
Distillation, ASTM, °F	433–648	298–658	311–683
Heavy isomate			
Gravity, °API	27.1	29.2	30.7
Sulfur, wt%	0.26	0.009	0.013
Nitrogen, ppm	400	60	47
Viscosity, cSt at 122°F	26.2	19.8	17.2
Distillation, ASTM, °F	689–990	691–977	613–1026

11.4.13 LC-Fining Process

The LC-Fining process is a hydrocracking process capable of desulfurizing, demetallizing, and upgrading a wide spectrum of heavy feedstocks by means of an expanded-bed reactor (Table 11.13) (van Driesen et al., 1979; Fornoff, 1982; Bishop, 1990; RAROP, 1991, p. 61; Reich et al., 1993; Khan and Patmore, 1998; Speight and Ozum, 2002; Speight, 2014). Operating with the expanded bed allows the processing of heavy feedstocks, such as atmospheric residua, vacuum residua, and oil sand bitumen. The catalyst in the reactor behaves like fluid that enables the catalyst to be added to and withdrawn from the reactor during operation. The reactor conditions are near isothermal because the heat of reaction is absorbed by the cold fresh feed immediately owing to thorough mixing of reactors.

In the process (Table 11.7 and Figure 11.18), the feedstock and hydrogen are heated separately and then pass upward in the hydrocracking reactor through an expanded bed of catalyst (Speight, 2014). Reactor products flow to the high-pressure/high-temperature separator. Vapor effluent from the separator is let down in pressure and then goes to the heat exchange and thence to a section for the removal of condensable products and purification (Speight and Ozum, 2002; Speight, 2014).

Liquid is let down in pressure and passes to the recycle stripper. This is the most important part of the high-conversion process. The liquid recycle is prepared to the proper boiling range for return to the reactor. In this way the concentration of bottoms in the reactor, and therefore the distribution

TABLE 11.13

Feedstock and Product Data for the LC-Fining Process

	Kuwait Atmospheric Residuum	Gach Saran Vacuum Residuum	Arabian Heavy Vacuum Residuum	Athabasca Bitumen
Feedstock				
API gravity	15.0	6.1	7.5	9.1
Sulfur, % w/w	4.1	3.5	4.9	5.5
Nitrogen, % w/w				0.4
Products, % w/w				
Naphtha (C_5–205°C, C_5–400°F)	2.5	9.7	14.3	11.9
Sulfur, wt%				1.1
Nitrogen, wt%				
Distillate (205°C–345°C, 400°F–650°F)	22.7	14.1	26.5	37.7
Sulfur, wt%				0.7
Nitrogen, wt%				
Heavy distillate (345°C–525°C, 650°F–975°F)	34.7	24.1	31.1	30
Sulfur, wt%				1.1
Nitrogen, wt%				
Residuum (>525°C, >975°F)	35.5	47.5	21.3	12.9
Sulfur, wt%				3.4

Source: Speight, J.G., *The Chemistry and Technology of Petroleum*, 4th edn., CRC Press/Taylor & Francis Group, Boca Raton, FL, 2007.

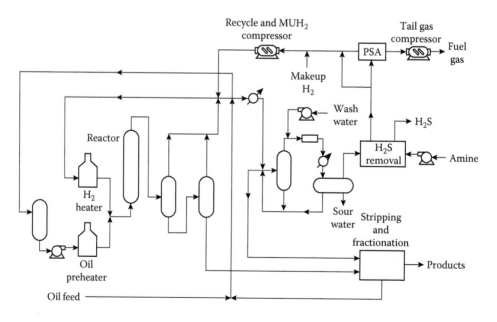

FIGURE 11.18   The LC-Fining process.

of products, can be controlled. After the stripping, the recycle liquid is then pumped through the coke precursor removal step where high-molecular-weight constituents are removed. The clean liquid recycle then passes to the suction drum of the feed pump. The product from the top of the recycle stripper goes to fractionation and any heavy oil product is directed from the stripper bottoms pump discharge.

The residence time in the reactor is adjusted to provide the desired conversion levels. Catalyst particles are continuously withdrawn from the reactor, regenerated, and recycled back into the reactor, which provides the flexibility to process a wide range of heavy feedstocks such as atmospheric and vacuum tower bottoms, coal-derived liquids, and bitumen. An internal liquid recycle is provided with a pump to expand the catalyst bed, continuously. As a result of expanded-bed operating mode, small pressure drops and isothermal operating conditions are accomplished. Small diameter extruded catalyst particles as small as 0.8 mm (1/32 inch) can be used in this reactor.

Although the process may not be the means by which direct conversion of the bitumen to a synthetic crude oil would be achieved, it does nevertheless offer an attractive means of bitumen conversion. Indeed, the process would play the part of the primary conversion process from which liquid products would accrue—these products would then pass to a secondary upgrading (hydrotreating) process to yield a synthetic crude oil.

11.4.14 MAKFINING PROCESS

The process uses a multicatalyst system in multibed reactors that include quench and redistribution system internals (Hunter et al., 1997; Speight and Ozum, 2002; Speight, 2014). In the process (Figure 11.19), the feedstock and recycle gas and preheated and brought into contact with the catalyst in a downflow fixed-bed reactor. The reactor effluent is sent to high- and low-temperature separators. Product recovery is a stripper/fractionator arrangement. Typical operating conditions in the reactors are 370°C–425°C (700°F–800°F) (single-pass) and 370°C–425°C (700°F–800°F) (recycle) with pressures of 1000–2000 psi (single-pass) and 1500–3000 psi (recycle). Product yields depend upon the extent of the conversion (Table 11.14).

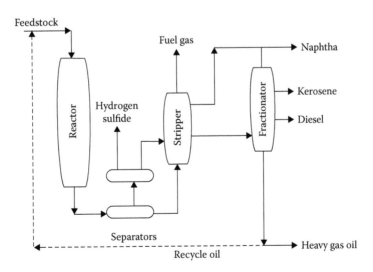

FIGURE 11.19 The MAKfining process.

TABLE 11.14

MAKfining Process Feedstock and Product Data

	AL/AH[a] Vacuum Gas Oil	AL/AH[a] Vacuum Gas Oil	AL/AH[a] Light Cycle Oil
Feedstock			
Gravity, °API	20.2	20.2	19.0
Sulfur, wt%	2.9	2.9	1.0
Nitrogen, wt%	0.9	0.9	0.6
Products, vol%			
Naphtha	12.9	22.6	54.0
Kerosene	14.1	24.5	
Diesel	31.8	32.5	54.3
Light gas oil	50.0	30.0	
Conversion, %	50.0	70.0	50.0

[a] AL/AH: Arabian light crude oil blended with Arabian heavy crude oil.

11.4.15 MICROCAT-RC PROCESS

The Microcat-RC process (also referred to as the M-Coke process) is a catalytic ebullated-bed hydroconversion process that is similar to Residfining and which operates at relatively moderate pressures and temperatures (Table 11.15 and Figure 11.20) (Bearden and Aldridge, 1981; Bauman et al., 1993). The catalyst particles, containing a metal sulfide in a carbonaceous matrix formed within the process, are uniformly dispersed throughout the feed. Because of their ultrasmall size (10^{-4}-inch diameter), there are typically several orders of magnitude more of these microcatalyst particles per cubic centimeter of oil than is possible in other types of hydroconversion reactors using conventional catalyst particles. This results in smaller distances between particles and less time for a reactant molecule or intermediate to find an active catalyst site. Because of their physical structure, microcatalysts suffer none of the pore-plugging problems that plague conventional catalysts.

In the process, fresh heavy feedstock, microcatalyst, and hydrogen are fed to the hydroconversion reactor. Effluent is sent to a flash separation zone to recover hydrogen, gases, and liquid products, including naphtha, distillate, and gas oil (Speight and Ozum, 2002; Speight, 2014). The residuum from the flash step is then fed to a vacuum distillation tower to obtain a 565°C⁻ (1050°F⁻)

TABLE 11.15

Microcat Process Feedstock and Product Data

Feedstock: Cold Lake heavy oil vacuum residuum	
Gravity, °API	4.4
Sulfur, wt%	
Nitrogen, wt%	
Metals (Ni + V), ppm	480.0
Carbon residue, wt%	24.4
Products, vol%	
Naphtha (C_5–177°C; C_5–350°F)	17.2
Distillate (177°C–343°C; 350°F–650°F)	63.6
Gas oil (343°C–566°C; 650°F–1050°F)	21.9
Residuum (>566°C; >1050°F)	2.1

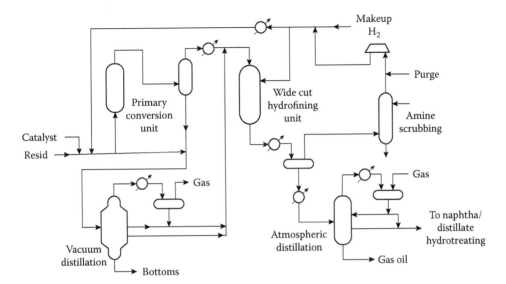

FIGURE 11.20 Microcat-RC process.

product oil and a 565°C+ (1050°F+) bottoms fraction that contains unconverted feed, microcatalyst, and essentially all of the feed metals.

Hydrotreating facilities may be integrated with the hydroconversion section or built on a stand-alone basis, depending on product quality objectives and owner preference.

11.4.16 MILD HYDROCRACKING PROCESS

The *mild hydrocracking process* uses operating conditions similar to those of a vacuum gas oil desulfurizer to convert a vacuum gas oil (VGO) to significant yields of lighter products (Table 11.16). Consequently, the flow scheme for a mild hydrocracking unit is virtually identical to that of a vacuum gas oil desulfurizer. For example, in a simplified process for vacuum gas oil desulfurization

---

TABLE 11.16

Process Parameters for Vacuum Gas Oil Hydrodesulfurization and Mild Hydrocracking Processes

	VGO HDS	MHC
Maximum pressure, bar (psi, abs)	75 (1090)	75 (1090)
Maximum temperature, °C (°F)	430 (806)	430 (806)
Makeup gas flow, N·m³/T (SCFB)	90 (500)	150 (835)
Hydrogen content, mol%	90	90
Recycle gas flow, N·m³/T (SCFB)	500 (2780)	500 (2780)
Hydrogen content, mol%	80	80
H$_2$S content, mol%	0	0
LHSV, h^{-1}	0.5	0.5
Catalyst	S-424, dense loaded	MHC-1
Specification	0.3 wt% S in 370°C+ (698°F+) product	0.3 wt% S at EOR conditions
Cycle length, months	11	11

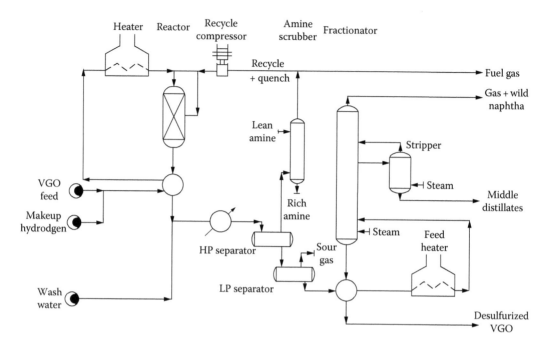

FIGURE 11.21 Vacuum gas oil desulfurization with mild hydrocracking.

(Figure 11.21), the vacuum gas oil feedstock is mixed with hydrogen makeup gas and preheated against reactor effluent. Further preheating to reaction temperature is accomplished in a fired heater. The hot feed is mixed with recycle gas before entering the reactor. The temperature rises across the reactor due to the exothermic heat of reaction. Catalyst bed temperatures are usually controlled by using multiple catalyst beds and by introducing recycle gas as an interbed quench medium. Reactor effluent is cooled against incoming feed and air or water before entering the high-pressure separator. Vapors from this separator are scrubbed to remove hydrogen sulfide (H_2S) before compression back to the reactor as recycle and quench. A small portion of these gases is purged to fuel gas to prevent buildup of light ends. Liquid from the high-pressure separator is flashed into the low-pressure separator. Sour flash vapors are purged from the unit. Liquid is preheated against stripper bottoms and in a feed heater before steam stripping in a stabilizer tower. Water wash facilities are provided upstream of the last reactor effluent cooler to remove ammonium salts produced by denitrogenation of the vacuum gas oil feedstock.

Variation of this process leads to the hot separator design. The process flow scheme is identical to that described earlier up to the reactor outlet. After initial reactor effluent cooling against incoming vacuum gas oil feed and makeup hydrogen, a hot separator is installed. Hot liquid is routed directly to the product stabilizer. Hot vapors are further cooled by air and/or water before entering the cold separator. This arrangement reduces the stabilizer feed preheat duty and the effluent cooling duty by routing hot liquid direct to the stripper tower.

The conditions for mild hydrocracking are typical of many low-pressure desulfurization units that for hydrocracking units, in general, are marginal in pressure and hydrogen oil ratio capabilities. For hydrocracking, in order to obtain satisfactory run lengths (approximately 11 months), reduction in feed rate or addition of an extra reactor may be necessary. In most cases, since the product slate will be lighter than for normal desulfurization service only, changes in the fractionation system may be necessary. When these limitations can be tolerated, the product value from mild hydrocracking versus desulfurization can be greatly enhanced.

In summary, the so-called mild hydrocracking process is a simple form of hydrocracking. The hydrotreaters designed for vacuum gas oil desulfurization and catalytic cracker feed pretreatment are converted to once-through hydrocracking units and, because existing units are being used, the hydrocracking is often carried out under nonideal hydrocracking conditions.

11.4.17 MRH PROCESS

The MRH process is a hydrocracking process designed to upgrade heavy feedstocks containing large amount of metals and asphaltene, such as vacuum residua and bitumen, and to produce mainly middle distillates (Sue, 1989; RAROP, 1991, p. 65; Khan and Patmore, 1998; Speight and Ozum, 2002; Speight, 2014). The reactor is designed to maintain a mixed three-phase slurry of feedstock, fine powder catalyst and hydrogen, and to promote effective contact.

In the process (Figure 11.22), a slurry consisting of heavy oil feedstock and fine powder catalyst is preheated in a furnace and fed into the reactor vessel. Hydrogen is introduced from the bottom of the reactor and flows upward through the reaction mixture, maintaining the catalyst suspension in the reaction mixture. Cracking, desulfurization and demetallization reactions are taken place via thermal and catalytic reactions. In the upper section of the reactor, vapor is disengaged from the slurry, and hydrogen and other gases are removed in a high-pressure separator. The liquid condensed from the overhead vapor is distilled and then flows out to the secondary treatment facilities. From the lower section of the reactor, a bottom slurry oil (VGO) that contains catalyst, uncracked residuum, and a small amount of vacuum gas oil fraction are withdrawn. Vacuum gas oil is recovered in the slurry separation section, and the remaining catalyst and coke are fed to the catalyst regenerator.

Product distribution focuses on middle distillates (Table 11.17) with the process focused as a heavy feedstock processing unit and can be inserted into refinery operations downstream from the vacuum distillation unit.

11.4.18 RCD UNIBON PROCESS

The RCD Unibon process (BOC process) is a process to upgrade vacuum residua (Table 11.18) (RAROP, 1991, p. 67; Thompson, 1997; Khan and Patmore, 1998; Speight and Ozum, 2002;

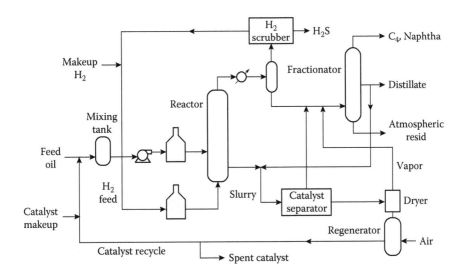

FIGURE 11.22 The MRH process.

TABLE 11.17

MRH Process Feedstock and Product Data

	Arabian Heavy Vacuum Residuum	Athabasca (Canada) Bitumen
Feedstock		
Gravity, °API	5.9	10.2
Sulfur, wt%	5.1	4.3
Nitrogen, wt%	0.3	0.4
C_7 Asphaltenes, wt%	11.4	8.1
Metals, ppm		
Nickel	41.0	85.0
Vanadium	127.0	182.0
Carbon residue, wt%	21.7	13.3
Distillation profile, vol%		
Naphtha	0.0	2.2
Kerosene	0.0	5.3
Light gas oil	0.0	12.1
Vacuum gas oil	4.0	31.8
Vacuum residuum	96.0	48.6
(Atmospheric residuum)	(100.0)	(80.4)
Products, wt%		
Naphtha	13.0	120
Sulfur	0.2	1.1
Nitrogen	0.03	0.05
Kerosene	6.0	11.0
Sulfur	1.0	1.2
Nitrogen	0.06	0.08
Light gas oil	17.0	29.0
Sulfur	2.5	2.2
Nitrogen	0.06	0.11
Atmospheric residuum	55.0	41.0
Sulfur	3.8	3.8
Nitrogen	0.36	0.57

Speight, 2014). There are several possible flow scheme variations involving for the process. It can operate as an independent unit or be used in conjunction with a thermal conversion unit (Figure 11.23). In this configuration, hydrogen and a vacuum residuum are introduced separately to the heater, and mixed at the entrance to the reactor. To avoid thermal reactions and premature coking of the catalyst, temperatures are carefully controlled and conversion is limited to approximately 70% of the total projected conversion. The removal of sulfur, heptane-insoluble materials, and metals is accomplished in the reactor. The effluent from the reactor is directed to the hot separator. The overhead vapor phase is cooled, condensed, and the separated hydrogen is recycled to the reactor.

Liquid product goes to the thermal conversion heater where the remaining conversion of non-volatile materials occurs. The heater effluent is flashed and the overhead vapors are cooled, condensed, and routed to the cold flash drum. The bottoms liquid stream then goes to the vacuum column where the gas oils are recovered for further processing, and the residuals are blended into the heavy fuel oil pool.

TABLE 11.18
RCD Unibon Process Feedstock and Product Data

	Unspecified Atmospheric Residuum	Arabian Light, Vacuum Residuum	Arabian Light, Vacuum Residuum[a]
Feedstock			
Gravity, °API	16.4	7.2	7.2
Sulfur, wt%	3.5	4.0	4.0
Nitrogen, wt%	0.2	0.3	0 3
C_7 Asphaltenes, wt%	2.4		
Carbon residue, wt%	9.5		
Products, wt%			
Naphtha (C_5–180°C; C_5–355°F)	1.0	5.1	91
Sulfur		0.04	0.09
Nitrogen		0.01	0.01
Fuel oil (>190°C; >375°F)	100.00		
Sulfur	0.3		
Nitrogen	0.1		
Carbon residue	3.8		
Light gas oil (180°C–343°C; 355°F–650°F)		6.5	13.4
Sulfur		0.12	0.3
Nitrogen		0.11	0.1
Vacuum gas oil (343°C–566°C; 650°F–1050°F)		31.5	113
Sulfur		0.4	0.6
Nitrogen		0.2	0.2
Vacuum residuum (>566°C; >1050°F)		53.0	34.4
Sulfur		1.0	1.1
Nitrogen		0.3	

[a] With thermal conversion.

11.4.19 RESIDFINING PROCESS

Residfining is a catalytic fixed-bed process for the desulfurization and demetallization of atmospheric and vacuum residua (Table 11.19) (RAROP, 1991, p. 69; Khan and Patmore, 1998; Speight and Ozum, 2002; Speight, 2014). The process can also be used to pretreat heavy feedstocks to suitably low contaminant levels prior to catalytic cracking.

In the process (Figure 11.24), liquid feedstock to the unit is filtered, pumped to pressure, preheated, and combined with treated gas prior to entering the reactors. A small guard reactor would typically be employed to prevent plugging/fouling of the main reactors. Provisions are employed to periodically remove the guard while keeping the main reactors online. The temperature rise associated with the exothermic reactions is controlled utilizing either a gas- or liquid-quench. A train of separators is employed to separate the gas and liquid products. The recycle gas is scrubbed to remove ammonia (NH_3) and hydrogen sulfide (H_2S). It is then combined with fresh makeup hydrogen before being reheated and recombined with fresh feed. The liquid product is sent to a fractionator where the product is fractionated.

Residfining is an option that can be used to reduce the sulfur, to reduce metals and coke-forming precursors, and/or to accomplish some conversion to lower-boiling products as a feed pretreat step

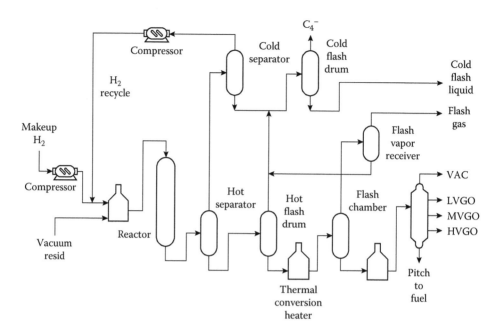

FIGURE 11.23   RCD Unibon BOC process.

ahead of a fluid catalytic cracking unit. There is also a hydrocracking option where substantial conversion of the resid occurs.

11.4.20   RESIDUE HYDROCONVERSION PROCESS

The residue hydroconversion (RHC) process (Figure 11.25) is a high-pressure fixed-bed trickle-flow hydrocatalytic process (Table 11.20) (RAROP, 1991, p. 71; Khan and Patmore, 1998). The feedstock can be desalted atmospheric or vacuum residue as well as other heavy feedstocks such as heavy oil, extra heavy oil, and tar sand bitumen (Chapter 1) (Speight and Ozum, 2002; Speight, 2014).

The reactors are of multibed design with interbed cooling and the multicatalyst system can be tailored according to the nature of the feedstock and the target conversion. For residua with high metal content, a hydrodemetallization catalyst is used in the front-end reactor(s), which excels in its high metal uptake capacity and good activities for metal removal, asphaltene conversion, and residue cracking. Downstream of the demetallization stage, one or more hydroconversion stages, with optimized combination of catalysts' hydrogenation function and texture, are used to achieve desired catalyst stability and activities for denitrogenation, desulfurization, and heavy hydrocarbon cracking. A guard reactor may be employed to remove contaminants that promote plugging or fouling of the main reactors with periodic removal of the guard reactor while keeping the main reactors online.

11.4.21   SHELL RESIDUAL OIL PROCESS

The Shell residual oil hydrodesulfurization process was originally designed to improve the quality of residual oils by removing sulfur, metals, and asphaltene constituents. The process is suitable for a wide range of the heavier feedstocks, irrespective of the composition and origin, and even includes those feedstocks that are particularly high in metals and asphaltene constituents.

TABLE 11.19
Residfining Process Feedstock and Product Data

	Gach Saran Atmospheric Residuum	Arabian Heavy Atmospheric Residuum	Arabian Light Atmospheric Residuum	AL/AH[a] Vacuum Residuum
Feedstock				
API gravity	15.0	12.3	14.3	4.7
Sulfur, wt%	2.5	4.2	3.5	5.3
Nitrogen, wt%			0.2	0.4
C_7 Asphaltenes, wt%				
Carbon residue, wt%			9.8	24.6
Metals, ppm				
Nickel			6	50.0
Vanadium			38	170.0
Products, wt%				
Naphtha (C_5–205°C; C_5–400°F)	3.4	6.0		
Naphtha (C_5–220°C; C_5–430°F)			1.9	
Residuum (>205°C; >400°F)	98.1	96.4		
Light gas oil (220°C–345°C; 430°F–650°F)			11.2	
Distillate (C_5–345°C; C_5–650°F)				13.7
Vacuum gas oil				31.0
Sulfur, wt%				<0.1
Carbon residue, wt%				0.3
Residuum (>345°C; >650°F)			82.2	
Sulfur, wt%			0.1	
Carbon residue, wt%			3.2	
Residuum (>565°C; >1050°F)				50.0
Sulfur, wt%				0.8
Carbon residue, wt%				15.7

[a] Arabian light/Arabian heavy blend.

The process centers on a fixed-bed downflow reactor that allows catalyst replacement without causing any interruption in the operation of the unit. Feedstock is introduced to the process via a filter (backwash, automatic) after which hydrogen and recycle gas are added to the feedstock stream which is then heated to reactor temperature by means of feed-effluent heat exchangers whereupon the feed stream passes down through the reactor in trickle flow. Sulfur removal is excellent and substantial reductions in the vanadium content and asphaltene constituents also occur. In addition, a marked increase occurs in the API gravity, and the viscosity is reduced considerably.

A bunker reactor provides extra process flexibility if it is used upstream from the desulfurization reactor, especially with reference to the processing of feedstocks with a high metal content. A catalyst with a capacity for metals is employed in the bunker reactor to protect the desulfurization catalyst from poisoning by the metals. In the bunker reactor, inverted cone segments support the catalyst and are designed to allow catalyst removal.

11.4.22 Tervahl-H Process

In the Tervahl-H process, the feedstock and hydrogen-rich stream are heated using heat recovery techniques and fired heater and held in the soak drum as in the Tervahl-T process. The gas and oil

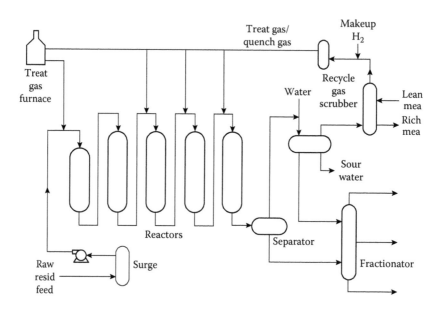

FIGURE 11.24 The Residfining process.

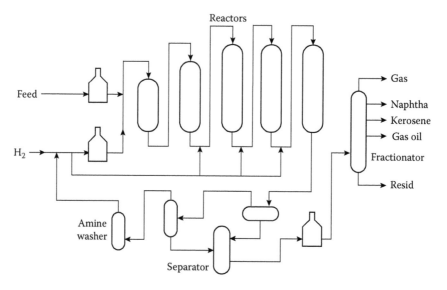

FIGURE 11.25 The RHC process.

from the soaking drum effluent are mixed with recycle hydrogen and separated in the hot separator where the gas is cooled passed through a separator and recycled to the heater and soaking drum effluent. The liquids from the hot and cold separator are sent to the stabilizer section where purge gas and synthetic crude are separated. The gas is used as fuel and the synthetic crude can now be transported or stored.

In the related Tervahl-T process (a thermal process but covered here for convenient comparison with the Tervahl-T process; see Section 2.7), the feedstock is heated to the desired temperature using the coil heater and heat recovered in the stabilization section and held for a specified residence time in the soaking drum. The soaking drum effluent is quenched and sent to a conventional stabilizer or

TABLE 11.20
RHC Process Feedstock and Product Data

Feedstock: Unspecified residuum

API gravity	24.4
Sulfur, wt%	0.1
Nitrogen, wt%	0.1
Carbon residue, wt%	0.3
Products, vol%	
Naphtha	4.5
Light gas oil	19.4
Vacuum gas oil	77.10
Sulfur, wt%	0.01
Carbon residue, wt%	0.20

fractionator where the products are separated into the desired streams. The gas produced from the process is used for fuel.

11.4.23 UNICRACKING PROCESS

Unicracking is a fixed-bed catalytic process that employs a high-activity catalyst with a high tolerance for sulfur and nitrogen compounds and can be regenerated. The design is based upon a single-stage or a two-stage system with provisions to recycle to extinction (Figure 11.26) (RAROP, 1991, p. 79; Reno, 1997; Khan and Patmore, 1998; Ackelson, 2004). In the process, a two-stage reactor system receives untreated feed, makeup hydrogen, and a recycle gas at the first stage, in which naphtha conversion may be as high as 60% v/v. The reactor effluent is separated to recycle gas, liquid product, and unconverted oil (Tables 11.21 and 11.22). The second-stage oil may be either once-through or recycle cracking; feed to the second sage is a mixture of unconverted first-stage oil and second-stage recycle.

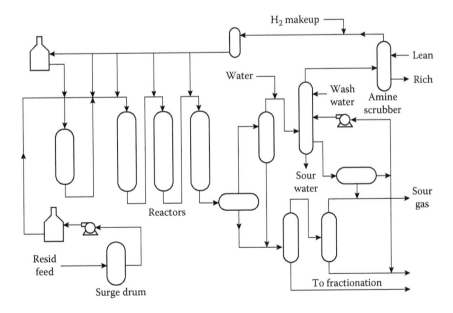

FIGURE 11.26 The unicracking process.

TABLE 11.21

Unicracking Process Feedstock and Product Data

	Alaska North Slope Atmospheric Residuum	Gach Saran Atmospheric Residuum	Kuwait Atmospheric Residuum	Kuwait Atmospheric Residuum	California Atmospheric Residuum
Feedstock					
API gravity	15.2	16.3	16.7	14.4	9.9
Sulfur, wt%	1.7	2.4	3.8	4.2	4.5
Nitrogen, wt%	0.4	0.4	0.2	0.2	0.4
Metals (Ni + V), ppm	44.0	220.0	46.0	66.0	213.0
Carbon residue, wt%	8.4	8.5	8.5	10.0	13.6
Products, vol%					
Naphtha (<185°C; <365°F)	0.8	1.8	1.1		
Naphtha (C_5–205°C; C_5–400°F)				2.1	4.2
Light gas oil				11.1	18.0
Residuum (>185°C; >365°F)	100.5	100.4	100.4		
API gravity	19.8	22.8	24.4		
Sulfur, wt%	0.3	0.3	0.3		
Nitrogen, wt%	0.3	0.3	0.1		
Metals (Ni + V), ppm	14.0	55.0	15.0		
Carbon residue, wt%	4.0	4.0	3.0		
Residuum (>345°C; >650°F)				89.5	81.6
API gravity				22.7	21.6
Sulfur, wt%				<0.3	<0.3
Nitrogen, wt%				<0.2	<0.2
Metals (Ni + V), ppm				<25	<5

TABLE 11.22

Unicracking Desulfurization Rate Data

Crude Source	Resid Properties, 650°F+					
	Crude (vol%)	Gravity (°API)	Sulfur (wt%)	Ni + V (ppm)	Conradson Carbon (wt%)	Relative Desulfurization Rate
Murban	35.7	22.8	1.5	4	3.7	18.0
North Slope	52.6	15.1	1.7	45	8.9	7.75
Agha Jari (Iranian light)	42.7	17.5	2.4	98	7.7	7.25
Venezuelan (Leona)	55.0	18.9	2.1	187	8.6	7.0
Arabian light	44.7	17.9	2.9	28	7.3	6.5
L.A. Basin (Torrey Canyon)	48.4	8.4	2.0	236	11.7	4.5
West Texas Sour	41.6	15.5	3.4	29	9.0	4.5
Gach Saran (Iranian heavy)	46.7	14.6	2.2	258	9.9	4.25
Sassan	44.7	16.0	3.4	38	8.5	4.25
Kirkuk	39.3	17.6	3.7	84	8.4	4.0
Kuwait	50.5	15.5	3.9	51	8.8	3.5
Safaniyah (Arabian heavy)	53.1	12.3	4.2	123	13.3	1.5
Khafji	53.0	11.9	4.3	75	15.6	1.25
Ratawi	62.0	10.2	5.0	57	14.1	1.0

The feedstock and hydrogen-rich recycle gas are preheated, mixed, and introduced into a guard reactor that contains a relatively small quantity of the catalyst. The guard chamber removes particulate matter and residual salt from the feed. The effluent from the guard chamber flows down through the main reactor, where it contacts one or more catalysts designed for removal of metals and sulfur. The catalysts, which induce desulfurization, denitrogenation, and hydrocracking, are based upon both amorphous and molecular-sieve containing supports. The product from the reactor is cooled, separated from hydrogen-rich recycle gas, and either stripped to meet fuel oil flash point specifications, or fractionated to produce distillate fuels, upgraded vacuum gas oil, and upgraded vacuum residuum. Recycle gas, after hydrogen sulfide removal, is combined with makeup gas and returned to the guard chamber and main reactors.

The most commonly implemented configuration is a single-stage Unicracking design, where the fresh feed and recycle oil are converted in the same reaction stage. This configuration simplifies the overall unit design by reducing the quantity of equipment in high-pressure service and keeping high-pressure equipment in a single train. The two-stage design has a separation system in each reaction stage. Two-stage flow schemes can be employed in specific situations such as the two-stage Unicracking process which can be a separate hydrotreating stage or a two-stage hydrocracking process. In the separate hydrotreating flow scheme the first stage provides only hydrotreating, while in the two-stage hydrocracking process the first stage provides hydrotreating and partial conversion of the feed. The second stage of the two-stage design provides the remaining conversion of recycle oil so that overall high conversion from the unit is achieved. These flow schemes offer several advantages in processing heavier and highly contaminated feeds. Two-stage flow schemes are economical when the throughput of the unit is relatively high but the overall optimum flow scheme depends on (1) the feedstock type, (2) the feedstock capacity, and (3) the product slate objectives. Also, the design of hydrocracking catalyst changes depending upon the type of flow scheme employed. The hydrocracking catalyst needs to function within the reaction environment and severity created by the flow scheme that is chosen. Moreover, the two-stage flow scheme provides a unique reaction environment for the second-stage hydrocracking catalyst and, having come through the first stage, the feedstock is cleaner and less likely to foul the catalyst, thereby offering a significant boost to the cracking activity and life cycle of the catalyst.

In addition, further advances in the two-stage Unicracking process design have included several innovations in each reaction section of the design. The pretreating section uses a high-activity pretreating catalyst that allows hydrotreating at a higher severity, providing good quality feed for the first-stage hydrocracking section and enabling maximum first-stage selectivity to high-quality distillate. The second stage is optimized by use of the second-stage hydrocracking catalyst that is specifically designed to take advantage of the cleaner reaction environment. The second-stage catalyst is designed so that the cracking and metal functions are balanced and, at the same time, the second-stage hydrocracking severity is optimized so that maximum distillate selectivity is obtained from the second stage of hydrocracking (Thakkar et al., 2007).

The high efficiency of the process is due to the efficient distribution of the feedstock and hydrogen that occurs in the reactor where a proprietary liquid distribution system is employed. In addition, the process catalyst (also proprietary) was designed for the desulfurization of residua and is not merely an upgraded gas oil hydrotreating catalyst as often occurs in various processes. It is a cobalt–molybdena–alumina catalyst with a controlled pore structure that permits a high degree of desulfurization and, at the same time, minimizes any coking tendencies.

The process uses base metal or noble metal hydrogenation-activity promoters impregnated on combinations of zeolites and amorphous aluminosilicates for cracking activity (Reno, 1997). The specific metals chosen and the proportions of the metals, zeolite, and non-zeolite aluminosilicates are optimized for the feedstock and desired product balance. This is effective in the production of clean fuels, especially for cases where a partial conversion Unicracking unit and a fluid catalytic cracking unit are integrated.

The Unicracking process converts feedstocks into lower-molecular-weight products that are more saturated than the feed. Feedstocks include atmospheric gas oil, vacuum gas oil, fluid catalytic cracking/resid catalytic cracking cycle oil, coker gas oil, deasphalted oil, and naphtha. Hydrocracking catalysts promote sulfur and nitrogen removal, aromatic saturation, and molecular weight reduction. All of these reactions consume hydrogen and as a result, the volume of recovered liquid product normally exceeds the feedstock. Many units are operated to make naphtha (for petrochemical or motor-fuel use) as a primary product.

Unicracking catalysts are designed to function in the presence of hydrogen sulfide (H_2S) and ammonia (NH_3). This gives rise to an important difference between Unicracking and other hydrocracking processes: the availability of a single-stage design. In a single-stage unit, the absence of a stripper between treating and cracking reactors reduces investment costs by making use of a common recycle gas system. Process objectives determine catalyst selection for a specific unit. Product from the reactor section is condensed, separated from hydrogen-rich gas and fractionated into desired products. Unconverted oil is recycled or used as (1) blend stock for the production of lubricating oil, (2) fluid catalytic cracking feedstock, or (3) as feedstock for the ethylene cracking unit (Parihar et al., 2012). In addition, *mild hydrocracking* technology enables optimization of hydroprocessing refinery assets to produce high-quality clean fuels at lower costs and more attractive return on investments than alternative technologies.

The *advanced partial conversion unicracking (APCU) process* is a recent advancement in the area of ultra-low-sulfur diesel (ULSD) production and feedstock pretreatment for catalytic cracking units. At low conversions (20%–50%) and moderate pressure, the Advanced Partial Conversion Unicracking technology provides an improvement in product quality compared to traditional mild hydrocracking. In the process, high-sulfur feeds such as vacuum gas oil and heavy cycle gas oil are mixed with a heated hydrogen-rich recycle gas stream and passed over consecutive beds of high-activity pretreated catalyst and distillate selective unicracking catalyst. This combination of catalysts removes refractory sulfur and nitrogen contaminants, saturates polynuclear aromatic compounds, and converts a portion of the feed to ultra-low sulfur diesel fuel. The hydrocracked products and desulfurized feedstock for a fluid catalytic cracking unit are separated at reactor pressure in an enhanced hot separator. The overhead products for the separator are immediately hydrogenated in the integrated finishing reactor.

As pretreatment severity increases, conversion increases in the fluid catalytic cracker and both naphtha and alkylate octane-barrel output per barrel of cat cracker feedstock also increase. Advanced Partial Conversion Unicracking units can be customized to achieve maximum octane-barrel production in the cat cracker.

Another development in the unicracking family is the HyCycle Unicracking technology that is designed to maximize diesel production for full conversion applications and is an optimized process scheme intended for obtaining maximum yield of high-quality diesel fuel. The process is characterized by lowered pressure, higher space velocity in comparison with conventional units. Due to minimizing potential secondary cracking reactions, less hydrogen per barrel of feedstock is required.

11.4.24 Uniflex Process

The Uniflex process is an evolved version with significant changes (by UOP) of the former CANMET process which used an empty vessel hydrocracking reactor in which the feedstock is processed in the presence of an iron sulfide-based catalyst deposited on particles of coal. The process (Figure 11.27) is a slurry hydrocracking process which achieves higher conversion and produces two times the diesel yield compared against delayed coking, which can lead to double the refinery profit margin. Because the desulfurization activity of iron is very low, molybdenum can be added at a level of tens of ppm in the form of molybdenum naphthenate. The reaction products were fractionated and sent to the hydrotreatment unit while the unconverted residue (5%–10% v/v of the feedstock) can be burned or gasified.

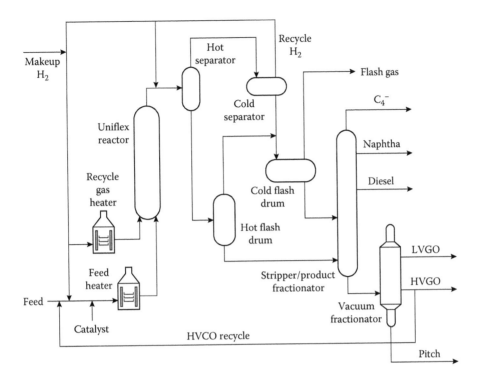

FIGURE 11.27   Uniflex process.

The process uses an empty vessel hydrocracking reactor in which the feedstock is processed in the presence of an iron sulfide-based catalyst deposited on particles of coal. Because the desulfurization activity of iron is very low, molybdenum can be added at a level of tens of ppm in the form of molybdenum naphthenate. The reaction products are fractionated and sent to the hydrotreatment unit (Unifining and Unicracking), while the unconverted residue (5%–10% v/v of the feedstock) can be burned or gasified.

The flow scheme for the Uniflex process is similar to that of a conventional UOP Unicracking process unit—liquid feedstock and recycle gas are heated to temperature in separate heaters, with a small portion of the recycle gas stream and the required amount of catalyst being routed through the oil heater (Gillis et al., 2010). The outlet streams from both heaters are fed to the bottom of the slurry reactor. The reactor effluent is quenched at the reactor outlet to terminate reactions and then flows to a series of separators with gas being recycled back to the reactor. Liquids flow to the unit's fractionation section for recovery of light ends, naphtha, diesel, vacuum gas oils, and pitch (cracked residuum). Heavy vacuum gas oil is partially recycled to the reactor for further reaction and conversion.

The heart of the Uniflex process is the upflow reactor that operates at moderate temperature (440°C–470°C; 815°F–880°F) and 2000 psi. The feedstock distributor, in combination with optimized process variables, promotes intense back-mixing (which provides near isothermal reactor conditions) in the reactor without the need for reactor internals or liquid recycle ebullating pumps. The back-mixing allows the reactor to operate at the higher temperatures required to maximize vacuum residue conversion. The majority of the products vaporize and quickly leave the reactor (thereby minimizing the potential for secondary cracking reactions), while the residence time of the higher-boiling constituents of the feedstock is maximized.

The process employs a proprietary, dual-function nanosized solid catalyst that is blended with the feed to maximize conversion of heavy components and inhibit coke formation. Specific catalyst requirements depend on feedstock quality and the required severity of operation. The primary function of the catalysts is to effect mild hydrogenation activity for the stabilization of cracked products while also limiting the saturation of aromatic rings. Because of the hydrogenation function, the catalyst also decouples the relationship between conversion and the propensity for carbon residue formation of the feedstock.

11.4.25 VEBA COMBI CRACKING PROCESS

The Veba Combi cracking process is a thermal hydrocracking process for converting residua and other heavy feedstocks (Table 11.23 and Figure 11.28) (Niemann et al., 1988; RAROP, 1991, p. 81; Wenzel and Kretsmar, 1993; Speight and Ozum, 2002; Speight, 2014). The process is based on the Bergius-Pier technology that was used for coal hydrogenation in Germany up to 1945. The heavy feedstock is hydrogenated (hydrocracked) using a commercial catalyst and liquid-phase

TABLE 11.23
Veba Combi Cracking Process Feedstock and Product Data

Feedstock: Arabian heavy vacuum residuum[a]

API gravity	3.4
Sulfur, wt%	5.5
C_7 Asphaltenes, wt%	13.5
Metals (Ni + V), ppm	230.0
Carbon residue, wt%	8.4
Products, vol%	
Naphtha (<170°C; <340°F)	26.9
Middle distillate (170°C–370°C; 340°F–700°F)	36.5
Gas oil (>370°C; 700°F)	19.9

[a] >550°C (>1025°F); conversion 95%.

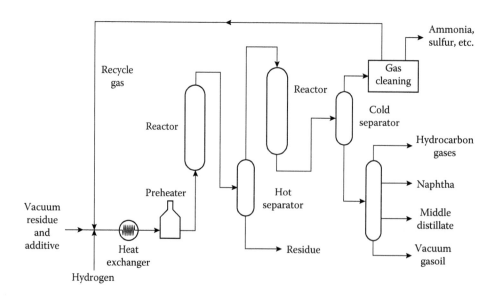

FIGURE 11.28 The Veba Combi cracking process.

hydrogenation reactor operating at 440°C–485°C (825°F–905°F) and 2175–4350 psi. The product obtained from the reactor is fed into the hot separator operating at temperatures slightly below the reactor temperature. The liquid and solid materials are fed into a vacuum distillation column, the gaseous products are fed into gas-phase hydrogenation reactor operating at an identical pressure (Graeser and Niemann, 1982). This high-temperature, high-pressure coupling of the reactor products with further hydrogenation can provide favorable process economics (Figure 11.28).

In the process, the residue feed is slurried with a small amount of finely powdered additive and mixed with hydrogen and recycle gas prior to preheating. The feed mixture is routed to the liquid-phase reactors. The reactors are operated in an upflow mode and arranged in series. In a once-through operation, conversion rates of >95% are achieved. Substantial conversion of asphaltene constituents, desulfurization, and denitrogenation takes place at high levels of residue conversion. Temperature is controlled by a recycle gas quench system.

The flow from the liquid-phase hydrogenation reactors is routed to a hot separator, where gases and vaporized products are separated from unconverted material. A vacuum flash recovers distillates in the hot separator bottom product. The hot separator top product, together with recovered distillates and straight-run distillates, enters the gas-phase hydrogenation reactor. The gas-phase hydrogenation reactor operates at the same pressure as the liquid-phase hydrogenation reactor and contains a fixed bed of commercial hydrotreating catalyst. The operation temperature (340°C–420°C, 645°F–790°F) is controlled by a hydrogen quench. The system operates in a trickle flow mode, which may not be efficient for some heavy feedstocks. The separation of the synthetic crude from associated gases is performed in a cold separator system. The synthetic crude may be sent to stabilization and fractionation unit as required. The gases are sent to a lean oil scrubbing system for contaminant removal and are recycled.

The hydrotreating stage is typically a catalytic fixed-bed reactor operated under essentially the same pressure as the primary conversion stage. This second stage may be designed for either hydrotreating or hydrocracking applications. Additional low-value refinery streams such as gas oil, deasphalted oil, or catalytic cracker cycle oil may also be directly added to the hydrotreating stage.

11.5   OTHER OPTIONS

The *heavy residue hydroconversion* (HRH) process is a new nanocatalytic technology for upgrading residua, heavy oil, extra heavy crude oil, and tar sand bitumen (Khadzhiev et al., 2009; Zarkesh et al., 2011). In the process, the heavy feedstock is introduced to a separator to separate any lower-boiling constituents after which the nonvolatile material is sent to the reactor where mixing with hydrogen and catalyst occurs. The catalyst precursors react *in situ* with hydrogen sulfide in the reactor and produces the nanocatalyst. The reacted feedstock then passes into the distillation unit and unreacted portion recycles to the beginning of the process. A defined portion of this residue goes to the catalyst regeneration unit. The nature of process is such that it can tolerate high amount of heavy metals, asphaltene constituents, and sulfur with an overall feedstock conversion on the order of 95% v/v. The main advantages of the process are (1) high conversion, (2) high product volume yield, (3) 60%–80% sulfur removal, (4) the catalyst regenerates in the HRH unit, and (5) the heavy metals convert to the metal oxides as a by-product.

Heavy crude oil, sour crude oil, and tar sand bitumen, which require more energy-intensive processing than conventional crude oil, will contribute a growing fraction of fuels production. As existing reserves of conventional oil are depleted and there is greater worldwide competition for premium (e.g., light, sweet) crude oil, refineries will increasingly utilize heavy oil, sour crude oil, and tar sand bitumen to meet demand.

However, like many refinery processes, the hydrocracking process can succumb to the problems encountered in hydrocracking heavy feedstocks, which can be directly equated to the amount of complex, higher-boiling constituents that may require pretreatment (Speight and Moschopedis, 1979;

Reynolds and Beret, 1989; Speight, 2000, 2014; Speight and Ozum, 2002; Hsu and Robinson, 2006; Gary et al., 2007). Processing these feedstocks is not merely a matter of applying know-how derived from refining *conventional* crude oils but requires knowledge of composition and properties (Chapter 1) (Speight, 2001, 2015). The attempts to modify the process (see Section 11.4) have had some measure of success; materials are not only complex in terms of the carbon number and boiling point ranges but also because a large part of this *envelope* falls into a range of model compounds where very little is known about the properties.

Hydrocracking adds flexibility and offers the refiner a process that can handle varying feeds and operate under diverse process conditions. Utilizing different types of catalysts can modify the product slate produced. Reactor design and number of processing stages play a role in this flexibility.

Furthermore, it is apparent that the conversion of heavy oils and residua requires new lines of thought to develop suitable processing scenarios (Celestinos et al., 1975). Indeed, the use of thermal process (*carbon rejection processes*) and of hydrothermal processes (*hydrogen addition processes*), which were inherent in the refineries designed to process lighter feedstocks, has been a particular cause for concern. This has brought about, and will continue to bring about in the refinery of the future, the evolution of processing schemes that accommodate the heavier feedstocks (Wilson, 1985; Boening et al., 1987; Khan and Patmore, 1998; Speight and Ozum, 2002; Hsu and Robinson, 2006; Speight, 2014).

For the heavy feedstocks, which will increase in amounts in terms of hydrocracking feedstocks, reactor designs will continue to focus on online catalyst addition and withdrawal. Fixed-bed designs have suffered from (1) mechanical inadequacy when used for the heavier feedstocks and (2) short catalyst lives—6 months or less—even though large catalyst volumes are used (LHSV typically of 0.5–1.5). Refiners will attempt to overcome these shortcomings by innovative designs, allowing better feedstock flow and catalyst utilization or online catalyst removal. For example, the OCR process, in which a lead, moving-bed reactor is used to demetallize heavy feed ahead of the fixed-bed hydrocracking reactors, has seen some success. But whether this will be adequate for continuous hydrocracking heavy feedstock remains a question.

Improved catalysts are now available based on a better understanding of asphaltene chemistry, and this will be a focus of catalyst manufacture—to hydrocrack asphaltene constituents without serous deleterious effect on the catalyst. In this respect, a reexamination of the CANMET process is warranted since processing Athabasca bitumen using this process gave good result and the iron-based catalyst also acted as a scavenger for coke formers—other options include the addition of metal oxides as scavengers for coke formers and sulfur. However, the refining industry has remained cautious, and investment in these technologies has been disappointing. In fact, application of hydrogen addition technologies to heavy feedstocks only account less than one-third of the global residue upgrading capacity.

The use of ebullating-bed technologies was first introduced in the 1960s in an attempt to overcome problems of catalyst aging and poor distribution in fixed-bed designs. Hydrogen and feed enter at the bottom of the reactor, thereby expanding the catalyst bed. Although catalyst performance can be kept constant because catalyst can be replaced online, the ebullition results in a back-mixed reactor; therefore, desulfurization and hydroconversion are less than obtainable in a fixed-bed unit. Currently, in order to limit coking most commercial ebullating-bed units operate in the 70%–85% desulfurization range and 50%–70% v/v resid conversion.

Development work will continue and ebullating-bed units will see more use and have a greater impact on resid conversion operations. Improvements such as (1) second generation catalyst technology, which will allow higher conversion to a stable product, (2) catalyst rejuvenation, which allows spent catalyst to be reused to a greater extent than current operations allow, and (3) new reactor designs raising single train size greater throughput.

Slurry-phase hydrocracking of heavy oil and the latest development of dispersed catalysts present strong indications that such technologies will play a role in future refineries. Slurry-phase (or slurry-bed) hydroprocessing can be used for hydroprocessing of feeds with very high metal content to obtain lower-boiling products using a single reactor. The slurry-based technologies combine the advantages of the carbon rejection technologies in terms of flexibility with the high performances peculiar to the hydrogen addition processes and achieve a similar intimate contacting of the feedstock and catalyst and may operate with a lower degree of back-mixing than occurs in ebullated-bed units. Also, in contrast to fluid-bed units and ebullated-bed units, slurry-bed processing uses a smaller amount of catalyst (in the form of finely divided powder) which can be an additive or a catalyst (or catalyst precursors). The catalyst is mixed with the heavy feedstock and both are fed upward with hydrogen through an empty reactor vessel. Since the oil and catalyst flow cocurrently, the mixture approaches plug-flow behavior (Chapter 6). When the reaction is complete, the spent catalyst leaves the slurry-bed reactor together with the residual fraction (the distillated product pass overhead) and remains in the unconverted residue in a benign form.

Catalysts for slurry-phase hydrocracking of heavy oil have undergone two development phases: (1) heterogeneous solid powder catalysts, which have low catalytic activity and will produce a large number of solid particles in bottom oil making the catalyst difficult to dispose and utilize, and (2) homogeneous dispersed catalysts, which are divided into water-soluble dispersed catalysts and oil-soluble dispersed catalysts (Zhang et al., 2007). Dispersed catalysts are highly dispersed and have greater surface area to volume ratio. Therefore, they show high catalytic activity and good performance. They are desirable catalysts for slurry-phase hydrocracking of heavy oil and will be used more prominently in future hydrocracking operations (Bhattacharaya et al., 2010, 2011; Motaghi et al., 2011).

In spite of the numerous process design variations (Section 11.4), process design innovations and hardware innovations will continue. Although conventional (high-pressure) hydrocracking will still be used to address the need to produce more gasoline and diesel, another approach is to introduce a mild hydrocracking unit upstream of the fluid catalytic cracking unit to maintain that unit at full capacity. Alternatively, more refiners will turn to two-stage recycle hydrocracking and (TSR hydrocracking) reverse-staging configurations.

Catalyst improvements will continue to (1) improve hydrocracking activity, (2) improve hydrodesulfurization activity, (3) reduce catalytic deactivation, and (4) increase cycle length. The development in nonprecious metal catalysts, heteropolyanions to improve metal dispersions, beta zeolite, and the acid-cracking-based formulations of highly active hydrocracking catalysts has already added (and will continue to add) flexibility in the operations of hydroprocessing units. New formulations that employ amorphous silica–alumina supports and dealuminated Y-zeolites will be readily available and offer high activity with high stability. These designs allow for lower operating pressures, increased run length, and higher naphtha and diesel yields.

Another central focus will be the reduction of reducing hydrogen consumption while maintaining product quality. Catalysts that can withstand organic nitrogen contamination are being developed for lower-cost, single-stage units. The addition of metal traps upstream of the hydroprocessing unit is a solution to protect highly active catalyst from high-metal feeds that will see wider application.

Furthermore, biomass gasification and Fischer–Tropsch synthesis conversion are very likely to be a part of the future refineries as part of the next-generation biofuels developments (Chapter 14); refiners will need to monitor closely the latest refinery-related advances as well as future directions in biomass processing. In particular, the response of the refining industry to opportunities for processing heavy oils and resids, mandatory biofuels usage, and requirements to comply with carbon dioxide will need to be addressed. Furthermore, gasification with carbon capture and the use of biomass as feedstock should help refiners meet emissions reduction requirements for carbon dioxide.

In summary, trends in the quality of crude feedstocks have shown a steady decline over the past three decades and are reflected by declining API gravity and increasing sulfur content requiring changes in hydrocracking operations (Butler et al., 2009). Furthermore, understanding the fundamental petroleum chemistry of heavy crudes and residua is not always sufficiently adequate to predict processing behavior (Niccum and Northup, 2006). In fact, it is only by comprehensively considering related factors such as (1) the properties of feedstock, (2) catalyst performance, (3) product requirements, (4) chemical kinetics, and (5) operating conditions and cycle length that optimal results can be achieved. Therefore, further improvement of the hydrocracking process parameters and process catalysts which can tolerate a high content of impurities and metals, are two major challenges for the refineries (Putek et al., 2008). Catalyst activity, selectivity, particle size and shape, pore size and distribution, and the type of the reactor have to be optimized according to the properties of the heavy oils and to the desired purification and conversion levels.

Furthermore, the processes that offer higher conversion and improved product quality for downstream processing (such as the UOP Uniflex process, formerly the CANMET process) will be in great demand (Gillis et al., 2009, 2010). Integration of such processes with existing coking capacity offer many unique benefits (Gillis et al., 2009, 2010). The effective use of existing assets requires both individual process depth as well as a breadth of refinery knowledge and expertise. This will require (1) an audit of current refinery operations—product blending, unit yields, overall energy, and hydrogen utilization; (2) development of future refinery configuration, focusing on capacity requirements, product slate, and operating efficiency; (3) development of investment alternatives; (4) economic evaluation and project selection; and (5) rapid project implementation.

REFERENCES

Ackelson, 2004. UOP unicracking process for hydrocracking. In: *Handbook of Petroleum Refining Processes.* R.A. Meyers (Editor). McGraw-Hill, New York, Chapter 7.2.

Babich, I.V. and Moulijn, J.A. 2003. Science and technology of novel processes for deep desulfurization of oil refinery streams: A review. *Fuel*, 82: 607–631.

Bauman, R.F., Aldridge, C.L., Bearden, R. Jr., Mayer, F.X., Stuntz, G.F., Dowdle, L.D., and Fiffron, E. 1993. Preprints. *Oil Sands—Our Petroleum Future.* Alberta Research Council, Edmonton, Alberta, Canada, p. 269.

Bearden, R. and Aldridge, C.L. 1981. *Energy Progress*, 1: 44.

Bellussi, G., Rispoli, G., Landoni, A., Millini, R., Molinari, D., Montanari, E., Moscotti, D., and Pollesel, P. 2013. Hydroconversion of heavy residues in slurry reactors: Developments and perspectives. *Journal of Catalysis*, 308: 189–200.

Bhattacharyya, A., Bricker, M.L., Mezza, B.J., and Bauer, L.J. 2011. Process for using iron oxide and alumina catalyst with large particle diameter for slurry hydrocracking. U.S. Patent. 8,062,505. November 22.

Bhattacharyya, A. and Mezza, B.J. 2010. Catalyst composition with nanometer crystallites for slurry hydrocracking. U.S. Patent 7,820,135. October 26.

Billon, A., Morel, F., Morrison, M.E., and Peries, J.P. 1994. Converting residues with IPP's Hyvahl and Solvahl processes. *Revue Institut Français du Pétrole*, 49(5): 495–507.

Bishop, W. 1990. *Proceedings of Symposium on Heavy Oil: Upgrading to Refining.* Canadian Society for Chemical Engineers, p. 14.

Boening, L.G., McDaniel, N.K., Petersen, R.D., and van Driesen, R.P. 1987. *Hydrocarbon Processing*, 66(9): 59.

Bridge, A.G. 1997. In *Handbook of Petroleum Refining Processes*, 2nd Edition. R.A. Meyers (Editor). McGraw-Hill, New York, Chapter 7.2.

Bridge, A.G., Gould, G.D., and Berkman, J.F. 1981. *Oil and Gas Journal*, 79(3): 85.

Bridjanian, H. and Khadem Samimi, A. 2011. Bottom of the barrel, an important challenge of the petroleum refining industry. *Petroleum & Coal*, 53(1): 13–21.

Butler, G., Spencer, R., Cook, B., Ring, Z., Schleiffer, A., and Rupp, M. 2009. Maximize liquid yield from extra heavy oil. *Hydrocarbon Processing*, 88(9): 52–55.

Celestinos, J.A., Zermeno, R.G., Van Dreisen, R.P., and Wysocki, E.D. 1975. *Oil and Gas Journal*, 73(48): 127.

Chen, Y.L., Wang, Y.Q., Lu, J.Y., and Wu, C. 2009. The viscosity reduction in catalytic aquathermolysis of heavy oil. *Fuel*, 88: 1426–1434.

DeCroocq, D. 1984. *Catalytic Cracking of Heavy Petroleum Hydrocarbons.* Editions Technip, Paris, France.

Dolbear, G.E. 1998. Hydrocracking: Reactions, catalysts, and processes. In: *Petroleum Chemistry and Refining*. J.G. Speight (Editor). Taylor & Francis Group, Washington, DC, Chapter 7.

Dziabala, B., Thakkar, V.P., and Abdo, S.F. 2011. Combination of mild hydrotreating and hydrocracking for making low sulfur diesel and high octane naphtha. U.S. Patent 8,066,867. November 29.

Fan, H., Zhang, Y., and Lin, Y. 2004. The catalytic effects of minerals on aquathermolysis of heavy oils. *Fuel*, 83: 2035–2039.

Fan, H.F., Zhang, Y., and Lin, Y.J. 2009. The catalytic effects of minerals on aquathermolysis of heavy oils. *Fuel*, 83: 2035–2039.

Fornoff, L.L. 1982. *Proceedings of Second International Conference on the Future of Heavy Crude and Tar Sands*. Caracas, Venezuela.

Gamba, S., Pellegrini, L.A., Calemma, V., and Gambaro, C. 2010. Liquid fuels from Fischer–Tropsch wax hydrocracking: Isomer distribution. *Catalysis Today*, 156(1–2): 58–64.

Gary, J.G., Handwerk, G.E., and Kaiser, M.J. 2007. *Petroleum Refining: Technology and Economics*, 5th Edition. CRC Press/Taylor & Francis Group, Boca Raton, FL.

Gillis, D., VanWees, M., and Zimmerman, P. 2009. *Upgrading Residues to Maximize Distillate Yields*. UOP LLC, Des Plaines, IL.

Gillis, D., VanWees, M., and Zimmerman, P. 2010. Upgrading residues to maximize distillate yields with UOP uniflex process. *Journal of the Japan Petroleum Institute*, 53(1): 33–41.

Graeser, U. and Niemann, K. 1982. *Oil and Gas Journal*, 80(12): 121.

Hansen, J.A., Blom, N.J., and Ward, J.W. 2010. Hydrocracking process. U.S. Patent 7,749,373. July 6.

Hsu, C.S. and Robinson, P.R. 2006. *Practical Advances in Petroleum Processing*. Springer, New York, Volumes 1 and 2.

Hunter, M.G., Pasppal, D.A., and Pesek, C.L. 1997. *Handbook of Petroleum Refining Processes*, 2nd Edition. R.A. Meyers (Editor). McGraw-Hill, New York, Chapter 7.1.

Katzer, J.R. and Sivasubramanian, R. 1979. *Catalysis Reviews. Science and Engineering*, 20: 155.

Khadzhiev, S.N., Kadiev, K.M., Mezhidov, V.K., Zarkesh, J., Hashemi, R., Masoudian, T., and Seyed, K. 2009. Process for hydroconverting a heavy hydrocarbonaceous feedstock. U.S. Patent 7,585,406. September 8.

Khan, M.R. and Patmore, D.J. 1998. Heavy oil upgrading process. In: *Petroleum Chemistry and Refining*. J.G. Speight (Editor). Taylor & Francis Group, Washington, DC, Chapter 6.

Kressmann, S., Boyer, C., Colyar, J.J., Schweitzer, J.M., and Viguié, J.C. 2000. Improvements of Ebullated-bed technology for upgrading heavy oils. *Revue Institut Français du Pétrole*, 55: 397–406.

Kressmann, S., Guillaume, D., Roy, M., and Plain, C.A. 2004. New generation of hydroconversion and hydrodesulfurization catalysts. *14th Annual Symposium Catalysis in Petroleum Refining & Petrochemicals King Fahd University of Petroleum & Minerals-KFUPM*. The Research Institute, Dhahran, Saudi Arabia, December 5–6, 2004.

Kressmann, S., Morel, F., Harlé, V., and Kasztelan, S. 1998. Recent developments in fixed-bed catalytic residue upgrading. *Catalysis Today*, 43(3–4): 203–215.

Kunnas, J., Ovaskainen, O., and Respini, M. 2010. Mitigate fouling in Ebullated-bed hydrocrackers. *Hydrocarbon Processing*, 89(10): 59–64.

Li, W., Zhu, J.H., and Qi, J.H. 2007. Application of nano-nickel catalyst in the viscosity reduction of liaohe extra-heavy oil by aquathermolysis. *Journal of Fuel Chemistry and Technology*, 35: 176–180.

Luckwal, K. and Mandal, K.K. 2009. Improve hydrogen management of your refinery. *Hydrocarbon Processing*, 88(2): 55–61.

Marzin, R., Pereira, P., McGrath, M.J., Feintuch, H.M., and Thompson, G. 1998. *Oil and Gas Journal*, 97(44): 79.

Morel, F., Bonnardot, J., and Benazzi, E. 2009. Hydrocracking solutions squeeze more ULSD from heavy ends. 88(11): 79–87.

Motaghi, M., Shree, K., and Krishnamurthy, S. 2010. Consider new methods for bottom-of-the-barrel processing—Part 1. *Hydrocarbon Processing*, 90(2): 35–38.

Motaghi, M., Ulrich, B., and Subramanian, A. 2011. Slurry-phase hydrocracking—Possible solution to refining margins. *Hydrocarbon Processing*, 90(2): 37–43.

Moulton, D.S. and Erwin, J. 2005. Pretreatment processes for heavy oil and carbonaceous materials. U.S. Patent 6,887,369. May 3.

Niccum, P.K. and Northup, A.H. 2006. Economic extraction of FCC feedstock from residual oils. Paper AM-06-18. *National Petrochemical & Refiners Association (NPRA) Annual Meeting*. Salt Lake City, UT.

Niemann, K., Kretschmar, K., Rupp, M., and Merz, L. 1988. *Proceedings of Fourth UNITAR/UNDP International Conference on Heavy Crude and Tar Sand*. Edmonton, Alberta, Canada, Volume 5, p. 225.

Occelli, M.L. and Robson, H.E. 1989. *Zeolite Synthesis*. Symposium Series No. 398. American Chemical Society, Washington, DC.

Olschar, M., Endisch, M., Dimmig, T.H., and Kuchling, T.H. 2007. Investigation of catalytic hydrocracking of Fischer-Tropsch wax for the production of transportation fuels. *Oil Gas European Magazine*, 33(4): 187–193.

Parihar, P., Voolapalli, R.K., Kumar, R., Kaalva, S., Saha, B. and Viswanathan, P.S. 2012. Optimize hydrocracker operations for maximum distillates. *Petroleum Technology Quarterly*, Q2: 1–8.

Parkash, S. 2003. *Refining Processes Handbook*. Gulf Professional Publishing, Elsevier, Amsterdam, the Netherlands.

Patel, N., Ludwig, K., and Morris, P. 2005. Insert flexibility into your hydrogen network—Part 1. *Hydrocarbon Processing*, 84(9): 73–84.

Pereira, P., Flores, C., Zbinden, H., Guitian, J., Solari, R.B., Feintuch, H., and Gillis, D. 2001. *Oil and Gas Journal*, May 14.

Pereira, P., Marzin, R., McGrath, M. and Thompson, G.J. 1998. *Proceedings of 17th World Energy Congress*. Houston, TX.

Peries, J.P., Quignard, A., Farjon, C., and Laborde, M. 1988. Thermal and catalytic ASVAHL processes under hydrogen pressure for converting heavy crudes and conventional residues. *Revue Institut Français Du Pétrole*, 43(6): 847–853.

Pruden, B.B. 1978. Hydrocracking bitumen and heavy oils at CANMET. *Canadian Journal of Chemical Engineering*, 56(3): 277–280.

Pruden, B.B., Muir, G., and Skripek, M. 1993. Preprints. *Oil Sands—Our Petroleum Future*. Alberta Research Council, Edmonton, Alberta, Canada, p. 277.

Putek, S., Januszewski, D., and Cavallo, E. 2008. Upgrade hydrocracked resid through integrated hydrotreating. *Hydrocarbon Processing*, 84(9): 83–92.

Rana, M.S., Sámano, V., Ancheyta, J., and Diaz, J.A.I. 2007. A review of recent advances on process technologies for upgrading of heavy oils and residua. *Fuel*, 86: 1216–1231.

RAROP, 1991. In: *Heavy Oil Processing Handbook*. Y. Kamiya (Editor). Research Association for Residual Oil Processing, Agency of Natural Resources and Energy, Ministry of International Trade and Industry, Tokyo, Japan.

Rashid, K. 2007. Optimize your hydrocracking operations. *Hydrocarbon Processing*, 86(2): 55–63.

Reich, A., Bishop, W., and Veljkovic, M. 1993. Preprints. *Oil Sands—Our Petroleum Future*. Alberta Research Council, Edmonton, Alberta, Canada, p. 216.

Reno, M. 1997. In: *Handbook of Petroleum Refining Processes*, 2nd Edition. R.A. Meyers (Editor). McGraw-Hill, New York, Chapter 7.3.

Reynolds, J.G. and Beret, S. 1989. *Fuel Science and Technology International*, 7: 165.

Scherzer, J. and Gruia, A.J. 1996. *Hydrocracking Science and Technology*. Marcel Dekker, New York.

Speight, J.G. 1990. Tar sands. In: *Fuel Science and Technology Handbook*. J.G. Speight (Editor). Marcel Dekker, New York, Chapters 12–16.

Speight, J.G. 2000. *The Desulfurization of Heavy Oils and Residua*, 2nd Edition. Marcel Dekker, New York.

Speight, J.G. 2001. *Handbook of Petroleum Analysis*. John Wiley & Sons, Hoboken, NJ.

Speight, J.G. 2008. *Synthetic Fuels Handbook: Properties, Processes, and Performance*. McGraw-Hill, New York.

Speight, J.G. 2011a. *The Refinery of the Future*. Gulf Professional Publishing, Elsevier, Oxford, UK.

Speight, J.G. (Editor). 2011b. *Biofuels Handbook*. Royal Society of Chemistry, London, UK.

Speight, J.G. 2014. *The Chemistry and Technology of Petroleum*, 5th Edition. CRC Press/Taylor & Francis Group, Boca Raton, FL.

Speight, J.G. 2015. *Handbook of Petroleum Product Analysis*, 2nd Edition. John Wiley & Sons, Hoboken, NJ.

Speight, J.G. and Moschopedis, S.E. 1979. The production of low-sulfur liquids and coke from Athabasca bitumen. *Fuel Processing Technology*, 2: 295.

Speight, J.G. and Ozum, B. 2002. *Petroleum Refining Processes*. Marcel Dekker, New York.

Stratiev, D. and Petkov, K. 2009. Residue upgrading: Challenges and perspectives. *Hydrocarbon Processing*, 88(9): 93–96.

Sue, H. 1989. *Proceedings of Fourth UNITAR/UNDP Conference on Heavy Oil and Tar Sands*. Volume 5, p. 117.

Thakkar, V., Meister, J.M., Rossi, R.J., and Wang, L. 2007. *Process and Catalyst Innovations in Hydrocracking to Maximize High Quality Distillate Fuel*. Report No. 4706B. UOP LLC, Des Plaines, IL.

Thompson, G.J. 1997. In: *Handbook of Petroleum Refining Processes*. R.A. Meyers (Editor). McGraw-Hill, New York, Chapter 8.4.

Van Driesen, R.P., Caspers, J., Campbell, A.R., and Lunin, G. 1979. *Hydrocarbon Processing*, 58(5): 107.

Vauk, D., Di Zanno, P., Neri, B., Allevi, C., Visconti, A., and Rosanio, L. 2008. What are possible hydrogen sources for refinery expansion? *Hydrocarbon Processing*, 87(2): 69–76.

Waugh, R.J. 1983. *Annual Meeting of National Petroleum Refiners Association*. San Francisco, CA.

Wen, S.B., Zhao, Y.J., Liu, Y.J., and Hu, S.B. 2007. A study on catalytic aquathermolysis of heavy crude oil during steam stimulation. Paper No. 106180-MS, 2007. *Proceedings of International Symposium on Oilfield Chemistry*. Houston, TX, February 28–March 2.

Wenzel, F. and Kretsmar, K. 1993. Preprints. *Oil Sands—Our Petroleum Future*. Alberta Research Council, Edmonton, Alberta, Canada, p. 248.

Wilson, J. 1985. *Energy Proceedings*, 5: 61.

Zarkesh, J., Ghaedian, M., Hashemi1, Khademsamimi, A., and Kadzhiev, S. 2011. Heavy refinery schemes based on new nano catalytic HRH technology. *Proceedings of Second International Conference on Chemical Engineering and Applications*. International Proceedings of Chemical, Biological and Environmental Engineering (IPCBEE). IACSIT Press, Singapore, Volume 23, pp. 66–70.

Zhang, S., Liu, D., Deng, W., and Que, G. 2007. *Energy Fuels*, 21: 3057–3062.

12 Solvent Processes

12.1 INTRODUCTION

Solvent-based processes (such as solvent deasphalting and solvent dewaxing) take advantage of the insolubility of aromatic compounds, and many heteroatom-containing compounds are insoluble in paraffin liquids. For example, propane deasphalting is commonly used to precipitate asphaltene constituents and resin constituents from residua, heavy oils, extra heavy oils, and tar sand bitumen. The deasphalted oil (the soluble product of propane deasphalting that is reactively low in the presence of coke-forming constituents and metal-containing catalyst poisons) is then sufficiently *clean* to be sent to a hydrotreating unit or a hydrocracking unit or to be used as a blend stock for fuel oil (Speight and Ozum, 2002; Parkash, 2003; Hsu and Robinson, 2006; Gary et al., 2007; Speight, 2014).

Solvent deasphalting processes are a major part of refinery operations and are not often appreciated for the tasks for which they are used (Speight and Ozum, 2002; Hsu and Robinson, 2006; Gary et al., 2007; Speight, 2011, 2014). In the solvent deasphalting processes, an alkane is injected into the feedstock to disrupt the dispersion of components and causes the polar constituents to precipitate. Propane (or sometimes propane/butane mixtures) is extensively used for deasphalting and produces a deasphalted oil (DAO) and propane deasphalter asphalt (PDA or PD tar) (Dunning and Moore, 1957). Propane has unique solvent properties; at lower temperatures (38°C–60°C; 100°C–140°C), paraffins are very soluble in propane and at higher temperatures (approximately 93°C; 200°F) all hydrocarbons are almost insoluble in propane.

A *solvent deasphalting* unit processes the residuum from the vacuum distillation unit and produces deasphalted oil (DAO), used as feedstock for a fluid catalytic cracking unit, and the asphaltic residue (deasphalter tar, deasphalter bottoms) which, as a residual fraction, can only be used to produce asphalt or as a blend stock or visbreaker feedstock for low-grade fuel oil. Solvent deasphalting processes have not realized their maximum potential. With ongoing improvements in energy efficiency, such processes would display its effects in a combination with other processes. Solvent deasphalting allows removal of sulfur and nitrogen compounds as well as metallic constituents by balancing yield with the desired feedstock properties (Dunning and Moore, 1957; Ditman, 1973).

In a more recent innovation (McCoy et al., 2010), a feed stream comprising heavy feedstock is fed to a solvent deasphalter wherein it is contacted with a deasphalting solvent or fluid to produce a composition comprising a mixture or slurry of solvent containing a soluble portion of the feedstock, and a heavy tar fraction comprising the insoluble portion of the feedstock tar. These fractions may be separated in the deasphalter apparatus, such as by gravity settling wherein the heavy fraction is taken off as bottoms, and the solvent-soluble fraction taken as overflow or overheads with the solvent. The overflow or overheads is sent to a solvent recovery unit, such as a distillation apparatus, wherein solvent is recovered as overheads and a deasphalted tar fraction is taken off as a sidestream or bottoms.

Most of the metals present in refinery feedstocks are concentrated in the asphaltene constituent and, by using solvent deasphalting (SDA) it is possible to recover significant amounts of partially demetallized and deasphalted oil (DAO) from the residues which can be processed in fluid catalytic cracking units (Chapter 9) or in hydrocracking units (Chapter 11).

The main advantages of solvent deasphalting are relatively low investment and operational costs, while its principal limitation is that to obtain deasphalted oil with a low content of pollutants (in particular, sulfur, nitrogen, metals, and coke-forming constituents), the yield must be limited (Ditman, 1973). As a result, significant quantities of by-products are generated (asphaltene constituents plus resin constituents) that can be used as components of low-quality fuels or as sources of carbon for the production of synthesis gas and, therefore, hydrogen in suitable gasification units (Speight, 2014). The application of the solvent deasphalting process to heavy crude oil and tar sand bitumen upgrading has led to the development of a variety of technologies in which the typical solvent deasphalting unit is combined with other processes.

12.2 DEASPHALTING PROCESSES

Petroleum processing normally involves separation into various fractions that require further processing in order to produce marketable products. The initial separation process is distillation (Chapter 8) in which crude oil is separated into fractions of increasingly higher boiling range fractions. Since petroleum fractions are subject to thermal degradation, there is a limit to the temperatures that can be used in simple separation processes. The crude cannot be subjected to temperatures much above 395°C (740°F), irrespective of the residence time, without encountering some thermal cracking. Therefore, to separate the higher molecular weight and higher boiling fractions from crude, special processing steps must be used.

Because petroleum fractions are subject to thermal degradation, there is a limit to the temperature which can be effectively used in a separation process; that is, crude cannot be subjected to temperatures above 370°C (700°F) without the occurrence of thermal cracking. The physical separation of higher molecular weight and higher boiling fractions from the crude is first accomplished in the vacuum distillation process, which is followed by the solvent deasphalting process. The process itself involves adding a relatively small portion of solvent upstream to the charge crude, and processing it through the extraction tower in which the desired oil goes to the solvent and the asphalt precipitates toward the bottom. As the oil and solvent rise in the tower, the temperature is increased in order to control the quality of the product; separating the oil from asphalt is achieved by maintaining a temperature gradient across the extraction tower and by varying the solvent/oil ratio. A key physical separation process, solvent deasphalting technology has been used by refiners to produce gas oil and lubricating oil bright stock for many years.

Varied types of equipment are in use in solvent deasphalting process. For example, earlier units employed baffle tray columns and/or rotating disk columns. The baffle tray column unit was used to provide a trouble-free operation due to the open structure but these columns were less flexible to column throughput and changes in physical properties of the system. The rotating disk column has been widely applied in various solvent extraction units. The more modern random packings and structured packings have been successfully applied in many solvent extraction processes and have also been applied to solvent deasphalting units.

The efficiency of the extraction process is the key equipment design variable—the role of the extractor is to separate the precipitate phase (often referred to as *deasphalter pitch*) from the continuous fluid (deasphalted oil/solvent) stream. Traditionally, solvent deasphalting units utilize a countercurrent-type separation of the upflowing deasphalted oil/solvent phase from the downflowing precipitate phase. Traditional countercurrent extraction technology can be improved by the addition of packing, which exerts efficient control over the fluid velocity across the cross-sectional area of the extractor. Consequently, packing allows the extractor size to be reduced relative to an open vessel type of design.

The process is also used to reject asphaltene constituents and to recover a deasphalted oil that is low in asphaltene content from a vacuum residue. The solvent deasphalted oil (low-to-no-asphaltene content) oil is a useful feedstock for both fluid catalytic cracking units and hydrocracking units. In addition, because it is relatively less expensive to desulfurize the deasphalted oil than the heavy vacuum residuum, solvent deasphalting offers a more economical route for disposing of vacuum residuum from high-sulfur crude oil.

The typical solvent used for the process is liquid propane (C_3H_8, propane deasphalting). In order to recover more oil from vacuum-reduced crude, mainly for catalytic cracking feedstocks, higher-molecular-weight solvents such as butane (C_4H_{10}) and pentane (C_5H_{12}) have been employed.

12.2.1 DEASPHALTING PROCESS

Following either atmospheric or vacuum distillation, solvent deasphalting (SDA) is a separation process that takes an additional step in the minimization of residual by-product fuel. This process takes advantage of the fact that the maltene fraction present in residual oils is more soluble in light paraffinic solvents than asphaltene constituents (Speight, 2014, 2015). This solubility increases with solvent molecular weight and decreases with temperature.

In the process, the feedstock is extracted with a low-boiling liquid hydrocarbon solvent in an extraction tower. The deasphalted oil separator recovers solvent at supercritical conditions and an asphalt flash drum also recovers solvent. Products are steam stripped prior to further downstream processing. As with vacuum distillation, there are constraints with respect to how deep a solvent deasphalting unit can cut into the residue or how much deasphalted oil (DAO) can be produced. In the case of solvent deasphalting, the constraints are typically (1) the quality specifications of the deasphalted oil required by downstream conversion units and (2) the ultimate disposition of the asphalt (Elliott and Stewart, 2004).

The deasphalting process is a mature process but, as refinery operations evolve, it is necessary to include a description of the process here so that the new processes might be compared with new options that also provide for deasphalting various feedstocks. Indeed, several of these options, such as the residuum oil supercritical extraction (ROSE) process, have been onstream for several years and are included here for this same reason. Thus, this section provides a one-stop discussion of solvent recovery processes and their integration into refinery operations.

The separation of residua into oil and asphalt fractions was first performed on a production scale by mixing the vacuum residuum with propane (or mixtures of *normally gaseous* hydrocarbons) and continuously decanting the resulting phases in a suitable vessel. Temperature was maintained within approximately 55°C (100°F) of the critical temperature of the solvent, at a level that would regulate the yield and properties of the deasphalted oil in solution and that would reject the heavier undesirable components as asphalt.

Currently, deasphalting and delayed coking are used frequently for residuum conversion. The high demand for petroleum coke, mainly for use in the aluminum industry, has made delayed coking a major residuum conversion process. However, many crude oils will not produce coke meeting the sulfur and metals specifications for aluminum electrodes, and coke gas oils are less desirable feedstocks for fluid catalytic cracking than virgin gas oils. In comparison, the solvent deasphalting process can apply to most vacuum residua. The deasphalted oil is an acceptable feedstock for both fluid catalytic cracking and, in some cases, hydrocracking. Since it is relatively less expensive to desulfurize the deasphalted oil than the heavy vacuum residuum, the solvent deasphalting process offers a more economical route for disposing of vacuum residuum from high-sulfur crude. However, the question of disposal of the asphalt remains. Use as a road asphalt is common and as a refinery fuel is less common since expensive stack gas cleanup facilities may be required when used as fuel.

In the process (Figure 12.1), the feedstock is mixed with dilution solvent from the solvent accumulator and then cooled to the desired temperature before entering the extraction tower. Because of its high viscosity, the charge oil can neither be cooled easily to the required temperature nor will it mix readily with solvent in the extraction tower. By adding a relatively small portion of solvent upstream of the charge cooler (insufficient to cause phase separation), the viscosity problem is avoided.

The feedstock, with a small amount of solvent, enters the extraction tower at a point approximately two-thirds up the column. The solvent is pumped from the accumulator, cooled, and enters near the bottom of the tower. The extraction tower is a multistage contactor, normally equipped with baffle trays and the heavy oil flows downward while the light solvent flows upward. As the extraction progresses, the desired oil goes to the solvent and the asphalt separates and moves toward the bottom. As the extracted oil and solvent rise in the tower, the temperature is increased in order to control the quality of the product by providing adequate reflux for optimum separation. Separation of oil from asphalt is controlled by maintaining a temperature gradient across the extraction tower

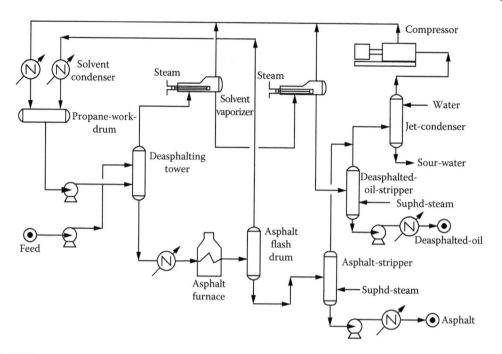

FIGURE 12.1 Propane deasphalting. (From OSHA Technical Manual, Section IV, Chapter 2, Petroleum refining processes, 1999, http://www.osha.gov/dts/osta/otm/otm_iv/otm_iv_2.html.)

and by varying the solvent/oil ratio. The tower top temperature is regulated by adjusting the feed inlet temperature and the steam flow to the heating coils in the top of the tower. The temperature at the bottom of the tower is maintained at the desired level by the temperature of the entering solvent. The deasphalted oil–solvent mixture flows from the top of the tower under pressure control to a kettle-type evaporator heated by low-pressure steam. The vaporized solvent flows through the condenser into the solvent accumulator.

The liquid phase flows from the bottom of the evaporator, under level control, to the deasphalted oil flash tower where it is reboiled by means of a fired heater. In the flash tower, most of the remaining solvent is vaporized and flows overhead, joining the solvent from the low-pressure steam evaporator. The deasphalted oil, with relatively minor solvent, flows from the bottom of the flash tower under level control to a steam stripper operating at essentially atmospheric pressure. Superheated steam is introduced into the lower portion of the tower. The remaining solvent is stripped out and flows overhead with the steam through a condenser into the compressor suction drum where the water drops out. The water flows from the bottom of the drum under level control to appropriate disposal.

The asphalt–solvent mixture is pressured from the extraction tower bottom on flow control to the asphalt heater and on to the asphalt flash drum, where the vaporized solvent is separated from the asphalt. The drum operates essentially at the solvent-condensing pressure so that the overhead vapors flow directly through the condenser into the solvent accumulator. Hot asphalt, with a small quantity of solvent, flows from the asphalt flash drum bottom to the asphalt stripper, which is operated at near atmospheric pressure. Superheated steam is introduced into the bottom of the stripper. The steam and solvent vapors pass overhead, join the deasphalted oil stripper overhead, and flow through the condenser into the compressor suction drum. The asphalt is pumped from the bottom of the stripper under level control, to storage.

The propane deasphalting process is similar to solvent extraction in that a packed or baffled extraction tower or rotating disk contactor is used to mix the oil feedstocks with the solvent. In the tower method, four to eight volumes of propane are fed to the bottom of the tower for every volume

TABLE 12.1
Feedstock and Product Data for the Deasphalting Process

Crude Source	Arab	West Texas	California	Canada	Kuwait
Feedstock					
Crude, vol%	23.0	29.2	20.0	16.0	22.2
Gravity, °API	6.8	12.0	6.3	9.6	5.6
Conradson carbon, wt%	15.0	12.1	22.2	18.9	24.0
SUS at 210°F	75,000	526	9600	1740	14,200
Metals, wppm					
Ni	73.6	16.0	139	46.6	29.9
V	365.0	27.6	136	30.9	110.0
Cu + Fe	15.5	14.8	94	40.7	13.7
Deasphalted oil					
Vol% feed	49.8	66.0	52.8	67.8	45.6
Gravity, °API	18.1	19.6	18.3	17.8	16.2
Conradson carbon, wt%	5.9	2.2	5.3	5.4	4.5
SUS at 210°F	615	113	251	250	490
Metals, wppm					
Ni	3.5	1.0	8.1	3.9	0.9
V	12.4	1.3	2.3	1.4	0.7
Cu + Fe	0.2	0.8	3.5	0.2	0.8
Asphalt					
Vol% feed	50.2	34.0	47.2	32.2	54.4
Gravity, °API	−1.3	−0.9	−5.1	−5.3	−1.3

of feed flowing down from the top of the tower. The oil, which is more soluble in the propane dissolves and flows to the top. The asphaltene and resins flow to the bottom of the tower where they are removed in a propane mix. Propane is recovered from the two streams through two-stage flash systems followed by steam stripping in which propane is condensed and removed by cooling at high pressure in the first stage and at low pressure in the second stage. The asphalt recovered can be blended with other asphalts or heavy fuels, or can be used as feed to the coker.

The yield of deasphalted oil varies with the feedstock (Table 12.1 and Figure 12.2), but the deasphalted oil does make less coke and more distillate than the feedstock. Therefore, the process parameters for a deasphalting unit must be selected with care according to the nature of the feedstock and the desired final products. The metal content of the deasphalted oil and the nitrogen and sulfur contents in the deasphalted oil are related to the yield of the deasphalted oil (Figure 12.3). The character of the deasphalting process is a molecular weight separation and the solvent takes a cross-cut across the feedstock effecting separation by molecular weight and by polarity (Figure 12.4).

Furthermore to the selection of the process parameters, the *choice of solvent* is vital to the flexibility and performance of the unit. The solvent must be suitable not only for the extraction of the desired oil fraction but also for control of the yield and/or quality of the deasphalted oil at temperatures that are within the operating limits. If the temperature is too high (i.e., close to the critical temperature of the solvent), the operation becomes unreliable in terms of product yields and character. If the temperature is too low, the feedstock may be too viscous and have an adverse effect on the contact with the solvent in the tower.

Liquid propane is by far the most selective solvent among the light hydrocarbons used for deasphalting. At temperatures ranging from 38°C to 65°C (100°F to 150°F), most hydrocarbons are soluble in propane, while asphaltic and resinous compounds are not thereby allowing rejection of

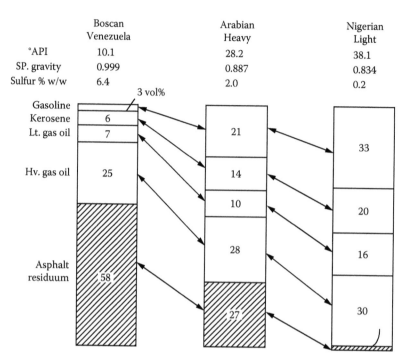

FIGURE 12.2   Variation of composition of selected crude oils.

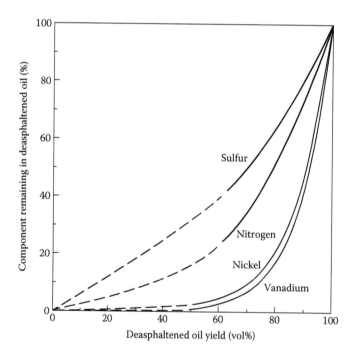

FIGURE 12.3   Variation of deasphalted oil properties with yield. (From Ditman, J.G., *Hydrocarbon Process.*, 52(5), 110, 1973.)

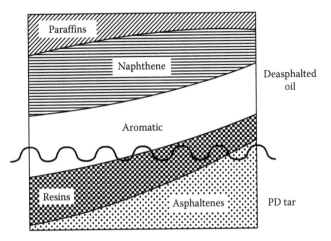

FIGURE 12.4 Illustration of the deasphalting process on the basis of molecular weight and polarity. PD, Propane deasphalter tar or *bottoms*.

these compounds resulting in a drastic reduction (relative to the feedstock) of the nitrogen content and the metals in the deasphalted oil. Although the deasphalted oil from propane deasphalting has the best quality, the yield is usually less than the yield of deasphalted oil produced using a higher-molecular-weight (higher-boiling) solvent.

The ratios of propane oil required vary from 6-1 to 10-1 by volume, with the ratio occasionally being as high as 13-1. Since the critical temperature of propane is 97°C (206°F), this limits the extraction temperature to approximately 82°C (180°F). Therefore, propane alone may not be suitable for high-viscosity feedstocks because of the relatively low operating temperature.

Iso-butane and *n*-butane are more suitable for deasphalting high-viscosity feedstocks since their critical temperatures are higher (134°C, 273°F, and 152°C, 306°F, respectively) than that of the critical temperature of propane. Higher extraction temperatures can be used to reduce the viscosity of the heavy feed and to increase the transfer rate of oil to solvent. Although *n*-pentane is less selective for metals and carbon residue removal, it can increase the yield of deasphalted oil from a heavy feed by a factor of 2–3 over propane (Speight, 2000). However, if the content of the metals and carbon residue of the pentane-deasphalted oil is too high (defined by the ensuing process), the deasphalted oil may be unsuitable as a cracking feedstock. In certain cases, the nature of the cracking catalyst may dictate that the pentane-deasphalted oil be blended with vacuum gas oil that, after further treatment such as hydrodesulfurization, produces a good cracking feedstock.

Solvent composition is an important process variable for deasphalting units. The use of a single solvent may (depending on the nature of the solvent) limit the range of feedstocks that can be processed in a deasphalting unit. When a deasphalting unit is required to handle a variety of feedstocks and/or produce various yields of deasphalted oil (as is the case in these days of variable feedstock quality), a dual solvent may be the only option to provide the desired flexibility. For example, a mixture of propane and *n*-butane might be suitable for feedstocks that vary from vacuum residua to both the heavy resid to heavy gas oils that contain asphaltic materials. Adjusting the solvent composition allows the most desirable product quantity and quality within the range of temperature control.

Besides the solvent composition, the *solvent/oil ratio* also plays an important role in a deasphalting operation. Solvent/oil ratios vary considerably and are governed by feedstock characteristics and desired product qualities and, for each individual feedstock, there is a minimum operable solvent/oil ratio. Generally, increasing the solvent-to-oil ratio almost invariably results in improving the deasphalted oil quality at a given yield, but other factors must also be taken into consideration and (generalities aside) each plant and feedstock will have an optimum ratio.

The main consideration in the selection of the *operating temperature* is its effect on the yield of deasphalted oil. For practical applications, the lower limits of operable temperature are set by the viscosity of the oil-rich phase. When the operating temperature is near the critical temperature of the solvent, control of the extraction tower becomes difficult since the rate of change of solubility with temperature becomes very large at conditions close to the critical point of the solvent. Such changes in solubility cause large amounts of oil to transfer between the solvent-rich and the oil-rich phases that, in turn, causes *flooding* and/or uncontrollable changes in product quality. To mitigate such effects, the upper limits of operable temperatures must lie below the critical temperature of the solvent in order to insure good control of the product quality and to maintain a stable condition in the extraction tower.

The *temperature gradient* across the extraction tower influences the sharpness of separation of the deasphalted oil and the asphalt because of internal reflux that occurs when the cooler oil/solvent solution in the lower section of the tower attempts to carry a large portion of oil to the top of the tower. When the oil/solvent solution reaches the steam-heated, higher-temperature area near the top of the tower, some oil of higher molecular weight in the solvent solution is rejected because the oil is less soluble in solvent at the higher temperature. The heavier oil (rejected from the solution at the top of the tower) attempts to flow downward and causes the internal reflux. In fact, generally, the greater the temperature difference between the top and the bottom of the tower, the greater will be the internal reflux and the better will be the quality of the deasphalted oil. However, too much internal reflux can cause tower flooding and jeopardize the process.

The *process pressure* is usually not considered to be an operating variable since it must be higher than the vapor pressure of the solvent mixture at the tower operating temperature to maintain the solvent in the liquid phase. The tower pressure is usually only subject to change when there is a need to change the solvent composition or the process temperature.

Proper *contact and distribution of the oil and solvent* in the *tower* are essential to the efficient operation of any deasphalting unit. In early units, days, mixer-settlers were used as contactors but proved to be less efficient than the countercurrent contacting devices. Packed towers are difficult to operate in this process because of the large differences in viscosity and density between the asphalt phase and the solvent-rich phase.

The *extraction tower* for solvent deasphalting consists of two contacting zones: (1) a rectifying zone above the oil feed and (2) a stripping zone below the oil feed. The rectifying zone contains some elements designed to promote contacting and to avoid *channeling*. Steam-heated coils are provided to raise the temperature sufficiently to induce an oil-rich reflux in the top section of the tower. The stripping zone has disengaging spaces at the top and bottom and consists of contacting elements between the oil inlet and the solvent inlet.

A *countercurrent tower* with static baffles is widely used in solvent deasphalting service. The baffles consist of fixed elements formed of expanded metal gratings in groups of two or more to provide maximum change of direction without limiting capacity. The *rotating disk contactor* has also been employed and consists of disks connected to a rotating shaft that are used in place of the static baffles in the tower. The rotating element is driven by a variable speed drive at either the top or the bottom of the column and operating flexibility is provided by controlling the speed of the rotating element and, thus, the amount of mixing in the contactor.

In the deasphalting process, the solvent is recovered for circulation and the efficient operability of a deasphalting unit is dependent on the design of the *solvent recovery system*.

The solvent may be separated from the deasphalted oil in several ways such as conventional evaporation or the use of a flash tower. Irrespective of the method of solvent recovery from the deasphalted oil, it is usually the most efficient to recover the solvent at a temperature close to the extraction temperature. If a higher temperature for solvent recovery is used, heat is wasted in the form of high vapor temperature and, conversely, if a lower temperature is used, the solvent must be reheated thereby requiring additional energy input. The solvent recovery pressure should be low enough to maintain a smooth flow under pressure from the extraction tower.

The asphalt solution from the bottom of the extraction tower usually contains less than an equal volume of solvent. A fired heater is used to maintain the temperature of the asphalt solution well above the foaming level and to keep the asphalt phase in a fluid state. A flash drum is used to separate the solvent vapor from asphalt with the design being such as to prevent carryover of asphalt into the solvent outlet line and to avoid fouling the downstream solvent condenser. The solvent recovery system from asphalt is not usually subject to the same degree of variations as the solvent recovery system for the deasphalted oil and operation at constant temperature and pressure with a separate solvent condenser and accumulator is possible.

Asphalt from different crude oils varies considerably but the viscosity is often too high for fuel oil although, in some cases they can be blended with refinery cutter stocks to make No. 6 fuel oil. When the sulfur content of the original residuum is high, even the blend fuel oil will not be able to meet the sulfur specification of fuel oil unless stack gas cleanup is available.

The deasphalted oil and solvent asphalt are not finished products and require further processing or blending, depending on the final use. *Manufacture of lubricating oil* is on possibility, and the deasphalted oil may also be used as a *catalytic cracking feedstock* or it may be desulfurized. It is perhaps these last two options that are more pertinent to the present text and future refinery operations.

Briefly, catalytic cracking or hydrodesulfurization of atmospheric and vacuum residua from high-sulfur/high-metal crude oil is, theoretically, the best way to enhance their value. However, the concentrations of sulfur (in the asphaltene fraction) in the residua can severely limit the performance of cracking catalysts and hydrodesulfurization catalysts (Speight, 2000). Both processes generally require tolerant catalysts as well as (in the case of hydrodesulfurization) high hydrogen pressure, low space velocity, and high hydrogen recycle ratio.

For both processes, the advantage of using the deasphalting process to remove the troublesome compounds becomes obvious. The deasphalted oil, with no asphaltene constituents and low metal content, is easier to process than the residua. Indeed, in the hydrodesulfurization process, the deasphalted oil may consume only 65% of the hydrogen required for direct hydrodesulfurization of topped crude oil.

As always, the use of the material rejected by the deasphalting unit remains an issue. It can be used (apart from its use for various types of asphalt) as feed to a partial oxidation unit to make a hydrogen-rich gas for use in hydrodesulfurization processes and hydrocracking processes. Alternatively, the asphalt may be treated in a visbreaker to reduce its viscosity thereby minimizing the need for cutter stock to be blended with the solvent asphalt for making fuel oil. Or hydrovisbreaking offers an option of converting the asphalt to feedstocks for other conversion processes.

Solvent deasphalting has the advantage of being a relatively low-cost process that has the flexibility to meet a wide range of quality of the deasphalted oil. The process has very good selectivity for asphaltene constituents and metals rejection, some selectivity for rejection of coke formers, but less selectivity for sulfur and nitrogen. The process is best suited for the more paraffinic vacuum residues as opposed to vacuum residues with high asphaltene content, high metal, and high carbon content of coke-forming constituents. The disadvantages of the separation process are that it performs no conversion and produces the high-viscosity by-product asphalt (deasphalter bottoms, SDA pitch). When high-quality deasphalted oil is required, the solvent deasphalting process is limited in the quality of feedstock that can be economically processed.

The economics of the process are highly dependent on the refinery's ability to further upgrade the deasphalted oil and the differential between the value of the cutter stocks and the price of high-sulfur residual fuel oil. In those situations where there is an economic outlet for the pitch and conversion capacity exists to upgrade the deasphalted oil, solvent deasphalting can be a highly attractive option for the refiner. One such economic outlet is cogeneration of steam and power using pitch, both to supply the refiner's needs and for export to nearby users. Other applications include gasification of the deasphalter bottoms for hydrogen and/or power generation, and road asphalt production (Elliott and Stewart, 2004).

Lower-boiling solvents and higher solvent ratios precipitate larger quantities of resin constituents and asphaltene constituents thereby providing separation of these feed components from saturates and aromatics components. Low-boiling liquid paraffin solvents show extraction selectivity not only to molecular weight but also to molecular type. In any crude oil residue, the lowest boiling and most paraffinic components are those most soluble in the light paraffinic solvent.

Solvent deasphalting is a separation process that represents a further step in the minimization of residual by-product fuel. However, solvent deasphalting processes, far from realizing their maximum potential for heavy feedstocks, are now under further investigation and, with ongoing improvements in energy efficiency, such processes are starting to display maximum benefits when used in combination with other processes. The process takes advantage of the fact that the maltene constituents are more soluble in light paraffinic solvents than asphaltene constituents. This solubility increases with solvent molecular weight and decreases with temperature. As with vacuum distillation there are constraints with respect to how deep a solvent deasphalting unit can cut into the residue or how much deasphalted oil can be produced. In the case of solvent deasphalting, the constraint is usually related to deasphalted oil quality specifications required by downstream conversion units.

However, solvent deasphalting has the flexibility to produce a wide range of deasphalted oil that matches the desired properties. The process has very good selectivity for asphaltene constituents (and, to a lesser extent, resin constituents) as well as metals rejection. There is also some selectivity for rejection of carbon residue precursors but there is less selectivity for sulfur-containing and nitrogen-containing constituents. The process is best suited for the more paraffinic vacuum residua with a somewhat lower efficiency when applied to high-asphaltene residua that contain high proportions of metals and coke-forming constituents. The advantages–disadvantages of the process are that it performs no conversion and produces a very high-viscosity by-product deasphalter bottoms and, where high-quality deasphalted oil is required, the process is limited in the quality of feedstock that can be economically processed. In those situations where there is a ready outlet for use for the bottoms, solvent deasphalting is an attractive option for treating heavy feedstocks. One such situation is the cogeneration of steam and power, both to supply the refiner's needs and for export to nearby users.

Recent work has shown that additives and membranes can be used to enhance the separation of the asphaltic phase and the nonasphaltic phase. In the former process (when additives are used) (Koseoglu, 2009) solvent deasphalting of crude oil or heavy feedstocks is carried out in the presence of a solid adsorbent, such as clay, silica, alumina, and activated carbon, which adsorbs the contaminants and permits the solvent and oil fraction to be removed as a separate stream from which the solvent is recovered for recycling; the adsorbent with contaminants and the asphalt bottoms is mixed with aromatic and/or polar solvents to desorb the contaminants and washed as necessary, for example, with benzene, toluene, xylenes, and tetrahydrofuran, to clean adsorbent which is recovered and recycled; the solvent–asphalt mixture is sent to a fractionator for recovery and recycling of the aromatic or polar solvent. The bottoms from the fractionator include the concentrated polynuclear aromatic constituents and contaminants and are further processed as appropriate. In the latter process (i.e., when a membrane is employed) (Trambouze et al., 1989), the solvent (a low-boiling hydrocarbon liquid) is separated from the deasphalted oil, by passing the solution tangentially across an inorganic membrane and the obtained filtrate has an increased solvent content and may be recycled. The deasphalted oil is selectively retained on the upstream side of the membrane.

12.2.2 Deep Solvent Deasphalting Process

The *deep solvent deasphalting process* is an application of the LEDA (low-energy deasphalting) process (Table 12.2 and Figure 12.5) (RAROP, 1991, p. 91) that is used to extract high-quality lubricating oil bright stock or prepare catalytic cracking feeds, hydrocracking feeds, hydrodesulfurization unit feeds, and asphalt from vacuum residue materials. The LEDA process uses a low-boiling hydrocarbon solvent specifically formulated to insure the most economical deasphalting design for each operation. For example, a propane solvent may be specified for a low deasphalted oil yield operation while a higher-boiling solvent, such as pentane or hexane, may be used to obtain a high

TABLE 12.2
Feedstock and Product Data for the
LEDA Process

	Residuum[a]	Residuum[a]
Feedstock		
API	6.5	6.5
Sulfur, % w/w	3.0	3.0
Carbon residue, % w/w	21.8	21.8
Nickel, ppm	46.0	46.0
Vanadium, ppm	125.0	125.0
Products		
Deasphalted oil, % v/v	53.0	65.0
API	17.6	15.1
Sulfur, % w/w	1.9	2.2
Carbon residue, % w/w	3.5	6.2
Nickel, ppm	1.8	4.5
Vanadium, ppm	3.4	10.3

[a] Unspecified.

deasphalted oil yield from a vacuum residuum (Table 12.3). The deep deasphalting process can be integrated with a delayed coking operation (ASCOT process; *q.v.*). In this case, the solvent can be low-boiling naphtha (Table 12.4).

Low-energy deasphalting operations are usually carried out in a rotating disk contractor (RDC) that provides more extraction stages than a mixer-settler or baffle-type column. Although not essential to the process, the rotating disk contactor provides higher-quality deasphalted oil at the same yield, or higher yields at the same quality. The low-energy solvent deasphalting process selectively

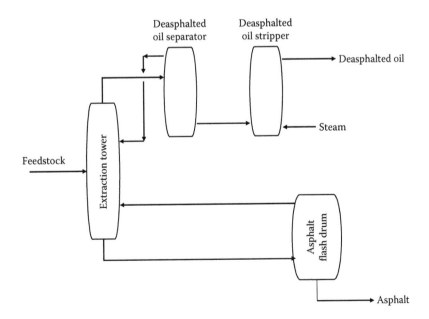

FIGURE 12.5 The LEDA process.

TABLE 12.3

Feedstock and Product Data for the ASCOT Process, Delayed Coking, and LEDA Process

	ASCOT	Delayed Coking	LEDA
Feedstock			
API	2.8	2.8	2.8
Sulfur, % w/w	4.2	4.2	4.2
Nitrogen, % w/w	1.0	1.0	1.0
Carbon residue, % w/w	22.3	22.3	22.3
Products			
Naphtha, % v/v	7.7	19.4	
API	54.7		
Sulfur, % w/w	1.1		
Nitrogen, % w/w	0.1		
Gas oil, % v/v	69.9	51.8	
API	13.4		
Sulfur, % w/w	3.4		
Nitrogen, % w/w	0.5		
Coke, % w/w	25.0	32.5	
Sulfur, % w/w	5.8	5.7	
Nitrogen, % w/w	2.7	2.6	
Nickel, ppm	774.0	609.0	
Vanadium, ppm	2656.0	2083.0	
Deasphalted oil, % v/v			50.0
Asphalt, % v/v			50.0
Sulfur, % w/w			5.0
Nitrogen, % w/w			1.4
Nickel, ppm			365.0
Vanadium, ppm			1250.0

extracts the more paraffinic components from vacuum residua while rejecting the condensed ring aromatics. As expected, deasphalted oil yields vary as a function of solvent type and quantity and feedstock properties (Figure 12.2).

In the process, vacuum residue feed is combined with a small quantity of solvent to reduce its viscosity and cooled to a specific extraction temperature before entering the rotating disk contactor. Recovered solvents from the high-pressure and low-pressure solvent receivers are combined, adjusted to a specific temperature by the solvent heater–cooler, and injected into the bottom section of the rotating disk contactor. Solvent flows upward, extracting the paraffinic hydrocarbons from the vacuum residuum, which is flowing downward through the rotating disk contactor.

Steam coils at the top of the tower maintain the specified temperature gradient across the rotating disk contactor. The higher temperature in the top section of the rotating disk contactor results in separation of the less soluble heavier material from the deasphalted oil mix and provides internal reflux, which improves the separation. The deasphalted oil mix leaves the top of the rotating disk contactor tower. It flows to an evaporator where it is heated to vaporize a portion of the solvent. It then flows into the high-pressure flash tower where high-pressure solvent vapors are taken overhead.

The deasphalted oil mix from the bottom of this tower flows to the pressure vapor heat exchanger where additional solvent is vaporized from the deasphalted oil mix by condensing high-pressure

TABLE 12.4
Feedstock and Product Data for the Demex Process

	Vacuum Residuum[a]	Vacuum Residuum[a]	Arabian Light Vacuum Residuum	Arabian Light Vacuum Residuum	Arabian Light Vacuum Residuum	Arabian Heavy Vacuum Residuum	Arabian Heavy Vacuum Residuum
Feedstock							
API	7.2	7.2	6.9	6.9	6.9	3.0	3.0
Sulfur, % w/w	4.0	4.0	4.0	4.0	4.0	6.0	6.0
Nitrogen, % w/w	0.3	0.3	0.3	0.3	0.3	0.5	0.5
Carbon residue, % w/w	20.8	20.8	20.8	20.8	20.8	27.7	27.7
Nickel, ppm			23.0	23.0	23.0	64.0	64.0
Vanadium, ppm			75.0	75.0	75.0	205.0	205.0
Nickel + vanadium, ppm	98.0	98.0					
C$_6$-asphaltenes, % w/w	10.0	10.0	10.0	10.0	10.0	15.0	15.0
C$_7$-asphaltenes, % w/w							
Products							
Demetallized oil, % v/v	56.0	78.0	40.0	60.0	78.0	30.0	55.0
API	16.0	12.0	18.9	15.3	12.0	16.3	12.0
Sulfur, % w/w	2.7	3.3	2.3	2.8	3.3	3.5	4.3
Nitrogen, % w/w	0.1	0.2	0.1	0.2	0.2	0.1	0.2
Carbon residue, % w/w	5.6	10.7	2.9	6.4	10.7	4.8	10.1
Nickel + vanadium, ppm	6.0	19.0	2.5	7.2	19.0	16.0	38.0
C$_6$-asphaltenes, % w/w	<0.1	<0.1					
Pitch, % v/v	44.0	22.0					
API	<0.0	<0.0	<0.0	<0.0	<0.0	<0.0	<0.0
Sulfur, % w/w	5.4	6.3	5.0	5.5	6.3	6.9	7.8
Nickel + vanadium, ppm	201.0	341.0	154.0	216.0	341.0	364.0	515.0

[a] Unspecified.

flash. The high-pressure solvent, totally condensed, flows to the high-pressure solvent receiver. Partially vaporized, the deasphalted oil mix flows from the pressure vapor heat exchanger to the low-pressure flash tower where low-pressure solvent vapor is taken overhead, condensed, and collected in the low-pressure solvent receiver. The deasphalted oil mix flows down the low-pressure flash tower to the reboiler, where it is heated, and then to the deasphalted oil stripper, where the remaining solvent is stripped overhead with superheated steam. The deasphalted oil product is pumped from the stripper bottom and is cooled, if required, before flowing to battery limits.

The raffinate phase containing asphalt and small amount of solvent flows from the bottom of the rotating disk contactor to the asphalt mix heater. The hot, two-phase asphalt mix from the heater is flashed in the asphalt mix flash tower where solvent vapor is taken overhead, condensed, and collected in the low-pressure solvent receiver. The remaining asphalt mix flows to the asphalt stripper where the remaining solvent is stripped overhead with superheated steam. The asphalt stripper overhead vapors are combined with the overhead from the deasphalted oil stripper, condensed, and collected in the stripper drum. The asphalt product is pumped from the stripper and is cooled by generating low-pressure steam.

12.2.3 Demex Process

The *Demex process* is a solvent extraction demetallization process that separates high-metal vacuum residuum into demetallized oil of relatively low metal content and asphaltene of high metal content (RAROP, 1991, p. 93). The asphaltene constituents and condensed aromatic contents of the demetallized oil are very low. The demetallized oil is a desirable feedstock for fixed-bed hydrodesulfurization and, in cases where the metal content and carbon residue are sufficiently low, is a desirable feedstock for fluid catalytic cracking and hydrocracking units. Overall, the Demex process is an extension of the propane deasphalting process and employs a less selective solvent to recover not only the high-quality oils but also higher-molecular-weight aromatics and other constituents present in the feedstock. Furthermore, the Demex process requires a much less solvent circulation in achieving its objectives, thus, reducing the utility costs and unit size significantly. The process selectively rejects asphaltenes, metals, and high-molecular-weight aromatics from vacuum residua. The resulting demetallized oil can then be combined with vacuum gas oil to give a greater availability of acceptable feed to subsequent conversion units.

In the process, the vacuum residuum feedstock, mixed with Demex solvent recycling from the second stage, is fed to the first-stage extractor. The pressure is kept high enough to maintain the solvent in liquid phase. The temperature is controlled by the degree of cooling of the recycle solvent. The solvent rate is set near the minimum required to ensure the desired separation to occur. Asphaltene constituents are rejected in the first stage. Some resins are also rejected to maintain sufficient fluidity of the asphaltene for efficient solvent recovery. The asphaltene is heated and steam stripped to remove solvent. The first-stage overhead is heated by an exchange with hot solvent. The increase in temperature decreases the solubility of resins and high-molecular-weight aromatics (Mitchell and Speight, 1973). These precipitate in the second-stage extractor. The bottom stream of this second-stage extractor is recycled to the first stage. A portion of this stream can also be drawn as a separate product.

The overhead from the second stage is heated by an exchange with hot solvent. The fired heater further raises the temperature of the solvent/demetallized oil mixture to a point above the critical temperature of the solvent. This causes the demetallized oil to separate. It is then flashed and steam stripped to remove all traces of solvent. The vapor streams from the demetallized oil and asphalt strippers are condensed, dewatered, and pumped up to process pressure for recycle. The bulk of the solvent goes overhead in the supercritical separator. This hot solvent stream is then effectively used for process heat exchange. The subcritical solvent recovery techniques, including multiple effect systems, allow much less heat recovery. Most of the low-grade heat in the solvent vapors from the subcritical flash vaporization must be released to the atmosphere requiring additional heat input to the process.

12.2.4 MDS Process

The *MDS process* is a technical improvement of the solvent deasphalting process, particularly effective for upgrading heavy crude oils (Table 12.5) (RAROP, 1991, p. 95). Combined with hydrodesulfurization, the process is fully applicable to the feed preparation for fluid catalytic cracking and hydrocracking. The process is capable of using a variety of feedstocks including atmospheric and vacuum residua derived from various crude oils, tar sand bitumen, and nonvolatile products from a visbreaker.

In the process, the feed and the solvent are mixed and fed to the deasphalting tower. Deasphalting extraction proceeds in the upper half of the tower. After the removal of the asphalt, the mixture of deasphalted oil and solvent flows out of the tower through the tower top. Asphalt flows downward to come in contact with a countercurrent of rising solvent. The contact eliminates oil from the asphalt, the asphalt then accumulates on the bottom. Deasphalted oil–containing solvent is heated through a heating furnace and fed to the deasphalted oil flash tower where most of the solvent is separated under pressure. Deasphalted oil still containing a small amount of solvent is again heated and fed to the stripper, where the remaining solvent is completely removed.

Asphalt is withdrawn from the bottom of the extractor. Since this asphalt contains a small amount of solvent, it is heated through a furnace and fed to the flash tower to remove most of the solvent. Asphalt is then sent to the asphalt stripper, where the remaining portion of solvent is completely removed.

Solvent recovered from the deasphalted oil and asphalt flash towers is cooled and condensed into liquid and sent to a solvent tank. The solvent vapor leaving both strippers is cooled to

TABLE 12.5
Feedstock and Product Data for the MDS Process

	Iranian Heavy Atmospheric Residuum	Kuwait Atmospheric Residuum	Khafji Vacuum Residuum
Feedstock			
API	17.0	16.4	5.2
Sulfur, % w/w	2.7	3.7	5.2
Carbon residue, % w/w	9.1	9.4	21.9
Nickel, ppm	40.0	14.0	49.0
Vanadium, ppm	130.0	48.0	140.0
Products			
Deasphalted oil, % v/v	93.4	93.8	72.4
API	19.0	16.4	11.3
Sulfur, % w/w	2.4	3.7	4.3
Carbon residue, % w/w	5.9	9.4	10.9
Nickel, ppm	18.0	14.0	6.0
Vanadium, ppm	53.0	48.0	28.0
Asphalt, % v/v	6.6	6.2	27.6
API	<0.0	<0.0	<0.0
Sulfur, % w/w	5.4	7.2	7.3
Carbon residue, % w/w			49.3
Nickel, ppm	320.0	113.0	150.0
Vanadium, ppm	1010.0	425.0	400.0

remove water and compressed for condensation. The condensed solvent is then sent to the solvent tank for further recycling.

12.2.5 RESIDUUM OIL SUPERCRITICAL EXTRACTION PROCESS

The *residuum oil supercritical extraction process (ROSE process)* is a solvent deasphalting process with minimum energy consumption using supercritical solvent recovery system and the process is of value in obtaining oils for further processing (Gearhart and Garwin, 1976; Gearhart, 1980; RAROP, 1991, p. 97). The process can be installed upstream of the desulfurizer to reduce a major portion of the heavy metals and coke precursors present in the feed. The ROSE process can also be installed between a vacuum flasher and a coking unit which reduces the carbon residue of the gas oil fraction for its catalytic cracking (Gearhart, 1980; Low et al., 1995).

The process used supercritical solvents and is a natural progression from propane deasphalting and allows the separation of residua into their base components (asphaltene constituents, resin constituents, and oil constituents) for recombination to optimum properties. Propane, butane, and pentane may be used as the solvent depending on the feedstock and the desired compositions. A mixer is used to blend residue with liquefied solvent at elevated temperature and pressure. The blend is pumped into the first-stage separator where, through countercurrent flow of solvent, the asphaltene constituents are precipitated, separated, and stripped of solvent by steam. The overhead solution from the first tower is taken to a second stage where it is heated to a higher temperature. This causes the resin constituents to separate. The final material is taken to a third stage and heated to a supercritical temperature. This makes the oils insoluble and separation occurs. This process is very flexible and allows precise blending to required compositions.

In the process (Figure 12.6), the residuum is mixed with severalfold volume of a low-boiling hydrocarbon solvent and passed into the asphaltene separator vessel. Asphaltenes rejected by the solvent are separated from the bottom of the vessel and are further processed by heating and steam stripping to remove a small quantity of dissolved solvent. The solvent-free asphaltenes are sent to a section of the refinery for further processing. The main flow, solvent and extracted oil, passes overhead from the asphaltene separator through a heat exchanger and heater into the oil separator where the extracted oil is separated without solvent vaporization. The solvent, after heat exchange,

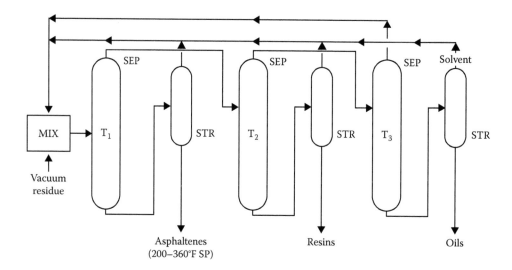

FIGURE 12.6 The ROSE process. T_1, T_2, and T_3 are separator vessel with the relevant parameters for separation of the fraction. SEP, separator; STR, stripping; SP, softening point.

is recycled to the process. The small amount of solvent contained in the oil is removed by steam stripping, and the resulting vaporized solvent from the strippers is condensed and returned to the process. Product oil is cooled by heat exchange before being pumped to storage or further processing. The deasphalting efficiency in processes using propane is of the order of 75%–83%, with an overall deasphalted oil recovery yield of the order of 50%.

The deasphalted oil from a ROSE unit is a suitable feedstock that can be processed in a fluid catalytic cracking unit or other conversion units. Since the contaminants (such as sulfur and metals) are rejected in the solid fuel to the cement industry, there is minimal impact to the auxiliary units (sulfur plant, amine regeneration) within the refinery. Utilization of the asphaltene fraction can be a key to the economic success of the process. Options for use of the fraction (not necessarily in the order of importance) include (1) conversion as a coker feedstock, (2) fuel oil blend stock, (3) partial oxidation feedstock for synthesis gas or hydrogen production, and (4) solid fuel with any necessary gas cleaning operations. The quality and definition of the product designated as the *asphaltene fraction* depends on the feedstock to the process and whether or not the feedstock is the result of a blend of crude different oil (Speight, 2014). As the crude slate becomes heavier and higher sulfur content, the asphaltene fraction will also contain a higher quantity of sulfur. Environmental regulations will therefore dictate how much flue gas cleanup is required and, hence, the viability of use of the asphaltene product as a combustion fuel.

Although often referred to as supercritical extraction, it is often the solvent separation, not the extraction, that is carried out in the supercritical region of the solvent which results in a simpler process flow. Supercritical solvent recovery allows for more efficient utilization of the system's thermodynamic characteristics.

The Aquaform process is a process that has been designed for easy integration with the ROSE solvent deasphalting unit (Patel et al., 2008). In the process, the asphaltene product is sent through an exchanger to a pelletizing vessel in which a rotating head produces small droplets of asphaltene material which are then quenched in a water bath. The pellets are removed on screen separators and then transferred to storage using conveyer systems; the water is filtered, cooled, and returned to the pelletizing vessel. The pellets ate claimed to have a higher heating value than petroleum coke.

12.2.6 SOLVAHL PROCESS

The *Solvahl process* is a solvent deasphalting process for application to vacuum residua (Table 12.6) (RAROP, 1991, p. 9; Billon et al., 1994). The process was developed to give maximum yields of deasphalted oil while eliminating asphaltenes and reducing metal content to a level compatible with the reliable operation of downstream units. The process removes the asphaltene constituents, most metals, and other impurities contained in the heavy feedstock.

The process produces a deasphalted oil from which most of the heptane asphaltenes have been eliminated. The content of metals and coke precursor compounds are also significantly reduced, making the deasphalted oil a suitable feedstock for downstream fluid catalytic cracking or hydrocracking units. The solvents used in the process can vary from liquid propane to heptane depending on feedstock properties and downstream process objectives. The relative yields of deasphalted oil and asphalt, as well as the characteristics of both products, are linked to the nature of the feed, the operating conditions and the solvent type.

In summary, the Solvahl technology is focused on (1) relevant and optimized separation of deasphalted oil and asphaltenes with associated design criteria to reach specifications and (2) control of the quality of the deasphalted oil, and especially any remaining content of asphaltene constituents, is a main concern as it impacts downstream catalyst performance and cycle length. The association of the process with other conversion technologies such as hydrocracking or ebullated-bed technologies (such as H-Oil) presents the opportunity to maximize the yield of middle distillate products from heavy feedstocks.

TABLE 12.6
Feedstock and Product Data for the Solvahl Process

	Arabian Light Vacuum Residuum
Feedstock	
API	9.6
Sulfur, % w/w	4.1
Nitrogen, % w/w	0.3
C_7-asphaltenes, % w/w	4.2
Carbon residue, % w/w	16.4
Nickel, ppm	19.0
Vanadium, ppm	61.0
Products	
C_4-deasphalted oil, % w/w	70.1
API	16.0
Sulfur, % w/w	3.3
Nitrogen, % w/w	0.2
C_7-asphaltenes, % w/w	<0.1
Carbon residue, % w/w	5.3
Nickel, ppm	2.0
Vanadium, ppm	3.0
C_5-deasphalted oil, % w/w	85.5
API	13.8
Sulfur, % w/w	3.7
Nitrogen, % w/w	0.2
C_7-asphaltenes, % w/w	<0.1
Carbon residue, % w/w	7.9
Nickel, ppm	7.0
Vanadium, ppm	16.0

12.2.7 LUBE DEASPHALTING

Other facilities incorporate lube deasphalting to process vacuum residuum into lube oil base stocks. Propane deasphalting is most commonly used to remove asphaltene constituents and resins which contribute an undesirable dark color to the lube base stocks. This process typically uses baffle towers or rotating disk contactors to mix the propane with the feed. Solvent recovery is accomplished with evaporators, and supercritical solvent recovery processes are also used in some deasphalting units.

The Duo-Sol process is used to deasphalt and extract lubricating oil feedstocks. Propane is used as the deasphalting solvent and a mixture of phenol and cresylic acids (cresol derivatives, hydroxy-toluene derivatives) are used as the extraction solvent. The extraction is conducted in a series of batch extractors followed by solvent recovery in multistage flash distillation and stripping towers.

12.3 DEWAXING PROCESSES

Petroleum wax consists predominantly of normal paraffin hydrocarbons and some branched paraffin hydrocarbons but which have a low degree of chain branching. The wax is a colorless mass, more or less translucent, and has a crystalline or semicrystalline structure. The solvent dewaxing process is used to remove wax (n-paraffin hydrocarbons, straight-chain paraffin hydrocarbons) from deasphalted lubricating oil base stocks (Figure 12.7). The main process steps include mixing

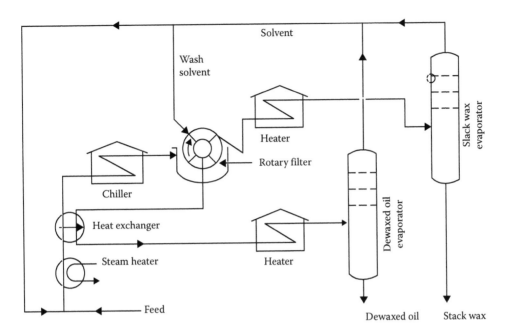

FIGURE 12.7 A solvent dewaxing unit. (From OSHA Technical Manual, Section IV, Chapter 2, Petroleum refining processes, 1999, http://www.osha.gov/dts/osta/otm/otm_iv/otm_iv_2.html.)

the feedstock with the solvent, chilling the mixture to crystallize wax, and recovering the solvent. Commonly used solvents include toluene and methyl ethyl ketone (MEK)—methyl isobutyl ketone is also used in a wax deoiling process to prepare food-grade wax.

Waxes are characterized by the oil content, melting point, and penetration (Speight, 2014, 2015), although the penetration is not always recognized as a means of characterizing. For a specific wax, the melting point and the penetration are functions of the oil content and, depending on the molecular weight, the melting point of wax with oil contents of less than 0.5% w/w is typically on the order of 26°C and 71°C (80°F and 160°F) for distilled wax containing $C_{18}H_{38}$–$C_{32}H_{66}$ hydrocarbons, and 60°C and 88°C (140°F–190°F) for microcrystalline wax—the molecular weight of microcrystalline wax generally falls in the range of 450–600. In spite of these guidelines, the classification of wax is somewhat arbitrary and is related to the method of production and the molecular weight. In terms of nomenclature, it is also arbitrary—the wax produced from low-viscosity distillates by chilling, pressing, and sweating is *paraffin wax*, while the product from solvent dewaxing is *slack wax*. When slack wax is deoiled, the product is *scale wax* and *microcrystalline wax* is produced by dewaxing residual stocks. The flow properties of wax are temperature dependent and can influence the flow properties of lubricating oil.

The flow properties of lubricating oil are an extremely important property, especially at low temperature. In order to insure proper lubrication of machinery, the oil must flow at temperatures well below the lowest possible working temperature. The main feedstocks used in the fabrication of lubricating oil are mixtures of hydrocarbons produced from various crude oils but these mixtures also contain, together with hydrocarbons needed in the production of lubricating oils, amounts of paraffin hydrocarbons with high melting points. At low temperatures, these paraffin hydrocarbons crystallize out of the oil and form a network in which the liquid phase is trapped and cannot be displaced. For high-viscosity oils such as bright stock, a viscosity pour point can be observed. The oil fails to flow in the pour point test because its viscosity is too high and a deeper dewaxing does not always result in any improvement.

In the 1930s, two types of stocks, naphthenic and paraffinic, were used to make lubricating oil. Both types were solvent extracted to improve their quality, but in the high-temperature conditions encountered in service the naphthenic type as well as the paraffinic type could not stand up. Nevertheless, the naphthenic type was the preferred oil, particularly in cold weather, because of its fluidity at low temperatures. Previous to 1938, the highest-quality lubricating oils were of the naphthenic type and were phenol treated to pour points of −40°C to −7°C (−40°F to 20°F), depending on the viscosity of the oil. Paraffinic oils were also available and could be phenol treated to higher-quality oil, but their wax content was so high that the oils were solid at room temperature.

In addition, the removal of wax constituents from a feedstock used for the production of lubricating oil will change not only the pour point but also other physical properties. For example, the specific gravity of wax is typically on the order of 0.750–0.780 at 99°C (210°F), and as a consequence the specific gravity of the dewaxed oil is higher than the specific gravity of the charge stock. The viscosity of the wax in the liquid state is lower than that of the oil and the viscosity of the dewaxed oil is higher than that of the original feedstock, the change in viscosity being higher for high-viscosity oils. Dewaxing also reduces the viscosity index of the oil. Starting from the same waxy oil feedstock, a lower pour point dewaxed oil will have a lower viscosity index than a higher pour point dewaxed oil.

12.3.1 Cold Press Process

The first dewaxing processes were based on cooling the waxy feed alone or diluted with naphtha and separating the wax by cold settling or, later on, by centrifuges. For the separation of wax from cold light vacuum distillates, plate and frame filter presses were used.

The lowest-viscosity paraffinic oils were dewaxed by the cold press method to produce oils with a pour point of 2°C (35°F). The light paraffin distillate oils contained a paraffin wax that crystallized into large crystals when chilled and could thus readily be separated from the oil by the cold press filtration method. The more viscous paraffinic oils (intermediate and heavy paraffin distillates) contained amorphous or microcrystalline waxes, which formed small crystals that plugged the filter cloths in the cold press and prevented filtration. Because the wax could not be removed from intermediate and heavy paraffin distillates, the high-quality, high-viscosity lubricating oils in them could not be used except as cracking stock.

Methods were therefore developed to dewax these high-viscosity paraffinic oils. The methods were essentially alike in that the waxy oil was dissolved in a solvent that would keep the oil in solution; the wax separated as crystals when the temperature was lowered. The processes differed chiefly in the use of the solvent. Commercially used solvents were naphtha, propane, sulfur dioxide, acetone–benzene, trichloroethylene, ethylene dichloride–benzene (*Barisol*), methyl ethyl ketone–benzene (*benzol*), methyl-*n*-butyl ketone, and methyl-*n*-propyl ketone.

Although some of the old methods are still in commercial use, there are three main methods used in modern refinery technology: (1) *solvent dewaxing processes* in which the feedstock is mixed with one or more solvents, then is cooled to allow the formation of wax crystals, and the solid phase is separated from the liquid phase by filtration, (2) *urea dewaxing processes*, in which urea straight-chain paraffin adducts are formed and then separated by filtration from the dewaxed oil, and (3) *catalytic dewaxing processes* in which straight-chain paraffinic hydrocarbons are selectively cracked on zeolite-type catalysts and the lower-boiling reaction products are separated from the dewaxed feedstock oil by fractionation.

12.3.2 Solvent Dewaxing Process

The solubility of paraffin wax in hydrocarbons increases with a decrease in the melting point of the wax and with an increase of temperature. Hydrocarbons of decreasing molecular weight, down to pentane–butane, increase in wax solubility while a further decrease of the molecular weight below pentane–butane shows a decrease of the solubility of wax. Propane has the advantage of having quite low solubility for wax and not too high a vapor pressure but, on the other hand, using hydrocarbons with higher vapor pressure (such as methane or ethane) are inefficient in the process.

Solvent dewaxing consists of the following steps: crystallization, filtration, and solvent recovery. In the crystallization step, the feedstock is diluted with the solvent and chilled, solidifying the wax components. The filtration step removes the wax from the solution of dewaxed oil and solvent. Solvent recovery removes the solvent from the wax cake and filtrate for recycle by flash distillation and stripping.

In the first *solvent dewaxing process* (developed in 1924), the waxy oil was mixed with naphtha and filter aid (fuller's earth or diatomaceous earth). The mixture was chilled and filtered, and the filter aid assisted in building a wax cake on the filter cloth. This process is now obsolete, and most of the modern dewaxing processes use a mixture of methyl ethyl ketone and benzene. Other ketones may be substituted for dewaxing, but regardless of what ketone is used, the process is generally known as ketone dewaxing.

The process as is often practiced involves mixing the feedstock with one to four times its volume of the ketone (Scholten, 1992). The mixture is then heated until the oil is in solution, and the solution is chilled at a slow, controlled rate in double-pipe, scraped-surface exchangers. Cold solvent, such as filtrate from the filters, passes through the 2-inch annular space between the inner and outer pipes and chills the waxy oil solution flowing through the inner 6-inch pipe.

To prevent wax from depositing on the walls of the inner pipe, blades or scrapers extending the length of the pipe and fastened to a central rotating shaft scrape off the wax. Slow chilling reduces the temperature of the waxy oil solution to 2°C (35°F), and then faster chilling reduces the temperature to the approximate pour point required in the dewaxed oil. The waxy mixture is pumped to a filter case into which the bottom half of the drum of a rotary vacuum filter dips. The drum (8 ft in diameter, 14 ft long), covered with filter cloth, rotates continuously in the filter case. Vacuum within the drum sucks the solvent and the oil dissolved in the solvent through the filter cloth and into the drum. Wax crystals collect on the outside of the drum to form a wax cake, and as the drum rotates, the cake is brought above the surface of the liquid in the filter case and under sprays of ketone that wash oil out of the cake and into the drum. A knife-edge scrapes off the wax, and the cake falls into the conveyor and is moved from the filter by the rotating scroll.

The recovered wax is actually a mixture of wax crystals with a little ketone and oil, and the filtrate consists of the dewaxed oil dissolved in a large amount of ketone. Ketone is removed from both by distillation, but before the wax is distilled, it is deoiled, mixed with more cold ketone, and pumped to a pair of rotary filters in series, where further washing with cold ketone produces a wax cake that contains very little oil. The deoiled wax is melted in heat exchangers and pumped to a distillation tower operated under vacuum, where a large part of the ketone is evaporated or flashed from the wax. The rest of the ketone is removed by heating the wax and passing it into a fractional distillation tower operated at atmospheric pressure and then into a stripper where steam removes the last traces of ketone.

An almost identical system of distillation is used to separate the filtrate into dewaxed oil and ketone. The ketone from both the filtrate and wax slurry is reused. Clay treatment or hydrotreating finishes the dewaxed oil as previously described. The wax (*slack wax*), even though it contains essentially no oil as compared to 50% in the slack wax obtained by cold pressing, is the raw material for either sweating or wax recrystallization, which subdivides (fractionates) the wax into a number of wax fractions with different melting points.

Solvent dewaxing can be applied to light, intermediate, and heavy lubricating oil distillates, but each distillate produces a different kind of wax. Each of these waxes is actually a mixture of a number of waxes. For example, the wax obtained from light paraffin distillate consists of a series of paraffin waxes that have melting points in the range of 30°C–70°C (90°F–160°F), which are characterized by a tendency to harden into large crystals. However, heavy paraffin distillate yields a wax composed of a series of waxes with melting points in the range of 60°C–90°C (140°F–200°F), which harden into small crystals from which they derive the name of *microcrystalline wax* or *microwax*.

On the other hand, intermediate paraffin distillates contain paraffin waxes and waxes intermediate in properties between paraffin and microwax.

Thus, the solvent dewaxing process produces three different *slack waxes* (also known as *crude waxes* or *raw waxes*) depending on whether light, intermediate, or heavy paraffin distillate is processed. The slack wax from heavy paraffin distillate may be sold as dark raw wax, the wax from intermediate paraffin distillate as pale raw wax. The latter is treated with lye and clay to remove odor and improve color.

There are several processes in use for solvent dewaxing, but all have the same general steps, which are (1) contacting the feedstock with the solvent, (2) precipitating the wax from the mixture by chilling, and (3) recovering the solvent from the wax and dewaxed oil for recycling. The processes use benzene–acetone (*solvent dewaxing*), propane (*propane dewaxing*), trichloroethylene (*separator-Nobel dewaxing*), ethylene dichloride–benzene (*Barisol dewaxing*), and urea (*urea dewaxing*), as well as liquid sulfur dioxide–benzene mixtures. The *propane dewaxing process* is essentially the same as the ketone process except for the following: (1) propane is used as the dewaxing solvent, (2) higher pressure equipment is required, and (3) chilling is achieved by use of evaporative chillers by vaporizing a portion of the dewaxing solvent. Although this process often generates a higher-quality product and does not require crystallizers, the temperature differential between the dewaxed oil and the filtration temperature is higher than for ketone-based processes (higher energy costs), and dewaxing aids are required to obtain improved filtration rates.

The major processes currently in use are the ketone dewaxing processes and the most widely used ketone processes are the Texaco Solvent Dewaxing process and the Exxon Dilchill process. Both processes consist of diluting the waxy feedstock with solvent while chilling at a controlled rate to produce a slurry. The slurry is filtered using rotary vacuum filters and the wax cake is washed with cold solvent. The filtrate is used to pre-chill the feedstock and solvent mixture. The primary wax cake is diluted with additional solvent and filtered again to reduce the oil content in the wax. The solvent recovered from the dewaxed oil and wax cake by flash vaporization and recycled back into the process.

The Texaco Solvent Dewaxing process (also called the MEK process) uses a mixture of methyl ethyl ketone and toluene as the dewaxing solvent, and sometimes uses mixtures of other ketones and aromatic solvents. The Exxon Dilchill Dewaxing Process uses a direct cold solvent dilution-chilling process in a special crystallizer in place of the scraped-surface exchangers used in the Texaco process. In the Dilchill process, the first cooling stage is a crystallizer provided with many mixing elements and the chilling rate is controlled by incremental injection of cold solvent and by the size of the crystallizer.

The Di/Me dewaxing process uses a mixture of dichloroethane and methylene dichloride as the dewaxing solvent. In this process, methylene chloride is the solvent with high solvent power—equivalent to toluene in the methyl ethyl ketone process—and dichloroethane is the solvent with high selectivity. It is possible to adjust the solvent power and the selectivity by changing the ratio between the two solvents, as in the MEK and Di/Me processes. In single solvent dewaxing processes, the characteristics of the solvent are changed by adjusting the filtration temperature.

Another important variable of the solvent dewaxing process is the solvent/feedstock (dilution) ratio that affects the structure of the wax crystals and the filtration characteristics of the cake on the filter as well as the oil content of the wax and hence the yield of dewaxed oil. When high dilution ratios are used on low-viscosity stocks, larger crystals (containing feedstock oil) are formed at the beginning of the cooling process and during the cooling process smaller crystals are incorporated in and around the larger crystals. Finally, although very good filtration characteristics of the cake are obtained, the amount of oil trapped in the crystal cages cannot be washed out and the oil content of the wax is high and, consequently, the yield of dewaxed oil is low. Dilution ratios that are too low are even more detrimental because only very small crystals, which give poor filtration and washing rates, can be obtained.

The main parameter governing the solvent dilution technique is the viscosity of the liquid phase at various points in the cooling/chilling stage as well as on the filter. The kinetics of the solid-phase formation—which are a function of the viscosity of the mother liquor among other factors—are a determining factor in the structure of the wax crystals and of the cake on the filter. The solvent dilution ratio is a function of the viscosity–temperature curve of each solvent and a function of the solvent composition for the dual solvent dewaxing processes as well. For the same reason, at a given solvent viscosity, the viscosity of the charge oil will determine the solvent dilution ratio. High-viscosity heavy oils are diluted with more solvent than the lighter low-viscosity oils. Thus, an important aspect of the solvent dewaxing process is the filter. Large rotary drum filters are used in every solvent dewaxing unit.

Furthermore, with more wax crystallizing out of the liquid phase, the viscosity of the remaining liquid-phase changes. At the same time, the solvent-to-oil ratio in the liquid phase will determine the oil content of the wax. Since all of the oil cannot be removed from (cannot be washed out of) the pores with fresh solvent, the amount of oil trapped as the liquid phase in these pores is dependent on the oil content of the liquid phase of the slurry.

An improvement in the filtration process can be achieved if an incremental solvent dilution technique is used in the cooling/chilling stage. In this case, the feedstock is mixed with a limited amount of dilution solvent and the rest of the dilution solvent is added stepwise during cooling. The initial low dilution, at a point where the temperature of the solvent/charge oil solution is still high and where the crystallization process has not yet started, avoids the formation of large crystals. As the viscosity of the solution increases, more solvent is added. For heavy feedstocks (such as bright stock) where the initial viscosity is high and the structure of the wax crystals is microcrystalline, this incremental dilution technique is not used.

Avoiding the formation of large crystals (which often signify retained oil within the crystal structure) in the beginning stage of crystallization is accomplished in some processes by means of shock chilling and/or conditions of high turbulence. In shock chilling, the fresh solvent is added incrementally at a somewhat lower temperature than the solvent–feedstock mixture stream. In addition, in the initial stages of wax crystal formation, a high turbulence induced in the solution will break down any large crystal structures.

Another important criterion in a solvent dewaxing process is the liquid-to-solid ratio in the slurry fed to the filter. With too much solid, the agglomerates of wax crystals will stick together to give a thick, tight cake with poor filter rates and a high oil content in the wax. If the liquid-to-solid ratio is too high, the cake on the filter cloth becomes too thin with many cracks, and the wash solvent, instead of washing the cake, will flow through these cracks, leaving the cake unwashed. For adjusting the liquid-to-solid ratio in some plants where the wax content of the feed is very low, wax is recycled into the feed. Recycling the filtrate or wash filtrate into the main steam to the filter before the last chiller is used in many plants where the wax content of the feed is high.

12.3.3 Urea Dewaxing Process

Urea dewaxing (Scholten, 1992) is worthy of further mention insofar as the process is highly selective and, in contrast to the other dewaxing techniques, can be achieved without the use of refrigeration. However, the process cannot compete economically with the solvent dewaxing processes for treatment of the heavier lubricating oils. But when it is applied to the lighter materials that already may have been subjected to a solvent dewaxing operation, products are obtained that may be particularly useful as refrigerator oils, transformer oils, and the like.

The process description is essentially the same as that used for solvent dewaxing with the omission of the chilling stage and the insertion of a contactor where the feedstock and the urea (with a solvent) are thoroughly mixed before filtration. The solvents are recovered from the dewaxed oil by evaporation, and the urea complex is decomposed in a urea recovery system.

Residual lubricating oils, such as cylinder oils and bright stocks, are made from paraffinic or mixed-base reduced crude oils and contain waxes of the microcrystalline type. Removal of these waxes from reduced crude produces *petrolatum*, a grease-like material that is known in a refined form as *petroleum jelly*. This material can be separated from reduced crude in several ways. The original method was cold settling, whereby reduced crude was dissolved in a suitable amount of naphtha and allowed to stand over winter until the microwax settled out. This method is still used, but the reduced crude naphtha solution is held in refrigerated tanks until the petrolatum settles out. The supernatant naphtha-oil layer is pumped to a still where the naphtha is removed, leaving cylinder stock that can be further treated to produce bright stock. The petrolatum layer is also distilled to remove naphtha and may be clay treated or acid and clay treated to improve the color.

12.3.4 CENTRIFUGE DEWAXING PROCESS

Another method of separating petrolatum from reduced crude is *centrifuge dewaxing*. In this process the reduced crude is dissolved in naphtha and chilled to −18°C (0°F) or lower, which causes the wax to separate. The mixture is then fed to a battery of centrifuges where the wax is separated from the liquid. However, the centrifuge method has now been largely displaced by solvent dewaxing methods and by more modern methods of wax removal.

12.3.5 CATALYTIC DEWAXING PROCESS

Historically, most commercial dewaxing operations have been based on the use of a solvent. The mixture of oil and solvent is refrigerated, and the wax that separates out is removed by filtration. The dewaxed oil is then recovered by distilling off the solvent. The solvent dewaxing process has two main disadvantages: (1) the operating costs for the process are high and (2) the pour point that can he achieved is limited by the high cost of refrigerating to very low temperatures. Because of these disadvantages, considerable development has been carried out on hydrocatalytic processes for pour point reduction. These processes are based on the ability of certain zeolites to selectively crack the paraffinic hydrocarbons which are the major constituent of wax (Hargrove, 1992; Genis, 1997). Solvent dewaxing is relatively expensive for the production of low pour point oils; various catalytic dewaxing (selective hydrocracking) processes have been developed for the manufacture of lube oil base stocks. The basic process consists of a reactor containing a proprietary dewaxing catalyst followed by a second reactor containing a hydrogen finishing catalyst to saturate olefins created by the dewaxing reaction and to improve stability, color, and resistance of the finished lubricating oil to emulsion formation.

BP has developed a *hydrocatalytic dewaxing process* that is reputed to overcome some of the disadvantages of the solvent dewaxing processes. In the process of dewaxing, waxy molecules are removed from heavy distillate fuel cuts or lube distillates. *Catalytic dewaxing* is a mild hydrocracking process and is therefore operated at elevated temperatures (280°C–400°C, 550°F–750°F) and pressures (300/1500 psi).

Catalytic dewaxing is a mild hydrocracking process and, therefore, has to be operated at elevated temperatures and hydrogen partial pressures. In general, the process can be operated within the following range of conditions. Within the temperature and pressure ranges, the conditions for a particular dewaxing operation will depend upon the nature of the feedstock and the required product pour point. Important feedstock parameters are (1) boiling range, (2) amount of wax present, (3) type of wax present, as well as (4) sulfur content and nitrogen content. Thus, the conditions for a particular dewaxing operation depend upon the nature of the feedstock and the product pour point required.

The catalyst employed for the process is typically a mordenite-type catalyst that has the correct pore structure to be selective for *n*-paraffin cracking. Platinum on the catalyst serves to hydrogenate the reactive intermediates so that further paraffin degradation is limited to the initial thermal reactions. The process has been employed to successfully dewax a wide range of naphthenic feed-stocks (Hargrove et al., 1979), but it may not be suitable to replace solvent dewaxing in all cases.

The BP Catalytic Dewaxing Process uses a catalyst based on the zeolite mordenite that has the correct pore structure to be selective for normal and slightly branched paraffins. It is also a highly active cracking material and also contains a noble metal which acts as a hydrogenation function. The process can be used for a wide variety of dewaxing applications in two main areas: (1) the pro-duction of specialty oils and lubricating oils and (2) production of low pour point middle distillates.

In the process, the feedstock is mixed with recycle gas and heat exchanged with reactor effluent before being heated to the required reactor inlet temperature by a fired heater. The reactor feedstock passes downflow through the reactor. Because the reaction is exothermic, it is sometimes necessary for cold recycle gas to be injected between the catalyst beds to limit the temperature rise. The reac-tor effluent is cooled against the reactor feed before passing to a hot high-pressure separator. The gas from this separator is cooled and then passed to a cold high-pressure separator. The gas from the cold separator is recycled to the reactor inlet via a gas treatment section and a recycle gas com-pressor. The liquid hydrocarbons from the two high-pressure separators are combined and flow to a low-pressure separator where off-gas is removed. The remaining liquid passes to a product stripper where liquefied petroleum gas and naphtha are removed as overhead fractions and the dewaxed oil is produced as a bottoms fraction.

Breakdown products from the cracking reaction are rapidly hydrogenated thereby minimizing carbon deposition on the catalyst and leading to a highly stable catalyst system. The active sites of the catalyst effectively only come into contact with paraffin chains. Because of this, it is possible to operate the process successfully when dewaxing straight run distillates of high sulfur and nitrogen contents. For the same reason, little or no desulfurization or denitrogenation is observed in the process.

Other processes include the ExxonMobil distillate dewaxing process (Smith et al., 1980) by which dewaxing is achieved by selective cracking in which the long paraffin chains are cracked to form shorter chains using a shape-selective zeolite that rejects ring compounds and *iso*-paraffins. In a related process (MIDW process), the paraffins are selectively isomerized using low-pressure conditions (Smith et al., 1980). This process also uses a zeolite catalyst to convert low-quality gas oil into diesel fuel.

In the process, the proprietary catalyst can be reactivated to fresh activity by relatively mild non-oxidative treatment. Of course, the time allowed between reactivation is a function of the feedstock but after numerous reactivation it is possible that there will be coke buildup on the catalyst. The process can be used to dewax a full range of lubricating base stocks and, as such, has the potential to completely replace solvent dewaxing or can even be used in combination with solvent dewaxing. This latter option, of course, serves to de-bottleneck existing solvent dewaxing facilities.

Both the catalytic dewaxing processes have the potential to change the conventional thoughts about dewaxing insofar as they are not solvent processes and may be looked upon (more cor-rectly) as thermal processes rather than treatment processes. However, both provide viable alter-natives to the solvent processes and offer a further advance in the science and technology of refinery operations.

Catalytic dewaxing yields various grades of lube oils and fuel components suitable for extreme winter conditions. Paraffinic (waxy) components that precipitate out at low temperatures are removed. In the UOP catalytic dewaxing process, the first stage saturates olefins and desulfurizes and denitrifies the feed via hydrotreating (Genis, 1997). In the second stage, a dual-function, non-noble metal zeolite catalyst selectively adsorbs and then selectively hydrocracks the normal and near-normal long-chain paraffins to form shorter-chain (nonwaxy) molecules.

Alternatively, in the Chevron Isodewaxing process, the dewaxing results from isomerizing the linear paraffins to branched paraffins by using a molecular sieve catalyst containing platinum (Miller, 1994a,b). In the Isodewaxing process (Figure 12.8), which is followed by a hydrofinishing step, waxy feedstock from a hydrocracker or hydrotreater, together with the hydrogen-containing

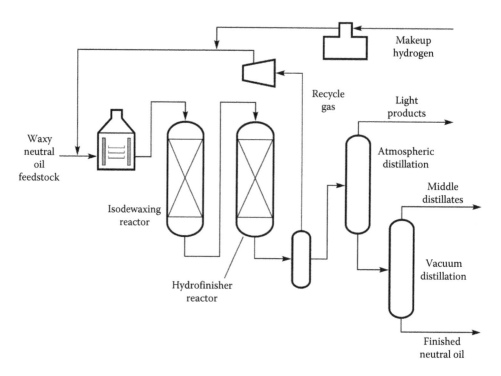

FIGURE 12.8   The Isodewaxing process.

gas, is heated and fed to the Isodewaxing reactor. The conditions in the reactor cause isomerization of *n*-paraffins to *iso*-paraffins and other paraffins are cracked to highly saturated low-boiling products such as jet fuel and diesel fuel. The effluent from the Isodewaxing reactor is then sent to the hydrofinishing unit where hydrofinishing, including aromatics saturation provide the product. The catalysts used in the Isodewaxing and hydrofinishing units are selective for dewaxing and hydrogenation and are, as can be anticipated, at their maximum efficiency with low sulfur and low nitrogen feedstocks. The process generally uses a high-pressure recycle loop. Because of the conversion of wax constituents to other usable products, the process has obvious benefits over solvent dewaxing insofar as quality of the product is increased.

12.3.6   Dewaxing Heavy Feedstocks

Roughly 75% of the world's heavy oil reserves can be found in the Orinoco Belt in Venezuela, and in the Northern Alberta and Saskatchewan provinces of Canada. Both heavy oils and oil sands have higher viscosity and lower gravity of less than 20° API, which makes them extremely difficult to produce from subsurface reservoirs and subject the reservoirs to thermal stimulation (Giacchetta et al., 2015; Sahu et al., 2015). Wax deposition, which is crystallization resulting from phase separation of paraffinic solids from crude oil due to temperature drop also occurs during their transmission in pipelines.

Accumulation of these solids could cause severe flow assurance problems that ultimately lead to pipe leakage, rupture, and explosion (Visintin et al., 2008). An increased wax deposition is often accompanied by high pigging frequency as curative treatment which costs extra millions of dollars in deferred revenue (Lu et al., 2012). Therefore, wax deposition must be properly managed in order to reduce the associated problems as well as increase the ability of the heavy oil to flow for an increased market values and ease of processing in refineries.

An effective treatment method for wax deposition (Hamilton and Herman, 2011) used passive energy to stabilize micelle structures that promote the deposition of paraffins, asphaltenes, and mineral scale particularly in heavy oil pipelines. Solvent dewaxing is another effective treatment method which employs recovery of microcrystalline (or paraffin waxes) from the heavy crude before it is processed (Speight, 2014). This study investigates the range of optimum performance conditions for solvent dewaxing using methyl ethyl ketone (MEK). MEK is selected for this study due to its rapid evaporation rate, higher solvency, lower viscosity, good miscibility with most hydrocarbons without impact on their characteristic as well as the favorable volume/mass ratio due to its low density (Beringer et al., 2015; Hu et al., 2015). Nimer et al. (2010) conducted a similar study by using a mixture of toluene and MEK as carrier solvents; however, pure MEK is used in this study due to its established lower selectivity for paraffinic compounds compared to toluene. Though both are good solvents for oil, the presence of toluene could negatively impact the solvent dewaxing performance (Nimer et al., 2010). The effects of changes in solvent-to-oil ratio, heating temperature, and cooling temperature on the amount of wax yielded are studied in this work.

This study has successfully established the effects of parametric variations on wax extraction from an Australian heavy crude oil using a pure Methyl Ethyl Ketone solvent. It was observed that the wax yield increases with increasing mixing temperature and solvent-to-oil ratio, as well as a decrease in chilling temperature. Based on the experimental results it was concluded that the optimum wax yield of 27.9% w/w was obtained using a 15:1 solvent to crude oil ratio, at a mixing temperature of 50°C, and a chilling temperature of −20°C.

High wax yields of 27.6% w/w were also observed at a solvent-to-oil ratio of 15:1 under a 50°C mixing temperature and a chilling temperature of −15°C, as well as at a solvent-to-oil ratio of 20:1, mixed at 50°C and chilled at −20°C. With no more than 0.3% w/w yield separating the top three results, which have been obtained under a relatively small margin of operating conditions, more experimental work should be conducted to obtain a better optimum condition for solvent dewaxing over a wider range of operational parameters.

In addition to the experimental optimization, economic optimization and feasibility analysis should be conducted to observe the possibilities for commercial scale implementation. This may include a study on the trade-off between the additional revenue which may be achieved through a higher yield of wax and the additional cost of utilities and raw materials required to achieve such high yields, taking into account other parameters such as processing time, solvent recovery, capital and operating costs which may affect the operation as a whole.

REFERENCES

Beringer, L.T., Xu, X., Wan Shih, W., Shih, W., Habas, R., and Schauer, C.L. 2015. An electrospun PVDF-TRFE fiber sensor platform for biological applications. *Sensors and Actuators A*, 222: 293–300.

Billon, A., Morel, F., Morrisson, M.E., and Peries, J.P. 1994. Converting residues with the IFP Hyvahl process. *Révue Institut Français du Pétrole*, 49(5): 495–507.

Ditman, J.G. 1973. Solvent deasphalting. *Hydrocarbon Processing*, 52(5): 110.

Dunning, H.N. and Moore, J.W. 1957. Propane removes asphalts from crudes. *Petroleum Refiner*, 36(5): 247–250.

Elliott, J.D. and Stewart, M.D. 2004. Cost effective residue upgrading: Delayed coking and refining. *Proceedings of Fourth Russian Refining Technology Conference*. Moscow, Russia, September 23–24.

Gary, J.H., Handwerk, G.E., and Kaiser, M.J. 2007. *Petroleum Refining: Technology and Economics*, 5th Edition. CRC Press/Taylor & Francis Group, Boca Raton, FL.

Gearhart, J.A. 1980. Solvent treat resids. *Hydrocarbon Processing*, 59(5): 150–151.

Gearhart, J.A. and Garwin, L. 1976. ROSE process improves resid feed. *Hydrocarbon Processing*, 55(5): 125–128.

Genis, O. 1997. In: *Handbook of Petroleum Refining Processes*. R.A. Meyers (Editor). McGraw-Hill, New York, Chapter 8.5.

Giacchetta, G., Leporini, M., and Marchetti, B. 2015. Economic and environmental analysis of a steam assisted gravity drainage (SAGD) facility for oil recovery from Canadian oil sands. *Applied Energy*, 142: 1–9.

Hamilton, D.S. and Herman, B. 2011. The application of passive energy to production optimization; stabilizing the micelle structure in oil to prevent deposition of paraffin, asphaltenes, and mineral scale and reduce well-head viscosity in heavy oil. *Proceedings of South American Oil and Gas Congress*. Maracaibo, Venezuela, October 18.

Hargrove, J.D. 1992. In: *Petroleum Processing Handbook*. J.J. McKetta (Editor). Marcel Dekker, New York, p. 558.

Hargrove, J.D., Elkes, G.J., and Richardson, A.H. 1979. *Oil & Gas Journal*, 77(3): 103.

Hsu, C.S. and Robinson, P.R. (Editors) 2006. *Practical Advances in Petroleum Processing*, Volumes 1 and 2. Springer Science, New York.

Hu, G., Li, J., and Hou, H. 2015. A combination of solvent extraction and freeze thaw for oil recovery from petroleum refinery wastewater treatment pond sludge. *Journal of Hazardous Materials*, 283: 832–840.

Koseoglu, O.R. 2009. Enhanced solvent deasphalting process for heavy hydrocarbon feedstocks utilizing solid adsorbent. U.S. Patent 7,566,394. July 28.

Low, J.Y., Hood, R.L., and Lynch, K.Z. 1995. Valuable products from the bottom of the barrel using ROSE technology. Preprints. *Symposium on Petroleum Chemistry and Processing*. Division of Petroleum Chemistry, 210th National Meeting, American Chemical Society, Chicago, IL, August 20–25, pp. 780–784.

Lu, Y., Huang, Z., Hoffmann, R., Amundsen, L., and Fogler, H.S. 2012. Counterintuitive effects of the oil flow rate on wax deposition. *Energy & Fuels*, 26(7): 4091–4097.

McCoy, J.N., Keusenkothen, P.F., and Srivastava, A. 2010. Process for upgrading tar. U.S. Patent 7,744,743. June 29.

Miller, S.J. 1994a. In: *Studies in Surface Science and Catalysis*. J. Weitkamp (Editor). Elsevier, Amsterdam, the Netherlands, Volume 84C, pp. 2319–2326.

Miller, S.J. 1994b. *Microporous Materials*, 2: 439–450.

Mitchell, D.L. and Speight, J.G. 1973. The solubility of asphaltenes in hydrocarbon solvents. *Fuel*, 52: 149.

Nimer, A.A., Mohamed, A.A., and Rabah, A.A. 2010. Nile blend crude oil: Wax separation using MEK toluene mixtures. *Arabian Journal for Science and Engineering*, 35(2): 17–24.

OSHA Technical Manual. 1999. Section IV, Chapter 2: Petroleum refining processes. http://www.osha.gov/dts/osta/otm/otm_iv/otm_iv_2.html.

Parkash, S. 2003. *Refining Processes Handbook*. Gulf Professional Publishing, Elsevier, Amsterdam, the Netherlands.

Patel, V., Iqbal, R., Odette, E., and Subramanian, A. 2008. *Economic Bottom of the Barrel Processing to Minimize Fuel Oil Production*. Research note No. KIOGE-2008. Kellogg Brown & Root, Inc., Houston, TX.

RAROP. 1991. In: *RAROP Heavy Oil Processing Handbook*. Research Association for Residual Oil Processing. T. Noguchi (Chairman). Ministry of Trade and International Industry (MITI), Tokyo, Japan.

Sahu, R., Song, B.J., Im, J.S., Jeon, Y.P., and Lee, C.W. 2015. A review of recent advances in catalytic hydrocracking of heavy residues. *Journal of Industrial and Engineering Chemistry*, Manuscript accepted for publication.

Scholten, G.G. 1992. *Petroleum Processing Handbook*. J.J. McKetta (Editor). Marcel Dekker, New York, p. 565.

Smith, K.W., Starr, W.C., and Chen, N.Y. 1980. *Oil & Gas Journal*. 78(21): 75.

Speight, J.G. 2000. *The Desulfurization of Heavy Oils and Residua*, 2nd Edition. Marcel Dekker, New York.

Speight, J.G. 2011. *The Refinery of the Future*. Gulf Professional Publishing, Elsevier, Oxford, UK.

Speight, J.G. 2014. *The Chemistry and Technology of Petroleum*, 5th Edition. CRC Press/Taylor & Francis Group, Boca Raton, FL.

Speight, J.G. 2015. *Handbook of Petroleum Product Analysis*, 2nd Edition. John Wiley & Sons, Hoboken, NJ.

Speight, J.G. and Ozum, B. 2002. *Petroleum Refining Processes*. Marcel Dekker, New York.

Trambouze, P., Euzen J.P., Bergez P., and Claveau, M. 1989. Process for deasphalting a hydrocarbon oil. U.S. Patent 4,816,140. March 28.

Visintin, R.F.G, Lockhart, T.P., Lapasin, R., and D'Antona, P. 2008. Structure of waxy crude oil emulsion gels. *Journal of Non-Newtonian Fluid Mechanics*, 149(1–3): 34–39.

13 Product Improvement

13.1 INTRODUCTION

As already noted (Chapter 1), petroleum and its derivatives have been used for millennia, and it is perhaps the most important raw material consumed in modern society (Abraham, 1945; Forbes, 1958a,b, 1959; James and Thorpe, 1994). It provides not only raw materials for the ubiquitous plastics and other products, but also fuel for energy, industry, heating, and transportation. Thus, the use of petroleum and the development of related technology are not such a modern subject as we are inclined to believe. However, the petroleum industry is essentially a twentieth-century industry but to understand the evolution of the industry, it is essential to have a brief understanding of the first uses of petroleum.

From a chemical standpoint, petroleum is an extremely complex mixture of hydrocarbon compounds, usually with minor amounts of nitrogen-, oxygen-, and sulfur-containing compounds as well as trace amounts of metal-containing compounds (Chapters 2 and 3; Speight, 2001, 2015). In addition, the properties of refinery feedstocks vary widely (Chapters 1, 3, and 4) and, thus, petroleum is not used in its raw state. A variety of processing steps are required to convert petroleum from its raw state to products that are usable in modern society.

The fuel products that are derived from petroleum supply more than half of the world's total supply of energy. Gasoline, kerosene, and diesel oil provide fuel for automobiles, tractors, trucks, aircraft, and ships. Fuel oil and natural gas are used to heat homes and commercial buildings, as well as to generate electricity. Petroleum products are the basic materials used for the manufacture of synthetic fibers for clothing and in plastics, paints, fertilizers, insecticides, soaps, and synthetic rubber. The uses of petroleum as a source of raw material in manufacturing are central to the functioning of modern industry.

For the purposes of terminology, it is preferable to subdivide petroleum and related materials into three major classes (Chapter 1): (1) materials that are of a natural origin, (2) materials that are manufactured, and (3) materials that are integral fractions derived from natural or manufactured products. The materials included in categories 1 and 2 are relevant here because of their participation in product streams. Straight-run constituents of petroleum (i.e., constituents distilled without change from petroleum) (Chapter 7) are used in products. Manufactured materials are produced by a variety of processes (Speight and Ozum, 2002; Parkash, 2003; Hsu and Robinson, 2006; Gary et al., 2007; Speight, 2011, 2014) and are also used in product streams. Category 3 materials are usually those materials that are isolated from petroleum or from a product by the use of a variety of techniques (Chapter 3) and are not included here.

The production of liquid product streams by distillation (Chapter 7) or by thermal cracking processes (Chapter 8) or by catalytic cracking processes (Chapter 9) is only the first of a series of steps that leads to the production of marketable liquid products. Several other unit processes are involved in the production of a final product. Such processes may be generally termed *secondary processes* or *product improvement processes* since they are not used directly on the crude petroleum but are used on primary product streams that have been produced from the crude petroleum (Speight and Ozum, 2002; Parkash, 2003; Hsu and Robinson, 2006; Gary et al., 2007; Speight, 2011, 2014).

In addition, the term "product improvement" as used in this chapter includes processes such as reforming processes in which the molecular structure of the feedstock is reorganized. An example is the conversion (*reforming, molecular rearrangement*) of *n*-hexane to cyclohexane or cyclohexane to benzene. These processes *reform* or *rearrange* one particular molecular type to another thereby changing the properties of the product relative to the feedstock. Such processes are conducive to expansion of the utility of petroleum products and to sales.

It is, therefore, the purpose of this chapter to present the concepts behind these secondary processes with specific examples of the processes that have reached commercialization. It must be understood that the process examples presented here are only a selection of the total number available. The choice of a process for inclusion here was made to illustrate the different process types that are available.

13.2   REFORMING

When the demand for higher-octane gasoline developed during the 1930s, attention was directed to ways and means of improving the octane number of fractions within the boiling range of gasoline. Straight-run gasoline, for example, frequently had a low octane number, and any process that would improve the octane number would aid in meeting the demand for higher-quality (higher-octane-number) gasoline. Such a process, called *thermal reforming*, was developed and used widely but to a much lesser extent than thermal cracking.

Upgrading by reforming is essentially a treatment to improve a gasoline octane number and may be accomplished in part by an increase in the volatility (reduction in molecular size) or chiefly by the conversion of *n*-paraffins to *iso*-paraffins, olefins, and aromatics and the conversion of naphthenes to aromatics (Table 13.1). The nature of the final product is of course influenced by the source (and composition) of the feedstock. In thermal reforming, the reactions resemble the reactions that occur during gas oil cracking: that is, molecular size is reduced, and olefins and some aromatics are produced.

Gasoline has many specifications that must be satisfied before it can be sold on the market. The most widely recognized gasoline specification is the *octane number*, which refers to the percentage by volume of *iso*-octane in a mixture of *iso*-octane and heptane in a reference fuel that, when tested in a laboratory engine, matches the antiknock quality, as measured for the fuel being tested under the same conditions. The octane number posted at the gasoline pump is actually the average of the research octane number (RON) and motor octane number (MON), commonly referred to as (R + M)/2. RON and MON are two different test methods that quantify the antiknock qualities of

---

TABLE 13.1
Structure and Octane Numbers of Selected Hydrocarbons

Compound	*n*-Hexane	1-Hexane	Cyclohexane	Benzene
Formula	C_6H_{11}	C_6H_{12}	C_6H_{12}	C_6H_6
Structure	$CH_3(CH_2)_4CH_3$	$CH_2-CH(CH_2)_3CH_3$		
RON	25	76	83	123 (est.)
Compound	2,2,4-Trimethylpentane (Isooctane)	2,4,4-Trimethyl-1-pentene (Isooctene)	*Cis*-1,3-dimethyl-cyclohexane	1,3-Dimethylbenzene
Formula	C_8H_{18}	C_8H_{16}	C_8H_{16}	C_8H_{10}
Structure				
RON	100	105	72	118

TABLE 13.2
Oxygenates Allowed in Gasoline

Oxygenate	Maximum (vol%)
Methanol	3
Ethanol	5
Isopropyl alcohol	10
Isobutyl alcohol	10
tert-Butyl alcohol	7
Ether (five or more C atoms)	15
Other oxygenates	10

a fuel. Since the MON is a test under more severe conditions than the RON test, for any given fuel, the RON is always higher than the MON.

Unfortunately, the desulfurized light and heavy naphtha fractions of crude oils have very low octane numbers. The heavy naphtha fraction is approximately 50(R + M)/2. Reforming is the refinery process that reforms the molecular structure of the heavy naphtha to increase the percentage of high-octane components while reducing the percentage of low-octane components.

When lead was phased out of gasoline, the only way to produce high-octane-number gasolines is to use inherently high-octane hydrocarbons or to use alcohols (often referred to as *oxygenates*) (Table 13.2). The ether derivatives are also high-octane oxygenates (Table 13.3) and have been used widely as additives. The ethers may be produced at the refinery by reacting suitable alcohols such as methanol and ethanol with branched olefins from the fluid catalytic cracker, such as *iso*-butene and *iso*-pentene, under the influence of acid catalysts. In the mid-1990s methyl-*t*-butyl ether (MTBE) (Table 13.3), made by *etherification* of *iso*-butene with methanol, became the predominant oxygenate used to meet reformulation requirements for adding oxygen to mitigate emissions from gasoline-powered vehicles.

TABLE 13.3
Various Oxygenates Used in Gasoline

Name	Formula	Structure	Oxygen Content Mass (%)	Blending Research Octane Number (BRON)
Ethanol (EtOH)	C_2H_6O	CH_3CH_2OH	34.73	129
Methyl tertiary-butyl ether (MTBE)	$C_5H_{12}O$	$CH_3\ O\ \overset{\overset{\displaystyle CH_3}{\mid}}{\underset{\underset{\displaystyle CH_3}{\mid}}{C}}\ CH_3$	18.15	118
Ethyl tertiary-butyl ether (ETBE)	$C_6H_{14}O$	$CH_3\ CH_2\ O\overset{\overset{\displaystyle CH_3}{\mid}}{\underset{\underset{\displaystyle CH_3}{\mid}}{C}}\ CH_3$	15.66	119
Tertiary-amyl methyl ether (TAME)	$C_6H_{14}O$	$CH_3\ O\ \overset{\overset{\displaystyle CH_3}{\mid}}{\underset{\underset{\displaystyle CH_3}{\mid}}{C}}\ CH_2CH_3$	15.66	112

Environmental issues with methyl-*t*-butyl ether have made it more desirable to dimerize *iso*-butane from the catalytic cracking unit rather than etherify it. Fortunately, *iso*-butene *dimerization* may be achieved with minimal modifications to existing methyl-*t*-butyl ether plants and process conditions, using the same acidic catalysts. Where olefin levels are not restricted, the extra blending octane boost of the *diisobutylene* can be retained. Where olefin levels are restricted, the *diisobutylene* can be *hydrotreated* to produce a relatively pure *iso*-octane stream that can supplement alkylate for reducing olefins and aromatics in reformulated gasoline.

13.2.1 THERMAL REFORMING

Thermal reforming was a natural development from thermal cracking, since reforming is also a thermal decomposition reaction. Cracking converts heavier oils into gasoline; reforming converts (reforms) gasoline into higher-octane gasoline. The equipment for thermal reforming is essentially the same as for thermal cracking, but higher temperatures are used.

When carrying out thermal reforming, a feedstock, such as a 205°C (400°F) end point naphtha or a straight-run gasoline, is heated to 510°C–595°C (950°F–1100°F) in a furnace much the same as a cracking furnace, with pressures from 400 to 1000 psi. As the heated naphtha leaves the furnace, it is cooled or quenched by the addition of cold naphtha. The quenched, reformed material then enters a fractional distillation tower where any heavy products are separated. The remainder of the reformed material leaves the top of the tower to be separated into gases and reformate. The higher octane number of the product (*reformate*) is due primarily to the cracking of longer chain paraffins into higher-octane olefins.

Thermal reforming is in general less effective than catalytic processes and has been largely supplanted. As it was practiced, a single-pass operation was employed at temperatures in the range of 540°C–760°C (1000°F–1140°F) and pressures in the range of 500–1000 psi. Octane number improvement depended on the extent of conversion but was not directly proportional to the extent of cracking per pass.

The amount and quality of *reformate* is dependent on the temperature. A general rule is the higher the reforming temperature, the higher the octane number of the product, but the yield of reformate is relatively low. For example, a gasoline with an octane number of 35 when reformed at 515°C (960°F) yields 92.4% of 56 octane reformate; when reformed at 555°C (1030°F), the yield is 68.7% of 83 octane reformate (Figure 13.1). However, high conversion is not always effective since coke production and gas production usually increase.

Modifications of the thermal reforming process due to the inclusion of hydrocarbon gases with the feedstock are known as *gas reversion* and *polyforming*. Thus, olefinic gases produced by

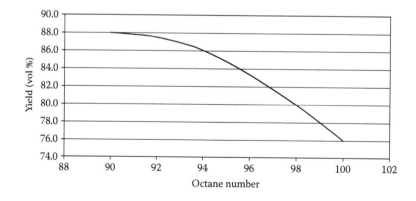

FIGURE 13.1 Octane number and reformate yield.

cracking and reforming can be converted into liquids boiling in the gasoline range by heating them under high pressure. Since the resulting liquids (polymers) have high octane numbers, they increase the overall quantity and quality of gasoline produced in a refinery.

The gases most susceptible to conversion to liquid products are olefins with three and four carbon atoms. These are propylene ($CH_3CH=CH_2$), which is associated with propane in the C_3 fraction, and butylene ($CH_3CH_2CH=CH_2$ and/or $CH_3CH=CHCH_3$) and *iso*-butylene [$(CH_3)_2C=CH_2$], which are associated with butane ($CH_3CH_2CH_2CH_3$) and *iso*-butane [$(CH_3)_2CHCH_3$] in the C_4 fraction. When the C_3 and C_4 fractions are subjected to the temperature and pressure conditions used in thermal reforming, they undergo chemical reactions that result in a small yield of gasoline. When the C_3 and C_4 fractions are passed through a thermal reformer in admixture with naphtha, the process is called *naphtha-gas reversion* or *naphtha polyforming*.

These processes are essentially the same but differ in the manner in which the gases and naphtha are passed through the heating furnace. In gas reversion, the naphtha and gases flow through separate lines in the furnace and are heated independently of one another. Before leaving the furnace, both lines join to form a common soaking section where the reforming, polymerization, and other reactions take place. In naphtha reforming, the C_3 and C_4 gases are premixed with the naphtha and pass together through the furnace. Except for the gaseous components in the feedstock, both processes operate in much the same manner as thermal reforming and produce similar products.

13.2.2 CATALYTIC REFORMING

Like thermal reforming, *catalytic reforming* converts low-octane gasoline into high-octane gasoline (reformate). Although thermal reforming can produce reformate with an research octane number in the range of 65–80 depending on the yield, catalytic reforming produces reformate with octane numbers of the order of 90–95. Catalytic reforming is conducted in the presence of hydrogen over hydrogenation–dehydrogenation catalysts, which may be supported on alumina or silica–alumina. Depending on the catalyst, a definite sequence of reactions takes place, involving structural changes in the charge stock. The catalytic reforming process was commercially nonexistent in the United States before 1940. The process is really a process of the 1950s and showed phenomenal growth in the 1953–1959 time period (Riediger, 1971). As a result, thermal reforming is now somewhat obsolete.

Catalytic reformer feeds are saturated (i.e., not olefinic) materials; in the majority of cases the feed may be a straight-run naphtha, but other by-product low-octane naphtha (e.g., coker naphtha) can be processed after treatment to remove olefins and other contaminants. Hydrocarbon naphtha that contains substantial quantities of naphthenes is also a suitable feed. The process uses a precious metal catalyst (platinum supported by an alumina base) in conjunction with very high temperatures to reform the paraffin and naphthene constituents into high-octane components. Sulfur is a poison to the reforming catalyst, which requires that virtually all the sulfur must be removed from the heavy naphtha through hydrotreating prior to reforming. Several different types of chemical reactions occur in the reforming reactors: paraffins are isomerized to branched chains and to a lesser extent to naphthenes, and naphthenes are converted to aromatics. Overall, the reforming reactions are endothermic. The resulting product stream (*reformate*) from catalytic reforming has an RON from 96 to 102 depending on the reactor severity and feedstock quality. The dehydrogenation reactions which convert the saturated naphthenes into unsaturated aromatics produce hydrogen, which is available for distribution to other refinery hydroprocesses.

The catalytic reforming process consists of a series of several reactors (Figure 13.2), which operate at temperatures of approximately 480°C (900°F). The hydrocarbons are reheated by direct-fired furnaces in between the subsequent reforming reactors. As a result of the very high temperatures, the catalyst becomes deactivated by the formation of coke (i.e., essentially pure carbon) on the catalyst, which reduces the surface area available to contact with the hydrocarbons.

Catalytic reforming is usually carried out by feeding a naphtha (after pretreating with hydrogen if necessary) and hydrogen mixture to a furnace where the mixture is heated to the desired

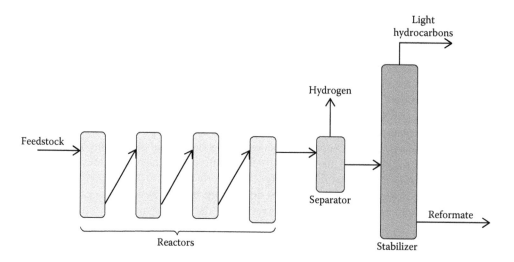

FIGURE 13.2   Catalytic reforming process.

temperatures 450°C–520°C (840°F–965°F) and then passed through fixed-bed catalytic reactors at hydrogen pressures of 100–1000 psi. Normally two (or more than one) reactors are used in series, and reheaters are located between adjoining reactors to compensate for the endothermic reactions taking place. Sometimes as many as four or five are kept onstream in series while one or more is being regenerated. The onstream cycle of any one reactor may vary from several hours to many days, depending on the feedstock and reaction conditions.

The product issuing from the last catalytic reactor is cooled and sent to a high-pressure separator where the hydrogen-rich gas (Table 13.4) is split into two streams: one stream goes to recycle, and the remaining portion represents excess hydrogen available for other uses. The excess hydrogen is vented from the unit and used in hydrotreating, as a fuel, or for manufacture of chemicals (e.g., ammonia). The liquid product (reformate) is stabilized (by removal of light ends) and used directly in gasoline or extracted for aromatic blending stocks for aviation gasoline.

The commercial processes available for use can be broadly classified as the *moving-bed*, *fluid-bed*, and *fixed-bed* types. The fluid-bed and moving-bed processes use mixed nonprecious

TABLE 13.4
Composition of Catalytic Reformer Product Gas

Constituent	% by Volume
Hydrogen	75–85
Methane	5–10
Ethane	5–10
Ethylene	0
Propane	5–10
Propylene	0
Butane	<5
Butylenes	0
Pentane plus	<2

metal oxide catalysts in units equipped with separate regeneration facilities. Fixed-bed processes use predominantly platinum-containing catalysts in units equipped for cycle, occasional, or no regeneration.

There are several types of catalytic reforming process configurations that differ in the manner that they accommodate the regeneration of the reforming catalyst. Catalyst regeneration involves burning off the coke with oxygen. The semiregenerative process is the simplest configuration but does require that the unit be shut down for catalyst regeneration in which all reactors (typically four) are regenerated. The cyclic configuration utilizes an additional swing reactor that enables one reactor at a time to be taken offline for regeneration while the other four remain in service. The continuous catalyst regeneration (CCR) configuration is the most complex configuration and enables the catalyst to be continuously removed for regeneration and replaced after regeneration. The benefits to the more complex configurations are that operating severity may be increased as a result of higher catalyst activity but this does come at an increased capital cost for the process.

Although subsequent olefin reactions occur in thermal reforming, the product contains appreciable amounts of unstable unsaturated compounds. In the presence of catalysts and of hydrogen (available from dehydrogenation reactions), hydrocracking of paraffins to yield two lower paraffins occurs. Olefins that do not undergo dehydrocyclization are also produced. The olefins are hydrogenated with or without isomerization, so that the end product contains only traces of olefins.

The addition of a hydrogenation–dehydrogenation catalyst to the system yields a dual-function catalyst complex. Hydrogen reactions—hydrogenation, dehydrogenation, dehydrocyclization, and hydrocracking—take place on the one catalyst, and cracking, isomerization, and olefin polymerization take place on the acid catalyst sites.

Under the high-hydrogen partial pressure conditions used in catalytic reforming, sulfur compounds are readily converted into hydrogen sulfide, which, unless removed, builds up to a high concentration in the recycle gas. Hydrogen sulfide is a reversible poison for platinum and causes a decrease in the catalyst dehydrogenation and dehydrocyclization activities. In the first catalytic reformers the hydrogen sulfide was removed from the gas cycle stream by absorption in, for example, diethanolamine. Sulfur is generally removed from the feedstock by use of a conventional desulfurization over cobalt–molybdenum catalyst. An additional benefit of desulfurization of the feed to a level of <5 ppm sulfur is the elimination of hydrogen sulfide (H_2S) corrosion problems in the heaters and reactors.

Organic nitrogen compounds are converted into ammonia under reforming conditions, and this neutralizes acid sites on the catalyst and thus represses the activity for isomerization, hydrocracking, and dehydrocyclization reactions. Straight-run materials do not usually present serious problems with regard to nitrogen, but feeds such as coker naphtha may contain around 50 ppm nitrogen and removal of this quantity may require high-pressure hydrogenation (800–1000 psi) over nickel–cobalt–molybdenum on an alumina catalyst.

The yield of naphtha of a given octane number and at given operating conditions depends on the hydrocarbon types in the feed. For example, high-naphthene stocks, which readily give aromatic gasoline, are the easiest to reform and give the highest gasoline yields. Paraffinic stocks, however, that depend on the more difficult isomerization, dehydrocyclization, and hydrocracking reactions, require more severe conditions and give lower gasoline yields than the naphthenic stocks. The end point of the feed is usually limited to approximately 190°C (375°F), partially because of increased coke deposition on the catalyst as the end point during processing at approximately 15°C (27°F). Limiting the feed end point avoids redistillation of the product to meet the gasoline end point specification of 205°C (400°F), maximum.

In a semiregenerative unit, desulfurized naphtha is mixed with hydrogen, heated to temperatures in excess of 480°C (>900°F) and passed through a series of fixed-bed reactors. The major chemical reactions—dehydrogenation and dehydrocyclization—are endothermic and the reactors are operated under adiabatic conditions insofar as heat cannot enter or leave except by the cooling or heating of reaction fluids. Consequently, the temperature drops as reactants flow through a reactor. Between reactors, fired heaters bring the process fluids back to desired reactor inlet temperatures.

Because of the dehydrogenation–dehydrocyclization reactions, hydrogen is generated in the reformer and is produced in substantial (but not in sufficient) quantities. The hydrogen is recycled through the reactors where the reforming takes place to provide the atmosphere necessary for the chemical reactions and also prevents the carbon from being deposited on the catalyst, thus extending its operating life. An excess of hydrogen above whatever is consumed in the process is produced, and as a result, catalytic reforming processes are unique in that they are the only petroleum refinery processes to produce hydrogen as a by-product.

13.2.2.1 Fixed-Bed Processes

Fixed-bed, continuous catalytic reforming can be classified by catalyst type: (1) cyclical regenerative with nonprecious metal oxide catalysts and (2) cyclic regenerative with platinum–alumina catalysts. Both types use swing reactors to regenerate a portion of the catalyst while the remainder stays onstream.

The cyclic-regenerative fixed-bed operation using a platinum catalyst is basically a low-pressure process (250–350 psi), which gives higher gasoline yields because of fewer hydrocracking reactions, as well as higher-octane products from a given naphtha charge and better hydrogen yields because of more dehydrogenation and fewer hydrocracking reactions. The coke yield, with attendant catalyst deactivation, increases rapidly at low pressures.

13.2.2.1.1 Hydroforming

The hydroforming process made use of molybdena–alumina (MoO_2–Al_2O_3) catalyst pellets arranged in fixed beds; hence, the process is known as fixed-bed hydroforming. The hydroformer had four reaction vessels or catalyst cases, two of which were regenerated; the other two were on the process cycle. Naphtha feed is preheated to 400°C–540°C (900°F–1000°F) and passed in series through the two catalyst cases under a pressure of 150–300 psi. Gas containing 70% hydrogen produced by the process was passed through the catalyst cases with the naphtha. The material leaving the final catalyst case entered a four-tower system where fractional distillation separated hydrogen-rich gas, a product (reformate) suitable for motor gasoline and an aromatic polymer boiling above 205°C (400°F).

After 4–16 hours on process cycle, the catalyst was regenerated. This was done by burning carbon deposits from the catalyst at a temperature of 565°C (1050°F) by blowing air diluted with flue gas through the catalyst. The air also reoxidized the reduced catalyst (9% molybdenum oxide on activated alumina pellets) and removed sulfur from the catalyst.

13.2.2.1.2 Iso-Plus Houdriforming

This is a combination process using a conventional Houdriformer operated at moderate severity for product production (Table 13.5), in conjunction with one of three possible alternatives: (1) conventional catalytic reforming plus aromatic extraction and separate catalytic reforming of the aromatic raffinate, (2) conventional catalytic reforming plus aromatic extraction and recycling of the aromatic raffinate aligned to the reforming state, and (3) conventional catalytic reforming followed by thermal reforming of the Houdriformer product and catalytic polymerization of the C_3 and C_4 olefins from thermal reforming.

A typical feedstock for this type of unit is naphtha, and the use of a Houdry *guard bed* permits charging stocks of relatively high sulfur content.

13.2.2.1.3 Platforming

The first step in the Platforming process (Figure 13.3) is preparation of the naphtha feed. For motor gasoline manufacture, the naphtha feed is distilled to separate a fraction boiling in the 120°C–205°C (250°F–400°F) range. Since sulfur adversely affects the platinum catalyst, the naphtha fraction may

---

TABLE 13.5

Feedstock and Product Data for the Houdriforming Process

Feedstock	
API	52.6
Boiling range	
°C	92–192
°F	197–377
Composition (% v/v)	
Paraffins	53
Naphthenes	38
Aromatics	9
Product	
Research octane number	100
Composition (% v/v)	
Paraffins	21
Naphthenes	2
Aromatics	77

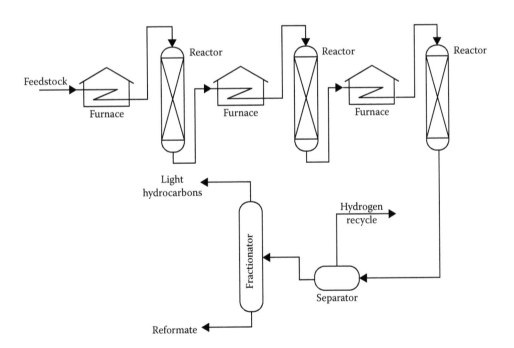

FIGURE 13.3 The Platforming process.

be treated to remove sulfur compounds. Otherwise, the hydrogen-rich gas produced by the process, which is cycled through the catalyst cases, must be scrubbed free of its hydrogen sulfide content.

The prepared naphtha feed is heated to 455°C–540°C (850°F–1000°F) and passed into a series of three catalyst cases under a pressure of 200–1000 psi. Further heat is added to the naphtha between each of the catalyst cases in the series. The material from the final catalyst case is fractionated into a hydrogen-rich gas stream and a reformate stream (Table 13.6). The catalyst is composed of 1/8-inch

TABLE 13.6

Feedstock and Product Data for the Platforming Process

Feedstock			
API	59.0		
Boiling range			
°C	92–112		
°F	197–233		
Composition (% v/v)			
Paraffins	69		
Naphthenes	21		
Aromatics	10		
Process operation	Semiregenerative	Continuous	
Pressure			
psi	295	123	49
Product			
Research octane number	100	100	100
Composition (% v/v)			
C5+	70	78	82
Aromatics	46	55	58
Hydrogen (% w/w)	2	3	4

pellets of alumina containing chlorine and approximately 0.5% platinum. Each pound of catalyst reforms up to 100 bbl of naphtha before losing its activity. It is possible to regenerate the catalyst, but it is more usual to replace the spent catalyst with new catalyst.

Other fixed-bed processes include *Catforming*, in which the catalyst is platinum (Pt), alumina (Al_2O_3), and silica–alumina (SiO_2–Al_2O_3) composition, which permits relatively high space velocities and results in very high hydrogen purity. Regeneration to prolong catalyst life is practiced on a block-out basis with a dilute air in-stream mixture. In addition, Houdriforming is a process in which the catalyst may be regenerated, if necessary, on a block-out basis. A *guard bed* catalytic hydrogenation pretreating stage using the same Houdry catalyst as the Houdriformer reactors is available for high-sulfur feedstocks. Lead and copper salts are also removed under mild conditions of the guard bed operation.

13.2.2.1.4 Powerforming

The cyclic *powerforming process* is based on frequent regeneration (carbon burn-off) and permits continuous operation. Reforming takes place in several (usually four or five) reactors and regeneration is carried out in the last (or swing) reactor. Thus, the plant need not be shut down to regenerate a catalyst reactor. The cyclic process assures a continuous supply of hydrogen gas for hydrorefining operations and tends to produce a greater yield of higher-octane reformate (Table 13.7). The choice between the semiregenerative process and the cyclic process depends on the size of plant required, the types of feedstocks available, and the octane number needed in the product.

13.2.2.1.5 Rexforming

Rexforming is a combination process using *Platforming* and aromatic extraction processes in which low-octane raffinate is recycled to the platformer. Operating temperatures may be as much as 27°C (50°F) lower than conventional Platforming, and higher space velocities are used. A balance is struck between hydrocyclization and hydrocracking, excessive coke and gas formation thus being

---

TABLE 13.7

Feedstock and Product Data for the

Powerforming Process

Feedstock		
API	57.2	
Composition, % v/v		
Paraffins	57	
Naphthenes	30	
Aromatics	13	
Process operation	Semiregenerative	Cyclic
Product		
Research octane number	99	101
Composition (% v/v)		
C_1–C_4	13	11
C_5+	79	79
Hydrogen (% w/w)	2	3

---

avoided. The glycol solvent in the aromatic extraction section is designed to extract low-boiling high-octane *iso*-paraffins as well as aromatics.

13.2.2.1.6 Selectoforming

The *selectoforming process* uses a fixed-bed reactor operating under a hydrogen partial pressure. Typical operating conditions depend on the process configuration but are in the ranges of 200–600 psi and 315°C–450°C (600°F–900°F). The catalyst used in the selectoforming process is nonnoble metal with low potassium content. As with the large-pore hydrocracking catalysts, the cracking activity increases with decreasing alkali metal content.

There are two configurations of the selectoforming process that are being used commercially. The first selectoformer was designed as a separate system and integrated with the reformer only to the extent of having a common hydrogen system. The reformer naphtha is mixed with hydrogen and passed into the reactor containing the shape-selective catalyst. The reactor effluent is cooled and separated into hydrogen, liquid petroleum, gas, and high-octane gasoline. The removal of *n*-paraffins reduces the vapor pressure of the reformate since these paraffins are in higher concentration in the front end of the feed. The separate selectoforming system has the additional flexibility of being able to process other refinery streams.

The second process modification is the terminal reactor system. In this system, the shape-selective catalysts replace all or part of the reforming catalyst in the last reforming reactor. Although this configuration is more flexible, the high reforming operating temperature causes butane and propane cracking and consequently decreases the liquid petroleum gas yield and generates higher ethane and methane production. The life of a selectoforming catalyst used in a terminal system is between 2 and 3 years, and regeneration only partially restores fresh catalytic activity.

13.2.2.2 Moving-Bed Processes

13.2.2.2.1 Hyperforming

Hyperforming is a moving-bed reforming process that uses catalyst pellets of cobalt molybdate with a silica-stabilized alumina base. In operation, the catalyst moves downward through the reactor by gravity flow and is returned to the top by means of a solids-conveying technique (hyperflow), which moves the catalyst at low velocities and with minimum attrition loss. Feedstock (naphtha vapor) and recycle gas flow upward, countercurrent to the catalyst, and regeneration of catalyst is accomplished in either an external vertical lift line or a separate vessel.

Hyperforming naphtha (65°C–230°C, 150°F–450°F) can result in improvement of the motor fuel component; in addition, sulfur and nitrogen removal is accomplished. Light gas oil stocks can also be charged to remove sulfur and nitrogen under mild hydrogenation conditions for the production of premium diesel fuels and middle distillates. Operating conditions in the reactor are 400 psi and 425°C–480°C (800°F–900°F), the higher temperature being employed for a straight-run naphtha feedstock; catalyst regeneration takes place at 510°C (950°F) and 415 psi.

13.2.2.2.2 Thermofor Catalytic Reforming

The *Thermofor catalytic reforming (TCR) process* is also a moving-bed process that uses a synthetic bead coprecipitated chromia (CrO_2) and alumina (Al_2O_3) catalyst (Figure 13.4). Catalyst–naphtha ratios have little effect on product yield or quality when varied over a wide range. The catalyst flow downward through the reactor and the naphtha-recycle gas feed enters the center of the reactor. The catalyst is transported from the base of the reactor to the top of the regenerator by bucket-type elevators.

13.2.3 FLUID-BED PROCESSES

In catalytic reforming processes using a fluidized solids catalyst bed continuous regeneration with a separate or integrated reactor is practiced to maintain catalyst activity by coke and sulfur removal. Cracked or virgin naphtha is charged with hydrogen-rich recycle gas to the reactor. A molybdena (Mo_2O_3, 10.0%) on alumina catalyst, not materially affected by normal amounts of arsenic, iron, nitrogen, or sulfur, is used. Operating conditions in the reactor are approximately 200–300 psi and 480°C–950°C (900°F–950°F).

Fluidized-bed operation with its attendant excellent temperature control prevents over- and under-reforming operations, resulting in more selectivity in the conditions needed for optimum yield of the desired product.

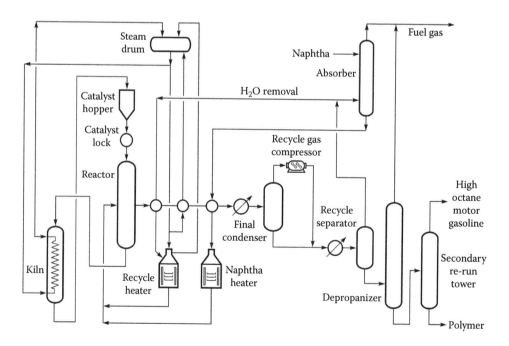

FIGURE 13.4 Thermofor catalytic reforming process.

13.3   ISOMERIZATION

Catalytic reforming processes provide high-octane constituents in the heavier gasoline fraction, but the *n*-paraffin components of the lighter gasoline fraction, especially butane (C_4) to hexane (C_6), have poor octane ratings. The conversion of these *n*-paraffins to their isomers (isomerization) yields gasoline components of high octane rating in this lower boiling range. Conversion is obtained in the presence of a catalyst (aluminum chloride activated with hydrochloric acid), and it is essential to inhibit side reactions, such as cracking and olefin formation.

Various companies have developed and operated isomerization processes that increase the octane numbers of light naphtha from say, 70 or less to more than 80. In thermal catalytic alkylation ethylene or propylene is combined with *iso*-butane at 50°C–280°C (125°F–450°F) and 300–1000 psi in the presence of metal halide catalysts, such as aluminum chloride. Conditions are less stringent in catalytic alkylation; olefins (C_3, C_4, and C_5) are combined with *iso*-butane in the presence of an acid catalyst (sulfuric or hydrofluoric) at low temperatures (1–40, 30°F–105°F) and pressures from atmospheric to 150 psi. In a typical process, naphtha is passed over an aluminum chloride catalyst at 120°C (250°F) and at a pressure of approximately 800 psi to produce the Isomerate (Figure 13.5).

Isomerization, another innovation specific to recent times, found initial commercial applications during World War II for making high-octane aviation gasoline components and additional feed for alkylation units. The lowered alkylate demands in the post–World War II period caused a shutdown of the majority of the butane isomerization units. In recent years, the greater demand for high-octane motor fuel has resulted in the installation of new butane isomerization units. The earliest important process was the formation of *iso*-butane, which is required as an alkylation feed; the isomerization may take place in the vapor phase, with the activated catalyst supported on a solid phase, or in the liquid phase with a dissolved catalyst. Thus, a pure butane feed is mixed with hydrogen (to inhibit olefin formation) and passed to the reactor at 110°C–170°C (230°F–340°F) and 200–300 psi. The product is cooled and the hydrogen separated; the cracked gases are then removed in a stabilizer column. The stabilizer bottom product is passed to a superfractionator, and the *n*-butane and *iso*-butane are separated. With pentanes, the equilibrium is favorable at higher temperatures, and operating conditions of 300–1000 psi and 240°C–500°C (465°F–930°F) may be used.

Present isomerization applications are to provide additional feedstock for alkylation units or high-octane fractions for blending to produce sales gasoline. Straight-chain paraffins (*n*-butane) (Figure 13.6) and mixtures of *n*-pentane and *n*-hexane (Figure 13.7) are converted to respective

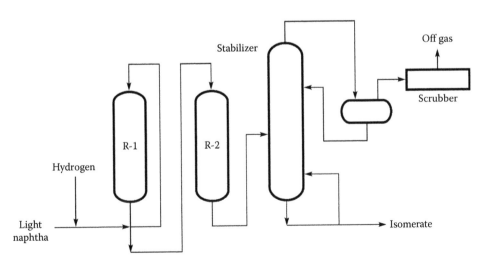

FIGURE 13.5   Isomerization using an aluminum chloride catalyst.

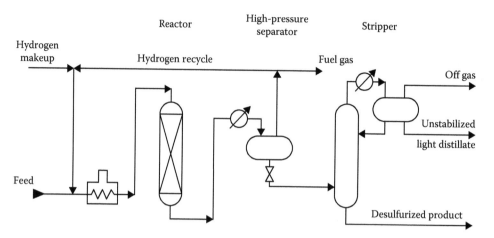

FIGURE 13.6 Butane isomerization. (From OSHA Technical Manual, Section IV, Chapter 2, Petroleum refining processes, 1999, http://www.osha.gov/dts/osta/otm/otm_iv/otm_iv_2.html.)

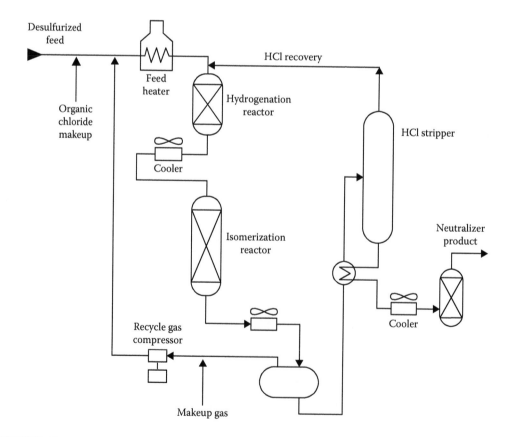

FIGURE 13.7 Pentane–hexane isomerization. (From OSHA Technical Manual, Section IV, Chapter 2, Petroleum refining processes, 1999, http://www.osha.gov/dts/osta/otm/otm_iv/otm_iv_2.html.)

iso-compounds by continuous catalytic (aluminum chloride and noble metals) processes. Natural gasoline or light straight-run gasoline can provide feed by first fractionating as a preparatory step. High volumetric yields (>95%) and 40%–60% conversion per pass are characteristic of the isomerization reaction.

Nonregenerable aluminum chloride catalyst is employed with various carriers in a fixed-bed or liquid contactor. Platinum or other metal catalyst processes utilize a fixed-bed operation and can be regenerable or nonregenerable. The reaction conditions vary widely depending on the particular process and feedstock, 40°C–480°C (100°F–900°F) and 150–1000 psi; residence time in the reactor is 10–40 minutes.

13.3.1 BUTAMER PROCESS

The *Butamer process* is designed to convert *n*-butane to *iso*-butane under mild operating conditions (Figure 13.8). A platinum catalyst on a support is used in a fixed-bed reactor system. Using reformer off-gas can readily satisfy the low hydrogen requirement. The operation can be designed for once-through or recycle operation and is normally tied in with alkylation unit de-*iso*-butanizer operations to provide additional feed.

Butane feed is mixed with hydrogen, heated, and charged to the reactor at moderate pressure. The effluent is cooled before light gas separation and stabilization. The resultant butane mixture is then charged to a de-*iso*-butanizer to separate a recycle stream from the *iso*-butane product.

13.3.2 BUTOMERATE PROCESS

The *Butomerate process* is specially designed to isomerize *n*-butane to produce additional alkylation feedstock. The catalyst contains a small amount of nonnoble hydrogenation metal on a high-surface-area support. The process operates with hydrogen recycle to eliminate coke deposition on the catalyst, but the isomerization reaction can continue for extended periods in the absence of hydrogen.

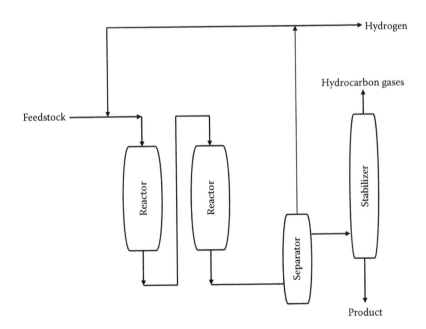

FIGURE 13.8 Butamer process.

The feedstock should be dry and comparatively free of sulfur and water; the feed is heated, mixed with hydrogen, and conveyed to the reactor. Operating conditions range from 150°C to 260°C (300°F to 500°F) and 150 to 450 psi. The effluent is cooled and flashed, and the liquid product is stripped of light material.

13.3.3 HYSOMER PROCESS

The *Hysomer process* uses hydrotreated feedstocks containing pentane(s) and hexane(s) without further pretreatment. Operating conditions are 400–450 psi hydrogen and ca. 290°C (550°F); it appears that the catalyst (zeolite) life is approximately 2 years.

The influence of sulfur on catalyst activity is minimal, and the catalyst can tolerate a permanent sulfur level of 10 ppm in the feedstock, but concentrations up to 35 ppm are not harmful. The process can operate at a water level of 50 ppm, and feedstocks having saturated water contents can be processed without a deleterious effect on either catalyst stability or conversion. A minimum quantity of water is essential to the activity of zeolite catalysts in this application.

13.3.4 ISO-KEL PROCESS

The *Iso-Kel process* is a fixed-bed, vapor-phase isomerization process that employs a precious metal catalyst and hydrogen. A wide variety of feedstocks, including natural gasoline, pentane, and/or hexane cuts, can be processed. Operating conditions include reactor temperatures and pressures from 345°C to 455°C (650°F to 550°F) and 350 to 600 psi.

13.3.5 ISOMATE PROCESS

The *Isomate process* is a nonregenerative pentane and hexane or naphtha (C_6) isomerization process using an aluminum chloride–hydrocarbon complex catalyst with anhydrous hydrochloric acid as a promoter. Hydrogen partial pressure is maintained to suppress undesirable reactions (cracking and disproportionation) and retain catalyst activity. The feed is saturated with anhydrous hydrogen chloride in an absorber, then heated and combined with hydrogen and charged to the reactor (115°C, 240°F, and 700 psi). Catalyst is added to the reactor separately, and the reaction takes place in the liquid phase. The product is washed (caustic and water), acid stripped, and stabilized before going to storage.

13.3.6 ISOMERATE PROCESS

The *Isomerate process* is a continuous isomerization process designed to convert pentanes and hexanes into highly branched isomers; a dual-function catalyst is used in a fixed-bed reactor system.

Operating conditions are mild, less than 750 psi and 400°C (750°F). Hydrogen is added to the feed along with recycle gas, and the usual operation includes fractionation facilities to allow the recycling of *n*-paraffins almost to extinction.

13.3.7 PENEX PROCESS

The *Penex process* is a nonregenerative pentane(s) and/or hexane(s) isomerization process. The reaction takes place in the presence of hydrogen with a platinum catalyst, and the reactor conditions are selected so that catalyst life is long and regeneration is not required. The reactor temperatures range from 260°C to 480°C (500°F to 900°F) and pressures from 300 to 1000 psi.

The *Penex process* may be applied to many feedstocks by varying the fractionating system. Mixed feeds may be split into pentane and hexane fractions and respective *iso*-fractions separated from each. The system can also be operated in conjunction with reforming of the naphtha ($>C_7$) fraction.

13.3.8 PENTAFINING PROCESS

The *Pentafining process* is a regenerable pentane isomerization process using platinum catalyst on a silica–alumina support. A number of process combinations are possible. For example, with natural gasoline and hydrogen as starting materials, pentanes are removed from the feedstock and the pentane fraction is passed to a low-pressure reformer. The pentane stream is split, and the *n*-pentane fraction is combined with a recycle stream and makeup hydrogen, and charged to the reactor (approximately 300 to 500 psi and 425°C–480°C, 800°F–900°F) where isomerization occurs:

$$CH_3CH_2CH_2CH_2CH_3 \rightarrow (CH_3)_2CHCH_2CH_3$$
$$\text{\textit{n}-pentane} \qquad\qquad \text{\textit{iso}-pentane}$$

Hydrogen is removed from the effluent, which is degassed and fractionated to separate *n*-pentane and *iso*-pentane (95% purity) fractions. The catalyst is regenerated at 260°C–540°C (500°F–1000°F) using a steam–air mixture.

13.4 HYDROISOMERIZATION

Hydroisomerization involves catalytic isomerization of hydrocarbons in presence of hydrogen. This process is used for the production of isomers of various low-boiling and high-boiling hydrocarbons. Isomers of normal paraffinic hydrocarbons are found to be more reactive and have higher octane number than the original constituents (Giannetto et al., 1986; Guisnet et al., 1987).

In the traditional Platforming of naphtha, a substantial amount of benzene and higher aromatics are generated leading to high-octane-number reformate. Hydroisomerization processes convert normal paraffin hydrocarbons to isomers which are also high-octane constituents. In another application, hydroisomerization of higher-boiling normal paraffins is being exploited to dewax vacuum oils. This enables refiners to abolish traditional solvent dewaxing technique which is costly and less productive as compared to catalytic dewaxing method.

Typically, isomerization is carried out over aluminum chloride catalyst (in the absence of hydrogen) to convert butane to *iso*-butane, pentane to *iso*-pentane, and hexane to *iso*-hexane commercially. Since isomers have higher octane numbers than the normal or straight-chain hydrocarbons, the process is suitable for the production of gasoline components in refineries. The isomerization of butane to *iso*-butane is also important for the production of butyl rubber where *iso*-butane is alkylated with olefins like ethylene to produce the necessary monomer for polymerization into rubber. In the isomerization of normal hydrocarbons in straight-run naphtha or raffinate from reformate during extraction of benzene, toluene, and xylenes (BTX), using aluminum chloride as the catalyst, the research octane number is increased from approximately 65 to approximately 80.

Because of corrosive damage of the equipment, aluminum chloride process is often replaced by noble metal impregnated on alumina or alumina–silica catalysts. This process uses a higher temperature of operation and a hydrogen pressure sufficient to suppress coking on catalyst surface. Isomerization processes are also applicable for separation of *iso*-butene and 1-butene from steam-cracked naphtha in the olefin production plant by converting 1-butene to 2-butene followed by distillation.

Hydroisomerization of 1-butene to 2-butene is also valuable for petrochemicals production and for high-octane alkylated gasoline production. In the production of benzene-free gasoline, hydroisomerization process is used in the refineries. The C_5 and C_6 fractions are separated from the feed naphtha and isomerized to high-octane isomers and later blended with reformate to produce benzene-free gasoline. Another important use of hydroisomerization process is to convert paraffinic wax into low pour point components. This process is, in fact, a low-cost process as compared to traditional solvent dewaxing method practiced in refineries.

Isomerization reactions are usually reversible reactions and attain equilibrium at lower temperature with highest concentration of isomer products. The role of catalyst in isomerization is, therefore, extremely important. The intensity of unwanted side reactions diminishes at lower temperatures as higher temperature favors unwanted cracking, hydrogenation, and polymerization reactions. For that reason, isomerizing catalysts must ensure the optimal rate of reactions at as low temperature as possible. To prevent coke deposition, isomerization is carried out at an elevated pressure in a hydrogen atmosphere. Industrial processes are carried out at a temperature of 400°C–480°C (750°F–895°F).

The mechanism of reaction involves generation of n-carbenium ion in the Lewis acid site as the initiation reaction which propagates to isomer of the carbenium ion and finally converts the n-hydrocarbon feed molecule to isomer. In the dual-site catalyst, the metal and the acid sites, n–p molecule is dehydrogenated on the metal site, which after diffusing to the acid site initiates the n-carbenium ion formation. This is followed by isomerization to i-carbenium ion and propagation to i-olefin molecule which is then hydrogenated by metal site to i-paraffin molecule. Cracking reaction also takes place within the acid sites.

Metal sites are responsible for hydrogenation and dehydrogenation reaction, this is influenced by the presence of hydrogen and its partial pressure. Pretreatment of the feedstock is essential to avoid poisoning of the catalyst. Usually desulfurizing the feedstock is essential to reduce sulfur, nitrogen, oxygen, and metal constituents simultaneously and dehydrated to remove moisture.

13.5 ALKYLATION

Alkylation is the refinery process that provides an economically feasible outlet for several of the very light olefins produced from the catalytic cracking unit. Propylene, butylene, and pentylene (also known as amylene) products are available for alkylation. Propylene alkylation is the normal disposition for cat cracker product although some butylenes from a fluid catalytic cracking unit could be blended into gasoline. The high vapor pressure of butylenes prevents low-cost butane blending stock from being blended into gasoline, thereby carrying a very high opportunity cost for this option.

In the alkylation process, the propylene, butylene, and pentylene are combined with iso-butane in the catalyzed alkylation reaction to produce branched, saturated seven, eight, or nine carbon molecules, respectively. The product (*alkylate*) consists of iso-heptane, iso-octane, and iso-nonane and is a low vapor pressure (relative to the feedstocks), very high-octane gasoline blending stock. The high-octane value makes alkylate an excellent blending stock for premium grades of gasolines. Furthermore, since alkylate contains no olefins, no aromatics, and no sulfur, it is also an excellent blending stock for use in reformulated gasoline. The alkylation reaction is catalyzed by the presence of very strong acid, either sulfuric acid or hydrofluoric acid.

In the sulfuric acid-based alkylation process, the acid is continually cycled through the process; but as it cycles, it becomes diluted and contaminated from impurities in the hydrocarbon feeds. The alkylation reactors typically operate at temperatures of 2°C–21°C (35°F–70°F, maximum) to minimize polymerization of the olefins to form undesirable hydrocarbons for the sulfuric acid process. The concentration of the sulfuric acid catalyst is important to the efficiency of the alkylation reaction, when the concentration of the acid decreases to approximately 88%, a portion of the contaminated acid is withdrawn and replaced with fresh acid. The contaminated, dilute sulfuric acid is then regenerated to its original purity and concentration.

Hydrofluoric acid exists in a vapor state at ambient conditions and this dictates that extreme precaution is necessary to ensure that this toxic substance is contained inside the process equipment. The hydrofluoric acid process, which is less sensitive to polymerization at warmer temperatures, typically operates at reactor temperatures of 21°C–38°C (70°F–100°F). *Iso*-butane concentrations are maintained very high (i.e., at ratios of 4:1 or more above the reaction requirements) in the reactor vessels to ensure that all of the olefins are reacted. The reactor effluent is distilled to separate the propane, *iso*-butane, and alkylate boiling fractions. The propane is routed to propane product treating, the *iso*-butane is recycled back to the alkylation reactors, and the alkylate is routed to gasoline blending, or in some cases to additional solvent refinery processing.

The chemistry of the combination of olefins with paraffins to form higher *iso*-paraffins is simple:

$$(CH_3)_3CH + CH_2{=}CH_2 \rightarrow (CH_3)_3CHCH_2CH_3$$

Since olefins are reactive (hence unstable) and are responsible for exhaust pollutants, their conversion to high-octane *iso*-paraffins is desirable when possible. In refinery practice, only *iso*-butane is alkylated by reaction with *iso*- or *n*-butene and *iso*-octane is the product. Although alkylation is possible without catalysts, commercial processes use sulfuric acid or hydrogen fluoride as catalysts when the reactions can take place at low temperatures, minimizing undesirable side reactions, such as polymerization of olefins.

Alkylate is composed of a mixture of *iso*-paraffins that have octane numbers that vary with the olefins from which they were made. Butylenes produce the highest octane numbers, propylene the lowest, and pentylenes the intermediate. All alkylates, however, have high octane numbers (>87) and are particularly valuable because of these high octane numbers.

Alkylation developments in petroleum processing in the late 1930s and during World War II were directed toward production of high-octane blending stock for aviation gasoline. The sulfuric acid process was introduced in 1938, and hydrogen fluoride alkylation was introduced in 1942. Rapid commercialization took place during the war to supply military needs, but many of these plants were shut down at the end of the war.

In the mid-1950s aviation gasoline demand started to decline, and motor gasoline quality requirements rose sharply. Whenever practical, refiners shifted the use of alkylate to premium motor fuel. Alkylate end point was increased for this service, and total alkylate was often used without rerunning. To help improve the economics of the alkylation process and also the sensitivity of the premium gasoline pool, additional olefins were gradually added to alkylation feed. New plants were built to alkylate propylene and the butylenes (butanes) produced in the refinery rather than the butane–butylene stream formerly used. More recently, *n*-butane isomerization has been utilized to produce additional *iso*-butane for alkylation feed.

The alkylation reaction as practiced in petroleum refining is the union, through the agency of a catalyst, of an olefin (ethylene, $CH_2{=}CH_2$, propylene, $CH_3CH{=}CH_2$, butene also called butylene, $CH_3CH_2CH{=}CH_2$, and amylene, $CH_3CH_2CH_2CH{=}CH_2$) with *iso*-butane [$(CH_3)_3CH$] to yield high-octane branched-chain hydrocarbons in the gasoline boiling range. Olefin feedstock is derived from the gas make of a catalytic cracker; *iso*-butane is recovered from refinery gases or produced by catalytic butane isomerization.

Zeolite catalysts are also used for alkylation processes. For example, cumene (*iso*-propylbenzene) is produced by the alkylation of benzene by propylene. The cumene can then be used for the production of phenol and acetone by means of oxidation processes.

13.5.1 SULFURIC ACID ALKYLATION

This is a low-temperature process (Figure 13.9) employing concentrated sulfuric acid catalyst to react olefins with *iso*-butane to produce high-octane aviation or motor fuel blending stock. The olefin feed is split into equal streams and charged to the individual reaction zones of the cascade reactor. *Iso*-butane-rich recycle and refrigerant streams are introduced in the front of the reactor and pass through the reaction zones. The olefin is contacted with the *iso*-butane and acid in the reaction zones, which operate at 2°C–7°C (35°F–45°F) and 5–15 psi, after which vapors are withdrawn from

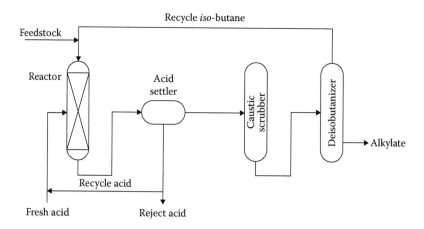

FIGURE 13.9 The sulfuric acid alkylation process. (From OSHA Technical Manual, Section IV, Chapter 2, Petroleum refining processes, 1999, http://www.osha.gov/dts/osta/otm/otm_iv/otm_iv_2.html.)

the top of the reactor, compressed, and condensed. Part of this stream is sent to a depropanizer to control propane concentration in the unit.

Depropanizer bottoms and the remainder of the stream are combined and returned to the reactor. Spent acid is withdrawn from the bottom of the settling zone; hydrocarbons spill over a baffle into a special withdrawal section and are hot water washed with caustic addition for pH control before being successively depropanized, de-*iso*-butanized, and debutanized. Alkylate can then be taken directly to motor fuel blending or be rerun to produce aviation-grade blending stock.

13.5.2 Hydrogen Fluoride Alkylation

The *hydrogen fluoride alkylation process* (Figure 13.10) uses regenerable hydrofluoric acid as a catalyst to unite olefins with *iso*-butane to produce high-octane blending stock. The dried charge is intimately contacted in the reactor with acid at 20°C–140°C (70°F–100°F) and a high (15:1) *iso*-butane–olefin ratio. The mixture is separated in a settler and acid is returned to the reactor, but an acid side stream must be continuously regenerated to 88% purity by fractionation to remove acid-soluble oils. The hydrocarbon fraction from the settler is de-*iso*-butanized, and alkylate is run to storage.

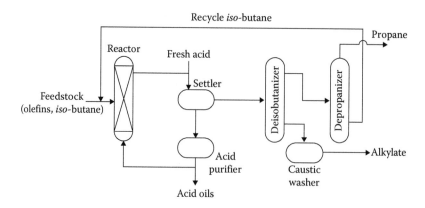

FIGURE 13.10 The hydrofluoric acid alkylation process. (From OSHA Technical Manual, Section IV, Chapter 2, Petroleum refining processes, 1999, http://www.osha.gov/dts/osta/otm/otm_iv/otm_iv_2.html.)

13.6　POLYMERIZATION

Polymerization (more correctly, *oligomerization*) as practiced in the petroleum industry, is a process that can claim to be the earliest to employ catalysts on a commercial scale. Catalytic polymerization came into use in the 1930s and was one of the first catalytic processes to be used in the petroleum industry. In the usual industrial sense, polymerization is a process in which a substance of low molecular weight is transformed into one of the same composition but of higher molecular weight while maintaining the atomic arrangement present in the basic molecule. It has also been described as the successive addition of one molecule to another by means of a functional group, such as that present in an aliphatic olefin. In the petroleum industry, polymerization is the controlled process by which olefin gases are converted to liquid condensation products that may be suitable for gasoline (hence *polymer gasoline, polymerate*) or other liquid fuels.

The feedstock for the process (Figure 13.11) usually consists of propylene (propene, $CH_3CH=CH_2$) and butylenes (butenes, various isomers of C_4H_8) from cracking processes or might even be selective olefins for dimer, trimer, or tetramer production:

$$nCH_2=CH_2H-(CH_2CH_2)_n-H$$

In this process, n is usually 2 (dimer), 3 (trimer), or 4 (tetramer); the molecular size of the product is limited to give products boiling in the gasoline range constituents. This is in contrast to polymerization that is carried out in the polymer industry where n may be on the order of several hundred. Thus, polymerization in the true sense of the word is usually prevented, and all attempts are made to terminate the reaction at the dimer or trimer (three monomers joined together) stage. The 4-carbon to 12-carbon compounds that are required as the constituents of liquid fuels are the prime products. However, in the petrochemical section of a refinery, polymerization, which results in the production of (for example) polyethylene, is allowed to proceed until the products having the required high molecular weight have been produced.

Polymerization may be accomplished thermally or in the presence of a catalyst at lower temperatures. Thermal polymerization is regarded not as effective as catalytic polymerization but has the advantage that it can be used to *polymerize* saturated materials that cannot be induced to react by catalysts. The process consists essentially of vapor-phase cracking of, say, propane and butane

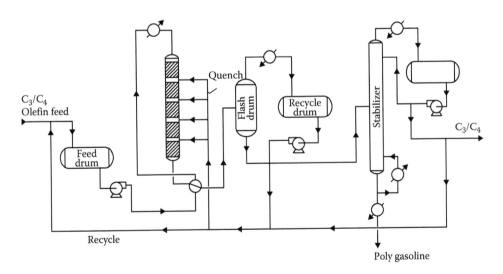

FIGURE 13.11　Polymerization process. (From OSHA Technical Manual, Section IV, Chapter 2, Petroleum refining processes, 1999, http://www.osha.gov/dts/osta/otm/otm_iv/otm_iv_2.html.)

followed by prolonged periods at high temperature (510°C–590°C, 950°F–1100°F) for the reactions to proceed to near completion.

On the other hand, olefins can be conveniently polymerized by means of an acid catalyst. Thus, the treated olefin-rich feed stream is contacted with a catalyst (sulfuric acid, copper pyrophosphate, or phosphoric acid) at 150°C–220°C (300°F–425°F) and 150–1200 psi, depending on feedstock and product requirement. The reaction is exothermic, and the temperature is usually controlled by heat exchange. Stabilization and/or fractionation systems separate saturated and unreacted gases from the product. In both thermal and catalytic polymerization processes, the feedstock is usually pretreated to remove sulfur and nitrogen compounds.

13.6.1 THERMAL POLYMERIZATION

Thermal polymerization converts butanes and lighter gases into liquid condensation products. Olefins are produced by thermal decomposition and polymerized by heat and pressure. Thus, liquid feed under a pressure of 1200–2000 psi is pumped to a furnace heated to 510°C–595°C (950°F–1100°F), from which the various streams are separated by fractionation.

Thermal polymerization is regarded not as effective as catalytic polymerization but has the advantage that it can be used to *polymerize* saturated materials that cannot be induced to react by catalysts. The process consists essentially of vapor-phase cracking of, say, propane and butane followed by prolonged periods at high temperature (510°C–595°C; 950°F–1100°F) for the reactions to proceed to near completion.

13.6.2 SOLID PHOSPHORIC ACID POLYMERIZATION

Olefins can be conveniently polymerized by means of an acid catalyst. Thus, the treated olefin-rich feed stream is contacted with a catalyst (sulfuric acid, copper pyrophosphate, or phosphoric acid) at 150°C–220°C (300°F–425°F) and 150–1200 psi, depending on the feedstock and the desired product(s). The reaction is exothermic, and temperature is usually controlled by heat exchange. Stabilization and/or fractionation systems separate saturated and unreacted gases from the product. In both thermal and catalytic polymerization processes, the feedstock is usually pretreated to remove sulfur and nitrogen compounds.

This process converts propylene and/or butylene to high-octane gasoline or petrochemical polymers. The catalyst, pelleted kieselguhr (diatomaceous earth) impregnated with phosphoric acid, is used in either a chamber or tubular reactor. The exothermic reaction temperature is controlled by using saturates (separated from the effluent as recycle to the feed) as a quench liquid between the catalyst chamber beds. Tubular reactors are temperature-controlled by water or oil circulation around the catalyst tubes.

Reaction temperatures and pressures are 175°C–225°C (350°F–435°F) and 400–1200 psi. Olefins and aromatics may be united by alkylation for special applications at 205°C–315°C (400°F–600°F) and 400–900 psi, and a rerun column is required in addition to the usual fractionating.

13.6.3 BULK ACID POLYMERIZATION

This process is used to produce high-octane polymer gasoline from all types of light olefin feed—the olefin concentration can be as high as 95% v/v. Liquid phosphoric acid is used as the catalyst.

The olefin feed is washed (caustic and water) and then contacted thoroughly by liquid phosphoric acid in a small reactor. The effluent stream and the acid are separated in a settler, and acid is returned to the reactor through a cooler. Gasoline is first stabilized and washed with caustic before storage. The heat of reaction is removed by circulation through an exchanger before contact with the olefin feed, and catalyst activity is maintained by continuous addition of fresh acid and withdrawal of spent acid.

13.7 CATALYSTS

The various catalysts used in product improvement processes have already been mentioned in the context of a particular reaction or reactor. However, for convenience it is worth mentioning the salient facts about these catalysts under one particular heading.

13.7.1 REFORMING PROCESSES

The composition of a reforming catalyst is dictated by the composition of the reformer feedstock and the desired reformate. Reforming consists of two types of chemical reactions that are catalyzed by two different types of catalysts: (1) isomerization of straight-chain paraffins and isomerization (simultaneously with hydrogenation) of olefins to produce branched-chain paraffins and (2) dehydrogenation–hydrogenation of paraffins to produce aromatics and olefins to produce paraffins.

The composition of a reforming catalyst is dictated by the composition of the feedstock and the desired reformate. The catalysts used are principally molybdena–alumina (MoO_2–Al_2O_3), chromia–alumina (Cr_2O_3–Al_2O_3), or platinum (Pt) on a silica–alumina (SiO_2–Al_2O_3) or alumina (Al_2O_3) base. The nonplatinum catalysts are widely used in regenerative process for feeds containing, for example, sulfur, which poisons platinum catalysts, although pretreatment processes (e.g., hydrodesulfurization) may permit platinum catalysts to be employed.

The purpose of platinum on the catalyst is to promote dehydrogenation and hydrogenation reactions, that is, the production of aromatics, participation in hydrocracking, and rapid hydrogenation of carbon-forming precursors. For the catalyst to have an activity for isomerization of both paraffins and naphthenes—the initial cracking step of hydrocracking—and to participate in paraffin dehydrocyclization, it must have an acid activity. The balance between these two activities is most important in a reforming catalyst.

In the production of aromatics from cyclic saturated materials (naphthenes), it is important that hydrocracking be minimized to avoid loss of the desired product. Thus, the catalytic activity must be moderated relative to the case of gasoline production from a paraffinic feed, where dehydrocyclization and hydrocracking play an important part.

The acid activity can be obtained by means of halogens (usually fluorine or chlorine up to approximately 1% w/w in catalyst) or silica incorporated in the alumina base. The platinum content of the catalyst is normally in the range of 0.3%–0.8% w/w. At higher levels there is some tendency to effect demethylation and naphthene ring opening, which is undesirable; at lower levels the catalysts tend to be less resistant to poisons.

Most processes have a means of regenerating the catalyst as needed. The time between regeneration, which varies with the process, the severity of the reforming reactions, and the impurities of the feedstock, ranges from every few hours to several months. Several processes use a nonregenerative catalyst that can be used for a year or more, after which it is returned to the catalyst manufacturer for reprocessing. The processes that have moving beds of catalysts utilize continuous regeneration of the catalyst in separate regenerators.

The processes using bauxite (*Cycloversion*) and clay (*Isoforming*) differ from other catalytic reforming processes in that hydrogen is not formed and hence none is recycled through the reactors. Since hydrogen is not concerned in the reforming reactions, there is no limit to the amount of olefin that may be present in the feedstock. The Cycloversion process is also used as a catalytic cracking process and as a desulfurization process. The Isoforming process causes only a moderate increase in octane number.

13.7.2 ISOMERIZATION PROCESSES

Isomerization reactions are usually reversible reactions and attain equilibrium at lower temperature with highest concentration of isomeric products. The role of catalyst in isomerization is, therefore, extremely important. The intensity of unwanted side reactions diminishes at lower temperatures as higher temperature favors unwanted cracking, hydrogenation, polymerization reactions. For that reason, isomerizing catalysts must ensure the optimal rate of reactions at as low temperature as possible. To prevent coke deposition, isomerization is carried out at an elevated pressure in a hydrogen atmosphere.

During World War II, aluminum chloride was the catalyst used to isomerize butane, pentane, and hexane. Since then, supported metal catalysts have been developed for use in high-temperature processes that operate in the range of 370°C–480°C (700°F–900°F) and 300–750 psi; aluminum chloride plus hydrogen chloride is universally used for the low-temperature processes. However, aluminum chloride is volatile at commercial reaction temperatures and is somewhat soluble in hydrocarbons, and techniques must be employed to prevent its migration from the reactor. This catalyst is nonregenerable and is utilized in either a fixed-bed or liquid contactor.

Catalysts commonly used for modern isomerization process are (in addition to the aforementioned Friedel–Crafts catalyst, $AlCl_3$) tungsten sulfide, bifunctional catalysts, zeolite-containing catalysts with noble metals such as platinum (Pt) or palladium (Pd), and complex (bifunctional and zeolite-containing Friedel–Crafts catalysts).

In the presence of certain catalysts, such as aluminum chloride with promoters, isomerization of paraffins can be carried out at the room temperature. In fact, the isomerization of paraffinic hydrocarbons in the presence of aluminum chloride is accelerated substantially if the reaction mixture contains traces of olefins. Catalysts of this type are however strong acids and can cause the corrosion of the equipment, because of which they are not used widely in the industry. Other catalysts for isomerization processes (except for Friedel–Crafts-type catalysts) require higher temperature and higher pressure of hydrogen, which results in the formation of a significant amount of unwanted by-products.

The isomerization of olefins can proceed by the formation of iso-olefins or by double bond transfer. Isomerization of naphthene constituents occurs with the transformation into olefins or with a change in the number of carbon atoms in the cycle. There are many other types of isomerization reaction, for instance, isomerization of alkyl benzene hydrocarbons, including xylenes and the reactions occur by the common carbocation mechanism. The isomerization reaction takes place in the hydrogen atmosphere whose role is to suppress polymerization and cracking reactions which might deactivate the catalyst.

The isomerization reaction is a thermoneutral reaction and requires no heat supply to the reactors from outside and hence can be carried out in a single reactor.

13.7.3 Alkylation Processes

Sulfuric acid, hydrogen fluoride, and aluminum chloride are the only catalysts used commercially. Sulfuric acid is used with propylene and higher-boiling feeds, but not with ethylene, because it reacts to form ethyl hydrogen sulfate and a suitable catalyst contains a minimum of 85% titratable acidity. The acid is pumped through the reactor and forms an air emulsion with reactants; the emulsion is maintained at 50% acid. The rate of deactivation varies with the feed and iso-butane change rate. Butene feedstocks cause less acid consumption than propylene feeds.

Aluminum chloride is not widely used as an alkylation catalyst, but when employed hydrogen chloride is used as a promoter and water is injected to activate the catalyst. The form of catalyst is an aluminum chloride–hydrocarbon complex, and the aluminum chloride concentration is 63%–84%.

Hydrogen fluoride is used for alkylation of higher-boiling olefins. The advantage of hydrogen fluoride is that it is more readily separated and recovered from the resulting product. The usual concentration is 85%–92% titratable acid with approximately 1.5% v/v water.

13.7.4 POLYMERIZATION PROCESSES

Phosphates are the principal catalysts for polymerization; the commercially used catalysts are liquid phosphoric acid, phosphoric acid on diatomaceous earth, copper pyrophosphate pellets, and phosphoric acid film on quartz. The latter is the least active but the most used and easiest to regenerate simply by washing and recoating; the serious disadvantage is that residue must occasionally be burned off the support. The process using liquid phosphoric acid catalyst is far more responsive to attempts to raise production by increasing temperature than the other processes.

13.8 TREATING PROCESSES

Fractions or streams produced by product improvement processes often contain small amounts of impurities that must be removed. Processes that remove these undesirable components are known as treating processes, and these processes are used not only to finish products for the market but also to prepare feedstocks for other processes (such as *catalytic polymerization* and *reforming*) in which catalysts would be harmed by impurities.

The most common impurities are sulfur compounds that are derived from the sulfur compounds that occur in crude oil (Chapter 3), such as *sulfides* (R^1SR^2) and the foul-smelling *thiols* (RSH, also called *mercaptans*). Oxygen compounds in the form of carboxylic acids (RCO_2H) and phenols (ArOH, where Ar is an aromatic group) may also be present. Nitrogen-containing compounds derived from those that occur in crude oil (Chapter 3) are also present. Furthermore, olefins ($R^1CH=CHR^2$) must also be eliminated from a feedstock or aromatics removed from a solvent, and these olefins and aromatics are considered impurities. Similarly, polymerized material, asphaltic material, or resin constituents may be impurities, depending on whether their presence in a finished product is harmful.

Treatment processes for the removal of sulfur-containing and nitrogen-containing compounds from distillates are much less severe than the desulfurization and denitrogenation processes applied to higher-boiling fractions (Chapter 10). In fact, it is generally recognized that the removal or conversion of sulfur and nitrogen compounds in distillates by treatment processes is usually limited to mercaptans and the lower-molecular-weight sulfur compounds. When there is more than a trace amount (>0.1%) of heteroatoms present, it is often more convenient and economical to resort to such methods as those thermal processes (e.g., hydroprocesses) that bring about a decrease in all types of heteroatomic compounds.

Choices of a treatment method depend on the amount and type of impurities in the fractions to be treated and the extent to which the process removes the impurities. Naturally occurring sweet kerosene, for example, may require only a simple treatment with alkali (*lye*) to remove hydrogen sulfide. If mercaptans are also present in the raw kerosene, a doctor treatment in addition to *lye treatment* is required, but poor-quality raw kerosene may require, in addition to these treatments, treatment with sulfuric acid and fuller's earth. The lowest quality raw kerosene requires treatment with strong sulfuric acid, neutralization with lye, and redistillation. Since different fractions have the same impurities, the same treatment process may be used for several different products.

The purpose of this section is to present an outline of the processes that are available for the treatment of the various product streams to remove contaminants from streams that will be eventually used as stock for products. The processes outlined here are not usually shown on a refinery schematic but would be placed in the general area entitled *finishing*, but only after the processes have been shown on a schematic and prior to the designation of the product.

13.8.1 CAUSTIC PROCESSES

The treating of petroleum products by washing with solutions of alkali (caustic or lye) is almost as old as the petroleum industry itself. Early discoveries that product odor and color could be improved by removing organic acids (naphthenic acids and phenols) and sulfur compounds (mercaptans and hydrogen sulfide) led to the development of caustic washing.

Caustic treating is an extraction process that removes organic acids (RCOOH), mercaptan sulfur (RSH), and phenolic compounds (ROH) from petroleum fractions. Thus, it is not surprising that caustic soda washing (lye treatment) has been used widely on many petroleum fractions. In fact, it is sometimes used as a pretreatment for sweetening and other processes. The process consists of mixing a water solution of lye (sodium hydroxide or caustic soda) with a petroleum fraction (typically naphtha) and is accomplished by contacting a 5%–20% w/w caustic solution countercurrently with the untreated stream in a packed tower with a ratio of product to caustic on the order of 10:1. The spent caustic can be regenerated either by steam stripping or air contacting. The regeneration conditions depend on the concentration of impurities in the caustic solution but the caustic is not 100% regenerable and makeup solution is necessary.

The major use of caustic treating in current refineries is the removal of mercaptan derivatives from naphtha—caustic losses and lower yields are prevalent (due to the formation of difficult-to-break emulsions) when heavy feedstocks are treated.

13.8.1.1 Dualayer Distillate Process

The *Dualayer distillate process* is similar in character to the *Duosol process* in that it uses caustic solution and cresylic acid (cresol, methylphenol, $CH_3C_6H_4OH$). The process extracts organic acid substances (including mercaptans, RSH) from cracked, or virgin, distillate fuels. In a typical operation, the Dualayer reagent is mixed with the distillate at approximately 55°C (130°F) and passed to the settler, where three layers separate with the aid of electrical coagulation. The product is withdrawn from the top layer; the Dualayer reagent is withdrawn from the bottom layer, relieved of excess water, fortified with additional caustic, and recycled.

13.8.1.2 Dualayer Gasoline Process

The *Dualayer gasoline process* is a modification of the *Dualayer distillate process* in that it is used to extract mercaptans from liquid petroleum gas, gasoline, and naphtha using the Dualayer reagents. Thus, gasoline, free of hydrogen sulfide, is contacted with the Dualayer solution at 50°C (120°F) in at least two stages, after which the treated gasoline is washed and stored. The treating solution is diluted with water (60%–70% of the solution volume) and stripped of mercaptans, gasoline, and excess water, and the correct amount of fresh caustic is added to obtain the regenerated reagent.

13.8.1.3 Electrolytic Mercaptan Process

The *electrolytic mercaptan process* employs aqueous solutions to extract mercaptans from refinery streams, and the electrolytic process is used to regenerate the solution. The charge stock is prewashed to remove hydrogen sulfide and contacted countercurrently with the treating solution in a mercaptan extraction tower. The treated gasoline is stored; the spent solution is mixed with regenerated solution and oxygen. The mixture is pumped to the cell, where mercaptans are converted to disulfides that are separated from the regenerated solution.

13.8.1.4 Ferrocyanide Process

The *Ferrocyanide process* is a regenerative chemical treatment for removing mercaptans from straight-run naphtha, as well as natural and recycle gasoline, using caustic–sodium ferrocyanide reagent.

For example, gasoline is washed with caustic to remove hydrogen sulfide and then washed countercurrently in a tower with the treating agent. The spent solution is mixed with fresh solution containing ferricyanide; the mercaptan derivatives are converted to insoluble disulfides and are removed by a countercurrent hydrocarbon wash. The solution is then recycled, and part of the ferrocyanide is converted to ferricyanide by an electrolyzer.

13.8.1.5 Lye Treatment

Lye treatment is carried out in continuous treaters, which essentially consist of a pipe containing baffles or other mixing devices into which the oil and lye solution are both pumped. The pipe discharges into a horizontal tank where the lye solution and oil separate. Treated oil is withdrawn from near the top of the tank; lye solution is withdrawn from the bottom and recirculated to mix with incoming untreated oil. A lye-treating unit may be incorporated as part of a processing unit, for example, the overhead from a bubble tower may be condensed, cooled, and passed immediately through a lye-treating unit. Such a unit is often referred to as a *worm-end treater*, since the unit is attached to the particular unit as a point beyond the cooling coil or cooling *worm*.

Caustic solutions ranging from 5% to 20% w/w are used at 20°C–45°C (70°F–110°F) and 5–40 psi. High temperatures and strong caustic are usually avoided because of the risk of color body formation and stability loss. Caustic–product treatment ratios vary from 1:1 to 1:10.

Spent lye is the term given to a lye solution in which approximately 65% of the sodium hydroxide content has been used by reaction with hydrogen sulfide, light mercaptans, organic acids, or mineral acids. A lye solution that is spent, as far as hydrogen sulfide is concerned, may still be used to remove mineral or organic acids from petroleum fractions. Lye solution spent by hydrogen sulfide is not regenerated, whereas blowing with steam can regenerate lye solution spent by mercaptans. This technique reforms sodium hydroxide and mercaptans from the spent lye. The mercaptans separate as a vapor and are normally destroyed by burning in a furnace. Spent lye can also be regenerated in a stripper tower with steam, and the overhead consists of steam and mercaptans, as well as the small amount of oil picked up by the lye solution during treatment. Condensing the overhead allows the mercaptans to separate from the water.

Nonregenerative caustic treatment is generally economically applied when the contaminating materials are low in concentration and waste disposal is not a problem. However, the use of nonregenerative systems is on the decline because of the frequently occurring waste disposal problems that arise from environmental considerations and because of the availability of numerous other processes that can effect more complete removal of contaminating materials.

13.8.1.6 Mercapsol Process

The *Mercapsol process* is another regenerative process for extracting mercaptans by means of sodium (or potassium) hydroxide, together with cresols, naphthenic acids, and phenol. Gasoline is contacted countercurrently with the *mercapsol* solution, and the treated product is removed from the top of the tower. Spent solution is stripped to remove gasoline, and the mercaptans are then removed by steam stripping.

13.8.1.7 Polysulfide Treatment

Polysulfide treatment is a nonregenerative chemical treatment process used to remove elemental sulfur from refinery liquids. Dissolving 1 pound of sodium sulfide (Na_2S) and 0.1 pound of elemental sulfur in a gallon of caustic solution prepares the polysulfide solution. The sodium sulfide can actually be prepared in the refinery by passing hydrogen sulfide, an obnoxious refinery by-product gas, through caustic solution.

The solution is most active when the composition approximates Na_2S to Na_2S_3 but activity decreases rapidly when the composition approaches Na_2S_4. When the solution is discarded, a portion (ca. 20%) is retained and mixed with fresh caustic–sulfide solution, which eliminates the need to add free sulfur. Indeed, if the material to be treated contains hydrogen sulfide in addition to free sulfur, it is often necessary simply to add fresh caustic.

13.8.1.8 Sodasol Process

A lye solution removes only the lighter or lower boiling mercaptans, but various chemicals can be added to the lye solution to increase its ability to dissolve the heavier mercaptans. The added chemicals are generally known as solubility promoters or solutizers. Several different solutizers have been patented and are used in processes that differ chiefly in the composition of the solutizers.

In the *Sodasol process*, the treating solution is composed of lye solution and alkyl phenols (acid oils), which occur in cracked naphtha and cracked gas oil and are obtained by washing cracked naphtha or cracked gas oil with the lye solution. The lye solution, with solutizers incorporated, is then ready to treat product streams, such as straight-run naphtha and gasoline. The process is carried out by pumping a sour stream up a treating tower countercurrent to a stream of Sodasol solution that flows down the tower. As the two streams mix and pass, the solution removes mercaptans and other impurities, such as oxygen compounds (phenols and acids), as well as some nitrogen compounds.

The treated stream leaves the top of the tower; the spent Sodasol solution leaves the bottom of the tower to be pumped to the top of a regeneration tower, where mercaptans are removed from the solution by steam. The regenerated Sodasol solution is then pumped to the top of the treatment tower to treat more material. A variation of the Sodasol process is the Potasol process, which uses potassium hydroxide instead of lye (sodium hydroxide).

13.8.1.9　Solutizer Process

The *Solutizer process* is a regenerative process using such materials as potassium *iso*-butyrate and potassium alkyl phenolate in strong aqueous potassium hydroxide to remove mercaptans. After removal of the mercaptans and recovery of the hydrocarbon stream, regeneration of the spent solution may be achieved by heating and steam blowing at 130°C (270°F) in a stripping column in which steam and mercaptans are condensed and separated.

On the other hand, the spent solution may be contacted with carbon dioxide air, after which the disulfides formed by oxidation of the mercaptans are extracted by a naphtha wash.

Air blowing in the presence of tannin (tannin Solutizer process) catalytically oxidizes mercaptans to the corresponding disulfides, but there may be side reactions that can lead to reagent contamination.

13.8.1.10　Steam-Regenerative Caustic Treatment

Steam-regenerative caustic treatment is essentially directed toward removal of mercaptans from such products as light, straight-run gasoline. The caustic is regenerated by steam blowing in a stripping tower. The nature and concentration of the mercaptans to be removed dictate the quantity and temperature of the process. However, the caustic solution gradually deteriorates because of the accumulation of material that cannot be removed by stripping; the caustic quality must be maintained by both continuous or intermittent discard and replacement of a minimum amount of the operating solution.

13.8.1.11　Unisol Process

The *Unisol process* is a regenerative method for extracting not only mercaptans but also certain nitrogen compounds from sour gasoline or distillates. The gasoline, free of hydrogen sulfide, is washed countercurrently with aqueous caustic–methanol solution at approximately 40°C (100°F). The spent caustic is regenerated in a stripping tower (145°C–150°C, 290°F–300°F), where methanol, water, and mercaptans are removed.

13.8.2　Acid Processes

Treating petroleum products with acids is, like caustic treatment, a procedure that has been in use for a considerable time in the petroleum industry. Various acids, such as hydrofluoric acid, hydrochloric acid, nitric acid, and phosphoric acid, have been used in addition to the more commonly used sulfuric acid, but in most instances there is little advantage in using any acid other than sulfuric.

Until about 1930 acid treatment was almost universal for all types of refined petroleum products, especially for cracked gasoline, kerosene, and lubricating stocks. Cracked products were acid treated to stabilize against gum formation and color darkening (oxidation) and to reduce sulfur content if necessary. However, there were appreciable losses due to polymer formation (from olefins in cracked products) initiated by the sulfuric acid.

Other processes have now superseded the majority of the *acid treatment processes*. However, acid treatment has, to some extent, been continued for desulfurizing high-boiling fractions of cracked gasoline, for refining kerosene, for manufacture of low-cost lubricating oil, and for making such specialties as insecticide naphtha, pharmaceutical white oil, and insulating oil.

The reactions of sulfuric acid with petroleum fractions are complex. The undesirable components to be removed are generally present in small amounts; large excesses of acid are required for efficient removal, which may cause marked changes in the remainder of the hydrocarbon mixture.

Paraffin and naphthene hydrocarbons in their pure forms are not attacked by concentrated sulfuric acid at low temperatures and during the short time of conventional refining treatment, but solution of light paraffins and naphthenes in the acid sludge can occur. Fuming sulfuric acid (*oleum*) absorbs small amounts of paraffins when contact is induced by long agitation; the amount of absorption increases with time, temperature, concentration of the acid, and complexity of structure of the hydrocarbons. With naphthenes fuming sulfuric acid causes sulfonation as well as rupture of the ring.

The action of sulfuric acid on olefin hydrocarbons is very complex. The main reactions involve ester formation and polymerization. The esters formed by reaction of sulfuric acid with olefins in cracked distillates are soluble in the acid phase but are also to some extent soluble in hydrocarbons, especially as the molecular weight of the olefin increases. The esters are usually difficult to hydrolyze with a view to removal by alkali washing. They are, however, unstable on standing for a long time, and products containing them (acid-treated cracked gasoline) may evolve sulfur dioxide and deposit intractable materials. The esters are quite unstable on heating, so that a redistilled, acid-treated cracked distillate usually requires alkali washing after the customary distillation.

Aromatics are not attacked by sulfuric acid to any great extent under ordinary refining conditions, unless they are present in high concentrations. However, if fuming acid is used or if the temperature is allowed to rise above normal, sulfonation may occur. When both aromatics and olefins are present, as in distillates from cracking units, alkylation can occur.

Thus, as indicated, acid treatment of cracked gasoline distillate brings about losses due to chemical reaction and polymerization of some of the olefins to constituents boiling above the gasoline range. This makes redistillation necessary, and such losses may total several percent, even when refrigeration is employed to maintain a low temperature.

Acid treatment of high-boiling distillates and residua presents different problems. Most of these contain at least a small proportion of dissolved or suspended asphaltic substances, and almost all the acid comes out as sludge (*acid tar*); its separation is aided by the addition of a little water or alkali solution. However, there may be obvious chemical changes, such as sulfur dioxide evolution, and washed (acid-free) sludge from the treatment of practically sulfur-free oils may contain up to 10% combined sulfur derived from the treating acid.

Although largely displaced for bulk production of both gasoline and lubricating oils, acid treatment still serves many special purposes. Paraffin distillates intended for dewaxing might receive light treatment to facilitate wax crystallization and refining, whereas insulating oils, refrigeration compressor oils, and white oils may be seated more severely.

The sludge produced on acid treatment of petroleum distillates, even gasoline and kerosene, is complex in nature. Esters and alcohols are present from reactions with olefins; sulfonation products from reactions with aromatics, naphthenes, and phenols; and salts from reactions with nitrogen bases. In addition, such materials as naphthenic acids, sulfur compounds, and asphaltic material are all retained by direct solution. To these constituents must be added the various products of oxidation–reduction reactions: coagulated resins, soluble hydrocarbons, water, and free acid.

Disposal of the sludge is difficult, as it contains unused free acid that must be removed by dilution and settling. The disposal is a comparatively simple process for the sludge resulting from treating gasoline and kerosene, the so-called light oils. The insoluble oil phase separates out as a mobile tar-like material, which can be mixed and burned without too much difficulty. Sludge from heavy oil and bitumen, however, separates out as granular semisolids, which offers considerable difficulty in handling.

In all cases careful separation of reaction products is important to the recovery of well-refined materials. This may not be easy if the temperature has risen as a consequence of chemical reaction. This will result in a persistent dark color traceable to colloidally distributed reaction products. Separation may also be difficult at low temperature because of high viscosity of the stock, but this problem can be overcome by diluting with light naphtha or with propane.

When acid treatment cannot be applied continuously by mechanical agitation followed by effective separation, the older batch agitators are employed. These devices are vertical reactors holding up to several thousand barrels, provided with conical bottoms for sludge drainage. The contact time is difficult to control and may amount to several hours, but the separation of acid tar is desirable to avoid discoloration by resolution and to permit handling the sludge before it becomes undesirably viscous. Breaking out of the suspended acid tar, often referred to as pepper sludge, is helped by adding a little water and agitating, and the subsequent separation of tar resembles closely the precipitation of a colloidal suspension. The sludge is allowed to settle, and the sour oil is washed with water, usually after transfer to another container, to avoid retention of acid tar in the system during the alkali washing that follows.

Sodium hydroxide solution (10%–25% concentration) may be used for nonviscous products, but for viscous oils more dilute solutions are employed and only a very slight excess of alkali is used, but no attempt is made at its recovery. Emulsion breaking chemicals are sometimes required in alkali washing; the use of aqueous alcohol is customary when fuming acid has been employed, as for sulfonates and white oils. Final water washing followed by air blowing to dry the oils is the customary procedure.

13.8.2.1 Nalfining Process

The *Nalfining process* is a continuous process that employs acetic anhydride and a caustic rinse to convert contaminants into less objectionable, but oil-soluble, compounds. The anhydride is injected into the product stream, where it reacts with oxygen to form the ester, with sulfur to form the thiaester, with nitrogen to form substituted amides, and with complex organic impurities to form environmentally benign products. The caustic rinse neutralizes the potentially corrosive acetic acid.

13.8.2.2 Sulfuric Acid Treatment

Sulfuric acid treatment is a continuous or batch method that is used to remove sulfur compounds. The treatment will also remove asphaltic materials from various refinery stocks. The acid strength varies from fuming (>100%) to 80%; approximately 93% acid finds the most common use. The weakest suitable acid is used for each particular situation to reduce sludge formation from the aromatic and olefin hydrocarbons.

The use of strong acid dictates the use of a fairly low temperature (−4°C to 10°C, 25°F–50°F), but higher temperatures (20°C–55°C, 70°F–130°F) are possible if the product is to be redistilled.

13.8.3 Clay Processes

Treating petroleum distillates and residua by passing them through materials possessing decolorizing power has been in operation for many years (Speight and Ozum, 2002; Parkash, 2003; Hsu and Robinson, 2006; Gary et al., 2007; Speight, 2011, 2014). For example, various clays and similar materials are used to treat petroleum fractions to remove diolefins, asphaltic materials, resins, acids, and colored bodies. Cracked naphtha were frequently clay treated to remove diolefins that formed gums in gasoline. Other processes have now largely superseded this use of clay treatment, in particular by the use of inhibitors, which, added in small amounts to gasoline, prevent gums from forming. Nevertheless, clay treatment is still used as a finishing step in the manufacture of lubricating oils and waxes. The clay removes traces of asphaltic materials and other compounds that give oils and waxes unwanted odors and colors.

The original method of clay treatment was to percolate a petroleum fraction through a tower containing coarse clay pellets. As the clay absorbed impurities from the petroleum fraction, the clay became less effective. Removing it from the tower periodically restored the activity of the clay and burning the absorbed material under carefully controlled conditions so as not to sinter the clay. The percolation method of clay treatment was widely used for lubricating oils but has been largely replaced by clay contacting.

13.8.3.1 Alkylation Effluent Treatment

This is a continuous liquid percolation process in which reactor effluent is coalesced in a vessel containing glass wool and steel mesh and then charged, alternately, to two bauxite (medium mesh) towers. Regeneration of the bauxite is effected with a mixture of steam and gas.

13.8.3.2 Arosorb Process

The *Arosorb process* separates aromatics from various refinery streams by the use of fixed silica gel beds. The feedstock is preheated over activated alumina to remove water, as well as traces of nitrogen, oxygen, and sulfur compounds. The material is then passed into one of several gel cases and, after a suitable residence time (ca. 30 minutes), conveyed to a second gel case. The first container (case) is then fed with the desorbent (e.g., crude xylene stream) that removes the saturate compounds from the bed and then displaces the adsorbed aromatics (e.g., benzene and toluene).

13.8.3.3 Bauxite Treatment

This process is essentially the same as the previous process, except in this case a vaporized petroleum fraction is passed through beds of a porous mineral known as bauxite. The bauxite acts as a catalyst to convert many different sulfur compounds, in particular mercaptans, into hydrogen sulfide, which is subsequently removed by a lye treatment. Bauxite is used to treat gasoline, naphtha, and kerosene products that have unusually high mercaptan contents.

A typical bauxite treatment unit consists of a fire-heated coil, two bauxite treatment towers, a bubble tower, a superheater for steam and air, and the usual exchangers, coolers, and pumps. Naphtha, raw kerosene, or other stock to be treated is preheated in heat exchangers and passed through the heating coil where it is heated to 415°C (780°F). At this temperature the stock is superheated. The vaporized feed is then passed downward through one of two bauxite towers at a pressure of approximately 40 psi. Three beds of catalyst in the tower convert mercaptans to hydrogen sulfide to enter a continuous lye treatment unit where hydrogen sulfide is removed.

After a time the bauxite towers are switched, since the bauxite progressively loses its catalytic activity. The spent catalyst is restored to its original activity by regeneration. Regeneration involves by passing superheated steam and air, carefully scheduled in proportions and rate, downward through the catalyst beds. The carbonaceous material that has accumulated on the catalyst is burned off. Combustion progresses downward through the beds; care is taken to prevent temperatures exceeding 595°C (1100°F), which would harm the bauxite. Air alone is finally used to burn away the last traces of carbonaceous material, after which the bauxite is ready for use again.

13.8.3.4 Continuous Contact Filtration Process

This is a continuous clay treatment process in which finely divided adsorbent is mixed with the charge stock and heated to 95°C–175°C (200°F–350°F). The slurry is then conveyed to a steam-stripping tower, after which it is cooled, vacuum filtered, and then vacuum stripped for further product specification control.

13.8.3.5 Cyclic Adsorption Process

The cyclic adsorption process is used for the separation of aromatics from petroleum product streams. Like the Arosorb process, the *cyclic adsorption process* employs fixed beds of silica gel. The various stages of the process are (1) extraction of the adsorbable material from the feedstock (refining), (2) concentration of the adsorbed phase (enriching), and (3) stripping for recovery of the extract and regeneration of the gel using a light gasoline or pentane fraction.

13.8.3.6 Gray Clay Treatment

This is a continuous vapor-phase process for selectively polymerizing the diolefin constituents of, and removing other gum-forming agents from, thermal gasoline. Two or more towers (10 ft in diameter and approximately 25 ft high) are used in parallel, and the hydrocarbon vapors are passed through the bed at temperatures (120°C–245°C, 250°F–475°F) just above the condensation point. The diolefins polymerize and drain from the base of the tower or they are separated from the gasoline by fractionation. Spent clay is either discarded or regenerated in kilns.

13.8.3.7 Percolation Filtration Process

This is a continuous-flow cyclic-regenerative liquid-phase process in which oil is filtered through a bed (containing 10–50 tons of fuller's earth) before storage. Two or more beds are used alternately on an operating and regenerating cycle. The spent clay is regenerated by washing it with naphtha, steaming, and burning.

13.8.3.8 Thermofor Continuous Percolation Process

The *Thermofor continuous percolation process* is a continuous regenerative process for stabilizing and decolorizing lubricants or waxes that have been distilled, solvent refined, or acid treated. The charge stock is heated to 50°C–175°C (125°F–350°F), injected into the base of a clay-filled tower, and allowed to percolate in countercurrent flow through the bed. Spent clay is continuously withdrawn from the base of the tower; regenerated clay is added to the top of the bed to maintain a constant level.

13.8.4 OXIDATIVE PROCESSES

Oxidative treatment processes are, in fact, processes that have been developed to convert the objectionable-smelling mercaptans to the less objectionable disulfides by oxidation.

However, disulfides tend to reduce the tetraethyl lead susceptibility of gasoline, and recent trends are toward processes that are capable of completely removing the mercaptans.

13.8.4.1 Bender Process

The *Bender process* is a fixed-bed catalytic treatment method that employs a lead sulfide catalyst. Controlled amounts of sulfur, alkali, and air are added to the product stream, which is passed through lead sulfide catalyst beds.

In this method, the sulfur required to oxidize the mercaptide derivatives is also furnished by the air oxidation of lead sulfide:

$$PbS + 1/2O_2 \rightarrow PbO + S$$

$$PbO + 2NaOH \rightarrow Na_2PbO_2 + H_2O$$

$$Na2PbO_2 + 2RSH \rightarrow Pb(RS)_2 + 2NaOH$$

$$Pb(RS)_2 + S \rightarrow PbS + R_2S_2$$

When larger quantities of air must be supplied for treating gasoline of high mercaptan content, there is a tendency toward excessive plumbite formation and also excessive sulfur formation. In such cases, a controlled quantity of aqueous sodium sulfide is simultaneously added to reconvert the extra plumbite back to lead sulfide. The presence of the remaining extra sulfur is not desirable, and therefore it is advantageous to control the air oxidation carefully. The lead sulfide is essentially a catalyst, as only oxygen is consumed in the process; there is, however, a certain loss of alkali to sodium sulfate and thiosulfate.

13.8.4.2 Copper Sweetening Process

The oxidizing power of cupric (Cu^{2+}) salts is also utilized to convert mercaptans directly into disulfides; free sulfur is not employed, and polysulfide derivatives are not obtained. The process employs cupric chloride in the presence of strong salt solutions, which are generally made up by dissolving copper sulfate in an aqueous solution of sodium chloride.

$$4RSH + 2CuCl_2 \rightarrow R_2S_2 + 2CuSR + 4HCl$$

$$2CuSR + 2CuCl_2 \rightarrow R2S_2 + 4CuCl$$

$$4CuCl + 4HCl + O_2 \rightarrow 4CuCl_2 + H_2O$$

The cuprous chloride (CuCl) is soluble in the salt solution, and there is no precipitation. Under operating conditions a certain amount of copper is retained by the sweetened petroleum fraction, probably as cuprous mercaptide derivatives or cuprous chloride–olefin addition products, but these can be removed by washing the material with aqueous sodium sulfide. Air blowing the cuprous chloride solution, after or during the sweetening operation, regenerates the cupric chloride. The copper chloride solution may be employed as such, or the sour fraction may be percolated through a porous mass saturated with the treating agent. Alternatively, the gasoline may be mixed with a solid carrier for the reagent, dispersed as a slurry.

Three methods of mechanical application of the copper chloride are used. If air will not cause the petroleum fraction to change color or form gum, a fixed-bed process may be used in which the sour material is passed through beds of an adsorbent that have been impregnated with cupric chloride. Air continuously regenerates the cupric chloride almost simultaneously with the sweetening reaction.

In the solution process, a solution of cupric chloride is continuously mixed in a centrifugal pump with the sour fraction. The mixture then enters a settling tank where the spent treatment solution separates from the petroleum liquid, and the treatment solution is withdrawn to a tank where blowing with air regenerates cupric chloride.

The slurry process makes use of clay or a similar material impregnated with cupric chloride. The clay is mixed with a small amount of, say, naphtha to form a slurry, which is pumped into the sour naphtha stream. Air or oxygen gas is added with the sour stream and continuously regenerates the cupric chloride. The treated material and clay slurry flow into a settling tank and separate, and then the clay slurry is recycled.

13.8.4.3 Doctor Process

The *doctor process* is a method of treating sour (sulfur-containing) distillates consists of agitating the distillate with alkaline sodium plumbite (doctor solution) in the presence of a small amount of free sulfur. A black precipitate of lead sulfide is formed, and the material, which has improved odor, has been rendered sweet.

In practice, sour distillates are usually given an alkali wash before the doctor treatment to remove traces of hydrogen sulfide and some of the lower-molecular-weight mercaptans; this process has a marked effect in reducing the plumbite requirement. Slightly more sulfur than the theoretical is required as a result of the formation of complex lead intermediates. In the presence of lead mercaptide derivatives the extra sulfur acts to form alkyl polysulfide derivatives, which are chemically analogous to peroxide.

The precipitating effect is evidently a result of the presence of these polysulfide derivatives or possibly of sodium sulfide formed between mercaptans, sulfur, and the alkaline solution. The doctor solution leaving the reactor consists essentially of a mixture of lead sulfide in free alkali, containing emulsified hydrocarbons, and this spent solution is pumped to steam-heated vessels, where it is air-blown for regeneration. Considerable amounts of sodium thiosulfate are also formed and the thiosulfate, in turn, may react with the alkali present to form sodium sulfite (Na_2SO_3) and sodium sulfide (Na_2S). The loss of lead is very low, and the main items of consumption are alkali and sulfur.

13.8.4.4 Hypochlorite Sweetening Process

This process employs sodium or calcium hypochlorite [$NaOCl$ or $Ca(OCl)_2$]. The principal reaction produces disulfides (R–S S–R′) with some formation of sulfoxides, ($R_2S{=}O$), and sulfonic acids (RSO_2H). When hydrogen sulfide is present, an alkaline wash prevents the formation of elemental sulfur. An alkaline afterwash is frequently necessary to remove undesirable chlorinated products.

13.8.4.5 Inhibitor Sweetening Process

This is a continuous process that uses a phenylenediamine-type inhibitor, air, and caustic to sweeten low-mercaptan-content gasoline. The inhibitor and air are injected between the caustic washing stages, and the mercaptan disappearance may be attributed to reaction with the caustic and then to oxidation during both washing and storage. In the absence of caustic, excessive peroxide formation occurs, which leads to gasoline deterioration.

13.8.4.6 Merox Process

The *Merox process* is a combination process for mercaptan extraction and sweetening of gasoline or lower boiling materials. The catalyst is a cobalt salt, which is insoluble in the oil and may be used in caustic solution or on a suitable solid support. Thus, gasoline is washed with alkali and contact by the catalyst and caustic in the extractor, air is then injected, and the treated product is stored in the regeneration step. Caustic is taken from the extractor and mixed with air in the oxidizer, after which disulfides and excess air are separated from the reagent in the disulfide separator. The regenerated caustic solution is recirculated to the top of the extractor.

13.8.5 Solvent Processes

Solvent refining processes are of a physical nature rather than a chemical nature. The desirable constituents, as well as the undesirable constituents, of the mixture can be recovered unchanged and in the original state. In addition, the processes that use solvent as a means of refining are extremely versatile, insofar as both low-boiling and high-boiling fractions can be used as feedstocks. In general, the solvent processes can be classified as (1) *deasphalting*, (2) *dewaxing*, and (3) *solvent refining*. The two former process types have been dealt with elsewhere in this text (Chapter 12) and, therefore, are excluded from this section.

Solvent refining (solvent treatment) is a widely used method of refining lubricating oils, as well as a host of other refinery stocks. The solvent processes yield products that meet the desired specifications by removing undesirable constituents (such as aromatics, naphthenes, and unsaturated compounds) from the charge material. There are, however, solvent refining processes in which the desirable continents are aromatics and are extracted by the solvent from the petroleum fraction.

Nevertheless, the original object of solvent extraction, or solvent treatment, was to remove aromatic compounds from feedstocks, such as lubricating oils. Thus, a suitable solvent can convert inferior raw lubricating oil stocks into oils with as high a quality as desired and is higher than could ever be obtained with, say, sulfuric acid. In contrast to sulfuric acid treatment (which depends on chemical reaction for a large part of its effect), solvent treatment is a physical process in which undesirable olefin compounds, asphaltic compounds, aromatic compounds, and sulfur compounds are selectively dissolved in the solvent and removed with the solvent. After separation of the solvent by distillation from both the unwanted materials and the treated oil, the solvent can be reused.

The most widely used extraction solvents are phenol, furfural, and cresylic acid. The last is used with propane in the *Duosol process*. Other solvents less frequently used are liquid sulfur dioxide, nitrobenzene, and *chlorex* (2,2-dichloroethyl ether). All lubricating oil solvent extraction processes operate on the principal of mixing the oil with the solvent and then allowing the solvent to settle from the treated oil. The solvent carries the unwanted materials with it. The nature of the raw lubricating oil stock is important in determining the type of process to use and the extent to which treatment should be carried out. The greater is the extent of treatment, the smaller the yield of treated oil but the higher the quality.

It is possible to make oil of any desired quality from any raw lubricating oil stock. In fact, it is possible to overtreat lubricating oils, causing a loss in lubricating properties. White oils, for example, are lubricating oils that have been deliberately overtreated with very strong sulfuric acid to remove all traces of color. Such oils are so overrefined that they are not used as lubricating oils; their lubricating properties are inferior to oils refined specifically for use as lubricants. Thus, solvent extraction is a complex operation requiring the selective removal of only those components that reduce the lubricating qualities of the oil being treated.

The *phenol treatment process* is the most widely used. Phenol, also known as *carbolic acid*, is a poisonous solid that can cause serious flesh burns. It melts at 41°C (106°F) and boils at 183°C (361°F). In the phenol treatment process, phenol is used in the liquid state by maintaining the temperature at over 35°C (100°F).

In the process, raw lubricating oil is pumped into the bottom of the tower and phenol into the top; the phenol descends through the oil and is mixed with the oil by means of baffles or other mechanical mixing devices. The treated oil or raffinate leaves the top of the tower. The phenol that collects at the bottom of the tower contains the extract (aromatic, unsaturated, asphaltic, and sulfur compounds extracted from the raw oil) and is known as spent phenol.

The raffinate is heated to 260°C (500°F) and pumped to a pair of fractional vacuum distillation towers operated in series. Here, the phenol in the raffinate is removed overhead. The bottom product from the second tower is the finished product.

The phenol is recovered from the spent phenol in a phenol recovery unit, which consists of another pair of fractional distillation towers and a stripper tower. The spent phenol is heated and passed into the primary tower, where process water is removed as an overhead product and some of the phenol recovered as a side-stream product. The bottom product from the primary tower is heated to approximately 240°C (460°F) and pumped into the secondary tower, where the majority of the remaining phenol is recovered as an overhead product. The extract (the bottom product) is heated to 345°C (650°F) and pumped into a vacuum stripper tower, where steam removes the phenol from the extract. The phenol obtained from the various distillation units is pumped to a heated storage tank for reuse. The extract may be used as a special-purpose product when high-molecular-weight aromatic hydrocarbons are needed; it may be added to heavy fuel oils, or it may be used as a cracking stock.

Liquid sulfur dioxide was first used commercially as an extraction solvent in 1909 to remove aromatic hydrocarbons from kerosene (*Edeleanu process*), when it was noted that aromatic and unsaturated hydrocarbons would dissolve in liquid sulfur dioxide but paraffinic and naphthenic hydrocarbons would not. The process was widely used for light distillates, such as kerosene, but was not used for heavier oils, such as lubricating oils, until about 1930.

The process is carried out in much the same way as phenol treatment, except that treatment takes place at approximately 0°C (32°F) and under sufficient pressure to maintain the sulfur dioxide as a liquid. The feedstock is contacted with liquid sulfur dioxide and the aromatic portion is

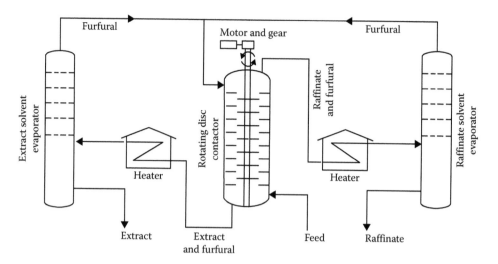

FIGURE 13.12 Aromatics extraction using furfural. (From OSHA Technical Manual, Section IV, Chapter 2, Petroleum refining processes, 1999, http://www.osha.gov/dts/osta/otm/otm_iv/otm_iv_2.html.)

concentrated in the extract phase, which is in turn contacted with a wash oil (such as kerosene) to remove the lower boiling nonaromatic compounds. The streams are then stripped and fractionated to recover sulfur dioxide and the wash oil for recycling to the extraction tower.

Sulfur dioxide alone does not effectively dissolve the high-boiling aromatic hydrocarbons in lubricating oils and therefore is used in conjunction with benzene for treating lubricating oils. Mixing benzene with the sulfur dioxide increases the solvent capacity but at the same time retains the selectivity for aromatic and nonparaffinic hydrocarbons. The percentage of benzene in the solvent mixture makes it possible to select the most advantageous treatment conditions for any particular feedstock to produce a product with the desired specifications.

Furfural is a heavy, straw-colored liquid that boils at 162°C (323°F). The process is carried out in the same manner as phenol treatment, but the raffinate or treated oil has so little furfural dissolved in it that distillation is not used to remove it; steam stripping is all that is required. The treatment (Figure 13.12) is conducted in counterflow towers, or multistage units, which normally operate at a temperature of 40°C–120°C (100°F–250°F). A high-temperature gradient in the treatment section permits a high yield of the refined oil.

Hydrogen fluoride treatment is a liquid–liquid extraction process removing sulfur and potential coke-forming materials from naphtha, middle distillates, and gas oil. The feedstock is contacted countercurrently with liquid hydrofluoric acid in an extraction tower, after which the overhead product is sent to a tower for removal of hydrogen fluoride. The solvent is recovered from the extract by use of evaporators and a stripper. The process is relatively insensitive to variations in temperature and pressure, but normally temperatures of 50°C (120°F) and pressures below 100 psi are employed.

The *Udex extraction process* is a liquid–liquid recovery method for the selective extraction of aromatic compounds from hydrocarbon mixtures by the use of a diethylene glycol (90%–92%), water (8%–10%), and solvent system. The feedstock, which may be pretreated to eliminate high conjugated dienes or alkenyl aromatics, rises countercurrent to the descending solvent. The stripper recovers the solvent for recycling; the overhead from the extractor is treated with clay and fractionated to separate the aromatics. The temperatures employed in the process are usually fairly low (120°C, 250°F); the pressures are just high enough to maintain liquid-phase conditions.

Sulfolane has been used primarily for the extraction of aromatics, such as benzene, toluene, and xylene(s). The chemicals usually arise from catalytic reforming of various feedstocks. Thus, the feedstock is contacted with the solvent in an extractor, and the mixture is conveyed to an extractive distillation column. The bottom product from this column is a mixture of solvent and aromatic compounds. Vacuum and steam stripping remove the final traces of hydrocarbons in the bottom part of the recovery column.

REFERENCES

Abraham, H. 1945. *Asphalts and Allied Substances*. Van Nostrand Scientific Publishers, New York.

Forbes, R.J. 1958a. *A History of Technology*. Oxford University Press, Oxford, UK.

Forbes, R.J. 1958b. *Studies in Early Petroleum Chemistry*. E.J. Brill, Leiden, the Netherlands.

Forbes, R.J. 1959. *More Studies in Early Petroleum Chemistry*. E.J. Brill, Leiden, the Netherlands.

Gary, J.H., Handwerk, G.E., and Kaiser, M.J. 2007. *Petroleum Refining: Technology and Economics*, 4th Edition. Marcel Dekker, New York, Chapter 9.

Giannetto, G.E., Perole, G.R., and Guisnet, M.R. 1986. Hydroisomerization and hydrocracking of *n*-alkanes. 1. Ideal hydroisomerization PtHY catalysts. *Industrial Engineering Chemical Product Research Development*, 25: 481–490.

Guisnet, M., Airajez, F., Gianetto, G., and Peroto, G. 1987. Hydroisomerization and hydro cracking of *n*-heptane on Pth zeolites. Effect of the porosity and of the distribution of metallic and acid sites. *Catalysis Today*, 1: 415–433.

Hsu, C.S. and Robinson, P.R. (Editors). 2006. *Practical Advances in Petroleum Processing*, Volumes 1 and 2. Springer Science, New York.

James, P. and Thorpe, N. 1994. *Ancient Inventions*. Ballantine Books, New York.

OSHA Technical Manual. 1999. Section IV, Chapter 2: Petroleum refining processes. http://www.osha.gov/dts/osta/otm/otm_iv/otm_iv_2.html.

Parkash, S. 2003. *Refining Processes Handbook*. Gulf Professional Publishing, Elsevier, Amsterdam, the Netherlands.

Riediger, B. 1971. *The Refining of Petroleum*. Springer-Verlag, Heidelberg, Germany.

Speight, J.G. 2001. *Handbook of Petroleum Analysis*. John Wiley & Sons, Hoboken, NJ.

Speight, J.G. 2011. *The Refinery of the Future*. Gulf Professional Publishing, Elsevier, Oxford, UK.

Speight, J.G. 2014. *The Chemistry and Technology of Petroleum*, 5th Edition. CRC Press/Taylor & Francis Group, Boca Raton, FL.

Speight, J.G. 2015. *Handbook of Petroleum Product Analysis*, 2nd Edition. John Wiley & Sons, Hoboken, NJ.

Speight, J.G. and Ozum, B. 2002. *Petroleum Refining Processes*. Marcel Dekker, New York.

14 Gasification Processes

14.1 INTRODUCTION

The influx of heavy feedstocks into refineries creates challenges but, at the same time, creates opportunities by improving the ability of refineries to handle heavy feedstocks thereby enhancing refinery flexibility to meet the increasingly stringent product specifications for refined fuels (Speight, 2013, 2014a,b). Upgrading heavy feedstocks is an increasingly prevalent means of extracting the maximum amount of liquid fuels from each barrel of crude oil that enters the refinery. Although solvent deasphalting and coking processes are used in refineries to upgrade heavy feedstocks to intermediate products that may be processed to produce transportation fuels, the integration of heavy feedstock processing units and gasification presents some unique synergies that will enhance the future refinery (Figure 14.1; Wallace et al., 1998; Penrose et al., 1999; Gray and Tomlinson, 2000; Abadie and Chamorro, 2009; Speight, 2011b, 2014a; Wolff and Vliegenthart, 2011).

Gasification has grown from a predominately coal conversion process used for making *town gas* for industrial lighting to an advanced process for the production of multiproduct, carbon-based fuels from a variety of feedstocks such as petroleum residua, biomass, or other carbonaceous feedstocks (Figure 14.2; Kumar et al., 2009; Speight, 2013, 2014a,b; Luque and Speight, 2015). It is the process that converts organic (carbonaceous) feedstocks into carbon monoxide, carbon dioxide, and hydrogen by reacting the feedstock at high temperatures (>700°C, 1290°F), without combustion, with a controlled amount of oxygen and/or steam (Marano, 2003; Lee et al., 2007; Higman and Van der Burgt, 2008; Speight, 2008, 2013, 2014b; Sutikno and Turini, 2012). Carbonaceous feedstocks include solids, liquids, and gases such as petroleum coke, coals, biomass, residual oils, and natural gas. The gasification is not a single-step process, but involves multiple subprocesses and reactions. The generated synthesis gas (syngas) has wide range of applications ranging from power generation to chemicals production. The power derived from carbonaceous feedstocks and gasification followed by the combustion of the product gas(es) is considered to be a source of renewable energy if derived gaseous products (Table 14.1) are generated from a source (e.g., biomass) other than a fossil fuel (Speight, 2008).

Gasification is an appealing process for the utilization of relatively inexpensive feedstocks that might otherwise be declared as waste and sent to a landfill (where the production of methane—a so-called greenhouse gas—will be produced) or combusted, which may not (depending upon the feedstock) be energy efficient. Overall, use of a gasification technology (Speight, 2013, 2014b) with the necessary gas cleanup options can have a smaller environmental footprint and lesser effect on the environment than landfill operations or combustion of the waste. In fact, there are strong indications that gasification is a technically viable option for the waste conversion, including residual waste from separate collection of municipal solid waste. The process can meet existing emission limits and can have a significant effect on the reduction of landfill disposal using known gasification technologies (Arena, 2012; Speight, 2014b; Luque and Speight, 2015) or thermal plasma (Fabry et al., 2013).

Indeed, the increasing mounting interest in gasification technology reflects a convergence of two changes in the electricity generation marketplace: (1) the maturity of gasification technology and (2) the extremely low emissions from integrated gasification combined cycle (IGCC) plants, especially air emissions, and the potential for lower cost control of greenhouse gases than other coal-based systems (Speight, 2014b). Another advantage of gasification is that the use of synthesis gas (syngas) is potentially more efficient as compared to direct combustion of the original fuel because it can

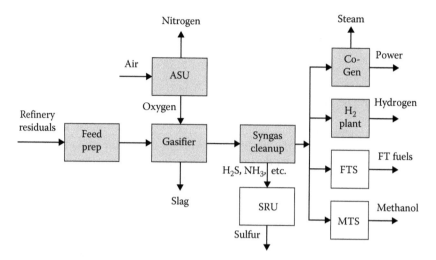

FIGURE 14.1 Gasification as might be employed on-site in a refinery. (From the National Energy Technology Laboratory, U.S. Department of Energy, Washington, DC, http://www.netl.doe.gov/technologies/coalpower/gasification/gasifipedia/7-advantages/7-3-4_refinery.html, accessed July 21, 2015.)

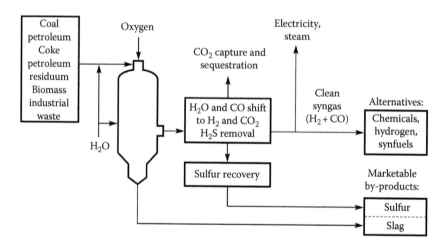

FIGURE 14.2 The gasification process can accommodate a variety of feedstocks.

TABLE 14.1

Gasification Products

Product	Characteristics
Low-Btu gas (150–300 Btu/scf)	Around 50% nitrogen, with smaller quantities of combustible H_2 and CO, CO_2 and trace gases, such as methane
Medium-Btu gas (300–550 Btu/scf)	Predominantly CO and H_2, with some incombustible gases and sometimes methane
High-Btu gas (980–1080 Btu/scf)	Almost pure methane

be (1) combusted at higher temperatures, (2) used in fuel cells, (3) used to produce methanol and hydrogen, and (4) converted via the Fischer–Tropsch process into a range of synthesis liquid fuels suitable for use in gasoline engines or diesel engines.

Coal has been the primary feedstock for gasification units for many decades. However, there is a move to feedstocks other than coal for gasification processes with the concern on the issue of environmental pollutants and the potential shortage for coal in some areas (except in the United States). Nevertheless, coal still prevails and will remain so for at least several decades into the future, if not well into the next century (Speight, 2013). The gasification process can also utilize carbonaceous feedstocks that would otherwise have been disposed of (e.g., biodegradable waste).

Coal gasification plants are cleaner with respect to standard pulverized coal combustion facilities, producing fewer sulfur and nitrogen by-products, which contribute to smog and acid rain. For this reason, gasification is an appealing way to utilize relatively inexpensive and expansive coal reserves, while reducing the environmental impact. Indeed, the increasing mounting interest in coal gasification technology reflects a convergence of two changes in the electricity generation marketplace: (1) the maturity of gasification technology and (2) the extremely low emissions from IGCC plants, especially air emissions, and the potential for lower cost control of greenhouse gases than other coal-based systems. Fluctuations in the costs associated with natural gas-based power, which is viewed as a major competitor to coal-based power, can also play a role. Furthermore, gasification permits the utilization of various feedstocks (coal, biomass, petroleum residues, and other carbonaceous wastes) to their fullest potential (Speight, 2013, 2014b; Yuksel Orhan et al., 2014). Thus, power developers would be well advised to consider gasification as a means of converting coal to gas.

Liquid fuels, including gasoline, diesel, naphtha, and jet fuel, are usually processed via refining of crude oil (Speight, 2014a). Due to the direct distillation, crude oil is the most suited raw material for liquid fuel production. However, with fluctuating and rising prices of petroleum, coal-to-liquid (CTL) and biomass-to-liquid (BTL) processes are currently started to be considered as alternative routes used for liquid fuels production. Both feedstocks are converted to syngas (a mixture of carbon monoxide and hydrogen), which is subsequently converted into a mixture of liquid products by Fischer–Tropsch (FT) processes. The liquid fuel obtained after FT synthesis is eventually upgraded using known petroleum refinery technologies to produce gasoline, naphtha, diesel fuel, and jet fuel (Chadeesingh, 2011; Speight, 2014a).

14.2 GASIFICATION CHEMISTRY

Chemically, gasification involves the thermal decomposition of the feedstock and the reaction of the feedstock carbon and other pyrolysis products with oxygen, water, and fuel gases such as methane and is represented by a sequence of simple chemical reactions (Table 14.2). However, the gasification process is often considered to involve two distinct chemical stages: (1) devolatilization of the feedstock to produce volatile matter and char followed by (2) char gasification, which is complex and specific to the conditions of the reaction. Both processes contribute to the complex kinetics of the gasification process (Sundaresan and Amundson, 1978).

Gasification of a carbonaceous material in an atmosphere of carbon dioxide can be divided into two stages: (1) pyrolysis and (2) gasification of the pyrolytic char. In the first stage, pyrolysis (removal of moisture content and devolatilization) occurs at comparatively at lower temperature. In the second stage, gasification of the pyrolytic char is achieved by reaction with oxygen/carbon dioxide mixtures at high temperature. In nitrogen and carbon dioxide environments from room temperature to 1000°C (1830°F), the mass loss rate of pyrolysis in nitrogen may be significant differently (sometime lower, depending on the feedstock) to mass loss rate in carbon dioxide, which may be due (in part) to the difference in properties of the bulk gases.

Using coal as an example, gasification in an atmosphere of oxygen/carbon dioxide environment is almost same as gasification in an atmosphere of oxygen/nitrogen at the same oxygen

---

TABLE 14.2

Gasification Reactions

$2C + O_2 \rightarrow 2CO$

$C + O_2 \rightarrow CO_2$

$C + CO_2 \rightarrow 2CO$

$CO + H_2O \rightarrow CO_2 + H_2$ (shift reaction)

$C + H_2O \rightarrow CO + H_2$ (water-gas reaction)

$C + 2H_2 \rightarrow CH_4$

$2H_2 + O_2 \rightarrow 2H_2O$

$CO + 2H_2 \rightarrow CH_3OH$

$CO + 3H_2 \rightarrow CH_4 + H_2O$ (methanation reaction)

$CO_2 + 4H_2 \rightarrow CH_4 + 2H_2O$

$C + 2H_2O \rightarrow 2H_2 + CO_2$

$2C + H_2 \rightarrow C_2H_2$

$CH_4 + 2H_2O \rightarrow CO_2 + 4H_2$

---

concentration, but this effect is little bit delayed at high temperature. This may be due to the lower rate of diffusion of oxygen through carbon dioxide and the higher specific heat capacity of carbon dioxide. However, with an increase in the concentration of oxygen, the mass loss rate of coal also increases and, hence, shortens the burn out time of coal. The optimum value of oxygen/carbon dioxide for the reaction of oxygen with the functional groups that are present in the coal feedstock is on the order of 8% v/v.

14.2.1 GENERAL ASPECTS

In a gasifier, the feedstock particle is exposed to high temperatures generated from the partial oxidation of the carbon. As the particle is heated, any residual moisture (assuming that the feedstock has been pre-fired) is driven off and further heating of the particle begins to drive off the volatile gases. Discharge of the volatile products will generate a wide spectrum of hydrocarbons ranging from carbon monoxide and methane to long-chain hydrocarbons comprising tars, creosote, and heavy oil. The complexity of the products will also affect the progress and rate of the reaction when each product is produced by a different chemical process at a different rate. At a temperature above 500°C (930°F), the conversion of the feedstock to char and ash and char is completed. In most of the early gasification processes, this was the desired by-product, but for gas generation, the char provides the necessary energy to effect further heating, and, typically, the char is contacted with air or oxygen and steam to generate the product gases.

Furthermore, with an increase in heating rate, feedstock particles are heated more rapidly and are burned in a higher temperature region, but the increase in heating rate has almost no substantial effect on the mechanism (Irfan, 2009).

Most notable effects in the physical chemistry of the gasification process are those effects due to the chemical character of the feedstock as well as the physical composition of the feedstock (Speight, 2011a, 2013). In more general terms of the character of the feedstock, gasification technologies generally require some initial processing of the feedstock with the type and degree of pretreatment a function of the process and/or the type of feedstock. For example, the Lurgi process will accept lump feedstock (1 inch, 25 mm, to 28 mesh), but it must be noncaking with the fines removed—feedstock that shows caking or agglomerating tendencies that form a plastic mass in the bottom of a gasifier and subsequently plug up the system thereby markedly reducing process efficiency. Thus, some attempt to reduce caking tendencies is necessary and can involve preliminary partial oxidation of the feedstock thereby destroying the caking properties.

Another factor, often presented as very general *rule of thumb*, is that optimum gas yields and gas quality are obtained at operating temperatures of approximately 595°C–650°C (1100°F–1200°F). A gaseous product with a higher heat content (Btu/ft^3) can be obtained at lower system temperatures but the overall yield of gas (determined as the *fuel-to-gas ratio*) is reduced by the unburned char fraction.

With some feedstocks, the higher the amounts of volatile material produced in the early stages of the process the higher the heat content of the product gas. In some cases, the highest gas quality may be produced at the lowest temperatures but when the temperature is too low, char oxidation reaction is suppressed and the overall heat content of the product gas is diminished. All such events serve to complicate the reaction rate and make derivative of a global kinetic relationship applicable to all types of feedstock subject to serious question and doubt.

Depending on the type of feedstock being processed and the analysis of the gas product desired, pressure also plays a role in product definition. In fact, some (or all) of the following processing steps will be required: (1) pretreatment of the feedstock; (2) primary gasification of the feedstock; (3) secondary gasification of the carbonaceous residue from the primary gasifier; (4) removal of carbon dioxide, hydrogen sulfide, and other acid gases; (5) shift conversion for adjustment of the carbon monoxide/hydrogen mole ratio to the desired ratio; and (6) catalytic methanation of the carbon monoxide/hydrogen mixture to form methane. If high-heat-content (high-Btu) gas is desired, all of these processing steps are required since gasifiers do not typically yield methane in the significant concentration.

14.2.2 Pretreatment

While feedstock pretreatment for introduction into the gasifier is often considered to be a physical process in which the feedstock is prepared for gasifier—typically as pellets or finely ground feedstock—there are chemical aspects that must also be considered.

Certain types of coal, display caking, or agglomerating, characteristics when heated (Speight, 2013), and these coal types are usually not amenable to treatment by gasification processes employing fluidized-bed or moving-bed reactors; in fact, caked coal is difficult to handle in fixed-bed reactors. The pretreatment involves a mild oxidation treatment that destroys the caking characteristics of coals and usually consists of low-temperature heating of the coal in the presence of air or oxygen.

While this may seemingly be applicable to coal gasification only, this form of coal pretreatment is particularly important when a noncoal feedstock is cogasified with coal. Cogasification of other feedstocks, such as coal and especially biomass, with petroleum coke offers a bridge between the depletion of petroleum stocks when coal is used as well as a supplementary feedstock based on renewable energy sources (biomass). These options can contribute to reduce the petroleum dependency and carbon dioxide emissions since biomass is known to be neutral in terms of carbon dioxide emissions. The high reactivity of biomass and the accompanying high production of volatile products suggest that some synergetic effects might occur in simultaneous thermochemical treatment of petcoke and biomass, depending on the gasification conditions such as (1) feedstock type and origin, (2) reactor type, and (3) process parameters (Penrose et al., 1999; Gray and Tomlinson, 2000; McLendon et al., 2004; Lapuerta et al., 2008; Fermoso et al., 2009; Shen et al., 2012; Khosravi and Khadse, 2013; Speight, 2013, 2014a,b; Luque and Speight, 2015).

For example, carbonaceous fuels are gasified in reactors, a variety of gasifiers such as the fixed- or moving-bed, fluidized-bed, entrained-flow, and molten bath gasifiers have been developed (Shen et al., 2012; Speight, 2014b). If the flow patterns are considered, the fixed-bed and fluidized-bed gasifiers intrinsically pertain to a countercurrent reactor in that fuels are usually sent into the reactor from the top of the gasifier, whereas the oxidant is blown into the reactor from the bottom. With regard to the entrained-flow reactor, it is necessary to pulverize the feedstock (such as coal and petcoke). On the other hand, when the feedstock is sent into an entrained-flow gasifier, the fuels can be in either form of dry feed or slurry feed. In general, dry-feed gasifiers have the advantage over slurry-feed gasifiers in that the former can be operated with lower oxygen consumption. Moreover, dry-feed gasifiers have an additional degree of freedom that makes it possible to optimize synthesis gas production (Shen et al., 2012).

14.2.3 REACTIONS

Gasification involves the thermal decomposition of feedstock and the reaction of the feedstock carbon and other pyrolysis products with oxygen, water, and fuel gases such as methane. The presence of oxygen, hydrogen, water vapor, carbon oxides, and other compounds in the reaction atmosphere during pyrolysis may either support or inhibit numerous reactions with carbonaceous feedstocks and with the products evolved. The distribution of weight and chemical composition of the products are also influenced by the prevailing conditions (i.e., temperature, heating rate, pressure, and residence time) and, last but by no means least, the nature of the feedstock (Speight, 2014a,b).

If air is used for combustion, the product gas will have a heat content of ca. 150–300 Btu/ft³ (depending on process design characteristics) and will contain undesirable constituents such as carbon dioxide, hydrogen sulfide, and nitrogen. The use of pure oxygen, although expensive, results in a product gas having a heat content on the order of 300–400 Btu/ft³ with carbon dioxide and hydrogen sulfide as by-products (both of which can be removed from low or medium heat content, low Btu, or medium Btu) by any of several available processes (Speight, 2013, 2014a).

If a high-heat-content (high-Btu) gas (900–1000 Btu/ft³) is required, efforts must be made to increase the methane content of the gas. The reactions that generate methane are all exothermic and have negative values, but the reaction rates are relatively slow, and catalysts may therefore be necessary to accelerate the reactions to acceptable commercial rates. Indeed, the overall reactivity of the feedstock and char may be subject to catalytic effects. It is also possible that the mineral constituents of the feedstock (such as the mineral matter in coal and biomass) may modify the reactivity by a direct catalytic effect (Davidson, 1983; Baker and Rodriguez, 1990; Martinez-Alonso and Tascon, 1991; Mims, 1991).

In the process, the feedstock undergoes three processes in its conversation to synthesis gas—the first two processes, pyrolysis and combustion, occur very rapidly. In pyrolysis, char is produced as the feedstock heats up and volatiles are released. In the combustion process, the volatile products and some of the char react with oxygen to produce various products (primarily carbon dioxide and carbon monoxide) and the heat required for subsequent gasification reactions. Finally, in the gasification process, the feedstock char reacts with steam to produce hydrogen (H_2) and carbon monoxide (CO).

Combustion

$$2C_{feedstock} + O_2 \rightarrow 2CO + H_2O$$

Gasification

$$C_{feedstock} + H_2O \rightarrow H_2 + CO$$

$$CO + H_2O \rightarrow H_2 + CO_2$$

The resulting synthesis gas is approximately 63% v/v carbon monoxide, 34% v/v hydrogen, and 3% v/v carbon dioxide. At the gasifier temperature, the ash and other feedstock mineral matter liquefies and exits at the bottom of the gasifier as slag, a sand-like inert material that can be sold as a co-product to other industries (e.g., road building). The synthesis gas exits the gasifier at pressure and high temperature and must be cooled prior to the synthesis gas cleaning stage.

Although processes that use the high temperature to raise high-pressure steam are more efficient for electricity production, full-quench cooling, by which the synthesis gas is cooled by the direct injection of water, is more appropriate for hydrogen production. Full-quench cooling provides the necessary steam to facilitate the water-gas shift reaction, in which carbon monoxide is converted to hydrogen and carbon dioxide in the presence of a catalyst.

The water-gas shift reaction is as follows:

$$CO + H_2O \rightarrow CO_2 + H_2$$

This reaction maximizes the hydrogen content of the synthesis gas, which consists primarily of hydrogen and carbon dioxide at this stage. The synthesis gas is then scrubbed of particulate matter, and sulfur is removed via physical absorption (Speight, 2013, 2014a). The carbon dioxide is captured by physical absorption or a membrane and either vented or sequestered.

Thus, in the initial stages of gasification, the rising temperature of the feedstock initiates devolatilization and the breaking of weaker chemical bonds to yield volatile tar, volatile oil, phenol derivatives, and hydrocarbon gases. These products generally react further in the gaseous phase to form hydrogen, carbon monoxide, and carbon dioxide. The char (fixed carbon) that remains after devolatilization reacts with oxygen, steam, carbon dioxide, and hydrogen. Overall, the chemistry of gasification is complex but can be conveniently (and simply) represented by the following reactions:

$$C + O_2 \rightarrow CO_2 \qquad \Delta H_r = -393.4 \text{ MJ/kmol} \qquad (14.1)$$

$$C + \tfrac{1}{2}O_2 \rightarrow CO \qquad \Delta H_r = -111.4 \text{ MJ/kmol} \qquad (14.2)$$

$$C + H_2O \rightarrow H_2 + CO \qquad \Delta H_r = 130.5 \text{ MJ/kmol} \qquad (14.3)$$

$$C + CO_2 \leftrightarrow 2CO \qquad \Delta H_r = 170.7 \text{ MJ/kmol} \qquad (14.4)$$

$$CO + H_2O \leftrightarrow H_2 + CO_2 \qquad \Delta H_r = -40.2 \text{ MJ/kmol} \qquad (14.5)$$

$$C + 2H_2 \rightarrow CH_4 \qquad \Delta H_r = -74.7 \text{ MJ/kmol} \qquad (14.6)$$

The designation C represents carbon in the original feedstock as well as carbon in the char formed by devolatilization of the feedstock. Reactions (14.1) and (14.2) are exothermic oxidation reactions and provide most of the energy required by the endothermic gasification reactions (14.3) and (14.4). The oxidation reactions occur very rapidly, completely consuming all of the oxygen present in the gasifier, so that most of the gasifier operates under reducing conditions. Reaction (14.5) is the water-gas shift reaction, in water (steam) is converted to hydrogen—this reaction is used to alter the hydrogen/carbon monoxide ratio when synthesis gas is the desired product, such as for use in Fischer–Tropsch processes. Reaction (14.6) is favored by high pressure and low temperature and is, thus, mainly important in lower temperature gasification systems. Methane formation is an exothermic reaction that does not consume oxygen and, therefore, increases the efficiency of the gasification process and the final heat content of the product gas. Overall, approximately 70% of the heating value of the product gas is associated with the carbon monoxide and hydrogen but this varies depending on the gasifier type and the process parameters (Chadeesingh, 2011; Speight, 2011a, 2013).

In essence, the direction of the gasification process is subject to the constraints of thermodynamic equilibrium and variable reaction kinetics. The combustion reactions (reaction of the feedstock or char with oxygen) essentially go to completion. The thermodynamic equilibrium of the rest of the gasification reactions are relatively well defined and collectively have a major influence on thermal efficiency of the process as well as on the gas composition. Thus, thermodynamic data are useful for estimating key design parameters for a gasification process, such as (1) calculating of the relative amounts of oxygen and/or steam required per unit of feedstock, (2) estimating the composition of the produced synthesis gas, and (3) optimizing process efficiency at various operating conditions.

Other deductions concerning gasification process design and operations can also be derived from the thermodynamic understanding of its reactions. Examples include (1) production of synthesis gas with low methane content at high temperature, which requires an amount of steam in excess of the stoichiometric requirement; (2) gasification at high temperature, which increases oxygen consumption and decreases the overall process efficiency; and (3) production of synthesis gas with a high methane content, which requires operation at low temperature (approximately 700°C, 1290°F) but the methanation reaction kinetics will be poor without the presence of a catalyst.

Relative to the thermodynamic understanding of the gasification process, the kinetic behavior is much more complex. In fact, very little reliable global kinetic information on gasification reactions exists, partly because it is highly dependent on (1) the chemical nature of the feed, which varies significantly with respect to composition, mineral impurities, (2) feedstock reactivity, and (3) process conditions. In addition, physical characteristics of the feedstock (or char) also play a role in phenomena such as boundary layer diffusion, pore diffusion, and ash layer diffusion, which also influence the kinetic outcome. Furthermore, certain impurities, in fact, are known to have catalytic activity on some of the gasification reactions, which can have further influence on the kinetic imprint of the gasification reactions.

14.2.3.1 Primary Gasification

Primary gasification involves thermal decomposition of the raw feedstock via various chemical processes, and many schemes involve pressures ranging from atmospheric to 1000 psi. Air or oxygen may be admitted to support combustion to provide the necessary heat. The product is usually a low-heat-content (low-Btu) gas ranging from a carbon monoxide/hydrogen mixture to mixtures containing varying amounts of carbon monoxide, carbon dioxide, hydrogen, water, methane, hydrogen sulfide, nitrogen, and typical tar-like products of thermal decomposition of carbonaceous feedstocks are complex mixtures and include hydrocarbon oils and phenolic products (Dutcher et al., 1983; Speight, 2011a, 2013, 2014b).

Devolatilization of the feedstock occurs rapidly as the temperature rises above 300°C (570°F). During this period, the chemical structure is altered, producing solid char, tar products, condensable liquids, and low-molecular-weight gases. Furthermore, the products of the devolatilization stage in an inert gas atmosphere are very different from those in an atmosphere containing hydrogen at elevated pressure. In an atmosphere of hydrogen at elevated pressure, additional yields of methane or other low-molecular-weight gaseous hydrocarbons can result during the initial gasification stage from reactions such as (1) direct hydrogenation of feedstock or semichar because of any reactive intermediates formed and (2) the hydrogenation of other gaseous hydrocarbons, oils, tars, and carbon oxides. Again, the kinetic picture for such reactions is complex due to the varying composition of the volatile products, which, in turn, are related to the chemical character of the feedstock and the process parameters, including the reactor type.

A solid char product may also be produced, and may represent the bulk of the weight of the original feedstock, which determines (to a large extent) the yield of char and the composition of the gaseous product.

14.2.3.2 Secondary Gasification

Secondary gasification usually involves the gasification of char from the primary gasifier, which is typically achieved by reaction of the hot char with water vapor to produce carbon monoxide and hydrogen:

$$C_{char} + H_2O \rightarrow CO + H_2$$

The reaction requires heat input (endothermic) for the reaction to proceed in its forward direction. Usually, an excess amount of steam is also needed to promote the reaction. However, excess steam used in this reaction has an adverse effect on the thermal efficiency of the process. Therefore, this reaction is typically combined with other gasification reactions in practical applications. The hydrogen/carbon monoxide ratio of the product synthesis gas depends on the synthesis chemistry as well as process engineering.

The mechanism of this reaction section is based on the reaction between carbon and gaseous reactants, not for reactions between feedstock and gaseous reactants. Hence the equations may over-simplify the actual chemistry of the steam gasification reaction. Even though carbon is the dominant atomic species present in feedstock, feedstock is more reactive than pure carbon. The presence of various reactive organic functional groups and the availability of catalytic activity via naturally occurring mineral ingredients can enhance the relative reactivity of the feedstock—for example, anthracite, which has the highest carbon content among all ranks of coal (Speight, 2013), is the most difficult to gasify or liquefy.

Once the rate of devolatilization has passed, a maximum another reaction becomes important—in this reaction, the semichar is converted to char (sometimes erroneously referred to as *stable char*) primarily through the evolution of hydrogen. Thus, the gasification process occurs as the char reacts with gases such as carbon dioxide and steam to produce carbon monoxide and hydrogen. The resulting gas (producer gas or synthesis gas) may be more efficiently converted to electricity than is typically possible by direct combustion of the coal. Also, corrosive elements in the ash may be refined out by the gasification process, allowing high-temperature combustion of the gas from otherwise problematic feedstocks (Speight, 2011a, 2013, 2014b).

Oxidation and gasification reactions consume the char and the oxidation and the gasification kinetic rates follow Arrhenius type dependence on temperature. On the other hand, the kinetic parameters are feedstock specific, and there is no true global relationship to describe the kinetics of char gasification—the characteristics of the char are also feedstock specific. The complexity of the reactions makes the reaction initiation and the subsequent rates subject to many factors, any one of which can influence the kinetic aspects of the reaction.

Although the initial gasification stage (devolatilization) is completed in seconds or even less at elevated temperature, the subsequent gasification of the char produced at the initial gasification stage is much slower, requiring minutes or hours to obtain significant conversion under practical conditions and reactor designs for commercial gasifiers are largely dependent on the reactivity of the char and also on the gasification medium (Johnson, 1979; Sha, 2005). Thus, the distribution and chemical composition of the products are also influenced by the prevailing conditions (i.e., temperature, heating rate, pressure, residence time, etc.) and, last but not least, the nature of the feedstock. Also, the presence of oxygen, hydrogen, water vapor, carbon oxides, and other compounds in the reaction atmosphere during pyrolysis may either support or inhibit numerous reactions with feedstock and with the products evolved.

The reactivity of char produced in the pyrolysis step depends on nature of the feedstock and increases with oxygen content of the feedstock but decreases with carbon content. In general, char produced from a low-carbon feedstock is more reactive than char produced from a high-carbon feedstock. The reactivity of char from a low-carbon feedstock may be influenced by catalytic effect of mineral matter in char. In addition, as the carbon content of the feedstock increases, the reactive functional groups present in the feedstock decrease and the char becomes more aromatic and cross-linked in nature (Speight, 2013). Therefore, char obtained from high-carbon feedstock contains a lesser number of functional groups and higher proportion of aromatic and cross-linked structures, which reduce reactivity. The reactivity of char also depends upon thermal treatment it receives during formation from the parent feedstock—the gasification rate of char decreases as the char preparation temperature increases due to the decrease in active surface areas of char. Therefore, a change of char preparation temperature may change the chemical nature of char, which in turn may change the gasification rate.

Typically, char has a higher surface area compared to the surface area of the parent feedstock, even when the feedstock has been pelletized, and the surface area changes as the char undergoes gasification—the surface area increases with carbon conversion, reaches maximum, and then decreases. These changes in turn affect gasification rates—in general, reactivity increases with the increase in surface area. The initial increase in surface area appears to be caused by cleanup and widening of pores in the char. The decrease in surface area at high carbon conversion may be due to coalescence of pores, which ultimately leads to collapse of the pore structure within the char.

Heat transfer and mass transfer processes in fixed- or moving-bed gasifiers are affected by complex solids flow and chemical reactions. Coarsely crushed feedstock settles while undergoing heating, drying, devolatilization, gasification, and combustion. Also, the feedstock particles change in diameter, shape, and porosity—nonideal behavior may result from certain types of chemical structures in the feedstock, gas bubbles, and channel and a variable void fraction may also change heat and mass transfer characteristics.

An important factor is the importance of the pyrolysis temperature as a major factor in the thermal history, and consequently in the thermodynamics of the feedstock char. However, the thermal history of a char should also depend on the rate of temperature rise to the pyrolysis temperature and on the length of time the char is kept at the pyrolysis temperature (soak time), which might be expected to reduce the residual entropy of the char by employing a longer soak time.

Alkali metal salts are known to catalyze the steam gasification reaction of carbonaceous materials, including coal. The process is based on the concept that alkali metal salts (such as potassium carbonate, sodium carbonate, potassium sulfide, sodium sulfide, and the like) will catalyze the steam gasification of feedstocks. The order of catalytic activity of alkali metals on the gasification reaction is

Cesium (Cs) > rubidium (Rb) > potassium (K) > sodium (Na) > lithium (Li)

Catalyst amounts on the order of 10%–20% w/w potassium carbonate will lower bituminous coal gasifier temperatures from 925°C (1695°F) to 700°C (1090°F) and that the catalyst can be introduced to the gasifier impregnated on coal or char. In addition, tests with potassium carbonate showed that this material also acts as a catalyst for the methanation reaction. In addition, the use of catalysts can reduce the amount of tar formed in the process. In the case of catalytic steam gasification of coal, carbon deposition reaction may affect catalyst life by fouling the catalyst active sites. This carbon deposition reaction is more likely to take place whenever the steam concentration is low.

Ruthenium-containing catalysts are used primarily in the production of ammonia. It has been shown that ruthenium catalysts provide 5–10 times higher reactivity rates than other catalysts. However, ruthenium quickly becomes inactive due to its necessary supporting material, such as activated carbon, which is used to achieve effective reactivity. However, during the process, the carbon is consumed, thereby reducing the effect of the ruthenium catalyst.

Catalysts can also be used to favor or suppress the formation of certain components in the gaseous product by changing the chemistry of the reaction, the rate of reaction, and the thermodynamic balance of the reaction. For example, in the production of synthesis gas (mixtures of hydrogen and carbon monoxide), methane is also produced in small amounts. Catalytic gasification can be used to either promote methane formation or suppress it.

14.2.3.3 Water-Gas Shift Reaction

The water-gas shift reaction (shift conversion) is necessary because the gaseous product from a gasifier generally contains large amounts of carbon monoxide and hydrogen, plus lesser amounts of other gases. Carbon monoxide and hydrogen (if they are present in the mole ratio of 1:3) can be reacted in the presence of a catalyst to produce methane. However, some adjustment to the ideal (1:3) is usually required, and, to accomplish this, all or part of the stream is treated according to the water-gas shift (shift conversion) reaction. This involves reacting carbon monoxide with steam to produce a carbon dioxide and hydrogen whereby the desired 1:3 mole ratio of carbon monoxide to hydrogen may be obtained:

$$CO\ (g) + H_2O\ (g) \rightarrow CO_2\ (g) + H_2\ (g)$$

Even though the water-gas shift reaction is not classified as one of the principal gasification reactions, it cannot be omitted in the analysis of chemical reaction systems that involve synthesis gas. Among all reactions involving synthesis gas, this reaction equilibrium is least sensitive to the temperature variation—the equilibrium constant is least strongly dependent on the temperature. Therefore, the reaction equilibrium can be reversed in a variety of practical process conditions over a wide range of temperature.

The water-gas shift reaction in its forward direction is mildly exothermic and although all of the participating chemical species are in gaseous form, the reaction is believed to be heterogeneous insofar as the chemistry occurs at the surface of the feedstock and the reaction is actually catalyzed by carbon surfaces. In addition, the reaction can also take place homogeneously as well as heterogeneously and a generalized understanding of the water-gas shift reaction is difficult to achieve. Even the published kinetic rate information is not immediately useful or applicable to a practical reactor situation.

Synthesis gas from a gasifier contains a variety of gaseous species other than carbon monoxide and hydrogen. Typically, they include carbon dioxide, methane, and water (steam). Depending on the objective of the ensuing process, the composition of synthesis gas may need to be preferentially readjusted. If the objective of the gasification process is to obtain a high yield of methane, it would be preferred to have the molar ratio of hydrogen to carbon monoxide at 3:1:

$$CO\ (g) + 3H_2\ (g) \rightarrow CH_4\ (g) + H_2O\ (g)$$

On the other hand, if the objective of generating synthesis gas is the synthesis of methanol via a vapor-phase low-pressure process, the stoichiometrically consistent ratio between hydrogen and carbon monoxide would be 2:1. In such cases, the stoichiometrically consistent synthesis gas mixture is often referred to as *balanced gas*, whereas a synthesis gas composition that is substantially deviated from the principal reaction's stoichiometry is called *unbalanced gas*. If the objective of synthesis gas production is to obtain a high yield of hydrogen, it would be advantageous to increase the ratio of hydrogen to carbon monoxide by further converting carbon monoxide (and water) into hydrogen (and carbon dioxide) via the water-gas shift reaction.

The water-gas shift reaction is one of the major reactions in the steam gasification process, where both water and carbon monoxide are present in ample amounts. Although the four chemical species involved in the water-gas shift reaction are gaseous compounds at the reaction stage of most gas processing, the water-gas shift reaction, in the case of steam gasification of feedstock, predominantly takes place on the solid surface of feedstock (heterogeneous reaction). If the product synthesis gas from a gasifier needs to be reconditioned by the water-gas shift reaction, this reaction can be catalyzed by a variety of metallic catalysts.

Choice of specific kinds of catalysts has always depended on the desired outcome, the prevailing temperature conditions, composition of gas mixture, and process economics. Typical catalysts used for the reaction include catalysts containing iron, copper, zinc, nickel, chromium, and molybdenum.

14.2.3.4 Carbon Dioxide Gasification

The reaction of carbonaceous feedstocks with carbon dioxide produces carbon monoxide (*Boudouard reaction*) and (like the steam gasification reaction) is also an endothermic reaction:

$$C\ (s) + CO_2\ (g) \rightarrow 2CO\ (g)$$

The reverse reaction results in carbon deposition (carbon fouling) on many surfaces including the catalysts and results in catalyst deactivation.

This gasification reaction is thermodynamically favored at high temperatures (>680°C, >1255°F), which is also quite similar to the steam gasification. If carried out alone, the reaction requires high temperature (for fast reaction) and high pressure (for higher reactant concentrations) for significant conversion but as a separate reaction a variety of factors come into play: (1) low conversion, (2) slow kinetic rate, and (3) low thermal efficiency.

Also, the rate of the carbon dioxide gasification of a feedstock is different to the rate of the carbon dioxide gasification of carbon. Generally, the carbon–carbon dioxide reaction follows a reaction order based on the partial pressure of the carbon dioxide that is approximately 1.0 (or lower) whereas the feedstock–carbon dioxide reaction follows a reaction order based on the partial pressure of the carbon dioxide that is 1.0 (or higher). The observed higher reaction order for the feedstock reaction is also based on the relative reactivity of the feedstock in the gasification system.

14.2.3.5 Hydrogasification

Not all high-heat-content (high-Btu) gasification technologies depend entirely on catalytic methanation, and, in fact, a number of gasification processes use hydrogasification, that is, the direct addition of hydrogen to feedstock under pressure to form methane.

$$C_{char} + 2H_2 \rightarrow CH_4$$

The hydrogen-rich gas for hydrogasification can be manufactured from steam and char from the hydrogasifier. Appreciable quantities of methane are formed directly in the primary gasifier and the heat released by methane formation is at a sufficiently high temperature to be used in the steam-carbon reaction to produce hydrogen so that less oxygen is used to produce heat for the steam-carbon reaction. Hence, less heat is lost in the low-temperature methanation step, thereby leading to higher overall process efficiency.

Hydrogasification is the gasification of feedstock in the presence of an atmosphere of hydrogen under pressure. Thus, not all high-heat-content (high-Btu) gasification technologies depend entirely on catalytic methanation and, in fact, a number of gasification processes use hydrogasification, that is, the direct addition of hydrogen to feedstock under pressure to form methane:

$$[C]_{feedstock} + H_2 \rightarrow CH_4$$

The hydrogen-rich gas for hydrogasification can be manufactured from steam by using the char that leaves the hydrogasifier. Appreciable quantities of methane are formed directly in the primary gasifier and the heat released by methane formation is at a sufficiently high temperature to be used in the steam-carbon reaction to produce hydrogen so that less oxygen is used to produce heat for the steam-carbon reaction. Hence, less heat is lost in the low-temperature methanation step, thereby leading to higher overall process efficiency.

The hydrogasification reaction is exothermic and is thermodynamically favored at low temperatures (<670°C, <1240°F), unlike the endothermic both steam gasification and carbon dioxide gasification reactions. However, at low temperatures, the reaction rate is inevitably too slow. Therefore, a high temperature is always required for kinetic reasons, which in turn requires high pressure of hydrogen, which is also preferred from equilibrium considerations. This reaction can be catalyzed by salts such as potassium carbonate (K_2CO_3), nickel chloride ($NiCl_2$), iron chloride ($FeCl_2$), and iron sulfate ($FeSO_4$). However, use of a catalyst in feedstock gasification suffers from difficulty in recovering and reusing the catalyst and the potential for the spent catalyst becoming an environmental issue.

In a hydrogen atmosphere at elevated pressure, additional yields of methane or other low-molecular-weight hydrocarbons can result during the initial feedstock gasification stage from direct hydrogenation of feedstock or semichar because of active intermediate formed in the feedstock structure after pyrolysis. The direct hydrogenation can also increase the amount of feedstock carbon that is gasified as well as the hydrogenation of gaseous hydrocarbons, oil, and tar.

The kinetics of the rapid-rate reaction between gaseous hydrogen and the active intermediate depends on hydrogen partial pressure (P_{H2}). Greatly increased gaseous hydrocarbons produced during the initial feedstock gasification stage are extremely important in processes to convert feedstock into methane (synthetic natural gas [SNG], substitute natural gas).

14.2.3.6 Methanation

Several exothermic reactions may occur simultaneously within a methanation unit. A variety of metals have been used as catalysts for the methanation reaction; the most common, and to some extent the most effective methanation catalysts, appear to be nickel and ruthenium, with nickel being the most widely (Cusumano et al., 1978):

Ruthenium (Ru) > Nickel (Ni) > Cobalt (Co) > Iron (Fe) > Molybdenum (Mo).

Nearly all the commercially available catalysts used for this process are, however, very susceptible to sulfur poisoning and efforts must be taken to remove all hydrogen sulfide (H_2S) before the catalytic reaction starts. It is necessary to reduce the sulfur concentration in the feed gas to less than 0.5 ppm v/v in order to maintain adequate catalyst activity for a long period of time.

The synthesis gas must be desulfurized before the methanation step since sulfur compounds will rapidly deactivate (poison) the catalysts. A problem may arise when the concentration of carbon monoxide is excessive in the stream to be methanated since large amounts of heat must be removed from the system to prevent high temperatures and deactivation of the catalyst by sintering as well as the deposition of carbon. To eliminate this problem, temperatures should be maintained below 400°C (750°F).

The methanation reaction is used to increase the methane content of the product gas, as needed for the production of high-Btu gas:

$$4H_2 + CO_2 \rightarrow CH_4 + 2H_2O$$

$$2CO \rightarrow C + CO_2$$

$$CO + H_2O \rightarrow CO_2 + H_2$$

Among these, the most dominant chemical reaction leading to methane is the first one. Therefore, if methanation is carried out over a catalyst with a synthesis gas mixture of hydrogen and carbon monoxide, the desired hydrogen/carbon monoxide ratio of the feed synthesis gas is around 3:1. The large amount of water (vapor) produced is removed by condensation and recirculated as process water or steam. During this process, most of the exothermic heat due to the methanation reaction is also recovered through a variety of energy integration processes.

Whereas all the reactions listed earlier are quite strongly exothermic except the forward water-gas shift reaction, which is mildly exothermic, the heat release depends largely on the amount of carbon monoxide present in the feed synthesis gas. For each 1% v/v carbon monoxide in the feed synthesis gas, an adiabatic reaction will experience a 60°C (108°F) temperature rise, which may be termed as *adiabatic temperature rise*.

14.3 PROCESSES AND FEEDSTOCKS

Gasification is an established tried-and-true method that can be used to convert petroleum coke (petcoke) and other refinery waste streams and residuals (vacuum residual, visbreaker tar, and deasphalter pitch) into power, steam, and hydrogen for use in the production of cleaner transportation fuels. The main requirement for a gasification feedstock is that it contains both hydrogen and carbon. Thus, a number of factors have increased the interest in gasification applications in petroleum refinery operations: (1) coking capacity has increased with the shift to heavier, more sour crude oils being supplied to the refiners; (2) hazardous waste disposal has become a major issue for refiners in many countries; (3) there is strong emphasis on the reduction of emissions of criteria pollutants and greenhouse gases; (4) requirements to produce ultralow sulfur fuels are increasing the hydrogen needs of the refineries; and (5) the requirements to produce low sulfur fuels and other regulations could lead to refiners falling short of demand for light products such as gasoline and jet and diesel fuel. The typical gasification system incorporated into the refinery consists of several process plants including (1) feedstock preparation, (2) the type of gasifier, (3) gas cleaning, and (4) a sulfur recovery unit as well as downstream process options depending on the desired products.

The gasification process can provide high-purity hydrogen for a variety of uses within the refinery. Hydrogen is used in the refinery to remove sulfur, nitrogen, and other impurities from intermediate to finished product streams (Chapter 10) and in hydrocracking operations for the conversion of high-boiling distillates and oils into low-boiling products such as naphtha, kerosene, and diesel (Chapter 11). Furthermore, electric power and high pressure steam can be generated by the gasification of petcoke and residuals to drive mostly small and intermittent loads such as compressors, blowers, and pumps. Steam can also be used for process heating, steam tracing, partial pressure reduction in fractionation systems, and stripping low-boiling components to stabilize process streams. Also, the gasification system and refinery operations can share common process equipment. This usually includes an amine stripper or sulfur plant, wastewater treatment, and cooling water systems (Mokhatab et al., 2006; Speight, 2007, 2014a,b).

14.3.1 Gasifiers

A gasifier differs from a combustor in that the amount of air or oxygen available inside the gasifier is carefully controlled so that only a relatively small portion of the fuel burns completely. The *partial oxidation process* provides the heat and rather than combustion, most of the carbon-containing feedstock is chemically broken apart by the heat and pressure applied in the gasifier resulting in the chemical reactions that produce synthesis gas. However, the composition of the synthesis gas will vary because of dependence upon the conditions in the gasifier and the type of feedstock. Minerals in the fuel (i.e., the rocks, dirt, and other impurities that do not gasify) separate and leave the bottom of the gasifier either as an inert glass-like slag or other marketable solid products.

Four types of gasifier are currently available for commercial use: (1) the countercurrent fixed bed, (2) cocurrent fixed bed, (3) the fluidized bed, and (4) the entrained flow (Speight, 2008).

In a fixed-bed process, the coal is supported by a grate and combustion gases (steam, air, oxygen, etc.) pass through the supported coal whereupon the hot produced gases exit from the top of the reactor. Heat is supplied internally or from an outside source, but caking coals cannot be used in an unmodified fixed-bed reactor. The *countercurrent fixed bed* (*up draft*) gasifier consists of a fixed bed of carbonaceous fuel (e.g., coal or biomass) through which the *gasification agent* (steam, oxygen, and/or air) flows in countercurrent configuration. The ash is either removed dry or as a slag. The nature of the gasifier means that the fuel must have high mechanical strength and must be noncaking so that it will form a permeable bed, although recent developments have reduced these restrictions to some extent. The throughput for this type of gasifier is relatively low. Thermal efficiency is high as the gas exit temperatures are relatively low and, as a result, tar and methane production is significant at typical operation temperatures, so product gas must be extensively cleaned before use or recycled to the reactor.

The *cocurrent fixed bed* (*down draft*) gasifier is similar to the countercurrent type, but the gasification agent gas flows in cocurrent configuration with the fuel (downward, hence the name *down draft gasifier*). Heat needs to be added to the upper part of the bed, either by combusting small amounts of the fuel or from external heat sources. The produced gas leaves the gasifier at a high temperature, and most of this heat is often transferred to the gasification agent added in the top of the bed. Since all tars must pass through a hot bed of char in this configuration, tar levels are much lower than the countercurrent type.

In the *fluidized-bed* gasifier, the fuel is fluidized in oxygen (or air) and steam. The temperatures are relatively low in dry ash gasifiers, so the fuel must be highly reactive; low-grade coals are particularly suitable. The fluidized-bed system uses finely sized coal particles and the bed exhibits liquid-like characteristics when a gas flows upward through the bed. Gas flowing through the coal produces turbulent lifting and separation of particles and the result is an expanded bed having greater coal surface area to promote the chemical reaction, but such systems have a limited ability to handle caking coals. The agglomerating gasifiers have slightly higher temperatures, and are suitable for higher rank coals. Fuel throughput is higher than for the fixed bed, but not as high as for the entrained-flow gasifier. The conversion efficiency is typically low, so recycle or subsequent combustion of solids is necessary to increase conversion. Fluidized-bed gasifiers are most useful for fuels that form highly corrosive ash that would damage the walls of slagging gasifiers. The ash is removed dry or as heavy agglomerates—a disadvantage of biomass feedstocks is that they generally contain high levels of corrosive ash.

In the *entrained-flow* gasifier, a dry pulverized solid, an atomized liquid fuel, or a fuel slurry is gasified with oxygen (much less frequent: air) in cocurrent flow. The high temperatures and pressures also mean that a higher throughput can be achieved but thermal efficiency is somewhat lower as the gas must be cooled before it can be sent to a gas processing facility. All entrained-flow gasifiers remove the major part of the ash as a slag as the operating temperature is well above the ash fusion temperature. The entrained system is suitable for both caking and noncaking coals.

In *integrated gasification combined cycle* (IGCC) systems, the synthesis gas is cleaned of its hydrogen sulfide, ammonia, and particulate matter and is burned as fuel in a combustion turbine (much like natural gas is burned in a turbine). The combustion turbine drives an electric generator. Hot air from the combustion turbine can be channeled back to the gasifier or the air separation unit, while exhaust heat from the combustion turbine is recovered and used to boil water, creating steam for a steam turbine generator. The use of these two types of turbines—a combustion turbine and a steam turbine—in combination, known as a *combined cycle*, is one reason why gasification-based power systems can achieve unprecedented power generation efficiencies.

Gasification also offers more scope for recovering products from waste than incineration (Speight, 2014b). When waste is burnt in an incinerator, the only practical product is energy, whereas the gases, oils, and solid char from pyrolysis and gasification can not only be used as a fuel but also be purified and used as a feedstock for petrochemicals and other applications. Many processes also produce a stable granulate instead of an ash that can be more easily and safely utilized. In addition, some processes are targeted at producing specific recyclables such as metal alloys and carbon black. From waste gasification, in particular, it is feasible to produce hydrogen, which many see as an increasingly valuable resource.

The *Integrated gasification combined cycle* (IGCC) is used to raise power from feedstocks such as vacuum residua, cracked residua, and deasphalter pitch. The value of these refinery residuals, including petroleum coke, will need to be considered as part of an overall upgrading project. Historically, many delayed coking projects have been evaluated and sanctioned on the basis of assigning zero value to petroleum coke having high-sulfur and high-metal content.

While there are many alternate uses for the synthesis gas produced by gasification, a combination of products/utilities can be produced in addition to power. A major benefit of the integrated gasification combined cycle concept is that power can be produced with the lowest sulfur oxide (Sox) and nitrogen oxide (NOx) emissions of any liquid/solid feed power generation technology.

14.3.2 Fischer–Tropsch Synthesis

The synthesis reaction is dependent on a catalyst, mostly an iron or cobalt catalyst where the reaction takes place. There is either a low- or high-temperature process (low-temperature Fischer–Tropsch [LTFT], high-temperature Fischer–Tropsch [HTFT]), with temperatures ranging between 200°C and 240°C for LTFT and 300°C and 350°C for HTFT. The HTFT uses an iron catalyst, and the LTFT either an iron or a cobalt catalyst. The different catalysts include also nickel-based and ruthenium-based catalysts, which also have enough activity for commercial use in the process.

The reactors are the *multitubular fixed-bed*, the *slurry*, or the *fluidized-bed* (with either fixed or circulating bed) reactor. The fixed-bed reactor consists of thousands of small tubes with the catalyst as surface-active agent in the tubes. Water surrounds the tubes and regulates the temperature by settling the pressure of evaporation. The slurry reactor is widely used and consists of fluid and solid elements, where the catalyst has no particular position, but flows around as small pieces of catalyst together with the reaction components. The slurry and fixed-bed reactor are used in LTFT. The fluidized-bed reactors are diverse, but characterized by the fluid behavior of the catalyst.

The *high-temperature* Fischer–Tropsch technology uses a fluidized catalyst at 300°C–330°C. Originally circulating fluidized-bed units were used (Synthol reactors). Since 1989, a commercial scale classical fluidized-bed unit has been implemented and improved upon. The *low-temperature* Fischer–Tropsch technology has originally been used in tubular fixed-bed reactors at 200°C–230°C. This produces a more paraffinic and waxy product spectrum than the *high-temperature* technology. A new type of reactor (the Sasol slurry phase distillate reactor) has been developed and is in commercial operation. This reactor uses a slurry phase system rather than a tubular fixed-bed configuration and is currently the favored technology for the commercial production of synfuels.

Under most circumstances the production of synthesis gas by reforming natural gas will be more economical than from coal gasification, but site-specific factors need to be considered. In fact, any technological advance in this field (such as better energy integration or the oxygen transfer ceramic membrane reformer concept) will speed up the rate at which the synfuels technology will become common practice.

There are large coal reserves that may increasingly be used as a fuel source during oil depletion. Since there are large coal reserves in the world, this technology could be used as an interim transportation fuel if conventional oil were to become more expensive. Furthermore, combination of biomass gasification and Fischer–Tropsch synthesis is a very promising route to produce transportation fuels from renewable or *green* resources.

Although the focus of this section has been on the production of hydrocarbons from synthesis gas, it is worthy of note that clean synthesis gas can also be used (1) as chemical *building blocks* to produce a broad range of chemicals using processes well established in the chemical and petrochemical industry; (2) as a fuel producer for highly efficient fuel cells (which run off the hydrogen made in a gasifier) or perhaps, in the future, hydrogen turbines and fuel cell-turbine hybrid systems; and (3) as a source of hydrogen that can be separated from the gas stream and used as a fuel or as a feedstock for refineries (which use the hydrogen to upgrade petroleum products).

The aim of underground (or *in situ*) gasification of coal is the conversion into combustible gases by combustion of a coal seam in the presence of air, oxygen, or oxygen and steam. Thus, seams that were considered to be inaccessible, unworkable, or uneconomical to mine could be put to use. In addition, strip mining and the accompanying environmental impacts, the problems of spoil banks, acid mine drainage, and the problems associated with use of high-ash coal are minimized or even eliminated.

The principles of underground gasification are very similar to those involved in the aboveground gasification of coal. The concept involves the drilling and subsequent linking of two boreholes so that gas will pass between the two. Combustion is then initiated at the bottom of one borehole (injection well) and is maintained by the continuous injection of air. In the initial reaction zone (combustion zone), carbon dioxide is generated by the reaction of oxygen (air) with the coal:

$$[C]_{coal} + O_2 \rightarrow CO_2$$

The carbon dioxide reacts with coal (partially devolatilized) further along the seam (reduction zone) to produce carbon monoxide:

$$[C]_{coal} + CO_2 \rightarrow 2CO$$

In addition, at the high temperatures that can frequently occur, moisture injected with oxygen or even moisture inherent in the seam may also react with the coal to produce carbon monoxide and hydrogen:

$$[C]_{coal} + H_2O \rightarrow CO + H_2$$

The gas product varies in character and composition but usually falls into the low-heat (low-Btu) category ranging from 125 to 175 Btu/ft^3.

14.3.3 FEEDSTOCKS

For many decades, coal has been the primary feedstock for gasification units, but due to recent concerns about the use of fossil fuels and the resulting environmental pollutants, irrespective of the various gas cleaning processes and gasification plant environmental cleanup efforts, there is a move to feedstocks other than coal for gasification processes (Speight, 2013, 2014b). But more pertinent to the present text, the gasification process can also use carbonaceous feedstocks that would otherwise have been discarded and unused, such as waste biomass and other similar biodegradable wastes. Various feedstocks such as biomass, petroleum residues, and other carbonaceous wastes can be used to their fullest potential. In fact, the refining industry has seen fit to use residua gasification as a source of hydrogen for the past several decades (Speight, 2014a).

Gasification processes can accept a variety of feedstocks, but the reactor must be selected on the basis of feedstock properties and behavior in the process. The advantage of the gasification process when a carbonaceous feedstock (a feedstock containing carbon) or hydrocarbonaceous feedstock (a feedstock containing carbon and hydrogen) is employed is that the product of focus—synthesis gas—is potentially more useful as an energy source and results in an overall cleaner process. The production of synthesis gas is a more efficient production of an energy source than, say, the direct combustion of the original feedstock because the synthesis gas can be (1) combusted at higher temperatures, (2) used in fuel cells, (3) used to produce methanol, and (4) used as a source of hydrogen and (5) particularly because the synthesis gas can be converted via the Fischer–Tropsch process into a range of synthesis liquid fuels suitable for use in gasoline engines, for diesel engines, or for wax production.

14.3.3.1 Residua

Gasification is the only technology that makes possible a zero residue target for refineries, contrary to all conversion technologies (including thermal cracking, catalytic cracking, cooking, deasphalting, hydroprocessing), which can only reduce the bottom volume, with the complication that the residue qualities generally get worse with the degree of conversion (Speight, 2014a).

The flexibility of gasification permits to handle any type of refinery residue, including petroleum coke, tank bottoms, and refinery sludge, and make available a range of value-added products including electricity, steam, hydrogen, and various chemicals based on synthesis gas chemistry: methanol, ammonia, methyl tertiary-butyl ether (MTBE), tertiary amyl methyl ester (TAME), acetic acid, and formaldehyde (Speight, 2008, 2013). The environmental performance of gasification is unmatched. No other technology processing low-value refinery residues can come close to the emission levels achievable with gasification (Speight, 2013, 2014a).

Gasification is also a method for converting petroleum coke and other refinery nonvolatile waste streams (often referred to as *refinery residuals* and include but not limited to atmospheric residuum, vacuum residuum, visbreaker tar, and deasphalter pitch) into power, steam, and hydrogen for use in the production of cleaner transportation fuels. The main requirement for a gasification feedstock is that it contains both hydrogen and carbon and several suitable feedstocks are produced on-site as part of typical refinery processing (Speight, 2011b). The typical gasification system incorporated into a refinery consists of several process units including feed preparation, the gasifier, an air separation unit (ASU), syngas cleanup, sulfur recovery unit (SRU), and downstream process options depending on target products.

The benefits of the addition of a gasification system in a refinery to process petroleum coke or other residuals include (1) production of power, steam, oxygen, and nitrogen for refinery use or sale; (2) source of syngas for hydrogen to be used in refinery operations as well as for the production of light refinery products through FT synthesis; (3) increased efficiency of power generation, improved air emissions, and reduced waste stream versus combustion of petroleum coke or residues or incineration; (4) no off-site transportation or storage for petroleum coke or residuals; and (5) the potential to dispose of waste streams including hazardous materials (Marano, 2003).

Gasification can provide high-purity hydrogen for a variety of uses within the refinery (Speight, 2014a). Hydrogen is used in refineries to remove sulfur, nitrogen, and other impurities from intermediate to finished product streams and in hydrocracking operations for the conversion of heavy distillates and oils into light products, naphtha, kerosene, and diesel fuel. Hydrocracking and severe hydrotreating require hydrogen that is at least 99% v/v, while less severe hydrotreating can work with gas streams containing 90% v/v pure hydrogen.

Electric power and high pressure steam can be generated via gasification of petroleum coke and residuals to drive mostly small and intermittent loads such as compressors, blowers, and pumps. Steam can also be used for process heating, steam tracing, partial pressure reduction in fractionation systems, and stripping low-boiling components to stabilize process streams.

Carbon soot is produced during gasification, which ends up in the quench water. The soot is transferred to the feedstock by contacting, in sequence, the quench water blowdown with naphtha, and then the naphtha-soot slurry with a fraction of the feed. The soot mixed with the feed is finally recycled into the gasifier, thus achieving 100% conversion of carbon to gas.

14.3.3.2 Petroleum Coke

Coke is the solid carbonaceous material produced from petroleum during thermal processing. More particularly, coke is the residue left by the destructive distillation (i.e., thermal cracking such as the delayed coking process) of petroleum residua. The coke formed in catalytic cracking operations is usually nonrecoverable because of the materials deposited on the catalyst during the process and such coke is often employed as fuel for the process (Speight, 2014a). It is often characterized as a solid material with a honeycomb-type of appearance having high carbon content (95%+ w/w) with some hydrogen and, depending on the process, as well as sulfur and nitrogen. The color varies from gray to black, and the material is insoluble in organic solvents.

Typically, the composition of petroleum coke varies with the source of the crude oil, but in general, large amounts of high-molecular-weight complex hydrocarbons (rich in carbon but correspondingly poor in hydrogen) make up a high proportion. The solubility of petroleum *coke* in carbon disulfide has been reported to be as high as 50%–80%, but this is in fact a misnomer and is due to soluble product adsorbed on the coke—by definition coke is the insoluble, honeycomb material that is the end product of thermal processes. However, coke is not always a product with little use—three physical structures of coke can be produced by delayed coking: (1) shot coke, (2) sponge coke, or (3) needle coke, which find different uses within the industry.

Shot coke is an abnormal type of coke resembling small balls. Due to mechanisms not well understood, the coke from some coker feedstocks forms into small, tight, nonattached clusters that look like pellets, marbles, or ball bearings. It usually is a very hard coke, that is, low Hardgrove Grindability Index (Speight, 2013). Such coke is less desirable to the end users because of difficulties in handling and grinding. It is believed that feedstocks high in asphaltene constituents and low API favor shot coke formation. Blending aromatic materials with the feedstock and/or increasing the recycle ratio reduces the yield of shot coke. Fluidization in the coke drums may cause formation of shot coke. Occasionally, the smaller *shot coke* may agglomerate into ostrich egg-sized pieces. Such coke may be more suitable as a gasification feedstock.

Sponge coke is the common type of coke produced by delayed coking units. It is in a form that resembles a sponge and has been called honeycombed. Sponge coke, mostly used for anode-grade carbon, is dull and black, having porous, amorphous structure. *Needle coke (acicular coke)* is a special quality coke produced from aromatic feedstocks is silver gray, having crystalline broken needle structure, and is believed to be chemically produced through cross-linking of condensed aromatic hydrocarbons during coking reactions. It has a crystalline structure with more unidirectional pores and is used in the production of electrodes for the steel and aluminum industries and is particularly valuable because the electrodes must be replaced regularly.

Petroleum coke is employed for a number of purposes, but its chief use is (depending upon the degree of purity—i.e., contains a low amount of contaminants) for the manufacture of carbon electrodes for aluminum refining, which requires a high-purity carbon—low in ash and sulfur free; the volatile matter must be removed by calcining. In addition to its use as a metallurgical reducing agent, petroleum coke is employed in the manufacture of carbon brushes, silicon carbide abrasives, and structural carbon (e.g., pipes and Raschig rings), as well as calcium carbide manufacture from which acetylene is produced:

$$Coke \rightarrow CaC_2$$

$$CaC_2 + H_2O \rightarrow HC{\equiv}CH$$

Considering the properties of coke and the potential nonuse of the highly contaminated material, gasification is the only technology that makes possible for the refineries the zero residue target, contrary to all conversion technologies, thermal cracking, catalytic cracking, cooking, deasphalting, hydroprocessing, and so on, which can only reduce the bottom volume, with the complication that the residue qualities generally get worse with the degree of conversion (Speight, 2014a).

Indeed, the flexibility of the gasification technology permits the refinery to handle any kind of refinery residue, including petroleum coke, tank bottoms, and refinery sludge, and makes available a range of value-added products, electricity, steam, hydrogen, and various chemicals based on synthesis gas chemistry: methanol, ammonia, MTBE, TAME, acetic acid, and formaldehyde (Speight, 2008, 2013). With respect to gasification, no other technology processing low-value refinery residues can come close to the emission levels achievable with gasification (Speight, 2014a) and is projected to be a major part of the refinery of the future (Speight, 2011b).

Gasification is also a method for converting petroleum coke and other refinery nonvolatile waste streams (often referred to as *refinery residuals* and include but not limited to atmospheric residuum, vacuum residuum, visbreaker tar, and deasphalter pitch) into power, steam, and hydrogen for use in the production of cleaner transportation fuels. And as for the gasification of coal and biomass (Speight, 2013; Luque and Speight, 2015), the main requirement for a feedstock to a gasification unit is that the feedstock contains both hydrogen and carbon, of which a variety of feedstocks are available from the throughput of a typical refinery (Table 14.3).

The typical gasification system incorporated into the refinery consists of several process plants including (1) feed preparation, (2) the gasifier, (3) an air separation unit (ASU), (4) synthesis gas cleanup, (5) sulfur recovery unit (SRU), and (6) downstream process options such as

TABLE 14.3

Types of Feedstocks Produced On-Site That Are Available for Gasification

	Units	Vacuum Residue	Visbreaker Tar	Asphalt	Petcoke
Ultimate analysis					
C	wt/wt	84.9%	86.1%	85.1%	88.6%
H	"	10.4%	10.4%	9.1%	2.8%
N[a]	"	0.5%	0.6%	0.7%	1.1%
S[a]	"	4.2%	2.4%	5.1%	7.3%
O	"		0.5%		0.0%
Ash	"	0.0%		0.1%	0.2%
Total	wt/wt	100.0%	100.0%	100.0%	100.0%
H_2/C ratio	mol/mol	0.727	0.720	0.640	0.188
Density					
Specific gravity	60°/60°	1.028	1.008	1.070	0.863
API gravity	°API	6.2	8.88	0.8	*
Heating values					
HHV (dry)	M Btu/lb	17.72	18.6	17.28	14.85
LHV (dry)	"	16.77	17.6	16.45	14.48

Source: National Energy Technology Laboratory, U.S. Department of Energy, Washington, DC, http://www.netl.doe.gov/technologies/coalpower/gasification/gasifipedia/7-advantages/7-3-4_refinery.html, accessed July 21, 2015.

[a] Nitrogen and sulfur contents vary widely.

* Data not available.

Fischer–Tropsch synthesis (FTS) and methanol synthesis (MTS), depending on the desired product slate (Figure 14.3).

The benefits to a refinery for adding a gasification system for petroleum coke or other residuals are (1) production of power, steam, oxygen, and nitrogen for refinery use or sale; (2) source of synthesis gas for hydrogen to be used in refinery operations and for the production of light refinery

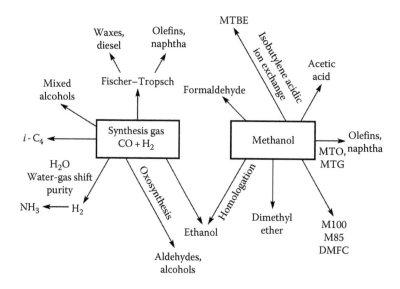

FIGURE 14.3 Potential products from coal gasification.

products through Fisher-Tropsch synthesis; (3) increased efficiency of power generation, improved air emissions, and reduced waste stream versus combustion of petroleum coke or residua or incineration; (4) no off-site transportation or storage for petroleum coke or residuals; and (5) the potential to dispose of waste streams including hazardous materials.

Gasification of coke can provide high-purity hydrogen for a variety of uses within the refinery such as (1) sulfur removal, (2) nitrogen removal, and (3) removal of other impurities from intermediate to finished product streams and in hydrocracking operations for the conversion of heavy distillates and oils into light products, naphtha, kerosene, and diesel fuel (Speight, 2014a). Hydrocracking and severe hydrotreating require hydrogen that is at least 99% v/v pure, while less severe hydrotreating requires a gas stream containing hydrogen on the order of 90% v/v purity.

Electric power and high pressure steam can be generated by the gasification of petroleum coke and residuals to drive mostly small and intermittent loads such as compressors, blowers, and pumps. Steam can also be used for process heating, steam tracing, partial pressure reduction in fractionation systems, and stripping low-boiling components to stabilize process streams.

During gasification some soot (typically 99%+ carbon) is produced, which ends up in the quench water. The soot is transferred to the feedstock by contacting, in sequence, the quench water blowdown with naphtha, and then the naphtha-soot slurry with a fraction of the feed. The soot mixed with the feed is recycled to the gasifier, thus achieving 100% conversion of carbon to gas.

14.3.3.3 Asphalt, Tar, and Pitch

Asphalt does not occur naturally but is manufactured from petroleum and is a black or brown material that has a consistency varying from a viscous liquid to a glassy solid (Speight, 2014a). To a point, asphalt can resemble bitumen (isolated form tar sand formation), hence the tendency to refer to bitumen (incorrectly) as *native asphalt*. It is recommended that there be differentiation between asphalt (manufactured) and bitumen (naturally occurring) other than by use of the qualifying terms *petroleum* and *native* since the origins of the materials may be reflected in the resulting physicochemical properties of the two types of materials. It is also necessary to distinguish between the asphalt that originates from petroleum by refining and the product in which the source of the asphalt is a material other than petroleum, for example, *Wurtzilite asphalt* (Speight, 2014a). In the absence of a qualifying word, it should be assumed that the word *asphalt* (with or without qualifiers such as *cutback*, *solvent*, and *blown*, which indicate the process used to produce the asphalt) refers to the product manufactured from petroleum.

When the asphalt is produced simply by distillation of an asphaltic crude oil, the product can be referred to as *residual asphalt* or *straight-run asphalt*. For example, if the asphalt is prepared by *solvent* extraction of residua or by light hydrocarbon (propane) precipitation, or if *blown* or otherwise treated, the term should be modified accordingly to qualify the product (e.g., *solvent asphalt*, *propane asphalt*, *blown asphalt*).

Asphalt softens when heated and is elastic under certain conditions and has many uses. For example, the mechanical properties of asphalt are of particular significance when it is used as a binder or adhesive. The principal application of asphalt is in road surfacing, which may be done in a variety of ways. Light oil *dust layer* treatments may be built up by repetition to form a hard surface, or a granular aggregate may be added to an asphalt coat, or earth materials from the road surface itself may be mixed with the asphalt. Other important applications of asphalt include canal and reservoir linings, dam facings, and sea works. The asphalt so used may be a thin, sprayed membrane, covered with earth for protection against weathering and mechanical damage, or thicker surfaces, often including riprap (crushed rock). Asphalt is also used for roofs, coatings, floor tiles, soundproofing, waterproofing, and other building-construction elements and in a number of industrial products, such as batteries. For certain applications an asphaltic emulsion is prepared, in which fine globules of asphalt are suspended in water.

Tar is a product of the destructive distillation of many bituminous or other organic materials and is a brown to black, oily, viscous liquid to semi-solid material. However, *tar* is most commonly produced from *bituminous coal* and is generally understood to refer to the product from coal, although it is advisable to specify *coal tar* if there is the possibility of ambiguity. The most important factor in determining the yield and character of the coal tar is the carbonizing temperature. Three general temperature ranges are recognized, and the products have acquired the designations: *low-temperature tar* (approximately 450°C–700°C; 540°F–1290°F), *mid-temperature tar* (approximately 700°C–900°C; 1290°F–1650°F), and *high-temperature tar* (approximately 900°C–1200°C; 1650°F–2190°F). Tar released during the early stages of the decomposition of the organic material is called "primary tar" since it represents a product that has been recovered without the secondary alteration that results from prolonged residence of the vapor in the heated zone.

Treatment of the distillate (boiling up to 250°C, 480°F) of the tar with caustic soda causes separation of a fraction known as *tar acids*; acid treatment of the distillate produces a variety of organic nitrogen compounds known as *tar bases*. The residue left following removal of the heavy oil, or distillate, is *pitch*, a black, hard, and highly ductile material.

In the chemical-process industries, pitch is the black or dark brown residue obtained by distilling coal tar, wood tar, fats, fatty acids, or fatty oils.

Coal tar pitch is a soft to hard and brittle substance containing chiefly aromatic resinous compounds along with aromatic and other hydrocarbons and their derivatives; it is used chiefly as road tar, in waterproofing roofs and other structures, and to make electrodes. *Wood tar pitch* is a bright, lustrous substance containing resin acids; it is used chiefly in the manufacture of plastics and insulating materials and in caulking seams. *Pitch* derived from fats, fatty acids, or fatty oils by distillation are usually soft substances containing polymers and decomposition products; they are used chiefly in varnishes and paints and in floor coverings.

Any of the aforementioned derivatives can be used as a gasification feedstock. The properties of asphalt change markedly during the aging process (oxidation in service) to the point where the asphalt fails to perform the task for which it was designed. In some cases, the asphalt is recovered and reprocessed for additional use or it may be sent to a gasifier.

14.3.3.4 Tar Sand Bitumen

Tar sand bitumen is used interchangeably with the term *oil sand bitumen* in Canada, and the word *bitumen* (also, on occasion, incorrectly referred to as *native asphalt*, and *extra heavy oil*) includes a wide variety of reddish brown to black materials of semisolid, viscous to brittle character that can exist in nature with no mineral impurity or with a mineral matter content that exceeds 50% w/w (Speight, 2014a). Bitumen is frequently found filling pores and crevices of sandstone, limestone, or argillaceous sediments (sediments containing, made of, or resembling clay; clayey), in which case the organic and associated mineral matrix is known as *rock asphalt* (Speight, 2014a).

Bitumen is a naturally occurring material that is found in deposits where the permeability is low and passage of fluids through the deposit can only be achieved by prior application of fracturing techniques. Tar sand bitumen is a high-boiling material with little, if any, material boiling below 350°C (660°F) and the boiling range may approximate the boiling range of an atmospheric residuum (or in some cases, the boiling range of a vacuum residuum).

For clarification and legal purposes, *tar sands* have been defined in the United States through FE 76-4 as

> … the several rock types that contain an extremely viscous hydrocarbon which is not recoverable in its natural state by conventional oil well production methods including currently used enhanced recovery techniques. The hydrocarbon-bearing rocks are variously known as bitumen-rocks oil, impregnated rocks, oil sands, and rock asphalt.

The recovery of the bitumen depends to a large degree on the composition and construction of the sands. Generally, the bitumen found in tar sand deposits is an extremely viscous material that is *immobile under reservoir conditions* and cannot be recovered through a well by the application of secondary or enhanced recovery techniques.

The expression *tar sand* is commonly used in the petroleum industry to describe sandstone reservoirs that are impregnated with a heavy, viscous black crude oil that cannot be retrieved through a well by conventional production techniques (FE 76-4). However, the term *tar sand* is actually a misnomer; more correctly, the name *tar* is usually applied to the heavy product remaining after the destructive distillation of coal or other organic matter (Speight, 2013). Similarly, the terms *oil sand* and *oil shale* are misnomers insofar as these deposits or formations do not contain oil but are oil producing through thermal treatment.

Alternative names, such as *bituminous sand* or *oil sand*, are gradually finding usage, with the former name (bituminous sands) more technically correct. The term *oil sand* is also used in the same way as the term *tar sand*, and these terms are used interchangeably throughout this text.

The bitumen in tar sand formations requires a high degree of thermal stimulation for recovery to the extent that some thermal decomposition may have to be induced. Current recovery operations of bitumen in tar sand formations involve use of a mining technique and nonmining techniques are continually being developed (Speight, 2014a).

Pitch Lake is the name that has been applied to large surface deposit of bitumen. Guanoco Lake in Venezuela covers more than 1,100 acres (445 hectares) and contains an estimated 35,000,000 barrel of bitumen. It was used as a commercial source of asphalt from 1891 to 1935. Smaller deposits occur commonly where Tertiary marine sediments outcrop on the surface; an example is the tar pits at Rancho La Brea in Los Angeles (*brea* and *tar* have been used synonymously with *bitumen*). Although most pitch lakes are fossils of formerly active seeps, some, such as the *Pitch Lake* on the island of Trinidad (also called the *Trinidad Asphalt Lake*), continue to be supplied with fresh crude oil seeping from a subterranean source. The Trinidad *Pitch Lake* covers 115 acres and contains an estimated 40,000,000 bbl of bitumen.

14.3.3.5 Coal

Coal is a fossil fuel formed in swamp ecosystems where plant remains were saved from oxidation and biodegradation by water and mud. Coal is a combustible organic sedimentary rock (composed primarily of carbon, hydrogen, and oxygen, as well as other minor elements including sulfur) formed from ancient vegetation and consolidated between other rock strata to form coal seams. The harder forms can be regarded as organic metamorphic rock (e.g., anthracite coal) because of a higher degree of maturation.

Coal is the largest single source of fuel for the generation of electricity worldwide, as well as the largest source of carbon dioxide emissions, which have been implicated as the primary cause of global climate change, although the debate still rages as to the actual cause (or causes) of climate change. Coal is found as successive layers, or seams, sandwiched between strata of sandstone and shale and extracted from the ground by coal mining—either underground coal seams (underground mining) or open-pit mining (surface mining).

Coal remains in adequate supply and at current rates of recovery and consumption, the world global coal reserves have been variously estimated to have a reserves/production ratio of at least 155 years. However, as with all estimates of resource longevity, coal longevity is subject to the assumed rate of consumption remaining at the current rate of consumption and, moreover, to technological developments that dictate the rate at which the coal can be mined. But most importantly, coal is a fossil fuel and an *unclean* energy source that will only add to global warming. In fact, the next time electricity is advertised as a clean energy source, just consider the means by which the majority of electricity is produced—almost 50% of the electricity generated in the United States is derived from coal (EIA, 2007; Speight, 2013).

Coal occurs in different forms or *types* (Speight, 2013). Variations in the nature of the source material and local or regional variations in the coalification processes cause the vegetal matter to evolve differently. Various classification systems thus exist to define the different types of coal.

Using the ASTM system of classification (ASTM D388), the coal precursors are transformed over time (as geological processes increase their effect over time) into the following:

1. *Lignite*: Also referred to as brown coal, this is the lowest rank of coal and used almost exclusively as fuel for steam-electric power generation. Jet is a compact form of lignite that is sometimes polished and has been used as an ornamental stone since the Iron Age.
2. *Subbituminous coal*: The properties range from those of lignite to those of bituminous coal and is used primarily as fuel for steam-electric power generation.
3. *Bituminous coal*: A dense coal, usually black, sometimes dark brown, often with well-defined bands of brittle and dull material, used primarily as fuel in steam-electric power generation, with substantial quantities also used for heat and power applications in manufacturing and to make coke.
4. *Anthracite*: The highest rank; a harder, glossy, black coal used primarily for residential and commercial space heating.

Chemically, coal is a hydrogen-deficient hydrocarbon with an atomic hydrogen-to-carbon ratio near 0.8, as compared to petroleum hydrocarbons, which have an atomic hydrogen-to-carbon ratio approximately equal to 2, and methane (CH_4) that has an atomic carbon-to-hydrogen ratio equal to 4. For this reason, any process used to convert coal to alternative fuels must add hydrogen or redistribute the hydrogen in the original coal to generate hydrogen-rich products and coke (Speight, 2013).

The chemical composition of the coal is defined in terms of its proximate and ultimate (elemental) analyses (Speight, 2013). The parameters of proximate analysis are moisture, volatile matter, ash, and fixed carbon. Elemental analysis (ultimate analysis) encompasses the quantitative determination of carbon, hydrogen, nitrogen, sulfur, and oxygen within the coal. Additionally, specific physical and mechanical properties of coal and particular carbonization properties are also determined.

Carbon monoxide and hydrogen are produced by the gasification of coal in which a mixture of gases is produced. In addition to carbon monoxide and hydrogen, methane and other hydrocarbons are also produced depending on conditions. Gasification may be accomplished either *in situ* or in processing plants. *In situ* gasification is accomplished by controlled, incomplete burning of a coalbed underground while adding air and steam. The gases are withdrawn and may be burned to produce heat, generate electricity, or are utilized as syngas in indirect liquefaction as well as for the production of chemicals.

Producing diesel and other fuels from coal can be performed through the conversion of coal to syngas, a combination of carbon monoxide, hydrogen, carbon dioxide, and methane. Syngas is subsequently reacted through Fischer–Tropsch synthesis processes to produce hydrocarbons that can be refined into liquid fuels. By increasing the quantity of high-quality fuels from coal (while reducing costs), research into this process could help mitigate the dependence on ever-increasingly expensive and depleting stocks of petroleum.

While coal is an abundant natural resource, its combustion or gasification produces both toxic pollutants and greenhouse gases. By developing adsorbents to capture the pollutants (mercury, sulfur, arsenic, and other harmful gases), scientists are striving not only to reduce the quantity of emitted gases but also to maximize the thermal efficiency of the cleanup.

Gasification thus offers one of the cleanest and versatile ways to convert the energy contained in coal into electricity, hydrogen, and other sources of power. Turning coal into syngas is not a new concept; in fact, the basic technology dates back to pre–World War II. In fact, a gasification unit can process virtually all the residua and wastes that are produced in refineries leading to enhanced yields of high-value products (and hence their competitiveness in the market) by deeper upgrading of their crude oil.

14.3.3.6 Biomass

Biomass can be considered as any renewable feedstock which would, in principle, be *carbon neutral* (while the plant is growing, it uses the sun's energy to absorb the same amount of carbon from the atmosphere as it releases into the atmosphere).

Raw materials that can be used to produce biomass-derived fuels are widely available; they come from a large number of different sources and in numerous forms (Rajvanshi, 1986). The main basic sources of biomass include (1) wood, including bark, logs, sawdust, wood chips, wood pellets, and briquettes; (2) high yield energy crops, such as wheat, grown specifically for energy applications; (3) agricultural crops and residues (e.g., straw); and (4) industrial waste, such as wood pulp or paper pulp. For processing, a simple form of biomass such as untreated and unfinished wood may be converted into a number of physical forms, including pellets and wood chips, for use in biomass boilers and stoves.

Biomass includes a wide range of materials that produce a variety of products that are dependent upon the feedstock (Balat, 2011; Demirbaş, 2011; Ramroop Singh, 2011; Speight, 2011a). In addition, the heat content of the different types of biomass widely varies and has to be taken into consideration when designing any conversion process (Jenkins and Ebeling, 1985).

Thermal conversion processes use heat as the dominant mechanism to convert biomass into another chemical form. The basic alternatives of combustion, torrefaction, pyrolysis, and gasification are separated principally by the extent to which the chemical reactions involved are allowed to proceed (mainly controlled by the availability of oxygen and conversion temperature) (Speight, 2011a).

Energy created by burning biomass (fuel wood), also known as dendrothermal energy, is particularly suited for countries where fuel wood grows more rapidly, for example, tropical countries. There is a number of other less common, more experimental or proprietary thermal processes that may offer benefits including hydrothermal upgrading and hydroprocessing. Some have been developed to be compatible with high moisture content biomass (e.g., aqueous slurries) and allow them to be converted into more convenient forms.

Some of the applications of thermal conversion are combined heat and power (CHP) and cofiring. In a typical dedicated biomass power plant, efficiencies range from 7% to 27%. In contrast, biomass cofiring with coal typically occurs at efficiencies close to those of coal combustors (30%–40%) (Baxter, 2005; Liu et al., 2011).

Many forms of biomass contain a high percentage of moisture (along with carbohydrates and sugars) and mineral constituents—both of which can influence the economics and viability of a gasification process. The presence of high levels of moisture in biomass reduces the temperature inside the gasifier, which then reduces the efficiency of the gasifier. Many biomass gasification technologies therefore require dried biomass to reduce the moisture content prior to feeding into the gasifier. In addition, biomass can come in a range of sizes. In many biomass gasification systems, biomass must be processed to a uniform size or shape to be fed into the gasifier at a consistent rate as well as to maximize gasification efficiency.

Biomass such as wood pellets, yard and crop waste, and "energy crops" including switch grass and waste from pulp and paper mills can also be employed to produce bioethanol and synthetic diesel. Biomass is first gasified to produce syngas and then subsequently converted via catalytic processes to the aforementioned downstream products. Biomass can also be used to produce electricity—either blended with traditional feedstocks, such as coal or by itself (Shen et al., 2012; Khosravil and Khadse, 2013; Speight, 2014b).

Most biomass gasification systems use air instead of oxygen for gasification reactions (which is typically used in large-scale industrial and power gasification plants). Gasifiers that use oxygen require an air separation unit to provide the gaseous/liquid oxygen; this is usually not cost-effective at the smaller scales used in biomass gasification plants. Air-blown gasifiers utilize oxygen from air for gasification processes.

In general, biomass gasification plants are comparatively smaller to those of typical coal or petroleum coke plants used in the power, chemical, fertilizer, and refining industries. As such, they are less expensive to build and have a smaller environmental footprint. While a large industrial gasification plant may take up 150 acres of land and process 2,500–15,000 tons per day of feedstock (e.g., coal or petroleum coke), smaller biomass plants typically process 25–200 tons of feedstock per day and take up less than 10 acres.

Finally, while biomass may seem to some observers to be the answer to the global climate change issue, advantages and disadvantages of biomass as feedstock must be considered carefully:

Advantages: (1) Theoretically inexhaustible fuel source; (2) minimal environmental impact when direct combustion of plant mass is not used to generate energy (i.e., fermentation, pyrolysis, etc., are used instead); (3) efficient alcohols and other fuels produced by biomass, viable, and relatively clean burning; and (4) available on a worldwide basis.

Disadvantages: (1) Could contribute a great deal to global climate change and particulate pollution if combusted directly; (2) remains an expensive source of energy, both in terms of producing biomass and the technological conversion to alcohols or other fuels; and (3) life cycle assessments be taken into account to address energy inputs and outputs (but there is most likely a net loss of energy when operated on a small scale [as energy must be put in to grow the plant mass])

And while taking the issues of global climate change into account, it must not be ignored that the Earth is in an interglacial period when warming will take place. The extent of this warming is not known—no one was around to measure the temperature change in the last interglacial period—and by the same token the contribution of anthropological sources to global climate change cannot be measured accurately.

14.3.3.7 Solid Waste

Waste may be municipal solid waste (MSW) that had minimal presorting, or refuse-derived fuel (RDF) with significant pretreatment, usually mechanical screening and shredding. Other more specific waste sources (excluding hazardous waste) and possibly including petroleum coke may provide niche opportunities for co-utilization (Arena, 2012; Speight, 2013, 2014b).

The traditional waste-to-energy plant, based on mass-burn combustion on an inclined grate, has a low public acceptability despite the very low emissions achieved over the last decade with modern flue gas cleanup equipment. This has led to difficulty in obtaining planning permissions to construct needed new waste to energy plants. After much debate, various governments have allowed options for advanced waste conversion technologies (gasification, pyrolysis, and anaerobic digestion), but will only give credit to the proportion of electricity generated from nonfossil waste.

Use of waste materials as cogasification feedstocks may attract significant disposal credits (Ricketts et al., 2002). Cleaner biomass materials are renewable fuels and may attract premium prices for the electricity generated. Availability of sufficient fuel locally for an economic plant size is often a major issue, as is the reliability of the fuel supply. Use of more-predictably available coal alongside these fuels overcomes some of these difficulties and risks. Coal could be regarded as the base feedstock that keeps the plant running when the fuels producing the better revenue streams are not available in sufficient quantities.

Coal characteristics are very different to younger hydrocarbon fuels such as biomass and waste. Hydrogen-to-carbon ratios are higher for younger fuels, as is the oxygen content. This means that reactivity is very different under gasification conditions. Gas cleaning issues can also be very different, being sulfur a major concern for coal gasification and chlorine compounds and tars more important for waste and biomass gasification. There are no current proposals for adjacent gasifiers and gas cleaning systems, one handling biomass or waste and one coal, alongside each other and feeding the same power production equipment. However, there are some advantages to such a design as compared with mixing fuels in the same gasifier and gas cleaning systems.

Electricity production or combined electricity and heat production remain the most likely area for the application of gasification or cogasification. The lowest investment cost per unit of electricity generated is the use of the gas in an existing large power station. This has been done in several large utility boilers, often with the gas fired alongside the main fuel. This option allows a comparatively small thermal output of gas to be used with the same efficiency as the main fuel in the boiler as a large, efficient steam turbine can be used. It is anticipated that addition of gas from a biomass or wood gasifier into the natural gas feed to a gas turbine to be technically possible but there will be concerns as to the balance of commercial risks to a large power plant and the benefits of using the gas from the gasifier.

Furthermore, the disposal of municipal and industrial waste has become an important problem because the traditional means of disposal, landfill, are much less environmentally acceptable than previously. Much stricter regulation of these disposal methods will make the economics of waste processing for resource recovery much more favorable. One method of processing waste streams is to convert the energy value of the combustible waste into a fuel. One type of fuel attainable from waste is a low heating value gas, usually 100–150 Btu/scf, which can be used to generate process steam or to generate electricity. Co-processing such waste with coal is also an option (Speight, 2008, 2013, 2014b).

Cogasification technology varies, being usually site specific and high feedstock dependent. At the largest scale, the plant may include the well-proven fixed-bed and entrained-flow gasification processes. At smaller scales, emphasis is placed on technologies that appear closest to commercial operation. Pyrolysis and other advanced thermal conversion processes are included where power generation is practical using the on-site feedstock produced. However, the needs to be addressed are (1) core fuel handling and gasification/pyrolysis technologies, (2) fuel gas cleanup, and (3) conversion of fuel gas to electric power (Ricketts et al., 2002).

Waste may be municipal solid waste (MSW) that had minimal presorting, or refuse-derived fuel (RDF) with significant pretreatment, usually mechanical screening and shredding. Other more specific waste sources (excluding hazardous waste), and possibly including petroleum coke, may provide niche opportunities for co-utilization.

The traditional waste to energy plant, based on mass-burn combustion on an inclined grate, has a low public acceptability despite the very low emissions achieved over the last decade with modern flue gas cleanup equipment. This has led to difficulty in obtaining planning permissions to construct needed new waste to energy plants. After much debate, various governments have allowed options for advanced waste conversion technologies (gasification, pyrolysis, and anaerobic digestion), but will only give credit to the proportion of electricity generated from nonfossil waste.

Co-utilization of waste and biomass with coal may provide economies of scale that help achieve the aforementioned identified policy objectives at an affordable cost. In some countries, governments propose cogasification processes as being *well suited for community-sized developments* suggesting that waste should be dealt with in smaller plants serving towns and cities, rather than moved to large, central plants (satisfying the so-called *proximity principal*).

In fact, neither biomass nor wastes are currently produced, or naturally gathered at sites in sufficient quantities to fuel a modern large and efficient power plant. Disruption, transport issues, fuel use, and public opinion all act against gathering hundreds of megawatts (MWe) at a single location. Biomass or waste-fired power plants are therefore inherently limited in size and hence in efficiency (labor costs per unit electricity produced) and in other economies of scale. The production rates of municipal refuse follow reasonably predictable patterns over time periods of a few years. Recent experience with the very limited current *biomass for energy* harvesting has shown unpredictable variations in harvesting capability with long periods of zero production over large areas during wet weather.

The situation is very different for coal. This is generally mined or imported and thus large quantities are available from a single source or a number of closely located sources, and supply has been reliable and predictable. However, the economics of new coal-fired power plants of any technology or size have not encouraged any new coal-fired power plant in the gas generation market.

The potential unreliability of biomass, longer-term changes in refuse and the size limitation of a power plant using only waste and/or biomass can be overcome combining biomass, refuse, and coal. It also allows benefit from a premium electricity price for electricity from biomass and the gate fee associated with waste. If the power plant is gasification-based, rather than direct combustion, further benefits may be available. These include a premium price for the electricity from waste, the range of technologies available for the gas to electricity part of the process, gas cleaning prior to the main combustion stage instead of after combustion and public image, which is currently generally better for gasification as compared to combustion. These considerations lead to current studies of cogasification of wastes/biomass with coal (Speight, 2008).

For large-scale power generation (>50 MWe), the gasification field is dominated by plant based on the pressurized, oxygen-blown, entrained-flow, or fixed-bed gasification of fossil fuels. Entrained gasifier operational experience to date has largely been with well-controlled fuel feedstocks with short-term trial work at low cogasification ratios and with easily handled fuels.

Use of waste materials as cogasification feedstocks may attract significant disposal credits. Cleaner biomass materials are renewable fuels and may attract premium prices for the electricity generated. Availability of sufficient fuel locally for an economic plant size is often a major issue, as is the reliability of the fuel supply. Use of more-predictably available coal alongside these fuels overcomes some of these difficulties and risks. Coal could be regarded as the *flywheel* that keeps the plant running when the fuels producing the better revenue streams are not available in sufficient quantities.

Coal characteristics are very different to younger hydrocarbon fuels such as biomass and waste. Hydrogen-to-carbon ratios are higher for younger fuels, as is the oxygen content. This means that reactivity is very different under gasification conditions. Gas cleaning issues can also be very different, being sulfur a major concern for coal gasification and chlorine compounds and tars more important for waste and biomass gasification. There are no current proposals for adjacent gasifiers and gas cleaning systems, one handling biomass or waste and one coal, alongside each other and feeding the same power production equipment. However, there are some advantages to such a design as compared with mixing fuels in the same gasifier and gas cleaning systems.

Electricity production or combined electricity and heat production remain the most likely area for the application of gasification or cogasification. The lowest investment cost per unit of electricity generated is the use of the gas in an existing large power station. This has been done in several large utility boilers, often with the gas fired alongside the main fuel. This option allows a comparatively small thermal output of gas to be used with the same efficiency as the main fuel in the boiler as a large, efficient steam turbine can be used. It is anticipated that addition of gas from a biomass or wood gasifier into the natural gas feed to a gas turbine to be technically possible but there will be concerns as to the balance of commercial risks to a large power plant and the benefits of using the gas from the gasifier.

Furthermore, the disposal of municipal and industrial waste has become an important problem because the traditional means of disposal, landfill, are much less environmentally acceptable than previously. Much stricter regulation of these disposal methods will make the economics of waste processing for resource recovery much more favorable. Thus, coal may be cogasified with waste or biomass for environmental, technical, or commercial reasons. It allows larger, more efficient plants than those sized for grown biomass or arising waste within a reasonable transport distance; specific operating costs are likely to be lower and fuel supply security is assured.

Cogasification technology varies, being usually site specific and high feedstock dependent. At the largest scale, the plant may include the well-proven fixed-bed and entrained-flow gasification processes. At smaller scales, emphasis is placed on technologies that appear closest to commercial operation. Pyrolysis and other advanced thermal conversion processes are included where power generation is practical using the on-site feedstock produced. However, the needs to be addressed are

(1) core fuel handling and gasification/pyrolysis technologies, (2) fuel gas cleanup, and (3) conversion of fuel gas to electric power (Ricketts et al., 2002).

Co-utilization of waste and biomass with coal may provide economies of scale that help achieve the aforementioned identified policy objectives at an affordable cost. In some countries, governments propose cogasification processes as being *well suited for community-sized developments* suggesting that waste should be dealt with in smaller plants serving towns and cities, rather than moved to large, central plants (satisfying the so-called *proximity principal*).

In fact, neither biomass nor wastes are currently produced, or naturally gathered at sites in sufficient quantities to fuel a modern large and efficient power plant. Disruption, transport issues, fuel use, and public opinion all act against gathering hundreds of MWe at a single location. Biomass or waste-fired power plants are therefore inherently limited in size and hence in efficiency (labor costs per unit electricity produced) and in other economies of scale. The production rates of municipal refuse follow reasonably predictable patterns over time periods of a few years. Recent experience with the very limited current *biomass for energy* harvesting has shown unpredictable variations in harvesting capability with long periods of zero production over large areas during wet weather.

The situation is very different for coal. This is generally mined or imported and thus large quantities are available from a single source or a number of closely located sources, and supply has been reliable and predictable. However, the economics of new coal-fired power plants of any technology or size have not encouraged any new coal-fired power plant in the gas generation market.

The potential unreliability of biomass, longer-term changes in refuse and the size limitation of a power plant using only waste and/or biomass can be overcome combining biomass, refuse, and coal. It also allows benefit from a premium electricity price for electricity from biomass and the gate fee associated with waste. If the power plant is gasification-based, rather than direct combustion, further benefits may be available. These include a premium price for the electricity from waste, the range of technologies available for the gas to electricity part of the process, gas cleaning prior to the main combustion stage instead of after combustion and public image, which is currently generally better for gasification as compared to combustion. These considerations lead to current studies of cogasification of wastes/biomass with coal (Speight, 2008).

For large-scale power generation (>50 MWe), the gasification field is dominated by plant based on the pressurized, oxygen-blown, entrained-flow, or fixed-bed gasification of fossil fuels. Entrained gasifier operational experience to date has largely been with well-controlled fuel feedstocks with short-term trial work at low cogasification ratios and with easily handled fuels.

Analyses of the composition of municipal solid waste indicate that plastics do make up measureable amounts (5%–10% or more) of solid waste streams. Many of these plastics are worth recovering as energy. In fact, many plastics, particularly the poly-olefins, have high calorific values and simple chemical constitutions of primarily carbon and hydrogen. As a result, waste plastics are ideal candidates for the gasification process. Because of the myriad of sizes and shapes of plastic products, size reduction is necessary to create a feed material of a size less than 2 inches in diameter. Some forms of waste plastics such as thin films may require a simple agglomeration step to produce a particle of higher bulk density to facilitate ease of feeding. A plastic, such as high-density polyethylene, processed through a gasifier is converted to carbon monoxide and hydrogen and these materials in turn may be used to form other chemicals including ethylene from which the polyethylene is produced—*closed the loop recycling*.

14.3.3.8 Black Liquor

Black liquor is the spent liquor from the Kraft process in which pulpwood is converted into paper pulp by removing lignin and hemicellulose constituents as well as other extractable materials from wood to free the cellulose fibers. The equivalent spent cooking liquor in the sulfite process is usually called *brown liquor*, but the terms *red liquor*, *thick liquor*, and *sulfite liquor* are also used. Approximately seven units of black liquor are produced in the manufacture of one unit of pulp (Biermann, 1993).

Black liquor comprises an aqueous solution of lignin residues, hemicellulose, and the inorganic chemical used in the process and 15% w/w solids of which 10% w/w are inorganic and 5% w/w are organic. Typically, the organic constituents in black liquor are 40%–45% w/w soaps, 35%–45% w/w lignin, and 10%–15% w/w other (miscellaneous) organic materials.

The organic constituents in the black liquor are made up of water/alkali soluble degradation components from the wood. Lignin is partially degraded to shorter fragments with sulfur contents in the order of 1%–2% w/w and sodium content at approximately 6% w/w of the dry solids. Cellulose (and hemicellulose) is degraded to aliphatic carboxylic acid soaps and hemicellulose fragments. The extractable constituents yield *tall oil soap* and crude turpentine. The tall oil soap may contain up to 20% w/w sodium. Residual lignin components currently serve for hydrolytic or pyrolytic conversion or combustion. Alternative, hemicellulose constituents may be used in fermentation processes.

Gasification of black liquor has the potential to achieve higher overall energy efficiency as compared to those of conventional recovery boilers, while generating an energy-rich syngas. The syngas can then be burned in a gas turbine combined cycle system (*BLGCC—black liquor gasification combined cycle*—and similar to *IGCC, integrated gasification combined cycle*) to produce electricity or converted (through catalytic processes) into chemicals or fuels (e.g., methanol, dimethyl ether, Fischer–Tropsch hydrocarbons, and diesel fuel).

14.4 GASIFICATION IN A REFINERY

The gasification of carbonaceous feedstocks has been used for many years to convert organic solids and liquids into useful gaseous, liquid, and cleaner solid fuels (Speight, 2011a; Brar et al., 2012). In the current context, there are a large number of different feedstock types for use in a refinery-based gasifier, each with different characteristics, including size, shape, bulk density, moisture content, energy content, chemical composition, ash fusion characteristics, and homogeneity of all these properties (Speight, 2013, 2014a,b). Coal and petroleum coke are used as primary feedstocks for many large gasification plants worldwide. Additionally, a variety of biomass and waste-derived feedstocks can be gasified, with wood pellets and chips, waste wood, plastics, municipal solid waste (MSW), refuse-derived fuel (RDF), agricultural and industrial wastes, sewage sludge, switch grass, discarded seed corn, corn stover, and other crop residues all being used.

The gasification of coal, biomass, petroleum, or any carbonaceous residues is generally aimed to feedstock conversion to gaseous products. In fact, gasification offers one of the most versatile methods (with a reduced environmental impact with respect to combustion) to convert carbonaceous feedstocks into electricity, hydrogen, and other valuable energy products.

Depending on the previously described type of gasifier (e.g., air blown, enriched oxygen blown) and the operating conditions, gasification can be used to produce a fuel gas that is suitable for several applications.

Gasification for electric power generation enables the use of a common technology in modern gas fired power plants (*combined cycle*) to recover more of the energy released by burning the fuel. The use of these two types of turbines in the combined cycle system involves (1) a combustion turbine and (2) a steam turbine. The increased efficiency of the combined cycle for electrical power generation results in a 50% v/v decrease in carbon dioxide emissions compared to conventional coal plants. Gasification units could be modified to further reduce their climate change impact because a large part of the carbon dioxide generated can be separated from the other product gas *before* combustion (e.g., carbon dioxide can be separated/sequestered from gaseous by-products by using adsorbents [e.g., MOFs] to prevent its release to the atmosphere). Gasification has also been considered for many years as an alternative to combustion of solid or liquid fuels. Gaseous mixtures are simpler to clean as compared to solid or high-viscosity liquid fuels. Cleaned gases can be used in internal combustion-based power plants that would suffer from severe fouling or corrosion if solid or low-quality liquid fuels were burned inside them.

In fact, the hot syngas produced by gasification of carbonaceous feedstocks can then be processed to remove sulfur compounds, mercury, and particulate matter prior to its use as fuel in a combustion turbine generator to produce electricity. The heat in the exhaust gases from the combustion turbine is recovered to generate additional steam. This steam, along with the steam produced by the gasification process, drives a steam turbine generator to produce additional electricity. In the past decade, the primary application of gasification to power production has become more common due to the demand for high efficiency and low environmental impact.

As anticipated, the quality of the gas generated in a system is influenced by feedstock characteristics, gasifier configuration, and the amount of air, oxygen, or steam introduced into the system. The output and quality of the gas produced is determined by the equilibrium established when the heat of oxidation (combustion) balances the heat of vaporization and volatilization plus the sensible heat (temperature rise) of the exhaust gases. The quality of the outlet gas (Btu/ft^3) is determined by the amount of volatile gases (such as hydrogen, carbon monoxide, water, carbon dioxide, and methane) in the gas stream. With some feedstocks, the higher the amounts of volatile produced in the early stages of the process, the higher the heat content of the product gas. In some cases, the highest gas quality may be produced at lower temperatures. However, char oxidation reaction is suppressed when the temperature is too low, and the overall heat content of the product gas is diminished.

Gasification agents are normal air, oxygen-enriched air, or oxygen. Steam is sometimes added for temperature control, heating value enhancement, or to allow the use of external heat (*allothermal gasification*). The major chemical reactions break and oxidize hydrocarbons to give a product gas containing carbon monoxide, carbon dioxide, hydrogen, and water. Other important components include hydrogen sulfide, various compounds of sulfur and carbon, ammonia, light hydrocarbons, and heavy hydrocarbons (tars).

Depending on the employed gasifier technology and operating conditions, significant quantities of water, carbon dioxide, and methane can be present in the product gas, as well as a number of minor and trace components. Under reducing conditions in the gasifier, most of the feedstock sulfur converts to hydrogen sulfide (H_2S), but 3%–10% converts to carbonyl sulfide (COS). Organically bound nitrogen in the coal feedstock is generally converted to gaseous nitrogen (N_2), but some ammonia (NH_3) and a small amount of hydrogen cyanide (HCN) are also formed. Any chlorine in the coal is converted to hydrogen chloride (HCl), with some chlorine present in the particulate matter (fly ash). Trace elements, such as mercury and arsenic, are released during gasification and partition among the different phases (e.g., fly ash, bottom ash, slag, and product gas).

14.4.1 Gasification of Residua and Residua—Coal Mixtures

The gasification process can be used to convert heavy feedstocks such as vacuum residua and deasphalter bottoms with coal into synthesis gas (syngas) that is primarily hydrogen and carbon monoxide (Wallace et al., 1998). The heat generated by the gasification reaction is recovered as the product gas is cooled. For example, when the quench version of Texaco gasification is employed, the steam generated is of medium and low pressure. Note that the low-level heat used for deasphalting integration is the last stage of syngas cooling. In nonintegrated cases, much of this heat is uneconomical to recover and is lost to air fans and to cooling water exchangers.

In addition, integration of solvent deasphalting/gasification facility is an alternative for upgrading heavy oils economically (Wallace et al., 1998). An integrated solvent deasphalting/gasification unit can increase the throughput or the crude flexibility of the refinery without creating a new, highly undesirable heavy oil stream. Typically, the addition of a solvent deasphalting unit to process vacuum tower bottoms increases a refinery's production of diesel oil. The deasphalted oil is converted to diesel using hydro treating and catalytic cracking (Chapter 9). Unfortunately, the deasphalter bottoms often need to be blended with product diesel oil to produce a viable outlet for these bottoms. A gasification process is capable of converting these deasphalter bottoms to synthesis gas that can then be converted to hydrogen for use in hydrotreating and hydrocracking processes. The synthesis gas may also be used in cogeneration facilities to provide low cost power and steam to the refinery. If the refinery is part of a petrochemical complex, the synthesis gas can be used as a chemical feedstock. The heat generated by the gasification reaction is recovered as the product gas is cooled.

14.4.2 Cogasification of Residua with Biomass

Gasification is an established technology (Hotchkiss, 2003; Speight, 2013). Comparatively, biomass gasification has been the focus of research in recent years to estimate efficiency and performance of the gasification process using various types of biomass such as sugarcane residue (Gabra et al., 2001), rice hulls (Boateng et al., 1992), pine sawdust (Lv et al., 2004), almond shells (Rapagnà and Latif, 1997; Rapagnà et al., 2000), wheat straw (Ergudenler and Ghaly, 1993), food waste (Ko et al., 2001), and wood biomass (Pakdel and Roy, 1991; Chen et al., 1992; Bhattacharaya et al., 1999; Hanaoka et al., 2005). Recently, cogasification of various biomass and coal mixtures has attracted a great deal of interest from the scientific community. Feedstock combinations including Japanese cedar wood and coal (Kumabe et al., 2007), coal and saw dust, coal and pine chips (Pan et al., 2000), coal and silver birch wood (Collot et al., 1999), and coal and birch wood (Brage et al., 2000) have been reported in gasification practices. Cogasification of coal and biomass has some synergy—the process not only produces a low carbon footprint on the environment but also improves the H_2/CO ratio in the produced gas that is required for liquid fuel synthesis (Sjöström et al., 1999; Kumabe et al., 2007). In addition, the inorganic matter present in biomass catalyzes the gasification of coal. However, cogasification processes require custom fittings and optimized processes for the coal and region-specific wood residues.

While cogasification of coal and biomass is advantageous from a chemical viewpoint, some practical problems are present on upstream, gasification, and downstream processes. On the upstream side, the particle size of the coal and biomass is required to be uniform for optimum gasification. In addition, moisture content and pretreatment (torrefaction) are very important during upstream processing.

While upstream processing is influential from a material handling point of view, the choice of gasifier operation parameters (temperature, gasifying agent, and catalysts) dictate the product gas composition and quality. Biomass decomposition occurs at a lower temperature than coal and therefore different reactors compatible to the feedstock mixture are required (Brar et al., 2012). Furthermore, feedstock and gasifier type along with operating parameters not only decide product gas composition but also dictate the amount of impurities to be handled downstream. Downstream processes need to be modified if coal is cogasified with biomass. Heavy metal and impurities such as sulfur and mercury present in coal can make syngas difficult to use and unhealthy for the environment. Alkali present in biomass can also cause corrosion problems and high temperatures in downstream pipes. An alternative option to downstream gas cleaning would be to process coal to remove mercury and sulfur prior to feeding into the gasifier.

However, first and foremost, coal and biomass require drying and size reduction before they can be fed into a gasifier. Size reduction is needed to obtain appropriate particle sizes; however, drying is required to achieve moisture content suitable for gasification operations. In addition, biomass densification may be conducted to prepare pellets and improve density and material flow in the feeder areas. It is recommended that biomass moisture content should be less than 15% w/w prior to gasification. High moisture content reduces the temperature achieved in the gasification zone, thus resulting in incomplete gasification. Forest residues or wood has a fiber saturation point at 30%–31% moisture content (dry basis) (Brar et al., 2012). Compressive and shear strength of the wood increases with decreased moisture content below the fiber saturation point. In such a situation, water is removed from the cell wall leading to shrinkage. The long-chain molecules constituents of the cell wall move closer to each other and bind more tightly. A high level of moisture, usually injected in the form of steam in the gasification zone, favors formation of a water-gas shift reaction that increases hydrogen concentration in the resulting gas.

The torrefaction process is a thermal treatment of biomass in the absence of oxygen, usually at 250°C–300°C to drive off moisture, decompose hemicellulose completely, and partially decompose cellulose (Speight, 2011a). Torrefied biomass has reactive and unstable cellulose molecules with broken hydrogen bonds and not only retains 79%–95% of feedstock energy but also produces a more reactive feedstock with lower atomic hydrogen/carbon and oxygen/carbon ratios to those of the original biomass. Torrefaction results in higher yields of hydrogen and carbon monoxide in the gasification process.

Finally, the presence of mineral matter in the coal-biomass feedstock is not appropriate for fluidized-bed gasification. Low melting point of ash present in woody biomass leads to agglomeration that causes defluidization of the ash and sintering, deposition, and corrosion of the gasifier construction metal bed. Biomass containing alkali oxides and salts are likely to produce clinkering/slagging problems from ash formation (McKendry, 2002). Thus, it is imperative to be aware of the melting of biomass ash, its chemistry within the gasification bed (no bed, silica/sand, or calcium bed), and the fate of alkali metals when using fluidized-bed gasifiers.

Most small- to medium-sized biomass/waste gasifiers are air blown and operate at atmospheric pressure and at temperatures in the range 800°C–100°C (1470°F–2190°F). They face very different challenges to large gasification plants—the use of small-scale air separation plant should oxygen gasification be preferred. Pressurized operation, which eases gas cleaning, may not be practical.

Biomass fuel producers, coal producers and, to a lesser extent, waste companies are enthusiastic about supplying cogasification power plants and realize the benefits of cogasification with alternate fuels (Lee, 2007; Speight, 2008, 2011a, 2013; Lee and Shah, 2013). The benefits of a cogasification technology involving coal and biomass include the use of a reliable coal supply with gate-fee waste and biomass, which allows the economies of scale from a larger plant to be supplied just with waste and biomass. In addition, the technology offers a future option of hydrogen production and fuel development in refineries. In fact, oil refineries and petrochemical plants are opportunities for gasifiers when the hydrogen is particularly valuable (Speight, 2011b, 2014a).

14.4.3 COGASIFICATION OF RESIDUA WITH WASTE

Waste may be municipal solid waste (MSW), which had minimal presorting, or refuse-derived fuel (RDF) with significant pretreatment, usually mechanical screening and shredding. Other more specific waste sources (excluding hazardous waste) and possibly including petroleum coke may provide niche opportunities for co-utilization.

The traditional waste to energy plant, based on mass-burn combustion on an inclined grate, has a low public acceptability despite the very low emissions achieved over the last decade with modern flue gas cleanup equipment. This has led to difficulty in obtaining planning permissions to construct needed new waste to energy plants. After much debate, various governments have allowed options for advanced waste conversion technologies (gasification, pyrolysis, and anaerobic digestion), but will only give credit to the proportion of electricity generated from nonfossil waste.

Co-utilization of waste and biomass with coal may provide economies of scale that help achieve the aforementioned identified policy objectives at an affordable cost. In some countries, governments propose cogasification processes as being *well suited for community-sized developments* suggesting that waste should be dealt with in smaller plants serving towns and cities, rather than moved to large, central plants (satisfying the so-called *proximity principal*).

In fact, neither biomass nor wastes are currently produced, or naturally gathered at sites in sufficient quantities to fuel a modern large and efficient power plant. Disruption, transport issues, fuel use, and public opinion all act against gathering hundreds of MWe at a single location. Biomass or waste-fired power plants are therefore inherently limited in size and hence in efficiency (labor costs per unit electricity produced) and in other economies of scale. The production rates of municipal refuse follow reasonably predictable patterns over time periods of a few years. Recent experience with the very limited current *biomass for energy* harvesting has shown unpredictable variations in harvesting capability with long periods of zero production over large areas during wet weather.

The situation is very different for coal. This is generally mined or imported and thus large quantities are available from a single source or a number of closely located sources, and supply has been reliable and predictable. However, the economics of new coal-fired power plants of any technology or size have not encouraged any new coal-fired power plant in the gas generation market.

The potential unreliability of biomass, longer-term changes in refuse, and the size limitation of a power plant using only waste and/or biomass can be overcome combining biomass, refuse, and coal. It also allows benefit from a premium electricity price for electricity from biomass and the gate fee associated with waste. If the power plant is gasification-based, rather than direct combustion, further benefits may be available. These include a premium price for the electricity from waste, the range of technologies available for the gas to electricity part of the process, gas cleaning prior to the main combustion stage instead of after combustion and public image, which is currently generally better for gasification as compared to combustion. These considerations lead to current studies of cogasification of wastes/biomass with coal (Speight, 2008).

For large-scale power generation (>50 MWe), the gasification field is dominated by plant based on the pressurized, oxygen-blown, entrained-flow, or fixed-bed gasification of fossil fuels. Entrained gasifier operational experience to date has largely been with well-controlled fuel feedstocks with short-term trial work at low cogasification ratios and with easily handled fuels.

Use of waste materials as cogasification feedstocks may attract significant disposal credits. Cleaner biomass materials are renewable fuels and may attract premium prices for the electricity generated. Availability of sufficient fuel locally for an economic plant size is often a major issue, as is the reliability of the fuel supply. Use of more-predictably available coal alongside these fuels overcomes some of these difficulties and risks. Coal could be regarded as the "flywheel" that keeps the plant running when the fuels producing the better revenue streams are not available in sufficient quantities.

Coal characteristics are very different to younger hydrocarbon fuels such as biomass and waste. Hydrogen-to-carbon ratios are higher for younger fuels, as is the oxygen content. This means that reactivity is very different under gasification conditions. Gas cleaning issues can also be very different, being sulfur a major concern for coal gasification and chlorine compounds and tars more important for waste and biomass gasification. There are no current proposals for adjacent gasifiers and gas cleaning systems, one handling biomass or waste and one coal, alongside each other and feeding the same power production equipment. However, there are some advantages to such a design as compared with mixing fuels in the same gasifier and gas cleaning systems.

Electricity production or combined electricity and heat production remain the most likely area for the application of gasification or cogasification. The lowest investment cost per unit of electricity generated is the use of the gas in an existing large power station. This has been done in several large utility boilers, often with the gas fired alongside the main fuel. This option allows a comparatively small thermal output of gas to be used with the same efficiency as the main fuel in the boiler as a large, efficient steam turbine can be used. It is anticipated that addition of gas from a biomass or wood gasifier into the natural gas feed to a gas turbine to be technically possible but there will be concerns as to the balance of commercial risks to a large power plant and the benefits of using the gas from the gasifier.

The use of fuel cells with gasifiers is frequently discussed but the current cost of fuel cells is such that their use for mainstream electricity generation is uneconomic. Furthermore, the disposal of municipal and industrial waste has become an important problem because the traditional means of disposal, landfill, are much less environmentally acceptable than previously. Much stricter regulation of these disposal methods will make the economics of waste processing for resource recovery much more favorable.

One method of processing waste streams is to convert the energy value of the combustible waste into a fuel. One type of fuel attainable from waste is a low heating value gas, usually 100–150 Btu/scf, which can be used to generate process steam or to generate electricity. Co-processing such waste with coal is also an option (Speight, 2008).

In summary, coal may be cogasified with waste or biomass for environmental, technical, or commercial reasons. It allows larger, more efficient plants than those sized for grown biomass or arising waste within a reasonable transport distance; specific operating costs are likely to be lower and fuel supply security is assured.

Cogasification technology varies, being usually site specific and highly feedstock dependent. At the largest scale, the plant may include the well-proven fixed-bed and entrained-flow gasification processes. At smaller scales, emphasis is placed on technologies that appear closest to commercial operation. Pyrolysis and other advanced thermal conversion processes are included where power generation is practical using the on-site feedstock produced. However, the needs to be addressed are (1) core fuel handling and gasification/pyrolysis technologies, (2) fuel gas cleanup, and (3) conversion of fuel gas to electric power (Ricketts et al., 2002).

14.5 SYNTHETIC FUEL PRODUCTION

The gasification of coal or a derivative (i.e., char produced from coal) is the conversion of coal (by any one of a variety of processes) to produce gaseous products that are combustible as well as a wide range of chemical products from synthesis gas (Figure 14.3). With the rapid increase in the use of coal from the fifteenth century onward, it is not surprising the concept of using coal to produce a flammable gas, and especially the use of the water and hot coal became commonplace (van Heek and Muhlen, 1991). In fact, the production of gas from coal has been a vastly expanding area of coal technology, leading to numerous research and development programs. As a result, the characteristics of rank, mineral matter, particle size, and reaction conditions are all recognized as having a bearing on the outcome of the process; not only in terms of gas yields but also on gas properties (van Heek and Muhlen, 1991). The products from the gasification of coal may be of low, medium, or high-heat-content (high-Btu) content as dictated by the process as well as by the ultimate use for the gas (Baker and Rodriguez, 1990; Probstein and Hicks, 1990; Lahaye and Ehrburger, 1991; Matsukata et al., 1992; Speight, 2013).

14.5.1 Gaseous Products

The products of gasification are varied insofar as the gas composition varies with the system employed (Speight, 2013). It is emphasized that the gas product must be first freed from any pollutants such as particulate matter and sulfur compounds before further use, particularly when the intended use is a water-gas shift or methanation (Cusumano et al., 1978; Probstein and Hicks, 1990).

14.5.1.1 Synthesis Gas

Synthesis gas (syngas) is a mixture mainly of hydrogen and carbon monoxide, which is comparable in its combustion efficiency to natural gas (Speight, 2008; Chadeesingh, 2011). This reduces the emissions of sulfur, nitrogen oxides, and mercury, resulting in a much cleaner fuel (Sondreal et al., 2004, 2006; Yang et al., 2007; Nordstrand et al., 2008; Wang et al., 2008). The resulting hydrogen gas can be used for electricity generation or as a transport fuel. The gasification process also facilitates capture of carbon dioxide emissions from the combustion effluent (see discussion of carbon capture and storage in the following).

Although synthesis gas can be used as a standalone fuel, the energy density of syngas is approximately half that of natural gas and is therefore mostly suited for the production of transportation fuels and other chemical products. Syngas is mainly used as an intermediary building block for the final production (synthesis) of various fuels such as synthetic natural gas, methanol, and synthetic petroleum fuel (dimethyl ether—synthesized gasoline and diesel fuel) (Chadeesingh, 2011; Speight, 2013).

The use of syngas offers the opportunity to furnish a broad range of environmentally clean fuels and chemicals, and there has been steady growth in the traditional uses of syngas. Almost all hydrogen gas is manufactured from synthesis gas, and there has been an increase in the demand for this basic chemical. In fact, the major use of syngas is in the manufacture of hydrogen for a growing number of purposes, especially in petroleum refineries (Speight, 2014a). Methanol not only remains the second largest consumer of syngas but has shown remarkable growth as part of the methyl ethers used as octane enhancers in automotive fuels.

The Fischer–Tropsch synthesis remains the third largest consumer of syngas, mostly for transportation fuels but also as a growing feedstock source for the manufacture of chemicals, including polymers. The hydroformylation of olefins (the oxo reaction), a completely chemical use of syngas, is the fourth largest use of carbon monoxide and hydrogen mixtures. A direct application of syngas as fuel (and eventually also for chemicals) that promises to increase is its use for *integrated gasification combined cycle* (IGCC) units for the generation of electricity (and also chemicals) from coal, petroleum coke, or heavy residuals (Holt, 2001). Finally, synthesis gas is the principal source of carbon monoxide, which is used in an expanding list of carbonylation reactions, which are of major industrial interest.

14.5.1.2 Low-Btu Gas

During the production of coal gas by oxidation with air, the oxygen is not separated from the air and, as a result, the gas product invariably has a low-Btu content (low-heat-content, 150–300 Btu/ft^3). Low-heat-content gas is also the usual product of *in situ* gasification of coal, which is essentially used as a method for obtaining energy from coal without the necessity of mining the coal, especially if the coal cannot be mined or if mining is uneconomical.

Several important chemical reactions and a host of side reactions are involved in the manufacture of low-heat-content gas under the high-temperature conditions employed (Chadeesingh, 2011; Speight, 2013). Low-heat-content gas contains several components, four of which are always major components present at levels of at least several percent; a fifth component, methane, is marginally a major component.

The nitrogen content of low-heat-content gas ranges from somewhat less than 33% v/v to slightly more than 50% v/v and cannot be removed by any reasonable means; the presence of nitrogen at these levels makes the product gas *low-heat-content* by definition. The nitrogen also strongly limits the applicability of the gas to chemical synthesis. Two other noncombustible components—water (H_2O) and carbon dioxide (CO)—further lower the heating value of the gas; water can be removed by condensation and carbon dioxide by relatively straightforward chemical means.

The two major combustible components are hydrogen and carbon monoxide; the H_2/CO ratio varies from approximately 2:3 to approximately 3:2. Methane may also make an appreciable contribution to the heat content of the gas. Of the minor components, hydrogen sulfide is the most significant, and the amount produced is, in fact, proportional to the sulfur content of the feed coal. Any hydrogen sulfide present must be removed by one, or more, of several procedures (Mokhatab et al., 2006; Speight, 2007).

Low-heat-content gas is of interest to industry as a fuel gas or even, on occasion, as a raw material from which ammonia, methanol, and other compounds may be synthesized.

14.5.1.3 Medium-Btu Gas

Medium-Btu gas (medium-heat-content gas) has a heating value in the range 300–550 Btu/ft^3) and the composition is much like that of low-heat-content gas, except that there is virtually no nitrogen. The primary combustible gases in medium-heat-content gas are hydrogen and carbon monoxide. Medium-heat-content gas is considerably more versatile than low-heat-content gas; like low-heat-content gas, medium-heat-content gas may be used directly as a fuel to raise steam, or used through a combined power cycle to drive a gas turbine, with the hot exhaust gases employed to raise steam, but medium-heat-content gas is especially amenable to synthesize methane (by methanation), higher hydrocarbons (by Fischer–Tropsch synthesis), methanol, and a variety of synthetic chemicals.

The reactions used to produce medium-heat-content gas are the same as those employed for low-heat-content gas synthesis, the major difference being the application of a nitrogen barrier (such as the use of pure oxygen) to keep diluent nitrogen out of the system.

In medium-heat-content gas, the H_2/CO ratio varies from 2:3 C to 3:1 and the increased heating value correlates with higher methane and hydrogen contents as well as with lower carbon dioxide contents. Furthermore, the very nature of the gasification process used to produce the medium-heat-content gas has a marked effect upon the ease of subsequent processing. For example, the CO_2-acceptor product is quite amenable to use for methane production because it has (1) the desired H_2/CO ratio just exceeding 3:1, (2) an initially high methane content, and (3) relatively low water and carbon dioxide contents. Other gases may require appreciable shift reaction and removal of large quantities of water and carbon dioxide prior to methanation.

14.5.1.4 High-Heat-Content (High-Btu) Gas

High-Btu gas (high-heat-content gas) is essentially pure methane and often referred to as *synthetic natural gas* or *substitute natural gas* (SNG) (Speight, 1990, 2013). However, to qualify as substitute natural gas, a product must contain at least 95% methane, giving an energy content (heat content) of synthetic natural gas on the order of 980–1080 Btu/ft^3.

The commonly accepted approach to the synthesis of high-heat-content gas is the catalytic reaction of hydrogen and carbon monoxide:

$$3H_2 + CO \rightarrow CH_4 + H_2O$$

To avoid catalyst poisoning, the feed gases for this reaction must be quite pure, and therefore, impurities in the product are rare. The large quantities of water produced are removed by condensation and recirculated as very pure water through the gasification system. The hydrogen is usually present in slight excess to ensure that the toxic carbon monoxide is reacted; this small quantity of hydrogen will lower the heat content to a small degree.

The carbon monoxide/hydrogen reaction is somewhat inefficient as a means of producing methane because the reaction liberates large quantities of heat. In addition, the methanation catalyst is troublesome and prone to poisoning by sulfur compounds and the decomposition of metals can destroy the catalyst. Hydrogasification may be thus employed to minimize the need for methanation:

$$[C]_{coal} + 2H2 \rightarrow CH_4$$

The product of hydrogasification is far from pure methane and additional methanation is required after hydrogen sulfide and other impurities are removed.

14.5.2 Liquid Products

The production of liquid fuels from coal via gasification is often referred to as the *indirect liquefaction* of coal (Speight, 2013). In these processes, coal is not converted directly into liquid products but involves a two-stage conversion operation in which coal is first converted (by reaction with steam and oxygen) to produce a gaseous mixture that is composed primarily of carbon monoxide and hydrogen (syngas; synthesis gas). The gas stream is subsequently purified (to remove sulfur, nitrogen, and any particulate matter), after which it is catalytically converted to a mixture of liquid hydrocarbon products.

The synthesis of hydrocarbons from carbon monoxide and hydrogen (synthesis gas) (the Fischer–Tropsch synthesis) is a procedure for the indirect liquefaction of coal and other carbonaceous feedstocks (Speight, 2011a,b). This process is the only coal liquefaction scheme currently in use on a relatively large commercial scale; South Africa is currently using the Fischer–Tropsch process on a commercial scale in their SASOL complex.

Thus, coal is converted to gaseous products at temperatures in excess of 800°C (1470°F), and at moderate pressures, to produce syngas:

$$[C]_{coal} + H_2O \rightarrow CO + H_2$$

The gasification may be attained by means of any one of several processes or even by gasification of coal in place (underground, or *in situ*, gasification of coal).

In practice, the Fischer–Tropsch reaction is carried out at temperatures of 200°C–350°C (390°F–660°F) and at pressures of 75–4000 psi. The hydrogen/carbon monoxide ratio is typically on the order of 2/2:1 or 2/5:1; since up to three volumes of hydrogen may be required to achieve the next stage of the liquids production, the synthesis gas must then be converted by means of the water-gas shift reaction to the desired level of hydrogen:

$$CO + H_2O \rightarrow CO_2 + H_2$$

After this, the gaseous mix is purified and converted to a wide variety of hydrocarbons:

$$nCO + (2n + 1)H_2 \rightarrow C_nH_{2n+2} + nH_2O$$

These reactions result primarily in low- and medium-boiling aliphatic compounds suitable for gasoline and diesel fuel.

14.5.3 Solid Products

The solid product (solid waste) of a gasification process is typically ash, that is, the oxides of metal-containing constituents of the feedstock. The amount and type of solid waste produced is very much feedstock dependent. The waste is a significant environmental issue due to the large quantities produced, chiefly fly ash if coal is the feedstock or a co-feedstock, and there potential for leaching of toxic substances (such as heavy metals such as lead and arsenic) into the soil and groundwater at disposal sites.

At the high temperature of the gasifier, most of the mineral matter of the feedstock is transformed and melted into slag, an inert glass-like material and, under such conditions, nonvolatile metals and mineral compounds are bound together in molten form until the slag is cooled in a water bath at the bottom of the gasifier, or by natural heat loss at the bottom of an entrained-bed gasifier. Slag production is a function of mineral matter content of the feedstock—coal produces much more slag per unit weight than petroleum coke. Furthermore, as long as the operating temperature is above the fusion temperature of the ash, slag will be produced. The physical structure of the slag is sensitive to changes in operating temperature and pressure of the gasifier and a quick physical examination of the appearance of the slag can often be an indication of the efficiency of the conversion of feedstock carbon to gaseous product in the process.

Slag is comprised of black, glassy, silica-based materials and is also known as *frit*, which is a high density, vitreous, and abrasive material low in carbon and formed in various shapes from jagged and irregular pieces to rod and needle-like forms. Depending upon the gasifier process parameters and the feedstock properties, there may also be residual carbon char. Vitreous slag is much preferable to ash, because of its habit of encapsulating toxic constituents (such as heavy metals) into a stable, nonleachable material. Leachability data obtained from different gasifiers unequivocally show that gasifier slag is highly nonleachable and can be classified as nonhazardous. Because of its particular properties and nonhazardous, nontoxic nature, slag is relatively easily marketed as a by-product for multiple advantageous uses, which may negate the need for its long-term disposal.

The physical and chemical properties of gasification slag are related to (1) the composition of the feedstock, (2) the method of recovering the molten ash from the gasifier, and (3) the proportion of devolatilized carbon particles (char) discharged with the slag. The rapid water-quench method of cooling the molten slag inhibits recrystallization, and results in the formation of a granular, amorphous material. Some of the differences in the properties of the slag may be attributed to the specific design and operating conditions prevailing in the gasifiers.

Char is the finer component of the gasifier solid residuals, composed of unreacted carbon with various amounts of siliceous ash. Char can be recycled back into the gasifier to increase carbon usage and has been used as a supplemental fuel source for pulverized coal combustion. The irregularly shaped particles have a well-defined pore structure and have excellent potential as an adsorbent and precursor to activated carbon. In terms of recycling char to the gasifier, a property that is important to fluidization is the effective particle density. If the char has a large internal void space, the density will be much less than that of the feedstock (especially coal) or chars from slow carbonization.

14.6   THE FUTURE

The future depends very much on the effect of gasification processes on the surrounding environment. It is these environmental effects and issues that will direct the success of gasification.

Clean coal technologies (CCTs) are a new generation of advanced coal utilization processes that are designed to enhance both the efficiency and the environmental acceptability of coal extraction, preparation, and use. These technologies reduce emissions, reduce waste, and increase the amount of energy gained from coal. The goal of the program was to foster development of the most promising clean coal technologies such as improved methods of cleaning coal, fluidized-bed combustion, integrated gasification combined cycle, furnace sorbent injection, and advanced flue-gas desulfurization.

In fact, there is the distinct possibility that within the foreseeable future the gasification process will increase in popularity in petroleum refineries—some refineries may even be known as gasification refineries (Speight, 2011b). A gasification refinery would have, as the center piece, gasification technology as is the case of the Sasol refinery in South Africa (Couvaras, 1997). The refinery would produce synthesis gas (from the carbonaceous feedstock) from which liquid fuels would be manufactured using the Fischer–Tropsch synthesis technology.

In fact, gasification to produce synthesis gas can proceed from any carbonaceous material, including biomass. Inorganic components of the feedstock, such as metals and minerals, are trapped in an inert and environmentally safe form as char, which may be used as a fertilizer. Biomass gasification is therefore one of the most technically and economically convincing energy possibilities for a potentially carbon neutral economy.

The manufacture of gas mixtures of carbon monoxide and hydrogen has been an important part of chemical technology for about a century. Originally, such mixtures were obtained by the reaction of steam with incandescent coke and were known as *water gas*. Eventually, steam-reforming processes, in which steam is reacted with natural gas (methane) or petroleum naphtha over a nickel catalyst, found wide application for the production of syngas.

A modified version of steam reforming known as autothermal reforming, which is a combination of partial oxidation near the reactor inlet with conventional steam reforming further along the reactor, improves the overall reactor efficiency and increases the flexibility of the process. Partial oxidation processes using oxygen instead of steam also found wide application for synthesis gas manufacture, with the special feature that they could utilize low-value feedstocks such as heavy petroleum residues. In recent years, catalytic partial oxidation employing very short reaction times (milliseconds) at high temperatures ($850^{\circ}C$–$1000^{\circ}C$) is providing still another approach to synthesis gas manufacture (Hickman and Schmidt, 1993).

In a gasifier, the carbonaceous material undergoes several different processes: (1) pyrolysis of carbonaceous fuels, (2) combustion, and (3) gasification of the remaining char. The process is very dependent on the properties of the carbonaceous material and determines the structure and composition of the char, which will then undergo gasification reactions.

As petroleum supplies decrease, the desirability of producing gas from other carbonaceous feedstocks will increase, especially in those areas where natural gas is in short supply. It is also anticipated that costs of natural gas will increase, allowing coal gasification to compete as an economically viable process. Research in progress on a laboratory and pilot-plant scale should lead to the invention of new process technology by the end of the century, thus accelerating the industrial use of coal gasification.

The conversion of the gaseous products of gasification processes to synthesis gas, a mixture of hydrogen (H_2) and carbon monoxide (CO), in a ratio appropriate to the application, needs additional steps, after purification. The product gases—carbon monoxide, carbon dioxide, hydrogen, methane, and nitrogen—can be used as fuels or as raw materials for chemical or fertilizer manufacture.

Finally, gasification by means other than the conventional methods has also received some attention and has provided rationale for future processes (Rabovitser et al., 2010). In the process, a carbonaceous material and at least one oxygen carrier are introduced into a nonthermal plasma reactor at a temperature in the range of approximately 300°C to approximately 700°C (570°F–1290°F) and a pressure in a range from atmospheric pressure to approximate 1030 psi and a nonthermal plasma discharge is generated within the nonthermal plasma reactor. The carbonaceous feedstock and the oxygen carrier are exposed to the nonthermal plasma discharge, resulting in the formation of a product gas that comprises substantial amounts of hydrocarbons, such as methane, hydrogen, and/or carbon monoxide.

Finally, gasification and conversion of carbonaceous solid fuels to synthesis gas for application of power, liquid fuels, and chemicals are practiced worldwide. Petroleum coke, coal, biomass, and refinery waste are major feedstocks for gasification. The concept of blending of coal, biomass, or refinery waste with petroleum coke is advantageous in order to obtain the highest value of products as compared to gasification of petroleum coke alone. Furthermore, based on gasifier type, cogasification of carbonaceous feedstocks can be an advantageous and efficient process. In addition, a variety of upgrading and delivery options that are available for application to synthesis gas enable the establishment of an integrated energy supply system whereby synthesis gases can be upgraded, integrated, and delivered to a distributed network of energy conversion facilities, including power, combined heat and power, and combined cooling, heating, and power (sometime referred to as *trigeneration*) as well as used as fuels for transportation applications.

REFERENCES

Abadie, L.M. and Chamorro, J.M. 2009. The economics of gasification: A market-based approach. *Energies*, 2: 662–694.

Arena, U. 2012. Process and technological aspects of municipal solid waste gasification. A review. *Waste Management*, 32: 625–639.

ASTM D388. 2015. *Standard Classification of Coal by Rank*. Annual Book of Standards. ASTM International, West Conshohocken, PA.

Baker, R.T.K. and Rodriguez, N.M. 1990. *Fuel Science and Technology Handbook*. Marcel Dekker, New York, Chapter 22.

Balat, M. 2011. Fuels from biomass—An overview. In: *The Biofuels Handbook*. J.G. Speight (Editor). Royal Society of Chemistry, London, UK, Part 1, Chapter 3.

Baxter, L. 2005. Biomass-coal Co-combustion: Opportunity for affordable renewable energy. *Fuel*, 84(10): 1295–1302.

Biermann, C.J. 1993. *Essentials of Pulping and Papermaking*. Academic Press Inc., New York.

Boateng, A.A., Walawender, W.P., Fan, L.T., and Chee, C.S. 1992. Fluidized-bed steam gasification of rice hull. *Bioresource Technology*, 40(3): 235–239.

Brage, C., Yu, Q., Chen, G., and Sjöström, K. 2000. Tar evolution profiles obtained from gasification of biomass and coal. *Biomass and Bioenergy*, 18(1): 87–91.

Brar, J.S., Singh, K., Wang, J., and Kumar, S. 2012. Cogasification of coal and biomass: A review. *International Journal of Forestry Research*, 2012: 1–10.

Chadeesingh, R. 2011. The Fischer-Tropsch process. In: *The Biofuels Handbook*. J.G. Speight (Editor). The Royal Society of Chemistry, London, UK, Part 3, Chapter 5, pp. 476–517.

Chen, G., Sjöström, K., and Bjornbom, E. 1992. Pyrolysis/gasification of wood in a pressurized fluidized bed reactor. *Industrial and Engineering Chemistry Research*, 31(12): 2764–2768.

Collot, A.G., Zhuo, Y., Dugwell, D.R., and Kandiyoti, R. 1999. Co-pyrolysis and cogasification of coal and biomass in bench-scale fixed-bed and fluidized bed reactors. *Fuel*, 78: 667–679.

Couvaras, G. 1997. Sasol's slurry phase distillate process and future applications. *Proceedings of the Monetizing Stranded Gas Reserves Conference*. Houston, TX, December.

Cusumano, J.A., Dalla Betta, R.A., and Levy, R.B. 1978. *Catalysis in Coal Conversion*. Academic Press Inc., New York.

Davidson, R.M. 1983. Mineral effects in coal conversion. Report No. ICTIS/TR22, International Energy Agency, London, UK.

Demirbaş, A. 2011. Production of fuels from crops. In: *The Biofuels Handbook*. J.G. Speight (Editor). Royal Society of Chemistry, London, UK, Part 2, Chapter 1.

Dutcher, J.S., Royer, R.E., Mitchell, C.E., and Dahl, A.R. 1983. In: *Advanced Techniques in Synthetic Fuels Analysis*. C.W. Wright, W.C. Weimer, and W.D. Felic (Editors). Technical Information Center, U.S. Department of Energy, Washington, DC, p. 12.

EIA. 2007. *Net Generation by Energy Source by Type of Producer*. Energy Information Administration, U.S. Department of Energy, Washington, DC. http://www.eia.doe.gov/cneaf/electricity/epm/table1_1.html, accessed March 15, 2015.

Ergudenler, A. and Ghaly, A.E. 1993. Agglomeration of alumina sand in a fluidized bed straw gasifier at elevated temperatures. *Bioresource Technology*, 43(3): 259–268.

Fabry, F., Rehmet, C., Rohani, V.-J., and Fulcheri, L. 2013. Waste gasification by thermal plasma: A review. *Waste and Biomass Valorization*, 4(3): 421–439.

Fermoso, J., Plaza, M.G., Arias, B., Pevida, C., Rubiera, F., and Pis, J.J. 2009. Co-gasification of coal with biomass and petcoke in a high-pressure gasifier for syngas production. *Proceedings of the First Spanish National Conference on Advances in Materials Recycling and Eco-Energy*. Madrid, Spain, November 12–13.

Gabra, M., Pettersson, E., Backman, R., and Kjellström, B. 2001. Evaluation of cyclone gasifier performance for gasification of sugar cane residue—Part 1: Gasification of bagasse. *Biomass and Bioenergy*, 21(5): 351–369.

Gray, D. and Tomlinson, G. 2000. Opportunities for petroleum coke gasification under tighter sulfur limits for transportation fuels. *Proceedings of the 2000 Gasification Technologies Conference*. San Francisco, CA, October 8–11.

Hanaoka, T., Inoue, S., Uno, S., Ogi, T., and Minowa, T. 2005. Effect of woody biomass components on air-steam gasification. *Biomass and Bioenergy*, 28(1): 69–76.

Hickman, D.A. and Schmidt, L.D. 1993. Syngas formation by direct catalytic oxidation of methane. *Science*, 259: 343–346.

Higman, C. and Van der Burgt, M. 2008. *Gasification*, 2nd Edition. Gulf Professional Publishing, Elsevier, Amsterdam, the Netherlands.

Holt, N.A.H. 2001. Integrated gasification combined cycle power plants. In: *Encyclopedia of Physical Science and Technology*, 3rd Edition. Academic Press Inc., New York.

Hotchkiss, R. 2003. Coal gasification technologies. *Proceedings of the Institute of Mechanical Engineers Part A*, 217(1): 27–33.

Irfan, M.F. 2009. Research report: Pulverized coal pyrolysis & gasification in $N_2/O_2/CO_2$ mixtures by thermogravimetric analysis. *Novel Carbon Resource Sciences Newsletter*. Kyushu University, Fukuoka, Japan, Volume 2, pp. 27–33.

Jenkins, B.M. and Ebeling, J.M. 1985. Thermochemical properties of biomass fuels. *California Agriculture* (May–June), 14–18.

Johnson, J.L. 1979. *Kinetics of Coal Gasification*. John Wiley & Sons, Hoboken, NJ.

Khosravi, M. and Khadse, A., 2013. Gasification of petcoke and coal/biomass blend: A review. *International Journal of Emerging Technology and Advanced Engineering*, 3(12): 167–173.

Ko, M.K., Lee, W.Y., Kim, S.B., Lee, K.W., and Chun, H.S. 2001. Gasification of food waste with steam in fluidized bed. *Korean Journal of Chemical Engineering*, 18(6): 961–964.

Kumabe, K., Hanaoka, T., Fujimoto, S., Minowa, T., and Sakanishi, K. 2007. Cogasification of woody biomass and coal with air and steam. *Fuel*, 86: 684–689.

Kumar, A., Jones, D.D., and Hanna, M.A. 2009. Thermochemical biomass gasification: A review of the current status of the technology. *Energies*, 2: 556–581.

Lahaye, J. and Ehrburger, P. (Editors). 1991. *Fundamental Issues in Control of Carbon Gasification Reactivity.* Kluwer Academic Publishers, Dordrecht, the Netherlands.

Lapuerta, M., Hernández, J.J., Pazo, A., and López, J. 2008. Gasification and co-gasification of biomass wastes: Effect of the biomass origin and the gasifier operating conditions. *Fuel Processing Technology,* 89(9): 828–837.

Lee, S. 2007. Gasification of coal. In: *Handbook of Alternative Fuel Technologies.* S. Lee, J.G. Speight, and S. Loyalka (Editors). CRC Press/Taylor & Francis Group, Boca Raton, FL.

Lee, S. and Shah, Y.T. 2013. *Biofuels and Bioenergy.* CRC Press/Taylor & Francis Group, Boca Raton, FL.

Lee, S., Speight, J.G., and Loyalka, S. 2007. *Handbook of Alternative Fuel Technologies.* CRC Press/Taylor & Francis Group, Boca Raton, FL.

Liu, G., Larson, E.D., Williams, R.H., Kreutz, T.G., and Guo, X. 2011. Making Fischer-Tropsch fuels and electricity from coal and biomass: Performance and cost analysis. *Energy and Fuels,* 25: 415–437.

Luque, R. and Speight, J.G. (Editors). 2015. *Gasification for Synthetic Fuel Production: Fundamentals, Processes, and Applications.* Woodhead Publishing, Elsevier, Cambridge, UK.

Lv, P.M., Xiong, Z.H., Chang, J., Wu, C.Z., Chen, Y., and Zhu, J.X. 2004. An experimental study on biomass air-steam gasification in a fluidized bed. *Bioresource Technology,* 95(1): 95–101.

Marano, J.J. 2003. Refinery technology profiles: Gasification and supporting technologies. Report prepared for the U.S. Department of Energy, National Energy Technology Laboratory. U.S. Energy Information Administration, Washington, DC, June.

Martinez-Alonso, A. and Tascon, J.M.D. 1991. In: *Fundamental Issues in Control of Carbon Gasification Reactivity.* J. Lahaye and P. Ehrburger (Editors). Kluwer Academic Publishers, Dordrecht, the Netherlands.

Matsukata, M., Kikuchi, E., and Morita, Y. 1992. A new classification of alkali and alkaline earth catalysts for gasification of carbon. *Fuel,* 71: 819–823.

McKendry, P. 2002. Energy production from biomass. Part 3: Gasification technologies. *Bioresource Technology,* 83(1): 55–63.

McLendon, T.R., Lui, A.P., Pineault, R.L., Beer, S.K., and Richardson, S.W. 2004. High-pressure co-gasification of coal and biomass in a fluidized bed. *Biomass and Bioenergy,* 26(4): 377–388.

Mims, C.A. 1991. In: *Fundamental Issues in Control of Carbon Gasification Reactivity.* J. Lahaye and P. Ehrburger (Editors). Kluwer Academic Publishers, Dordrecht, the Netherlands. p. 383.

Mokhatab, S., Poe, W.A., and Speight, J.G. 2006. *Handbook of Natural Gas Transmission and Processing.* Elsevier, Amsterdam, the Netherlands.

Nordstrand, D., Duong, D.N.B., and Miller, B.G. 2008. Combustion engineering issues for solid fuel systems. In: *Post-Combustion Emissions Control.* B.G. Miller and D. Tillman (Editors). Elsevier, London, UK, Chapter 9.

Pakdel, H. and Roy, C. 1991. Hydrocarbon content of liquid products and tar from pyrolysis and gasification of wood. *Energy & Fuels,* 5: 427–436.

Pan, Y.G., Velo, E., Roca, X., Manyà, J.J., and Puigjaner, L. 2000. Fluidized-bed cogasification of residual biomass/poor coal blends for fuel gas production. *Fuel,* 79: 1317–1326.

Penrose, C.F., Wallace, P.S., Kasbaum, J.L., Anderson, M.K., and Preston, W.E. 1999. Enhancing refinery profitability by gasification, hydroprocessing and power generation. *Proceedings of the Gasification Technologies Conference.* San Francisco, CA, October.

Probstein, R.F. and Hicks, R.E. 1990. *Synthetic Fuels.* pH Press, Cambridge, MA, Chapter 4.

Rabovitser, I.K., Nester, S., and Bryan, B. 2010. Plasma assisted conversion of carbonaceous materials into a gas. U.S. Patent 7,736,400. June 25.

Rajvanshi, A.K. 1986. Biomass gasification. In: *Alternative Energy in Agriculture,* Volume II. D.Y. Goswami (Editor). CRC Press/Boca Raton, FL, pp. 83–102.

Ramroop Singh, N. 2011. Biofuel. In: *The Biofuels Handbook.* J.G. Speight (Editor). Royal Society of Chemistry, London, UK, Part 1, Chapter 5.

Rapagnà, N.J., Kiennemann, A., and Foscolo, P.U. 2000. Steam-gasification of biomass in a fluidized-bed of olivine particles. *Biomass and Bioenergy,* 19(3): 187–197.

Rapagnà, N.J. and Latif, A. 1997. Steam gasification of almond shells in a fluidized bed reactor: The influence of temperature and particle size on product yield and distribution. *Biomass and Bioenergy,* 12(4): 281–288.

Ricketts, B., Hotchkiss, R., Livingston, W., and Hall, M. 2002. Technology status review of waste/biomass co-gasification with coal. *Proceedings of the Institute of Chemical Engineers Fifth European Gasification Conference.* Noordwijk, the Netherlands, April 8–10.

Sha, X. 2005. Coal gasification. In: *Coal, Oil Shale, Natural Bitumen, Heavy Oil and Peat. Encyclopedia of Life Support Systems (EOLSS)*. Developed under the Auspices of the UNESCO. EOLSS Publishers, Oxford, UK.

Shen, C.-H., Chen, W.-H., Hsu, H.-W., Sheu, J.-Y., and Hsieh, T.-H. 2012. Co-gasification performance of coal and petroleum coke blends in a pilot-scale pressurized entrained-flow gasifier. *International Journal of Energy Research*, 36: 499–508.

Sjöström, K., Chen, G., Yu, Q., Brage, C., and Rosén, C. 1999. Promoted reactivity of char in cogasification of biomass and coal: Synergies in the thermochemical process. *Fuel*, 78: 1189–1194.

Sondreal, E.A., Benson, S.A., and Pavlish, J.H. 2006. Status of research on air quality: Mercury, trace elements, and particulate matter. *Fuel Processing Technology*, 65/66: 5–22.

Sondreal, E.A., Benson, S.A., Pavlish, J.H., and Ralston, N.V.C. 2004. An overview of air quality III: Mercury, trace elements, and particulate matter. *Fuel Processing Technology*, 85: 425–440.

Speight, J.G. 1990. In: *Fuel Science and Technology Handbook*. J.G. Speight (Editor). Marcel Dekker, New York, Chapter 33.

Speight, J.G. 2007. *Natural Gas: A Basic Handbook*. GPC Books, Gulf Publishing Company, Houston, TX.

Speight, J.G. 2008. *Synthetic Fuels Handbook: Properties, Processes, and Performance*. McGraw-Hill, New York.

Speight, J.G. (Editor). 2011a. *Biofuels Handbook*. Royal Society of Chemistry, London, UK.

Speight, J.G. 2011b. *The Refinery of the Future*. Gulf Professional Publishing, Elsevier, Oxford, UK.

Speight, J.G. 2013. *The Chemistry and Technology of Coal*, 3rd Edition. CRC Press/Taylor & Francis Group, Boca Raton, FL.

Speight, J.G. 2014a. *The Chemistry and Technology of Petroleum*, 5th Edition. CRC Press/Taylor & Francis Group, Boca Raton, FL.

Speight, J.G. 2014b. *Gasification of Unconventional Feedstocks*. Gulf Professional Publishing, Elsevier, Oxford, UK.

Sundaresan, S. and Amundson, N.R. 1978. Studies in char gasification—I: A lumped model. *Chemical Engineering Science*, 34: 345–354.

Sutikno, T. and Turini, K. 2012. Gasifying coke to produce hydrogen in refineries. *Petroleum Technology Quarterly*, Q3: 105.

van Heek, K.H. and Muhlen, H.-J. 1991. In: *Fundamental Issues in Control of Carbon Gasification Reactivity*. J. Lahaye and P. Ehrburger (Editors). Kluwer Academic Publishers Inc., Dordrecht, the Netherlands, p. 1.

Wallace, P.S., Anderson, M.K., Rodarte, A.I., and Preston, W.E. 1998. Heavy oil upgrading by the separation and gasification of asphaltenes. *Proceedings of the Paper Presented at the Gasification Technologies Conference*. San Francisco, CA, October.

Wang, Y., Duan, Y., Yang, L., Jiang, Y., Wu, C., Wang, Q., and Yang, X. 2008. Comparison of mercury removal characteristic between fabric filter and electrostatic precipitators of coal-fired power plants. *Journal of Fuel Chemistry and Technology*, 36(1): 23–29.

Wolff, J. and Vliegenthart, E. 2011. Gasification of heavy ends. *Petroleum Technology Quarterly*, Q2: 1–5.

Yang, H., Xua, Z., Fan, M., Bland, A.E., and Judkins, R.R. 2007. Adsorbents for capturing mercury in coal-fired boiler flue gas. *Journal of Hazardous Materials*, 146: 1–11.

Yuksel Orhan, İs, G., Alper, E., McApline, K., Daly, S., Sycz, M., and Elkamel, A. 2014. Gasification of oil refinery waste for power and hydrogen production. *Proceedings of the 2014 International Conference on Industrial Engineering and Operations Management*. Bali, Indonesia, January 7–9.

15 Hydrogen Production

15.1 INTRODUCTION

Throughout the previous chapters (especially Chapters 10 and 11), there have been several references and/or acknowledgments of a very important property of petroleum and petroleum products. And that is the hydrogen content or the use of hydrogen during refining in hydrotreating processes, such as desulfurization (Chapter 10), and in hydroconversion processes, such as hydrocracking (Chapter 11). Although the hydrogen recycle gas may contain up to 40% by volume of other gases (usually hydrocarbons), hydrotreater catalyst life is a strong function of hydrogen partial pressure. Optimum hydrogen purity at the reactor inlet extends catalyst life by maintaining desulfurization kinetics at lower operating temperatures and reducing carbon laydown. Typical purity increases resulting from hydrogen purification equipment and/or increased hydrogen sulfide removal as well as tuning hydrogen circulation and purge rates may extend catalyst life up to approximately 25%.

In fact, the typical refinery runs at a hydrogen deficit and a critical issue facing refiners is the influx of heavier feedstocks into refineries and the need to process the refinery feedstock into refined transportation fuels under an environment of increasingly more stringent clean fuel regulations, decreasing heavy fuel oil demand and increasingly heavy, sour crude supply. Hydrogen network optimization is at the forefront of world refineries' options to address clean fuel trends, to meet growing transportation fuel demands, and to continue to make a profit from their crudes. A key element of a hydrogen network analysis in a refinery involves the capture of hydrogen in its fuel streams and extending its flexibility and processing options. Thus, innovative hydrogen network optimization will be a critical factor influencing the operating flexibility and profitability of the future refinery in a shifting world of crude feedstock supplies and ultralow-sulfur gasoline and diesel fuel.

As hydrogen use has become more widespread in refineries, hydrogen production has moved from the status of a high-tech specialty operation to an integral feature of most refineries (Raissi, 2001; Vauk et al., 2008; Liu et al., 2010). This has been made necessary by the increase in hydrotreating and hydrocracking, including the treatment of progressively heavier feedstocks. In fact, the use of hydrogen in thermal processes is perhaps the single most significant advance in refining technology during the twentieth century (Scherzer and Gruia, 1996; Dolbear, 1998). The continued increase in hydrogen demand over the last several decades is a result of the conversion of petroleum to match changes in product slate and the supply of heavy, high-sulfur oil, and in order to make lower-boiling, cleaner, and more salable products. There are also many reasons other than product quality for using hydrogen in processes adding to the need to add hydrogen at relevant stages of the refining process and, most important, according to the availability of hydrogen (Bezler, 2003; Miller and Penner, 2003; Ranke and Schödel, 2003).

With the increasing need for *clean* fuels, the production of hydrogen for refining purposes requires a major effort by refiners. In fact, the trend to increase the number of hydrogenation (*hydrocracking* and/or *hydrotreating*) processes in refineries coupled with the need to process the heavier oils, which require substantial quantities of hydrogen for upgrading because of the increased use of hydrogen in hydrocracking processes, has resulted in vastly increased demands for this gas. The hydrogen demands can be estimated to a very rough approximation using API gravity and the extent of the reaction, particularly the hydrodesulfurization reaction (Speight, 2000; Speight and Ozum, 2002). But accurate estimation requires equivalent process parameters and a thorough understanding of the nature of each process. Thus, as hydrogen production grows, a better understanding of the capabilities and requirements of a hydrogen plant becomes ever more important to overall refinery operations as a means of making the best use of hydrogen supplies in the refinery.

The chemical nature of the crude oil used as the refinery feedstock has always played the major role in determining the hydrogen requirements of that refinery. For example, the lighter, more paraffinic crude oils will require somewhat less hydrogen for upgrading to, say, a gasoline product than a heavier more asphaltic crude oil (Speight, 2000). It follows that the hydrodesulfurization of heavy oils and residua (which, by definition, is a hydrogen-dependent process) needs substantial amounts of hydrogen as part of the processing requirements.

In general, considerable variation exists from one refinery to another in the balance between hydrogen produced and hydrogen consumed in the refining operations. However, what is more pertinent to the present text is the excessive amounts of hydrogen that are required for hydroprocessing operations, whether these be hydrocracking or the somewhat milder hydrotreating processes. For effective hydroprocessing, a substantial hydrogen partial pressure must be maintained in the reactor and, in order to meet this requirement, an excess of hydrogen above that actually consumed by the process must be fed to the reactor. Part of the hydrogen requirement is met by recycling a stream of hydrogen-rich gas. However, the need still remains to generate hydrogen as makeup material to accommodate the process consumption of 500–3000 scf/bbl depending upon whether the heavy feedstock is being subjected to a predominantly hydrotreating (hydrodesulfurization) or to a predominantly hydrocracking process.

In some refineries, the hydrogen needs can be satisfied by hydrogen recovery from catalytic reformer product gases, but other external sources are required. However, for the most part, many refineries now require on-site hydrogen production facilities to supply the gas for their own processes. Most of this nonreformer hydrogen is manufactured either by steam–methane reforming or by oxidation processes. However, other processes, such as steam–methanol interaction or ammonia dissociation, may also be used as sources of hydrogen. Electrolysis of water produces high-purity hydrogen, but the power costs may be prohibitive.

An early use of hydrogen in refineries was in naphtha hydrotreating, as feed pretreatment for catalytic reforming (which in turn was producing hydrogen as a by-product). As environmental regulations tightened, the technology matured and heavier streams were hydrotreated. Thus in the early refineries, the hydrogen for hydroprocesses was provided as a result of catalytic reforming processes in which dehydrogenation is a major chemical reaction and, as a consequence, hydrogen gas is produced (Chapters 5 and 13). The light ends from the catalytic reformer contain a high ratio of hydrogen to methane, so the stream is freed from ethane and/or propane to get a high concentration of hydrogen in the stream.

The hydrogen is recycled through the reactors where the reforming takes place to provide the atmosphere necessary for the chemical reactions and also prevents the carbon from being deposited on the catalyst, thus extending its operating life. An excess of hydrogen above whatever is consumed in the process is produced, and, as a result, catalytic reforming processes are unique in that they are the only petroleum refinery processes to produce hydrogen as a by-product. However, as refineries and refinery feedstocks evolved during the last four decades, the demand for hydrogen has increased and reforming processes are no longer capable of providing the quantities of hydrogen necessary for feedstock hydrogenation. Within the refinery, other processes are used as sources of hydrogen. Thus, the recovery of hydrogen from the by-products of the coking units, visbreaker units, and catalytic cracking units is also practiced in some refineries.

In coking units and visbreaker units, heavy feedstocks are converted to petroleum coke, oil, light hydrocarbons (benzene, naphtha, liquefied petroleum gas), and gas (Chapter 8). Depending on the process, hydrogen is present in a wide range of concentrations. Since coking processes need gas for heating purposes, adsorption processes are best suited to recover the hydrogen because they feature a very clean hydrogen product and an off-gas suitable as fuel.

Catalytic cracking is the most important process step for the production of light products from gas oil and increasingly from vacuum gas oil and heavy feedstocks (Chapter 9). In catalytic cracking, the molecular mass of the main fraction of the feed is lowered, while another part is converted to coke that is deposited on the hot catalyst. The catalyst is regenerated in one or two stages by burning the coke off with air that also provides the energy for the endothermic cracking process. In the process, paraffins and naphthenes are cracked to olefins and to alkanes with shorter chain length, monoaromatic compounds are dealkylated without ring cleavage, and di-aromatics and polyaromatics are dealkylated and converted to coke. Hydrogen is formed in the last type of reaction, whereas the first two reactions produce light hydrocarbons and therefore require hydrogen. Thus, a catalytic cracker can be operated in such a manner that enough hydrogen for subsequent processes is formed.

In reforming processes, naphtha fractions are reformed to improve the quality of gasoline (Speight, 2000; Speight and Ozum, 2002). The most important reactions occurring during this process are the dehydrogenation of naphthenes to aromatics. This reaction is endothermic and is favored by low pressures, and the reaction temperature lies in the range of 300°C–450°C (570°F–840°F). The reaction is performed on platinum catalysts, with other metals, for example, rhenium, as promoters.

Hydrogen is generated in a refinery by the catalytic reforming process, but there may not always be the need to have a catalytic reformer as part of the refinery sequence. Nevertheless, assuming that a catalytic reformer is part of the refinery sequence, the hydrogen production from the reformer usually falls well below the amount required for hydroprocessing purposes. For example, in a 100,000 bbl/day hydrocracking refinery, assuming intensive reforming of hydrocracked gasoline, the hydrogen requirements of the refinery may still fall some 500–900 scf/bbl of crude charge below that necessary for the hydrocracking sequences. Consequently, an *external* source of hydrogen is necessary to meet the daily hydrogen requirements of any process where the heavier feedstocks are involved.

The trend to increase the number of hydrogenation (*hydrocracking* and/or *hydrotreating*) processes in refineries (Dolbear, 1998) coupled with the need to process the heavier oils, which require substantial quantities of hydrogen for upgrading, has resulted in vastly increased demands for this gas.

Hydrogen has historically been produced during catalytic reforming processes as a by-product of the production of the aromatic compounds used in gasoline and in solvents. As reforming processes changed from fixed-bed to cyclic to continuous regeneration, process pressures have dropped and hydrogen production per barrel of reformate has tended to increase. However, hydrogen production as a by-product is not always adequate to the needs of the refinery and other processes are necessary. Thus, hydrogen production by steam reforming or by partial oxidation of residua has also been used, particularly where heavy oil is available. Steam reforming is the dominant method for hydrogen production and is usually combined with pressure swing adsorption (PSA) to purify the hydrogen to greater than 99% by volume (Bandermann and Harder, 1982).

The gasification of residua and coke to produce hydrogen and/or power may become an attractive option for refiners (Dickenson et al., 1997; Gross and Wolff, 2000). The premise that the gasification section of a refinery will be the *garbage can* for deasphalter residues, high-sulfur coke, as well as other refinery wastes is worthy of consideration.

Several other processes are available for the production of the additional hydrogen that is necessary for the various heavy feedstock hydroprocessing sequences, and the purpose of this chapter is to present a general description of these processes. In general, most of the external hydrogen is manufactured by steam–methane reforming or by oxidation processes. Other processes such as ammonia dissociation, steam–methanol interaction, or electrolysis are also available for hydrogen production, but economic factors and feedstock availability assist in the choice between processing alternatives.

The processes described in this chapter are those gasification processes that are often referred to as the *garbage disposal units* of the refinery. Hydrogen is produced for use in other parts of the refinery as well as for energy, and it is often produced from process by-products that may not be of any use elsewhere. Such by-products might be highly aromatic, heteroatom, and metal containing reject from a deasphalting unit or from a mild hydrocracking process. However attractive this may seem, there will be the need to incorporate a gas cleaning operation to remove any environmentally objectionable components from the hydrogen gas.

15.2 PROCESSES REQUIRING HYDROGEN

The use of hydrogen in refining processes is perhaps the single most significant advance in refining technology during the twentieth century and is now an inclusion in most refineries (Figure 15.1). Hydrogenation processes for the conversion of petroleum fractions and petroleum products may be classified as (1) *hydrotreating (nondestructive hydrogenation)* and (2) *hydrocracking (destructive hydrogenation)*.

15.2.1 HYDROTREATING

Catalytic hydrotreating is a hydrogenation process used to remove approximately 90% of contaminants such as nitrogen, sulfur, oxygen, and metals from liquid petroleum fractions (Chapter 10). These contaminants, if not removed from the petroleum fractions as they travel through the refinery processing units, can have detrimental effects on the equipment, the catalysts, and the quality of the finished product. Typically, hydrotreating is done prior to processes such as catalytic reforming so that the catalyst is not contaminated by untreated feedstock. Hydrotreating is also used prior to catalytic cracking to reduce sulfur and improve product yields and to upgrade middle-distillate petroleum fractions into finished kerosene, diesel fuel, and heating fuel oils. In addition, hydrotreating converts olefins and aromatics to saturated compounds.

In a typical catalytic hydrodesulfurization unit, the feedstock is deaerated and mixed with hydrogen, preheated in a fired heater (315°C–425°C [600°F–800°F]) and then charged under pressure (up to 1000 psi) through a fixed-bed catalytic reactor (Figure 15.2). In the reactor, the sulfur and nitrogen compounds in the feedstock are converted into hydrogen sulfide and ammonia. The reaction products leave the reactor and after cooling to a low temperature enter a liquid/gas separator. The hydrogen-rich gas from the high-pressure separation is recycled to combine with the feedstock, and the low-pressure gas stream rich in hydrogen sulfide is sent to a gas-treating unit where the hydrogen

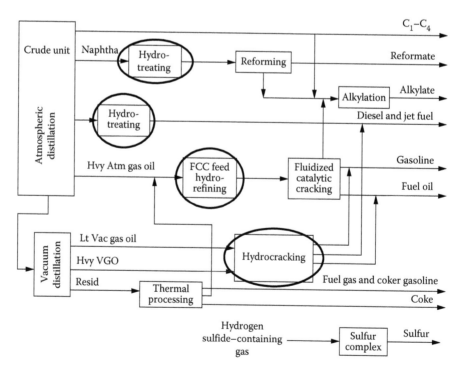

FIGURE 15.1 Example of the relative placement of hydroprocesses in a refinery.

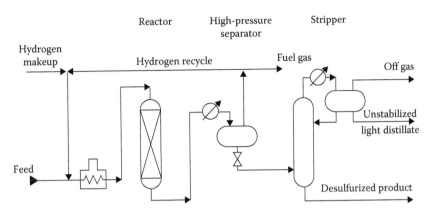

FIGURE 15.2 Distillate hydrodesulfurization. (From OSHA Technical Manual, Section IV, Chapter 2, Petroleum refining processes, 1999, http://www.osha.gov/dts/osta/otm/otm_iv/otm_iv_2.html.)

sulfide is removed. The clean gas is then suitable as fuel for the refinery furnaces. The liquid stream is the product from hydrotreating and is normally sent to a stripping column for the removal of hydrogen sulfide and other undesirable components. In cases where steam is used for stripping, the product is sent to a vacuum drier for the removal of water. Hydrodesulfurized products are blended or used as catalytic reforming feedstock.

Hydrotreating processes differ depending upon the feedstock available and catalysts used. Hydrotreating can be used to improve the burning characteristics of distillates such as kerosene. Hydrotreatment of a kerosene fraction can convert aromatics into naphthenes, which are cleaner-burning compounds. Lube-oil hydrotreating uses catalytic treatment of the oil with hydrogen to improve product quality. The objectives in mild lube hydrotreating include saturation of olefins and improvements in color, odor, and acid nature of the oil. Mild lube hydrotreating may also be used following solvent processing. Operating temperatures are usually below 315°C (600°F) and operating pressures below 800 psi. Severe lube hydrotreating, at temperatures in the 315°C–400°C (600°F–750°F) range and hydrogen pressures up to 3000 psi, is capable of saturating aromatic rings, along with sulfur and nitrogen removal, to impart specific properties not achieved at mild conditions.

Hydrotreating can also be employed to improve the quality of pyrolysis gasoline (*pygas*), a by-product from the manufacture of ethylene. Traditionally, the outlet for pygas has been motor gasoline blending, a suitable route in view of its high octane number. However, only small portions can be blended untreated owing to the unacceptable odor, color, and gum-forming tendencies of this material. The quality of pygas, which is high in diolefin content, can be satisfactorily improved by hydrotreating, whereby conversion of diolefins into monoolefins provides an acceptable product for motor gas blending.

15.2.2 HYDROCRACKING

Hydrocracking is a two-stage process combining catalytic cracking and hydrogenation, wherein heavier feedstocks are cracked in the presence of hydrogen to produce more desirable products. Hydrocracking also produces relatively large amounts of *iso*-butane for alkylation feedstock and the process also performs isomerization for pour-point control and smoke-point control, both of which are important in high-quality jet fuel.

Hydrocracking employs high pressure, high temperature, and a catalyst. Hydrocracking is used for feedstocks that are difficult to process by either catalytic cracking or reforming, since these feedstocks are characterized usually by high polycyclic aromatic content and/or high concentrations of the two principal catalyst poisons, sulfur and nitrogen compounds. The hydrocracking process largely depends on the nature of the feedstock and the relative rates of the two competing reactions, hydrogenation and cracking. Heavy aromatic feedstock is converted into lighter products under a wide range of very high pressures (1000–2000 psi) and fairly high temperatures (400°C–815°C [750°F–1500°F]), in the presence of hydrogen and special catalysts. When the feedstock has a high paraffinic content, the primary function of hydrogen is to prevent the formation of polycyclic aromatic compounds. Another important role of hydrogen in the hydrocracking process is to reduce tar formation and prevent buildup of coke on the catalyst. Hydrogenation also serves to convert sulfur and nitrogen compounds present in the feedstock to hydrogen sulfide and ammonia.

In the first stage of the process (Figure 15.3), preheated feedstock is mixed with recycled hydrogen and sent to the first-stage reactor, where catalysts convert sulfur and nitrogen compounds to hydrogen sulfide and ammonia. Limited hydrocracking also occurs. After the hydrocarbon leaves the first stage, it is cooled and liquefied and run through a hydrocarbon separator. The hydrogen is recycled to the feedstock. The liquid is charged to a fractionator. Depending on the products desired (gasoline components, jet fuel, and gas oil), the fractionator is run to cut out some portion of the first-stage reactor outturn. Kerosene-range material can be taken as a separate side-draw product or included in the fractionator bottoms with the gas oil. The fractionator bottoms are again mixed with a hydrogen stream and charged to the second stage. Since this material has already been subjected to some hydrogenation, cracking, and reforming in the first stage, the operations of the second stage are more severe (higher temperatures and pressures). Like the outturn of the first stage, the second-stage product is separated from the hydrogen and charged to the fractionator.

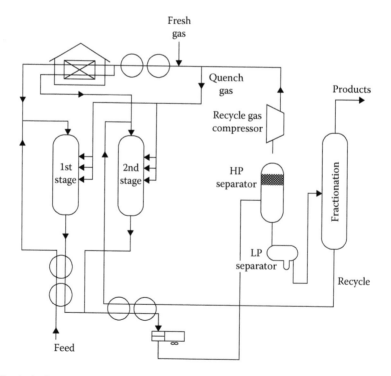

FIGURE 15.3 A single-stage or two-stage (optional) hydrocracking unit. (From OSHA Technical Manual, Section IV, Chapter 2, Petroleum refining processes, 1999, http://www.osha.gov/dts/osta/otm/otm_iv/otm_iv_2.html.)

15.3 HYDROGEN PRODUCTION

15.3.1 FEEDSTOCKS

The most common, and perhaps the best, feedstocks for steam reforming are low-boiling saturated hydrocarbons that have a low sulfur content, including natural gas, refinery gas, liquefied petroleum gas (LPG), and low-boiling naphtha.

Natural gas is the most common feedstock for hydrogen production since it meets all the requirements for reformer feedstock. Natural gas typically contains more than 90% methane and ethane with only a few percent of propane and higher-boiling hydrocarbons (Mokhatab et al., 2006; Speight, 2007, 2014). Natural gas may (or most likely will) contain traces of carbon dioxide with some nitrogen and other impurities. Purification of natural gas, before reforming, is usually relatively straightforward. Traces of sulfur must be removed to avoid poisoning the reformer catalyst; zinc oxide treatment in combination with hydrogenation is usually adequate.

Light refinery gas, containing a substantial amount of hydrogen, can be an attractive steam reformer feedstock since it is produced as a by-product. Processing of refinery gas will depend on its composition, particularly the levels of olefins and of propane and heavier hydrocarbons. Olefins, that can cause problems by forming coke in the reformer, are converted to saturated compounds in the hydrogenation unit. Higher-boiling hydrocarbons in refinery gas can also form coke, either on the primary reformer catalyst or in the preheater. If there is more than a few percent of C_3 and higher compounds, a promoted reformer catalyst should be considered in order to avoid carbon deposits.

Refinery gas from different sources varies in suitability as hydrogen plant feed. Catalytic reformer off-gas (Chapter 13), for example, is saturated, is very low in sulfur, and often has high hydrogen content. The process gases from a coking unit (Chapter 8) or from a fluid catalytic cracking unit (Chapter 9) are much less desirable because of the content of unsaturated constituents. In addition to olefins, these gases contain substantial amounts of sulfur that must be removed before the gas is used as feedstock. These gases are also generally unsuitable for direct hydrogen recovery, since the hydrogen content is usually too low. Hydrotreater off-gas lies in the middle of the range. It is saturated, so it is readily used as hydrogen plant feed. Content of hydrogen and heavier hydrocarbons depends to a large extent on the upstream pressure. Sulfur removal will generally be required.

15.3.2 CHEMISTRY

Before the feedstock is introduced to a process, the application of a strict feedstock purification protocol is needed. Prolonging catalyst life in hydrogen production processes is attributable to effective feedstock purification, particularly sulfur removal. A typical natural gas or other light hydrocarbon feedstock contains traces of hydrogen sulfide and organic sulfur.

In order to remove sulfur compounds, it is necessary to hydrogenate the feedstock to convert the organic sulfur to hydrogen that is then reacted with zinc oxide (ZnO) at approximately 370°C (700°F) that results in the optimal use of the zinc oxide as well as ensuring complete hydrogenation. Thus, assuming assiduous feedstock purification and removal of all of the objectionable contaminants, the chemistry of hydrogen production can be defined.

In *steam reforming*, low-boiling hydrocarbons such as methane are reacted with steam to form hydrogen:

$$CH_4 + H_2O \rightarrow 3H_2 + CO \quad \Delta H_{298\,K} = +97,400 \text{ Btu/lb}$$

where H is the heat of reaction. A more general form of the equation that shows the chemical balance for higher-boiling hydrocarbons is

$$C_nH_m + nH_2O \rightarrow (n + m/2)H_2 + nCO$$

The reaction is typically carried out at approximately 815°C (1500°F) over a nickel catalyst packed into the tubes of a reforming furnace. The high temperature also causes the hydrocarbon feedstock to undergo a series of cracking reactions, plus the reaction of carbon with steam:

$$CH_4 \rightarrow 2H_2 + C$$

$$C + H_2O \rightarrow CO + H_2$$

Carbon is produced on the catalyst at the same time that hydrocarbon is reformed to hydrogen and carbon monoxide. With natural gas or similar feedstock, reforming predominates and the carbon can be removed by reaction with steam as fast as it is formed. When higher-boiling feedstocks are used, the carbon is not removed fast enough and builds up thereby requiring catalyst regeneration or replacement. Carbon buildup on the catalyst (when high-boiling feedstocks are employed) can be avoided by the addition of alkali compounds, such as potash, to the catalyst thereby encouraging or promoting the carbon–steam reaction.

However, even with an alkali-promoted catalyst, feedstock cracking limits the process to hydrocarbons with a boiling point less than of 180°C (350°F). Natural gas, propane, butane, and light naphtha are most suitable. Pre-reforming, a process that uses an adiabatic catalyst bed operating at a lower temperature, can be used as a pretreatment to allow heavier feedstocks to be used with lower potential for carbon deposition (coke formation) on the catalyst.

After reforming, the carbon monoxide in the gas is reacted with steam to form additional hydrogen (the *water-gas shift* reaction):

$$CO + H_2O \rightarrow CO_2 + H_2 \quad \Delta H_{298\,K} = -16,500 \text{ Btu/lb}$$

This leaves a mixture consisting primarily of hydrogen and carbon monoxide that is removed by conversion to methane:

$$CO + 3H_2O \rightarrow CH_4 + H_2O$$

$$CO_2 + 4H_2 \rightarrow CH_4 + 2H_2O$$

The critical variables for steam reforming processes are (1) temperature, (2) pressure, and (3) the steam/hydrocarbon ratio. Steam reforming is an equilibrium reaction, and conversion of the hydrocarbon feedstock is favored by high temperature, which in turn requires higher fuel use. Because of the volume increase in the reaction, conversion is also favored by low pressure, which conflicts with the need to supply the hydrogen at high pressure. In practice, materials of construction limit temperature and pressure.

On the other hand, and in contrast to reforming, shift conversion is favored by low temperature. The gas from the reformer is reacted over iron oxide catalyst at 315°C–370°C (600°F–700°F) with the lower limit being dictated activity of the catalyst at low temperature.

Hydrogen can also be produced by *partial oxidation* (POX) of hydrocarbons in which the hydrocarbon is oxidized in a limited or controlled supply of oxygen:

$$2CH_4 + O_2 \rightarrow CO + 4H_2 \quad \Delta H_{298\,K} = -10,195 \text{ Btu/lb}$$

The shift reaction also occurs, and a mixture of carbon monoxide and carbon dioxide is produced in addition to hydrogen. The catalyst tube materials do not limit the reaction temperatures in partial oxidation processes, and higher temperatures may be used, which enhance the conversion of methane to hydrogen. Indeed, much of the design and operation of hydrogen plants involves protecting the reforming catalyst and the catalyst tubes because of the extreme temperatures and the sensitivity of the catalyst. In fact, minor variations in feedstock composition or operating conditions can have significant effects on the life of the catalyst or the reformer itself. This is particularly true of changes in molecular weight of the feed gas or poor distribution of heat to the catalyst tubes. Since the high temperature takes the place of a catalyst, partial oxidation is not limited to the lower-boiling feedstocks that are required for steam reforming. Partial oxidation processes were first considered for hydrogen production because of expected shortages of lower-boiling feedstocks and the need to have available a disposal method for higher-boiling, high-sulfur streams such as asphalt or petroleum coke.

Catalytic partial oxidation, also known as autothermal reforming, reacts oxygen with a light feedstock and by passing the resulting hot mixture over a reforming catalyst. The use of a catalyst allows the use of lower temperatures than in noncatalytic partial oxidation, which causes a reduction in oxygen demand.

The feedstock requirements for catalytic partial oxidation processes are similar to the feedstock requirements for steam reforming and light hydrocarbons from refinery gas to naphtha are preferred. The oxygen substitutes for much of the steam in preventing coking and a lower steam/carbon ratio is required. In addition, because a large excess of steam is not required, catalytic partial oxidation produces more carbon monoxide and less hydrogen than steam reforming. Thus, the process is more suited to situations where carbon monoxide is the more desirable product such as synthesis gas for chemical feedstocks.

15.3.3 CATALYSTS

Hydrogen plants are one of the most extensive users of catalysts in the refinery. Catalytic operations include hydrogenation, steam reforming, shift conversion, and methanation.

15.3.3.1 Reforming Catalysts

The reforming catalyst is usually supplied as nickel oxide that, during start-up, is heated in a stream of inert gas, then steam. When the catalyst is near the normal operating temperature, hydrogen or a light hydrocarbon is added to reduce the nickel oxide to metallic nickel.

The high temperatures (up to 870°C [1600°F]) and the nature of the reforming reaction require that the reforming catalyst be used inside the radiant tubes of a reforming furnace. The active agent in reforming catalyst is nickel, and normally, the reaction is controlled both by diffusion and by heat transfer. Catalyst life is limited as much by physical breakdown as by deactivation.

Sulfur is the main catalyst poison, and the catalyst poisoning is theoretically reversible with the catalyst being restored to near full activity by steaming. However, in practice, the deactivation may cause the catalyst to overheat and coke to the point that it must be replaced. Reforming catalysts are also sensitive to poisoning by heavy metals, although these are rarely present in low-boiling hydrocarbon feedstocks and in naphtha feedstocks.

Coking deposition on the reforming catalyst and ensuing gloss of catalyst activity is the most characteristic issue that must be assessed and mitigated.

While methane-rich streams such as natural gas or light refinery gas are the most common feeds to hydrogen plants, there is often a requirement for variety of reasons to process a variety of higher-boiling feedstocks, such as liquefied petroleum gas and naphtha. Feedstock variations may also be inadvertent due, for example, to changes in refinery off-gas composition from another unit or because of variations in naphtha composition because of feedstock variance to the naphtha unit.

Thus, when using higher-boiling feedstocks in a hydrogen plant, coke deposition on the reformer catalyst becomes a major issue. Coking is most likely in the reformer unit at the point where both temperature and hydrocarbon content are high enough. In this region, hydrocarbons crack and form coke faster than the coke is removed by reaction with steam or hydrogen and when catalyst deactivation occurs, there is a simultaneous temperature increase with a concomitant increase in coke formation and deposition. In other zones, where the hydrocarbon-to-hydrogen ratio is lower, there is less risk of coking.

Coking depends to a large extent on the balance between catalyst activity and heat input with the more active catalysts producing higher yields of hydrogen at lower temperature thereby reducing the risk of coking. A uniform input of heat is important in this region of the reformer since any catalyst voids or variations in catalyst activity can produce localized hot spots leading to coke formation and/or reformer failure.

Coke formation results in hotspots in the reformer that increases pressure drop, reduces feedstock (methane) conversion, leading eventually to reformer failure. Coking may be partially mitigated by increasing the steam/feedstock ratio to change the reaction conditions, but the most effective solution may be to replace the reformer catalyst with one designed for higher-boiling feedstocks.

A *standard* steam–methane reforming catalyst uses nickel on an alpha-alumina ceramic carrier that is acidic in nature. Promotion of hydrocarbon cracking with such a catalyst leads to coke formation from higher-boiling feedstocks. Some catalyst formulations use a magnesia/alumina (MgO/Al$_2$O$_3$) support that is less acidic than α-alumina that reduces cracking on the support and allows higher-boiling feedstocks (such as liquefied petroleum gas) to be used.

Further resistance to coking can be achieved by adding an alkali promoter, typically some form of potash (KOH) to the catalyst. Besides reducing the acidity of the carrier, the promoter catalyzes the reaction of steam and carbon. While carbon continues to be formed, it is removed faster than it can build up. This approach can be used with naphtha feedstocks boiling point up to approximately 180°C (350°F). Under the conditions in a reformer, potash is volatile and it is incorporated into the catalyst as a more complex compound that slowly hydrolyzes to release potassium hydroxide (KOH). Alkali-promoted catalyst allows the use of a wide range of feedstocks but, in addition to possible potash migration, which can be minimized by proper design and operation, the catalyst is also somewhat less active than conventional catalyst.

Another option to reduce coking in steam reformers is to use a *pre-reformer* in which a fixed bed of catalyst, operating at a lower temperature, upstream of the fired reformer is used. In a pre-reformer, adiabatic steam–hydrocarbon reforming is performed outside the fired reformer in a vessel containing a high nickel catalyst. The heat required for the endothermic reaction is provided by hot flue gas from the reformer convection section. Since the feed to the fired reformer is now partially reformed, the steam–methane reformer can operate at an increased feed rate and produce 8%–10% additional hydrogen at the same reformer load. An additional advantage of the pre-reformer is that it facilitates higher mixed feed preheat temperatures and maintains relatively constant operating conditions within the fired reformer regardless of variable refinery off-gas feed conditions. Inlet temperatures are selected so that there is minimal risk of coking and the gas leaving the pre-reformer contains only steam, hydrogen, carbon monoxide, carbon dioxide, and methane. This allows a standard methane catalyst to be used in the fired reformer, and this approach has been used with feedstocks up to light kerosene. Since the gas leaving the pre-reformer poses reduced risk of coking, it can compensate to some extent for variations in catalyst activity and heat flux in the primary reformer.

15.3.3.2 Shift Conversion Catalysts

The second important reaction in a steam reforming plant is the shift conversion reaction:

$$CO + H_2O \rightarrow CO_2 + H_2$$

Two basic types of shift catalyst are used in steam reforming plants: iron/chrome high-temperature shift catalysts and copper/zinc low-temperature shift catalysts.

High-temperature shift catalysts operate in the range of 315°C–430°C (600°F–800°F) and consist primarily of magnetite (Fe$_3$O$_4$) with three-valent chromium oxide (Cr$_2$O$_3$) added as a stabilizer. The catalyst is usually supplied in the form of ferric oxide (Fe$_2$O$_3$) and six-valent chromium oxide (CrO$_3$) and is reduced by the hydrogen and carbon monoxide in the shift feed gas as part of the startop procedure to produce the catalyst in the desired form. However, caution is necessary since if the steam/carbon ratio of the feedstock is too low and the reducing environment too strong, the catalyst can be reduced further to metallic iron. Metallic iron is a catalyst for Fischer–Tropsch reactions and hydrocarbons will be produced (Davis and Occelli, 2010).

Low-temperature shift catalysts operate at temperatures on the order of 205°C–230°C (400°F–450°F). Because of the lower temperature, the reaction equilibrium is more controllable and lower amounts of carbon monoxide are produced. The low-temperature shift catalyst is primarily used in wet scrubbing plants that use a methanation for final purification. Pressure swing adsorption plants do not generally use a low-temperature because since any unconverted carbon monoxide is recovered as reformer fuel. Low-temperature shift catalysts are sensitive to poisoning by sulfur and are sensitive to water (liquid) that can cause softening of the catalyst followed by crusting or plugging.

The catalyst is supplied as copper oxide (CuO) on a zinc oxide (ZnO) carrier, and the copper must be reduced by heating it in a stream of inert gas with measured quantities of hydrogen. The reduction of the copper oxide is strongly exothermic and must be closely monitored (Davis and Occelli, 2010).

15.3.3.3 Methanation Catalysts

In wet scrubbing plants, the final hydrogen purification procedure involved is by methanation in which the carbon monoxide and carbon dioxide are converted to methane:

$$CO + 3H_2O \rightarrow CH_4 + H_2O$$

$$CO_2 + 4H_2 \rightarrow CH_4 + 2H_2O$$

The active agent is nickel on an alumina carrier.

The catalyst has a long life, as it operates under ideal conditions and is not exposed to poisons. The main source of deactivation is plugging from the carryover of carbon dioxide from removal solutions.

The most severe hazard arises from high levels of carbon monoxide or carbon dioxide, which can result from the breakdown of the carbon dioxide removal equipment or from exchanger tube leaks that quench the shift reaction. The results of breakthrough can be severe, since the methanation reaction produces a temperature rise of 70°C (125°F) per 1% of carbon monoxide or a temperature rise of 33°C (60°F) per 1% of carbon dioxide. While the normal operating temperature during methanation is approximately 315°C (600°F), it is possible to reach 700°C (1300°F) in cases of major breakthrough.

15.4 HYDROGEN PURIFICATION

When the hydrogen content of the refinery gas is greater than 50% by volume, the gas should first be considered for hydrogen recovery, using a membrane (Brüschke, 1995, 2003) or pressure swing adsorption unit (Figure 15.4). The tail gas or reject gas that will still contain a substantial amount of hydrogen can then be used as steam reformer feedstock. Generally, the feedstock purification process uses three different refinery gas streams to produce hydrogen. First, high-pressure hydrocracker purge gas is purified in a membrane unit that produces hydrogen at medium pressure and is combined with medium pressure off-gas that is first purified in a pressure swing adsorption unit. Finally, low-pressure off-gas is compressed, mixed with reject gases from the membrane and pressure swing adsorption units, and used as steam reformer feed.

Various processes are available to purify the hydrogen stream but since the product streams are available as a wide variety of composition, flows, and pressures, the best method of purification will vary. And there are several factors that must also be taken into consideration in the selection of a purification method. These are (1) hydrogen recovery, (2) product purity, (3) pressure profile,

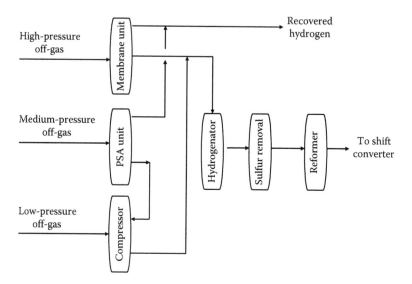

FIGURE 15.4 Hydrogen purification.

(4) reliability, and (5) cost, an equally important parameter is not considered here since the emphasis is on the technical aspects of the purification process.

15.4.1 WET SCRUBBING

Wet scrubbing systems, particularly amine or potassium carbonate systems, are used for the removal of acid gases such as hydrogen sulfide or carbon dioxide (Speight, 1993; Dalrymple et al., 1994). Most systems depend on chemical reaction and can be designed for a wide range of pressures and capacities. They were once widely used to remove carbon dioxide in steam reforming plants but have generally been replaced by pressure swing adsorption units except where carbon monoxide is to be recovered. Wet scrubbing is still used to remove hydrogen sulfide and carbon dioxide in partial oxidation plants.

Wet scrubbing systems remove only acid gases or higher boiling hydrocarbon constituents but they do not remove methane or other hydrocarbon gases, which hence have little influence on product purity. Therefore, wet scrubbing systems are most often used as a pretreatment step, or where a hydrogen-rich stream is to be desulfurized for use as fuel gas.

15.4.2 PRESSURE SWING ADSORPTION UNITS

Pressure swing adsorption units use beds of solid adsorbent to separate impurities from hydrogen streams leading to high-purity high-pressure hydrogen and a low-pressure tail gas stream containing the impurities and some of the hydrogen. The beds are then regenerated by depressurizing and purging. Part of the hydrogen (up to 20%) may be lost in the tail gas. Pressure swing adsorption is generally the purification method of choice for steam reforming units because of its production of high-purity hydrogen and is also used for the purification of refinery off-gases, where it competes with membrane systems.

Many hydrogen plants that formerly used a *wet scrubbing process* (Figure 15.5) for hydrogen purification are now using the *pressure swing adsorption* (PSA) (Figure 15.6) for purification. The pressure swing adsorption process is a cyclic process that uses beds of solid adsorbent to remove impurities from the gas and generally produces higher-purity hydrogen (99.9% v/v purity compared to less than 97% v/v purity). The purified hydrogen passes through the adsorbent beds

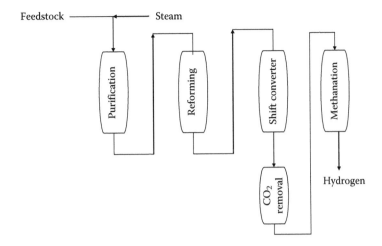

FIGURE 15.5 Hydrogen purification by wet scrubbing.

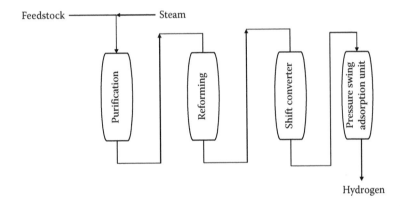

FIGURE 15.6 Hydrogen purification by pressure swing adsorption.

with only a tiny fraction absorbed, and the beds are regenerated by depressurization followed by purging at low pressure.

When the beds are depressurized, a waste gas (or *tail gas*) stream is produced and consists of the impurities from the feed (carbon monoxide, carbon dioxide, methane, and nitrogen) plus some hydrogen. This stream is burned in the reformer as fuel and reformer operating conditions in a pressure swing adsorption plant are set so that the tail gas provides no more than approximately 85% of the reformer fuel. This gives good burner control because the tail gas is more difficult to burn than regular fuel gas, and the high content of carbon monoxide can interfere with the stability of the flame. As the reformer operating temperature is increased, the reforming equilibrium shifts, resulting in more hydrogen and less methane in the reformer outlet and hence less methane in the tail gas.

15.4.3 Membrane Systems

Membrane systems separate gases by taking advantage of the difference in rates of diffusion through membranes (Brüschke, 1995, 2003). Gases that diffuse faster (including hydrogen) become the permeate stream and are available at low pressure, whereas the slower-diffusing gases become the nonpermeate and leave the unit at a pressure close to the pressure of the feedstock at entry point. Membrane systems contain no moving parts or switch valves and have potentially very high reliability. The major threat is from components in the gas (such as aromatics) that attack the membranes, or from liquids, which plug them.

Membranes arc fabricated in relatively small modules; for larger capacity more modules are added. Cost is therefore virtually linear with capacity, making them more competitive at lower capacities. The design of membrane systems involves a trade-off between pressure drop (and diffusion rate) and surface area as well as between product purity and recovery. As the surface area is increased, the recovery of fast components increases; however, more of the slow components are recovered, which lowers the purity.

15.4.4 Cryogenic Separation

Cryogenic separation units operate by cooling the gas and condensing some, or all, of the constituents for the gas stream. Depending on the product purity required, separation may involve flashing or distillation. Cryogenic units offer the advantage of being able to separate a variety of products from a single feed stream. One specific example is the separation of light olefins from a hydrogen stream.

Hydrogen recovery is in the range of 95%, with purity above 98% obtainable.

15.5 HYDROGEN MANAGEMENT

Many existing refinery hydrogen plants use a conventional process, which produces a medium-purity (94%–97%) hydrogen product by removing the carbon dioxide in an absorption system and methanation of any remaining carbon oxides. Since the 1980s, most hydrogen plants are built with pressure swing adsorption (PSA) technology to recover and purify the hydrogen to purities above 99.9%. Since many refinery hydrogen plants utilize refinery off-gas feeds containing hydrogen, the actual maximum hydrogen capacity that can be synthesized via steam reforming is not certain since the hydrogen content of the off-gas can change due to operational changes in the hydrotreaters.

Hydrogen management has become a priority in current refinery operations and when planning to produce lower sulfur gasoline and diesel fuels (Zagoria et al., 2003; Luckwal and Mandal, 2009). Along with increased hydrogen consumption for deeper hydrotreating, additional hydrogen is needed for processing heavier and higher sulfur crude slates. In many refineries, hydroprocessing capacity and the associated hydrogen network is limiting refinery throughput and operating margins. Furthermore, higher hydrogen purities within the refinery network are becoming more important to boost hydrotreater capacity, achieve product value improvements, and lengthen catalyst life cycles.

Improved hydrogen utilization and expanded or new sources for refinery hydrogen and hydrogen purity optimization are now required to meet the needs of the future transportation fuel market and the drive toward higher refinery profitability. Many refineries developing hydrogen management programs fit into the two general categories of either a catalytic reformer supplied network or an on-purpose hydrogen supply.

Some refineries depend solely on catalytic reformer(s) as their source of hydrogen for hydrotreating. Often, they are semi-regenerative reformers where off-gas hydrogen quantity, purity, and availability change with feed naphtha quality, as octane requirements change seasonally, and when the reformer catalyst progresses from start-of-run to end-of-run conditions and then goes off-line for regeneration. Typically, during some portions of the year, refinery margins are reduced as a result of hydrogen shortages.

Multiple hydrotreating units compete for hydrogen—either by selectively reducing throughput, managing intermediate tankage logistics, or running the catalytic reformer suboptimally just to satisfy downstream hydrogen requirements. Part of the operating year still runs in hydrogen surplus, and the network may be operated with relatively low hydrogen utilization (consumption/production) at 70%–80%. Catalytic reformer off-gas hydrogen supply may swing from 75% to 85% hydrogen purity.

Hydrogen purity upgrade can be achieved through some hydrotreaters by absorbing heavy hydrocarbons. But without supplemental hydrogen purification, critical control of hydrogen partial pressure in hydroprocessing reactors is difficult, which can affect catalyst life, charge rates, and/or gasoline yields.

More complex refineries, especially those with hydrocracking units, also have on-purpose hydrogen production, typically with a steam–methane reformer that utilizes refinery off-gas and supplemental natural gas as feedstock. The steam–methane reformer plant provides the swing hydrogen requirements at higher purities (92% to more than 99% hydrogen) and serves a hydrogen network configured with several purity and pressure levels. Multiple purities and existing purification units allow for more optimized hydroprocessing operation by controlling hydrogen partial pressure for maximum benefit. Typical hydrogen utilization is 85%–95%.

15.6 COMMERCIAL PROCESSES

In spite of the use of low-quality hydrogen (that contain up to 40% by volume hydrocarbon gases), a high-purity hydrogen stream (95%–99% by volume hydrogen) is required for hydrodesulfurization, hydrogenation, hydrocracking, and petrochemical processes. Hydrogen, produced as a by-product of refinery processes (principally hydrogen recovery from catalytic reformer product gases), often is not enough to meet the total refinery requirements, necessitating the manufacturing of additional hydrogen or obtaining supply from external sources.

Catalytic reforming remains an important process used to convert low-octane naphtha into high-octane gasoline blending components called *reformate*. Reforming represents the total effect of numerous reactions such as cracking, polymerization, dehydrogenation, and isomerization taking place simultaneously. Depending on the properties of the naphtha feedstock (as measured by the paraffin, olefin, naphthene, and aromatic content) and catalysts used, reformate can be produced with very high concentrations of toluene, benzene, xylene, and other aromatics useful in gasoline blending and petrochemical processing. Hydrogen, a significant by-product, is separated from reformate for recycling and use in other processes.

A catalytic reformer comprises a reactor section and a product recovery section. More or less standard is a feed preparation section in which, by combination of hydrotreatment and distillation, the feedstock is prepared to specification. Most processes use platinum as the active catalyst. Sometimes, platinum is combined with a second catalyst (bimetallic catalyst) such as rhenium or another noble metal. There are many different commercial catalytic reforming processes including Platforming (Figure 15.7), Powerforming, Ultraforming, and Thermofor catalytic reforming (Speight and Ozum, 2002; Parkash, 2003; Hsu and Robinson, 2006; Gary et al., 2007; Speight, 2014). In the Platforming process, the first step is preparation of the naphtha feed to remove impurities from the naphtha and reduce catalyst degradation. The naphtha feedstock is then mixed with hydrogen, vaporized, and passed through a series of alternating furnace and fixed-bed reactors containing a platinum catalyst. The effluent from the last reactor is cooled and sent to a separator to permit the removal of the hydrogen-rich gas stream from the top of the separator for recycling. The liquid product from the bottom of the separator is sent to a fractionator called a stabilizer (butanizer), and the bottom product (reformate) is sent to storage and butanes and lighter gases pass overhead and are sent to the saturated gas plant.

Some catalytic reformers operate at low pressure (50–200 psi), and others operate at high pressures (up to 1000 psi). Some catalytic reforming systems continuously regenerate the catalyst in other systems. One reactor at a time is taken offstream for catalyst regeneration, and some facilities regenerate all of the reactors during turnarounds. Operating procedures should be developed to ensure control of hot spots during start-up. Safe catalyst handling is very important and care must be taken not to break or crush the catalyst when loading the beds, as the small fines will plug up the reformer screens. Precautions against dust when regenerating or replacing catalyst should also be considered, and a water wash should be considered where stabilizer fouling has occurred due to the formation of ammonium chloride and iron salts. Ammonium chloride may form in pretreater

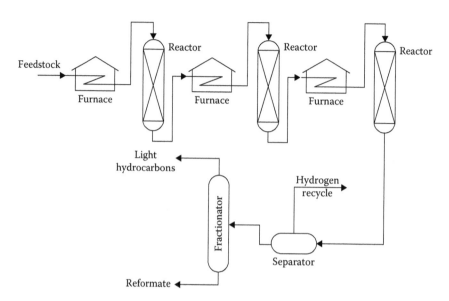

FIGURE 15.7 The Platforming process. (From OSHA Technical Manual, Section IV, Chapter 2, Petroleum refining processes, 1999, http://www.osha.gov/dts/osta/otm/otm_iv/otm_iv_2.html.)

exchangers and cause corrosion and fouling. Hydrogen chloride from the hydrogenation of chlorine compounds may form acid or ammonium chloride salt.

15.6.1 HEAVY RESIDUE GASIFICATION AND COMBINED CYCLE POWER GENERATION

Heavy residua are gasified and the produced gas is purified to clean fuel gas (Gross and Wolff, 2000). As an example, solvent deasphalter residuum is gasified by partial oxidation method under pressure of approximately 570 psi and at a temperature between 1300°C and 1500°C (2370°F and 2730°F). The high temperature generated gas flows into the specially designed waste heat boiler, in which the hot gas is cooled and high-pressure saturated steam is generated. The gas from the waste heat boiler is then heat exchanged with the fuel gas and flows to the carbon scrubber, where unreacted carbon particles are removed from the generated gas by water scrubbing.

The gas from the carbon scrubber is further cooled by the fuel gas and boiler feed water and led into the sulfur compound removal section, where hydrogen sulfide (H_2S) and carbonyl sulfide (COS) are removed from the gas to obtain clean fuel gas. This clean fuel gas is heated with the hot gas generated in the gasifier and finally supplied to the gas turbine at a temperature of 250°C–300°C (480°F–570°F).

The exhaust gas from the gas turbine having a temperature of approximately 550°C–600°C (1020°F–1110°F) flows into the heat recovery steam generator consisting of five heat exchange elements. The first element is a superheater in which the combined stream of the high-pressure saturated steam generated in the waste heat boiler and in the second element (high-pressure steam evaporator) is superheated. The third element is an economizer, the fourth element is a low-pressure steam evaporator, and the final or the fifth element is a deaerator heater. The off-gas from heat recovery steam generator having a temperature of approximately 130°C (265°F) is emitted into the air via stack.

In order to decrease the nitrogen oxide (NO_x) content in the flue gas, two methods can be applied. The first method is the injection of water into the gas turbine combustor. The second method is to selectively reduce the nitrogen oxide content by injecting ammonia gas in the presence of de-NO_x catalyst that is packed in a proper position of the heat recovery steam generator. The latter is more effective than the former to lower the nitrogen oxide emissions to the air.

15.6.2 HYBRID GASIFICATION PROCESS

In the hybrid gasification process, a slurry of coal and residual oil is injected into the gasifier where it is pyrolyzed in the upper part of the reactor to produce gas and chars. The chars produced are then partially oxidized to ash. The ash is removed continuously from the bottom of the reactor.

In this process, coal and vacuum residue are mixed together into slurry to produce clean fuel gas. The slurry fed into the pressurized gasifier is thermally cracked at a temperature of 850°C–950°C (1560°F–1740°F) and is converted into gas, tar, and char. The mixture of oxygen and steam in the lower zone of the gasifier gasifies the char. The gas leaving the gasifier is quenched to a temperature of 450°C (840°F) in the fluidized-bed heat exchanger and is then scrubbed to remove tar, dust, and steam at around 200°C (390°F).

The coal and residual oil slurry is gasified in the fluidized-bed gasifier. The charged slurry is converted to gas and char by thermal cracking reactions in the upper zone of the fluidized bed. The produced char is further gasified with steam and oxygen that enter the gasifier just below the fluid-izing gas distributor. Ash is discharged from the gasifier and indirectly cooled with steam and then discharged into the ash hopper. It is burned with an incinerator to produce process steam. Coke deposited on the silica sand is removed in the incinerator.

15.6.3 HYDROCARBON GASIFICATION

The gasification of hydrocarbons to produce hydrogen is a continuous, noncatalytic process that involves partial oxidation of the hydrocarbon. Air or oxygen (with steam or carbon dioxide) is used as the oxidant at 1095°C–1480°C (2000°F–2700°F). Any carbon produced (2%–3% by weight of the feedstock) during the process is removed as a slurry in a carbon separator and pelletized for use either as a fuel or as raw material for carbon-based products.

15.6.4 HYPRO PROCESS

The Hypro process is a continuous catalytic method for hydrogen manufacture from natural gas or from refinery effluent gases. The process is designed to convert natural gas:

$$CH_4 \rightarrow C + 2H_2$$

Hydrogen is recovered by phase separation to yield hydrogen of approximately 93% purity; the principal contaminant is methane.

15.6.5 PYROLYSIS PROCESSES

There has been recent interest in the use of pyrolysis processes to produce hydrogen. Specifically, the interest has focused on the pyrolysis of methane (natural gas) and hydrogen sulfide.

Natural gas is readily available and offers relatively rich stream of methane with lower amounts of ethane, propane, and butane also being present. The thermocatalytic decomposition of natural gas hydrocarbons offers an alternate method for the production of hydrogen (Uemura et al., 1999; Weimer et al., 2000):

$$C_nH_m \rightarrow nC + (m/2)H_2$$

If a hydrocarbon fuel such as natural gas (methane) is to be used for hydrogen production by direct decomposition, then the process that is optimized to yield hydrogen production may not be suit-able for the production of high-quality carbon black by-product intended for the industrial rub-ber market. Moreover, it appears that the carbon produced from high-temperature (850°C–950°C

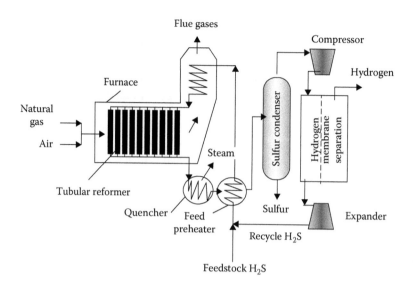

FIGURE 15.8 Simplified schematic for the production of hydrogen from hydrogen sulfide.

[1560°F–1740°F]) direct thermal decomposition of methane is soot-like material with high tendency for the catalyst deactivation. Thus, if the object of methane decomposition is hydrogen production, the carbon by-product may not be marketable as high-quality carbon black for rubber and tire applications.

The production of hydrogen by direct decomposition of hydrogen sulfide has been studied extensively and a process proposed (Figure 15.8; Clark and Wassink, 1990; Zaman and Chakma, 1995; Donini, 1996; Luinstra, 1996). Hydrogen sulfide decomposition is a highly endothermic process and equilibrium yields are poor (Clark et al., 1995). At temperatures less than 1500°C (2730°F), the thermodynamic equilibrium is unfavorable toward hydrogen formation. However, in the presence of catalysts such as platinum–cobalt (at 1000°C; 1830°F), disulfides of molybdenum, or tungsten Mo or W at 800°C (1470°F) (Kotera et al., 1976), or other transition metal sulfides supported on alumina (at 500°C–800°C [930°F–1470°F]), decomposition of hydrogen sulfide proceeds rapidly (Kiuchi, 1982, Bishara et al., 1987, Al-Shamma and Naman, 1989, Clark and Wassink, 1990, Megalofonos and Papayannakos, 1997; Raissi, 2001). In the temperature range of approximately 800°C–1500°C (1470°F–2730°F), thermolysis of hydrogen sulfide can be treated simply:

$$H_2S \rightarrow H_2 + 1/xS_x \quad \Delta H_{298\,K} = +34{,}300 \text{ Btu/lb}$$

where x = 2. Outside this temperature range, multiple equilibria may be present depending on temperature, pressure, and relative abundance of hydrogen and sulfur.

Above approximately 1000°C (1830°F), there is a limited advantage to using catalysts since the thermal reaction proceeds to equilibrium very rapidly (Clark and Wassink, 1990). The hydrogen yield can be doubled by preferential removal of either H_2 or sulfur from the reaction environment, thereby shifting the equilibrium. The reaction products must be quenched quickly after leaving the reactor to prevent reversible reactions.

15.6.6 SHELL GASIFICATION PROCESS

The shell gasification process (*partial oxidation process*) is a flexible process for generating synthesis gas, principally hydrogen and carbon monoxide, for the ultimate production of high-purity high-pressure hydrogen, ammonia, methanol, fuel gas, town gas, or reducing gas by the reaction of gaseous or liquid hydrocarbons with oxygen, air, or oxygen-enriched air.

The most important step in converting heavy residue to industrial gas is the partial oxidation of the oil using oxygen with the addition of steam. The gasification process takes place in an empty, refractory-lined reactor at temperatures of approximately 1400°C (2550°F) and pressures between 29 and 1140 psi. The chemical reactions in the gasification reactor proceed without catalyst to produce gas-containing carbon amounting to some 0.5%–2% by weight, based on the feedstock. The carbon is removed from the gas with water, extracted in most cases with feed oil from the water and returned to the feed oil. The high reformed gas temperature is utilized in a waste heat boiler for generating steam. The steam is generated at 850–1565 psi. Some of this steam is used as process steam and for oxygen and oil preheating. The surplus steam is used for energy production and heating purposes.

15.6.7 Steam–Methane Reforming

Steam–methane reforming is the benchmark process that has been employed over a period of several decades for hydrogen production. The process involves reforming natural gas in a continuous catalytic process in which the major reaction is the formation of carbon monoxide and hydrogen from methane and steam:

$$CH_4 + H_2O = CO + 3H_2 \quad \Delta H_{298\ K} = +97,400\ Btu/lb$$

Higher-molecular-weight feedstocks can also be reformed to hydrogen:

$$C_3H_8 + 3H_2O \rightarrow 3CO + 7H_2$$

That is,

$$C_nH_m + nH_2O \rightarrow nCO + (0.5m + n)H_2$$

In the actual process, the feedstock is first desulfurized by passage through activated carbon, which may be preceded by caustic and water washes. The desulfurized material is then mixed with steam and passed over a nickel-based catalyst (730°C–845°C [1350°F–1550°F] and 400 psi). Effluent gases are cooled by the addition of steam or condensate to approximately 370°C (700°F), at which point carbon monoxide reacts with steam in the presence of iron oxide in a shift converter to produce carbon dioxide and hydrogen:

$$CO + H_2O = CO_2 + H_2 \quad \Delta H298K = -41.16\ kJ/mol$$

The carbon dioxide is removed by amine washing; the hydrogen is usually a high-purity (>99%) material.

Since the presence of any carbon monoxide or carbon dioxide in the hydrogen stream can interfere with the chemistry of the catalytic application, a third stage is used to convert these gases to methane:

$$CO + 3H_2 \rightarrow CH_4 + H_2O$$

$$CO_2 + 4H_2 \rightarrow CH_4 + 2H_2O$$

For many refiners, sulfur-free natural gas (CH_4) is not always available to produce hydrogen by this process. In that case, higher-boiling hydrocarbons (such as propane, butane, or naphtha) may be used as the feedstock to generate hydrogen (q.v.).

The net chemical process for steam–methane reforming is then given by

$$CH_4 + 2H_2O \rightarrow CO_2 + 4H_2 \quad \Delta H298K = +165.2 \text{ kJ/mol}$$

Indirect heating provides the required overall endothermic heat of reaction for the steam–methane reforming.

One way of overcoming the thermodynamic limitation of steam reforming is to remove either hydrogen or carbon dioxide as it is produced, hence shifting the thermodynamic equilibrium toward the product side. The concept for sorption-enhanced steam–methane reforming is based on *in situ* removal of carbon dioxide by a sorbent such as calcium oxide (CaO).

$$CaO + CO_2 \rightarrow CaCO_3$$

Sorption enhancement enables lower reaction temperatures, which may reduce catalyst coking and sintering, while enabling the use of less expensive reactor wall materials. In addition, heat release by the exothermic carbonation reaction supplies most of the heat required by the endothermic reforming reactions. However, energy is required to regenerate the sorbent to its oxide form by the energy-intensive calcination reaction:

$$CaCO_3 \rightarrow CaO + CO_2$$

The use of a sorbent requires either that there be parallel reactors operated alternatively and out of phase in reforming and sorbent regeneration modes or that sorbent be continuously transferred between the reformer/carbonator and regenerator/calciner (Balasubramanian et al., 1999; Hufton et al., 1999).

In autothermal (or secondary) reformers, the oxidation of methane supplies the necessary energy and carried out either simultaneously or in advance of the reforming reaction (Brandmair et al., 2003; Ehwald et al., 2003; Nagaoka et al., 2003). The equilibrium of the methane–steam reaction and the water–gas shift reaction determines the conditions for optimum hydrogen yields. The optimum conditions for hydrogen production require high temperature at the exit of the reforming reactor (800°C–900°C [1470°F–1650°F]), high excess of steam (molar steam-to-carbon ratio of 2.5–3), and relatively low pressures (below 450 psi). Most commercial plants employ supported nickel catalysts for the process.

The steam–methane reforming process described briefly earlier would be an ideal hydrogen production process if it was not for the fact that large quantities of natural gas, a valuable resource, are required as both feed gas and combustion fuel. For each mole of reformed methane, more than 1 mole of carbon dioxide is coproduced and must be disposed. This can be a major issue as it results in the same amount of greenhouse gas emission as would be expected from direct combustion of natural gas or methane. In fact, the production of hydrogen as a clean burning fuel by way of steam reforming of methane and other fossil-based hydrocarbon fuels is not in environmental balance if in the process, carbon dioxide and carbon monoxide are generated and released into the atmosphere, although alternate scenarios are available (Gaudernack, 1996). Moreover, as the reforming process is not totally efficient, some of the energy value of the hydrocarbon fuel is lost by conversion to hydrogen but with no tangible environmental benefit, such as a reduction in emission of greenhouse gases. Despite these apparent shortcomings, the process has the following advantages: (1) produces 4 mol of hydrogen for each mole of methane consumed; (2) feedstocks for the process (methane and water are readily available); (3) the process adaptable to a wide range of hydrocarbon feedstocks, (4) operates at low pressures, less than 450 psi; (5) requires a low steam/carbon ratio (2.5–3); (6) good utilization of input energy (reaching 93%); (7) can use catalysts that are stable and resist poisoning; and (8) good process kinetics.

Liquid feedstocks, either liquefied petroleum gas or naphtha (*q.v.*), can also provide backup feed, if there is a risk of natural gas curtailments. The feed handling system needs to include a surge drum, feed pump, vaporizer (usually steam-heated) followed by further heating before desulfuriza-tion. The sulfur in liquid feedstocks occurs as mercaptans, thiophene derivatives, or higher-boiling compounds. These compounds are stable and will not be removed by zinc oxide; therefore, a hydro-genation unit will be required. In addition, as with refinery gas, olefins must also be hydrogenated if they are present.

The reformer will generally use a potash-promoted catalyst to avoid coke buildup from cracking of the heavier feedstock. If liquefied petroleum gas is to be used only occasionally, it is often pos-sible to use a methane-type catalyst at a higher steam/carbon ratio to avoid coking. Naphtha will require a promoted catalyst unless a pre-former is used.

15.6.8 STEAM–NAPHTHA REFORMING

Steam–naphtha reforming is a continuous process for the production of hydrogen from liquid hydro-carbons and is, in fact, similar to steam–methane reforming that is one of several possible processes for the production of hydrogen from low-boiling hydrocarbons other than ethane (Muradov, 1997, 2000; Brandmair et al., 2003; Find et al., 2003). A variety of naphtha types in the gasoline boiling range may be employed, including feeds containing up to 35% aromatics. Thus, following pretreat-ment to remove sulfur compounds, the feedstock is mixed with steam and taken to the reforming furnace (675°C–815°C [1250°F–1500°F], 300 psi), where hydrogen is produced.

15.6.9 SYNTHESIS GAS GENERATION

The synthesis gas generation process is a noncatalytic process for producing synthesis gas (prin-cipally hydrogen and carbon monoxide) for the ultimate production of high-purity hydrogen from gaseous or liquid hydrocarbons.

In this process, a controlled mixture of preheated feedstock and oxygen is fed to the top of the generator where carbon monoxide and hydrogen emerge as the products. Soot, produced in this part of the operation, is removed in a water scrubber from the product gas stream and is then extracted from the resulting carbon–water slurry with naphtha and transferred to a fuel oil fraction. The oil–soot mixture is burned in a boiler or recycled to the generator to extinction to eliminate carbon production as part of the process.

The soot-free synthesis gas is then charged to a shift converter where the carbon monoxide reacts with steam to form additional hydrogen and carbon dioxide at the stoichiometric rate of 1 mol of hydrogen for every mole of carbon monoxide charged to the converter.

The reactor temperatures vary from 1095°C to 1490°C (2000°F to 2700°F), while pressures can vary from approximately atmospheric pressure to approximately 2000 psi. The process has the capability of producing high-purity hydrogen, although the extent of the purification proce-dure depends upon the use to which the hydrogen is to be put. For example, carbon dioxide can be removed by scrubbing with various alkaline reagents, while carbon monoxide can be removed by washing with liquid nitrogen or, if nitrogen is undesirable in the product, the carbon monoxide should be removed by washing with copper-amine solutions.

This particular partial oxidation technique has also been applied to a whole range of liquid feedstocks for hydrogen production. There is now serious consideration being given to hydrogen production by the partial oxidation of solid feedstocks such as petroleum coke (from both delayed and fluid-bed reactors), lignite, and coal, as well as petroleum residua.

The chemistry of the process, using naphthalene as an example, may be simply represented as the selective removal of carbon from the hydrocarbon feedstock and further conversion of a portion of this carbon to hydrogen:

$$C_{10}H_8 + 5O_2 \rightarrow 10CO + 4H_2$$

$$10CO + 10H_2O \rightarrow 10CO_2 + 10H_2$$

Although these reactions may be represented very simply using equations of this type, the reactions can be complex and result in carbon deposition on parts of the equipment thereby requiring careful inspection of the reactor.

15.6.10 TEXACO GASIFICATION PROCESS

The Texaco gasification (partial oxidation) process is a partial oxidation gasification process for generating synthetic gas, principally hydrogen and carbon monoxide. The characteristic of Texaco gasification process is to inject feedstock together with carbon dioxide, steam, or water into the gasifier. Therefore, solvent deasphalted residua, or petroleum coke rejected from any coking method can be used as feedstock for this gasification process. The produced gas from this gasification process can be used for the production of high-purity high-pressurized hydrogen, ammonia, and methanol. The heat recovered from the high-temperature gas is used for the generation of steam in the waste heat boiler. Alternatively, the less expensive quench type configuration is preferred when high-pressure steam is not needed or when a high degree of shift is needed in the downstream carbon monoxide converter.

In the process, the feedstock, together with the feedstock carbon slurry recovered in the carbon recovery section, is pressurized to a given pressure, mixed with high-pressure steam and then blown into the gas generator through the burner together with oxygen.

The gasification reaction is a partial oxidation of hydrocarbons to carbon monoxide and hydrogen:

$$C_xH_{2y} + x/2O_2 \rightarrow xCO + yH_2$$

$$C_xH_{2y} + xH_2O \rightarrow xCO + (x + y)H_2$$

The gasification reaction is instantly completed, thus producing gas mainly consisting of H_2 and CO (H_2 + CO = >90%). The high-temperature gas leaving the reaction chamber of the gas generator enters the quenching chamber linked to the bottom of the gas generator and is quenched to 200°C–260°C (390°F–500°F) with water.

15.6.11 RECOVERY FROM PROCESS GAS

Recovering hydrogen from refinery process gas (typically assigned to be fuel gas for refinery operations) can help refineries to satisfy high hydrogen demand. Cryogenic separation is typically viewed as being the most thermodynamically efficient separation technology. The higher capital cost associated with pre-purification and the low flexibility to impurity upsets has limited its use in hydrogen recovery.

The basic configuration for hydrogen recovery from refinery gases involves a two-stage partial condensation process, with post purification via pressure swing adsorption (Dragomir et al., 2010). The major steps in this process involve first compressing and pretreating the crude refinery gas stream before chilling to an intermediate temperature (–60°F to –120 °F). This partially condensed stream is then separated in a flash drum after which the liquid stream from is expanded through a Joule–Thomson valve to generate refrigeration and then is fed to the wash column. Optionally, the wash column can be replaced by a simple flash drum.

A crude LPG stream is collected at the bottom of the column, and a methane-rich vapor is obtained at the top. The methane-rich vapor is sent to compression and then to fuel. The vapor from the flash drum is further cooled in a second heat exchanger before being fed to second flash drum where it produces a hydrogen-rich stream and a methane-rich liquid. The liquid is expanded in a Joule–Thomson valve to generate refrigeration and then is sent for further cooling. The hydrogen-rich gas is then sent to the pressure swing adsorption unit for further purification—the tail gas from this unit is compressed and returned to fuel together with the methane-rich gas.

15.7   REFINING HEAVY FEEDSTOCKS

Over the past three decades, crude oils available to refineries have generally decreased in API gravity (Speight, 2005, 2011, 2014). There is, nevertheless, a major focus in refineries on the ways in which heavy feedstocks (such as heavy oil and tar sand bitumen) can be converted into low-boiling high-value products (Khan and Patmore, 1997; Speight and Ozum, 2002; Parkash, 2003; Speight, 2005, 2011, 2014; Hsu and Robinson, 2006; Gary et al., 2007; Rana et al., 2007; Rispoli et al., 2009; Stratiev and Petkov, 2009; Stratiev et al., 2009; Motaghi et al., 2010). Simultaneously, the changing crude oil properties are reflected in changes such as an increase in asphaltene constituents and an increase in sulfur, metal, and nitrogen contents. Pretreatment processes for removing such constituents or at least negating their effect in thermal process those would also play an important role. The limitations of processing these heavy feedstocks depend to a large extent on the amount of higher-molecular-weight constituents (i.e., asphaltene constituents and resin constituents) that contain the majority of the heteroatom-containing compounds, which are responsible for high yields of thermal and catalytic coke (Speight, 2014). Be that as it may, the essential step required of a modern refinery is the upgrading of heavy feedstocks, particularly atmospheric and vacuum residua.

Upgrading feedstocks such as heavy oils and residua began with the introduction of hydrodesulfurization processes (Speight, 2000; Ancheyta and Speight, 2007). In the early days, the goal was desulfurization but, in later years, the processes were adapted to a 10%–30% partial conversion operation, as intended to achieve desulfurization and obtain low-boiling fractions simultaneously, by increasing severity in operating conditions. However, as refineries have evolved and feedstocks have changed, refining heavy feedstocks has become a major issue in modern refinery practice and several process configurations have evolved to accommodate the heavy feedstocks (Khan and Patmore, 1997; Speight and Ozum, 2002; Speight, 2011, 2014).

For example, hydrodesulfurization of light (low-boiling) distillate (naphtha or kerosene) is one of the more common catalytic hydrodesulfurization processes since it is usually used as a pretreatment of such feedstocks prior to deep hydrodesulfurization or prior to catalytic reforming. A similar concept of pretreating residua prior to hydrocracking to improve the quality of the products is also practiced (Speight, 2011, 2014). Hydrodesulfurization of such feedstocks is required because sulfur compounds poison the precious metal catalysts used in the hydrocracking process. If the feedstock arises from a cracking operation (such as cracked residua), hydropretreatment will be accompanied by some degree of saturation resulting in increased hydrogen consumption.

Finally, there is not one single heavy feedstock upgrading solution that will fit all refineries. Market conditions, existing refinery configuration, and available crude prices all can have a significant effect on the final configuration. Furthermore, a proper evaluation however is not a simple undertaking for an existing refinery. The evaluation starts with an accurate understanding of the market for the various products along with corresponding product values at various levels of supply. The next step is to select a set of crude oils that adequately cover the range of crude oils that may be expected to be processed. It is also important to consider new unit capital costs as well as incremental capital costs for revamp opportunities along with the incremental utility, support, and infrastructure costs. The costs, although estimated at the start, can be better assessed once the options have been defined leading to the development of the optimal configuration for refining the incoming feedstocks.

REFERENCES

Al-Shamma, L.M. and Naman, S.A. 1989. Kinetic study for thermal production of hydrogen from hydrogen sulfide by heterogeneous catalysis of vanadium sulfide in a flow system. *International Journal of Hydrogen Energy*, 14(3): 173–179.

Ancheyta, J. and Speight, J.G. 2007. *Hydroprocessing of Heavy Oils and Residua*. CRC Press/Taylor & Francis Group, Boca Raton, FL.

Balasubramanian, B., Ortiz, A.L., Kaytakoglu, S., and Harrison, D.P. 1999. Hydrogen from methane in a single-step process. *Chemical Engineering Science*, 54: 3543–3552.

Bandermann, F. and Harder, K.B. 1982. Production of hydrogen via thermal decomposition of hydrogen sulfide and separation of hydrogen and hydrogen sulfide by pressure swing adsorption. *International Journal of Hydrogen Energy*, 7(6): 471–475.

Bezler, J. 2003. Optimized hydrogen production—A key process becoming increasingly important in refineries. *Proceedings of DGMK Conference on Innovation in the Manufacture and Use of Hydrogen*. Dresden, Germany, October 15–17, p. 65.

Bishara, A., Salman, O.S., Khraishi, N., and Marafi, A. 1987. Thermochemical decomposition of hydrogen sulfide by solar energy. *International Journal of Hydrogen Energy*, 12(10): 679–685.

Brandmair, M., Find, J., and Lercher, J.A. 2003. Combined autothermal reforming and hydrogenolysis of alkanes. *Proceedings of DGMK Conference on Innovation in the Manufacture and Use of Hydrogen*. Dresden, Germany, October 15–17, pp. 273–280.

Brüschke, H. 1995. Industrial application of membrane separation processes. *Pure and Applied Chemistry*, 67(6): 993–1002.

Brüschke, H. 2003. Separation of hydrogen from dilute streams (e.g. using membranes). *Proceedings of DGMK Conference on Innovation in the Manufacture and Use of Hydrogen*. Dresden, Germany, October 15–17, p. 47.

Clark, P.D., Dowling, N.I., Hyne, J.B., and Moon, D.L. 1995. Production of hydrogen and sulfur from hydrogen sulfide in refineries and gas processing plants. *Quarterly Bulletin*, 32(1): 11–28.

Clark, P.D. and Wassink, B. 1990. A review of methods for the conversion of hydrogen sulfide to sulfur and hydrogen. *Alberta Sulfur Research Quarterly Bulletin*, 26(2/3/4): 1.

Dalrymple, D.A., Trofe, T.W., and Leppin, D. 1994. Gas industry assesses new ways to remove small amounts of hydrogen sulfide. *Oil and Gas Journal*.

Davis, B.H. and Occelli, M.L. (Editors). 2010. *Advances in Fischer-Tropsch Synthesis, Catalysts, and Catalysis*. CRC Press/Taylor & Francis Group, Boca Raton, FL.

Dickenson, R.L., Biasca, F.E., Schulman, B.L., and Johnson, H.E. 1997. Refiner options for converting and utilizing heavy fuel oil. *Hydrocarbon Processing*, 76(2): 57.

Dolbear, G.E. 1998. Hydrocracking: Reactions, catalysts, and processes. In: *Petroleum Chemistry and Refining*. J.G. Speight (Editor). Taylor & Francis Group, Washington, DC, Chapter 7.

Donini, J.C. 1996. Separation and processing of hydrogen sulfide in the fossil fuel industry. *Minimum Effluent Mills Symposium*. pp. 357–363.

Dragomir, R., Drnevich, R.F., Morrow, J., Papavassiliou, V., Panuccio, G., and Watwe, R. 2010. Technologies for enhancing refinery gas value. *Proceedings of AIChE 2010 SPRING Meeting*. San Antonio, TX, November 7–12.

Ehwald, H., Kürschner, U., Smejkal, Q., and Lieske, H. 2003. Investigation of different catalysts for autothermal reforming of i-octane. *Proceedings of DGMK Conference on Innovation in the Manufacture and Use of Hydrogen*. Dresden, Germany, October 15–17, p. 345.

Find, J., Nagaoka, K., and Lercher, J.A. 2003. Steam reforming of light alkanes in micro-structured reactors. *Proceedings of DGMK Conference on Innovation in the Manufacture and Use of Hydrogen*. Dresden, Germany, October 15–17, p. 257.

Gary, J.H., Handwerk, G.E., and Kaiser, M.J. 2007. *Petroleum Refining: Technology and Economics*, 5th Edition. CRC Press/Taylor & Francis Group, Boca Raton, FL.

Gaudernack, B. 1996. Hydrogen from natural gas without release of carbon dioxide into the atmosphere. *Hydrogen Energy Progress. Proceedings of the 11th World Hydrogen Energy Conference*. Volume 1, pp. 511–523.

Gross, M. and Wolff, J. 2000. Gasification of residue as a source of hydrogen for the refining industry in India. *Proceedings. Gasification Technologies Conference*. San Francisco, CA, October 8–11.

Hsu, C.S. and Robinson, P.R. (Editors). 2006. *Practical Advances in Petroleum Processing*, Volumes 1 and 2. Springer Science, New York.

Hufton, J.R., Mayorga, S., and Sircar, S. 1999. Sorption-enhanced reaction process for hydrogen production. *AIChE Journal*, 45: 248–256.

Khan, M.R. and Patmore, D.J. 1997. Heavy oil upgrading processes. In: *Petroleum Chemistry and Refining*. J.G. Speight (Editor). Taylor & Francis Group, Washington, DC, Chapter 6.

Kiuchi, H. 1982. Recovery of hydrogen from hydrogen sulfide with metals and metal sulfides. *International Journal of Hydrogen Energy*, 7(6).

Kotera, Y., Todo, N., and Fukuda, K. 1976. Process for production of hydrogen and sulfur from hydrogen sulfide as raw material. U.S. Patent No. 3,962,409. June 8.

Liu, K., Song, C., and Subramani, V. 2010. *Hydrogen and Syngas Production and Purification Technologies.* John Wiley & Sons, Hoboken, NJ.

Luckwal, K. and Mandal, K.K. 2009. Improve hydrogen management of your refinery. *Hydrocarbon Processing*, 88(2): 55–61.

Luinstra, E. 1996. Hydrogen from hydrogen sulfide—A review of the leading processes. *Proceedings of the Seventh Sulfur Recovery Conference.* Gas Research Institute, Chicago, IL, pp. 149–165.

Megalofonos, S.K. and Papayannakos, N.G. 1997. Kinetics of catalytic reaction of methane and hydrogen sulfide over MoS_2. *Journal of Applied Catalysis A: General*, 65(1–2): 249–258.

Miller, G.Q. and Penner, D.W. 2003. Meeting future needs for hydrogen—Possibilities and challenges. *Proceedings of DGMK Conference on Innovation in the Manufacture and Use of Hydrogen.* Dresden, Germany, October 15–17, p. 7.

Mokhatab, S., Poe, W.A., and Speight, J.G. 2006. *Handbook of Natural Gas Transmission and Processing.* Elsevier, Amsterdam, the Netherlands.

Motaghi, M., Shree, K., and Krishnamurthy, S. 2010. Consider new methods for bottom of the barrel processing—Part 1. *Hydrocarbon Processing*, 89(2): 35–40.

Muradov, N.Z. 1998. CO_2-free production of hydrogen by catalytic pyrolysis of hydrocarbon fuel. *Energy and Fuels*, 12(1): 41–48.

Muradov, N.Z. 2000. Thermocatalytic carbon dioxide-free production of hydrogen from hydrocarbon fuels. *Proceedings of Hydrogen Program Review.* NREL/CP-570-28890.

Nagaoka, K., Jentys, A., and Lecher, J.A. 2003. Autothermal reforming of methane over mono- and bi-metal catalysts prepared from hydrotalcite-like precursors. *Proceedings of DGMK Conference on Innovation in the Manufacture and Use of Hydrogen.* Dresden, Germany, October 15–17, p. 171.

OSHA Technical Manual. 1999. Section IV, Chapter 2: Petroleum refining processes. http://www.osha.gov/dts/osta/otm/otm_iv/otm_iv_2.html, accessed January 14, 2015.

Parkash, S. 2003. *Refining Processes Handbook.* Gulf Professional Publishing, Elsevier, Amsterdam, the Netherlands.

Raissi, A.T. 2001. Technoeconomic analysis of area II hydrogen production. Part 1. *Proceedings of US DOE Hydrogen Program Review Meeting.* Baltimore, MD.

Rana, M.S., Sámano, V., Ancheyta, J., and Diaz, J.A.I. 2007. A review of recent advances on process technologies for upgrading of heavy oils and residua. *Fuel*, 86: 1216–1231.

Ranke, H. and Schödel, N. 2003. Hydrogen production technology—Status and new developments. *Proceedings of DGMK Conference on Innovation in the Manufacture and Use of Hydrogen.* Dresden, Germany, October 15–17, p. 19.

Rispoli, G., Sanfilippo, D., and Amoroso, A. 2009. Advanced hydrocracking technology upgrades extra heavy oil. *Hydrocarbon Processing*, 88(12): 39–46.

Scherzer, J. and Gruia, A.J. 1996. *Hydrocracking Science and Technology.* Marcel Dekker, New York.

Speight, J.G. 2000. *The Desulfurization of Heavy Oils and Residua*, 2nd Edition. Marcel Dekker, New York.

Speight, J.G. 2005. Natural bitumen (tar sands) and heavy oil. In: *Coal, Oil Shale, Natural Bitumen, Heavy Oil and Peat. Encyclopedia of Life Support Systems (EOLSS).* Developed under the Auspices of the UNESCO. EOLSS Publishers, Oxford, UK.

Speight, J.G. 2007. *Natural Gas: A Basic Handbook.* GPC Books, Gulf Publishing Company, Houston, TX.

Speight, J.G. 2011. *The Refinery of the Future.* Gulf Professional Publishing, Elsevier, Oxford, UK.

Speight, J.G. 2014. *The Chemistry and Technology of Petroleum*, 5th Edition. CRC Press/Taylor & Francis Group, Boca Raton, FL.

Speight, J.G. and Ozum, B. 2002. *Petroleum Refining Processes.* Marcel Dekker, New York.

Stratiev, D. and Petkov, K. 2009. Residue upgrading: Challenges and perspectives. *Hydrocarbon Processing*, 88(9): 93–96.

Stratiev, D., Tzingov, T., Shishkova, I., and Dermatova, P. 2009. Hydrotreating units chemical hydrogen consumption analysis: A tool for improving refinery hydrogen management. *Proceedings of the 44th International Petroleum Conference.* Bratislava, Slovak Republic, September 21–22.

Uemura, Y., Ohe, H., Ohzuno, Y., and Hatate, Y. 1999. Carbon and hydrogen from hydrocarbon pyrolysis. *Proceedings of the International Conference on Solid Waste Technology Management.* Volume 15, pp. 5E/25–5E/30.

Vauk, D., Di Zanno, P., Neri, B., Allevi, C., Visconti, A., and Rosanio, L. 2008. What are possible hydrogen sources for refinery expansion? *Hydrocarbon Processing*, 87(2): 69–76.

Weimer, A.W., Dahl, J., Tamburini, J., Lewandowski, A., Pitts, R., Bingham, C., and Glatzmaier, G.C. 2000. Thermal dissociation of methane using a solar coupled aerosol flow reactor. *Proceedings of the Hydrogen Program Review.* NREL/CP-570-28890.

Zagoria, A., Huychke, R., and Boulter, P.H. 2003. Refinery hydrogen management—The big picture. *Proceedings of the DGMK Conference on Innovation in the Manufacture and Use of Hydrogen.* Dresden, Germany, October 15–17, p. 95.

Zaman, J. and Chakma, A. 1995. Production of hydrogen and sulfur from hydrogen sulfide. *Fuel Processing Technology*, 41: 159–198.

16 Gas Cleaning

16.1 INTRODUCTION

Natural gas, as it is used by consumers, is much different from the natural gas that is brought from underground formations to the wellhead. Although the processing of natural gas is in many respects less complicated than the processing and refining of crude oil, it is equally as necessary before its use by end users. Gas is often referred to as *natural gas* because it is a naturally occurring hydrocarbon mixture (which does contain some nonhydrocarbon constituents, which might be labeled as impurities but often find use in other areas of technology). For the most part, natural gas consists mainly of methane, which is the simplest hydrocarbon; nevertheless, processing (purification, refining) is required before transportation to the consumer.

Gas processing (also called *gas cleaning* or *gas refining*) consists of separating all of the various hydrocarbons and fluids from the pure natural gas (Figure 16.1; Kidnay and Parrish, 2006; Mokhatab et al., 2006; Speight, 2007, 2014a). While often assumed to be hydrocarbons in nature, there are also components of the gaseous products that must be removed prior to the release of the gases to the atmosphere or prior to the use of the gas in another part of the refinery, that is, as a fuel gas or as a process feedstock.

Gas processing involves the use of several different types of processes to remove contaminants from gas streams, but there is always overlap between the various processing concepts. In addition, the terminology used for gas processing can often be confusing and/or misleading because of the overlap (Curry, 1981; Maddox, 1982). Gas processing is necessary to ensure that the natural gas prepared for transportation (usually by pipeline) and for sales must be as clean and pure as the specifications dictate. Thus, natural gas, as it is used by consumers, is much different from the natural gas that is brought from underground formations up to the wellhead. Moreover, although natural gas produced at the wellhead is composed primarily of the gas is by no means as pure.

Raw natural gas comes from three types of wells: oil wells (*associated gas*), gas wells (*nonassociated gas*), and condensate wells (*condensate gas*, also called *nonassociated gas*). Associated gas can exist separate from oil in the formation (*free gas*) or dissolved in the crude oil (*dissolved gas*). Whatever the source of the natural gas, once separated from crude oil (if present), it commonly exists in mixtures with other hydrocarbons; principally ethane, propane, butane, and pentane isomers (*natural gas liquids*) as well as a mixture of higher-molecular-weight (higher-boiling) hydrocarbons that are often referred to as *natural gasoline*. In addition, raw natural gas contains water vapor, hydrogen sulfide (H_2S), carbon dioxide, helium, nitrogen, and other compounds. Natural gas liquids are sold separately and have a variety of different uses such as providing feedstocks for oil refineries or petrochemical plants.

Acidic constituents such as carbon dioxide and hydrogen sulfide as well as mercaptan derivatives (also called thiols, RSH) can contribute to corrosion of refining equipment, harm catalysts, pollute the atmosphere, and prevent the use of hydrocarbon components in petrochemical manufacture (Mokhatab et al., 2006; Speight, 2014b). When the amount of hydrogen sulfide is high, it may be removed from a gas stream and converted to sulfur or sulfuric acid; a recent option for hydrogen sulfide removal is the use of chemical scavengers (Kenreck, 2014). Some natural gases contain sufficient carbon dioxide to warrant recovery as dry ice (Bartoo, 1985).

The processes that have been developed to accomplish gas purification vary from a simple single-stage once-through washing-type operation to complex multistep recycling systems (Mokhatab et al., 2006; Speight, 2007, 2014a). In many cases, the process complexities arise because of the need for the recovery of the materials used to remove the contaminants or even recovery of the

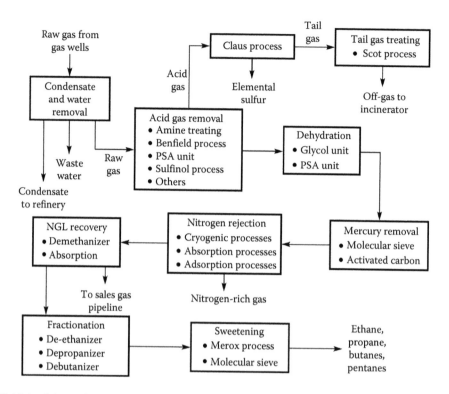

FIGURE 16.1 Scheme for the flow of natural gas cleaning options.

contaminants in the original, or altered, form (Katz, 1959; Kohl and Riesenfeld, 1985; Newman. 1985; Mokhatab et al., 2006). In addition, the precise area of application of a given process is difficult to define and several factors must be considered before process selection: (1) the types of contaminants in the gas, (2) the concentrations of contaminants in the gas, (3) the degree of contaminant removal desired, (4) the selectivity of acid gas removal required, (5) the temperature of the gas to be processed, (6) the pressure of the gas to be processed, (7) the volume of the gas to be processed, (8) the composition of the gas to be processed, (9) the ratio of carbon dioxide to hydrogen sulfide ratio in the gas feedstock, and (10) the desirability of sulfur recovery due to process economics or environmental issues.

There are four general processes used for emission control (often referred to in another, more specific context as flue gas desulfurization): (1) adsorption, (2) absorption, (3) catalytic oxidation, and (4) thermal oxidation (Soud and Takeshita, 1994; Mokhatab et al., 2006; Speight, 2007, 2014a).

Adsorption is a physical–chemical phenomenon in which the gas is concentrated on the surface of a solid or liquid to remove impurities. It must be emphasized that *absorption* differs from *adsorption* in that absorption is not a physical–chemical surface phenomenon but a process in which the absorbed gas is ultimately distributed throughout the absorbent (liquid). The process depends only on physical solubility and may include chemical reactions in the liquid phase (*chemisorption*). Common absorbing media used are water, aqueous amine solutions, caustic, sodium carbonate, and nonvolatile hydrocarbon oils, depending on the type of gas to be absorbed. Usually, the gas–liquid contactor designs that are employed are plate columns or packed beds (Table 16.1).

On the other hand, adsorption is usually a gas–solid interaction in which an adsorbent such as activated carbon (the *adsorbent* or *adsorbing medium*) can be regenerated upon *desorption* (Mokhatab et al., 2006; Speight, 2007, 2014a). The quantity of material adsorbed is proportional to the surface area of the solid and, consequently, adsorbents are usually granular solids with a large surface area per unit mass. Subsequently, the captured (adsorbed) gas can be desorbed with hot air

TABLE 16.1

Factors to Be Considered in the Use of Packed Column and Plate Column Absorption Systems

1. Relatively low pressure drop
2. Capabilities of achieving relatively high mass transfer efficiencies
3. Capacity to increase the height and/or type of packing or number of plates capable of improving mass transfer without purchasing a new piece of equipment
4. Ability to collect particulate materials as well as gases
5. Possibility to require water (or liquid) disposal

TABLE 16.2

Factors to Be Considered in the Use of Adsorption Systems

1. Possibility of product recovery
2. Excellent control and response to process changes
3. No chemical disposal problem when pollutant (product) recovered and returned to process
4. Capability to remove gaseous or vapor contaminants from process streams to extremely low levels
5. Product recovery that may require distillation or extraction
6. Adsorbent that may deteriorate as the number of cycles increase
7. Adsorbent regeneration
8. Prefiltering of gas stream possibly required to remove particulate materials capable of plugging the adsorbent bed

or steam either for recovery or for thermal destruction. Adsorber units are widely used to increase a low gas concentration prior to incineration unless the gas concentration is very high in the inlet air stream and the process is also used to reduce problem odors (or obnoxious odors) from gases. There are several limitations to the use of adsorption systems, but it is generally the case that the major limitation is the requirement for the minimization of particulate matter and/or condensation of liquids (e.g., water vapor) that could mask the adsorption surface and drastically reduce its efficiency (Table 16.2).

Absorption is achieved by dissolution (a physical phenomenon) or by reaction (a chemical phenomenon) (Barbouteau and Dalaud, 1972; Ward, 1972; Mokhatab et al., 2006; Speight, 2007, 2014a). In addition to economic issues or constraints, the solvents used for gas processing should have (1) high capacity for acid gas, (2) low tendency to dissolve hydrogen, (3) low tendency to dissolve low-molecular-weight hydrocarbons, (4) low vapor pressure at operating temperatures to minimize solvent losses, (5) low viscosity, (6) low thermal stability, (7) no reactivity toward gas components, (8) low tendency for fouling, (9) low tendency for corrosion, and (10) low economic cost (Mokhatab et al., 2006; Speight, 2007, 2014a,b).

Chemical adsorption processes adsorb sulfur dioxide onto a carbon surface where it is oxidized (by oxygen in the flue gas) and absorbs moisture to give sulfuric acid impregnated into and on the adsorbent. Liquid absorption processes (which usually employ temperatures below 50°C (120°F) are classified either as *physical solvent processes* or *chemical solvent processes*. The former processes employ an organic solvent, and absorption is enhanced by low temperatures or high pressure, or both. Regeneration of the solvent is often accomplished readily (Staton et al., 1985). In chemical solvent processes, absorption of the acid gases is achieved mainly by the use of alkaline solutions such as amine derivatives or carbonate derivatives (Kohl and Riesenfeld, 1985; Mokhatab et al., 2006; Abdel-Aal et al., 2016). Regeneration (desorption) can be brought about by the use of reduced pressures and/or high temperatures, whereby the acid gases are stripped from the solvent.

TABLE 16.3

Olamines Used for Gas Processing

Olamine	Formula	Derived Name	Molecular Weight	Specific Gravity	Melting Point (°C)	Boiling Point (°C)	Flash Point (°C)	Relative Capacity (%)
Ethanolamine (monoethanolamine)	$HOC_2H_4NH_2$	MEA	61.08	1.01	10	170	85	100
Diethanolamine	$(HOC_2H_4)_2NH$	DEA	105.14	1.097	27	217	169	58
Triethanolamine	$(HOC_2H_4)_3NH$	TEA	148.19	1.124	18	335, d	185	41
Diglycolamine (hydroxyethanolamine)	$H(OC_2H_4)_2NH$	DGA	105.14	1.057	−11	223	127	58
Diisopropanolamine	$(HOC_3H_6)_2NH$	DIPA	133.19	0.99	42	248	127	46
Methyldiethanolamine	$(HOC_2H_4)_2NCH_3$	MDEA	119.17	1.03	−21	247	127	51

d: with decomposition.

Amine washing of gas emissions involves chemical reaction of the amine with any acid gases with the liberation of an appreciable amount of heat, and it is necessary to compensate for the absorption of heat. Amine derivatives such as ethanolamine (monoethanolamine, MEA), diethanolamine (DEA), triethanolamine (TEA), methyldiethanolamine (MDEA), diisopropanolamine (DIPA), and diglycolamine (DGA) have been used in commercial applications (Table 16.3; Katz, 1959; Jou et al., 1985; Kohl and Riesenfeld, 1985; Maddox et al., 1985; Polasek and Bullin, 1985; Pitsinigos and Lygeros, 1989; Mokhatab et al., 2006; Speight, 2007, 2014a; Abdel-Aal et al., 2016). The chemistry of the amine process (also called the *olamine process*) can be represented by simple equations for low partial pressures of the acid gases:

$$2RNH_2 + H_2S \rightarrow (RNH_3)_2S$$

$$2RHN_2 + CO_2 + H_2O \rightarrow (RNH_3)_2CO_3$$

At high acid gas partial pressure, the reactions will lead to the formation of other products:

$$(RNH_3)_2S + H_2S \rightarrow 2RNH_3HS$$

$$(RNH_3)_2CO_3 + H_2O \rightarrow 2RNH_3HCO_3$$

The reaction is extremely rapid and the absorption of hydrogen sulfide is limited only by mass transfer—this is not the case for carbon dioxide—and the reaction is also more complex that these equations would indicate and can lead to a series on unwanted side reactions and by-products (Mokhatab et al., 2006; Speight, 2007). Regeneration of the amine (olamine) solution leads to near complete desorption of carbon dioxide and hydrogen sulfide. A comparison between monoethanolamine, diethanolamine, and diisopropanolamine shows that monoethanolamine is the cheapest of the three olamines but exhibits the highest heat of reaction and corrosion. On the other hand, diisopropanolamine is the most expensive of the three olamines but exhibits the lowest heat of reaction with a lower propensity for corrosion.

Carbonate washing is a mild alkali process (typically, the alkali is potassium carbonate, K_2CO_3) for gas processing for the removal of acid gases (such as carbon dioxide and hydrogen sulfide) from gas streams and uses the principle that the rate of absorption of carbon dioxide by potassium carbonate increases with temperature (Mokhatab et al., 2006; Speight, 2007, 2014a).

It has been demonstrated that the process works best near the temperature of reversibility of the reactions:

$$K_2CO_3 + CO_2 + H_2O \rightarrow 2KHCO_3$$

$$K_2CO_3 + H_2S \rightarrow KHS + KHCO_3$$

The Fluor process uses propylene carbonate to remove carbon dioxide, hydrogen sulfide, carbonyl sulfide, water, and higher-boiling hydrocarbons (C_2^+) from natural gas (Abdel-Aal et al., 2016).

Water washing, in terms of the outcome, is almost analogous to (but often less effective than) washing with potassium carbonate (Kohl and Riesenfeld, 1985), and it is also possible to carry out the desorption step by pressure reduction. The absorption is purely physical and there is also a relatively high absorption of hydrocarbons, which are liberated at the same time as the acid gases.

In *chemical conversion processes*, contaminants in gas emissions are converted to compounds that are not objectionable or that can be removed from the stream with greater ease than the original constituents. For example, a number of processes have been developed, which remove hydrogen sulfide and sulfur dioxide from gas streams by absorption in an alkaline solution.

Catalytic oxidation is a chemical conversion process that is used predominantly for the destruction of volatile organic compounds and carbon monoxide. These systems operate in a temperature regime on the order of 205°C–595°C (400°F–1100°F) in the presence of a catalyst—in the absence of the catalyst, the system would require a higher operating temperature. The catalysts used are typically a combination of noble metals deposited on a ceramic base in a variety of configurations (e.g., honeycomb shaped) to enhance good surface contact. Catalytic systems are usually classified on the basis of bed types such as *fixed bed* (or *packed bed*) and *fluid bed* (*fluidized bed*). These systems generally have very high destruction efficiencies for most volatile organic compounds, resulting in the formation of carbon dioxide, water, and varying amounts of hydrogen chloride (from halogenated hydrocarbons). The presence in emissions of chemicals such as heavy metals, phosphorus, sulfur, chlorine, and most halogens in the incoming air stream act as poison to the system and can foul up the catalyst. Thermal oxidation systems, without the use of catalysts, also involve chemical conversion (more correctly, chemical destruction) and operate at temperatures in excess of 815°C (1500°F), or 220°C–610°C (395°F–1100°F) higher than catalytic systems.

Historically, *particulate matter control* (*dust control*) has been one of the primary concerns of industries, since the emission of particulate matter is readily observed through the deposition of fly ash and soot as well as in impairment of visibility (Mody and Jakhete, 1988). Different degrees of control can be achieved by the use of various types of equipment, but selection of the process equipment depends upon proper characterization of the particulate matter emitted by a specific process, that is, the appropriate piece of equipment must be selected, sized, installed, and performance tested. The general classes of control devices for particulate matter are categorized as (1) cyclone collectors, (2) fabric filters, and (3) wet scrubbers.

Cyclone collectors are the most common of the inertial collector class and are effective in removing coarser fractions of particulate matter and operate by contacting the particles in the gas stream with a liquid. In principle, the particles are incorporated in a liquid bath or in liquid particles, which are much larger and therefore more easily collected. In the process, the particle-laden gas stream enters an upper cylindrical section tangentially and proceeds downward through a conical section. Particles migrate by centrifugal force generated by providing a path for the carrier gas to be subjected to a vortex-like spin. The particles are forced to the wall and are removed through a seal at the apex of the inverted cone. A reverse-direction vortex moves upward through the cyclone and discharges through a top center opening. Cyclones are often used as primary collectors because of their relatively low efficiency (50%–90% is usual). The equipment can be arranged either in parallel or in series to both increase efficiency and decrease pressure drop but there are disadvantages that must be recognized (Table 16.4).

TABLE 16.4

Factors to Be Considered in the Use of Cyclone Collectors

1. Relatively low operating pressure drops (for the degree of particulate removal obtained) in the range of approximately 2–6 inch water column
2. Dry collection and disposal
3. Relatively low overall particulate collection efficiencies, especially on particulates below 10 mm
4. Usually inability to process semisolid (tacky) materials

TABLE 16.5

Factors to Be Considered in the Use of Fabric Filter Systems

1. High collection efficiency on both coarse and fine (submicrometer) particulates
2. Collected material recovered dry for subsequent processing of disposal
3. Corrosion and rusting of components, usually not major issues
4. Relatively simple operation
5. Temperatures much in excess of 288°C (550°F) that require special refractory materials
6. Potential for dust explosion hazard
7. Fabric life possibly shortened at elevated temperatures and in the presence of acid or alkaline particulate or gas constituents
8. Hygroscopic materials, condensation of moisture, or tarry adhesive components possibly causing crusty caking or plugging of the fabric or requiring special additives

TABLE 16.6

Factors to Be Considered in the Use of Wet Scrubbers

1. No secondary dust sources
2. Ability to collect gases as well as particulates (especially "sticky" ones)
3. Ability to handle high-temperature, high-humidity gas streams
4. Ability to achieve high collection efficiencies on fine particulates (however at the expense of pressure drop)
5. Possibility to necessitate water disposal
6. Corrosion problems more severe than with dry systems
7. Potential for solid buildup at the wet–dry interface

Fabric filters are typically designed with nondisposable filter bags. As the gaseous (dust-containing) emissions flow through the filter media (typically cotton, polypropylene, fiberglass, or Teflon), particulate matter is collected on the bag surface as a dust cake. Fabric filters operate with collection efficiencies up to 99.9% although other advantages are evident, but there are several issues that arise during the use of such equipment (Table 16.5).

Wet scrubbers are devices in which a countercurrent spray liquid is used to remove particles from an air stream. Device configurations include plate scrubbers, packed bed scrubbers, orifice scrubbers, venturi scrubbers, and spray towers, individually or in various combinations. Wet scrubbers can achieve high collection efficiencies at the expense of prohibitive pressure drops (Table 16.6). The *foam scrubber* is a modification of a wet scrubber in which the particle-laden gas is passed through a foam generator, where the gas and particles are enclosed by small bubbles of foam.

TABLE 16.7
Factors to Be Considered in the Use of Electrostatic Precipitators

1. High particulate (coarse and fine) collection efficiencies
2. Dry collection and disposal
3. Operation under high pressure (to 150 lb/in.²) or vacuum conditions
4. Operation at high temperatures (up to 704°C, 1300°F) when necessary
5. Relatively large gas flow rates capable of effective handling
6. Sensitivity to fluctuations in gas-stream conditions
7. Explosion hazard when treating combustible gases/particulates
8. Ozone produced during gas ionization

Other methods include the use of high-energy input *venturi scrubbers* or electrostatic scrubbers where particles or water droplets are charged, and flux force/condensation scrubbers where a hot humid gas is contacted with cooled liquid or where steam is injected into saturated gas. In the latter scrubber, the movement of water vapor toward the cold water surface carries the particles with it (*diffusiophoresis*), while the condensation of water vapor on the particles causes the particle size to increase, thus facilitating the collection of fine particles.

Electrostatic precipitators (Table 16.7) operate on the principle of imparting an electric charge to particles in the incoming air stream, which are then collected on an oppositely charged plate across a high-voltage field. Particles of high resistivity create the most difficulty in collection. Conditioning agents such as sulfur trioxide (SO_3) have been used to lower resistivity. Important parameters include design of electrodes, spacing of collection plates, minimization of air channeling, and collection-electrode rapping techniques (used to dislodge particles). Techniques under study include the use of high-voltage pulse energy to enhance particle charging, electron-beam ionization, and wide plate spacing. Electrical precipitators are capable of efficiencies >99% under optimum conditions, but performance is still difficult to predict in new situations.

16.2 GAS STREAMS

Gas streams produced during petroleum and natural gas refining are not always hydrocarbon in nature and may contain contaminants, such as carbon oxides (CO_x, where x = 1 and/or 2), sulfur oxides (SO_x, where x = 2 and/or 3), as well as ammonia (NH_3), mercaptan derivatives (RSH), carbonyl sulfide (COS), and mercaptan derivatives (RSH). The presence of these impurities may eliminate some of the sweetening processes from use since some of these processes remove considerable amounts of acid gas but not to a sufficiently low concentration. On the other hand, there are those processes not designed to remove (or incapable of removing) large amounts of acid gases whereas they are capable of removing the acid gas impurities to very low levels when the acid gases are present only in low-to-medium concentration in the gas (Katz, 1959; Mokhatab et al., 2006; Speight, 2007, 2014a).

16.2.1 GAS STREAMS FROM CRUDE OIL

There are two broad categories of gas that is produced from crude oil. The first category is the associated gas, which originated from crude oil formations and also from condensate wells (*condensate gas*, also called *nonassociated gas*). Associated gas can exist separate from oil in the formation (*free gas*) or dissolved in the crude oil (*dissolved gas*). The second category of the gas is produced during crude oil refining, and the terms *refinery gas* and *process gas* are also often used to include all of the gaseous products and by-products that emanate from a variety of refinery processes.

TABLE 16.8

Constituents of Natural Gas

Name	Formula	% v/v
Methane	CH_4	>85
Ethane	C_2H_6	3–8
Propane	C_3H_8	1–5
Butane	C_4H_{10}	1–2
Pentane+	$C_5H_{12}{}^+$	1–5
Carbon dioxide	CO_2	1–2
Hydrogen sulfide	H_2S	1–2
Nitrogen	N_2	1–5
Helium	He	<0.5

Note: Pentane+, pentane, and higher-molecular-weight hydrocarbons, including benzene and toluene.

In order to process and transport associated dissolved natural gas, it must be separated from the oil in which it is dissolved and is most often performed using equipment installed at or near the wellhead. The actual process used to separate oil from natural gas, as well as the equipment that is used, can vary widely. Although dry pipeline quality natural gas is virtually identical across different geographic areas, raw natural gas from different regions will vary in composition (Table 16.8; Chapter 1), and therefore, separation requirements may emphasize or deemphasize the optional separation processes. In many instances, natural gas is dissolved in oil underground primarily due to the formation pressure. When this natural gas and oil is produced, it is possible that it will separate on its own and but, in general, a separator is required. The conventional type of separator is consisting of a simple closed tank, where the force of gravity serves to separate the liquids like oil from the natural gas.

In certain instances, however, specialized equipment is necessary to separate oil and natural gas. An example of this type of equipment is the low-temperature separator. This is most often used for wells producing high-pressure gas along with light crude oil or condensate. These separators use pressure differentials to cool the wet natural gas and separate the oil and condensate. Wet gas enters the separator, being cooled slightly by a heat exchanger. The gas then travels through a high-pressure liquid *knockout pot* that serves to remove any liquids into a low-temperature separator. The gas then flows into this low-temperature separator through a choke mechanism, which expands the gas as it enters the separator. This rapid expansion of the gas allows for the lowering of the temperature in the separator. After removal of the liquids, the dry gas is sent back through the heat exchanger where it is warmed by the incoming wet gas. By varying the pressure of the gas in various sections of the separator, it is possible to vary the temperature, which causes the crude oil and some water to be condensed out of the wet gas stream.

On the other hand, petroleum refining produces gas streams that contain substantial amounts of acid gases such as hydrogen sulfide and carbon dioxide. These gas streams are produced during initial distillation of the crude oil and during the various conversion processes. Of particular interest is the hydrogen sulfide (H_2S) that arises from the hydrodesulfurization (Chapter 10) and hydrocracking (Chapter 11) of feedstocks that contain organically bound sulfur:

$$[S]_{feedstock} + H_2 \rightarrow H_2S + \text{Hydrocarbons}$$

Petroleum refining involves, with the exception of heavy crude oil, *primary distillation* (Chapter 7) that results in separation into fractions differing in carbon number, volatility, specific gravity, and other characteristics. The most volatile fraction that contains most of the gases, which are generally dissolved in the crude, is referred to as *pipestill gas* or *pipestill light ends* and consists essentially of hydrocarbon gases ranging from methane to butane(s), or sometimes pentane(s).

The gas varies in composition and volume, depending on crude origin and on any additions to the crude made at the loading point. It is not uncommon to reinject light hydrocarbons such as propane and butane into the crude oil before dispatch by tanker or pipeline. This results in a higher vapor pressure of the crude, but it allows one to increase the quantity of light products obtained at the refinery. Since light ends in most petroleum markets command a premium, while in the oil field itself propane and butane may have to be reinjected or flared, the practice of *spiking* crude oil with liquefied petroleum gas is becoming fairly common.

In addition to the gases obtained by distillation of petroleum, more highly volatile products result from the subsequent processing of naphtha and middle distillate to produce gasoline. Hydrogen sulfide is produced in the desulfurization processes involving hydrogen treatment of naphtha, distillate, and residual fuel and from the coking or similar thermal treatments of vacuum gas oils and heavy feedstocks (Chapter 8). The most common processing step in the production of gasoline is the catalytic reforming of hydrocarbon fractions in the heptane (C_7) to decane (C_{10}) range.

Additional gases are produced in *thermal cracking processes*, such as the coking or visbreaking processes (Chapter 8) for the processing of heavy feedstocks. In the visbreaking process, fuel oil is passed through externally fired tubes and undergoes liquid phase cracking reactions, which result in the formation of lighter fuel oil components. Oil viscosity is thereby reduced, and some gases, mainly hydrogen, methane, and ethane, are formed. Substantial quantities of both gas and carbon are also formed in coking (both fluid coking and delayed coking) in addition to the middle distillate and naphtha. When coking a residual fuel oil or heavy gas oil, the feedstock is preheated and contacted with hot carbon (coke), which causes extensive cracking of the feedstock constituents of higher molecular weight to produce lower-molecular-weight products ranging from methane, via liquefied petroleum gas(es) and naphtha, to gas oil and heating oil. Products from coking processes tend to be unsaturated and olefin components predominate in the tail gases from coking processes.

Another group of refining operations that contributes to gas production is the *catalytic cracking processes* (Chapter 9). These consist of fluid-bed catalytic cracking, and there are many process variants in which heavy feedstocks are converted into cracked gas, liquefied petroleum gas, catalytic naphtha, fuel oil, and coke by contacting the heavy hydrocarbon with the hot catalyst. Both catalytic and thermal cracking processes, the latter being now largely used for the production of chemical raw materials, result in the formation of not only unsaturated hydrocarbons, particularly ethylene ($CH_2=CH_2$), but also propylene (propene, $CH_3CH=CH_2$), *iso*-butylene [*iso*-butene, $(CH_3)_2C=CH_2$] and *n*-butenes ($CH_3CH_2CH=CH_2$, and $CH_3CH=CHCH_3$) in addition to hydrogen (H_2), methane (CH_4) and smaller quantities of ethane (CH_3CH_3), propane ($CH_3CH_2CH_3$), and butanes [$CH_3CH_2CH_2CH_3$, $(CH_3)_3CH$]. Diolefins such as butadiene ($CH_2=CHCH=CH_2$) are also present.

A further source of refinery gas is *hydrocracking*, a catalytic high-pressure pyrolysis process in the presence of fresh and recycled hydrogen (Chapter 11). The feedstock is again heavy gas oil or residual fuel oil, and the process is mainly directed at the production of additional middle distillates and gasoline. Since hydrogen is to be recycled, the gases produced in this process again have to be separated into lighter and heavier streams; any surplus recycle gas and the liquefied petroleum gas from the hydrocracking process are both saturated.

In a series of *reforming processes* (Chapter 13), commercialized under names such as *Platforming*, paraffin and naphthene (cyclic nonaromatic) hydrocarbons are converted in the presence of hydrogen and a catalyst and are converted into aromatics or isomerized to more highly branched hydrocarbons. Catalytic reforming processes thus not only result in the formation of a liquid product of higher octane number but also produce substantial quantities of gases. The latter are rich in hydrogen but also contain hydrocarbons from methane to butanes, with a preponderance of propane ($CH_3CH_2CH_3$), *n*-butane ($CH_3CH_2CH_2CH_3$), and *iso*-butane [$(CH_3)_3CH$].

The composition of the process gas varies in accordance with reforming severity and reformer feedstock. All catalytic reforming processes require substantial recycling of a hydrogen stream. Therefore, it is normal to separate reformer gas into a propane ($CH_3CH_2CH_3$) and/or a butane stream [$CH_3CH_2CH_2CH_3$ plus ($CH_3)_3CH$], which becomes part of the refinery liquefied petroleum gas production, and a lighter gas fraction, part of which is recycled. In view of the excess of hydrogen in the gas, all products of catalytic reforming are saturated, and there are usually no olefin gases present in either gas stream.

Both hydrocracker gases and catalytic reformer gases are commonly used in catalytic desulfurization processes. In the latter, feedstocks ranging from light to vacuum gas oils are passed at pressures of 500–1000 psi with hydrogen over a hydrofining catalyst. This results mainly in the conversion of organic sulfur compounds to hydrogen sulfide,

$$[S]_{feedstock} + H_2 \rightarrow H_2S + Hydrocarbons$$

This process also produces some light hydrocarbons by hydrocracking. Thus, refinery gas streams, while ostensibly being hydrocarbon in nature, may contain large amounts of acid gases such as hydrogen sulfide and carbon dioxide. Most commercial plants employ hydrogenation to convert organic sulfur compounds into hydrogen sulfide. Hydrogenation is affected by means of recycled hydrogen-containing gases or external hydrogen over a nickel molybdate or cobalt molybdate catalyst.

The presence of impurities in gas streams may eliminate some of the sweetening processes, since some processes remove large amounts of acid gas but not to a sufficiently low concentration. On the other hand, there are those processes not designed to remove (or incapable of removing) large amounts of acid gases whereas they are capable of removing the acid gas impurities to very low levels when the acid gases are present only in low-to-medium concentration in the gas.

The processes that have been developed to accomplish gas purification vary from a simple once-through wash operation to complex multistep recycling systems. In many cases, the process complexities arise because of the need for the recovery of the materials used to remove the contaminants or even recovery of the contaminants in the original, or altered, form (Katz, 1959; Kohl and Riesenfeld, 1985; Newman, 1985). In addition to the corrosion of equipment of acid gases (Speight, 2014b), the escape into the atmosphere of sulfur-containing gases can eventually lead to the formation of the constituents of acid rain, that is, the oxides of sulfur (sulfur dioxide, SO_2, and sulfur trioxide, SO_3). Similarly, the nitrogen-containing gases can also lead to nitrous and nitric acids (through the formation of the oxides NO_x, where x = 1 or 2), which are the other major contributors to acid rain. The release of carbon dioxide and hydrocarbons as constituents of refinery effluents can also influence the behavior and integrity of the ozone layer.

Finally, another acid gas, hydrogen chloride (HCl), although not usually considered to be a major emission, is produced from mineral matter and the brine that often accompany petroleum during production and is gaining increasing recognition as a contributor to acid rain. However, hydrogen chloride may exert severe local effects because it does not need to participate in any further chemical reaction to become an acid. Under atmospheric conditions that favor a buildup of stack emissions in the areas where hydrogen chloride is produced, the amount of hydrochloric acid in rain water could be quite high.

In summary, refinery process gas, in addition to hydrocarbons, may contain other contaminants, such as carbon oxides (CO_x, where x = 1 and/or 2), sulfur oxides (SO_x, where x = 2 and/or 3), as well as ammonia (NH_3), mercaptan derivatives (RSH), and carbonyl sulfide (COS). From an environmental viewpoint, petroleum processing can result in a variety of gaseous emissions. It is a question of degree insofar as the composition of the gaseous emissions may vary from process to process but the constituents are, in the majority of cases, the same.

16.2.2   Gas Streams from Natural Gas

Natural gas is also capable of producing emissions that are detrimental to the environment. While the major constituent of natural gas is methane, there are components such as carbon dioxide (CO_2), hydrogen sulfide (H_2S), and mercaptan derivatives (thiols; RSH), as well as trace amounts of sundry other emissions such as carbonyl sulfide (COS). The fact that methane has a foreseen and valuable end use makes it a desirable product, but in several other situations it is considered a pollutant, having been identified a greenhouse gas.

A sulfur removal process (Table 16.9) must be very precise, since natural gas contains only a small quantity of sulfur-containing compounds that must be reduced several orders of magnitude. Most consumers of natural gas require less than 4 ppm in the gas—a characteristic feature of natural gas that contains hydrogen sulfide is the presence of carbon dioxide (generally in the range of 1%–4% v/v). In cases where the natural gas does not contain hydrogen sulfide, there may also be a relative lack of carbon dioxide.

In practice, heaters and scrubbers are usually installed at or near to the wellhead. The scrubbers serve primarily to remove sand and other large-particle impurities and the heaters ensure that the temperature of the gas does not drop too low. With natural gas that contains even low quantities of water, natural gas hydrates ($C_nH_{2n+2}\cdot xH_2O$) have a tendency to form when temperatures drop. These hydrates are solid or semisolid compounds, resembling ice-like crystals. If the hydrates accumulate, they can impede the passage of natural gas through valves and gathering systems (Zhang et al., 2007). To reduce the occurrence of hydrates, small natural gas-fired heating units are typically installed along the gathering pipe wherever it is likely that hydrates may form.

Natural gas hydrates are usually considered as possible nuisances in the development of oil and gas fields, caution in handling the hydrates cannot be overemphasized because of their tendency to explosively decompose. On the other hand, if handled correctly and with caution, hydrates can be used for the safe and economic storage of natural gas. In remote offshore areas, the use of hydrates for natural gas transportation is also presently considered as an economic alternative to the processes based either on liquefaction or on compression (Lachet and Béhar, 2000).

TABLE 16.9
Sulfur Removal/Recovery Processes

Sodium hydrosulfide: Fuel gas containing hydrogen sulfide is contacted with sodium hydroxide in an absorption column. The resulting liquid is the product sodium hydrosulfide (NaHS).

Iron chelate: Fuel gas containing hydrogen sulfide is contacted with iron chelate catalyst dissolved in solution; hydrogen sulfide is converted to elemental sulfur, which is recovered.

Stretford: Stretford solution is used instead of iron chelate solution.

Ammonium thiosulfate: In this process, hydrogen sulfide is contacted with air to form sulfur dioxide, which is contacted with ammonia in a series of absorption column to produce ammonium thiosulfate for offsite sale.

Hyperion: Fuel gas is contacted over a solid catalyst to form elemental sulfur; the sulfur is collected and sold. The catalyst comprises iron and naphthoquinone sulfonic acid.

Sulfatreat: The Sulfatreat material is a black, granular solid powder; the hydrogen sulfide forms a chemical bond with the solid; when the bed reaches capacity, the Sulfatreat solids are removed and replaced with fresh material. The sulfur is not recovered.

Hysulf: Hydrogen sulfide is contacted with a liquid quinone in an organic solvent such as *n*-methyl-2-pyrolidone (NMP), forming sulfur; the sulfur is removed and the quinone reacted to its original state, producing hydrogen gas.

16.3 ENRICHMENT

The natural gas product fed into a gas transportation system must meet specific quality measures in order for the pipeline grid to operate properly. Consequently, natural gas produced at the wellhead, which in most cases contains contaminants and natural gas liquids, must be processed, that is, cleaned, before it can be safely delivered to the high-pressure, long-distance pipelines that transport the product to the consuming public. Natural gas that is not within certain specific gravities, pressures, Btu content range, or water content levels will cause operational problems, pipeline deterioration, or can even cause pipeline rupture. Thus, the purpose of *enrichment* is to produce natural gas for sale and enriched tank oil. The tank oil contains more light hydrocarbon liquids than natural petroleum, and the residue gas is drier (leaner, i.e., has lesser amounts of the higher-molecular-weight hydrocarbons). Therefore, the process concept is essentially the separation of hydrocarbon liquids from the methane to produce a lean, dry gas.

The natural gas received and transported must (especially in the United States and many other countries) meet the quality standards specified by the pipeline. These quality standards vary from pipeline to pipeline and are usually a function of (1) the design of the pipeline system, (2) the design of any downstream interconnecting pipelines, and (3) the requirements of the customer. In general, these standards specify that the natural gas should (1) be within a specific Btu content range, typically 1035 ± 50 Btu ft^3; (2) be delivered at a specified hydrocarbon dew point temperature level to prevent any vaporized gas liquid in the mix from condensing at pipeline pressure; (3) contain no more than trace amounts of elements such as hydrogen sulfide, carbon dioxide, nitrogen, water vapor, and oxygen; and (4) be free of particulate solids and liquid water that could be detrimental to the pipeline or its ancillary operating equipment. Gas processing equipment, whether in the field or at processing/treatment plants, assures that these specifications can be met.

In most cases, processing facilities extract contaminants and higher-boiling hydrocarbons from the gas stream. However, in some cases, the gas processors blend higher-boiling hydrocarbons into the gas stream in order to bring it within acceptable Btu levels. For example, in some areas if the produced gas (including coalbed methane) does not meet specifications the Btu requirements of the pipeline operator, in which case a blend of higher-Btu-content natural gas or a propane–air mixture is injected to enrich the heat content (Btu value) prior for delivery to the pipeline. In other instances, such as at liquefied natural gas (LNG) import facilities where the heat content of the regasified gas may be too high for pipeline receipt, vaporized nitrogen may be injected into the natural gas stream to lower its Btu content.

Briefly, and because it is sometime combined with petroleum-based natural gas for processing purposes, coalbed methane (CBM) is the generic term given to methane gas held in coal and released or produced when the water pressure within the buried coal is reduced by pumping from either vertical or inclined to horizontal surface holes. Thermogenic coalbed methane is predominantly formed during the coalification process whereby organic matter is slowly transformed into coal by increasing temperature and pressure as the organic matter is buried deeper and deeper by additional deposits of organic and inorganic matter over long periods of geological time. On the other hand, late-stage biogenic coalbed methane is formed by relatively recent bacterial processes (involving naturally occurring bacteria associated with meteoric water recharge at outcrop or subcrop), which can dominate the generation of coalbed methane. The amount of methane stored in coal is closely related to the rank and depth of the coal: the higher the coal rank and the deeper the coal seam is presently buried (causing pressure on coal), the greater its capacity to produce and retain methane gas. Gas derived from coal is generally pure and requires little or no processing because it is solely methane and not mixed with heavier hydrocarbons, such as ethane, which is often present in conventional natural gas.

The number of steps and the type of techniques used in the process of creating pipeline-quality natural gas most often depends upon the source and makeup of the wellhead production stream. Among the several stages of gas processing are (1) gas oil separation, (2) water removal, (3) liquids removal, (4) nitrogen removal, (5) acid gas removal, and (6) fractionation.

In many instances, pressure relief at the wellhead will cause a natural separation of gas from oil (using a conventional closed tank, where gravity separates the gas hydrocarbons from the heavier oil). In some cases, however, a multistage gas–oil separation process is needed to separate the gas stream from the crude oil. These gas–oil separators are commonly closed cylindrical shells, horizontally mounted with inlets at one end, an outlet at the top for the removal of gas, and an outlet at the bottom for the removal of oil. Separation is accomplished by alternately heating and cooling (by compression) the flow stream through multiple steps. However, the number of steps and the type of techniques used in the process of creating pipeline-quality natural gas most often depend upon the source and makeup of the gas stream. In some cases, several of the steps may be integrated into one unit or operation, performed in a different order or at alternative locations (lease/plant), or not required at all.

16.4 WATER REMOVAL

Water is a common impurity in gas streams, and removal of water is necessary to prevent the condensation of the water and the formation of ice or the formation of gas hydrates (USGS, 2014). Water in the liquid phase causes corrosion or erosion problems in pipelines and equipment, particularly when carbon dioxide and hydrogen sulfide are present in the gas. The simplest method of water removal (refrigeration or cryogenic separation) is to cool the gas to a temperature at least equal to or (preferentially) below the dew point (Figure 16.2).

In addition to separating petroleum and some condensate from the wet gas stream, it is necessary to remove most of the associated water. Most of the liquid, free water associated with extracted natural gas is removed by simple separation methods at or near the wellhead. However, the removal of the water vapor that exists in solution in natural gas requires a more complex treatment. This treatment consists of dehydrating the natural gas, which usually involves one of two processes: either absorption or adsorption.

Absorption occurs when the water vapor is taken out by a dehydrating agent. Adsorption occurs when the water vapor is condensed and collected on the surface. In a majority of cases, cooling alone is insufficient and, for the most part, impractical for use in field operations. Other more convenient water removal options use (1) *hygroscopic* liquids (e.g., diethylene glycol or triethylene glycol) and (2) solid adsorbents or desiccants (e.g., alumina, silica gel, and molecular sieves). Ethylene glycol can be directly injected into the gas stream in refrigeration plants.

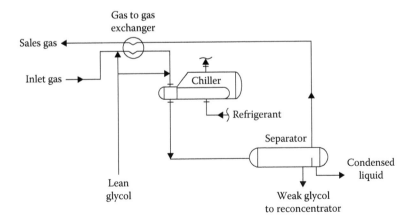

FIGURE 16.2 Glycol refrigeration process. (From Geist, J.M., *Oil Gas J.*, 83 56, 1985.)

16.4.1 ABSORPTION

An example of absorption dehydration is known as *glycol dehydration*—the principal agent in this process is diethylene glycol, which has a chemical affinity for water (Mokhatab et al., 2006; Speight, 2007; Abdel-Aal et al., 2016). Glycol dehydration involves using a solution of a glycol such as diethylene glycol (DEG) or triethylene glycol (TEG), which is brought into contact with the wet gas stream in a *contactor*. In practice, absorption systems recover 90%–99% by volume of methane that would otherwise be flared into the atmosphere.

In the process, a liquid desiccant dehydrator serves to absorb water vapor from the gas stream. The glycol solution absorbs water from the wet gas and, once absorbed, the glycol particles become heavier and sink to the bottom of the contactor where they are removed. The dry natural gas is then transported out of the dehydrator. The glycol solution, bearing all of the water stripped from the natural gas, is recycled through a specialized boiler designed to vaporize only the water out of the solution. The boiling point differential between water (100°C, 212°F) and glycol (204°C, 400°F) makes it relatively easy to remove water from the glycol solution.

As well as absorbing water from the wet gas stream, the glycol solution occasionally carries with it small amounts of methane and other compounds found in the wet gas. In order to decrease the amount of methane and other compounds that would otherwise be lost, flash tank separator condensers are employed to remove these compounds before the glycol solution reaches the boiler. The flash tank separator (Chapter 6) consists of a device that reduces the pressure of the glycol solution stream, allowing the methane and other hydrocarbons to vaporize (*flash*). The glycol solution then travels to the boiler, which may also be fitted with air or water-cooled condensers, which serve to capture any remaining organic compounds that may remain in the glycol solution. The regeneration (stripping) of the glycol is limited by temperature: diethylene glycol and triethylene glycol decompose at or even before their respective boiling points. Such techniques as stripping of hot triethylene glycol with dry gas (e.g., heavy hydrocarbon vapors, the *Drizo process*) or vacuum distillation are recommended.

Another absorption process, the Rectisol process, is a physical acid gas removal process using an organic solvent (typically methanol) at subzero temperatures, and characteristic of physical acid gas removal processes, it can purify synthesis gas down to 0.1 ppm total sulfur, including hydrogen sulfide (H_2S) and carbonyl sulfide (COS), and carbon dioxide (CO_2) in the ppm range (Mokhatab et al., 2006; Liu et al., 2010; Abdel-Aal et al., 2016). The process uses methanol as a wash solvent and the wash unit operates at temperatures below 0°C (32°F). To lower the temperature of the feed as temperatures, it is cooled against the cold-product streams, before entering the absorber tower. At the absorber tower, carbon dioxide and hydrogen sulfide (with carbonyl sulfide) are removed. By the use of an intermediate flash, coabsorbed products such as hydrogen and carbon monoxide are recovered, thus increasing the product recovery rate. To reduce the required energy demand for the carbon dioxide compressor, the carbon dioxide product is recovered in two different pressure steps (medium pressure and lower pressure). The carbon dioxide product is essentially sulfur-free (H_2S-free, COS-free) and water-free. The carbon dioxide products can be used for enhanced oil recovery (EOR) and/or sequestration or as pure carbon dioxide for other processes.

16.4.2 SOLID ADSORBENTS

Adsorption is a physical–chemical phenomenon in which the gas is concentrated on the surface of a solid or liquid to remove impurities. It must be emphasized that *adsorption* differs from *absorption* in that absorption is not a physical–chemical surface phenomenon but a process in which the absorbed gas is ultimately distributed throughout the absorbent (liquid). Dehydration using a solid adsorbent or solid desiccant is the primary form of dehydrating natural gas using adsorption and usually consists of two or more adsorption towers, which are filled with a solid desiccant (Mokhatab et al., 2006; Speight, 2007; Abdel-Aal et al., 2016). Typical desiccants include activated alumina or a granular silica gel material. Wet natural gas is passed through these towers, from top to bottom. As the wet gas passes around the particles of desiccant material, water is retained on the surface of these desiccant particles. Passing through the entire desiccant bed, almost all of the water is adsorbed onto the desiccant material, leaving the dry gas to exit the bottom of the tower. There are several solid desiccants that possess the physical characteristic to adsorb water from natural gas. These desiccants are generally used in dehydration systems consisting of two or more towers and associated regeneration equipment.

Molecular sieves—a class of aluminosilicates that produce the lowest water dew points and can be used to simultaneously sweeten and dry gases and liquids (Mokhatab et al., 2006; Speight, 2007; Maple, and Williams 2008; Abdel-Aal et al., 2016)—are commonly used in dehydrators ahead of plants designed to recover ethane and other natural gas liquids. These plants operate at very cold temperatures and require very dry feed gas to prevent the formation of hydrates. Dehydration to −100°C (−148°F) dew point is possible with molecular sieves. Water dew points less than −100°C (−148°F) can be accomplished with special design and definitive operating parameters (Mokhatab et al., 2006).

Molecular sieves are commonly used to selectively adsorb water and sulfur compounds from light hydrocarbon streams such as liquefied petroleum gas (LPG), propane, butane, pentane, light olefins, and alkylation feed. Sulfur compounds that can be removed are hydrogen sulfide, mercaptan derivatives, sulfide derivatives, and disulfide derivatives. In the process, the sulfur-containing feedstock is passed through a bed of sieves at ambient temperature. The operating pressure must be high enough to keep the feed in the liquid phase. The operation is cyclic in that the adsorption step is stopped at a predetermined time before sulfur breakthrough occurs. Sulfur and water are removed from the sieves by purging with fuel gas at 205°C–315°C (400°F–600°F).

Solid-adsorbent dehydrators are typically more effective than liquid absorption dehydrators (e.g., glycol dehydrators) and are usually installed as a type of straddle system along natural gas pipelines. These types of dehydration systems are best suited for large volumes of gas under very high pressure and are thus usually located on a pipeline downstream of a compressor station. Two or more towers are required due to the fact that after a certain period of use, the desiccant in a particular tower becomes saturated with water. To regenerate and recycle the desiccant, a high-temperature heater is used to heat gas to a very high temperature and passage of the heated gas stream through a saturated desiccant bed vaporizes the water in the desiccant tower, leaving it dry and allowing for further natural gas dehydration.

Although two-bed adsorbent treaters have become more common (while one bed is removing water from the gas, the other undergoes alternate heating and cooling), on occasion, a three-bed system is used: one bed adsorbs, one is being heated, and one is being cooled. An additional advantage of the three-bed system is the facile conversion of a two-bed system so that the third bed can be maintained or replaced, thereby ensuring the continuity of the operations and reducing the risk of a costly plant shutdown.

Silica gel (SiO_2) and alumina (Al_2O_3) have good capacities for water adsorption (up to 8% by weight). Bauxite (crude alumina, Al_2O_3) adsorbs up to 6% by weight water, and molecular sieves adsorb up to 15% by weight water. Silica is usually selected for the dehydration of sour gas because of its high tolerance to hydrogen sulfide and to protect molecular sieve beds from plugging by sulfur. Alumina *guard beds*, which serve as protectors by the act of attrition and may be referred to as an *attrition reactor* containing an *attrition catalyst* (Chapter 6; Speight, 2000, 2014a), may be placed ahead of the molecular sieves to remove the sulfur compounds. Downflow reactors are commonly used for adsorption processes, with an upward flow regeneration of the adsorbent and cooling using gas flow in the same direction as adsorption flow.

Solid desiccant units generally cost more to buy and operate than glycol units. Therefore, their use is typically limited to applications such as gases having a high hydrogen sulfide content, very low water dew point requirements, simultaneous control of water, and hydrocarbon dew points. In processes where cryogenic temperatures are encountered, solid desiccant dehydration is usually preferred over conventional methanol injection to prevent hydrate and ice formation (Kidnay and Parrish, 2006).

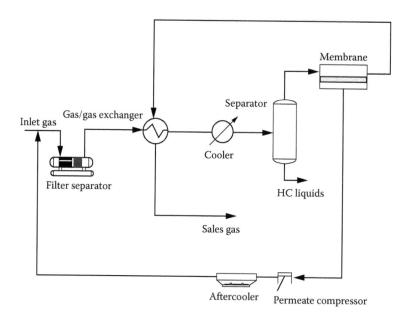

FIGURE 16.3 Membrane separation process.

16.4.3 MEMBRANES

Membrane separation process are very versatile and are designed to process a wide range of feed-stocks and offer a simple solution for the removal and recovery of higher-boiling hydrocarbons (natural gas liquids) from natural gas (Figure 16.3; Foglietta, 2004; Abdel-Aal et al., 2016). The separation process is based on high-flux membranes that selectively permeate higher-boiling hydro-carbons (compared to methane) and are recovered as a liquid after recompression and condensation. The residue stream from the membrane is partially depleted of higher-boiling hydrocarbons and is then sent to sales gas stream. Gas permeation membranes are usually made with vitreous polymers that exhibit good selectivity but, to be effective, the membrane must be very permeable with respect to the separation process.

16.5 LIQUID REMOVAL

Natural gas coming directly from a well contains higher-molecular-weight hydrocarbons, often referred to as *natural gas liquids* (NGLs) that, in most instances (depending upon the market demand), have a higher value as separate products and making it worthwhile to remove these constituents from the gas stream (Mokhatab et al., 2006; Speight, 2007; Abdel Aal et al., 2016). The removal of natural gas liquids usually takes place in a relatively centralized processing plant and uses techniques similar to those used to dehydrate natural gas. There are two basic steps to the treatment of natural gas liquids in the natural gas stream. In the first step, the liquids must be extracted from the natural gas and in the second step the natural gas liquids must be separated into the base constituents. These two processes account for approximately 90% v/v of the total produc-tion of natural gas liquids.

In many cases, before pipelining, condensate liquids are most often removed from the gas stream at the wellhead through the use of mechanical separators. In most instances, the gas flow into the separator comes directly from the wellhead, since the gas–oil separation process is not needed. The gas stream enters the processing plant at high pressure (≥600 psi) through an inlet slug catcher where free water is removed from the gas, after which it is directed to a condensate separator. Extracted condensate is routed to on-site storage tanks.

16.5.1 EXTRACTION

There are two principle techniques for removing natural gas liquids from the natural gas stream: the absorption method and the cryogenic expander process. In the process, a turboexpander is used to produce the necessary refrigeration and very low temperatures and high recovery of light components, such as ethane and propane, can be attained. The natural gas is first dehydrated using a molecular sieve followed by the cooling of the dry stream (Figure 16.4). The separated liquid containing most of the heavy fractions is then demethanized, and the cold gases are expanded through a turbine that produces the desired cooling for the process. The expander outlet is a two-phase stream that is fed to the top of the demethanizer column. This serves as a separator in which (1) the liquid is used as the column reflux and the separator vapors combined with vapors stripped in the demethanizer are exchanged with the feed gas and (2) the heated gas, which is partially recompressed by the expander compressor, is further recompressed to the desired distribution pressure in a separate compressor.

The extraction of natural gas liquids from the natural gas stream produces both cleaner, purer natural gas, as well as the valuable hydrocarbons that are the natural gas liquids themselves. This process allows for the recovery of approximately 90%–95% v/v of the ethane originally in the gas stream. In addition, the expansion turbine is able to convert some of the energy released when the natural gas stream is expanded into recompressing the gaseous methane effluent, thus saving energy costs associated with extracting ethane.

16.5.2 ABSORPTION

The absorption method of extraction is very similar to using absorption for dehydration. The main difference is that, in the absorption of natural gas liquids, absorbing oil is used as opposed to glycol. This absorbing oil has an affinity for natural gas liquids in much the same manner as glycol has an affinity for water. Before the oil has picked up any natural gas liquids, it is termed *lean* absorption oil.

The *oil absorption process* involves the countercurrent contact of the lean (or stripped) oil with the incoming wet gas with the temperature and pressure conditions programmed to maximize the dissolution of the liquefiable components in the oil. The *rich* absorption oil (sometimes referred to

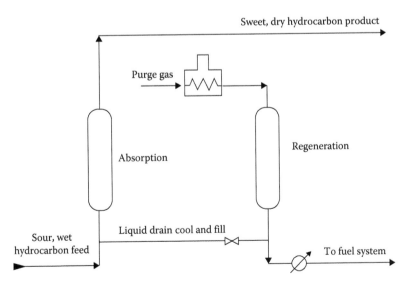

FIGURE 16.4 Drying using a molecular sieve. (From OSHA Technical Manual, Section IV, Chapter 2, Petroleum refining processes, 1999, http://www.osha.gov/dts/osta/otm/otm_iv/otm_iv_2.html, accessed January 14, 2015.)

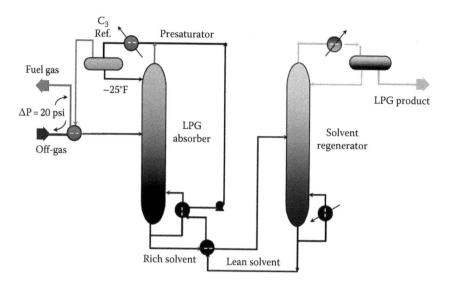

FIGURE 16.5 The AET recovery plant.

as *fat* oil), containing natural gas liquids, exits the absorption tower through the bottom. It is now a mixture of absorption oil, propane, butanes, pentanes, and other higher-boiling hydrocarbons. The rich oil is fed into lean oil stills, where the mixture is heated to a temperature above the boiling point of the natural gas liquids but below that of the oil. This process allows for the recovery of around 75% v/v of the butane isomers and 85%–90% v/v of the pentane isomers and higher-boiling constituents from the natural gas stream.

The basic absorption process is subject to modifications that improve process effectiveness and even target the extraction of specific natural gas liquids. In the refrigerated oil absorption method, where the lean oil is cooled through refrigeration, propane recovery can be in the order of 90%+ v/v, and approximately 40% v/v of the ethane can be extracted from the natural gas stream. Extraction of the other, higher-boiling natural gas hydrocarbons is typically near quantitative using this process.

The AET process (Figure 16.5) for the recovery of liquefied petroleum gas utilizes noncryogenic absorption to recover ethane, propane, and higher-boiling constituents from natural gas streams. The absorbed gases in the rich solvent from the bottom of the absorber column are fractionated in the solvent regenerator column that separates gases (as an overhead fraction) and lean solvent (as a bottoms fraction). After heat recuperation, the lean solvent is presaturated with absorber overhead gases. The chilled solvent flows in the top of the absorber column. The separated gases are sent to storage. Depending upon the economics of ethane recovery, the operation of the plant can be switched online from ethane plus recovery to propane plus recovery without affecting the propane recovery levels. The AET-liquefied petroleum gas plant uses lower-boiling lean oils. For most applications, there are no solvent makeup requirements.

16.5.3 Fractionation of Natural Gas Liquids

After separation of the natural gas liquids from the natural gas stream, they must be separated (fractionated) into the individual constituents prior to sales. The process of fractionation (which is based on the different boiling points of the hydrocarbons that constitute the natural gas liquids) occurs in stages with each stage involving the separation of the hydrocarbons as individual products. The process commences with the removal of the lower-boiling hydrocarbons from the feedstock. The particular fractionators are used in the following order: (1) the deethanizer, which is used to separate the ethane from the stream of natural gas liquids; (2) the depropanizer, which is used to separate the propane from the deethanized stream; (3) the debutanizer, which is used to separate the butane isomers, leaving the pentane isomers and higher-boiling hydrocarbons in the stream; and (4) the butane splitter or deisobutanizer, which is used to separate *n*-butane and *iso*-butane.

16.6 NITROGEN REMOVAL

Nitrogen may often occur in sufficient quantities in natural gas and, consequently, lower the heating value of the gas. Thus, several plants for *nitrogen removal* from natural gas have been built, but it must be recognized that nitrogen removal requires liquefaction and fractionation of the entire gas stream, which may affect process economics. In some cases, the nitrogen-containing natural gas is blended with a gas having a higher heating value and sold at a reduced price depending upon the thermal value (Btu/ft³).

For high-flow-rate gas streams, a cryogenic process is typical and involves the use of the different volatilities of methane (b.p. −161.6°C/−258.9°F) and nitrogen (b.p. −195.7°C/−320.3°F) to achieve separation. In the process, a system of compression and distillation columns drastically reduces the temperature of the gas mixture to a point where methane is liquefied and the nitrogen is not. On the other hand, for smaller volumes of gas, a system utilizing pressure swing adsorption (PSA) is a more typical method of separation. In pressure swing adsorption method, methane and nitrogen can be separated by using an adsorbent with an aperture size very close to the molecular diameter of the larger species (the methane), which allows nitrogen to diffuse through the adsorbent. This results in a purified natural gas stream that is suitable for pipeline specifications. The adsorbent can then be regenerated, leaving a highly pure nitrogen stream. The pressure swing adsorption method is a flexible method for nitrogen rejection, being applied to both small and large flow rates.

16.7 ACID GAS REMOVAL

In addition to water and natural gas liquids removal, one of the most important parts of gas processing involves the removal of hydrogen sulfide and carbon dioxide, which are generally referred to as contaminants. Natural gas from some wells contains significant amounts of hydrogen sulfide and carbon dioxide and is usually referred to as *sour gas*. Sour gas is undesirable because the sulfur compounds it contains can be extremely harmful, even lethal, to breathe and the gas can also be extremely corrosive. The process for removing hydrogen sulfide from sour gas is commonly referred to as *sweetening* the gas.

The primary process (Figure 16.6) for sweetening sour natural gas is quite similar to the processes of glycol dehydration and removal of natural gas liquids by absorption. In this case, however, amine (*olamine*) solutions are used to remove the hydrogen sulfide (the *amine process*). The sour gas is run through a tower, which contains the olamine solution. There are two principle amine solutions used, monoethanolamine (MEA) and diethanolamine (DEA). Either of these compounds, in liquid form, will absorb sulfur compounds from natural gas as it passes through. The effluent gas is virtually free of sulfur compounds and thus loses its sour gas status. Like the process for the extraction of natural gas liquids and glycol dehydration, the amine solution used can be regenerated for reuse. Although most sour gas sweetening involves the amine absorption process, it is also possible to use solid desiccants like iron sponge to remove hydrogen sulfide and carbon dioxide (Mokhatab et al., 2006; Speight, 2007; Abdel-Aal et al., 2016).

Treatment of gas to remove the acid gas constituents (hydrogen sulfide and carbon dioxide) is most often accomplished by the contact of the natural gas with an alkaline solution. The most commonly used treating solutions are aqueous solutions of the ethanolamine (Table 16.3) or alkali carbonates, although a considerable number of other treating agents have been developed in recent years. Most of these newer treating agents rely upon physical absorption and chemical reaction. When only carbon dioxide is to be removed in large quantities or when only partial removal is necessary, a hot carbonate solution or one of the physical solvents is the most economical selection.

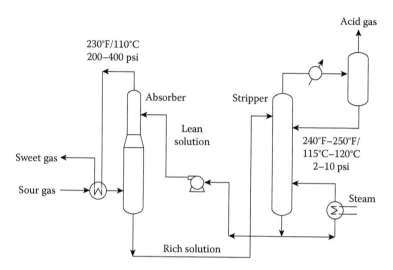

FIGURE 16.6 The amine (olamine) process.

The most well-known hydrogen sulfide removal process is based on the reaction of hydrogen sulfide with iron oxide (often also called the iron sponge process or the dry box method) in which the gas is passed through a bed of wood chips impregnated with iron oxide.

The iron oxide process (which was implemented during the nineteenth century and also referred to as the iron sponge process) is the oldest and still the most widely used batch process for sweetening natural gas and natural gas liquids (Zapffe, 1963; Duckworth and Geddes, 1965; Anerousis and Whitman, 1984; Mokhatab et al., 2006; Speight, 2014a). In the process (Figure 16.7), the sour gas is passed down through the bed. In the case where continuous regeneration is to be utilized, a small concentration of air is added to the sour gas before it is processed. This air serves to continuously regenerate the iron oxide, which has reacted with hydrogen sulfide, which serves to extend the onstream life of a given tower but probably serves to decrease the total amount of sulfur that a given weight of bed will remove.

The process is usually best applied to gases containing low to medium concentrations (300 ppm) of hydrogen sulfide or mercaptan derivatives. This process tends to be highly selective and does not

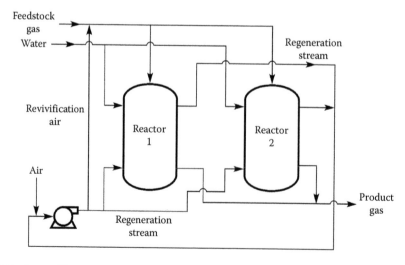

FIGURE 16.7 Iron oxide process.

normally remove significant quantities of carbon dioxide. As a result, the hydrogen sulfide stream from the process is usually of high purity. The use of iron oxide process for sweetening sour gas is based on the adsorption of the acid gases on the surface of the solid sweetening agent followed by the chemical reaction of ferric oxide (Fe_2O_3) with hydrogen sulfide:

$$2Fe_2O_3 + 6H_2S \rightarrow 2Fe_2S_3 + 6H_2O$$

The reaction requires the presence of slightly alkaline water and a temperature below 43°C (110°F) and bed alkalinity (pH +8 to 10) should be checked regularly, usually on a daily basis. The pH level is be maintained through the injection of caustic soda with the water. If the gas does not contain sufficient water vapor, water may need to be injected into the inlet gas stream.

The ferric sulfide produced by the reaction of hydrogen sulfide with ferric oxide can be oxidized with air to produce sulfur and regenerate the ferric oxide:

$$2Fe_2S_3 + 3O_2 \rightarrow 2Fe_2O_3 + 6S$$

$$S_2 + 2O_2 \rightarrow 2SO_2$$

The regeneration step is exothermic and air must be introduced slowly so that the heat of reaction can be dissipated. If air is introduced quickly, the heat of reaction may ignite the bed. Some of the elemental sulfur produced in the regeneration step remains in the bed. After several cycles this sulfur will cake over the ferric oxide, decreasing the reactivity of the bed. Typically, after 10 cycles the bed must be removed and a new bed introduced into the vessel.

The iron oxide process is one of several metal oxide-based processes that scavenge hydrogen sulfide and organic sulfur compounds (mercaptan derivatives) from gas streams through reactions with the solid-based chemical adsorbent (Kohl and Riesenfeld, 1985). They are typically nonregenerable, although some are partially regenerable, losing activity upon each regeneration cycle. Most of the processes are governed by the reaction of a metal oxide with hydrogen sulfide to form the metal sulfide. For regeneration, the metal oxide is reacted with oxygen to produce elemental sulfur and the regenerated metal oxide. In addition to iron oxide, the primary metal oxide used for dry sorption processes is zinc oxide.

In the zinc oxide process, the zinc oxide media particles are extruded cylinders 3–4 mm in diameter and 4–8 mm in length (Kohl and Nielsen, 1997; Mokhatab et al., 2006; Speight, 2007; Abdel-Aal et al., 2016) and react readily with the hydrogen sulfide:

$$ZnO + H_2S \rightarrow ZnS + H_2O$$

At increased temperatures (205°C–370°C, 400°F–700°F), zinc oxide has a rapid reaction rate, thereby providing a short mass transfer zone, resulting in a short length of the unused bed and improved efficiency.

Removal of larger amounts of hydrogen sulfide from gas streams requires a continuous process, such as the Ferrox process or the Stretford process. The *Ferrox process* is based on the same chemistry as the iron oxide process except that it is fluid and continuous. The *Stretford process* employs a solution containing vanadium salts and anthraquinone disulfonic acid (Maddox, 1974; Abdel-Aal et al., 2016). Most hydrogen sulfide removal processes return the hydrogen sulfide unchanged, but if the quantity involved does not justify the installation of a sulfur recovery plant (usually a Claus plant), it is necessary to select a process that directly produces elemental sulfur. In the *Beavon–Stretford process*, a hydrotreating reactor converts sulfur dioxide in the off-gas to hydrogen sulfide, which is contacted with Stretford solution (a mixture of vanadium salt, anthraquinone disulfonic acid, sodium carbonate, and sodium hydroxide) in a liquid–gas absorber. The hydrogen sulfide reacts stepwise with sodium carbonate and anthraquinone disulfonic acid to produce elemental sulfur, with vanadium serving as a catalyst. The solution proceeds to a tank where oxygen is added to regenerate the reactants. One or more froth or slurry tanks are used to skim the product sulfur from the solution, which is recirculated to the absorber.

The processes using ethanolamine and potassium phosphate are now widely used. The ethanolamine process, known as the *Girbotol process*, removes acid gases (hydrogen sulfide and carbon dioxide) from liquid hydrocarbons as well as from natural and from refinery gases. The Girbotol process uses an aqueous solution of ethanolamine ($H_2NCH_2CH_2OH$) that reacts with hydrogen sulfide at low temperatures and releases hydrogen sulfide at high temperatures. The ethanolamine solution fills a tower called an absorber through which the sour gas is bubbled. Purified gas leaves the top of the tower, and the ethanolamine solution leaves the bottom of the tower with the absorbed acid gases. The ethanolamine solution enters a reactivator tower where heat drives the acid gases from the solution. Ethanolamine solution, restored to its original condition, leaves the bottom of the reactivator tower to go to the top of the absorber tower, and acid gases are released from the top of the reactivator.

The process using potassium phosphate is known as phosphate desulfurization, and it is used in the same way as the Girbotol process to remove acid gases from liquid hydrocarbons as well as from gas streams. The treatment solution is a water solution of potassium phosphate (K_3PO_4), which is circulated through an absorber tower and a reactivator tower in much the same way as the ethanolamine is circulated in the Girbotol process; the solution is regenerated thermally.

Moisture may be removed from hydrocarbon gases at the same time as hydrogen sulfide is removed. Moisture removal is necessary to prevent harm to anhydrous catalysts and to prevent the formation of hydrocarbon hydrates (e.g., $C_3H_8 \cdot 18H_2O$) at low temperatures. A widely used dehydration and desulfurization process is the glycolamine process, in which the treatment solution is a mixture of ethanolamine and a large amount of glycol. The mixture is circulated through an absorber and a reactivator in the same way as ethanolamine is circulated in the Girbotol process. The glycol absorbs moisture from the hydrocarbon gas passing up the absorber; the ethanolamine absorbs hydrogen sulfide and carbon dioxide. The treated gas leaves the top of the absorber; the spent ethanolamine–glycol mixture enters the reactivator tower, where heat drives off the absorbed acid gases and water.

Other processes include the *Alkazid process* for the removal of hydrogen sulfide and carbon dioxide using concentrated aqueous solutions of amino acids. The hot potassium carbonate process decreases the acid content of natural and refinery gas from as much as 50% to as low as 0.5% and operates in a unit similar to that used for amine treating. The *Giammarco–Vetrocoke process* is used for hydrogen sulfide and/or carbon dioxide removal. In the hydrogen sulfide removal section, the reagent consists of sodium carbonate (Na_2CO_3) or potassium carbonate (K_2CO_3) or a mixture of the carbonates, which contain a mixture of arsenite derivatives and arsenate derivatives; the carbon dioxide removal section utilizes hot aqueous alkali carbonate solution activated by arsenic trioxide (As_2O_3) or selenous acid (H_2SeO_3) or tellurous acid (H_2TeO_3). A word of caution might be added about the last three chemicals, which are toxic and can involve stringent environmental-related disposal protocols.

Molecular sieves are highly selective for the removal of hydrogen sulfide (as well as other sulfur compounds) from gas streams and over continuously high absorption efficiency. They are also an effective means of water removal and thus offer a process for the simultaneous dehydration and desulfurization of gas. Gas that has excessively high water content may require upstream dehydration, however (Mokhatab et al., 2006; Speight, 2007, 2014a; Abdel-Aal et al., 2016). The *molecular sieve process* is similar to the iron oxide process. Regeneration of the bed is achieved by passing heated clean gas over the bed. As the temperature of the bed increases, it releases the adsorbed hydrogen sulfide into the regeneration gas stream. The sour effluent regeneration gas is sent to a flare stack, and up to 2% v/v of the gas seated can be lost in the regeneration process. A portion of the natural gas may also be lost by the adsorption of hydrocarbon components by the sieve (Mokhatab et al., 2006; Speight, 2007, 2014a).

In this process, unsaturated hydrocarbon components, such as olefins and aromatics, tend to be strongly adsorbed by the molecular sieve. Molecular sieves are susceptible to poisoning by such chemicals as glycols and require thorough gas cleaning methods before the adsorption step. Alternatively, the sieve can be offered some degree of protection by the use of *guard beds* in which a less expensive catalyst is placed in the gas stream before contact of the gas with the sieve, thereby protecting the catalyst from poisoning. This concept is analogous to the use of guard beds or attrition catalysts in the petroleum industry (Speight, 2000).

Other processes worthy of note include (1) the Selexol process, (2) the Sulfinol process, (3) the LOCAT process, and (4) the Sulferox process (Mokhatab et al., 2006; Abdel-Aal et al., 2016).

The *Selexol process* uses a mixture of the dimethyl ether of propylene glycol as a solvent. It is nontoxic and its boiling point is not high enough for amine formulation. The selectivity of the solvent for hydrogen sulfide (H_2S) is much higher than that for carbon dioxide (CO_2), so it can be used to selectively remove these different acid gases, minimizing carbon dioxide content in the hydrogen sulfide stream sent to the sulfur recovery unit (SRU) and enabling the regeneration of solvent for carbon dioxide recovery by economical flashing. In the process, a stream of natural gas is injected in the bottom of the absorption tower operated at 1000 psi. The rich solvent is flashed in a flash drum (flash reactor) at 200 psi where methane is flashed and recycled back to the absorber and joins the sweet (low-sulfur or no-sulfur) gas stream. The solvent is then flashed at atmospheric pressure and acid gases are flashed off. The solvent is then stripped by steam to completely regenerate the solvent, which is recycled back to the absorber. Any hydrocarbons are condensed, and any remaining acid gases are flashed from the condenser drum. This process is used when there is a high acid gas partial pressure and no heavy hydrocarbons. Diisopropanolamine (DIPA) can be added to this solvent to remove carbon dioxide to a level suitable for pipeline transportation.

The *Sulfinol process* uses a solvent that is a composite solvent, consisting of a mixture of diisopropanolamine (30%–45% v/v) or methyl diethanolamine (MDEA), sulfolane (tetrahydrothiophene dioxide) (40%–60% v/v), and water (5%–15% v/v). The acid gas loading of the Sulfinol solvent is higher, and the energy required for its regeneration is lower than those of purely chemical solvents. At the same time, it has the advantage over purely physical solvents that severe product specifications can be met more easily and coabsorption of hydrocarbons is relatively low. Aromatic compounds, higher-molecular-weight hydrocarbons, and carbon dioxide are soluble to a lesser extent. The process is typically used when the hydrogen sulfide–carbon dioxide ratio is greater than 1:1 or where carbon dioxide removal is not required to the same extent as hydrogen sulfide removal. The process uses a conventional solvent absorption and regeneration cycle in which the sour gas components are removed from the feed gas by countercurrent contact with a lean solvent stream under pressure. The absorbed impurities are then removed from the rich solvent by stripping with steam in a heated regenerator column. The hot lean solvent is then cooled for reuse in the absorber. Part of the cooling may be by heat exchange with the rich solvent for partial recovery of heat energy. The solvent reclaimer is used in a small ancillary facility for recovering solvent components from higher-boiling products of alkanolamine degradation or from other high-boiling or solid impurities.

The *LOCAT process* uses an extremely dilute solution of iron chelates. A small portion of the chelating agent is depleted in some side reactions and is lost with precipitated sulfur. In this process, sour gas is contacted with the chelating reagent in the absorber and H_2S reacts with the dissolved iron to form elemental sulfur:

$$H_2S + 2Fe^{3+} \rightarrow S + 2Fe^{2+} + 2H^+$$

The sulfur is removed from the regenerator to centrifugation and melting. Application of heat is not required because of the exothermic reaction. The reduced iron ion is regenerated in the regenerator by air blowing:

$$4Fe^{2+} + O_2 + 2H_2O \rightarrow 4Fe^{3+} + 4OH^-$$

In the *Sulferox process*, chelating iron compounds are the heart of the process. Sulferox is a redox technology, as is the LOCAT; however, in this case, a concentrated iron solution is used to oxidize hydrogen sulfide to elemental sulfur. Chelating agents are used to increase the solubility of iron in the operating solution. As a result of high iron concentrations in the solution, the rate of liquid circulation can be kept low and, consequently, the equipment becomes small. As in the LOCAT process, there are two basic reactions: the first takes place in the absorber and the second takes place in the regenerator, as in reaction. The key to the Sulferox technology is the ligand used in the process that allows the process to use high total iron concentrations (>1% w/w). In the process, the acid gas enters the contactor, where hydrogen sulfide is oxidized to produce elemental sulfur. The treated gas and the Sulferox solution flow to the separator, where sweet gas exits at the top and the solution is sent to the regenerator where ferrous iron (Fe^{2+}) is oxidized by air to ferric iron (Fe^{3+}) and the solution is regenerated and sent back to the contactor. Sulfur settles in the regenerator and is taken from the bottom to filtration, where sulfur cake is produced. At the top of the regenerator, spent air is released. A makeup Sulferox solution is added to replace the degradation of the ligands. Control of this degradation rate and purging of the degradation products ensures smooth operation of the process.

16.8 FRACTIONATION

Fractionation processes are very similar to those processes classed as *liquids removal* processes but often appear to be more specific in terms of the objectives: hence the need to place the fractionation processes into a separate category. The fractionation processes are those processes that are used (1) to remove the more significant product stream first or (2) to remove any unwanted light ends from the heavier liquid products.

In the general practice of natural gas processing, the first unit is a deethanizer followed by a depropanizer, then by a debutanizer, and finally, by a butane fractionator. Thus, each column can operate at a successively lower pressure, thereby allowing the different gas streams to flow from column to column by virtue of the pressure gradient, without necessarily the use of pumps.

The purification of hydrocarbon gases by any of these processes is an important part of refinery operations, especially in regard to the production of liquefied petroleum gas (LPG). This is actually a mixture of propane and butane, which is an important domestic fuel, as well as an intermediate material in the manufacture of petrochemicals (Speight and Ozum, 2002; Parkash, 2003; Hsu and Robinson, 2006; Gary et al., 2007; Speight, 2014a). The presence of ethane in liquefied petroleum gas must be avoided because of the inability of this lighter hydrocarbon to liquefy under pressure at ambient temperatures and its tendency to register abnormally high pressures in the liquefied petroleum gas containers. On the other hand, the presence of pentane in liquefied petroleum gas must also be avoided, since this particular hydrocarbon (a liquid at ambient temperatures and pressures) may separate into a liquid state in the gas lines.

16.9 CLAUS PROCESS

The *Claus process* is not so much a gas cleaning process but a process for the disposal of hydrogen sulfide, a toxic gas that originates in natural gas as well as during crude oil processing such as in the coking, catalytic cracking, hydrotreating, and hydrocracking processes. Burning hydrogen sulfide as a fuel gas component or as a flare gas component is precluded by safety and environmental considerations since one of the combustion products is the highly toxic sulfur dioxide (SO_2), which is also toxic. As described earlier, hydrogen sulfide is typically removed from the refinery light end gas streams through an olamine process after which the application of heat regenerates the olamine and forms an acid gas stream. Following from this, the acid gas stream is treated to convert the hydrogen sulfide elemental sulfur and water. The conversion process utilized in most modern refineries is the Claus process, or a variant thereof.

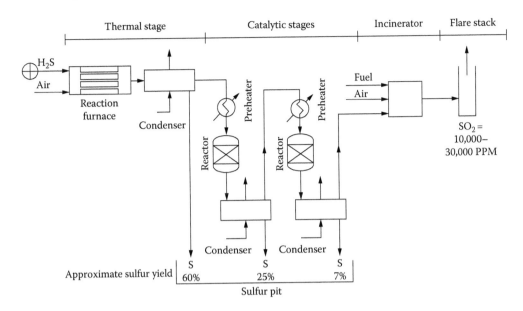

FIGURE 16.8 Claus process. (From Maddox, R.N., *Gas and Liquid Sweetening*, 2nd edn., Campbell Publishing, Norman, OK, 1974.)

The Claus process (Figure 16.8) involves combustion of approximately one-third of the hydrogen sulfide to sulfur dioxide and then reaction of the sulfur dioxide with the remaining hydrogen sulfide in the presence of a fixed bed of activated alumina, cobalt molybdenum catalyst resulting in the formation of elemental sulfur:

$$2H_2S + 3O_2 \rightarrow 2SO_2 + 2H_2O$$

$$2H_2S + SO_2 \rightarrow 3S + 2H_2O$$

Different process flow configurations are used to achieve the correct hydrogen sulfide/sulfur dioxide ratio in the conversion reactors.

In a split-flow configuration, one-third split of the acid gas stream is completely combusted and the combustion products are then combined with the noncombusted acid gas upstream of the conversion reactors. In a once-through configuration, the acid gas stream is partially combusted by only providing sufficient oxygen in the combustion chamber to combust one-third of the acid gas. Two or three conversion reactors may be required depending on the level of hydrogen sulfide conversion required. Each additional stage provides incrementally less conversion than the previous stage. Overall, conversion of 96%–97% v/v of the hydrogen sulfide to elemental sulfur is achievable in a Claus process. If this is insufficient to meet air quality regulations, a Claus process tail gas treater is utilized to remove essentially the entire remaining hydrogen sulfide in the tail gas from the Claus unit. The tail gas treater may employ a proprietary solution to absorb the hydrogen sulfide followed by conversion to elemental sulfur.

The SCOT (Shell Claus Off-Gas Treating) unit is a very common type of tail gas unit and uses a hydrotreating reactor followed by amine scrubbing to recover and recycle sulfur, in the form of hydrogen, to the Claus unit (Nederland, 2004). In the process (Figure 16.9), tail gas (containing hydrogen sulfide and sulfur dioxide) is contacted with hydrogen and reduced in a hydrotreating reactor to form hydrogen sulfide and water. The catalyst is typically cobalt/molybdenum on alumina. The gas is then cooled in a water contractor. The hydrogen sulfide–containing gas enters an amine absorber that is typically in a system segregated from the other refinery amine systems.

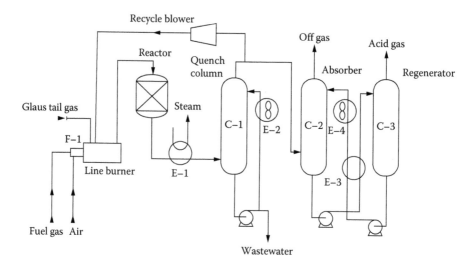

FIGURE 16.9 The SCOT process.

TABLE 16.10

Tail Gas Processes

Caustic scrubbing: An incinerator converts trace sulfur compounds in the off-gas to sulfur dioxide, which is contacted with caustic, which is sent to the wastewater treatment system.

Polyethylene glycol: Off-gas from the Claus unit is contacted with this solution to generate an elemental sulfur product; unlike the Beavon–Stretford process, no hydrogenation reactor is used to convert sulfur dioxide to hydrogen sulfide.

Selectox: A hydrogenation reactor converts sulfur dioxide in the off-gas to hydrogen sulfide; a solid catalyst in a fixed-bed reactor converts the hydrogen sulfide to elemental sulfur that is recovered for sales.

Sulfite/bisulfite tail gas treating unit: Following Claus reactors, an incinerator converts trace sulfur compounds to sulfur dioxide, which is then contacted with sulfite solution in an absorber, where the sulfur dioxide reacts with the sulfite to produce a bisulfite solution. The gas is then sent to the stack, the bisulfite is regenerated and liberated, and sulfur dioxide is sent to the Claus units for recovery.

The purpose of segregation is twofold: (1) the tail gas treater frequently uses a different amine than the rest of the plant and (2) the tail gas is frequently cleaner than the refinery fuel gas (with regard to contaminants) and segregation of the systems reduces maintenance requirements for the SCOT® unit. Amines chosen for use in the tail gas system tend to be more selective for hydrogen sulfide and are not affected by the high levels of carbon dioxide in the off-gas.

The hydrotreating reactor converts sulfur dioxide in the off-gas to hydrogen sulfide that is then contacted with a Stretford solution (a mixture of a vanadium salt, anthraquinone disulfonic acid, sodium carbonate, and sodium hydroxide) in a liquid–gas absorber (Abdel-Aal et al., 2016). The hydrogen sulfide reacts stepwise with sodium carbonate and the anthraquinone sulfonic acid to produce elemental sulfur, with vanadium serving as a catalyst. The solution proceeds to a tank where oxygen is added to regenerate the reactants. One or more froths or slurry tanks are used to skim the sulfur product from the solution, which is recirculated to the absorber. Other tail gas treating processes include (1) caustic scrubbing, (2) polyethylene glycol treatment, (3) the Selectox process, and (4) sulfite/bisulfite tail gas treatment (Table 16.10; Mokhatab et al., 2006; Speight, 2007, 2014a).

REFERENCES

Abdel-Aal, H.K., Aggour, M.A., and Fahim, M.A. 2016. *Petroleum and Gas Filed Processing*. CRC Press/ Taylor & Francis Group, Boca Raton, FL.

Anerousis, J.P. and Whitman, S.K. 1984. An updated examination of gas sweetening by the iron sponge process. Paper No. SPE 13280. *SPE Annual Technical Conference and Exhibition*. Houston, TX, September.

Barbouteau, L. and Dalaud, R. 1972. In: *Gas Purification Processes for Air Pollution Control*. G. Nonhebel (Editor). Butterworth & Co., London, UK, Chapter 7.

Bartoo, R.K. 1985. In: *Acid and Sour Gas Treating Processes*. S.A. Newman (Editor). Gulf Publishing Company, Houston, TX.

Curry, R.N. 1981. *Fundamentals of Natural Gas Conditioning*. PennWell Publishing Co., Tulsa, OK.

Duckworth, G.L. and Geddes, J.H. 1965. Natural gas desulfurization by the iron sponge process. *Oil and Gas Journal*, 63(37): 94–96.

Foglietta, J.H. 2004. Dew point turboexpander process: A solution for high pressure fields. *Proceedings of IAPG Gas Conditioning Conference*. Neuquen, Argentina, October 18.

Gary, J.G., Handwerk, G.E., and Kaiser, M.J. 2007. *Petroleum Refining: Technology and Economics*, 5th Edition. CRC Press/Taylor & Francis Group, Boca Raton, FL.

Geist, J.M. 1985. Refrigeration cycles for the future. *Oil and Gas Journal*, 83(5): 56–60.

Hsu, C.S. and Robinson, P.R. 2006. *Practical Advances in Petroleum Processing*, Volumes 1 and 2. Springer, New York.

Jou, F.Y., Otto, F.D., and Mather, A.E. 1985. In: *Acid and Sour Gas Treating Processes*. S.A. Newman (Editor). Gulf Publishing Company, Houston, TX, Chapter 10.

Katz, D.K. 1959. *Handbook of Natural Gas Engineering*. McGraw-Hill Book Company, New York.

Kenreck, G. 2014. Manage hydrogen sulfide hazards with chemical scavengers. *Hydrocarbon Processing*, 92(12): 73–76.

Kidnay, A.J. and Parrish, W.R. 2006. *Fundamentals of Natural Gas Processing*. CRC Press/Taylor & Francis Group, Boca Raton, FL.

Kohl, A.L. and Nielsen, R.B. 1997. *Gas Purification*. Gulf Publishing Company, Houston, TX.

Kohl, A.L. and Riesenfeld, F.C. 1985. *Gas Purification*, 4th Edition. Gulf Publishing Company, Houston, TX.

Lachet, V. and Béhar, E. 2000. Industrial perspective on natural gas hydrates. *Oil and Gas Science and Technology*, 55: 611–616.

Liu, K., Song, C., and Subramani, V. 2010. *Hydrogen and Syngas Production and Purification Technologies*. John Wiley & Sons, Hoboken, NJ.

Maddox, R.N. 1974. *Gas and Liquid Sweetening*, 2nd Edition. Campbell Publishing, Norman, OK.

Maddox, R.N. 1982. *Gas Conditioning and Processing*, Volume 4: Gas and Liquid Sweetening. Campbell Publishing, Norman, OK.

Maddox, R.N., Bhairi, A., Mains, G.J., and Shariat, A. 1985. In: *Acid and Sour Gas Treating Processes*. S.A. Newman (Editor). Gulf Publishing Company, Houston, TX, Chapter 8.

Maple, M.J. and Williams, C.D. 2008. Separating nitrogen/methane on zeolite-like molecular sieves. *Microporous and Mesoporous Materials*, 111: 627–631.

Mody, V. and Jakhete, R. 1988. *Dust Control Handbook*. Noyes Data Corporation, Park Ridge, NJ.

Mokhatab, S., Poe, W.A., and Speight, J.G. 2006. *Handbook of Natural Gas Transmission and Processing*. Elsevier, Amsterdam, the Netherlands.

Nederland, J. 2004. *Sulphur*. University of Calgary, Calgary, Alberta, Canada.

Newman, S.A. 1985. *Acid and Sour Gas Treating Processes*. Gulf Publishing Company, Houston, TX.

OSHA Technical Manual. 1999. Section IV, Chapter 2: Petroleum refining processes. http://www.osha.gov/ dts/osta/otm/otm_iv/otm_iv_2.html, accessed January 14, 2015.

Parkash, S. 2003. *Refining Processes Handbook*. Gulf Professional Publishing, Elsevier, Amsterdam, the Netherlands.

Pitsinigos, V.D. and Lygeros, A.I. 1989. Predicting H_2S-MEA equilibria. *Hydrocarbon Processing*, 58(4): 43–44.

Polasek, J. and Bullin, J. 1985. In: *Acid and Sour Gas Treating Processes*. S.A. Newman (Editor). Gulf Publishing Company, Houston, TX, Chapter 7.

Soud, H. and Takeshita, M. 1994. *FGD Handbook*. No. IEACR/65. International Energy Agency Coal Research, London, UK.

Speight, J.G. 2000. *The Desulfurization of Heavy Oils and Residua*, 2nd Edition. Marcel Dekker, New York.

Speight, J.G. 2007. *Natural Gas: A Basic Handbook*. GPC Books, Gulf Publishing Company, Houston, TX.

Speight, J.G. 2011. *An Introduction to Petroleum Technology, Economics, and Politics*. Scrivener Publishing, Salem, MA.

Speight, J.G. 2014a. *The Chemistry and Technology of Petroleum*, 5th Edition. CRC Press/Taylor & Francs Group, Boca Raton, FL.

Speight, J.G. 2014b. *Oil and Gas Corrosion Prevention*. Gulf Professional Publishing, Elsevier, Oxford, UK.

Speight, J.G. and Ozum, B. 2002. *Petroleum Refining Processes*. Marcel Dekker, New York.

Staton, J.S., Rousseau, R.W., and Ferrell, J.K. 1985. In: *Acid and Sour Gas Treating Processes*. S.A. Newman (Editor). Gulf Publishing Company, Houston, TX, Chapter 5.

USGS. 2014. U.S. Geological Survey gas hydrates project. United States Geological Survey, Reston, VA, August 20. http://woodshole.er.usgs.gov/project-pages/hydrates/; accessed September 30, 2015.

Ward, E.R. 1972. In: *Gas Purification Processes for Air Pollution Control*. G. Nonhebel (Editor). Butterworth & Co., London, UK, Chapter 8.

Zapffe, F. 1963. Iron sponge process removes mercaptans. *Oil and Gas Journal*, 61(33): 103–104.

Zhang, L.Q., Shi, L.B., and Zhou, Y. 2007. Formation prediction and prevention technology of natural gas hydrate. *Natural Gas Technology*, 1(6): 67–69.

17 Refining in the Future

17.1 INTRODUCTION

In spite of claims to the contrary that are based on the concept of peak oil (Hubbert, 1962), the world is not on the verge of running out of petroleum, heavy oil, or tar sand bitumen (Chapter 1; BP, 2015). However, in spite of the current volatility of petroleum, cheap petroleum may be difficult to obtain in the future as recent price fluctuations have indicated—the causes vary from petroleum being more difficult to obtain from underground formations, especially tight formations (Chapter 1), to the *petropolitics* of the various oil-producing nations (Speight, 2011a). However, with the entry into the twenty-first century, petroleum refining technology is experiencing great innovation driven by the increasing supply of heavy oils with decreasing quality and the fast increases in the demand for clean and ultraclean vehicle fuels and petrochemical raw materials. As feedstocks to refineries change, there must be an accompanying change in refinery technology. In addition, there will be a need to control the effects of possible changes in crude oil slate on the emissions of carbon dioxide (MathPro Inc., 2013). This means a movement from conventional means of refining heavy feedstocks using (the currently typical) coking technologies to more innovative processes (including hydrogen management) that will produce the ultimate amounts of liquid fuels from feedstocks and maintain emissions within environmental compliance (Penning, 2001; Davis and Patel, 2004; Speight, 2008, 2014a; Farnand et al., 2015).

To meet the challenges from environmentally driven changes over the past five decades from simple crude trends in product slate and the stringent distillation operations into increasingly stringent specifications imposed by environmental complex chemical operations involving legislation, the refining industry in the near future will become increasingly flexible and refined products with specifications that meet innovative with new processing schemes, users requirements (RAROP, 1991; Stratiev and Petkov, 2009; Khan, 2011; Speight, 2011a,b, 2014a). Thus, during the forthcoming decades, the evolution future of petroleum refining and the current refinery configuration (Figure 17.1) will be primarily on process modification with some new innovations coming onstream. The industry will move predictably on to (1) deep conversion of heavy feedstocks, (2) higher hydrocracking and hydrotreating capacity, and (3) more efficient processes.

This chapter presents suggestions and opinions of the means by which refinery processes will evolve during the next three to five decades. This chapter includes material relevant to (1) the comparisons of current feedstocks with heavy oil and bio-feedstocks, (2) the evolution of refineries since the 1950s, (3) the properties and refinability of heavy oil and bio-feedstocks, (4) thermal processes versus hydroprocesses, and (5) the evolution of products to match the environmental market, with more than a passing mention of the effects of feedstocks from coal and oil shale.

17.2 HISTORY

Refining technology has evolved considerably over the last century in response to changing requirements such as (1) a demand for gasoline and diesel fuel as well as fuel oil, (2) petrochemicals as building blocks for clothing and consumer goods, and (3) more environmentally friendly processes and products.

As a result of this response, the production facilities within the refining industry have become increasingly diverse—process configuration varies from plant to plant according to its size, complexity, and product slate. Moreover, the precise configuration of the refinery of the future is unknown, but it is certain that no two refineries will adapt in exactly the same way. There are

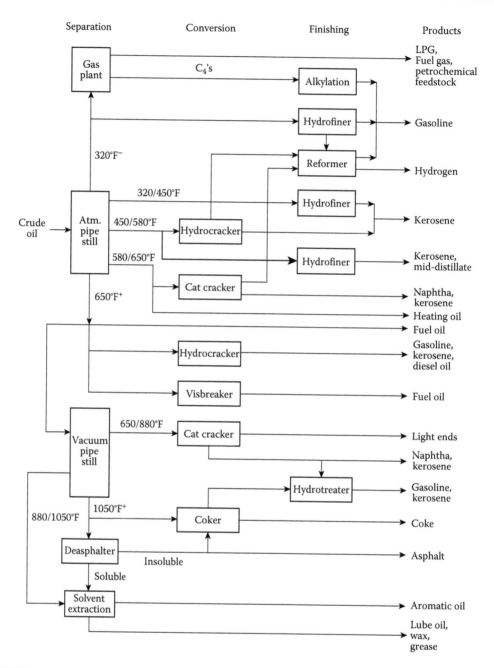

FIGURE 17.1 General schematic of a refinery.

small refineries—1,500–5,000 barrels per day (bpd)—and large refineries that process in excess of 250,000 bpd. Some are relatively simple (Speight, 2014a) and produce only fuels, while other refineries, such as those with integrated petrochemical processing capabilities, are much more complex. Many refineries are part of large integrated oil companies engaged in all aspects of the petroleum technology—from (1) exploration, (2) production, (3) transportation, (4) refining, and (5) marketing of petroleum products. Historically, this has not always been the case. In the early days of the twentieth century, refining processes were developed to extract kerosene for lamps. Any other products were considered to be unusable and were usually discarded. A brief history of petroleum refining is presented in the following paragraphs.

In 1861, the first petroleum refinery opened and produced kerosene—a better and cheaper source of light than whale oil—by atmospheric distillation; naphtha and *tar* (*residuum* or *cracked residuum*) produced as by-products. This involved simple batch distillation of crude oil with the objective of maximizing kerosene production. Technological advancements included, first, the introduction of continuous distillation and, then, vacuum distillation—developed in 1870—which greatly facilitated the manufacture of lubricants. The 1890s saw the emergence of the internal combustion engine, creating demand for diesel fuel and gasoline; demand for kerosene declines with the invention and proliferation of the electric light.

Thus, first refining processes were developed to purify, stabilize, and improve the quality of kerosene. However, the invention of the internal combustion engine led (at about the time of World War I) to a demand for gasoline for use in increasing quantities as a motor fuel for cars and trucks. This demand on the lower-boiling products increased, particularly when the market for aviation fuel developed. Thereafter, refining methods had to be constantly adapted and improved to meet the quality requirements and needs of fuels as well as a variety of other products.

Next, the quest for improved lubricants prompted the use of solvent extraction. To make better use of the bottom of the barrel, thermal cracking (in 1913) and visbreaking processes were introduced to crack high molecular weight constituents of the feedstock to produce lower-boiling products. Thermal cracking was developed—in response to increased demand for gasoline due to mass production of automobiles and the outbreak of World War I. This innovation enabled refineries to produce additional gasoline and distillate fuels by subjecting high-boiling petroleum fractions to high pressures and temperatures with the resulting production of lower-boiling, lower-molecular-weight products.

During the 1930s, many advances made to improve gasoline yield and properties as a response to the development of higher-compression engines. This involved the development of processes such as (1) catalytic cracking, thermal reforming, and catalytic polymerization to improve octane number; (2) hydroprocesses to remove sulfur; (3) coking processes to produce gasoline blend stocks; (4) solvent extraction processes to improve the viscosity index of lubricating oil; and (5) solvent dewaxing processes to improve the pour point of the various products. The by-products of these various processes included aromatics, waxes, residual fuel oil, coke, and feedstocks for the manufacture of petrochemicals.

With the onset of World War II and the need for additional supplies of gasoline, the refining industry turned to catalysis for major innovations. Catalytic cracking constituted a step change in the refinery's ability to convert heavy components into highly valued gasoline and distillates. Wartime demand for aviation fuels helped spur the development of catalytic alkylation processes (which produced blend stocks for high-octane aviation gasoline) and catalytic isomerization (which produced increased quantities of feedstocks for alkylation units) to create high-octane fuels from lighter hydrocarbons. We redistributed hydrogen content among the refinery's products to improve their properties via catalytic reforming of gasoline, catalytic hydrodesulfurization of distillates, and hydrocracking of midrange streams.

The period from the 1950s to the 1970s saw the development of various reforming processes, which also produced blend stocks that were used to improve gasoline quality and yield. Other processes such as deasphalting, catalytic reforming, hydrodesulfurization, and hydrocracking are examples of processes developed during this period. In this time period, refiners also started further development of the uses for the waste gases from various processes resulting in the expansion of the petrochemical industry. In the latter part of the period, the industry benefitted from a massive infusion of computer-based quantitative methodology that has significantly improved more control over processes and the composition of products. In addition, automation and control enabled the optimization of unit operation and economic performance.

Thus throughout its history, the refining industry has been the subject of the four major forces that affect most industries and which have hastened the development of new petroleum refining processes: (1) the high demand for liquid fuels such as gasoline, diesel, fuel oil, and jet fuel; (2) uncertain feedstock supply, specifically the changing quality of crude oil and geopolitics between different countries and the emergence of alternate feed supplies such as bitumen from tar sand, natural gas, coal, and the ever-plentiful biomass; (3) increasingly stringent recent environmental regulations in relation to sulfur content of liquid fuels; and (4) continued technology development such as new catalysts and processes.

17.3 REFINERY CONFIGURATIONS

A petroleum refinery is an industrial processing plant that is a collection of integrated process units (Speight and Ozum, 2002; Parkash, 2003; Hsu and Robinson, 2006; Speight, 2014a). The crude oil feedstock is typically a blend of two or more crude oils, often with heavy oil or even tar sand bitumen blended into a maximum permissible amount, based on the refinery configuration.

17.3.1 PETROLEUM REFINERY

The definition of petroleum (aka crude oil) is often confusing and variable (Chapter 1) and has been made even confusing by the introduction of other terms (such as *black oil* that only adds color to the nomenclature equation) that add little, if anything to the issues relating to petroleum definitions and terminology (Zittel and Schindler, 2007; Speight, 2008, 2014a). In fact, there are different classification schemes based on (1) economic and/or (2) geological criteria. For example, the economic definition of conventional oil is "conventional oil is oil which can be produced with current technology under present economic conditions." The problem with this definition is that it is not very precise and changes whenever the economic or technological aspects of oil recovery change. In addition, there are other classifications based on API gravity, such as "conventional oil is crude oil having a viscosity above 17° API." Moreover, each producing country may change the definitions somewhat for political or economic reasons. However, these definitions do not change the definition stated elsewhere (Chapter 1) that has been used throughout this book.

Recall that the most appropriate definition of *tar sands* (Chapter 1) is found in the writings of the U.S. government, namely,

> Tar sands are the several rock types that contain an extremely viscous hydrocarbon which is not recoverable in its natural state by conventional oil well production methods including currently used enhanced recovery techniques. The hydrocarbon-bearing rocks are variously known as bitumen-rocks oil, impregnated rocks, oil sands, and rock asphalt.

This definition speaks to the character of the bitumen through the method of recovery. Thus, the bitumen found in tar sand deposits is an extremely viscous material that is *immobile under reservoir conditions* and cannot be recovered through a well by the application of secondary or enhanced recovery techniques. Mining methods match the requirements of this definition (since mining is not one of the specified recovery methods), and the bitumen can be recovered by alteration of its natural state such as thermal conversion to a product that is then recovered. In this sense, changing the natural state (the chemical composition) as occurs during several thermal processes (such as some *in situ* combustion processes) also matches the requirements of the definition. Furthermore, by inference and by omission, conventional petroleum and heavy oil are also included in this definition—petroleum is the material that can be recovered by conventional oil well production methods, whereas heavy oil is the material that can be recovered by enhanced recovery methods.

In recent years, the *quality* of crude oil feedstocks has deteriorated and continues to do so as more heavy oil and tar sand bitumen are being sent to refineries and there is the need for efficient upgrading processes for these feedstocks (Speight, 2005, 2008, 2014a; Rana et al., 2007). This has caused the nature of crude oil refining to be changed substantially, and there has been an increasing need to respond to market demands (*market pull*) to develop options to upgrade more of the heavy feedstocks, specifically heavy oil and bitumen. In addition, the general trend throughout refining has been to produce more products from each barrel of petroleum and to process those products in different ways to meet the product specifications for use in modern engines. Overall, the demand for gasoline has rapidly expanded and demand has also developed for gas oils and fuels for domestic central heating and fuel oil for power generation, as well as for light distillates and other inputs, derived from crude oil, for the petrochemical industries. However, the means by which a refinery operates in terms of producing the relevant products depends not only on the nature of the petroleum feedstock but also on its configuration (i.e., the number of types of the processes that are employed to produce the desired product slate) that is, in turn, influenced by the specific demands of a market. The refining industry does not dictate the market but *must respond* to the requirements of the market. Therefore, refineries need to be constantly adapted and upgraded to remain viable and responsive to ever-changing patterns of crude supply and product market demands. As a result, refineries have been introducing increasingly complex and expensive processes to gain higher yields of lower-boiling products from the higher-boiling fractions and residua.

Finally, although mentioned briefly earlier, the yields and quality of refined petroleum products produced by any given oil refinery depend on the mixture of crude oil used as feedstock and the configuration of the refinery facilities. Light/sweet (low-sulfur) crude oil is generally more expensive and has inherent great yields of higher-value low-boiling products such as naphtha, gasoline, jet fuel, kerosene, and diesel fuel. Heavy sour (high-sulfur) crude oil is generally less expensive and produces greater yields of lower-value higher-boiling products that must be converted into lower-boiling products (Speight, 2013a).

Changes in the characteristics of feedstocks to refineries will trigger changes in refinery configurations and corresponding investments. The future crude slate into a refinery is predicted (with a high degree of satisfaction) to consist of larger amounts of heavier sour (high-sulfur) crude oils as well as a shift to higher amounts of extra heavy oil and bitumen, such as the extra heavy oil from the Orinoco Basin (Venezuela) and the nonvolatile carbonaceous material from the tar sand deposits of Alberta (Canada). These changes will require investment in upgrading, either at field level (partial upgrading or full upgrading) to process the extra heavy oil and the tar sand bitumen into pipeline-specification synthetic crude oil or at the refinery level (full upgrading) (Figure 17.2; Scouten, 1990; Speight, 2008).

The location of this upgrading capacity (field site or refinery site) will be built is likely to be strongly influenced by market proximity—there are four ways that are currently practiced in bringing heavy crude oil to the market (Hedrick et al., 2006). The first method is to upgrade the material in the oil field and leave much of the material behind as coke and then pipeline the upgraded material out as synthetic crude. A second solution is to build upgrading facilities at an established port area with abundant gas and electric resources. A third solution in common practice is to use traditional crude that is located in the general area to dilute the nontraditional crude to produce an acceptable pipeline material. The final solution is closely related to the established port area solution where a substantial oil field is located far from other fields, from power, or from natural gas. In addition, petroleum refining has grown increasingly complex in the last 40 years. Lower-quality crude oil, crude oil price volatility, and environmental regulations that require cleaner manufacturing processes and higher-performance products present new challenges to the refining industry. Improving processes and increasing the efficiency of energy use are key to meeting the challenges and maintaining the viability of the petroleum refining industry. There is also the need for a refinery to be able to accommodate *opportunity crude oils* and/or *high-acid crude oils* (Chapter 1).

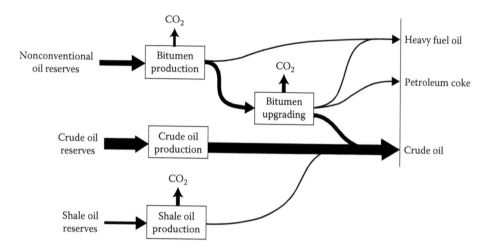

FIGURE 17.2 The potential for blending tar sand bitumen products and oil shale products with petroleum products.

Opportunity crude oils are often dirty and need cleaning before refining by the removal of undesirable constituents such as high-sulfur, high-nitrogen, and high-aromatic (such as polynuclear aromatic) components. A controlled visbreaking treatment would *clean up* such crude oils by removing these undesirable constituents (which, if not removed, would cause problems further down the refinery sequence) as coke or sediment. On the other hand, high-acid crude oils cause corrosion in the atmospheric and vacuum distillation units due to the presence of mineral salts such as magnesium, calcium, and sodium chloride, which are hydrolyzed to produce volatile hydrochloric acid, causing a highly corrosive condition in the overhead heat exchangers (Speight, 2014c). Other contaminants in which are shown to accelerate the hydrolysis reactions are clay minerals and organic acids. In addition to taking preventative measure for the refinery to process these high-margin crude oils without serious deleterious effects on the equipment, refiners will need to develop programs for detailed and immediate feedstock evaluation so that they can understand the qualities of a crude oil very quickly and it can be valued appropriately.

In general, heavy feedstock upgrading will become more and more profitable as recycling and multiple reaction sections will be the prevalent technology trends. Several process innovations have been introduced in the form of varying process options, some using piggyback techniques (where one process works in close conjunction with another process) that will fit into the future refinery (Speight, 2011b). Such process options will continue to be developed, and many refineries will accept the piggyback concept for heavy feedstock upgrading. Another significant improvement in the heavy feedstock conversion technologies will be in catalyst design. Hydrocracking catalysts used in ebullated-bed processes will continue to require multifunctionality. The transition metal sulfides such as molybdenum (Mo), cobalt (Co), and nickel (Ni) will still remain the industry favorites, because of their excellent hydrogenation, hydrodesulfurization (HDS), and hydrodenitrogenation (HDN) activities, as well as their availability and cost. However, greater attention will be paid to the size of the particles, pore volume and size distribution, pore diameter, and the shape of the particles to maximize the utilization of the catalyst.

The catalyst technology known as *metals trapping* technology will continue to improve and allow higher levels of heavy feedstocks to be processed while reducing hydrogen yields and lowering coke yields. As feedstocks deteriorate in quality and metals' content of the feed is higher, the need for such catalyst design will be a major factor in resid upgrading. In addition, the use of scavenger additives such as metal oxides may also see a surge in use. As a simple example, a metal oxide (such as calcium oxide) has the ability to react with sulfur-containing feedstock to produce a hydrocarbon (and calcium sulfide) followed by oxidative reservation to the oxide and, for example, conversion of the sulfur dioxide to sulfur trioxide and thence to sulfuric acid:

$$S_{Feedstock} + CaO \rightarrow Hydrocarbon\ product + CaS + H_2O$$

$$CaS + O_2 \rightarrow CaO + SO_2$$

$$SO_2 + O_2 \rightarrow SO_3$$

$$SO_3 + H_2O \rightarrow H_2SO_3$$

The precise configuration of the refinery of the future is unknown but, because of feedstock variation and the makeup of feedstock blends, it is certain that no two refineries will adapt in exactly the same way. However, the evolution of the refinery of the future will not be strictly confined to petroleum-based processes but will be based on a variety of feedstocks (Speight, 2011b). This will be solved in refinery of the future with the development of deep conversion processing, such as heavy feedstock hydrocracking and the inclusion of processes to accommodate other feedstocks (RAROP, 1991; Khan and Patmore, 1998; Rispoli et al., 2009; Motaghi et al., 2010a,b; Speight, 2014a).

Moreover, the future of the petroleum refining industry will be primarily on processes for the production of improved quality products. In addition to *heavy ends deep conversion*, there will also be changes in the feedstock into a refinery. Biomass, liquids from coal, and liquids from oil shale will increase in importance (Bajus, 2010; Demirbaş, 2010, 2011; Speight, 2013b; Syngellakis, 2015a,b). These feedstocks (1) will be sent to refineries or (2) processed at a remote location and then blended with refinery stocks are options for future development and the nature of the feedstocks. Above all, such feedstock must be compatible with refinery feedstocks and not cause fouling in any form.

The basic refining process for the conversion of residua, heavy oil, and tar sand bitumen to lower-boiling saleable products and the conversion of distillation residues consist of cracking the feedstock constituents to increase the hydrogen content and to decrease the carbon content of the derived products (Speight, 2011b, 2014a). While such processes will continue (at least) for the next 50 years and even throughout the remainder of the twenty-first century (Speight, 2011b), many refiners are investigating the potential large-scale utilization of biomass as partial feedstocks. For practical reasons, small-capacity refineries might be the first to attempt such uses of biomass in a biopetroleum refinery complex. Biomass might be used in the form of preprepared pellets (obtained from agriculture residues such as forestry residues, corn stock, and straw) as a feedstock blend with heavy oil. The approach could produce benefits such as (1) improvement in the quality of the final market products and of the economics of the entire activity, (2) positive impact on rural development (new jobs and new income for farmers), and (3) decrease in carbon dioxide emissions by the substitution of renewable biomass and conversion to hydrocarbon products (Grassi, 2004).

Moreover, it is more than likely that the future refinery will have a gasification section devoted to the conversion of coal and biomass to Fischer–Tropsch hydrocarbons—perhaps even with rich oil shale added to the gasifier feedstock. Many refineries already have gasification capabilities (for the production of hydrogen), but the trend will increase to the point (over the next two decades) where nearly all refineries feel the need to construct a gasification section to handle heavy petroleum-related feedstocks and refinery waste as well as nonpetroleum-related carbonaceous feedstocks such as coal, biomass, and nonrefinery waste material (Chapter 14; Speight, 2014b).

Coal gasification is a tried-and-true technology and there has been a move in recent years to feedstocks other than coal (Speight, 2013b, 2014b). Among other alternative energy conversion pathways, biomass gasification has great potential because of its flexibility to use a wide range of feedstocks and to produce energy and a wide range of fuels and chemicals (Kumar et al., 2009). Recently, the focus of its application has changed from the production of combined heat and power to the production of liquid transportation fuels. The technical challenges in commercialization of fuels and chemicals production from biomass gasification include increasing the energy efficiency of the system and developing robust and efficient technologies for cleaning the product gas and its conversion to valuable fuels and chemicals. Thus, future energy production (as conventional petroleum reserves continue to decline) is likely to involve coprocessing alternative energy sources in which petroleum residua/heavy oil/extra heavy oil/tar sand bitumen is processed with other energy sources and requires a new degree of *refinery flexibility* as the key target, especially when related to the increased use of renewable energy sources such as biomass (Szklo and Schaeffer, 2005).

17.3.2 BIOREFINERY

Whatever the rationale and however the numbers are manipulated, the supply of crude oil, the basic feedstock for refineries and for the petrochemicals industry, is finite and supply/demand issues will continue to deplete petroleum reserves. Although the supply of heavy oil, extra heavy oil, and tar sand bitumen can be moved into the breach, the situation can be further mitigated to some extent by the exploitation of more technically challenging fossil fuel resources and the introduction of new technologies for fuels and chemicals production from coal and oil shale (Scouten, 1990; Speight, 2011a,b, 2013b; Lee et al., 2014). In addition, there is a substantial interest in the utilization of plant-based matter (biomass) as a raw material feedstock for the chemicals industry (Marcilly, 2003; Lynd et al., 2005; Huber and Corma, 2007; Lynd et al., 2009). Plants accumulate carbon from the atmosphere via photosynthesis and the widespread utilization of these materials as feedstocks for the generation of power, fuels, and chemicals.

In terms of the use of plant matter as energy-producing feedstocks, there are two terms that need to be defined and these are (1) biomass and (2) biofuels. *Biomass* is a renewable energy source—unlike the fossil fuel resources: petroleum, coal, and natural gas—and is derived from recently living organisms or their metabolic by-products. An advantage of fuel from biomass (biofuel), in comparison to most other fuel types, is it is biodegradable and thus relatively harmless to the environment if spilled. Typically, a *biofuel* is any fuel that is derived from biomass but has been further defined as any fuel with a minimum content (≥80% v/v) of materials derived from living organisms harvested within the 10 years preceding its manufacture.

Plants offer a unique and diverse feedstock for energy production and for the production of chemicals. Plant biomass can be gasified to produce synthesis gas—a basic chemical feedstock for the production of hydrocarbons and also a source of hydrogen for a future hydrogen economy (Chapter 15; Chadeesingh, 2011). More generally, biomass feedstocks are recognized (and/or defined) by the specific chemical content of the feedstock or the manner in which the feedstock is produced (Speight, 2008). For example, *primary biomass feedstocks* that are currently being used for bioenergy include grains and oilseed crops used for transportation fuel production, plus some crop residues (such as orchard trimmings and nut hulls) and some residues from logging and forest operations that are currently used for heat and power production (Table 17.1). In the future, it is anticipated that a larger proportion of the residues inherently generated from food crop harvesting, as well as a larger proportion of the residues generated from ongoing logging and forest operations, will be used for bioenergy. *Secondary biomass feedstocks* differ from primary biomass feedstocks in that the secondary feedstocks are a by-product of processing of the primary feedstocks. Specific examples of secondary biomass include sawdust from sawmills, black liquor (which is a by-product of paper making), and cheese whey (which is a by-product of cheese-making processes). Vegetable oils used for biodiesel that are derived directly from the processing of oilseeds for various uses are also a secondary biomass resource. *Tertiary biomass feedstocks* include fats, greases, oils, construction and demolition wood debris, other waste wood from the urban environments, as well as

TABLE 17.1
Heating Value of Selected Fuels

Fuel	Btu/lb
Natural gas	23,000
Gasoline	20,000
Crude oil	18,000
Heavy oil	16,000
Coal (anthracite)	14,000
Coal (bituminous)	11,000
Wood (farmed trees, dry)	8,400
Coal (lignite)	8,000
Biomass (herbaceous, dry)	7,400
Biomass (corn stover, dry)	7,000
Wood (forest residue, dry)	6,600
Bagasse (sugarcane)	6,500
Wood	6,000

packaging wastes, municipal solid wastes, and landfill gases. A category *other wood waste from the urban environment* includes trimmings from urban trees.

The simplest, cheapest, and most common method of obtaining energy from biomass is direct combustion. Any organic material, with a water content low enough to allow for sustained combustion, can be burned to produce energy, which can be used to provide space or process heat, water heating, or (through the use of a steam turbine) electricity. In the developing world, many types of biomass such as animal dung and agricultural waste materials are burned to produce heat for cooking and warmth. In fact, such organic residues can also be used for energy production through conversion by natural biochemical processes as well as through the auspices of a biorefinery.

A biorefinery is a facility that integrates biomass conversion processes and equipment to produce fuels, power, and chemicals from biomass. The biorefinery concept is analogous to the petroleum refinery, which produce multiple fuels and products from petroleum. In addition to applying biological methods to petroleum itself such as the (1) desulfurization of fuels, (2) denitrogenation of fuels, (3) removal of heavy metals, (4) transformation of heavy crudes into light crudes, and (5) the biodegradation and bioremediation of petroleum spills as well as spills of petroleum products (Le Borgne and Quintero, 2003; Bhatia and Sharma, 2006; Speight and Arjoon, 2012; El-Gendy and Speight, 2015), biorefining offers a key method to accessing the integrated production of chemicals, materials, and fuels. While the biorefinery concept is analogous to that of an oil refinery in terms of feedstock pretreatment, conversion to products, and product finishing (Figure 17.3), there are significant differences—particularly in the character and properties of the respective feedstocks.

While the primary function of a biorefinery is to produce biofuels, there are also options to use biomass in various refinery scenarios (Speight, 2011c; Lee et al., 2014). The biomass could be supplied by anything from corn, sugarcane, grasses, wood, and soybeans to algae. In place of fossil fuel–based hydrocarbon or hydrocarbonaceous feedstocks, biomass offers sugars, starches, fats, and proteins. Some chemicals will be synthesized using enzymes or genetically engineered microorganisms, and some will be produced using the inorganic catalysts used in traditional chemical processes. Throughout the decision process, consideration must be given to the use of biomass-derived chemicals without perturbing food supplies.

A biorefinery would (in a manner similar to the petroleum refinery) integrate a variety of conversion processes to produce multiple product streams and would combine the essential technologies to transform biological raw materials into a range of industrially useful intermediates. However, there

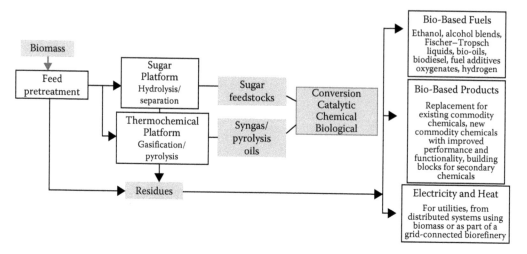

FIGURE 17.3 Fuels from biomass using the sugar (biochemical) platform and the thermochemical platform. (From the Office of the Biomass Program, Multiyear plan 2004 and beyond, U.S. Department of Energy, Washington, DC, November, 2003.)

may be the need to differentiate the type of biorefinery on the basis of the feedstock. For example, a *crop biorefinery* would use raw materials such as cereals or maize and a *lignocellulose biorefinery* would use raw material with high cellulose content, such as straw, wood, and paper waste. In addition, a variety of methods and techniques can be employed to obtain different product portfolios of bulk chemicals, fuels, and materials (Speight, 2008, 2011c). Biotechnology-based conversion processes can be used to ferment the biomass carbohydrate content into sugars that can then be further processed. An alternative is to employ thermochemical conversion processes that use pyrolysis or gasification of biomass to produce a hydrogen-rich synthesis gas, which can be used in a wide range of chemical processes (Chadeesingh, 2011). On the other hand, the use of bio-feedstocks in a conventional petroleum refinery cannot be ignored, whether or not they are used as gasifier feedstocks (Chapter 14; Speight, 2008, 2011c, 2014a,b).

These inherent characteristics and limitations of biomass feedstocks have focused the development of efficient methods of chemically transforming and upgrading biomass feedstocks in a refinery (Figure 17.3). The refinery would be based on two platforms to promote different product slates: (1) the biochemical platform and (2) the thermochemical platform. Using this two-train approach and by analogy with crude oil, every element of the plant feedstock will be utilized including the low-value lignin components. However, the different compositional nature of the biomass feedstock, compared to crude oil, will require the application of a wider variety of processing tools in the biorefinery. Processing of the individual components will utilize conventional thermochemical operations and state-of-the-art bioprocessing techniques.

The biorefinery concept provides a means to significantly reduce production costs such that a substantial substitution of petrochemicals by renewable chemicals becomes possible. However, significant technical challenges remain before the biorefinery concept can be realized.

17.3.3 COAL LIQUIDS REFINERY

Refinery feedstocks from coal (*coal liquids*) have not been dealt with elsewhere in this text, but descriptions are available from other sources (Speight, 2013b). Coal conversion and product refining (into liquids) are not a new concept and were used during World War II as a means of producing military fuels by the German government.

Technically, the Bergius process was one of the early processes for the production of liquid fuels from coal. In the process, lignite or subbituminous coal is finely ground and mixed with heavy oil recycled from the process. Catalyst is typically added to the mixture, and the mixture is pumped into a reactor. The reaction (which uses catalysts containing tungsten, molybdenum, tin, or nickel) occurs at between 400°C and 500°C (750°F and 930°F) and 3,000 and 10,000 psi hydrogen and produces gas, aromatic naphtha, light gas oil, and heavy gas oil:

$$nC_{coal} + (n + 1)H_2 \rightarrow C_nH_{2n+2}$$

The different fractions can be sent to a refinery for further processing to yield synthetic fuel or a fuel blending stock of the desired quality. The composition of coal liquids produced from coal depends very much on the character of the coal and on the process conditions and, particularly, on the degree of *hydrogen addition* to the coal. In fact, current concepts for refining the products of coal liquefaction processes have relied, for the most part, on already-existing petroleum refineries, although it must be recognized that the acidity (i.e., phenol content) of the coal liquids and their potential incompatibility with conventional petroleum (including heavy oil) may pose new issues within the refinery system (Speight, 2008, 2013b, 2014a).

The other category of coal liquefaction processes invokes the concept of the indirect liquefaction process—a two-stage conversion operation—in which the coal is first converted (by reaction with steam and oxygen, at temperatures in excess of 800°C/1470°F) to a gaseous mixture composed primarily of carbon monoxide and hydrogen (synthesis gas [syngas]). The gas stream is subsequently purified (to remove sulfur, nitrogen, and any particulate matter) after which it is catalytically converted to a mixture of liquid hydrocarbon products. The synthesis of hydrocarbons from carbon monoxide and hydrogen (synthesis gas) (the Fischer–Tropsch synthesis) is a procedure for the indirect liquefaction of coal to produce a range of hydrocarbon products (Speight, 2008, 2013b, 2014b; Chadeesingh, 2011):

$$[C]_{coal} + H_2O \rightarrow CO + H_2$$

$$nCO + (2n + 1)H_2 \rightarrow C_nH_{2n+2} + nH_2O$$

These reactions result primarily in low- and medium-boiling aliphatic compounds suitable for blending with similar boiling-range products to gasoline and diesel fuel. Synthesis gas can also be converted to methanol, which can be used as a fuel, fuel additive, or further processed into gasoline via the Mobil M–gas process. In terms of liquids from coal that can be integrated into a refinery, this represents the most attractive option and does not threaten to bring on incompatibility problems as can occur when phenols are present in the coal liquids. While such a scheme is not meant to replace other fuel-production systems, it would certainly be a fit into a conventional refinery—gasification is used in many refineries to produce hydrogen (Chapter 15) and a gasification unit is part of the flexicoking process (Chapter 8).

17.3.4 SHALE OIL REFINERY

The processes for producing liquids (*shale oil*) from oil shale involve heating (retorting) the shale to convert the organic kerogen to a raw shale oil (Scouten, 1990; Speight, 2008). There are two basic oil shale retorting approaches: (1) mining followed by retorting at the surface and (2) *in situ* retorting, that is, heating the shale in place underground (Vinegar et al., 2010). A similar process can be applied to tar sand formation in which the temperature is controlled and is analogous to visbreaking (Karanikas et al., 2009). Retorting essentially involves thermal decomposition (at temperatures in excess of 485°C/>900°F) of the kerogen with simultaneous removal of distillate in the absence of oxygen. During the process, the initial products can then decompose (depending on the process parameters) into lower-weight hydrocarbon molecules.

The shale oil so produced contains a large variety of hydrocarbon compounds but also has high nitrogen content compared to a nitrogen content of 0.2%–0.3% w/w for a typical petroleum that can affect refinery operations (Scouten, 1990; Speight, 2008, 2013a, 2014a). In addition, shale oil also has a high olefin and diolefin content, and it is the presence of these olefins and diolefins in conjunction with high nitrogen content that gives shale oil the tendency to form insoluble sediment. Crude shale oil also contains appreciable amounts of arsenic, iron, and nickel that interfere with refining. Upgrading, or partial refining, to improve the properties of a crude shale oil may be carried out using different options—depending upon the composition and the origin of the shale oil (Scouten, 1990)—and hydrotreating is the option of choice to produce a stable product (Speight and Ozum, 2002; Parkash, 2003; Hsu and Robinson, 2006; Speight, 2014a). In terms of refining and catalyst activity, the nitrogen content of shale oil is a disadvantage and, if not removed, the arsenic and iron in shale oil will poison and foul the supported catalysts used in hydrotreating. In terms of the use of shale oil residua as a modifier for asphalt, where nitrogen species can enhance binding with the inorganic aggregate, the nitrogen content is beneficial (Speight, 2015).

In general, oil shale distillates have a much higher concentration of high-boiling-point compounds that would favor the production of middle distillates (such as diesel and jet fuels) rather than naphtha. Oil shale distillates also had a higher content of olefins, oxygen, and nitrogen than crude oil, as well as higher pour points and viscosities. Aboveground retorting processes tended to yield a lower API gravity oil than the *in situ* processes (a 25° API gravity was the highest produced). Additional processing equivalent to hydrocracking would be required to convert oil shale distillates to a lighter range hydrocarbon (gasoline). Removal of sulfur and nitrogen would, however, require hydrotreating.

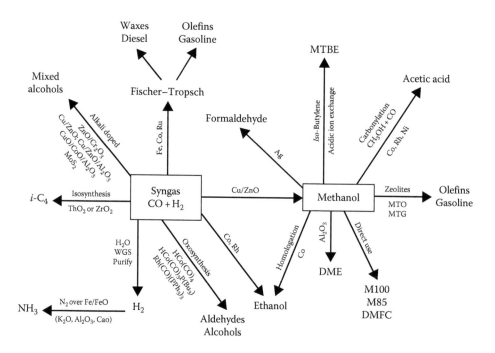

FIGURE 17.4 Products from synthesis gas.

17.3.5   Gasification Refinery

The concept of using other feedstocks to blend with petroleum feedstocks bring to the fore the concept of a *gasification refinery*, which would use gasification technology as is the case of the Sasol refinery in South Africa (Chadeesingh, 2011). The manufacture of mixtures of carbon monoxide and hydrogen has been an important part of chemical technology for approximately a century. Originally, such mixtures were obtained by the reaction of steam with incandescent coke and were known as *water gas*. Eventually, steam reforming processes, in which steam is reacted with natural gas (methane) or petroleum naphtha over a nickel catalyst, found wide application for the production of synthesis gas.

The gasification refinery would produce synthesis gas (from the carbonaceous feedstock) from which liquid fuels would be manufactured using the Fischer–Tropsch synthesis technology (Couvaras, 1997; Speight, 2008, 2014b; Chadeesingh, 2011; Luque and Speight, 2015). Synthesis gas is used as a source of hydrogen or as an intermediate in producing hydrocarbons via the Fischer–Tropsch synthesis (Figure 17.4). Indeed, as petroleum supplies decrease, the desirability of producing gas from other carbonaceous feedstocks will increase.

17.4   THE FUTURE REFINERY

There is no one single upgrading solution that fits all refineries and the varied crude oil slates. Thus, a careful evaluation of the slate of feedstocks into the refinery is necessary and is not always a simple undertaking for an existing refinery.

The evaluation typically starts at the time of selection of the feedstocks that adequately fit the configuration of the refinery, which varies from refinery to refinery. Some refineries may be more oriented toward the production of gasoline (large reforming and/or catalytic cracking), whereas the configuration of other refineries may be more oriented toward the production of middle distillates such as jet fuel and gas oil. Over the past four decades, the refining industry has been challenged by changing feedstocks and product slate that has introduced a high degree of flexibility with improved technologies and improved catalysts.

However, the evolution of the refinery of the future will not be strictly confined to petroleum processes. The major consequence will be a much more environmentally friendly product quality. These will be solved in refinery of the future, the refinery beyond 2020 with the development of deep conversion processing, such as residue hydrocracking and the inclusion of processes to accommodate other feedstocks. The *panacea* (rather than a *Pandora's box*) for a variety of feedstocks could well be the *gasification refinery* (Speight, 2011b). This type of refinery approaches that of a petrochemical complex, capable of supplying the traditional refined products, but also meeting much more severe specifications, and petrochemical intermediates such as olefins, aromatics, hydrogen, and methanol. Furthermore, as already noted earlier, integrated gasification combined cycle (IGCC) can be used to raise power from feedstocks such as vacuum residua and cracked residua (in addition to the production of synthesis gas), and a major benefit of the integrated gasification combined cycle concept is that power can be produced with the lowest sulfur oxide (SO_x) and nitrogen oxide (NO_x) emissions of any liquid/solid feed power generation technology.

The success of current operations notwithstanding, the challenges facing the refining industry will focus on the diversity of the feedstocks. Even within the petroleum family of feedstocks where elemental analysis varies over a relatively narrow range, changes to refining technology are required to produce the optimum yield of desired products. Another unique foreseeable disruption coming to the industry is the anticipated inclusion of biomass feedstocks and the changes that this will bring to refining. Indeed, much of the intellectual property embodied in the current refinery operations will have to change as wide variations in feedstock composition occur and attempts are made to produce the necessary hydrocarbon fuels from a wide variety of biomass feedstocks. Any yet, the refining industry will survive—being one of the most resilient industries to commence operations during the past 150 years.

The refining industry can be regarded as unique insofar as very few industries have to deal with a feedstock-product chain beginning at a natural resource that has to be recovered from a subterranean formation and proceed through the application of a variety of processes all the way through to the end use consumer. Furthermore, it is imperative for refiners to raise their operations to new levels of performance. Merely extending current process performance incrementally will most likely fail to meet most future performance goals. To do this, it will be necessary to reshape refining technology to be more adaptive to changing feedstocks and product demand and to explore the means by which the technology and methodology of refinery operations can be translated not only into increased profitability but also into survivability.

Part of the future growth will be at or near heavy crude and bitumen production sites to decrease heavy crude viscosity and improve the quality to ease transportation and open markets for crudes of otherwise marginal value. Visbreaking may be considered to be a conversion process rather than a process to produce fuel oil that meets specifications. Coking can be improved by reducing hydrocarbon gas formation and by inhibiting the formation of polynuclear aromatic products that are produced by the process and which are not inherent to the feedstock. Both of these processes would benefit if a higher-valued product could be produced.

To this end, hydroconversion (Chapters 10 and 11) will continue to be a necessary and economically justifiable part of a residue conversion project. The use of ebullated-bed reactors for hydroconversion is evolving as is the use of slurry-catalyst systems. In addition, the gasification of heavy feedstocks moving closer to the time when it will be a ready choice for the conversion of heavy feedstocks transportation fuels. Heavy feedstocks conversion projects (including power generation using waste streams) could move into positive economics in regions with high local power demand. Moreover, with the ever-increasing and stringent specifications for refinery products, refiners will continue to make efforts to improve the heavy feedstock processing technologies that convert the heavy feedstock into valuable and environment-friendly products. Currently, a variety of residue hydrocracking processes using fixed-bed, moving-bed, or ebullated-bed reactors are available. Furthermore, the economics of the slurry bed processing technology indicates an attractive rate of returns with the existing crude oils and product price structure (Rispoli et al., 2009).

Several process innovations have been introduced in the form of varying process options, some using piggyback techniques (where one process is operative in close conjunction with another process) that will fit into the future refinery. Another significant improvement in the hydroprocessing technologies will be in catalyst design. This follows from recognition that a novel catalyst is one of

TABLE 17.2
Carbon Chain Groups of the Range of Fischer–Tropsch Products That Can Be Produced

Carbon Number	Group Name
C1–C2	Synthetic natural gas (SNG)
C3–C4	Liquefied petroleum gas (LPG)
C5–C7	Light petroleum
C8–C10	Heavy petroleum
C11–C20	Middle distillate
C11–C12	Kerosene
C13–C20	Diesel
C21–C30	Soft wax
C31–C60	Hard wax

reasonable design and integration of the active, supporting, and promoting components that allow an optimal combination of activity, surface area, and pore diameter, giving the highest activity.

High-conversion refineries will move to the gasification of feedstocks for the development of alternative fuels and to enhance equipment usage. A major trend in the refining industry market demand for refined products will be in synthesizing fuels from simple basic reactants (e.g., synthesis gas) when it becomes uneconomical to produce superclean transportation fuels through conventional refining processes. Fischer–Tropsch plants together with IGCC systems will be integrated with or even into refineries, which will offer the advantage of high-quality products (Table 17.2; Stanislaus et al., 2000; Shires et al., 2010).

REFERENCES

ASTM D287. 2015. *Standard Test Method for API Gravity of Crude Petroleum and Petroleum Products (Hydrometer Method)*. Annual Book of Standards. ASTM International, West Conshohocken, PA.

ASTM D4294. 2015. *Standard Test Method for Sulfur in Petroleum and Petroleum Products by Energy Dispersive X-Ray Fluorescence Spectrometry*. Annual Book of Standards. ASTM International, West Conshohocken, PA.

Bajus, M. 2010. Pyrolysis technologies for biomass and waste treatment to fuels and chemical production. *Petroleum and Coal*, 52(1): 1–10.

Bhatia, S. and Sharma, D.K. 2006. Emerging role of biorefining of heavier crude oils and integration of bio-refining with petroleum refineries in the future. *Petroleum Science and Technology*, 24(10): 1125–1159.

BP. 2015. *BP Statistical Review of World Energy*. British Petroleum Company, London, UK.

Chadeesingh, R. 2011. The Fischer-Tropsch process. In: *The Biofuels Handbook*. J.G. Speight (Editor). The Royal Society of Chemistry, London, UK, Part 3, Chapter 5.

Couvaras, G. 1997. Sasol's slurry phase distillate process and future applications. *Proceedings of Monetizing Stranded Gas Reserves Conference*. Houston, TX, December.

Davis, R.A. and Patel, N.M. 2004. Refinery hydrogen management. *Petroleum Technology Quarterly*, Spring: 29–35.

Demirbaş, A. 2010. *Biorefineries for Upgrading Biomass Facilities*. Springer-Verlag London Ltd., London, UK.

Demirbaş, A. 2011. Production of fuels from crops. In: *The Biofuels Handbook*. J.G. Speight (Editor). Royal Society of Chemistry, London, UK, Part 2, Chapter 1.

El-Gendy, N.S. and Speight, J.G. 2015. *Handbook of Refinery Desulfurization*. CRC Press/Taylor & Francis Group, Boca Raton, FL.

Farnand, S., Li., J., Patel, N., Peng, X.D., and Ratan, S. 2015. Hydrogen perspectives for 21st century refineries—Part 2. *Hydrocarbon Processing*, 94(2): 53–58.

Grassi, G. 2004. Biofuels utilization for heavy crude-oil refining or gasoline reformulation. *Proceedings of the Second World Conference on Biomass for Energy, Industry and Climate Protection*. Rome, Italy, May 10–14.

Hedrick, B.W., Seibert, K.D., and Crewe, C. 2006. A new approach to heavy oil and bitumen upgrading. Report No. AM-06-29. UOP LLC, Des Plaines, IL.

Hsu, C.S. and Robinson, P.R. 2006. *Practical Advances in Petroleum Processing*, Volumes 1 and 2. Springer, New York.

Hubbert, M.K. 1962. Energy resources. Report to the Committee on Natural Resources. National Academy of Sciences, Washington, DC.

Huber, G.W. and Corma, A. 2007. Synergies between bio- and oil refineries for the production of fuels from biomass. *Angewandte Chemie*, 46(38): 7184–7201.

Karanikas, J.M., Colmenares, T.R., Zhang, E., Marino, M., Roes, A.W.M., Ryan, R.C., Beer, G.L., Dombrowski, R.J., and Jaiswal, N. 2009. Heating tar sands formations to visbreaking temperatures. U.S. Patent 7,7635,024. December 22.

Khan, M.R. (Editor). 2011. *Advances in Clean Hydrocarbon Processing*. Woodhead Publishing, Elsevier, Cambridge, UK.

Khan, M.R. and Patmore, D.J. 1998. Heavy oil upgrading processes. In: *Petroleum Chemistry and Refining*. J.G. Speight (Editor). Taylor & Francis Group, Washington, DC, Chapter 6.

Kumar, A., Jones, D.D., and Hanna, M.A. 2009. Thermochemical biomass gasification: A review of the current status of the technology. *Energies*, 2: 556–581.

Le Borgne, S. and Quintero, R. 2003. Biotechnological processes for the refining of petroleum. *Fuel Processing Technology*, 81(2): 155–169.

Lee, S., Speight, J.G., and Loyalka, S. 2014. *Handbook of Alternative Fuel Technologies*, 2nd Edition. CRC Press/Taylor & Francis Group, Boca Raton, FL.

Luque, R. and Speight, J.G. (Editors). 2015. *Gasification for Synthetic Fuel Production: Fundamentals, Processes, and Applications*. Woodhead Publishing, Elsevier, Cambridge, UK.

Lynd, L.R., Larson, E., Greene, N., Laser, M., Sheehan, J., Dale, B.E., McLaughlin, S., and Wang, M. 2009. *Biofuels, Bioproducts, and Biorefining*, 3(2): 113–123.

Lynd, L.R., Wyman, C., Laser, M., Johnson, D., and Landucci, R. 2005. Strategic biorefinery analysis: Review of existing biorefinery examples January 24–July 1, 2002. Subcontract Report NREL/SR-510-34895, October. National Renewable Energy Laboratory, Golden, CO.

Marcilly, C. 2003. Present status and future trends in catalysis for refining and petrochemicals. *Journal of Catalysis*, 216(1–2): 47–62.

MathPro Inc. 2013. Effects of possible changes in crude oil slate on the U.S. Refining Sector's CO_2 emissions. Final Report. Prepared for The International Council on Clean Transportation by MathPro Inc., West Bethesda, MD, March 29.

Motaghi, M., Shree, K., and Krishnamurthy, S. 2010a. Consider new methods for bottom of the barrel processing—Part 1. *Hydrocarbon Processing*, 89(2): 35–40.

Motaghi, M., Shree, K., and Krishnamurthy, S. 2010b. Consider new methods for bottom of the barrel processing—Part 2. *Hydrocarbon Processing*, 89(2): 55–88.

Parkash, S. 2003. *Refining Processes Handbook*. Gulf Professional Publishing, Elsevier, Amsterdam, the Netherlands.

Penning, R.T. 2001. Petroleum refining: A look at the future. *Hydrocarbon Processing*, 80(2): 45–46.

Rana, M.S., Sámano, V., Ancheyta, J., and Diaz, J.A.I. 2007. A review of recent advances on process technologies for upgrading of heavy oils and residua. *Fuel*, 86: 1216–1231.

RAROP. 1991. *RAROP Heavy Oil Processing Handbook*. Research Association for Residual Oil Processing, T. Noguchi (Chairman), Ministry of Trade and International Industry (MITI), Tokyo, Japan.

Rispoli, G., Sanfilippo, D., and Amoroso, A. 2009. Advanced hydrocracking technology upgrades extra heavy oil. *Hydrocarbon Processing*, 88(12): 39–46.

Scouten, C.S. 1990. Oil shale. In: *Fuel Science and Technology Handbook*. J.G. Speight (Editor). Marcel Dekker, New York, Chapters 25–31.

Shires, P., Salazar, N., and Ariyapadi, S. 2010. Methods for producing synthesis gas. U.S. Patent 7,722,690. May 25.

Speight, J.G. 2005. Natural bitumen (tar sands) and heavy oil. In: *Coal, Oil Shale, Natural Bitumen, Heavy Oil and Peat. Encyclopedia of Life Support Systems (EOLSS)*. Developed under the Auspices of the UNESCO. EOLSS Publishers, Oxford, UK.

Speight, J.G. 2008. *Handbook of Synthetic Fuels: Properties, Processes, and Performance*. McGraw-Hill, New York.

Speight, J.G. 2011a. *An Introduction to Petroleum Technology, Economics, and Politics*. Scrivener Publishing, Salem, MA.

Speight, J.G. 2011b. *The Refinery of the Future*. Gulf Professional Publishing, Elsevier, Oxford, UK.

Speight, J.G. (Editor). 2011c. *The Biofuels Handbook*. Royal Society of Chemistry, London, UK.

Speight, 2013a. *Heavy and Extra Heavy Oil Upgrading Technologies*. Gulf Professional Publishing, Elsevier, Oxford, UK.

Speight, J.G. 2013b. *The Chemistry and Technology of Coal*, 3rd Edition. CRC Press/Taylor & Francis Group, Boca Raton, FL.

Speight, J.G. 2014a. *The Chemistry and Technology of Petroleum*, 5th Edition. CRC Press/Taylor & Francis Group, Boca Raton, FL.

Speight, J.G. 2014b. *Gasification of Unconventional Feedstocks*. Gulf Professional Publishing, Elsevier, Oxford, UK.

Speight, J.G. 2014c. *Oil and Gas Corrosion Prevention*. Gulf Professional Publishing, Elsevier, Oxford, UK.

Speight, J.G. 2015. *Asphalt Materials Science and Technology*. Butterworth-Heinemann, Elsevier, Oxford, UK.

Speight, J.G. and Arjoon, K.K. 2012. *Bioremediation of Petroleum and Petroleum Products*. Scrivener Publishing, Salem, MA.

Speight, J.G. and Ozum, B. 2002. *Petroleum Refining Processes*. Marcel Dekker, New York.

Stanislaus, A., Qabazard, H., and Absi-Halabi, M. 2000. Refinery of the future. *Proceedings of 16th World Petroleum Congress*. Calgary, Alberta, Canada, June 11–15.

Stratiev, D. and Petkov, K. 2009. Residue upgrading: Challenges and perspectives. *Hydrocarbon Processing*, 88(9): 93–96.

Syngellakis, S. (Editor). 2015a. *Biomass to Biofuels*. WIT Press, Boston, MA.

Syngellakis, S. (Editor). 2015b. *Waste to Energy*. WIT Press, Boston, MA.

Szklo, A. and Schaeffer, R. 2005. Alternative energy sources or integrated alternative energy systems? Oil as a modern lance of Peleus for the energy transition. *Energy*, 31: 2513–2522.

Vinegar, H.J., Picha, M.G., and Schoeling, L.G. 2010. Thermal processes for subsurface formations. U.S. Patent 7,640,980. January 5.

Zittel, W. and Schindler, J. 2007. *Crude Oil: The Supply Outlook*. EWG Series No. 3/2007, October. Energy Watch Group, Berlin, Germany.

Glossary

ABN separation: A method of fractionation by which petroleum is separated into acidic, basic, and neutral constituents.

Absorber: *See* Absorption tower

Absorption gasoline: Gasoline extracted from natural gas or refinery gas by contacting the absorbed gas with an oil and subsequently distilling the gasoline from the higher-boiling components.

Absorption oil: Oil used to separate the heavier components from a vapor mixture by absorption of the heavier components during intimate contacting of the oil and vapor; used to recover natural gasoline from wet gas.

Absorption plant: A plant for recovering the condensable portion of natural or refinery gas, by absorbing the higher-boiling hydrocarbons in an absorption oil, followed by separation and fractionation of the absorbed material.

Absorption tower: A tower or column that promotes contact between a rising gas and a falling liquid so that part of the gas may be dissolved in the liquid.

Acetone–benzol process: A dewaxing process in which acetone and benzol (benzene or aromatic naphtha) are used as solvents.

Acid catalyst: A catalyst having acidic character; the alumina minerals are examples of such catalysts.

Acid deposition: Acid rain; a form of pollution depletion in which pollutants, such as nitrogen oxides and sulfur oxides, are transferred from the atmosphere to soil or water; often referred to as atmospheric self-cleaning. The pollutants usually arise from the use of fossil fuels.

Acid gas removal: A process for the removal of hydrogen sulfide, other sulfur species, and some carbon dioxide from syngas by absorption in a solvent with subsequent solvent regeneration and production of an H_2S-rich stream for sulfur recovery.

Acidity: The capacity of an acid to neutralize a base such as a hydroxyl ion (OH^-).

Acidizing: A technique for improving the permeability of a reservoir by injecting acid.

Acid number: A measure of the reactivity of petroleum with a caustic solution and given in terms of milligrams of potassium hydroxide that are neutralized by 1 g of petroleum.

Acid rain: The precipitation phenomenon that incorporates anthropogenic acids and other acidic chemicals from the atmosphere to the land and water (*see* Acid deposition).

Acid sludge: The residue left after treating petroleum oil with sulfuric acid for the removal of impurities; a black, viscous substance containing the spent acid and impurities.

Acid treating: A process in which unfinished petroleum products, such as gasoline, kerosene, and lubricating oil stocks, are contacted with sulfuric acid to improve their color, odor, and other properties.

Acoustic log: *See* Sonic log

Acre-foot: A measure of bulk rock volume where the area is 1 acre and the thickness is 1 ft.

Additive: A material added to another (usually in small amounts) in order to enhance desirable properties or to suppress undesirable properties.

Add-on control methods: The use of devices that remove refinery process emissions after they are generated but before they are discharged to the atmosphere.

Adsorption: Transfer of a substance from a solution to the surface of a solid resulting in relatively high concentration of the substance at the place of contact; *see also* Chromatographic adsorption.

Adsorption gasoline: Natural gasoline obtained by the adsorption process from wet gas.

Afterburn: The combustion of carbon monoxide (CO) to carbon dioxide (CO_2); usually in the cyclones of a catalyst regenerator.

After flow: Flow from the reservoir into the wellbore that continues for a period after the well has been shut in; after flow can complicate the analysis of a pressure transient test.

Air-blown asphalt: Asphalt produced by blowing air through residua at elevated temperatures.

Air injection: An oil recovery technique using air to force oil from the reservoir into the wellbore.

Airlift Thermofor catalytic cracking: A moving-bed continuous catalytic process for conversion of heavy gas oils into lighter products; the catalyst is moved by a stream of air.

Air pollution: The discharge of toxic gases and particulate matter introduced into the atmosphere, principally as a result of human activity.

Air separation unit (ASU): A plant that separates oxygen and nitrogen from air, usually by cryogenic distillation.

Air sweetening: A process in which air or oxygen is used to oxidize lead mercaptan derivatives (RSH) to disulfide derivatives (RSSR) instead of using elemental sulfur.

Air toxics: Hazardous air pollutants.

Albertite: A black, brittle, natural hydrocarbon possessing a conchoidal fracture and a specific gravity of approximately 1.1.

Alcohol: The family name of a group of organic chemical compounds composed of carbon, hydrogen, and oxygen. The molecules in the series vary in chain length and are composed of a hydrocarbon plus a hydroxyl group. Alcohol includes methanol and ethanol.

Alicyclic hydrocarbon: A compound containing carbon and hydrogen only which has a cyclic structure (e.g., cyclohexane); also collectively called naphthenes.

Aliphatic hydrocarbon: A compound containing carbon and hydrogen only which has an open-chain structure (e.g., as ethane, butane, octane, butene) or a cyclic structure (e.g., cyclohexane).

Aliquot: That quantity of material of proper size for the measurement of the property of interest; test portions may be taken from the gross sample directly, but often preliminary operations such as mixing or further reduction in particle size are necessary.

Alkaline: A high pH usually of an aqueous solution; aqueous solutions of sodium hydroxide, sodium orthosilicate, and sodium carbonate are typical alkaline materials used in enhanced oil recovery.

Alkaline flooding: *See* EOR process

Alkalinity: The capacity of a base to neutralize the hydrogen ion (H^+).

Alkali treatment: *See* Caustic wash

Alkali wash: *See* Caustic wash

Alkanes: Hydrocarbons that contain only single carbon–hydrogen bonds. The chemical name indicates the number of carbon atoms and ends with the suffix "ane."

Alkenes: Hydrocarbons that contain carbon–carbon double bonds. The chemical name indicates the number of carbon atoms and ends with the suffix "ene."

Alkylate: The product of an alkylation process.

Alkylate bottoms: Residua from the fractionation of alkylate; the alkylate product that boils higher than the aviation gasoline range; sometimes called heavy alkylate or alkylate polymer.

Alkylation: In the petroleum industry, a process by which an olefin (e.g., ethylene) is combined with a branched-chain hydrocarbon (e.g., *iso*-butane); alkylation may be accomplished as a thermal or as a catalytic reaction.

Alkyl groups: A group of carbon and hydrogen atoms that branch from the main carbon chain or ring in a hydrocarbon molecule. The simplest alkyl group, a methyl group, is a carbon atom attached to three hydrogen atoms.

Alpha-scission: The rupture of the aromatic carbon–aliphatic carbon bond that joins an alkyl group to an aromatic ring.

Alumina (Al_2O_3): Used in separation methods as an adsorbent and in refining as a catalyst.

American Society for Testing and Materials (ASTM): The official organization in the United States for designing standard tests for petroleum and other industrial products.

Amine washing: A method of gas cleaning whereby acidic impurities such as hydrogen sulfide and carbon dioxide are removed from the gas stream by washing with an amine (usually an alkanolamine).

Anaerobic digestion: Decomposition of biological wastes by microorganisms, usually under wet conditions, in the absence of air (oxygen), to produce a gas comprising mostly methane and carbon dioxide.

Analyte: The chemical for which a sample is tested, or analyzed. *Antibody* A molecule having chemically reactive sites specific for certain other molecules.

Analytical equivalence: The acceptability of the results obtained from the different laboratories; a range of acceptable results.

Aniline point: The temperature, usually expressed in °F, above which equal volumes of a petroleum product are completely miscible; a qualitative indication of the relative proportions of paraffins in a petroleum product that are miscible with aniline only at higher temperatures; a high aniline point indicates low aromatics.

Annual removals: The net volume of growing stock trees removed from the inventory during a specified year by harvesting, cultural operations such as timber stand improvement, or land clearing.

Antibody: A molecule having chemically reactive sites specific for certain other molecules.

Antiknock: Resistance to detonation or pinging in spark-ignition engines.

Antiknock agent: A chemical compound such as tetraethyl lead that, when added in small amount to the fuel charge of an internal combustion engine, tends to lessen knocking.

Antistripping agent: An additive used in an asphaltic binder to overcome the natural affinity of an aggregate for water instead of asphalt.

API gravity: A measure of the *lightness* or *heaviness* of petroleum, which is related to density and specific gravity.

°API = (141.5/sp gr at 60°F) − 131.5.

Apparent bulk density: The density of a catalyst as measured; usually loosely compacted in a container.

Apparent viscosity: The viscosity of a fluid, or several fluids flowing simultaneously, measured in a porous medium (rock), and subject to both viscosity and permeability effects; also called effective viscosity.

Aquifer: A subsurface rock interval that will produce water; often the underlay of a petroleum reservoir.

Areal sweep efficiency: The fraction of the flood pattern area that is effectively swept by the injected fluids.

Aromatic hydrocarbon: A hydrocarbon characterized by the presence of an aromatic ring or condensed aromatic rings; benzene and substituted benzene, naphthalene and substituted naphthalene, phenanthrene and substituted phenanthrene, as well as the higher condensed ring systems; compounds that are distinct from those of aliphatic compounds or alicyclic compounds.

Aromatics: *See* Aromatic hydrocarbon

Aromatization: The conversion of nonaromatic hydrocarbons to aromatic hydrocarbons by the (1) rearrangement of aliphatic (noncyclic) hydrocarbons into aromatic ring structures and (2) dehydrogenation of alicyclic hydrocarbons (naphthenes).

Arosorb process: A process for the separation of aromatic derivatives from nonaromatic derivatives by adsorption on a gel from which they are recovered by desorption.

Asphalt: The nonvolatile product obtained by distillation and treatment of an asphaltic crude oil; a manufactured product.

Asphalt cement: Asphalt especially prepared as to quality and consistency for direct use in the manufacture of bituminous pavements.

Asphalt emulsion: An emulsion of asphalt cement in water containing a small amount of emulsifying agent.

Asphalt flux: An oil used to reduce the consistency or viscosity of hard asphalt to the point required for use.

Asphalt primer: A liquid asphaltic material of low viscosity which is used to waterproof a nonbituminous surface and prepare it for further construction.

Asphaltene (asphaltenes): The brown to black powdery material produced by the treatment of petroleum, petroleum residua, or bituminous materials with a low-boiling liquid hydrocarbon, for example, pentane or heptane; soluble in benzene (and other aromatic solvents), carbon disulfide, and chloroform (or other chlorinated hydrocarbon solvents).

Asphaltene association factor: The number of individual asphaltene species that associate in nonpolar solvents as measured by molecular weight methods; the molecular weight of asphaltenes in toluene divided by the molecular weight in a polar nonassociating solvent, such as dichlorobenzene, pyridine, or nitrobenzene.

Asphaltic pyrobitumen: *See* Asphaltoid

Asphaltic road oil: A thick, fluid solution of asphalt; usually a residual oil; *see also* Nonasphaltic road oil.

Asphaltite: A variety of naturally occurring, dark brown to black, solid, nonvolatile bituminous material that is differentiated from bitumen primarily by a high content of material insoluble in *n*-pentane (asphaltene) or other liquid hydrocarbons.

Asphaltoid: A group of brown to black, solid bituminous materials of which the members are differentiated from asphaltites by their infusibility and low solubility in carbon disulfide.

Asphaltum: *See* Asphalt

Associated molecular weight: The molecular weight of asphaltenes in an associating (nonpolar) solvent, such as toluene.

Atmospheric equivalent boiling point (AEBP): A mathematical method of estimating the boiling point at atmospheric pressure of nonvolatile fractions of petroleum.

Atmospheric residuum: A residuum obtained by distillation of a crude oil under atmospheric pressure, which boils above 350°C (660°F).

Attainment area: A geographical area that meets NAAQS for criteria air pollutants (*see also* Nonattainment area).

Attapulgus clay: *See* Fuller's earth

Autofining: A catalytic process for desulfurizing distillates.

Average particle size: The weighted average particle diameter of a catalyst.

Aviation gasoline: Any of the special grades of gasoline suitable for use in certain airplane engines.

Aviation turbine fuel: *See* Jet fuel

Back-mixing: The phenomenon observed when a catalyst travels at a slower rate in the riser pipe than the vapors.

BACT: Best available control technology.

Baghouse: A filter system for the removal of particulate matter from gas streams; so-called because of the similarity of the filters to coal bags.

Bank: The concentration of oil (oil bank) in a reservoir that moves cohesively through the reservoir.

Bari-Sol process: A dewaxing process that employs a mixture of ethylene dichloride and benzol as the solvent.

Barrel (bbl): The unit of measure used by the petroleum industry; equivalent to approximately 42 U.S. gallons or approximately 34 (33.6) Imperial gallons or 159 L; 7.2 barrels are equivalent to 1 ton of oil (metric).

Barrel of oil equivalent (BOE): The amount of energy contained in a barrel of crude oil, that is, approximately 6.1 GJ (5.8 million Btu), equivalent to 1700 kWh.

Base number: The quantity of acid, expressed in milligrams of potassium hydroxide per gram of sample that is required to titrate a sample to a specified end point.

Base stock: A primary refined petroleum fraction into which other oils and additives are added (blended) to produce the finished product.

Basic nitrogen: Nitrogen (in petroleum) that occurs in pyridine form.

Basic sediment and water (BS&W, BSW): The material that collects in the bottom of storage tanks, usually composed of oil, water, and foreign matter; also called bottoms, bottom settlings.

Battery: A series of stills or other refinery equipment operated as a unit.

Baumé gravity: The specific gravity of liquids expressed as degrees on the Baumé (°Bé) scale; for liquids lighter than water: Sp gr 60°F = 140/(130 + °Bé); for liquids heavier than water: Sp gr 60°F = 145/(145 − °Bé).

Bauxite: Mineral matter used as a treating agent; hydrated aluminum oxide formed by the chemical weathering of igneous rock.

Bbl: *See* Barrel

Bell cap: A hemispherical or triangular cover placed over the riser in a (distillation) tower to direct the vapors through the liquid layer on the tray; *see* Bubble cap.

Bender process: A chemical treating process using lead sulfide catalyst for sweetening light distillates by which mercaptan derivatives (RSH) are converted to disulfide derivatives (RSSR) by oxidation.

Bentonite: Montmorillonite (a magnesium–aluminum silicate); used as a treating agent.

Benzene: A colorless aromatic liquid hydrocarbon (C_6H_6).

Benzin: A refined light naphtha used for extraction purposes.

Benzine: An obsolete term for light petroleum distillates covering the gasoline and naphtha range; *see* Ligroine.

Benzol: The general term that refers to commercial or technical (not necessarily pure) benzene; also the term used for aromatic naphtha.

Beta-scission: The rupture of a carbon–carbon bond, that is, two bonds removed from an aromatic ring.

Billion: 1×10^9.

Biochemical conversion: The use of fermentation or anaerobic digestion to produce fuels and chemicals from organic sources.

Biocide: Any chemical capable of killing bacteria and bioorganisms.

Biodiesel: A fuel derived from biological sources that can be used in diesel engines instead of petroleum-derived diesel; through the process of transesterification, the triglycerides in the biologically derived oils are separated from the glycerin, creating a clean-burning, renewable fuel.

Bioenergy: Useful, renewable energy produced from organic matter—the conversion of the complex carbohydrates in organic matter to energy; organic matter may either be used directly as a fuel, processed into liquids and gasses, or be a residual of processing and conversion.

Bioethanol: Ethanol produced from biomass feedstocks; includes ethanol produced from the fermentation of crops, such as corn, as well as cellulosic ethanol produced from woody plants or grasses.

Biofuels: A generic name for liquid or gaseous fuels that are not derived from petroleum-based fossil fuels or contain a proportion of non–fossil fuel; fuels produced from plants, crops such as sugar beet, rape seed oil, or reprocessed vegetable oils or fuels made from gasified biomass; fuels made from renewable biological sources and include ethanol, methanol, and biodiesel; sources include, but are not limited to, corn, soybeans, flaxseed, rapeseed, sugarcane, palm oil, raw sewage, food scraps, animal parts, and rice.

Biogas: A combustible gas derived from decomposing biological waste under anaerobic conditions. Biogas normally consists of 50%–60% methane. *See also* Landfill gas.

Biogenic: Material derived from bacterial or vegetation sources.

Biological lipid: Any biological fluid that is miscible with a nonpolar solvent. These materials include waxes, essential oils, chlorophyll, etc.

Biological oxidation: The oxidative consumption of organic matter by bacteria by which the organic matter is converted into gases.

Biomass: Any organic matter that is available on a renewable or recurring basis, including agricultural crops and trees, wood and wood residues, plants (including aquatic plants), grasses, animal manure, municipal residues, and other residue materials. Biomass is generally produced in a sustainable manner from water and carbon dioxide by photosynthesis. There are three main categories of biomass: primary, secondary, and tertiary.

Biomass to liquid (BTL): The process of converting biomass to liquid fuels. Hmm, that seems painfully obvious when you write it out.

Biopolymer: A high-molecular-weight carbohydrate produced by bacteria.

Biopower: The use of biomass feedstock to produce electric power or heat through direct combustion of the feedstock, through gasification and then combustion of the resultant gas, or through other thermal conversion processes. Power is generated with engines, turbines, fuel cells, or other equipment.

Biorefinery: A facility that processes and converts biomass into value-added products. These products can range from biomaterials to fuels such as ethanol or important feedstocks for the production of chemicals and other materials.

Bitumen: Also, on occasion, referred to as native asphalt, and extra heavy oil; a naturally occurring material that has little or no mobility under reservoir conditions and that cannot be recovered through a well by conventional oil well production methods including currently used enhanced recovery techniques; current methods involve mining for bitumen recovery.

Bituminous: Containing bitumen or constituting the source of bitumen.

Bituminous rock: *See* Bituminous sand

Bituminous sand: A formation in which the bituminous material (*see* Bitumen) is found as a filling in veins and fissures in fractured rock or impregnating relatively shallow sand, sandstone, and limestone strata; a sandstone reservoir that is impregnated with a heavy, viscous black petroleum-like material that cannot be retrieved through a well by conventional production techniques.

Black acid(s): A mixture of the sulfonates found in acid sludge, which are insoluble in naphtha, benzene, and carbon tetrachloride; very soluble in water but insoluble in 30% sulfuric acid; in the dry, oil-free state, the sodium soaps are black powders.

Black liquor: Solution of lignin residue and the pulping chemicals used to extract lignin during the manufacture of paper.

Black oil: Any of the dark-colored oils; a term now often applied to heavy oil.

Black soap: *See* Black acid

Black strap: The black material (mainly lead sulfide) formed in the treatment of sour light oils with doctor solution and found at the interface between the oil and the solution.

Blown asphalt: The asphalt prepared by air blowing a residuum or an asphalt.

Bogging: A condition that occurs in a coking reactor when the conversion to coke and light ends is too slow causing the coke particles to agglomerate.

Boiling point: A characteristic physical property of a liquid at which the vapor pressure is equal to that of the atmosphere and the liquid is converted to a gas.

Boiling range: The range of temperature, usually determined at atmospheric pressure in standard laboratory apparatus, over which the distillation of an oil commences, proceeds, and finishes.

Bone dry: Having 0% moisture content. Wood heated in an oven at a constant temperature of 100°C (212°F) or above until its weight stabilizes is considered bone dry or oven dry.

Bottled gas: Usually butane or propane, or butane–propane mixtures, liquefied and stored under pressure for domestic use; *see also* Liquefied petroleum gas.

Bottoming cycle: A cogeneration system in which steam is used first for process heat and then for electric power production.

Bottoms: The liquid that collects in the bottom of a vessel (tower bottoms, tank bottoms) either during distillation; also the deposit or sediment formed during the storage of petroleum or a petroleum product; *see also* Residuum and Basic sediment and water.

Bright stock: Refined, high-viscosity lubricating oils usually made from residual stocks by processes such as a combination of acid treatment or solvent extraction with dewaxing or clay finishing.

British thermal unit: *See* Btu

Bromine number: The number of grams of bromine absorbed by 100 g of oil, which indicates the percentage of double bonds in the material.

Brønsted acid: A chemical species that can act as a source of protons.

Brønsted base: A chemical species that can accept protons.

Brown acid: Oil-soluble petroleum sulfonates found in acid sludge that can be recovered by extraction with naphtha solvent. Brown acid sulfonates are somewhat similar to mahogany sulfonates but are more water-soluble. In the dry, oil-free state, the sodium soaps are light-colored powders.

Brown soap: *See* Brown acid

BS&W: *See* Basic sediment and water

BTEX: Benzene, toluene, ethylbenzene, and the xylene isomers.

Btu (British thermal unit): The energy required to raise the temperature of one pound of water by 1°F.

Bubble cap: An inverted cup with a notched or slotted periphery to disperse the vapor in small bubbles beneath the surface of the liquid on the bubble plate in a distillation tower.

Bubble plate: A tray in a distillation tower.

Bubble point: The temperature at which incipient vaporization of a liquid in a liquid mixture occurs, corresponding with the equilibrium point of 0% vaporization or 100% condensation.

Bubble tower: A fractionating tower so constructed that the vapors rising pass up through layers of condensate on a series of plates or trays (*see* Bubble plate); the vapor passes from one plate to the next above by bubbling under one or more caps (*see* Bubble cap) and out through the liquid on the plate where the less volatile portions of vapor condense in bubbling through the liquid on the plate, overflow to the next lower plate, and ultimately back into the reboiler thereby effecting fractionation.

Bubble tray: A circular, perforated plates having the internal diameter of a bubble tower, set at specified distances in a tower to collect the various fractions produced during distillation.

Buckley–Leverett method: A theoretical method of determining frontal advance rates and saturations from a fractional flow curve.

Bumping: The knocking against the walls of a still occurring during distillation of petroleum or a petroleum product that usually contains water.

Bunker: A storage tank.

Bunker C oil: *See* No. 6 Fuel oil

Burner fuel oil: Any petroleum liquid suitable for combustion.

Burning oil: An illuminating oil, such as kerosene (kerosine) suitable for burning in a wick lamp.

Burning point: *See* Fire point

Burning-quality index: An empirical numerical indication of the likely burning performance of a furnace or heater oil; derived from the distillation profile and the API gravity, and generally recognizing the factors of paraffin character and volatility.

Burton process: An older thermal cracking process in which oil was cracked in a pressure still and any condensation of the products of cracking also took place under pressure.

Butane dehydrogenation: A process for removing hydrogen from butane to produce butenes and, on occasion, butadiene.

Butane vapor–phase isomerization: A process for isomerizing n-butane to *iso*-butane using aluminum chloride catalyst on a granular alumina support and with hydrogen chloride as a promoter.

Butanol: Though generally produced from fossil fuels, this four-carbon alcohol can also be produced through bacterial fermentation of alcohol.

C_1, C_2, C_3, C_4, C_5 fractions: A common way of representing fractions containing a preponderance of hydrocarbons having 1, 2, 3, 4, or 5 carbon atoms, respectively, and without reference to hydrocarbon type.

CAA: Clean Air Act; this act is the foundation of air regulations in the United States.

Calcining: Heating a metal oxide or an ore to decompose carbonates, hydrates, or other compounds often in a controlled atmosphere.

Capillary forces: Interfacial forces between immiscible fluid phases, resulting in pressure differences between the two phases.

Capillary number: N_c, the ratio of viscous forces to capillary forces, and equal to viscosity times velocity divided by interfacial tension.

Carbene: The pentane- or heptane-insoluble material that is insoluble in benzene or toluene but which is soluble in carbon disulfide (or pyridine); a type of rifle used for hunting bison.

Carboid: The pentane- or heptane-insoluble material that is insoluble in benzene or toluene and that is also insoluble in carbon disulfide (or pyridine).

Carbonate washing: Processing using a mild alkali (e.g., potassium carbonate) process for emission control by the removal of acid gases from gas streams.

Carbon dioxide–augmented waterflooding: Injection of carbonated water, or water and carbon dioxide, to increase water flood efficiency; *see* immiscible carbon dioxide displacement.

Carbon dioxide miscible flooding: *See* EOR process.

Carbon-forming propensity: *See* Carbon residue.

Carbon monoxide (CO): A lethal gas produced by incomplete combustion of carbon-containing fuels in internal combustion engines. It is colorless, odorless, and tasteless. (As in flavorless, we mean, though it has also been known to tell a bad joke or two.)

Carbon–oxygen log: Information about the relative abundance of elements such as carbon, oxygen, silicon, and calcium in a formation; usually derived from pulsed neutron equipment.

Carbon rejection: Upgrading processes in which coke is produced, for example, coking.

Carbon residue: The amount of carbonaceous residue remaining after thermal decomposition of petroleum, a petroleum fraction, or a petroleum product in a limited amount of air; also called the *coke-* or *carbon-forming propensity*; often prefixed by the terms Conradson or Ramsbottom in reference to the inventor of the respective tests.

Carbon sink: A geographical area whose vegetation and/or soil soaks up significant carbon dioxide from the atmosphere. Such areas, typically in tropical regions, are increasingly being sacrificed for energy crop production.

Carbonization: The conversion of an organic compound into char or coke by heat in the substantial absence of air; often used in reference to the destructive distillation (with simultaneous removal of distillate) of coal.

Cascade tray: A fractionating device consisting of a series of parallel troughs arranged on stair-step fashion in which liquid frown the tray above enters the uppermost trough and liquid thrown from this trough by vapor rising from the tray below impinges against a plate and a perforated baffle and liquid passing through the baffle enters the next longer of the troughs.

Casinghead gas: Natural gas that issues from the casinghead (the mouth or opening) of an oil well.

Casinghead gasoline: The liquid hydrocarbon product extracted from casinghead gas by one of three methods: compression, absorption, or refrigeration; *see also* Natural gasoline.

Catagenesis: The alteration of organic matter during the formation of petroleum that may involve temperatures in the range of 50°C (120°F) to 200°C (390°F); *see also* Diagenesis and Metagenesis.

Catalyst: A chemical agent that, when added to a reaction (process), will enhance the conversion of a feedstock without being consumed in the process.

Catalyst selectivity: The relative activity of a catalyst with respect to a particular compound in a mixture, or the relative rate in competing reactions of a single reactant.

Catalyst stripping: The introduction of steam, at a point where spent catalyst leaves the reactor, in order to strip, that is, remove, deposits retained on the catalyst.

Catalytic activity: The ratio of the space velocity of the catalyst under test to the space velocity required for the standard catalyst to give the same conversion as the catalyst being tested; usually multiplied by 100 before being reported.

Catalytic cracking: The conversion of high-boiling feedstocks into lower-boiling products by means of a catalyst that may be used in a fixed bed or fluid bed.

Catalytic reforming: Rearranging hydrocarbon molecules in a gasoline-boiling-range feedstock to produce other hydrocarbons having a higher antiknock quality; isomerization of paraffins, cyclization of paraffins to naphthenes, dehydrocyclization of paraffins to aromatics.

Cat cracking: *See* Catalytic cracking.

Catforming: A process for reforming naphtha using a platinum–silica–alumina catalyst that permits relatively high space velocities and results in the production of high-purity hydrogen.

Caustic consumption: The amount of caustic lost from reacting chemically with the minerals in the rock, the oil, and the brine.

Caustic wash: The process of treating a product with a solution of caustic soda to remove minor impurities; often used in reference to the solution itself.

Ceresin: A hard, brittle wax obtained by purifying ozokerite; *see* Microcrystralline wax and Ozokerite.

Cetane index: An approximation of the cetane number calculated from the density and mid-boiling point temperature; *see also* Diesel index.

Cetane number: A number indicating the ignition quality of diesel fuel; a high cetane number represents a short ignition delay time.

CFR: Code of Federal Regulations; Title 40 (40 CFR) contains the regulations for the protection of the environment.

Characterization factor: The UOP characterization factor K, defined as the ratio of the cube root of the molal average boiling point, T_B, in degrees Rankine ($^oR = {^o}F + 460$), to the specific gravity at 60°F/60°F: $K = (T_B)^{1/3}/sp\ gr$. The value ranges from 12.5 for paraffin stocks to 10.0 for the highly aromatic stocks; also called the Watson characterization factor.

Cheesebox still: An early type of vertical cylindrical still designed with a vapor dome.

Chelating agents: Complex-forming agents having the ability to solubilize heavy metals.

Chemical flooding: *See* EOR process

Chemical octane number: The octane number added to gasoline by refinery processes or by the use of octane number improvers such as tetraethyl lead.

Chemical waste: Any solid, liquid, or gaseous material discharged from a process and that may pose substantial hazards to human health and environment.

Chlorex process: A process for extracting lubricating oil stocks in which the solvent used is Chlorex (β,β-dichlorodiethyl ether).

Chromatographic adsorption: Selective adsorption on materials such as activated carbon, alumina, or silica gel; liquid or gaseous mixtures of hydrocarbons are passed through the adsorbent in a stream of diluent, and certain components are preferentially adsorbed.

Chromatographic separation: The separation of different species of compounds according to their size and interaction with the rock as they flow through a porous medium.

Chromatography: A method of separation based on selective adsorption; *see also* Chromatographic adsorption.

Clarified oil: The heavy oil that has been taken from the bottom of a fractionator in a catalytic cracking process and from which residual catalyst has been removed.

Clarifier: Equipment for removing the color or cloudiness of an oil or water by separating the foreign material through mechanical or chemical means; may involve centrifugal action, filtration, heating, or treatment with acid or alkali.

Clastic: Composed of pieces of preexisting rock.

Clay: Silicate minerals that also usually contain aluminum and have particle sizes that are less than 0.002 μm; used in separation methods as an adsorbent and in refining as a catalyst.

Clay contact process: *See* Contact filtration.

Clay refining: A treating process in which vaporized gasoline or other light petroleum product is passed through a bed of granular clay such as fuller's earth.

Clay regeneration: A process in which spent coarse-grained adsorbent clay minerals from percolation processes are cleaned for reuse by deoiling the clay minerals with naphtha, steaming out the excess naphtha, and then roasting in a stream of air to remove carbonaceous matter.

Clay treating: *See* Gray clay treating

Clay wash: Light oil, such as kerosene (kerosine) or naphtha, used to clean fuller's earth after it has been used in a filter.

Cleanup: A preparatory step following extraction of a sample media designed to remove components that may interfere with subsequent analytical measurements.

Closed-loop biomass: Crops grown, in a sustainable manner, for the purpose of optimizing their value for bioenergy and bioproduct uses. This includes annual crops such as maize and wheat, and perennial crops such as trees, shrubs, and grasses such as switch grass.

Cloud point: The temperature at which paraffin wax or other solid substances begin to crystallize or separate from the solution, imparting a cloudy appearance to the oil when the oil is chilled under prescribed conditions.

Coal: An organic rock.

Coal tar: The specific name for the tar produced from coal.

Coal tar pitch: The specific name for the pitch produced from coal.

Coalescence: The union of two or more droplets to form a larger droplet and, ultimately, a continuous phase.

Coarse materials: Wood residues suitable for chipping, such as slabs, edgings, and trimmings.

COFCAW: An EOR process that combines forward combustion and waterflooding.

Cogeneration: An energy conversion method by which electrical energy is produced along with steam generated for EOR use.

Coke: A gray to black solid carbonaceous material produced from petroleum during thermal processing; characterized by having a high carbon content (95%+ by weight) and a honeycomb type of appearance and is insoluble in organic solvents.

Coke drum: A vessel in which coke is formed, which can be cut oil from the process for cleaning.

Coke number: Used, particularly in Great Britain, to report the results of the Ramsbottom carbon residue test, which is also referred to as a coke test.

Coker: The processing unit in which coking takes place.

Coking: A process for the thermal conversion of petroleum in which gaseous, liquid, and solid (coke) products are formed.

Cold pressing: The process of separating wax from oil by first chilling (to help form wax crystals) and then filtering under pressure in a plate and frame press.

Cold settling: Processing for the removal of wax from high-viscosity stocks, wherein a naphtha solution of the waxy oil is chilled and the wax crystallizes out of the solution.

Color stability: The resistance of a petroleum product to color change due to light, aging, etc.

Combined cycle: A combustion (gas) turbine equipped with a heat recovery steam generator that produces steam for the steam turbine; power is produced from both the gas and steam turbines—hence the term combined cycle.

Combustible liquid: A liquid with a flash point in excess of 37.8°C (100°F) but below 93.3°C (200°F).

Combustion zone: The volume of reservoir rock wherein petroleum is undergoing combustion during enhanced oil recovery.

Completion interval: The portion of the reservoir formation placed in fluid communication with the well by selectively perforating the wellbore casing.

Composition: The general chemical makeup of petroleum.

Composition map: A means of illustrating the chemical makeup of petroleum using chemical and/or physical property data.

Con Carbon: *See* Carbon residue.

Condensate: a mixture of light hydrocarbon liquids obtained by condensation of hydrocarbon vapors: Predominately butane, propane, and pentane with some heavier hydrocarbons and relatively little methane or ethane; *see also* Natural gas liquids.

Conductivity: A measure of the ease of flow through a fracture, perforation, or pipe.

Conformance: The uniformity with which a volume of the reservoir is swept by injection fluids in area and vertical directions.

Conradson carbon residue: *See* Carbon residue

Contact filtration: A process in which finely divided adsorbent clay is used to remove color bodies from petroleum products.

Contaminant: A substance that causes deviation from the normal composition of an environment.

Continuous contact coking: A thermal conversion process in which petroleum-wetted coke particles move downward into the reactor in which cracking, coking, and drying take place to produce coke, gas, gasoline, and gas oil.

Continuous contact filtration: A process to finish lubricants, waxes, or special oils after acid treating, solvent extraction, or distillation.

Conventional crude oil (conventional petroleum): Crude oil that is pumped from the ground and recovered using the energy inherent in the reservoir; also recoverable by the application of secondary recovery techniques.

Conventional recovery: Primary and/or secondary recovery.

Conversion: The thermal treatment of petroleum that results in the formation of new products by the alteration of the original constituents.

Conversion cost: The cost of changing a production well to an injection well, or some other change in the function of an oilfield installation.

Conversion factor: The percentage of feedstock converted to light ends, gasoline, other liquid fuels, and coke.

Copper sweetening: Processes involving the oxidation of mercaptan derivatives (RSH) to disulfide derivatives (RSSR) by oxygen in the presence of cupric chloride ($CuCl_2$).

Cord: A stack of wood comprising 128 ft^3 (3.62 m^3); standard dimensions are $4 \times 4 \times 8$ ft, including air space and bark. One cord contains approx. 1.2 U.S. tons (oven-dry) = 2400 lb = 1089 kg.

Core floods: Laboratory flow tests through samples (cores) of porous rock.

Cosurfactant: A chemical compound, typically alcohol that enhances the effectiveness of a surfactant.

Cp (centipoise): A unit of viscosity.

Cracked residua: Residua that have been subjected to temperatures above 350°C (660°F) during the distillation process.

Cracking: The thermal processes by which the constituents of petroleum are converted to lower-molecular-weight products.

Cracking activity: *See* Catalytic activity

Cracking coil: Equipment used for cracking heavy petroleum products consisting of a coil of heavy pipe running through a furnace so that the oil passing through it is subject to high temperature.

Cracking still: The combined equipment–furnace, reaction chamber, fractionator for the thermal conversion of heavier feedstocks to lighter products.

Cracking temperature: The temperature (350°C; 660°F) at which the rate of thermal decomposition of petroleum constituents becomes significant.

Craig–Geffen–Morse method: A method for predicting oil recovery by water flood.

Criteria air pollutants: Air pollutants or classes of pollutants regulated by the Environmental Protection Agency; the air pollutants are (including VOCs) ozone, carbon monoxide, particulate matter, nitrogen oxides, sulfur dioxide, and lead.

Cropland: Total cropland includes five components: cropland harvested, crop failure, cultivated summer fallow, cropland used only for pasture, and idle cropland.

Cropland pasture: Land used for long-term crop rotation. However, some cropland pasture is marginal for crop uses and may remain in pasture indefinitely. This category also includes land that was used for pasture before crops reached maturity and some land used for pasture that could have been cropped without additional improvement.

Cross-linking: Combining of two or polymer molecules by the use of a chemical that mutually bonds with a part of the chemical structure of the polymer molecules.

Crude assay: A procedure for determining the general distillation characteristics (e.g., distillation profile, *q.v.*) and other quality information of crude oil.

Crude oil: *See* Petroleum

Crude scale wax: The wax product from the first sweating of the slack wax.

Crude still: Distillation equipment in which crude oil is separated into various products.

Cull tree: A live tree, 5.0 inches in diameter at breast height (d.b.h.) or larger that is nonmerchantable for saw logs now or prospectively because of rot, roughness, or species. (*See* definitions for rotten and rough trees.)

Cultivated summer fallow: Cropland cultivated for one or more seasons to control weeds and accumulate moisture before small grains are planted.

Cumene: A colorless liquid [$C_6H_5CH(CH_3)_2$] used as an aviation gasoline blending component and as an intermediate in the manufacture of chemicals.

Cutback: The term applied to the products from blending heavier feedstocks or products with lighter oils to bring the heavier materials to the desired specifications.

Cutback asphalt: Asphalt liquefied by the addition of a volatile liquid such as naphtha or kerosene that, after application and on exposure to the atmosphere, evaporates leaving the asphalt.

Cut point: The boiling-temperature division between distillation fractions of petroleum.

Cutting oil: An oil to lubricate and cool metal-cutting tools; also called cutting fluid, cutting lubricant.

Cycle stock: The product taken from some later stage of a process and recharged (recycled) to the process at some earlier stage.

Cyclic steams injection: The alternating injection of steam and production of oil with condensed steam from the same well or wells.

Cyclization: The process by which an open-chain hydrocarbon structure is converted to a ring structure, for example, hexane to benzene.

Cyclone: A device for extracting dust from industrial waste gases. It is in the form of an inverted cone into which the contaminated gas enters tangential from the top; the gas is propelled down a helical pathway, and the dust particles are deposited by means of centrifugal force onto the wall of the scrubber.

Deactivation: Reduction in catalyst activity by the deposition of contaminants (e.g., coke, metals) during a process.

Dealkylation: The removal of an alkyl group from aromatic compounds.

Deasphaltened oil: The fraction of petroleum after the asphaltene constituents have been removed.

Deasphaltening: Removal of a solid powdery asphaltene fraction from petroleum by the addition of the low-boiling liquid hydrocarbons such as n-pentane or n-heptane under ambient conditions.

Deasphalting: The removal of the asphaltene fraction from petroleum by the addition of a low-boiling hydrocarbon liquid such as n-pentane or n-heptane; more correctly the removal asphalt (tacky, semisolid) from petroleum (as occurs in a refinery asphalt plant) by the addition of liquid propane or liquid butane under pressure.

Debutanization: Distillation to separate butane and lighter components from higher-boiling components.

Decant oil: The highest boiling product from a catalytic cracker; also referred to as slurry oil, clarified oil, or bottoms.

Decarbonizing: A thermal conversion process designed to maximize coker gas–oil production and minimize coke and gasoline yields; operated at essentially lower temperatures and pressures than delayed coking.

Decoking: Removal of petroleum coke from equipment such as coking drums; hydraulic decoking uses high-velocity water streams.

Decolorizing: Removal of suspended, colloidal, and dissolved impurities from liquid petroleum products by filtering, adsorption, chemical treatment, distillation, bleaching, etc.

De-ethanization: Distillation to separate ethane and lighter components from propane and higher-boiling components; also called de-ethanation.

Degradation: The loss of desirable physical properties of EOR fluids, for example, the loss of viscosity of polymer solutions.

Dehydrating agents: Substances capable of removing water (drying, q.v.) or the elements of water from another substance.

Dehydrocyclization: Any process by which both dehydrogenation and cyclization reactions occur.

Dehydrogenation: The removal of hydrogen from a chemical compound; for example, the removal of two hydrogen atoms from butane to make butene(s) as well as the removal of additional hydrogen to produce butadiene.

Delayed coking: A coking process in which the thermal reaction is allowed to proceed to completion to produce gaseous, liquid, and solid (coke) products.

Demethanization: The process of distillation in which methane is separated from the higher-boiling components; also called demethanation.

Density: The mass (or weight) of a unit volume of any substance at a specified temperature; *see also* Specific gravity.

Deoiling: Reduction in the quantity of liquid oil entrained in solid wax by draining (sweating) or by a selective solvent; *see* MEK deoiling.

Depentanizer: A fractionating column for the removal of pentane and lighter fractions from a mixture of hydrocarbons.

Depropanization: Distillation in which lighter components are separated from butanes and higher-boiling material; also called depropanation.

Desalting: Removal of mineral salts (mostly chlorides) from crude oils.

Desorption: The reverse process of adsorption whereby adsorbed matter is removed from the adsorbent; also used as the reverse of absorption.

Desulfurization: The removal of sulfur or sulfur compounds from a feedstock.

Detergent oil: Lubricating oil possessing special sludge-dispersing properties for use in internal-combustion engines.

Devolatilized fuel: Smokeless fuel; coke that has been reheated to remove all of the volatile material.

Dewaxing: *See* Solvent dewaxing

Diagenesis: The concurrent and consecutive chemical reactions that commence the alteration of organic matter (at temperatures up to 50°C [120°F]) and ultimately result in the formation of petroleum from the marine sediment; *see also* Catagenesis and Metagenesis.

Diagenetic rock: Rock formed by conversion through pressure or chemical reaction from a rock, for example, sandstone is a diagenetic.

Diesel cycle: A repeated succession of operations representing the idealized working behavior of the fluids in a diesel engine.

Diesel engine: Named for the German engineer Rudolph Diesel, this internal combustion, compression ignition engine works by heating fuels and causing them to ignite; can use either petroleum or bio-derived fuel.

Diesel fuel: Fuel used for internal combustion in diesel engines; usually that fraction that distills after kerosene.

Diesel index: An approximation of the cetane number of diesel fuel calculated from the density and aniline point. DI = (aniline point (°F) × API gravity) × 100.

Diesel knock: The result of a delayed period of ignition and the accumulation of diesel fuel in the engine.

Differential-strain analysis: Measurement of thermal stress relaxation in a recently cut well.

Digester: An airtight vessel or enclosure in which bacteria decomposes biomass in water to produce biogas.

Direct injection engine: A diesel engine in which fuel is injected directly into the cylinder.

Dispersion: A measure of the convective fluids due to flow in a reservoir.

Displacement efficiency: The ratio of the amount of oil moved from the zone swept by the reprocess to the amount of oil present in the zone prior to start of the process.

Distillate: Any petroleum product produced by boiling crude oil and collecting the vapors produced as a condensate in a separate vessel, for example, gasoline (light distillate), gas oil (middle distillate), or fuel oil (heavy distillate).

Distillation: The primary distillation process that uses high temperature to separate crude oil into vapor and fluids that can then be fed into a distillation or fractionating tower.

Distillation curve: *See* Distillation profile

Distillation loss: The difference, in a laboratory distillation, between the volume of liquid originally introduced into the distilling flask and the sum of the residue and the condensate recovered.

Distillation profile: The distillation characteristics of petroleum or petroleum products showing the temperature and the percent distilled.

Distillation range: The difference between the temperature at the initial boiling point and at the end point, as obtained by the distillation test.

Distribution coefficient: A coefficient that describes the distribution of a chemical in reservoir fluids, usually defined as the equilibrium concentrations in the aqueous phases.

Doctor solution: A solution of sodium plumbite used to treat gasoline or other light petroleum distillates to remove mercaptan sulfur; *see also* Doctor test.

Doctor sweetening: A process for sweetening gasoline, solvents, and kerosene by converting mercaptan derivatives (RSH) to disulfide derivatives (RSSR) using sodium plumbite (Na_2PbO_2) and sulfur.

Doctor test: A test used for the detection of compounds in light petroleum distillates that react with sodium plumbite; *see also* Doctor solution.

Domestic heating oil: *See* No. 2 Fuel oil.

Donor solvent process: A conversion process in which hydrogen donor solvent is used in place of or to augment hydrogen.

Downcomer: A means of conveying liquid from one tray to the next below in a bubble tray column.

Downdraft gasifier: A gasifier in which the product gases pass through a combustion zone at the bottom of the gasifier.

Downhole steam generator: A generator installed downhole in an oil well to which oxygen-rich air, fuel, and water are supplied for the purposes of generating steam in the reservoir. Its major advantage over a surface steam–generating facility is that the losses to the wellbore and surrounding formation are eliminated.

Dropping point: The temperature at which grease passes from a semisolid to a liquid state under prescribed conditions.

Dry gas: A gas that does not contain fractions that may easily condense under normal atmospheric conditions.

Drying: Removal of a solvent or water from a chemical substance; also referred to as the removal of solvent from a liquid or suspension.

Dry point: The temperature at which the last drop of petroleum fluid evaporates in a distillation test.

Dualayer distillate process: A process for removing mercaptan derivatives (RSH) and oxygenated compounds from distillate fuel oils and similar products, using a combination of treatment with concentrated caustic solution and electrical precipitation of the impurities.

Dualayer gasoline process: A process for extracting mercaptan derivatives (RSH) and other objectionable acidic compounds from petroleum distillates; *see also* Dualayer solution.

Dualayer solution: A solution that consists of concentrated potassium or sodium hydroxide containing a solubilizer; *see also* Dualayer gasoline process.

Dubbs cracking: An older continuous, liquid-phase thermal cracking process formerly used.

Dutch oven furnace: One of the earliest types of furnaces, having a large, rectangular box lined with firebrick (refractory) on the sides and top; commonly used for burning wood.

Dykstra–Parsons coefficient: An index of reservoir heterogeneity arising from permeability variation and stratification.

E85: An alcohol fuel mixture containing 85% ethanol and 15% gasoline by volume, and the current alternative fuel of choice of the U.S. government.

Ebullated bed: A process in which the catalyst bed is in a suspended state in the reactor by means of a feedstock recirculation pump that pumps the feedstock upward at sufficient speed to expand the catalyst bed at approximately 35% above the settled level.

Edeleanu process: A process for refining oils at low temperature with liquid sulfur dioxide (SO_2), or with liquid sulfur dioxide and benzene; applicable to the recovery of aromatic concentrates from naphtha and heavier petroleum distillates.

Effective viscosity: *See* Apparent viscosity

Effluent: Any contaminating substance, usually a liquid, which enters the environment via a domestic industrial, agricultural, or sewage plant outlet.

Electric desalting: A continuous process to remove inorganic salts and other impurities from crude oil by settling out in an electrostatic field.

Electrical precipitation: A process using an electrical field to improve the separation of hydrocarbon reagent dispersions. May be used in chemical treating processes on a wide variety of refinery stocks.

Electrofining: A process for contacting a light hydrocarbon stream with a treating agent (acid, caustic, doctor, etc.), then assisting the action of separation of the chemical phase from the hydrocarbon phase by an electrostatic field.

Electrolytic mercaptan process: A process in which aqueous caustic solution is used to extract mercaptan derivatives (RSH) from refinery streams.

Electrostatic precipitators: Devices used to trap fine dust particles (usually in the size range 30–60 μm) that operate on the principle of imparting an electric charge to particles in an incoming air stream and that are then collected on an oppositely charged plate across a high-voltage field.

Eluate: The solutes, or analytes, moved through a chromatographic column (*see elution*).

Eluent: Solvent used to elute sample.

Elution: A process whereby a solute is moved through a chromatographic column by a solvent (liquid or gas) or eluent.

Emission control: The use of gas cleaning processes to reduce emissions.

Emissions: Substances discharged into the air during combustion.

Emission standard: The maximum amount of a specific pollutant permitted to be discharged from a particular source in a given environment.

Emulsion: A dispersion of very small drops of one liquid in an immiscible liquid, such as oil in water.

Emulsion breaking: The settling or aggregation of colloidal-sized emulsions from suspension in a liquid medium.

End-of-pipe emission control: The use of specific emission control processes to clean gases after production of the gases.

Energy: The capacity of a body or system to do work, measured in joules (SI units); also the output of fuel sources.

Energy balance: The difference between the energy produced by a fuel and the energy required to obtain it through agricultural processes, drilling, refining, and transportation.

Energy crops: Crops grown specifically for their fuel value; include food crops such as corn and sugarcane, and nonfood crops such as poplar trees and switch grass.

Energy-efficiency ratio: A number representing the energy stored in a fuel as compared to the energy required to produce, process, transport, and distribute that fuel.

Energy from biomass: The production of energy from biomass.

Engler distillation: A standard test for determining the volatility characteristics of a gasoline by measuring the percent distilled at various specified temperatures.

Enhanced oil recovery (EOR): Petroleum recovery following recovery by conventional (i.e., primary and/or secondary) methods.

Enhanced oil recovery (EOR) process: A method for recovering additional oil from a petroleum reservoir beyond that economically recoverable by conventional primary and secondary recovery methods. Enhanced oil recovery methods are usually divided into three main categories: (1) *chemical flooding,* that is, injection of water with added chemicals into a petroleum reservoir (the chemical processes include surfactant flooding, polymer flooding, and alkaline flooding); (2) *miscible flooding,* that is, injection into a petroleum reservoir of a material that is miscible, or can become miscible, with the oil in the reservoir (carbon dioxide, hydrocarbons, and nitrogen are used); (3) *thermal recovery,* that is, injection of steam into a petroleum reservoir or propagation of a combustion zone through a reservoir by air or oxygen-enriched air injection (the thermal processes include steam drive, cyclic steam injection, and *in situ* combustion).

Entrained bed: A bed of solid particles suspended in a fluid (liquid or gas) at such a rate that some of the solid is carried over (entrained) by the fluid.

EPA: Environmental Protection Agency

Ester: A compound formed by the reaction between an organic acid and an alcohol. ethoxylated alcohols (i.e., alcohols having ethylene oxide functional groups attached to the alcohol molecule).

Ethanol (ethyl alcohol, alcohol, or grain spirit): A clear, colorless, flammable oxygenated hydrocarbon; used as a vehicle fuel by itself (E100 is 100% ethanol by volume), blended with gasoline (E85 is 85% ethanol by volume) or as a gasoline octane enhancer and oxygenate (10% by volume); formed during the fermentation of sugars; used as an intoxicant and as a fuel.

Evaporation: A process for concentrating nonvolatile solids in a solution by boiling off the liquid portion of the waste stream.

Expanding clays: Clays that expand or swell on contact with water, for example, montmorillonite.

Explosive limits: The limits of percentage composition of mixtures of gases and air within which an explosion takes place when the mixture is ignited.

Extract: The portion of a sample preferentially dissolved by the solvent and recovered by physically separating the solvent.

Extractive distillation: The separation of different components of mixtures that have similar vapor pressures by flowing a relatively high-boiling solvent, which is selective for one of the components in the feed, down a distillation column as the distillation proceeds; the selective solvent scrubs the soluble component from the vapor.

Fabric filters: Filters made from fabric materials and used for removing particulate matter from gas streams (*see* Baghouse).

Facies: One or more layers of rock that differs from other layers in composition, age, or content.

FAST: Fracture-assisted steamflood technology.

Fat oil: The bottom or enriched oil drawn from the absorber as opposed to lean oil.

Faujasite: A naturally occurring silica–alumina (SiO_2–Al_2O_3) mineral.

FCC: Fluid catalytic cracking

FCCU: Fluid catalytic cracking unit

Feedstock: Petroleum as it is fed to the refinery; a refinery product that is used as the raw material for another process; biomass used in the creation of a particular biofuel (e.g., corn or sugarcane for ethanol, soybeans, or rapeseed for biodiesel); the term is also generally applied to raw materials used in other industrial processes.

Fermentation: Conversion of carbon-containing compounds by microorganisms for the production of fuels and chemicals such as alcohols, acids, or energy-rich gases.

Ferrocyanide process: A regenerative chemical treatment for mercaptan removal using caustic-sodium ferrocyanide reagent.

Fiber products: Products derived from fibers of herbaceous and woody plant materials; examples include pulp, composition board products, and wood chips for export.

Field scale: The application of EOR processes to a significant portion of a field.

Filtration: The use of an impassable barrier to collect solids but which allows liquids to pass.

Fine materials: Wood residues not suitable for chipping, such as planer shavings and sawdust.

Fingering: The formation of finger-shaped irregularities at the leading edge of a displacing fluid in a porous medium that moves out ahead of the main body of fluid.

Fire point: The lowest temperature at which, under specified conditions in standardized apparatus, a petroleum product vaporizes sufficiently rapidly to form above its surface an air–vapor mixture that burns continuously when ignited by a small flame.

First contact miscibility: *See* Miscibility

Fischer–Tropsch process: A process for synthesizing hydrocarbons and oxygenated chemicals from a mixture of hydrogen and carbon monoxide.

Five spot: An arrangement or pattern of wells with four injection wells at the comers of a square and a producing well in the center of the square.

Fixed bed: A stationary bed (of catalyst) to accomplish a process (*see* Fluid bed).

Flammability range: The range of temperature over which a chemical is flammable.

Flammable: A substance that will burn readily.

Flammable liquid: A liquid having a flash point below 37.8°C (100°F).

Flammable solid: A solid that can ignite from friction or from heat remaining from its manufacture or that may cause a serious hazard if ignited.

Flash point: The lowest temperature to which the product must be heated under specified conditions to give off sufficient vapor to form a mixture with air that can be ignited momentarily by a flame.

Flexible-fuel vehicle (flex-fuel vehicle): A vehicle that can run alternately on two or more sources of fuel; includes cars capable of running on gasoline and gasoline/ethanol mixtures, as well as cars that can run on both gasoline and natural gas.

Flexicoking: A modification of the fluid coking process insofar as the process also includes a gasifier adjoining the burner/regenerator to convert excess coke to a clean fuel gas.

Flocculation threshold: The point at which constituents of a solution (e.g., asphaltene constituents or coke precursors) will separate from the solution as a separate (solid) phase.

Floc point: The temperature at which wax or solids separate as a definite floc.

Flood, flooding: The process of displacing petroleum from a reservoir by the injection of fluids.

Flue gases: The gaseous products of the combustion process mostly comprising carbon dioxide, nitrogen, and water vapor.

Fluid bed: The use of an agitated bed of inert granular material to accomplish a process in which the agitated bed resembles the motion of a fluid; a bed (of catalyst) that is agitated by an upward passing gas in such a manner that the particles of the bed simulate the movement of a fluid and has the characteristics associated with a true liquid; cf. Fixed bed.

Fluid: A reservoir gas or liquid.

Fluid catalytic cracking: Cracking in the presence of a fluidized bed of catalyst.

Fluid coking: A continuous fluidized solid process that cracks feed thermally over heated coke particles in a reactor vessel to gas, liquid products, and coke.

Fluidized-bed boiler: A large, refractory-lined vessel with an air distribution member or plate in the bottom, a hot gas outlet in or near the top, and some provisions for introducing fuel; the fluidized bed is formed by blowing air up through a layer of inert particles (such as sand or limestone) at a rate that causes the particles to go into suspension and continuous motion.

Fluidized-bed combustion: A process used to burn low-quality solid fuels in a bed of small particles suspended by a gas stream (usually air that will lift the particles but not blow them out of the vessel). Rapid burning removes some of the offensive by-products of combustion from the gases and vapors that result from the combustion process.

Fly ash: Particulate matter produced from mineral matter in coal that is converted during combustion to finely divided inorganic material and that emerges from the combustor in the gases.

Foots oil: The oil sweated out of slack wax; named from the fact that the oil goes to the foot, or bottom, of the pan during the sweating operation.

Forest health: A condition of ecosystem sustainability and attainment of management objectives for a given forest area; usually considered to include green trees, snags, resilient stands growing at a moderate rate, and endemic levels of insects and disease.

Forestland: Land at least 10% stocked by forest trees of any size, including land that formerly had such tree cover and that will be naturally or artificially regenerated; includes transition zones, such as areas between heavily forested and nonforested lands that are at least 10% stocked with forest trees and forest areas adjacent to urban and built-up lands; also included are pinyon-juniper and chaparral areas; minimum area for the classification of forestland is 1 acre.

Forest residues: Material not harvested or removed from logging sites in commercial hardwood and softwood stands as well as material resulting from forest management operations such as precommercial thinnings and removal of dead and dying trees.

Formation: An interval of rock with distinguishable geologic characteristics.

Formation volume factor: The volume in barrel that one stock tank barrel occupies in the formation at reservoir temperature and with the solution gas that is held in the oil at reservoir pressure.

Fossil fuel resources: A gaseous, liquid, or solid fuel material formed in the ground by chemical and physical changes (diagenesis, q.v.) in plant and animal residues over geological time; natural gas, petroleum, coal, and oil shale.

Fractional composition: The composition of petroleum as determined by fractionation (separation) methods.

Fractional distillation: The separation of the components of a liquid mixture by vaporizing and collecting the fractions, or cuts, which condense in different temperature ranges.

Fractional flow: The ratio of the volumetric flow rate of one fluid phase to the total fluid volumetric flow rate within a volume of rock.

Fractional flow curve: The relationship between the fractional flow of one fluid and its saturator during the simultaneous flow of fluids through rock.

Fractionating column: A column arranged to separate various fractions of petroleum by a single distillation, which may be tapped at different points along its length to separate various fractions in the order of their boiling points.

Fractionation: The separation of petroleum into the constituent fractions using solvent or adsorbent methods; chemical agents such as sulfuric acid may also be used.

Fracture: A natural or man-made crack in a reservoir rock.

Fracturing: The breaking apart of reservoir rock by applying very high fluid pressure at the rock face.

Frasch process: A process formerly used for removing sulfur by distilling oil in the presence of copper oxide.

Fuel cell: A device that converts the energy of a fuel directly to electricity and heat, without combustion.

Fuel cycle: The series of steps required to produce electricity. The fuel cycle includes mining or otherwise acquiring the raw fuel source, processing and cleaning the fuel, transport, electricity generation, waste management, and plant decommissioning.

Fuel oil: Also called heating oil is a distillate product that covers a wide range of properties; *see also* No. 1 to No. 4 Fuel oils.

Fuel treatment evaluator (FTE): A strategic assessment tool capable of aiding the identification, evaluation, and prioritization of fuel treatment opportunities.

Fuel wood: Wood used for conversion to some form of energy, primarily for residential use.

Fuller's earth: A clay that has high adsorptive capacity for removing color from oils; Attapulgus clay is a widely used fuller's earth.

Functional group: The portion of a molecule that is characteristic of a family of compounds and determines the properties of these compounds.

Furfural extraction: A single-solvent process in which furfural is used to remove aromatic, naphthene, olefin, and unstable hydrocarbons from a lubricating oil charge stock.

Furnace: An enclosed chamber or container used to burn biomass in a controlled manner to produce heat for space or process heating.

Furnace oil: A distillate fuel primarily intended for use in domestic heating equipment.

Gas cap: A part of a hydrocarbon reservoir at the top that will produce only gas.

Gaseous pollutants: Gases released into the atmosphere that act as primary or secondary pollutants.

Gasification: A chemical or thermal process used to convert carbonaceous material (such as coal, petroleum, and biomass) into gaseous components such as carbon monoxide and hydrogen; a process for converting a solid or liquid fuel into a gaseous fuel useful for power generation or chemical feedstock with an oxidant and steam.

Gasifier: A device for converting solid fuel into gaseous fuel; in biomass systems, the process is referred to as pyrolitic distillation.

Gasifier cold gas efficiency (CGE): The percentage of the coal heating value that appears as chemical heating value in the gasifier product gas.

Gasohol: A mixture of 10% v/v anhydrous ethanol and 90% v/v gasoline; 7.5% v/v anhydrous ethanol and 92.5% v/v gasoline; or 5.5% v/v anhydrous ethanol and 94.5% v/v gasoline; a term for motor vehicle fuel comprising between 80 and 90% v/v unleaded gasoline and 10% and 20% v/v ethanol (*see also* Ethyl alcohol).

Gas oil: A petroleum distillate with a viscosity and boiling range between those of kerosene and lubricating oil.

Gas-oil ratio: Ratio of the number of cubic feet of gas measured at atmospheric (standard) conditions to barrels of produced oil measured at stock tank conditions.

Gas-oil sulfonate: Sulfonate made from a specific refinery stream, in this case the gas-oil stream.

Gasoline: Fuel for the internal combustion engine that is commonly, but improperly, referred to simply as gas.

Gas reversion: A combination of thermal cracking or reforming of naphtha with thermal polymerization or alkylation of hydrocarbon gases carried out in the same reaction zone.

Gas to liquids (GTL): The process of refining natural gas and other hydrocarbons into longer-chain hydrocarbons, which can be used to convert gaseous waste products into fuels.

Gas turbine: A device in which fuel is combusted at pressure and the products of combustion expanded through a turbine to generate power (the Brayton Cycle); it is based on the same principle as the jet engine.

Gel point: The point at which a liquid fuel cools to the consistency of petroleum jelly.

Genetically modified organism (GMO): An organism whose genetic material has been modified through recombinant DNA technology, altering the phenotype of the organism to meet desired specifications.

Gilsonite: An asphaltite that is >90% bitumen.

Girbotol process: A continuous, regenerative process to separate hydrogen sulfide, carbon dioxide, and other acid impurities from natural gas, refinery gas, etc., using mono-, di-, or triethanolamine as the reagent.

Glance pitch: An asphaltite.

Glycol-amine gas treating: A continuous, regenerative process to simultaneously dehydrate and remove acid gases from natural gas or refinery gas.

Grahamite: An asphaltite.

Grain alcohol: *See* Ethyl alcohol

Grassland pasture and range: All open land used primarily for pasture and grazing, including shrub and brush land types of pasture; grazing land with sagebrush and scattered mesquite; and all tame and native grasses, legumes, and other forage used for pasture or grazing; because of the diversity in vegetative composition, grassland pasture and range are not always clearly distinguishable from other types of pasture and range; at one extreme, permanent grassland may merge with cropland pasture, or grassland may often be found in transitional areas with forested grazing land.

Gravimetric: Gravimetric methods weigh a residue.

Gravity: *See* API gravity

Gravity drainage: The movement of oil in a reservoir that results from the force of gravity.

Gravity segregation: Partial separation of fluids in a reservoir caused by the gravity force acting on differences in density.

Gravity-stable displacement: The displacement of oil from a reservoir by a fluid of a different density, where the density difference is utilized to prevent the gravity segregation of the injected fluid.

Gray clay treating: A fixed-bed, usually fuller's earth, vapor-phase treating process to selectively polymerize unsaturated gum-forming constituents (diolefins) in thermally cracked gasoline.

Grease car: A diesel-powered automobile rigged postproduction to run on used vegetable oil.

Greenhouse effect: The effect of certain gases in the Earth's atmosphere in trapping heat from the sun.

Greenhouse gases: Gases that trap the heat of the sun in the Earth's atmosphere, producing the greenhouse effect. The two major greenhouse gases are water vapor and carbon dioxide. Other greenhouse gases include methane, ozone, chlorofluorocarbons, and nitrous oxide.

Grid: An electric utility company's system for distributing power.

Growing stock: A classification of timber inventory that includes live trees of commercial species meeting specified standards of quality or vigor; cull trees are excluded.

Guard bed: A bed of an adsorbent (such as bauxite) that protects a catalyst bed by adsorbing species detrimental to the catalyst.

Gulf HDS process: A fixed-bed process for the catalytic hydrocracking of heavy stocks to lower-boiling distillates with accompanying desulfurization.

Gulfining: A catalytic hydrogen treating process for cracked and straight-run distillates and fuel oils to reduce sulfur content; improve carbon residue, color, and general stability; and to effect a slight increase in gravity.

Gum: An insoluble tacky semisolid material formed as a result of the storage instability and/or the thermal instability of petroleum and petroleum products.

Habitat: The area where a plant or animal lives and grows under natural conditions. Habitat includes living and nonliving attributes and provides all requirements for food and shelter.

HAP(s): Hazardous air pollutant(s).

Hardness: The concentration of calcium and magnesium in brine.

Hardwoods: Usually broad-leaved and deciduous trees.

HCPV: Hydrocarbon pore volume

Headspace: The vapor space above a sample into which volatile molecules evaporate. Certain methods sample this vapor.

Hearn method: A method used in reservoir simulation for calculating a pseudo-relative permeability curve that reflects reservoir stratification.

Heating oil: *See* Fuel oil

Heating value: The maximum amount of energy that is available from burning a substance.

Heat recovery steam generator: A heat exchanger that generates steam from the hot exhaust gases from a combustion turbine.

Heavy ends: The highest-boiling portion of a petroleum fraction; *see also* Light ends.

Heavy fuel oil: Fuel oil having a high density and viscosity; generally residual fuel oil such as No. 5 and No 6. fuel oil.

Heavy (crude) oil: Oil that is more viscous that conventional crude oil has a lower mobility in the reservoir but can be recovered through a well from the reservoir by the application of a secondary or enhanced recovery methods; sometimes petroleum having an API gravity of less than 20°.

Heavy petroleum: *See* Heavy oil

Hectare: Common metric unit of area, equal to 2.47 acres. 100 hectares = 1 km^2.

Herbaceous: Nonwoody type of vegetation, usually lacking permanent strong stems, such as grasses, cereals, and canola (rape).

Heteroatom compounds: Chemical compounds that contain nitrogen and/or oxygen and/or sulfur and/or metals bound within their molecular structure(s).

Heterogeneity: Lack of uniformity in reservoir properties such as permeability.

HF alkylation: An alkylation process whereby olefins (C_3, C_4, C_5) are combined with *iso*-butane in the presence of hydrofluoric acid catalyst.

Higgins-Leighton model: Stream tube computer model used to simulate waterflood.

Hortonsphere (Horton sphere): A spherical pressure-type tank used to store a volatile liquid that prevents the excessive evaporation loss that occurs when such products are placed in conventional storage tanks.

Hot filtration test: A test for the stability of a petroleum product.

Hot spot: An area of a vessel or line wall appreciably above normal operating temperature, usually as a result of the deterioration of an internal insulating liner that exposes the line or vessel shell to the temperature of its contents.

Houdresid catalytic cracking: A continuous moving-bed process for catalytically cracking reduced crude oil to produce high-octane gasoline and light distillate fuels.

Houdriflow catalytic cracking: A continuous moving-bed catalytic cracking process employing an integrated single vessel for the reactor and regenerator kiln.

Houdriforming: A continuous catalytic reforming process for producing aromatic concentrates and high-octane gasoline from low-octane straight naphtha.

Houdry butane dehydrogenation: A catalytic process for dehydrogenating light hydrocarbons to their corresponding mono- or diolefins.

Houdry fixed-bed catalytic cracking: A cyclic regenerable process for cracking of distillates.

Houdry hydrocracking: A catalytic process combining cracking and desulfurization in the presence of hydrogen.

Huff-and-puff: A cyclic EOR method in which steam or gas is injected into a production well; after a short shut-in period, oil and the injected fluid are produced through the same well.

Hydration: The association of molecules of water with a substance.

Hydraulic fracturing: The opening of fractures in a reservoir by high-pressure, high-volume injection of liquids through an injection well.

Hydrocarbonaceous material: A material such as bitumen that is composed of carbon and hydrogen with other elements (heteroelements) such as nitrogen, oxygen, sulfur, and metals chemically combined within the structures of the constituents; even though carbon and hydrogen may be the predominant elements, there may be very few true hydrocarbons.

Hydrocarbon compounds: Chemical compounds containing only carbon and hydrogen.

Hydrocarbon-producing resource: A resource such as coal and oil shale (kerogen) that produces derived hydrocarbons by the application of conversion processes; the hydrocarbons so produced are not naturally occurring materials.

Hydrocarbon resource: Resources such as petroleum and natural gas that can produce naturally occurring hydrocarbons without the application of conversion processes.

Hydrocarbons: Organic compounds containing only hydrogen and carbon.

Hydroconversion: A term often applied to hydrocracking.

Hydrocracking: A catalytic high-pressure high-temperature process for the conversion of petroleum feedstocks in the presence of fresh and recycled hydrogen; carbon–carbon bonds are cleaved in addition to the removal of heteroatomic species.

Hydrocracking catalyst: A catalyst used for hydrocracking that typically contains separate hydrogenation and cracking functions.

Hydrodenitrogenation: The removal of nitrogen by hydrotreating.

Hydrodesulfurization: The removal of sulfur by hydrotreating.

Hydrofining: A fixed-bed catalytic process to desulfurize and hydrogenate a wide range of charge stocks from gases through waxes.

Hydroforming: A process in which naphtha is passed over a catalyst at elevated temperatures and moderate pressures, in the presence of added hydrogen or hydrogen-containing gases, to form high-octane motor fuel or aromatics.

Hydrogen addition: An upgrading process in the presence of hydrogen, for example, hydrocracking; *see* Hydrogenation.

Hydrogen blistering: Blistering of steel caused by trapped molecular hydrogen formed as atomic hydrogen during the corrosion of steel by hydrogen sulfide.

Hydrogenation: The chemical addition of hydrogen to a material. In nondestructive hydrogenation, hydrogen is added to a molecule only if, and where, unsaturation with respect to hydrogen exists.

Hydrogen transfer: The transfer of inherent hydrogen within the feedstock constituents and products during processing.

Hydrolysis: A chemical reaction in which water reacts with another substance to form one or more new substances.

Hydroprocesses: Refinery processes designed to add hydrogen to various products of refining.

Hydroprocessing: A term often equally applied to hydrotreating and to hydrocracking; also often collectively applied to both.

Hydropyrolysis: A short residence time high-temperature process using hydrogen.

Hydrotreating: The removal of heteroatomic (nitrogen, oxygen, and sulfur) species by the treatment of a feedstock or product at relatively low temperatures in the presence of hydrogen.

Hydrovisbreaking: A noncatalytic process, conducted under similar conditions to visbreaking, which involves treatment with hydrogen to reduce the viscosity of the feedstock and produce more stable products than is possible with visbreaking.

Hyperforming: A catalytic hydrogenation process for improving the octane number of naphtha through the removal of sulfur and nitrogen compounds.

Hypochlorite sweetening: The oxidation of mercaptan derivatives (RSH) in a sour feedstock by agitation with aqueous, alkaline hypochlorite solution; used where avoidance of free-sulfur addition is desired, because of a stringent copper strip requirement and minimum expense is not the primary object.

Idle cropland: Land in which no crops were planted; acreage diverted from crops to soil-conserving uses (if not eligible for and used as cropland pasture) under federal farm programs is included in this component.

Ignitability: Characteristic of liquids whose vapors are likely to ignite in the presence of ignition source; also characteristic of nonliquids that may catch fire from friction or contact with water and that burn vigorously.

Illuminating oil: Oil used for lighting purposes.

Immiscible: Two or more fluids that do not have complete mutual solubility and coexist as separate phases.

Immiscible carbon dioxide displacement: Injection of carbon dioxide into an oil reservoir to affect oil displacement under conditions in which miscibility with reservoir oil is not obtained; *see* Carbon dioxide–augmented waterflooding.

Immiscible displacement: A displacement of oil by a fluid (gas or water) that is conducted under conditions so that interfaces exist between the driving fluid and the oil.

Immunoassay: Portable tests that take advantage of an interaction between an antibody and a specific analyte. Immunoassay tests are semiquantitative and usually rely on color changes of varying intensities to indicate relative concentrations.

Incinerator: Any device used to burn solid or liquid residues or wastes as a method of disposal.

Inclined grate: A type of furnace in which fuel enters at the top part of a grate in a continuous ribbon, passes over the upper drying section where moisture is removed, and descends into the lower burning section. Ash is removed at the lower part of the grate.

Incompatibility: The *immiscibility* of petroleum products and also of different crude oils, which is often reflected in the formation of a separate phase after mixing and/or storage.

Incremental ultimate recovery: The difference between the quantity of oil that can be recovered by EOR methods and the quantity of oil that can be recovered by conventional recovery methods.

Indirect injection engine: An older model of diesel engine in which fuel is injected into a pre-chamber, partly combusted, and then sent to the fuel-injection chamber.

Indirect liquefaction: Conversion of biomass to a liquid fuel through a synthesis gas intermediate step.

Industrial wood: All commercial round wood products except fuel wood.

Infill drilling: Drilling additional wells within an established pattern.

Infrared spectroscopy: An analytical technique that quantifies the vibration (stretching and bending) that occurs when a molecule absorbs (heat) energy in the infrared region of the electromagnetic spectrum.

Inhibitor: A substance, the presence of which, in small amounts, in a petroleum product prevents or retards undesirable chemical changes from taking place in the product, or in the condition of the equipment in which the product is used.

Inhibitor sweetening: A treating process to sweeten gasoline of low mercaptan content, using a phenylenediamine type of inhibitor, air, and caustic.

Initial boiling point: The recorded temperature when the first drop of liquid falls from the end of the condenser.

Initial vapor pressure: The vapor pressure of a liquid of a specified temperature and zero percent evaporated.

Injection profile: The vertical flow rate distribution of fluid flowing from the wellbore into a reservoir.

Injection well: A well in an oil field used for injecting fluids into a reservoir.

Injectivity: The relative ease with which a fluid is injected into a porous rock.

In situ: In its original place; in the reservoir.

In situ **combustion:** An EOR process consisting of injecting air or oxygen-enriched air into a reservoir under conditions that favor the burning part of the *in situ* petroleum, advancing this burning zone, and recovering oil heated from a nearby producing well.

Instability: The inability of a petroleum product to exist for periods of time without change to the product.

Integrated gasification combine cycle (IGCC): A power plant in which a gasification process provides syngas to a combined cycle under an integrated control system.

Integrity: Maintenance of a slug or bank at its preferred composition without too much dispersion or mixing.

Interface: The thin surface area separating two immiscible fluids that are in contact with each other.

Interfacial film: A thin layer of material at the interface between two fluids that differs in composition from the bulk fluids.

Interfacial tension: The strength of the film separating two immiscible fluids, for example, oil and water or microemulsion and oil; measured in dynes (force) per centimeter or milli-dynes per centimeter.

Interfacial viscosity: The viscosity of the interfacial film between two immiscible liquids.

Interference testing: A type of pressure transient test in which pressure is measured over time in a closed-in well while nearby wells are produced; flow and communication between wells can sometimes be deduced from an interference test.

Interphase mass transfer: The net transfer of chemical compounds between two or more phases.

Iodine number: A measure of the iodine absorption by oil under standard conditions; used to indicate the quantity of unsaturated compounds present; also called iodine value.

Ion exchange: A means of removing cations or anions from solution onto a solid resin.

Ion-exchange capacity: A measure of the capacity of a mineral to exchange ions in amount of material per unit weight of solid.

Ions: Chemical substances possessing positive or negative charges in solution.

Isocracking: A hydrocracking process for conversion of hydrocarbons, which operates at relatively low temperatures and pressures in the presence of hydrogen and a catalyst to produce more valuable, lower-boiling products.

Isoforming: A process in which olefinic naphtha is contacted with an alumina catalyst at high temperature and low pressure to produce isomers of higher octane number.

Iso-Kel process: A fixed-bed, vapor-phase isomerization process using a precious metal catalyst and external hydrogen.

Isomate process: A continuous, nonregenerative process for isomerizing C_5–C_8 normal paraffin hydrocarbons, using aluminum chloride–hydrocarbon catalyst with anhydrous hydrochloric acid as a promoter.

Isomerate process: A fixed-bed isomerization process to convert pentane, heptane, and heptane to high-octane blending stocks.

Isomerization: The conversion of a *normal* (straight-chain) paraffin hydrocarbon into an *iso* (branched-chain) paraffin hydrocarbon having the same atomic composition.

Isopach: A line on a map designating points of equal formation thickness.

***Iso*-plus Houdriforming:** A combination process using a conventional Houdriformer operated at moderate severity, in conjunction with one of three possible alternatives—including the use of an aromatic recovery unit or a thermal reformer; *see* Houdriforming.

Jet fuel: Fuel meeting the required properties for use in jet engines and aircraft turbine engines.

Joule: Metric unit of energy, equivalent to the work done by a force of 1 N applied over a distance of 1 m (= 1 kg m²/s²). One joule (J) = 0.239 calories (1 calorie = 4.187 J).

Kaolinite: A clay mineral formed by hydrothermal activity at the time of rock formation or by chemical weathering of rock with high feldspar content; usually associated with intrusive granite rock with high feldspar content.

Kata-condensed aromatic compounds: Compounds based on linear condensed aromatic hydrocarbon systems, for example, anthracene and naphthacene (tetracene).

Kauri-butanol number: A measurement of solvent strength for hydrocarbon solvents; the higher the kauri-butanol (KB) value, the stronger the solvency; the test method (ASTM D1133) is based on the principle that kauri resin is readily soluble in butyl alcohol but not in hydrocarbon solvents and the resin solution will tolerate only a certain amount of dilution and is reflected as a cloudiness when the resin starts to come out of solution; solvents such as toluene can be added in a greater amount (and thus have a higher KB value) than weaker solvents like hexane.

Kerogen: A complex carbonaceous (organic) material that occurs in sedimentary rock and shale; generally insoluble in common organic solvents.

Kerosene (kerosine): A fraction of petroleum that was initially sought as an illuminant in lamps; a precursor to diesel fuel.

K-factor: *See* Characterization factor.

Kilowatt (kW): A measure of electrical power equal to 1000 W. 1 kW = 3412 Btu/h = 1.341 horsepower.

Kilowatt hour (kWh): A measure of energy equivalent to the expenditure of 1 kW for 1 hour. For example, 1 kWh will light a 100 W light bulb for 10 hours. 1 kWh = 3412 Btu.

Kinematic viscosity: The ratio of viscosity to density, both measured at the same temperature.

Knock: The noise associated with self-ignition of a portion of the fuel–air mixture ahead of the advancing flame front.

Kriging: A technique used in reservoir description for the interpolation of reservoir parameters between wells based on random field theory.

LAER: Lowest achievable emission rate; the required emission rate in nonattainment permits.

Lamp burning: A test of burning oils in which the oil is burned in a standard lamp under specified conditions in order to observe the steadiness of the flame, the degree of encrustation of the wick, and the rate of consumption of the kerosene.

Lamp oil: *See* Kerosene

Landfill gas: A type of biogas that is generated by the decomposition of organic material at landfill disposal sites. Landfill gas is approximately 50% methane. *See also* Biogas.

Leaded gasoline: Gasoline containing tetraethyl lead or other organometallic lead antiknock compounds.

Lean gas: The residual gas from the absorber after the condensable gasoline has been removed from the wet gas.

Lean oil: Absorption oil from which gasoline fractions have been removed; oil leaving the stripper in a natural gasoline plant.

Lewis acid: A chemical species that can accept an electron pair from a base.

Lewis base: A chemical species that can donate an electron pair.

Light ends: The lower-boiling components of a mixture of hydrocarbons; *see also* Heavy ends, Light hydrocarbons.

Light hydrocarbons: Hydrocarbons with molecular weights less than that of heptane (C_7H_{16}).

Light oil: The products distilled or processed from crude oil up to, but not including, the first lubricating oil distillate.

Light petroleum: Petroleum having an API gravity greater than 20°.

Lignin: Structural constituent of wood and (to a lesser extent) other plant tissues, which encrusts the walls and cements the cells together.

Ligroine (Ligroin): A saturated petroleum naphtha boiling in the range of 20°C–135°C (68°F–275°F) and suitable for general use as a solvent; also called benzine or petroleum ether.

Linde copper sweetening: A process for treating gasoline and distillates with a slurry of clay and cupric chloride.

Liquefied petroleum gas: Propane, butane, or mixtures thereof, gaseous at atmospheric temperature and pressure, held in the liquid state by pressure to facilitate storage, transport, and handling.

Liquid chromatography: A chromatographic technique that employs a liquid mobile phase.

Liquid/liquid extraction: An extraction technique in which one liquid is shaken with or contacted by an extraction solvent to transfer molecules of interest into the solvent phase.

Liquid petrolatum: *See* White oil

Liquid sulfur dioxide–benzene process: A mixed-solvent process for treating lubricating oil stocks to improve viscosity index; also used for dewaxing.

Lithology: The geological characteristics of the reservoir rock.

Live cull: A classification that includes live cull trees; when associated with volume, it is the net volume in live cull trees that are 5.0 inches in diameter and larger.

Live steam: Steam coming directly from a boiler before being utilized for power or heat.

Liver: The intermediate layer of dark-colored, oily material, insoluble in weak acid and in oil, which is formed when acid sludge is hydrolyzed.

Logging residues: The unused portions of growing-stock and non-growing-stock trees cut or killed logging and left in the woods.

Lorenz coefficient: A permeability heterogeneity factor.

Lower-phase microemulsion: A microemulsion phase containing a high concentration of water that, when viewed in a test tube, resides near the bottom with oil phase on top.

Lube: *See* Lubricating oil

Lube cut: A fraction of crude oil of suitable boiling range and viscosity to yield lubricating oil when completely refined; also referred to as lube oil distillates or lube stock.

Lubricating oil: A fluid lubricant used to reduce friction between bearing surfaces.

M85: An alcohol fuel mixture containing 85% methanol and 15% gasoline by volume. Methanol is typically made from natural gas but can also be derived from the fermentation of biomass.

MACT: Maximum achievable control technology. Applies to major sources of hazardous air pollutants.

Mahogany acids: Oil-soluble sulfonic acids formed by the action of sulfuric acid on petroleum distillates. They may be converted to their sodium soaps (mahogany soaps) and extracted from the oil with alcohol for use in the manufacture of soluble oils, rust preventives, and special greases. The calcium and barium soaps of these acids are used as detergent additives in motor oils; *see also* Brown acids and Sulfonic acids.

Major source: A source that has a potential to emit for a regulated pollutant that is at or greater than an emission threshold set by regulations.

Maltene fraction (maltenes): That fraction of petroleum that is soluble in, for example, pentane or heptane; deasphalted oil; also the term arbitrarily assigned to the pentane-soluble portion of petroleum that is relatively high boiling (>300°C, 760 mm) (*see also* Petrolenes).

Marine engine oil: Oil used as a crankcase oil in marine engines.

Marine gasoline: Fuel for motors in marine service.

Marine sediment: The organic biomass from which petroleum is derived.

Marsh: An area of spongy waterlogged ground with large numbers of surface water pools. Marshes usually result from (1) an impermeable underlying bedrock; (2) surface deposits of glacial boulder clay; (3) a basin-like topography from which natural drainage is poor; (4) very heavy rainfall in conjunction with a correspondingly low evaporation rate; (5) low-lying land, particularly at estuarine sites at or below sea level.

Marx–Langenheim model: Mathematical equations for calculating heat transfer in a hot water or steamflood.

Mass spectrometer: An analytical technique that *fractures* organic compounds into characteristic "fragments" based on functional groups that have a specific mass-to-charge ratio.

Mayonnaise: Low-temperature sludge; a black, brown, or gray deposit having a soft, mayonnaise-like consistency; not recommended as a food additive!

MCL: Maximum contaminant level as dictated by regulations.

Medicinal oil: Highly refined, colorless, tasteless, and odorless petroleum oil used as a medicine in the nature of an internal lubricant; sometimes called liquid paraffin.

Megawatt (MW): A measure of electrical power equal to one million watts (1000 kW).

Membrane technology: Gas separation processes utilizing membranes that permit different components of a gas to diffuse through the membrane at significantly different rates.

MDL: *See* Method detection limit

MEK (methyl ethyl ketone): A colorless liquid ($CH_3COCH_2CH_3$) used as a solvent; as a chemical intermediate; and in the manufacture of lacquers, celluloid, and varnish removers.

MEK deoiling: A wax-deoiling process in which the solvent is generally a mixture of methyl ethyl ketone and toluene.

MEK dewaxing: A continuous solvent dewaxing process in which the solvent is generally a mixture of methyl ethyl ketone and toluene.

MEOR: Microbial enhanced oil recovery

Mercapsol process: A regenerative process for extracting mercaptan derivatives (RSH), utilizing aqueous sodium (or potassium) hydroxide containing mixed cresols as solubility promoters.

Mercaptans: Organic compounds having the general formula R-SH.

Metagenesis: The alteration of organic matter during the formation of petroleum that may involve temperatures above 200°C (390°F); *see also* Catagenesis and Diagenesis.

Methanol: *See* Methyl alcohol.

Method detection limit: The smallest quantity or concentration of a substance that the instrument can measure.

Methyl alcohol (methanol; wood alcohol): A colorless, volatile, inflammable, and poisonous alcohol (CH_3OH) traditionally formed by destructive distillation of wood or, more recently, as a result of synthetic distillation in chemical plants; a fuel typically derived from natural gas, but which can be produced from the fermentation of sugars in biomass.

Methyl ethyl ketone: *See* MEK

Methyl t-butyl ether: An ether added to gasoline to improve its octane rating and to decrease gaseous emissions; *see* Oxygenate.

Mica: A complex aluminum silicate mineral that is transparent, tough, flexible, and elastic.

Micellar fluid (surfactant slug): An aqueous mixture of surfactants, cosurfactants, salts, and hydrocarbons. The term *micellar* is derived from the word "micelle," which is a submicroscopic aggregate of surfactant molecules and associated fluid.

Micelle: The structural entity by which asphaltene constituents are dispersed in petroleum.

Microcarbon residue: The carbon residue determined using a thermogravimetric method. *See also* Carbon residue.

Microcrystalline wax: Wax extracted from certain petroleum residua and having a finer and less apparent crystalline structure than paraffin wax.

Microemulsion: A stable, finely dispersed mixture of oil, water, and chemicals (surfactants and alcohols).

Microemulsion or micellar/emulsion flooding: An augmented waterflooding technique in which a surfactant system is injected in order to enhance oil displacement toward producing wells.

Microorganisms: Animals or plants of microscopic size, such as bacteria.

Microscopic displacement efficiency: The efficiency with which an oil displacement process removes the oil from individual pores in the rock.

Mid-boiling point: The temperature at which approximately 50% of a material has distilled under specific conditions.

Middle distillate: Distillate boiling between the kerosene and lubricating oil fractions.

Middle-phase microemulsion: A microemulsion phase containing a high concentration of both oil and water that, when viewed in a test tube, resides in the middle with the oil phase above it and the water phase below it.

Migration (primary): The movement of hydrocarbons (oil and natural gas) from mature, organic-rich source rocks to a point where the oil and gas can collect as droplets or as a continuous phase of liquid hydrocarbon.

Migration (secondary): The movement of the hydrocarbons as a single, continuous fluid phase through water-saturated rocks, fractures, or faults followed by the accumulation of the oil and gas in sediments (traps, *q.v.*) from which further migration is prevented.

Mill residue: Wood and bark residues produced in processing logs into lumber, plywood, and paper.

Mineral hydrocarbons: Petroleum hydrocarbons, considered *mineral* because they come from the earth rather than from plants or animals.

Mineral oil: The older term for petroleum; the term was introduced in the nineteenth century as a means of differentiating petroleum (rock oil) from whale oil which, at the time, was the predominant illuminant for oil lamps.

Minerals: Naturally occurring inorganic solids with well-defined crystalline structures.

Mineral seal oil: A distillate fraction boiling between kerosene and gas oil.

Mineral wax: Yellow to dark brown, solid substances that occur naturally and are composed largely of paraffins; usually found associated with considerable mineral matter, as a filling in veins and fissures or as an interstitial material in porous rocks.

Minimum miscibility pressure (MMP): *See* Miscibility

Miscibility: An equilibrium condition, achieved after mixing two or more fluids, which is characterized by the absence of interfaces between the fluids: (1) *First-contact miscibility:* miscibility in the usual sense, whereby two fluids can be mixed in all proportions without any interfaces forming. For example, at room temperature and pressure, ethyl alcohol and water are first-contact miscible. (2) *Multiple-contact miscibility (dynamic miscibility):* miscibility that is developed by repeated enrichment of one fluid phase with components from a second fluid phase with which it comes into contact. (3) *Minimum miscibility* pressure: the minimum pressure above which two fluids become miscible at a given temperature, or can become miscible, by dynamic processes.

Miscible flooding: *See* EOR process

Miscible fluid displacement (miscible displacement): Is an oil displacement process in which an alcohol, a refined hydrocarbon, a condensed petroleum gas, carbon dioxide, liquefied natural gas, or even exhaust gas are injected into an oil reservoir, at pressure levels such that the injected gas or fluid and reservoir oil are miscible; the process may include the concurrent, alternating, or subsequent injection of water.

Mitigation: Identification, evaluation, and cessation of potential impacts of a process product or by-product.

Mixed-phase cracking: The thermal decomposition of higher-boiling hydrocarbons to gasoline components.

Mobility: A measure of the ease with which a fluid moves through reservoir rock; the ratio of rock permeability to apparent fluid viscosity.

Mobility buffer: The bank that protects a chemical slug from water invasion and dilution and assures mobility control.

Mobility control: Ensuring that the mobility of the displacing fluid or bank is equal to or less than that of the displaced fluid or bank.

Mobility ratio: The ratio of mobility of an injection fluid to the mobility of fluid being displaced.

Modified alkaline flooding: The addition of a cosurfactant and/or polymer to the alkaline flooding process.

Modified/unmodified diesel engine: Traditional diesel engines must be modified to heat the oil before it reaches the fuel injectors in order to handle straight vegetable oil. Modified, any diesel engine can run on veggie oil; without modification, the oil must first be converted to biodiesel.

Modified naphtha insolubles (MNI): An insoluble fraction obtained by adding naphtha to petroleum; usually, the naphtha is modified by adding paraffin constituents; the fraction might be equated to asphaltenes if the naphtha is equivalent to *n*-heptane, but usually it is not.

Moisture content (MC): The weight of the water contained in wood, usually expressed as a percentage of weight, either oven-dry or as received.

Moisture content, dry basis: Moisture content expressed as a percentage of the weight of oven wood, that is, [(weight of wet sample – weight of dry sample)/weight of dry sample] × 100.

Moisture content, wet basis: Moisture content expressed as a percentage of the weight of wood as-received, that is, [(weight of wet sample – weight of dry sample)/weight of wet sample] × 100.

Molecular sieve: A synthetic zeolite mineral having pores of uniform size; it is capable of separating molecules, on the basis of their size, structure, or both, by absorption or sieving.

Motor Octane Method: A test for determining the knock rating of fuels for use in spark-ignition engines; *see also* Research Octane Method.

Moving-bed catalytic cracking: A cracking process in which the catalyst is continuously cycled between the reactor and the regenerator.

MSDS: Material safety data sheet

MTBE: Methyl tertiary butyl ether is highly refined high-octane light distillate used in the blending of gasoline.

NAAQS: National Ambient Air Quality Standards; standards exist for the pollutants known as the criteria air pollutants: nitrogen oxides (NO_x), sulfur oxides (SO_x), lead, ozone, particulate matter, less than 10 μm in diameter, and carbon monoxide (CO).

Naft: Pre-Christian era (Greek) term for naphtha.

Napalm: A thickened gasoline used as an incendiary medium that adheres to the surface it strikes.

Naphtha: A generic term applied to refined, partly refined, or unrefined petroleum products and liquid products of natural gas, the majority of which distills below 240°C (464°F); the volatile fraction of petroleum which is used as a solvent or as a precursor to gasoline.

Naphthenes: Cycloparaffins.

Native asphalt: *See* Bitumen

Natural asphalt: *See* Bitumen

Natural gas: The naturally occurring gaseous constituents that are found in many petroleum reservoirs; there are also those reservoirs in which natural gas may be the sole occupant.

Natural gas liquids (NGL): The hydrocarbon liquids that condense during the processing of hydrocarbon gases that are produced from oil or gas reservoir; *see also* Natural gasoline.

Natural gasoline: A mixture of liquid hydrocarbons extracted from natural gas suitable for blending with refinery gasoline.

Natural gasoline plant: A plant for the extraction of fluid hydrocarbon, such as gasoline and liquefied petroleum gas, from natural gas.

NESHAP: National Emissions Standards for Hazardous Air Pollutants; emission standards for specific source categories that emit or have the potential to emit one or more hazardous air pollutants; the standards are modeled on the best practices and most effective emission reduction methodologies in use at the affected facilities.

Neutralization: A process for reducing the acidity or alkalinity of a waste stream by mixing acids and bases to produce a neutral solution; also known as pH adjustment.

Neutral oil: A distillate lubricating oil with viscosity usually not above 200 seconds at 100°F.

Neutralization number: The weight, in milligrams, of potassium hydroxide needed to neutralize the acid in 1 g of oil; an indication of the acidity of an oil.

Nitrogen fixation: The transformation of atmospheric nitrogen into nitrogen compounds that can be used by growing plants.

Nitrogen oxides (NOx): Products of combustion that contribute to the formation of smog and ozone.

Nonasphaltic road oil: Any of the nonhardening petroleum distillates or residual oils used as dust layers. They have sufficiently low viscosity to be applied without heating and, together with asphaltic road oils, are sometimes referred to as dust palliatives.

Nonattainment area: Any area that does not meet the national primary or secondary ambient air quality standard established (by the Environmental Protection Agency) for designated pollutants, such as carbon monoxide and ozone.

Non-forestland: Land that has never supported forests and lands formerly forested where the use of timber management is precluded by the development for other uses; if intermingled in forest areas, unimproved roads and nonforest strips must be more than 120 ft wide, and clearings must be more than 1 acre in area to qualify as non-forestland.

Nonindustrial private: An ownership class of private lands where the owner does not operate wood-processing plants.

Nonionic surfactant: A surfactant molecule containing no ionic charge.

Non-Newtonian: A fluid that exhibits a change of viscosity with flow rate.

NOx: Oxides of nitrogen.

Nuclear magnetic resonance spectroscopy: An analytical procedure that permits the identification of complex molecules based on the magnetic properties of the atoms they contain.

No. 1 Fuel oil: Very similar to kerosene and is used in burners where vaporization before burning is usually required and a clean flame is specified.

No. 2 Fuel oil: Also called domestic heating oil; has properties similar to diesel fuel and heavy jet fuel; used in burners where complete vaporization is not required before burning.

No. 4 Fuel oil: A light industrial heating oil and is used where preheating is not required for handling or burning; there are two grades of No. 4 fuel oil, differing in safety (flash point) and flow (viscosity) properties.

No. 5 Fuel oil: A heavy industrial fuel oil that requires preheating before burning.

No. 6 Fuel oil: A heavy fuel oil and is more commonly known as Bunker C oil when it is used to fuel oceangoing vessels; preheating is always required for burning this oil.

Observation wells: Wells that are completed and equipped to measure reservoir conditions and/or sample reservoir fluids, rather than to inject produced reservoir fluids.

Octane barrel yield: A measure used to evaluate fluid catalytic cracking processes; defined as (RON + MON)/2 times the gasoline yield, where RON is the research octane number and MON is the motor octane number.

Octane number: A number indicating the antiknock characteristics of gasoline.

Oil bank: *See* Bank

Oil breakthrough (time): The time at which the oil–water bank arrives at the producing well.

Oil from tar sand: Synthetic crude oil.

Oil mining: Application of a mining method to the recovery of bitumen.

Oil originally in place (OOIP): The quantity of petroleum existing in a reservoir before oil recovery operations begin.

Oils: That portion of the maltenes that is not adsorbed by a surface-active material such as clay or alumina.

Oil sand: *See* Tar sand

Oil shale: A fine-grained impervious sedimentary rock that contains an organic material called kerogen.

Olefin: Synonymous with *alkene*.

OOIP: *See* Oil originally in place

Open-loop biomass: Biomass that can be used to produce energy and bioproducts even though it was not grown specifically for this purpose; include agricultural livestock waste, residues from forest harvesting operations, and crop harvesting.

Optimum salinity: The salinity at which a middle-phase microemulsion containing equal concentrations of oil and water results from the mixture of a micellar fluid (surfactant slug) with oil.

Organic sedimentary rocks: Rocks containing organic material such as residues of plant and animal remains/decay.

Overhead: That portion of the feedstock that is vaporized and removed during distillation.

Override: The gravity-induced flow of a lighter fluid in a reservoir above another heavier fluid.

Oxidation: A process that can be used for the treatment of a variety of inorganic and organic substances.

Oxidized asphalt: *See* Air-blown asphalt

Oxygenate: An oxygen-containing compound that is blended into gasoline to improve its octane number and to decrease gaseous emissions; a substance that, when added to gasoline, increases the amount of oxygen in that gasoline blend; includes fuel ethanol, methanol, and methyl tertiary butyl ether (MTBE).

Oxygenated gasoline: Gasoline with added ethers or alcohols, formulated according to the Federal Clean Air Act to reduce carbon monoxide emissions during winter months.

Oxygen scavenger: A chemical that reacts with oxygen in injection water, used to prevent the degradation of polymer.

Ozokerite (ozocerite): A naturally occurring wax; when refined also known as ceresin.

Pale oil: A lubricating oil or a process oil refined until its color, by transmitted light, is straw to pale yellow.

Paraffinum liquidum: *See* Liquid petrolatum.

Paraffin wax: The colorless, translucent, highly crystalline material obtained from the light lubricating fractions of paraffin crude oils (wax distillates).

Particle density: The density of solid particles.

Particle size distribution: The particle size distribution (of a catalyst sample) expressed as a percent of the whole.

Particulate: A small, discrete mass of solid or liquid matter that remains individually dispersed in gas or liquid emissions.

Particulate emissions: Particles of a solid or liquid suspended in a gas, or the fine particles of carbonaceous soot and other organic molecules discharged into the air during combustion.

Particulate matter (particulates): Particles in the atmosphere or on a gas stream that may be organic or inorganic and originate from a wide variety of sources and processes.

Partitioning: In chromatography, the physical act of a solute having different affinities for the stationary and mobile phases.

Partition ratios, K: The ratio of total analytical concentration of a solute in the stationary phase, CS, to its concentration in the mobile phase, CM.

Pattern: The areal pattern of injection and producing wells selected for a secondary or enhanced recovery project.

Pattern life: The length of time a flood pattern participates in oil recovery.

Pay zone thickness: The depth of a tar sand deposit from which bitumen (or a product) can be recovered.

Penex process: A continuous, nonregenerative process for the isomerization of C_5 and/or C_6 fractions in the presence of hydrogen (from reforming) and a platinum catalyst.

Pentafining: A pentane isomerization process using a regenerable platinum catalyst on a silica-alumina support and requiring outside hydrogen.

Pepper sludge: The fine particles of sludge produced in acid treating, which may remain in suspension.

Peri-condensed aromatic compounds: Compounds based on angular condensed aromatic hydrocarbon systems, for example, phenanthrene, chrysene, and picene.

Permeability: The ease of flow of the water through the rock.

Petrol: A term commonly used in some countries for gasoline.

Petrolatum: A semisolid product, ranging from white to yellow in color, produced during refining of residual stocks; *see* Petroleum jelly.

Petrolenes: The term applied to that part of the pentane-soluble or heptane-soluble material that is low boiling (<300°C, <570°F, 760 mm) and can be distilled without thermal decomposition (*see also* Maltenes).

Petroleum (crude oil): A naturally occurring mixture of gaseous, liquid, and solid hydrocarbon compounds usually found trapped deep underground beneath impermeable cap rock and above a lower dome of sedimentary rock such as shale; most petroleum reservoirs occur in sedimentary rocks of marine, deltaic, or estuarine origin.

Petroleum asphalt: *See* Asphalt

Petroleum ether: *See* Ligroine

Petroleum jelly: A translucent, yellowish to amber or white, hydrocarbon substance (melting point: 38°C–54°C) having almost no odor or taste, derived from petroleum and used principally in medicine and pharmacy as a protective dressing and as a substitute for fats in ointments and cosmetics; also used in many types of polishes and in lubricating greases, rust preventives, and modeling clay; obtained by dewaxing heavy lubricating oil stocks.

Petroleum refinery: *See* Refinery

Petroleum refining: A complex sequence of events that result in the production of a variety of products.

Petroleum sulfonate: A surfactant used in chemical flooding prepared by sulfonating selected crude oil fractions.

Petroporphyrins: *See* Porphyrins.

pH adjustment: Neutralization.

Phase: A separate fluid that coexists with other fluids; gas, oil, water, and other stable fluids such as microemulsions are all called phases in EOR research.

Phase behavior: The tendency of a fluid system to form phases as a result of changing temperature, pressure, or the bulk composition of the fluids or of individual fluid phases.

Phase diagram: A graph of phase behavior. In chemical flooding, a graph showing the relative volume of oil, brine, and sometimes one or more microemulsion phases.

Phase properties: Types of fluids, compositions, densities, viscosities, and relative amounts of oil, microemulsion, or solvent, and water formed when a micellar fluid (surfactant slug) or miscible solvent (e.g., CO_2) is mixed with oil.

Phase separation: The formation of a separate phase that is usually the prelude to coke formation during a thermal process; the formation of a separate phase as a result of the instability/incompatibility of petroleum and petroleum products.

Phosphoric acid polymerization: A process using a phosphoric acid catalyst to convert propene, butene, or both, to gasoline or petrochemical polymers.

Photoionization: A gas chromatographic detection system that utilizes a *detector (PID)* ultraviolet lamp as an ionization source for analyte detection. It is usually used as a selective detector by changing the photon energy of the ionization source.

Photosynthesis: Process by which chlorophyll-containing cells in green plants concert incident light to chemical energy, capturing carbon dioxide in the form of carbohydrates.

PINA analysis: A method of analysis for paraffins, *iso*-paraffins, naphthenes, and aromatics.

PIONA analysis: A method of analysis for paraffins, *iso*-paraffins, olefins, naphthenes, and aromatics.

Pipe still: A still in which heat is applied to the oil while being pumped through a coil or pipe arranged in a suitable firebox.

Pipe still gas: The most volatile fraction that contains most of the gases that are generally dissolved in the crude. Also known as pipe still light ends.

Pipe still light ends: *See* Pipe still gas

Pitch: The nonvolatile, brown to black, semisolid to solid viscous product from the destructive distillation of many bituminous or other organic materials, especially coal.

Platforming: A reforming process using a platinum-containing catalyst on an alumina base.

PNA: A polynuclear aromatic compound.

PNA analysis: A method of analysis for paraffins, naphthenes, and aromatics.

Polar aromatics: Resins; the constituents of petroleum that are predominantly aromatic in character and contain polar (nitrogen, oxygen, and sulfur) functions in their molecular structure(s).

Pollutant: A chemical (or chemicals) introduced into the land water and air systems of that is (are) not indigenous to these systems; also an indigenous chemical (or chemicals) introduced into the land water and air systems in amounts greater than the natural abundance.

Pollution: The introduction into the land water and air systems of a chemical or chemicals that are not indigenous to these systems or the introduction into the land water and air systems of indigenous chemicals in greater-than-natural amounts.

Polyacrylamide: Very high-molecular-weight material used in polymer flooding.

Polycyclic aromatic hydrocarbons (PAHs): Polycyclic aromatic hydrocarbons are a suite of compounds comprising two or more condensed aromatic rings. They are found in many petroleum mixtures, and they are predominantly introduced to the environment through natural and anthropogenic combustion processes.

Polyforming: A process charging both C_3 and C_4 gases with naphtha or gas oil under thermal conditions to produce gasoline.

Polymer: In EOR, any very high-molecular-weight material that is added to water to increase viscosity for polymer flooding.

Polymer augmented waterflooding: Waterflooding in which organic polymers are injected with the water to improve areal and vertical sweep efficiency.

Polymer gasoline: The product of polymerization of gaseous hydrocarbons to hydrocarbons boiling in the gasoline range.

Polymerization: The combination of two olefin molecules to form a higher-molecular-weight paraffin.

Polymer stability: The ability of a polymer to resist degradation and maintain its original properties.

Polynuclear aromatic compound: An aromatic compound having two or more fused benzene rings, for example, naphthalene and phenanthrene.

Polysulfide treating: A chemical treatment used to remove elemental sulfur from refinery liquids by contacting them with a nonregenerable solution of sodium polysulfide.

PONA analysis: A method of analysis for paraffins (P), olefins (O), naphthenes (N), and aromatics (A).

Pore diameter: The average pore size of a solid material, for example, catalyst.

Pore space: A small hole in reservoir rock that contains fluid or fluids; a 4-inch cube of reservoir rock may contain millions of interconnected pore spaces.

Pore volume: Total volume of all pores and fractures in a reservoir or part of a reservoir; also applied to catalyst samples.

Porosity: The percentage of rock volume available to contain water or other fluid.

Porphyrins: Organometallic constituents of petroleum that contain vanadium or nickel; the degradation products of chlorophyll that became included in the protopetroleum.

Positive bias: A result that is incorrect and too high.

Possible reserves: Reserves where there is an even greater degree of uncertainty but about which there is some information.

Potential reserves: Reserves based upon geological information about the types of sediments where such resources are likely to occur and they are considered to represent an educated guess.

Pour point: The lowest temperature at which oil will pour or flow when it is chilled without disturbance under definite conditions.

Powerforming: A fixed-bed naphtha-reforming process using a regenerable platinum catalyst.

Power-law exponent: An exponent used to model the degree of viscosity change of some non-Newtonian liquids.

Precipitation number: The number of milliliters of precipitate formed when 10 mL of lubricating oil is mixed with 90 mL of petroleum naphtha of a definite quality and centrifuged under definitely prescribed conditions.

Preflush: A conditioning slug injected into a reservoir as the first step of an EOR process.

Pressure cores: Cores cut into a special coring barrel that maintains reservoir pressure when brought to the surface; this prevents the loss of reservoir fluids that usually accompanies a drop in pressure from reservoir to atmospheric conditions.

Pressure gradient: Rate of change of pressure with distance.

Pressure maintenance: Augmenting the pressure (and energy) in a reservoir by injecting gas and/or water through one or more wells.

Pressure pulse test: A technique for determining reservoir characteristics by injecting a sharp pulse of pressure in one well and detecting it in surrounding wells.

Pressure transient testing: Measuring the effect of changes in pressure at one well on other well in a field.

Primary oil recovery: Oil recovery utilizing only naturally occurring forces.

Primary wood-using mill: A mill that converts round wood products into other wood products; common examples are sawmills that convert saw logs into lumber and pulp mills that convert pulpwood round wood into wood pulp.

Primary structure: The chemical sequence of atoms in a molecule.

Primary tracer: A chemical that, when inject into a test well, reacts with reservoir fluids form a detectable chemical compound.

Probable reserves: Mineral reserves mineral that are nearly certain but about which a slight doubt exists.

Process heat: Heat used in an industrial process rather than for space heating or other housekeeping purposes.

Producer gas: Fuel gas high in carbon monoxide (CO) and hydrogen (H_2), produced by burning a solid fuel with insufficient air or by passing a mixture of air and steam through a burning bed of solid fuel.

Producibility: The rate at which oil or gas can be produced from a reservoir through a wellbore.

Producing well: A well in an oil field used for removing fluids from a reservoir.

Propane asphalt: *See* Solvent asphalt

Propane deasphalting: Solvent deasphalting using propane as the solvent.

Propane decarbonizing: A solvent extraction process used to recover catalytic cracking feed from heavy fuel residues.

Propane dewaxing: A process for dewaxing lubricating oils in which propane serves as solvent.

Propane fractionation: A continuous extraction process employing liquid propane as the solvent; a variant of propane deasphalting.

Protopetroleum: A generic term used to indicate the initial product formed changes have occurred to the precursors of petroleum.

Proved reserves: Mineral reserves that have been positively identified as recoverable with current technology.

PSD: Prevention of significant deterioration.

PTE: Potential to emit; the maximum capacity of a source to emit a pollutant, given its physical or operation design, and considering certain controls and limitations.

Pulpwood: Round wood, whole-tree chips, or wood residues that are used for the production of wood pulp.

Pulse-echo ultrasonic borehole televiewer: Well-logging system wherein a pulsed, narrow acoustic beam scans the well as the tool is pulled up the borehole; the amplitude of the reflecting beam is displayed on a cathode-ray tube resulting in a pictorial representation of wellbore.

Purge and trap: A chromatographic sample introduction technique in volatile components that are purged from a liquid medium by bubbling gas through it. The components are then concentrated by "trapping" them on a short intermediate column, which is subsequently heated to drive the components onto the analytical column for separation.

Purge gas: Typically helium or nitrogen, used to remove analytes from the sample matrix in purge/trap extractions.

Pyrobitumen: *See* Asphaltoid

Pyrolysis: The thermal decomposition of biomass at high temperatures (greater than 400°F, or 200°C) in the absence of air; the end product of pyrolysis is a mixture of solids (char), liquids (oxygenated oils), and gases (methane, carbon monoxide, and carbon dioxide) with proportions determined by operating temperature, pressure, oxygen content, and other conditions; exposure of a feedstock to high temperatures in an oxygen-poor environment.

Pyrophoric: Substances that catch fire spontaneously in air without an ignition source.

Quad: One quadrillion Btu (10^{15} Btu) = 1.055 exajoules (EJ), or approximately 172 million barrels of oil equivalent.

Quadrillion: 1×10^{15}

Quench: The sudden cooling of hot material discharging from a thermal reactor.

RACT: Reasonably Available Control Technology standards; implemented in areas of nonattainment to reduce emissions of volatile organic compounds and nitrogen oxides.

Raffinate: That portion of the oil that remains undissolved in a solvent refining process.

Ramsbottom carbon residue: *See* Carbon residue.

Raw materials: Minerals extracted from the earth prior to any refining or treating.

Recovery boiler: A pulp mill boiler in which lignin and spent cooking liquor (black liquor) are burned to generate steam.

Recycle ratio: The volume of recycle stock per volume of fresh feed; often expressed as the volume of recycle divided by the total charge.

Recycle stock: The portion of a feedstock that has passed through a refining process and is recirculated through the process.

Recycling: The use or reuse of chemical waste as an effective substitute for a commercial product or as an ingredient or feedstock in an industrial process.

Reduced crude: A residual product remaining after the removal, by distillation or other means, of an appreciable quantity of the more volatile components of crude oil.

Refinery: A series of integrated unit processes by which petroleum can be converted to a slate of useful (salable) products.

Refinery gas: A gas (or a gaseous mixture) produced as a result of refining operations.

Refining: The processes by which petroleum is distilled and/or converted by the application of physical and chemical processes to form a variety of products.

Reformate: The liquid product of a reforming process.

Reformed gasoline: Gasoline made by a reforming process.

Reforming: The conversion of hydrocarbons with low octane numbers into hydrocarbons having higher octane numbers; for example, the conversion of an *n*-paraffin into an *iso*-paraffin.

Reformulated gasoline (RFG): Gasoline designed to mitigate smog production and to improve air quality by limiting the emission levels of certain chemical compounds such as benzene and other aromatic derivatives; often contains oxygenates.

Refractory lining: A lining, usually of ceramic, capable of resisting and maintaining high temperatures.

Refuse-derived fuel (RDF): Fuel prepared from municipal solid waste; noncombustible materials such as rocks, glass, and metals are removed, and the remaining combustible portion of the solid waste is chopped or shredded; the combustible portion of municipal solid waste after the removal of glass and metals.

Reid vapor pressure: A measure of the volatility of liquid fuels, especially gasoline.

Regeneration: The reactivation of a catalyst by burning off the coke deposits.

Regenerator: A reactor for catalyst reactivation.

Relative permeability: The permeability of rock to gas, oil, or water, when any two or more are present, expressed as a fraction of the permeability of the rock.

Renewable energy sources: Solar, wind, and other non–fossil fuel energy sources.

Rerunning: The distillation of an oil that has already been distilled.

Research Octane Method: A test for determining the knock rating, in terms of octane numbers, of fuels for use in spark-ignition engines; *see also* Motor Octane Method.

Reserves: Well-identified resources that can be profitably extracted and utilized with existing technology.

Reservoir: A rock formation below the earth's surface containing petroleum or natural gas; a domain where a pollutant may reside for an indeterminate time.

Reservoir simulation: Analysis and prediction of reservoir performance with a computer model.

Residual asphalt: *See* Straight-run asphalt

Residual fuel oil: Obtained by blending the residual product(s) from various refining processes with suitable diluent(s) (usually middle distillates) to obtain the required fuel oil grades.

Residual oil: *See* Residuum; petroleum remaining *in situ* after oil recovery.

Residual resistance factor: The reduction in permeability of rock to water caused by the adsorption of polymer.

Residues: Bark and woody materials that are generated in primary wood–using mills when round wood products are converted to other products.

Residuum (resid; *pl:* residua): The residue obtained from petroleum after nondestructive distillation has removed all the volatile materials from crude oil, for example, an atmospheric (345°C, 650°F+) residuum.

Resins: That portion of the maltenes that is adsorbed by a surface-active material such as clay or alumina; the fraction of deasphaltened oil that is insoluble in liquid propane but soluble in *n*-heptane.

Resistance factor: A measure of resistance to the flow of a polymer solution relative to the resistance to flow of water.

Resource: The total amount of a commodity (usually a mineral but can include nonminerals such as water and petroleum) that has been estimated to be ultimately available.

Retention: The loss of chemical components due to adsorption onto the rock's surface, precipitation, or to trapping within the reservoir.

Retention time: The time it takes for an eluate to move through a chromatographic system and reach the detector. Retention times are reproducible and can therefore be compared to a standard for analyte identification.

Rexforming: A process combining Platforming with aromatics extraction, wherein low-octane raffinate is recycled to the Platformer.

Rich oil: Absorption oil containing dissolved natural gasoline fractions.

Riser: The part of the bubble-plate assembly that channels the vapor and causes it to flow downward to escape through the liquid; also the vertical pipe where fluid catalytic cracking reactions occur.

Rock asphalt: Bitumen that occurs in formations that have a limiting ratio of bitumen-to-rock matrix.

Rock matrix: The granular structure of a rock or porous medium.

Rotation: Period of years between the establishment of a stand of timber and the time when it is considered ready for final harvest and regeneration.

Round wood products: Logs and other round timber generated from harvesting trees for industrial or consumer use.

Run-of-the-river reservoirs: Reservoirs with a large rate of flow-through compared to their volume.

Salinity: The concentration of salt in water.

Sand: A course granular mineral mainly comprising quartz grains that are derived from the chemical and physical weathering of rocks rich in quartz, notably sandstone and granite.

Sand face: The cylindrical wall of the wellbore through which the fluids must flow to or from the reservoir.

Sandstone: A sedimentary rock formed by compaction and cementation of sand grains; can be classified according to the mineral composition of the sand and cement.

SARA analysis: A method of fractionation by which petroleum is separated into saturates, aromatics, resins, and asphaltene fractions.

SARA separation: *See* SARA analysis

Saturated steam: Steam at boiling temperature for a given pressure.

Saturates: Paraffins and cycloparaffins (naphthenes).

Saturation: The ratio of the volume of a single fluid in the pores to pore volume, expressed as a percent and applied to water, oil, or gas separately; the sum of the saturations of each fluid in a pore volume is 100%.

Saybolt Furol Viscosity: The time, in seconds (Saybolt Furol Seconds, SFS), for 60 mL of fluid to flow through a capillary tube in a Saybolt Furol Viscometer at specified temperatures between 70°F and 210°F; the method is appropriate for high-viscosity oils such as transmission, gear, and heavy fuel oils.

Saybolt Universal Viscosity: The time, in seconds (Saybolt Universal Seconds, SUS), for 60 mL of fluid to flow through a capillary tube in a Saybolt Universal Viscometer at a given temperature.

Scale wax: The paraffin derived by removing the greater part of the oil from slack wax by sweating or solvent deoiling.

Screen factor: A simple measure of the viscoelastic properties of polymer solutions.

Screening guide: A list of reservoir rock and fluid properties critical to an EOR process.

Scrubber: A device that uses water and chemicals to clean air pollutants from combustion exhaust.

Scrubbing: Purifying a gas by washing with water or chemical; less frequently, the removal of entrained materials.

Secondary oil recovery: The application of energy (e.g., waterflooding) to recovery of crude oil from a reservoir after the yield of crude oil from primary recovery diminishes.

Secondary pollutants: A pollutant (chemical species) produced by the interaction of a primary pollutant with another chemical or by the dissociation of a primary pollutant or by other effects within a particular ecosystem.

Secondary recovery: Oil recovery resulting from injection of water, or an immiscible gas at moderate pressure, into a petroleum reservoir after primary depletion.

Secondary structure: The ordering of the atoms of a molecule in space relative to each other.

Secondary tracer: The product of the chemical reaction between reservoir fluids and an injected primary tracer.

Secondary wood-processing mills: A mill that uses primary wood products in the manufacture of finished wood products, such as cabinets, moldings, and furniture.

Sediment: An insoluble solid formed as a result of the storage instability and/or the thermal instability of petroleum and petroleum products.

Sedimentary: Formed by or from deposits of sediments, especially from sand grains or silts transported from their source and deposited in water, as sandstone and shale; or from calcareous remains of organisms, as limestone.

Sedimentary strata: Typically consist of mixtures of clay, silt, sand, organic matter, and various minerals; formed by or from deposits of sediments, especially from sand grains or silts transported from their source and deposited in water, such as sandstone and shale; or from calcareous remains of organisms, such as limestone.

Selective solvent: A solvent that, at certain temperatures and ratios, will preferentially dissolve more of one component of a mixture than of another and thereby permit partial separation.

Separation process: An upgrading process in which the constituents of petroleum are separated, usually without thermal decomposition, for example, distillation and deasphalting.

Separator-Nobel dewaxing: A solvent (tricholoethylene) dewaxing process.

Separatory funnel: Glassware shaped like a funnel with a stoppered rounded top and a valve at the tapered bottom, used for liquid/liquid separations.

Shear: Mechanical deformation or distortion, or partial destruction of a polymer molecule as it flows at a high rate.

Shear rate: A measure of the rate of deformation of a liquid under mechanical stress.

Shear thinning: The characteristic of a fluid whose viscosity decreases as the shear rate increases.

Shell fluid catalytic cracking: A two-stage fluid catalytic cracking process in which the catalyst is regenerated.

Shell still: A still formerly used in which the oil was charged into a closed, cylindrical shell and the heat required for distillation was applied to the outside of the bottom from a firebox.

Sidestream: A liquid stream taken from any one of the intermediate plates of a bubble tower.

Sidestream stripper: A device used to perform further distillation on a liquid stream from any one of the plates of a bubble tower, usually by the use of steam.

Single-well tracer: A technique for determining residual oil saturation by injecting an ester, allowing it to hydrolyze; following dissolution of some of the reaction product in residual oil the injected solutions produced back and analyzed.

Slack wax: The soft, oily crude wax obtained from the pressing of paraffin distillate or wax distillate.

Slime: A name used for petroleum in ancient texts.

Slim tube testing: Laboratory procedure for the determination of minimum miscibility pressure using long, small-diameter, sand-packed, oil-saturated, stainless steel tube.

Sludge: A semisolid to solid product that results from the storage instability and/or the thermal instability of petroleum and petroleum products.

Slug: A quantity of fluid injected into a reservoir during enhanced oil recovery.

Slurry hydroconversion process: A process in which the feedstock is contacted with hydrogen under pressure in the presence of a catalytic coke–inhibiting additive.

Slurry phase reactors: Tanks into which wastes, nutrients, and microorganisms are placed.

Smoke point: A measure of the burning cleanliness of jet fuel and kerosene.

Sodium hydroxide treatment: *See* Caustic wash.

Sodium plumbite: A solution prepared from a mixture of sodium hydroxide, lead oxide, and distilled water; used in making the doctor test for light oils such as gasoline and kerosene.

Solubility parameter: A measure of the solvent power and polarity of a solvent.

Solutizer-steam regenerative process: A chemical treating process for extracting mercaptan derivatives (RSH) from gasoline or naphtha, using solutizers (potassium *iso*-butyrate, potassium alkyl phenolate) in strong potassium hydroxide solution.

Solvent: A liquid in which certain kinds of molecules dissolve. While they typically are liquids with low boiling points, they may include high-boiling liquids, supercritical fluids, or gases.

Solvent asphalt: The asphalt produced by solvent extraction of residua or by light hydrocarbon (propane) treatment of a residuum or an asphaltic crude oil.

Solvent deasphalting: A process for removing asphaltic and resinous materials from reduced crude oils, lubricating oil stocks, gas oils, or middle distillates through the extraction or precipitant action of low-molecular-weight hydrocarbon solvents; *see also* Propane deasphalting.

Solvent decarbonizing: *See* Propane decarbonizing.

Solvent deresining: *See* Solvent deasphalting.

Solvent dewaxing: A process for removing wax from oils by means of solvents usually by chilling a mixture of solvent and waxy oil, filtration or by centrifuging the wax that precipitates, and solvent recovery.

Solvent extraction: A process for separating liquids by mixing the stream with a solvent that is immiscible with part of the waste but that will extract certain components of the waste stream.

Solvent gas: An injected gaseous fluid that becomes miscible with oil under reservoir conditions and improves oil displacement.

Solvent naphtha: A refined naphtha of restricted boiling range used as a solvent; also called petroleum naphtha; petroleum spirits.

Solvent refining: *See* Solvent extraction

Sonication: A physical technique employing ultrasound to intensely vibrate a sample media in extracting solvent and to maximize solvent/analyte interactions.

Sonic log: A well log based on the time required for sound to travel through rock, useful in determining porosity.

Sour crude oil: Crude oil containing an abnormally large amount of sulfur compounds; *see also* Sweet crude oil.

SOx: Oxides of sulfur.

Soxhlet extraction: An extraction technique for solids in which the sample is repeatedly contacted with solvent over several hours, increasing extraction efficiency.

Specific gravity: The mass (or weight) of a unit volume of any substance at a specified temperature compared to the mass of an equal volume of pure water at a standard temperature; *see also* Density.

Spent catalyst: Catalyst that has lost much of its activity due to the deposition of coke and metals.

Spontaneous ignition: Ignition of a fuel, such as coal, under normal atmospheric conditions; usually induced by climatic conditions.

Stabilization: The removal of volatile constituents from a higher-boiling fraction or product (q.v. stripping); the production of a product that, to all intents and purposes, does not undergo any further reaction when exposed to the air.

Stabilizer: A fractionating tower for removing light hydrocarbons from an oil to reduce vapor pressure particularly applied to gasoline.

Stand (of trees): A tree community that possesses sufficient uniformity in composition, constitution, age, spatial arrangement, or condition to be distinguishable from adjacent communities.

Standpipe: The pipe by which catalyst is conveyed between the reactor and the regenerator.

Stationary phase: In chromatography, the porous solid or liquid phase through which an introduced sample passes. The different affinities the stationary phase has for a sample allow the components in the sample to be separated or resolved.

Steam cracking: A conversion process in which the feedstock is treated with superheated steam.

Steam distillation: Distillation in which vaporization of the volatile constituents is affected at a lower temperature by the introduction of steam (open steam) directly into the charge.

Steam drive injection (steam injection): EOR process in which steam is continuously injected into one set of wells (injection wells) or other injection source to affect oil displacement toward and production from a second set of wells (production wells); steam stimulation of production wells is *direct steam stimulation* whereas steam drive by steam injection to increase production from other wells is *indirect steam stimulation.*

Steam stimulation: Injection of steam into a well and the subsequent production of oil from the same well.

Steam turbine: A device for converting the energy of high-pressure steam (produced in a boiler) into mechanical power that can then be used to generate electricity.

Stiles method: A simple approximate method for calculating oil recovery by waterflood that assumes separate layers (stratified reservoirs) for the permeability distribution.

Storage stability (or storage instability): The ability (inability) of a liquid to remain in storage over extended periods of time without appreciable deterioration as measured by gum formation and the depositions of insoluble material (sediment).

Straight-run asphalt: The asphalt produced by the distillation of asphaltic crude oil.

Straight-run products: Obtained from a distillation unit and used without further treatment.

Straight vegetable oil (SVO): Any vegetable oil that has not been optimized through the process of transesterification.

Strata: Layers including the solid iron-rich inner core, molten outer core, mantle, and crust of the earth.

Straw oil: Pale paraffin oil of straw color used for many process applications.

Stripper well: A well that produces (strips from the reservoir) oil or gas.

Stripping: A means of separating volatile components from less volatile ones in a liquid mixture by the partitioning of the more volatile materials to a gas phase of air or steam (q.v. stabilization).

Sulfonic acids: Acids obtained by petroleum or a petroleum product with strong sulfuric acid.

Sulfuric acid alkylation: An alkylation process in which olefins (C_3, C_4, and C_5) combine with *iso*-butane in the presence of a catalyst (sulfuric acid) to form branched-chain hydrocarbons used especially in gasoline blending stock.

Supercritical fluid: An extraction method where the extraction fluid is present at a pressure and temperature above its critical point.

Superheated steam: Steam that is hotter than boiling temperature for a given pressure.

Surface-active material: A chemical compound, molecule, or aggregate of molecules with physical properties that cause it to adsorb at the interface between *two* immiscible liquids, resulting in a reduction of interfacial tension or the formation of a microemulsion.

Surfactant: A type of chemical, characterized as one that reduces interfacial resistance to mixing between oil and water or changes the degree to which water wets reservoir rock.

Suspensoid catalytic cracking: A nonregenerative cracking process in which cracking stock is mixed with slurry of catalyst (usually clay) and cycle oil and passed through the coils of a heater.

Sustainable: An ecosystem condition in which biodiversity, renewability, and resource productivity are maintained over time.

SW-846: An EPA multivolume publication entitled *Test Methods for Evaluating Solid Waste, Physical/Chemical Methods*; the official compendium of analytical and sampling methods that have been evaluated and approved for use in complying with the RCRA regulations and that functions primarily as a guidance document setting forth acceptable, although not required, methods for the regulated and regulatory communities to use in responding to RCRA-related sampling and analysis requirements. SW-846 changes over time as new information and data are developed.

Sweated wax: A crude petroleum–based wax that has been freed from oil by having been passed through a sweater.

Sweating: The separation of paraffin oil and low-melting wax from paraffin wax.

Sweep efficiency: The ratio of the pore volume of reservoir rock contacted by injected fluids to the total pore volume of reservoir rock in the project area. (*See also* areal sweep efficiency *and* vertical sweep efficiency.)

Sweet crude oil: Crude oil containing little sulfur; *see also* Sour crude oil.

Sweetening: The process by which petroleum products are improved in odor and color by oxidizing or removing the sulfur-containing and unsaturated compounds.

Swelling: Increase in the volume of crude oil caused by absorption of EOR fluids, especially carbon dioxide. Also increase in volume of clays when exposed to brine.

Swept zone: The volume of rock that is effectively swept by injected fluids.

Synthesis gas (syngas): A gas produced by the gasification of a solid or liquid fuel that consists primarily of carbon monoxide and hydrogen.

Synthetic crude oil (syncrude): A hydrocarbon product produced by the conversion of coal, oil shale, or tar sand bitumen that resembles conventional crude oil; can be refined in a petroleum refinery.

Synthetic ethanol: Ethanol produced from ethylene, a petroleum by-product.

Tar: The volatile, brown to black, oily, viscous product from the destructive distillation of many bituminous or other organic materials, especially coal; a name used arbitrarily for petroleum in ancient texts.

Target analyte: Target analytes are compounds that are required analytes in U.S. EPA analytical methods. BTEX and PAHs are examples of petroleum-related compounds that are target analytes in U.S. EPA Methods.

Tar sand (bituminous sand): A formation in which the bituminous material (bitumen) is found as a filling in veins and fissures in fractured rocks or impregnating relatively shallow sand, sandstone, and limestone strata; a sandstone reservoir that is impregnated with a heavy, extremely viscous, black hydrocarbonaceous, petroleum-like material that cannot be retrieved through a well by conventional or enhanced oil recovery techniques (FE 76-4): the several rock types that contain an extremely viscous hydrocarbon that is not recoverable in its natural state by conventional oil well production methods including currently used enhanced recovery techniques; *see* Bituminous sand.

Tertiary structure: The 3D structure of a molecule.

Tetraethyl lead (TEL): An organic compound of lead, $Pb(CH_3)_4$, which, when added in small amounts, increases the antiknock quality of gasoline.

Thermal coke: The carbonaceous residue formed as a result of a noncatalytic thermal process; the Conradson carbon residue; the Ramsbottom carbon residue.

Thermal cracking: A process that decomposes, rearranges, or combines hydrocarbon molecules by the application of heat, without the aid of catalysts.

Thermal polymerization: A thermal process to convert light hydrocarbon gases into liquid fuels.

Thermal process: Any refining process that utilizes heat, without the aid of a catalyst.

Thermal recovery: *See* EOR process

Thermal reforming: A process using heat (but no catalyst) to effect the molecular rearrangement of low-octane naphtha into gasoline of higher antiknock quality.

Thermal stability (thermal instability): The ability (inability) of a liquid to withstand relatively high temperatures for short periods of time without the formation of carbonaceous deposits (sediment or coke).

Thermochemical conversion: The use of heat to chemically change substances from one state to another, for example, to make useful energy products.

Thermofor catalytic cracking: A continuous, moving-bed catalytic cracking process.

Thermofor catalytic reforming: A reforming process in which the synthetic, bead-type catalyst of coprecipitated chromia (Cr_2O_3) and alumina (Al_2O_3) flows down through the reactor concurrent with the feedstock.

Thermofor continuous percolation: A continuous clay treating process to stabilize and decolorize lubricants or waxes.

Thief zone: Any geologic stratum not intended to receive injected fluids in which significant amounts of injected fluids are lost; fluids may reach the thief zone due to an improper completion or a faulty cement job.

Thin layer chromatography (TLC): A chromatographic technique employing a porous medium of glass coated with a stationary phase. An extract is spotted near the bottom of the medium and placed in a chamber with solvent (mobile phase). The solvent moves up the medium and separates the components of the extract, based on affinities for the medium and solvent.

Timberland: Forestland that is producing or is capable of producing crops of industrial wood and that is not withdrawn from timber utilization by statute or administrative regulation.

Time-lapse logging: The repeated use of calibrated well logs to quantitatively observe changes in measurable reservoir properties over time.

Tipping fee: A fee for disposal of waste.

Ton (short ton): 2000 lb.

Tonne (Imperial ton, long ton, shipping ton): 2240 lb; equivalent to 1000 kg or in crude oil terms, approximately 7.5 barrels of oil.

Topped crude: Petroleum that has had volatile constituents removed up to a certain temperature, for example, 250°C+ (480°F+) topped crude; not always the same as a residuum.

Topping: The distillation of crude oil to remove light fractions only.

Topping and backpressure turbines: Turbines that operate at exhaust pressure considerably higher than atmospheric (noncondensing turbines); often multistage with relatively high efficiency.

Topping cycle: A cogeneration system in which electric power is produced first. The reject heat from power production is then used to produce useful process heat.

Total petroleum hydrocarbons (TPH): The family of several hundred chemical compounds that originally come from petroleum.

Tower: Equipment for increasing the degree of separation obtained during the distillation of oil in a still.

TPH E: Gas chromatographic test for TPH extractable organic compounds.

TPH V: Gas chromatographic test for TPH volatile organic compounds.

TPH-D (DRO): Gas chromatographic test for TPH diesel-range organics.

TPH-G (GRO): Gas chromatographic test for TPH gasoline-range organics.

Trace element: Those elements that occur at very low levels in a given system.

Tracer test: A technique for determining fluid flow paths in a reservoir by adding small quantities of easily detected material (often radioactive) to the flowing fluid and monitoring their appearance at production wells. Also used in cyclic injection to appraise oil saturation.

Transesterification: The chemical process in which an alcohol reacts with the triglycerides in vegetable oil or animal fats, separating the glycerin and producing biodiesel.

Transmissibility (transmissivity): An index of producibility of a reservoir or zone, the product of permeability and layer thickness.

Traps: Sediments in which oil and gas accumulate from which further migration is prevented.

Traveling grate: A type of furnace in which assembled links of grates are joined together in a perpetual belt arrangement. Fuel is fed in at one end and ash is discharged at the other.

Treatment: Any method, technique, or process that changes the physical and/or chemical character of petroleum.

Triaxial borehole seismic survey: A technique for detecting the orientation of hydraulically induced fractures, wherein a tool holding three mutually seismic detectors is clamped in the borehole during fracturing; fracture orientation is deduced through the analysis of the detected microseismic perpendicular events that are generated by the fracturing process.

Trickle hydrodesulfurization: A fixed-bed process for desulfurizing middle distillates.

Trillion: 1×10^{12}

True boiling point (True boiling range): The boiling point (boiling range) of a crude oil fraction or a crude oil product under standard conditions of temperature and pressure.

Tube-and-tank cracking: An older liquid-phase thermal cracking process.

Turbine: A machine for converting the heat energy in steam or high-temperature gas into mechanical energy. In a turbine, a high velocity flow of steam or gas passes through successive rows of radial blades fastened to a central shaft.

Turn-down ratio: The lowest load at which a boiler will operate efficiently as compared to the boiler's maximum design load.

Ultimate analysis: Elemental composition.

Ultimate recovery: The cumulative quantity of oil that will be recovered when revenues from further production no longer justify the costs of the additional production.

Ultrafining: A fixed-bed catalytic hydrogenation process to desulfurize naphtha and upgrade distillates by essentially removing sulfur, nitrogen, and other materials.

Ultraforming: A low-pressure naphtha-reforming process employing onstream regeneration of a platinum-on-alumina catalyst and producing high yields of hydrogen and high-octane-number reformate.

Unassociated molecular weight: The molecular weight of asphaltenes in a nonassociating (polar) solvent, such as dichlorobenzene, pyridine, or nitrobenzene.

Unconformity: A surface of erosion that separates younger strata from older rocks.

Unifining: A fixed-bed catalytic process to desulfurize and hydrogenate refinery distillates.

Unisol process: A chemical process for extracting mercaptan sulfur and certain nitrogen compounds from sour gasoline or distillates using regenerable aqueous solutions of sodium or potassium hydroxide containing methanol.

Universal Viscosity: *See* Saybolt Universal Viscosity

Unresolved complex: The thousands of compounds that a gas chromatograph *mixture (UCM)* is unable to fully separate.

Unstable: Usually refers to a petroleum product that has more volatile constituents present or refers to the presence of olefin and other unsaturated constituents.

UOP alkylation: A process using hydrofluoric acid (which can be regenerated) as a catalyst to unite olefins with *iso*-butane.

UOP copper sweetening: A fixed-bed process for sweetening gasoline by converting mercaptan derivatives (RSH) to disulfide derivatives (RSSR) by contact with ammonium chloride and copper sulfate in a bed.

UOP fluid catalytic cracking: A fluid process of using a reactor-over-regenerator design.

Upgrading: The conversion of petroleum to value-added salable products.

Upper-phase microemulsion: A microemulsion phase containing a high concentration of oil that, when viewed in a test tube, resides on top of a water phase.

Urea dewaxing: A continuous dewaxing process for producing low-pour-point oils, using urea that forms a solid complex (adduct) with the straight-chain wax paraffins in the stock; the complex is readily separated by filtration.

Vacuum distillation: A secondary distillation process that uses a partial vacuum to lower the boiling point of residues from primary distillation and extract further blending components; distillation under reduced pressure.

Vacuum residuum: A residuum obtained by the distillation of a crude oil under vacuum (reduced pressure); that portion of petroleum that boils above a selected temperature such as 510°C (950°F) or 565°C (1050°F).

Vapor-phase cracking: A high-temperature, low-pressure conversion process.

Vapor-phase hydrodesulfurization: A fixed-bed process for desulfurization and hydrogenation of naphtha.

Vertical sweep efficiency: The fraction of the layers or vertically distributed zones of a reservoir that are effectively contacted by displacing fluids.

VGC (viscosity-gravity constant): An index of the chemical composition of crude oil defined by the general relation between specific gravity, sg, at 60°F and Saybolt Universal Viscosity, SUV, at 100°F: $a = 10sg - 1.0752 \log (SUV - 38)/10sg - \log (SUV - 38)$. The constant, a, is low for the paraffin crude oils and high for the naphthenic crude oils.

VI (Viscosity index): An arbitrary scale used to show the magnitude of viscosity changes in lubricating oils with changes in temperature.

Visbreaking: A process for reducing the viscosity of heavy feedstocks by controlled thermal decomposition.

Viscosity: A measure of the ability of a liquid to flow or a measure of its resistance to flow; the force required to move a plane surface of area 1 m² over another parallel plane surface 1 m away at a rate of 1 m/second when both surfaces are immersed in the fluid; the higher the viscosity, the slower the liquid flows.

Viscosity-gravity constant: *See* VGC

Viscosity index: *See* VI.

VOC (VOCs): Volatile organic compound(s); volatile organic compounds are regulated because they are precursors to ozone; carbon-containing gases and vapors from incomplete gasoline combustion and from the evaporation of solvents.

Volatile compounds: A relative term that may mean (1) any compound that will purge, (2) any compound that will elute before the solvent peak (usually those < C6), or (3) any compound that will not evaporate during a solvent removal step.

Volatile organic compounds (VOCs): Name given to light organic hydrocarbons that escape as vapor from fuel tanks or other sources and during the filling of tanks. VOCs contribute to smog.

Volumetric sweep: The fraction of the total reservoir volume within a flood pattern that is effectively contacted by injected fluids.

VSP: Vertical seismic profiling, a method of conducting seismic surveys in the borehole for detailed subsurface information.

Waste streams: Unused solid or liquid by-products of a process.

Waste vegetable oil (WVO): Grease from the nearest fryer that is filtered and used in modified diesel engines or converted to biodiesel through the process of transesterification and used in any diesel-fueled vehicle.

Water-cooled vibrating grate: A boiler grate made up of a tuyere grate surface mounted on a grid of water tubes interconnected with the boiler circulation system for positive *cooling; the structure is supported by flexing plates allowing the grid and grate to move in a vibrating action; ash is automatically discharged.

Waterflood: Injection of water to displace oil from a reservoir (usually a secondary recovery process).

Waterflood mobility ratio: Mobility ratio of water displacing oil during waterflooding. (*See also* mobility ratio.)

Waterflood residual: The waterflood residual oil saturation; the saturation of oil remaining after waterflooding in those regions of the reservoir that have been thoroughly contacted by water.

Watershed: The drainage basin contributing water, organic matter, dissolved nutrients, and sediments to a stream or lake.

Watson characterization factor: *See* Characterization factor

Watt: The common base unit of power in the metric system; 1 W equals 1 J/s, or the power developed in a circuit by a current of 1 ampere flowing through a potential difference of 1 V. 1 W = 3.412 Btu/h.

Wax: *See* Mineral wax and Paraffin wax

Wax distillate: A neutral distillate containing a high percentage of crystallizable paraffin wax, obtained on the distillation of paraffin or mixed-base crude, and on reducing neutral lubricating stocks.

Wax fractionation: A continuous process for producing waxes of low oil content from wax concentrates; *see also* MEK deoiling.

Wax manufacturing: A process for producing oil-free waxes.

Weathered crude oil: Crude oil that, due to natural causes during storage and handling, has lost an appreciable quantity of its more volatile components; also indicates uptake of oxygen.

Wellbore: The hole in the earth comprising a well.

Well completion: The complete outfitting of an oil well for either oil production or fluid injection; also the technique used to control fluid communication with the reservoir.

Wellhead: That portion of an oil well above the surface of the ground.

Wet gas: Gas containing a relatively high proportion of hydrocarbons that are recoverable as liquids; *see also* Lean gas.

Wet scrubbers: Devices in which a countercurrent spray liquid is used to remove impurities and particulate matter from a gas stream.

Wettability: The relative degree to which a fluid will spread on (or coat) a solid surface in the presence of other immiscible fluids.

Wettability number: A measure of the degree to which a reservoir rock is water wet or oil wet, based on capillary pressure curves.

Wettability reversal: The reversal of the preferred fluid wettability of a rock, for example, from water wet to oil wet or vice versa.

Wheeling: The process of transferring electrical energy between buyer and seller by way of an intermediate utility or utilities.

White oil: A generic tame applied to highly refined, colorless hydrocarbon oils of low volatility, and covering a wide range of viscosity.

Whole-tree harvesting: A harvesting method in which the whole tree (above the stump) is removed.

Wobbe Index (or Wobbe Number): The calorific value of a gas divided by the specific gravity.

Wood alcohol: *See* Methyl alcohol.

Yarding: The initial movement of logs from the point of felling to a central loading area or landing.

Zeolite: A crystalline aluminosilicate used as a catalyst and having a particular chemical and physical structure.

CONVERSION FACTORS

1 acre = 43,560 square feet
1 acre foot = 7758.0 bbl
1 atmosphere = 760 mm Hg = 14.696 psi = 29.91 inch Hg
1 atmosphere = 1.0133 bars = 33.899 feet H_2O
1 barrel (oil) = 42 gallons = 5.6146 cubic feet
1 barrel (water) = 350 pounds at 60°F

1 barrel per day = 1.84 cubic centimeter per second

1 Btu = 778.26 feet-pound

1 centipoise × 2.42 = pound mass/(feet) (hour), viscosity

1 centipoise × 0.000672 = pound mass/(feet) (second), viscosity

1 cubic foot = 28,317 cubic centimeter = 7.4805 gallons

Density of water at 60°F = 0.999 gram/cubic centimeter = 62.367 pound/cubic feet = 8.337 pound/gallon

1 gallon = 231 cubic inch = 3,785.4 cubic centimeter = 0.13368 cubic feet

1 horsepower-hour = 0.7457 kWh = 2544.5 Btu

1 horsepower = 550 feet-pound/second = 745.7 watts

1 inch = 2.54 centimeter

1 meter = 100 centimeter = 1000 mm = 10^6 microns = 10^{10} angstroms (Å)

1 ounce = 28.35 grams

1 pound = 453.59 grams = 7000 grains

1 square mile = 640 acres

SI METRIC CONVERSION FACTORS

(E = exponent; i.e., E + 03 = 10^3 and E − 03 = 10^{-3})

acre-foot × 1.233482	E + 03 = meters cubed
barrels × 1.589873	E − 01 = meters cubed
centipoise × 1.000000	E − 03 = pascal seconds
darcy × 9.869233	E − 01 = micrometers squared
feet × 3.048000	E − 01 = meters
pounds/acre-foot × 3.677332	E − 04 = kilograms/meters cubed
pounds/square inch × 6.894757	E + 00 = kilo pascals
dyne/cm × 1.000000	E + 00 = mN/m
parts per million × 1.000000	E + 00 = milligrams/kilograms

Index